TELETRAFFIC ISSUES
in an Advanced Information Society

ITC-11

STUDIES IN TELECOMMUNICATION
VOLUME 5

International Teletraffic Congress
International Advisory Council
The Development and Application of Teletraffic Theory

The first International Teletraffic Congress, entitled 'On the Application of the Theory of Probability in Telephone Engineering and Administration', took place in 1955 in Copenhagen. Every 3 years since then, specialists from tele-administrations, industry, and universities have gathered to present new methodologies and applications of the theory of teletraffic and teleplanning. Their audience is composed of experts and users, and the main focus is on issues of telecommunication traffic, as they affect customer service, and efficient telecommunication equipment loading, with special emphasis on probabilistic and other mathematical handling of traffic problems.

The International Advisory Council – with representatives from each past Congress country and the two future Congress countries – is responsible for the Congresses. Up till now, the following have served on the Council:

NORTH-HOLLAND – AMSTERDAM • NEW YORK • OXFORD • TOKYO

TELETRAFFIC ISSUES

in an Advanced Information Society

ITC-11

Proceedings of the Eleventh International Teletraffic Congress
Kyoto, Japan, September 4-11, 1985

Edited by:

Minoru AKIYAMA

Department of Electrical Engineering
The University of Tokyo
Tokyo, Japan

Part 2

1985

NORTH-HOLLAND – AMSTERDAM • NEW YORK • OXFORD • TOKYO

ISBN Part 1: 0 444 87917 X
ISBN Part 2: 0 444 87918 8
ISBN Set: 0 444 87919 6

Published by:
ELSEVIER SCIENCE PUBLISHERS B.V.
P.O. BOX 1991
1000 BZ AMSTERDAM
THE NETHERLANDS

Sole distributors for the U.S.A. and Canada:
ELSEVIER SCIENCE PUBLISHING COMPANY, INC.
52 VANDERBILT AVENUE
NEW YORK, N.Y. 10017
U.S.A.

Legal Notice:
All opinions expressed in these proceedings are those of the authors and are not binding on the International Advisory Council of the International Teletraffic Congress.

PRINTED IN THE NETHERLANDS

CONTENTS

PART 1

SESSION 1.2: OPENING SESSION

SESSION 1.3: ISDN AND NEW SERVICES I

Chairperson: C.W. Pratt (Australia)
Vice-Chairperson: J. Matsumoto (Japan)

SESSION 1.4: TRAFFIC MODELS AND METHODOLOGIES

Chairperson: B. Wallstroem (Sweden)
Vice-Chairperson: J. Matsuda (Japan)

SESSION 2.1A: ISDN AND NEW SERVICES II

Chairperson: H. Terada (Japan)
Vice-Chairperson: D.R. Mansfield (Canada)

SESSION 2.1B: OVERLOAD CONTROL

Chairperson: N.W. MacFadyen (United Kingdom)
Vice-Chairperson: T. Hoshi (Japan)

SESSION 2.3A: NETWORK PLANNING AND DESIGN I
Chairperson: S. Katz (U.S.A.)
Vice-Chairperson: T. Takemura (Japan)

SESSION 2.3B: SATELLITE AND RADIO SYSTEMS I
Chairperson: A. Myskja (Norway)
Vice-Chairperson: H. Okada (Japan)

SESSION 2.4A: NETWORK PLANNING AND DESIGN II
Chairperson: V.B. Iversen (Denmark)
Vice-Chairperson: T. Yamaguchi (Japan)

SESSION 3.1B: TRAFFIC ADMINISTRATION

Chairperson: M. Bonatti (Italy)
Vice-Chairperson: T. Takenaka (Japan)

SESSION 3.2A: QUEUEING SYSTEMS II

Chairperson: R.G. Schehrer (F.R.G.)
Vice-Chairperson: Y. Takahashi (Japan)

SESSION 3.2B: TRAFFIC MEASUREMENT

Chairperson: K. Bø (Norway)
Vice-Chairperson: S. Iisaku (Japan)

MEETING S3.1J: PERFORMANCE EVALUATION AND TRAFFIC THEORY

MEETING S3.1K: ISDN AND TRAFFIC STUDIES

MEETING S3.2K: LAN AND SATELLITE

PART 2

SESSION 4.1A: LOCAL AREA NETWORKS I

Chairperson: A. Descloux (U.S.A.)
Vice-Chairperson: N. Shiratori (Japan)

SESSION 4.1B: CCITT RELATED PROBLEMS

Chairperson: G. Gosztony (Hungary)
Vice-Chairperson: H. Murata (Japan)

SESSION 4.4B: SWITCHING SYSTEMS AND MODELING II
Chairperson: G. Miranda (Italy)
Vice-Chairperson: Y. Takahashi (Japan)

SESSION 5.1A: ALTERNATE ROUTING
Chairperson: H. Akimaru (Japan)
Vice-Chairperson: R.J. Harris (Australia)

SESSION 5.1B: SWITCHING SYSTEMS AND MODELING III
Chairperson: H. Hofstetter (F.R.G.)
Vice-Chairperson: Y. Sakita (Japan)

SESSION 5.3A: DATA COMMUNICATION NETWORKS II
Chairperson: C.E.M. Pearce (Australia)
Vice-Chairperson: K. Endo (Japan)

SESSION 5.3B: REPEATED CALL
Chairperson: D.J. Songhurst (U.K.)
Vice-Chairperson: Y. Tanaka (Japan)

SESSION 5.4A: DATA COMMUNICATION NETWORKS III
Chairperson: O. Gonzalez Soto (Spain)
Vice-Chairperson: H. Miyahara (Japan)

SESSION 6.3: INTERNATIONAL COOPERATION IN TRAFFIC ENGINEERING

Chairperson: M. Kato (Japan)

MEETING S4.1K: SWITCHING SYSTEMS AND MODELING

MEETING S4.2K: TRAFFIC STUDIES ON NEW SERVICES

MEETING S5.1K: TRAFFIC SIMULATION AND THEORY

MEETING S5.2K: OVERLOAD CONTROL AND TRANSIENT ANALYSIS

CLOSING ADDRESS

APPENDIX: INDEX OF PAPERS PRESENTED AT INTERNATIONAL CONGRESSES 1–10

TELETRAFFIC ISSUES in an Advanced Information Society
ITC-11
Minoru Akiyama (Editor)
Elsevier Science Publishers B.V. (North-Holland)

Statistics of Mixed Data Traffic on a Local Area Network

W. T. MARSHALL and S. P. MORGAN

AT&T Bell Laboratories
Murray Hill, New Jersey, U.S.A.

ABSTRACT

We have analyzed a week's worth of data traffic on a DATAKIT* Virtual Circuit Switch network at AT&T Bell Laboratories. The network includes 5 nodes connecting 22 host computers and 226 terminals, with trunks to nodes elsewhere at Bell Laboratories. Users are predominantly researchers using the UNIX† operating system via teletypewriter terminals and diskless work stations at 9.6 kb/s. Comparable fractions of the traffic are generated by terminal-to-host calls, by indirect logins, by interactive remote command executions, and by host-to-host file transfers. We display histograms representing the distributions of interarrival times and call lengths associated with the various types of calls, and the distributions of transmission bursts in individual calls. We characterize typical distributions by their means and coefficients of variation, and propose a model for time-sharing traffic which depends on a relatively small number of parameters and statistical distributions.

1. INTRODUCTION

Few detailed measurements of data traffic on local area networks have been published. Little appears to be known about the statistics of such traffic, beyond the conventional wisdom that data traffic is bursty and that the capabilities of networks, terminals, and hosts are evolving so fast that every case is different. However, even a snapshot of a particular system, if carefully interpreted, can give some guidance to network designers, and can suggest a model for traffic measurements on other systems.

We report measurements of a week's worth of traffic on a DATAKIT* Virtual Circuit Switch network at AT&T Bell Laboratories. The host computers (DEC VAX-11/780's and 750's) run the UNIX† time-shared operating system and are used by researchers for tasks such as prototype software development, graphics, text editing, and numerical computation, as well as data storage. Terminals run at an access speed of 9.6 kb/s and include screen teletypewriters as well as diskless work stations such as the TELETYPE§ 5620 Dot-Mapped Display. A UNIX/DATAKIT user can connect to a host directly from a terminal, or indirectly through one or more other hosts. He or she can execute a single interactive command on a remote host and return automatically to the original host. In addition, users can routinely cause files to be transferred from one host to another.

Data traffic measurements can be either user-oriented or network-oriented. Examples of user-oriented measurements would be timestamped records of successive characters sent from and received by a particular terminal, or the character counts and transmission times of host-to-host file transfers. Examples of network-oriented measurements would be counts of the numbers and lengths of packets passing various points in the network during given intervals of time. The relationship between user-level traffic and network-level traffic depends on the network architecture and protocols.

The easiest traffic data to obtain are mean values, such as number of characters transmitted per hour. However, for predictions of detailed network behavior much more complete load statistics are needed. If the interarrival times of messages are independent and exponentially distributed, and if the message lengths are also independent and exponentially distributed, then a large number of theoretical results are available. If one or both of the distributions are not exponential, the queueing problem must be treated by more complicated analysis, approximation, or simulation. Measurements of real traffic are therefore of substantial importance.

User-oriented traffic measurements including statistical distributions as well as mean values began with Fuchs and Jackson's measurements [1] of low-speed, half-duplex terminal traffic in the late 1960's. In a series of papers, summarized in 1981, Pawlita [2] has reported the statistics of terminal traffic generated by a number of different user populations using half-duplex terminals at speeds from 200 b/s to 4800 b/s. However, there are no published measurements for full-duplex terminals, for work stations, or for speeds of 9600 b/s.

Network-oriented traffic measurements include Shoch and Hupp's measurements [3] on the original Ethernet at Xerox PARC. These authors give histograms of packet length and interpacket arrival time, and they estimate the ratio of total overhead bits to user data bits on their network. However they do not break their traffic down in as much detail as we propose to do here.

Section 2 of this paper describes the configuration of our network and the nature of the per-call data collected by software monitors in each switching node. Sections 3 and 4 show qualitative features of the switching and packet traffic as a function of hour of day. Section 5 includes holding-time and packet-count histograms for different types of calls, and also samples of detailed character counts and timings for some individual calls.

In Section 6 we propose a traffic model for the Bell Laboratories DATAKIT population. We argue that our traffic can be represented as consisting of terminal-like calls and file transfer calls. We characterize the statistical distributions that are relevant to our model by their means $\bar{x}$ and coefficients of variation C_x, where C_x is the ratio of the standard deviation to the mean. These parameters can go directly into two-moment approximations for the behavior of queueing networks with non-Poisson arrival processes and nonexponential service time distributions [4]. Furthermore, if one does not wish to use the means and standard deviations measured for the Bell Laboratories population, it is apparent what measurements would be required to get the corresponding parameters for a different population.

* DATAKIT is a trademark of AT&T.

† UNIX is a trademark of AT&T Bell Laboratories.

§ TELETYPE is a trademark of Teletype Corporation.

In this paper we do not fit specific functions to the empirical statistical distributions. Most of the empirical distributions have C_x substantially greater than unity; that is, they have longer tails than an exponential distribution. Some of the histograms, especially those corresponding to message interarrival times, look as if they could be well fitted by a lognormal distribution or a mixture of two lognormals; others, especially those having to do with work-station traffic, are irregular or represent samples too small to be definitive. More sophisticated statistical analyses of data traffic may be desirable in the future, but for the present it appears most important to publish the existing measurements, and to encourage the measurement of other systems.

2. NETWORK CONFIGURATION AND MEASUREMENTS

A DATAKIT network [5] consists of terminals and hosts connected to one or more virtual circuit switches or *nodes*. The network that we studied is shown in Fig. 1. The nodes are connected to each other and to other nodes at the same location by 1.7 Mb/s trunks. They are connected to nodes at other Bell Laboratories locations by 56 kb/s trunks. During the reference week, the network of Fig. 1 supported 22 hosts, 226 terminals, 10 dial-in lines, and 3 dial-out lines.

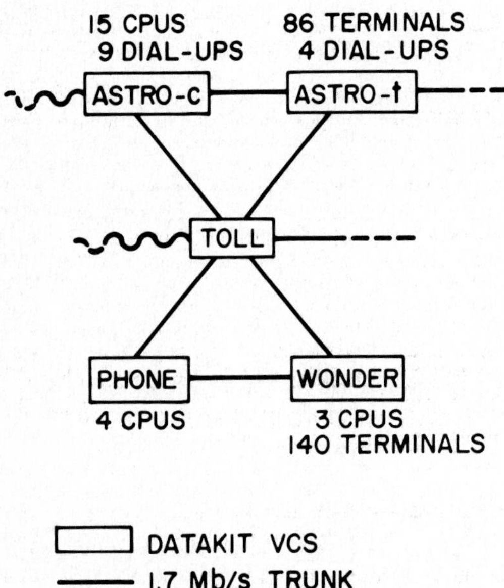

Fig. 1 Network configuration.

Raw traffic data are collected as follows. A program running in each node controller collects all call setup and takedown records, as well as cumulative packet counts for each call in progress approximately every 10 minutes. (A DATAKIT packet is 16 bytes long. For slow input devices, a packet will contain fewer than 16 bytes of data and will be padded with null bytes.) A program running on one of the hosts calls up each controller and receives the data collected.

The output includes three types of records, each timestamped to the nearest second. Call-setup records identify the circuit endpoint of the originator and display the "dialstring". The dialstring contains the destination name and service requested, and the source name including the user identification of the caller. Packet-count records identify two circuit endpoints and give the cumulative packet count in each direction. Call takedown records identify two circuit endpoints and give the total packet count in each direction. A complete record of a call consists of one call setup record, zero or more packet-count records, and one call takedown record, all of which can be associated with each other by the circuit endpoints.

The present study is based on the week of November 7-13, 1983. During this time no special precautions were taken except to hold the hardware configuration constant, so that we could associate each circuit endpoint with a unique terminal, host, or trunk. In addition, because the dialstring for host-to-host calls includes the command name, it was possible to distinguish the characteristics of traffic generated by different types of commands. We inspected every call placed during the week and established the following classification.

1. Terminal calls. These calls connect a terminal directly to a host computer. Teletypewriter and work-station calls were separated into two subcategories.
2. Remote login calls. These calls between hosts logically attach a terminal to a destination, and are expected to have characteristics similar to terminal calls.
3. Interactive remote command executions. These host-to-host calls are characterized by short holding times and the transfer of small numbers of packets.
4. Host-to-host file transfers. These include not only obvious file-transfer commands but also commands that send a file to a host that serves an output device such as a printer.

3. SWITCHING TRAFFIC

Fig. 2 shows 10-minute average switching rates during a typical day on one of the controllers. A "switch" is either a setup or a takedown; that is, 100 completed calls would correspond to 200 switches. The figure, although quite spiky, shows the familiar double-humped shape with a lunch dip. Fig. 3 shows hourly averages of the number of virtual circuits connected through the controller of Fig. 2.

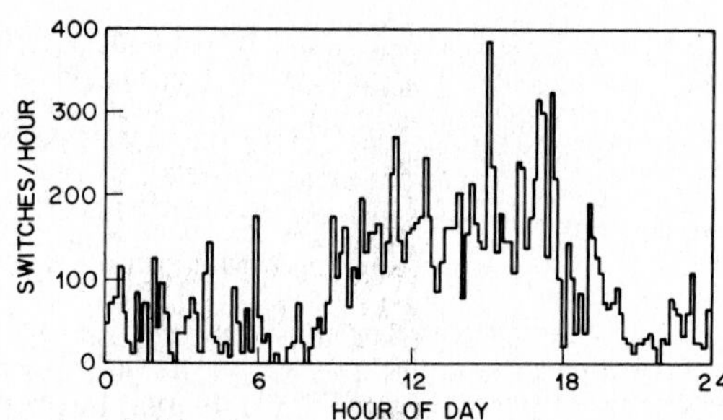

Fig. 2 10-minute average switching rates.

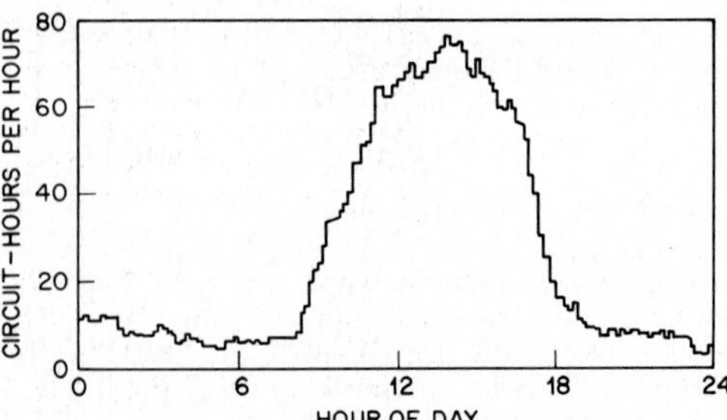

Fig. 3 Average circuit occupancy.

The network controllers are very lightly loaded. During the busiest 10 minutes of the week, the busiest controller is utilizing less than 4% of its processing capacity, and a similarly small fraction of its memory.

A further observation, which results from a more detailed breakdown of our data, is that the ratio of host-to-host to terminal-to-host calls is much larger than the ratio of host-to-host to terminal-to-host circuits. The reason is that terminal-to-host calls typically hold a circuit for a much longer time than host-to-host interactions.

Fig. 4 is a log-histogram of interarrival times of switching requests at a typical controller. The plot represents merged data for the 5 busiest hours of the week. All of the log-histograms for switching arrivals, including total calls and terminal-to-host calls separately, tend to be bimodal. There is no numerical difficulty in fitting a mixture of two lognormals to the empirical distributions. However, we have not yet made any attempt at interpretation.

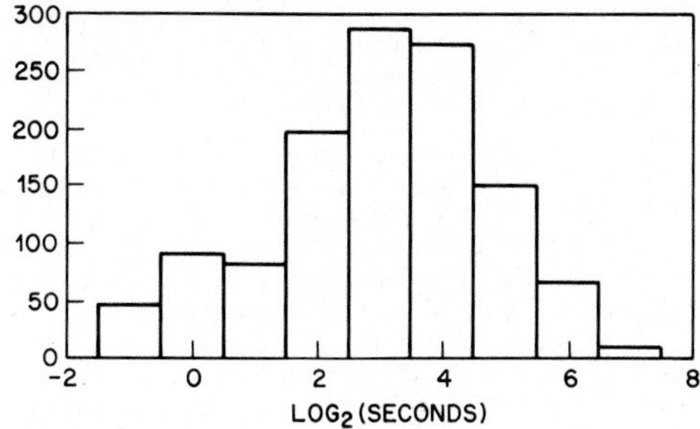

Fig. 4 Interswitch arrival distribution.

4. PACKET TRAFFIC

Fig. 5 is a bar plot of the total one-way packet traffic on a particular trunk during a typical day (10-minute averages). No two such plots are alike in detail; spikes corresponding to file transmissions can and do occur at any time of the day or night. We observed, however, that sustained average rates in excess of 400 packets per second for as long as 10 minutes were extremely rare. Such a rate is only about 4% of the capacity of a 1.7 Mb/s trunk.

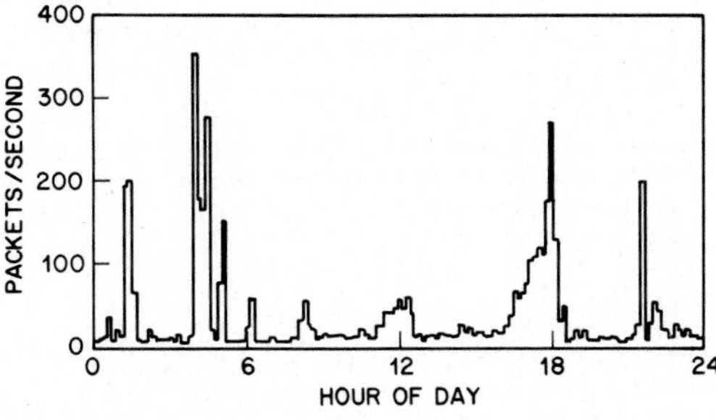

Fig. 5 Trunk packet traffic.

The traffic due to terminal calls alone, which might be of interest in an environment where users interact with a single host and rarely transfer files, is much less spiky than traffic including file transfers. Terminal packet traffic shows the characteristic two-humped profile with a lunch dip, and the number of packets sent from terminal to host is roughly half the number of packets sent from host to terminal.

5. CHARACTERISTICS OF DIFFERENT TYPES OF CALLS

As described in Section 2, setup and takedown times are recorded for each call, and cumulative packet counts for each continuing call are recorded by a polling program at approximately 10-minute intervals. For terminal calls and remote logins, frequently there are polling intervals during which no packets are transmitted. We can join together all the contiguous polling intervals in which any packets are transmitted and call these the active segments of a given call. If polls were taken more frequently, one could vary the definition of active segment; but it is not clear how much additional insight would be gained.

From the present data, for each call we know the lengths of the active segments and the total holding time, as well as the packet counts in each direction for each active segment. From these data we can calculate the mean number of packets transmitted in each direction per active second, as well as the ratio of mean active seconds to mean holding seconds per call. The results are shown approximately in Table 1.

TABLE 1. MEAN VALUES OF CALL PARAMETERS

	T'type-writer	Work station	Remote login	Remote exec	File transfer
Holding time	150 min	150 min	45 min	120 sec	50 sec
Active segment	30 min	45 min	20 min	-	-
Active/ holding ratio	0.4	0.6	0.6	-	-
Terminal packets/ active second	0.8	1.1	1.5	1.3	12⁺
Host packets/ active second	1.7	1.9	2.4	2.5	45*

* Forward direction + Backward direction

A. Terminal traffic

Holding times average about 2.5 hours, while the average active segment is about 30 minutes for teletypewriters and 45 minutes for work stations. (The difference in mean active segments may be related to the fact that some work-station users run a program that continually displays the load on the host. Such a work station would never show an inactive interval by our definition.) Typical log-histograms of holding times and active segments are shown in Figs. 6 and 7. The log-histogram for holding time falls off sharply above 2^{15} seconds (= 9.1 hours), suggesting that many people turn their terminals off only at the end of the day. The ratio of total active time to holding time is between 40 and 60%. Perhaps in a more structured environment, the ratio of active time to holding time would be higher.

The average packet rate from host to terminal during an active segment is a little less than 2 packets per second, while the average packet rate from terminal to host is about 1 packet per second. Interactions via remote login are similar to but more concentrated than interactions between a terminal and its immediate host (shorter active and holding times, but more packets per second). Interactive remote command executions differ from terminal-to-host calls in that the mean holding times are much shorter, being only a couple of minutes, and the mean packet rates per second are a little larger than for terminal-to-host calls.

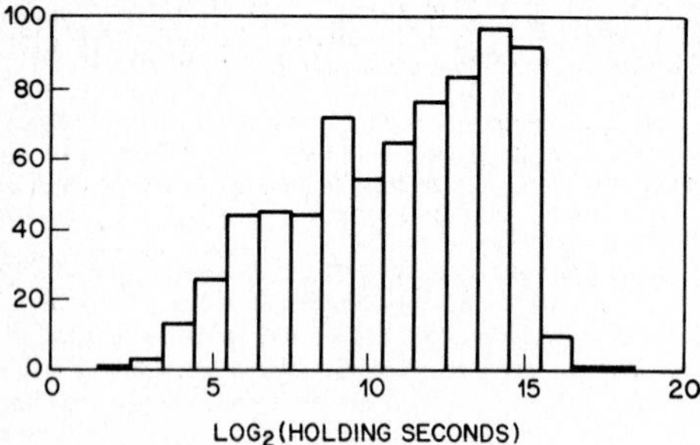

Fig. 6 Work-station holding times.

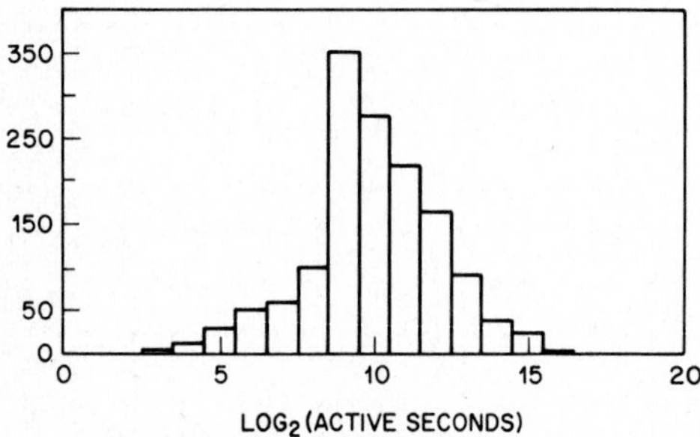

Fig. 7 Work-station active segments.

To understand the relationship between character and packet traffic in terminal calls, we used a hardware monitor to observe the character traffic produced by users during individual terminal sessions. The monitor records every character transmitted in either direction between the terminal and the network, timestamped to the nearest millisecond. Altogether, about 25 different technical users were monitored for a total of 30 hours on teletypewriters and 70 hours on work stations. In addition, we monitored three professional typists for a total of about 10 hours of word processing using teletypewriter terminals on another network at Bell Laboratories. Although these samples are not really large enough for statistical analysis, the data do indicate significant differences among the different types of terminals and users.

We find that a technical user of a teletypewriter terminal generates about 1 character per active second from the keyboard, whereas a professional typist engaged in word processing may average 2 characters/second. In a full-duplex system the total number of characters transmitted from the host is greater than the total number of characters transmitted from the terminal, since each character from the terminal is echoed by the host before it is displayed on the screen, and presumably the user wants to see some "real" output in addition to the echoes. The actual ratio of characters from host to characters from terminal depends critically on what the user is doing; the value of this ratio is probably the quantity that is determined with least accuracy by our small sample. We have observed single-session character ratios of over 35:1 for teletypewriters in technical use, with perhaps half of that (15:1 to 20:1) being a representative average, as compared to 3:1 for teletypewriters used for word processing. The larger ratios for technical users may suggest, inter alia, that 9.6 kb/s access lines encourage users to search sizeable files by scrolling them over the screen.

The situation for work stations is more complex. A work station contains its own microprocessor, which controls multiplexing (windowing) on the screen. When the terminal is in multiplexed mode, keyboard activity sends bursts of varying length to the host, and the host responds with bursts of varying length, but not on an individual keystroke basis. It turns out, however, that total daily character counts between a work station and a host are dominated by downloads of programs from the host into the terminal, and by transfers of bitmaps from the terminal screen to hard-copy devices such as laser printers. The end result is that overall character rates between a work station and a host are 2 to 3 times higher than between a teletypewriter in technical use and a host. We have observed average rates as high as nearly 15 characters per active second from work station to host and 3 times as much from host to work station, again depending strongly on what the user is doing.

The observation that a teletypewriter generates about 1 character per active second agrees with measurements by Fuchs and Jackson [1] and by Pawlita [2]. The average number of characters returned to the teletypewriter by the host depended for them, as it does for us, on what the users were doing. No previous results have been published for work-station terminals.

The flow of packets during a terminal session was not recorded directly but was calculated from the flow of characters by applying the DATAKIT packetizing algorithm, as follows: A "packetizing clock" ticks every 16 2/3 ms (60 times per second). A packet is begun when the first character arrives, and is closed after the 16th character or the third clock tick, whichever arrives first. If the packet does not contain 16 characters by the third tick, the remaining bytes are padded with nulls. For terminal calls, the ratio of packets-from-host to packets-from-terminal is smaller than the ratio of characters, since the host sends to the terminal at line speed or nearly so, and thus there are more characters on average in a packet from the host than in a packet from the terminal. For the same reason, the markedly higher character rates of work stations are reflected in only modestly higher packet rates.

As might have been expected, packet rates from the monitored sessions are comparable to but somewhat higher than the network averages. This is understandable because the monitored sessions were relatively short and users were aware of the monitoring. The network data are probably more representative of long-term averages.

Detailed analyses of user-computer interactions have generally followed the model introduced by Fuchs and Jackson [1] for low-speed, half-duplex teletypewriter terminals. In this model, each time-sharing session is described as a sequence of contiguous, nonoverlapping dialog segments identified as user think time, user input time, compute time, and computer output time. When characters are being transmitted, the variability of sending rate is described by breaking up the transmission into "user bursts" or "computer bursts" separated by interburst intervals. In full-duplex transmission, characters can be sent simultaneously from terminal to host and from host to terminal, and automatic separation of a session into a sequence of disjoint intervals defined as for the half-duplex model is difficult. We have accordingly considered the streams of characters from terminal to host and from host to terminal to be separate processes.

It is still useful to treat each process as a series of bursts separated by interburst intervals. The distribution of interarrival intervals between characters in any one stream contains relatively few intervals in the range between 10 and 50 ms. A human user does not strike successive keys as close together as 50 ms, while if a work station sends several bytes as a result of one keystroke, they go at access line speed (1 character per millisecond). Similarly, host responses go at line speed, with pauses due to timesharing which generally exceed 50 ms. In the present work we have defined a burst from either the terminal or the host as a sequence of characters separated by intervals of not more than 16 ms. This definition is obviously somewhat arbitrary.

User bursts from asynchronous teletypewriters are invariably single characters. By far the largest number of host bursts are one or two characters, representing single-character echoes or carriage-return-line-feed combinations. Accordingly we separate host bursts into one- or two-character echoes and longer responses.

For the TELETYPE 5620 work station, user characters are buffered in the terminal and transmitted to the host a line at a time. The line is acknowledged but not echoed. We have simply lumped the terminal-to-host and host-to-terminal transmissions into two sets of burst-length and interarrival-time log-histograms, as shown in Fig. 8. In this figure, the spikes in the interarrival distributions at 2000 ms represent a program which reports the load on the system via a 9-byte burst every 2 seconds whether anything else is going on at the terminal or not. In the traffic model described in Section 6, we have attempted to subtract out this component of our work station traffic.

In all of the user sessions that we recorded, user interarrival times look roughly lognormal, as do echo interarrival times for teletypewriter terminals. Lognormal interarrival distributions appear in the observations of Fuchs and Jackson [1]; Pawlita [2] fits mixtures of exponentials to his empirical distributions. "Echo" times for work stations are more irregularly distributed. Host response lengths and interresponse times are also somewhat irregular, but seem frequently to be bimodal on a logarithmic scale. Substantially more data would be required in order to determine whether there are significant regularities in the statistics of computer response bursts.

B. File traffic

Host-to-host file transfers differ from terminal-like calls in that file transfer calls hold for shorter times (less than a minute on the average), and the mean packet rates are more than an order of magnitude greater than the rates for terminal-like calls.

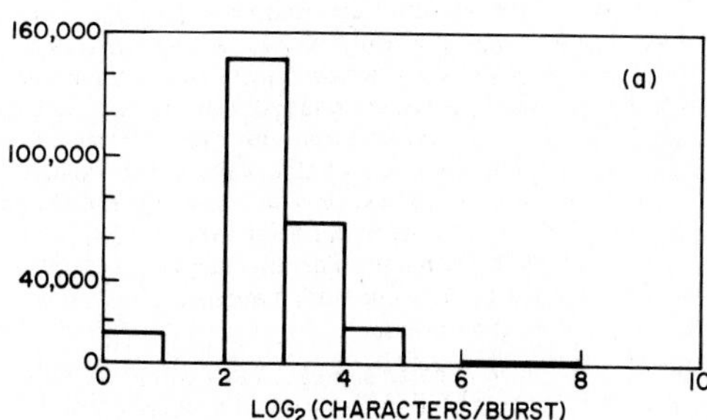

Fig. 8a Characters/burst from work station.

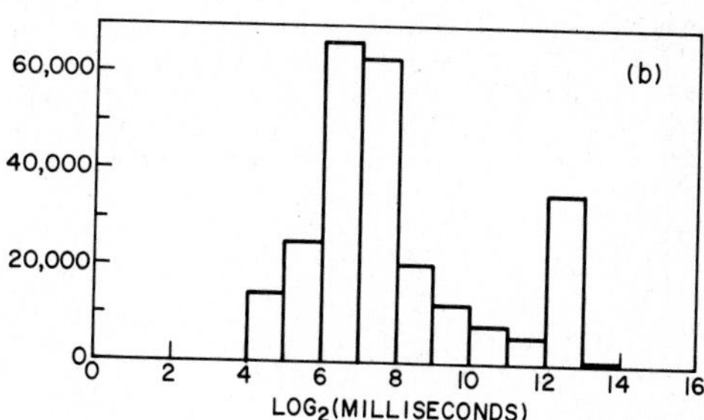

Fig. 8b Interburst intervals from work station.

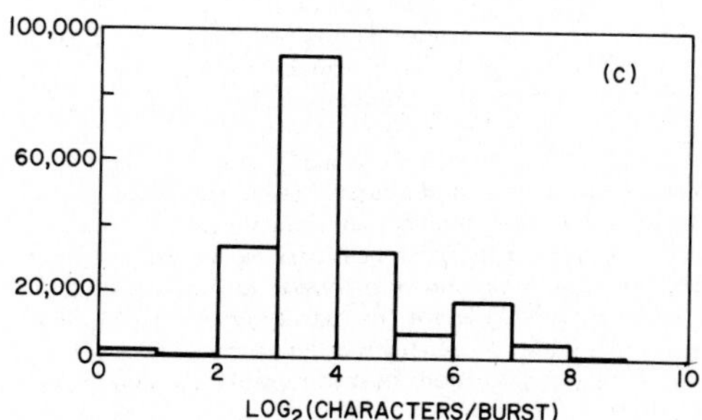

Fig. 8c Characters/response burst from host.

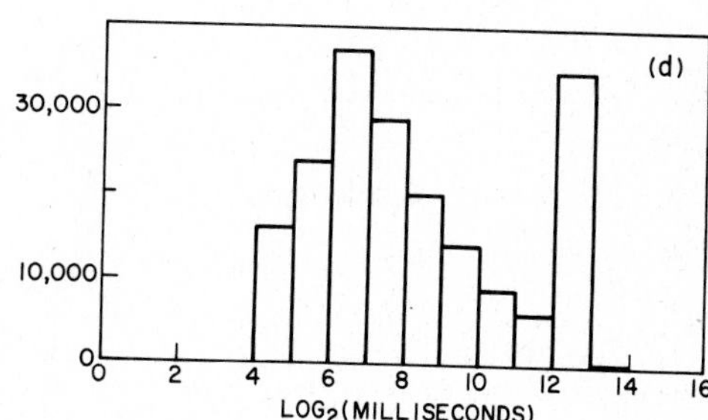

Fig. 8d Interresponse intervals from host.

In addition, file transfers involve a window flow control mechanism that leads to a substantial reverse flow of packets during each transfer. In Table 1, the reverse flow is about a quarter of the forward flow.

The effective limiting speed of file transfers on a local area network is almost never set by the raw transmission speed of the network, but rather by the host's operating system and the hardware interface between the host and the network. At the time these traffic data were taken, almost all of the hosts were connected to the network via DEC DR11-C interfaces. At the present writing, almost all of the interfaces are DEC KMC11-B microcomputers. Measurements of maximum disk-to-disk file transfer rates between otherwise idle hosts on our network were made for both interfaces. We obtained about 720 packets/second (7700 characters/second) for DR11 interfaces, and 5900 packets/second (75,000 characters/second) for KMC11 interfaces, with slightly higher rates for program-to-program transfers that bypass the UNIX file system overhead. The character-to-packet ratios would be lower for short files because of startup overhead.

Log-histograms of the distribution of packet counts of file-transfer calls and the distribution of busy-hour interarrival times are shown in Fig. 9. It would be convenient, if one were modeling a network with a different set of file transfer protocols, to know something about the distribution of file-transfer requests from users. Unfortunately one cannot get information about individual user requests from DATAKIT call records, because a single file-transfer call may involve the transfer of several files. Existing UNIX system software captures user-level data for certain classes of file transfers, but not for all file transfers.

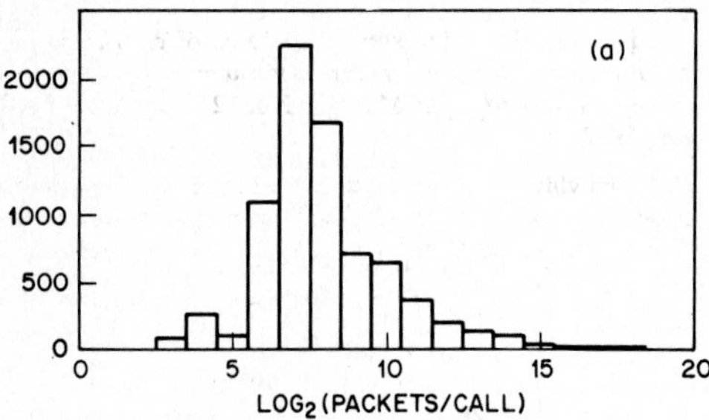

Fig. 9a Packets per file-transfer call.

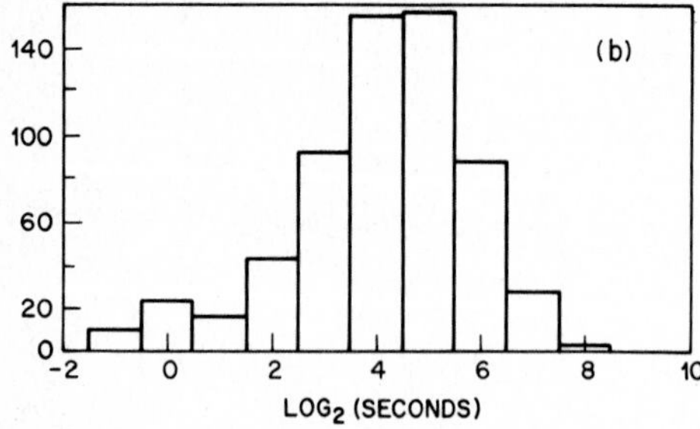

Fig. 9b File-call interarrival times.

6. TRAFFIC MODEL

The traffic generated by a typical user consists of the traffic between the user's own terminal and the host into which that terminal is connected, plus all the host-to-host traffic that the primary call generates. The approximate percentages of traffic are shown in Table 2. On the average one terminal call generates approximately 8 host-to-host calls, and one terminal packet generates somewhat more than 1.5 host-to-host packets.

TABLE 2. TRAFFIC PERCENTAGES

	Calls	Packets
Terminal calls	11	37
Remote logins	12	20
Remote executions	28	3
File transfers	49	40
Total	100	100

Table 2 does not show the relative numbers of user characters transmitted by the different types of calls, because the network measurements only recorded packets. However, we estimate that the average packet contains 8 bytes of user data (perhaps a little more than this for terminal traffic, and a little less for files). Since one trunk frame encapsulates 3 packets in 530 bits, this suggests that only $(3 \times 8 \times 8)/530 = 36\%$ of the bits on the local trunk represent user data. This estimate may be compared with Shoch and Hupp's estimate [3] that about 69% of the bits on their Ethernet represent user data, while about 31% encompass all forms of overhead. It should be noted, however, that null bytes are stripped from DATAKIT packets before being transmitted over long-distance trunks (Fig. 1), where bandwidth may be at more of a premium.

We now construct a per-user description of the traffic on our network. Suppose that N terminals are available for connection to a given node. According to our observations, the maximum number that will be simultaneously connected to a host at any time during the week is about $0.5N$. Furthermore, the maximum number of simultaneously existing host-to-host circuits is about one-half the maximum number of simultaneously existing terminal-to-host circuits. Since the instantaneous number of circuits does not fluctuate much about its local mean, one would feel safe if the circuit capacity of the switch were, say, twice the maximum number of simultaneous terminal connections expected during the week.

So far as the switching load is concerned, we see by looking at peak 10-minute averages that each simultaneously connected port generates terminal-to-host switching transitions at an average rate from 2 to 3 per hour. If we add in the switching rate for host-to-host calls, the total rate in the busiest 10 minutes corresponds to 10 to 20 switches per hour per connected terminal. This suggests that the controllers need to be able to handle a statistical distribution of 20 switches per hour per simultaneously connected terminal.

A calculation of congestion in the controller could make the following assumptions. A single setup or takedown operation on a controller takes about 0.1 s. A log-histogram of an interswitch arrival time distribution is shown in Fig. 4. If one is willing to use two-moment approximations for queueing delays, one can employ the following parameters for the interarrival and service time distributions:

$$\bar{t} = 3600/n_s \text{ seconds}, \qquad C_t = 1.25,$$

$$\bar{x} = 0.1 \text{ second}, \qquad C_x = 0,$$

where n_s is the average number of switches per hour.

Turning now to the modeling of packet flow, we can assume that on the average half the number of terminal circuits that are up are active. For each packet generated by a terminal circuit, roughly one-half of a terminal-circuit-like packet is generated on remote login and/or interactive remote command execution circuits, and about one file-transfer packet (Table 2).

We model a terminal-like circuit in the following way. An active asynchronous terminal sends a burst of x characters after t seconds to the host, where x and t are random variates with specified means and coefficients of variation. The host sends back a stream of "echoes" and a stream of longer responses. Our current best assessment of the parameters describing teletypewriter and work-station traffic on the Bell Laboratories DATAKIT network is shown in Table 3. Conversion of these character streams into equivalent packet streams depends, of course, on the packetizing algorithm associated with the given network.

TABLE 3. CHARACTER-BURST PARAMETERS

		$\bar{x}$ char	C_x	$\bar{t}$ sec	C_t	$\bar{x}/\bar{t}$ char/sec	$\sum \bar{x}/\bar{t}$ char/sec
Work station	From terminal	12	1.5	1	3.5	12	44
	From host	40	10	1.25	3	32	
Teletypewriter	From terminal	1	0	1	4	1	18
	Host echo	1	0	1	4	1	
	Host response	128	2	8	3	16	

Finally, we need a model for file transfer calls. On our network, the total number of packets transferred in file-transfer calls is approximately equal to the total number of packets transferred in terminal-to-host calls, but file transfer packets go in much larger bursts. A pessimistic estimate of the congestion due to file transfer calls would be obtained by assuming that the distribution of forward packet counts follows the log-histogram of Fig. 9a (unimodal with a long tail), with a mean length of 20000 user characters (2000 packets) per forward transfer, and a mean "delivery" rate from the sending host of, say, 4000 packets/s. Each forward transmission would be paired with a backward transmission of one-quarter the length — that is, 500 packets including no user characters. The intervals between transmissions could be distributed according to the plot in Fig. 9b, with mean adjusted to achieve the desired overall average rate of file transfers. If there are N characters per second on the average, then

$$t = 20000/N \text{ seconds}, \quad C_t = 1,$$

$$\bar{x} = 20000 \text{ characters}, \quad C_x = 10.$$

The file-transfer assumptions are conservative on two counts, so far as congestion is concerned. In the first place, a single file-transfer call may involve the transfer of a number of short files rather than a single long file; we had no way of recording individual user files during the packet-count measurements. Secondly, although a file transfer between KMC11 host interfaces can run at 6000 packets/second under ideal conditions, in practice file transfers run substantially slower because the hosts are being timeshared.

Finally, it should be emphasized that in none of the distributions of interarrival times have we investigated possible correlations between successive arrivals. If the arrival process is not a renewal process (successive intervals independently and identically distributed), the results of a conventional queueing-theory analysis may be seriously in error. This matter deserves further inquiry.

7. CONCLUSIONS

We have obtained a reasonably detailed statistical description of mixed data traffic, in terms of terminal-like calls and file transfer calls, by combining user-level and network-level measurements on a given network. The resulting model should be useful in predicting the response of networks having various different architectures to similar kinds of traffic.

Models of teletypewriter traffic are inadequate to describe the volume and statistical characteristics of traffic generated by work-station terminals. We have made some measurements of traffic due to diskless work stations. Measurements are also needed on networks of semiautonomous work stations, which have their own disk storage combined with substantial internal processing capability.

Statistical distributions associated with data traffic have coefficients of variation substantially greater than unity (up to 10 in the present study). It is important to learn how such large coefficients of variation affect the performance of queueing networks.

ACKNOWLEDGMENTS

A. E. Kaplan made the file-transfer packet flow measurements, and E. J. Sitar did the measurements of terminal character traffic.

REFERENCES

[1] E. Fuchs and P. E. Jackson, "Estimates of distributions of random variables for certain computer communications models," Comm. ACM **13**, No. 12 (December 1970), pp. 752-757.

[2] P. F. Pawlita, "Traffic measurements in data networks, recent measurement results, and some implications," IEEE Trans. Comm. **COM-29**, No. 4 (April 1981), pp. 525-535.

[3] J. F. Shoch and J. A. Hupp, "Measured performance of an Ethernet local network," Comm. ACM **23**, No. 12 (December 1980), pp. 711-721.

[4] W. Whitt, "The Queueing Network Analyzer," B.S.T.J. **62**, No. 9 (November 1983), pp. 2779-2815, and attached references.

[5] A. G. Fraser, "Towards a universal data transport system," IEEE Journal on Selected Areas in Communications **SAC-1**, No. 5 (November 1983), pp. 803-816.

TELETRAFFIC ISSUES in an Advanced Information Society
ITC-11
Minoru Akiyama (Editor)
Elsevier Science Publishers B.V. (North-Holland)

OPTIMAL DELAYS FOR RETRANSMISSION IN MULTI-ACCESS COMMUNICATION SYSTEMS

Frits C. SCHOUTE

Philips' Telecommunicatie Industrie, Hilversum, Netherlands
University of California, Berkeley, USA.

ABSTRACT

In the context of a small unslotted carrier sense multi-access system we want to answer the following questions: a) Does it make a difference whether retransmission delays are drawn from an uniform or an exponential distribution? b) Is there a simple way to implement exponentially distributed delays? and c) Should one double the mean delay on subsequent retransmissions of the same packet (binary backoff)? We find the following:

a) A very simple two station simulation model shows an advantage for exponentially distributed delays over uniformly distributed delays.
b) In fixed point binary notation, the bits of an exponential random variable are independent binary random variables. This leads to an algorithm that is simple to implement on a microprocessor without logarithm capabilities.
c) An analytical model which gives an adequate approximation of a multi-access system with states (a next state is entered each time that transmission of a packet needs to be rescheduled) can be used to optimize the mean delay for each state.

1. INTRODUCTION

Ever since the implementation of the Aloha system at the University of Hawaii [1], similar forms of multi-access communication systems are enjoying great interest in academia for many of the challenging theoretical questions and in practice for the flexibility of implementation. One of the well known earlier publications that addresses the issue of control of retransmissions is [2]. It is for me impossible to give even a rudimentary overview of what has been published on this subject since. Rather, I would like to address some related questions, which have been raised by implementors and for which, to my knowledge, the answers are still unknown.

The multi-access systems addressed here are small-scale and simple. Small-scale means e.g. a mobile radio system with a few tens of users, an office system with, say, ten stations or maybe a home system with, say, five stations. Simple means that one has very limited amounts of hardware and software. Typically such a system is a non-slotted carrier sense multi-access system. The questions that we want to answer for such a system are: a) Does it make difference whether retransmission delays are drawn from an uniform or an exponential distribution? b) Is there a simple way to implement exponentially distributed delays? and c) Should one double the mean delay on subsequent retransmissions of the same packet (binary backoff)?

Correspondingly the paper has the following sections:
In section 2 we introduce a simple model of two stations or users that share a multi-access channel. Transmissions are to be rescheduled after a random delay when packets of the two users collide or when one of the users senses the channel buzy. For a given set of parameters we compare, using discrete event simulation, the throughput of the channel for uniformly distributed and exponentially distributed retransmission delays.
In section 3 we show that in fixed point binary notation, the bits of a exponential random variate can be generated as the outcomes of independent Bernouli experiments and for the less significant bits the success-probability is ½. This observation is used for an algorithm for exponential variates that can easily be implemented on a microprocessor.
In section 4 we consider a multi-access system where each station, when rescheduling transmission of a packet, enters a new state. Each state can have its own mean retransmission delay. The complexity of optimization of mean retransmission delays by simulation becomes prohibitive. We shall derive an approximating analytical model for which one can use standard optimization routines.

2. UNIFORM vs. EXPONENTIAL DISTRIBUTION

In a non-slotted carrier sense multi-access communication system collisions occur if two or more stations have scheduled their start of transmission less than Δ apart. We assume Δ to be a constant that incorporates processing delay between sensing the channel idle and start of transmission, as well as propagation delay. If a collision occurs then the start of transmission is rescheduled with a random retransmission delay. Rescheduling also takes place if the channel is sensed busy before an intended start of transmission. To investigate a possible difference in performance between using uniformly distributed retransmission delays or exponentially distributed retransmission delays, we consider an idealized model of a two station multi-access channel. The performance measure to be maximized, η, is the utilization of the channel by successful transmissions.

To specify the model further, let us assume that the transmission time of one packet is constant and equal to one unit of time, that each successful transmission is followed by an exponentially distributed idle time V, and that there are only 2 stations When a transmission collides or when the channel is sensed busy before the intended transmission takes place, it will be rescheduled $\Delta+D$ after the original (intended) start of transmission. Here D is the random retransmission delay and Δ is added because packet transmission may be going on already this long before a collision occurs. At any point in time we let T_1 give the time of the first upcomming start of transmission and T_2 the time at which the other user intends to start transmission. Note that the indices do not correspond to the number of the station but to the ***order*** of the scheduled transmissions. We use the symbol T_1^* and T_2^* to denote the next scheduled transmission of the station that was first and second, respectively. Depending on the relative distance of T_1 and T_2 we can distinguish the following three cases:

coll	$T_2 \le T_1+\Delta$	$T_1^n := T_1+\Delta+D$ $T_2^n := T_2+\Delta+D$
busy	$T_1+\Delta < T_2 \le T_1+1$	$T_1^n := T_1+1+V$ $T_2^n := T_2+\Delta+D^\dagger$
succ	$T_2 > T_1+1$	$T_1^n := T_1+1+V$ $T_2^n := T_2$

In the last two cases (*busy* and *succ*) the first station has a successful transmission, and we collect one point. The utilization of the channel, η, is the number of points collected per unit of time.

For given Δ and mean idle time $E\{V\}$, η depends on the probability distribution of the retransmission delay. We analyzed this dependence with a small simulation program which, each cycle, updates the values of T_1 and T_2 according to

$$T_1 := \min(T_1^n, T_2^n); \qquad T_2 := \max(T_1^n, T_2^n)$$

The symbol ":=" is to be read as "gets the value". The values of T_1^n and T_2^n for the next cycle are again determined depending on the case *coll*, *busy* or *succ* from the list above.

Figure 1 summarizes the the difference in utilization of the channel for uniform and exponential distribution of retransmission delay as function of mean retransmission delay. For those graphs we have taken Δ=0.2 and $E\{V\}$=1. The maximum attainable utilization of the channel is larger with exponentially distributed retransmission delays than with uniformly distributed delays.

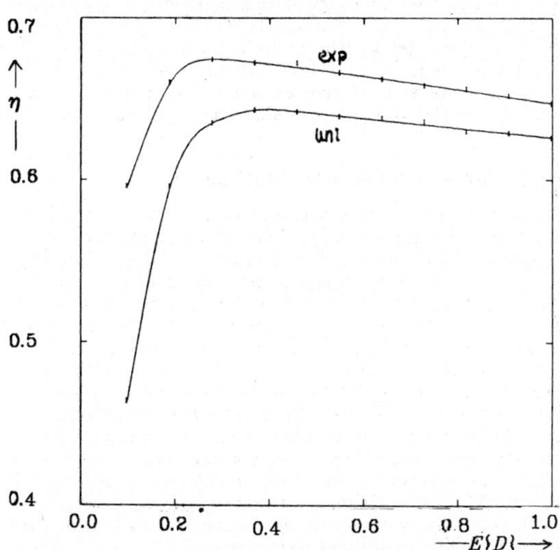

Fig. 1 Channel utilization as function of retransmission delay.

3. GENERATING EXPONENTIAL RANDOM VARIABLES

Suppose one has available a random number generator that generates U, uniformly distributed in the interval (0,1). Then, an easy way to generate an exponential random variable, X, is to compute $X = -ln(U)$ [3]. However when implementing a multi-access algorithm on a micro-processor, the small advantage of the exponential distribution as outlined in the previous section may not be worth the expense of adding code to compute the natural logarithm. Therefore we want to show a way to generate exponential variables that is well suited for efficient implementation on a microprocessor.

† Note that we can not have $T_2^n < T_1+1$, therefore we actually have in the simulation that $\Delta+D$ is added as often as needed (most of the time: once) to get $T_2^n \ge T_1$

Consider an exponential random variable X with mean 1 (to change the mean, just multiply X with the desired mean). In binary fixed point notation X can be written as

$$X = \cdots b_2 b_1 b_0 \,.\, b_{-1} b_{-2} b_{-3} \cdots \tag{1}$$

where bit b_i indicates whether the term 2^i is in X. Define $p_i = \Pr\{b_i=1\}$. In appendix A we give an expression for p_i and we show that b_i and b_j $(i \ne j)$ are independent random variables. Before we compile a table of bit probabilities a choice has to be made as to which accuracy these probabilities will be specified. For our application 16 bit accuracy is more than sufficient. The probability that bit 4, or a more significant bit, is 1 is negligable ($<2^{-23}$). Therefore we start the table with p_3 and specify this probability with 16 bit accuracy. To match this accuracy for less significant bits we specify p_j with $13+j$ bit accuracy ($j=-12, \cdots, 4$)

i	p_i	p_i binary
3	0.00034	.000000000001011
2	0.01799	.000001001001101
1	0.11920	.00011110100001
0	0.26894	.0100010011011
-1	0.37754	.011000001010
-2	0.43782	.01110000001
-3	0.46879	.01111
-4	0.48438	.011111
-5	0.49219	.0111111
-6	0.49609	.1
-7	0.49805	.1
-8	0.49902	.1
-9	0.49951	.1
-10	0.49976	.1
-11	0.49988	.1
-12	0.49994	.1

Table 1. Probability p_i that bit i of an exponential random variable (with mean 1) is 1. The right hand column gives p_i in binary fixed point notatation with $13+i$ bit accuracy. Trailing zeros are omitted.

Suppose that we have a source of random bits, e.g. from a Tausworthe generator [4], in the form of a coded function with the name RanBit. Upon each call, the function RanBit returns a "1" with probability ½ or else a "0". As an example of how a bit of X is generated, we give below the flow diagram of the code to generate b_{-3}. Each line corresponds to one digit of p_{-3} binary.

line	decision	assignment
0		b_{-3}:=0 ⟵
1	RanBit=1? — yes ⟶ (no ↓)	
2	RanBit=1? — yes ⟶ (no ↓)	b_{-3}:=1
3	RanBit=1? — yes ⟶ (no ↓)	b_{-3}:=1
4	RanBit=1? — yes ⟶ (no ↓)	b_{-3}:=1
5	RanBit=1? — yes ⟶ (no ↓)	b_{-3}:=1

To understand how this works note that b_{-3} will only get the value "1" if the assignment of line 2 or 3 or 4 or is executed. The probability that this happens is $\frac{1}{4}+\frac{1}{8}+\frac{1}{16}+\frac{1}{32}$ = 0.46875, which is the 13−3 bit approximation of p_{-4}. For b_3 downto b_{-5} we have similar flow diagrams, and for the low order bits we only need to copy random bits, i.e. b_i = RanBit ($i=-6, \cdots, -12$). These flow diagrams are most suitable for implementation in machine code. What makes the code efficient is its linearity (no loops) and the fact the 105 bits of the column 'p_i binary' are directly expressed in the code. Although it seems that for the high order bits there are many calls to RanBit, it can be seen from the example flow diagram that as soon as RanBit returns a "1", no more calls to RanBit are made for that bit of X. So the average number of calls to RanBit is less than 2 for each of the 16 bits of X.

4. OPTIMAL MEAN DELAY FOR EACH STATE

In some systems, like Ethernet, rescheduling of a transmission puts the station into a next state. Here we shall address the question of the optimal mean delay for each state. The model is more elaborate than the model of section 2, because we want to consider the case of $n>2$ stations and moreover each station can be in one of m states. So, even when ignoring the fact that the holding time of each state is not exactly exponential, we still have that the size of the state space is of the order m^n. Given the complex constraint that only one station at a time can be successfully transmitting, there is not much hope for an exact *analytical* model. A *simulation* model that allows for different mean retransmission delays for each state, makes optimization by simulation, as we did in section 2, not very feasible. The approach that we want to take in this section is to derive an analytical model that is simplified in the sense that we ignore most necessary conditions for each step in the derivation. We verify the analytical model by comparing it with the outcome of a more realistic simulation model. The analytical model can then be used for optimization of mean retransmission delays.

For future reference we list below the most important variables used in the analytical model.

n	number of stations
m	number of states
v	mean idle time after transmission
Δ	min. separation of start of transm. for *succ*
1	transmission time of packet
d_i	mean retransmission delay state i
h_i	$=d_i+\Delta$ $(i>0)$ mean time spent in state i
h_0	$=v+1$ mean time spent in state 0
f_i	probability station is in state i
p_i	*succ* probability from state i

In the state diagram of figure 2, each state corresponds to a possible state of a station. State 0 comprises the (successful) transmission time and the exponentially distributed idle time, V, of a station.

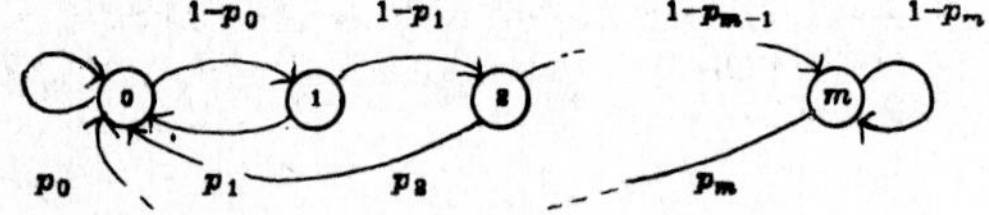

Fig. 2 State transition diagram for a station.

Let h_i be the mean time that a station spends in state i and f_i be the probability that a station is in state i at an arbitrary point in time, then we get from equating the probabilities of an upward transition and a downward transition across an imaginary boundary between state i and $i+1$:

$$\frac{(1-p_i)f_i}{h_i}=\sum_{j=i+1}^{m}\frac{p_j f_j}{h_j} \tag{2}$$

If we assume the success probabilities, p_j, are given, then f_i $i=0,\cdots m$ can be determined from the normalizing condition

$$\sum_{i=0}^{m} f_i = 1 \tag{3}$$

The next step is to derive the success probabilities p_i assuming that f_i is given. At an arbitrary point in time a planned transmission will not be rescheduled if the channel is idle and if there is no interference from an other transmission:

$$\Pr\{succ\} = \Pr\{\text{no interference}\}\,\Pr\{\text{channel idle}\} \tag{4}$$

The rate of transmission of a station in state i (i.e. the probability per unit of time that a start of transmission is planned) is $1/h_i$. Therefore the weighted rate is:

$$r=\sum_{i=0}^{m}\frac{f_i}{h_i} \tag{5}$$

A station will have no interference if the $n-1$ other stations have not planned their start of transmission within Δ of its start of transmission, hence:

$$\Pr\{\text{no interference}\} = e^{-2\Delta(n-1)r} \tag{6}$$

A station spends on average a fraction $1/(v+1)$ of state 0 in transmitting. Thus the probability that no *other* station is transmitting at an arbitrary point in time is:

$$\Pr\{\text{channel idle}\} = 1-\frac{n-1}{v+1}f_0 \tag{7}$$

An unsuccessful attempt that is repeated again after a very short time is likely to be unsuccessful again. We found that the "steady state" value $\Pr\{succ\}$ was approached exponentially with time constant $\Delta/2$. This gives us:

$$p_i = \Pr\{succ\}(1-e^{2h_i/\Delta}) \tag{8}$$

The iterative procedure uses equations (2)-(3) to compute f_i, then, with f_i given, it uses (4)-(8) to compute p_i. In the next iteration we start again at (2) with the new values of p_i. For those values of d_i that correspond to a stable system, we experienced rapid convergence of the iterative procedure to a solution to the set of equations (2)-(8).

The performance that we want to maximize is the number of successful transmissions per unit of time. Each successful transmission corresponds to one visit to state 0 followed by zero or more visits to higher states; we call this a cycle. Hence maximizing successful transmissions corresponds to maximizing the number of cycles per unit of time, which corresponds to minimizing the mean cycle time, t_{cycl}. Since each cycle has one visit to state 0, it is not hard to see that $f_0 t_{cycl} = h_0$, or,

$$t_{cycl} = \frac{h_0}{f_0} \tag{9}$$

Two cases of retransmission delays in state i are of special interest to us: (i): $d_{i+1}=2d_i$, corresponding to 'binary backoff' which is done in the widely implemented Ethernet and (ii): $d_i=constant$ which turned out to minimize t_{cycl}. For a system with $n=5$ stations and $m=8$ states and a mean idle time $v=8$ packet lengths we simulated and computed the quantities of (2)-(9). The input for d_i was (i): $d_1=0.1,\cdots,d_8=12.8$ (ii): $d_i=0.5$ $(i=1,\cdots,8)$ In figure 3 we show the computed values and confidence intervals of the simulated value of t_{cycl}, for 'binary backoff' and constant retransmission delay.

The approximate model is not exact in predicting the outcome of the simulation. However, varying the values for the retransmission delays, we found that the changes in t_{cycl} and other quantities of the analytical model paralleled the changes observed in the simulation model. Therefore the analytical model is very useful when searching for optimal retransmission delays. The search was done by numerical optimization. The outcome: to have the same retransmission delay for all states, was not what we had initially expected, but not so suprising afterwards, if one considers the equal roles that state 1 through 8 play in the model.

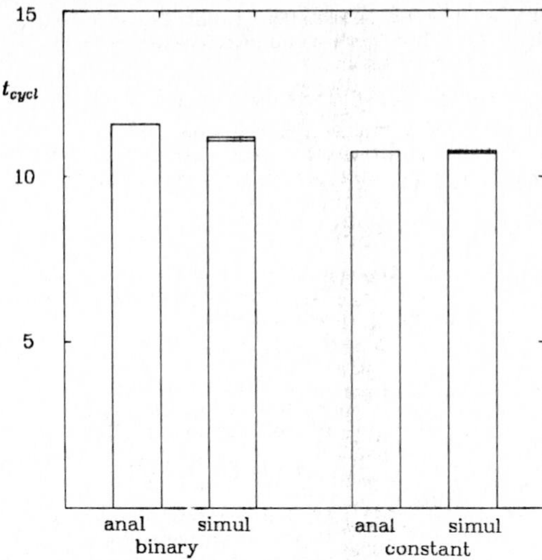

Fig. 3 Mean time between successful transmissions of the same station, t_{cycl}, for 'binary backoff' and constant retransmission delays.

5. CONCLUSIONS.

Investigation of questions on retransmission delays that were posed for a small unslotted carrier sense multi-access communication system led to the following results.

a) A higher throughput can be obtained with retransmission delays from the exponential distribution than with uniformly distributed delays.

b) Constructing an exponential random variable, from a source of random bits, can be done very well on a micro-processor without logarithm capabilities. From the point of view of programming such a routine in machine language, the code is straightforward and efficient.

c) From the point of view of maximizing throughput in a system with a given (small) number of stations and given equal mean idle time (think time) for each user, constant retransmission delays is better than binary backoff. However, we did not address aspects of stability of the system. Instability would show up in the analytical model as oscillations in the iterative procedure of section 4.1. Further research is still needed to know which set of retransmission delays makes the system robust.

REFERENCES

[1] N. Abramson, " The ALOHA system - another alternative for computer communications", *1970 Fall Joint Comput. Conf., AFIPS Conf. Proc.*, vol 37, 1970, pp. 281-285.

[2] L. Kleinrock and S. Lam, "Packet switching in a Multiaccess Broadcast Channel: Performance Evaluation", *IEEE Trans. on Comm.*, vol. COM-23, April 1975, pp. 410-423.

[3] G.S. Fishman, *Principles of Discrete Event Simulation*, Wiley, 1978

[4] P. Bratley, B. Fox and L. Schrage, *A Guide to Simulation*, Springer, 1983

APPENDIX A

Let X be an exponential random variable with mean 1. In fixed point binary notation it is written as

$$X = \cdots b_2 b_1 b_0 \,.\, b_{-1} b_{-2} b_{-3} \cdots$$

Bit i has the value "1" if X is in one of the intervals $[k\,2^{i+1}+2^i, (k+1)2^{i+1})$ $k=0,1,2,\cdots$, hence by integrating the exponential density over those intervals we get

$$\Pr\{b_i=1\} = p_i = \sum_{k=0}^{\infty} \int_{k\,2^{i+1}+2^i}^{(k+1)2^{i+1}} e^{-x}\,dx$$

Using the notation $e_i=e^{-2^i}$ (note that $e_{i+1}=e_i^2$) this can be worked out as

$$p_i = \sum_{k=0}^{\infty} (e_{i+1}^k e_i - e_{i+1}^{k+1}) = \frac{e_i - e_{i+1}}{1-e_{i+1}} = \frac{e_i}{1+e_i}$$

The next thing we want to show is that b_i is statistically independent of $b_{i+1}, b_{i+2}, \cdots$ When $b_{i+1}, b_{i+2}, \cdots$ are given this means that we know in which interval of size 2^{i+1} the random variable X is. Let this be the interval $[k\,2^{i+1}+2^i, (k+1)2^{i+1})$. Now the conditional probability

$$\Pr\{b_i=1 \mid b_{i+1}, b_{i+2}, \cdots\} = \frac{\int_{k\,2^{i+1}+2^i}^{(k+1)2^{i+1}} e^{-x}\,dx}{\int_{k\,2^{i+1}}^{(k+1)2^{i+1}} e^{-x}\,dx}$$

$$= \frac{e_{i+1}^k e_i - e_{i+1}^{k+1}}{e_{i+1}^k - e_{i+1}^{k+1}} = \frac{e_i}{1+e_i} = \Pr\{b_i=1\}$$

turns out to be equal to the unconditional probability from which we may conclude independence of the binary random variables b_i and b_j for $i \neq j$.

TELETRAFFIC ISSUES in an Advanced Information Society
ITC-11
Minoru Akiyama (Editor)
Elsevier Science Publishers B.V. (North-Holland)

DYNAMIC CONGESTION CONTROL IN INTERCONNECTED LOCAL AREA NETWORKS

Takeshi NISHIDA* Masayuki MURATA** Hideo MIYAHARA* Kensuke TAKASHIMA*

* Osaka University, Department of Information Sciences
** IBM Japan,Ltd., Science Institute

ABSTRACT

We propose the congestion control for interconnected local area networks (LANs), named ESP (Emigrant Suppressing with P-persistent CSMA/CD) scheme. It is shown through the simulation that this scheme can relieve the performance degradation of the congested LAN by controlling the internet-traffic which enter into it, and that it can prevent the congestion of a LAN by limiting the internet-traffic which may cause the congestion of that LAN. Moreover, the method for realizing the dynamic control with this scheme is shown, and it is verified through the numerical examples that this works well.

1. INTRODUCTION

Various types of interconnected LAN can be considered: tandem (Fig.1(a)), hierarchical (Fig.1(b)). Each network is connected through a gateway having various functions[1][2], e.g., protocol conversion, address filtering, routing, etc.

In a single LAN, the channel access schemes themselves have the functions to avoid the congestion by staggering the instance of transmission from each terminal (e.g., the backoff algorithm in CSMA/CD and the token handling method in token passing). Therefore internal network congestion can be relieved to a certain extent. In an interconnected system, however, if a LAN is overloaded, it becomes to be more congested due to traffic from other LANs. Further, since the buffer of the gateway connected with that LAN becomes full, the terminals in the other LANs cannot send the messages to the desired terminals in the congested LAN. Thus, they must retransmit these messages repeatedly, which causes increase of load in those LANs. This congestion propagates to other LANs (so-called "backpressure effect"), resulting in the system deadlock. Therefore, a control mechanism is required in order to prevent and to relieve the congestion.

Bux and Grillo[3] study the flow control in interconnected LANs with token ring access method. They use the window scheme to accomplish flow control. We deal with the congestion control in the interconnection of bus type LANs, e.g., Ethernet[4]. In this paper, we propose the ESP (Emigrant Suppressing with P-persistent CSMA/CD) scheme to control the internet-traffic flow for the purpose of avoiding congestion of a LAN, or to prevent the congestion of the LAN into which a great deal of internet-traffic enter. This scheme uses the property of p-persistent CSMA/CD[5] in which the channel throughput is controlled by changing the value of p.

2. ESP SCHEME

2.1 Interconnected LAN

For simplicity, all terminals are classified into two groups throughout the paper; the one is that the traffic from it is transmitted to the terminals connected with the same LAN, and the other is that it is transmitted to the other LAN. They are called the inner and outer terminal, respectively. The traffic generated from each terminal is called inner and outer traffic, respectively. The model of interconnected LANs used in this paper is shown in Fig.2. The detail description of the model is as follows.

(1) Both inner and outer terminals are connected with LAN_1, and only the inner terminals are connected with LAN_2. That is, traffic flow through the gateway is unidirectional (LAN_1 --> LAN_2).

(2) Each terminal and the gateway transmit their packets according to p-persistent CSMA/CD. The value of p is fixed at 1 at all inner terminals and the gateway.

(3) The number of inner terminals in LAN_1 and LAN_2 is denoted by M_{in1} and M_{in2} respectively, and the number of outer terminals is denoted by M_{out1}. Inner terminals in LAN_1 and LAN_2, and outer terminals in LAN_1 generate the packet according to a Poisson process with a mean of q_{in1}, q_{in2} and q_{out1}, respectively. If we define the offered load as the ratio between the sum of traffic generated by all terminals and the channel capacity, the offered load of inner of LAN_1 and LAN_2 and outer traffic becomes

$$\lambda_{in1}=M_{in1}q_{in1}T_{in1}, \tag{2.1}$$

$$\lambda_{in2}=M_{in2}q_{in2}T_{in2}, \tag{2.2}$$

$$\lambda_{out1}=M_{out1}q_{out1}T_{out1}, \tag{2.3}$$

where T_{in1}, T_{in2} and T_{out1} are respective one packet transmission time.

(4) All terminals possess one packet buffer. The number of packets which can be stored in the gateway is fixed to a certain amount.

(5) The rescheduling delay of a collided packet is exponentially distributed with a mean of r (i.e., exponential backoff).

(6) The gateway does not accept any more packets if the number of packets in its exceeds the capacity of buffer, the gateway transmits the

jam signal and forces to destroy the incomming packet.

(7) If the gateway succeedes the one packet transmission, it continues its transmission until the gateway buffer becomes empty (exhaustive service).

In this section, we show that the performance of each LAN degrades if there are no controls. The system parameters are shown in Table.1. The value of p in the outer terminals is fixed at 1. Fig.3 shows the throughput of inner traffic of LAN_2 as a function of λ_{in2}, where λ_{out1} is taken as a parameter. The throuhput of inner traffic of LAN_2 is less affected by the outer traffic of LAN_1 within the usage where λ_{in2} is less than 0.4. The throughput, however, becomes smaller compared with the one of λ_{out1}=0, if it becomes larger than 0.4. The throughput becomes to be less than 0.2 when λ_{in2}=1.0 even if there is no incomming traffic. In this case, although the congestion control may be considered to be useless, through applying some congestion control scheme, it becomes possible to make the time of congestion shorter than that without control.

Next we show how the performance of LAN_2 degrades as the outer traffic increases. The load of inner traffic in LAN_2 is fixed at 0.7. Figs.4 and 5 show the throughput and average packet delay vs. λ_{out1}. The average packet delay is defined as the time duration from the generation of packet until its arrival at the destination terminal. Thus, the packet delay of outer traffic is sum of the time duration from the generation until when the packet is transmitted successfully (transmission success of outer traffic means that it does not collide with the other packets and does not meet the buffer overflow of the gateway) and the time duration from its arrival at the gateway until the gateway succeeds the transmission of packet. The increase of λ_{out1} results in the drastic degradation of the performance of inner traffic in LAN_2, and the throughput of outer traffic in LAN_1 does not increase so much. The throuhgput of inner traffic in LAN_2 increases when λ_{out1} becomes more than 0.5. This is because the entering flow of outer traffic in LAN_2 becomes small for the frequent collisions in LAN_1. Moreover, the throuhgput of inner traffic in LAN_1 becomes small. This is not only because λ_{out1} itself increases, but also because the load by the retransmitted traffic increases for the buffer overflow in the gateway. Thus, if no congestion control scheme is applied, LAN_2 which has been operated without congestion would be in congestion state because of increase of load caused by the frequent retransmission of outer traffic. This "back-pressure effect" of congestion may result in the system deadlock. Thus, any control mechanism is required in order to prevent and to eliminate the network congestion.

2.2 ESP scheme

The ESP scheme is based on the property of p-persistent CSMA/CD protocol in which the throughput can be controlled by changing the value of p. That is, the outer terminals are able to control their traffic to be transmitted to another LAN by choosing the appropriate value of p according to the degree of congestion. The ESP scheme can prevent or eliminate the network congestion, because the inner traffic in the congested LAN can be transmitted with higher priority over the entering outer traffic, or because it prevents a great deal of outer traffic which will cause the congestion from entering a LAN.

3. PERFORMANCE EVALUATION OF ESP SCHEME

We study the availability of the ESP scheme through the numerical results obtained by simulation. The system parameters are the same as those in Table.1. Figs.6 and 7 show the load-throughput and the load-delay curves for different values of p when λ_{out1}=0.3 and λ_{in1}=0.3. In Fig.6, the curves of λ_{out1}=0 correspondes to the case that there is no flow entering into LAN_2 from other LAN, and the curves of p=1 correspondes to the case of no control. The characteristics of throughput of the inner traffic in LAN_2 are improved by making the value of p smaller, while the parameter of outer traffic of LAN_1 becomes worse. There is the trade-off relation between the throughput of inner traffic in LAN_2 and that of outer one in LAN_1. Thus, the performance criteria which resolves this problem of trade-off must be considered. We will discuss the performance criteria in the later chapter. From these figures, it can be seen that the ESP scheme can improve the performance of inner traffic in LAN_2, compared with the no control (p=1) (e.g., if we set p=0.001, the throughput of inner traffic is assured to be 85% of the one in the case where λ_{out1}=0, at λ_{in2}=1.0).

Figs.8 and 9 show the load-throughput and the load-delay curves when λ_{in2}=0.9 and λ_{in1}=0.3, where the load of outer traffic of LAN_1 is changed. It can be also seen that the control of λ_{out1} by changing the value of p avoids the throughput degradation of inner traffic in LAN_2 and improves the performance of inner traffic in LAN_1.

4. DYNAMIC CONGESTION CONTROL

4.1 Dynamic control

In this control, the gateway is assumed to have the function of monitoring traffic condition of LAN_2. The gateway determines the value of p in accordance with that condition, and informs the value of p to all terminals in other LANs. If an improper value of p is chosen, the outer traffic may not be transmitted at all. Hence, we define the performance criteria that maximizes the throughput of the outer traffic with keeping the one of inner traffic to some desired value. The value of p is chosen so as to satisfy the performance criteria. The dynamic congestion control with ESP scheme is summarized as follows (we use the network model of Fig.2 for explanation).

Every T_p (msec.), the gateway

(1) estimates the load of outer traffic ($\lambda_{out1}(t)$) and inner traffic of LAN_2 ($\lambda_{in2}(t)$),

(2) determines the value of p in accordance with the outer and inner traffic estimated in (1) to satisfy the performance criteria, and

(3) informs that value to the outer terminals in LAN_1, if $|p(t)-p(t-T_p)|>\alpha$, where α is the arbitrary positive constant.

The method of how to estimate the load is discussed in Sec.4.3.

4.2 The performance criteria

Here we consider the general network model shown in Fig.10. Two kinds of traffic we assumed to be transmitted in LAN_A; the one is that generated from the terminals in LAN_A (intranet-traffic) and the other is that enters into LAN_A through the gateway (internet-traffic). The load of each traffic in LAN_A is λ_{in} and λ'_{out}, respectively. The load of internet-traffic generated by the all terminals outside LAN_A is λ_{out} ($\lambda'_{out} \le \lambda_{out}$). Denote the throughput of intranet-traffic by $S_{in}(\lambda_{in},\lambda'_{out})$. We define the following performance criteria:

> maximizes λ_{out}
> subject to
> $$S(\lambda_{in},\lambda'_{out}) \ge b(\lambda_{in},\lambda_{out})S(\lambda_{in},0), \qquad (4.1.1)$$
> where b is the function of both λ_{in} and λ_{out}, and $0 \le b \le 1$.

If b is constant, the intranet-traffic in LAN_A is assured to get (100·b)% throughput of the one which is achieved in case $\lambda'_{out}=0$, regardless with λ_{out}. If we set

$$b=\lambda_{in}/(\lambda_{in}+\gamma\lambda_{out}), \qquad (4.1.2)$$

where γ is an arbitrary positive constant. λ_{out} can be taken into consideration in this performance criteria. That is, λ_{out} becomes larger, we can suppress the achievable throughput of intranet-traffic in accordance with the magnitude of λ_{out}.

4.3 The estimation of load

We return the network model of Fig.2. As stated in Sec.4.1, the gateway must estimate the load of inner traffic (λ_{in2}) and outer traffic (λ_{out1}). At first we discuss the method of estimation of $\lambda_{in2}(t)$. The following two measurable data are used for this purpose.

i) The load of gateway traffic at time t: $\lambda'_{out1}(t)$

ii) The average waiting time of packet in the gateway buffer: W_p

The waiting time of packet is defined as the time lapse from its arrival until its transmission success. The gateway gathers these data during a given time interval (we refer it a sampling time and denote by T_s) and takes the average value of them.

If λ'_{out1} is fixed and the buffer size is not so small, it can be expected that there exists a one-to-one correspondence between λ_{in2} and W_p. The relation between λ_{in2} and W_p in the steady state is shown in Fig.11. The system parameters used in this chapter are the same as those in Sec.2.1. From this figure, it can be prospected that λ_{in2} is estimated by W_p. Since W_p is also dependent on λ'_{out}, λ_{in2} is estimated by the combination of λ'_{out} and W_p. In order to make a real-time estimation in the dynamic control, it is necessary to have the table in which the relation among λ'_{out}, W_p and λ_{in2} is specified in advance in the gateway. (The table can be obtained by simulation or analysis.) In case of the small sized buffer, W_p becomes to be saturated, when λ_{in2} is higher. In this case, the buffer overflow rate must be also used in the estimation of λ_{in2}.

The following four information are necessary for the gateway in order to estimate λ_{out1}, the gateway must obtain the following four information: (1)$\lambda_{in1}(t)$, (2)$\lambda_{in2}(t)$, (3)the current value of p, and (4)λ'_{out}. The gateway can monitor (3) and (4), and estimate (2) by using the method mentioned above. As for (1), it can be estimated easily because it is almost equal to the throughput if 1) is small. Since the throughput is the rate of successful transmission of packet, it can be measured by counting the number of packets in which both source and destination address are LAN_1.

We call the time interval that the gateway refreshes the value of p as an adjustment time (T_p), ($T_s \ge T_p$). That is, every T_p (msec.), the gateway calculates the average value of W_p and X'_{out} from the data collected during the time interval T_s (msec.) ahead. Fig.12 illustrates the relation between T_s and T_p. We call this (T_s,T_p)-control in ESP.

4.4 NUMERICAL EXAMPLES AND DISCUSSION

In the reminder of this section, we set b=0.8 in the performance criteria. All results are obtained by simulation, and the system parameters are same as those in Sec.2.1. Fig.13 shows the load-throughput curves, where the value of p is taken as a parameter. $S_{in}(\lambda_{in2},0)$ and $0.8 \cdot S_{in}(\lambda_{in2},0)$ are also shown in this figure. Table.2 shows the relation between λ_{in2} and p satisfying the performance criteria. Though the congestion control is not needed at λ_{in2} which is less than 0.55, the value of p must be decreased as λ_{in2} increases, e.g., p=0.005 in λ_{in2}=0.65, p=0.0015 in λ_{in2}=0.7. The gateway estimates the load, determines the value of p by reference to this table, and informs this p to the outer terminals in LAN_1.

We evaluate the performance of dynamic control with ESP scheme. Executing the simulation, λ_{in1} and λ_{out1} are assumed to be known, i.e., only λ_{in2} is estimated, and the delay incurred by the transmission of p is ignored.

In the simulation model hereafter, λ_{in2} is increased from 0.3 to 0.7 at time 2000 (mesc.) and is returned to an original value at time 4000 (msec.). Fig.14 shows (a) the value of p, (b) the throughput of inner traffic of LAN_2, and (c) the throughput of outer traffic of LAN_1 in (500,100)-control in ESP. The curves of no control are also shown. Though the throughput of inner traffic of LAN_2 degrades temporarily in ESP scheme when the LAN_2 is congested, it increases by decreasing the value of p and reaches the value for satisfying the performance criteria ($0.8 \cdot S_{in}(\lambda_{in2},\lambda'_{out})$=0.51). Without control, the performance of inner traffic of LAN_2 degrades extremely.

We study the effect of sampling time to the results of control (the adjustment time is fixed at 100 msec.). Fig.15 shows the comparison between T_s=100(msec.) and 500(msec.). The curve of T_s=100 shows more real time response to the change of λ_{in2} than that of T_s=500, but it varies widely. We show the average and variance of estimated λ_{in2} in Table.3 in order to examine the relation between the sampling time and the estimated λ_{in2}. In this calculation, we set λ_{in2}=0.7, T_p=500(msec.) and T_s=100, 300, 500(msec.). As the sampling time is made larger, the variance of the estimated λ_{in2} becomes smaller. Though the control with the smaller sampling time reacts rapidly to the traffic change, it may result in the incorrect estimation. On the other hand, though the reaction of the control with the longer sampling time is slow, it estimates correctly. As a conclusion, if the load changes drastically, the shorter sampling time is better for the real time control, on the other hand, if the variation of the load is not so widely, the longer sampling time is better for the accurate control.

5. CONCLUSION

We proposed the dynamic congestion control, named ESP scheme, in the interconnected LANs. We applied this scheme to the LANs which use p-persistent CSMA/CD as the channel access protocol. This scheme controls the internet-traffic entering the congested LAN by adjusting the value of p in p-persistent CSMA/CD in accordance with the degree of congestion. Since most of the functions of the scheme are implemented on the gateway, the terminal only changes the value of p according to the command from the gateway. Through the various numerical examples, it was shown that the ESP can relieve and avoid the congestion, and prevent the "backpressure" effect of congestion.

The research of interconnection of LAN with long-haul networks has been progressed. In general, the speed of the long-haul network is much slower than that of LAN (the latter is several thousands faster than the former). This mismatching of speed may result in the buffer overflow in the gateway connecting the LAN with the long-haul network, even if the long-haul network is not so congested. The ESP scheme can also applied to this case by the following procedure: The gateway monitors both the traffic condition and the state of buffer, chooses the appropriate value of p, and informs this to the terminals in the LAN. In this case, however, the performance criteria defined in the paper may not be applied, because the buffer overflow occurs in the gateway even if the long-haul network is not so congested. Thus, a new performance criteria must be considered.

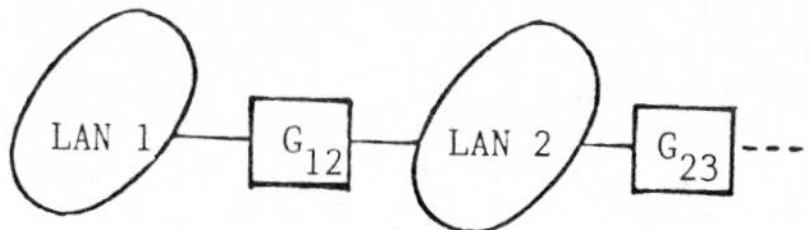

(a) Tandem connection model

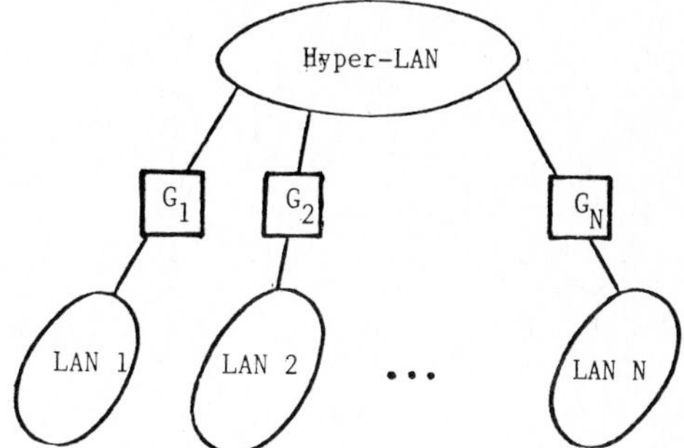

(b) Hierachical connection model

Fig.1 Various types of interconnected LANs

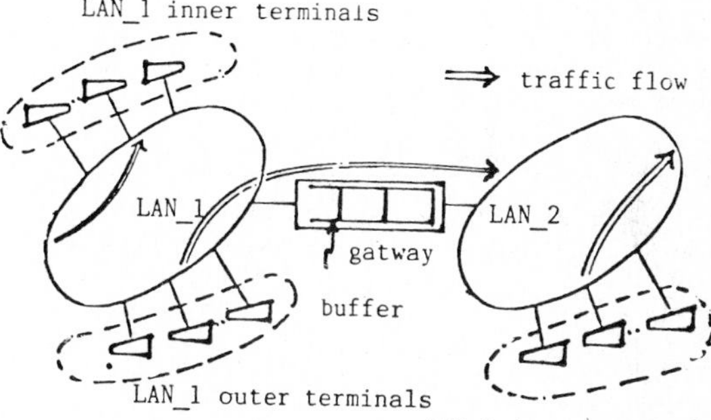

Fig.2 Interconnected LAN model

REFERENCES

[1] V.G. Cerf & P. Kirstein, "Issues in packet-network interconnection," Proc. IEEE, vol.66, pp.1386-1408, Nov. 1978.

[2] D.R. Boggs, J.F. Shoh, E.A. Taft & R. Metcalfe, "Pup:An internetwork architecture," IEEE Trans. Commun., vol.COM- 28, no.4, 1980.

[3] W. Bux & D. Grillo, "End-to-end performance in local-area networks of interconnected rings," INFOCOM'84, pp.60-68, 1984.

[4] R. Metcalfe & D. Boggs, "Ethernet:Distributed packet switching for local computer networks," Comm. Ass. Comput. Mach., vol.21, Dec. 1978.

[5] F. A. Tobagi & V. B. Hunt, "Performance analysis of carrier sence multiple access with collision ditection," Comput. Networks, vol.4, pp.245-259, 1980.

parameter	LAN_1	LAN_2
max. propagation delay (usec.)	5	5
channel capacity (Mbps)	10	10
jam length	0	0
retrans. ratio	0.1	0.1
bus length (km)	1	1
number of term.	12(inner),12(outer)	24
buffer size (gateway)	30	

Table.1 System parameters in interconnected LANs

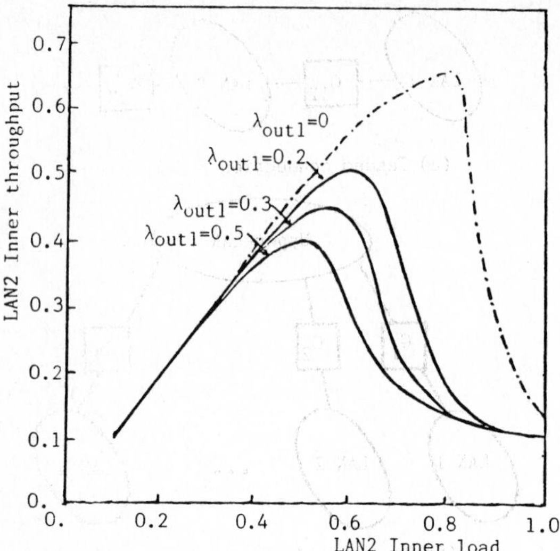

Fig.3 LAN_2 inner load vs. throughput (without control)

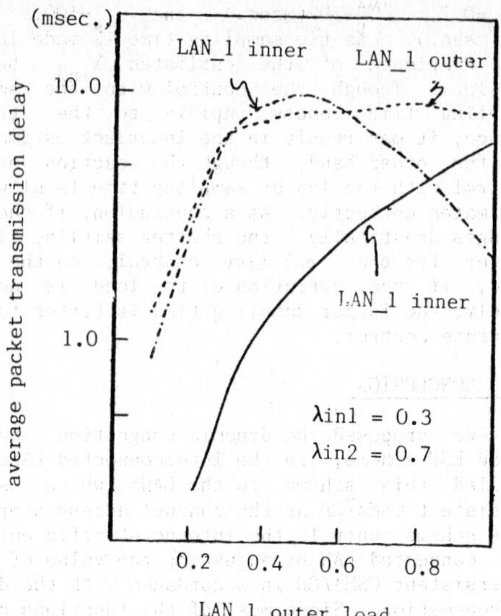

Fig.5 LAN_1 outer load vs. delay (without control)

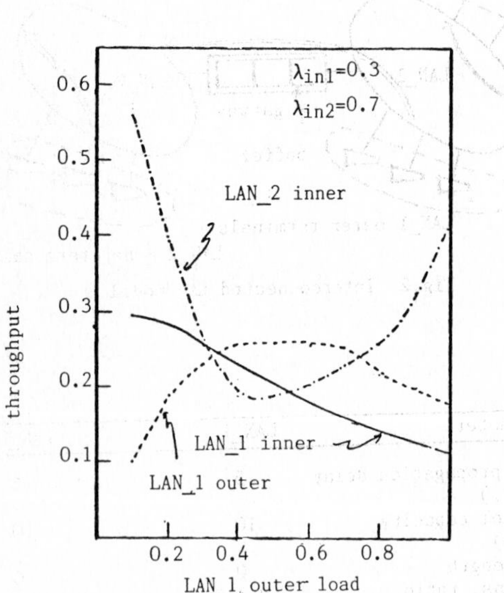

Fig.4 LAN_1 outer load vs. throughput (without control)

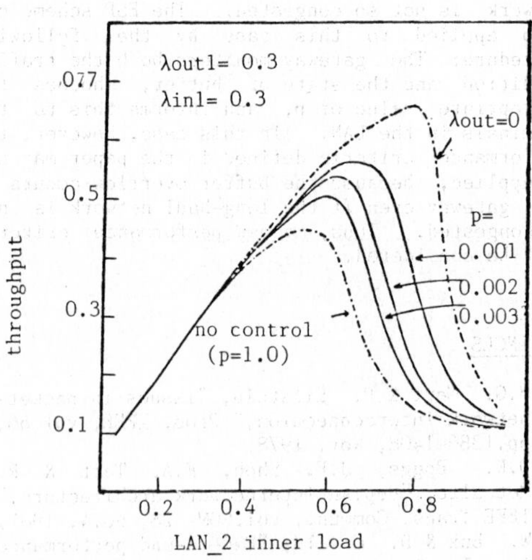

Fig.6 LAN_2 inner load vs. throughput (with ESP scheme)

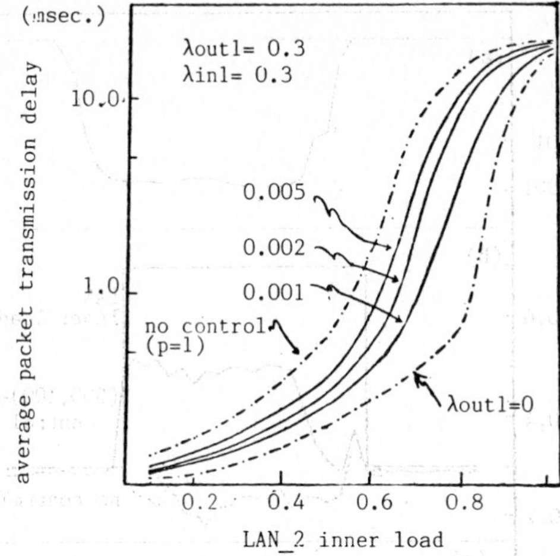

Fig.7 LAN_2 inner load vs. delay (with ESP scheme)

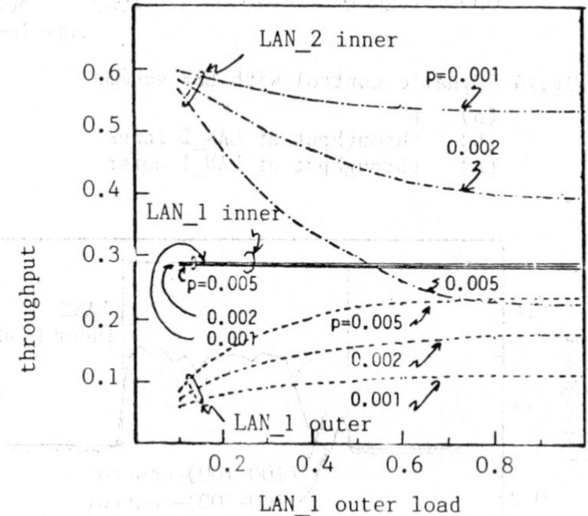

Fig.8 LAN_1 outer load vs. throughput (with ESP scheme)

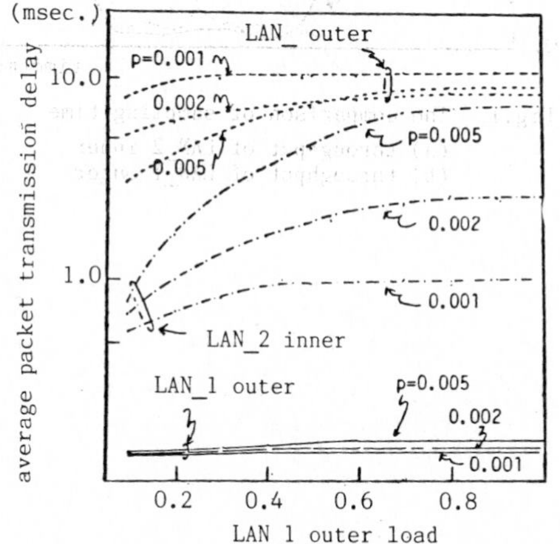

Fig.9 LAN_1 outer load vs. delay (with ESP scheme)

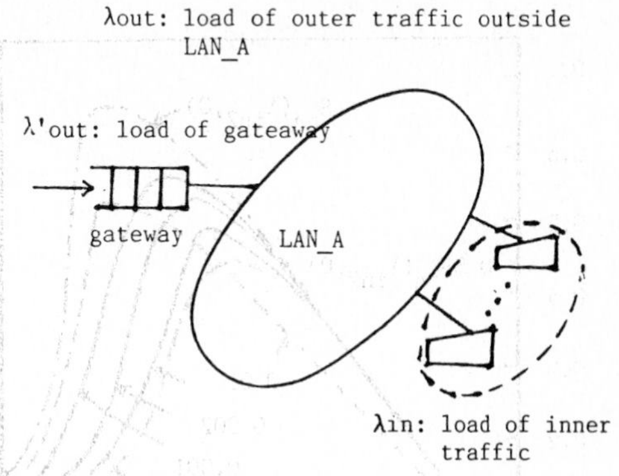

Fig.10 The network model for performance crteria

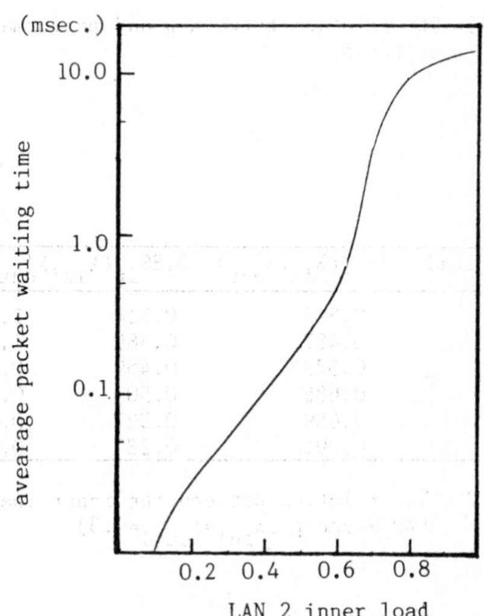

Fig.11 The average waiting time of gateway

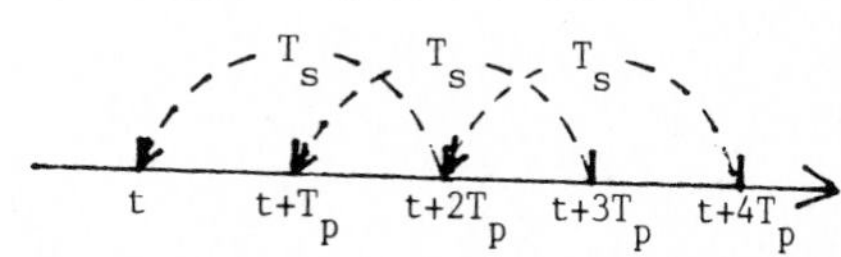

Fig.12 The ralation between T_s and T_p

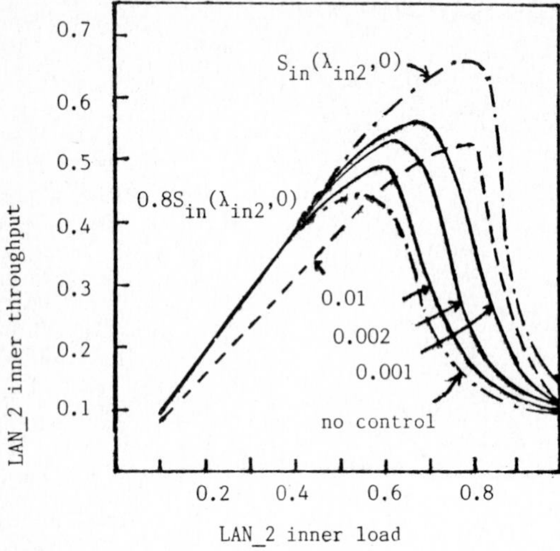

Fig.13 The relation between p and performance criteria

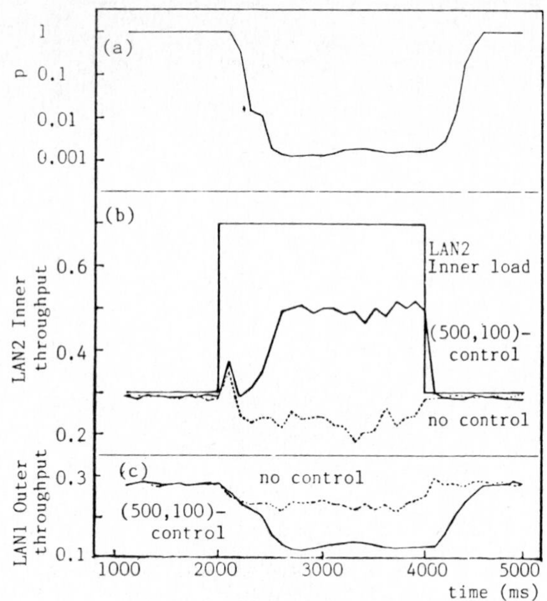

Fig.14 Dynamic control with ESP scheme

(a) p
(b) throughput of LAN_2 inner
(c) throughput of LAN_1 outer

LAN_2 load	$S_{in}(\lambda_{in},\lambda'_{out})$	$0.8S_{in}(\lambda_{in},\lambda'_{out})$	p
0.4	0.393	0.314	1.0
0.5	0.487	0.389	1.0
0.6	0.573	0.459	0.03
0.7	0.636	0.509	0.0015
0.8	0.659	0.527	0.00034
0.9	0.291	0.233	0.0003

Table.2 The relation between the inner load in LAN_2 and p ($\lambda_{in1}=\lambda_{out1}=0.3$)

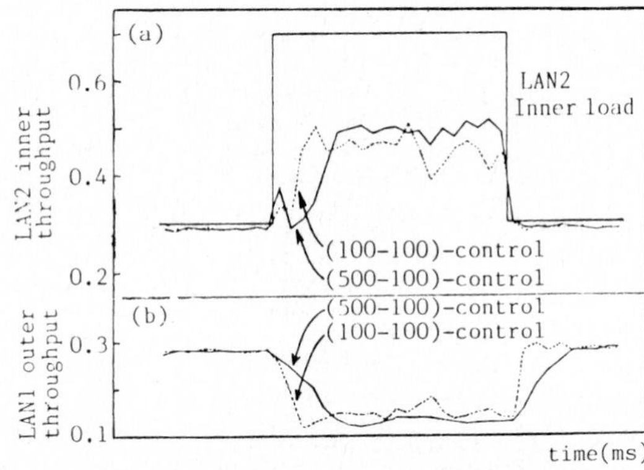

Fig.15 The comparison of sampling time

(a) throughput of LAN_2 inner
(b) throughput of LAN_1 outer

sampling time (msec.)	average	variance
100	0.696	0.00507
200	0.702	0.00323
300	0.705	0.00199

Table.3 The sampling time vs. average and variance of estimated load

TELETRAFFIC ISSUES in an Advanced Information Society
ITC-11
Minoru Akiyama (Editor)
Elsevier Science Publishers B.V. (North-Holland)

Performance Evaluation of Medium-Access Control Protocols in Local Area Networks

Hiromi Okada

Kobe University

Nada, Kobe, JAPAN 565

Abstract This paper presents a comparative performance evaluation of medium-access control (MAC) protocols in LANs. This paper deals with three representative MAC protocols; Token-ring, Token-bus, and CSMA/CD. First, this paper defines a unit-period as the time basis for analysis and obtains the probability distribution on the length of this unit-period. By considering the beginning of this unit-period as the imbedded Markov point, this paper makes it possible to analyze the network performance of the three MAC protocols, stochastically. Then, this paper compares the analytical results to the simulation results to verify the validity of the analysis. Finally, this paper shows the performance comparison among the three MAC protocols for various kinds of LAN configurations and discusses the suitable situations of them.

1. Introduction

There have been the increasing needs for local area networks (LANs) in the computer communications environment including office automation, factory automation and so on. A great number of LANs with various configuration schemes have been developed and also developing. These numerous LANs are generally classified by the medium access control (MAC) protocol. Three major schemes of them are Token-ring, Token-bus and CSMA/CD. These three schemes already have been treated for the standardization of LAN by several standard organizations.

When users construct LANs, they must select the eligible schemes for their own environment from at least the three standardized schemes. So, It is necessary to present the comparative evaluation of manifold features of LAN configurations from various points of view at the initial stage. Further, it is also necessary to present the detail performance analysis for the selected configuration scheme of LAN at the designing stage. Hence, in the further researches of LANs, one of the most important problems is to evaluate and compare the performance of these network schemes.

Up to now, a number of works for evaluating and comparing the network performance of LANs have been presented [1,2,3,4,5]. Stuck[1] evaluated the throughput characteristics with some simple network models. While the computation of this evaluation is simple enough, there are some problems on the accuracy of the analysis. Bux[2], Cherukuri et al [3] presented detail analyses on the performance evaluation of networks. However, these papers quoted some results on the network analysis from other papers and applied them to suitable MAC protocols. They adopt the results of Kohnheim and Meiser[6] to Token-ring and the results of Lam[7] to CSMA/CD. Thus, the differences in the analytical methods are left as problems.

The objective of this paper is to present a comparative performance evaluation of the representative MAC protocols in LANs, i.e. Token-ring, Token-bus, and CSMA/CD by an identical analysis. This paper defines a unit-period consisting of an access time and a packet transmission time as the time basis for analysis. This paper derives the probability distribution on the length of this unit-period for the three MAC protocols. Especially, for CSMA/CD, the back-off behavior depending on the number of packet collisions is taken into consideration. By considering the beginning of the unit-period as the imbedded Markov point and introducing some approximation, this paper makes it possible to obtain the throughput-delay performance of the three MAC protocols.

Further, this paper shows some quantitative characteristics of the three MAC protocols on typical network configurations by numerical analysis. First, this paper compares the analytical results to the simulation results to verify the validity of the approximated analysis. Secondly, this paper compares the three MAC protocols from various points of view including channel capacity, network distance, number of nodes and so on, and discusses the suitable situations of the three MAC protocols and their potential features. By extending the analytical method presented in this paper, we can obtain the performance of various kinds of LAN schemes including hybrid schemes of Token-passing and CSMA/CD, some schemes with transmission priorities, and so on.

2. Model

A model of LANs for analysis is described in this section. The model consists of four parts. The first part (1)-(4) is the network configuration and the input traffic property in LANs. This is common to all MAC protocols. The second part 5.1)- 5.6), the third part 6.1)- 6.6), and the fourth part 7.1)- 7.5) are specific properties of Token-ring, Token-bus, and CSMA/CD, respectively.

(1) The network configuration
channel capacity C bps

number of nodes N
maximum distance (or cable length) D km
buffer size in each node 1

(2) In analysis, the time axis is divided into slots with a constant length τ. The slot length depends on the MAC protocol.

(3) The packet arrival process has the Poisson distribution. The arrival rate per sec is λ. Thus, the arrival rate per slot σ is $\sigma=\lambda\cdot\tau$.

(4) Packets have a constant length, Lp bits. Then, the packet length per slot Lps is $Lps = \lceil Lp / \tau C \rceil$. Where, the operator $\lceil \ \rceil$ means the minimum integer that is greater than or equal to the content.

(5) Token-ring

5.1) The distance between any adjacent nodes is identical to D / N Km.

5.2) The repeat delay at each node is B bits.

5.3) The length of the free-token is Hr bits. Then the length of the free token per slot Hrs is $Hrs = \lceil Hr / \tau C \rceil$.

5.4) The position of the free-token in the network is considered to be probabilistic.

5.5) The slot length of Token-ring τ is defined as the propagation delay between adjacent nodes. It includes the repeat delay B.

$$\tau = D/(2\times 10^5 \times N) + B/C \tag{1}$$

5.6) The state of the network is expressed by the number of ready nodes. Here, a ready node means a node ready to transmit a packet.

(6) Token-bus

6.1) The distance between any adjacent nodes is identical to D/(N-1) Km.

6.2) The order of passing the free-token (explicit token) is a count-up scheme of the node index, except the node N. That is, (1)-(2)-(3)-....-(N-1)-(N)-(1) as shown in Fig.1. This is the best case of token passing situations in Token-bus.

6.3) The position of the free-token in the network is considered to be probabilistic.

6.4) The switch time from receiving mode to transmitting mode is t sec.

6.5) The length of the free-token is Hb bits. Thus, the length per slot is $Hbs = \lceil Hb / \tau C \rceil$.

6.6) The slot length of token-bus τ is the average propagation delay of the free-token. It includes the switch time t.

$$\tau = \frac{1}{N}\left\{\frac{(N-1)D}{2\times 10^5 \times (N-1)} + \frac{D}{2\times 10^5}\right\} + \frac{Hb}{C} + t$$
$$= \frac{D}{10^5 \times N} + \frac{Hb}{C} + t \tag{2}$$

(7) CSMA/CD

7.1) The slot length of CSMA/CD τ is defined as the propagation delay from end to end in the network.

$$\tau = D/(2\times 10^5) \tag{3}$$

7.2) The state of the network is expressed by 2-tuple vector (i,m), where i is the number of ready nodes and m is the number of packet collisions.

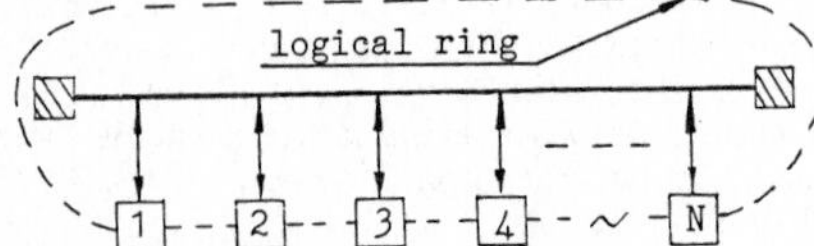

Fig.1 Logical ring in Token-bus network

7.3) When a packet collision occurs, the confusing status on the channel due to the packet collision is dissolved in a slot.

7.4) The number of packet collisions m coordinates to each node and is updated as follows;
a) to add 1 whenever a packet collision is observed in the network, and
b) to reset 0 whenever a successful transmission of packet is observed in the network.

7.5) If a ready node senses the channel idle at the state (i,m), then this node transmits a packet with probability Pd(i,m). Pd(i,m) is depending on the Back-off protocol. Pd(i,m) for BEB (Binary Exponential Back-off) protocol is the following.

$$Pd(i,m) \begin{cases} = 1 & n = 0 \\ = 1/(2 \cdot 2^{m-1}) & n \geq 1 \end{cases} \tag{4}$$

3. Analysis

First, we define a unit-period as the time-basis for analysis (it is depicted in Fig.2).

[Definition] **A unit-period is a time interval between a pair of the end times of successively transmitted packets, except the interval when there is no ready node in the network.**

Here, the meaning of the end time depends on the network topology and so depends on MAC protocol. In a bus network(CSMA/CD or Token-bus), the end time means the time when the end of the transmitted packet propagates to the far end of the network. The darken triangle in Fig.2 shows the propagation of the last bit of a transmitted packet. On the other hand, in a ring network, the propagation of the end of packet transmission and the starting of the next packet transmission may simultaneously occurs for some specific situations. We can not identify these two events clearly on the time axis. So, we consider the time of releasing the free-token as the end of the packet transmission in Token-ring.

Generally, a unit-period consists of an access time and an packet transmission time as shown in Fig.2. Here, an access time is an interval from the end of the latest packet transmission to the start of the following packet transmission. We exclude the time duration when there is no ready node in the network from the access time. We call this time as a null time. Thus, an access time includes the token-passing time for both Token-ring and Token-bus, and includes some idle times, collision times and the subsequent times to recover the channel from the collision state for CSMA/CD. In other words, it is a time required for some specific node to get the right to transmit a packet on the channel. A packet transmission time is a time to transmit a packet

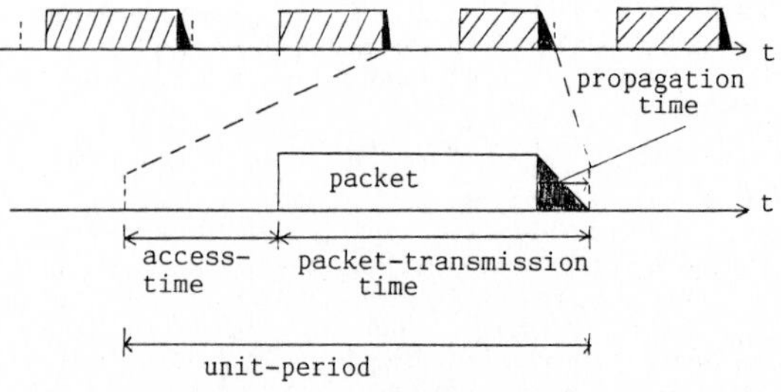

Fig.2 Structure of unit-period

completely to transmission medium. A packet transmission time includes a propagation time in Token-bus and CSMA/CD. In Token-ring, it includes a time required to release the free-token.

The analysis for evaluating the performances of LANs is divided into two parts. In the first part, we obtain the probability distributions on the access time and the packet transmission time for the three MAC protocols. In the second part, we develop a Markov analysis by considering the beginning of the unit-period as the imbedded Markov point. Further, we derive some basic characteristics (average values on throughput, number of ready nodes, and delay) from this Markov analysis.

3.1 Probability distribution on length of unit-period

The unit-period consists of two, i.e. the access time and the packet transmission time. Here, we define three probabilities Punit(k/i), Pacc(k/i), and Ppt(k/i), under the condition that the state of the network just before the beginning of the unit-period is i (we call this condition simply the condition C(i)).

Punit(k/i); probability that a unit-period is k slots.

Pacc(k/i) ; probability that an access time is k slots.

Ppt(k/i) ; probability that a packet transmission time is k.

Further, we define the probability generating functions of them Gunit(z/i), Gacc(z/i), and Gpt(z/i), respectively.

The access time is substantially depending on MAC protocols. On the other hand, the packet transmission time is almost identical for all protocols. The access time and the packet transmission time are independent of each other.

$$\mathrm{Gunit}(z/i) = \mathrm{Gacc}(z/i)\ \mathrm{Gpt}(z/i) \qquad (5)$$

Then, we derive all of them for the three MAC protocols.

(1) Token-ring

In this MAC protocol, there is only one node that has the transmission right (the free-token) and can transmit a packet in the contention-free scheme. When he completely transmits his packet, he releases the free-token and a unit-period ceases. If there are some ready nodes in the network, a new unit-period begins successively. Otherwise, the next unit-period will begin when there occur some ready nodes in the network.

So, we can consider that the access time is the time interval from the very beginning of the new unit-period to the instant that a node acquires the free-token. From the model 5-4) mentioned before, Pacc(k/i) is derived combinatorially.

$$\mathrm{Pacc}(k/i) = \frac{{}_{N-k-1}C_{i-1}}{{}_{N-1}C_{i}} \qquad (i \neq 0) \qquad (6)$$

$$\mathrm{Gacc}(z/i) = \sum_{k=1}^{N-i} \mathrm{Pacc}(k/i)\cdot z^{i} \qquad (i \neq 0) \qquad (7)$$

The packet transmission time is depending on both the packet length and the control scheme to release the free-token. In this paper, we only deal with a constant packet-length and the single-token scheme *1). In the single-token scheme, a node with the token releases it when the following two conditions are satisfied.

1) The node transmits his packet completely to the transmission medium.
2) The node receives a part of its packet from the receiving port after a ring propagation latency and confirms its address in the source address field of the receiving packet.

Then, the probability generating function on the packet-transmission time is

$$\mathrm{Gpt}(Z/i) = \begin{cases} Z^{N} & \mathrm{Hrs} + \mathrm{Lps} \leq N \\ Z^{(\mathrm{Hrs} + \mathrm{Lps})} & \mathrm{Hrs} + \mathrm{Lps} > N \end{cases} \qquad (8)$$

In this equation, the propagation delay from the sending node to the destination node does not included. Thus, the following correction term should be added to the delay performance of Token-ring in the next subsection (Eq.(31)).

$$\mathrm{Dc} = \begin{cases} (\mathrm{Hrs} + \mathrm{Lps})/\ \mathrm{Lps} & \mathrm{Hrs+Lps} \leq N \\ N\ /\ \mathrm{Lps} & \mathrm{Hrs+Lps} > N \end{cases} \qquad (9)$$

(2) Token-bus

In Token-bus, a logical ring is composed in a bus network and the control token is passed on the node-by-node basis. So, the control scheme of Token-bus is almost similar to Token-ring. In this analysis, the difference between them is included in the definition of each slot. Hence, Pacc(k/i) and Gacc(z/i) of Token-bus is the same as those of Token-ring. Gpt(z/i) is the following.

$$\mathrm{Gpt}(z/i) = z^{\mathrm{Lps} + \mathrm{Cp}} \qquad (10)$$

In this equation, Cp is the maximum propagation time of the end of packet. So, Token-bus does not need the correction term.

(3) CSMA/CD

In this MAC protocol, the state of the network is described as (i,m), where i is the number of ready nodes and m is the number of the packet collisions, as mentioned in the Model. First, we define four probabilities to deal with the channel situation. Under the condition that the state of the network is (i,m), the probability Ps(i,m) that a node successfully transmits a packet in a slot, the probability Pi(i,m) that the channel is idle in a slot, the probability Pc(i,m) that there is a packet collision on the channel in a slot, and the probability Pa(i,m) that there is a new arrival in a slot are, respectively.

$$\mathrm{Ps}(i,m) = i \cdot \mathrm{Pd}(i,m)\ \{\ 1 - \mathrm{Pd}(i,m)\ \}^{i-1} \qquad (11)$$

$$\mathrm{Pi}(i,m) = \{\ 1 - \mathrm{Pd}(i,m)\ \}^{i} \qquad (12)$$

$$\mathrm{Pc}(i,m) = 1 - \mathrm{Ps}(i,m) - \mathrm{Pi}(i,m) \qquad (13)$$

$$\mathrm{Pa}(i,m) = 1 - e^{-(N-i)\cdot\sigma} \qquad (14)$$

Let the probability that an access time is just k slots, on condition that the state of network is (i,m), be Pacc(k/i,m). Then Pacc(k/i,m) satisfies the following Proposition.

---

*1) Generally, token-passing control schemes are classified into three, multi-token, single-token (single-token multi-packet), and single-packet (single-token single packet). IEEE 802 committee recommends the single-token scheme.

[Proposition]

$$
Pacc(k/i,m) = \begin{cases} Ps(i,m) & k = 1 \\ Pi(i,m)\{1 - Pa(i,m)\}\,Pacc(k-1/i,m) & \\ \quad + Pi(i,m)\,Pa(i,m)\,Pacc(k-1/i+1,m) & \\ \quad + Pc(i,m)\,\{1 - Pa(i,m)\}\,Pacc(k-1/i,m+1) & \\ \quad + Pc(i,m)\,Pa(i,m)\,Pacc(k-1/i+1,m+1) & k \geq 2 \end{cases} \tag{15}
$$

The above difference equation gives the correlation between the probabilistic behavior of the first and the second slots of the unit-period. Here, we omit the proof of this Proposition. Then, the probability generating function of Pacc(k/i,m) satisfies the following difference equation.

$$
\begin{aligned} Gacc(z/i,m) = {} & \frac{z}{1-Pi(i,m)\{1 - Pa(i,m)\}\cdot z} \\ & \times [\, Pi(i,m)\,Pa(i,m)\,Gacc(z/i+1,m) \\ & + Pc(i,m)\{1 - Pa(i,m)\}\,Gacc(z/i,m+1) \\ & + Pc(i,m)\,Pa(i,m)\,Gacc(z/i+1,m+1)\,] \end{aligned} \tag{16}
$$

Gacc(z,i,m) has the following boundary condition.

$$
\begin{aligned} &1) \quad i > N \quad Gacc(z/i,m) = 0 \\ &2) \quad m > M \quad Gacc(z/i,m) = Gacc(z/i,M) \end{aligned} \tag{17}
$$

Where M is the maximum number of collisions for increasing the back-off interval at the packet collision in CSMA/CD. From these conditions above, Gacc(z/i,m) is recursively obtained for any values of i and m. The probability generating function of the packet-transmission time Gpt(z/i,m) is

$$
Gpt(z/i,m) = z^{Lps+1} \tag{18}
$$

From the Model 6-3), we assume the initial value of **m** at the beginning of each unit-period be 0. Hereafter, we will express the functions Gacc(z/i,0) and Gpt(z/i,0) of CSMA/CD simply as Gacc(z/i) and Gpt(z/i), respectively.

3.2 Markov analysis

In this subsection, a Markov analysis is developed to obtain the network performance of the three MAC protocols. Let us consider the beginning of the unit-period as the imbedded Markov point and the number of ready nodes in the network at this point as the state of the network. Further, the following approximation is introduced.

[Approximation] **The state transition in the unit-period depends on only the length of the unit-period and the packet arrival rate.**

Thus, the analytical procedure of this second part is not depending on MAC control protocol but on the length of the unit-period. If the unit-period begins with the state i and has the length of k slots, then the conditional one step state transition probability Pt(j/i,k) that the state at the beginning of the next unit-period is j, is

$$
Pt(j/i,k) = \begin{cases} 0 & (0 \leq j \leq i-2) \\ {}_{N-i}C_{j-i+1}(1-e^{-\sigma k})^{j-i+1}\, e^{-\sigma k(N-j-1)} & (i-1 \leq j \leq N,\ i \neq 0) \\ Pt(j/1,k) & (i = 0) \end{cases} \tag{19}
$$

Where N is the total number of nodes in the network and σ is the packet arrival rate per slot-node. As mentioned at the definition, the null times when there is no ready node in the network are excluded from the unit-period. Thus, Pt(i/0,k)=Pt(i/1,k).

Then the one step state transition probability Pt(j/i) that the state of the next unit-period is j, on the condition C(i) (0 < i,j < N-1), is obtained as follows.

$$
Pt(j/i) = \sum_{k=1}^{\infty} Pt(j/i,k)\,Punit(k/i)
$$

$$
= \begin{cases} 0 & (0 \leq j \leq i-2) \\ {}_{N-i}C_{j-i+1} \displaystyle\sum_{r=0}^{j-i+1} {}_{j-i+1}C_r(-1)^r \times Gunit(e^{-\sigma(N+r-j-1)}/i) & (i-1 \leq j \leq N,\ i \neq 0) \\ Pt(j/1) & (i = 0) \end{cases} \tag{20}
$$

The state sequence consists of an aperiodic Markov chain with finite states. Thus, the steady state probability distribution $\Pi = \{\pi(i)\}$ (i=0,1,2,...,N-1) exists.

$$
\Pi = \Pi \cdot \mathbf{Pt} \tag{21}
$$

Where **Pt** is the one-step state transition probability matrix. Here, we obtain the average length of the unit-period with the initial state i, Tunit(i), and the average of Tunit(i) on i, Tunit.

$$
Tunit(i) = \lim_{z \to 1} \frac{d}{dz} Gunit(z/i) \tag{22}
$$

$$
Tunit = \sum_{i=0}^{N-1} \pi(i) \cdot Tunit(i) \tag{23}
$$

Then, the throughput S is obtained by

$$
S = Lps \,/\, \left(Tunit + \pi(0) \cdot \frac{1}{1-\exp(-\sigma N)} \right) \tag{24}
$$

Now, we obtain the average number of ready nodes Q in the network at an arbitrary time instant. As described above, we have the steady state probability distribution π(i). However, this probability is concerned to only the imbedded Markov points. So, the expectation of π(i) is not equal to Q. In the following part of this chapter, we will derive Q.

First, we define the conditional probability P(m,n/i,k,j) that there is **m** ready nodes at the **n**-th slot in the unit-period, on condition that the state just before the beginning of the unit-period is i, the length of the unit-period is **k** slots, and the state just after the end of the unit-period is j (hereafter, we call this condition simply the condition C(i,k,j)). Then, P(m,n/i,k,j) is as follows.

$$
P(m,n/i,k,j) = \begin{cases} \dfrac{{}_{N-i}C_{m-i}(1-e^{-\sigma n})^{m-i}(e^{-\sigma n})^{N-m} \cdot {}_{N-m}C_{j-m+1}}{{}_{N-i}C_{j-i+1}(1-e^{-\sigma k})^{j-i+1}} \times \dfrac{(1-e^{-\sigma(k-n)})^{j-m+1}(e^{-\sigma(k-n)})^{N-j-1}}{(e^{-\sigma k})^{N-j-1}} & (1 \leq i \leq N-1,\ j \geq i-1) \\ P(m,n/1,k,j) & (i = 0) \end{cases} \tag{25}
$$

Then, under the condition C(i,k,j), the average number of ready nodes at the n-th slot in the unit-period Q(n,i,k,j) is

$$Q(n,i,j,k) = \sum_{m=i}^{j+1} m \cdot P(m,n/i,k,j)$$
$$= i + (j-i+1) \cdot \frac{1-e^{-\sigma n}}{1-e^{-\sigma k}} \quad (26)$$

Under the condition C(i,k,j), the average accumulated number of ready nodes in the unit-period A(i,k,j) is

$$A(i,k,j) = \sum_{n=1}^{k} Q(n,i,k,j) \quad (27)$$
$$= k(i + \frac{j-i+1}{1-e^{-\sigma k}}) - (j-i+1) \cdot \frac{e^{-\sigma}}{1-e^{-\sigma}}$$

So, under the condition C(i) and the condition that the length of the unit-period is k, the average accumulated number of ready nodes in the unit-period A(i,k) is as follows.

$$A(i,k) = \sum_{j=i-1}^{N-1} P(j/i,k) \cdot A(i,k,j)$$
$$= k \cdot i + (N-i) \frac{e^{-\sigma}}{1-e^{-\sigma}} (1-e^{-k\sigma}) \quad (28)$$

Hence, under the condition C(i), the average accumulated number of ready nodes in the unit-period A(i) is as follows.

$$A(i) = \sum_{k=1}^{\infty} Punit(k/i) \cdot A(i,k) \quad (29)$$
$$= N \cdot Tunit - (N-i) \cdot \frac{e^{-\sigma}}{1-e^{-\sigma}} (1- Gunit(e^{-\sigma}/i))$$

From the definition of the unit-period, it is easily shown that A(0) = A(1). Hence, the average number of ready nodes Q is

$$Q = (\sum_{i=1}^{N-1} \pi(i) \cdot A(i))/(Tunit + \pi(0) \cdot \frac{1}{1-e^{-\sigma N}}) \quad (30)$$

Finally, the packet delay (normalized by one packet transmission time without any overheads)

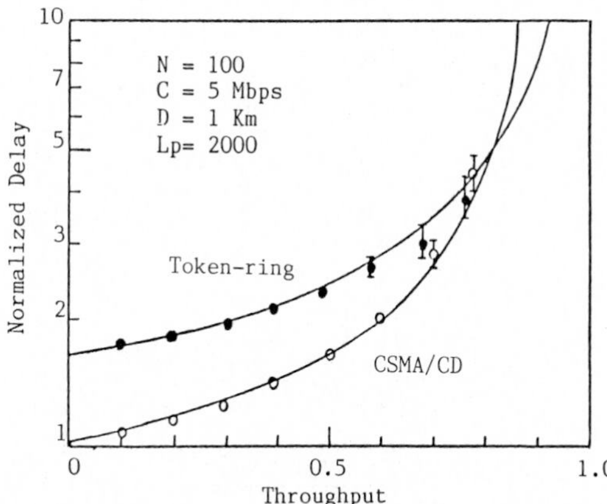

Fig.3 Comparison between analytical results and simulation results on throughput-delay

Dp is obtained from Eqs(24) and (30) by applying the Little's Formula, as follows.

$$Dp = Q / S \quad (31)$$

It should be noted that the correction term Dc (Eq.(9)) must be added to the above equation (31), in Token-ring.

4. Numerical analysis

We show numerical results for typical LAN configurations. We consider the basic parameters for the network configuration and MAC protocols as follows;

1) network parameter
channel capacity C = 5 Mbps
number of nodes N = 100
maximum distance D = 1 Km
packet length Lp = 2000 bits
2) Token-ring (Single token)
token header length Hr = 24 bits
repeat delay at node B = 8 bits
3) Token-bus
token length Hb = 96 bits
switch delay t = 50 nsec
4) CSMA/CD (BEB Back-off protocol)
maximum length of contention to control M = 10

First, we compare the analytical results to the results of the computer simulation to check the validity of the approximated analysis. The simulation is a continuous-time event-to-event typed simulation and so it does not introduce the concept of the time slot (Model 2)). Additionally, the assumption 7.4) and 7.5) of the model is released and the Ethernet-like scheme is introduced for the back-off control of CSMA/CD. The computer simulation is performed by the single-run-method [12], where the number of sample packets at each run is 10000 except the initial run and every 500 packets composes one sample point for 95 % confidence interval.

Fig.3 shows the comparison of the results by the analysis and the simulation. Here, we only deal with Token-ring and CSMA/CD. The delay performance is normalized by one packet transmission time Lp/C. We can see that both results for two MAC protocols coordinate each other for lightly loaded and moderately loaded situations. In heavily loaded situation (almost saturated area), there is a slight difference between two results. However, the analytical results are included in the 95 % confidence intervals of the simulation results for these saturated area. Hence, we can conclude that the approximated analysis presented in this paper is valid for almost all loaded situations of the typical LAN configuration.

Then, we show the performance comparison among the three MAC protocols for various kinds of LAN configurations. Fig.4 shows the comparison on the throughput-delay characteristics for three cases of the channel capacity, C = 1, 5, 20 Mbps. First, we consider the case of C = 5 Mbps as the basic situation. In lightly and moderately loaded situations, CSMA/CD is the most desirable on the delay. On the contrary, Token-ring is preferable on both the delay and the throughput (the maximum throughput) to the others in heavily loaded situation. The performances of Token-ring and CSMA/CD are crossing around the heavily loaded situation. The critical point for the superiority on the performance fundamentally depends on the network configurations. Token-bus has the worst performance for almost all situations, especially on the delay. This is due to the overhead for passing the explicit token.

Now, we consider the other cases of the channel capacity. While the performance of both Token-ring and Token-bus are hardly affected by the variation of the channel capacity C, CSAM/CD is quite sensitive to the variation of C. For the small value of the channel capacity (C = 1 MBPS), CSMA/CD is entirely superior to Token-ring and also Token-bus. For the large value of the channel capacity (C = 20 Mbps), Token-ring relatively becomes to be preferable to CSMA/CD, because CSMA/CD inherently has the capacity limit due to the Carrier Sense effect. When we consider the larger capacity than 20 Mbps, this tendency will become more notable.

Here, we have to note on the value of the packet delay. As mentioned before, the packet delay is measured on the time basis normalized by one packet transmission time Lp/C. The one packet transmission time depends on the packet length and the channel capacity. Hence, it should be noted that when we consider the absolute value of delay, the unit delay (Dp = 1) for some value of the channel capacity C is not equal to the one for another value of C. For instance, Dp = 1 means 2 msec in the case of C = 1 Mbps and Lp= 2000 bits, while Dp = 1 means 0.4 msec in the case of C = 5 Mbps and Lp = 2000 bits.

Fig.5 shows the performance comparison on the throughput-delay characteristics for the three cases of the network distance D, D = 0.1, 1, 10 Km. When the network distance increases from 1 Km to 10 Km, it is observed that the performance of CSMA/CD is drastically degraded. The maximum throughput of CSMA/CD for D = 10 Km is only 40 % of the channel capacity. On the other hand, when the network distance decreases to 0.1 Km, the performance of CSMA/CD is improved to almost the perfect scheduling scheme. However, the variation of the network distance hardly affects to the performance of Token-ring and Token-bus. The reasons of this insensitiveness are different from each other for two schemes. In Token-ring, when the network distance increases and one round-trip propagation delay becomes larger than the one packet transmission time, the performance will be degraded, significantly. Until the network distance is less than the critical value, the performance is almost insensitive to the variation of the distance. In Token-bus, the overhead due to the explicit token is substantially large and so the overhead due to the network distance is almost negligible.

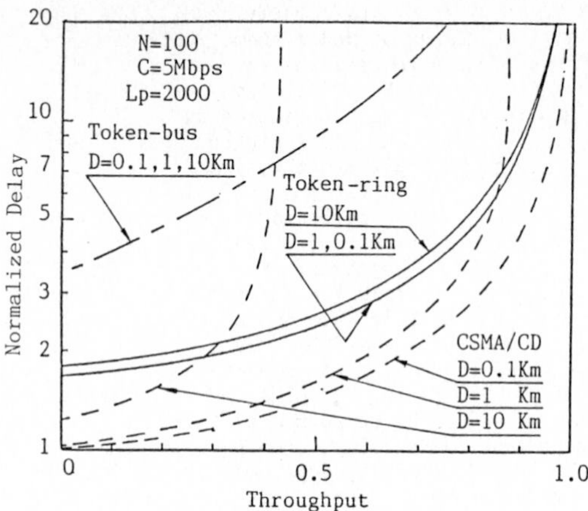

Fig.5 Throughput-delay characteristics with varying network distance D

Fig.6 shows the performance comparison on the throughput-delay characteristics for three cases of the number of nodes N, N = 10, 100, 200. As far as the traffic intensity and/or traffic load is kept the same, CSMA/CD shows almost the equivalent performance for the increasing the number of nodes from 100 to 200. This tendency will be preserved for larger values of N. However, the performances of Token-ring and

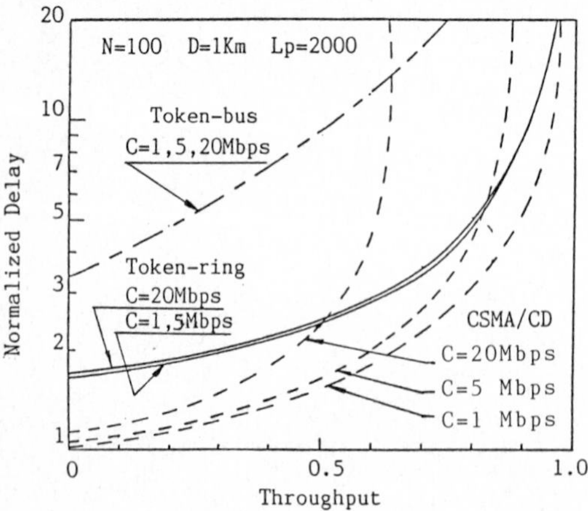

Fig.4 Throughput-delay characteristics with varying channel capacity C

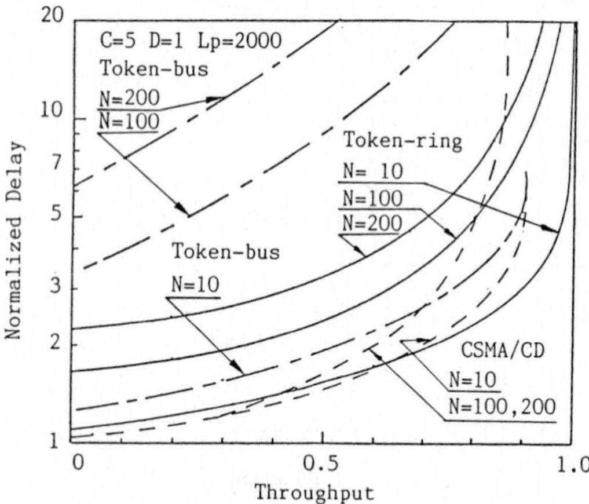

Fig.6 Throughput-delay characteristics with varying number of nodes N

Token-bus are significantly degraded by increasing the number of nodes from 100 to 200 due to the overhead to pass the token. For the small size network (N = 10), the three schemes show almost similar characteristics on the throughput-delay. Token-ring is a little preferable to the others for the maximum throughput ability.

Fig.7 shows the performance comparison on the throughput-delay characteristics for three cases of the packet length Lp, Lp = 500, 2000, 10000. When the packet length increases, the performances of the three protocols are improved. For the case of Lp = 10000, there is scarcely any difference on the throughput-delay performance between Token-ring and CSMA/CD. For the case of Lp = 500, the performance of all schemes are degraded. Especially, the performance of Token-ring is drastically degraded, comparing to the other schemes. This is due to the reason that the round-trip propagation delay becomes larger than the packet transmission delay.

As a result, we can describe as follows.

1) **Token-ring** As far as the round-trip propagation delay on the ring network is smaller than the one packet transmission time, Token-ring shows the excellent performance on the throughput-delay. This scheme is the most suitable to large channel capacity, large network distance, and small user size.

2) **Token-bus** From the viewpoint of the network performance, Token-bus has the least potential. This scheme has some advantageous features which can not be evaluated quantitatively, e.g. to utilize off-the-shelf CATV facility, to be easily adaptable to broadband situations, and so on. Thus, this scheme will be adopted to such specific environment.

3)**CSMA/CD** This scheme shows the low delay ability for lightly and moderately loaded situations. This scheme is suitable to networks with large user size where the aggregate traffic load is not so high. When the network distance is small enough, this scheme show the notable performance almost similar to the perfect schedule.

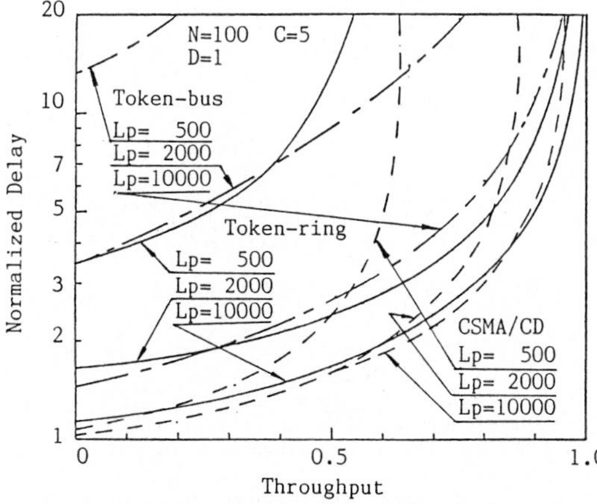

Fig.7 Throughput-delay characteristics with varying packet length Lp

5. Conclusion

The performance evaluation of various LAN schemes is one of the most important problems in the researches of LAN. The objective of this paper is to present an analytical method for evaluating three MAC protocols of LANs, Token-ring, Token-bus, and CSMA/CD. This paper introduced the concept of the unit-period that consists of an access time and one packet transmission time. By using some approximation, this paper has developed the imbedded Markov analysis on the basis of the unit-period. In order to verify the validity of the approximation, this paper compared the analytical results to the simulation results.

This paper has shown the performance comparison among the three MAC protocols for typical network configurations. By the numerical analysis, this paper has discussed the suitable situations for each medium-access control protocols in the LAN environment. We can easily obtain the competitive boundary on the network performance for various kinds of LAN configurations. Further, by extending the analytical method, we can obtain the performance of various kinds of LAN schemes including hybrid schemes of Token-passing and CSMA/CD, some schemes with transmission priorities, and so on.

Acknowledgment The author would like to appreciate the discussion of Prof. Yoshihiro Tsunoda of Kobe University and Prof. Yoshiro Nakanishi of Osaka University. This research is partially supported by the National Science Grant of the Japanese Government in 1984.

References

1) B.Stuck,"Calculating the maximum mean data rate in local area networks", IEEE Computer, May 1983.
2) W.Bux,"Local-area subnetworks: a performance comparison",IEEE Trans. Com., Oct. 1981.
3) R.Cherukuri, L.Li, and L.Louis," Evaluation of Token Passing Schemes in Local Area Networks", Proc. of Computer Networking Symp.'82, 1982.
4) W.Stallings," Local Network performance", IEEE Commun. Magazine, Vol.22, No.2, Feb. 1984.
5) H.Okada, T.Yamamoto, Y.Nomura, Y.Nakanishi, "Comparative evaluation of Token-ring and CSMA/CD Medium-Access Control Protocols in LAN Configurations", Proc. of Computer Networking Symp.'84, 1984.
6) A.Kohnheim, B.Meiser,"Waiting lines and times in a system with polling", JACM, Vol.21, 1974.
7) S.Lam, " A carrier sense multiple access protocol for local networks", Computer Networks, Vol.4, No.1, 1980.
8) F.Tobagi, V.Hunt," Performance analysis of carrier sense multiple access with collision detection",Computer Networks, Vol.4 No.5, 1980.
9) M.Reiser, " Performance Evaluation of Data Communication Systems", Proc. of IEEE, Vol.70, No.2, Feb. 1982.
10)J.Kurose, M.Schwartz, and Y.Yemini," Multiple-Access Protocols and Time-Constrained Communication",Computing Surveys, Vol.16, No.1, 1984.
11)IEEE Project 802, " Local Area Network Standards", Draft E, 1983.
12)H.Kobayashi, "Modeling and Analysis" Addison-Wesley Pub.Co. 1978.

TELETRAFFIC ISSUES in an Advanced Information Society
ITC-11
Minoru Akiyama (Editor)
Elsevier Science Publishers B.V. (North-Holland)

AN INTEGRATED CIRCUIT/PACKET SWITCHING LOCAL AREA NETWORK - PERFORMANCE ANALYSIS AND COMPARISON OF STRATEGIES

Ernst-Heinrich GOELDNER

Institute of Communications Switching and Data Technics
University of Stuttgart, Fed. Rep. of Germany

ABSTRACT

A new inhouse communication system is suggested integrating circuit switched services with variable bitrate as well as packet switched services with variable throughput rate. The advantages of LANs and PBXes will be combined in the new architecture. In this paper several strategies to achieve the integration of both switching principles on a ring system are presented. Performance evaluation is done by means of maximum PS-throughput calculation as well as simulations of the detailed system, resulting in mean waiting times for PS-messages under different offered PS- and CS-traffic load conditions.

1 INTRODUCTION

Currently, two mainstream developments characterize the field of inhouse communication:

- o Introduction of pure packet switched (PS) local area networks (LAN) for computer application, and
- o Introduction of full-digital private branch exchange (PBX) systems for circuit switched (CS) voice traffic.

The PBX systems will be upgraded to include also circuit switched text and data traffic, matching the specifications given by the new public networks (ISDN) with transmission over one ore more basic 64 kbps channels (B-Channels) and a separate 16 kbps channel for frame-oriented signalling, low-speed packet switched data and teleaction information (D-Channel) on existing subscriber lines. Several developments are directed to bridge these separated networks through gateways to provide an arbitrary connection of any terminal and end system, and to connect these private networks to the public networks.

This technique maintains the dominating features for each class of application: Circuit-switching for stream-type voice communication and packet switching for brust-type data communication. But the rapid development within the area of terminal equipment, integrated work stations, and office automation requires multifunctional terminals and a universal network interface for both circuit- and packet switched services with variable bandwidth and throughput rate, respectively. These requirements cannot be fulfilled momentarily, neither by known LANs, nor by digital PBXes, since

- o LANs are often not suitable for real-time applications (e.g. voice) due to the load-dependent and therefore randomly distributed waiting times.
- o PBXes are normally star-shaped and have only narrow-band subscriber loops which give a limitation of the maximum bandwidth. Therefore, bursty traffic with high-speed transmission rates cannot be carried with adequate small delays.

Bridging of distinct networks may then result into throughput bottlenecks, large buffer equipment, delays as well as high protocol conversion overhead. For this reason, many concepts have been proposed to overcome these drawbacks by integration of circuit and packet switching into one network.

2 INTEGRATED INHOUSE NETWORK CONCEPT

2.1 Network Structure

Even in future, most terminals will be of the ISDN-type for CS-voice, -text, and -data. They all will be connected to centralized ISDN-PBXes through the existing subscriber loops. Higher communication requirements are usually concentrated to much smaller spatial areas, such as a department for research or development, or a university institute with functionally higher end systems, such as work stations, department computers, data bases, and graphic equipment to use simultaneously voice, data and graphic communication.

Accomplishing these new requirements we suggest a new inhouse communication system [1] which provides

- o synchronous transmission for circuit switched voice and circuit switched data (e.g. mass data transfer) with a variable and adaptable bandwidth

as well as

- o asynchronous transmission of data packets, similar to the well known LANs with a high throughput rate.

The structure of the new system is shown in Fig.1 and deals with the following aspects:

- o Small LANs (SLAN) on a ring basis for the real integration of circuit and packet switched traffic. For voice synchronous transmission with fixed time

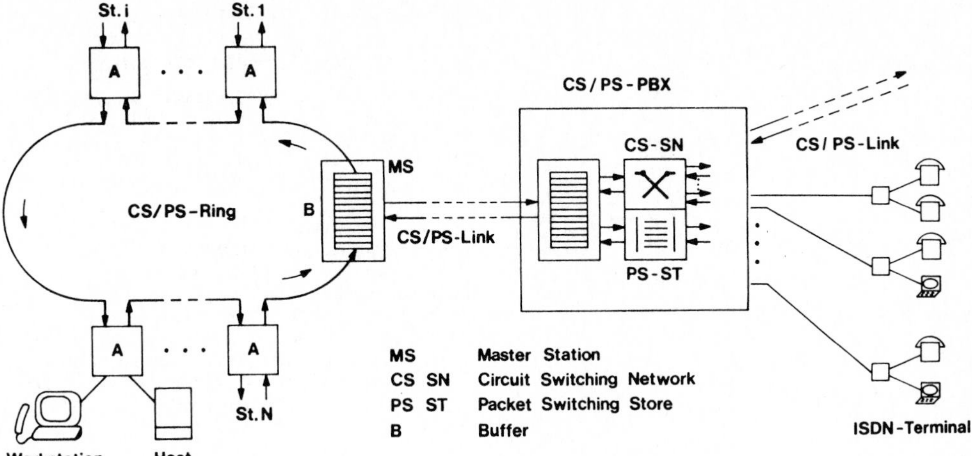

Fig. 1 Basic structure of the integrated inhouse network

slot allocation is provided, whereas for data synchronous as well as asynchronous (i.e. packet switched) transmission can be chosen. Both services allow a variable bandwidth allocation or a variable throughput rate by window flow control, respectively.

o Interconnection of several SLANs (e.g. within a plant) with a new type of PBXes for integrated circuit and packet switching. Within larger inhouse areas several integrated PBXes may be interconnected through integrated CS/PS-links in a mesh-type structure.

o The dominating ISDN-terminals are also connected to the CS/PS-PBXes in the usual star-configuration through the existing subscriber loops, using the standardized ISDN interfaces.

This structure reveals a number of advantages as

- Limitation of distributed functions to very small areas (SLAN)
- Handling of mass traffic, operational and maintenance functions by centralized nodes
- Maintaining of the adequate service-specific switching principles (CS or PS)
- Matching of specific grade of service criteria as throughput and delay for PS-services and blocking for CS-services
- Concentrating of all traffic to foreign exchanges or public networks to one gateway
- Imbedding of the new structure into the existing infrastructure.

2.2 Integrated Ring-System

An integrated, distributed ring-system (SLAN) has been developed to connect the multi-functional terminals, providing CS and PS on demand, cf. [2,3]. Several end systems are connected to one ring access station, which operates as a cluster controller to keep the costs small for the decentralized logic.

Basis for the CS/PS-integrated ring and the CS/PS-link between the integrated systems is a synchronous pulse frame with fixed length. This frame is partitioned into equal sized time slots, similar to the well-known PCM-frame. One time slot is able to carry one CS-channel with 64 kbps transmission rate, where the same time slot can be used for both transmission directions providing fullduplex connectivity. Allocation of time slots to new CS-calls is done by means of a signalling procedure at call establishment. To achieve short delays and independence from the PS-traffic, a separate signalling channel in one ore more time slots is necessary.

One station in the ring generates the pulse frame and buffers the frame, compensating for different propagation delays. This station is also suited to provide gateway-functions, gaining access from the ring to the other parts of the inhouse network and to the public networks. This station is called the Master Station and may also be responsible for managing the CS-calls.

This paper focuses only on the CS/PS-integration in a distributed ring-system. The integrated links have been subject of many papers [5] - [8] and will therefore not be discussed here.

3 HYBRID SWITCHING PRINCIPLES

The principles of integration can be classified into 3 main categories:

- Carrying all packet switched data over circuit switched channels (pure CS)

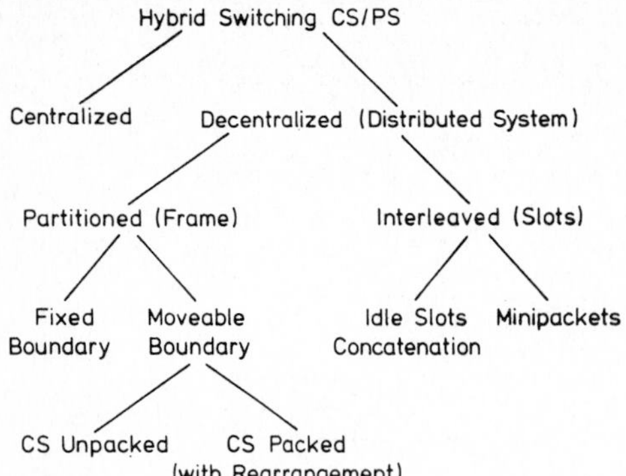

Fig. 2 Hybrid switching principles

- Packetizing all circuit switched information (also voice) and all-packetized communication in the network (pure PS)

- Sharing the transmission capacity and providing CS and PS (hybrid switching)

Several possibilities are known to integrate PS-traffic into the synchronous pulse frame. Fig.2 gives an overview for the following taxonomy of these principles.

3.1 Partitioned Frame

This principle divides the pulse frame into two parts for CS- and for PS-traffic. The boundary between the two parts may be fixed, i.e. it provides a fixed bandwidth for each traffic type. Each part of the pulse frame can be seen as an independent system and, therefore, the management of each traffic type can be easily implemented. However, the drawback in this case is that empty time slots within the CS-part of the frame cannot be used by the PS-traffic and vice versa. An optimal utilization of the system is not possible under varying traffic loads.

An inprovement of the ring utilization can be achieved by moving the boundary between the CS- and PS-part as a result of the momentarily CS-occupancy pattern. PS-traffic is carried in the second part of the pulse frame, beginning immediately after the CS-connection with the largest time slot number. Two further distinctions can be made, depending on the use of unoccupied time slots <u>within</u> the CS-occupancy pattern:

Packed and Unpacked

Packed means, that existing CS-calls become rearranged after clearing down a CS-call. All occupied time slots are shifted to the beginning of the pulse frame to avoid empty and unused slots within the CS-part. However, for rearrangement of CS-calls is not practical in a distributed system, an unused time slot may occur after clearing down of any of the existing CS-calls exept the last one used (unpacked). This additional empty bandwidth cannot be used for PS-transmission, and a total utilization is not always possible. Besides this, implementation problems may arise from the management of this moveable boundary, and from moving the boundary into the PS-part. Fig.3 illustrates both cases of the partitioned frame with fixed and movable boundary.

Within the pulse frame, the only information being required additionaly is

- One channel for CS-signalling and

- One bit every CS-time slot to indicate whether CS-data are still valid or not.

However, several unused time slots in the CS-part may waste useful bandwidth.

3.2 Interleaved Slots

The second way of hybrid switching on a ring-system is based on a slotted frame and interleaving of CS- and PS-data. A CS-call may occupy any empty time slot within the whole pulse frame as long as the maximum number of allowable CS-channels is not exceeded. All other time slots can be used for PS-traffic. These "idle" slots can be considered as being concatenated to one remaining PS-channel.

This principle needs an additional flag within every time slot to distinguish the occupied CS-slots from slots available for PS. In this case, the total ring utilization is guaranteed, but a similar problem arises as in case of the moveable boundary scheme:
Once a new CS-connection has to be set up, the PS-transmission must be interrupted immediately at the slot boundary and delayed without any loss or disturbance of data.

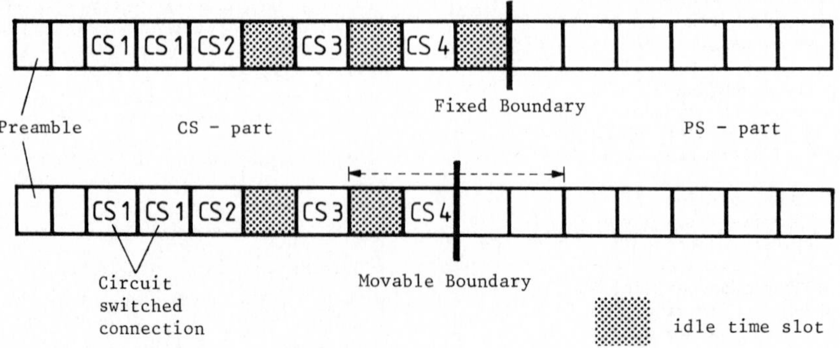

Fig. 3 Frame structure: partitioned frame fixed / movable boundary

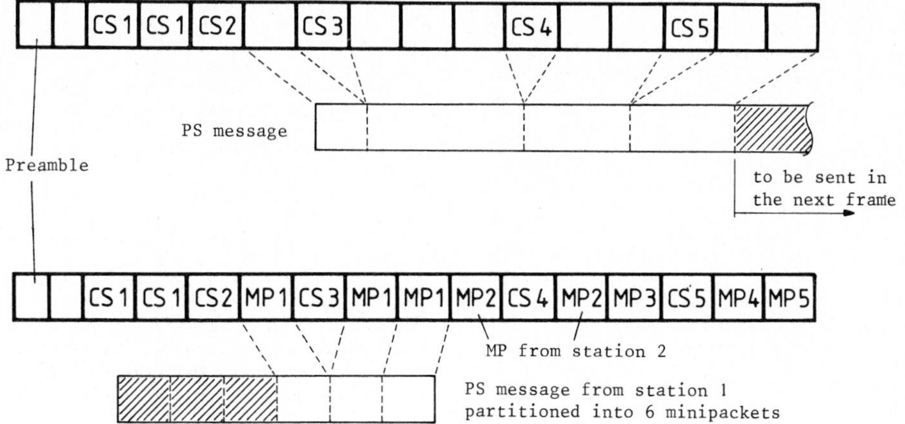

Fig. 4 Frame structure: interleaved slotted frame idle slot concatenation / subdivision of messages into minipackets

The access to the PS-part of the pulse frame has to be controlled by a suited protocol for all three principles mentioned above. To guarantee a high throughput and a stable behavior for high loads, the distributed token-passing scheme is the most adequate media access protocol.

The fourth and last principle discussed here uses the same strategy for CS-traffic as before. But the remaining bandwidth for PS-traffic is not given to one station as in case of the token-passing protocol.

Messages delivered from a terminal at the ring-station will first be partitioned into equal-sized and individually addressed mini-packets (MP). Each minipacket fits exactly into one time slot. Every time slot is marked by two bits, indicating whether this time slot is available for PS and whether this PS-slot is empty or not.
Therefore, a station being prepared to send PS-data within minipackets watches the slots passing by and inserts the minipackets into unused slots. A station detecting its own address in the header of a minipacket copies the contents into its buffer, sets the time slot empty immediately or may even insert an own mimipacket to be sent. That means that empty slots are used on demand by the sending station, whereas the receiver is responsible for clearing of the slot. This method allows an excessive ring utilization since one slot may carry more than one minipacket within one cycle, depending only on the sender-receiver relations for communication. In Fig.4 the two implementation choices for the interleaved accesses are illustrated.

Every minipacket contains the addresses of the receiving and sending ring-station. Additionally, several minipackets containing parts of one message can be reordered and easily restored by an extention of the addressfield, the so called service indicator. Priority schemes are also possible and - as another advantage - the CS-signalling can be done by minipackets with high priority. Therefore, no separate CS-signalling channel is necessary.

The additional overhead for addressing of minipackets seems to be relatively large. In our laboratory implementation, 16 bits are used for addressing and 48 bits are available for user-data.
On the other hand, one time slot may be used by more than one minipacket within one frame cycle and the improving of the ring utilization may compensate or even overcompensate for the additional overhead.

4 ANALYTICAL PERFORMANCE EVALUATION

4.1 Overview

The first performance aspect is the maximum throughput achievable for each hybrid switching principle.

For CS-traffic takes priority against PS-traffic, the throughput of the CS-traffic will not be affected significantly by PS. On the contrary, the maximum PS-traffic is directly depending on the monentarily CS-traffic.

The fixed boundary scheme is easy to analyze. Each traffic type makes use of its own bandwidth without any affect to the rest system.

The movable boundary with rearranging of existing CS-calls and the idle slot concatenation principles are quite similar for maximum throughput considerations. Both principles allow the total use of the residual bandwidth for PS. Only the necessity of different overhead-bits may cause small differences in the maximum PS-throughput. The fixed and the movable boundary schemes have been discussed in several papers [6] - [13]; more references can be found therein.

The access to the PS-bandwidth is controlled by a normal token-passing protocol. A station detecting the token is allowed to send its messages. The messages are copied by the receiving station but will be removed only at the sender. Therefore, the maximum utilization of the ring is independent of the routing matrix.

4.2 Performance Evaluation for MP-Protocol

Opposit to this, the maximum PS-utilization of a minipacket ring-system depends on the routing matrix and on the traffic rates generated by each station. The maximum PS-throughput will be calculated as follows.

In Fig.5 one ring-station is shown, operating under the minipacket protocol.

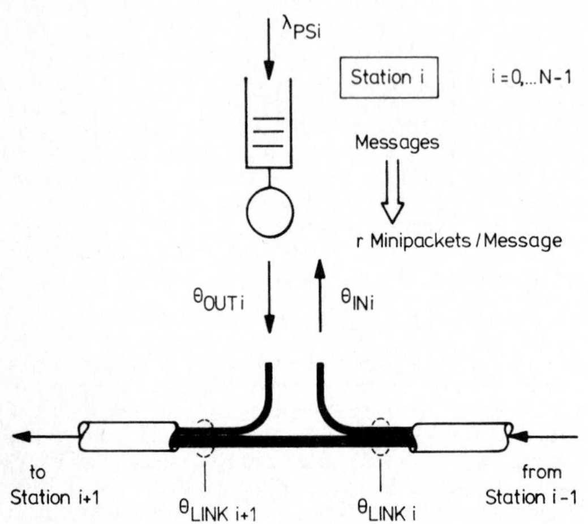

Fig. 5 Traffic parameters for minipacket protocol

N : total number of ring-stations

p_{ij} : probability, that one minipacket is sent from station i to station j

with $\sum_{j}^{N-1} p_{ij} = 1$, $p_{ii} = 0$

Θ_{OUTi} : total MP-traffic rate from station i to all other stations

Θ_{INi} : total MP-traffic rate from all stations to station i

Θ_{mi} : MP-traffic rate from station m to station i,i+1,...m+N-1 (modulo N)

Θ_{LINKi} : MP-traffic rate on the link between station i-1 and station i

$$\Theta_{LINKi} = \sum_{m=0}^{N-1} \Theta_{mi} = \sum_{m=0}^{N-1} \left(\Theta_{OUTm} \sum_{\substack{k=i \\ (\mathrm{mod}\ N)}}^{k=m+N-1} p_{mk} \right) \quad (1)$$

Calculation of the maximum throughput of a minipacket ring-system:

n_{PS} : mean number of time slots available for PS

n : total number of time slots within a pulse frame

Y_{CS} : mean number of time slots occupied by CS CS-traffic is transmitted FDX having only an influence to the total link capacity.

$$n_{PS} = n - Y_{CS}$$

Max. minipacket-utilization:

$$\max.(\Theta_{LINKi}) = n_{PS} \quad i=0,1,..,N-1 \quad (2)$$

Be λ_{PS} the total arrival rate of PS-messages, r the number of MPs per message, and c_i the relative traffic part contributed by station i. Then

$$\lambda_{PSi} = c_i * \lambda_{PS} \quad (3)$$
$$\Theta_{OUTi} = r * \lambda_{PSi}$$

The maximum PS-traffic rate λ_{PS} can be easily computed by using the equations (1) to (3).

Furtheron, a factor α can be defined to characterize the systems capacity, based on the communications relations:

$$\alpha_i = \frac{\lambda_{PS}}{\Theta_{LINKi}/r} \quad \text{for all links } i-1,i \quad (4)$$

If $\alpha_i = \alpha$ for all $i = 0,..,N-1$ the ring is called symmetrically loaded; then

$$\sum_{j=0}^{N-1} p_{ij} * \lambda_{PSi} = \sum_{j=0}^{N-1} p_{ji} * \lambda_{PSj} \quad (5)$$

<u>Example:</u>

Parameters:

10 Mbps transmission rate
1 ms pulse frame duration
146 time slots with 64 kbps each for CS or MP
64 bits overhead per frame and
4 bits overhead per time slot
16 bits for addessfield per MP
48 bits for userdata per MP
22 MP per message
10 stations

In this example, a symmetrically loaded ring is considered with unbalanced station load. Two stations (0,3) send and receive MPs at a higher rate than the residual stations. This is expresed by the routing matrix and the relative traffic amounts.

$$p_{ij} = \frac{1}{9}; \quad i = 0,3, \quad j = 0,..,9, \quad i \neq j,$$
$$p_{i0} = p_{i3} = \frac{5}{17}; \quad i = 1,2,4,..,9,$$
$$p_{ij} = \frac{1}{17}; \quad i,j = 1,2,4,..,9, \quad i \neq j,$$
$$c_0 = c_3 = \frac{45}{226};$$
$$c_i = \frac{17}{226}; \quad i = 1,2,4,..,9.$$

From equations (1), (3) and (4) it follows $\alpha_i = 2$ for all $i = 0,1,..,9$.

This can be interpreted that every time slot used for PS carries two minipackets every cycle, or, with respect to the overhead, this ring-system carries in total 14.016 Mbps for PS net rate.

Assuming, that 50% traffic is circuit switched (i.e.,73 FDX-connections with 64 kbps transmission rate each), the maximum PS-arrival rate to the whole system is

$$\lambda_{PS} = 6\ 636.4 \text{ messages/sec.}$$

Several other systems have been analysed, but if equation (5) is fulfilled, in most realistic cases α is approximately 2. The factor α is bounded by

$$\frac{N}{N-1} < \alpha < N$$

Especially, in case of a complete symmetrically loaded system, the factor $\alpha = 2$ holdes exactly.

Fig.6 shows the maximum PS-arrival rate as a function of the relative CS-traffic load. The idle slot concatenation or movable boundary principle is compared with the minipacket protocol for seveal values of α.

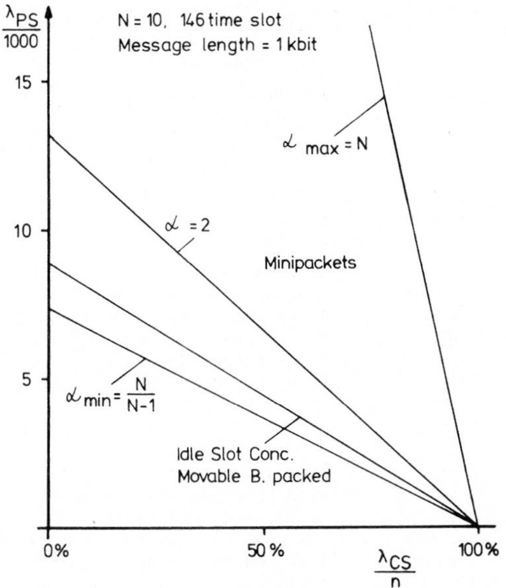

Fig. 6 Maximum PS-traffic rates vs. CS-traffic rates

The analytical delay analysis is based on an equivalent M/G/1 delay system with service interrupts. The delay depends heavily on the statistics of the interrupt periodes. The analytical delay evaluation has not yet been fully validated and will be reported in a later paper. Results on the delay performance are obtained by simulation, see chapter 5.

5 SIMULATION

5.1 Simulation Technique

The second performance aspect is the mean waiting time for PS-messages. These results have been derived from a simulation model of the ring system. The simulation method is based on the event-by-event-simulation: due to the diversity of temporal and spatial events the method had to be extended to include such effects.

The circulating frame is the only synchronism within the system. The stations are operating simultaneously on the ring, but accessing a single slot strictly after the preeceding station. This can be modeled exactly by a station, which works alone on the frame for one whole frame period having its own time/event-schedule. The arrival events at a single station are independent from events at all other stations and, therefore, they can be generated for a larger interval in the future. After finishing all operations on the frame, it is passed to the next station which starts working at the correct arrival time. The simulation system time runs along the pulse frame within a station and jumps than back to that time, the frame reaches the consecutive ring station.

5.2 Simulation Results

Waiting times for PS-messages derived from simulation runs are shown in Fig.7. A ring-system operating under the minipacket protocol is compared with an implementation of the fixed boundary. In both cases, a system with 5 symmetrically loaded ring-stations has been simulated. All other system parameters are identical to the example in chapter 4.2. Results from systems with more symmetrically loaded stations will be quite similar.

The waiting times are shown as a function of the total offered PS-load, based on a constant message length of 1024 bits or 22 minipackets, respectively. As a parameter, the CS-loads 30%, 50%, and 70% (in average 43.8, 73, and 102.2 simultaneous FDX-connections, using 64 kbps bandwidth each) are presented. To carry these CS-loads with suitable small loss on the fixed boundary system, the boundary between CS- and PS-part has been defined to 50, 80, and 110 time slots for CS and 1, 2, and 3 additional time slots for CS-signalling, respectively.

The CS-occupations of a time slot (e.g. telephone call) are very long compared with the short occupations of a slot for a single PS-message. Therefore, the simulation results for the minipacket protocol involve relatively large 95%-confidence intervals in spite of long simulation times. The lines have been drawn only for interpolation - dashed representing the fixed boundary's performance, bold for the minipacket version.

In the fixed boundary system a station holding the token sends all PS-messages which are in the buffer (exhaustive service). This leads to a better utilization of the PS-bandwidth, but also to higher delays in a very low loaded system. The waiting times in a minipacket system are much lower, but the processing times for partitioning a message in several minipackets at the sender and for reassembling the message at the receiver should be also taken into consideration.

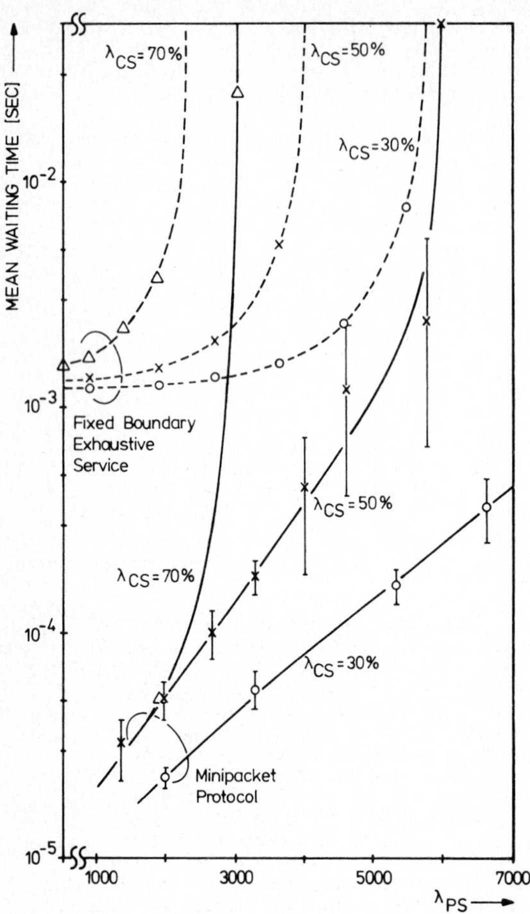

Fig. 7 Mean waiting times for PS-messages

6 CONCLUSION

New services, new multi-functional workstations and new communication requirements are claiming for new network-solutions, even in the private field. A new CS/PS integrated inhouse communication concept has been presented. It is based on integrated rings, providing CS with variable bandwidth and PS with high throughput-rate on demand. Different possibilties to implement these features on a synchronously circulating pulse frame have been discussed. One part of the performance evaluation is the maximum throughput calculation, especially for the minipacket protocol, showing that the relatively high overhead, necessary for addressing the MPs will be overcompensated for most realistic systems. An exact event-by-event simulation is the second tool, used for more results to evaluate the different protocols. The mean waiting times for PS-messages show also that the minipacket protocol, due to the higher maximum throughput, will be the more efficiently one.

7 ACKNOWLEDGEMENTS

The author would like to thank Prof. Dr.-Ing. P.J. Kuehn for his continuing interest and for supporting this project. The helpful discussions with W. Weiss in the design-phase and the programming efforts of J. Rothenburg are also appreciated.

8 REFERENCES

[1] E.- H. Goeldner, P.J. Kuehn, "Integration of Voice and Data in the Local Area," Proc. Data Communication in the ISDN-Era, Tel Aviv, Israel, pp. 103-117, 1985.

[2] C. Fruchard, J. Dejean, "A Hybrid Switched Open Network for Voice and Data Services," Proc. XI Int. Switching Symposium, Florence, (ISS), session 42-B, paper 2, 1984.

[3] J. Eberspaecher, "Optisches Lokales Netz fuer Sprache und Daten," Proc. Telematica, Stuttgart, pp. 224-233, 1984.

[4] M.W. Crozier, R.N. Pandya, R. Doshi, "Integrating Voice and Data in a Switching Node - A Comparison of Strategies," Proc. 10th Int. Teletraffic Congress, (ITC), Montreal, paper 1.1-4, 1983.

[5] M. Ross, O.A.Mowafi, "Perforamnce Analysis of Hybrid Switching Concepts for Integrated Voice/Data Communications," IEEE Trans. Com, vol. COM-30, no. 5, pp. 1072-1087, 1982.

[6] K. Kuemmerle, "Multiplexer Performance for Integrated Line- and Packet-Switched Traffic," Proc. Int. Conf. on Comp.Communication (ICCC), Stockholm, pp. 507-515, 1974.

[7] E. Arthurs, B. Stuck, "Traffic Analysis for Integrated Digital Time-Division Link Level Multiplexing of Synchronous and Asynchronous Message Streams," IEEE Journal on Selected Areas in Comm., vol. SAC-1, no.6, 1983.

[8] I. Gitman, W.-N. Hsieh, B.J. Occhiogrosso, "Analysis and Design of Hybrid Switched Networks," IEEE Trans. Com., vol. COM-29, pp. 1290-1300, 1981.

[9] M.J. Fischer, T.C. Harris, "A Model for Evaluating the Performance of an Integrated Circiut- and Packet-Switched Multiplex Structure," IEEE Trans. Com, vol. COM-24, pp. 195-202, 1976.

[10] B. Maglaris, M. Schwartz, "Performance Evaluation of a Variable Frame Multiplexer for Integrated Switched Networks," IEEE Trans. Com, vol. COM-29, no. 6, pp. 800-807, 1981.

[11] R.H. Kwong, A. Leon-Garcia, "Performance Analysis of an Integrated Hybrid-Switched Multiplex Structure," Performance Evaluation vol. 4, pp. 81-91, 1984.

[12] A.G. Konheim, R.L. Pickholtz, "Analysis of Integrated Voice/Data Multiplexing," IEEE Trans. Com., vol.COM-32, no. 2, pp. 140-147, 1984.

[13] W. Hilal, M.T. Liu, "Local Area Networks Supporting Speech Traffic," Computer Networks, vol. 8, pp.325-337, 1984.

TELETRAFFIC ISSUES in an Advanced Information Society
ITC-11
Minoru Akiyama (Editor)
Elsevier Science Publishers B.V. (North-Holland)

CHANGES IN TRAFFIC PARAMETERS DUE TO INTRODUCTION OF TIME DEPENDENT TARIFFS

Rune BÄCKSTRÖM and Paul DAHLSTRÖM

Swedish Telecommunications Administration
Farsta, Sweden

ABSTRACT

This paper presents a study on changes in some traffical parameters due to bringing time dependent tariffs into use in Sweden. Nationwide measurements referring to both traffic and calls have been executed during a period of two years before and two years after the change-over.

From these measurements a lot of measuring parameters have been formed and analysed according to changes in their values. As the measurements comprises a stringent limited time period, changes caused only by the change-over could be detected.

Our conclusions are that the daily traffic profile which already earlier was rather even has been further smoothed out a little and that this depends on move of calls between time tariffs more than changes in holding times. Differences in parameter values follow the expected directions but their sizes have been more difficult to predict in several cases.

MOTIVES FOR REORGANIZATION OF THE TARIFF SYSTEM

Earlier Sweden has had telephone tariffs based only on distance but in January 1983 a new tariff system was adopted, where the interval between the meter pulses also depends upon the time in the day when the call is made. The main motives for the reorganization were:

- An attempt to smoothe down the differences in the traffic intensity during both the daily hours and days of the week and in that case investments will be postponed to a later time owing to a more efficient use of the available traffic capacity.
- An attempt to improve the relationship between the tariffs and the real costs for different types of traffic.

THE TARIFF SYSTEM

Before the change-over the tariff system consisted of nine tariff classes for national automatical traffic. Connected with the change-over the three lowest classes were replaced by a new one. The other classes kept their puls-intervals without changes during 1200-1800 monday-friday (normal tariff), were shortened about 25 % during 0800-1200 monday-friday (high tariff) and were lengthened about 33 % during 1800-0800 monday-friday and all weekend (low tariff). The number of classes were reduced from nine to seven but were simultaneously differentiated so that a total number of 18 tariff intervals are in use. (In some cases a low tariff interval in a higher class coincides with a high tariff interval in a lower class).

MEASURING PLAN

Before the change-over nationwide traffic measurements were done in order to study the traffic behaviour as well as the call distributions. Initially measurements were done at some 100 exchanges in the size of 200-40000 subscriber lines. Later on the number of exchanges have been reduced to about 60, depending on the high claims for accuracy and completeness in the results. Every selected exchange have been measured twice a year (spring/autumn) and the measurement time comprises 14 days between 0600-2400 every day. These measurements started in spring 1981 and therefore at most comprise 8 measured weeks per exchange.

After the change-over corresponding measurements have been done in the same exchanges and a complete result would also consist of 8 weeks. Because the measurements had continued under the same conditions after the change-over, the results are directly comparable. The measurements have now been finished (autumn 1984) but will eventually be resumed in the future as a tool to study the effects of coming change-overs.

MEASUREMENT EQUIPMENTS

Traffic measurements have been executed using two different types of equipment, TM800 and TM-T1/T2. TM800 is an equipment fixed connected in an exchange and has a capacity of max 24000 devices (800 groups of 30 devices). TM-T1/T2 is a transportable equipment with a capacity of max 9600 devices (320 groups of 30 devices). The traffic values have been collected every hour by a central computer which called up the equipments using switched lines and asked for transmission of the stored data.

The registration of the calls have been done by an equipment (TAL-F) with a capacity of 16 inlets. Each inlet consists of 4 wires which will give a possibility to record the starting time, holding time, dialled number, conversation time and number of meter pulses. The result outputs are done on 3M-cartridges.

MEASURING POINTS

In order to obtain the totally generated traffic in an exchange, traffic measurements have been done on all the subscriber stages. TAL-F has been connected to 16 of the most used outgoing lines from the stages. This means that the recording of calls is a sample of the total traffic distribution. (Fig. 1)

RESULT PROCESSING

At the end of each measurement period the data are processed in 4 steps.

Step 1 handles the recorded data from TAL-F and analyses the dialled number using a data file containing all destination codes in use for the exchange in order to supply each call with a correct tariff. A summary per hour is done regarding the number of call attempts, total holding time, number of conversations, total conversation time and number of meter pulses per tariff.

Step 2 handles the traffic values from TM800 and TM-T1/T2. If more than one equipment are used a summary of their values are made.

Step 3 calculates a coefficient per hour based on the relation between measured traffic (TM800 and TM-T1/T2) and its corresponding value from TAL-F. This coefficients are used to transform the data resulting from step 1 to comprise the whole exchange. Certain limitation in this transforming have been done in order to achieve greatest possible accuracy in the results.

Step 4 forms the final summary. Traffic profiles, -volumes and matrices of tariffs are calculated per time and distance tariffs. One of the distance tariffs contains international, service and unidentified traffic and is not used in the analysis. A flowing chart over the result processing is shown in fig. 2.

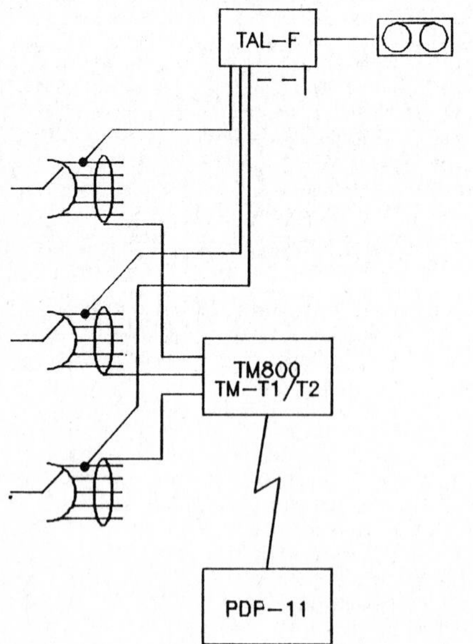

Figure 1

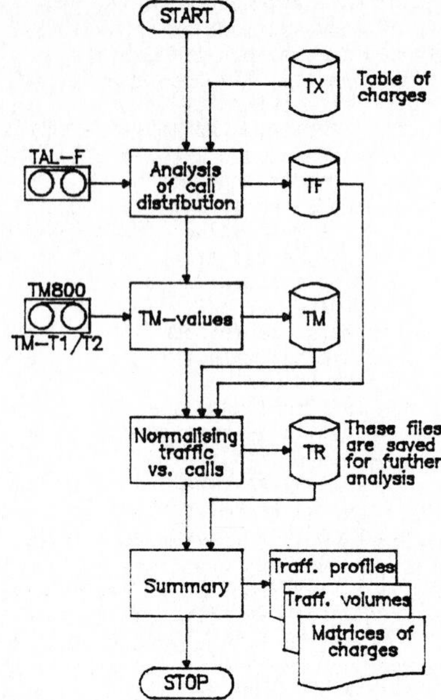

Figure 2

CRITERIA FOR ANALYSING

The results, which are to participate in the analysis, must consist of almost perfect measured weeks from each exchange. An almost perfect measured week is defined as a week which has not lost more than 3 hours during high tariff time or in total 8 hours. In order not to effect volume values lost hours have been given as probable values as possible. Furthermore to satisfy the criteria mentioned above each exchange must supply two or more accepted weeks both before and after the change-over to participate in the analysis. These criteria are reducing the available exchanges to a number of 38.

These 38 exchanges represent in total 159100 subscriber lines in year 1981/-82 and 164500 in year 1983/-84.

In order to form comparable results from exchanges of different sizes, a normalization must be done. Therefore we have chosen to transform the result to be based on 1000 subscriber lines. This is applicated on all volumes and traffic values for each exchange.

In this way the increase (3.4 %) in number of subscriber lines, between the years mentioned above and representing the time before and after the change-over, does not effect the comparison between the corresponding values.

To show the size of the sample we can mention that the total number of subscriber lines are just above 5 millions (1984).

PARAMETERS USED IN THE ANALYSIS

From the weekly results the following primary measurement parameters are formed:

- Mean value of traffic per hour, working-day and weekend respectively

- Number of call attempts per time tariff
- Number of conversations per time tariff
- Total holding time per time tariff
- Total conversation time per time tariff
- Number of meter pulses per time tariff.

For each of the selected exchanges the parameter values are accumulated in groups according to the measurement period (before/after change-over). By doing this each exchange will get a mean value for every parameter, both before and after, and we have got a base for simple comparisons.

From the primary parameters the following secondary parameters are formed:

- Traffic volume per time tariff
- Total traffic volume
- Busy hour traffic (TCBH)
- Busy hour traffic as part of daily volume
- Traffic level related to TCBH-value per hour, working-day and weekend respectively
- Traffic flow and -level sorted and accumulated per hour, working-day and weekend respectively
- Parts of call attempts per time tariff
- Mean holding time per time tariff
- Mean conversation time per time tariff
- Mean number of pulses/call per time tariff
- Answer ratio per time tariff
- Efficiency ratio per time tariff.

Besides weighted total values are formed for the time tariff dependent parameters.

METHODS OF ANALYSIS

At the analysis of the changes caused by the change-over very simple methods have been used. From the result from each exchange, every parameter form a difference

$$d_i = a_i - b_i$$

where a and b are the parameter values after and before the change-over respectively. Taking the same parameter differences from all the exchanges, the mean value and standard deviation

$$m = \sum_{i=1}^{n} d_i / n$$

$$s = \sqrt{\left(\sum_{i=1}^{n} d_i^2 - n \cdot m^2\right) / (n - 1)}$$

where n is the number of exchanges, are calculated.

In order to confirm if significant changes have occurred we have used a singlesided t-test with the expression

$$s \cdot t_\alpha(f)/\sqrt{n}$$

where α is the significance level chosen to 10 %, and f is the number of degrees of freedom, n - 1, in the t-distribution.

The absolute mean values for each parameter are also calculated

$$\bar{a} = \sum_{i=1}^{n} a_i / n$$

$$\bar{b} = \bar{a} - m$$

By doing this we have created a way to express also relative changes both for mean values and for significant changes, sized at 2 % and 5 %.

TRAFFICAL ANALYSIS

- Busy hour traffic
  The busy hour is defined by TCBH for every exchange, both before and after. The traffic value belongs to the busy hour even if the latter has been moved.
  A slight decrease, even though not significant, is shown in table 1.
- Daily traffic profile
  Related to every point of time, working-day and weekend respectively, traffic values are shown.
  Even after the change-over, the most heavy traffic hours remain in the mornings of the working-day as can be seen in figure 3. However it shall be noticed that the profile has been smoothed out and that the differences in a very conspicuous way follows the time tariff intervals of the working-day, as can be seen in the lower part of the figure.
  Figure 4 shows that weekend-traffic has increased throughout the whole day. However, the traffic volume of the weekend has not reached the same level as the working-day.
  Of the hours surrounding the tariff changes an obvious change can be seen in two cases, partly a decrease during the first hour in high tariff and partly an increase during the first hour in low tariff. Otherwise traffic values are not affected.
  (Figure 3)
- Traffic volumes per time tariff
  The volumes are measured in erlang hours per week. Low tariff refers to both working-day and weekend.
  From table 1 an observable decrease in high tariff volume can be seen. The increase in low tariff volume, significant + 5 %, more than compensate for the loss described above. Normal tariff volume has not been affected at all.
- Total traffic volume
  The volume is measured in erlang hours per week.
  A slight increase, however not significant, is shown in table 1.
- Busy hour traffic as part of daily volume
  The ratio is made of the busy hour traffic part of the traffic volume in an average working-day.
  From table 1 a significant, - 2 %, lower ratio can be seen. This means an improved use of the traffic capacity.

Tariff	mean b	mean a	diff	quota	sig
Busy hour traffic (TCBH)					
	24.98	24.19	-0.79	0.969	-
Traffic volumes per time tariff					
High	434.3	413.7	-20.6	0.953	-2%
Normal	580.6	581.3	+0.7	1.001	-
Low	745.9	804.7	+58.9	1.079	+5%
Total traffic volume					
	1760.7	1799.7	+39.0	1.022	-
Busy hour traffic as part of daily volume					
	0.0879	0.0841	-.0038	0.957	-2%

Table 1 All the 38 exchanges

- Sorted and accumulated traffic level
  For every exchange the hourly traffic values have been sorted in decreasing size, working-day and weekend respectively, and related to the busy hour traffic.
  By sequently adding the sorted values each succeeding hour will get an accumulated value showing the number of busy hour volumes up to that number of hours. This will be a measure for the evenness of the daily profile.
  The values described above are illustrated in figures 5 and 6 for working-day and weekend. For the working-day the difference is significant, + 2 %, from 8 hours and upwards and for weekends, + 5 %, throughout all values.

CALL ANALYSIS

- Number of call attempts (table 2)
  The values refer to the number of call attempts per subscriber line and week and show a slight increase in total, significant + 2 %, during low tariff.
- Number of conversations (table 2)
  The values refer to the number of conversations per subscriber line and week and show significant, + 2 %, increases during both low tariff and totally.
- Parts of call attempt volume (table 2)
  The values refer to parts belonging to corresponding time tariffs and related to the total volume. A significant movement, - 2 % and + 2 %, of the values from high tariff towards low tariff can be seen.

- Mean holding time (table 3)
  The values are given in seconds per call and show a negligible decrease.
- Mean conversation time (table 3)
  The values are given in seconds per conversation and show a small decrease, however significant, - 2 %, during high tariff. Some increase for nonlocal conversations during low tariff occurs.

Tariff	mean b	mean a	diff	quota	sig
Number of call attempts					
High	7.87	7.71	-0.16	0.979	-
Normal	11.41	11.66	+0.25	1.022	-
Low	12.11	12.96	+0.85	1.070	+2%
Total	31.39	32.33	+0.94	1.030	-
Number of conversations					
High	5.06	5.07	+0.01	1.001	-
Normal	7.10	7.31	+0.21	1.030	-
Low	7.40	7.90	+0.50	1.067	+2%
Total	19.57	20.28	+0.72	1.037	+2%
Parts of call attempt volume					
High	0.251	0.238	-0.013	0.950	-2%
Normal	0.363	0.360	-0.003	0.991	-
Low	0.386	0.402	+0.016	1.041	+2%

Table 2 All the 38 exchanges

- Number of meter pulses per conversation (table 3)
  The values are given in number of pulses. Of course the change-over has caused great changes in the number of meter pulses during high and low tariff respectively. Because the local conversations form the greater part (75 %) of the total, the values differ less than the corresponding changes in time intervals.
  The total decrease is significant, - 2 %.

Tariff	mean b	mean a	diff	quota	sig
Mean holding time					
High	194.1	188.0	-6.1	0.969	-
Normal	179.1	175.4	-3.7	0.979	-
Low	218.5	219.9	+1.4	1.006	-
Total	197.9	196.1	-1.8	0.991	-
Mean conversation time					
High	252.7	239.7	-13.0	0.949	-2%
Normal	235.6	230.6	-5.0	0.979	-
Low	304.9	309.2	+4.3	1.014	-
Total	265.9	263.2	-2.7	0.990	-
Number of pulses per conversation					
High	4.46	4.99	+0.53	1.118	+5%
Normal	4.08	3.98	-0.10	0.975	-
Low	4.71	3.86	-0.85	0.819	-5%
Total	4.42	4.19	-0.23	0.948	-2%

Table 3 All the 38 exchanges

- Answer ratio (table 4)
  The values show the number of conversations divided by the number of call attempts and no visible changes can be seen.
- Efficiency ratio (table 4)
  The values show the total conversation time divided by the total holding time. These values too are unaffected by the change-over.

Tariff	mean b	mean a	diff	quota	sig
Answer ratio					
High	0.643	0.657	+0.014	1.022	-
Normal	0.623	0.628	+0.005	1.008	-
Low	0.612	0.611	-0.001	0.998	-
Total	0.624	0.628	+0.004	1.007	-
Efficiency ratio					
High	0.835	0.836	+0.001	1.002	-
Normal	0.818	0.824	+0.006	1.008	-
Low	0.852	0.858	+0.006	1.007	-
Total	0.837	0.842	+0.006	1.007	-

Table 4 All the 38 exchanges

DISTANCE TARIFFS

As we earlier mentioned the Swedish tariff system in respect to distances has not been changed mutually, except for local calls. These were before the change-over charged for one meter puls independent of duration of the call, but have now a pulse interval of 6 minutes (high and normal tariff) or 12 minutes (low tariff). Since this change is less important from an economic point of view, though most discussed in mass-media, the presented results up to now refer to differences and values wholly based on time tariffs.

However we have analysed the call parameters also in respect to three distance classes and would mention a few of these results. The results for number of call attempts have changed slightly and show small increases for all the three classes, but in parts of volume there is a small move, 0.6 %, from the class of local calls to the longer distance classes. Also the number of conversations have increased slightly in all the classes.

The mean conversation times are unaffected on the whole with a decrease for local calls with 1.5 % as the greatest relative change all over the distance classes. On the other hand there are great differences in the absolute mean values and these differences tend to increase. This depends on that already before the change-over the local calls during "high tariff" were the shortest calls and the long distance calls during "low tariff" had the longest duration. The mean values now vary from 225 seconds for the former to 470 seconds for the latter. Obviously this confirms earlier observations that the duration of calls increases with the distance.

Number of meter pulses per conversation show decreasing mean values, except for local calls, in consequence to partly greater lowering in low tariff than raising in high tariff and partly certain move of calls from high to low tariff. The average local call costs during high tariff 1.24 meter pulses and during low tariff 1.10 meter pulses.

EXCHANGES WITH DIFFERENT TRAFFICAL BEHAVIOUR

Already before the change-over there were quite a number of exchanges that had their most busy hours in the afternoon and in the evening. They are found in suburbs, decided domestic areas and small villages. Out of traffical view it is naturally not desirable that these exchanges move a great part of their total traffic volume to low tariff time.

In order to analyse this category we have out of our sample chosen the exchanges that before the change-over had the greatest parts of their traffic volumes during "low tariff". By setting a limit for that parameter to 43 % we got for analysis a selection of 16 exchanges that may represent this category. As can be seen from figure 7 they are not decided "low tariff exchanges" since also for this group the resulting busy hour occurs in the morning.

Table 5 shows the same parameters as table 1 for the whole sample and figures 7 and 8 correspond to figures 3 and 4 respectively. The traffical changes seem to be rather similar to them presented above and most important the volume during low tariff show less increase than above.

Tariff	mean b	mean a	diff	quota	sig
Busy hour traffic (TCBH)					
	24.08	23.33	-0.75	0.969	-
Traffic volumes per time tariff					
High	413.6	388.7	-24.9	0.940	-2%
Normal	561.6	556.3	-5.3	0.991	-
Low	803.9	847.7	+43.8	1.055	+2%
Total traffic volume					
	1779.0	1792.7	+13.6	1.008	-
Busy hour traffic as part of daily volume					
	0.0856	0.0831	-0.0025	0.971	-

Table 5 The 16 "low tariff" exchanges

CONCLUSIONS

The traffical analysis shows a desired move of traffic volume from high to low tariff time. Though the changes are limited we can notice that a decrease in busy hour traffic occurs simultaneous with a slight increase in total traffic volume. Of course the changes mentioned above form a favourable combination leading to a more efficient use of the traffical capacity. However we ought to be aware of that an even traffic profile consisting of mean values during a week conceals day to day variations. Since these individual variations are much greater in the evening hours than in the morning, a mean value profile with the busy hour in the morning is preferable.

The call analysis shows some increase in the number of conversations, especially during low tariff time. On the other hand the mean conversation time has decreased slightly. Certainly the trend of the conversation times is as could be expected between the time tariffs mutually, but the increase during low tariff time is perhaps less than expected. An explanation to that is that the subscribers have not in any further extension used the opportunity to lengthen their calls. It could be that they even before the cost reduction regarded the charges for the conversations as not too expensive. International comparisons also shows that Sweden has low telephone costs and that this fact can explain the minor changes in the conversation times.

Generated traffic per 1000 subscr. and hour, working-days

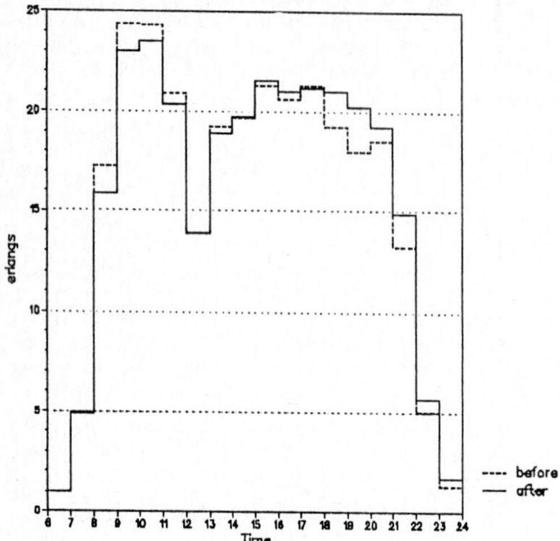

Difference

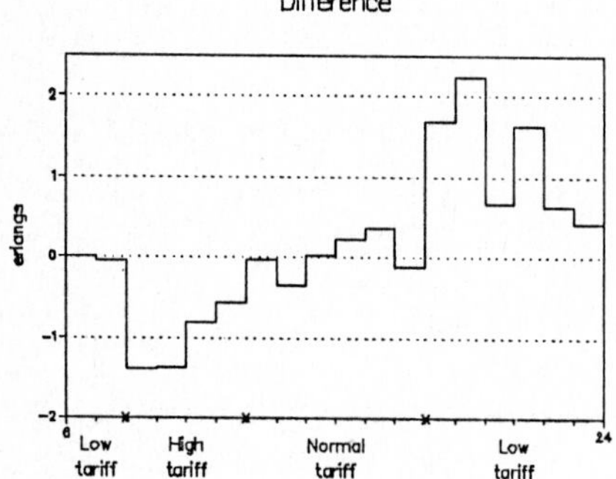

Figure 3 All the 38 exchanges

Generated traffic per 1000 subscr. and hour, weekends

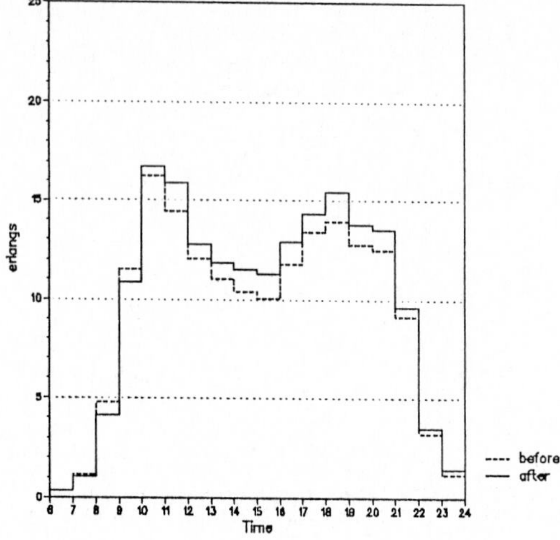

Difference

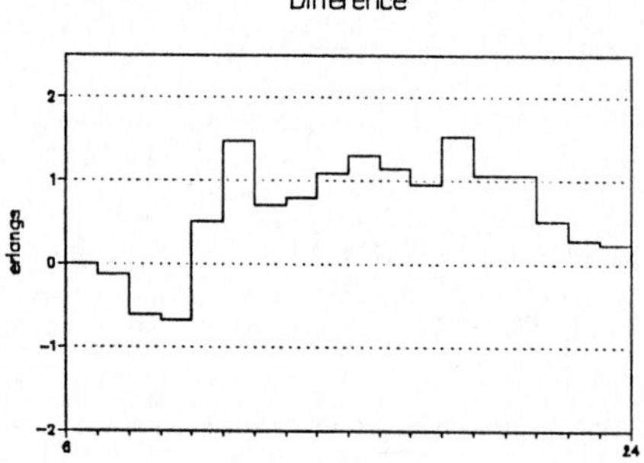

Figure 4 All the 38 exchanges

Accumulated traffic volume, working-days

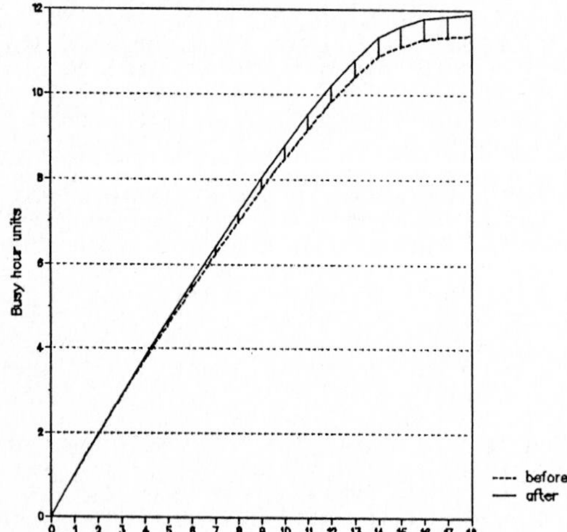

Figure 5 All the 38 exchanges

Accumulated traffic volume, weekends

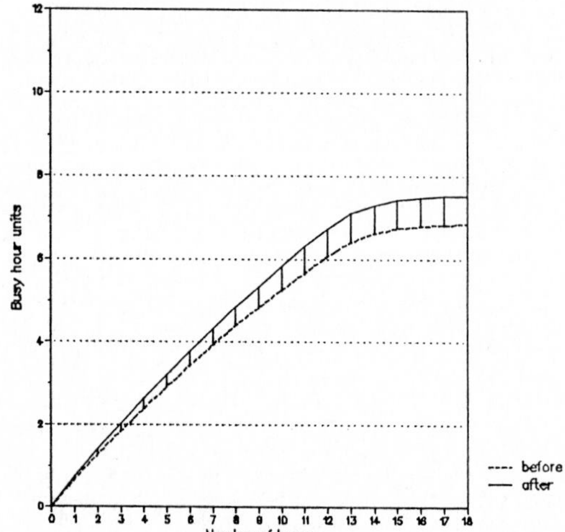

Figure 6 All the 38 exchanges

Generated traffic per 1000 subscr. and hour, working–days

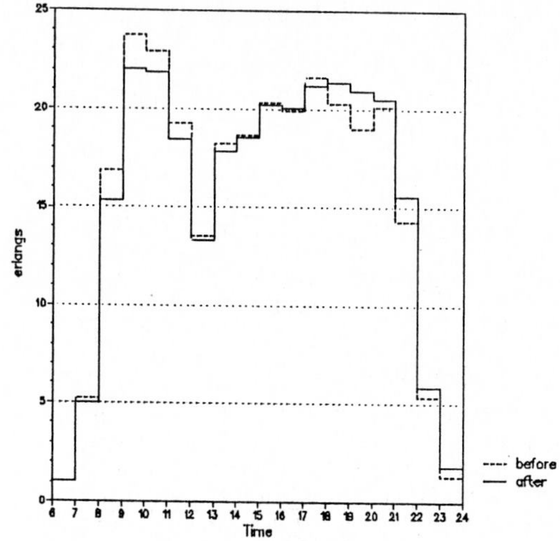

Difference

2
1
0
−1
−2
erlangs
6
24
Low tariff
High tariff
Normal tariff
Low tariff

Figure 7 The 16 "low tariff" exchanges

Generated traffic per 1000 subscr. and hour, weekends

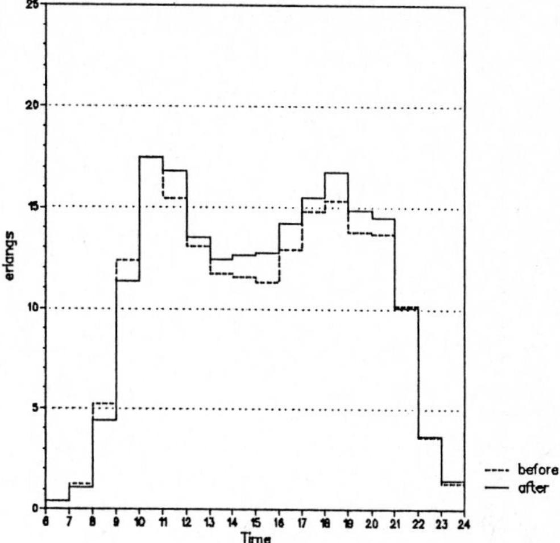

Difference

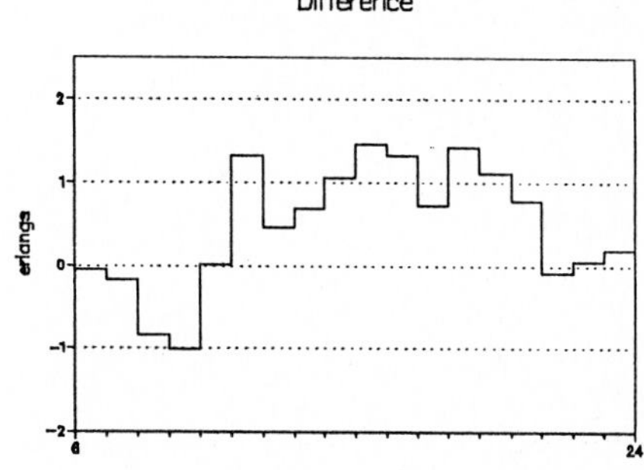

Figure 8 The 16 "low tariff" exchanges

TELETRAFFIC ISSUES in an Advanced Information Society
ITC-11
Minoru Akiyama (Editor)
Elsevier Science Publishers B.V. (North-Holland)

ECONOMY AND SERVICE ASPECTS OF DIFFERENT DESIGNS OF ALTERNATE ROUTING NETWORKS

Per LINDBERG, Krister NIVERT and Barbro SAGERHOLM

Swedish Telecommunications Administration
Farsta, Sweden

ABSTRACT

The concept of cluster grade of service is introduced and a design algorithm for cluster engineering is presented. Protection trunk groups are discussed and an algorithm for their optimal dimensioning is given. The two design criteria cluster grade of service and final choice grade of service are used in optimizing a three level 234 node hierarchical network. Resulting network cost and end-to-end grade of service are compared, and possible improvement using end-to-end blocking constraints in the optimization are investigated.

1. INTRODUCTION

Automatic operation was early introduced in the Swedish long distance network. The electromechanical switching machines were provided with common control, and offered extensive facilities for overflow to other trunk groups, in case of blocking situations. A design principle was called for adopting the idea of alternate routing and taking the non poisson properties of overflow traffic into account.

A traditional hierarchical network already existed with a starshaped network of backbone trunk groups and the odd direct trunk group where these were justified. It was immediately observed that alternate routing allowed direct trunk groups to be implemented even for small traffic parcels depending on the interaction between less utilized cheaper direct circuits and better utilized but more expensive backbone paths. The trade off between high usage and overflow paths was recognized as an important target for network design, and cost effectiveness has henceforth been intimately related to alternate routing network design methods, implying the sense of optimization techniques in the application of cost minimization under Grade of Service constraints.

2. THE 'SWEDISH METHOD'

A network design method of this kind generally consists of two parts, one designated to estimate a good balance between direct and overflow circuits, the other to dimension the overflow path so that a prescribed GOS standard is obtained. The method developed in Sweden during the -50's by Tånge [1] takes a start in a fully provided single trunk group. Direct high usage trunk groups are then justified by comparing the cost per erlang for carrying traffic on the last circuit of a direct trunk group and the first circuit in the overflow path. The latter is subsequently reduced by a number of circuits carrying the same amount of traffic as the directs overflowing on them. This step may be repeated any number of times, 'weighting' high usage trunk groups out of 'detours' as the Swedish expressions would appear in English, allowing calculations of hierarchical networks with any number of tiers.

The dimensioning stage in this algorithm is based on calculational methods for gradings presented by Berkeley [2] requiring only one parameter viz. the mean of traffic offered in this case all traffic parcels offered to the grading. The correspondence in alternate routing networks is the cluster consisting of a final and all trunk groups overflowing on it. Thus the concept of cluster grade of service was adopted as a criterion for alternate routing network design in Sweden. Tests in the electromechanical artificial traffic machine revealed, that although the GOS was fairly evenly distributed over traffic parcels first offered to high usage trunk groups, the traffic first routed to the final sometimes faced an unacceptable poor GOS. To overcome this a service protection group was introduced for this traffic. To simply increase the final slightly to give first routed traffic acceptable GOS may prove cheaper but the protection group is considered to have some advantages in overload situations.

This 'Swedish Method' has some obvious advantages in situations when all engineering is carried out manually by using tables and diagrams. It was accepted and recommended by CCITT for the study periods 1956-1968 [3] and was used in its original form even as computerised in Sweden until some years ago.

3. THE EQUIVALENT RANDOM THEORY

During the 50's Wilkinson developed his well known Equivalent Random Theory [4] and this result was presented internationally at the same time as the 'Swedish Method'. Applying ERT one would use the mean and variance of the traffic offered to the final as basis for its sizing. Thus it may seem natural to relate the GOS standard to the traffic on the final.

A dimensioning method for trunk groups offered overflow traffic, using the ERT, relating the GOS standard to the final, and fixing it at 1 % to ascertain reasonable service for first routed traffic to the final, was adopted by CCITT in parallel with the 'Swedish Method' and is now the only still in recommendation [5]. As it is a two parameter method it is naturally superior in accuracy but also incurs more calculation work, even if the solution of a system of equations is avoided by use of tabulated results.

Comparisons with the 'Swedish Method' revealed that ERT in combination with 1 % GOS on finals required more circuits on the finals, but combined with 1 % GOS on the cluster gave smaller finals than the 'Swedish Method', typically 2 to 5 circuits depending on the size and structure of the cluster, even if the need for protection trunk groups was accounted for.

4. NON STATIONARY TRAFFIC

The fact that the ERT approach allowed for a meaner dimensioning of finals caused of course an increasing interest in applying ERT in practical engineering. Such investigations were however discouraged when Neal presented updated engineering tables taking day-to-day variations into account [6]. It was found that after extending the finals to allow for unexpected high peakedness in the traffic offered attributed to non stationary properties they were in general of the same size as those obtained by the 'Swedish Method'. Hence the interest in ERT as a 'cheaper' method faded.

The apparent property of the 'Swedish Method' to take day-to-day variations into account was however quite coincidental and varied a lot for individual trunk groups. Some asset was to be found in the fact that the basis for engineering was first offered traffic to the cluster rather than overflow traffic offered to the final. This was however not a sufficient answer to the problem as it was seen that even if the sizing of finals was on average correct, some were overdimensioned and other were underdimensioned. A more accurate dimensioning method was asked for as well as a method to establish GOS standards for non stationary traffic. The latter problem was treated by Gunnarsson et. el. at ITC 10 [7] and is not dealt with here. The result is in general a GOS standard depending on the traffic offered, so the more traffic the lower congestion.

5. NEW ANALYTHICAL AND NUMERICAL METHODS

The adoption of more complex routing patterns, some of them not strictly hierarchical, especially in connection with digitalisation of switching centres, disabled the method introduced by Tånge. It was recognised that, to facilitate analytical treatment of the situation that appeared in the network, at least two parameters had to be handled, and not only for the overflow but also for carried traffic. The investigations leading up to full analytical treatment of overflow arrangements were presented by Lindberger at ITC 10 [8].

The Hayward [9] approach to cater for peakedness was originally proposed for numerical treatment. It was however found wanting in accuracy in extreme situations which frequently enough appear in real life networks. The 'Hayward Method' is retained for manual calculations as a reasonable compromise between calculational effort and accuracy. For computer calculations however another trade off between calculational effort and accuracy was looked for.

The ERT was accurate enough to avoid the identified problems but the calculational work involved in the optimisation of large networks needed approximations with controlled accuracy. The solution to this problem is presented in Annex A.

6. CLUSTER ENGINEERING

A new design method for alternate routing networks with a more general applicability was investigated to replace the 'Tånge Method'. The solution is in principle a cost minimisation of subnetworks under GOS constraints. By repeated treatment of the same subnetwork in an iterative procedure an improved total network cost is obtained. The GOS standard is still related to the total traffic offered to the cluster, thus the method is called cluster engineering.

The following basic steps may be seen in the calculation algorithm.

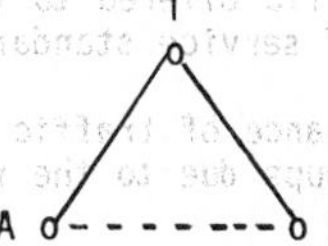

Fig. 1

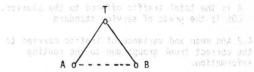

1. Subdivide the network in triangular subnetworks like fig. 1. Using the routing information establish a calculation sequence so that a trunk group is not calculated until all trunk groups overflowing on it and all trunk groups carrying traffic offered to it are calculated.

This is a necessary condition that restricts the set of networks to be treated but still allows a fair variety of not strictly hierarchical routings interesting in practical networks.

2. Obtain the number of circuits n_1 in trunk group AB:

2.1 m_1 and v_1 are mean and variance for traffic offered to AB. Calculate using 'ERT' the fictitious trunk group's A;x and obtain the optimum number of circuits n_1 in trunk group AB so that:

$$A \cdot \frac{dE(x + n_1, A)}{d(x + n_1)} = h$$

using the well known Rapp approximation for the optimal occupancy:

$$h = \varepsilon[1 - .3(1 - \varepsilon^2)]$$

where

$$\varepsilon = \frac{C_1}{C_2 + C_3}$$

C_i being the cost per circuit in trunk groups AB, AT and TB respectively.

2.2 Subdivide both mean and variance of the carried and overflow traffic AB into traffic groups, using the method presented by Lindberger [8]. Each group shall have an unique routing.

2.3 Add each traffic group to the trunk group corresponding to the routing information.

3. Repeat step 2 following the sequence established in 1 until all trunk groups with an overflow possibility are calculated.

4. Dimension the final trunk groups to meet the cluster grade GOS standard.

4.1 m and v are mean and variance of traffic offered to the final. Calculate using 'ERT' A_f and x of the fictitious group yielding m and v and obtain the number of circuits n in the final so that

$$A_f \cdot E(x + n, A_f) = A \cdot GOS$$

where

A is the total traffic offered to the cluster.
GOS is the grade of service standard

4.2 Add mean and variance of traffic carried to the correct trunk groups due to the routing information.

5. Repeat 4 until all finals are dimensioned.

6. Interrupt if solution is obtained. Criteria for interruption may be that no changes in number of circuits occur or that network cost does not improve satisfactorily.

7. Obtain optimum number of circuits n_1 in trunk group AB.

7.1 m and v are mean and variance for traffic offered to trunk group AB. Calculate using 'ERT' A_1 and x_1 for fictitious trunk group yielding m and v.

7.2 Calculate

$$m_1' = A_1 \cdot E(x + n_1', A_1)$$

and corresponding v_1'
n_1' beeing old solution to n_1.

7.3 Calculate $m_1'' = A_1 \cdot E(x_1 + n_1' + 1, A_1)$

and corresponding v_1''

7.4 m_2 and v_2 are mean and variance of traffic offered to trunk group AT in previous solution. Calculate $m_2' = m_2 - m_1' + m_1''$

$$v_2' = v_2 - v_1' + v_1''$$

7.5 Calculate using 'ERT' A_2 and x_2 for fictitious group yielding m_2' and v_2' and obtain number of circuits n_2' in trunk group AT so that:

$$A_2 \cdot E(x_2 + n_2', A_2) = M_2$$

where

M_2 is overflow from trunk group AT in previous iteration.

7.6 Repeat 7.4 and 7.5 to obtain n_3' for trunk group TB.

7.7 Calculate

$$h = \frac{C_1}{\dfrac{C_2(n_2 - n_2')}{m_2 - m_2'} + \dfrac{C_3(n_3 - n_3')}{m_3 - m_3'}}$$

7.8 Calculate new n_1 so that

$$A \cdot \frac{dE(x_1 + n_1, A_1)}{d(x_1 + n_1)} = h$$

7.9 Update mean and variance of traffic offered to correct trunk groups due to the routing information using new mean and variance of traffic carried and overflow traffic for trunk group AB.

8. Repeat 7 following the sequence in 1 until all trunk groups with an overflow possibility are calculated.

9. Go to 4.

7. PROTECTION TRUNK GROUPS

When calculations are interrupted, normally one iteration after the initiation with the Rapp approximation, a network with economically justified trunk group sizes is obtained. In this network some trunk groups carry both overflow traffic and traffic parcels first routed to them. In the latter case a substantially higher blocking probability were to be expected if no counteractive measures were taken.

The solution adopted in the Swedish network is protection trunk groups to which first routed traffic is offered before its overflow is merged with the rest of overflows and offered to the next trunk group in sequence. This method is implemented due to its relatively simple technical application in electromechanical switching machines and that it is not inferior to other methods of protection in failure free situations. As the circuits in the protection trunk group are similar to those in the overflow trunk group a special algorithm to optimize the size is required.

The reason for the protection trunk group is to limit the congestion for first routed traffic and the problem may be formulated:

Find the size of the protection trunk group that together with the overflow trunk group yields the lowest cost and gives first routed traffic a congestion of not more than k times the average congestion in the cluster.

A solution to this problem is included in [8] but its essence will be recapitulated here in connection with fig. 2 which depicts a small cluster.

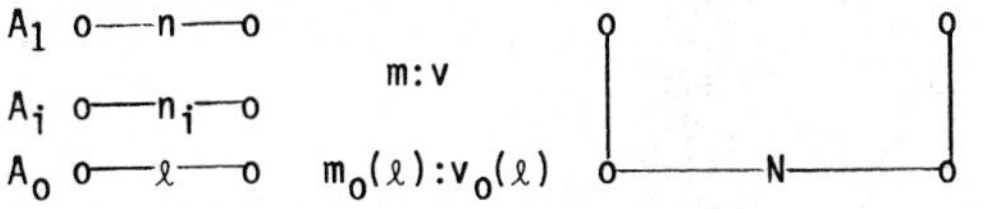

A_1 o—n—o
A_i o—n_i—o   m:v
A_0 o—ℓ—o $m_0(\ell):v_0(\ell)$ o—N—o

Fig. 2

The overflow traffic has mean m and variance v. A is the sum of first routed traffic entering the cluster on all levels and A_0 is the first routed traffic entering on this level. Note that the size of the overflow group may not be known yet. The sizing of the protection trunk group ℓ so that:

$$v_0(\ell) = \frac{v \cdot k \cdot A_0}{A - k \cdot A_0}$$

will give first routed traffic at this level a congestion of maximum k times the average congestion in the cluster.

The protection trunk group may fall short of a lower limit for implementation. In this case the grade of service for first routed traffic is protected by sizing the overflow trunk group so that the time congestion in the latter is less thank k times the average call congestion in the cluster. For the time congestion the approximation given by Fredericks [10] is used and the problem is defined as:

Find n so that:

$$B_T \leq \frac{2A_F \cdot E(x+n,A_F)}{A_F - x - n + \sqrt{(A_F - x - n)^2 + 4A_F \cdot n}} \leq k \cdot B$$

where A_F and x are the Wilkinson parameters for the fictitious group.

By implementing protection trunk groups on all levels in the network the variation of individual blockings within a cluster may be controlled and kept in a narrow band. In practice however protection trunk groups are mostly used only for finals so a practical value for k seems to be 2 allowing for a twice higher blocking than average in the worst cases.

8. FINAL CHOICE BLOCKING

The GOS standard recommended by CCITT [5] is based on the blocking probability in the final and differs from the cluster GOS in step 4 of the algorithm where

$$A_f \cdot E(x + n, A_f) = m \cdot GOS$$

would replace the engineering equation.

9 END-TO-END BLOCKING

With end-to-end blocking constraints the objective of step 4 in the algorithm may be rephrased as:

Minimize the sum of cost of the finals subject to the constraint

$$B_r \leq B_{max}$$

for all relations r.

A solution to this problem was presented at ITC 10 by COST-201 [11].

10. TEST NETWORK

The three design methods are applied to optimize a real size network. It is realistic in the sense that it is a full scale representation of a fully digitalised Swedish national network. The test network is a three tier network with 234 switching nodes. There are four transit nodes on the highest level and 16 transit nodes on the intermediate level. There are 470 hierarchical finals (two nodes are dedicated to incoming traffic respectively) all unidirectional and the network is offered 45800 erlang. Subscribers are connected to exchanges on all levels so traffic may originate and terminate also on transit nodes.

All possible direct high usage trunk groups are allowed provided they are economically justified and exceeds a lower limit of 10 circuits. The same lower limit is used for protection trunk groups if applicable. The routing principle is hierarchical far to near and each traffic parcel is given two alternatives and is thereafter offered to a final. High usage to own top level transit is used only for traffic within its transit area. These rules have been the same for all cases.

With cluster GOS and final trunk group blocking respectively an optimum balance between high usage and final trunk groups has been established yielding for each method the lowest network cost. Thus the high usage part of the network is not identical in both cases. In cluster GOS the protection trunk groups have been introduced to limit blocking probability for first routed traffic to twice average cluster blocking on all levels.

For the two high usage networks obtained with cluster GOS and final trunk group blocking an optimization of the final trunk groups is made with end-to-end blocking constraint giving the same average EEB.

11. RESULTS

TABLE I
FINAL CHOICE BLOCKING

TRUNK-GROUPS		CIRC.	COST
FC	470	118322	865.8

TRAFFIC CARR. 44690
LOST 1100
TOTAL 45789

AVERAGE EEB 2.40%
AV.CLUSTER GOS 1.00%

Table I reports a network with only backbone routes, dimensioned for 1 % GOS. As there is no overflow traffic in this network the cluster is the final and the engineering algorithm is quite trivial. Diagram I shows typical peaks for 1 %, 2 % and 4 % congestion depending on the number of trunk groups in a path.

AV. EEB %	NO OF REL	TRAFFIC	
.97	95	2299	****
1.14	302	6422	***********
1.95	458	2540	****
2.13	3778	15063	*************************
2.86	7115	8621	**************
3.16	6040	4144	*******
3.71	13431	5268	*********
4.23	5560	1234	**
4.58	970	197	

DIAGRAM I

Tables II and III report networks dimensioned to meet cluster grade of service and final choice grade of service standards respectively.

TABLE II
CLUSTER GRADE OF SERVICE

TRUNK-GROUPS		CIRC.	COST
HU P	83	2298	22.5
HU	1515	34238	327.9
FC P	120	6422	50.0
FC	470	40218	229.7
SUM	2188	83176	630.1

TRAFFIC CARR. 44808
LOST 981
TOT 45789

AVERAGE EEB 2.14%
AV. CLUSTER GOS .89%

TABLE III
FINAL CHOICE BLOCKING

TRUNK-GROUPS		CIRC.	COST
HU	1533	36951	355.2
FC	470	48702	295.6
SUM	2003	85653	650.8

TRAFFIC CARR. 45384
LOST 405
TOT 45788

AVERAGE EEB .88%
AV. CLUSTER GOS .37%

As can be seen from TABLE II above 83 out of the 1515 high usage trunk groups and 120 out of the 470 final choice trunk groups are provided with protection for first routed traffic. TABLE III shows 18 more high usage trunk groups but a total of 185 less. The sum of circuits is however 2477 higher in the latter case and the total network cost is 3.3 % higher for a network optimised for final choice grade of service than for cluster grade of service relative to the costs that are affected by the engineering method. The cheaper network has a 2.14 % average end-to-end blocking and carries 575 erlang less during the network busy hour. In the CCITT recommendation an only route is dimensioned to give 1 % congestion, but if one or more high usage trunk groups are introduced the resulting GOS will decrease as can be seen as the difference in cluster GOS.

AV EEB %	NO OF REL	TRAFFIC	
0.30	135	1539	******
0.71	579	8163	**************************
1.25	944	6437	********************
1.76	2418	6953	*********************
2.25	4539	6265	*******************
2.74	6192	6383	********************
3.24	5919	3910	************
3.72	5748	2882	*********
4.22	3768	1437	****
4.73	2464	754	**
5.42	2919	635	**
6.39	1131	247	*
7.44	454	83	
8.49	205	42	
9.48	86	19	

Cluster grade of service
DIAGRAM II

AV EEB %	NO OF REL	TRAFFIC	
0.24	898	10966	***************
0.76	7507	18652	*************************
1.22	13179	9006	************
1.71	11466	6545	*********
2.17	4054	565	*
2.61	580	50	
3.18	60	3	

Final choice grade of service
DIAGRAM III

The final choice grade of service standard will give a more generously dimensioned network if 1 % standard is applied. If a 2.5 % standard is used for this case a result as given in table IV and diagram IV is obtained, which is quite similar to the cluster GOS result as far as network cost and average EEB is concerned.

TABLE IV
FINAL CHOICE GRADE OF SERVICE

TRUNK-GROUPS		CIRC.	COST
HU	1519	36465	350.3
FC	470	46129	279.9
SUM	1989	82594	630.2

TRAFFIC	CARR.	44766
	LOST	1023
	TOT	45789

AVERAGE EEB 2.23%
AV.CLUSTER GOS .94%

AV EEB %	NO OF REL.	TRAFFIC	
.37	274	3378	********
.73	556	6971	*****************
1.22	683	4726	************
1.78	1740	3897	**********
2.28	4930	10154	*************************
2.72	5043	4558	***********
3.25	5399	3264	********
3.76	5631	3327	********
4.24	5585	3410	********
4.67	3986	1657	****
5.23	2331	300	*
5.69	1179	118	

Final choice grade of service = 2.5 %
DIAGRAM IV

In no of the studied cases has optimization under EEB constraints, been able to improve the results that is to create a cheaper network with at least the same traffic carrying capacity.

12. CONCLUSIONS

Two design criteria has been investigated and compared regarding network economy and grade of service. Cluster grade of service with 1 % blocking for the traffic offered to the cluster gives a mean EEB closer to the original result with only backbone routes. Final choice grade of service with 2.5 % blocking in the final gives an almost identical result. This seems to be a better grade of service standard to be recommended in connection with final choice grade of service. Cluster engineering may have an advantage in its control of grade of service for first routed traffic. This advantage is however impossible to evaluate generally, as it is closely related to traffic charging principles.

Network optimization under EEB constraints has turned out to be extremely CPU-time consuming without giving any better results than traditional design methods.

Annex A. SOME BASIC FORMULAS USED

The optimization approach used is to subdivide the problem into the optimization of a large number of simple triangles. Each of those can be optimized by using standard functions for the blocking of overflow traffic. In a real size network there will be a large number of triangles and in order to get a short total run time, each must be performed quickly. Therefore some new basic formulas have been developed to speed up the calculations.

One often used function is the Erlang formula. This is efficiently calculated by use of different methods in different ranges. For a large combination of parameters the fastest computation method has been found to be the continued fraction expansion. This will for trunk groups from a size of 50 and upwards give an essential reduction of computation time compared to the ordinary recurrence formula.

One crucial problem is the handling of overflow traffic. The adopted model is Wilkinson's ERT. If the equivalent traffic is determined by some standard inversion algorithm such as Newton-Raphson, it will require several time-spending calls of the Erlang formula. The approximative value given by Rapp [13] was not considered as precise enough, since it degenerates when the variance is much larger than the mean value, as will occur for the overflow from large trunk groups. Also an modification suggested by Farmer and Kaufman [12] was found to be insufficient. This problem was solved by developing that modification to a new and much more precise formula, which was determined after extensive numerical tests.

$$A = z(m + (2 + \beta^{\gamma})(z - 1))$$

where

$$\gamma = (2.36z - 2.17)\log\left(\frac{z - 1}{m(z + 1.5)} + 1\right)$$

and

$$\beta = \frac{z}{1.5m + 2z - 1.3}$$

The formula has been thoroughly tested for $0.36 \leq A \leq 1600$ and for blockings down to 10^{-6}. Quite often the error of Rapp's approximation is 30 times larger than that of the new formula. Also this has its best precision at high blockings. For low blockings it still gives values about the correct magnitude.

Another subproblem frequently encountered is the trunk inverse of Erlangs formula. This is for example used for the dimensioning of the final trunk groups. Although the real network allows integer sizes only, it is important to be able to get accurate real values for the evaluation of the optimization techniques. The problem can be solved by standard inversion algorithms, but the adopted interpolation method gives accurate solutions in a much shorter time.

The interpolation formula is of the type exp(P) where P is a second-degree polynomial. It is used over any interval of unit length, and is determined such that the derivative satisfies the differentiated recurrence formula. When the wanted blocking is b the Erlang recurrence is used to find a size x with

$$E_x(A) \geq b \geq E_{x+1}(A)$$

then set

$$B = E_x(A)$$

$$G = \log(E_{x+1}(A)/B)$$

$$H = (ABG + 1)/(2x + 2 + AB)$$

An accurate interpolation for $0 \leq y \leq 1$ is now given by

$$E_{x+y}(A) \approx B\, e^{(H + G)x - Hx^2}$$

For the wanted application it is simply inverted

$$y = \frac{1}{2H}(H + G + \sqrt{(H + G)^2 - 4H\log\frac{b}{B}})$$

The interpolation formula has been tested for a large range of variables. It was everywhere found to have an absolute error less than 0.001 circuits. Also the error in offered traffic is less than this value. Since the formula is so robust it is applied in all ranges. Another formula, an generalisation of Rapp's interpolation, has been proposed by Sanders [14]. This gives a higher precision at high blockings but a lower when the blocking is small. For a medium size traffic the formulas are comparable at a blocking of about 5 %.

During the optimization iterations only directed trunk groups are used. The procedure allows however the introduction of both-way trunk groups after the main optimization loop. This can be done at all levels of the network. The method used is to substitute the two directed trunk groups between a pair of nodes with an optimal combination giving the same overflowing traffic. The problem to find the optimal combination here should be more simple than the general triangle optimisation. However it was found that the obtained solution was rather sensitive to the model used. Simple optimization criteria as comparing the cost per erlang on the last directed circuit with the first both-way was not found to be satisfactory. When considering the precision obtained by Rapp's approximation for the optimal occupancy in the triangular case, we felt that a similar formula could be obtained here. After extensive numerical tests the following formula was determined for the optimal occupancy of the last directed trunk.

$$h = (2C - 1)(1 - 0.7616(1 - C)^{.2285})A^{-.3071}$$

C is the cost proportion of directed to both-way circuits $\frac{1}{2} \leq C \leq 1$ and A is the offered (equivalent) traffic. The size of the directed trunk group is given by the integer giving the occupancy closest to the value of the formula.

REFERENCES

[1] Tånge I., 'Optimum methods for determining routes and number of lines in a telephone network with alternate traffic facilities.' TELE No 2 (1959) English ed.

[2] Berkeley G.S., 'Traffic and trunking principles in automatic telephony.' London 1934. (1949).

[3] CCITT. 'Recommendation E.93' Blue Book pp 202-203 (1964).

[4] Wilkinson R.I., 'Theories for toll traffic engineering in the U.S.A.' B.S.T.J. 35 pp 421-514 (1956).

[5] CCITT. 'Recommendation E.521, E.522' Yellow Book pp 57-70 (1980).

[6] Hill D.W., Neal S., 'The traffic capacity of a probability-engineered trunk group.' B.S.T.J. (September 1976).

[7] Nivert K., Gunnarsson R., Sjöström L.E., 'Optimum distribution of congestion in a national trunk network. An economical evaluation as a base for GOS-standards.' Proc ITC 10 Montreal 1983.

[8] Lindberger K., 'Simple approximations of overflow system quantities for additional demands in the optimization.' Proc ITC 10 Montreal 1983.

[9] Fredericks A.A., 'Congestion in blocking systems. A simple approximation technique.' B.S.T.J. 59 No 6 pp 805-827 (1980).

[10] Fredericks A.A., 'Approximating parcel blocking via state dependent birth rates.' Proc ITC 10 Montreal 1983.

[11] Dressler J., Gomes J.A.C., Mantel R.J., Mepuis J.M., Sara E.J., 'COST 201: A European research project. A flexible procedure for minimizing the cost of a switched network, taking into account mixed technologies and end-to-end blocking constraints.' Proc ITC 10 Montreal 1983.

[12] Farmer R.F., Kaufman I., 'On the numerical evaluation of some basic traffic formulæ.' Networks 8:2, pp 153-186 (1978).

[13] Rapp Y., 'Planning of junction network in a multi-exchange area 1. General principles.' Ericsson Technics 1, pp 77-130 (1964).

[14] Sanders B., 'Comments on 'Calculation of some functions arising in problems of queueing and communications traffic'.' IEEE Trans. on communications 28:6, pp 906-907 (June 1980).

TELETRAFFIC ISSUES in an Advanced Information Society
ITC-11
Minoru Akiyama (Editor)
Elsevier Science Publishers B.V. (North-Holland)

ADAPTIVE, TARIFF DEPENDENT TRAFFIC ROUTING AND NETWORK MANAGEMENT IN MULTI-SERVICE TELECOMMUNICATIONS NETWORKS

Edmund SZYBICKI

Geneve, Switzerland

ABSTRACT

The paper presents an adaptive and revenue driven traffic routing and network management system for multi-service networks, where the customers are served based on differentiated tariffs and/or differentiated grade-of-service, as is the case in the inter-city and in the integrated services networks.

The system has a global view of the network and based on this view it routes traffic depending on the actual network status and such, that maximum possible revenues are achieved. This is realized by routing traffic in relation to distance and service sensitive tariffs.

Two routing algorithms are studied. One is based on tariffs and applied to toll networks. The other one is based on differentiated grade-of-service and applicable to integrated services networks.

Performance of networks using the system is demonstrated by utilizing the adaptive routing operating system as a simulator (emulator).

1. INTRODUCTION.

The telephone networks of today are as a rule dimensioned for a constant traffic figure, which in a certain way represents the busy season, busy hour traffic in the network. For a given grade-of-service a fixed amount of network equipment is installed. Further, the networks are designed for the fixed-hierarchical routing, allowing traffic between switch nodes to be routed in a predetermined manner. In reality, however the traffic is subject to variations. It may exceed the average level in some part of the network, at the same time as in another part,the traffic may be below the average level. The present network design leads to a situation, where a static network is expected to handle a rather dynamic traffic process. In normal traffic situations the network can provide an acceptable grade-of-service. During overload, however, the network if uncontrolled, can no longer provide adequate service to the customers, resulting in irritations, reduced traffic handled and reduced revenues. The same applies to situations where network equipment is out-of-service.

In order to avoid these problems, telephone administrations perform network management, which however, is based on more or less manual interventions. The disadvantage is that the procedure is tedious and so it may be a matter of hours to execute changes. Meantime the situation may have changed again and the performed servicing may have no effect. In fact, a late intervention may even have a disturbing effect.

The Stored Program Control used in telephone systems today, makes it possible to introduce a more intelligent traffic routing and network management system into the networks. Such a system has been proposed by the author in Ref./1/ for metropolitan networks. There are, however assential differences between metropolitan and multi-service networks, e.g. the inter-city and integrated services networks. In the metropolitan networks, where flat rate is used, the main objective of the adaptive routing was to maximize the traffic throughput. It would not be quite correct to use the same routing criteria in multi-service networks, where the service is provided based on differentiated tariffs or based on differentiated grade-of-service.

This paper is an extension of work presented by the author in Ref./1/ at the ICC81, Denver, Col., 1981. The Denver paper was dealing with metropolitan applications. This paper presents an adaptive routing and automatic network management system for multi-service networks. Traffic studies performed on networks using the system demonstrate superior performance as compared to the presently used fixed-hierarchical networks.

2. GENERAL.

The studies presented in this paper cover the two network alternatives:

1) inter-city networks,
2) integrated services networks.

In the first case, the network handles traffic applying different tariffs. In the second case, the traffic is handled based on differentiated tariffs and using different priorities for the different services, which results in differentiated grade-of-service. It is assumed that the different parameters are properly related, such that high priority calls should have better grade-of-service and are subject to higher tariffs than other priorities.

The system presented here performs the following functions:

1) adaptive, revenue driven traffic routing,
2) automatic network management,
3) on-demand network management.

The objective of the system is to route traffic such, that the total revenues are as high as possible, subject to grade-of-service constraints. The traffic routing is performed based on:

a) traffic indicators, obtained from network observations,
b) data base information, such as tariffs, priority assignments and grade-of-service assignments.

The automatic network management is based on:

c) out-of-service indicators for switches and trunk groups,
d) load indicators for switches and signalling devices.

In addition, some of the interventions are performed on demand. This is done by means of a special System Management and Intervention Language (SMIL), which allows scheduling of actions on-line and at any point of time.

In the case of inter-city networks, the system routes traffic maximizing the revenues and providing more or less uniform grade-of-service.

In the case of integrated services networks, the routing control is performed such, that

* the assigned grade-of-service for the different priorities is satisfied at the engineering traffic level;
* during overload, the first priority traffic is protected by successively canceling the low priority traffic;
* at underload, the low priority traffic is successively allowed higher network access.

Provided that priorities, grade-of-service and tariffs are properly assigned, highest possible revenues will be automatically achieved.

In both cases, the following network management functions can be performed:

* blocking of calls at origin if no free path to destination exists;
* automatic by-pass of out-of-service and overloaded network equipment;
* on-demand blocking of calls to restricted destinations;
* on-demand reservation of path to a destination for a given point of time;
* on-demand management of two-way trunks;
* on-demand modification of network data, such as network definition, tariffs, priority assignments, grade of-service assignments, etc.

The on-demand interventions are performed by means of SMIL, which is modular and ca easily be extended. Similarily, the data bases can be extended to accomodate other functions, such as subscriber supervision data, central number data, etc. The system facilities can also be used for a more general network control, e.g. transfer of signalling information between the nodes.

3. DESCRIPTION OF THE ADAPTIVE ROUTING AND NETWORK MANAGEMENT SYSTEM.

3.1 Basic Principles.

The system is described assuming all switches to be of SPC type. The system can also be used in mixed, marker and SPC node networks. In that case, however, calls originating in a marker node must connect first to one of the SPC nodes in order to benefit from the adaptive routing system.

The general configuration of the system is similar to that presented in Ref./1/, see also Fig. 1. New principles are presented in this paper as regards its function, specifically applicable to the multi-service networks.

The system consists of a central computer and data links providing the physical connections between the computer and the switch nodes of the network. The computer has a global network view and based on this information it provides the switch nodes with routing instructions.

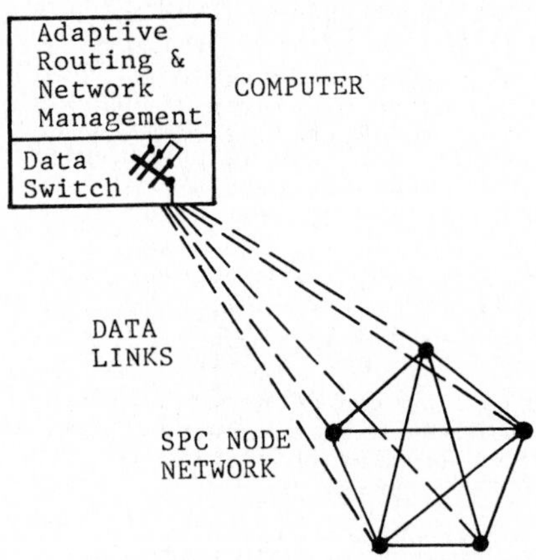

Fig.1. Adaptive Routing and Network Management. System Configuration.

The global view is obtained by means of network scanning and is updated with certain frequency. During the scanning, information is collected on the actual traffic and out-of-service status in the trunk groups and the switch nodes. The observed traffic indicators are:

* no. of busy trunks in each group,
* traffic status in the receiver and sender groups of the switch nodes,
* status of CPU load in each node,
* call intensities for different traffic streams,
* holding time statistics.

The out-of-service indicators cover:

* individual trunks out-of-service,
* trunk groups out-of-service,
* switch nodes out-of-service.

The scanning interval depends on the actual application. However, the system must pick up the quick processes related to load variations in the CPU and in the signalling devices. Therefore, in general a scanning interval of 5-10 seconds can be envisaged.

Apart from the on-line observed indicators, the system is using a data base information, which covers the following parameters:

* tariffs,
* priority assignments,
* grade-of-service assignments,
* destination restrictions,
* network definition,
* trunk type definition.

The system defines the routing scheme for each pair of nodes, taking into account the current values of traffic and out-of-service indicators, as well as the current information in data bases. Instructions for routing are then properly distributed to the different switch nodes. The process is repeated every scanning interval. The central computer can be provided with a data switch, which apart from the routing information can also transfer signalling information between the nodes, Ref./2/.

3.2. System Management and Intervention Language, SMIL.

SMIL has been designed to facilitate on-line network management interventions. It is a coding system, accommodating at present 25 codes, but is modular and can easily be extended. SMIL allows scheduling of interventions as well as activation of interventions during the operations. It can be used, for instance, for scheduling of systematic interventions for the day, or for on-line activation of interventions for immediate reaction, such as:

* modification of network definition,
* modification of scanning interval,
* modification of restricted areas, etc.

3.3. Path search Principles.

The path search is performed by the central computer for each originating node to each destination. It is based on conditional selection using the following three choices:

1. Direct trunk group; 1:st choice.
2. Two-link overflow; 2:nd choice.
3. Three-link overflow; 3:rd choice.

The three-link choice is used to assure an acceptable grade-of-service also to nodes, to which there is no two-link path, e.g.:

* nodes, which due to low traffic and high trunk costs have acces to one, or only few other nodes;
* nodes, which due to out-of-service, extensions, or the current traffic situation can be reached from one, or only few other nodes.

The use of three-link alternative has a favorable effect on the node-to-node congestion.

3.4. Number of Indicated Paths.

If a free path to a destination does not exist the system indicates congestion. On the other hand an indicated free path can be snatched away during the scanning interval. The incident of snatching can be reduced by indicating several free paths to be used alternatively, should the principal path be snatched away. There is however a limit, at which an additional path indicated, will not result in a substantial decrease of congestion. Fig.2 shows the general trend, which is valid for most practical cases. Consequently, the system operates based on two path indication, if existing, each of them using a maximum of 3 links, see Fig.3.

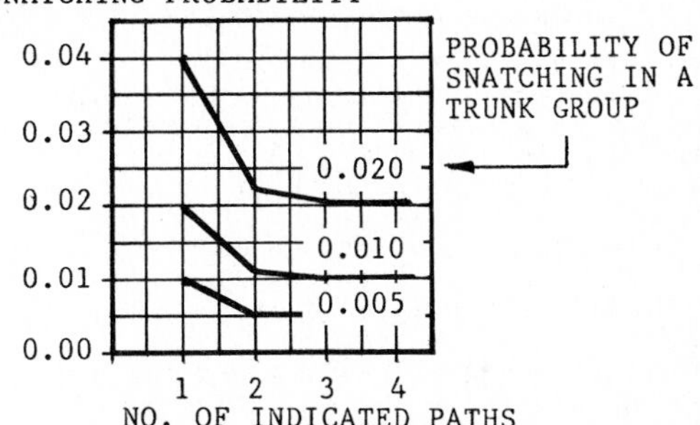

Fig.2 Optimal Number of Paths.

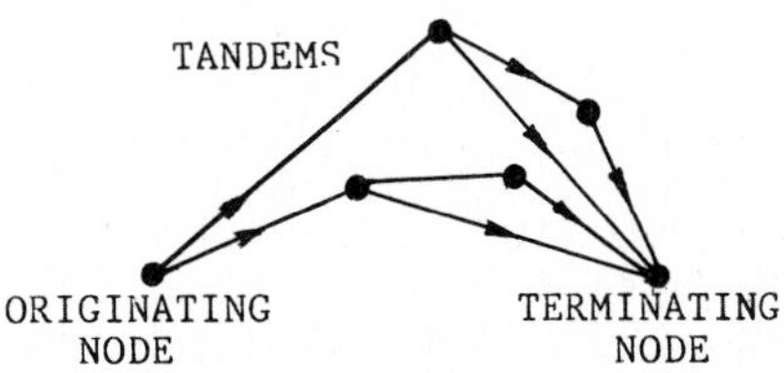

Fig.3 Number of Overflow Choices.

3.5. Control of tariff Dependent Routing.

Detailed description of this routing scheme is given in Ref./3/.

An overflow call occupies two, or three links in series, hindering in this way the respective links to handle direct traffic. The principle of this routing scheme is to allow overflow traffic to enter the network only if the expected revenue from it is greater than the sum of revenues expected from handling direct traffic by the respective links.

Set A_1 =traffic offered to the 1:st link,
A_2 =ditto, 2:nd link,
N_1 =no. of trunks in the 1:st link,
N_2 =ditto, 2:nd link,
s =average holding time,

y_{ij} =first offered call intensity from node i to j,
c_{ij} =tariff per time unit for calls from i to j,
h =duration of overflow call,
t =tandem node number,
$B(N_1)$=blocking for link 1,
$B(N_2)$=ditto, link 2.

During the duration of overflow call the expected total revenue from the 2 links is

$$h \cdot c_{ij} + h \cdot y_{it} \cdot s \cdot c_{it} \cdot \left[1 - B(N_1 - 1)\right] + h \cdot y_{tj} \cdot s \cdot c_{tj} \cdot \left[1 - B(N_2 - 1)\right]; \tag{1}$$

The expected total revenue from the 2 links in case the overflow call is rejected is:

$$h \cdot y_{it} \cdot s \cdot c_{it} \cdot \left[1 - B(N_1)\right] + h \cdot y_{tj} \cdot s \cdot c_{tj} \cdot \left[1 - B(N_2)\right]; \tag{2}$$

Consequently, the overflow call will be set up if:

$$c_{ij} > c_{it} \cdot A_1 \cdot \left[B(N_1 - 1) - B(N_1)\right] + c_{tj} \cdot A_2 \cdot \left[B(N_2 - 1) - B(N_2)\right]; \tag{3}$$

and will be rejected otherwise.

The corresponding routing condition for three link overflow is as follows:

$$c_{ij} > c_{it} \cdot A_1 \cdot \left[B(N_1 - 1) - B(N_1)\right] + c_{tr} \cdot A_2 \cdot \left[B(N_2 - 1) - B(N_2)\right] + c_{rj} \cdot A_3 \cdot \left[B(N_3 - 1) - B(N_3)\right]; \tag{4}$$

The described routing scheme provides an uniform grade-of-service. However, should the network handle different services using different tariffs, the high tariff services will receive better grade-of-service.

3.6. Control of Priorities.

In the case of integrated services networks the different services might be handled based on priorities. It is fair to assume that non-preemptive priorities must be used. In the system presented here, the priorities are controlled by providing high network access to high priority calls and reduced network access to lower priority calls. For two priorities, the following recursive formulas have been obtained by means of equations of states, for the calculation of congestion, see also Ref./4/ and /5/:

$$E_{k,\nu} = \frac{A \cdot E_{k-1,\nu-1}}{k + A \cdot E_{k-1,\nu-1}}; \tag{5}$$

$$E'_{k,\nu} = \frac{k \cdot E'_{k-1,\nu-1}}{k + A \cdot E_{k-1,\nu-1}} + E_{k,\nu}; \tag{6}$$

$$E_{n,0} = E'_{n,0} = E_n(A + A_1); \quad k,\nu = n+1, 1; \; \cdots - N, N-n;$$

where N =tot. no. of trunks in the group =network access for priority 1.
n =max. no. of trunks in the group to be used for priority 2 calls.
A =offered priority 1 traffic.
A_1=offered priority 2 traffic.
E =blocking for priority 1 calls.
E' =blocking for priority 2 calls.

The formulas assume full availability group and pure chance traffic. For given traffics and blockings E and E', the number of trunks N and network access n can be calculated. During the network operations the parameter n can be kept constant or adaptively modified.

3.7. Path Search Algorithms.

3.7.1 General.

The path search is performed for each node to each destination and provides the following routing indications:

* two 2-link paths over two different tandems;
* if no 2-link path exists, a 3-link path is searched and if free, indicated;
* if no free path to a destination exists, network congestion is indicated.

As a rule, different services will have different routing instructions. Only in some cases the same instruction may be used for two or more services. The path search is based on conditional selection; a free path indication is issued only if a complete path between two nodes exists. The out-of-service and overloaded nodes and trunk groups are excluded from the path search.

Definitions:

N_{ij} =tot.no.trunks in group ij; can be used for priority 1 calls;
n_{ij} =no.of trunks in group ij that can be used for priority 2 calls;
ν_{ij} =no.of busy trunks in group ij;
A_{ij} =1:st offered traffic from i to j;
$B(N_{ij})$=blocking in trunk group ij;
c_{ij} =tariff per time unit for calls from i to j;
$q_{1\mu}$ =availability indicator for node μ;
$q_{2\mu}$ =load indicator for CPU in node μ;
$q_{3\mu}$ =load indicator for receivers in node μ;
$q_{4\mu}$ =load indicator for senders in node μ;
$q_{5\mu}$ =load indicator for the connecting network in node μ;
t =tandem node number;
G_{ij} =characteristic value for trunk group between node i and j.

The value of the indicator $q_{\nu\mu} = 1$ if the device is available for service and $q_{\nu\mu} = 0$ if the device is out-of-service.
The path search is performed only if the availability indicators for destination node j satisfy the condition:

$$Q_j = \prod_{s=1}^{5} q_{sj} = 1 \tag{7}$$

Similarily, node t can be used as tandem only if it satisfies the condition:

$$Q_t = \prod_{s=1}^{5} q_{st} = 1 \tag{8}$$

3.7.2 Inter-City Networks. Tariff Dependent Routing.

The selection of a 2-link overflow path for connections from i to j is based on the actual values G_{it} and G_{tj}. These are obtained from the following formulas:

$$\left.\begin{aligned} G_{it} &= N_{it} - \nu_{it} \\ G_{tj} &= N_{tj} - \nu_{tj} \end{aligned}\right\} \text{ if } Q_t=1 \qquad (9)$$

$$G_{it} = G_{tj} = 0 \qquad \text{if } Q_t=0; \quad t=1,2--- \qquad (10)$$

Not all tandems t provide a profitable overflow path. Only those t can be used, which satisfy the conditions:

$$V_t = \min(G_{it}; G_{tj}); \qquad (11)$$

$$c_{ij} > c_{it} \cdot A_{it} \cdot [B(N_{it}-1) - B(N_{it})] + c_{tj} \cdot A_{tj} \cdot [B(N_{tj}-1) - B(N_{tj})]; \qquad (12)$$

Two tandems are selected at random from the group of candidates, t=1,2,---. If no 2-link path exists, a 3-link path is selected in a similar way. In that case, the tandem candidates r and t must satisfy the conditions

$$t \neq j \quad \text{and} \quad r \neq i \qquad (13)$$

In general, this routing scheme provides an uniform grade-of-service. However, should the network handle different type of services using different tariffs, the high tariff services will receive a better grade-of-service.

3.7.3 Metropolitan Applications.

The routing scheme described in Section 3.7.2 can also be used for networks applying flat rate. In that case it is necessary to set:

$$c_{ij} = c_{it} = c_{tj} = 1;$$

A 2-link overflow path can be used if:

$$1 > A_{it} \cdot [B(N_{it}-1) - B(N_{it})] + A_{tj} \cdot [B(N_{tj}-1) - B(N_{tj})]; \qquad (14)$$

In this case an overflow call is allowed to enter the network if it is expected to contribute to the increase of the handled traffic. Else, it is more profitable to use the respective links for handling direct traffic.

3.7.4 Integrated Services Networks. Priority Differentiated Routing.

In this paper two different service types using different priorities are assumed to be handled by the network. Each of the priorities needs a separate path search. The following procedure is applied:

* path search is performed only if the destination node satisfies the condition:

$$Q_j = \prod_{s=1}^{5} q_{sj} = 1 \qquad (15)$$

  else, the calls are rejected at the origin,
* for priority 1 calls the parameters G_{it} and G_{tj} are calculated from:

$$\left.\begin{aligned} G_{it} &= N_{it} - \nu_{it} - P_{it} \\ G_{tj} &= N_{tj} - \nu_{tj} - P_{tj} \end{aligned}\right\} \text{ if } Q_t = 1 \qquad (16)$$

$$G_{it} = G_{tj} = 0 \qquad \text{if } Q_t = 0 \qquad (17)$$

A 2-link path is selected using those tandems t, for which

$$Q_t = \prod_{s=1}^{5} q_{st} = 1 \qquad (18)$$

$$V_t = \min(G_{it}; G_{tj}) > 0 \qquad (19)$$

If several tandems satisfy the condition, two are selected at random. If no free, 2-link path exists, a 3-link path is selected. In that case only those tandems r and t can be used, which satisfy the condition:

$$t \neq j \quad \text{and} \quad r \neq i \qquad (20)$$

* for priority 2 calls, the parameters G_{it} and G_{tj} are calculated from:

$$\left.\begin{aligned} G_{it} &= n_{it} - \nu_{it} - P_{it} \\ G_{tj} &= n_{tj} - \nu_{tj} - P_{tj} \end{aligned}\right\} \text{ if } Q_t = 1 \qquad (21)$$

$$G_{it} = G_{tj} = 0 \qquad \text{if } Q_t = 0 \qquad (22)$$

  The 2-link and 3-link path search is then carried out in the same way as for priority 1 calls.

The parameter $P_{s\mu}$ is used to protect the network from overload. For calculation of the state protection parameter, see Ref./5/. The network access parameter n, for priority 2 calls can be calculated from formulas (5) and (6).

4. OPERATING SYSTEM.

Operating software for the described system has been developed and tested, including the three routing algorithms described in Section 3.7.2, 3.7.3 and 3.7.4. For the development of programs, VS FORTRAN has been used. A test performed on IBM 3081 indicated that for a network of 100 nodes the system required in average, approximately 0.4 seconds to carry out the scheduled tasks during a scanning interval of 10 seconds.

5. STUDIES PERFORMED.

5.1 General.

The basic objective of the study is to demonstrate the performance of a network operating under the control of the adaptive routing and network management system. The study covered the following items:

* comparison of the fixed-hierarchical routing and the 2-link version of adaptive routing;
* performance testing of an inter-city network operating under the control of the tariff dependent routing;
* performance testing of the same inter-city network, which was using the priority differentiated routing;
* performance comparison of the fixed-hierarchical routing and the priority differentiated routing in the case of switch out-of-service;
* network performance when the adaptive routing system was out-of service.

5.2 Study Tool. Using the Adaptive Routing Operating System as a Simulator,

A computer aided simulator has been developed for the study. It is based on the adaptive routing operating system, where some additional modules have been included in order to have it operational in batch environment and to obtain the required statistics. The software is modular so that by exchanging modules different routing and network management strategies can be tested. At present the following routing modules exist:

* direct routing,
* fixed routing,
* 2-link adaptive routing,
* 3-link tariff dependent routing,
* 3-link priority control routing,
* 3-link adaptive routing for flat rate applications.

The simulator can also be used for testing of the required adaptive routing switch modifications. The general simulator configuration is shown in Fig. 4.

SIMULATOR
OPERATING SYSTEM
Adaptive Routing
Network Management
INTERFACES
STATISTICS
NETWORK DEFINITION
TRAFFIC GENERATOR
DATA LINKS
REAL TRAFFIC

Fig.4 Simulator Configuration.

5.3 Given Conditions.

The study network is a small inter-city network, which has been modified by excluding node relations with negligible traffic interest and so that certain network problems can be studied. The network has the fixed routing design as shown in Fig. 5. Tables 1 and 2 show the traffic and the number of trunks matrices, respectively. The same traffics and the same network configuration has been used also in the case of adaptive routing. However, since the fixed routing sizing is not optimal for adaptive routing, some trunk groups, in this case, have been redimensioned. The total number of trunks was 2671 in the case of adaptive routing and 2675 in the case of fixed routing. Two service types were used with the same probability of occurrence. Priority 1 was used for service 1 and priority 2 for service 2. The tariff ratio (service 1)/(service 2) was as 2/1.

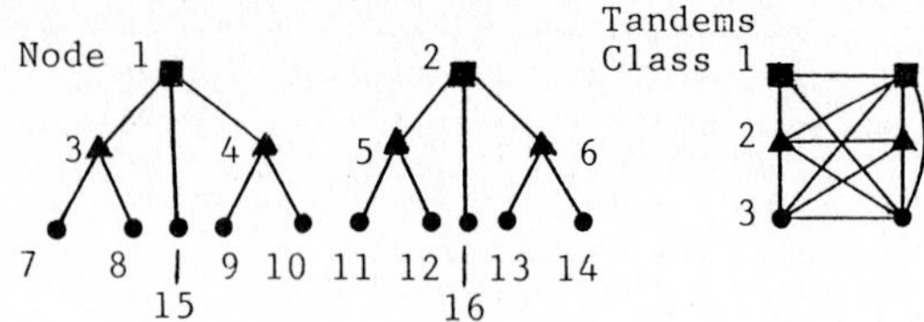

Fig.5 Homing Arrangement and Hierarchy for the Fixed Routing Network.

TABLE 1. Node-To-Node Traffic Offered, erl.

From \ To	1	2	3	4	5	6	7	8	9	10	11	12	13	14	15	16
1	-	-	3	2	-	-	2	2	3	1	-	-	-	-	-	-
2	-	-	-	-	3	2	-	-	-	-	2	2	3	1	-	-
3	3	-	-	15	10	15	12	14	12	3	12	14	12	1	8	8
4	2	-	15	-	15	22	17	21	17	7	17	21	17	2	13	13
5	-	3	10	15	-	15	12	14	12	1	12	14	12	3	8	8
6	-	2	15	22	15	-	17	21	17	2	17	21	17	7	13	13
7	2	-	12	17	12	17	-	16	13	1	13	16	13	2	10	10
8	2	-	14	21	14	21	16	-	16	2	16	20	16	2	12	12
9	3	-	12	17	12	17	13	16	-	1	13	16	13	2	10	10
10	1	-	3	7	1	2	1	2	1	-	2	2	2	-	2	2
11	-	2	12	17	12	17	13	16	13	2	-	16	13	1	10	10
12	-	2	14	21	14	21	16	20	16	2	16	-	16	2	12	12
13	-	3	12	17	12	17	13	16	13	2	13	16	-	1	10	10
14	-	1	1	2	3	7	2	2	2	-	1	2	1	-	2	2

Total traffic offered = 1744 erl.

TABLE 2. Fixed Routing Network. Number of Trunks.

	1	2	3	4	5	6	7	8	9	10	11	12	13	14	15	16
1	-	-	47	39	29	17	-	-	-	-	-	-	-	-	123	-
2	-	-	29	17	47	39	-	-	-	-	-	-	-	-	-	123
3	47	29	-	46	21	43	79	93	36	-	35	40	23	-	-	-
4	39	17	46	-	43	50	45	55	92	80	46	54	46	-	-	-
5	29	47	21	43	-	46	35	40	23	-	79	93	36	-	-	-
6	17	39	43	50	46	-	46	54	46	-	45	55	92	80	-	-
7	-	-	79	45	35	46	-	36	26	-	27	33	27	-	-	-
8	-	-	93	55	40	54	36	-	32	-	33	40	33	-	-	-
9	-	-	36	92	23	46	26	32	-	-	27	33	27	-	-	-
10	-	-	-	80	-	-	-	-	-	-	-	-	-	-	-	-
11	-	-	35	46	79	45	27	33	27	-	-	36	26	-	-	-
12	-	-	40	54	93	55	33	40	33	-	36	-	32	-	-	-
13	-	-	23	46	36	92	27	33	27	-	26	32	-	-	-	-
14	-	-	-	-	-	80	-	-	-	-	-	-	-	-	-	-

Two-way trunks; tot. no. of trunks = 2675

6. STUDY RESULTS AND CONCLUSIONS.

The results are shown in Tables 3-7. Table 3 gives the comparison between the fixed-hierarchical and the 2-link adaptive routing. It is evident that there are no physical 2-link connections between nodes 7 through 14 and nodes 15 & 16. Hence, in the case of 2-link adaptive routing the corresponding calls experience 100% node-to-node congestion. For all other cases, the adaptive routing provides a better node-to-node grade-of-service than the fixed-hierarchical routing.

TABLE 3. Node-To-Node Congestion.

From Nodes	Adaptive Routing To Nodes		Fixed Routing To Nodes	
	1-14	15,16	1-14	15,16
1- 6	0.018	0.029	0.038	0.300
7-14		1.000		0.278

Table 4 shows comparison between the two adaptive routing schemes, the priority differentiated and the tariff dependent routing. They show comparable quality. The tariff dependent results in a lower probability of snatching, which indicates that the routing algorithm provides an efficient control of overflow traffic. The priority differentiated routing uses state protection to control overflow traffic.

TABLE 4. Comparison of Adaptive Routing Schemes.

3-LINK ROUTING TYPE	NETWORK CONGESTION	PROBABILITY OF SNATCHING	REVENUES K$
PRIORITY CONTROL	0.0114 ±0.0034	0.0103	183.1
TARIFF RELATED	0.0108 ±0.0033	0.0040	183.2

Table 5 shows comparison between the fixed-hierarchical and the priority differentiated, adaptive routing. For the same traffic offered, the adaptive routing (Alt.1) provides much better grade-of-service than the fixed routing. For increased traffic (Alt.2), the adaptive routing gives approx. 7% higher revenues and still results in lower congestion for priority 1 calls than the fixed routing. Comparing the two alternatives (1 and 2), it is evident that the priority driven routing provides a rather good control of priorities. For increased traffic the scheme has increased the revenues of priority 1 services by 6% and reduced at the same time the priority 2 revenues by 2.5%. Service 1 calls have obviously received the highest priority.

TABLE 5. Comparison of the Fixed-Hierarchical and Adaptive Routing Schemes.

ITEM	SERVICE	FIXED ROUTING	3-LINK ADAPTIVE ROUTING WITH PRIORITIES	
			ALT. 1	ALT. 2
CONGESTION	1	0.057	0.006	0.021
	2	0.057	0.017	0.146
	ALL	0,057	0.011	0.083
CONFIDENCE INTERVAL *	ALL	±0.006	±0.004	±0.018
REVENUES K$	1	119.6	123.9	131.7
	2	58.1	59.2	57.7
	ALL	177.7	183.1	189.4

ALT.1: 1744 erl. ALT.2: 1924 erl.

* 90%

Tables 6 and 7 show results related to out-of-service situations. Table 7 indicates that during the period of 9 minutes when the adaptive routing system was out-of-function the network was still providing a better grade-of-service than the fixed routing. During the absence of adaptive routing the network is routing traffic according to the last issued instructions, constituting in this case a dispersed, double sector tandem routing. This is an efficient and simple way of backing up the adaptive routing during temporary break down situations.

TABLE 6. Comparison of Performance when Node 5 is Out-of-Service.

ITEM	SERVICE	FIXED ROUTING	3-LINK ADAPTIVE ROUTING WITH PRIORITIES
CONGESTION	1	0.098	0.011
	2	0.093	0.041
	ALL	0.095	0.026
REVENUES K$	1	98.5	105.1
	2	48.1	49.3
	ALL	146.7	154.3

TABLE 7. Network Congestion when Adaptive Routing Is Out-of-Service.

NETWORK TIME Minutes	3-LINK ADAPTIVE ROUTING WITH PRIORITIES Out-of-service	3-LINK ADAPTIVE ROUTING WITH PRIORITIES In service	FIXED ROUTING
0-24	-	0.007	0.047
24-33*	0.023	0.012	0.057
33-51	0.013	0.012	0.062

* Adaptive Routing was out-of-service 9 minutes between the 24:th and the 33:rd minute.

7. BENEFITS.

The following benefits can be listed:

* efficient control of priorities,
* increased revenues through revenue dependent routing,
* automatic by-pass of overloaded and out-of-service equipment,
* automatic and on-line network management interventions.

8. REFERENCES.

/1/ Szybicki,E.:"Calculation of Congestion in Trunk Networks Provided with Adaptive Traffic Routing and Network Management System",ICC81,Denver,Col.

/2/ Hamrin,P.,Rasmusson,G.:"Signalling Networks", TELE 2/81.

/3/ Szybicki,E.:"Algorithm for the Control of Tariff Differentiated Traffic in Toll Networks", August 1984.

/4/ Roberts,J.W.:"Teletraffic Models for the Telecom 1 Integrated Service Network", ITC 10, 1983.

/5/ Szybicki,E.:"Adaptive Control of Non-Preemptive Priority Calls in Integrated Services Networks", October, 1984.

TELETRAFFIC ISSUES in an Advanced Information Society
ITC-11
Minoru Akiyama (Editor)
Elsevier Science Publishers B.V. (North-Holland)

A SOCIO–ECONOMIC MODEL EXPLAINING THE TELECOMMUNICATION DEMAND

Lars ENGVALL

Technical Cooperation Department, ITU
Geneva, Switzerland

ABSTRACT

The need of developing a methodology for estimating the demand of public services that also takes into consideration socio-economic factors is today recognized as of great importance. The author presents in this paper a model for forecasting the need of telephone services in households. This group of users, usually the residential category, ultimatley reaches 70-80 % of the total number of all main lines of a country.

The paper explains how socio-economic sample surveys are used, and the modifications of such that will help telephone planners in forecasting future demand. With the procedure explained in the paper it is possible to evaluate the potential demand of various user categories and surveyed strata of the country.

The importance tariff policies have on the users possibility to afford the services is discussed as well as how a "new" tariff policy could stimulate a more equitable, socially desired, use of the telephone services in the developing countries in particular.

1. INTRODUCTION

Today we often hear it stated that telecommunication services should be treated as an integral part of economic and social development.

The relatively low priority accorded to telecommunications in many countries is being investigated the world over in order to demonstrate the negative impact an insufficient development of telecommunication services will have on the development of other sectors of the society. Such investigations have been inlcuded in the ITU/OECD report "Telecommunications for Development".[1]

The "Independent Commission for World Wide Telecommunications Development", created by the Plenipotentiary Conference of ITU (1982, Nairobi) has recently presented their Report "The Missing Link".[2] Among the many recommendations made in the report is the establishment of a Centre for Telecommunications Development, which would include, among others, a Development Policy Unit, who´s "main function would be to collect information about telecommunications policies and experience, including experience of the role of telecommunications in economic and social development throughout the world, and to make the results available to developing countries to help them formulate policies for the evolution of their own networks".

The need of developing a methodology for estimating (forecasting) the demand of public services, here particularly devoted to the telecommuniation sector, but otherwise valid for many other sectors, such as radio and TV broadcasting, has been particularly felt by the author during activities in several parts of the world on ITU assignments. It has often required forecasting telephone development in countries having scarce and incomplete data. Usually the telephone services were only available in the most important urban areas of the countries. The rural areas in the developing world often represent 70-80 % of the population. Although the urbanization process goes on rapidly the rural population, will by year 2000, still be over 50 %. The possibility of forecasting the telephone service requirement in rural areas is therefore particularly difficult due to the scarcity of statistical data.

This paper presents a model to use for forecasting telephone requirements at the household level. This group of users, mainly of the so-called residential category, potentially represent 70-80 % of the total (future) telephone users. In the early phase of telephone development of a country, however, the business/official user group dominates. For this group of users other methods of forecasting must be applied.

The development pattern of the residential user group relative to the total number of main telephone lines (DEL) is illustrated by Figures 1 and 2 below.

JAPAN Development trends for business and residential users.

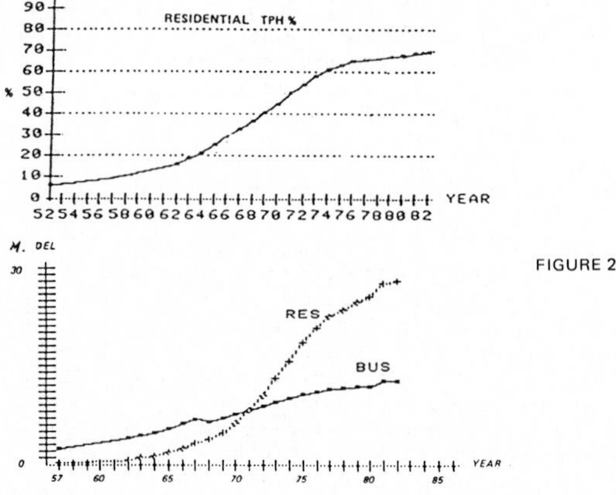

FIGURE 1

FIGURE 2

2. USE OF SOCIO-ECONOMIC SAMPLE SURVEYS

Most countries have undertaken, or are preparing, sample surveys of a socio-economic character. They may have many names, household expenditure surveys, socio-economic surveys, family budget surveys, etc. The fact that several UN agencies, such as the International Labour Organization, ILO, and the UN-sponsored National Household Survey Capability Programme, NHSCP, has given guidance on how to undertake household sample surveys, has been extremely useful. Thereby the procedures for data collection and analysis have been improved and as more and more countries follow similar procedures, it allows international comparisons to be made.

Since sample surveys have rarely been used by telecommunication planners/researchers, the statistical authorities of the countries are usually not analysing the eventually existing data on telephone possession by households. Thus, the possible use of such surveys for studying the telephone (or other public services) penetration among households depends on what data was collected. An investigation must therefore start by checking the questionnaires used at the time of the data collection. It is often found that a large amount of data was collected, and is stored in the data base, but has not been analysed or tabulated in the printed documents from the survey. Having investigated what data is available in the sample survey data base, and the data that has been published, it can be specified what additional tabulations from the data base should be made. It will almost certainly involve rerunning the data tapes with new instructions. Such work will have to be made locally in the countries with assistance from the responsible governmental authorities.

The possession of a telephone by the household is usually listed as "durable goods", together with e.g. radio, TV, refrigerator, motorcar, bicycle, etc. If this is the case the penetration (density) of telephones among households at various income or expenditure levels can be studied.

If the data base is suitably large the investigations can be specialized to study the penetration in urban, rural or regional areas, or by professional category of the household (referring to the head of the household), educational level, etc. Such detailed investigations have been made in a number of countries. As an example, some particular results of the study made from the sample survey of Thailand in 1975/76 is used to illustrate the methodology. [3]

3. FUTURE POSSIBILITY OF STUDYING THE USAGE OF THE TELEPHONE

The data available from present sample surveys will only allow us to analyse the possession of telephones, and not how much or for what purpose they are used. To do this we would need to merge telecommunication data with socio-economic sample survey data. This could be achieved if, in the questionnaires used for the interviews of the households, the telephone number is recorded (for such households possessing a telephone!). By obtaining statistical data from the telecommunication authorities that refer to such (surveyed) telephone users, details about user category, number of local, trunk and/or international calls made (per month or year), revenue collected for such calls, etc. could be investigated and related to households according to income (expenditure) groups, urban or rural users, professional categories, etc. Such information will considerably improve our understanding of the users habits and give guidance for future planning of the telephone services. It will also allow us to investigate the influence tariff charges have on both the possession of telephones and the use of this service.

The author has, so far, obtained an understanding on the importance of such data from the statistical departments of three countries in which new sample surveys are under way, which will, when completed, allow this detailed analysis of the use of telephone to be made.

4. THE MATHEMATICAL MODEL

The model describes how to calculate the demand for the studied services, e.g. telephone, television and radio. The most important relationships are :

- the income (or expenditure) frequency function of the households;
- the density function of the service considered, related to income/household (or expenditure/household).

These relationships can be analysed for various strata of the survey, urban, rural, various regions/districts of the country, ethnic groups of the population. For any of these groups the investigations can be further detailed by investigating how the income frequency functions and density functions for various professional categories (of the head of the HH), educational level, age etc. are related.

The characteristic functions for these relationships are illustrated by Figure 3 below :

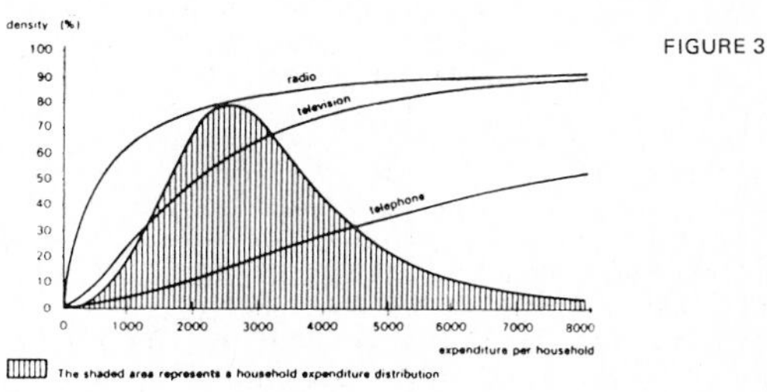

Figure 3 : Household expenditure distribution and density of services

Income/Expenditure functions

Data to analyse for such functions are generally directly available from the published sample survey reports, or on request from the Government's statistical department.

The income/expenditure data are presented according to selected income/expenditure groups. By using the accumulated distribution of the percentage distribution data, the mathermatical treatment can be clearly defined at the upper limit of the income/expenditure groups.

The following formula for the accumulated distribution F(x) [4] has proved to represent the actual data in almost all cases studied at a regression correlation coefficient R 99 %.

$$F(x) = \frac{100}{(1 + \exp(A)*x^{B})}$$

x is the income (expenditure) per month and per household

(Note that B is always negative!)

The actual income (expenditure) frequency function is then obtained by taking the first derivation of the above formula, i.e. dF/dx.

$$F'(x) = \frac{-100*B*\exp(A)*x^{B-1}}{(1 + \exp(A)*x^{B})^{2}}$$

Service density functions

A density function of a mathematical expression that is similar to the above-mentioned accumulated distribution function has shown good correlation with actual data. We call this density function D(x) :

$$D(x)_{(Serv)} = \frac{100}{(1 + \exp(A)*x^{B})}$$

where D(x) is the density in % of households at the income (expenditure) level "x" possessing the service (Serv) indicated, and

x is the income (expenditure) per month and per household.

The constants A and B are determined from a regression analysis as for the accumulated household´s income (expenditure) functions.

Demand calculation

Knowing the frequency function of the income (expenditure) per household and the density function of the service under consideration, the demand of the service can be calculated.

The mathematical formula for calculating this demand, called U, is :

$$U = \frac{1}{100} \int_{0}^{\infty} F'(x)*D(x)\, dx$$

where U is the percentage average demand, at the studied year, and for the household group, region, etc. concerned;

F´(x) is the income (expenditure) frequency function;

D(x) is the density function for the service considered.

When the sample sizes of a survey are sufficiently large, more detailed investigations can be made. For a particular region, urban or rural area, etc. the income and density functions can be evaluated for each professional category group.

The mathematical expression to be used will be more detailed than for the above U function.

$$U(i) = \frac{1}{100} \sum_{q} \int_{0}^{\infty} v(i)_{(q)}*D(i,x)_{(q)}*F'(i,x)_{(q)}*dx$$

where, (i) : Regional, Rural, Urban, Metropolis, etc.
(q) : Professional category, Educational level, etc.
v(i) : Proportion of Category (q) in the area (i)

$$\sum_{q} v(i)_{(q)} = 1 \text{ , for each } (i)$$

By using data from several sample surveys the time dependence of the factors "A" and "B" in the formulae can be analyzed, providing a possibility for forecasting the trends of the functions, particularly as the trend on the F(x), and consequently the F´(x), are important to determinate.

With the procedure explained it is possible to evaluate potential demand of various surveyed strata. The potential demand of telephone services in the rural areas is particularly revealed to be large, and it is also possible to learn from the analysis how different groups of households (based on professional category, educational level, age etc. of the head of the household) are interested in having a telephone.

5. EXAMPLES FROM ANALYSIS OF THAILAND DATA (1975/76) [5]

The socio-economic survey undertaken in Thailand in 1975/76 was studied in detail. As the particular data about the telephone, TV and radio services, we are interested in analyzing are not available from the published data, it was necessary to rerun the data tape according to programmed instructions.

The sample survey covered 11514 households throughout the Kingdom, which, since the total number of households is 7903000, represents 0.115 %. The samples are subdivided by regions and Bangkok Metropolitan area. The samples are further stratified into three community types, i.e. municipal areas (MA), sanitary districts (SD) and villages (V). In the analysis we have usually taken the SD and V together since they represent the so-called rural areas of Thailand.

After investigating which sample areas (villages, etc) were not within reach of TV and/or telephone services, the resulting data bases for the various services were determined.

The analysis was made for all regions and for MA and SD+V for all services. In addition, further detailed studies were made for particular groups of households (HH) distinguishing between them according to, professional category, educational level, age, referring to the head of the household.

Many other criteria can be tried, according to the available data records of the sample survey in question. Of the above the "professional category" seems to be the more useful.

HH Distribution according to Expenditure

The expenditure distribution in 1975/76 in Thailand for Bangkok and the Provinces (all regions outside Bangkok Metropolis area) are shown in the figure below, Figure 4.

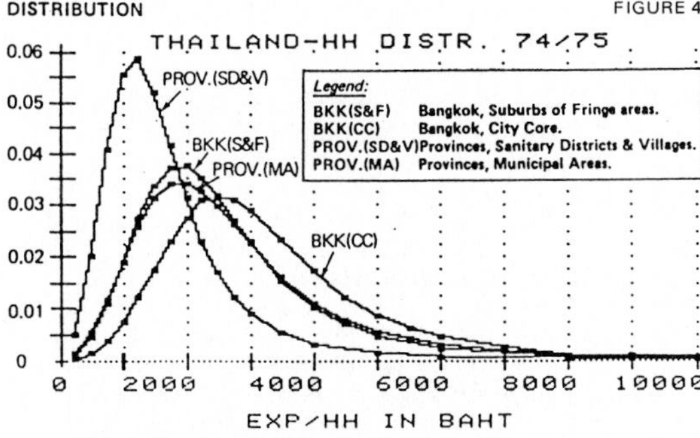

This figure shows the typical situation that in the urban areas the expenditure (and the income) is higher than in the rural areas. The form of the distribution curves also shows the typical changing form resulting from increasing income (or inflation) as a function of time. In our particular example we notice that the expenditure function for the urban areas in the provinces is almost identical to the situation in the Bangkok suburbs and fringe areas.

Penetration of Services in Thailand (as per 1975/76)

The density functions give the penetration situation. These functions have been calculated from the data, and Figure 5 below shows the data for Municipal Areas (MA). Separate curves for Bangkok (City Core) and the Provinces (together) are given. Similar calculations have also been made for the Sanitary Districts and Villages (SD+V). In the latter case we found the curves for TV and Telephone at a lower level.

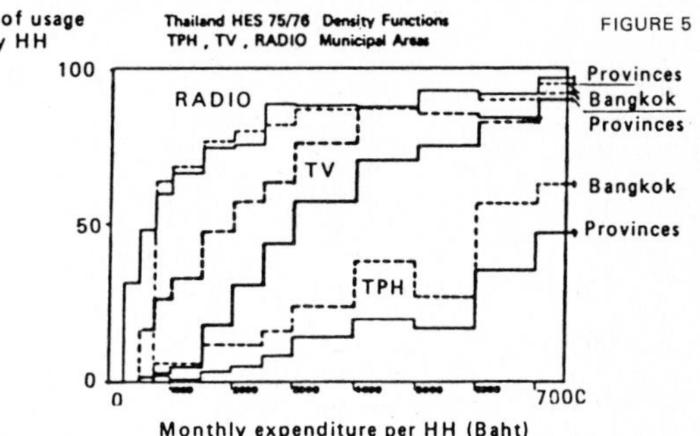

The fact that the users in urban and rural areas are composed of different user professional categories, who have different density functions, explains the differences in the overall density functions for "Bangkok" and "Provinces" in Figure 5.

Table 1 below illustrates the composition of professional categories amongst the sample households for the Bangkok area and the Northern Region. The expenditure situation for the various professional/categories in the Northern Region is illustrated in Figure 6.

TABLE 1

Graph No.	Category	Northern	%	Bangkok	%
1 + 2 + 3	Farming	1235	55.7	268	12.1
4	Production & Construction W.	129	5.8	391	17.7
5	Clerical , Sales & Service W.	198	8.9	582	26.4
6	Non-farming entrepreneurs	415	18.7	665	30.1
7	Prof. Admin. & Techn. W.	141	6.4	187	8.6
-	Economically inactive	101	4.6	114	5.2
	Σ	2219	100	2207	100

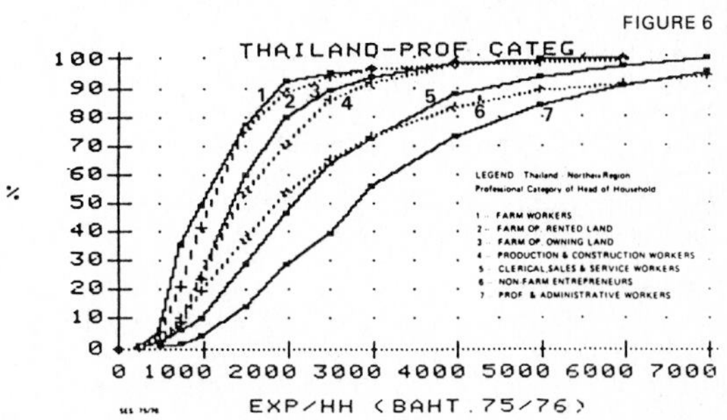

Knowing the expenditure distribution functions and the corresponding service density functions the actual demand, i.e. the number of users which would have afforded the services at the actual (1975/76) tariff charges and if the networks covered the whole kingdom, can be estimated.

The overall estimate of potential users (residential) for Thailand in 1975/76 were calculated to be about three times those actually provided, as detailed in Table 2 below.

Table 2

Area	Existing DEL	Est. Demand DEL
Bangkok	101800	150000
Provinces	11636	200000
Total :	113436	350000

6. TARIFF POLICIES VS DEMAND

It is not intended in this paper to develop a tariff model. The material so far obtained is insufficient. It is expected that the continued research on these matters will give data on the influence tariffs have on user categories at various income (expenditure) levels, etc.

However, it seems obvious that in order to be able to satisfy the socio-economic demand for telephone services a fundamental change in the present tariff structures in many of the developing countries must occur. The most obvious change will have to come in the access charges, which in most developing low telephone density countries are so high that they virtually block all middle and low income households from having a telephone.

The other tariff components, the rental charges and the unit call (meter-pulse) charges must be studied carefully and adapted to the aims of the countries, safeguarding a justified return of investments.

The unit charges, and particularly the STD charging plan for a country, must be reviewed. It is often seen that due to high STD charges the revenue from subscribers in the various regions of a country is considerably higher than the average revenue in the capital. A balance will have to be found so that telephone users in the distant regions are not "penalized" by excessive STD rates.

The ultimate goal of a public service must be that its use should be equally charged wherever it is used in the country.

Like the charges for a kWh electricity consumption or postal charges per mail are usually unique in the country, the charges per time unit of using the telephone services ought to be uniform throughout the country. This would eliminate the distance factor making the cost only dependent on the time used.

With the cost effective digital transmission systems now emerging the marginal costs for circuits on the main arteries becomes very low, which will pave the way for such fundamental changes on the "traditional" charges. It will help in linking people closer together, a possibility now within reach.

7. REFERENCES

[1] ITU/OECD. Report "Telecommunications for Development", World Communication Year, June 1983.

[2] Report of the Independent Commission for World-Wide Telecommunications Development, "The Missing Link", December 1984.

[3] Thailand : Socio-Economic Survey, 1975/76.
Documentation in 5 volumes :
- Greater Bangkok Metropolis Area;
- Northern Region;
- Northeastern Region;
- Central Region;
- Southern Region

National Statistical Office, Office of the Prime Minister, Bangkok, Thailand

[4] The F(x) formula can be further improved by exponentiating the function,

$$F(x) = \frac{100}{(1 + \exp(A)*x^{B})^{C}}$$

By trying various "C-values" a better fit of the function can be obtained. However, the basic function F(x), with C=1, has usually shown such excellent results that any further "refinements" have not been considered justified.

[5] Lars Engvall, "A socio-economic study on the usage of telephone, television and radio services", Report TRITA-TTDS-8402, Royal Institute of Technology, Stockholm, May 1984.

[6] Lars Engvall, "Telephone (TV and radio) usage in developing countries", ITC-10 paper #4, Session 4.3.

[7] Lars Engvall, "Etude socio-économique de l´usage des services téléphoniques, de télévision et de radiodiffusion dans les pays en développement", Le Bulletin de l´IDATE, No. 16, Juillet Juillet 1984.

TELETRAFFIC ISSUES in an Advanced Information Society
ITC-11
Minoru Akiyama (Editor)
Elsevier Science Publishers B.V. (North-Holland)

TRAFFIC ENGINEERING IN DIFFERENTIATED RATE PERIOD

Tito CERASOLI - Alberto MORENO (*) - Paulo BORGES

Empresa Brasileira de Telecomunicações (EMBRATEL)
(*) Telecomunicações do Rio de Janeiro (TELERJ)

ABSTRACT

The Brazilian Telecomunications System has been using since 1980 a 50% charge reduction in long distance calls between 8 p.m. and 8 a.m.

This has resulted in an expressive concentration of interest in calls between 8 p.m. and 9 p.m. which led to call congestions.

This scene brings about a problem to the traffic engineering in Brazil.

Which grade of service level should be used in trunk groups with busy hour localized in differentiated rate period?

This study was then developed to provide a support in a decision about a grade of service that would assure a good service to the subscribers and at the same time make possible the economic investiment in the plant.

1. INTRODUCTION

The economic analysis is based on the comparison between revenue loss due to a non increase in the trunk group versus the increasing cost, in circuit per circuit basis, during the equipment life time scope.

The revenue loss calculation must consider the seasonal traffic variations around the annual representative traffic value (VRA).

Loss value determination which is fundamental in this calculation is obtained with simulation support and will be described in next chapter.

In the increasing costs only the equipment costs (transmission and switching) are computed considering the manpower costs as marginal, not changing the results.

The study consists of the following steps:

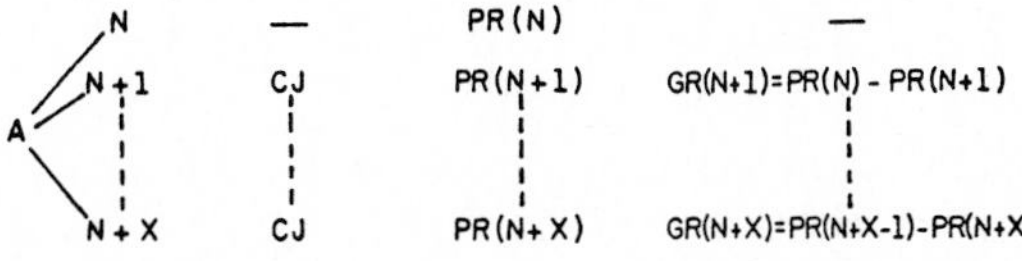

Fig. 1 - Additional Revenue Calculation Flow

N - number of existent circuits.

CJ - costs of one circuit (transmission+ switching).

PR - Annual revenue loss.

GR - Annual additional revenue.

This means that the investment of one additional circuit to N existing will imply an additional revenue GR(N+1) that can be a good economic business decision or not, based on the return rate (TIR) of this investment, calculated in the equipment life time scope.

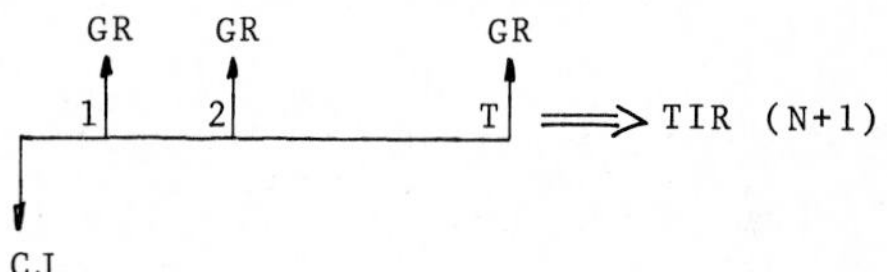

Fig. 2 - Cash Flow Diagram.

Where T means life time.

The analysis is repeated with the addition of one circuit at a time until the TIR becomes less than a minimum payment level (attractivity rate - TA), when the last additional circuit will not be economically a good bussiness.

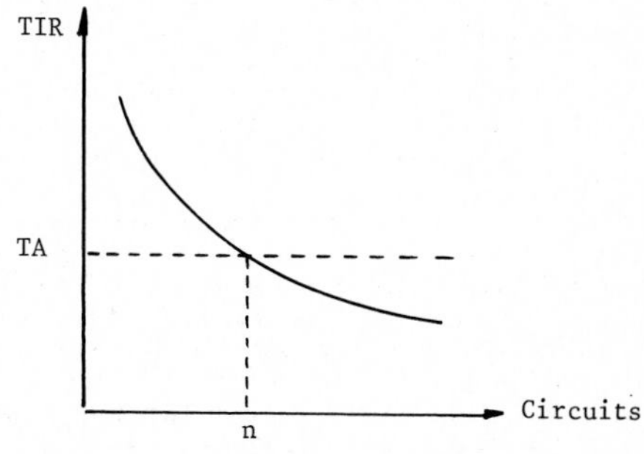

Fig. 3 - TIR Comportment

Where n means economic circuit number.

2. THEORYTICAL DEVELOPMENT

2.1 Valuation of Losses

To evaluate the call congestion, necessary to an economical valuation, it was firstly attempted a formulation on an analitical Markovian model.

This idea was put aside, since the simulation that had to be made to attend the analysed situation within this kind of model, could mask the results, and invalid the study.

It was then choosen a model which was improved as the study developed. At its final configuration, the simulated system corresponds to the following diagram.

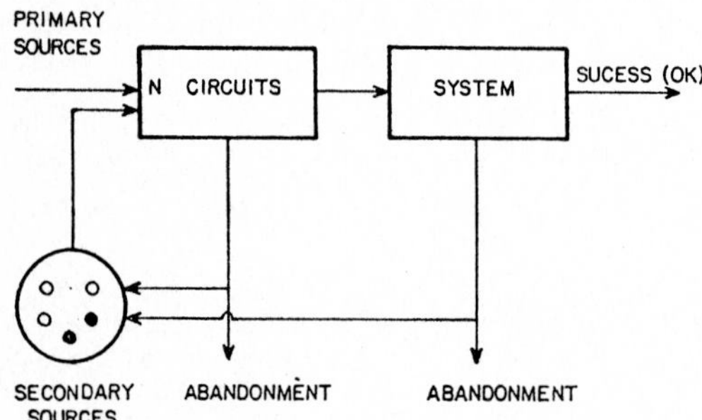

Fig. 4 - Simulated System Diagram

Calls can be proceeded from a primary source, if it is a first attempt, or from a secondary source, if it is a repeated call. The calls which find idle circuits are completed sucessfully and occupy these circuits during a randomic time interval.

If the circuits are congested or the call can't be completed, there are two possible situations: the call abandonment or its repetition due to the customers retrial (perseverance), after a randomic time interval.

The calls which return to the system due to the customers perseverance ocuppy a limited number of positions in the secundary sources to make new attempts. However, this number can be considered so greather as we want.

Along the simulation development the following hipothesis were considered:

- . full availability.
- . sequencial search of idle circuits.
- . exponentially distributed time interval between consecutive primary calls.
- . constant time interval to repeat a call that returns due to congestion.
- . exponentially distributed time interval to repeat a call that returns because the system is not OK.
- . the ratio of calls that returns to the system due to congestion or because they cannot be completed by the system is constant (f1 e f2 respectively).
- . the well succeeded calls occupy the circuits during a time interval equal to the sum of a constant (time of switching) and a randomic value exponencially distributed (conversation interval time).
- . the calls that dont't find congestion but are not completed by the system occupy the circuits during a constant time of switching.

The main variables involved in the simulation process are the following:

- TTC_i - instant time when the $(i+1)^{th}$ primary call arrives in the system.
- TS_i - service time of the $(i+1)^{th}$ call that seized the circuit (primary or repeated).
- TTR_i - instant time when the $(i+1)^{th}$ repeated call arrives in the system.
- KC - number of congestions.
- KR - number of repeated calls.
- KO - number of primary calls generated.
- TOC - time interval during which all circuits are busy.
- CONG - ratio between the number of congestions and the number of primary or secondary offered calls (lost of calls).
- CONGT - ratio between the total seizure time (TOC) and the total observation interval (time congestion).

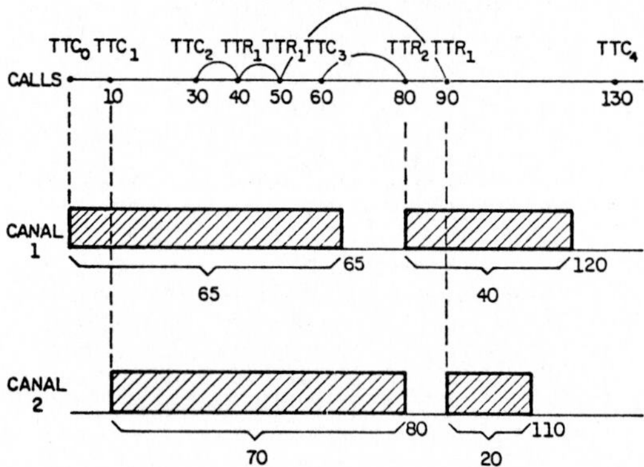

Fig. 5 - Simulation of a telephonic system with repeated call attempts

The first and second primary calls seize the idle circuits during service times respectively equal to 65 and 70 seconds. The third primary call arrives at TTC_2=30 sec., finds all circuits busy and returns three times at instants TTR_1 respectively equals to 40, 50 and 90 sec., when it finds an idle circuit and sizes it for 20 sec.

In TTC_3=60 sec. the fourth primary call arrives, doesn't find an idle circuit and returns once in TTR_2=80 sec., when it seizes an idle circuit for 40 sec.

Finally, the fifth primary calls arrive at TTC_4=130 sec. and seizes the idle circuit.

In accordance with the above conditions we can now write:

KO = 5
KR = 4
KC = 4
TOC = (65-10)+(110-90)= 75 seconds

The call congestion and time congestion can be estimated as follows:

CONG = KC/(KO+KR)=4/(4+5)=0.444
CONGT = TOC/observation interval
CONGT = 75/130=0.577

Naturally the precision of the above valuation improves as the observation time of the simulation or the number of generated calls grows.

During the development of the simulation it was necessary to study the following questions to achieve good results within an acceptable precision, with a minimum computational effort.

a - how to generate randomic numbers?
b - how many simulated calls would garantee the wanted precision?
c - how to garantee that the precision was achieved with the least number of simulated calls?

Whith these answers:

a - To generate randomic numbers uniformly distributed it was used the congruential method. The routine RANDU, developed by IBM, and presented in [1] was attached to the computer programm, codified in PL/I. To the generation of randomic variables with exponencial distribution it was employed the Inverse Transformation method [2]
b - To settle the number of calls that had to be simulated it was applied the following theorem presented in [3]
Theorem - for a process tending asymptotically to a Markovian process or μ-Markovian, the relative reliable interval, with significance level of 95%, is given by.

$$\Delta p = 2\sqrt{\frac{\lambda}{B}}, \text{ where}$$

p - call congestion
Δp - reliability interval to call congestion
B - number of times that the congested event was observed

$$\lambda = \frac{S}{p} \cdot VAR(p), \text{ where}$$

S - number of simulated calls
VAR(p) - variance of p

c - The variance reduction technich employed was the anti-thetic variable method. So during the processing of a simulation to a certain number of calls S we worked in two phases. In the first to S/2 calls we applied a set of uniform randomics and in the second S/2 calls we applied the set of complementary randomics. The result applied was taken as the average results of the two mentioned phases [4].

2.2 Daily Revenue Loss Calculation (PR_{DIA})

Considering that the daily revenue loss was only computed for the busy hour revenue loss, without considering the other hours, the following diagram can be used:

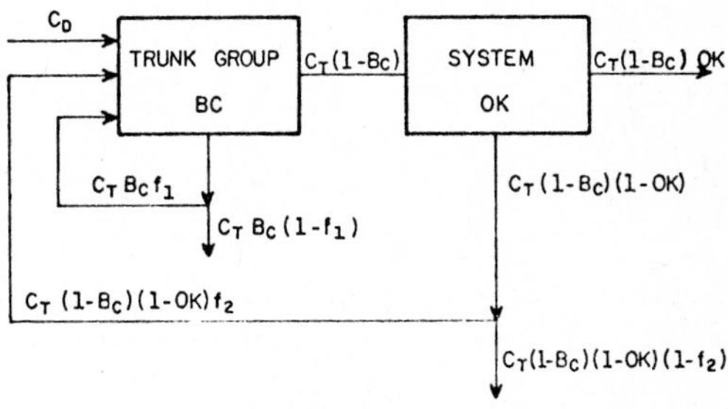

Fig. 6 - System Call Flow

C_D - Busy hour basic calls.

C_T - Busy hour total calls.

B_C - Busy hour lost call rate.

OK - Busy hour completed call rate.

f_1 - Busy hour probability of retrial for blocked calls loss.

f2 - Busy hour probability of retrial for non completed calls due to line busy or absent subscriber.

$$RP = [C_T B_C(1-f_1) + C_T(1-OK)(1-f_2)]\ TC.DM$$

$$RA = C_T(1-B_C)OK\ TC.DM$$

Where:

R_P - Busy hour revenue loss.

RA - Busy hour revenue.

TC - Mean time conversation.

DM - Average revenue per minute.

PRD - Revenue loss rate.

$$PRD = \frac{RP}{RA} = \frac{B_C(1-f_1)+(1-B_C)(1-OK)(1-f_2)}{(1-B_C)\ OK}$$

Small variations in the involved factors do not affect the reliability of this relation.

If the revenue loss were directly calculated, the result wouldn't have the same reliability as the one with the relation applied to a separated revenue calculation, described below:

$$RA = A_C.(\%\ TRAF.CONV.).60.DM$$

A_C - Carried traffic

$$A_C = \frac{A_0(1-B_T)}{(1-B_T.f_1)}$$

A_0 - Busy hour offered traffic

B_T - Time congestion

% TRAF.CONV.- conversation traffic percentage

$$\%\ TRAF.CONV. = \frac{OK.T_C}{T_X+OK.T_C}$$

T_X - Mean switching time.

$$RP_{(DIA)} = RP_{HMM} = PRD.RA_{HMM}$$

$$RP_{DIA} = \left[\frac{(1-f_1)\,B_C + (1-B_C)(1-OK)(1-f_2)}{(1-B_C)\,OK}\right] \times \frac{A_0\,(1-B_T)\,OK.TC.60.DM}{(1-B_T\,f_1)(TX+OK.TC)}$$

2.3 Annual Revenue Loss Calculation

The daily revenue loss during the year must be considered in this calculation.

This was made considering the mean frequency distribution of the traffic daily values in relation to the annual representative traffic values (VRA) for the network in Brazil, as showed in the following table, where NU means the number of days in which the traffic is greather than the VRA.

Table 1 - Daily traffic values distributon

% VRA	NU	% VRA	NU
130	1	90	10
120	1	89	10
115	1	88	10
110	3	87	8
105	7	86	8
100	22	85	6
99	9	84	5
98	10	83	6
97	11	82	4
96	12	81	4
95	13	80	5
94	13	75	13
93	13	70	9
92	12	60	7
91	12		

For each traffic distribution value all necessary parameters to revenue loss calculation are determined. When all values accumulated we have the annual revenue loss.

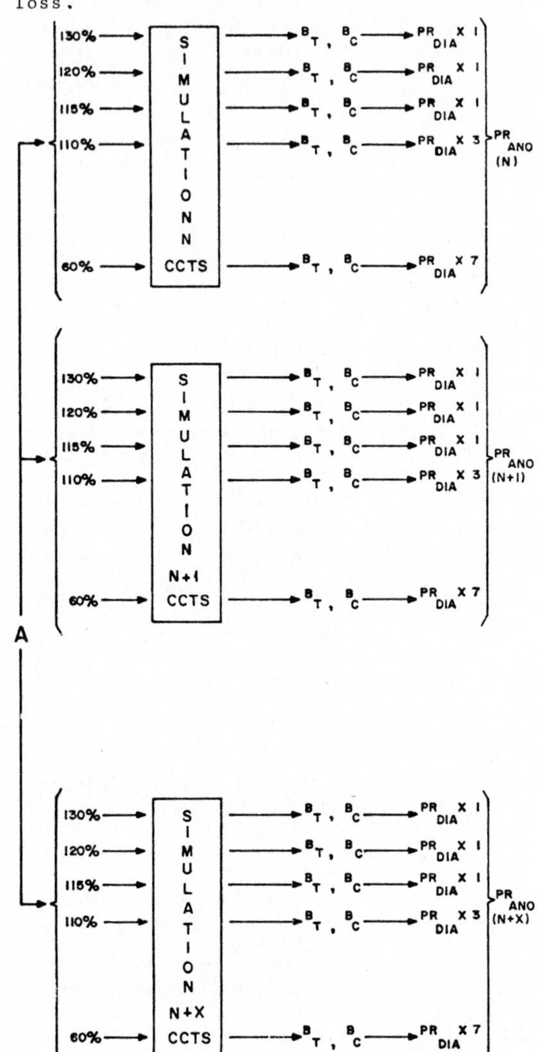

Fig. 7 - Annual Revenue Loss Calculation

2.4 Annual Additional Revenue Calculation (GR)

This will be obtained from the difference between adjacents circuits revenue loss.

N — PR_{ANO} (N)

N+1 — PR_{ANO} (N+1) - GR_{ANO}(N+1) = PR_{ANO}(N) - PR_{ANO}(N+1)

N+X — PR_{ANO} (N+X) - GR_{ANO} (N+X) = PR_{ANO}(N+X-1) - PR_{ANO}(N+X)

Fig. 8 - Annual Additional Revenue Calculation

2.5 Return Rate Calculation (TIR)

The economic viability in the circuit will be decided by the Return Rate (TIR), conjugating the additional circuit cost and the additional revenue with this additional circuit. The calculation considers the circuit life time scope.

$$TIR = i \left| \sum_{t=0}^{T} \frac{F_X}{(1+i)^t} = 0 \right.$$

F_X - cash flow component.

T - life time scope.

3. RESULTS

3.1 Input Data

The following input data were used in the study.

T - Life time scope - 13 years

T_C - Mean time conversation - 300 sec.

T_X - Mean time switching - 30 sec.

OK - Completed calls rate - 61.7%

DM - Mean revenue per minute - Cr$180

Circuit Costs (Switching + Transmission)
- Local-Toll Ticket Circuit
  Cr$7 000 000 (US$6 700)
- Toll-Toll Circuit
  Cr$4 760 000

3.2 Simulation Results

From the measures obtained in practice we have the following parameters:
- perseverance due to congestion(f_1) = 82%
- perseverance due to the fact that the system is not OK (f_2) = 72.66%
- mean time of conversation(T_C) = 300 sec.
- mean time of switching(T_X) = 30 sec.
- completed calls rate (OK) = 61.60%

Besides the above parameters we need the time interval between the arrivals of two primary calls as well as that one between secondary calls. Since direct measures of these parameters are not available in practice, they were obtained as follows:

- time interval between the arrivals of primary calls:
  $(T_C.OK+T_X)/(A(1-f_2(1-OK)))$, where:

  $(T_C.OK+T_C)$ - mean holding time

  A - total offered traffic, including the traffic due to repeatead calls (supposing there is not congestion).

  $A(1-f_2(1-OK))$ - traffic from primary sources

- time interval between the arrivals of repeated calls:
  It was considered as a fraction of the time interval between the arrivals of primary calls.

Some typical results obtained from the simulation process for the above parameters are following presented:

A	N	Call Congestion	Time Congestion
20	18	.4884	.3579
	19	.4341	.3125
	20	.3809	.2609
30	40	.0415	.0250
	41	.0275	.0166
	42	.0159	.0105

A - offered traffic
N - number of circuits

These results correspond to a 20 000 calls simulation. The relative precision of the call congestion can be evaluated from the theorem already stated.

From the determination of variance of $\hat{p}$ (VAR($\hat{p}$)), in the simulation process we have obtained λ=4.

$$\frac{\Delta p}{p} = 2\sqrt{\frac{\lambda}{B}} = 4\sqrt{\frac{1}{S.\hat{p}}} \quad \text{for a significance level of 95\%}$$

For a call congestion .4884 (N=18 and A=20) the relative precision of 20 000 calls is:

$$\frac{\Delta p}{p} = 4\sqrt{\frac{1}{20000 \times 0.04884}} = 4.03\%$$

3.3 Economic Study Example

The chosen example was 100 Erlangs offered to 99 and 100 circuits, leading to the following results:

VRA	A	NU	N	BC	BT	PR (Cr$)	N	BC	BT	PR (Cr$)
100	130	1	99	.5576	.5410	797 681	100	.5474	.5244	787 799
	120	1		.4672	.4477	621 128		.4562	.4340	609 696
	115	1		.4103	.3968	531 839		.3959	.3787	519 043
	110	3		.3422	.3259	1 344 480		.3226	.3150	1 288 106
	105	7		.2622	.2566	2 553 736		.2428	.2380	2 458 846
	100	22		.1750	.1733	6 433 077		.1634	.1576	6 297 695
	99	9		.1636	.1578	2 551 591		.1500	.1472	2 479 249
	98	10		.1503	.1470	2 729 168		.1358	.1349	2 647 242
	97	11		.1354	.1361	2 877 862		.1211	.1215	2 795 336
	96	12		.1209	.1216	3 016 390		.1077	.1044	2 940 336
	95	13		.1081	.1047	3 154 931		.0949	.0924	3 068 231
	94	13		.0948	.0946	3 032 430		.0848	.0835	2 970 798
	93	13		.0854	.0842	2 942 791		.0744	.0740	2 874 244
	92	12		.0747	.0728	2 627 854		.0633	.0669	2 559 265
	91	12		.0635	.0659	2 533 720		.0525	.0541	2 475 337
	90	10		.0522	.0529	2 039 496		.0434	.0461	1 599 476
	89	10		.0431	.0448	1 976 717		.0377	.0377	1 954 179
	88	10		.0377	.0384	1 932 366		.0305	.0311	1 902 186
	87	8		.0308	.0310	1 505 671		.0253	.0251	1 487 753
	86	8		.0251	.0264	1 469 197		.0207	.0213	1 455 332
	85	6		.0207	.0216	1 078 690		.0165	.0181	1 068 403
	84	5		.0160	.0177	878 851		.0135	.0140	874 343
	83	6		.0135	.0136	1 036 870		.0104	.0180	1 026 945
	82	4		.0099	.0114	677 105		.0080	.0083	674 516
	81	4		.0083	.0080	666 880		.0064	.0069	663 853
	80	5		.0060	.0060	818 981		.0050	.0050	817 141
	75	13		.0020	.0020	1 978 361		.0013	.0007	1 975 653
	70	5		.0005	.0005	1 274 003		.0001	.0002	1 272 807
100	60	7	99	.0000	.0000	1 575 555	100	.0000	.0000	1 575 555
TOTAL						56 657 421				55 519 368

The additional annual revenue with one additional circuit to 99 existing will be Cr$1 138 056 and the investment Cr$7 000 000 (Local-Toll circuit) or Cr$4 760 000 (Toll-Toll circuit).

The return rate of this investment, in 13 wears, will be.

Local-Toll circuit - 12.2%
Toll -Toll circuit - 21.6%

The adjacent circuits analysis have the following results:

TRAF	CIRC	B%	REVENUE (Cr$) LOSS	REVENUE (Cr$) ADDITIONAL	TIR (%) LC-TR	TIR (%) TR-TR
100	99	8.2	56657421	-	-	-
	100	7.6	55519365	1138056	12.2	21.6
	101	7.0	54386548	1132817	12.1	21.5
	102	6.4	53303310	1083238	11.1	20.3
	103	5.8	52419412	883898	7.1	15.2
	104	5.3	51 611 259	808153	5.4	13.1
100	105	4.8	50859267	751992	4.1	11.5

Considering 12% as a minimum payment level (Attractivity Rate), the minimum loss for 100 Erlangs offered traffic will be:

Local-Toll Group Trunk - 7.0%
Toll -Toll Group Trunk - 5.3%

3.4 Overall Results

The exemplified procedure in 3.3 was made for 10 Erlangs, with the following results:

VRA	Local-Toll Group Trunk	Toll-Toll Group Trunk
10	12.0%	8.4%
100	7.0%	5.3%

4. MAXIMUM LIMIT LOSS

The loss values obtained consider only economic aspects and not the service quality.

The service quality consideration was introduced according to the CCITT 541 recommendation (white Book - VI).

"It is experience of Administrations That an acceptable automatic service on a final circuit group cannot be maintained if the traffic loadind on the group exceeds a level corresponding to a calculated Erlang grade of service of 10%.

Beyond this traffic loading, and especially owing to the cumulative effect of repeat attempt calls, the service rapidly deteriorates."

Conjugating this recommendation with the economic analysis results, we have the following table:

VRA	Local-Toll Group Trunk	Toll-Toll Group Trunk
10	10%	8.4%
100	7%	5.3%

5. CONSIDERATIONS

The obtained results cannot be assumed as definitive. The study development must continue under many different approaches summaryed below:

A - Improvement of the Data Collection

The used process must be object of an intense analysis through a more comprehensive measurement process.

B - Model Validity Check

The assumed hipotesys must be tested through a statistic process to be made through a specific experiment.

C - Simulation Model

It will ocasionaly be necessary to change some hipotesis to improve the model. In this item it would be logical to mention the subscriber behaviour when submitted to a hard congestion situation.
It is a reasonable supposition that the retrial probability (perseverance) is not constant with direct variatons in relation to the trunk group congestion. This fact brings about an important problem to studied - the calls loss control limit determination that saves the system from congestions above certain limits without economic considerations.

D - Investment Costs

The investment costs should consider a broader possibility set, for example, the investment only in switching equipments if transmission equipments are available or vice-versa.

E - Conclusion

This study considered only economic aspects and doesn't intend to obtained definitive dimentioning values, showing that trunk groups with greater traffic capacity must be privilegies.

Nowadays the trunk groups with differenciated busy hour rate in the Brazilian network are dimentioned with 10% loss, specially owing to investment limitations.
The study will continue, including traffic demands in normal rate hours, better costs investment analysis and more considerations regarding the service quality.

REFERENCES

[1] Schmidt and Taylor - Simulation and Analysis of Indestril Systems - R.D., Inc. - Illinois - 1970

[2] Newman T.G. e Odell P.L. - TL Generation of Randon Variats - Griffuis Statistical Monograph & Courses - London - 1971

[3] Le Gall, P. - Les Methods di Simulation - Dunod Paris - 1968

[4] Moreno A.O. e Beranger J.E.S. - Simulação - Aplicação ao Tráfego Telefônico - Revista TELEBRÁS - Dezembro de 1984 - BRASIL

TELETRAFFIC ISSUES in an Advanced Information Society
ITC-11
Minoru Akiyama (Editor)
Elsevier Science Publishers B.V. (North-Holland)

Mean Waiting Time for a Token Ring with Nodal Dependent Overheads

Michael J. Ferguson

INRS-Télécommunications

Montréal, Canada

Abstract

This paper obtains the generating function and bounds on the first moment of the waiting time for a token ring system with nodal dependent overheads and *exhaustive* service. Extensions to *gated* service, although not given, are almost immediate. The particular form of nodal dependent overhead studied can be interpreted as an additional time required to seize the token when the node has a message to send, compared to the time required when just passing it on. Eisenberg's [1] result on the dependence of the waiting time on the intervisit time is extended, and the new relation between the intervisit time and the *terminal service time* of Ferguson and Aminetzah [2] is established. The resulting generating function depends upon the probability of *no-message* when the token arrives at a node. This in turn requires the solution of the complete generating function, a formidable task, to determine even the moments.

1. INTRODUCTION

This paper determines the generating function, bounds, and approximations to the waiting time for an *exhaustive* service token ring† with arbitrary nodal dependent overheads, arrival rates and message service distributions. This is the first paper to include *nodal* dependent overheads. An example of nodal dependent overhead, and the one that will be assumed for illustration purposes in this paper, is that difference in overhead caused by token passing, when a node has a message and when it does not. This overhead, to be designated as *token-acquisition* overhead, includes the length of the token and any other necessary ancilliary processing. A recent paper by Ferguson and Aminetzah [2] established a procedure for computing exact results for the mean waiting time in a token ring with arbitrary internodal delays, arrival rates, and service times. The major feature of the approach in that paper was the need, when there were N nodes, for only N^2 rather than the N^3 equations required in previous work (see for example [1, 4, 5] among others).

The analysis of an *exhaustive* service token ring in Ferguson and Aminetzah [2] is extended in Section 2 to include nodal dependent overhead. As in previous analyses, for example [2, 1, 4], the waiting time distribution is seen to be determined by the intervisit time, which in turn can be obtained by the joint distribution of the *terminal service time* first used by Aminetzah [2, 6]. The key difference, and difficulty, is that generating function for the *terminal service time* has a boundary condition involving the probability of *no-message* at the node when the token arrives.

Eisenberg [1] was the first to relate the waiting time distribution to the intervisit time. His expression is extended in Section 2 to include the effects of the *token-acquisition* overhead. The result, Eq. (27), is perhaps not too surprising but the implications are. The equivalent mean waiting time expression is shown as in Eq. (28). The methodology is similar to that used in [2, 6].

The *waiting* time of the token ring system is upper bounded by **either** appending a *token-acquisition* overhead to every transmission of a specific node **or** by appending it to the overhead of passing a token to the next node independent of whether the node has a message to send. For N nodes, there evidently exists 2^N systems each of which is an upper bound to the specific one in question. Since none of these systems have nodal dependent overheads, each is amenable to the exact analysis in Ferguson and Aminetzah [2]. The choice of system to use for the bound is made, not by solving the exact system, but rather by finding the system that "minimizes" the simple waiting time conservation equation found in [2] (see Eqs. (29) and (30) below). This discussion is found in Section 3.

A lower bound on the mean *waiting* time, given a knowledge of the **actual** *no-message* probabilities, is obtained in Section 4. The system used to obtain this bound randomly adds the *token-acquisition* overhead to the internodal delay according to the actual *no-message* probabilities. Since the *no-message* probabilities must be estimated these results actually lead to the various approximations discussed in Section 5. Section 6 discusses numerical results, with simulations, for symmetric traffic and a system where there is one dominant node with 90% of the traffic.

There are no other approximations or bounds in the literature applicable to rings with nodal dependent overheads. In fact those approximations that do exist for the "simple" case of no overhead, for example Bux [7], are quite inaccurate for asymmetric nodal traffics (see [2] for a discussion).

2. Basic Analytic Considerations

The queuing system under consideration is shown in Figure 1. There are N terminals or nodes. The queue capacity at each node is infinite. The arrival process at each node is independent and Poisson with arrival rate λ_i. The length of the messages and hence the service time has density $P_{s_i}(\cdot)$ and generating function $S_i(x)$. All message service times are independent. The service time is made up of a fixed size header of length h, and a message of some arbitrary density that may be nodal dependent.

The fixed overhead for passing the token from the ith to $(i+1)^{st}$ node is independent of any of the queue parameters, including whether there is any message at all in the queue, and is designated by d_i. The corresponding density is $P_{d_i}(\cdot)$ and generating function is $D_i(x)$. If there is a message at the node, an additional *token-acquisition* overhead is incurred. This will have the density $P_{a_i}(\cdot)$ and generating function $A_i(x)$. If there are any new message arrivals during this *token-acquisition*, they are simply added to the queue. The token is passed on when there are no messages left in the queue (*exhaustive* service). The *terminal service time* is the sum of the internodal delay from the previous node, the *token-acquisition* overhead if there are waiting messages, plus the time of n_k busy periods if there were n_k messages after the token seizure. The *terminal service time*, for probabilistic purposes, may be viewed as the sum of $n_k \geq 1$ busy periods where n_k is number of messages at token arrival plus a *token-acquisition* induced busy period. The queue is known to be empty when the token departs.

† A good taxonomy of ring systems is found in Penney [3].

The method of attack, as in [2, 6], is to first derive the joint generating function for the *terminal service time* during the kth visit to a node, and to relate this to the intervisit time. The intervisit time, including the complications due to the *token-acquisition* overhead, is then related to the waiting time.

The intervisit time, v_k at the kth visit to some node in terms of the *terminal service time* is

$$v_k = \sum_{i=1}^{N-1} \theta_{k-i} + d_{k-1} \qquad (1).$$

The intervisit times are determined by the last $N-1$ *terminal service times*. The *terminal service time* N-vector is defined as

$$\vartheta_k = [\theta_k, \theta_{k+1}, \ldots, \theta_{k+(N-1)}].$$

Note that ϑ_k and ϑ_{k+1} have $(N-1)$ values in common. The probability density is defined as $P_{\vartheta_k}(\cdot)$ and the generating function is defined as

$$\begin{aligned} G_{\vartheta_k}(\mathbf{x}) &\triangleq E(e^{-\sum_{i=0}^{N-1} x_i\theta_{k+i}}) \\ &= \int_0^\infty \cdots \int_0^\infty e^{-\sum_{i=0}^{N-1} x_i\theta_{k+i}} \\ &\qquad \times P_{\vartheta_k}(\theta_k, \ldots, \theta_{k+(N-1)}) \prod_{i=0}^{N-1} d\theta_{k+i} \end{aligned} \qquad (2)$$

where $\Re[x_i] \geq 0$.

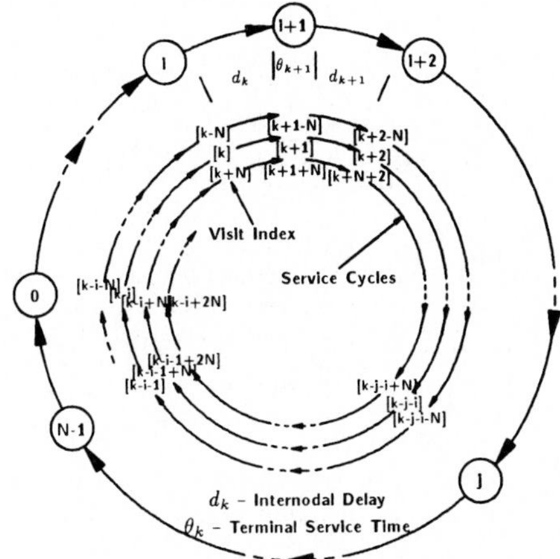

Fig. 1 The Basic Queuing Model

2.1 Generating Function and Moments for joint *terminal service time*

To derive a recursion for $G_{\vartheta_k}(\mathbf{x})$ first note that if the server arrives to see n_k messages in the queue, the busy period is the sum of n_k *simple* busy periods of an $M/G/1$ queue with arrival rate λ_k and service density $P_{s_k}(\cdot)$ plus the busy period, for the same type of queue, induced by arrivals during the *token-acquisition* overhead.

Let the *simple* busy period density be defined as $P_{b_k}(\cdot)$ and the corresponding generating function as $B_k(\cdot)$, then it is well known (see, for example [8]) that

$$B_k(x) = S_k(x + \lambda_k(1 - B_k(x))) \qquad (3)$$

In the sequel the products $\lambda_k\bar{b}_k$ and $\lambda_k\overline{b^2}_k$ frequently arise. The traffic intensity ρ_k and the *normalized second moment of the service time*, γ_k^2 are defined as

$$\rho_k \triangleq \lambda_k\bar{s}_k \qquad \gamma_k^2 \triangleq \frac{\overline{s_k^2}}{\bar{s}_k}. \qquad (4)$$

Using these definitions, and the well known results for the busy period of an M/G/1 queue,

$$\lambda_k\bar{b}_k = \frac{\rho_k}{1-\rho_k} \qquad \lambda_k\overline{b^2}_k = \frac{\rho_k\gamma_k^2}{(1-\rho_k)^3}. \qquad (5)$$

Generating Function and Moments for the *token-acquisition-induced busy period*

The *token-acquisition-induced busy period* will be designated with a subscript *ab*. Thus $a_{ab||k}$ is the length of the *token-acquisition* induced busy period, $A_{ab||k}(\cdot)$ is its generating function and $P_{ab||k}(\cdot)$ is its density. Thus $A_{ab||k}(\cdot)$ is derived by using standard queueing arguments.*

$$A_{ab||k}(x) = A_k(x + \lambda_k(1 - B_k(x))) \qquad (6)$$

The moments of the *token-acquisition-induced busy period* are found by differentiating Eq. (6) by x and setting to $x = 0$. The result, noting Eq. (4) and (5), for the first moment is

$$\bar{a}_{ab||k} = \frac{\bar{a}_k}{1-\rho_k} \qquad (7)$$

and for the second moment is

$$\overline{a^2}_{ab||k} = \frac{\overline{a^2}_k}{(1-\rho_k)^2} + \frac{\bar{a}_k\rho_k\gamma_k^2}{(1-\rho_k)^3} \qquad (8)$$

Generating Function for the *terminal service time*

Thus the generating function of the total time spent in service, given that the server met n_k messages on arrival is just $B_k^{n_k}(x)A_{ab||k}(x)$. The generating function of the *terminal service time*, θ_k, given n_k is just $B_k^{n_k}(x)A_{ab||k}(x)D_{k-1}(x)$. To find the unconditional generating function it is noted that n_k, given the time the server has been away, *i.e.* the intervisit time v_k, is Poisson with parameter $\lambda_k v_k$ where v_k is given in Eq. (1). Averaging with respect to the θ_{k-i} in v_k gives the result.

If $P^{*(n)}(t)$ is defined as the n fold convolution of $P(\cdot)$ evaluated at t, the fundamental probability equation is easily seen to be

$$\begin{aligned} P_{\vartheta_{k+1}}(\vartheta_{k+1}) = &\int_0^\infty P_{\vartheta_k}(\vartheta_k)d\theta_k \int_0^{\theta_{k+N}} P_{d_{k+N-1}}(t)dt \\ &\times \left(1 + \sum_{n_{k+N}=1}^{\infty} \left(P_{b_{k+N}}^{*(n_{k+N})} * P_{ab||k+N}\right)(\theta_k - t)\right) \\ &\times \frac{1}{n_{k+N}!}\left(\lambda_{k+N}\left(t + \sum_{i=1}^{N-1}\theta_{k+i}\right)\right)^{n_{k+N}} \\ &\times e^{-\lambda_{k+N}\left(t+\sum_{i=1}^{N-1}\theta_{k+i}\right)} \end{aligned} \qquad (9)$$

Multiplying the left hand side of Eq. (9) by $\exp\left(-\sum_{i=1}^{N} x_i\theta_{k+i}\right)$ and integrating gives $G_{\vartheta_{k+1}}(\mathbf{x})$. Multiplying by the same thing on the right-hand side of Eq. (9) leaves a sum over n_{k+N}, and integrations over $\theta_k, \ldots, \theta_{k+N}$ and t. To complete the recursion, integrate over θ_{k+N}, sum over n_{k+N}, integrate over t, and then finally integrate over $\theta_k, \ldots, \theta_{k+N-1}$. The result, noting the definition again in Eq. (2) is

$$\begin{aligned} G_{\vartheta_{k+1}}(x_1, \ldots, x_N) = &A_{ab||k+N}(x_N)D_{k+N-1}(x_N + y_{kN}) \\ &\times G_{\vartheta_k}(0, x_1 + y_{kN}, \ldots, x_{N-1} + y_{kN}) \\ &+ (1 - A_{ab||k+N}(x_N))\, D_{k+N-1}(x_N + \lambda_k) \\ &\times G_{\vartheta_k}(0, x_1 + \lambda_k, \ldots, x_{N-1} + \lambda_k) \end{aligned} \qquad (10)$$

---

* Details of these standard queuing arguments may be found in any good introductory queuing text such as Kleinrock [8] or in the ring context in Aminetzah [6].

where y_{kN} is defined as

$$y_{kN} \triangleq \lambda_k(1 - B_k(x_N)). \tag{11}$$

Note that the second summand has the unfortunate property of not being a known value when $\mathbf{x}$ is set to 0. This means that the equations for the moments depend upon the detailed form of $G_{\vartheta_k}(\cdot)$. From the definition of the intervisit time (see Eq. (1)) and the definition of $G_{\vartheta_k}(\cdot)$, it is evident that the generating function of the intervisit time, $V_{k+N}(x)$, is

$$V_{k+N}(x) = G_{\vartheta_k}(0, x, \ldots, x)D_{k+N-1}(x), \tag{12}$$

Thus $G_{\vartheta_k}(0, \lambda_k, \ldots, \lambda_k)D_{k+N-1}(\lambda_k)$ in Eq. (10) is the generating function of v_{k+N} evaluated at λ_k. Since the arrival process at the node of visit k is Poisson with rate λ_k, $e^{-\lambda_k v_k}$ is the probability of no arrivals in v_k and the expectation of this is $V_k(\lambda_k)$, which is thus the marginal probability of there being no customers at the node when the token arrives.

Moments of *terminal service time* and intervisit time

The first moments of the intervisit time and *terminal service time*, assuming $V_k(\lambda_k)$ is known, can be obtained by differentiating Eq. (10) with respect to x_N and setting $\mathbf{x} = 0$. Noting that $\bar{a}_{ab||k}$ is the mean of *token-acquisition-induced busy period*, $\bar{\theta}_{k+N}$ can be written as

$$\bar{\theta}_{k+N} = \bar{a}_{ab||k}(1 - V_k(\lambda_k)) + \frac{\bar{d}_{k-1}}{1-\rho_k} + \frac{\rho_k}{1-\rho_k}\sum_{i=1}^{N-1}\bar{\theta}_{k+i} \tag{13}$$

Inserting Eq. (7) for $\bar{a}_{ab||k}$ gives

$$\bar{\theta}_{k+N} = \frac{\bar{d}_{k-1} + \bar{a}_k(1 - V_k(\lambda_k))}{1-\rho_k} + \frac{\rho_k}{1-\rho_k}\sum_{i=1}^{N-1}\bar{\theta}_{k+i} \tag{14}$$

Inserting a $\bar{\theta}_{k+N}$ in the right-hand sum gives

$$\bar{\theta}_{k+N} = (\bar{d}_{k-1} + \bar{a}_k(1 - V_k(\lambda_k))) + \rho_k\sum_{i=1}^{N}\bar{\theta}_{k+i} \tag{15}$$

If ϕ_j is defined as the *terminal service time* an **node** j (rather than with respect to the visit k), then it is evident that the sum in Eq. (15) is independent of j and is just the mean cycle time, $\bar{c}$. Thus

$$\bar{\phi}_j = \bar{a}_j(1 - V_j(\lambda_j)) + \bar{d}_{j-1} + \rho_j\bar{c}. \tag{16}$$

Summing over j and solving for $\bar{c}$ gives

$$\bar{c} = \frac{1}{1-\rho}\left(\bar{d} + \sum_{j=1}^{N}\bar{a}_j(1 - V_j(\lambda_j))\right) \tag{17}$$

where ρ, the total traffic intensity, is defined as $\rho \triangleq \sum_{j=1}^{N}\rho_j$ and $\bar{d} \triangleq \sum_{j=1}^{N}\bar{d}_j$. The intervisit time is defined by Eq. (1) and easily found using Eqs. (16) and (17) as

$$\bar{v}_j = \left(\frac{1-\rho_j}{1-\rho}\right)\left(\bar{d} + \sum_{\ell=1}^{N}\bar{a}_\ell(1 - V_\ell(\lambda_\ell))\right) - \bar{a}_j(1 - V_j(\lambda_j)) \tag{18}$$

The form of the cycle mean in Eq. (17) shows that the *token-acquisition* overhead serves to increase the effective ring delay. Further discussion of the relation to random adding of *token-acquisition* overhead is found in Section 4. An explicit knowledge of $V_\ell(\lambda_\ell)$ is required to compute these means.

The computation of the second moments is conceptually similar to that of the first moments. Differentiating Eq. (10) twice, once with respect to x_N and a second time with respect to x_j gives an equation in $\overline{\theta_{k+N}\theta_{k+j}}$. Like the equations for the first moments these equations depend upon more than just the moments themselves. In fact in order to solve them it is necessary to compute $\overline{\theta_{k+j}e^{-\lambda_k v_{k+N}}}$. This is clearly even more difficult than the estimating of $V_k(\lambda_k) = \overline{e^{-\lambda_k v_{k+N}}}$ required for the first moments. These equations, although messy, serve as the basis of the computations. They were obtained by hand and confirmed using MACSYMA [9]. The results are displayed below.

$$\begin{aligned}\overline{\theta^2}_{k+N} = {} & \overline{a^2}_{ab||k+N}(1 - V_{k+N}(\lambda_k)) + \overline{d^2}_{k+N-1}\left(\frac{\rho_k}{1-\rho_k}\right) \\ & + 2\bar{a}_{ab||k+N}\left(\overline{d_{k+N-1}(1 - e^{-\lambda_k v_{k+N}})} + \frac{\rho_k}{1-\rho_k}\bar{d}_{k+N-1}\right) \\ & + \frac{\rho_k\bar{d}_{k+N-1}\gamma_k^2}{(1-\rho_k)^3} \\ & + \left(\frac{2\rho_k}{1-\rho_k}\left(\bar{a}_{ab||k+N} + \frac{\bar{d}_{k+N-1}}{1-\rho_k}\right) + \frac{\rho_k\gamma_k^2}{(1-\rho_k)^3}\right)\sum_{i=1}^{N-1}\bar{\theta}_{k+i} \\ & + \left(\frac{\rho_k}{1-\rho_k}\right)^2\sum_{i=1}^{N-1}\sum_{\ell=1}^{N-1}\overline{\theta_{k+i}\theta_{k+\ell}}\end{aligned} \tag{19a}$$

and

$$\begin{aligned}\overline{\theta_{k+j}\theta_{k+N}} = {} & \bar{a}_{ab||k+N}\overline{\theta_{k+j}(1 - e^{-\lambda_k v_{k+N}})} + \frac{\bar{d}_{k+N-1}}{1-\rho_k}\bar{\theta}_{k+j} \\ & + \frac{\rho_k}{1-\rho_k}\sum_{i=1}^{N-1}\overline{\theta_{k+j}\theta_{k+i}}\end{aligned} \tag{19b}$$

These equations are exact for **any** form of nodal dependent overhead that depends on whether there is message to be served. It can be specialized to the case considered in this paper by substituting in the forms for $\bar{a}_{ab||k+N}$ and $\overline{a^2}_{ab||k+N}$ in Eqs. (7) and (8) respectively. After substituting and some intentional rearrangement, the result is

$$\begin{aligned}\overline{\theta^2}_{k+N} = {} & \frac{\overline{d^2}_{k+N-1} + (\overline{a^2}_{k+N} + 2\bar{d}_{k+N-1}\bar{a}_{k+N})(1 - V_{k+N}(\lambda_k))}{(1-\rho_k)^2} \\ & + (\bar{d}_{k+N-1} + \bar{a}_{k+N}(1 - V_{k+N}(\lambda_k)))\frac{\rho_k\gamma_k^2}{(1-\rho_k)^3} \\ & + \boxed{\left(\frac{2\bar{a}_{k+N}}{(1-\rho_k)^2}\right)\left(\overline{d_{k+N-1}(1 - e^{-\lambda_k v_{k+N}})} + \rho_k\overline{d_{k+N-1}e^{-\lambda_k v_{k+N}}}\right)} \\ & + \left(\frac{2\rho_k}{(1-\rho_k)^2}(\bar{d}_{k+N-1} + \bar{a}_{k+N}(1 - V_{k+N}(\lambda_k)) + \boxed{V_{k+N}(\lambda_k)\bar{a}_{k+N}}) + \frac{\rho_k\gamma_k^2}{(1-\rho_k)^3}\right)\sum_{i=1}^{N-1}\bar{\theta}_{k+i} \\ & + \left(\frac{\rho_k}{1-\rho_k}\right)^2\sum_{i=1}^{N-1}\sum_{\ell=1}^{N-1}\overline{\theta_{k+i}\theta_{k+\ell}}\end{aligned} \tag{20a}$$

and

$$\begin{aligned}\overline{\theta_{k+j}\theta_{k+N}} = {} & \frac{\overline{\theta_{k+j}}}{1-\rho_k}(\bar{d}_{k+N-1} + \bar{a}_{k+N}(1 - V_{k+N}(\lambda_k))) \\ & + \boxed{\frac{\bar{a}_{k+N}}{1-\rho_k}\left(\overline{\theta_{k+j}(1 - e^{-\lambda_k v_{k+N}})} + \bar{\theta}_{k+j}\overline{(1 - e^{-\lambda_k v_{k+N}})}\right)} \\ & + \frac{\rho_k}{1-\rho_k}\sum_{i=1}^{N-1}\overline{\theta_{k+j}\theta_{k+i}}\end{aligned} \tag{20b}$$

These equations have been written in this way to show the similarity to the form that would result if the *token-acquisition* overhead were added randomly to the node according to the probability $(1 - V_{k+N}(\lambda_k))$. The boxed terms in Eq. (20) are due to the actual nonrandom addition of the *token-acquisition* overhead. These equations will be discussed further in Section 4.

2.2 The Generating Function and Moments for the Waiting Time

The relation between the waiting time and the intervisit time was first derived by Eisenberg [1] for the case of no nodal overhead. The expression derived here is an extension and follows the same methodology as that found in Aminetzah [6]. The basic idea is to identify a test message as having exactly ℓ messages served **after** it. This allows expressions for the generating function of this number to be related to both the waiting time and independently to the total number served in a *terminal service time*. The pair of expressions allow for a solution of the waiting time generating function.

First note that the generating function, $H_k(x)$, of the number of messages served in a *simple* busy time of an $M/G/1$ queue is given by the solution of

$$H_k(x) = e^{-x} S_k\left(\lambda_k(1 - H_k(x))\right) \tag{21}$$

Since the arrival process at visit k is Poisson with rate λ_k, each message arrival in the intervisit time v_k leads to a *simple* busy period. In addition, when there is at least one arrival, there is a *token-acquisition-induced busy period*. Thus the generating function of the total number served in the *terminal service time* at visit k, $N_k(x)$ is

$$\begin{aligned} N_k(x) = {} & V_k(\lambda_k(1 - H_k(x)))A_k(\lambda_k(1 - H_k(x))) \\ & + V_k(\lambda_k)\left((1 - A_k(\lambda_k(1 - H_k(x))))\right) \end{aligned} \tag{22}$$

This form is **not** what would be expected for a *randomly added token-acquisition overhead*. The average number served in a *terminal service time* will be required later. This is obtained by differentiating $N_k(x)$ and setting to zero. The result, $\overline{n}_k$ is

$$\overline{n}_k = \frac{\lambda_k\left(\overline{v}_k + (1 - V_k(\lambda_k))\overline{a}_k\right)}{1 - \rho_k}. \tag{23}$$

The first relation for the number served after the test message is found by noting that all the arrivals during the test message waiting and service time will generate *simple* busy periods. Letting $L_k(x)$ be defined as the generating function of the number of messages served after the test message and $W_k(x)$ as the generating function of the waiting time, then using similar arguments as above gives

$$L_k(x) = W_k(\lambda_k(1 - H_k(x)))S(\lambda_k(1 - H_k(x))). \tag{24}$$

The second expression for $L_k(x)$ is determined by noting that for every *terminal service time* there can be only one message that has exactly ℓ messages served after it, and that this can occur only if there are at least $\ell + 1$ messages served. In any set of M *terminal service times*, the total number of these that have at least $\ell + 1$ messages is about $M\,P_{n||k}(n_k \geq \ell + 1)$. The average number served is $M\overline{n}_k$. The test message is any one of these with equal probability. Thus, using the weak law of large numbers,

$$P_{L||k}(\ell) = \lim_{M\to\infty} \frac{M\,P_{n||k}(n_k \geq \ell + 1)}{M\overline{n}_k} = \frac{P_{n||k}(n_k \geq \ell + 1)}{\overline{n}_k} \tag{25}$$

Using this and Eq. (23) for $\overline{n}_k$, gives the second expression for the generating function $L_k(x)$ as

$$L_k(x) = \frac{1 - \rho_k}{\lambda_k\left(\overline{v}_k + (1 - V_k(\lambda_k))\overline{a}_k\right)} \frac{N_k(x) - 1}{e^{-x} - 1}. \tag{26}$$

Solving Eqs. (24) and (26) for $W_k(x)$ and simplifying using Eq. (21) to reduce $(e^{-x} - 1)$ gives

$$\begin{aligned} W_k(x) = {} & \frac{1 - \rho_k}{\left(\overline{v}_k + (1 - V_k(\lambda_k))\overline{a}_k\right)} \\ & \times \frac{1 - \left((V_k(x)A_k(x) + V_k(\lambda_k)(1 - A_k(x))\right)}{(x - \lambda_k(1 - S_k(x)))} \quad 0 \leq x \leq \lambda_k. \end{aligned} \tag{27}$$

If $A_k(x) = 1$, corresponding to a zero time *token-acquisition* overhead, Eq. (27) reduces to that derived by Eisenberg [1].

The mean waiting time is found by differentiating Eq. (27) and gives

$$\overline{w}_k = \frac{\left(\overline{v_k^2} + 2\overline{v}_k\overline{a}_k + (1 - V_k(\lambda_k))\overline{a_k^2}\right)}{2\left(\overline{v}_k + \overline{a}_k\right)} + \frac{\rho_k}{2(1 - \rho_k)} \frac{\overline{s_k^2}}{\overline{s}_k}. \tag{28}$$

3. A set of "upper" bounds

The basic idea for the upper bound is to note that on a sample function basis, and for any specific node, if the *token-acquisition* overhead is added to either, every message **or** the internodal token delay, the modified system stochastically overbounds the delays in the original system. Since the *token-acquisition* overheads are always added, there is no nodal dependent overhead and the exact equations in [2] are applicable. When there are N nodes there are evidently 2^N possible upper bounds. For N small it is an easy matter to explore all of them by computing the exact waiting time for each system. The best could be determined, rather arbitrarily, by either using the average mean or the *intensity weighted mean*. The *intensity weighted mean*, is defined

$$\overline{w}^\rho = \sum_{j=1}^{N} \frac{\rho_j}{\rho} \overline{w}^j \tag{29}$$

This has the advantage of being independent of the nodal placements on the ring and having the simple form, for no nodal dependent overhead (see [2]),

$$\overline{w}_\rho = \sum_{i=1}^{N} \frac{\rho_i(1 - \rho_i)}{2\rho(1 - \rho)}\overline{d} + \frac{\sigma_d^2}{2\overline{d}} + \sum_{i=1}^{N} \frac{\rho_i}{2(1 - \rho)} \frac{\overline{s_i^2}}{\overline{s}_i} \tag{30}$$

A particular *boundary* state, β, is a mix of *token-acquisition* overheads appended to data messages at some nodes and to the internodal delay at other nodes is *feasible* if the resultant total traffic intensity $\rho_\beta < 1$. Evidence, but not proof, for the unimodality of $\overline{w}_\rho$ over β was that a *greedy* algorithm, that started from a feasible *boundary* state and progressed to the nearest state (one node altered) with the greatest reduction in $\overline{w}_\rho$ always resulted in the same optimum, no matter the starting state.

The mean waiting time is computed according to the simple form first introduced by Eisenberg [1] and is the form obtained when Eq. (28) has $\overline{a}_k$ and $\overline{a^2}_k$ both 0. The intervisit mean is computed according to Eq. (18) with the $V(\lambda_k)$ being set to zero when the *token-acquisition* is added to the internodal delay.

4. A Lower Bound on the Mean Waiting Time

This section derives lower bound for the mean waiting time, **given that the** *no-message*, $V_j(\lambda_j)$ **are known exactly**. The system that adds *token-acquisition* overheads **randomly** to the internodal delay according to this probability, results in **a lower** bound on the mean waiting time for all nodes. This implies that the correlation between the *token-acquisition* overhead and the message in the queue leads to worse performance. Similar results have been noticed for the even stronger assumption of full nodal independence in ring systems, both exhaustive and not (see for instance [7, 10, 11]).

The argument proceeds in several stages. The first is to show that the mean intervisit time, suitably interpreted, is identical for both systems. The second is to show that the second moment of the intervisit time is lower bounded by the random *token-acquisition* overhead system. The third is to relate these to the mean *waiting* time, $\overline{w}_k$ in Eq. (28) and to find the resulting conditions on the intervisit moments.

4.1 Mean Intervisit Relations

If $v_{ra||k}$ is defined as the intervisit time for the *randomly added token-acquisition overhead system* then the equivalent of Eq. (1) in terms of the *terminal service time* may be written as

$$\begin{aligned} v_{ra||k} &= \sum_{i=1}^{N-1} \theta_{k-i} + d_{k-1} + a_{ra||k} \\ &= v_{dra||k} + a_{ra||k} \end{aligned} \tag{31}$$

where $v_{dra||k}$ is implicitly defined and $a_{ra||k}$ is the *token-acquisition* overhead added with probability $(1 - V_k(\lambda_k))$. The intervisit time of Eq. (18) corresponds to $v_{dra||k}$. Since the $a_{ra||k}$ is being added to the internodal travel time, the intervisit tme of the *randomly added token-acquisition overhead system* is just $v_{ra||k}$. From Ferguson and Aminetzah [2] or Eq. (18) with $\bar{a}_j = 0$, it is easy to show, with respect to node j, that

$$\begin{aligned}\bar{v}_{ra||j} &= \bar{v}_{dra||j} + \bar{a}_j(1 - V_j(\lambda_j)) \\ &= \left(\frac{1-\rho_j}{1-\rho}\right)\left(\bar{d} + \sum_{\ell=1}^{N} \bar{a}_\ell (1 - V_\ell(\lambda_\ell))\right)\end{aligned} \tag{32}$$

Comparing Eq. (32) with Eq. (18) shows that $\bar{v}_{dra||j} = \bar{v}_j$. Thus the intervisit means are identical in the two systems.

4.2 Relation of intervisit second moments

From Eq. (31), it is evident that

$$\overline{v^2}_{ra||j} = \overline{v^2}_{dra||j} + 2\bar{a}_j(1 - V_j(\lambda_j))\bar{v}_{dra||j} + \overline{a^2}_j(1 - V_j(\lambda_j)) \tag{33}$$

Solutions of the *randomly added token-acquisition overhead give* $\bar{v}_{ra||j}$ and $\overline{v^2}_{ra||j}$. Eq. (32) and (33) allow them to be easily related to $\overline{v^2}_{dra||j}$ and $\bar{v}_{dra||j}$. In this section it will be shown that $\overline{v^2}_{dra||j} \leq \overline{v^2}_j$. Since

$$\overline{v^2}_{k+N} = \sum_{i=1}^{N-1}\sum_{\ell=1}^{N-1} \overline{\theta_{k+i}\theta_{k+\ell}} + 2\bar{d}_{k+N-1}\sum_{\ell=1}^{N-1}\bar{\theta}_{k+\ell} + \overline{d^2}_{k+N+1} \tag{34}$$

where the $\sum_{i=1}^{N-1}\sum_{\ell=1}^{N-1}\overline{\theta_{k+i}\theta_{k+\ell}}$ satisfy Eq. (20). From the definition of $v_{dra||k}$ in Eq. (31), it is evident that $\overline{v^2}_{dra||k+N}$ has the same form as $\overline{v^2}_{k+N}$ in Eq. (34). In addition since $\sum_{\ell=1}^{N-1}\bar{\theta}_{k+\ell}$ and d_{k+N-1} are identical, the only difference in them occurs from the first term. Thus it is only necessary to consider the first term.

It will now be shown that $\sum_{i=1}^{N-1}\sum_{\ell=1}^{N-1}\overline{\theta_{k+i}\theta_{k+\ell}}$ is smaller when the *token-acquisition* overhead is added randomly than when it is not. Note that Eq. (20), deletion of the boxed terms, would be the solution for the *randomly added token-acquisition overhead system*. Since all of the random variables involved are positive **and** positively correlated, it is easy to show that the boxed terms are all positive. This observation implies that $\overline{v^2}_{dra||j} \leq \overline{v^2}_j$.

4.3 Relation to mean *waiting* time

The mean *waiting* time, $\bar{w}_k$ in Eq. (28), depends on the internodal delays and *token-acquisition* overheads only through the ratio

$$\frac{\left(\overline{v_k^2} + 2\bar{v}_k\bar{a}_k + \overline{a_k^2}\right)}{2\left(\bar{v}_k + \bar{a}_k\right)}. \tag{35}$$

Since $\overline{v^2}_k \geq \overline{v^2}_{dra||k}$ and $\bar{v}_k = \bar{v}_{dra||k}$, (35) can be written as

$$\begin{aligned}\frac{\left(\overline{v_k^2} + 2\bar{v}_k\bar{a}_k + \overline{a_k^2}\right)}{2\left(\bar{v}_k + \bar{a}_k\right)} &\geq \frac{\left(\overline{v^2}_{dra||k} + 2\bar{v}_{dra||k}\bar{a}_k + \overline{a_k^2}\right)}{2\left(\bar{v}_{dra||k} + \bar{a}_k\right)} \\ &= \frac{\left(\overline{v^2}_{ra||k} + 2\left(\bar{v}_{ra||k} - \bar{a}_k(1 - V_k(\lambda_k))\right)\bar{a}_kV_k(\lambda_k) + \overline{a_k^2}V_k(\lambda_k)\right)}{2\left(\bar{v}_{ra||k} + \bar{a}_kV_k(\lambda_k)\right)}\end{aligned} \tag{36}$$

The last expression is computable from the solution of the exact equations for the intervisit moments of the *randomly added token-acquisition overhead*. The expression evidently is a lower bound on the waiting time when substituted into Eq. (28) for $\bar{w}_k$. It should be remembered, though, that this bound depends on the knowledge of the exact $V_k(\lambda_k)$.

5. Approximations

5.1 A "Lower-Bound" Approximation

The "lower-bound" approximations are based on the *randomly added token-acquisition overhead along* with an estimate of the *no-message*, $V_j(\lambda_j)$. Since the boxed term in Eq. (20a) is computable assuming knowledge of the system parameters and $V_j(\lambda_j)$, it is added when solving "no-overhead" exact equations in [2]. The boxed term in Eq. (20b) requires a second approximation as it involves the correlation of θ_j and v_k. For very high and very low ρ, this term is approximately 0. This is discussed below.

no-message Estimation

The approximation for the *no-message* probabilities in this section tends to be too high, although it is quite accurate for low traffic. An (upper) estimate probability of *at-least-one-message* at node j is $(1 - V_j(\lambda_j)) = \overline{1 - e^{-\lambda_j v_j}}$. Any lower bound on v_j results in a lower bound on $(1 - V_j(\lambda_j))$. If the token is seized at a node, there is at least one message sent. A lower bound on the *terminal service time*, θ_j, designated as $\tilde{\theta}_j$, and hence on $\tilde{v}_j$ determined from these, is

$$\tilde{\theta}_j = \begin{cases} d_{j-1} & \text{with probability } \overline{e^{-\lambda_j v_j}} \\ d_{j-1} + a_{j-1} + s_{j-1} & \text{with probability } \overline{1 - e^{-\lambda_j v_j}} \end{cases} \tag{37}$$

Now since the set $e^{-\lambda_j\theta_j}$ are positive, positively correlated random variables,

$$\begin{aligned}V_j(\lambda_j) = \overline{e^{-\lambda_j v_j}} &\geq \overline{e^{-\lambda_j d_{j-1}}} \prod_{\substack{\ell=1 \\ \ell \neq j}}^{N} \overline{e^{-\lambda_j \theta_\ell}} \\ &\leq \overline{e^{-\lambda_j d_{j-1}}} \prod_{\substack{\ell=1 \\ \ell \neq j}}^{N} \overline{e^{-\lambda_j \tilde{\theta}_\ell}}\end{aligned} \tag{38}$$

Note that the second inequality is reversed. This is unfortunate and unavoidable. It is what prevents the obtaining of a bound. The "upper" approximation $\tilde{V}_{u||j}(\lambda_j)$ for $V_j(\lambda_j)$ is obtained by replacing the latter by the former in Eq. (37) and using it to evaluate $\overline{e^{-\lambda_\ell \tilde{\theta}_\ell}}$ in Eq. (38). The result is that $\tilde{V}_{u||j}(\lambda_j)$ is found by iteratively solving

$$\tilde{V}_{u||j}(\lambda_j) = D(\lambda_j) \prod_{\substack{\ell=1 \\ \ell \neq j}}^{N} \left(\tilde{V}_{u||\ell}(\lambda_\ell) + (1 - \tilde{V}_{u||\ell}(\lambda_\ell))A_\ell(\lambda_j)S_\ell(\lambda_j)\right) \tag{39}$$

$\theta_i\theta_j$ Correlation Term

The boxed term in Eq. (20b) is evidently positive, and difficult to estimate. Numerical work showed that both the convergence and the final results were quite sensitive to this approximation. Two approximations are reported. The first, or "lower-bound" just sets this term to zero, and uses Eq. (39) for the *no-message*.

The second approximation starts with a Taylor series expansion of the boxed term that gives

$$\begin{aligned}&\frac{\bar{a}_{k+N}}{1-\rho_k}\left(\overline{\theta_{k+j}(1 - e^{-\lambda_k v_{k+N}})} + \bar{\theta}_{k+j}\overline{(1 - e^{-\lambda_k v_{k+N}})}\right) \\ &\quad = \frac{\bar{a}_{k+N}}{1-\rho_k}\sum_{i=1}^{N-1}\overline{(\theta_{k+j} - \bar{\theta}_{k+j})(\theta_{k+i} - \bar{\theta}_{k+i})} \\ &\quad\quad + \langle \textit{third order terms} \rangle\end{aligned} \tag{40}$$

The $\langle$*third order terms*$\rangle$ are negative and the first term overestimates the effect of the correlation. It is easiest to see the effect of this form if the central moment version of Eq. (20b) is written by using Eq. (14) for θ_{k+N} and Eq. (40) above, to give

$$\begin{aligned}&\overline{(\theta_{k+j} - \bar{\theta}_{k+j})(\theta_{k+N} - \bar{\theta}_{k+N})} \\ &\quad = \frac{\rho_k + \lambda_k\bar{a}_{k+N}}{1-\rho_k}\sum_{i=1}^{N-1}\overline{(\theta_{k+j} - \bar{\theta}_{k+j})(\theta_{k+i} - \bar{\theta}_{k+i})} \\ &\quad\quad + \langle \textit{third order terms} \rangle\end{aligned} \tag{41}$$

The net effect of the approximation, ignoring the <*third order terms*> is the addition of the $\lambda_k\bar{a}_{k+N}$ term. As noted above, this term **overestimates** the correlation. In fact numerical work showed that this addition, even when the *no-message* probabilities are underestimated, results in nonconvergence of the equations.

From physical considerations, it is clear that the correlation becomes very small when ρ is either close to 0 or 1. It is also clear that any modification of this term should not involve the second moments that are being computed, or another set of iterations will be required. The function $\rho(1-\rho)$ has the feature of being 0 at both extremes. Numerical studies showed that the replacing of $\lambda_k \bar{a}_{k+N}$ by $\lambda_k \bar{a}_{k+N} \rho(1-\rho)$ did not cause convergence problems but still underestimated the mean waiting time. The actual approximation reported below replaces $\lambda_k \bar{a}_{k+N}$ in Eq. (41) by $2\lambda_k \bar{a}_{k+N} \rho(1-\rho)$. There is no really convincing argument for this, but it does give reasonable results.

6. Numerical results and comparisons

Along with a simulation, the following approximations are reported.

U This is the formulation of Section 3.

L This is the *randomly added token-acquisition overhead with* the $V_j(\lambda_j)$ approximated by Eq. (39), the boxed term of Eq. (20a) added, and the boxed term of Eq. (20b) set to zero.

M This is the *randomly added token-acquisition overhead with* the $V_j(\lambda_j)$ approximated by Eq. (39), the boxed term of Eq. (20a) added, and the boxed term of Eq. (20b) approximated as reported in the previous section.

The mean waiting times are reported in Figs. 2-6. The following assumptions are made for the numerical results.

- There are 20 nodes.
- The arrival processes are Poisson. Three arrival patterns are reported — symmetric; single node with 50% of traffic and remainder symmetric; single node with 90% of traffic and remainder symmetric.
- The message consists of a fixed length header of 0.5 and an exponential body of 1.0. Although only the mean and variance of the messages are important for the "no-overhead" equations, the exact form of the message distribution is used in Eq. (39). This is realistically high.
- The total delay is 0.1, 10.0. The variance of the delay is zero. This means that the inter-node delay is 0.005, and 0.5. A delay of 1.0 has accuracies and performance about midway between the two extremes. Note that this is **two** orders of magnitude!
- The *token-acquisition* overhead is 0.5, the same as the header size, and constant.
- The results are reported for the *big* node with the most traffic and the *small* nodes. The reason for collapsing the other nodes is that the mean waiting times are quite close, typically varying by only a few percent at most, and this allows for greater statistical accuracy for the simulations.
- The simulations consisted of either 200,000 served messages or 400,000 cycles, whichever came first. Observation of the means as a function of the cycle times indicated that these values had most likely converged.

The parameters were chosen to cover a reasonably wide range of values and, hopefully, extremes. It should be remembered that the equations used are **exact** when there is no *token-acquisition* overhead.

A sampling of results is reported in Figures 2-4. The mean waiting time prediction is quite accurate for low traffic, or **large non-queue dependent** walking times. Interestingly, the predictions appear to be more accurate with nonhomogeneous traffic rather than homogeneous. When a homogeneous ring goes into *momentary* saturation, due to statistical fluctuations of the input traffic, all the nodes tend to be busy simultaneously.

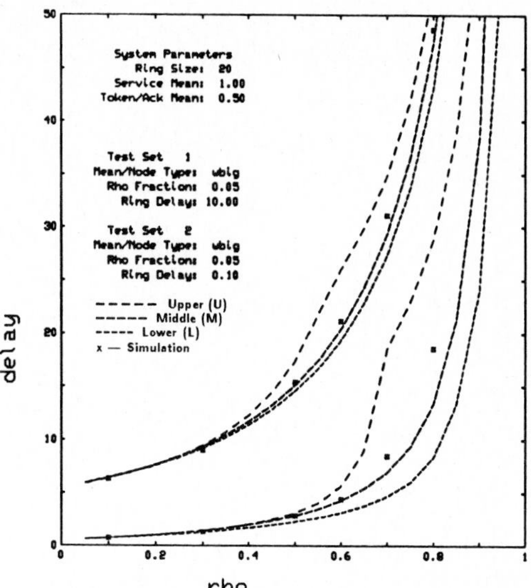

Fig. 2 The Mean Waiting Time for Symmetric Traffic

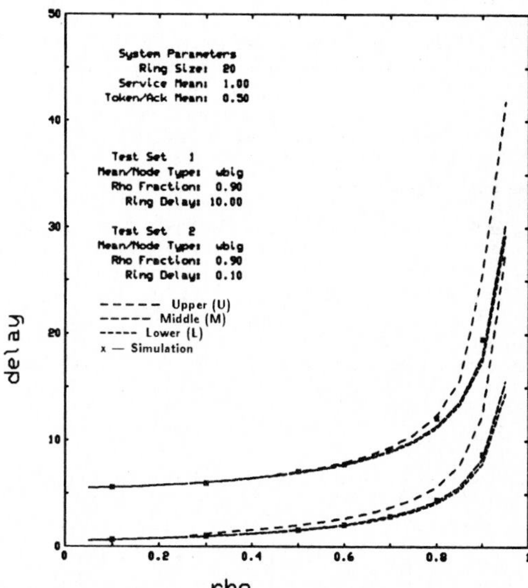

Fig. 3 The Mean Waiting Time at the High Traffic Node - 90% Traffic

The general conclusions are as follows:

- Ring systems are very sensitive to the correlation of nodal *terminal service time*. The equations here capture the non-queue dependent form exactly and lead to much more accurate results in the case of *token-acquisition* overhead. (see Ferguson and Aminetzah, [2] for more details). When the *token-acquisition* overhead is present, it represents a significant amount of time, and is noticed by other nodes when it happens. This in turn causes more *token-acquisition* over-

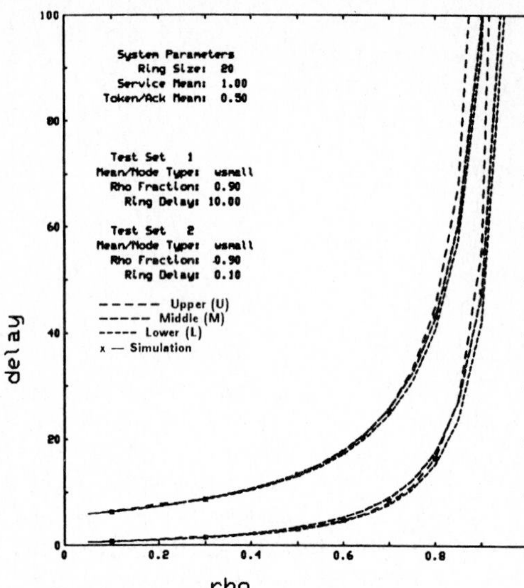

Fig. 4 The Mean Waiting Time at a Low Traffic Node - 90% Traffic

heads.

- For internodal and ***token-acquisition*** overheads of roughly the same magnitude ($\bar{d} = 10$), the **U** and **L** are quite tight and **M** is within a few percent of the simulated values. For internodal delays .01 of the *token-acquisition* overhead, the bounds are looser but the **M** approximation is within about 15% for $\rho < .5$ rising to about 25% for $\rho = .8$.

- The upper bound **U** is nonconvex, due to the changing of the optimum mix of adding the overhead to a message or a walking time. This is indeed a bound and can be used for worst case analyses. In fact the convex hull of the results is probably, but not provably, a bound. **U** has the advantage of depending on only second order statistics. The accuracy of **U** depends upon the traffic patterns total traffic ρ. It can be within a few percent for $\rho < 0.3$. However, as the figures show, it can also be unduly pessimistic. However, for very high traffic levels, the true behaviour approaches this bound.

- The "lower bound" **L**, appears in most cases to be an actual lower bound. The inaccuracies in this bound are due to both the underestimate of the *no-message* probabilities and the dropping of the correlation term. Of the two, the correlation appears to be the most important. For example, in the symmetric case, with $\bar{d} = .1$, the *no-message* probabilities are negligibly different than the simulation results up to about $\rho = .8$, yet the lower bound is out by about a factor of 2 at $\rho = .8$. At $\rho = .9$ the difference between the *no-message* probabilities is 27% but the change in the mean waiting time is only from 23.6 to 30.9 (the simulation value is 52.1). This suggests that the correlation is more important than an accurate estimate of the *no-message* probabilities.

- The **M** approximation tends to underestimate the mean delay for low and high ρ and overestimate it for medium ρ. However, in general it is within about 15% of the simulated value for all $\rho < .5$ and ranges in error from a few percent to about 15% for $\rho < .9$ and $\bar{d} = 10$.

- **M** is actually slightly higher than the **U** in Fig. 6 for some ρ. However, both are within a few per cent of the simulated results. However, it does show that this approximation sometimes overestimates

7. Conclusions and Open Problems

Although there are exact results available for the *exhaustive* and *gated* rings when the token overhead is independent of the existence of messages in the nodes, the addition of *token-acquisition* overhead considerably complicates the problem. The results here are quite robust and reasonably accurate, given the probable knowledge of traffic patterns. Extensions of these results to *gated* rings are straightforward and the accuracies are expected to be similar. Extensions to *nonexhaustive* rings are more difficult.

The accuracy of the approximation could be improved. To be consistent, this must be done without having to solve for higher moments. It will probably take some comparison with simulation to determine the optimum forms.

References

1 M. Eisenberg, " Queues with periodic service and changeover time", *Oper. Res.*, **Vol. 20**, *pp* 440-451, 1972.

2 M.J. Ferguson and Y. Aminetzah, " Exact Results for Non-Symmetric Token Ring Systems", *IEEE Trans. on Comm.*, **COM-33 No. 3**, Mar. 1985.

3 B.K. Penney, and A.A. Baghdadi, " Survey of computer communications loop networks: Parts 1 and 2", *Computer Communications*, **Vol. 2, No. 4 and 5**, *pp* 165-180 and 224-241, 1979.

4 A.G. Konheim and B. Meister, " Polling in a multidrop communication system: Waiting line analysis", *2nd Symposium on Problems in Optimization of Communication Networks and Teletraffic*, , *pp* 124-129, Oct. 1971.

5 I. Rubin and L.F. DeMoraes, " Polling schemes for local communication networks", *Conf. Rec. of ICC'81*, , *pp* 33.5.1-33.5.7, 1981.

6 Y. Aminetzah, " An Exact Approach to the Polling System", *PhD Thesis*, March 1975, McGill University

7 W. Bux and H.L. Truong, " Token-ring performance: Mean delay approximation ", *Proceedings of 10th International Teletraffic Congress*, **ITC-10**, *pp* 3.3.3-1/7, June 1983.

8 L. Kleinrock, **Queueing Systems Volume 1: Theory**, John Wiley & Sons, Inc., New York 1975,

9 , **MACSYMA Reference Manual**, The Mathlab Group, Laboratory for Computer Science, Mass. Inst. of Tech., Cambridge, Mass. Jan. 83,

10 P.J. Kuehn, " Multiqueue systems with nonexhaustive cyclic service", *Bell System Tech. Jour.*, **Vol. 58, No. 3**, *pp* 671-698, March 1979.

11 O. Hashida, " Analysis of Multiqueue", *Review of the Electr. Comm. Labs, Nippon Telegraph ad Telephone*, **Vol. 20**, *pp* 189-199, 1972.

TELETRAFFIC ISSUES in an Advanced Information Society
ITC-11
Minoru Akiyama (Editor)
Elsevier Science Publishers B.V. (North-Holland)

ETHERNET TRANSMISSION DELAY DISTRIBUTION: AN ANALYTIC MODEL

Kiyoshi YONEDA

Toshiba Corp. Systems and Software Engrg. Div.
Tokyo, Japan

ABSTRACT

An analytic model is presented for calculating Ethernet transmission delay distribution, not confined to a few of its representative values. An ether state model allowing arbitrary packet size distribution is obtained on the way, which yields a closed form throughput formula. The model is elementary and straightforward, using a discrete Markov chain and standard manipulation of distributions. Comparison with a more complex ether state model shows almost identical throughput characteristics in the range of practical interest. A sample computation is compared with a simulation of one bit accuracy.

1. INTRODUCTION

How realtime is Ethernet? An effort to answer this boils down to finding the ether (or channel) acquisition delay, viz. the time required to start a succssful packet transmission after the packet is passed to the transceiver. Since Ethernet makes use of randomness, the delay tends to incorporate a large variability. While most users are satisfied to know the mean delay, some demand the coefficient of variation, while others want to know quantiles. For instance, the 99% point in delay distribution is sometimes used as a measure of acoustic quality in packet voice communication; 99.7% delay seems to be a customary index in process control. All such performance indices may be produced if the entire delay distribution is known. The distribution has traditionally been made available through discrete event simulation: in order to observe, say, three packets exceeding 99.7% delay, a sample of size one thousand is required on the average.

An analytic model to compute the ether acquisition delay distribution is provided in this paper, not confined to just a few of distribution's representative values. As a byproduct, a simple model of the ether state is introduced which allows arbitrary packet size distributions; its solution gives a closed form throughput formula.

10[Mb/s] Ethernet specifications are provided in [1]. Its data link layer protocol is summarized in Figure 1 for convenience. This protocol is a well-known variation in a 1-persistent carrier sense multiple access with collision detection (CSMA/CD). A somewhat more suggestive term, persistent listen while talk (PLWT) [2], will be preferred throughout. "Persistent" refers to the property where a deferred station starts talking as soon as the present talker is over. Note that, in case there are two or more stations in deferrence, a collision will inevitably take place at the time the talking station goes quiet and all those in deferrence begin to talk.

The backoff strategy employed by Ethernet is called truncated binary exponential backoff (TBEB). The source of information used in this strategy, concerning the ether congestion, is restricted to a particular station's experience with respect to a particular packet.

The author is aware of six different analytic models of PLWT [3-8], each built for a different purpose. They may be classified into two categories: finite and infinite population models.

The infinite population models [3,4] circumvent the difficulty in dealing with complex backoff strategies, such as TBEB, by assuming that the stations as a whole become ready at random to access the ether. Under this assumption, [3] derives the Laplace transform of delay distribution for an

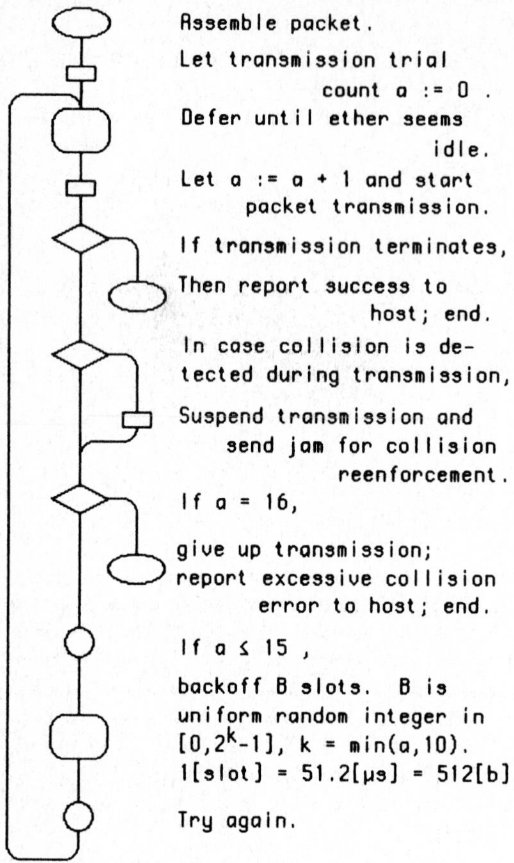

Assemble packet.
Let transmission trial count a := 0 .
Defer until ether seems idle.
Let a := a + 1 and start packet transmission.
If transmission terminates,
Then report success to host; end.
In case collision is detected during transmission,
Suspend transmission and send jam for collision reenforcement.
If a = 16,
give up transmission; report excessive collision error to host; end.
If a ≤ 15 ,
backoff B slots. B is uniform random integer in [0,2^k-1], k = min(a,10).
1[slot] = 51.2[μs] = 512[b].
Try again.

Figure 1 Ethernet data link protocol.

optimal backoff strategy. The result is not directly applicable to the Ethernet's strategy, which is suboptimal. [4] is concerned mostly with throughput.

Unlike the infinite population models, the finite population models [5-8] do not allow circumventing considerations on backoff strategies, due to the need to describe the state transition for individual stations. Since Ethernet employs TBEB, a model with the same strategy would suit the present purpose most naturally. [5,6] assume not TBEB but exponentially distributed backoff time. [7,8] can accomodate to a version of BEB, which differs from the Ethernet's in that, after a cycle consisting of a nontransmission period and a transmission period, all stations restart transmission trial count from scratch.

The overall structure of the model to be introduced is explained in 2. The model decomposes into two parts: a simple infinite population ether model and a station model attached to it. The ether state model is described in 3. The station model, which receives parameters from the ether model, is found in 4. Remarks in 5., suggesting improvements in the LWT strategy, concludes the paper.

2. OVERALL MODEL STRUCTURE

The model belongs to the class of infinite population models avoiding the backoff strategy formulation, provided that all stations are considered as a whole. However, when a single station is considered, the model must produce delay distribution. This is made possible by decomposition of the system into two submodels, as shown in Figure 2: one concerned with the ether state representing the behavior of collective stations, and the other concerned with the delay distribution describing the behavior of a particular station. The dichotomy is such that an individual cannot affect the rest of the world, while the rest of the world can affect an individual.

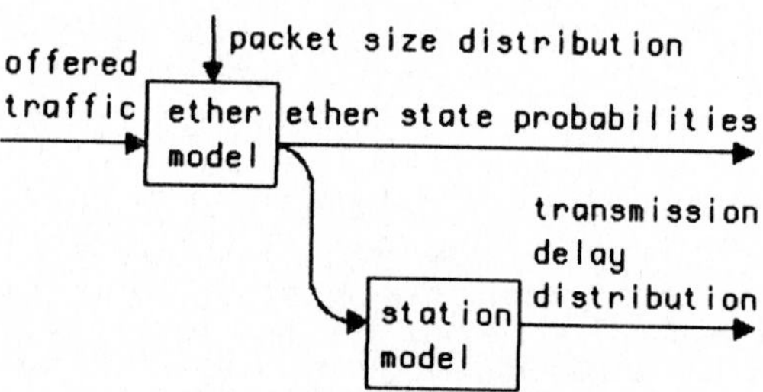

Figure 2 Overall model structure.

Given offered traffic and packet size distribution, the ether model transforms them into the ether's steady state probabilities. Some performance measures, such as throughput, follow from them. The state probabilities are then fed to the station model which, in turn, produces delay distribution. The ether model is by discrete Markov chain; the station model is by convolutions and weighted averages of distributions: both are elementary and straightforward.

3. ETHER MODEL

Time is slotted into 51.2 [µs] (= 512 [b]), which Ethernet adopts as the backoff unit time. A slot is slightly larger than double the end to end signal propagation delay. A transmission is assumed to start only at the beginning of a slot. During a time slot, the ether is in one of three possible states: idle, carrying information, or in collision. The ether is idle when there is no station transmitting signal; carrying information if exactly one is transmitting; in collision if two or more are transmitting. The stations as a whole are assumed to become ready at random to transmit a packet. The Ethernet transceiver holds at most one packet to transmit at a time.

Let symbols be as follow, where ":=" stands for "is defined by":

M := maximum packet size = 24 [slots] .

g := offered traffic [packets/slot]; $0 \le g$.

$P[.]$:= probability; $0 \le P[.] \le 1$.

r_i := rate the number of size i [slot] packets occupy among all;

$0 \le r \le 1$, $\Sigma_{1 \le i \le M}\, r_i = 1$.

x_0 := P[ether is idle] .

x_{ij} := P[ether is transmitting the j-th part of a size i packet] .

x_{M+1} := P[ether is in collision] .

Figure 3 describes the state transition diagram. To avoid cluttering, only size h and k [slot] packets among sizes 1 to M are illustrated in the diagram.

Since $x_{i1} = \ldots = x_{ii}$, let

$$x_i := x_{i1} = \ldots = x_{ii} \;;\; 1 \le i \le M .$$

The probability conservation is then

$$x_0 + \Sigma_{1 \le i \le M}\, i\, x_i + x_{M+1} = 1 .$$

The (right to left) state transition is:

$$x_0 = e^{-g} x_0 + \Sigma_{1 \le j \le M}\, e^{-jg} x_j + e^{-g} x_{M+1} \;;$$

$$x_i = g\, e^{-g} r_i\, x_0 + \Sigma_{1 \le j \le M}\, j\, g\, e^{-jg} r_i\, x_j + g\, e^{-g} r_i\, x_{M+1} \,,\quad 1 \le i \le M \;;$$

$$x_{M+1} = \{ 1 - (1 + g) e^{-g} \} x_0 + \Sigma_{1 \le j \le M} \{ 1 - (1 + jg) e^{-jg} \} x_j + \{ 1 - (1 + g) e^{-g} \} x_{M+1} .$$

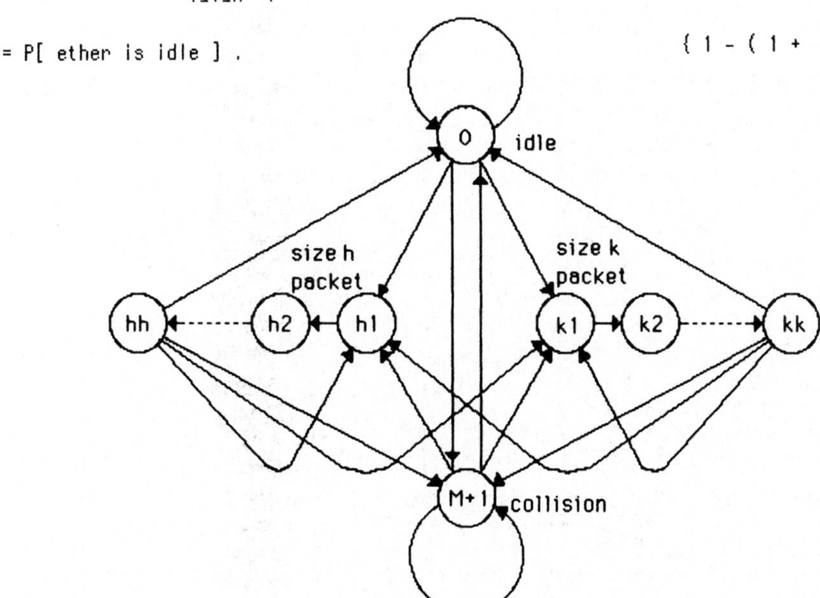

Figure 3 State transition diagram.

The first equation means that the present state is idle, if there is neither new arrival, nor continuation of a previously carried packet. Likewise, the second equation tells that, if there is exactly one station to talk now, then the ether enters the state of carrying a packet. The third captures the complement of the cases considered by the first two. The state transition, as described above, defines a regular Markov chain.

Set auxiliary variables as below, where the sums are over $1 \le i \le M$:

$$s \quad := \text{mean packet size} = \Sigma\, i\, r_i \;;$$

$$b \quad := \Sigma\, e^{-ig}\, r_i \;;$$

$$c \quad := \Sigma\, i\, g\, e^{-ig}\, r_i \;;$$

$$w \quad := 1 + s\, g\, e^{-g} - c \;.$$

The solution to the stationary equation is routinely verified to be

$$x_0 = e^{-g}\,(1 + g\, b - c) / w \;;$$

$$x_i = g\, e^{-g}\, r_i / w \;, \quad 1 \le i \le M \;;$$

$$x_{M+1} = \{ (1 - e^{-g})(1 - c) - g\, e^{-g}\, b \} / w \;.$$

Throughput follows from the solution.

$$S \quad := \text{throughput [slots / slot time]}$$

$$= P[\text{ether is transmitting information}]$$

$$= \Sigma_{1 \le i \le M}\, i\, x_i = s\, g\, e^{-g} / w \;.$$

In the special case of a single packet size, using a more popular notation G := offered traffic [packets / packet time] = sg and u := slot size [packet time] = $1/s$,

$$S = G\, e^{-uG} / (1 + G\, e^{-uG} - G\, e^{-G}) \;.$$

Returning to the case with multiple packet sizes and proceeding as noted in [9],

$$n \quad := \text{mean number of trials until success}$$

$$= \text{offered traffic / throughput} \quad \text{(both in [slots / slot time])}$$

$$= s\, g / S = w\, e^{g} \;.$$

$$p \quad := P[\text{a single trial turns out a success}]$$

$$= 1 / n = e^{-g} / w.$$

A comparison of three models with different time grains is sketched in Figure 3. The model presented above is the coarsest with grain = slot = 51.2[μs]. The infinite population model in [4] is intermediate with grain = minislot = slot/2 = 25.6[μs]. The simulation model is finest with grain = bit =

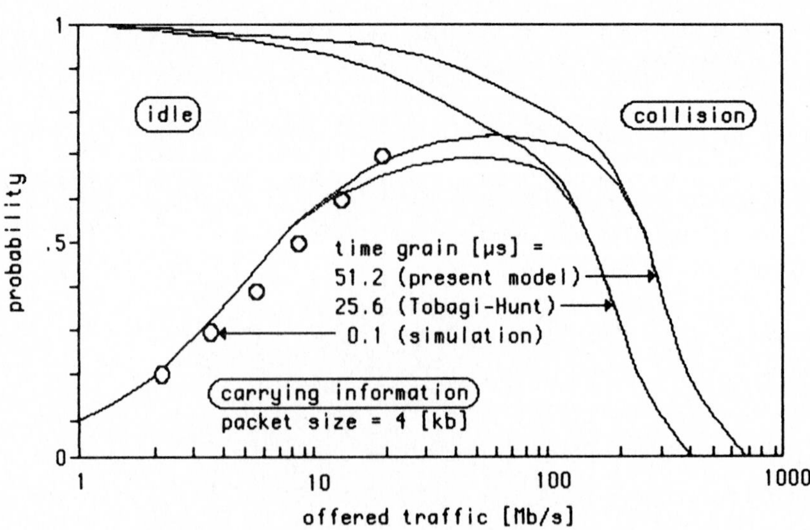

Figure 4 Sample ether state probabilities.

0.1[μs]: network diameter 2.5[km] (Ethernet maximum); 1 024 (Ethernet maximum) equally spaced stations; initial state with no packet in system; 5[s] initial data discarded; 100 000 sample packets. Packet size is fixed to 4[kb] because the minislot grain model does not allow mixed sizes. The right half of the graph has no practical meaning because of the excessive collision errors. In the noncritical range, the slot grain model throughput is practically identical to the minislot grain model's throughput. The bit grain model's throughput curve lacks the hump which is characteristic of the analytic PLWT models, looking more as if it were nonpersistent. Simulation results not presented here suggest that throughput's dependence on the number of stations is slight, if there are over a hundred of them.

4. STATION MODEL

Time is again slotted into the backoff unit time, as in the ether model. This time, the model is with respect to a particular station. Let

T := ether acquisition delay [slots]

= time to start a successful packet transmission after the transceiver has received a packet; $0 \le T$.

A := accesses [times]

= number of trials to start successful packet transmission; $1 \le a \le A$.

K_a := delay due to the a-th trial [slots]; $1 \le a \le A$.

Then,

$$T = \Sigma_{1 \le a \le A} K_a ;$$

$$K_a = \begin{cases} B_a + D_0 + c & 1 \le a \le A , \\ B_A + D_1 & a = A ; \end{cases}$$

where

B_a := backoff time due to the (a-1)-th trial, with $B_1 = 0$, $1 \le a \le 16$;

c := time required to detect and process collision

= 1 [slot] ;

D_0 := deferrence time when a trial resulted in failure [slots] ;

D_1 := deferrence time when a trial resulted in success [slots].

Note that B_a is defined as the backoff due to the (a-1)-th trial, rather than the a-th. This is equivalent to considering that a backoff B_a is performed before the a-th trial, rather than after the (a-1)-th failure. The quantities above, represented by capital letters (T, A, Ka, Ba, D_0 and D_1), are all taken to be random variables.

The distributions for individual random variables are now considered. Assuming the mutual independence of trials, the number of accesses A is geometrically distributed in the range $1 \le A \le 16$:

$$P[\ 1 \le a = A \le 16\ ] = p\, q^{a-1} \ ; \ q := 1 - p \ .$$

The remaining probability q^{16} will be dealt with separately. Since B_a, $1 \le a \le 16$, follows the uniform distribution in { 0, 1, ..., $2^{\min(a-1,10)}-1$ }, its probability generating function (pgf) is given by

$$B_a^*(z) := 1/b(a) + z/b(a) + \ldots + z^{b(a)-1}/b(a)$$

$$= (\ 1 - z^{b(a)}\ )/\{\ b(a)\ (\ 1 - z\ )\ \} \ ;$$

$$b(\ a\ ) := \min(\ a\ ,10\ ).$$

Since c = 1, its pgf is $C^*(\ z\ ) := z$. Let

D := deferrence time [slots] ,

which is not conditional to the trial result, failure or success. From the ether model,

$$P[\ 0 = D\ ] = x_0 + \Sigma_{1 \le i \le M} x_i + x_{M+1} ,$$

$$P[\ 1 \le d = D \le M\ ] = \Sigma_{d+1 \le i \le M} (\ i - d\ )\, x_i \ .$$

Consider D_0, the case in which the trial

failed. By Bayes's rule,

$$P[\,0=D \mid \text{failure}\,] = \{\, x_{M+1} + \Sigma_{1\le i\le M}\, x_i \,\} / q$$

$$P[\,1\le d=D\le M \mid \text{failure}\,] = \{\, \Sigma_{d+1\le i\le M}\, (i-d)\, x_i\, (1-e^{-ig}) \,\} / q.$$

Similarly, for D_1,

$$P[\,0=D \mid \text{success}\,] = x_0 / p\,,$$

$$P[\,1\le d=D\le M \mid \text{success}\,] = \{\, \Sigma_{d+1\le i\le M}\, (\, i - d\,)\, x_i\, e^{-ig} \,\} / p\,.$$

Hence, the pgfs for D_0 and D_1 are

$$D_0^*(\,z\,) = \{\, (\, x_{M+1} + \Sigma_{1\le i\le M}\, x_i \,) + \Sigma_{1\le d\le M}\, z^d\, \Sigma_{d+1\le i\le M}\, (i-d) x_i (1-e^{-ig}) \,\} / q\,;$$

$$D_1^*(\,z\,) = \{\, x_0 + \Sigma_{1\le d\le M}\, z^d\, \Sigma_{d+1\le i\le M}\, (i-d)\, x_i\, e^{-ig} \,\} / p.$$

Next, pgf for K_a will be determined from those for A, B, c and D. Assuming the necessary independence, the pgfs for K_a, $1\le a\le A-1$, and K_A are, respectively, $B_a^*(z)D_0^*(z)z$ and $B_A^*(z)D_1^*(z)$. If A were equal to a constant k, then the sum $T_k = \Sigma_{1\le a\le k}\, K_a$ would have a pgf

$$T_k^*(\,z\,) = \Pi_{1\le a\le k-1}\, \{\, B_a^*(\,z\,)\, D_0^*(\,z\,)\, z\,\}\, \{\, B_k^*(\,z\,)\, D_1^*(\,z\,)\,\}$$

$$= D_1^*(z)\, \{zD_0^*(z)\}^{k-1}\, \Pi_{1\le a\le k}\, B_a^*(z).$$

Since A is a random variable, T's distribution is a mixture of T_k's distributions, with pgf

$$T^*(\,z\,) = \Sigma_{1\le a\le 16}\, p\, q^{a-1}\, T_a^*(z) + q^{16} z^{"}$$

$$= p\, D_1^*(z)\; \Sigma_{1\le a\le 16}\, \{\, q\, z\, D_0^*(z)\,\}^{a-1}\, \Pi_{1\le k\le a}\, B_k^*(z) \;+\; q^{16} z^{"},$$

where the last term is for excessive collision error.

Actual computation is carried out by convolutions and weighted sums of distributions. Three arrays are required to reside in the main memory, the rest being stored in a secondary memory. With a minicomputer, it takes several minutes to compute. An example is illustrated in Figure 5; the simulation condition is as in the previous section, except that the sample size is 12 422. Distributions obtained by simulation tend to have longer tails.

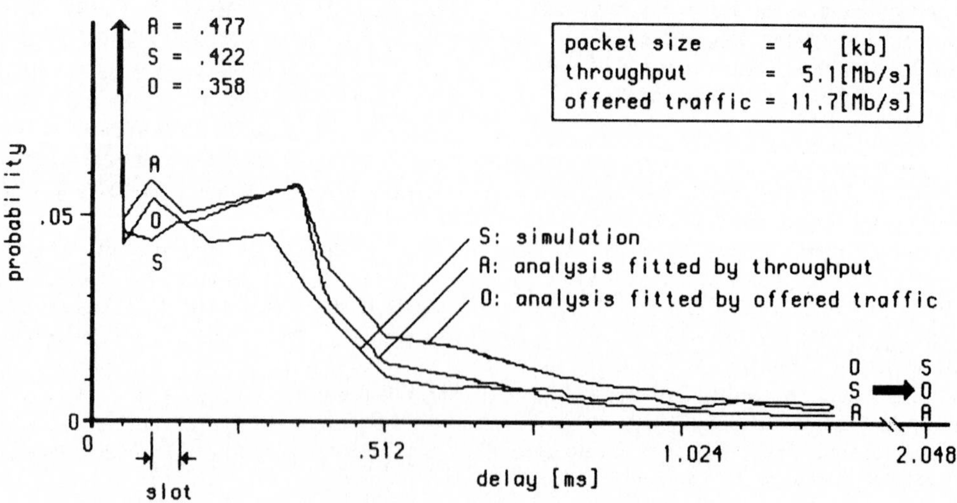

Figure 5 Sample delay comparison.

5. CONCLUDING REMARKS

Some of the difficulties encountered in modeling Ethernet are due to its following properties: persistence, TBEB, and the existence of the first trial which has to be treated separately from retrials. A nonpersistent protocol is easier to model than a persistent one since there is no need to consider deferrence. A backoff strategy employing more information than TBEB and closer to an optimal opens a possibility to regard P[some station succeeds] as being load independent. Also, since such a strategy will have a more stable backoff range than TBEB, the backoff time might be treated as a uniformly distributed random number over a fixed interval, given a constant load. Backoff performed before each trial rather than after failure renders unnecessary the distinction between a trial and a retrial.

Design of a protocol taking these in consideration not only facilitates modeling but also leads naturally to improved network performance. One such protocol may be found in [10]. Each station continuously observes the channel state and tries to adjust its backoff range so that the sum of idle time observed per collison is close to a preset constant. The protocol was designed for a 32[Mb/s] optical star network with 100 ports, 1[km] diameter, with packet priorities. An analysis method suitable for such protocols, prallel to but simpler than the present paper's, is described in [11]. It is interesting to note that the simulator for the network mentioned above was also simpler and considerably faster than Ethenet's, with the same one bit accuracy.

ACKNOWLEDGEMENTS

The author is indebted to Tomoo Kunikyo, Toshiba Information and Communication System Laboratory, for encouragement and the necessary familiarity with Ethernet. Akira Asai, a Tokyo University student, debugged the formulation, came up with many useful ideas, and did a large portion of the programming. Simulation is by Mutsumi Fujihara, Toshiba Research and Development Center.

REFERENCES

[1] DEC, Intel, Xerox; "The Ethernet: Data Link Layer and Physical Layer Specifications, Version 1.0"; 9/1980.

[2] C. E. Labarre, Analytic and Simulation Results for CSMA contention Protocols," MITRE Technical Report, MTR-3672, 9/1978.

[3] S. S. Lam, "A Carrier Sense Multiple Access Protocol for Local Networks," Computer Networks, Vol. 4, pp. 21-32, 1980.

[4] F. A. Tobagi and V. B. Hunt, "Performance Analysis of Carrier Sense Multiple Access with Collision Detection," Computer Networks, Vol. 4, pp.245-259, 1980.

[5] Y. Ishibashi and S. Tasaka, "Equilibrium Point Analysis of the 1-persistent CSMA-CD Protocol," IECE Report CS83-35, 69-76, 1983 (in Japanese).

[6] N. Shacham and V. B. Hunt, "Performance Evaluation of the CSMA/CD (1-persistent) Channel-Access Protocol in Common-Channel Local Networks," in P. C. Ravasio, et al, eds., "Local Computer Networks," pp.401-437, IFIP, 1981.

[7] Y. Nomura, H. Okada, and Y. Nakanishi, "Performance Evaluation of CSMA/CD with Various Back-Off Protocols," Trans. IECE of Japan, Vol. J67-D, No. 2, pp.184-191, 2/1984 (in Japanese).

[8] K. Tamaru and M. Tokoro, "Performance Analysis of Collision Control Algorithms and Acknowledgement Schemes in CSMA-CD," Trans. IECE of Japan, Vol. 65, No. 5, pp.527-534, 5/1982 (in Japanese).

[9] L. Kleinrock and F. A. Tobagi, "Packet Switching in Radio Channels: Part I -- Carrier Sense Multiple-Access Modes and Their Throughput-Delay Characteristics," IEEE Trans. on Comm., COM-23, 12, 1400-1416, 12/1975.

[10] K. Yoneda, M. Fujihara, and K. Oguchi, "Design of a Heavy Load Real Time CSMA/CD Protocol for 32Mb/s Star Configured Optical Local Area Network," IECE Conference, 3/1985 (in Japanese).

[11] K. Yoneda, "Percentile Calculation for the Infinite Population Nonpersistent CSMA/CD," to appear in Trans. IECE of Japan.

TELETRAFFIC ISSUES in an Advanced Information Society
ITC-11
Minoru Akiyama (Editor)
Elsevier Science Publishers B.V. (North-Holland)

PERFORMANCE EVALUATION OF PRIORITIZED TOKEN RING PROTOCOLS

Zaiming SHEN*, Shigeru MASUYAMA**, Shojiro MURO**, and Toshiharu HASEGAWA**

* Changchun Posts and Telecommunications Institute
Changchun, China

** Department of Applied Mathematics and Physics
Faculty of Engineering, Kyoto University
Kyoto, 606 Japan

ABSTRACT

In this paper, two types of new prioritized token ring access protocols in local area networks called R-PTR and WR-PTR are proposed, and their throughput-delay performance is evaluated through the theoretical analysis and simulation experiments. Our protocols are based on the standardizing protocol by IEEE 802.5 Project (denoted by IEEE-PTR), however, the stack mechanisms used in the IEEE-PTR are not employed and the schemes are rather simpler than the IEEE-PTR. In the R-PTR, the token is reserved for the next circulation by the highest prioritized packets waiting for transmission during the ongoing circulation of the token, and in the WR-PTR, such reservation scheme is not used and the priority level of the token is decided before the next circulation. The obtained results are compared with that of the IEEE-PTR, and it is shown that the performance of our proposed protocols is not inferior to that of the IEEE-PTR under almost all traffic load types.

1. INTRODUCTION

As the channel speed of local area networks (abbreviated LAN's) becomes higher and higher due to the use of optical fiber cables and the remarkable improvement of hardware integration, the development of technology of token ring access schemes has been prompted, and many token ring access protocols have already been proposed (see, e.g., [1], [3], [5], [8], [10]). Especially, for local area networks with integrated services where multifarious traffic (e.g., voice, data, image) is generated from various kinds of subscribers in the systems, several prioritized token ring access schemes (abbreviated PTR's) have been proposed up to the present (e.g., [5], [7], [8]).

In investigating the PTR's presented in [5] and [8] (note that both PTR's are considered to be the same protocol in principle, thus we referred to these two protocols as the IEEE-PTR), we face to the following two problems which motivate our studies of this paper:

(1) The IEEE-PTR is rather complex, and this would not be advantageous to implement the PTR scheme. As the access protocols of LAN's, simpler schemes are desired even if the performance is somewhat deteriorated comparing to the complex one with high performance.

(2) The IEEE-PTR uses two stacks at each station to control the priority level of the token, and the space consumed by these stacks becomes large in an unsound situation. This is also undesirable for implementation of the scheme. Especially, in a LAN with high channel speed, it would be necessary to realize the controller at each station as much compact as possible.

Based on these observations, we propose in section 2 two new PTR's without stacks. One is the PTR with reservation (called R-PTR) and the other is the PTR without reservation (called WR-PTR). For these two protocols, we further consider two variations according to the principle of packet transmission at each station i.e., gating and exhaustive transmission (see, e.g., [2], [4]).

Although token ring access protocols without priority have been analyzed (e.g., [1]) by applying the analysis methods of polling systems (e.g., [6], [9]), only a few attempts have been made so far to analyze PTR's (see, e.g., [7], [10]). The analysis by Nishida et al. [7] is for their proposed protocol whose scheme is fully different from the R-PTR or the IEEE-PTR. Although Yamamoto et al. [10] analyzed PTR's under a rather general framework, the model of their analysis doesn't directly treat the token priority determination mechanism of PTR's like the R-PTR. Thus, in section 3, we approximately analyze the average throughput-delay performance of the R-PTR by a queueing model with continuous-time and one-message buffer at each station, in which the decision mechanism of token priority of the R-PTR is explicitly reflected.

In section 4, we examine the accuracy of our approximation analysis of the R-PTR by simulation studies. Further, by the numerical results for the performance of the R-PTR and the extensive simulation studies on the IEEE-PTR and the WR-PTR, we evaluate their performance and prove that, compared with the IEEE-PTR, the performance of our proposed PTR's is not inferior to that of the IEEE-PTR under almost all traffic load environments.

Finally, we conclude this paper in section 5 with suggestions of some future research topics.

2. PROTOCOLS

2.1 R-PTR

In token ring access protocols, a particular short frame, i.e., a token circulates around the ring, and gives each station the right of data transmission if a free token is captured by the station. If a free token is captured by a station, say, station S, for data transmission, it becomes a busy token and is appended to the header part of the transmitted data packets, and the busy token is changed to a free token at station S after accomplishing the data transmission. In the R-PTR (also in the WR-PTR and the IEEE-PTR), the prioritized schemes are realized by assigning a priority to the free token, and a station can send the access waiting packets only if it captures a free token and the waiting packets have a priority not lower than that of the captured free token.

The basic idea behind the reservation of the R-PTR is that the circulating busy token picks up the information of the highest priority of all access waiting packets at each station when it visits the station, and gets the highest priority level p among the access waiting packets of all stations. When the new free token is created after the circulation of this busy token, the priority of the free token is set to be p. Principally, this scheme is achieved by just appending reservation bits to the busy token. However, we use a gimmick, i.e., the priority field of the busy token is also used to keep the highest priority level of the access waiting packets during a circulation. Consequently, the token size of the R-PTR become smaller than that of the IEEE-PTR.

Now, we describe the R-PTR protocol in details by using the token bit pattern. The configuration of the token bit pattern of R-PTR protocol is illustrated in Fig.1, where each component is as follows:

T is the token bit, where token is free if T=0, and token is busy if T=1;
M is the monitor bit, where the system is normal if M=0, and the system is abnormal if M=1;
PPP is the value of the current ring service priority level.

In the IEEE-PTR, another component RRR which is for the value of the reservation bits of busy token is provided. However, the component RRR is eliminated from the token of the R-PTR because of the above reason, and the component PPP is also used on behalf of the RRR when the token is busy.

Let Pm be the register to store the highest priority level of the packets of a station (say, S), where the packets with larger level numbers have higher priority, i.e., level 1 packets are the lowest prioritized packets. Then the protocol of R-PTR at the station S is described as follows:

(1) If station S captures the token satisfying T=0, i.e., free token, the values of PPP and Pm are compared. Then,
if PPP>Pm, station S sends the free token to the next station without updating the values of T and PPP;
otherwise, station S changes the captured free token to the busy token, i.e., T=1, and sets the value of PPP to be 0, then starts the transmission of the packet.

(2) If station S captures the busy token, it compares the value of PPP of the token with Pm to reserve the token. Then,
If PPP<Pm, set the value of PPP to be Pm;
otherwise; don't update the value of PPP.

(3) Assume that station S had generated the busy token and transmitted the packets, and the generated busy token has just returned to station S. Then, station S records the value of PPP of the token in register Rr, and the new free token with the following bit pattern is generated:

Fig.1 The bit pattern of the token of the R-PTR and the WR-PTR.

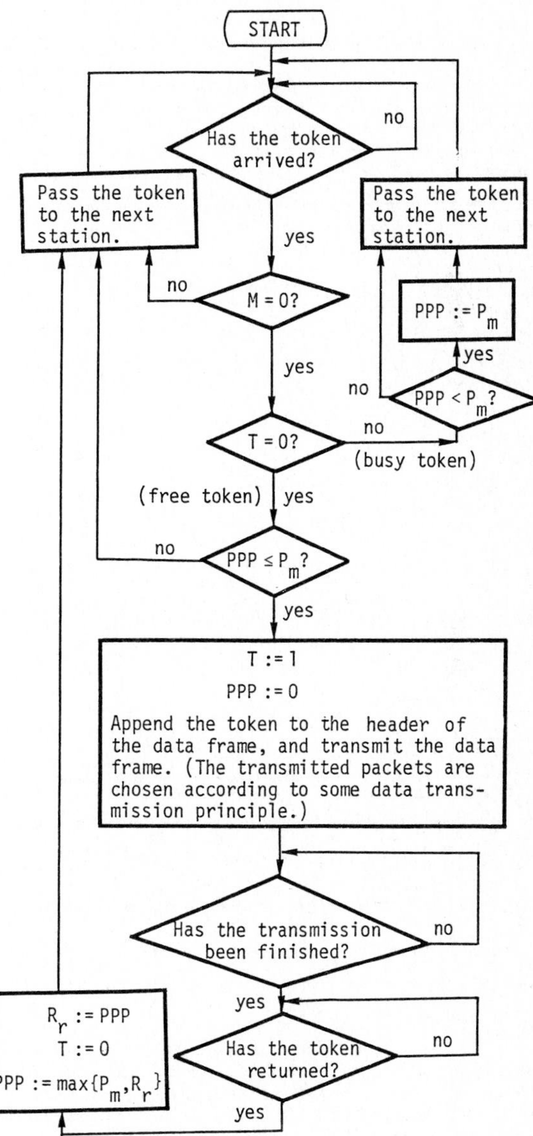

Fig.2 Behavior of a station under the R-PTR.

PPP=max(Pm,Rr), T=0, and M=0.

Note that the station transmitted the packets takes the responsibility to remove the transmitted packets from the ring. The algorithm of R-PTR is illustrated in Fig.2.

2.2 The Difference between R-PTR and IEEE-PTR

We would like to describe two essential differences between the R-PTR and the IEEE-PTR.

First, in the R-PTR, all the information gathered during a circulation of the busy token is fully utilized, while in the IEEE-PTR this information is partially utilized as explained in the following. In the IEEE-PTR [5], [8], the priority of a new free token is given by the maximum of the current token priority and the highest priority of access waiting packets known to the busy token during its one-round circulation.

Second, in the IEEE-PTR, two types of stacks are used in order to reduce the priority of the free token to the previous priority level in the case where no packet with the priority higher than or equal to that of the free token is found by the free token during its one-round circulation. The mechanism of manipulating these stacks makes the IEEE-PTR rather complex. On the contrary, in the R-PTR, such stacks are not used. In the IEEE-PTR, the priority level of the busy token is preserved even after its one-round circulation unless it is decreased by the stack manuplating mechanism, which may be beneficial to the packets with higher priority level as examplified in the following situation: Assume that a packet with the highest priority level arrives at station S just after the busy token with the highest priority is transferred to the next station from station S. Thus, the station S cannot reserve the highest priority level in the busy token for the next circulation of the token. In such a case, the possibility that station S will be able to transmit the packet with the highest priority in the next circulation in the IEEE-PTR is higher than that in the R-PTR, since the priority level of the token of the going circulation may be preserved in the next circulation in the IEEE-PTR. Including the above case, the fairness concerning an opportunity of transmission of each prioritized packet in both protocols should be taken into consideration. We examined many situations with respect to priority level of the packets waiting for transmission at each station, and concluded that there is nothing to choose between them about the fairness of transmission of each prioritized packet. Thus, it is considered that the R-PTR is rather advantageous comparing to the IEEE-PTR because of the simpler mechanism of each station as well as the smaller size of the memory for achieving almost the same performance as will be shown in section 4.

2.3 WR-PTR

In this subsection, we describe the protocol WR-PTR as well as the motivation of introducing this protocol. In a LAN in which very long packets like data files are mainly transmitted, a certain station may occupy the right of transmission of the packets for a long time if the channel speed is not so large comparing to the packet length. Then, the information on which the reservation schemes such as explained above are based is too old to utilize, and it becomes useless. Thus, it is useful to consider a prioritized scheme without reservation in such circumstances.

In the WR-PTR, whenever a new free token is created, its priority, i.e., the value of PPP, is always set to be the highest one. Thus the register Rr is also not necessary. If a station S recognizes that the free token created at station S returns without being captured by any other station (i.e., idle run), then station S decreases the priority level of the free token by one. Each station can recognize this situation idle run by its register Rt (this is not used in the IEEE-PTR).

It is considered that the WR-PTR makes good use of the priority mechanism in the LAN with properties explained above.

2.4 Several Variations of R-PTR

The protocols mentioned above are classified into several variations according to the principle of packet transmission at each station; namely, decision on which packets are transmitted among the packets with the same priority. We consider the following two popular principles:

(a) Gating principle [2]

The station transmits the packets in its queue with the priority relevant to be transmitted at the instance that the free token is captured.

(b) Exhaustive principle [2]

The station which captures the free token continues to transmit packets which have the priority high enough to be transmitted until the queue for this priority level packets becomes idle.

3. PERFORMANCE ANALYSIS OF THE R-PTR

In this section, we analyze the performance (i.e., average throughput and average transmission delay) of the R-PTR with gating transmission principle. The results for the exhaustive transmission principle have already been obtained. However, we only discuss on the gating model.

First, we introduce the following six assumptions (A1)-(A6) for simplicity of analysis.

(A1) The number of stations is N and two priority levels of packets are considered, i.e., level 2 (higher priority) packet and level 1 (lower priority) packet. Each station has two buffers with size 1 (packet) each of which is for level 1 packet or level 2 packet.

(A2) The state of the network is represented by a tuple (i,j,f), where i is the number of stations with a level 2 packet, j is the number of stations with a level 1 packet, and f is the level of a free token. It obviously holds that $0\leq i,j\leq N$ and f=1,2.

(A3) The size of a packet (both level 2 and 1) transmitted in the system is 1 (unit time), i.e., the channel time is normalized by the packet transmission time of a packet.

(A4) The level 2 (level 1) packets arrive at the system according to Poisson process with mean value s_2 (s_1) (packets per 1 (unit time)), respectively. The arrival of the level 2 packets and the level 1 packets are mutually independent.

(A5) The propagation delay between the nearest two stations is constant and is given by r (unit time). Further, the overhead for the free token to pass through a station without packet transmission is neglected.

(A6) Let us consider the following four time instants, where we assume that the free token visits station B next to station A.

tA: the instant that station A captures a free token,

tA': the instant that station A generates a free token,

tB: the instant that station B captures the free token generated by station A.

tB': the instant that station B generates a new free token.

The system state at instant tA' is determined by the system state at instant tA as well as the number of packets of each priority level arrived at the system between tA and tA' and the system state at tB is determined by the system state at tA' as well as the number of packets of each priority level arrived at the system between tA and tA'. Consequently, each instant when the station captures or generates a free token becomes the Markov renewal point.

Under these six assumptions, we analyze the delay and throughput of the R-PTR by means of usual Markovian model.

State Transition Probability

First, we obtain the state transition probability of two neighbouring Markov renewal points. Assume that the system state at tA' (tB) is given by (i,j,f) ((i',j',g)), respectively. Let us denote the state transition probability from tA' to tB by P(i',j',g/i,j,f). Since the level of a free token is not changed during the time interval between tA' to tB, g=f is satisfied. From the above assumptions, P(i',j',f/i,j,f) is given as follows:

$$P(i',j',f/i,j,f)={}_{N-i}C_{i'-i}(1-\exp(-s_2r))^{i'-i} \cdot(\exp(-s_2r))^{N-i'}\,{}_{N-j}C_{j'-j}(1-\exp(-s_1r))^{j'-j} \cdot(\exp(-s_1r))^{N-j'} \quad (i\le i'\le N,\ j\le j'\le N,\ f=1,2).$$

Next, we consider the following three probabilities at time instant tB with system state (i',j',f): (Case 1) station B has only level 1 packet; (Case 2) station B has only level 2 packet; (Case 3) station B has both level 1 and level 2 packets. Let us denote these three probabilities by P1(i',j'), P2(i',j'), and P3(i',j'), respectively. Then P1(i',j'), P2(i,',j'), and P3(i',j') are given as follows:

$$P1(i',j')=\begin{cases}({}_{N-1}C_{i'}\cdot{}_{N-1}C_{j'-1})/({}_{N}C_{i'}\cdot{}_{N}C_{j'}) & (j'\neq 0)\\ 0 & (j'=0),\end{cases}$$

$$P2(i',j')=\begin{cases}({}_{N-1}C_{i'-1}\cdot{}_{N-1}C_{j'})/({}_{N}C_{i'}\cdot{}_{N}C_{j'}) & (i'\neq 0)\\ 0 & (i'=0),\end{cases}$$

$$P3(i',j')=\begin{cases}({}_{N-1}C_{i'-1}\cdot{}_{N-1}C_{j'-1})/({}_{N}C_{i'}\cdot{}_{N}C_{j'}) & (i',j'\neq 0)\\ 0 & (i'=0 \text{ or } j'=0).\end{cases}$$

Now, we consider the state transition probability from instant tB to instant tB'. When the level of a free token is 2, each station sends a level 2 packet if it has a level 2 packet. On the other hand, when the level of a free token is 1, each station sends any level packets (a level 1 packet, a level 2 packet, or both level 1 and level 2 packets) if it has. Thus, there are three cases concerning the packet transmission from station B; namely, no packet transmission, one packet transmission, and two packets transmission. If no packet is transmitted, then tB'-tB is equal to 0 by assumption A5. If a packet is transmitted, then tB'-tB is given by Nr+1 (we shall denote this value q), and if two packets are transmitted, then tB'-tB becomes q+1. The state transition probability from tB to tB' is given as follows:

Case I. (f=2)

$$P(0,j',2/0,j',2)=1 \quad (\text{by (A5)}),$$

$$P(i',j',2/i',j',2)=1-P2(i',j')-P3(i',j') +(P2(i',j')+P3(i',j'))\,{}_{N-i'}C_1(1-\exp(-s_2q)) \cdot(\exp(-s_2q))^{N-i'-1}(\exp(-s_1q))^{N-j'}\,Res(2) \quad (i'\neq 0),$$

$$P(n,m,g/i',j',2)={}_{N-i'}C_{n-i'+1}(1-\exp(-s_2q))^{n-i'+1} \cdot(\exp(-s_2q))^{N-n-1}\,{}_{N-j'}C_{m-j'} \cdot(1-\exp(-s_1q))^{m-j'}(\exp(-s_1q))^{N-m} \cdot(P2(i',j')+P3(i',j'))Res(g)$$

$$(i'\neq 0,\ g=1,2,\ i'-1\le n\le N-1,\ j'\le m\le N, \text{ and } (n,m,g)\neq(i',j',2)), \quad (1)$$

where in eq. (1)

$$Res(2)=\begin{cases}1 & (i'\ge 2)\\ 1-\exp(-s_2rN(N+1)/2) & (i'=0,1),\end{cases}$$

$$Res(1)=\begin{cases}0 & (i'\ge 2)\\ \exp(-s_2rN(N+1)/2) & (i'=0,1).\end{cases}$$

Case II (f=1)

$$P(0,0,1/0,0,1)=1 \quad (\text{by (A5)}),$$

$$P(0,j',1/0,j',1)=1-P1(0,j')+P1(0,j')Res(1) \cdot(\exp(-s_2q))^{N}\,{}_{N-j'}C_1(1-\exp(-s_1q)) \cdot(\exp(-s_1q))^{N-j'-1} \quad (j'\neq 0),$$

$$P(i',0,1/i',0,1)=1-P2(i',0)+P2(i',0) \cdot(\exp(-s_1q))^{N}\,{}_{N-i'}C_1(1-\exp(-s_2q)) \cdot(\exp(-s_2q))^{N-i'-1}Res(1) \quad (i'\neq 0),$$

$$P(i',j',1/i',j',1)=1-P1(i',j')-P2(i',j') -P3(i',j')+\{P2(i',j')\,{}_{N-i'}C_1(1-\exp(-s_2q)) \cdot(\exp(-s_2q))^{N-i'-1}(\exp(-s_1q))^{N-j'} +P1(i',j')(\exp(-s_2q))^{N-i'}\,{}_{N-j'}C_1(1-\exp(-s_1q)) \cdot(\exp(-s_1q))^{N-j'-1} +P3(i',j')\,{}_{N-i'}C_1(1-\exp(-s_2(q+1))) \cdot(\exp(-s_2(q+1)))^{N-i'-1}\,{}_{N-j'}C_1(1-\exp(-s_1(q+1))) \cdot(\exp(-s_1(q+1)))^{N-j'-1}\}Res(1) \quad (i',j'\neq 0),$$

$$P(n,m,g/i',j',1)=\{{}_{N-i'}C_{n-i'+1}(1-\exp(-s_2q))^{n-i'+1}$$
$$\cdot(\exp(-s_2q))^{N-n-1}{}_{N-j'}C_{m-j'}(1-\exp(-s_1q))^{m-j'}$$
$$\cdot(\exp(-s_1q))^{N-m}P2(i',j')$$
$$+{}_{N-i'}C_{n-i'}(1-\exp(-s_2q))^{n-i'}(\exp(-s_2q))^{N-n}$$
$$\cdot{}_{N-j'}C_{m-j'+1}(1-\exp(-s_1q))^{m-j'+1}$$
$$\cdot(\exp(-s_1q))^{N-m-1}P1(i',j')$$
$$+P3(i',j'){}_{N-i'}C_{n-i'+1}(1-\exp(-s_2(q+1)))^{n-i'+1}$$
$$\cdot(\exp(-s_2(q+1)))^{N-n-1}$$
$$\cdot{}_{N-j'}C_{m-j'+1}(1-\exp(-s_1(q+1)))^{m-j'+1}$$
$$\cdot(\exp(-s_1(q+1)))^{N-m-1}\}Res(g)$$

$$(i',j'\neq 0,\ g=1,2,\ i'-1\leq n\leq N,\ j'-1\leq m\leq N,\ \text{and}\ (n,m,g)\neq(i',j',1)).$$

Steady State Probability

Here, we consider the steady state probability $\pi(n,m,g)$ of system state (n,m,g). $\pi(n,m,g)$ is given as follows:

$$\pi(n,m,g)=\sum_{0\leq i+j<2N,\ 0\leq i'+j'<2N,\ f=1,2} P(n,m,g/i',j',f)\cdot P(i',j',f/i,j,f)\pi(i,j,f).$$

Next, let $Pt(k/i,j,f)$ denote the probability that the transmission time (i.e., $tA'-tA$) is k when the system state (at instant tA) is (i,j,k). $Pt(k/i,j,k)$ is given as follows:

$$Pt(0/i,j,f)=\begin{cases}1-P2(i,j)-P3(i,j) & (f=2)\\ 1-P1(i,j)-P2(i,j)-P3(i,j) & (f=1),\end{cases}$$

$$Pt(Nr+1/i,j,f)=\begin{cases}P2(i,j)+P3(i,0j) & (f=2)\\ P1(i,j)+P2(i,j) & (f=1),\end{cases}$$

$$Pt(Nr+2/i,j,f)=\begin{cases}0 & (f=2)\\ P3(i,j) & (f=1),\end{cases}$$

$$Pt(k/i,j,f)=0\quad (k\neq 0,\ Nr+1,\ Nr+2).$$

Further, let us denote the probability generating function of $Pt(k/i,j,f)$ by $G(z/i,j,f)$. Then the average transmission time $F(i,j,f)$ is given by

$$F(i,j,f)=dG(z/i,j,f)/dz|_{z=1}.$$

Average Throughput

Now, the average throughput S1 (S2) of level 1 packet (level 2 packet) is given by the following equation:

$$Su=\sum_{0\leq i+j<2N,\ f=1,2}\pi(i,j,f)Ru(i,j,f)/\sum_{0\leq i+j<2N,\ f=1,2}\pi(i,j,f)F(i,j,f)\quad (u=1,2),\quad (2)$$

where in eq. (2) Ru is the probability that a level u packet is transmitted when the state is (i,j,f) and is given as follows:

$$R2(i,j,f)=P2(i,j)+P3(i,j),$$

$$R1(i,j,f)=(P1(i,j)+P3(i,j))Res(1).$$

Average Transmission Delay

Next, we consider the average number of packets in the system at arbitrary time. Let us denote the probability that the number of stations which has a level 2 packet (level 1 packet)) at the instant that k (unit time) has been elapsed after a free token was generated at the system state (i,j,f) is n (m) by $Pa2(n/i,j,f,k)$ $(Pa1(m/i,j,f,k))$, respectively. These two probabilities are given as follows:

$$Pa2(n/i,j,f,k)={}_{N-i}C_{n-i}\cdot(1-\exp(-s_2k))^{n-i}(\exp(-s_2k))^{N-n},$$

$$Pa1(m/i,j,f,k)={}_{N-j}C_{m-j}\cdot(1-\exp(-s_1k))^{m-j}(\exp(-s_1k))^{N-m}.$$

Then, the average number of level 1 packets (level 2 packets) in the system at the instant that k (unit time) has been elapsed after a free token was generated at the system state (i,j,f) (denoted $A1(i,j,f,k)$ $(A2(i,j,f,k))$, is respectively given by

$$A2(i,j,f,k)=\sum_{n=i}^{N} nPa2(n/i,j,f,k)$$
$$=N-(N-i)\exp(-s_2k),$$

$$A1(i,j,f,k)=\sum_{m=j}^{N} mPa1(m/i,j,f,k)$$
$$=N-(N-j)\exp(-s_1k).$$

Further, the accumulated average number of level 1 packets (level 2 packets) in the system from the generation of a free token at the system state (i,j,f) to the next generation of a free token (denoted $T2(i,j,f)$ $(T1(i,j,f))$ is respectively given by

$$T2(i,j,f)=\sum_{k=1}^{\infty}\{\sum_{x=1}^{k}A2(i,j,f,x)\}Pt(k/i,j,f)$$
$$=N\cdot F(i,j,f)-(N-i)\exp(-s_2)\cdot\{1-G(\exp(-s_2)/i,j,f)\}/(1-\exp(-s_2))\quad (i+j\neq 0),$$

$$T1(i,j,f)=\sum_{k=1}^{\infty}\{\sum_{x=1}^{k}A1(i,j,f,x)\}Pt(k/i,j,f)$$
$$=N\cdot F(i,j,f)-(N-j)\exp(-s_1)\cdot\{1-G(\exp(-s_1)/i,j,f)\}/(1-\exp(-s_1))\quad (i+j\neq 0),$$

$$Tu(0,0,1)=(1-\sum_{x=1}^{k}\exp(-s_2x))Tu(1,0,f)+(1-\sum_{x=1}^{k}\exp(-s_1x))Tu(0,1,f)\quad (u=1,2,\ i=0,\ j=0).\ (3)$$

Consequently, from eq. (3) the average number of level 2 packets (level 1 packets) in the system (denoted M2 (M1)) is respectively given as follows:

$$Mu = \sum_{0<i+j<2N,\ f=1,2} \pi(i,j,f)Tu(i,j,f) / \sum_{0<i+j<2N,\ f=1,2} \pi(i,j,f)F(i,j,f) \quad (u=1,2). \quad (4)$$

Finally, from eqs. (2), (4) and Little's formula, the average packet transmission delay Du (u=1,2) is given by

$$Du=Mu/Su \quad (u=1,2). \quad (5)$$

4. THE RESULTS OF ANALYSIS AND SIMULATION

In this section, we first examine the accuracy of the approximate analysis in section 3. Then, we compare the performance of the R-PTR and the IEEE-PTR. Throughout this section we will adopt the following system parameters of a LAN.

Channel speed:	32 (Mbits/sec.)
Delay between two stations:	2 (micro sec.)
Number of stations:	10
Highest priority level:	2

In Fig.3, we show the performance of the packet transmission delay time versus average packet interarrival time for the R-PTR with gating transmission principle, where the packet length is 1000 (bits). The solid (broken) lines show the numerical results obtained from the approximation analysis of section 3 (the results from simulation experiments), respectively. Although the difference between the results of approximation analysis and the simulation experiments is observed to some extent (i.e., at most 200 (micro sec.)), the results of the analysis well reflect the behavior obtained from the simulation experiments.

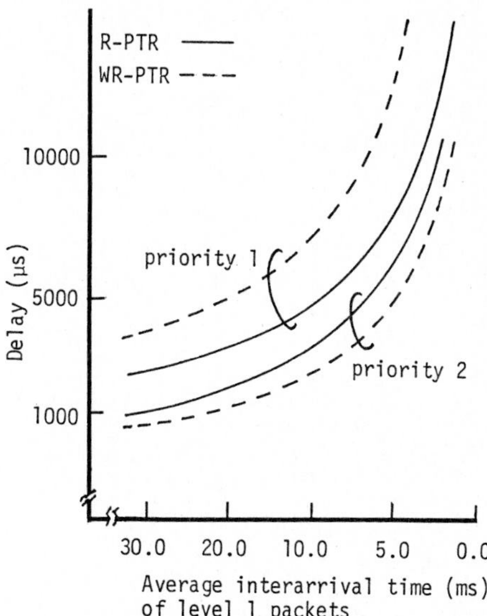

Fig.4 Comparison of the R-PTR and the WR-PTR by simulation experiments
(packet length: 8000 (bits), average interarrival time of level 2 packets: 10 (ms)).

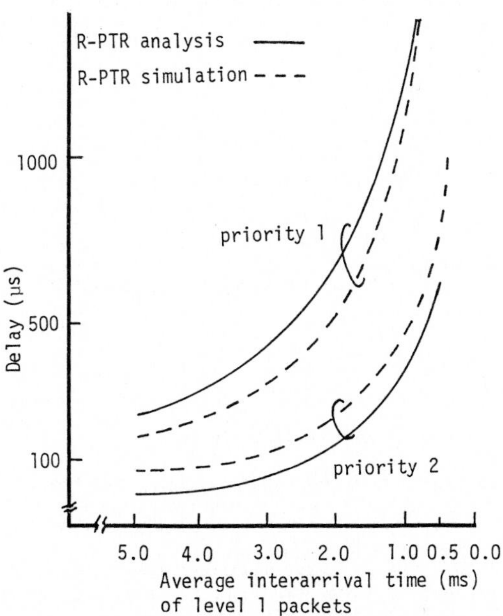

Fig.3 Comparison of the analysis and simulation results of the R-PTR (gating model)
(packet length: 1000 (bits), average interarrival time of level 2 packets: 1 (ms)).

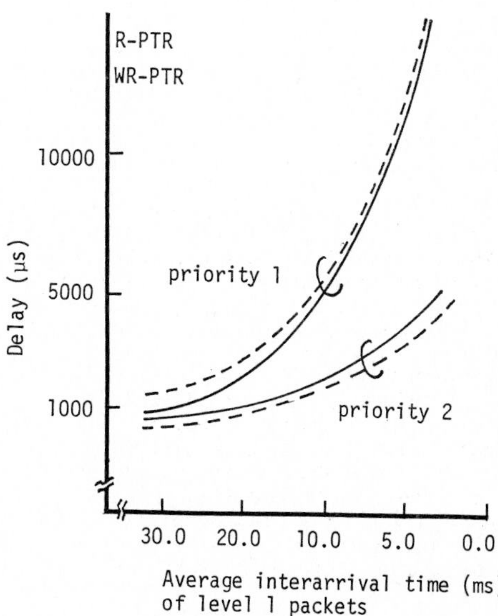

Fig.5 Comparison of the R-PTR and the WR-PTR by simulation experiments
(packet length: 16000 (bits), average interarrival time of level 2 packets: 10 (ms)).

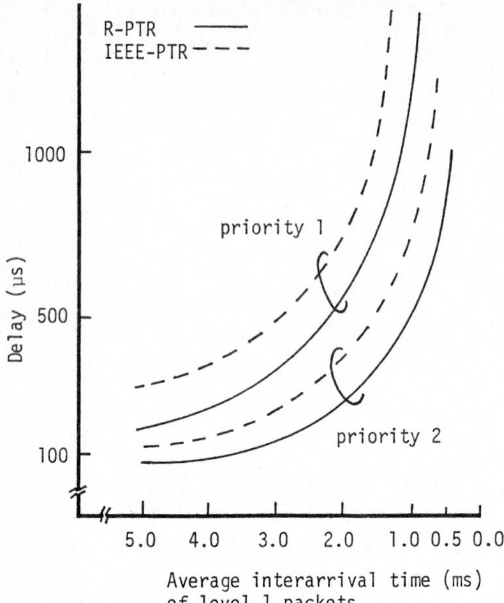

Fig.6 Comparison of the R-PTR and the IEEE-PTR by simulation experiments (packet length: 1000 (bits), average interarrival time of level 2 packets: 1 (ms)).

In Fig.4 and Fig.5, the packet transmission delay of the R-PTR and WR-PTR for different offered loads of lower priority packets is illustrated, where the exhaustive transmission principle is assumed and in Fig.4 (Fig.5) the length is 8000 (bits) (16000 (bits)), respectively. From these two figures, we can observe the following concerning the packet transmission delay in the case that the channel speed is 32 (Mbits/sec.): If the packet length is not so large (i.e., less than 10000 (bits)), the packet transmission delay of the R-PTR is smaller than that of the WR-PTR, while the WR-PTR has better transmission delay performance if the packet length becomes larger than 16000 (bits).

In Fig.6, we plot the simulation results of transmission delay of the R-PTR and IEEE-PTR under different offered loads of lower priority packets, where the exhaustive transmission is adopted since the IEEE-PTR employs the exhaustive principle with time-out mechanism. From Fig.6, it becomes clear that the packet transmission delay performance of the R-PTR is not inferior to that of the IEEE-PTR in almost all traffic load environments.

Consequently, it is conclusive that the R-PTR has simpler mechanism and better transmission delay performance as compared with the IEEE-PTR. Furthermore, the WR-PTR has also simple mechanism and shows good performance in a restricted traffic condition.

5. CONCLUSIONS

Based on the IEEE-PTR (prioritized token ring protocols standardized by the IEEE 802.5 Project), we proposed two prioritized token ring protocols; namely, the R-PTR and WR-PTR. Our proposed protocols are rather simpler than the IEEE-PTR. From our performance evaluation of the R-PTR and WR-PTR under the comparison with the IEEE-PTR, it is conclusive that, due to their simplicity and good performance, our proposed R-PTR and WR-PTR would be invaluable varieties of the prioritized token ring access protocols in LAN's.

To remedy the discrepancy observed between the result of analysis and that of the simulation experiments for the R-PTR, improvement of the approximation analysis will be required. The analysis of the IEEE-PTR by extending our approximation method is also one of the future research topics.

ACKNOWLEDGMENT

The authors wish to thank Mr. H. Asakura of Kyoto University for his several useful discussions of this work and his kind cooperation in supplementing their computational results. Thanks are due to Dr. M. Tokoro of Keio University and Dr. J. Shibata, Messrs. A. Toyooka and T. Murase of Sumitomo Electric Industries, Ltd. for their valuable comments. Finally, one of the authors, Z. Shen wishes to express his sincere appreciation to Prof. H. Mine of Kyoto University; Prof. T. Ibaraki of Toyohashi University of Technology, and Prof. H. Kise of Kyoto Institute of Technology for their encouragement since he came to Japan.

REFERENCES

[1] W. Bux, "Local-Area Subnetworks: A Performance Comparison", IEEE Trans. Commun., vol.COM-29, no.10, pp.1465-1473, 1981.
[2] R.B. Cooper and G. Murray, "Queues Served in Cyclic Order", Bell Syst. Tech. J., vol.48, no.3, pp.675-689, 1969.
[3] I.N. Dallas and E.B. Spratt (e.d.), "Ring Technology Local Area Networks", (Proc. IFIP WG6.4/Univ. of Kent Workshop on Ring Technology Based Local Area Networks, Kent, U.K., Sept. 1983), North-Holland, 1984.
[4] O. Hashida, "Gating Multiqueues Served in Cyclic Order", (in Japanese), IECEJ. vol.53-A, no.1, pp.43-50, 1970.
[5] "Draft Standard IEEE 802.5 Token Ring Access Method and Physical Layer Specification", IEEE 802.5 project, Feb. 1984.
[6] A.G. Konheim and B. Meister, " Waiting Lines and Times in a System with Polling", J.ACM, vol.21, no.3, pp.470-490, 1974.
[7] T. Nishida, M. Murata, H. Miyahara, and K. Takashima, "Prioritized Token Passing Method in Ring-Shaped Local Area Networks", Proc. ICC'83, pp.D2.3.1-D2.3.6, Boston, 1983.
[8] N. Strole, "A Local Communications Network Based on Interconnected Token-Access Rings: A Tutorial", IBM J. Res. Dev., vol.27, no.5, pp.481-496, Sept. 1983.
[9] H. Takagi and L. Kleinrock, "Analysis of Polling Systems", IBM JSI Res. Rep., TR87-0002, IBM Japan Science Institute, 1985.
[10] T. Yamamoto, H. Okada, and Y. Nakanishi, "Analysis of Token-Passing Ring Networks with Transmission Priority", (in Japanese), IECEJ., vol.J67-D, no.9, pp.989-996, 1984.

TELETRAFFIC ISSUES in an Advanced Information Society
ITC-11
Minoru Akiyama (Editor)
Elsevier Science Publishers B.V. (North-Holland)

END-TO-END TRANSPORT DELAYS IN LOCAL AREA NETWORKS

W.M. KIESEL

Institute of Communications Switching and Data Technics
University of Stuttgart
Stuttgart, Fed. Republic of Germany

ABSTRACT

Decentralized computer communication systems are based on local area networks (LAN) which connect various users to a common transmission channel. This paper presents a modified structure of a LAN, supported by a multi-access-protocol, to improve the utilization of the network interfaces, to combine the advantages of the contention and reservation protocols, and to achieve high efficiency and stability. The performance aspects of the access stations, like processor scheduling, cluster-internal connections, buffer limitations, and bidirectional traffic conditions are discussed by the models of the network access controller (NAC) and the channel access module (CAM). The analysis identify the main factors contributing to the end-to-end-transport-delays in local area networks. It will be shown that the system performance is more dominated by the interface processing and queueing rather than by the channel access delays.

1. INTRODUCTION

The increasing need of flexibility and reliability in computer based communication systems and the decreasing costs of hardware components, e.g. communication controllers, lead more and more to dezentralized solutions for such distributed systems. Local area networks (LAN) form the interconnection and transmission subsystems for these systems within a limited area.
Two basic interconnection topologies have shown most practical among all investigated networks: Ring systems, like the Cambridge Ring or others [1,2], are actually a series of connections between the consecutive access stations and use active interfaces to connect the stations to the network, whereas bus systems like the Ethernet [3] consist of a passive common transmission channel, onto which all access stations tap.

To provide an integral transport-system for open systems interconnections the access stations to the local area network have to perform most of those functions which are defined in the levels 1 - 4 of the extended ISO-OSI reference model, taking into account the special structure of local area networks.

These functions are:

- synchronization
- access management
- data encapsulation/decapsulation
- establishment and release of connections
- sequence control
- flow control
- error detection and error recovery
- segmenting and concatenation of data
- transfer of data units (packets,frames)
- supervisory functions
- access station management

The functions of level 2a are covered by multi-access-protocols which are used in the LAN to assign the channel capacity to the competing users. The CSMA/CD protocol standing for the contention protocols, used in bus structured systems, and the Token Passing Scheme as a reservation protocol, used in ring stuctured systems, are the mostly known access protocols for LAN's and they have been accepted by ISO and IEEE as standards [11,12]. These protocols have been analyzed in great detail [4,5,6,7] and compared to each other [5,6], but if the additional functions of the access stations mentioned above are taken into account, it is clearly seen that the implemented access method may contribute only in part to the general performance of the system.
Additional performance aspects are the processing overhead at the access stations to execute the higher protocol functions, the times spent for the interactions between the different protocol levels, the management of buffers and processors within the access station, and the speed mismatch between the transmission channel, the access stations, and the attached devices.
In this paper, some of these performance aspects, particularly those caused by the network interface structure are considered in detail, and it is shown that the system performance may be more influenced by the interface processing and queueing rather than by the channel access protocol and the channel speed.

Section 2 describes the system architecture which is subdivided into 3 main functional layers: peripherals, network access controllers (NAC), and the communication subsystem, consisting of the channel access modules (CAM) and the transmission channel.

In Section 3 the system is modelled and divided into submodels which are analyzed separately: the network access controller, the channel access module, and the access protocol.

Finally, some of the main factors are identified which contribute to the end-to-end-transport delays in local area networks.

2. SYSTEM ARCHITECTURE

The basic idea of local area networks is to provide a cost-effective communication system betweer computers, terminals, or I/O-devices within a spatially limited area. Under these conditions, decentralized structures become important using high speed carriers and integrated circuits. The demand of standardized interfaces and the need of higher protocol functions lead to a greater functionality of the network access points which results in higher costs. These costs can only be reduced by an effective use of the equipment, but this is contradicting to the large number of peripherals and their low traffic intensities.

To combine the main advantages of a common carrier local area network with the requirements of open systems interconnection, we conclude the following design requirements:

- a communication protocol with high throughput and low transfer times combining the advantages of contention and reservation protocols.
- a hybrid system structure with a decentralized communication subsystem and with network access stations clustering a limited number of peripheral devices, providing higher protocol functions for shared use, and guaranteeing economic use of the implemented resources.

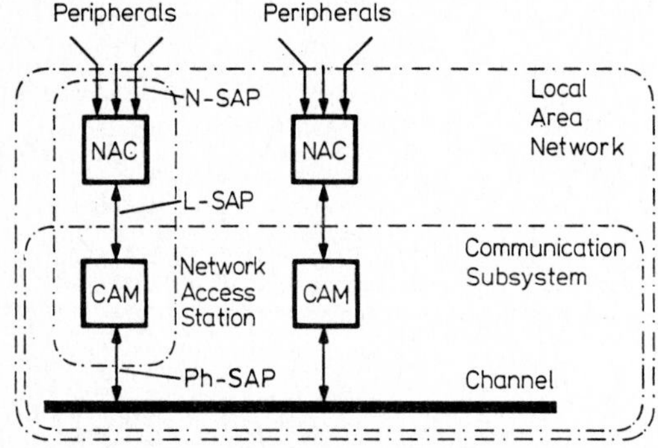

Fig. 1 System structure

NAC network access controller
CAM channel access module
N-SAP network service access point
L-SAP link service access point
Ph-SAP physical service access point

Fig.1 shows the basic system structure of the considered system. The peripheral devices are connected to the common transmission channel by the network access stations (NAS) in a clustered manner via the network service access points (N-SAP). The access stations are subdivided into two parts: the network access controller (NAC) and the channel access module (CAM).
The NAC provides basically the functions of levels 2b and 3 of the ISO-OSI-reference model; It consists of I/O-interface modules for the realization of terminal and host protocols, a microprocessor based control module for the call functions and for internal supervisory functions, a main data storage module and an interface module connecting the NAC to the communication subsystem, consisting of the CAM's and the common channel.
This interface represents the link service access point (L-SAP). The CAM performs the functions of layers 1 and 2a which have essentially to be processed at the speed of the common channel, e.g. synchronization, carrier sensing, collision detection, processing of frame control information, and intermediate buffering of frames. The CAM interfaces with the common channel by the physical service access point (Ph-SAP). Communication among the channel access modules runs under the control of a carrier-sense-multiple-access-protocol with collision detection and access conflict resolution by dynamic send priorities (CSMA-CD-DP) [8]. The send priorities are realized by staggered transmission delays which refer to the maximum propagation delay and the number of connected NAS's.

This 3-level structure (instead of the commonly discussed simple 2-level-structure) has the following advantages:

- Reduction of the number of access points
- Economic use of the common channel by minimizing the access control overhead
- Shared use of higher network protocol functions by the clustered peripherals.

3. MODELLING AND PERFORMANCE ANALYSIS

3.1 System Modelling

Besides the costs being involved by hardware and software implementation, the usefulness of a distributed system depends heavily on its throughput and delay performance. These characteristics are most sensitive to resource allocation, overhead, and traffic statistics.
The quantitative qualification is subject to modelling and performance analysis.

The main topics are:

- throughput and delay
- system response with respect to unbalanced load and dynamic overload
- identification of the most influencing system parameters
- optimization of system parameters.

Due to the complexity of the system, the complete model is subdivided into three submodels:

- the network access controller (NAC)
- the channel access module (CAM)
- the access protocol (CSMA-CD-DP).

These submodels are treated individually to identify the main influencing parameters.

3.2 The Network Access Controller

3.2.1 Modelling

Besides the call control functions, the main communication processing functions of the NAC are:

- receiving of data units from the peripherals
- transmission of data units to the peripherals
- packetizing/depacketizing of data
- buffering of data and packets
- address processing
- transmission of packets to the CAM
- receiving of packets from the CAM

These functions are the mostly time consuming ones and due to the bidirectional characteristic of the traffic, the internal structure of the NAC has to be considered carefully. The grouping of the functions into sending and receiving direction leads to the basic submodel of the NAC as a processor model with four server phases, see Fig 2.

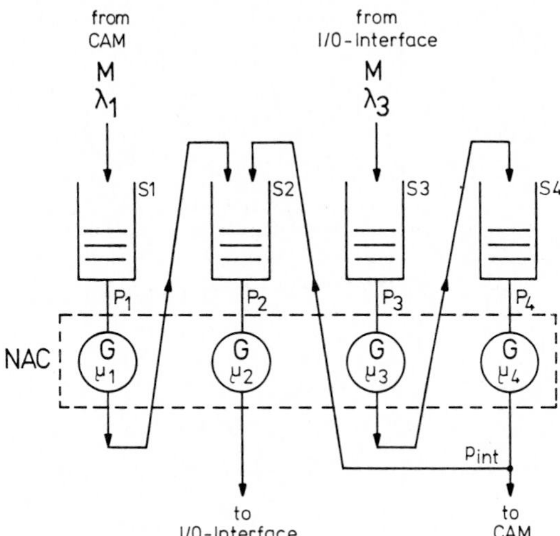

Fig. 2 Queueing model of the network access controller (NAC)
P_i: priority level of queue i

Phase 1 describes the transmission of received data units from the CAM-buffer to the NAC-data-buffer, including serial/parallel conversion, data decapsulation, address processing, sequence and flow control.
Phase 2 corresponds to those functions which are used for the transmission of data units to the different attached devices via the I/O-interfaces.
Phase 3 describes the transmission from the I/O-interfaces to the input buffer within the NAC. Collecting the data by the I/O-interfaces is assumed to run in parallel and is therefore covered by this phase. The completion of the arrivals of data units at the NAC-input-buffer is described by negative exponentially distributed interarrival times. This phase includes the time for data encapsulation and address processing, too.
Phase 4 deals with the transmission of ready data packets from the NAC-data-buffer into the CAM-buffer. This phase includes the parallel/serial conversion and the processing time for sequence and flow control.
Internal connections within the cluster of peripherals via the access station are modelled by the additional linkage from phase 4 to queue 2 and the branching probability p_{int}.

The priority schedule by which the different processor phases are activated will be discussed in Section 3.2.2 and the cluster-internal connections are considered in Section 3.2.4.

The analysis is carried out by a mean value analysis and workload considerations, depending on the actual priority schedule.

The most important results of this model are:

- the flow times through the NAC for sending and receiving direction
- the resulting queue lengths and, therefore, the determination of the NAC storage capacity.

The flow times through the NAC are defined as follows:

$T_{Rec.}$ = Waiting time in CAM-buffer (S1)
+ Service time in phase 1
+ Waiting time in NAC-data-buffer (S2)
+ Service time in phase 2

T_{Send} = Waiting time in input-buffer (S3)
+ Service time in phase 3
+ Waiting time in NAC-data-buffer (S4)
+ Service time in phase 4

The offered load is defined by:

$$\rho = \lambda_1/\mu_1 + \lambda_1/\mu_2 + \lambda_3/\mu_3 + \lambda_3/\mu_4$$

3.2.2 Processor Schedules

The first problem which arise in the structuring of the NAC and which is treated by the presented model is the scheduling of the processor phases.

To minimize the overhead, the scheduling of the phases follows a nonpreemptive priority scheme.

Giving highest priorities to the processing of foreign data units guarantees the fastest clearing of received data from the CAM.
However, if the NAC gets no packets through to the CAM-buffer, since there are always packets to receive, the station will waste its access rights and the station throughput will be degraded.
If priority is given to the sending direction, the CAM-send-buffer could be overloaded and packets sent by other stations and arriving at the considered station in sequence (back-to-back packets) will block the capacity of the common channel, due to the overloading of the CAM-receive-buffer. Therefore the channel throughput will be degraded.
Giving higher priorities to phases 1 and 3 raises the throughput of the I/O-interfaces and the CAM-buffers, but the NAC-data-buffer could be overloaded and the flow times are increasing. On the other hand, if phases 2 and 4 get the higher priorities, the necessary capacity of the NAC-data-buffer is minimized and the flow times through the NAC are decreasing.

3.2.3 Results

Tab.1 gives an overview of the resulting waiting times in each queue for some selected schedules. The formulae given in Tab.1 have been found by application of residual service times from renewal theory and Little's theorem, see, e.g. [10].

Fig. 3a,b displays results about the flow times through the NAC for the receiving and sending direction and for all priority schedules.
The arrival rates are symmetrical and the service times are exponentially distributed with mean 1.0 for each of the four phases. It is assumed that cluster-internal connections do not exist (p_{int} = 0.0).

P_1 P_2 P_3 P_4	$E[T_{W1}]$	$E[T_{W2}]$	$E[T_{W3}]$	$E[T_{W4}]$
2 1 4 3	$\frac{E[T_R] + \rho_1 h_2}{1-\rho_1-\rho_2}$	0	$\frac{E[T_R] + \rho_1 h_2 + \rho_3 h_4}{(1-\rho_1-\rho_2-\rho_3-\rho_4)(1-\rho_1-\rho_2)}$	$\frac{h_3(\rho_1+\rho_2)}{1-\rho_1-\rho_2}$
1 2 4 3	$\frac{E[T_R]}{1-\rho_1}$	$\frac{\rho_1+\rho_2}{1-\rho_1-\rho_2}\left[\frac{E[T_R]}{1-\rho_1} + h_1\right]$	$\frac{E[T_R] + \rho_1 h_2 + \rho_3 h_4}{(1-\rho_1-\rho_2-\rho_3-\rho_4)(1-\rho_1-\rho_2)}$	$\frac{h_3(\rho_1+\rho_2)}{1-\rho_1-\rho_2}$
3 1 4 2	$\frac{E[T_R] + \rho_1 h_2 + \rho_3 h_4}{1-\rho_1-\rho_2}$	0	$\frac{E[T_R] + \rho_1 h_2 + \rho_3 h_4}{(1-\rho_1-\rho_2-\rho_3-\rho_4)(1-\rho_1-\rho_2)}$	0

Tab. 1 Waiting times within the NAC-queues for selected processor schedules ($p_{int} = 0.0$)

$h_i = 1/\mu_i$ average service time — $\rho_i = \lambda_1/\mu_i$ $i = 1,2$ utilization phase 1,2

P_i priority level queue i — $\rho_j = \lambda_3/\mu_j$ $j = 3,4$ utilization phase 3,4

$E[T_R] = 0.5 \cdot \sum_{i=1}^{4} \rho_i h_i (1+c_{hi}^2)$ — c_{hi} service time coeff. of variation

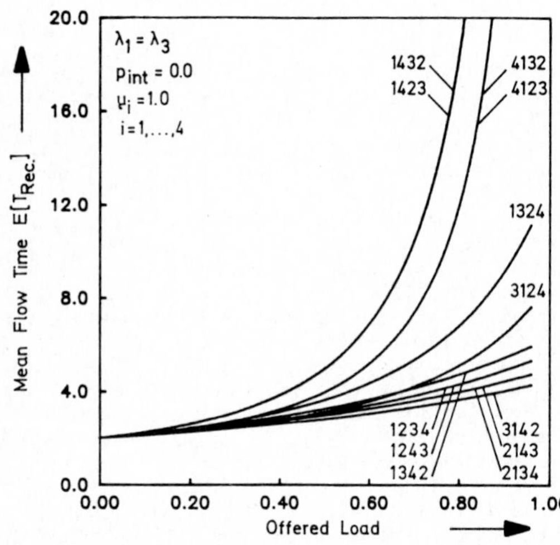

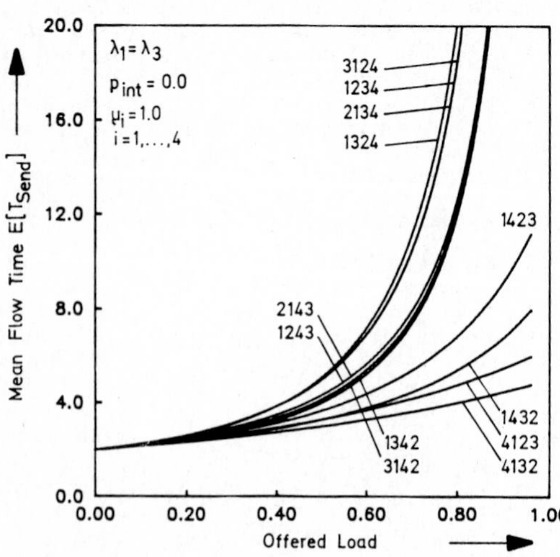

Fig. 3a,b Flow times vs. offered load NAC processor schedules

3.2.4 Cluster-Internal Connections

The clustering of peripheral devices to one network access station allows the connection of two devices without using the communication subsystem. Only the NAC to which these devices are attached, is involved in the connection.
Data units belonging to an internal connection arrive at the NAC-input-buffer and will be processed like all other data units. After recognition of the address information, the NAC-processor decides to route back these data units to the NAC-data-buffer used for ready data units which will be sent to the I/O-interfaces. The processing of connections among devices attached to the same NAC results in an additional processor load, because these connections use functions of the sending and receiving path within the NAC simultaneously.

Hence, the processor load is given by:

$$\rho_{tot} = \lambda_1/\mu_1 + \lambda_1/\mu_2 + \lambda_3/\mu_3 + \lambda_3/\mu_4 + p_{int}\lambda_3/\mu_2$$

Fig. 4a,b displays the flow times versus the offered load for a representative schedule (1234) and for different probabilities of cluster-internal connections. The arrival rates are again assumed to be symmetrical and the service times are exponentially distributed with mean 1.0 for each phase.
The results for $p_{int} = 0.0$ and $p_{int} = 1.0$ are the limiting cases of the model; for $p_{int} = 0.0$ there exists no cluster-internal connection and for $p_{int} = 1.0$ no data units are sent to the CAM.

Cluster-internal connections influence especially the sending direction of the NAC. However, the increasing of the flow times in the receiving direction are rather small.

3.2.5 Limited Buffer Capacity and Buffer Partitioning

The basic model of the NAC, given in Section 3.2.1 does not take into account the limitations of the NAC-data-buffers and gives only global criterions for the optimization of the buffer-partitioning.

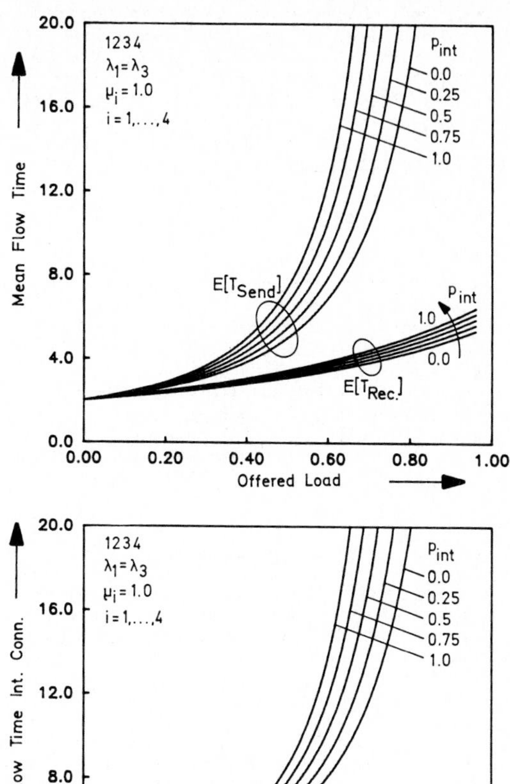

Fig. 4a,b Flow times vs. offered load
Cluster-internal connections

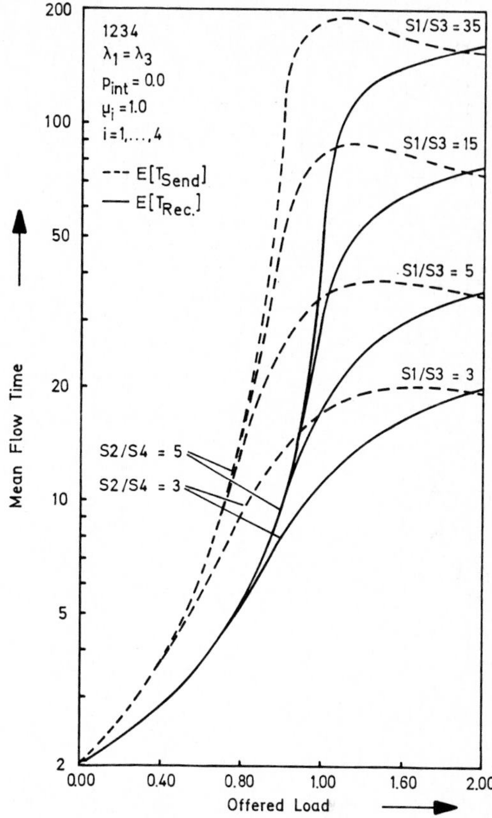

Fig. 5 Flow times vs. offered load
Limited NAC-buffer-capacities

Therefore, a model with limited queue lengths has been developed and analyzed for Markovian arrival and service processes. Additionally the processor schedule is extended to avoid internal blocking: Reaching a preset capacity limit of the storage, highest priority is given to the phases 2 and 4 to empty the buffers.
This model allows the dimensioning of the NAC-buffers, particularly for nonsymmetrical traffic characteristics. Fig. 5 shows the resulting flow times for sending and receiving direction for different buffer capacities. The arrival rates are assumed to be symmetrical, the service rates are equal for all phases with mean 1.0, and cluster-internal connections do not exist ($p_{int} = 0.0$).

Due to the limited buffer capacities, the blocking probabilities are rapidly increasing [9]. Therefore, the number of accepted data units is limited for increasing offered load and the flow times are bounded. The increase of the flow time for the receiving direction, compared to Fig. 3a, is the result of the processor schedule extension to avoid internal blocking.

3.3 The Channel Access Module

3.3.1 Modelling

Apart from processing the access protocol functions, the channel access module has to buffer incoming or outgoing packets to balance the speed mismatch between network access controller and channel. This is done by a double buffering mechanism for the sending and receiving direction, respectively. To consider the bidirectional behaviour of the channel access module the model shown in Fig. 6 was developed. The NAC is represented by the two transfer phases from and to the CAM-buffers. The common channel is modelled by two server phases standing for the transmission of departing and arriving data-frames.

3.3.2 Analysis

For Markovian arrival and service processes the performance analysis is based on establishing the multi-dimensional state-space and solving the resulting state-equations explicitly.
Results are obtained about the waiting times within the CAM, the flow times through the CAM, and the blocking probabilities of the CAM-buffers for bidirectional traffic assumptions.
Fig. 7 displays the flow times versus the offered channel load for different NAC processor speeds. It is clearly indicated that the speed mismatch between the channel and the NAC is the main factor contributing to the delays within the CAM.

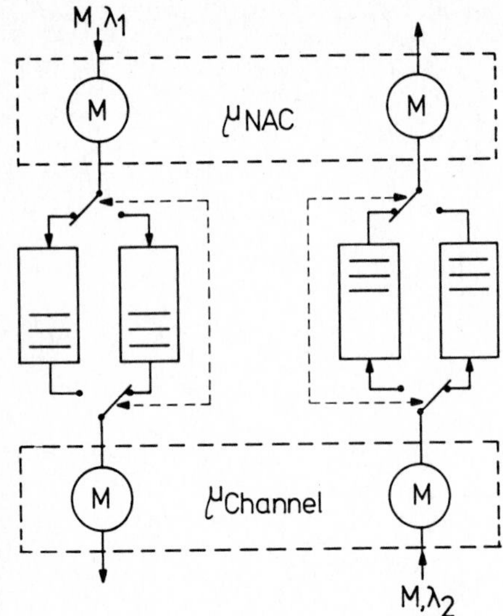

Fig. 6 Queueing model of the channel access module (CAM)

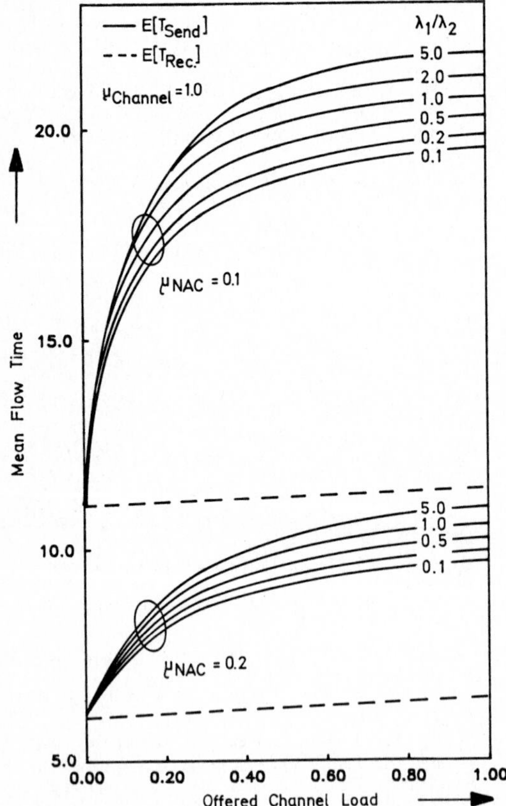

Fig. 7 Flow times vs. offered channel load Bidirectional double buffering model

An increase in the relation of the arrival rates for sending and receiving direction results also in rising flow times, if the NAC processor speed is fixed. The different flow times in sending and receiving direction result from the chosen schedule of the NAC-phases by giving higher priority to the arriving packets. Due to the limitation of the buffers, the number of accepted frames will be limited and, therefore, the waiting times are bounded, too.

3.3.3 Network Influence

As pointed out in [9], the next step in the analysis of the CAM is to consider the influence of the other stations. This can be done by an extension of the model treated in section 3.3.2. by adding a channel server phase for the transmission of packets, sent by other stations. The interaction of the channel phases is then given by a set of transition probabilities, taking into account the channel states, the traffic matrix, and the protocol dependencies (e.g. access right sequence).

3.4 The Access Protocol

The operation of the network underlies an extended CSMA-CD-protocol with dynamic priorities for the channel access using deterministically staggered transmission delays for each station. The transmission delays can be adapted to meet different requirements:

- fair access for all stations through cyclically changing of the transmission delays
- fixed prioritized access by fixed assignment of the transmission delays
- dynamic prioritized access through an adaptive assignment of transmission delays to single stations or a group of stations according to criteria of unbalanced load or overload.

The CSMA-CD-DP-protocol uses an immediate acknowledgement after each packet-transmission. The transmission delays are adjusted in each station after reception of the broadcasted acknowledgement according to the actual schedule of the transmission delays.

3.4.1 Modelling and Performance Analysis

The performance of the protocol has been investigated by means of simulation and mathematical analysis, allowing the study of allmost all cases of system parameters, process characteristics, and protocol options. Results of this analysis are reported in [8], considering especially the different protocol extensions and unbalanced load situations.

The mathematical analysis is based on the idea of collecting all waiting frames into a virtual queue and establishing a virtual service-time, depending on the mean number of waiting packets. Results can then be calculated for the resulting M/G/1-system. This analysis is reported in [8], too.

The results of the access protocol analysis are used in the end-to-end-transport-delay analysis for deriving the channel access delay and for the calculation of the access protocol overhead.

4. END-TO-END-TRANSPORT DELAY

The end-to-end delay of a message in the considered system, sent from one peripheral device to another, is composed of the following components:

- the flow time through the sending NAC
- the waiting time in the sending CAM
- the channel transmission time including the access protocol overhead
- the waiting time in the receiving CAM
- the flow time through the receiving NAC

The most influencing factors contributing to the end-to-end delay have been identified by the specific submodels:

- the processing and waiting times in the NAC depending on the schedule of the processor phases and the data-buffer dimensioning
- the intermediate buffering within the CAM under bidirectional traffic conditions
- the channel access delay and the protocol overhead.

If the parameters of the submodels are chosen in such a way that the interfaces between the submodels are represented adequately, the results of the submodels can be composed together to calculate the total end-to-end-transport delay.

4.1 Analysis Procedure

The calculation of the end-to-end-transport delay proceeds as follows:

STEP 1

Determination of the virtual channel transmission time for a particular activity at access protocol level. This calculation includes the channel characteristics (transmission speed, packet length distribution), and the network configuration (number of NAS's) and is performed by the protocol analysis.

STEP 2

Consideration of the two involved CAM's with respect to the traffic matrix and the resulting sending and receiving arrival rates at CAM level. The influence of the system is taken into account according to the extended CAM analysis [9] or the extended CAM-model, mentioned in Section 3.3.3. This step yields the resulting waiting times within the sending and receiving CAM, respectively.

STEP 3

Calculation of the flow times through the sending and receiving NAC, according to the model given in section 3.2. This step considers the actual processor schedules, the data-buffer-dimensioning and the arrival rates, found from the source-destination traffic matrix at peripheral level. The probability of internal connections for each NAC is determined by this traffic matrix, too.

CONCLUSION

A local area network with a 3-level structure has been presented. The access stations of this network have been analyzed in detail, separating them into two submodels, the network access controller and the channel access module.

The performance of the network access controller has been studied, considering the schedule of the processor phases, the occurence of cluster-internal connections, and buffer limitations.
The analysis of the CAM has been performed by extending the CAM-model to bidirectional traffic conditions.

These models reflect the main factors contributing to the end-to-end-transport delay in a local area network. As shown by the analysis, the structure of the access stations heavily influences the system performance. Therefore the delays within the NAS's have taken carefully into account. The analysis allows the conclusion that in many cases the system performance of a LAN is primarily given by the maximum channel throughput and the end-to-end-transport delay of data.

ACKNOWLEDGEMENTS

The author would like to thank Prof. P.J. Kuehn for helpful discussions.

REFERENCES

[1] Wilkes, M.V., Wheeler, D.J.: "The Cambridge Digital Communication Ring", Local Area Comm. Network Symposium, Boston, May 1979, pp. 47-61.

[2] Bux, W., Closs, F., Kuemmerle, K., Keller, H.J., Mueller, H.R.: "Architecture and Design of a Reliable Token-Ring Network", IEEE Selected Areas in Comm., Vol. SAC-1, No. 5, Nov. 1983, pp. 756-765.

[3] Metcalfe, R.M., Boggs, D.R.: "ETHERNET: Distributed Packet Switching for Local Computer Networks", Comm. ACM, Vol. 19, July 1976, pp. 395-403.

[4] Tobagi, F.A., Hunt, V.B.: "Performance Analysis of Carrier Sense Multiple Access with Collision Detection", Local Area Comm. Network Symposium, Boston, May 1979, pp. 217-245.

[5] Bux, W.: "Local Area Subnetworks: A Performance Comparision", IEEE Trans. Comm., Vol. COM-29, Oct. 1981, pp.1465-1473.

[6] Arthurs, E., Stuck, B.W.: "A Theoretical Performance Analysis of Polling and Carrier Sense Collision Detection Communication Systems", Int. Conf. on Performance of Data Communications Systems. and Their Applications, Paris, Sept. 1981.

[7] Tobagi, F.A.: "Multiaccess Protocols in Packet Communication Systems", IEEE Trans. Comm., Vol. COM-28, No. 4, April 1980 pp. 468-488.

[8] Kiesel, W.M., Kuehn, P.J.: "A New CSMA-CD Protocol for Local Area Networks with Dynamic Priorities and Low Collision Probability", IEEE Selected Areas in Comm., Vol. SAC-1, No. 5, Nov. 1983, pp. 869-876.

[9] Kiesel, W.M.: "End to End Delay in Local Area Networks", 10th Int. Teletraffic Congress, Montreal, 1983, pp. 3.1.5.1-3.1.5.7.

10] Herzog, U.: "Priority Models for Communication Processors Including System Overhead", 8th Int. Teletraffic Congress, Melbourne, 1976, pp. 623.1-623.7

11] International Organization for Standardization, ISO/TC97/SC6 Local Area Network Standardization

12] IEEE Project 802 Local Area Standard, Draft D, Dec. 1982.

TELETRAFFIC ISSUES in an Advanced Information Society
ITC-11
Minoru Akiyama (Editor)
Elsevier Science Publishers B.V. (North-Holland)

PERFORMANCE ANALYSIS OF MULTIBUS INTERCONNECTION NETWORKS IN DISTRIBUTED SYSTEMS

Thomas RAITH

Institute of Communications Switching and Data Technics
University of Stuttgart, Fed. Rep. of Germany

ABSTRACT

In distributed systems loosely coupled units (e.g. microprocessor-based control devices, peripheral processors etc.) often communicate with each other through a communication subsystem. The communication subsystem consists of a transmit and a receive buffer per unit which are connected via an interconnection network through an individual port. In the paper the performance evaluation of a communication subsystem with an interconnection network consisting of one or several high-speed busses is considered. The approximative analytical solution is based on approaches as decomposition methods, two moment matching and imbedded Markov chains. The approximation is validated by means of computer simulations. Numerical results were found to be in good agreement over a wide parameter range. The class of models considered in this paper arise in performance evaluation of switching systems with distributed control, token-ring local area networks, and distributed systems of multiple interconnected computers.

1. INTRODUCTION

Autonomous loosely coupled units in systems with distributed control often communicate with each other by message passing through a communication subsystem. Each of the units is considered as a processor with its local memory. The units operate according to principles of load and function sharing and communicate with each other by addressed messages via their transmit and receive buffers which are connected to the interconnection network through an individual port, see Fig.1. The ports are assumed to operate in a full duplex mode, i.e. a port can transmit and receive messages simultaneously. In the paper an interconnection network consisting of one or several high-speed busses is considered. In the case of multibus systems, the busses are assumed to work in parallel and independently, i.e. overtaking of bus grants can occur. The bus allocation to each particular unit is organized by means of a cyclic schedule. The transmit buffers are served in a nonexhaustive manner, that means per polling instant only one message will be transmitted.

The main subject of this paper is the performance investigation of the communication subsystem. The subsystem consists of the interconnection network, the transmit and receive buffers as well as the ports and may cause a performance degradation which depends mainly on the following blocking effects:

- transmit blocking due to port limitation, i.e. the transmit buffer can only be served by one bus at a time, therefore overtaking of bus grants may occur and the units are not able to use the full transmission capacity provided by the multibus interconnection network.
- receive blocking due to port limitation, i.e. the occupation of the receiver port is not possible due to the fact that the receiver port has been already occupied by another bus of the interconnection network.
- receive blocking due to buffer limitation, i.e. the transmission between units fails due to finite capacity of the receive buffer of the receiving unit.

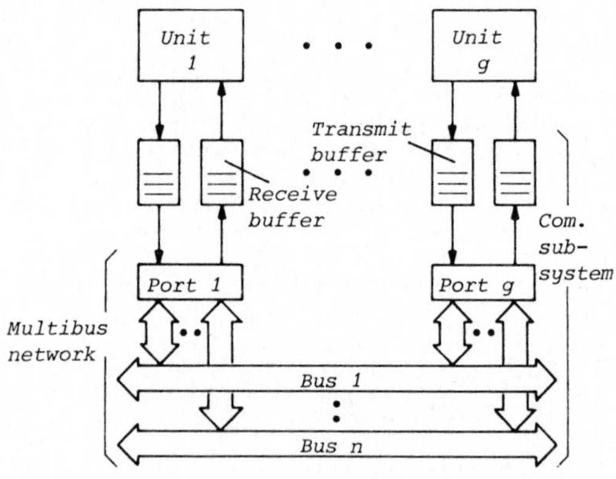

Fig. 1 *Structure of a communication subsystem with a multibus interconnection network*

All communication buffers are of finite capacity. Therefore, messages of a unit may be blocked in the transmit buffer due to its finite length. Blocking according to port limitation arises in interconnetion networks where several busses are working in parallel and interfer with each other. In the case of receive blocking, either the scheduler proceeds to the transmit buffer of the following unit and the message waits further on within its transmit buffer until the next bus grant occurs, or the bus remains occupied and the message waits in the transmit buffer until the blocked receive buffer gets idle. Thus, throughput and delay performance of the communication subsystem depend on its structure and bus scheduling mode as well as on parameters like arrival rates of messages, the physical transmission rates of the busses, transmit and receive buffer capacities, and the rate of emptying the receive buffer of the receiving units.

The paper aims at the modelling of a communication subsystem with a multibus interconnection network operating under several scheduling modes, the evaluation of its performance and a comparison between a single- and multibus interconnection network.

2. MODELLING

Modelling of the communication subsystem with a multibus interconnection network leads to a queueing model depicted in Fig. 2, which consists of a number g of finite transmit and receive buffers with the capacity S_i and R_i, respectively, and a number n of high-speed busses with transmission rate r_b. Each bus is allocated to one of the transmit buffers at a time. The allocation is done by a cyclic schedule with nonexhaustive service. Due to the parallel transmission capability and finite buffers , the blocking effects discussed above may occur. The arrival process of messages from the sending units are assumed to be Poissonian with the queue-individual rates $\lambda_1,\ldots,\lambda_g$ at the transmit buffers. Receive buffers are emptied at rates $\mu_1,\ldots,\mu_g$ with individual Markovian service times. The bus service time, i.e. the time to transmit a message from transmit buffer to receive buffer is considered to be generally distributed. After service of a transmit buffer the bus will be allocated by the scheduler to the succeeding unit to serve its transmit buffer, if there is at least one message waiting for service. If the transmit buffer is empty, the observed interscan period will be denoted as switchover time. This switchover time, which models all overheads spent and procedures performed by the scheduler to allocate the busses in a cyclic manner, is assumed to have an unit-individual general distribution function. In case of receive blocking two bus scheduling modes are considered :

i) the scheduler proceeds to the following transmit buffer and the message waits further on within its transmit buffer until the next bus grant occurs (bus repeat mode).

ii) the bus remains occupied and the message has to wait until the blocked receive buffer gets idle and the receiving unit is able to accept the message (bus wait mode).

Without receive buffer consideration the model in Fig. 2 corresponds to a multiserver polling system with nonexhaustive cyclic service and finite queue capacity. In the literature, multiqueue systems served by a single server have been subject of numerous investigations. An approximation technique introducing the method of conditional cycle times for cyclic queues with nonexhaustive service and general switchover time has been developed by Kuehn [3]. A survey on single server polling system analysis, where various system classes are considered, was provided by Takagi and Kleinrock [4]. In [2] Tran-Gia and Raith propose an approximative analysis method for a single server polling system with finite buffer capacity, i.e. an investigation on the effect of message blocking due to transmit buffer limitation. A previous study on the problem of multiqueue systems with multiple cyclic servers was performed by Morris and Wang [1]. They give a simple formulae for the mean sojourn time in the multiqueue system and consider several service disciplines of the queues but they do not consider any blocking effects as well as additional overhead (switchover times) caused by server overtaking.

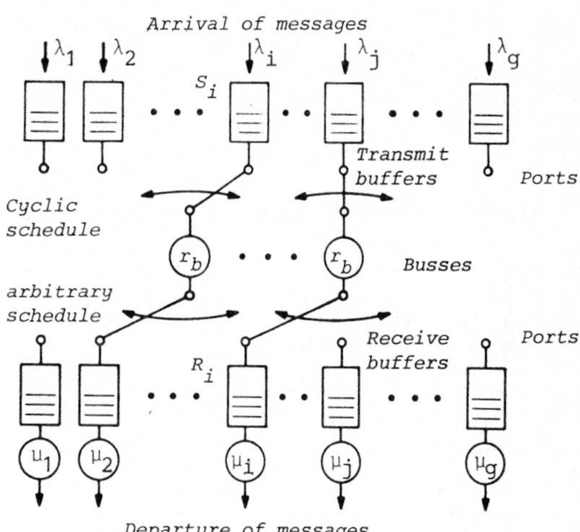

Fig. 2 *Model of the communication subsystem*

3. PERFORMANCE ANALYSIS

In this section, a numerical algorithm for an approximate analysis of a multiserver polling system with finite buffer capacity, i.e. the communication subsystem of Fig. 1, will be derived. Basically, the analysis follows the method presented in [2]. However, some modifications must be provided in order to take into account the port and memory blocking effect of the transmit and receive buffers. The main idea in the calculation method presented in this paper is to develop an alternating calculation algorithm to obtain values for the Markov chain state probabilities and the server intervisit times of a considered transmit buffer. The probability density function (pdf) of the intervisit time was approximated by a two moment technique proposed by Kuehn [5].

3.1 Markov Chain State Probabilities

A particular transmit buffer j is considered in the following, which is observed at scanning instants, i.e. instants where a bus grant occurs. Let t_n be the time of the the n-th scanning epoch and let $X^{(n)}(0^-)$ be the number of messages in this buffer at time t_n, just prior the n-th scanning epoch, then the Markov chain state probabilities are defined as

$$P^{(n)}_{k,j} = \Pr\{ X^{(n)}(0^-) = k \},\ k = 0,1,\ldots,S_j\ . \quad (3.1)$$

For ease of reading, the subscript j indicating the observed transmit buffer will be suppressed, e.g., the notation P_k will be used instead of $P_{k,j}$.

In order to calculate the transition probabilities of the Markov chain

$$p_{jk} = \Pr\{ X^{(n+1)}(0^-) = k \mid X^{(n)}(0^-) = j \} \qquad (3.2)$$

the state $X^{(n)}(t)$ of the transmit buffer at time t_n+t is observed. Considering the pure birth process in the buffer between two consecutive visits of an arbitrary server, the state probabilities at time t_n+t can be obtained as follows

$$P_k^{(n)}(t) = \Pr\{ X^{(n)}(t) = k \},\ k=0,1,\ldots,S_j. \qquad (3.3)$$

According to the consideration of conditional cycle times [3], the following random variables (r.v.) for the time between two succeeding server visits to buffer j are defined :

- T_{iv} r.v. for the intervisit time with respect to the observed transmit buffer j.
- T'_{iv} r.v. for the intervisit time, conditioning on an empty buffer at the previous scanning instant, that means without service of buffer j during the cycle.
- T''_{iv} r.v. for the intervisit time, conditioning on a non-empty buffer at the previous scanning instant (i.e., with service of buffer j during the intervisit time).

Some algebraic manipulations, see [2], yield to the set of Markov chain state equations (3.4), which are useful for the numerical calculation of the steady state probabilities $\{P_k\}$

$$P_k^{(n+1)} = P_0^{(n)} b'_k + \sum_{i=1}^{k+1} P_i^{(n)} b''_{k-i+1}, k=0,\ldots,S_j-1$$

$$P_{S_j}^{(n+1)} = P_0^{(n)} \sum_{i=S_j}^{\infty} b'_i + \sum_{i=1}^{S_j} P_i^{(n)} \sum_{m=S_j-i+1}^{\infty} b''_m, \qquad (3.4)$$

where the arrival probabilities, i.e. the probabilities for m arrivals during a conditional intervisit time of type T'_{iv} or T''_{iv} are

$$b'_m = \int_0^\infty a_m(t)\, f_{T'_{iv}}(t)\, dt$$

$$b''_m = \int_0^\infty a_m(t)\, f_{T''_{iv}}(t)\, dt . \qquad (3.5)$$

$a_m(t)$ corresponds to the probability that m Poisson arrivals during the time interval t occur. In order to calculate the arrival probabilities, the pdf of the conditional intervisit times have to be determined.

3.2 Cycle Time Segment Analysis

In this subsection some considerations where made to obtain the probability of blocking effects with respect to the bus scheduling modes discussed above. Based on these blocking effects a random variable $T_{E,j}$ for the time interval between the scanning epochs of transmit buffer j and (j+1) is defined, i.e. the cycle time segment corresponding to buffer j with respect to an arbitrary server.

The port blocking probability of a transmit or receive buffer j ($B_{PT,j}$, $B_{PR,j}$) corresponds to the probability that the servers interfer with each other at the port of the considered buffers (transmit- or receive buffer). Under the assumption that messages are equally distributed to the receive buffers, they are obtained as

$$B_{PT} = B_{PR} = \frac{(n-1)}{n} \lambda\ (1-B_{MT}) E[T_h]. \qquad (3.6)$$

The subscript j again is suppressed for ease of reading. The memory blocking probability of the receive buffer j ($B_{MR,j}$) corresponds to the probability that a bus transmission will be blocked due to finite capacity of the receive buffer and will be approximated by the blocking probability of a $M/M/1-R_j$ system. The memory blocking probability of a transmit buffer j ($B_{MT,j}$) corresponds to the probability that a message will be blocked due to finite capacity of the transmit buffer and will be obtained by the arbitrary time state probability $P_{S,j}$ of the considered transmit buffer. The probability of service of a transmit buffer where messages are waiting depends on the bus scheduling mode and is defined as $q_{s,j}$, i.e. the probability of additional overhead (switchover time) caused by server overtaking or receive buffer blocking.

3.2.1 Repeat Mode

According to bus scheduling mode "repeat" the probability of service of transmit buffer j ($q_{s,j}$), is obtained as (j being omitted) :

$$q_s = 1 - (B_{PT} + n\, B_{PR} + B_{MR}) . \qquad (3.7)$$

The Laplace-Stieltjes-Transform (LST) of the cycle time segment corresponding to buffer j of an arbitrary server ($T_{E,j}$) can be given as

$$\Phi_E(s) = \Phi_u(s)((1-P_0)\, q_s\, \Phi_h(s) + P_0). \qquad (3.8)$$

Thus, mean and variance of the cycle time segment are

$$E[T_E] = E[T_u] + (1-P_0)\, q_s\, E[T_h]$$
$$VAR[T_E] = VAR[T_u] + (1-P_0)\, q_s(VAR[T_h] + E[T_h]^2(1 - q_s(1-P_0))). \qquad (3.9)$$

3.2.2 Wait Mode

According to bus scheduling mode "wait" the probability of service transmit buffer j ($q_{s,j}$), is obtained as

$$q_s = 1 - (B_{PT} + B_{PR} + B_{MR}) . \qquad (3.10)$$

In this subsection the Laplace-Stieltjes-Transform (LST) of the cycle time segment ($T_{E,j}$) includes the forward recurrence time of bus service time and receive buffer service time. Based on the LST, again mean and variance of the cycle time segment can be calculated.

3.3 Conditional Intervisit Time Approximation

Under the assumption of independence between $T_{E,j}$, j=1,2,...,g, the LST of the conditional cycle times of an arbitrary server can be given as follows

$$\Phi_{C',j}(s) = \Phi_{uj}(s) \cdot \prod_{\substack{k=1\\k\neq j}}^{g} \Phi_{E,k}(s)$$

$$\Phi_{C'',j}(s) = \Phi_{uj}(s) \cdot \Phi_{hj}(s) \cdot \prod_{\substack{k=1\\k\neq j}}^{g} \Phi_{E,k}(s) \; . \tag{3.11}$$

Eqns. (3.11) yield the first two moments of the conditional cycle times, thus

$$E[T'_{C,j}] = E[T_{uj}] + \sum_{\substack{k=1\\k\neq j}}^{g} E[T_{E,k}]$$

$$VAR[T'_{C,j}] = VAR[T_{uj}] + \sum_{\substack{k=1\\k\neq j}}^{g} VAR[T_{E,k}]$$

$$E[T''_{C,j}] = E[T_{uj}] + E[T_{hj}] + \sum_{\substack{k=1\\k\neq j}}^{g} E[T_{E,k}] \tag{3.12}$$

$$VAR[T''_{C,j}] = VAR[T_{uj}] + VAR[T_{hj}] + \sum_{\substack{k=1\\k\neq j}}^{g} VAR[T_{E,k}] .$$

To obtain the resulting conditional intervisit times, the bus-individual conditional intervisit times are superimposed under the assumption of independent renewal processes [5]. Since only one bus can serve a queue at a time, overtaking has to be considered. Assuming a geometrical distribution for overtaking a transmit buffer by an arbitrary server, the first and second moment of the conditional intervisit times can be obtained. Based on these first two moments the calculation of the arrival probabilities (3.5) can be performed according to [2].

3.4 Calculation Algorithm for Markov Chain State Probabilities

Using the expressions for the Markov chain state probabilities and the conditional intervisit times obtained by composition of server cycle time processes where the server overtaking effect, i.e. transmit buffer blocking due to port limitation, are taken into account, a numerical algorithm is developed. Details of the algorithm are given in [2]. The main elements of the algorithm are :

- iteration of the Markov chain state probabilities and the intervisit time.
- during an iteration cycle the state probabilities of all buffers are determined in a cyclic manner; the calculation for each buffer is done according to eqn.(3.4).
- during an iteration cycle, depending on the actual state probabilities the conditional intervisit times are updated; these values will be used in the next iteration cycle.
- calculation of the arrival probabilities by means of a two-moment approximation of the intervisit time probability density function according to [5] in conjunction with a substitute process description.

3.5 Arbitrary Time State Probabilities

In order to calculate system characteristics, e.g. the blocking probability for messages or mean waiting time in a buffer, it is useful to obtain first the arbitrary time state probabilities (cf.[6]). Define $\{ P_k^* , k = 0,1,\ldots,S_j\}$ to be the arbitrary time state probabilities, i.e. the distribution of the number of messages in the considered buffer j at an arbitrary observation instant; the time interval from the last scanning epoch until this observation point is the backward recurrence time of the intervisit time with the probability density function (j beeing omitted)

$$f^{v}_{iv'}(t) = (1 - F_{iv'}(t)) \,/\, E[T'_{iv}]$$

and

$$f^{v}_{iv''}(t) = (1 - F_{iv''}(t)) \,/\, E[T''_{iv}] \; . \tag{3.13}$$

The arrival probabilities during the backward recurrence times T'^{v}_{iv} and T''^{v}_{iv} can be given as

$$b'^{*}_{m} = \int_0^\infty a_m(t)\, f^{v}_{iv'}(t)\; dt$$

and

$$b''^{*}_{m} = \int_0^\infty a_m(t)\, f^{v}_{iv''}(t)\; dt \; . \tag{3.14}$$

According to the two types of conditional intervisit times the probability that an outside observer sees an intervisit time of type T'_{iv} or T''_{iv}, respectively, corresponds to the two terms

$$P_0 \, E[T'_{iv}] \,/\, E[T_{iv}]$$

and

$$(1-P_0)\, E[T''_{iv}] \,/\, E[T_{iv}]$$

where

$$E[T_{iv}] = P_0\, E[T'_{iv}] + (1-P_0)\, E[T''_{iv}] \; .$$

Considering both types of conditional intervisit times and combining the above results, the arbitrary time state probabilities can be written as follows

$$P_k^* = \frac{E[T'_{iv}]}{E[T_{iv}]} P_0 b'^{*}_{k} + \frac{E[T''_{iv}]}{E[T_{iv}]} \sum_{i=1}^{k+1} P_i b''^{*}_{k-i+1} \qquad k=0,1,\ldots,S_j-1$$

and

$$P_{S_j}^* = \frac{E[T'_{iv}]}{E[T_{iv}]} P_0 \sum_{i=S_j}^{\infty} b'^{*}_{i} + \frac{E[T''_{iv}]}{E[T_{iv}]} \sum_{i=1}^{S_j} P_i \sum_{m=S_j-i+1}^{\infty} b''^{*}_{m} . \tag{3.15}$$

Analogous to the approximate calculation of the arrival probabilities in eqns. (3.5) using the substitute distribution function (cf. [5]) the arrival probabilities during the backward recurrence conditional intervisit times given by eqn. (3.14) can be determined.

3.6 System Characteristics

With the arbitrary time state probabilities the memory blocking probability for messages of transmit buffer j can be determined as

$$B_{MT,j} = P^*_{S_j} \; . \tag{3.16}$$

The mean delay in the transmit buffer j, referred to transmitted messages, is found from Little's law as

$$E[T_{wj}] = \frac{L_{T,j}}{\lambda_j (1-B_{MT,j})} \; , \tag{3.17}$$

where $L_{T,j}$ is the mean length of buffer j

$$L_{T,j} = \sum_{i=1}^{S_j} i \, P_i^* \; . \tag{3.18}$$

4. RESULTS

In the following, numerically obtained results will be presented and discussed for the case of a symmetrically loaded communication subsystem with a single- or double bus interconnection network, in order to illustrate the accuracy of the derived algorithm. The system consists of g = 8 units communicating over the communication subsystem. Each of the transmit and receive buffers have the same capacity $S_j = R_j = 10$. For the results presented, the time variables are standardized by $E[T_{hj}] = 1$, $j=1,2,\ldots,g$, i.e. the mean bus service time at transmit buffer j. The receive buffers are assumed to be emptied according to a Markovian service time with mean $g/n \cdot E[T_{hj}]$. The switchover time is chosen to be constant where $E[T_{uj}]/E[T_{hj}] = 0.5$.

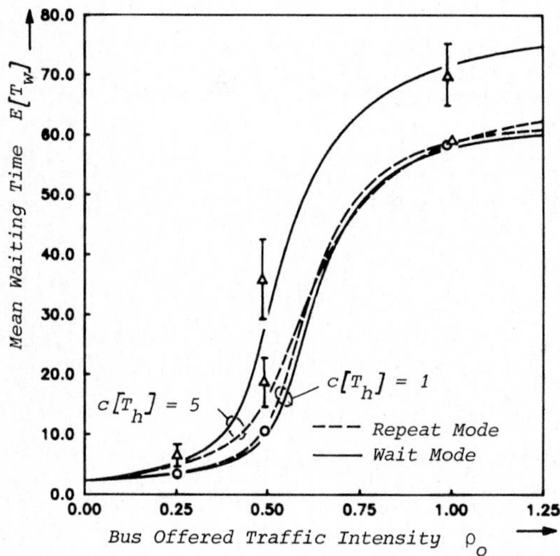

Fig. 3 Mean waiting time vs offered traffic

In order to validate the approximation, computer simulations are provided. The simulation results will be depicted with their 95 percent confidence intervals, where the circular symbol will be used for Markovian bus service times ($c[T_h] = 1$) and the triangle symbol for hyperexponential bus service times ($c[T_h] = 5$). The graphs will be drawn as function of the offered traffic per bus

$$\rho_0 = \sum_{j=1}^{g} \lambda_j E[T_{hj}] \tag{4.1}$$

The overall approximation accuracy for the given system parameters is in general less than 15 percent and depends strongly on the value of the service time coefficient of variation and the mean switchover time. The accuracy of the algorithm increases with increasing values of switchover time and decreasing values of the service time coefficient of variation. Results delivered by the presented method always show the same tendencies and phenomena as they are obtained by computer simulations.

4.1 Communication Subsystem with Double Bus Interconnection Network

In this subsection, an interconnection network of two identical busses is taken into account. The transmit buffer mean waiting time as well as its blocking probability for messages are shown as functions of the offered traffic intensity in Figs. 3 and 4, respectively, for different coefficients of variation for the service time and according to the various bus scheduling modes "wait" and "repeat".

In Fig. 3 a crossover effect of the waiting time characteristics can be recognized for the bus scheduling mode "repeat". Large values of the service time coefficient of variation and the scheduling mode "wait" lead to higher waiting times and blocking probabilities than mode "repeat" which is caused by a large forward recurrence time in case of waiting for the port to become free. For small values of the service time coefficient of variation (cf. $c[T_{hj}] = 1$) there is only a small difference between modes "wait" and "repeat", therefore, "repeat" seems to be the best strategy.

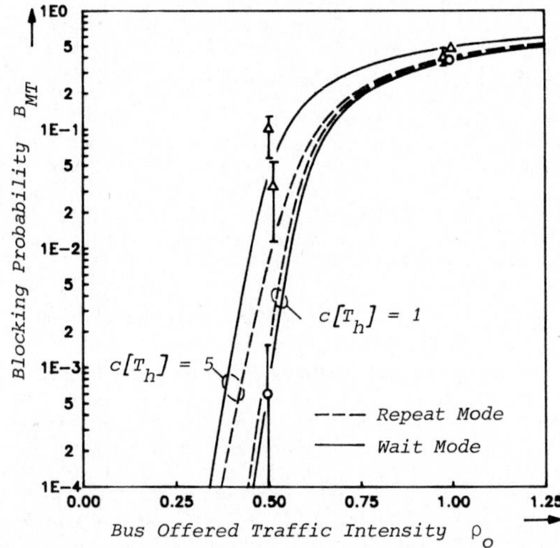

Fig. 4 Blocking probability vs offered traffic

4.2 Comparison Between Single and Double Bus Interconnection Networks

In a single bus interconnection network no server interference occurs, thus only memory blocking effects arise, but the mean intervisit times of the single server to the transmit buffers is much greater compared to the double bus interconnection network intervisit times. Therefore, for higher load a significant difference between the mean waiting time is obtained (cf. Fig. 5). In general, it can be observed, that the receive buffer memory blocking effect is very small, because the total bus service capacity is equal to the total service capacity of all receive buffers (i.e. sum of all receive buffer empty rates).

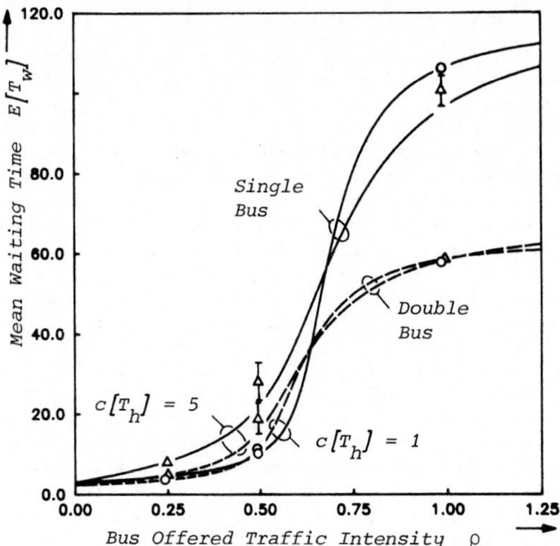

Fig. 5 *Mean waiting time vs offered traffic*

Fig. 6 shows that the double bus network compared to the single bus leads for higher load condition to worse blocking probabilities caused by additional overhead (switchover times) in case of blocking a bus transmission.

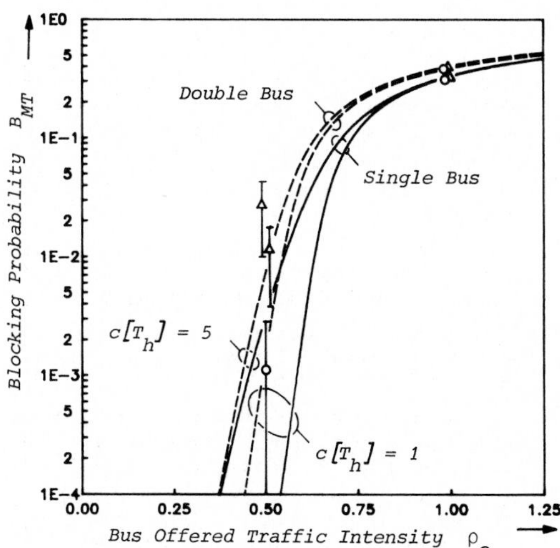

Fig. 6 *Blocking probability vs offered traffic*

The effect that service times with higher variances lead to shorter mean intervisit times in the range $0.5 < \rho_o < 0.75$ of traffic intensity as depicted in Fig. 7 can be explained considering the higher blocking probability by large $c[T_{hj}]$ (cf. Fig. 6). Fig. 7 shows also the two limiting cases of the mean intervisit time which correspond to low and overload traffic levels. The limiting intervisit times of the single bus interconnection network, see Fig. 7, are given either by the sum of all switchover times

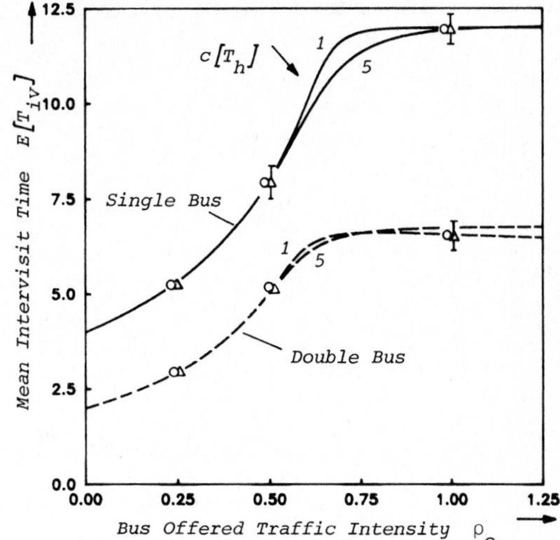

Fig. 7 *Mean intervisit time vs offered traffic*

(minimal intervisit $E[T_{iv}] = 4$) or the sum of switchover and service times of all transmit buffers (maximal intervisit $E[T_{iv}] = 12$). In principle, multibus intervisit times are obtained by the single bus intervisit times divided by the number of busses considered. However, for higher traffic the probability of service of a transmit buffer decreases; therefore, the mean intervisit time decreases, too.

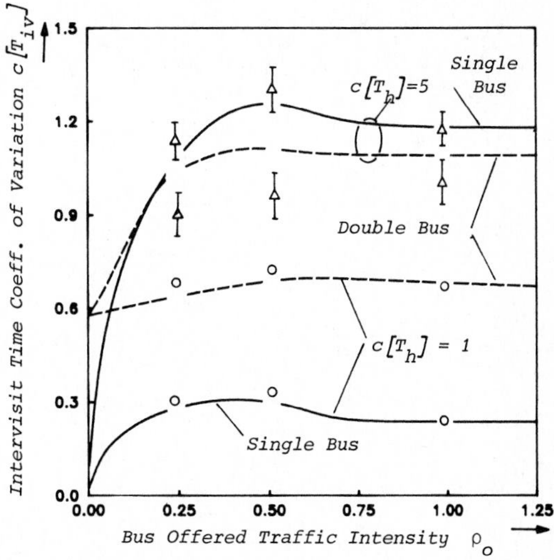

Fig. 8 *Intervisit time coefficient of variation vs offered traffic*

As depicted in Fig. 8, the cycle time coefficient of variation increases by increasing service time variation. If the traffic intensity approaches zero the intervisit time coefficient of variation of the single bus network starts from zero because the empty bus cycle is determined by the constant switchover times. The composition of two bus cycle time processes leads to the intervisit time coefficient of variation shown in Fig. 8. The simulation results show, that in case of higher service time variation the assumption of independece provide a higher calculated intervisit time coefficient of variation, i.e. the algorithm is less accurate.

Finally, the double bus network is considered under the assumption that one of the busses fails and the offered traffic stays equal, that means the residual bus is heavyly overloaded. Figs. 9 and 10 show, that for total traffic load larger than 0.25 the mean waiting time and blocking probabilitiy of the communication subsystem strongly increase.

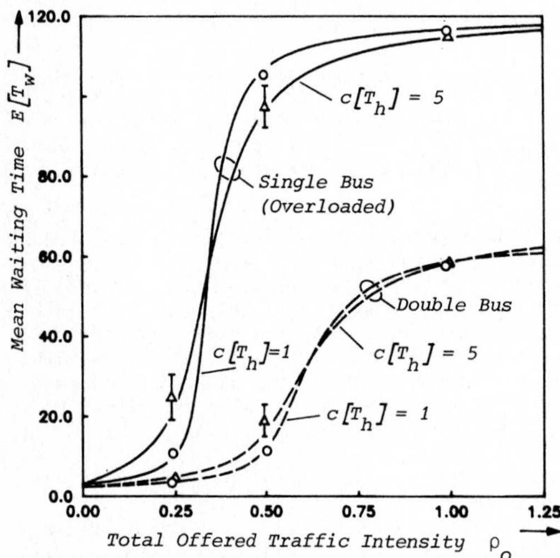

Fig. 9 Mean waiting time vs total offered traffic

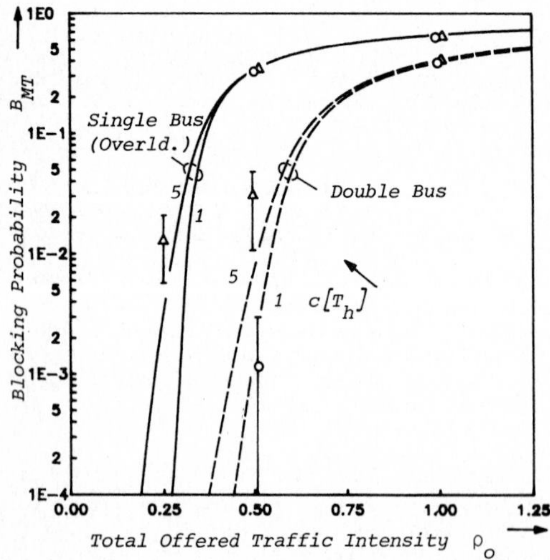

Fig. 10 Blocking probability vs total offered traffic

5. CONCLUSION

In the paper an approximative performance analysis for distributed systems with a communication subsystem consisting of transmit and receive buffers per unit and a multibus interconnection network is provided. The communication subsystem is modelled by means of a multi server polling system with finite buffer capacity and nonexhaustive cyclic service. An effective numerical algorithm is developed where different blocking effects are taken into account. Under consideration of two bus scheduling modes results for mean waiting time, blocking probability for messages etc. are derived. The accuracy of the presented algorithm is good over a wide range of parameters. This class of models can be applied to performance investigations of computer and communication systems, such as token-ring local area networks or systems of multiple interconnected computers with a distributed structure. Thus, the influence of structure, scheduling mode and parameters of the distributed system with a multibus interconnection network is analyzed and may be a basis for the decision process in the development and engineering of such systems.

ACKNOWLEDGEMENT

The author would like to express his thanks to Prof. P. Kuehn and to Dr. P. Tran-Gia for helpful discussions.

REFERENCES

[1] R.J.T. Morris, Y.T. Wang, "Some results for multiqueue systems with multiple cyclic servers", Proc. 2nd Int. Symp. on the Performance of Computer Communication Systems, Zurich, pp. 245-258, 1984.

[2] P. Tran-Gia, T. Raith, "Multiqueue systems with finite capacity and nonexhaustive cyclic service", to be published.

[3] P.J. Kuehn, "Multiqueue systems with nonexhaustive cyclic service", Bell Syst. Tech. J., vol. 58, pp. 671-699, 1979

[4] H. Takagi, L. Kleinrock, "Analysis of polling systems", Japan Science Institute Research Report, 1985.

[5] P.J. Kuehn, "Approximate analysis of general queueing networks by decomposition", IEEE Trans. Comm., COM-27, pp. 113-126, 1979.

[6] P. Tran-Gia, D.R. Manfield, "Performance analysis of delay optimization in distributed processing environments", Proc. 2nd Int. Symp. on Performance of Computer Communication Systems, Zurich, pp. 259-274, 1984.

TELETRAFFIC ISSUES in an Advanced Information Society
ITC-11
Minoru Akiyama (Editor)
Elsevier Science Publishers B.V. (North-Holland)

REDUCING LOST CALLS DUE TO THE CALLED SUBSCRIBER
AN EXPERIENCE IN BRAZIL

Roberto CALDEIRA and Ricardo NASCIMENTO

TELEBRÁS
Brasília, Brazil

ABSTRACT

It is well known that high busy line (BY) and dont't answer (DA) rates, lead to lower revenues and higher investment costs for the Telecommunication Companies, and poor quality of service for the users.

The BY and DA rates improvement campaign being carried out by TELEBRÁS, the brazilian Telecommunication Holding Company, and its 29 subsidiaries, since 1979, is briefly described. The main causes of the problem, the adopted solutions and the results obtained are presented.

So far, the results achieved show an improvement of about 8% on the DDD traffic completion rate, representing around US$ 40 million increase on the TELEBRÁS annual revenue.

1. INTRODUCTION

TELEBRÁS, the Government owned Telecommunication Administration, was created in 1972 to manage the Telecommunications services in Brazil. It is the holding company of TELEBRÁS System, which consists of 28 Telecommunication Operating Companies, responsible for operating the local and intrastate toll service, and EMBRATEL, the interstate toll and international services Operating Company.

Presently, the brazilian telephone network consists of 155 toll centers and 3100 local exchanges, to which are connected about 10.5 million telephones, spread over an area of 8.5 million Km² of the national territory.

In 1978 the quality of the DDD service was very poor. The call completion rate (OK) was 36.3% and the equipment blockage and failure rate (EBF) was 23.6%.

By that time, TELEBRÁS and its subsidiaries were already carrying out an EBF rate improvement campaign. The settled target was 6%. Prospective studies showed that, after this target had been reached (and it was actually met by 1981), the completion rate would be only 45.7%, which was considered unsatisfactory. The sum of the busy line (BY) and don't answer (DA) rates which, by that time, amounted to 34.5%, would tend to increase to 43.5%, due to the EBF rate reduction. This value was considered very high in comparison with the accepted international standards (about 25%).

High BY/DA rates are undesirable because they lead to:

- Lower revenues, as the call abandonment increases;
- Higher investment costs, as the number of ineffective attempts increases;
- Poorer quality of service for the subscribers.

2. PROBLEM DEFINITION

The main causes of the high BY/DA rates were:

1) very few high DDD incoming traffic subscribers, with an insufficient number of lines, responded for a large amount of BY/DA calls.
Real traffic measurements had shown that about 22% of all BY/DA calls during the peak hours (9:00 to 11:00 a.m.) were directed to only 0.08% of the subscribers. This meant that, as an average, only 8 subscribers in a 10.000 lines telephone exchange were responsible for 22% of all BY/DA incoming calls to that exchange. This amazing fact was the consequence of the existing high down-payment for the subscriber's line and the high cost of larger PABX'S, which led some subscribers to have an insufficient number of lines.

2) lack of automatic hunting facilities for many line groups connected to key systems.
The reasons were technical limitations of some local exchanges and unawareness of the benefits provided by the hunting facilities.

3) Erroneous publishing of telephone number of all individual lines belonging to a hunting group.
Sometimes this occurred in the telephone Directory itself and, more often, in the subscribers' publishings.

4) Insufficient number of operator positions in some PBX and, quite often, poor maintenance and operation procedures.

3. THE SOLUTION

Top management at TELEBRÁS and its subsidiaries decided to carry out a BY/DA rates reduction campaign, based on the following actions:

a) Pinpointing the high DDD incoming traffic subscribers with high BY/DA rates and correction of their service deficiency.
Based on AMA tapes data, monthly reports were generated listing all lines with more than 20 call attempts terminating on BY or DA condition, during the observation period (9.00 to 11:00 a.m., 2 days per month). These reports, released by EMBRATEL , are tailored to each Operating Company and lists only the called lines served by that Company.

All lines which appeared twice on a six-month period were considered as a problem. In such cases, an in-depth analysis should be made and a solution found. This, in general, requires the addition of new lines, expansion or replacement of PABX or key systems, etc..

b) implementing hunting facilities as much as

possible, for all subscribers having more than one line in a same address. This same action should also be applied to lower DDD traffic lines not listed on the monthly report.

c) Publishing the hunting number only.
This action should be applied directly to the Telephone Directories. The subscribers should also be oriented to do the same thing on its advertisements, pamphlets, name-cards, stationery, sales gadgets, etc...

d) Implementing the terminal hunting system, which performs the line hunting even when any individual line in the group is dialed.
This type of hunting makes the performance of the hunting process much less sensitive to erroneous Directory number publishing. In Brazil it is inherent to some switching equipment models but, for others, some circuit additions are required.

e) Improving PBX maintenance and operation procedures.
It was developed and implemented a more effective PBX maintenance service quality control system and intensive training courses were offered to the operators.

The analysis of the BY/DA rates, by itself, did not show the results of the campaign due to the existing correlation between the BY/DA and EBF rates. It was expected an increase in the BY/DA rates as a result of the reduction of the EBF rate. For this reason is was devised a new indicator called "Loss due to the Called Subscriber (LCS)", as follows,

$$LCS = \frac{BY + DA}{BY+DA+OK} \times 100\%$$

This indicator measures the loss probability (due to BY/DA), for those calls which have already reached the called line. It depends essentially on the status of the called line, being fairly independent of the public telephone network status (measured by the EBF rate).

The effectiveness of implementing the measures listed on b, c and d was demonstrated by field trials accomplished in three different cities (Montenegro, Cachoeira do Sul and Vitória). The experience was very simple: implementing hunting facilities as much as possible, including individual lines, and publishing the hunting number only.

The results were excellent. Figura 1 shows, as an example, the evolution of the number of hunting lines in Montenegro, a small city with a 2.000 lines exchange. The results obtained are depicted on figure 2. The OK and LCS rates variations observed during the third quarter of 1983 were due to the implementation of the terminal hunting facility.

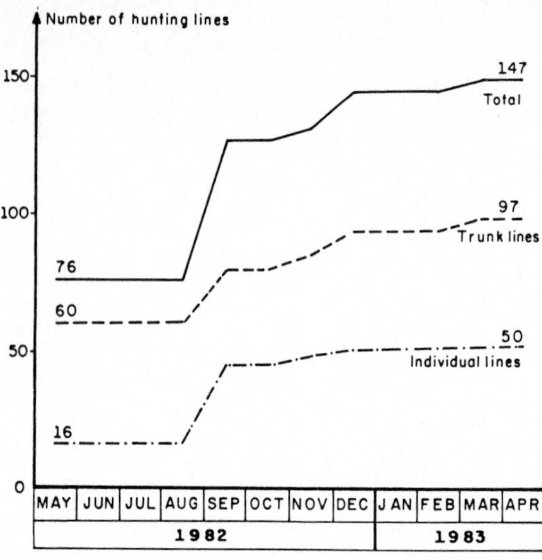

Fig. 1 – MONTENEGRO – HUNTING LINES EVOLUTION

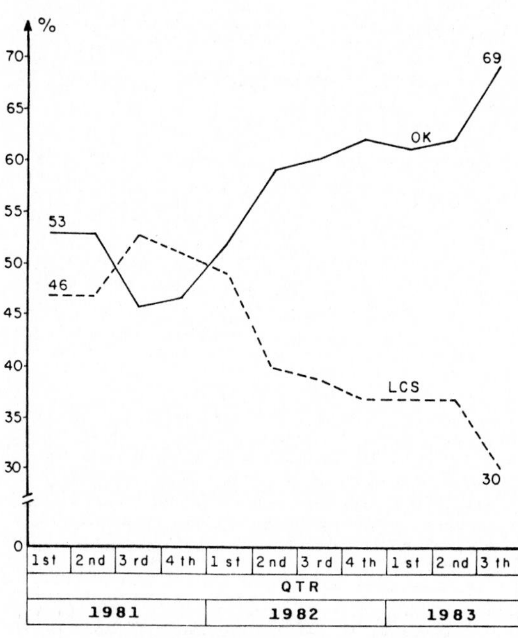

Fig.2 – CALL COMPLETION AND LCS RATES EVOLUTION (DDD TRAFFIC) – MONTENEGRO

4. CAMPAIGN COORDINATION

The successful accomplishment of those actions listed on part 3 requires the involvement of many people in different technical and operating areas of all 28 Operating Companies. An specific management struture was created, with the following key elements:

(a) An LCS manager in each one of the 28 Operating Companies, responsible for coordinating all the actions related to the campaign;

(b) a general supervisor in EMBRATEL, responsible for collecting the required data and releasing the monthly report to each Operating Company;

(c) a general manager in TELEBRÁS, responsible for the overall coordination of the campaign.

5. THE RESULTS

Up to now the results are excellent. The LCS index was reduced from 48.7% (1978) to 38.8% (1984), as shown on figure 3.

The evolution of the BY/DA and OK rates are shown on figures 4 and 5; one curve for the actual evolution and the other for the estimated values for LCS = 48.7%, i.e., if there were no

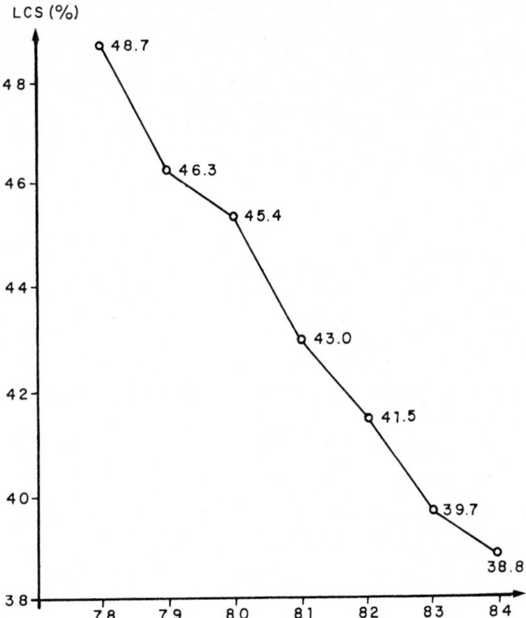

Fig. 3 REDUCTION ON LCS RATES

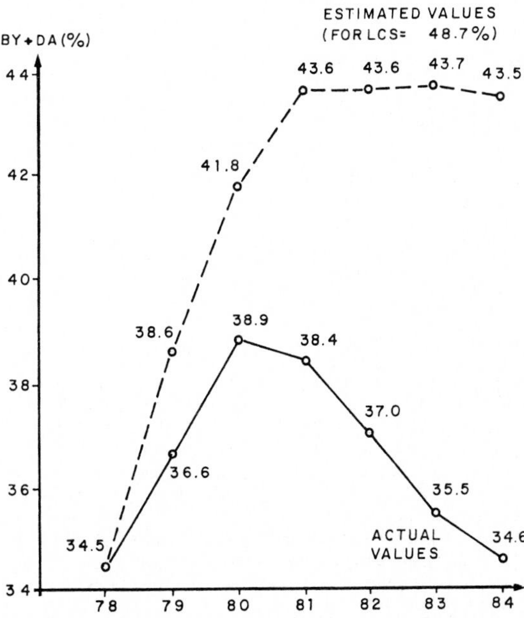

Fig. 4 ESTIMATED AND ACTUAL (BY + DA) RATE VALUES.

LCS reduction campaign. The latter represents the evolution of BY/DA and OK rates due only to the reduction of EBF rates from 23.6% (1978) to 6.8% (1984).

The result of the campaign is a reduction of 9.9% on the LCS rate, which led to an estimated increase of about 8% on the OK rate. It is also estimated that this improvement on the OK rate is worth about US$ 40 million/year (about 4.6% of

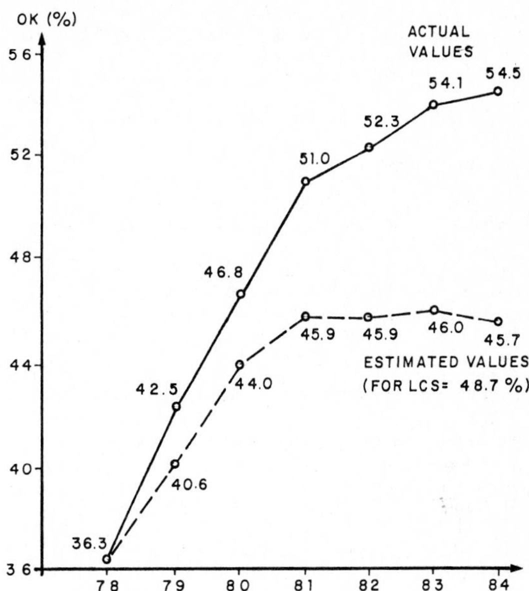

Fig. 5 ESTIMATED AND ACTUAL OK RATE VALUES

the TELEBRÁS System annual revenue from charged DDD calls). All those estimed values were calculated with the aid of the simplified theoretical model presented on the Appendix.

6. CONCLUSIONS

The improvement of call completion rate is a fundamental concern of all Telephone Operating Companies. Reducing the BY and DA rates, a hard and long lasting task, is an obligatory step and the most effective and rewarding way towards higher call completion rates compatible with international standards.

To identify and solve the originating problems of high BY/DA rates is a feasible task, even for Telecommunication Administrations which operate telephone networks with a high down-payment for the subscriber's line and with depressed demand (and so, high traffic/line), such as in Brazil.

Managing BY/DA rates is better accomplished with the help of the LCS indicator. The reduction on the LCS rate from 48.7% (1978) to 38.8% (1984), as a consequence of the LCS campaign started in 1979, represents an additional revenue of about US$ 4.00/telephone/year for the TELEBRÁS System. The tendency of the LCS rate is to go further down. The present target is to reach values below 35%.

Brazil is showing nowadays call completion rates around 55% for the international incoming traffic, measured by AT&T [2]. These figures are better than those shown by several developed countries.

The end result of this work is a better service quality for the subscribers, higher revenues and lower capital costs for the Telecommunication Administrations.

REFERENCES

[1] A. Elldin, "Approach to the Theoretical Description of Repeated Call Attempts", Ericsson

Technics Nº 3, 1967.

[2] "International Network Completion (IDDD) - From System to", AT&T monthly report (Jan. Nov./1984).

APPENDIX

A.1 INTRODUCTION

The call completion rate status of any telephone network can be defined by the following parameters: OK (completion rate), BY (busy line rate), DA (don't answer rate), EBF (equipment blockage and failure rate) and OT ("others" rate, which includes all unsuccessful calls not considered in the previous parameters). Those parameters are related to mutually exclusive events and their sum should be equal to one. An important additional auxiliary parameter, which was previously defined, is the LCS (loss due to the called subscriber). All those parameters, representing the status of any network, can allways be depicted on an abacus as the one shown on figure A.1.

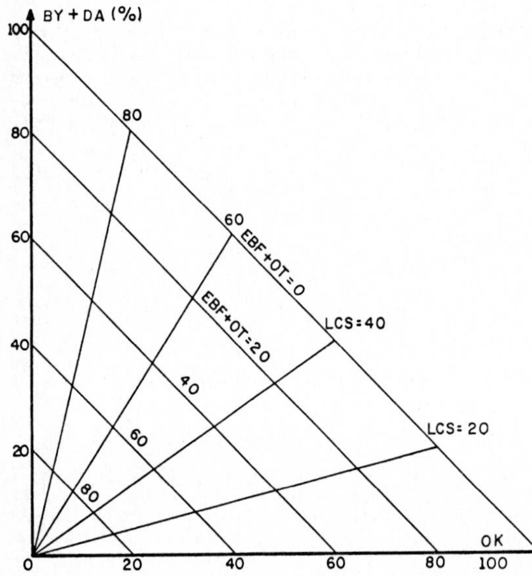

Fig.A.1 - NETWORK CALL COMPLETION STATUS ABACUS

Any network status change is allways due to an (EBF+OT) rate variation or to an LCS rate variation, or both.

It is expected that OK rate variations should lead to revenue variations, due to the variation on the probability that a series of attempts (of the same call intent) finally ends in a conversation.

A simplified theoretical model for estimating completion rate and revenue variations, as a result of LCS and (EBF+OT) rates variations, is given. The aim of the model is not to provide precise results but rough estimates, being rather a managerial and a decision support tool than a new scientific development.

A.2 ESTIMATING COMPLETION RATE VARIATIONS DUE TO EBF AND LCS RATES VARIATIONS

A.2.1 General Considerations

Figure A.2 gives an example of a generic network status change (from status A to status B).

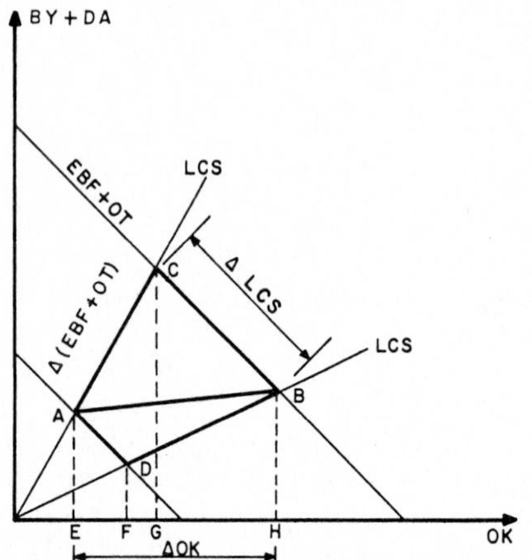

Fig.A.2-EXAMPLE OF A GENERIC NETWORK STATUS CHANGE

This change can be made by following any line connecting points A and B, but the net OK rate variation will be allways the same (ΔOK), and $\Delta OK = \Delta OK_E + \Delta OK_L$

ΔOK_E - OK rate variation due to the (EBF+OT) rate variation.

ΔOK_L - OK rate variation due to the LCS rate variation.

The values of ΔOK_E and ΔOK_L depend on the specific curve chosen to go from A to B. The abacus geometry (and equations (4) and (6) of this Appendix) shows that:

a) $|\Delta OK_E|$ is maximum when LCS in minimum and vice versa

b) $|\Delta OK_L|$ is maximum when (EBF+OT) is minimum and vice versa.

As a consequence, the maxima and minima values of ΔOK_E and ΔOK_L occur for curves ACB and ADB (v. figure A.2). The former curve implies allways in an (EBF+OT) rate variation (over the constant LCS rate line) in first place, followed by on LCS rate variation (performed over the constant EBF+OT rate line). On the other hand, for the ADB curve the LCS rate variation occurs first (over the constant EBF+OT line), followed by the (EBF+OT) rate variation (performed over the constant LCS rate line).

For the specific example shown on figure A.2 it is true that,

$$MIN\ [\Delta OK_E] = EG \qquad MAX\ [\Delta OK_E] = FH$$

$$MIN\ [\Delta OK_L] = EF \qquad MAX\ [\Delta OK_L] = GH$$

The maxima and minima values of ΔOK_E and ΔOK_L can be calculated as shown in parts A.2.2 and A.2.3, respectively.

A.2.2 OK Rate Variations Due to (EBF+OT) Rate Variations

Let $\overline{OK}$, $\overline{BY}$, $\overline{DA}$, $\overline{EBF}$, $\overline{OT}$ and $\overline{LCS}$, be the new parameters of a network after an (EBF+OT) rate variation, with no change on the value of the LCS rate (as in sections AC and DB, in figure A.2).

$$\overline{OK} = OK + \Delta OK_E$$

$$(\overline{BY} + \overline{DA}) = (BY + DA) + \Delta(BY + DA)$$

$$(\overline{EBF} + \overline{OT}) = (EBF+OT) + \Delta(EBF + OT)$$

$$\overline{LCS} = LCS \tag{1}$$

$$\overline{OK} + \overline{BY} + \overline{DA} + \overline{EBF} + \overline{OT} = OK+BY+DA+EBF+OT=1$$

$$\Delta(BY+DA) = - [\Delta OK_E + \Delta(EBF + OT)] \tag{2}$$

From equation (1), it follows that

$$\overline{LCS} = \frac{\overline{BY} + \overline{DA}}{\overline{BY}+\overline{DA}+\overline{OK}} = LCS = \frac{BY + DA}{BY+DA+OK}$$

$$(\overline{BY}+\overline{DA})\ (BY+DA+OK)=(\overline{BY}+\overline{DA}+\overline{OK})\ (BY+DA)$$

$$\overline{OK} = \frac{\overline{BY}+\overline{DA}}{BY+DA}\ OK$$

$$\overline{OK} = \frac{(BY+DA) + \Delta(BY+DA)}{BY+DA}\ OK$$

$$\overline{OK} = OK + \Delta OK_E = (1+\frac{\Delta(BY+DA)}{BY+DA})\ OK$$

$$\Delta OK_E = \frac{\Delta(BY+DA)}{BY+DA}\ OK \tag{3}$$

Using equations (2) and (3)

$$\Delta OK_E = - \frac{OK}{BY+DA+OK}\ \Delta(EBF+OT)=-(1-LCS)\Delta(EBF+OT) \tag{4}$$

From equations (2) and (4), it follows

$$\Delta(BY+DA) = - \frac{BY+DA}{BY+DA+OK}\Delta\ (EBF+OT) \tag{5}$$

A.2.3 OK Rate Variations Due to LCS Rate Variations

Let $\overline{OK}$, $\overline{BY}$, $\overline{DA}$, $\overline{EBF}$, $\overline{OT}$ and $\overline{LCS}$, be the new parameters of a network after an LCS rate variation, keeping (EBF+OT) rate as a constant (as in segments CB and AD, in figure A.2).

$$\overline{OK} = OK + \Delta OK_L$$

$$\overline{LCS} = LCS + \Delta LCS$$

$$(\overline{BY}+\overline{DA}) = (BY+DA) + \Delta(BY+DA)$$

$$(\overline{EBF} + \overline{OT}) = (EB + OT)$$

$$\overline{OK}+\overline{BY}+\overline{DA}+\overline{EBF}+\overline{OT}=OK+BY+DA+EBF+OT=1$$

$$\Delta OK_L = - \Delta(BY+DA)$$

$$\overline{LCS} = \frac{(\overline{BY}+\overline{DA})}{(\overline{BY}+\overline{DA})+\overline{OK}} = \frac{(BY+DA) + \Delta(BY+DA)}{(BY+DA) + \Delta(BY+DA)+OK+\Delta OK_L}$$

$$\overline{LCS} = \frac{BY+DA}{BY+DA+OK}+ \frac{\Delta(BY+DA)}{BY+DA+OK} = LCS + \Delta LCS$$

$$\Delta LCS = \frac{\Delta(BY+DA)}{BY+DA+OK} = - \frac{\Delta OK_L}{BY+DA+OK}$$

$$\Delta OK_L = - (BY+DA+OK).\Delta LCS=-[1-(EBF+OT)].\Delta LCS \tag{6}$$

A.3 REVENUE VARIATIONS DUE TO OK RATE VARIATIONS

The calculation of revenue variations due to OK rate variations will be performed with the aid of the parameter called "Success Rate (S)", given by

$$S = \frac{N_{OK}}{N_{CI}}$$

N_{OK} - number of completed (successful) calls.

N_{CI} - number of call intents (which equals the number of first attempts).

The number N_{CI} of call intents is a measure of the potential revenue from charged calls, i.e., it is the maximum number of calls that could be completed and charged. The number of completed calls (N_{OK}) is a measure of the effectivelly achieved revenue. Thus, the sucess rate measures the effectiveness of the Administration in generating the necessary revenues and is a key indicator for the whole Administration achievements.

To measure the number of call intents is a very difficult and laborious task. Nevertheless, the use of the theoretical model of call repeated attempts developed by Anders Elldin [1], make it easier to measure the success rate with a reasonable precision. This can be better understood with the aid of the simplified model of a telephone network shown on figure A.3.

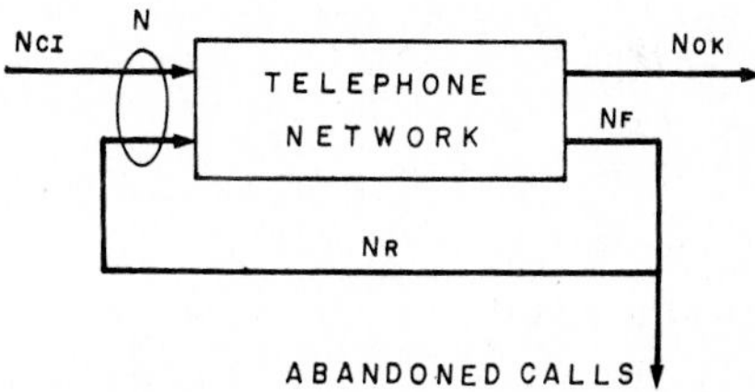

Fig. A.3- SIMPLIFIED MODEL OF A TELEPHONE NETWORK

N - total number of attempts
N_F - number of failed attempts
N_R - number of repeated attempts

$$N = N_{CI} + N_R$$

$$N_R = p.\ N_F$$

where p is the mean probability of renewing a call after the failure of the previous attempt. It is called perseverance.

$$N_F = N\ .\ (1 - OK)$$

$$N = N_{CI} + p\ .\ N_F = N_{CI} + p\ .\ N\ .\ (1 - OK)$$

$$N = \frac{N_{CI}}{1-p.(1-OK)}$$

$$S = \frac{N_{OK}}{N_{CI}} = \frac{OK.N}{N_{CI}} = \frac{OK}{1-p.(1-OK)} \tag{7}$$

The perseverance is in general a steady function in time. In the last three yearly real traffic measurements performed on the whole brazilian DDD network, the perseverance values varied only from

67.25% to 67.66%. Those measurements used a sample of about 3 million call attempts and took into account all call reattempts made during the whole day.

Thus, the perseverance can be considered as a constant over a certain period of time and measured less frequently. The success rate can be calculated using equation(7), as frequently as the OK rate is measured.

The success rate is a crescent monotonic function of the OK rate. Thus, increasing OK rates should lead to increasing revenues, due to the augment of the success rate (i.e., augment of the probability that a series of call attempts finally ends in a conversation). Quantifying those revenue variations is a task which will be done with the aid of an example of a generic network status change (figure A.4).

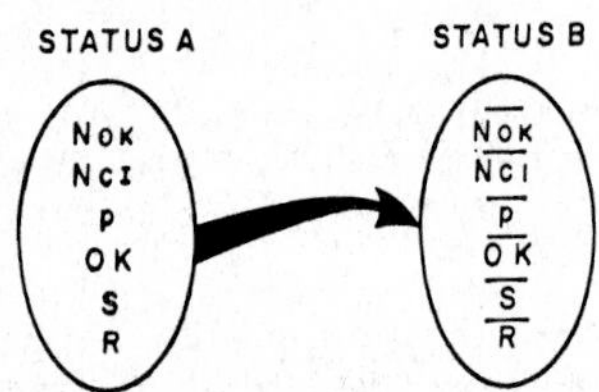

Fig. A.4 - EXAMPLE OF A GENERIC NETWORK STATUS CHANGE

For the sake of simplicity, it will be assumed that the OK rate variations won't significantly change the following parameters:

a) mean revenue per completed call. This means that there is no change on call durations and traffic distributions

($\frac{\overline{R}}{R} = \frac{\overline{N_{OK}}}{N_{OK}}$, where R and $\overline{R}$ are the revenues from charged calls).

b) number of call intents ($\overline{N_{CI}} = N_{CI}$)

c) perseverance ($\overline{p} = p$). This assumption was confirmed for the brazilian DDD network by real traffic measurements.

$$\frac{\overline{R}}{R} = \frac{\overline{N_{OK}}}{N_{OK}} = \frac{\overline{N_{OK}/N_{CI}}}{N_{OK}/N_{CI}} = \frac{\overline{S}}{S} \longrightarrow \overline{R} = \frac{\overline{S}}{S} R \tag{8}$$

$$\frac{\overline{R} - R}{\overline{R}} = \frac{\overline{S} - S}{\overline{S}} \longrightarrow \Delta R = \overline{R} - R = (1 - \frac{S}{\overline{S}}).\overline{R} \tag{9}$$

$$\frac{\overline{R} - R}{R} = \frac{\overline{S} - S}{S} \longrightarrow \Delta R = \overline{R} - R = (\frac{\overline{S}}{S} - 1).R \tag{10}$$

By the use of equations (7), (8), (9) and (10), it is now possible to quantify revenue variations due to OK rate variations.

A.4 COMPLETION RATE AND REVENUE VARIATIONS IN THE BRAZILIAN DDD NETWORK

The brazilian DDD network completion rate status has drastically changed in the period 1978-1984, as shown on Table A.1

The net reductions on the (EBF+OT) and LCS rates, in that time period, were 18.4% and 9.9%, respectively. The contributions ΔOK_E and ΔOK_L

TABLE A.1 - BRAZILIAN DDD NETWORK STATUS EVOLUTION (1978 TO 1984)

YEAR	OK	BY + DA	EBF + OT	LCS
1978	0.363	0.345	0.293	0.487
1979	0.425	0.366	0.209	0.463
1980	0.468	0.389	0.143	0.454
1981	0.510	0.384	0.106	0.430
1982	0.523	0.370	0.106	0.415
1983	0.541	0.355	0.104	0.397
1984	0.545	0.346	0.109	0.388

(corresponding to each one of those reductions) to the net increase of 18.2% on the OK rate, will be calculated as described on part A.2

The maxima and minima values of ΔOK_E and ΔOK_L occur when it is assumed that the change from status Ai to status Bi, corresponding to two consecutive years i and (i+1), is done over the curves ACB and ADB (v. figure A.2).

In the case of curve ACB, for each pair of points Ai and Bi, it follows that,

$$OK_{Ci} = OK_{Ai} + \Delta OK_{Ei}$$

$$R_{Ci} = R_{Ai} + \Delta R_{Ei}$$

$$OK_{Bi} = OK_{Ci} + \Delta OK_{Li}$$

$$R_{Bi} = R_{Ci} + \Delta R_{Li}$$

OK_{Ci} (R_{Ci}) - OK rate (revenue) corresponding to point C_i, which is the intermediate point between years i and (i+1).

OK_{Ai} (R_{Ai}) - OK rate (revenue) corresponding to point A_i.

OK_{Bi} (R_{Bi}) - OK rate (revenue) corresponding to point B_i.

ΔOK_{Ei} (ΔR_{Ei}) - OK rate (revenue) variation from year i to (i+1), due to the (EBF+OT) rate variation.

ΔOK_{Li} (ΔR_{Li}) - OK rate (revenue) variation from year i to (i+1), due to the LCS rate variation.

The values of OK_{Ai} and OK_{Bi} are known (table A.1) and the values of OK_{Ci}, R_{Ai}, R_{Bi} and R_{Ci} can be calculated by the use of equations (4), (7) and (8). The results are shown on Table A.2, where all revenue values are referred to the 1978 revenue, which was made equal to 1.000.

TABLE A.2 COMPLETION RATE AND REVENUE VARIATION-CURVE ACB

YEAR	OK		ΔOK		R		ΔR	
	$OK_{A,B}$	OK_C	ΔOK_E	ΔOK_L	$R_{A,B}$	R_C	ΔR_E	ΔR_L
1978	0.363		+0.043		1.000		+0.065	
		0.406				1.065		
1979	0.425		+0.036	+0.019	1.092		+0.047	+0.027
		0.461				1.139		
1980	0.468		-0.020	+0.007	1.148		+0.025	+0.009
		0.488				1.173		
1981	0.510		0	+0.022	1.199		0	+0.026
		0.510				1.199		
1982	0.523		+0.001	+0.013	1.214		+0.001	+0.015
		0.524				1.215		
1983	0.541		-0.003	+0.017	1.234		-0.004	+0.019
		0.538				1.230		
1984	0.545			+0.007	1.238			+0.008

The same calculations were made for curve ADB, for which it follows

$$OK_{Di} = OK_{Ai} + \Delta OK_{Li}$$

$$R_{Di} = R_{Ai} + \Delta R_{Li}$$

$$OK_{Bi} = OK_{Di} + \Delta OK_{Ei}$$

$$R_{Bi} = R_{Di} + \Delta R_{Ei}$$

The results of those calculations are shown on Table A.3

TABLE A.3 COMPLETION RATE AND REVENUE VARIATION–CURVE ADB

YEAR	OK		ΔOK		R		ΔR	
	$OK_{A,B}$	OK_D	ΔOK_L	ΔOK_E	$R_{A,B}$	R_D	ΔR_L	ΔR_E
1978	0.363		+0.017		1000		+0.026	
		0.380				1.026		
1979	0.425		+0.007	+0.045	1.092		+0.009	+0.066
		0.432				1.101		
1980	0.468		+0.021	+0.036	1.148		+0.026	+0.047
		0.489				1.174		
1981	0.510		+0.013	+0.021	1.199		+0.015	+0.025
		0.523				1.214		
1982	0.523		+0.016	0	1.214		+0.018	0
		0.539				1.232		
1983	0.541		+0.008	+0.002	1.234		+0.008	+0.002
		0.549				1.242		
1984	0.545			-0.004	1.238			-0.004

Tables A.2 and A.3 show the maxima and minima values of ΔOK_{Ei} and ΔOK_{Li}, for each pair of points (A_i, B_i). The net variations ΔOK_E and ΔOK_L, for the whole period 1978-1984, are restricted to the following limiting values,

$$MIN[\Delta OK_E] = \Sigma MIN\ [\Delta OK_{Ei}] = 0.043 + 0.036 + 0.020 + 0 + 0.001 - 0.004 = 0.096$$

$$MAX[\Delta OK_E] = \Sigma MAX\ [\Delta OK_{Ei}] = 0.045 + 0.036 + 0.021 + 0 + 0.002 - 0.003 = 0.101$$

$$MIN[\Delta OK_L] = \Sigma MIN\ [\Delta OK_{Li}] = 0.017 + 0.007 + 0.021 + 0.013 + 0.016 + 0.007 = 0.081$$

$$MAX[\Delta OK_L] = \Sigma MAX\ [\Delta OK_{Li}] = 0.019 + 0.007 + 0.022 + 0.013 + 0.017 + 0.008 = 0.086$$

Thus, as shown on Table A.1, the net OK rate increase was,

$$\Delta OK = \Delta OK_E + \Delta OK_L = 0.182$$

$$0.096 \leq \Delta OK_E \leq 0.101$$

$$0.081 \leq \Delta OK_L \leq 0.086$$

The reduction of 18.4% on the (EBF+OT) rate, that occurred between 1978 and 1984, led to an increase of 9.6% to 10.1% on the OK rate. This represents an average increase of 0.52% to o.55% on the OK rate for each 1% reduction on the (EBF+OT) rate.

In the same period, a reduction of 9.9% on the LCS rate led to an increase of 8.1% to 8.6% on the OK rate. This gives an average increase of 0.82% to 0.87% on the OK rate for each 1% reduction on the LCS rate. In general, LCS rate variations have stronger effects on the OK rate than (EBF+OT) rate variations.

By the use of equation (10), it follows that the increase of 18.2% on the OK rate that occurred from 1978 to 1984 led to a revenue increase

$$\Delta R = \overline{R} - R = (\frac{\overline{S}}{S} - 1).\ R = (\frac{0.784}{0.633} - 1).\ R = 0.238R$$

$$\Delta R = \Delta R_E + \Delta R_L = 0.238R$$

ΔR_E - net revenue variation due to the (EBT+OT) rate variation.

ΔR_L - net revenue variation due to the LCS variation.

The limiting values for ΔR_E and ΔR_L can be derived from tables A.2 and A.3, as it was done for the limiting values for ΔOK_E and ΔOK_L.

$$MIN[\Delta\ R_E] = \Sigma\ MIN\ [\Delta R_{Ei}] = 0.134$$

$$MAX[\Delta\ R_E] = \Sigma\ MAX\ [\Delta R_{Ei}] = 0.136$$

$$MIN[\Delta\ R_L] = \Sigma\ MIN\ [\Delta R_{Li}] = 0.102$$

$$MAX[\Delta\ R_L] = \Sigma\ MAX\ [\Delta R_{Li}] = 0.104$$

$$\Delta R = \Delta R_E + \Delta R_L = 0.238R \quad (11)$$

$$0.134R \leq \Delta R_E \leq 0.136R \quad (12)$$

$$0.102R \leq \Delta R_L \leq 0.104R \quad (13)$$

where R is the revenue projected back to 1978.

In order to translate the revenue variations into dollars, it is necessary to take into account that the network parameters values shown on Table A.1 refer to the test period (9:00 to 11:00 a.m., of working days). It will be assumed that they are also valid for the three afternoon peak hours of working days (2:00 to 5:00 p.m.). Those five hours of working days count for 45% of the monthly revenue from charged DDD calls.

$$\overline{R} = 0.45\ \overline{R}_T$$

$\overline{R}$ - revenue from charged DDD calls, in the five peak hours of working days (1984).

$\overline{R}_T$ - total revenue from charged DDD calls (1984).

In 1984 the TELEBRÁS System annual revenue from charged DDD calls was aproximately US$ 1.000 million.

$$\overline{R} = 0.45\ \overline{R}_T = \text{US\$ 450 million}$$

$$\overline{R} = R + \Delta R = 1.238\ R$$

$$R = \text{US\$ 363 million}$$

R - revenue from charged DDD calls, in the five peak hours of working days (1978), to which equations (11), (12) and (13) can be applied.

By the use of equations (11), (12) and (13)

$$\Delta R = 0.238R = 86.4 \text{ million}$$

$$48.6 \text{ million} \leq \Delta R_E \leq 49.4 \text{ million}$$

$$37.0 \text{ million} \leq \Delta R_L \leq 37.8 \text{ million}$$

Thus, it follows that the 9.9% LCS rate reduction that occurred from 1978 to 1984, led to an annual revenue increase varying from US$ 37.0 to US$ 37.8 million.

TELETRAFFIC ISSUES in an Advanced Information Society
ITC-11
Minoru Akiyama (Editor)
Elsevier Science Publishers B.V. (North-Holland)

AN APPROXIMATION FOR EVALUATING THE QUALITY OF SERVICE OF A TELEPHONE NETWORK IN FAILURE CONDITIONS

Elisa CAVALLERO, Alberto TONIETTI

CSELT - Centro Studi e Laboratori Telecomunicazioni S.p.A.
Via G. Reiss Romoli, 274 - 10148 TORINO (Italy)

ABSTRACT

An approximate method for computing the end-to-end blockings of a network in failure conditions is presented. The application of the method to a network dimensioning problem is also described. The tests performed on real networks for evaluating the influence of the approximations, show that the proposed method is accurate enough for engineering the network, taking reliability aspects into account.

1. INTRODUCTION

In the design of circuit switched communication networks, reliability and survivability considerations are of paramount importance. The constraints of an optimisation procedure should regard not only the grade of service in normal condition but also the network behaviour in abnormal states, i.e. overloads and failures. In this paper the failure situations only are considered. But the method presented can also be extended to evaluate overloaded networks.

According to the most recent trends, in normal condition the end-to-end blocking probabilities are used for characterizing the performance of the network. The same quantities can be adopted to represent the quality of service in any failure conditions.

The computation of end-to-end blockings for a real network, using simulations or accurate analytical models is a big task even for fast computers. So these methods, which can be used efficiently for evaluating the network performance in normal condition, are not suitable in the design phase or when it is necessary to examine a great number of failure configurations.

This paper presents a new approximate method for evaluating the quality of service of a network in failure states. The suggestion is to modify the set of parameters characterizing the performance of the network in normal condition, according to the circuit break-downs.

This method has been adopted within the project COST 201, a cooperation of 10 European countries with the aim ot developing computer based planning procedures to optimize the dimensioning of telecommunication networks [1, 2, 3].

2. THE EQUIVALENT TRUNK GROUP (ETG) APPROACH

The problem is to compute the end-to-end blockings in a network for many failure states.

The network behaviour in no failure condition can be evaluated precisely through simulation or analytic techniques. Considering any failure as a "small" perturbation of the normal state of the network, the end-to-end blockings in failure condition can be deduced in an approximate way. The approximation is based on the idea of taking advantage of the knowledge of the network behaviour in no failure condition, in order to evaluate the blockings in a failure state.

More precisely, the traffic carried by a route in normal condition is supposed to be carried by an equivalent number of circuits. When a failure is considered, the equivalent number of circuits is reduced as a function of the reduced sizes of the trunk groups of the route hit by the failure. Summing up the equivalent number of circuits of the routes used in each traffic relation, an equivalent trunk group is obtained. The loss probability of this trunk group can be used as an approximation of the end-to-end blocking.

Let us introduce the following notations:

A_r - traffic offered to relation r (Poisson)

N_{to} - number of circuits of trunk group t in no failure condition

N_{tf} - number of circuits of trunk group t in failure f

y_{ko} - traffic carried by route k in no failure condition

q_{ko} - number of circuits equivalent to route k in no failure condition

q_{kf} - number of circuits equivalent to route k in failure f

Q_{ro} - number of circuits equivalent to relation r in no failure condition

Q_{rf} - number of circuit equivalent to relation r in failure f

B_{ro} - end-to-end blocking of relation r in no failure condition

B_{rf} - end-to-end blocking of relation r in failure f

In order to show the use of ETG approach, the network of Fig. 1a is considered. The traffic offered to relation A-B is carried by the 3 routes shown in Fig. 1b.

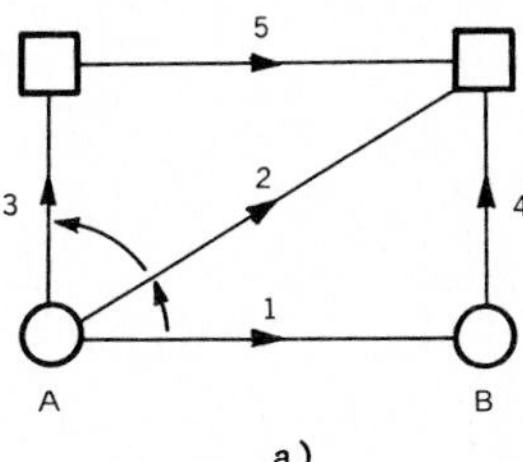

a)

route	trunk groups	carried traffic	equivalent circuits
1	1	y_{10}	q_{10}
2	2 4	y_{20}	q_{20}
3	3 5 4	y_{30}	q_{30}

b)

Fig. 1 - An example of the ETG approach

As an example, let us assume that failure f involves trunk groups 2 and 4. Being g(·) the relationship between the actual circuits and the equivalent ones, we assume:

$$q_{1f} = q_{10}$$

$$q_{2f} = g\ (q_{20},\ N_{20},\ N_{2f},\ N_{40},\ N_{4f})$$

$$q_{3f} = g\ (q_{30},\ N_{40},\ N_{4f})$$

In failure f, the relation A-B has the following number of equivalent circuits:

$$Q_{ABf} = q_{1f} + q_{2f} + q_{3f}$$

and the end-to-end blocking is simply expressed by the Erlang B formula:

$$B_{ABf} = E_{Q_{ABf}}\ (A_{AB})$$

3. COMPUTATION OF EQUIVALENT CIRCUITS IN NO FAILURE CONDITION

In no failure condition, the traffics carried by the routes can be computed by analytical models, such those presented in [4], [5], or by simulation.

Let Y_k be the traffic carried by the first k routes of relation r:

$$Y_k = \sum_{i=1}^{k} y_{io}$$

A trunk group carries the traffic Y_k if it has a number of circuits D_k satisfying the expression:

$$E_{D_k}\ (A_r) = 1 - Y_k/A_r$$

From the set D_k (k = 1, 2,...) the number of circtuis q_{ko} equivalent to the routes are easily computed:

$$q_{ko} = D_k - D_{k-1}$$

with $D_o = 0$.

As the routes are assumed to be each other indipendent, in no failure condition the relation r has a number of equivalent circuits equal to the sum of equivalent circuits of its routes:

$$Q_{ro} = \sum_{k} q_{ko}$$

The end-to-end blocking of relation r in no failure condition can be expressed using the Erlang B formula:

$$B_{ro} = E_{Q_{ro}}\ (A_r)$$

This formula gives the exact blocking, because the equivalent circuits of the routes are determined is such a way as they carry the actual traffics.

In case of failure, the same expressions are used:

$$Q_{rf} = \sum_{k} q_{kf} \tag{1}$$

$$B_{rf} = E_{Q_{rf}}\ (A_r) \tag{2}$$

4. COMPUTATION OF EQUIVALENT CIRCUITS IN FAILURE CONDITIONS

The crucial point in the ETG (Equivalent Trunk Group) approach is the relationship between the actual circuits and the equivalent ones in case of failure.

In order to answer to this problem an investigation on a single route is performed. The following assumptions are made:

i) there is only one trunk group in the route

ii) the offered traffics are not changed by the failures

iii) the offered traffics are Poisson distributed

The route k under examination offers a traffic stream A_{ko} to the trunk group t, which is globally offered a traffic A_{to}. The parameters $\beta (0 < \beta < 1)$ is the ratio of the two traffics:

$$\beta_{kt} = \frac{A_{ko}}{A_{to}} \tag{3}$$

The trunk group t has N_{to} circuits in normal condition. As all traffics are pure chance, the blocking of the trunk group t and the blocking of the route k are equal and given by the Erlang formula:

$$B_{ko} = E_{N_{to}}\ (A_{to})$$

The number of circuits q_{ko} equivalent to route k in no failure condition is given by the expression:

$$B_{ko} = E_{q_{ko}} (A_{ko})$$

Let us assume that the failure f reduces the size of trunk group t according to the factor δ_{tf} ($0 < \delta_{tf} < 1$):

$$\delta_{tf} = \frac{N_{tf}}{N_{to}}$$

The blocking of trunk group t and route k in failure f is:

$$B_{kf} = E_{N_{tf}} (A_{to})$$

This blocking probability corresponds to a number of equivalent circuits given by:

$$B_{kf} = E_{q_{kf}} (A_{ko})$$

So, for having the right blocking, in failure f the reduction in terms of equivalent circuits must be:

$$\gamma_{kf} = \frac{q_{kf}}{q_{ko}}$$

For evaluating the relationship between the actual circuits and the equivalent ones in case of failure, it is sufficient to compute the values of γ_{kf} as a function of δ_{kf} in various situations. Some numerical results, obtained varying the values of parameters N_{to}, B_{ko} and β_{kt}, are shown in Fig. 2.

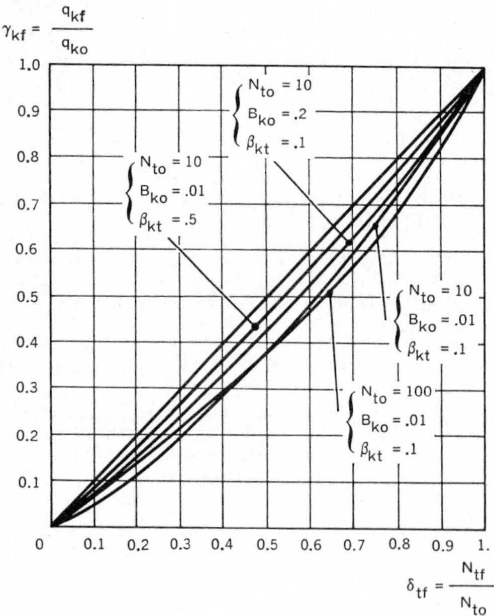

Fig. 2 - Ratio of equivalent circuits as a function of the ratio of actual circuits in failure condition

Similar results are obtained in cases of non Poisson traffics.

The curves of Fig. 2 can be approximated in a simple way by the function:

$$\frac{q_{kf}}{q_{ko}} = \left(\frac{N_{tf}}{N_{to}}\right)^{\lambda} \tag{4}$$

In Tab. 1 are shown the values of exponent λ giving the correct results for $\delta_{tf} = 0.5$.

N_{to}	10	10	10	100
B_{ko}	0.01	0.01	0.20	0.01
β_{kt}	0.1	0.5	0.1	0.1
λ	1.36	1.11	1.21	1.38

Tab. 1 - Best values for exponent λ ($\delta_{tf} = 0.5$)

When more than one trunk group fails in a route the expression (4) is extended in the following way:

$$\frac{q_{kf}}{q_{ko}} = \min_{t \in k} \left(\frac{N_{tf}}{N_{to}}\right)^{\lambda} \tag{5}$$

This means that the effect of a failure involving more than one trunk group in a route is assumed to be equivalent to a failure on the trunk group having the greatest percentage of circuit reduction.

5. THE VALUE OF EXPONENT λ

As it is shown in Tab. 1 the coefficient λ varies according to a set of parameters, the most important of which is the ratio β_{kt} between the traffic offered to the route and the total traffics offered to the trunk groups.

Moreover the expression (4) is not the only approximation implied in the ETG method. Other approximations regards the indipendence among the traffic relations, the indipendence among the routes, the expression (5) used to evaluate the effect of a failure involving more than one trunk group in a route,.... .

In order to test if a good choice of the exponent λ can improve the ETG approach with regards to all the assumptions implied in the ETG method, an investigation on some networks is performed.

With a program based on the algorithms presented in [5], the end-to-end blockings B_{rf} of all relation r in all failure f are computed. These blockings are attributed to Q_{rf} equivalent circuits such as

$$B_{rf} = E_{Q_{rf}} (A_r)$$

From (1) and (5), we obtain that the right value of exponent λ for relation r in failure f is given by

$$\sum_k q_{ko} \cdot \min_t \left(\frac{N_{tf}}{N_{to}}\right)^{\lambda_{rf}} = Q_{rf}$$

In this way a set of good λ values corrisponding to different situations are obtained.

It must be noticed that not for all the blockings it is possible to obtain the right values, choosing a suitable λ. This happens because, in some failure, some relations not using the failed trunk group have nevertheless blocking values different from the normal ones. The reason is that the traffic pattern is changed, against the assumption ii) in section 4.

From the numerical results obtained on some test networks it appears that the "optimal" λ_{rf} are rather stable with respect to the failures. The variations of λ_{rf} among the relations are stronger.

An attempt of correlating the λ values with the coefficients β introduced in section 4 is carried out. To this aim, the definition of β, given for a route composed by a single trunk group, is generalized to a traffic relation, according to the following formula:

$$\beta_r = \frac{\sum_{k \epsilon r} q_{ko} \min_t \beta_{kt}}{\sum_{k \epsilon r} q_{ko}}$$

where β_{kt} is given by (3).

β_r represents for relation r the degree of sharing the trunk groups with the other relations. The greater is β_r the more indipendent is the relation and the interference from extraneus traffics should be less. Decreasing values of λ_{rf} should correspond to increasing values of β_r. This is true only as a mean tendency; the single values of λ_{rf} are rather scattered and sometimes it happens that $\lambda_{r_1 f_1} < \lambda_{r_2 f_2}$ corresponds to $\beta_{r_1} < \beta_{r_2}$.

A linear regression based on the numerical results of some small and rather meshed test networks has led to the following expression:

$$\lambda_r = 1.87 - 0.87\ \beta_r \qquad (6)$$

The point ($\beta_r = 1$, $\lambda_r = 1$) is considered as fixed because in this case, the relation r is completly indipendent from the other ones.

In large networks the computation or the storage of coefficients β_r required by (6) can represent a problem. This can be avoided if the exponent λ is assumed to be constant for all relations. From the numerical results of the same test networks, we obtain

$$\lambda = \underset{f}{\text{mean}} \text{ of } \underset{r}{\text{mean}}\ \lambda_{rf} = 1.25 \qquad (7)$$

In Tab. 2 some parameters which can be used to characterize the quality of service of a network in failure conditions are shown. The approximate models based on (6) and (7) are compared with the exact one. The results regard one of the networks used for computing the (6) and (7).

The agreement between the approximate models and the exact one seems to be rather good. Model (6) is more accurate of model (7), because the former tries to take into account the peculiarities of each traffic relation.

QUALITY OF SERVICE	EXACT MODEL	MODEL (6)	MODEL (7)
$\underset{f}{\text{mean}}$ of $\underset{r}{\text{mean}}$ B_{rf}	0.276	0.295	0.302
$\underset{f}{\text{mean}}$ of $\underset{r}{\max}$ B_{rf}	0.595	0.578	0.532
$\underset{f}{\max}$ of $\underset{r}{\max}$ B_{rf}	0.812	0.801	0.798

Tab. 2 - Evaluation of models based on formulae (6) and (7) with respect to Quality of Service parameters.

The results would have been less close to exact model if a network not used for deducing (6) and (7) were tested. The constants appearing in (6) and (7) depend, indeed, by the characteristics of the network. An attempt of correlating the λ values with the general features of the network (mean number of traffic relation using a trunk group, mean number of trunk groups involved in a failure and so on) is actually under study.

6. AN APPLICATION OF ETG APPROACH TO NETWORK DIMENSIONING

In order to show how the ETG approach can be used in planning procedures let us consider the following problem:

> The switching network has been dimensioned in no failure condition, the trunk groups have been routed on the transmission media, we want to know how many circuits should be protected by a stand-by network in order to meet a quality of service constraint. The constraint regards the maximum end-to-end blocking allowed in failure states.

But, in any failure, this constraint can be satisfied imposing different degree of protection to the various trunk groups. The best we can do from the cost point of view, is to minimize the total number of circuits to be protected in each failure situation.

Adopting the ETG approximation this problem can be approached in the following way.

To guarantee, in failure f condition, an end-to-end blocking not greater than a threshold $\overline{B}$, the number of equivalent circuits of relation r must be greater than the value $\overline{Q}_r$ given by:

$$\overline{B} = E_{\overline{Q}_r}(A_r)$$

So the constraint on the blocking is expressed in terms of equivalent circuits. If these circuits are not sufficient, it is necessary to increase their number with an amount:

$$\Delta Q_{rf} = \overline{Q}_r - Q_{rf}$$

In general this can be done in different ways, depending from the subdivision of needed capacity among the various routes used by the relation:

$$\Delta Q_{rf} = \sum_k \Delta q_{kf} \qquad (8)$$

In order to increase the traffic carrying capacity of a route, it is necessary to increase the number of operating circuits of failed trunk-groups, adding some stand-by capacity. Being ΔN_{tf} the stand-by capacity of trunk group t in failure f, from (5) we have:

$$\Delta q_{kf} = q_{ko} \min_{t \epsilon k} \left\{ \frac{N_{tf} + \Delta N_{tf}}{N_{to}} \right\}^{\lambda} - q_{kf} \qquad (9)$$

The (8) and (9) represents a constraint in our problem which is difficult to handle because of the minimum function. An approximate solution can be obtained assuming that the stand-by capacities of all failed trunk groups used in route k are such as to reach exactly the value Δq_{kf} of the equivalent circuits. This is sensible because, according to ETG approach, it is useless to put more capacity in one trunk group than in the others belonging to the same route.

Following this simplification, from (9) we have:

$$\Delta N_{tf} = N_{to} \cdot \left(\frac{q_{kf} + \Delta q_{kf}}{q_{ko}} \right)^{1/\lambda} - N_{tf} \qquad (10)$$

and setting

$$\alpha_{kt} = \frac{N_{to}}{q_{ko}^{1/\lambda}} \qquad (11)$$

$$x_k = q_{kf} + \Delta q_{kf} \qquad (12)$$

we obtain

$$\Delta N_{tf} = \alpha_{kt} \, x_k^{1/\lambda} - N_{tf} \qquad (13)$$

The quantity to be minimized is

$$\sum_t \Delta N_{tf} = \sum_k \sum_{t \epsilon k} \alpha_{kt} \, x_k^{\gamma}$$

neglecting the term N_{tf} which is constant and summing the trunk groups route by route.

Set

$$\alpha_k = \sum_{t \epsilon k} \alpha_{kt} \qquad (14)$$

the problem becomes, in mathematical term:

$$\min \sum_k \alpha_k \, x_k^{1/\lambda} \qquad (15)$$

subject to

$$\sum_k x_k = \Delta Q_{rf} + \sum_k q_{kf} \qquad (16)$$

$$x_k \geq q_{kf} \qquad (17)$$

$$x_k \leq q_{ko} \qquad (18)$$

The (15) implies the minimization of stand-by requirements. The variables of the problem are the equivalent capacities x_k of the routes. The (16) expresses the quality-of-service constraint. The (17) and (18) give the boundaries of the x_k's.

The problem has always a feasible solution provided that the end-to-end blockings in no failure condition are less than the allowed blocking B in failure.

As it is $\lambda \geq 1$, the curves $\alpha_k \, x_k^{1/\lambda}$ never intersect each other. The solution of the problem (15) ÷ (18) is obtained by ordering the values α_k. The algorithm is the following:

1. set $X = \sum_k q_{kf}$
2. for each route k without failed trunk group, set $\alpha_k = \infty$
3. while $X < \Delta Q_{rf} + \sum_k q_{kf}$, choose

   $\tilde{k} = \min_k \alpha_k$ and set

   $q = \min \left\{ \Delta Q_{rf} + \sum_k q_{kf} - X, \; q_{\tilde{k}o} - q_{\tilde{k}f} \right\}$

   $\Delta q_{\tilde{k}f} = q$

   $x = X + q$

   $\alpha_{\tilde{k}} = \infty$

At the end of the algorithm the quantities Δq_{kf} are known and the stand-by capacities are obtained through (10).

The solution adopted to computed the stand-by capacities requires essentially the indipendence of the routes. If the same trunk group is used in more than one route, these are no longer indipendent. Two situations are distinguished:

CASE I: more than one route is in failure, the same trunk group is common to all the failed routes and no other trunk groups fails. As every route in failure uses the same failed trunk group $\tilde{t}$ they can be aggregated together to constitute a fictitions route $\tilde{k}$ with an equivalent number of circuits equal to the sum of equivalent circuits of the failed routes using $\tilde{t}$. This fictitions route $\tilde{k}$ can be treated as a real one according to the proposed procedure.

CASE II: more than one route is in failure, some trunk groups are common to the failed routes, some others not. The proposed solution is to adopt the simplified approach, neglecting the fact that some "anomalous" trunk groups are used in more than one route and to set the stand-by capacities of these trunk groups at the maximum number computed for each route using them.

More details can be found in [6].

So far the traffic relations have been considered one at a time. But, what happens if a trunk group, used in more than one relation, requires different numbers of stand-by circuits for protecting the different relations? The suggested solution is to take the maximum of stand-by requirements. This approach is in line with ETG approximation, which assumes a fixed sharing of the trunk group capacity among the traffic relations using it.

7. CONCLUSIONS

The ETG approach has been tested on real networks (up to 600 switching nodes) and it results to be accurate enough in network dimensioning problems. As the proposed method requires reduced computer times, it can be considered an useful tool for the design of large networks under reliability constraints.

ACKNOWLEDGMENTS

The contribution of all colleagues involved in COST 201 project is greatly appreciated.

REFERENCES

[1] K. Nivert, N. Noort: "COST 201: A European research project. Methods for planning and optimization of telecommunication networks" -II International Network Planning Symposium, Brighton, 1983

[2] J. Dressler, J. Gomes, R. Mantel, J.M. Mepuis, E. Sara: "COST 201: A European research project. A flexible procedure for minimizing the cost of a switched network taking into account mixed technologies and end-to-end blocking constraints" - X International Teletraffic Congress, Montreal, 1983

[3] P. Lindberg, U. Mocci, A. Tonietti: "COST 201: A European research project. A procedure for minimizing the cost of a transmission network under service availability constraints in failure conditions" - X International Tele traffic Congress, Montreal, 1983

[4] S. Katz: "Statistical performance analysis of a switched communications network" - V International Teletraffic Congress, New York, 1967

[5] M. Buttò, G. Colombo, A. Tonietti: "On point-to-point losses in communications networks" -VIII International Teletraffic Congress, Melbourne, 1976

[6] E. Cavallero, A. Tonietti: COST 201: Procedure Manual - Module SBRM.

TELETRAFFIC ISSUES in an Advanced Information Society
ITC-11
Minoru Akiyama (Editor)
Elsevier Science Publishers B.V. (North-Holland)

MODELS FOR AN EFFECTIVE DEFINITION OF END-TO-END GOS PARAMETERS AND FOR THEIR REPARTITION IN IDN NETWORKS

Mario BONATTI, Claudio BARBUIO, Giuliano CAPPELLINI, Ugo PADULOSI

Central Research Laboratories, ITALTEL
Settimo Milanese (Milano), Italy

ABSTRACT

A manufacturing company point of view on End-to-End and on Link-by-Link GOS specification is presented. Simple reference simulation models for subscriber behaviour, trunking network and exchanges are developed to help proposals for national and international specifications.
Statistical, traffic and economical criteria are presented as guidelines in the choice of GOS specification sets.

1. INTRODUCTION

1.1. Position

The interest of a manufacturing company in GOS studies arises from the impact that GOS standards have on the switching families (a switching family being a coherent range of switching systems with different capacities and capabilities). This impact can be analyzed in terms of:

- impact on each switching system (architecture, dimensioning, introduction, growth, overlay, replacement.)

- overall impact on a switching family (architecture, global number of lines served.)

Modern switching systems

- can co-operate in adopting advanced network management techniques to cope with overloads and/or faults, both through the identification of abnormal conditions, through their control and through an extensive tracing of call servicing.

- have, in general, blocking switching network and control structures introducing non negligible waiting times.

- have an high sensitivity to overloads: the throughput increases when the offered traffic increases, until a maximum capacity has been reached, then it can drop dramatically.

1.2 Problem definition

On the basis of an extensive review of GOS studies (from Moe up to now) [1] the following conclusions are reached:

- The common interest of the three parts involvolved in switching business (operating companies, subscribers and manufacturing companies) would be to set GOS standards for exchanges at such levels that a global economical criterion will be kept at a minimum trend.

- the choice of GOS parameters (end-to-end, link-by-link and node-to-node) and of their optimal values wouldn't be confined to traditional traffic engineering aspects

- the optimisation would be a dynamical one (i.e. extended over a period of y years) and not a static one (i.e. limited to year by year snapshots).

The economical criterion used for many years has been the cost of the network in terms of capital expenditures needed to face the network growth. This criterion can be improved taking into account the operation costs (particularly maintenance) and the rearrangement costs incurred in order to profit from technological progress.

But, in our opinion, a more global criterion is needed (cfr. Littlechild [2], Freidenfelds [3]) that should consider also

- the revenue losses of operating companies due to the end-to-end grade of service values and to the ineffective occupation of network resources by unsuccessful attempts

- the revenue losses of subscribers due to the "waiting" or to the "abandon" in obtaining required services.

The optimal investment sequence, corresponding to the optimum trend of the chosen Global Economical Criterion (GEC) would be sensitive to particular dependance of GOS parameters on "loads" offered and network availabilities.

The more significative combinations of load conditions and of availability conditions will be named "reference cases"

The optimal values of GOS parameters would be also sensitive to the network management techniques adopted during the planning period. A network management evolution scenario is needed.

The problem of optimal values of GOS parameters can now be formalized in the following way

P: find MIN(GEC)
versus node-to-node and link-by-link service level variables
given
a network management scenario
a reference case set C

subject to
a set of end-to-end service level constraints defined on C
over a y years period
in a repeated call attempt environment.

The formulation of P will be compared with the present state of the art.

The problem

P1: A1) Find MIN(total cost in circuits)
versus node-to-node number of circuits
subject to a last choice traffic service level constraint
stationarity conditions
poisson sources
lost calls cleared
negligible link-by-link losses and waiting times

B1) Reference Cases: normal loads, full network availability
monodirectional trunks
trunk modularity = 1
no layout diversity

has been solved, for HNDR (Hierarchical Non Dynamical Routing), both in the static and in the dynamical form.

Partial solutions of the problem

P2: A2) = A1
B2) Reference case: normal load, partial network availability,
both mono- and bi-directional trunks
trunk modularity 1, 6, 30, ...
layout diversity

have been presented for some HNDR network with service protections, only in the static form.

The State of the art evolves towards an unified study of the following aspects:

1) non-negligible link-by-link losses and waiting times
2) repeated call attempts
3) advanced network management techniques

and towards the adoption of a Global Economical Criterion.

1.3 The Specification Problem

In the formalization of the problem P a reference has been made to the set of reference cases and to the set of end-to-end service level constraints. The second set is obviously defined on the first, but in the sense of probability theory. As shown, for example, by Jacobaeus/Elldin [4], Molnar [5] and Erke/Rahko [6], this definition is not simple.

The real cases (traffic environments and availability configurations) fluctuate around benchmark cases.

The values of performability parameters for the real cases fluctuate around the values of performability parameters for the benchmark cases.

Sometimes the distributions are narrow, sometimes they are very broad.

Furthermore the set of end-to-end service level constraints must be developed as multipoint GOS specifications according to, among others, Wright [7], Horn [8] and Kodaira/Harada [9] and must take into account the composition of blocking and waiting effect, as pointed out by Katz [10] (inadequately handled call attempts in CCITT XI ?).

The problem of reference cases definition and of multipoint GOS specification can be formalized as follows

S : Given

$\mathcal{U}$ the set of all realizations of end-to-end traffic matrix and of all network availability conditions, with a probability function defined on it,

$\Theta(u)$ the duration of the realization u

$\mathcal{R}_{\pi}(u)$ the set of all end-to-end relations whose probability of blocking doesn't exceed ϵ_{π}

$\mathcal{R}_{\beta}(u)$ the set of all end-to-end relations for which the probability that the end-to-end post-dialling delay exceeds t doesn't exceed ϵ_{β}

$\mathcal{F}(\mathcal{U})$ the probability family that an end-to-end relation belongs to $\mathcal{R}_{\pi}(u)$ and to $\mathcal{R}_{\beta}(u)$ during a period $T(\Theta)$

find

a set of reference cases $\mathcal{C}$ and a set of end-to-end service level constraints $\mathcal{S}$ defined on $\mathcal{C}$ such that $\mathcal{S}(\mathcal{C})$ would select a significant sample of $\mathcal{F}(\mathcal{U})$ from the point of view of a criterion K (for example: comparison between systems, dimensioning, provisoning criteria).

A practicable approach to this complex problem could be to formulate some work hypothesis as system qualification benchmarks (CCITT XI) or as criteria for the optimal utilization in the network (CCITT II).

Anyway the goal is to choose the minimum set of specifications rules (performability parameters values vs traffic environments and availability configurations) that allows a definition of the optimal utilization of the systems in the network.

1.4 This Paper

The problems P (1.2) and S (1.3) are very hard. On the basis of our experience this isn't the only reason of difficulties in GOS standards definition. Some of the difficulties met arise from the lack of reference models of offered loads (subscriber behaviour etc), of network (in fault conditions) and of switching machines.

In this paper we present some models (network models in § 2, subscriber behaviour models in § 3, exchange models in § 4), that we hope to be general enough not to privilege any particular architecture or any particular implementation, specific enough to be useful and simple enough to be practicable (few decisional variables).

In § 5 the structure of the hypothetical sets of specification rules is described, alongh with the structure of experiments undertaken in order to check the optimality power of each hypothetical set. The concept of Reference Connection, proposed by Molnar [5] and improved by Harvey/ Hill [11] has also been adopted to evaluate each hypothetical set.

Some provisional results are also reported (synthetical results will be presented at the Congress).

2. NETWORK MODELS

2.1 Network Topology

Network topology is modeled by the Incidence Matrix with exchanges, (or subscribers' groups) as nodes and trunks, (or subscriber's line groups) as arcs.

The Incidence Matrix is modified as follows: each entry corresponding to an existing arc has the following values

1 := for first level trunks (same area)

2 := for second level trunks (same region)

3 := for third level trunks (different region)

2.2 Network Resources (dimensioning)

Network dimensioning is modeled by a Capacity Matrix: to each oriented arc, identified by the pair (originatig node, terminating node) are assigned two values; namely the number of circuits of the monodirectional and of the bidirectional trunks.

2.3 Network Management

All the network management techniques based on local control may be modeled.

The routing is modeled by a multilevel matrix: the first plane shows the direct routings, the second plane the routings through an high level exchange, and so on. If the entry value is equal to the column index value, the routing is completed.

Hierarchical Non Dynamical Routings are modeled fixing the sequence of routing trials.

2.4 Subscribers' Groups Modeling

Each subscribers' group A(i) assigned to an exchange X is modeled as a remote concentrator without loss.

2.5 Network Decomposition

The modeling of a very large network in a repeated call attempts environment, the analysis of design alternatives and the selection of the best ones can be helped by a decomposition of the network and by an adequate recomposition of the results [12].

3. SUBSCRIBER'S BEHAVIOUR MODELS

3.1 Initial Call Attempts Generation

Interarrival rates between initial call attempts for each origin-destination pair of nodes (o,d), modeled by the matrix Λ(o,d), are independent stationary stochastic variables, distributed according to a negexp. The matrix Λ(o,d) is generated starting from the matrices of interarrival rates between initial call attempts for each source-destination pair of subscribers classes, through an assignement procedure of the subscribers to the nodes. Each source-destination pair originates a sequence of repeated attempts if the initial one hasn't been successful, whichever the reason.

3.2 Holding Times Generation

Holding times for each completed call after answer are independent stationary stochastic variables distributed according to a two exponential $(p.\exp(-t/\theta 1) + (1-p)\exp(-t/\theta 2)$, modeled by the following three matrices p(o,d),θ1(o,d),θ2(o,d), where (o,d) stands for an origin-destination pair of subscriber classes (Liu [13]).

Holding times on busy tone and on ringing tone are generated according to the experimental distribution of subscriber's patience versus these events, truncated by network time-outs.

Holding times on dialling and dialling habits are generated according to experimental distributions.

Holding times on any other reason failures are generated from a lognormal distribution with parametric values.

3.3 Subscriber Reattempt Behaviour

For each source-destination class of subscribers the following retrial characteristics are considered:

- attempt dispositions probabilities
  - = complete attempts and unsuccessful attempts due to: network congestion blocking, excessive dial-tone and post-dialling delay, network time-outs, network faults and busy called subscriber: modeled through real-time network and subscribers models.
  - = unanswered attempts: modeled through a truncation of ringing tone by subscriber impatience or by network time-outs.
  - = any other reasons: modeled by a lognormal random generation
- probabilities of the (k+1)-th attempt and
- retrial time intervals between the failure of the k-th attempt and the origin of the (k+1)-th attempt,
  - if the k-th attempt disposition is i(k),
    - the initial attempt disposition is i(1),
    - the total elapsed time after the initial attempt is T;

  modeled through marginal distributions, choosing the highest outcome.

3.4 Service Attributes

The following attributes can be assigned to each initial call attempts: local calls, district calls, intertoll calls.

4. EXCHANGE MODELS

The aim of the exchange models is to generate the blockings and the delays due to the interaction between the instantaneous traffic and the state of the exchange, the network surrounding the exchange being simulated by mono- or bi-directional trunks.
The choice of the models depends on:
- the search for a large independence from each particular implementation
- the capability to cope with different cases of overload and unavailability
- the capability of discrimination between traffic flows (internal, external outgoing, external incoming, external transit).

4.1 Switching Network

The switching network is modeled as shown in fig. 1.
The X1,X2,X3,X4 switches model the blocking on each flow; the general Call Intensity Function (A. Jensen) Λ(M-me) models the blocking on substages; the Wallstrom functions [14] W(N-ne) model the blocking in trunk stages and the blocking generated by the common switching resources is modeled by a transform from the pair (n incoming trunks and m subscribers busy) to the pair (ne, me). The introduction of ne in W and of me in Λ models the interaction between all the flows.
An Y switch for each trunk models the protectio of trunks against overloads and or faults.
Overload conditions are modeled increasing offered traffic.
Fault conditions may be modeled by "worsening" one or more of the Internal Loss Functions (X1, X2,X3,X4,Ys,Λ,W).

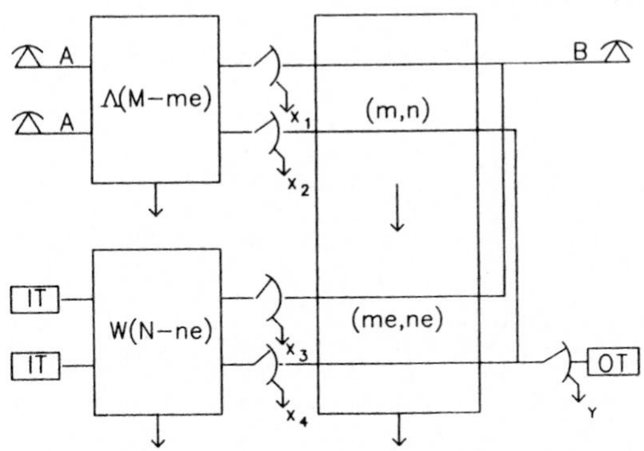

FIG.1 SWITCHING NETWORK MODELING

4.2 Control Structure

The control structure is modeled by the multi-queue one-server system of fig. 2.
The following classes of queues are modeled:

1) dialling tone queue
2) incoming respose queue
3) exchange-call-set-up queue (4 queues, 1 for each traffic flow)
4) through-connection queue (4 queues, 1 for each traffic flow)
5) connection output queue
6) clearing request queue
7) disconnection output queue

Priorities are assigned to the queues. Flow controls are modeled also on the queues, and they can generate a rejection of attempts and a subsequent busy tone (equivalent to busy tones from the switching network). Queue visiting disciplines are adopted to avoid the exhaustion of service on a high priority queue in case of unbalanced traffic conditions. The service times for items belonging to the different queues may be different.
Fault conditions are modeled by "worsening" the service time and the flow controls parameters.
Flow controls mechanisms are activated and deactivated on the basis of a progressive procedure (Gimpelson [15]).

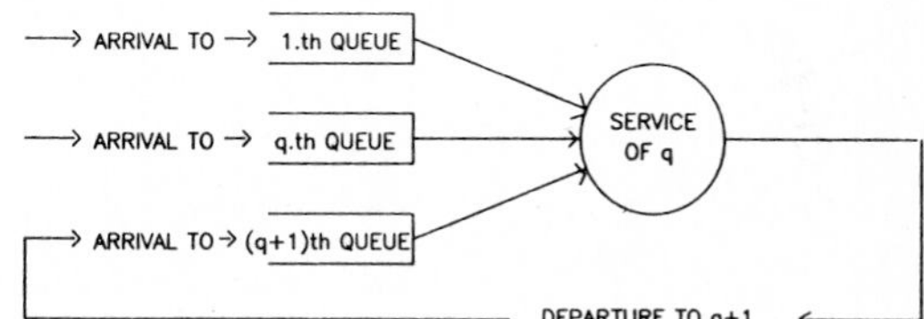

FIG.2 CONTROL STRUCTURE MODELING

5. DESIGN OF EXPERIMENTS

Networks of growing complexity are exhamined measuring and comparing the following variables for each origin-destination pair:

M1) end-to-end probability of loss
M2) end-to-end post-dialling delay
M3) revenue loss as sum of ineffective holding times and of times spent in trials or waiting by the subscribers
M4) contribution of arc and link congestions to the failure probability for the initial attempt, to the probability of abandon and to the number of attempts for each initial attempt
M5) resources capacities (cfr Machine Capacities in Gimpelson [15]) for nodes and trunks

for different reference cases and different assignement of performability parameters to the links and to the flows in the nodes.

5.1 Reference Cases

Given a traffic matrix, the following load conditions can be generated:

B0 := no overload = normal load
B1 := homogeneous diffuse overload
B2 := local overload
B3 := focused overload on one origin

B4 := focused overload on one destination
B5 := focused overload between an origin-destination pair

Given an assignment of performability parameters the following availability conditions can be generated

CN0 := full availability
CN1 := extreme value partial (evp) degradation of one terminating or origin function
CN2 := CN1 + B2 or B3 or B4
CN3 := evp degradation of a transit function
CN4 := CN3 + B2 or B3 or B4 or B5

regarding flows in the nodes, and

CL1 := evp degradation of first choice trunks
CL2 := evp degradation of intermediate choice trunks
CL3 := evp degradation of last choice trunks

regarding links

A case is a combination of one B, one CN and one CL condition. The C conditions can be generated by an adequate dimensioning of the trunks or an adequate choice of the node models.

5.2 Performability Parameters

Switching network blocking:
Bi := for internal flows
Be := for incoming flows
Bo := for outgoing flows
Bt := for transit flows

Control structure delays:
dtd := dial-tone delay
ird := incoming response delay
csudx := call-set-up delay for x flows
tcdx := through-connection delay for x flows (x=i,e,o,or t)

Maximum node capacity:
Cx := (completed-per-time-unit attempts)/(offered-per-time-unit attempts) in the node (x=i,e,o, or t)

Link blocking:
P(ik) := for the link between node i and node k
Plc := for the last choice trunks

Maximum link capacity: obvious

Arc blocking:
$\pi(ik)$:= sum of all quotas that will block an attempt between the node i and the node k
π_{lc} := sum of all quotas that will block a last choice attempt

Maximum arc capacity: obvious

5.3 Specification Hypothetical Set

In the traffic design of the UT LINE (a family of electronic digital exchanges that covers the capacity network needs up to 100K subscribers/60K circuits, and the capabilities needed in the range from the local to the intertoll exchanges) a set of national specifications has been used according to CCITT II and CCITT XI recommandations. To check the optimal utilization of the family, the hypothetical set illustrated in tables 1 and 2, has also been used and its optimality power has been evaluated.

CASE		A/A							
		LOCAL				REGIONAL			
		OF	XF	IF	RF	OF	XF	IF	RF
A	B0+CN0+CL0	0	0	0	0	0	0	0	0
B1	B1+CN0+CL0	4	4	4	4	4	4	4	4
B2	B2+CN0+CK0	ND	ND	ND	ND	ND	ND	ND	ND
B3	B3+CN0+CL0	25	4	4	4	10	8	4	4
B4	B4+CN0+CL0	4	4	20	20	4	8	10	10
B5	B5+CN0+CL0	25	6	20	20	10	15	10	10
C1	CN1+B5+CL0	25	6	25	25	NV	NV	NV	NV
C2	CN2+CL0	30	10	30	30	NV	NV	NV	NV
C3	CN3+a B1+CL0	NV	NV	NV	NV	6	6	6	6
C4	CN4+CL0	NV	NV	NV	NV	10	10	10	10

legenda
OF := OUTGOING FLOWS
XF := TRANSIT FLOWS
IF := INCOMING FLOWS
RF := INTERNAL FLOWS
ND := NOT DEFINED
NV := NOT VALID

TAB.1 REFERENCE CASES AND A/A REFERENCE VALUES

CASE	FLOW TYPE			
	OF	RF	IF	XF
A	1.5	2.	6.5	0.2
B1	3.	5.	1.5	3.
B3	12.	5.	1.5	5.
B4	3.	10.	3.	5.
B5	12.	10.	3.	5.
C1	25.	20.	20.	5.
C2	30.	30.	30.	5.
C3	4.	6.	2.	3.
C4	8.	8.	2.	5.

TAB.2 HYPOTHETICAL SPECIFICATION SET FOR BLOCKING REFERENCE VALUES

5.4 Data Collection Procedure

1 Given a HNDR network management technique without protection mechanisms

2 the initial GOS assignements

3 and an initial (B0,CN0,CL0) case matrix

4 find optimal network without repeated call attempts

5 simulate the networks with repeated call attempts and measure M1, M2, M3, M4, M5

6 reiterate 5 for different reference cases

7 reiterate from 5 to 6 with different network management techniques

8 reiterate from 4 to 7 with other initial matrices having the same total traffic

9 reiterate from 3 to 8 with different GOS assignments

10 reiterate from 3 to 9 with different initial matrices having a higher total traffic

5.5 Growing Complexity Network

Networks of growing complexity (fig. 3) are studied with and without decomposition to assess the decomposition methodology.

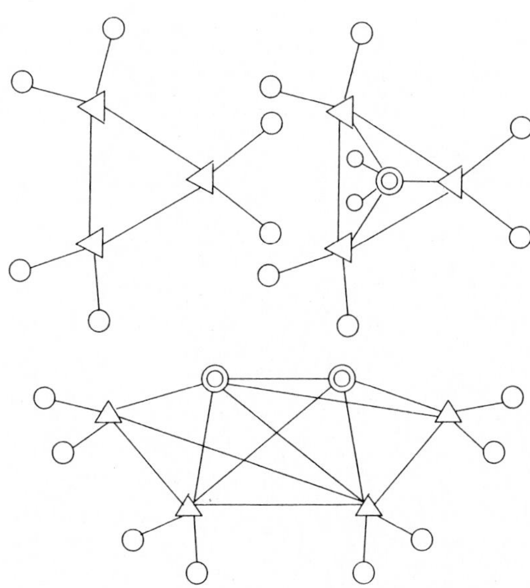

FIG.3 ANALYZED NETWORKS

6. RESULTS OF EXPERIMENTS

The measures are reorganized to obtain, for each reference case

- an Overall Gos/Sum Of Design.Vs.Mean Percent Of End Traffic chart (Harvey-Hill [11])(see fig. 4 in which the transition from dotted to broken lines is due to the repeated call attempt effect (RC), and the transition from broken to unbroken lines, to the GOS repartition between nodes and links (β)).

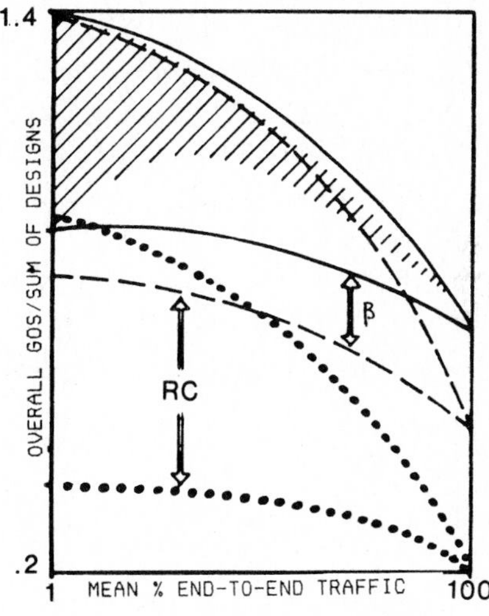

FIG.4 HARVEY-HILL CHART

- a GEC Sensitivity Map (see fig. 5), showing the sensitivity of minimum GEC (Δ%) .Vs. loss assignment, where loss assignment parameters are:

  1) ST := the end-to-end sum of design
  2) SL := the lower level network sum of design
  3) β := the mean ratio between link blocking and arc blocking for all node-to-node links of the lower level network (generalization of Gavassuti-Giacobbo charts: [16])

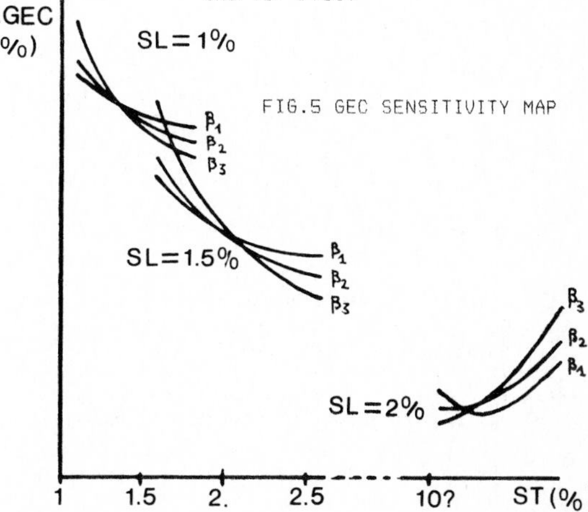

FIG.5 GEC SENSITIVITY MAP

REFERENCES

[1] M. Bonatti, "State of the art review on Gos-specification problems", Internal Memo ITALTEL/DCRS/PRS/0203/Feb.85 (contains also an ample bibliography).

[2] S.C. Littlechild, "Elements of Telecommunications Economics", Peter Peregrinns, London, 1979

[3] J. Freidenfelds, "Capacity Expansion: Analysis of Simple Models with Applications", North Holland, New York, 1981

[4] C. Jacobaeus, A. Elldin, "Telephone traffic theory. Present status and future trends", Ericsson Review no. 4, pp. 153-168, 1966

[5] I. Molnar, "An onlooker's reflections on recent grade-of-service deliberations by the International Telegraph and Telephone Consultative Committee (CCITT)", Proc. 7th Int. Teletraffic Congress, Stockholm, 1973

[6] T. Erke, K. Rahko, "Traffic measurements and the grade-of-service", Proc. 8th Int. Teletraffic Congress, Melbourne, 1976

[7] E.P.G. Wright, "Conclusions and recommandations of Working Party no. 5 of CCITT Study Group", Proc. 4th Int. Teletraffic Congress London, 1964

[8] R.W. Horn, "End-to-end connection probability. The next major engineering issue?", Proc. 9th Int. Teletraffic Congress, Torre Molinos, 1979

[9] K. Kodaiara, K. Ohara, "Traffic variation and grade-of-service specification in a telephone network", Rev. Electrical Communication Lab., Vol. 29, no. 3-4, pp. 134-144, March-April 1981

[10] S.S. Katz, "Improved traffic network administration process utilizing end-to-end service considerations", Proc. 9th Int. Teletraffic Congress, Torre Molinos, 1979

[11] C. Harvey, C.R. Hills, "Determining grades-of-service in a network", Proc. 9th Int. Teletraffic Congress, Torre Molinos, 1979

[12] C.Barbuio,G.Cappellini,"Decomposition methods in simulation analysis of complex telecommunication networks" Internal Memo ITALTEL/DCRS/PRS/0191/Jun.84

[13] K.S. Liu,"Direct-distance-Dialling Call Completion and customer retrial behaviour", Proc. 9th Int. Teletraffic Congress, Torre Molinos, 1979

[14] B. Wallstroem, "Congestion studies in telephone systems with overflow facilities", Ericsson Technics, Vol. 22, no. 3, 1967

[15] L.A. Gimpelson, "Network Management: Design and control of communications networks", Electrical Communication, Vol. 49, nO. 1, pp. 4-22, 1974

[16] M. Gavassuti, G. Giacobbo Scavo, "Distribution of the grade-of-loss in a long distance network", Proc. 9th Int. Teletraffic Congress, Torre Molinos, 1979

TELETRAFFIC ISSUES in an Advanced Information Society
ITC-11
Minoru Akiyama (Editor)
Elsevier Science Publishers B.V. (North-Holland)

TRAFFIC OFFERED GRADE OF SERVICE AND CALL COMPLETION RATIO IN A TOLL NETWORK VERSUS SUBSCRIBER RETRIAL BEHAVIOUR

G.GIACOBBO SCAVO, G.MIRANDA

SIP - Direzione Generale - Roma Italy

ABSTRACT

This paper focuses on the role of subscriber behaviour in the research of setting up suitable strategies in order to improve the network grade of service and revenue.

A study activity to evaluate the impact of repetead call attempts on the network has been carried out by the aid of an analytical and simulation model.

Moreover the input data for the simulation model and results of some case studies are explained in detail.

1. INTRODUCTION

The subscriber retrial behaviour has been extensively studied in the environment of traffic engineering. Indeed the aim of these surveys has been very often the determination of real offered traffic on the basis of carried one. The dimensioning standards and the well known network design method don't take into account of repeated attempts.

The goal of this paper is to deal with the impact of the subscriber behaviour on the network design, grade of service, call completion ratio and the revenue. As a consequence of the repeated attempts caused by the network congestion and mainly by the busy called party, offered traffic increases fictitiously affecting the call completion ratio and the revenue. Charge policy and some marketing oriented strategies could change the user behaviour, improving, at the same time, both the completion ratio and the revenue, avoiding to modify the network dimensioning and the relevant plant cost. As the user behaviour is depending both on the social, economic, human aspects peculiar of each country and on the network performance, the studies carried out on this subject cannot be easily standardized, it is suitable that each operating telephone company detects the behaviour of own users through appropriate observation campaigns and evaluates the corresponding results.

First of all in the paper an observation campaign conducted in order to investigate the subscriber behaviour is described.

The second step of the paper is relevant to a study conducted by the aid of a simulation method modelled in order to detect the parameters of the subscriber behaviour which have a sensible impact on the network grade of service.

The results of the simulation are compared with those obtained by classical analytical model in order to demonstrate the insufficiency of this last one and the necessity to define a new traffic model comprehensive of the subscriber retrial behaviour.

The third stage has just been the definition of this analytic model suitable to investigate the call congestion. The underlying theoretical basis for our approach was developed by Frederiks and Reisner.

The fourth stage, on the basis of:
- the observation campaign;
- the most important parameters pointed out by the simulation;
- the call congestion analytically determined,

evaluates the impact on plant costs, revenue and network grade of service of the marketing policies, whose aim is to modify user behaviour.

These evaluations are made undertaking a second simulation study comprehensive of both the network, as a whole, and the user behaviour (caller and called party) inserting one at time the fixed marketing policy.

2. MEASUREMENT RESULTS

All the operating telephone companies concentrate more and more a steady care in supervising the flow of telephone traffic by means of suitable measurement campaigns carried out in the SPC exchanges directly by the common control system and as relevant to the electromechanical ones by "ad hoc" additional measurement equipments.

The measurement data are gathered with the purpose of supporting the fundamental activities of traffic engineering, operation, network management, traffic administration.

The analysis of measurement methodologies and of the most suitable apparatus able to detect the traffic parameters is out of the goal of this paper.

Here it is only reported some data gathered by usual and special measurements relevant to subscriber behaviour, to call completion ratio and to the holding times quantitatively and concisely described by parameters to be used in a network simulation conducted with the goal to achieve suitable analyses of the grade of service, traffic offered and revenues versus subscriber behaviour. The measurement results are concerned with the toll traffic originated by some representative toll nodes and the incoming traffic terminating to a sample of 10 000 subscribers in a local exchange of Rome.

As relevant the first survey a total of 1 900 820 call attempts has been recorded whose disposition is:

a) call attempts failed for the behaviour of caller subscriber = 7%.
   This category encompasses:
   - abandon, i.e. caller hangs up before the end of dialing or does not wait the answer;
   - subscriber mishandling;
   - no such number;
   - customer dialed wrong number;
   - customer omitting an access code;

b) ineffective call attempts due to network problems including the exchange (BL) = 3%.
   This category encompasses:
   - failures such as: no ring, equipment irregularities;
   - network congestion:
   - exchange congestion;

c) call attempts failed due to the status of called subscriber = 36%.
   This category is splitted into:
   - called customer did not answer (NA) = 16%;
   - called customer busy (BY) = 20%;

d) successful call attempts (with answer)= 54%.

In the following instead of use the call completion ratio (equal to 0.54 in our case), it is used the inverse fraction i.e. the number of attempts per message (AM) (= 1.85 in our case).

In conclusion 90% of originated traffic reaches the called subscriber, so if we refer the percentage to the incoming traffic, BY becomes 22% and NA 18%.

As relevant to the holding times, the measurements have shown:
e) average conversation time = 190 s;
f) average holding time = 135 s;
g) average holding time for BY = 25 s;
h) average holding time for NA = 45 s.

As aforementioned, in order to investigate in detail the unsuccessful call attempts for user not in (NA), a special measurement has been undertaken sampling the traffic terminating to 10 000 customers. This survey has two basic aspects relevant to marketing strategy and grade of service (AM). In fact if an amount of subscribers were provided with simple equipments as phone answerers able to give the answer signal and to transmitt a short message, two goals would be achieved: first an improvement of the revenue, second a decrease of AM i.e. an improvement of call completion ratio.

The results achieved with this second measurement on the incoming traffic are in agreement with the previous one. In fact on a sample of about 1 milion of terminating call attempts on the 10 000 selected subscriber lines, 18% result ineffective because the user is not in.

The cumulative distribution of NA versus subscribers (the subscribers are directly sorted according to the relevant NA) is sketched in fig.1. Note that only 10% of subscribers cause 39% of unsuccessful call attempts, and 18% of subscribers 50% of unsuccessful call attempts.

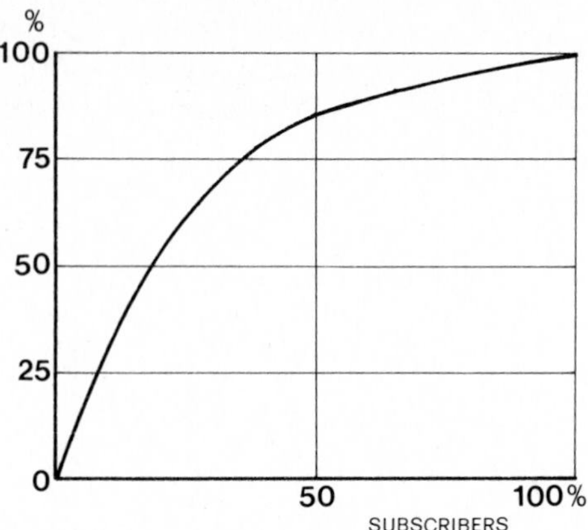

Fig.1-Cumulative distribution of NA-Calls versus subscribers.

3. SIMULATION - BASIC TRAFFIC PARAMETERS

The phenomenon of repeated call attempts represents still a very complicated problem and not yet completely understood. Various theoretical approaches and simulation studies have been carried out since the early fifties. An analytical approach without tempting a simplified outline of the telephone system with repeated attempts seems to be very difficult. The simulation study presented in this paragraph has the aim to offer a contribution in evaluating the impact of repeated attempts on the telephone system and in determining few but basic parameters able to define the traffic model. In this way it will be possible to set up a simplified analytical model avoiding the complexity of the state equations and also to define a further more sophisticated simulation model able to satisfy wider objectives like to evaluate different marketing policies versus ineffective call attempts due to called subscriber behaviour.

The hypotesis of Erlang traffic model, disregarding the phenomenon of repeated attempts, becomes always less and less able to keep up with the real traffic conditions as more sophisticated facilities which allow a fast and retrial access to network are provided to the users. Therefore this calls for setting up new analytical approaches and simulation models able to confirm the theoretical result.

3.1. Simulation model

The system considered in the simulation whose logical scheme is reported in fig.2, is represented by a full available trunk group to which the traffic generated by a great number of subscribers is offered. The connection of A-subscriber with B-subscriber requires only a free circuit of the trunk group. If all the circuits of the trunk group are busy the call attempt will result unsuccessful and further attempts may be generated.

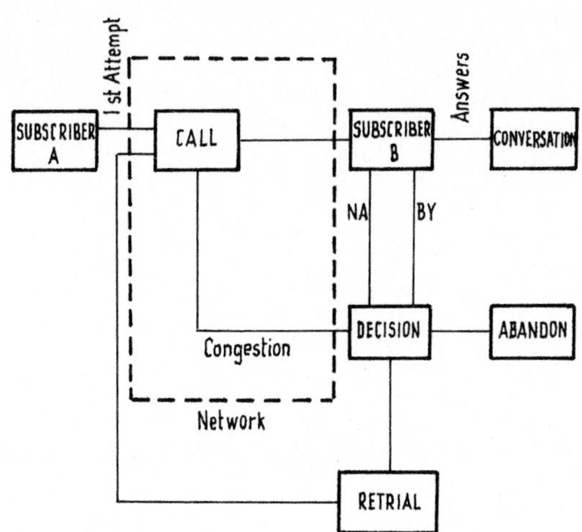

Fig.2 - First simulation scheme.

Let F the probability to find no idle circuit in the trunk group (network congestion), BY and NA the probability repectively to find the called subscriber busy or not in, the probability to have a successful call attempt will be: 1-F-BY-NA.

The blocked call attempts due to congestion and called subscriber status don't disappear completly but they re-present to the system according to a perseverance function. The grade of perseverance is defined [1] as

$$H(x) = A(x+1)/B(x)$$

where:

- A(x+1) is the number of attempts that produced an (x+1)-th attempt;
- B(x) is the number of 1-st attempts that produced an x-th unsuccessful attempt.

It has been derived qualitatively in the measurement campaign that the subscriber's perseverance depends strongly on the cause of failure but also increases systematically with the attempt rank for a given cause of failure. As we couldn't quantify the perseverance functions we have adopted those presented by Roberts [2] . Likewise for the functions for repetition times of the blocked calls, BY-calls and NA-calls, in the following we use, failing a measurement verification, those reported by Evers [3] .

3.2. Simulation results

The main results achieved by the simulation and that will be the basic inputs for the analytical models, are outlined as follows:

a) the call congestion is not substantially modified adopting for each cause of failure the average value of the perseverance function instead of the value relevant each attempt rank;

b) there are no meaningful differencies assuming the repetition times for the BY and NA repeated retrials are very long compared with the holding time. This means that the repeated call-attempts caused by the called subscriber status don't modify the traffic distribution that remain of Poisson type with an higher average value;

c) on the contrary the congestion probability is strongly affected by the distribution of the repetition time of the call-attempts blocked for network congestion.

These three basic conclusions are sketched in fig.3 where the call congestion versus traffic offered is reported for a trunk group of 10 circuits.

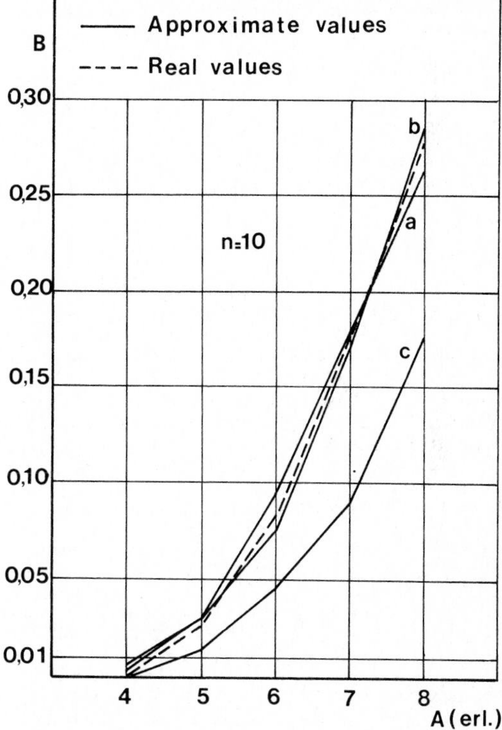

Fig.3-Call congestion versus offered traffic.

The dotted line represents the call congestion according to the real traffic condition. The curve a represents the call congestion assuming the average perseverance. The curve b represents the call congestion considering as "fresh" traffic the BY and NA repeated call attempts. The curve c represents the call congestion assuming for the call attempts blocked by the system the same hypotesis relevant to curve b.

Note that the curves a and b fit very well the real one while the curve c doesn't.

Therefore for the repeated calls it is necessary to assume a more complex but more real interarrival time distribution defined as follows:

$$R(t) = A[1-\exp(-t/a)] + (1-A)[1-\exp(-t/b)] \quad (1)$$

which represents the probability to have a repeated attempt after a time <t.

In fig.4 the function $G(t)=1-R(t)$ for the BL,BY and NA call attempts is reported.

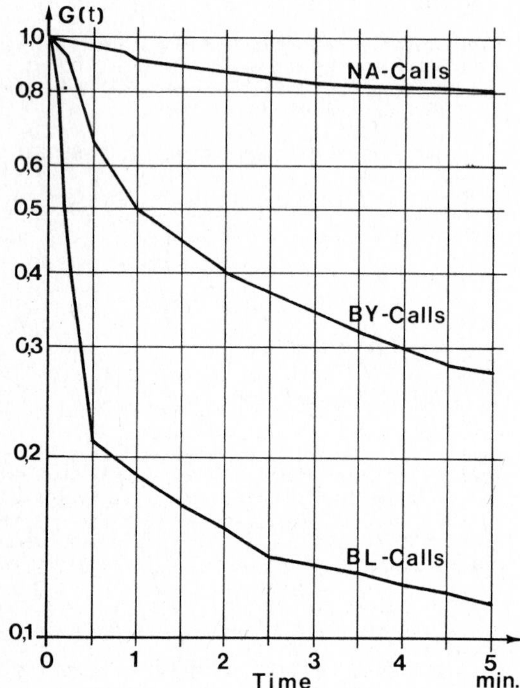

Fig.4-Arrival distribution for repeated attempts.

The parameters A, a and b appearing in the function (1) (this function has already been proposed by Evers [4] and Liu [5]) have the following meanings:

- A is proportional to the percentage of BL-repeated call attempts which represent themselves almost immediately;
- a represents the average value of the time repetition interval for the BL-repeated call attempts which represent themselves immediately;
- b represents the average value of the time repetition interval for the call attempts which don't represent themselves immediately.

As usual the a-value is about some seconds while b-value is about thousands of seconds.

In the following it will be shown the function (1), assuming some suitable hypoteses, depends only by two parameters: A and a.

4. ANALYTICAL MODEL

The results achieved by the simulation have allowed to define some basic simplified parameters which have been used to set up an analytical model. As regard the unsuccessful call attempts caused by the called subscriber status (BY and NA), the simulation has proved that it is possible to assume for them a very long inter-arrival time. This means hence that there is no difference between "fresh" calls and those BY and NA repeated call-attempts. Therefore, except the effects of system congestion, to the network a pure chance traffic indipendent from the trunks status is offered. Moreover the perseverance rate may be considered independent by the attempt rank. Finally, if F and H denote respectively the failure probability and the perseverance rate, the following value:

$$C = H_{BY} \cdot F_{BY} + H_{NA} \cdot F_{NA}$$

named <u>repetition constant</u>, represents the probability that a fresh call-attempt generates a repeated attempt caused by the called subscriber status. This C parameter is, as shown later, the sole parameter necessary to take into account of the BY and NA call attempts in the traffic model.

Let:

- l_i = the average birth rate (for fresh+repeated call attempts) when there are i circuits busy;
- l_o = the Poisson input rate;
- S_i^o = the average retrial intensity for BL-call attempts when there are i circuits busy;
- U_i = the average retrial intensity for BY and NA call attempts when there are i circuits busy.

There is the following relation:

$$l_i = l_o + S_i + U_i \qquad (2)$$

If l denotes the intensity of the total traffic offered to the system, we obtain:

$$l = \sum_{i=o}^{n} l_i \; P_i$$

where:

- n is the number of circuits of the trunk group;
- P_i is the probability that i circuits are busy at a time t.

Noting that it is feasible to assume as "fresh" the BY and NA repeated call attempts, as the simulation proved, we obtain:

$$U_i = l.(1-B).C$$

where B is the congestion of the trunk group evaluated with Erlang-B formula.

Note that U_i results indipendent from the number of busy trunks.

Introducing in the formula due to Fredericks and Reisner [6] the expression:

$$S_i = Bl_n H/P_i \int P_{ni}(t)dG(t)$$

where:

- P_{ni} is the probability that there are n busy circuits at the time t=o and i busy circuits at the time t;
- R(t) is the interarrival distribution function reported in (1);

and noting that the value of b is great enough to assume:

$$1/b \int P_{ni}(t)\exp(-t/b)dt = P_i$$

We have:

$$S_i = Bl_n H/P_i \{A/a\ L[P_{ni}(t/a)] + (1-A)P_i\}$$

where L denotes the Laplace transform.

Finally the equation (2) becomes:

$$l_i = l_o + l(1-B)C + S_i \qquad (3)$$

Using the Fredericks and Reisner iteration scheme it is possible from (3) calculate l_i and P_i and then the call congestion

$$B_c = l_n P_n / \sum_{i=0}^{n} l_i \cdot P_i \qquad (4)$$

The call congestion for the "fresh" call attempts derived from (4), is:

$$B_p = l_o P_n / \sum_{i=0}^{n} l_o \cdot P_i = P_n$$

Note that the call congestion for the fresh call-attempts is equal to the time-congestion.

For comparison in fig.5 the call congestion values versus traffic offered calculated by simulation, by the proposed formula (4) and by Erlang-B formula are reported.

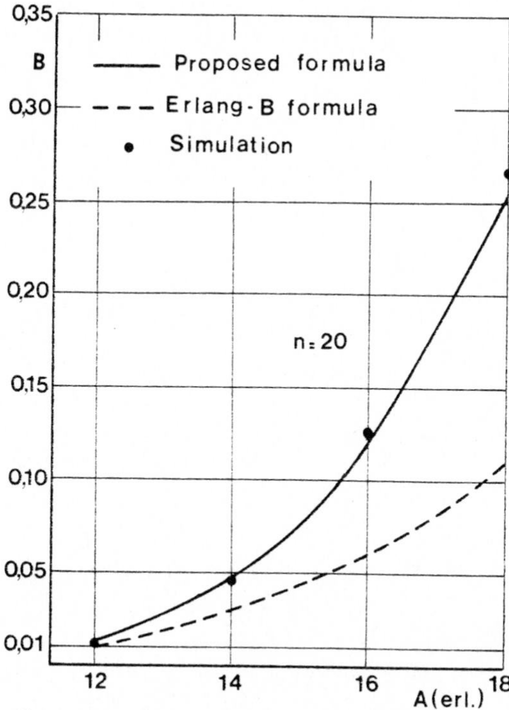

Fig.5-Call congestion versus offered traffic.

Note that the proposed analytical model well approximates the simulation results, while the Erlang-B formula yields values of call congestion too optimistic; this was easily foreseeable because the relevant traffic model doesn't take into account the repeated attempts.

The conclusions that can be pointed out are the following:

- the classical Erlang traffic model in case of not negligeable network congestion is inadeguate to represent the real traffic flow;

- in the study of repeated call attempts it is necessary to take into account the failure causes because the caller subscriber behaviour is strongly dependent by these ones. In fact the time interarrival distribution (see fig.4) is quite different according to the failure causes;

- the failures caused by the telephone system worse the network performance, because the retrials occurr immediately when the system is still in congestion.

5. NETWORK PERFORMANCE VERSUS CALLED SUBSCRIBER BEHAVIOUR

As usual the network is dimensioned in order to carry out the "fresh" traffic offered by the users, but really the network is also loaded by repeated call attempts generated by the caller user according to his perseverance rate. Even in normal traffic conditions the system presents a loss probability sufficient to provoke a not negligeable repeated attempt process. This process is, in its turn, more and more stressed by the called subscriber status.

The analytical approach doesn't allow to evaluate the overall grade of service of the network in real traffic conditions that is in presence of repeated call attempts, failures and overload phenomena. Therefore a second more sophisticated simulation method has been set up in order to take into account all the links between two users.

The Italian telephone network (serving about 15 million subscribers) is presently structured on five switching levels. The whole territory is subdivided into 231 District areas of univocal numbering and identified by different areas codes; the local area consists of 2 levels and the toll network consists of three hierarchical levels on 231 District Centers (CD), 21 Compartment Centers (CC) and 2 higher-level Transit Centers (TC). In the adopted simulation scheme (reported in fig.6) the connection of the user to the relevant toll center (CD) consists in one trunk group for the caller and an other one for the called user.

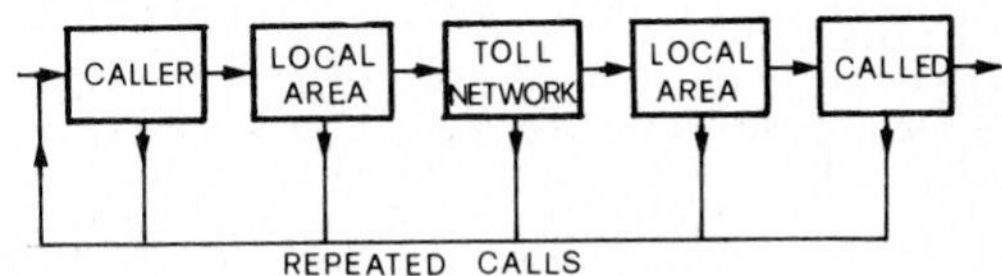

Fig.6 - Second simulation scheme.

This means to concentrate the loss probability for the outgoing and incoming toll traffic in one exchange and one trunk group. An other simplification has been to consider as a whole the toll network. The simulation has taken into account of:

- caller subscriber behaviour;

- failures;

- network congestion;

- called subscriber behaviour.

The traffic parameters relevant to average holding and conversation time, to the successful and unsuccessful call attempts have been fixed according to the values gathered by the measurement campaign. The perseverance function, as aforementioned, has been derived by Roberts and as far as the interarrival retrials the function (1) reported in the analytical model has been used.

5.1. Grade of Service Analysis.

First of all the relation between the parameter AM - number of attempts for message, i.e. the inverse of completion ratio-and the average point-to-point network loss probability has been investigated.

The results are sketched in fig.7, in which is reported AM versus the point-to-point loss probability (B). Note that really there is no correlation between B and AM; indeed in the B-range 1.5%÷2.5%, AM doesn't practically vary changing from 1.58 to 1.60.

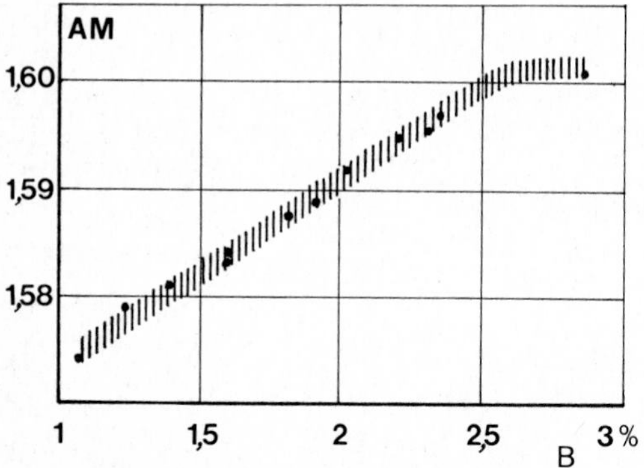

Fig.7 - AM-parameter versus point-to-point blocking probability (B).

On the other hand the network cost has a tight and sensible relation versus the point-to-point loss probability. In fact note that the results-obtained by using a network design optimization method [7] - sketched in fig.8 - show that the drop of the point-to-point loss probability B from 2.5% to 1.5% causes an additional network cost of about 5% without improving, as already seen, the call completion ratio. As far as this last point more interesting results can be achieved acting on the called subscriber behaviour; in fact this causes the 78% of the unseccessful attempts while the network congestion and failures represent only 7% (the remaining 15% is due to caller subscriber behaviour).

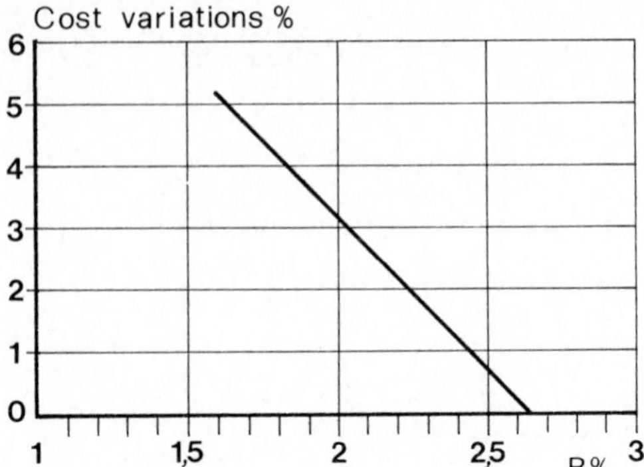

Fig.8-Network cost variations versus point-to-point blocking probability.

5.2. Called subscriber behaviour analysis

In order to take into account the called subscriber behaviour a variational analysis of offered traffic, grade of service, expressed by AM parameter, versus the BY and NA probability has been carried out. The approach used has been to maintain alternatively constant one of the two values (BY or NA) and to detect the influence of the other one on the inverse of call completion ratio and on the offered traffic, expressed in number of messages (MN) and conversation time (CT).

The modification of NA probability may be achieved installing to the subscriber lines simple phone-answerers able to transmitt a short message.

The effects of such strategy are outlined in fig.9 and 10 which respectively yield the variation of the inverse of completion ratio (AM) and traffic offered (CT, MN) versus the probability that the called user is not in (NA) for a fixed value of the probability that called subscriber is busy (BY=20%).

The following considerations can be drawn:

- AM-parameter is quite sensible versus NA probability. As an example a decrease of NA from 20% to 10% provokes an improvement of AM from 1.77 to 1.54;

- the number of completed calls (MN) improves of about 5.7%, while the average conversation time (CT) has a more limited variation of about 1.2%;

- the carried traffic doesn't register any real variation; in fact the holding time elapsing between the start of ringing tone and the abandon (in case of called not in) is substituted by the lenght of the phone answerer message.

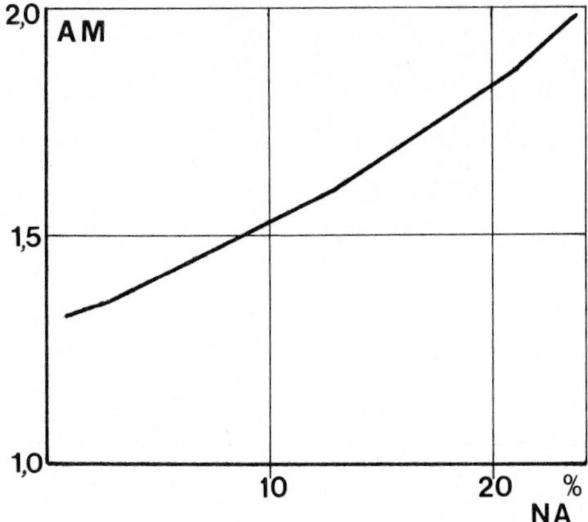

Fig.9-AM-parameter versus NA-probability.

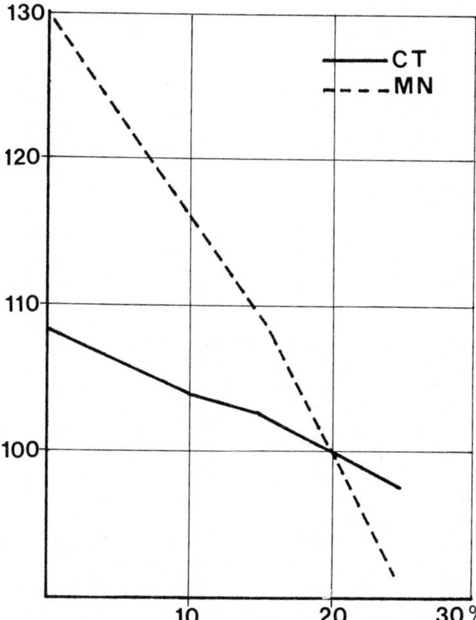

Fig.10-CT,MN parameters versus NA-probability

After all that a decrease of NA probability has a favourable impact because it provokes an improvement of call completion ratio and a relevant augmentation of the revenue without requiring a re-design of the network and hence further extra-costs.

In order to maximize the benefit/cost ratio, therefore, it is convenient a diffusion of authomatic phone answerers according to the results achieved by the measurements and outlined in fig.1. The number of phone answerers versus the average value of NA probability can be easily derived from fig.1. As an example to take back the NA probability from the usual value of 16% to 10% it is necessary to provide with phone answerers the 17% of the users. Such users have to be chosen not at random but among those who more contribute to the phenomenon.

An other analysis with the aid of the same simulation method has been carried out to evaluate the influence of BY probability. The improvement of AM and the increase of CT versus BY (maintaining constant NA=16%) are roughly equal to those just described and relevant to NA variations. On the contrary the variation of MN is smaller. Also in this case the carried traffic and hence the design traffic remain practically constant.

In order to modify BY, the subscriber lines with high busy rate can be augmented. Studies [8] have proved that adding only 3% of the lines, the busy rate decrease of about 50%.

In conclusion for a telephone operating company it is more suitable to set up strategies relevant to subscriber behaviour than to adopt more stringent design criteria in order to improve the call completion ratio and to alleviate the effect of repeated call attempts.

In this manner moreover it is possible to limit the calls which cannot be delayed and hence are definitively lost. An estimate of this variable is very difficult, Liu [5] evaluates in about 10% this amount.

Technological advances and competition for new revenue opportunities in the telecommunications industries have stimulated an increasing rate of introduction of new network services. The knowledges of behaviour subscriber will be more and more useful especially regarding the economical evaluation of the diffusion of such services.

REFERENCES

1. R.Kerebel, "Results of Observations of Unsuccessful Telephone Call Attempts in the Network of Paris" - Commutation et Electronique 26 (1969), pp.95-112.
2. J.W.Roberts, "Recent Observations of Subscriber Behaviour", 9th International Teletraffic Congress, Torremolinos, 1979.
3. R.Evers, "A Survey of Subscriber Behaviour Including Repeated Call Attempts - Result of Measurements in Two PABX's", 6th Human Factors Simulation Symposium, Stockholm, 1972.
4. R.Evers, "Measurement of Subscriber Reaction to Unsuccessful Call Attempts and the Influence of Reasons of Failure", 7th International Teletraffic Congress, Stockholm, 1973.
5. K.S.Liu, "Direct-Distance-Dialing Call Completion and Customer Retrial Behaviour", 9th International Teletraffic Congress, Torremolinos, 1979.
6. A.A.Fredericks and G.A.Reisner, "Approximations to Stochastic Service Systems, with an Application to a Retrial Model", Bell System Technical Journal, March 1979.
7. M.Gavassuti, G.Giacobbo Scavo, U.Trimarco, "A Computer Procedure for Optimal Dimensioning of Evolving Toll Networks under Constraints" 10th International Teletraffic Congress, Montreal 1983.
8. J.Biot and J.Massant, "On the Observation and augmentation of Subscriber Lines with High Probability of Being Busy", 9th International Teletraffic Congress, Torremolinos, 1979.

TELETRAFFIC ISSUES in an Advanced Information Society
ITC-11
Minoru Akiyama (Editor)
Elsevier Science Publishers B.V. (North-Holland)

EQUIVALENT PATH APPROACH FOR CIRCUIT SWITCHED NETWORKS ANALYSIS

Zbigniew DZIONG

Institute of Telecommunications, Technical University of Warsaw
Warsaw, Poland

ABSTRACT

The paper presents a new method for determining node to node grade of service for circuit switched networks with alternative routing. It is based on decomposition of the overall network analysis into a set of "path analysis problems" solved by means of the Equivalent Path Method. This approach takes into account correlation between occupancies of links carrying the same traffic stream. The method is valid for both hierarchical and non-hierarchical networks with conditional or step by step selection.

1. INTRODUCTION

The main problem of the performance analysis of a circuit switched network is the determining of node to node grade of service (NNGOS) for each origin-destination (O-D) pair. This problem was treated by several authors [1] - [10]. Usual analysis techniques decompose the overall network analysis into a set of link analysis problems. As a consequence, dependence of states of subsequent links carrying common traffic stream is neglected. It was shown (for the case of a single path) [11]-[13] that this approach results in considerable errors in GOS evaluation, i.e that the above dependence is an important factor. This observation motivated the development of the network analysis method proposed in the paper. It employs decomposition of the overall analysis into a set of "path analysis problems". Each of these subproblems is solved by means of the Equivalent Path Method (EPM) the idea of which was introduced in the authors previous paper [13]. The EPM treats all paths links and streams totally and simultaneously thus the correlation between links carrying common traffic streams on a path is taken into account.

The traffic model of the presented method is based on the following widely used assumptions:

- streams offered to origin-destination node pairs are Poissonian,
- traffic streams are well described by their first two moments, namely the mean (M) and variance (V),
- call holding times satisfy a negative exponential distribution,
- the network is in statistical equilibrium,
- call set up times are negligible,
- blocked calls are cleared and do not return,
- no congestion is encountered in the nodes.

The following Section presents a modified version of EPM. The modification lies in a new analytical model based on Bernoulli-Poisson-Pascal distribution; the main idea of the EPM is preserved. Section 3 considers the utilisation of EPM to determine NNGOS values for whole networks. In Section 4 the EPM-based approach is compared (numerically) with simulation and some other methods.

2. EQUIVALENT PATH METHOD (EPM)

2.1. The idea

We recall in short the basic idea of EPM [13]. Consider a path between origin (O) and destination (D) nodes which is offered a traffic stream (further refered to as O-D traffic stream) with parameters $d=(M,V)$. An example of a path traffic structure is depicted in Fig.1; N_i denotes i-th link dimension and $\hat{d}_j=(\hat{M}_j, \hat{V}_j)$ denotes parameters of j-th background traffic stream offered to a part of the path .

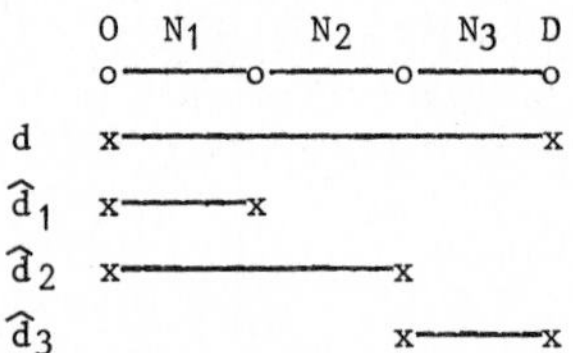

Fig. 1 Example of a path traffic structure.

To simplify notation let us introduce

$\underline{N}=(N_1,\ldots,N_i,\ldots,N_l)$,

$\underline{\hat{d}}=(\hat{d}_1,\ldots,\hat{d}_j,\ldots,\hat{d}_u)$

where: l - number of links, u - number of background streams.

The PATH-problem is formulated as follows:

Given $d,\underline{\hat{d}},\underline{N}$,
find parameters of traffic carried on and rejected from the path; $\overline{d}=(\overline{M},\overline{V})$, $\tilde{d}=(\tilde{M},\tilde{V})$ respectively.

Let X and X_i denote the number of O-D connections and the number of background connections carried on i-th link at any time, respectively. Now consider the path's state as "seen" by an arbitrary arrival from the O-D stream. The situation for the example from Fig.1 is depicted in Fig.2a. Notice that the arrival "sees" the path as if it was onelink (a "pipe") of dimension $N_{min}=MIN(N_1,N_2,N_3)$ with x trunks occupied by O-D traffic and x_t trunks occupied by background traffic. This "view through the pipe" is shown in Fig.2b.

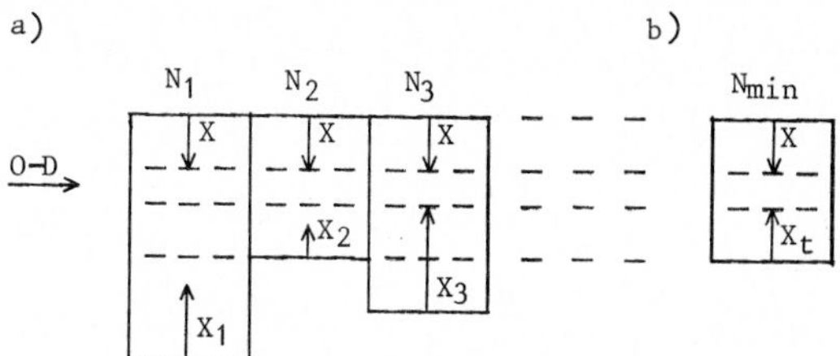

Fig.2 "View" through the path.

It is obvious that

$$x_t = MAX(x_1 - \Delta N_1, x_2 - \Delta N_2, x_3 - \Delta N_3) \tag{1}$$

where $\Delta N_i = N_i - N_{min}$

Previous remarks lead to the conclusion that instead of the entire path we may analyse an equivalent link of dimension $N_e = N_{min}$ which is offered the O-D stream and an equivalent background stream with parameters $d_e = (M_e, V_e)$. This equivalent path model is presented in Fig.3.

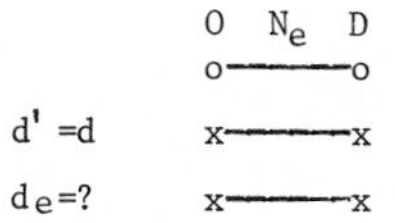

Fig.3 Equivalent path model.

Now the main problem is to determine d_e. Let X' and X_e denote the number of O-D connections and equivalent background connections at any time, respectively. To achieve exact equivalence the equivalent background stream should be such that

$$Pr(X=x) = Pr(X'=x') \tag{2}$$

for $x=x'$,

$$Pr(X_t=x_t) = Pr(X_e=x_e) \tag{3}$$

for $x_t=x_e$.

It is difficult to find exact solution satisfying this condition. To overcome this problem an approximation is used. It is based on the assumption that eqs (2), (3) approximately hold provided they are exact for $N_{min}=\infty$. This can be stated as follows

$$Pr(X_e=m) = Pr(MAX(X_1-\Delta N_1, \ldots, X_l-\Delta N_l)=m) \tag{4}$$

for $m = 1,2,\ldots$.

Further $Pr(X_e=x_e)$ is denoted $P_e(x_e)$. Notice that the distributions of X_i $(i=1,\ldots,l)$ and X_e for $N_{min}=\infty$ equal to the distributions of background traffic streams and equivalent background traffic stream , respectively (according to usual def. of offered traffic distribution). Thus d_e can be determined using eq.(4).

2.2. Analytical model

Consider the path provided O-D stream is neglected and all links are infinite $(N_{min}=\infty)$. Let Z_j denote the number of calls in service from the j-th background stream at any time. The path's state description may be of the two forms:

- vector $\underline{x}=(x_1,\ldots,x_i,\ldots,x_l)$,
- vector $\underline{z}=(z_1,\ldots,z_j,\ldots,z_u)$.

Denote steady-state probabilities as follows:

$$P_x(\underline{x}) = Pr(X_1=x_1,\ldots,X_i=x_i,\ldots,X_l=x_l),$$

$$P_z(\underline{z}) = Pr(Z_1=z_1,\ldots,Z_j=z_j,\ldots,Z_k=z_u).$$

The advantage of $\underline{Z}$ description lies in the fact that the Markov chain taking value from $\{\underline{z}\}$ may be modeled by a "Multidimensional Birth-Death Process" and thus has a product form solution [14]-[16]. Assume the birth rate for j-th background stream in state z_j $(\lambda_j(z_j))$ is defined according to the Bernoulli-Poisson-Pascal distribution. Additionally, to simplify the notation, assume the mean service time is equal 1. Thus, following [16]-[17],

$$\lambda_j(z_j) = \begin{cases} \alpha_j + \beta_j z_j \ , & \text{if } \alpha_j+\beta_j z_j > 0 \\ 0 & , \text{otherwise} \end{cases}$$

where: $\alpha_j = \hat{M}_j^2/\hat{V}_j$, $\beta_j = (\hat{V}_j - \hat{M}_j)/\hat{V}_j$.

As a consequence

$$z_j P_z(\underline{z}) = \lambda_j(z_j-1) P_z(\underline{z}-\underline{s}_j) \tag{5}$$

where $\underline{s}_j = (s_1,\ldots,s_i,\ldots,s_u)$,

$$s_i = \begin{cases} 1 & , \text{ if } i=j \\ 0 & , \text{ otherwise} \end{cases} .$$

Since we are looking for $P_e(x_e)$ we need $P_x(\underline{x})$. The latter can be obtained in two ways. The first is based on a simple realtion

$$x_i = \sum_j \delta_{ij} z_j \tag{6}$$

where $$\delta_{ij} = \begin{cases} 1 \ , & \text{if j-th stream uses i-th link} \\ 0 \ , & \text{otherwise} \end{cases} ,$$

so

$$P_x(\underline{x}) = \sum_{\underline{z} \in R(\underline{x})} P_z(\underline{z}) \tag{7}$$

where $R(\underline{x}) = \{\underline{z} : \bigwedge_i x_i = \sum_j \delta_{ij} z_j\}$.

The second is based on a recurrence formula which may be derived from eq.(5)

$$x_i P_x(\underline{x}) = \sum_j \delta_{ij} \bar{\lambda}_j (\underline{x} - \underline{\Delta}_j) P_x(\underline{x} - \underline{\Delta}_j) \tag{8}$$

where: $\underline{\Delta}_j = (\delta_{1j},\ldots,\delta_{lj})$,

$$\bar{\lambda}_j(\underline{x}-\underline{\Delta}_j) = \sum_{\underline{z}\in R(\underline{x})} \lambda_j(z_j-1) \frac{P_z(\underline{z}-\underline{s}_j)}{\sum_{\underline{z}\in R(\underline{x})} P_z(\underline{z}-\underline{s}_j)}$$

Notice that $\bar{\lambda}_j(\underline{x})$ can be treated as a mean value of the j-th stream's birth rate on condition the system is in state $\underline{x}$.

Unfortunately both ways are rather complex (except for Poissonian case: $\bar{\lambda}_j(\underline{x}) = \lambda_j$ - eq.(8)). Moreover transition from $P_x(\underline{x})$ to $P_e(x_e)$ using eq.(4) still requires a lot of computational effort. To overcome these difficulties we will use the following approach.

First consider the case $\underline{z}=\underline{x}$ (no overlaps of background streams) for which a recurrent relation (11) may be derived. To see that we define

$$\underline{g}(\underline{x})=(g_1,\dots,g_i,\dots,g_l)$$

where

$$g_i=\begin{cases}1 & \text{, if } x_i-\Delta N_i=\mathrm{MAX}(x_1-\Delta N_1,\dots,x_l-\Delta N_l)\\ 0 & \text{, otherwise}\end{cases} \quad .$$

this vector takes values from the set of patterns

$$\underline{c}_k=(c_{1k},\dots,c_{ik},\dots,c_{lk}), \qquad k=1,2,\dots,2^l-1$$

where c_{ik} equals 0 or 1 and $k=\sum_{i=1}^{l} c_{ik}2^{i-1}$.

Next we define the probability $Q_k(m)$ that $X_e=m$ on condition $X_i=m+\Delta N_i$ for $c_{ik}=1$, $X_i<m+\Delta N_i$ for $c_{ik}=0$:

$$Q_k(m)=\Pr(X_e=m\mid \underline{g}(\underline{x})=\underline{c}_k). \tag{9}$$

It is obvious that

$$P_e(m)=\sum_{k=1}^{2^l-1} Q_k(m). \tag{10}$$

Finally using eq .(5) the following recurrent relation is obtained:

$$Q_k(m)=\sum_{r=1}^{2^l-1} \varrho_{rk}Q_r(m-1)\prod_j\left[\frac{\lambda_j(m-1+\Delta N_j)}{m+\Delta N_j}\right]^{c_{jk}} \tag{11}$$

where $\varrho_{rk}=\begin{cases}1 & \text{, if } \underline{c}_r \boxtimes \underline{c}_k=\underline{c}_k\\ 0 & \text{, otherwise}\end{cases}$;

$\boxtimes$ - logical multiplication.

The initial values $Q_k(0)$ are given by

$$Q_k(0)=P_x(\underline{0})\left[\prod_j\left(\prod_{i=1}^{\Delta N_j}\frac{\lambda_j(i-1)}{i}\right)^{c_{jk}}\right]\times$$

$$\times\left[\prod_j\left(\sum_{s=1}^{\Delta N_j-1}\prod_{i=1}^{s}\frac{\lambda_j(i-1)}{i}\right)^{1-c_{jk}}\right] \tag{12}$$

where $P_x(\underline{0})=\Pr(X_1=0,\dots,X_l=0)$.

Now it is easy to find $P_e(X_e)$ from eqs (9),(10),(11),(12). Then we can find d_e by definition.

The presented procedure can be stated using function H:

$$d_e=H(\underline{\hat{d}},\underline{\Delta N})$$

where $\underline{\Delta N}=(\Delta N_1,\dots,\Delta N_l)$.

Since there is a trend to use at most two links in one alternate path, function H determines d_e for most cases of practical interest.

If overlaps of background streams exist d_e can be determined by recurrent use of function H. First consider "nice overlapping" of background streams. It means that if there is overlapping of i-th and j-th streams then one "covers" the other totally. This can be stated as follows:

$$\underline{\Delta}_j \boxtimes \underline{\Delta}_i \text{ equals } \underline{\Delta}_j \text{ or } \underline{\Delta}_i .$$

```
      O  N1      N2      N3  D
      o-------o-------o-------o
d̂1    x---------------x
d̂2    x-------x
d̂3            x-------x
d̂4                    x-------x
```

Fig.4 "Nice overlapping" of background streams.

For the example from Fig.4 the recurrent use of function H may be applied in the following manner:

$$d'_e=H((\hat{d}_2,\hat{d}_3),(\Delta N'_1,\Delta N'_2)),$$

$$d''_e=d'_e+\hat{d}_1=(M'_e+\hat{M}_1,V'_e+\hat{V}_1) \quad ,$$

$$d_e=H((d''_e,\hat{d}_4),(\Delta N''_e,\Delta N_3)) \quad ,$$

where $\Delta N'_i=N_i-\mathrm{MIN}(N_1,N_2)$; $i=1,2$

$\Delta N''_e=\mathrm{MIN}(N_1,N_2)-N_{min}$.

Now consider "nasty overlapping" of background streams cf. Fig.5a.

a)

```
      O  N1      N2      N3  D
      o-------o-------o-------o
d̂1    x---------------x
d̂2            x---------------x
```

b)

```
      O  N1      N2      N3  D
      o-------o-------o-------o
d̂'1   x-------x
d̂"1           x-------x
d̂2            x---------------x
```

Fig.5 "Nasty overlapping" of background streams.

In this case $\underline{\Delta}_j \boxtimes \underline{\Delta}_i$ does not equal $\underline{\Delta}_j$ or $\underline{\Delta}_i$ so we cannot use the procedure used for "nice overlapping". To overcome this problem we apply "cutting" of background streams. For the example from Fig. 5a "cutting" results in "nice overlapping" of background streams - see Fig.5b - and thus we may use function H recurrently just as before. One might expect that the error introduced by "cutting" can be minimized by taking values $\hat{d}'_i$ and $\hat{d}''_i$ which satisfy the following equations:

$$\hat{d}_i=H(\hat{d}'_i,\hat{d}''_i)$$

$$\hat{d}'_i=\hat{d}''_i$$

2.3. Equivalent path function

Provided parameters (d_e) of equivalent background traffic stream are determined the equivalent path model from Fig.3 can be analysed by means of several existing methods. It was verified that the most suitable for our purposes are the following:

- Hayward's approximation [18] determining parameters of total carried stream $(\overline{d}+\overline{d}_e)$ and total overflowing stream $(\tilde{d}+\tilde{d}_e)$. The advantage of this approach lies in that it treats uniformly cases with rough and smooth offerd trafic.
- Lindberg's approach [19] for determining individual stream blocking probability and then $\overline{M}$ and $\tilde{M}$ values. The advantage of this method lies in that it gives good results for wide range of blocking probabilities.
- The formulae presented in [20] for determining individual stream variances $\overline{V},\tilde{V}$:

$$\tilde{\theta}=\theta+p(\tilde{\theta}_t'-\theta) \tag{13}$$

$$\overline{\theta}=\theta+p(\overline{\theta}_t'-\theta) \tag{14}$$

where

$p=M/(M+M_e)$: O-D stream's share,

$\theta=V/M$: O-D stream's peakedness factor,

$\overline{\theta}_t',\tilde{\theta}_t'$: peakedness factors of total carried and rejected traffics in modified equivalent path model.

The modification of the equivalent path model consists in substituting V_e (variance of equivalent background stream) by a new value V_e' :

$V_e'=M_e\theta$ (i.e. $\theta_e'=\theta$) ;

other parameters remain unchanged. The validation of this approximation was tested in [20]. Notice that eqs.(13),(14) become exact for the Poissonian case $(\theta=\theta_e=1)$. This was proved in [21] and [20] respectively.

Summarising the discussion presented in this section we can introduce an Equivalent Path Function (EPM-function) of the following form

$$(\overline{d},\tilde{d})=f_e(d,\underline{\hat{d}},PTS)$$

where PTS denotes path traffic structure:

$$PTS=(\underline{N},\underline{S}),\quad \underline{S}=(\underline{\Delta}_1,\dots,\underline{\Delta}_u).$$

Thus the PATH-problem is solved.

3. NETWORK ANALYSIS

The NETWORK-problem is formulated as follows:

Given:

- network structure,
- set of routing schemes for each O-D pair,
- vector of link dimensions $\underline{N}$,
- vector of parameters of traffic streams offered to O-D pairs

$$\underline{A}=(A_1,\dots,A_n,\dots,A_r),$$

find vector of NNGOS values

$$\underline{B}=(B_1,\dots,B_n,\dots,B_r).$$

The routing scheme for each O-D pair is described by a path-loss sequence (PLS) according to the concept introduced in [6]. PLS consists of alternate paths and loss paths (loss paths are situated between alternate paths if step by step selection is applied). The first free path from a sequence is used. If it is a loss path then the call is lost, otherwise (alternate path) a connection is established.

The idea of the proposed network analysis is to decompose the overall analysis into a set of path analysis problems. Then a set of network equations is derived by means of the EPM-function. This set of equations consists of the three following subsets.

Equations determining parameters $(\overline{d}_i^n)$ of traffic carried on alternate paths.

These are simply obtained by applying the EPM-function to each alternate path:

$$(\overline{d}_i^n,\tilde{d}_i^n)=f_e(d_i^n,\underline{\hat{d}}_i^n,PTS_i^n) \tag{15}$$

where n — O-D pair index,
i — alternate path index.

In general, parameters of background streams $(\hat{d}_i^n)$ and offered streams are not given.

Equations determining parameters $(\hat{d}_{ij}^n)$ of background traffic streams.

Notice that the j-th background stream from PTS_i^n is a part of a traffic stream offered to an other alternate path determined by, say, PTS_1^m. Thus to gauge $\hat{d}_{ij}^n$ we have to divide PTS_1^m into two parts. The first (PTS') is common with PTS_i^n and the second (PTS") is separate from PTS_i^n. An example of this operation is shown in Fig.6.

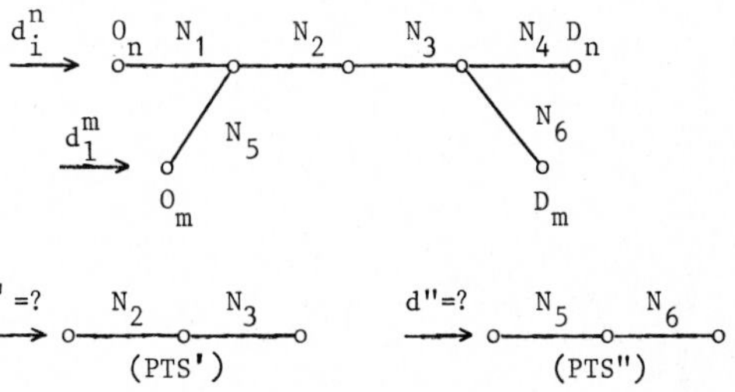

Fig.6 Path's division.

Then by determining parameters d' of traffic offered to PTS' we can find $\hat{d}_{ij}^n$ assuming

$$\hat{d}_{ij}^n=d' \tag{16}$$

It is reasonable to demand that parameters of traffic streams carried on the paths determined by PTS', PTS" and PTS_1^m are equal:

$$\overline{d}'=\overline{d}''=\overline{d}_1^m \tag{17}$$

Thus according to eq.(17) and applying the EPM-function to PTS' we have

$$(\overline{d}_1^m,\tilde{d}')=f_e(d',\underline{\hat{d}}',PTS') \tag{18}$$

To solve eq.(18) for d' we use

$$M'(k+1)=\frac{\overline{M}_1^m}{1-\tilde{M}'(k)/M'(k)},\quad (k)\text{-iteration index} \tag{19}$$

and aproximation

$$\theta'=\theta_1^m$$

which altogether constitute iterative procedure.

Equations determining parameters (d_i^n) of traffic offered to alternate path.

Traffic stream offered to the first alternate path (i=1) of each O-D routing scheme is equal to traffic offered to the O-D pair. Thus

$$d_1^n=(A_n,A_n) \tag{21}$$

If the considered alternate path is preceded by an other alternate path (conditional selection) then the traffic stream offered to it is equal to the traffic rejected from the preceding alternate path. Thus

$$d_i^n=\tilde{d}_{i-1}^n \tag{22}$$

where $\tilde{d}_{i-1}^n$ can be determined from equation of type (15). If there is at least one loss path between the considered one and its preceding alternate path (step by step selection) we have to divide the path traffic structure of the preceding alternate path (PTS_{i-1}^n) into two parts

- PTS'; rejected calls from PTS' are offered to the next alternate path,
- PTS"; rejected calls from PTS" are lost.

An example of a paths structure is depicted in Fig.7.

d_{i-1}^n → O_n — N_1 — N_2 — N_3 — D_n

N_1 — N_2 ; loss path

N_1 ; loss path

d_i^n → O_n — N_4 — N_5 — N_6 — D_n

$d'=?$ → N_1 (PTS') $d''=?$ → N_2 — N_3 (PTS")

Fig.7 Path's division (step by step selection).

Parameters of traffic offered to and rejected from PTS' and PTS" can be evaluated in the same manner as parameters of background streams (eqs (18),(19),(20)). However, note that all calls rejected when both PTS' and PTS" are congested are in fact rejected from PTS'. This is due to the fact that all calls offered to the PTS_{i-1}^n are first offered to PTS'. To take this into account we use the following approximation:

$$d_i^n=\tilde{d}'+(B'B''\tilde{M}'',\ B'B''\tilde{V}'') \tag{23}$$

where $B'=\tilde{M}'/M'$, $B''=\tilde{M}''/M''$

The derived set of network equations (15)-(23) can be stated as a function F:

$$(\underline{d},\underline{\bar{d}},\underline{\tilde{d}})=F(\underline{d},\underline{\bar{d}},\underline{\tilde{d}},\mathcal{PTS}) \tag{24}$$

where $\mathcal{PTS}$ - set of all PTS generated according to presented rules,

$\underline{d},\underline{\bar{d}},\underline{\tilde{d}}$ - vectors of parameters of traffics offered to, carried on and rejected from all PTS$\in\mathcal{PTS}$

This function may be easily solved using an iterative procedure. Finaly the NNGOS value for n-th O-D pair is given by

$$B_n=1-\frac{\sum_i \bar{M}_i^n}{A_n} \tag{25}$$

Thus the NETWORK-problem is solved.

4. MODEL VALIDATION

The presented model was compared with a simulation model. We have also compared the equivalent path approach with [10]- PIORO, [6]- LIN, [8]- MANF, [1]- KATZ, i.e. methods based on decemposition of the network analysis into a set of link analysis problems. The simulation results are given with a .05 confidence coefficient. Four network structures were tested.

Example 1 ; non-hierarchical, fully symmetrical, two-way trunks, 3 nodes, 3 O-D pairs (two-way). Each O-D pair has one direct and one alternate path consisting of two links. The parameters are as follows:

$N_k=20$, $k=1,2,3$,

$$A_n=\begin{cases}14, & \text{low blocking}\\ 17, & \text{overload}\end{cases} \quad , n=1,2,3.$$

The results are presented in Table 1.

Example 2 ; non-hierarchical, fully symetrical, two-way trunks, 5 nodes, 10 O-D pairs (two-way). Each O-D pair has one direct and three alternate paths consisting of two links. The parameters are as follows:

$N_k=20$, $k=1,2,\ldots,10$,

$$A_n=\begin{cases}14, & \text{low blocking}\\ 16, & \text{overload}\end{cases} \quad , n=1,2,\ldots,10.$$

The results are presented in Table 2.

Example 3 ; non-hierarchical, non-symetrical, one-way trunks, 5 nodes, 4 O-D pairs. This example was analysed in [8]. The structure of the network is depicted in Fig.8 where e_k denotes k-th link.

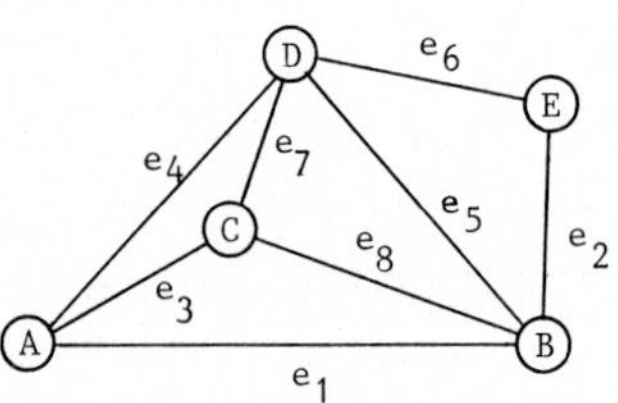

Fig.8 Network structure /Ex.3/.

Alternate path sequences for each O-D pair and parameters are as follows:

1. A-B : $e_1, e_3e_8, e_4e_5, e_4e_7e_8$; $A_1=20$
2. C-B : e_8 ; $A_2=8$
3. D-B : e_5, e_7e_8 ; $A_3=6$
4. E-B : $e_2, e_6e_5, e_6e_7e_8$; $A_4=16$

$N_1=N_2=N_7=12$, $N_3=N_5=8$, $N_4=N_6=10$, $N_8=24$

The results for two cases - conditional and step by step selection - are presented in Table 3.

Example 4 ; hierarchical, fully symetrical, two-way trunks, 4 local nodes, 1 transit node, 6 O-D pairs (two-way). Each O-D pair has one direct and one alternate path through the transit node.

Table 1 NNGOS values for Ex.1.

	low blocking	overload
SIMUL.	.0132 - .0176	.086 - 0.99
EPM	.0137	.093
PIORO	.0100	.116
LIN	.0054	.099

Table 2 NNGOS values for Ex.2.

	low blocking	overload
SIMUL.	.0025 - 0.0069	.073 - .108
EPM	.0030	.093
PIORO	.0022	.135
LIN	.0001	.123

Table 3 NNGOS values for Ex.3.

a) Conditional selection

O-D	A-B	C-B	D-B	E-B
SIMUL.	.043 - .049	.087 - .101	.068 - .078	.050 - .058
EPM	.041	.098	.105	.054
PIORO	.027	.105	.071	.039
LIN	.022	.094	.052	.025
MANF	.028	.080	.055	.046

b) Step by step selection

O-D	A-B	C-B	D-B	E-B
SIMUL.	.046 - .052	.081 - .093	.062 - .068	.048 - .056
EPM	.052	.091	.086	.049
PIORO	.042	.101	.064	.038
LIN	.035	.089	.047	.023
MANF	.037	.077	.051	.045

Table 4 NNGOS values for Ex.4.

	low blocking	overload
SIMUL.	.0090 - .0130	.056 - .067
EPM	.0113	.063
PIORO	.0121	.064
KATZ	.0125 - .0127	.065 - .070

The parameters are as follows:

$$N_k = \begin{cases} 17\,, & k=1,2,3,4 \;;\text{ high usage links} \\ 16\,, & k=5,\dots,10 \;;\text{ final links} \end{cases}$$

$$A_n = \begin{cases} 16\,, & \text{low blocking} \\ 19\,, & \text{overload} \end{cases} \qquad n=1,2,3.$$

The results are presented in Table 4.

The equivalent path approach was compared with simulation and PIORO-method for all examples. Moreover LIN-method(destined for non-hierachical networks) was used for exs 1,2,3 and KATZ-method (destined for hierarchical networks) was used for ex. 4. Additionally results given in [8] (MANF) are presented for ex.3.

The results show that the EPM approach gives NNGOS values very close to those obtained by simulation (in almost all cases they lie in the confidence interval). It is also important that in most cases the EPM approach gives better results than methods based on decomposition of the overall analysis into set of link analysis problems. This means that correlation between path links carrying common traffic streams is an important factor in evaluation of the NNGOS.

5. CONCLUSIONS

A new method for circuit switched networks analysis has been proposed. It is based on decomposition of the overall analysis into a set of path analysis problems. Each path problem is solved by means of the Equivalent Path Method (EPM). In this paper a new analytical model for EPM basing on BPP distribution has been presented. Comparison with simulation has indicated that the proposed approach gives good results for both hierarchical and non-hierarchical networks with conditional or step by step selection. Moreover comparisons with other methods have shown that in most cases the EPM approach gives better results. This is due to the fact that the presented method takes into account correlation between links carrying common traffic stream.

Note that the EPM features make it suitable for analysis of networks with dynamic routing. The author plans to present result on this subject soon.

REFERENCES

[1] S.S. Katz, "Statistical performance analysis of a switched communication network", Proc. 5th ITC, New York, 1967.

[2] M. Butto, G. Colombo, A. Tonietti, "On point to point losses in communications networks", Proc. 8th ITC, Melbourne, 1976.

[3] Ph.J. Deschamps, "Analytic approximation of blocking probabilities in circuit switched communication networks", IEEE Trans. on Com., 27, 603-606, 1979.

[4] K.S. Schneider, D. Minoli, "An algorithm for computing average loss probability in a circuit switched communication network", IEEE Trans. on Com., 28, 27-32, 1980.

[5] A. Kuczura, D. Bajaj, "A method of moments for the analysis of switched communication network's performance", IEEE Trans. on Com., 25, 185-193, 1977.

[6] P.M. Lin, B.J. Leon, C.R. Steward, "Analysis of circuit switched networks employing originating office control with spill-forward", IEEE Trans. on Com., 26, 754-765, 1978.

[7] D.R. Manfield, T. Downs, "On the one moment analysis of telephone traffic networks", IEEE Trans. on Com., 27, 1979.

[8] D.R. Manfield, T. Downs., "A method for the analysis of telephone traffic networks by decomposition", Proc. 9th ITC, Torremolinos, 1979.

[9] F. Le Gall, J. Bernussou, "An analytical formulation for grade of service determination in telephone networks", IEEE

Trans. on Com., 31, 420-424, 1983.
[10] M.P.Pioro, "A uniform approach to the analysis and optimisation of circuit switched communication networks", Proc. 10th ITC, Montreal, 1983.
[11] C.Harvey, C.R. Hills, "Determining grades of service in a network", Proc. 9th ITC, Torremolinos, 1979.
[12] D.R. Manfield, "Carried traffic in circuit switched networks", AEU, 360-368, september 1981.
[13] Z. Dziong, "A method for calculation of traffic carried on network paths", Proc. 10th ITC, Montreal, 1983.
[14] J.S. Kaufman, "Blocking in a shared resource environment", IEEE Trans. on Com., 29, 1981.
[15] J. Lubacz, W. Burakowski, "Multidimensional birth-death processes in the analysis of multi-user-class systems", Rozprawy Elektrotechniczne - PWN Warsaw, No3, 837-846, 1981.
[16] J.W. Roberts, "Teletraffic models for the Telecom 1 Integrated Services Network", Proc. 10th ITC, Montreal, 1983.
[17] L. Delbrouck, "A unified approximate evaluation of congestion functions for smooth and peaky traffics", IEEE Trans. on Com., 29, 85-91, 1981.
[18] A.A. Fredericks, "Congestion in blocking systems - a simple approximation technique", B.S.T.J., vol. 59, No 6, 1980.
[19] K.Lindberg,"Simple approximation of overflow system quantities for additional demands in the optimization", Proc. 10th ITC, Montreal, 1983.
[20] Z. Dziong, Phd Thesis, Technical University of Warsaw, 1980.
[21] A. Descloux, "On the components of overflow traffic", Unpublished work.

TELETRAFFIC ISSUES in an Advanced Information Society
ITC-11
Minoru Akiyama (Editor)
Elsevier Science Publishers B.V. (North-Holland)

A DECOMPOSITION METHOD FOR DNHR NETWORK DESIGN BASED ON DYNAMIC PROGRAMMING

Steven L. Dodd

AT&T Bell Laboratories
Holmdel, New Jersey, U.S.A.

Abstract: In this paper we discuss a decompostion method, which is based on dynamic programming, for determining the numbers of trunks and time-varying traffic routings in the AT&T Communications Dynamic Nonhierarchical Routing (DNHR) network. This new heuristic method, the Incremental Sequential Allocation Procedure (ISAP), produced network designs that compare favorably in some instances with those produced by an iterative DNHR network design method called the Unified Algorithm (UA). More work on the ISAP is warranted, and we conclude that the method embodied in the ISAP is worth further investigation for DNHR network design.

First we present an introduction and a formal statement of the DNHR design problem. Next we discuss the ISAP in detail, giving particular attention to a description of the dynamic program, and then compare ISAP-designed networks with those produced by the UA. Finally, we present our conclusions and some suggestions for further research.

1.0 INTRODUCTION

Since July, 1984 the AT&T Communications (AT&T-C) domestic network that connects Local Access and Transport Areas (interLATA network) has had 16 switches using time-sensitive traffic routing rules. For a given grade of service, Dynamic Nonhierarchical Routing (DNHR) utilizes network facilities more efficiently than conventional hierarchical routing. Since the estimated size of the 1990s AT&T-C interLATA network is more than 100 nodes, the projected savings from full implementation of DNHR are substantial [1].

Realizing the full monetary benefit from DNHR implementation depends, among other things, on the existence of engineering methods that produce a near-minimum-cost network design providing a specified grade of service for each point-to-point traffic item. (In the DNHR context the words "network design" mean the specification of the number of trunks and the time-varying traffic routings in the network.) At present the only available design method is the Unified Algorithm (UA) [2], which is a global, iterative heuristic method. The goal of this work was the development of a local design procedure; i.e. a procedure that provides a useful means of studying network changes under perturbations of a few point-to-point traffic loads or for studying network changes when the sizes of a few links in the network are changed. (These questions are some of the fundamental problems in DNHR network forecasting and servicing; e.g., see [3].) The result of our efforts is the Incremental Sequential Allocation Procedure (ISAP). In this paper we will describe the ISAP and compare its network designs with those produced by the UA.

The ISAP is a two-level decomposition heuristic method that is similar to techniques used in designing transportation networks [4]. The method divides the DNHR design problem into smaller optimization problems. Each small problem is solved at the lower level (Level 2) and a coordinating procedure (Level 1) uses the small-problem solutions to find a solution for the entire DNHR design problem (see, for example, [4],[5],[6], and [7].).

The purpose of the coordinating procedure, which is Level 1 of the ISAP optimization, is to determine which one of the set of the Level 2-proposed link sizes and traffic routings is the most economical. After producing a fixed ordering of all parcels, the ISAP will make M rounds through this ordered list, where M is an input parameter. When parcel i is examined on the kth round, $1 \le k \le M$, then, starting from a network that serves P_{k-1}% of parcel i's traffic, Level 2 of the ISAP uses a dynamic program [8] to generate a set of local minimum incremental cost network designs that serve with the appropriate grade of service in all design hours P_k%, $100 \ge P_k > P_{k-1}$, of parcel i's traffic. Level 1 chooses the Level 2 generated solution that minimizes total (global) network incremental cost incurred in order to serve in all design hours the traffic increment from parcel i.

Level 2 of ISAP is the level at which the local minimum incremental cost network designs are found. For a fixed parcel i and a fixed percentage (not depending on the value of i) of the load offered by that parcel, level 2 of the ISAP designs a network that will carry the incremental load on parcel i, in addition to all other loads it was previously designed to carry, with acceptable grades of service at each design hour. In the design process each design hour is considered in turn. (Therefore, for each parcel i examined during a round, the ISAP enters Level 2 H times, where H is the number of design hours.) A two state-variable three-stage dynamic program and supplemental sizing routine choose a route for the percentage of parcel i's traffic we wish to serve in that hour and insure that the network has sufficient capacity to carry the traffic with a particular grade of service. The sizes of those links offered parcel i's traffic may increase, decrease, or remain constant during this part of the procedure. If not all of the design hours for the parcel have been examined, the ISAP estimates the link blockings and link loads only on those links carrying traffic from the parcel. When all the design hours have been examined, the ISAP returns to Level 1.

After Level 1 chooses the Level 2 solution that minimizes total network incremental cost, the ISAP recalculates link loads and link blockings during all hours for all links in the network. Any blocking problems are corrected. Finally, if the network has been designed for the full load from all the parcels, the ISAP stops; otherwise, it chooses the next parcel in the list and continues.

The ISAP and UA differ in three important ways. First, the UA is an iterative method; the ISAP is a direct method and stops when all the point-to-point traffic items are allocated. Second, the UA separates the routing problem from the sizing problem; the ISAP determines the routing for a traffic item in a particular hour at the same time it allocates the capacity to carry that traffic. Third, inasmuch as the UA looks at all the traffic items in all hours at once, it is a global method; in contrast, the ISAP is local, examining traffic from a particular parcel in a particular hour.

2.0 THE DNHR DESIGN PROBLEM

This section contains a statement of the DNHR design problem. For completeness we now make the following definitions.

The words *switch* and *node* are equivalent and refer to the switches in the DNHR network. The trunk group between two nodes will be called a *link*. The *direct path* between the nodes A and B is the link joining A and B. A *two-link path* between the nodes A and B consists of the direct paths between the nodes A and C and the nodes C and B, where C is another node called the *via node*. (Paths with more than two links are not allowed in DNHR and, therefore, are not defined here.) An ordered set of paths is called a *route*. Note that these definitions imply that in a given path a switch can be used in only one of two possible ways: either traffic originates/terminates at the switch or the switch is a via node. In the discussion that follows, the direction of traffic flow between two nodes is not relevant; only the amount of traffic flowing between two nodes matters. If we call the traffic between the nodes A and B the *parcel* AB (this term is equivalent to *point-to-point item* AB), then we assume that only the parcel AB load is important. Finally, call the time of day during which the routing is fixed a *load set period*. The terms *hour*, *load set period*, and *design hour* will be used interchangeably.

2.1 The Objective Function

Our design goal is to minimize the total network cost over all possible routes for each parcel in every hour, i.e.,

$$\underset{\substack{\text{over the set of all}\\ \text{possible routings in}\\ \text{all load set periods}}}{\text{minimize}} \left\{ \sum_{\ell=1}^{L} C^{\ell} S^{\ell} \right\}, \qquad (2.1.1)$$

where L is the number of links, and C^{ℓ} and S^{ℓ} denote, respectively, the cost and size of the link ℓ.

2.2 Network Flow Equations

The network flow equations relate the load offered to each link and the parcel loads for every design hour. If link ℓ in path j is offered load from parcel i, the size of that offered load depends on the blocking on paths 1, 2,...,j - 1, and thus depends on the route for parcel i's traffic. Each path blocking depends, in turn, on the blockings on the link(s) comprising the path. We need the following definitions in order to express these relationships mathematically. Assume that $A_i(h)$ is the ith parcel load in hour h, $a^{\ell}(h)$ is the actual load offered to link ℓ in hour h, $\underset{\sim}{a}(h)$ is the row vector $[a^1(h), ..., a^L(h)]$, $e_{ij}(h)$ is the blocking of ith parcel on path j in hour h, $R_{ij}(h)$ is the amount of parcel i's traffic offered to path j in hour h, $b_{ij}^{\ell_k}(h)$ is the blocking of ith parcel on link ℓ_k, $k=1,2$, of path j in hour h (ℓ_2 does not exist for the direct path).

If we denote by $I_i(h)$ the number of paths in parcel i's route in hour h, then for h = 1,...,H, i = 1,...,R, the equations

$$R_{i1}(h) = A_i(h)$$

$$\text{and} \qquad (2.2.1)$$

$$R_{ij}(h) = R_{ij-1}(h)e_{ij-1}(h), \qquad j=2,...,I_i(h),$$

state that for h=1,...,H, all of the load from parcel i is offered to the first path in its route and that the overflow from path j-1 is offered to path j. Implicit in (2.2.1) is the assumption that the path blockings are independent.

Assuming that the blockings of the parcel on each link also are independent, we can write a simple expression for $e_{ij}(h)$ in terms of the blocking of the parcel on each link. We have

$$e_{ij}(h) = b_{ij}^{\ell_1}(h) + b_{ij}^{\ell_2}(h) - b_{ij}^{\ell_1}(h)b_{ij}^{\ell_2}(h) \quad \text{for j a 2-link path,} \qquad (2.2.2)$$

$$= b_{ij}^{\ell_1}(h) \quad \text{for j the direct path.}$$

To find the total load offered to a given link in hour h, we must sum the loads offered by each parcel to that link in hour h. Let $\rho(h)$ be the $R \times L$ matrix whose entries are the amount of traffic offered to link ℓ by parcel i in hour h. Note that the matrix $\rho(h)$ is well defined since any link can appear at most once in the route for parcel i. The load on any link ℓ in hour h is just the ℓth entry of the row vector $\underset{\sim}{a}(h)$ given by

$$\underset{\sim}{a}(h) = \underset{\sim}{1}\rho(h), \qquad (2.2.3)$$

where $\underset{\sim}{1}$ is the row vector of all 1s.

2.3 Sizing, Blocking, and Modularity Constraints

The DNHR network must be designed in order that the average blocking experienced by each parcel in a given hour is less than or equal to a prescribed maximum grade of service. Suppose that $B_i(h)$ is the grade of service for parcel i in hour h.

To demand that each parcel in each hour satisfies a particular grade of service means that for each parcel i and each hour h, the inequality

$$f_i(h) \le B_i(h), \qquad (2.3.1)$$

where $$f_i(h) = \prod_{j=1}^{I_i(h)} e_{ij}(h)$$

must be satisfied (c.f. 2.2.2).

Previously we defined $b_{ij}^{\ell_k}(h)$ as the blocking of the ith parcel on link ℓ_k, $k=1,2$, of path j in hour h and $b^{\ell}(h)$ as the average blocking on link ℓ in hour h. To insure that all the blockings we calculate represent probabilities, we also must include constraints which bound the $b^{\ell}(h)$ and $b_{ij}^{\ell_k}(h)$ between 0 and 1, inclusive. Therefore, for all $1 \le i \le R$, $1 \le j \le I_i(h)$, $1 \le h \le H$ and all links ℓ_k, $k = 1,2$ we require that

$$0 \le b_{ij}^{\ell_k}(h) \le 1,$$

$$\text{and for all links that} \quad 0 \le b^{\ell}(h) \le 1. \qquad (2.3.2)$$

The link size is the maximum over all hours h = 1,...,H, of a nonlinear function $\beta(a,z,b)$ that relates the hourly link load, $a^{\ell}(h)$, hourly traffic peakedness $z^{\ell}(h)$, and average hourly link blocking $b^{\ell}(h)$ to the link size S^{ℓ}:

$$S^{\ell} = \max_{h=1,...,H} \beta(a^{\ell}(h), b^{\ell}(h), z^{\ell}(h)), \quad h=1,...,H. \qquad (2.3.3)$$

In practice $\beta(.,.,.)$ is often not explicitly stated. The "hourly trunk quantities," of which the link size S^{ℓ} is the maximum, can be calculated by an application of the Equivalent Random Method and the inverse Erlang -B formula.

Trunks can be installed only in multiples of the module size m^{ℓ}. Therefore, we must also include the restriction that

$$S^{\ell} = 0, m^{\ell}, 2m^{\ell}, ... \qquad (2.3.4)$$

The constraint set is incomplete without equations that define the relationship between the average hourly link blocking $b^{\ell}(h)$ on link ℓ and the blockings experienced by all of the parcels that offer traffic to link ℓ (i.e., those blocking $b_{ij}^{\ell_k}(h)$ with $\ell_k=\ell$ and such that ℓ_k is in path j of parcel i's route in hour h). If a closed form for these equations exists, then it appears to involve all of the moments of the distributions of the call arrival processes to link ℓ as well as the moments of the distribution of all holding times. Developing the notation to carry these ideas requires too much space and sheds no light on the problem.

3.0 THE INCREMENTAL SEQUENTIAL ALLOCATION PROCEDURE

Figure 1 is a block diagram of the ISAP. In the following subsections we shall explain the function of each of the boxes in Figure 1 in more detail; the reasons underlying the routing and sizing block require the most attention since it is the heart of Level 2 and the most important part of the ISAP. In order to keep the discussion as simple as possible, we shall explain the

Level 2 computations (see section 3.5) before we explain the Level 1 computations (see section 3.7).

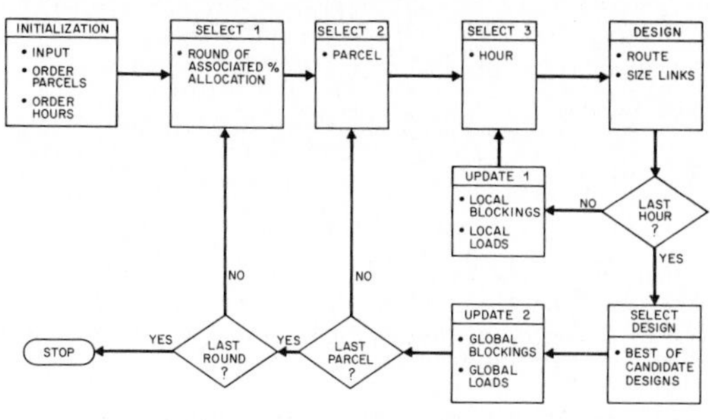

FIGURE 1 ISAP BLOCK DIAGRAM

3.1 *Initialization*

Initializing the computations proceeds as follows: set all link sizes to 0 and set both link blockings and the parcel blockings to 1. Other inputs are the cost per trunk for each link, the number of design hours, the module size for each link, the point-to-point loads in each hour, the point-to-point grade of service for each hour, switch locations and capacities, and the candidate paths for each point-to-point item. Moreover, the algorithm requires information about the traffic peakedness, day-to-day variation, and path restrictions due to transmission constraints (echo control restrictions, etc.).

The preceding information must be supplied to any network design procedure since it conveys the design criteria. The ISAP also requires the values of other parameters. One of these parameters is the positive integer M, which tells the routine how many times it must examine each parcel. During each look at parcel i, $i = 1,...,R$ (R is the number of parcels), the ISAP will allocate capacity to carry $P_k(i)$ % of parcel i's traffic, where $0 < P_k(i) \leq 100$, $k = 1,...,M$, and $P_M(i) = 100$. A vector $\underset{\sim}{U}$ of ten probabilities also is needed by the procedure. The entries in $\underset{\sim}{U}$ will serve as upper bounds on path blockings in the design step; we will discuss the values for the probabilities and how they are used in more detail in Section 3.5.2.

The last steps in the initialization procedure are to decide on an order in which to examine the parcels and, once a parcel is given, to choose an order in which to examine the design hours. For a given parcel, we achieved our best designs by examining the design hours in order from largest to smallest on the basis of hourly load for that parcel.

3.2 *Select 1*

Though the order in which parcels are to be considered is decided on in the Initialization step, the ISAP must keep a count of the number of rounds it has made through the ordered list of parcels. The box in Figure 1 labeled "Select 1" represents this control.

3.3 *Select 2*

The ISAP must keep a record of the parcels it has examined in the current round, the current parcel under consideration, and the percentage of the current parcel's traffic that the next network design must serve. The box in Figure 1 labeled "Select 2" represents this control.

3.4 *Select 3*

For a given parcel the ISAP must keep a record of the design hours it has examined as well as the current design hour under consideration. The box in Figure 1 labeled "Select 3" represents this control.

3.5 *Design*

Remember that we have said that the ISAP orders the parcels and then looks at each parcel with respect to that ordering. Each time a parcel is examined, we choose a route in each hour and network capacity to carry a larger percentage of that parcel's traffic than the network carried before. Given a round, a parcel, and an hour, the ISAP starts from the present routing and link sizes, which have been calculated to serve prescribed traffic with a particular grade of service, and determines routings and trunk augments/disconnects, if any, required to serve all of the traffic the network served before plus the traffic increment from the parcel in that hour.

The basis of the routing and sizing step in the ISAP is a two state-variable three-stage dynamic program and a supplemental sizing method. At the end of the final stage of the dynamic program, the ISAP has produced the first three paths in the route for the particular parcel and hour as well as the trunk augments/disconnects on the links in the paths. If the required grade of service (GOS) is met, the procedure examines a new hour and, possibly, a new parcel. If the GOS is not met, the ISAP uses the supplemental sizing routine to append more paths to the route until it is satisfied.

3.5.1 *The Dynamic Program*

Our procedure's most important component is a two state-variable three-stage dynamic program that is solved by backward recursion beginning at stage 3. At each stage n, n = 3,2,1, there are two decisions to make. We must decide on the nth path in the route at stage n, and we must choose the blocking on that path. Of course, these decisions in stage n depend on the decisions made in stage n+1, for if path j is the n+1th path in the route, then path j cannot be the nth path in the route. Similarly, particular path blockings at stage n+1 preclude the choice of other path blockings at stage n. Remember that the modularity conditions for trunk quantities imply that the possible choices for path blocking constitute a discrete space, not the interval [0,1]. For simplicity's sake, explicit reference to the design hour h under consideration is dropped.

Let the state of the system at stage n be given by the ordered pair (R_n^i, F_n^i) where R_n^i is the collection of the last 4-n, n = 3,2,1, paths in the route for parcel i and F_n^i gives the product of the path blockings on the last 4-n paths in this route. Also, suppose that P_i is the set of candidate paths for parcel i in hour h and that e_{ij} gives the blocking in hour h of parcel i on path j. Then the state transition functions that relate the state of the system in stage n with its state in stage n+1 are given by

$$R_n^i = d_n^1 o R_{n+1}^i , \quad n = 3,2,1, \tag{3.5.1}$$

$$F_n^i = d_n^2 \cdot F_{n+1}^i , \quad n = 3,2,1, \tag{3.5.2}$$

where $d_n^1 \in P_i$, $d_n^1 \notin R_{n+1}^i$ is the nth path in parcel i's route, $d_n^2 \in \underset{\sim}{U}$ is the blocking on this path, and the symbol "o" is the concatenation operator. The boundary conditions are defined by

$$R_4^i = \{ \}, \tag{3.5.3}$$

where { } denotes the empty set and

$$F_4^i = 1 . \tag{3.5.4}$$

Equation (3.5.1) says that the last 4-n paths in the route are obtained by picking the nth path in the route and appending the remaining paths. Equation (3.5.2) means that the blocking on the last 4-n paths in the route is obtained by multiplying the blocking chosen for the nth path, d_n^2, by the blocking on the remaining paths. (Note that the independence of path blockings is implicit here.)

The decisions d_n^1 and d_n^2 at each stage minimize incremental network cost given the state of the system. The minimum incremental network cost at stage n, $f_n^*(R_n^l,F_n^l)$, is the minimum of the sum of the one-stage incremental cost function $c(d_n^1,d_n^2)$ at stage n and the minimum incremental network cost at stage n+1, $f_{n+1}^*(R_{n+1}^l,F_{n+1}^l)$. Thus, we can write

$$f_n^*(R_n^l,F_n^l) = \min_{\substack{d_n^1 \in P_l/R_{n+1}^l \\ d_n^2 \in \underset{\sim}{U}}} \{c(d_n^1,d_n^2)+f_{n+1}^*(R_{n+1}^l,F_{n+1}^l)\}. \quad (3.5.5)$$

The minimum in the first variable is taken over the set $\{P_l/R_{n+1}^l\}$, which is nothing more than all the candidate paths except those already in the set R_{n+1}^l. The minimum in the second variable d_n^2 is taken over $\underset{\sim}{U}$, the discrete probability vector defined above, and the GOS requirement for parcel i and hour h are used in its computation.

In section 3.5.2 the solution to (3.5.5) will be illustrated by the example. We also will discuss how to evaluate the one-stage cost function.

Note that this formulation is completely general because it can be used, in principle at least, to find all $\bar{N}$ paths in a route. All of the previous development remains unchanged if references to three paths are replaced by references to $\bar{N}$ paths. However, since most of a parcel's traffic is offered to the first three paths, choosing and sizing more than three paths with the dynamic program uses too much computer time with little or no benefit. Section 3.5.3 explains how the other paths are chosen.

3.5.2 The Dynamic Program: An Explanation by Example

The following example of the two-state, three-stage dynamic program assumes that the network has 16 nodes, 10 design hours (load set periods), and that traffic from a parcel can be offered to the direct path or use any node other than the terminating node as a via node. Therefore, each parcel has fifteen paths in its list of candidate paths, but for simplicity we shall discuss only the first three paths in the list, (6,-), (2,33), and (3,45). The symbol "-" indicates that no link is present; thus, (6,-) is the direct link.

Suppose that the ISAP is in round 2 of its design, that our requirement is for the network to carry with the appropriate GOS 25% more traffic from every parcel at the end of round k than it carried at the end of round k-1, k=1,2,3,4, and that parcel 6 and hour 2 are the current parcel and hour under consideration. Therefore, at the beginning of the design step for parcel 6 and hour 2 we have a network that carries 25% of the parcel 6, hour 2 traffic plus traffic from other parcels and hours. At the end of this step the ISAP will have constructed a network serving with the prescribed GOS, 50% of the traffic from parcel 6 in hour 2 and the traffic it served before.

We specify a vector $\underset{\sim}{U}$ of ten probabilities. The elements of $\underset{\sim}{U}$ serve as upper bounds on path blockings in the dynamic program. Let $\underset{\sim}{U}$ = {.01,.05,.1,.2,.3,.4,.5,.6,.7,.8}. We will explain the use of this vector below.

To decide on the first path in parcel 6's route in hour 2, the ISAP computes a table like Table 1. Note that the entries in the top row of the table are the u_c, c = 1,...,10, and the left column lists the first three candidate paths. Assume that 45% of the parcel 6, hour 2 traffic is 5 erlangs. (The beginning of Section 3.5.3 explains why, at this point in the calculations, we offer 45% rather than 50% of the parcel load.) The number in the rth row and c^{th} column of the table is the minimum total number of trunks which must be added to the link(s) of path r to achieve a path blocking less than or equal to u_c, when the load offered to path r is 5 erlangs plus the load contributed in this hour by parcels other than parcel 6. (The values in these tables comprise the range of the one-stage cost function (see (3.5.5)) and, therefore, are used in the dynamic programming recursion.) If the path is a two-link path, note that the table entry gives the sum of the numbers of trunks added to both links since the optimization requires only this sum; a record of the number of trunks added to each link is kept for use in blocking calculations. We will assume that trunks cost 1 dollar, so that the entries in the table are the costs of augmenting/disconnecting trunks.

PATH \ u_c	.01	.05	.1	.2	.3	.4	.5	.6	.7	.8
(6,–)	6	5	5	4	3	3	2	2	0	-1
(2,33)	9	7	7	7	6	5	4	4	3	2
(3,45)	6	6	5	5	4	3	2	0	-1	-2

TABLE 1 FIRST - PATH MINIMUM MODULAR INCREMENTS TABLE

This table, which is called the first-path minimum modular increments table (MMIT), is constructed in a straightforward way with the Erlang-B function. In this example we ignore traffic peakedness and day-to-day variation. With the equivalent random method and appropriate inflation factors for overflow, the model could account for both of them. To calculate the entries for a two-link path, say (2,33), the ISAP chooses upper and lower bounds for the sizes of links 2 and 33, respectively, and within that domain it estimates a minimum cost configuration for the two-link path that gives the required path blocking u_c. If these proposed new link sizes result in blocking violations for other parcels or for this parcel in other hours, the links are augmented until the problems are solved. The modularity requirements are always enforced. (Undoubtedly the reader noticed that this method is suboptimal. The appropriate scheme is to calculate for each configuration of links 2 and 33 the augments required to correct any blocking problems on other traffic items. Though some method other than exhaustive search can be used here, checking more than one configuration appears to be too time consuming.) To calculate the entries for a one-link path, the ISAP determines the smallest modular size that gives a path blocking less than or equal to u_c, and then, if necessary, it determines modular increases in the link size in order to correct blocking problems for other parcels or for the current parcel in other hours.

Readers should note the negative entries in Table 1. They indicate that trunks can be removed from a link or links and still provide an acceptable path blocking for a given value u_c. Also observe that this table gives enough information to compute the minimum incremental cost one-path route for this parcel and this hour. Column 1 gives the minimum number of trunks that must be added to each path to guarantee that the blocking on that path is less than or equal to .01. We choose the increment with the smallest cost as our one-path route, which in this case, is the direct path (6, -).

Calculating the information needed to choose the second path is more complicated. First we must know the offered load to the second path, and of course that value depends on the first-path blocking. Once the offered load is known, the methods used in the first-path calculations are applicable. But we have a potential problem in determining the offered load.

Recall that the entries in the vector $\underset{\sim}{U}$ are upper bounds for the blocking on each path. Since the path blockings depend on both offered load and number of trunks, each entry in the first-path MMIT determines a path blocking, and it is conceivable that each of these path blockings is different from the other path blockings determined by a particular MMIT. Thus, for a 15×10 first-path MMIT, there can be 150 different path blockings and, consequently, 150 different second-path offered loads (second-path offered load = first-path offered load × first-path blocking). Producing the MMIT for each of these offered loads is computationally prohibitive. As a result we approximate these different second-path offered loads by assuming that the first-path blocking can be equal only to one of the elements of $\underset{\sim}{U}$. Once a first-path blocking in hour 2, $e_{i1}(2)$, is known for parcel i, we can compute an MMIT for the corresponding load. An example is given in Table 2.

FIRST PATH BLOCKING: $u_1 = e_{i1}(2) = .2$

PATH \ u_c	.01	.05	.1	.2	.3	.4	.5	.6	.7	.8
(6,–)	–	2	2	1	1	1	0	-1	-2	-3
(2,33)	–	3	3	2	2	1	0	0	-1	-2
(3,45)	–	1	1	1	0	0	-1	-2	-2	-3

TABLE 2 SECOND - PATH MINIMUM MODULAR INCREMENTS TABLE

Note that now we have enough information to compute a minimum cost two-path route satisfying the GOS, which is .01 in our discussion. For example, consider the two MMITs given in Tables 1 and 2. If $e_{i1}(2) = .2$ and $e_{i2}(2) = .05$, then any first path having blocking less than or equal to .2 and any second path having blocking less than or equal to .05 must, together, comprise a route which satisfies the GOS for this parcel in this hour. Suppose (2,33) is the second path in the route; the marginal cost of 3 trunks is given in Table 2 under the column headed ".05" opposite the row "(2,33)". By referring to Table 1, we see that the best choice for the first path, given $e_{i1}(2) = .2$, is (6,-) since the total cost with that option is $4 + 3 = 7$ (look under column .2 and row (6, -)), whereas choosing (3,45) costs $5 + 3 = 8$. If (3,45) is the second path, then choosing (6,-) for the first path is the best option since that incurs an incremental marginal cost of 5 versus the cost of 8 that results from the choice of (2,33) as the first path. If (6, -) is the second path, the best choice for the first path is (3,45), which costs 7. Clearly, if these were our only options, then {(6,–),(3,45)} provides the best possible two-path route satisfying the GOS for this parcel and this hour.

Computing the MMITs for the third path is nearly the same as for the second path. As before we must determine the offered load, which, in this case, is the load overflowing from the first two paths. Again, the elements of $\tilde{U}$ are the only permissible values for blockings on the first two paths; this restriction insures reasonable computation times.

The third stage MMITs, together with the one- and two-stage tables, are used to compute a minimum cost three-path route for the additional traffic. The details are similar to the above-described calculation for the two-path route satisfying the GOS. For each triple $(e_{i1}(2), e_{i2}(2), e_{i3}(2))$ we find the sequence of three paths which gives the minimum total incremental cost that must be incurred to provide the traffic with a grade of service less than or equal to the product $e_{i1}(2) \times e_{i2}(2) \times e_{i3}(2)$. Finding the minimum incremental cost three-path route satisfying the GOS reduces, then, to choosing the minimum incremental cost three-path route for which the preceding product is less than or equal to the GOS. This completes the description of the dynamic program.

At this point in the computations, we have available the MMITs from each stage of the dynamic program. From these tables we can identify minimum cost one-, two-, and three-path routes satisfying the GOS. Moreover, we can compute *any* optimal route of one, two, or, three paths providing a grade of service less than or equal to

$$\prod_{j=1}^{k} e_{ij}(2) \ , \qquad (3.5.2.1)$$

where $1 \leq k \leq 3$ is the number of paths in the route and the $e_{ij}(2)$, $j = 1,...,k$, are elements of $\tilde{U}$. The ability to calculate routes with a blocking given by (3.5.2.1) is important if we want to understand the cost effectiveness of offering traffic to routes with more than three paths. Section 3.5.3 discusses one way to order the remaining paths and to increment the network's capacity to serve with the requisite GOS the traffic overflowing from the first three paths.

3.5.3 *The Supplemental Routing/Sizing Routine*

As we said before, the ISAP calculates the routing and capacity for 45% hour 2 load in our example, yet our objective is to serve 50% of the parcel 6 hour 2 load at the completion of this step. There is a simple reason the procedure works this way. Intuitively we feel that the modularity requirements and our conservative sizing rules create spare capacity in the network. Therefore, the same numbers of trunks often will serve 50% of the traffic, with an acceptable GOS. Offering 50% of the load at first may create unnecessary augmentation. So the next thing that is done at this stage of the calculation is to check whether the route and capacity calculations produced by the dynamic program can serve the additional traffic.

When the route produced by the dynamic program gives a blocking on the first three paths greater than the GOS, the ISAP uses a simple scheme to find a route which does satisfy the GOS. The dynamic program provides SPACAP (SPAre CAPacity) with the current route, GOS, and the list of candidate paths for this parcel and hour. We shall explain the routine's operation using a simple example.

Let the parcel 6, hour 2 route calculated by the dynamic program be (6,-), (2,33), (3,45). Since we assumed that 45% of this parcel's hourly load is 5 erlangs, then 50% of it is 5.56 erlangs. Suppose that the blocking on the first three paths is equal to .03 if we offer 5.56 erlangs of traffic to the route, and we want a GOS of .01. Hence, we must economically *carry* $5.56 \times .02 = .111$ erlangs more load. First, starting from the network produced by the dynamic program, SPACAP configures the network to carry half of the remaining load; thus, for this example we configure the network to carry an additional .0556 erlangs of traffic. Under the constraint that the first three paths in the route are fixed, the same methods used to calculate the Minimum Marginal Increments Table are used in the calculation of a minimum incremental cost solution to the problem of serving the extra traffic. Either SPACAP finds a path not in the present route that can carry the traffic without augmentation and makes it part of the route, or it augments a path already in the route, or it augments a path not in the present route and makes it part of the route, whichever is the least expensive alternative. Then the process is repeated for the remaining half of the load. When SPACAP stops it has produced a route that serves parcel 6, hour 2 traffic with the appropriate GOS.

The reader may wonder why we offer .0556 erlangs twice to the candidate paths rather than offering .111 erlangs once. Unlike the dynamic program, SPACAP looks at only one path at a time, and, therefore, cannot recognize when appending two more paths to the route costs less than appending one path. Thus, offering larger increments will be more likely to force augmentation than offering smaller increments.

This completes the discussion of the design block in Figure 1. The next three sections will complete the presentation of the ISAP.

3.6 *Update 1*

After the design block chooses the best incremental network design for a fixed parcel and hour, the ISAP stores new estimates for the hourly link blockings, hourly parcel blockings, hourly link loads, and link sizes. These blocking estimates are simple approximations based on the Erlang-B formula with new hourly link loads and link sizes.

We update only local link and blocking information for the particular parcel rather than recalculating blockings for the entire network. Remember that the ISAP is a two-level decomposition and that the design block operates on the lower level by choosing good incremental solutions by parcel and hour. Our choice to augment or decrement a certain link affects traffic in all other hours. For a given parcel, changes in link ℓ that result in a small (or negative) incremental network cost in hour h_1 may force a very large incremental cost for the same parcel in another hour h_2.

Thus, the link sizes we choose at the end of the design block are, in fact, only candidate changes in the network. We do not recalculate all the network blockings until Level 1 finally chooses the link changes we wish to allow.

3.7 Select Design

To monitor accurately how single-hour choices affect other hours, we require a controlling objective function, i.e., an objective function that, for a given parcel, relates single-hour incremental costs to the total incremental network cost required to carry the parcel's traffic during all hours. Finding an explicit expression for this function seems difficult. Fortunately, a simple alternative coordinating procedure is available; it is Level 1 of the ISAP.

The MMITs and supplemental routing and sizing calculations for a particular parcel and hour provide information from which we can construct several different sets of routings and trunk augments/disconnects to serve the load with the appropriate GOS. The previous example shows how to calculate the least cost routing and augmentation strategy *without* taking into account this strategy's effect on the costs in other design hours for this parcel. Rather than calculating just the least cost strategy, the ISAP can determine different network configurations generated during the design step, and when all hours for a parcel have been examined, the procedure can choose the least cost alternative, with respect to the other design hours, from among those it has saved.

There are several ways in which we can select the network configurations to save. When the direct path proves in (the phrase "proves in" means that load above a minimum threshold is offered to the path), it is the cheapest path in the candidate list for almost all parcels. Hence, we should expect the direct path to be the first choice path in most hours and for most parcels, and it makes sense always to examine the case that the direct path is the first path in the route. Another network configuration we can store is the least incremental cost network when the first path in the route is not the direct path. Clearly the ISAP can order network alternatives and keep them up to the limits of storage and computation time. How many network alternatives need to be saved for large networks is an open issue. Our examples consider only three alternatives.

3.8 Update 2

When we have chosen the least incremental cost network for a particular parcel, the ISAP recomputes the hourly link loads, hourly link blockings, and hourly parcel blockings for the entire network. This means finding the solution to the nonlinear system of equations (2.2.1) - (2.2.4) and the Equivalent Random Method that, for a given routing, expresses the relationship between hourly link loads, link sizes, and hourly link blockings. Given the hourly link blockings calculated from the routing, one can determine for each design hour the blocking on each path in a parcel's route and thereby obtain the parcel blocking.

To solve this system of equations, which we call the network flow and blocking equations, the ISAP uses an iterative technique that assumes that the routing and link sizes are fixed; it begins by choosing a set of loads for the links in each hour h and assuming all the link blockings are unknown. The next step is to calculate new hourly link blockings for every link in the network with respect to these loads. The absolute values of the differences between the old hourly link blockings and new hourly link blockings are summed over all hours, and the sum is called Error. If the value of Error is less than a predetermined threshold, the method stops. Otherwise it calculates a new set of link offered loads; i.e., for each design hour $h = 1,...,H$, it finds a new set of loads (see 2.2.4) using the link blockings calculated above. The new hourly link blockings are used to find the values of $e_{ij}(h)$ as in (2.2.2). Once the $e_{ij}(h)$ are known, (2.2.1) gives the entries of the matrix $\underset{\sim}{p}(h)$, which, in turn, gives the new vector $\underset{\sim}{a}(h)$. Though we have no convergence proof for this method, it has never failed to converge in more than 50 iterations for any reasonable stopping criterion.

When the iterative method has converged, the ISAP verifies that the grade of service is met for each parcel in each hour. If the grade of service is violated for any parcel in any hour, the ISAP adds a trunk module to that (those) link(s) which is (are) the first path in the route of the parcel with the highest hourly blocking. Then we solve the above system of equations again and repeat the link augmentation where necessary to assure the appropriate hourly grade of service for each parcel. When the GOS is satisfied for all parcels in all hours, the ISAP determines whether all the traffic is served and stops if all of it is served. Otherwise, the ISAP chooses either the next hour or the next parcel and hour and continues.

This completes the discussion of the ISAP. In the next section we present our results and outline problems that this work suggests.

4.0 RESULTS

The following discussion presents a comparison between the design costs of the ISAP and the UA on small, multi-hour networks. We compared the design costs of the 3-node/2-hour, 4-node/2-hour, and 16-node/10-hour networks, respectively.

To assure an equitable comparison between the UA and the ISAP, we simplified the design problem as much as possible. Upper limits on switch capacity were ignored as were transmission constraints that limit via-route candidates. Thus, any node other than the originating node or the terminating node was allowed to be a via node. All traffic parcel peakednesses assume the value 1. Of course, the same network data were used as input to both algorithms. In addition, in the examples below, the ISAP passes through the ordered list of parcels for the 4-node and 16-node networks exactly once and designs the network to carry 100% of each parcel's traffic during each design hour. In the 3-node network the parcel list is traversed four times with 25% of every parcel's load allocated during each pass. The vector $\underset{\sim}{U}$ is {.01, .05, .1, .15, .2, .3, .4, .5, .6, .7} for the 3-node network and is {.01, .05, .1, .15, .2, .25, .3, .35, .4, .5} for the other examples. All networks were designed to provide parcel blocking in any design hour less than or equal to 1%.

For the 3-Node/2 hour network, each traffic parcel was 10 erlangs in both hours 1 and 2. Each trunk in a link cost 1 dollar, and the module size equaled 1. The UA produced a design that cost $47 and the ISAP produced a design that cost $46, which is a savings of 2.2% in the network design cost. The routing was identical in each design and was symmetric, as one expects. Neither design gave the symmetric trunk group sizes our intuition demands, but the ISAP design was more appealing in this respect than the UA design.

Refer to Figure 2 for the data on the 4-node/2-hour network. The left hand column of the uppermost table gives the link number and traffic parcel associated with the 4-node network depicted in the drawing. The parcel sizes are actual loads from measured data (see table at lower left), and the trunk group costs are true costs of the direct paths associated with the traffic parcels (see table at lower right). All module sizes have been set to 1, and any fractional trunk group sizes produced by the UA have been rounded to the next greater integer. The reader will notice that the ISAP and UA designs are different, yet they cost nearly the same; the ISAP is .1% less expensive. The cost results in Figure 2 are typical of ISAP designs for a very broad range of values for the vector $\underset{\sim}{U}$.

For AT&T-C the smallest practical problem of interest is the network with 16 switches and 10 load set periods since this is the size of the initial DNHR network. Our design problem, therefore, has 16 nodes and 10 hours. The data on which our tests were based are the data used to produce the 16-node network forecasts in December 1983. The ISAP designed network costs $56.4M, whereas the UA-designed network costs only $54.6M, a difference of 3.3%. In Section 5 we shall propose explanations for the ISAP's performance and suggest ways to improve it.

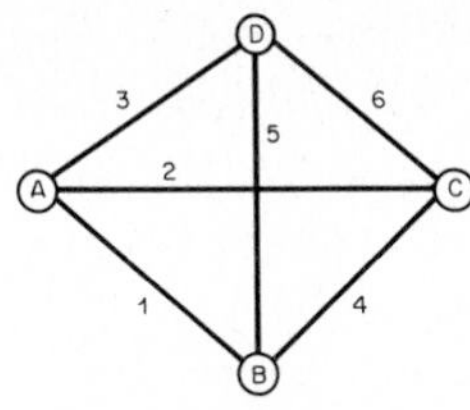

LINK/PARCEL	UA			ISAP		
	TRUNKS	ROUTE		TRUNKS	ROUTE	
		HOUR 1	HOUR 2		HOUR 1	HOUR 2
1/AB	110	(1,-)	(1, -) (3, 5) (2,4)	116	(1,-)	(1,-) (3,5)
2/AC	68	(2,-)	(2,-) (3, 6) (1, 4)	61	(2,-)	(2,-) (3,6)
3/AD	219	(3,-)	(3,-) (2, 6) (1, 5)	232	(3,-)	(3,-)
4/BC	168	(4,-) (5, 6)	(4,-) (5, 6)	183	(4,-) (5, 6)	(4,-)
5/BD	260	(5,-) (4, 6)	(5,-) (4, 6)	220	(5,-) (4, 6)	(5,-) (4, 6)
6/CD	621	(6,-)	(6,-) (4, 5)	634	(6,-)	(6,-)
TOTAL COST	4,380,118			4,374,270		

PARCEL	PARCEL SIZE (ERLANGS)	
	HOUR 1	HOUR 2
AB	73.27	117.79
AC	30.57	69.73
AD	115.07	188.77
BC	150.85	137.21
BD	235.08	230.68
CD	526.28	568.60

LINK	$/TRUNK
1	4805
2	4223
3	3499
4	2309
5	2986
6	2631

FIGURE 2 4-NODE/2-HOUR NETWORK

5.0 CONCLUSIONS

The performance of the ISAP and the UA is comparable for designing small networks, yet the example in the preceding section suggests that the UA is superior to the present form of the ISAP for designing networks that arise in practical problems. Numerical evidence suggests that there are two reasons for the ISAP's more costly 16-node/10-hour network design. Recall that the ISAP chooses cost-effective routes and link augments/disconnects under the constraints of its estimates of how these load and link-size perturbations affect blockings in the whole network without solving the network flow and blocking equations. The procedure's inability to estimate these blocking changes accurately contributes to its more expensive 16-node/10-hour design. Furthermore, the present implementation of the algorithm uses more computation time than the UA and will not permit more than one pass through the list of parcels for the 16-node/10-hour network. Examining a parcel more than once is the *only* way the ISAP can take advantage of global traffic non-coincidence in the network.

The introduction of a branch and bound scheme to conduct the searches for optimal routings and link augments/disconnects in the dynamic program (see Section 2.5.1) can save arithmetic operations in the ISAP. Broad applications of this idea can result in good run-time reductions. The method must use crude bounds on link blocking to establish a set of link sizes on which to search. Each different link size represents a different network configuration to evaluate. What is wanted at present is a way to eliminate most of these network configurations without having to solve the network flow and blocking equations. There is numerical evidence that when changing link sizes by adding (deleting) trunks results in a more expensive network, then adding (deleting) more trunks will only make the network even more expensive. This observation can be used to eliminate costly network configurations. Implicit in this discussion is the assumption that whenever the ISAP does examine a configuration, the network flow and blocking equations are solved iteratively.

The preceding assumption eliminates, in theory at least, the ISAP's problem with estimating changes in link blocking under hourly link load and link size perturbations. To make the ISAP competitive with the UA, though, we feel that run time reductions must be great enough to allow the ISAP to examine each parcel more than once. Evidence in [4] suggests that examining each parcel four times will give good results.

6.0 ACKNOWLEDEMENTS

A. N. Kashper, K. R. Krishnan, R. P. Murray, M. K. Nassar, and B. A. Whitaker served as a critical audience for my ideas during the development of the ISAP. S. L. Miller cheerfully did some computer studies when I asked. I wish to thank them all.

G. R. Ash made valuable comments on the early drafts of this paper.

References

[1] G. R. Ash, A. H. Kafker, and K. R. Krishnan, "Dynamic Routing for Inter-City Telephone Networks", *Proceedings of the Tenth International Teletraffic Congress*, Vol. 1, June, 1983.

[2] G. R. Ash, R. H. Cardwell, and R. P. Murray, "Design and Optimization of Networks with Dynamic Routing", *The Bell System Technical Journal*, Vol. 60, No. 8, October, 1981, pp. 1787-1820.

[3] G. R. Ash, A. H. Kafker, and K. R. Krishnan, "Servicing and Real-Time Control of Networks with Dynamic Routing", *The Bell System Technical Journal*, Vol. 60, No. 8, October, 1981, pp. 1821-1845.

[4] P. A. Steenbrink, *Optimization of Transport Networks*, New York: John Wiley & Sons,1974.

[5] P. J. Courtois, *Decomposability: Queueing and Computer System Applications*, New York: Academic Press, 1977.

[6] A. M. Geoffrion, "Generalized Benders Decomposition", *Journal of Optimization Theory and Applications*, Vol. 10, Nov. 4, 1972, pp. 237-260.

[7] L. S. Ladson, *Optimization Theory for Large Systems*, New York: MacMillan Publishing Co., Inc., 1970.

[8] R. Bellman, *Dynamic Programming*, Princeton, New Jersey: Princeton University Press, 1957.

TELETRAFFIC ISSUES in an Advanced Information Society
ITC-11
Minoru Akiyama (Editor)
Elsevier Science Publishers B.V. (North-Holland)

ANALYTICAL MODELING OF GTE TELENET DYNAMIC ROUTING

Michael EPELMAN* and Alexander GERSHT**

* Bentley College and GTE Laboratories, Inc.
** GTE Laboratories, Inc., 40 Sylvan Road, Waltham, MA 02254

ABSTRACT

An analytical model of equilibria in a virtual circuit network under the GTE Telenet isolated dynamic routing is presented. In the model the distributions of virtual circuits over the network links are obtained as solutions to equilibrium programming problem formulated for the network. Average network delay for the packet level is analyzed. The proposed approach is illustrated by simple examples where the results are obtained analytically. General numerical algorithm applicable to multinode networks is given.

I. INTRODUCTION

The purpose of this paper is to present a mathematical model of equilibrium regimes in the GTE Telenet network. Telenet is a packet-switched network operating in a virtual circuit mode [WEIR 82]. In such networks each session utilizes a single route until its termination. The route (virtual circuit) is set up by routing the first packet of every session, the call request packet, from the node of its origin through transit nodes towards the node of its destination. Decisions are made at every node as to which of the eligible links will carry a new virtual circuit. Telenet's routing rule dictates opening the circuit on the link that is currently least saturated. Thus, routing decisions are taken only on the basis of current information available locally in each node; this is *isolated adaptive* (or *dynamic*) routing [DAVI 79]. Mathematical modeling of networks under adaptive routing presents considerable difficulties [GERL 81, BERT 81, GEKL 76]. Is appears that classical queueing models are too detailed in their description, and lead to virtually intractable models. Ours is a macroscopic model in that it deals with average flows. However, the stochastic nature of the underlying processes constantly manifests itself in the model. In fact, the model owes its validity to the result obtained in [EPGE 84] which states that for large link capacities and proportionally large demand the random numbers of virtual circuits carried by the network links are well represented by the average loads. This result is based on the observation that the process of establishment and termination of virtual circuits is independent of the packet transport process mainly due to enormous difference between comparatively large virtual circuit life span and small packet transport time [GERS 82]. The implication of this is that the description of the equlibrium of the slower virtual circuit level is uninfluenced by the packet level. On the other hand, analysis of the faster packet level is conducted on the premise that the virtual circuit picture is static. Incidentally, all results pertaining to the virtual circuit level are applicable to circuit-switched networks.

To repeat, our purpose in this paper is to describe an existing routing rather than deduce an optimal routing from some global criterion. As a matter of fact, such global criterion is unlikely to exist for an isolated dynamic routing since in this case we are dealing with many decision making nodes, each pursuing a goal of its own. Not coincidentally, we describe the network equilibria via several interdependent optimization problems each representing a node of the network. Equilibrium programming [GAZA 81] is a ready-made tool for our purposes. A four-node example is given where equilibrium flows are found analytically for all possible input flows. In Appendix A we propose a numerical procedure applicable to multinode networks that finds equilibrium distribution of virtual circuits over the network links.

The layout of the paper is as follows: First, we describe the Telenet routing rule. Then routing variables are introduced and the network equilibrium is discussed for the virtual circuit level. We formulate conditions of network equilibrium in relation to the Telenet routing rule. Our next step is to devise an optimization problem for every node of the network. These nodal problems taken together constitute an equilibrium programming problem. We show that the Kuhn-Tucker conditions of optimality for the nodal problems are just another expression for the conditions of network equilibrium. Finally, on the packet level, we discuss the average network delay, one of the major performance characteristics of interest to practitioners.

II. TELENET ROUTING STRATEGY

As was observed, Telenet is a packet-switched network operating in the virtual circuit mode. A fixed end-to-end path is established through the network at session set-up time and all packets associated with that session follow this path. When the session is set up, route selection in the GTE Telenet network is executed on a node by node basis. When a node receives a call request packet, the first packet of every session, it searches its routing tables for a set of eligible outgoing links using the destination address as a search key. Links that belong to the minimum hop paths between the node and the call's destination are called primary links. Some of the remaining links are designated as secondary.

The current GTE Telenet routing rule [WEIR 80] prescribes opening of a new virtual circuit on a primary link that is least saturated at the moment. The saturation is evaluated by the ratio of the number of active sessions to link capacity. If the outgoing links are of equal capacity (which we will assume for simplicity) then the least saturated link is the one that carries the smallest number of virtual circuits. The secondary links are used only if all primary links at the node are fully saturated. If a call request reaches a node at which no primary or secondary links are available the virtual circuit associated with the call is cleared back to the previous node where the route search is resumed with the link of the blocked path excluded. If the blocking occurs at the origin node, the session is assumed to leave the network.

We see that the routing decisions are taken on the basis of *current* traffic information available *locally* in each node. In the taxonomy of routing, the above routing strategy is classified as *dynamic* and *isolated*.

III. MACRO DESCRIPTION

One very important aspect of the Telenet network operation is that the process of establishment and termination of virtual circuits is essentially independent of the packet transport process. The reason for this is twofold. Firstly, it is the count of the virtual circuits on the outgoing links and not the size of the packet queues that determines the routing decisions at the nodes of the network. Secondly, the average duration of a session is much longer than the average time required for the call request packet to travel from its origin to its destination and establish a virtual circuit. In other words, in a slower time frame of the virtual circuit birth-and-death process the effect of the packet transport is negligible. This allows us to consider the virtual circuit level separately and independently of the packet transport level. In fact, the main part of the proposed model is concerned with the virtual circuit level and describes a mathematical model for the equilibrium distributions of the virtual circuits over the links of the network for given inputs. All calculations pertaining to the data packet traffic can be carried out under the assumption that the virtual circuit picture remains static. This assumption is perfectly justified due to the earlier mentioned difference between the comparatively large virtual circuit life span and very fast data packet transport time.

In virtual circuit based networks the links are allowed to carry only a limited number of virtual circuits to avoid congestion in the packet queues. However, the "virtual capacity" of the links is usually large enough to play the role of a large parameter. Asymptotic analysis with respect to this large parameter shows that, for Poisson inputs and exponentially distributed holding time, the network attains an equilibrium under which the number of virtual circuits carried by each link deviates very little from its equilibrium value; more precisely, the variance (appropriately normalized) tends to zero with the increase in the link virtual capacity [EPGE 84].

In a well designed network the primary routes should be able to accommodate the average peak demand. We assume that this is true for all examples analyzed in this paper. Interestingly, one of the conclusions drawn from the asymptotic analysis mentioned above is that nonprimary routes are used only for service of very rare large fluctuations; in other words, the variability of the underlying processes does not force the network to use the nonprimary routes systematically provided there is enough capacity for the average demand.

Operating experience with the GTE Telenet network where links do have large virtual capacity has shown that "traffic flows in the network are extremely stable and predictable" [WEIR 82].

IV. ROUTING VARIABLES

We will now turn to the mathematical formulation of the model. Networks will be represented by undirected graphs. As an illustration, consider a four-node network in Figure 1.

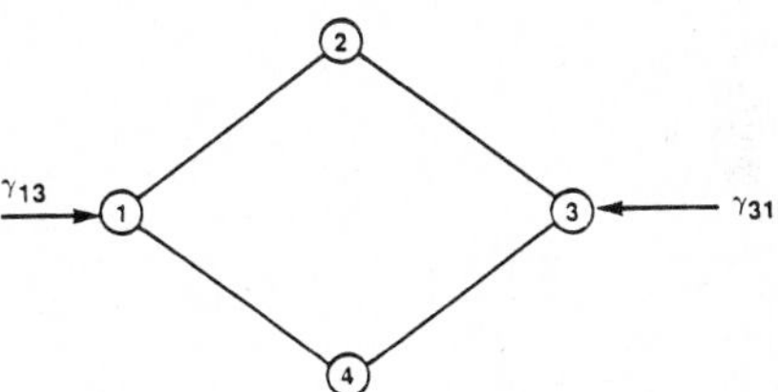

Fig. 1: Four Node Example

Let γ_{13} be the average number of sessions originating at node 1 and destined for node 3. The links (1, 2) and (1, 4) are on the primary (minimum hop) paths for the origin-destination pair {1,3} and the flow γ_{13} will be split between them in certain proportions say p_{12} and p_{14}, $p_{12} + p_{14} = 1$. We will call these proportions link routing variables. Clearly, these variables are always non-negative.

In node 3 the flow γ_{31} destined for node 1 is split between the primary links (3, 2) and (3, 4) in accordance with routing variables p_{32} and p_{34}. What values these variables are assigned is determined by the routing strategy and about it later. We assume for simplicity that there is no external input at nodes 2 and 4. Viewed from a node, the total flow on an outgoing link typically consists of two components: one shaped at the node and the other shaped elsewhere. For instance, at node 1 the total flow on the link (1, 2) is given by

$$f_{12} = p_{12}\gamma_{13} + p_{32}\gamma_{31} + \overset{\circ}{f}_{12} \qquad (1)$$

Here the component $p_{12}\gamma_{13}$ is the result of routing decisions made in node 1 while the component $p_{32}\gamma_{31}$ is forwarded on the link (1, 2) via the link (2, 3) by the node 3, and the flow $\overset{\circ}{f}_{12}$, although originating in node 1, is not affected by routing there since it is destined for node 2 and has only one primary way to go - the link (1, 2). Thus, in the node 1, $p_{12}\gamma_{13}$ is the controlled flow component and $p_{32}\gamma_{31} + \overset{\circ}{f}_{12}$ is the "imposed" component on the link (1, 2). Similarly, total flows on links (1, 4), 2, 3), and (3, 4) are

$$f_{14} = p_{14}\gamma_{13} + p_{34}\gamma_{31} + \mathring{f}_{14}$$

$$f_{32} = p_{32}\gamma_{31} + p_{12}\gamma_{13} + \mathring{f}_{32} \qquad (1)$$

$$f_{34} = p_{34}\gamma_{31} + p_{14}\gamma_{13} + \mathring{f}_{34}$$

respectively.

In general, routing variables are of macroscopic nature; they are time averaged portions of nodal loads diverted on appropriate links. However, very often our reasoning related to average quantities invokes instantaneous, dynamic considerations. We justify this frivolity by the previously mentioned smallness of the variance. Moreover, we will be calculating the equilibrium virtual circuit loads on the network links by finding appropriate values for routing variables.

V. NETWORK EQUILIBRIUM

As was observed earlier, the Telenet routing rule requires the node to open a new virtual circuit on the least loaded eligible link. Clearly, this tends to minimize the difference between the virtual circuit loads carried by the outgoing links. However, every node tries to even out the load without regard to the consequences of its actions for other nodes; every node pursues its own goal, often being at cross-purposes with its neighbor nodes. The network equilibrium occurs when each node has achieved its goal of best possible load balance on its outgoing links. An equivalent characterization of the network equilibrium is that at each node no flow utilizes a link unless the link's total flow is minimal among the flows on eligible links. This is an adaptation of Wardrop's definition of equilibrium for the case of isolated dynamic routing [WARD 52].

In our four-node example, let us assume that equilibrium flow configuration is such that at node 1 the total flow on link (1, 2) is strictly greater than the total flow on link (1, 4), $f_{12} > f_{14}$. Clearly, this implies that all of the flow γ_{13} is directed on the link (1, 4), i.e., $p_{12} = 0$ and $p_{14} = 1$, since otherwise the difference $f_{12} - f_{14}$ would have been only greater, which is contrary to the stated nodal goal. At the other node, any of the three situations are possible $f_{32} = f_{34}$, $f_{32} < f_{34}$ or $f_{32} > f_{34}$.

Assume for definiteness, that $f_{32} > f_{34}$. As in the case of node 1, this implies $p_{32} = 0$ (and hence $p_{34} = 1$) since $p_{32} > 0$ would be contrary to the goal of local routing at node 3 which is to minimize the difference between the total flows on links (3, 2) and (3, 4). Let us now combine the obtained values for the routing variables and the unequalities $f_{12} > f_{14}$ and $f_{32} > f_{34}$. In view of Eqs. (1), this gives

$$\mathring{f}_{12} + p_{12}\gamma_{13} + p_{32}\gamma_{13} > \mathring{f}_{14} + p_{14}\gamma_{13} + p_{34}\gamma_{31}$$

$$\mathring{f}_{32} + p_{32}\gamma_{13} + p_{12}\gamma_{31} > \mathring{f}_{34} + p_{34}\gamma_{31} + p_{14}\gamma_{13}$$

or, since $p_{12} = 0$, $p_{14} = 1$, $p_{32} = 0$, and $p_{34} = 1$,

$$\mathring{f}_{12} - \mathring{f}_{14} > \gamma_{13} + \gamma_{31}$$

$$\mathring{f}_{32} - \mathring{f}_{34} > \gamma_{13} + \gamma_{31} \qquad (2)$$

More important however is that the converse is true: inequalities (2) imply the equilibrium flow configuration with the routing $p_{12} = 0$, $p_{14} = 1$ and $p_{32} = 0$, $p_{34} = 1$. In other words, if the inputs γ_{13} and γ_{31} are not large enough to overcome the imbalances $\Delta_1 = \mathring{f}_{12} - \mathring{f}_{14}$ and $\Delta_3 = \mathring{f}_{32} - \mathring{f}_{34}$ due to the committed flow components, then the equilibrium flows correspond to the routing $p_{12} = 0$, $p_{14} = 1$ at node 1 and $p_{32} = 0$, $p_{34} = 1$ at node 3.

Altogether, there are nine regions in the space of inputs γ_{13} and γ_{31} and nodal imposed imbalances Δ_1 and Δ_3, each region being associated with a distinct equilibrium flow pattern. These are catalogued in Table 1. Interestingly, in eight out of nine cases, at least one node uses only one of the two available primary routes, i.e., fixed routing sets in at least partially (cf. [KLEI 76], p. 346). Equilibrium with complete balance at both nodes, $f_{12} = f_{14}$ and $f_{32} = f_{34}$, occurs only when the imposed nodal imbalances are equal, $\Delta_1 = \Delta_3$, condition unlikely to be satisfied. Finally, the presence of splitable inputs at nodes 2 and 4 would not affect the picture presented since the imbalances Δ_1 and Δ_3 are unaffected by these inputs. The way we analyzed our four-node network quickly becomes impractical for larger networks. The remainder of this paper is devoted to developing an alternate approach. First, however, we return to routing variables and give a general procedure that will allow us to avoid redundant variables.

Equilibrium Flow Configuration			Equilibrium Routing	
At Node 1	At Node 3	Parameter Relationship	At Node 1	At Node 3
$f_{12} = f_{14}$	$f_{32} = f_{34}$	$\Delta_1 = \Delta_3 \ (= \Delta)$ $\lvert\Delta\rvert \le \gamma_{13} + \gamma_{31}$	$p_{12} = \frac{1}{2} - \frac{\Delta}{2(\gamma_{13} + \gamma_{31})}$	$p_{32} = \frac{1}{2} - \frac{\Delta}{2(\gamma_{13} + \gamma_{31})}$
$f_{12} = f_{14}$	$f_{32} > f_{34}$	$\Delta_3 > \Delta_1$ $\gamma_{13} + \gamma_{31} \ge \Delta_1$ $-\gamma_{13} + \gamma_{31} \le \Delta_1$	$p_{12} = \frac{\gamma_{13} + \gamma_{31} - \Delta_1}{2\gamma_{13}}$	$p_{32} = 0$
$f_{12} = f_{14}$	$f_{32} < f_{34}$	$\Delta_3 < \Delta_1$ $\gamma_{13} + \gamma_{31} \ge -\Delta_1$ $\gamma_{13} - \gamma_{31} \ge \Delta_1$	$p_{12} = \frac{\gamma_{13} - \gamma_{31} - \Delta_1}{2\gamma_{13}}$	$p_{32} = 1$
$f_{12} > f_{14}$	$f_{32} = f_{34}$	$\Delta_1 > \Delta_3$ $\gamma_{13} + \gamma_{31} \geqslant \Delta_3$ $\gamma_{13} - \gamma_{31} \leqq \Delta_3$	$p_{12} = 0$	$p_{32} = \frac{\gamma_{13} + \gamma_{31} - \Delta_3}{2\gamma_{31}}$
$f_{12} > f_{14}$	$f_{32} > f_{34}$	$\gamma_{13} - \gamma_{31} < \Delta_1$ $\gamma_{13} - \gamma_{31} > \Delta_3$	$p_{12} = 0$	$p_{32} = 1$
$f_{12} > f_{14}$	$f_{32} > f_{34}$	$\gamma_{13} + \gamma_{31} < \Delta_1$ $\gamma_{13} + \gamma_{31} < \Delta_3$	$p_{12} = 0$	$p_{32} = 0$
$f_{12} < f_{14}$	$f_{32} = f_{34}$	$\Delta_1 < \Delta_3$ $-\gamma_{13} + \gamma_{31} \ge \Delta_3$ $\gamma_{13} + \gamma_{31} \ge -\Delta_3$	$p_{12} = 1$	$p_{32} = \frac{-\gamma_{13} + \gamma_{31} - \Delta_3}{2\gamma_{31}}$
$f_{12} < f_{14}$	$f_{32} < f_{34}$	$\gamma_{13} + \gamma_{31} < -\Delta_1$ $\gamma_{13} + \gamma_{31} < -\Delta_3$	$p_{12} = 1$	$p_{32} = 1$
$f_{12} < f_{14}$	$f_{32} > f_{34}$	$-\gamma_{13} + \gamma_{31} > \Delta_1$ $-\gamma_{13} + \gamma_{31} < \Delta_3$	$p_{12} = 1$	$p_{32} = 0$

TABLE 1

VI. SHARING LINKS

Consider now a more general network shown in Figure 2.

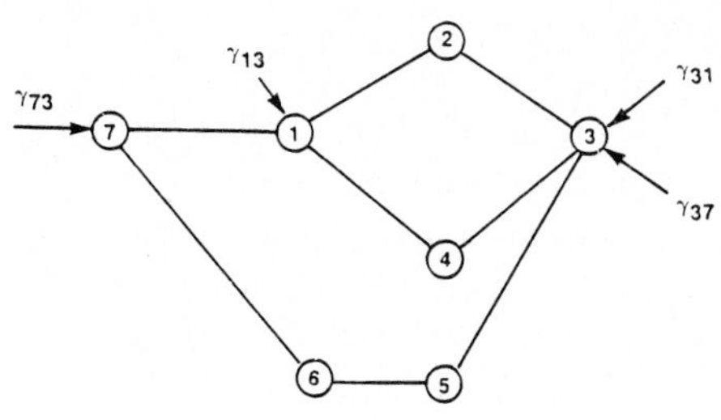

Fig. 2

At node 3 the flow γ_{31} destined for node 1 shares its primary links, (3, 2) and 3, 4), with the flow γ_{37} destined for node 7. The latter flow has one additional primary link, (3, 5). The fact that the two flows share some primary links affects the way routing variables are introduced at node 3. If p_1 and p_2 are the link routing variables for the flow γ_{31} and q_1, q_2, q_3 are the link routing variables for the flow γ_{37}, then there should be a relation between the variables serving different flow components on the same links. Indeed, suppose a call request destined for node 7 is being routed at node 3. Three primary links, (3, 2), (3, 4) and (3, 5), are eligible for the opening of a new virtual circuit. Assume for the moment that link (3, 5), the one that is not available to flow γ_{31}, is carrying more virtual circuits than either link (3, 2) or link (3, 4). Then, according to the Telenet routing rule, the choice will be made between links (3, 2) and (3, 4), the two least loaded links. This is exactly the choice we have got for the call requests destined for node 1. The outcome should not depend on the destination (whether it is node 1 or whether it is node 7) since only the current loads on the eligible links affect the choice. Hence, we will use the variables p_1 and p_2 for both the flow γ_{31} and the part of the flow γ_{37} served by the same links. Schematically, the assignment of routing variables in node 3 is shown in Figure 3. (A general procedure for assignment of routing variables to links has been developed.)

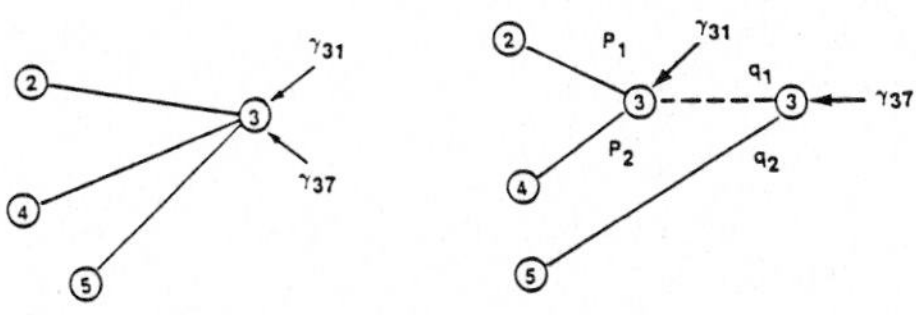

Fig. 3

Altogether we now have four routing variables instead of five that we would have if the connection between allocations of the two flows that share links were ignored. By taking into account this connection we made an important step towards "sufficiency" of the model. By that we mean exclusion of infeasible equilibria. In our four-node example the model admits a unique equilibrium except for the symmetric case of $\Delta_1 = \Delta_3$. Symmetry seems to unleash additional degrees of freedom in our model of network equilibrium. This is best illustrated by the two-node example shown in Figure 4.

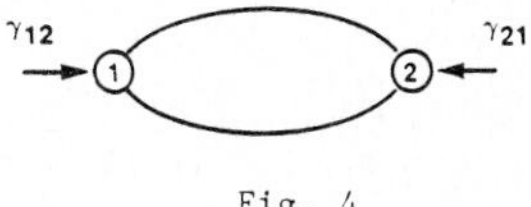

Fig. 4

This "network" has infinitely many equilibria -- any equal split of the total of the two inputs between the two links is an equilibrium. Which equilibria are feasible and which are not? This depends on the dynamics that results from the routing rule employed in the network. For the Telenet routing rule the equilibrium that will prevail should reflect the fact that at both nodes the decisions are made on the basis of the same information, the difference between the counts of virtual circuits currently open on the two links. Hence, we expect that at both nodes the input flow will be split equally between the two links. It is also confirmed by a probabilistic analysis that can be readily done in this simple case.

In general, the qualitative asymptotic analysis in [EPGE 84] shows that for a very wide class of routing strategies there is a unique equilibrium in the virtual circuit networks under unchanging inputs.

VII. EQUILIBRIUM PROGRAMMING DESCRIPTION

In this section we will give an equilibrium programming description of the network equilibrium flow configurations.

Equilibrium programming (EP) is a generalization of mathematical programming in which several optimization problems are solved simultaneously; the optimization problems are interdependent in that decision variables of one problem may enter as parameters into the goal function and/or constraints of other problems [GAZA 81]. As an analytical tool EP is innately suitable for modeling systems which comprise several decision-makers, each of whom optimizes a portion of the system.

We will formulate an EP problem whose solutions, called equilibria, satisfy the conditions of equilibrium flow configuration in a network under Telenet-type local routing. Before we do it in general form let us return to the network in Figure 1 for which we have found all equilibrium flow configurations for all inputs. It turns out that equilibrium flows in a network under the Telenet routing rule can be obtained as solutions to a multicomponent optimization problem. We have observed previously that in the Telenet network every node pursues its own goal of achieving the best possible balance of total flows on the outgoing links. Accordingly, every node is assigned a goal function to be optimized over the control variables at the disposal of this node. Thus, for node 1 of our four-node example, let

$$S_1 = - f_{12} \ln f_{12} - f_{14} \ln f_{14} \quad (3)$$

where

$$f_{12} = p_{12}\gamma_{13} + p_{32}\gamma_{31} + f^{\circ}_{12}$$

$$f_{14} = p_{14}\gamma_{13} + p_{34}\gamma_{31} + f^{\circ}_{14}$$

$$p_{12} + p_{14} = 1 \ , \ p_{12} \geqslant 0, \ p_{14} \geqslant 0$$

Expressions for f_{12} and f_{14} were obtained in Sec. IV; constraints on p_{12} and p_{14} are straightforward properties of link routing variables. The function S_1, called *nodal entropy*, depends on p_{12} and p_{14} via f_{12} and f_{14}. Similarly, the nodal entropy for node 3 is given by

$$S_3 = - f_{32} \ln f_{32} - f_{32} \ln f_{34} \quad (4)$$

where

$$f_{32} = p_{32}\gamma_{31} + p_{12}\gamma_{13} + f^{\circ}_{32}$$

$$f_{34} = p_{34}\gamma_{31} + p_{14}\gamma_{13} + f^{\circ}_{34}$$

$$p_{32} + p_{34} = 1, \ p_{32} \geqslant 0, \ p_{34} \geqslant 0$$

Observe that variables p_{32} and p_{34} play the role of parameters in the entropy S_1 while p_{12} and p_{14} are parameters in S_3. Our next step is to write out the Kuhn-Tucker conditions for the nodal entropies. For S_1 these conditions are

$$\partial S_1/\partial p_{12} = -\gamma_{13}(\ln f_{12} + 1) = \mu - \lambda_1$$

$$\partial S_1/\partial p_{14} = -\gamma_{13}(\ln f_{14} + 1) = \mu - \lambda_2$$

where $\lambda_1 \geqslant 0$, $\lambda_2 \geqslant 0$, and μ is real. After some algebra we get

$$f_{12} = e^{\mu+\lambda_1}, \ f_{14} = e^{\mu+\lambda_2}. \quad (5)$$

These equations are accompanied by the so-called complimentarity conditions

$$\lambda_1 p_{12} = 0, \ \lambda_2 p_{14} = 0, \ \lambda_1\lambda_2 = 0. \quad (6)$$

Similarly, the Kuhn-Tucker conditions for a maximum of S_3 are

$$f_{32} = e^{\nu+\zeta_1}, \ f_{34} = e^{\nu+\zeta_2} \quad (7)$$

where ν is real, $\zeta_1 \geqslant 0$ and $\zeta_2 \geqslant 0$, with the complimentarity conditions

$$\zeta_1 p_{32} = 0, \ \zeta_2 p_{34} = 0, \ \zeta_1\zeta_2 = 0. \quad (8)$$

The meaning of the Kuhn-Tucker conditions for S_1 is as follows: the routing p_{12}, p_{14} renders S_1 maximal if and only if there exist real μ and non-negative λ_1 and λ_2 such that Eqs. (5) and (6) hold. Assume now that p_{12} and p_{14} satisfy these equations and that $p_{12} > 0$. Then in view of (6), $\lambda_1 = 0$ and thus $f_{12} \leqslant f_{14}$. That is, if link (1, 2) is utilized by the flow γ_{13} then its total flow is less than or equal to the total flow on the other link. Conversely, let $f_{12} < f_{14}$; then, due to Eqs. (5), $\lambda_1 < \lambda_2$. Since $\lambda_1\lambda_2 = 0$, we have $\lambda_1 = 0$ and hence $\lambda_2 > 0$. Finally, $\lambda_2 > 0$ implies via Eqs. (6) that $p_{14} = 0$ and thus $p_{12} = 1$. That is, all of the flow γ_{13} is directed onto link (1, 2). To summarize the above argument, the nodal entropy S_1 is maximized if and only if the routing p_{12} and p_{14} is such that the flow γ_{13} utilizes only the least loaded link (unless the outgoing links carry equal total flows and then both links may be utilized). If the same situation prevails at node 3, then the routings p_{12}, p_{14} and p_{32}, p_{34} and the corresponding total link flows make up what we previously called equilibrium flow configuration (see Sec. V). Thus network equilibria can be obtained as solutions to simultaneous maximization of the nodal entropies. A proof of this statement for a general network under Telenet-type routing rule follows the argument we used for our four-node example and won't be given here.

In general, the equilibrium programming description of the equilibria of a packet-switched virtual circuit network with n nodes consists of n interdependent optimization problems. The optimization problems are all similarly structured: for node i the goal function has the form

$$S_i = -\sum_j f_j \ln f_j,$$

where summation goes over the outgoing links utilized by the flows routed at the node; constraints consist of the unity sums for every group of the routing variables (see Appendix A) and expressions for the total link flows f_j.

Several general methods available for solution of EP problems as well as important existence results can be found in [GAZA 81]. However, the simplicity of the constraints and the special structure of the goal functions in our network EP problems allow us to devise specialized solution procedures. One such procedure is outlined in Appendix B where it is used to find equilibrium virtual circuit configuration for the seven-node network shown in Figure 2.

VIII. PACKET LEVEL: DELAY ANALYSIS

Average network delay is a widely used measure of performance for packet-switched networks. It is the average delay endured by the average packet traversing the network. As has been observed in Sec. III, the packet transport process is evolving much faster than the birth-and-death process of the virtual circuits. The difference is so significant that a packet traveling through the network almost certainly will encounter no changes in the counts of virtual circuits on the links of its path. In equilibrium the data packets will be pushed through the network in accordance with the route map established for them on the virtual circuit level. This map reflects the equilibrium values of the routing variables that are obtained as a solution to the EP problem for given external inputs. In sum, for each equilibrium virtual circuit configuration, the packet level is modeled as a network under static routing. Each link in this network has two infinite buffer servers, one for each direction. We will assume that the average number of packets per session is large and equal in both directions. (Note here that the largeness of the virtual capacity of the link does not imply smallness of the variances of the data packet queue lengths.) Under this assumption it is not difficult to show that the packet flow is proportional to the virtual circuit flow. Specifically, if λ_{ij} is the packet flow on the link (i, j) then

$$\lambda_{ij} = \beta f_{ij} \tag{9}$$

and

$$\beta = \frac{NW}{c\tau} \tag{10}$$

where

c - the link capacity in bits/sec
τ - the session average holding time in sec
N - the average number of packets per session
W - the average packet length in bits

The formula for the average network delay, T, ([KLEI 76], pp. 322 ff) can be expressed as follows:

$$T = \frac{1}{\beta c \sum \gamma_{rs}} \sum \frac{\lambda_{ij}}{1-\lambda_{ij}} \tag{11}$$

Formula (11) is valid under the standard assumption of Poissonian interarrival times and the Kleinrock independence assumption of exponentially distributed packet length.

We will rewrite formula (11) in the form which lends itself readily to qualitative analysis:

$$T = \bar{n}\,\bar{t} \tag{12}$$

where

$$\bar{n} = \frac{\sum \lambda_{ij}}{\beta \sum \gamma_{rs}} = \frac{\sum f_{ij}}{\sum \gamma_{rs}} \tag{13}$$

is the average path length (by length we refer to the number of links encountered in the path), and

$$\bar{t} = \frac{W}{c\sum \lambda_{ij}} \sum \frac{\lambda_{ij}}{1-\lambda_{ij}} = \frac{W}{c\sum f_{ij}} \sum \frac{f_{ij}}{1-\beta f_{ij}} \tag{14}$$

is the average (over the network) link delay.

Recall that under the Telenet routing rule, minimum-hop paths are used as primary routes and they carry all of the traffic. Thus, the first factor, $\bar{n}$, in formula (12) is minimal. As far as the second factor, $\bar{t}$, is concerned it turns out that the Telenet routing rule together with the best load balance achieveable locally at the nodes of the network, simultaneously attains minimal average link delay over the same set of control parameters.

IX. CONCLUSIONS

In this paper we have presented a mathematical model of equilibria in a packet-switched network operating in the virtual circuit node under an isolated dynamic routing. The model is comprised of several interdependent maximization problems, one for each node of the network. Solving these problems simultaneously for given external demand gives the equilibrium distribution of virtual circuits over the links of the network. On the basis of this distribution, packet level performance characteristics such as the average network delay are calculated.

Generally speaking, we model a dynamic isolated routing by an equivalent static routing. In fact, the network in equilibrium behaves as if the static routing were employed. Moreover, experience as well as analytical and numerical analysis show that a single path is utilized by the sessions between most of the origin-destination pairs in the network. The temptation therefore is to use the cheaper static routing. One should keep in mind however that this would require a mechanism for routing adjustment in response to changing external demand. Transient processes should also be taken into account.

APPENDIX A

We will use the network in Figure 2 to outline an algorithm of gradient type that finds an equilibrium flow configuration by solving the EP problem formulated for the network.

In node 3 where two inputs are routed towards their respective destinations over two overlapping groups of primary links, a maximum of the nodal entropy

$$S_3 = -f_{32} \ln f_{32} - f_{34} \ln f_{34} - f_{35} \ln f_{35}$$

is sought subject to

$$p_1 + p_2 = 1\ ,\ p_1 \geq 0\ ,\ p_2 \geq 0$$

and

$$q_1 + q_2 = 1\ ,\ q_1 \geq 0\ ,\ q_2 \geq 0$$

Here the routing variables p_1, p_2 and q_1, q_2 are as in Figure 3 and f_{ij} are the total flows on the links with the corresponding indices. The expressions for the flows are

$$f_{35} = f^v_{35} + q_2\ \gamma_{37}$$

$$f_{34} = f^v_{34} + p_2\ (q_1\ \gamma_{37} + \gamma_{31})$$

$$f_{32} = f^v_{32} + p_1\ (q_1\ \gamma_{37} + \gamma_{31})$$

where f^v_{ij} are the components of the total flows f_{ij} that are not controlled in node 3. For instance, for link (3, 4), the component f^v_{34} may include the committed flow as well as the flow directed via link (1, 4) at node 1 and via link

(7, 1) at node 7. The expressions for nodal entropies S_1 and S_7 are constructed similarly. To describe the iterative procedure that finds equilibrium values of the routing variables we need the function

$$H(x) = \begin{cases} 1 \text{ if } x > 1 \\ x \text{ if } 0 \leqslant x \leqslant 1 \\ 0 \text{ if } x < 0 \end{cases}$$

that will help us to allow for the fact that a routing variable is never greater than one or less than zero.

The procedure is initialized by distributing the external inputs at the network nodes in accordance with some arbitrary set of values for the routing variables. The k+1 step of the procedure is given by

$$q_1^{k+1} = H[q_1^k + \Delta (f^k_{35} - \min\{f^k_{32}, f^k_{34}\})]$$

$$p_1^{k+1} = H[p_1^k + \Delta (f^k_{34} - f^k_{32})]$$

$$r_1^{k+1} = H[r_1^k + \Delta (f^k_{14} - f^k_{12})]$$

$$s_1^{k+1} = H[s_1^k + \Delta (f^k_{76} - f^k_{71})]$$

where r_1 and s_1 are routine variables for links (1, 2) and (7, 1), respectively, Δ is the step size, and the rest are as defined above. Strictly speaking, for the procedure to be called gradient it ought to include the derivatives of the optimized function with respect to the control variables. In our formulae instead we have the total link flows of which the derivatives are monotonic functions.

The intent of the procedure is at each iteration to move the flows closer to satisfying the Kuhn-Tucker conditions in all nodes of the network, i.e., to move them closer to the best balance locally achievable. The procedure simultaneously maximizes the nodal entropies S_1, S_3 and S_7 subject to the flow conservation constraints.

The simplicity of this example is in the fact that every group of routing variables consists of two variables that sum up to unity; hence only one of the two need to be traced.

In Figure 5-a are shown the inputs and committed link flows and in Figure 5-b the equilibrium flows obtained by the above procedure.

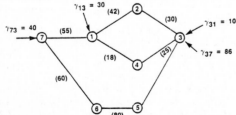

Fig. 5 a

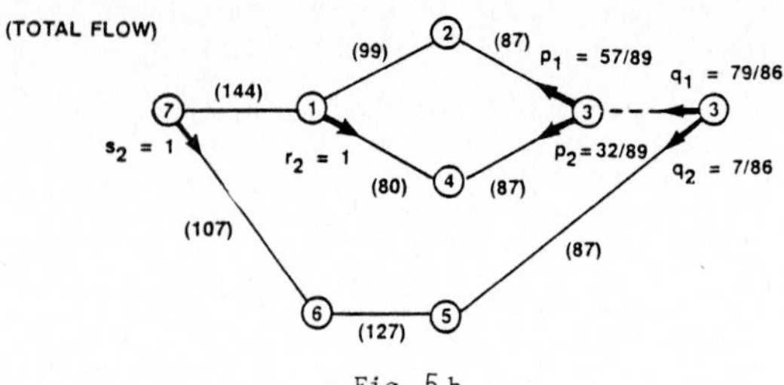

Fig. 5 b

In general case when variable groups may include more than two variables the following "cautious" procedure has been tested: at every iteration the redistribution is effected only between the activated link(s) carrying the largest total load and the link(s) carrying the smallest total load. Activated means that the routing variable of the link is greater than zero, i.e., the link carries the load that is controllable in the current node.

REFERENCES

[BERT 81] D. P. Bertsekas, "Notes on Optimal Routing and Flow Control for Communication Networks," LIDS Report R-1169, M.I.T., Cambridge, MA, December, 1981.

[DAVI 79] D. W. Davis, D. L. A. Barber, W. L. Price, and C. M. Solomonides, Computer Networks and Their Protocols, Wiley, 1979.

[EPGE 84] M. Epelman and A. Gersht, "Asymptotic Qualitative Analysis of a Link Imbedded into a Packet-Switched Virtual Circuit Network," Proceedings of INFOCOM-84, San-Francisco, 1984.

[GAZA 81] C. B. Garcia and W. I. Zangwill, Pathways to Solutions, Fixed Points, and Equilibria, Prentice-Hall, 1981.

[GEKL 77] M. Gerla and L. Kleinrock, "On the Topological Design of Distributed Computer Networks," IEEE Trans. on Commun. Vol. COM-25, pp. 48-60, Jan. 1977.

[GERL 81] M. Gerla, "Routing and Flow Control", in Protocols and Techniques for Data Communication Networks, F. Kuo (ed.), Prentice-Hall, 1981.

[GERS 82] A. Gersht, "Analytical Model of Dynamic Routing in Virtual Circuit Packet-Switched Network," Proceedings of ICCC-82, New York, 1982, pp. 76-80.

[KLEI 76] L. Kleinrock, Queueing Systems, Vol. 2, Wiley, 1976.

[WARD 52] J. G. Wardrop, "Some Theoretical Aspects of Road Traffic Research", Proceedings of the Institute of Civil Engineers, Part II, pp. 325-378, 1952.

[WEIR 80] D. Weir, J. Holmblad, and A. Rothberg, "An X.75 Based Network Architecture," Proceedings of ICCC-80, Atlanta, GA, 1980.

TELETRAFFIC ISSUES in an Advanced Information Society
ITC-11
Minoru Akiyama (Editor)
Elsevier Science Publishers B.V. (North-Holland)

DESIGN AND PERFORMANCE EVALUATION OF INTERNATIONAL TELEPHONE NETWORKS WITH DYNAMIC ROUTING

Yu WATANABE, Jun MATSUMOTO and Hiromichi MORI

Kokusai Denshin Denwa Co., Ltd. (KDD)
Tokyo, Japan

ABSTRACT

Dynamic or time varying routing schemes have been developed from present telephone network time fixed routing schemes by the recent rapid introduction of new processor control technology into telecommunication networks. In order to evaluate the benefit of such advanced routing schemes, it is necessary to first establish a dimensioning method for these networks. This paper presents a dimensioning method for networks with dynamic routing, and evaluates the merit of dynamic routing schemes for international networks. Under the dynamic routing scheme considered in this paper, the designations of high usage circuit groups and final circuit groups are changed hourly. A dimensioning method for multiple hour traffic demands with time varying routing patterns is proposed, and numerical examples are presented to illustrate the cost reduction effects of this dynamic routing in international telephone networks.

1. INTRODUCTION

Due to the remarkable progress of stored program controlled (SPC) exchanges and common channel signalling systems, highly sophisticated network controls are now available, such as dynamic/adaptive routings based on instantaneous network congestion levels and/or profiles of traffic demands. With these dynamic/adaptive routings, it is possible to improve network resource utilization, which eventually leads to decreases in network construction cost.

AT&T originated the DNHR (Dynamic Non-Hierarchical Routing) concept where network structure is completely non-hierarchical, i.e., where there is no distinction between high usage groups and final circuit groups, and routing patterns between each origin and destination pair are predetermined but dynamically changed hourly. Computer analysis of the effect of DNHR for a large network indicated that its usage would reduce network cost by 15 percent [1].

The fact that busy hour traffic demand does not appear at the same hour for all routes becomes an advantage in reducing network cost under the dynamic or time varying routing scheme. International networks can expect a greater network cost saving effect than domestic networks since traffic profiles for international networks are influenced by time differences, causing busy hour traffic to appear at different time periods for. each origin/ destination pair [2].

This paper presents a method for the design of dynamic routing networks, and evaluates the cost savings realized with dynamic routing in international networks. The dynamic routing scheme considered in this paper changes the designations of high usage and final circuit groups hourly, although there is a concept of high usage and final circuit groups. The routing patterns are predetermined according to traffic profiles in order to effectively utilize idle network capacity. The grade of service in this network is guaranteed by that of the final circuit groups, while minimum cost criteria are used to dimension high usage circuit groups.

For the design of networks using this dynamic routing scheme, circuit group dimensionings that take multiple hourly traffic demands into account are needed. Horn [3] proposed a multiple hour optimization method in which networks are sequentially optimized for each hourly traffic demand, and where hourly optimization results up through the previous hour are used as lower bounds for circuit group sizes. However, it is known that results are greatly dependent on the optimization sequence, especially for networks which have widely different busy hours in circuit groups, such as international networks.

In this paper, a design method is proposed for a dynamic routing scheme which does not depend upon an optimization sequence. The described dimensioning method treats multiple hour traffic demands simultaneously, and a near optimum solution is obtained regardless of the variance in peak periods. An example for a small telephone network is given to illustrate the benefits of dynamic routing in international networks.

2. DIMENSIONING METHOD

For single hour traffic demands, the well known marginal occupancy procedure for network dimensioning was established by Pratt [4], [5]. This paper proposes a multiple hour network dimensioning method in which single hour optimization results for all hourly traffic demands are used to produce an initial solution, which is then reduced by an iterative algorithm to increase network efficiency.

We will explain the proposed dimensioning method using the small 3-node sample network shown in Figure 1. The traffic demands for three hours are given in Table 1. In this example, we assume that the required grade of service is 1% blocking on final circuit groups.

The time varying routing patterns are determined in order to use idle network capacity effectively. In time period I, calls on route 1-2 overflow to alternate route 1-3-2. In other words. route 1-2 is operated as a high usage group, and routes 1-3 and 3-2 are operated as final groups. Routes 1-3 and 2-3 are operated as high usage groups in time periods II and III, respectively. The routing patterns for each time period are summarized in Figure 2.

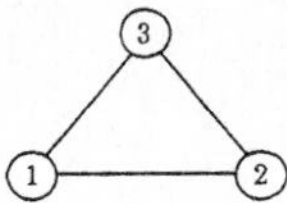

Fig. 1 Sample 3-node network

Table 1 Hourly traffic demand

Route \ Time period	I	II	III
1-2	10.0	3.5	4.8
1-3	5.0	11.0	4.1
2-3	7.2	7.9	16.0

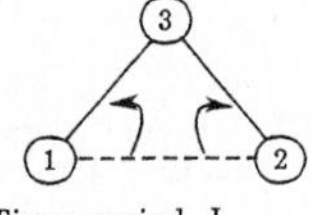

Time period I

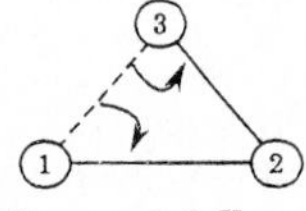

Time period II

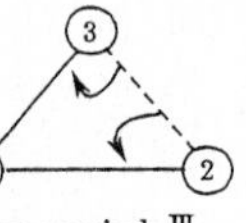

Time period III

——— final circuit group

- - - - - high usage circuit group

Fig. 2 Time varying routing patterns for sample network

2.1 Calculation of the initial solution

Step 1 Single hour optimizations are performed for each hourly traffic demand.

Step 2 The maximum circuit group size for each route is chosen as the initial solution, from the total of all single hour optimization results.

Using the marginal occupancy procedure, the sample network in Figure 1 can be dimensioned for separate hourly traffic demands. The circuit group sizes for each hour resultant from Step 1 are given in Table 2. From this result, the maximum circuit group sizes for routes 1-2, 1-3 and 2-3 are found to be 13, 14 and 20 trunks respectively, and these maximum circuit group sizes are chosen as the initial solution.

2.2 Reduction of initial solution

Maximum circuit group sizes of routes are reduced in order of decreasing network cost. Maximum circuit group sizes are optimized for single hour traffic demand and reduction of maximum circuit group sizes will cause increase in single hour network cost. Therefore, reduction

Table 2 Single hour dimensioning results

Route \ Time period	I	II	III	Maximum
1-2	13	10	12	13
1-3	12	14	11	14
2-3	15	16	20	20

of maximum circuit group sizes is made in order to minimize the increase of single hour network costs. Proposed algorithm for reduction is summarized in following 5 Steps.

Step 3 Routes are selected where decreases in maximum circuit group size will not cause increases in maximum circuit group sizes for other routes.

Step 4 Of the routes from Step 3, one route is determined by the following criteria:
Condition A: The route offers the largest reduction without a corresponding increase in maximum circuit group sizes for other routes.
Condition B: For routes which are selected by condition A, the route where the total incremental cost to decrease the maximum circuit group size is the smallest.

Step 5 The maximum circuit group size. for the route from Step 4 is reduced by adjusting the size of related circuit groups.

Step 6 The initial solution is replaced by the modified maximum circuit group size determined in Step 5.

Step 7 Steps 3 through 7 are repeated until no route meets the selection parameter of Step 3.

For the sample network in Figure 1, the circuit group size of route 1-2 is largest in time period I, that of route 1-3 in period II, and that of route 2-3 in period III, The circuit group sizes for the three routes do not show maximums in the same time periods, which means that the maximum circuit group size for any of them can be reduced without increasing the maximum circuit group size for any other. Consequently, all of the routes (1-2, 1-3, and 2-3) are selected by Step 3 criteria.

Next the conditions in Step 4 are estimated. To reduce the initial solution for route 1-2, circuit group sizes for routes 1-3 and 2-3 must be increased in time period I, Similarly, reductions in maximum circuit group size for route 1-3 or 2-3 would require circuit group size increases in time periods II and III for other routes. Table 3 shows the circuit group size increases required to decrease maximum circuit group sizes.

In time period I, route 1-3 and 2-3 circuit group sizes do not exceed the initial solutions when more than 10 trunks exist on route 1-2. However, reduction in route 1-2 circuit group size by more than 1 trunk is not effective. since the second maximum circuit group size of route 1-2 is 12 trunks, as shown in Table 2. The initial solution of 13 trunks for route 1-2 can therefore be reduced by 1 trunk.

Similarly, the initial solution of route 1-3 can be reduced by 2 trunks. If the initial solution for, route 2-3 is reduced to less than 18 trunks, the circuit group size for route 1-2 will exceed the maximum circuit group size of 13 trunks in time period III, and therefore route 2-3 reduction is also 2 trunks. The net result of Condition A of Step 4 is the selection of routes 1-3 and 2-3.

Condition B of Step 4 is considered next. In order to reduce either route 1-3 or 2-3 by 2 trunks, the overall circuit group size increase for the other routes is 3 trunks. This means that the incremental costs to reduce the initial solution of routes 1-3 and 2-3 are the same. However, if we regard circuit group size as a real number, 0.2 trunks are required to reduce the initial selection of route 1-3, while 0.11 trunks are required for route 2-3, Condition B therefore selects route 2-3 for the Step 4 result, as its circuit group size reduction can be accomplished with the minimum cost.

In Step 5, the maximum circuit group size of route 2-3 is reduced, and in Step 6 the initial solution is replaced by these modified circuit group sizes. The resultant circuit group sizes are shown in Table 4. This modified solution satisfies the blocking requirement of 1% on the final circuit group.

Returning then to Step 3, another route is selected for further reduction of circuit group sizes. The circuit group size of route 2-3 has already been reduced in the previous Step 5, and additional reduction is not possible. Route 1-2 maximum circuit group size appears in time periods I and III, but it is obvious that any decrease in route 1-2 circuit group size would cause an increase in route 2-3 circuit group size in time period III. Therefore. only route 1-3 circuit group size can be reduced, to an extent that does not cause maximum circuit group size increases for the other routes.

Consequently, route 1-3 is selected in Step 4. The possible reduction is 1 trunk, resulting in 13 trunks for routes 1-2 and 1-3, and 18 trunks for route 2-3. The circuit group sizes after Step 5 are shown in Table 5.

Following Step 5, all routes show maximum circuit group sizes in time period III, and further reductions are not possible. These circuit group sizes are therefore the final results, and the repetition process is completed. Figure 3 shows the circuit group size reductions in this example schematically.

In Step 4, Condition A is a criterion for large scale reduction, while Condition B is a criterion for large scale reduction in other routes. If the route which enables extremely large reduction is selected by Condition A, and the maximum possible reduction is made at once, reductions of other routes are restricted. The upper bound for immediate reduction in circuit sizes is therefore set.

In order to evaluate Conditions A and B of Step 4. it is possible to apply single hour design parameters which are used in marginal occupancy procedures.

In the sample network of Figure 1, all routes to be reduced are operated as a high usage groups, which means that the amount of final circuit group size adjustment required to decrease this high usage group size is uniquely determined. However, routes to be reduced are not necessarily operated as high usage groups. For circuit group size reduction of the final circuit group, circuit group sizes for many high usage groups must be adjusted. and the adjustment amounts among them cannot be determined uniquely. The assignment of circuit group size adjustment amount among high usage groups should therefore be

Table 3 Required increase of circuit group size

(1) For the reduction of circuit group size of route 1-2 (Time period I)

Route	Circuit group Sizes			
1-2	13	12	11	10
1-3	11.95	12.43	13.03	13.72
2-3	14.91	15.36	15.94	16.59
Total	39.86	39.79	39.97	40.31

(2) For the reduction of circuit group size of route 1-3 (Time period II)

Route	Circuit group Sizes			
1-3	14	13	12	11
1-2	9.96	10.48	11.12	11.84
2-3	15.95	16.41	16.99	17.65
Total	39.91	39.89	40.11	40.31

(3) For the reduction of circuit group size of route 2-3 (Time period III)

Route	Circuit group Sizes			
2-3	20	19	18	17
1-2	11.93	12.40	12.97	13.60
1-3	10.95	11.43	12.02	12.66
Total	42.88	42.83	42.99	43.26

Table 4 Circuit group sizes after first step 5

Time period / Route	I	II	III	Maximum
1-2	13	10	13	13
1-3	12	14	13	14
2-3	15	16	18	18

Table 5 Circuit group sizes after second step 5

Time period / Route	I	II	III	Maximum
1-2	13	11	13	13
1-3	12	13	13	13
2-3	15	17	18	18

optimized. This optimization is possible through the repetition of circuit group size dimensioning with increased final group circuit cost.

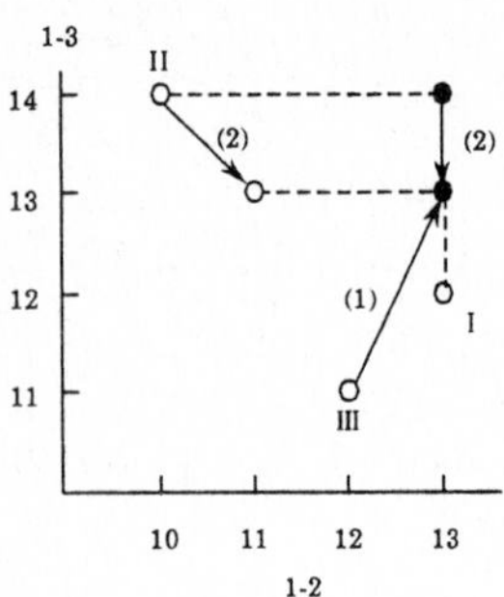

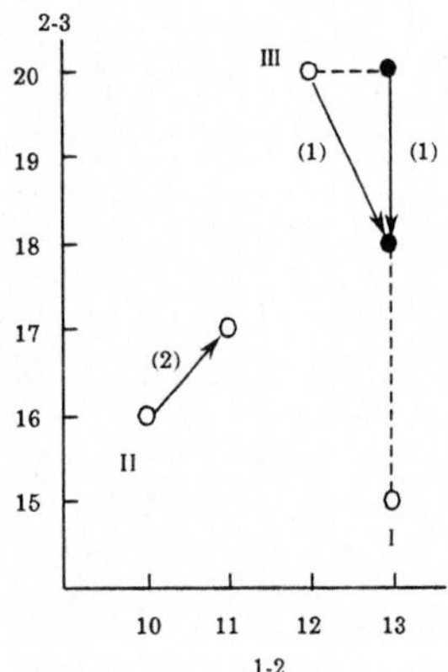

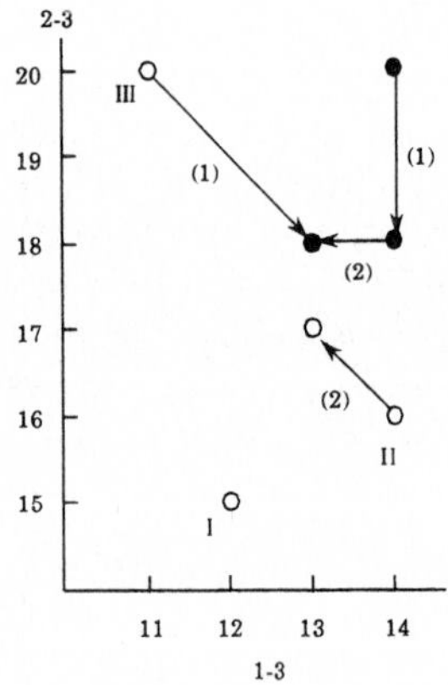

○ : Single hour dimesioning results
● : Initial solution
(1), (2): Number of repetitions of Step 5

Fig. 3 - Circuit group size reductions for sample network

3 NUMERICAL RESULT

A sample 4-node telephone network is shown in Figure 4. and the time difference in local standard time between the two nations connected by the route and the busy hour traffic demands are given. This network is a model of the international telephone network existing between Japan (Node 1), other asian nations (Node 2), Europe (Node 3), and North America (Node 4). The hourly traffic demands listed are calculated from the standard traffic profiles given in CCITT Recommendation E.523. Table 6 shows the traffic demands for 6 time periods. Traffic demands for the remaining time periods are small, and design results for the larger traffic demand of the listed 6 periods satisfies the service grade requirement for the remaining 18 periods as well.

Four routing patterns are considered for this network, as shown in Figure 5, and a routing pattern is selected according to hourly traffic demands that capitalizes on idle network capacity. Traffic demand to/from Node 3 is considerably small in time periods 9, 10, and 11 compared with the traffic demand in other time period, and traffic demand to/from Node 4 is considerably small in time period 17. In these time periods, therefore, routing patterns are used which call overflow to the idle node (patterns a and b).

In time periods 22 and 23, traffic demands on all routes are high, with traffic demand to/from routes 1 and 2 slightly lower. In these time periods, therefore, routing patterns c and d are considered. Time varying routing patterns for this sample network are summarized in Table 7,

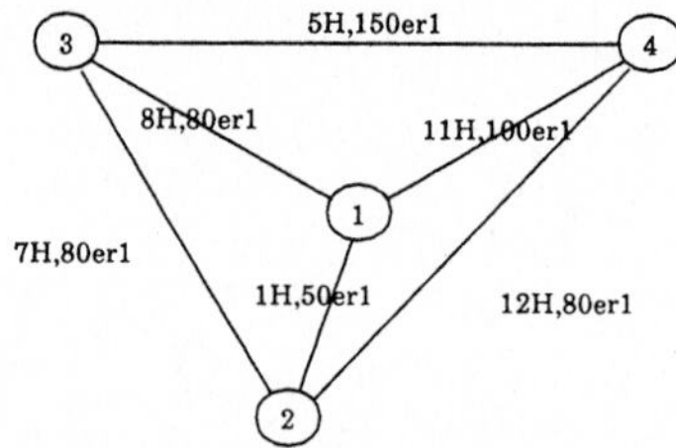

Fig. 4 Sample 4-node network

Table 6 Hourly traffic demand

Route \ Time period*	9	10	11	17	22	23
1-2	35.0	47.5	50.0	30.0	12.5	7.5
1-3	4.0	4.0	4.0	80.0	48.0	40.0
1-4	100.0	95.0	70.0	5.0	70.0	60.0
2-3	4.0	4.0	4.0	80.0	52.0	40.0
2-4	64.0	80.0	64.0	4.0	80.0	64.0
3-4	7.5	7.5	7.5	7.5	142.5	150.0
Total	214.5	237.5	199.5	206.5	405.0	361.5

* Local time in node 1

where routing pattern c and d are used in time period 22 and 23 for Case 1 and Case 2 respectively.

The reduction of the initial solutions for sample network using proposed dimensioning algorithm are shown in Table 8, where peakedness of overflow traffic is ignored and final circuit group blocking is assumed to be 1%. Total trunk requirements to serve the entire traffic demand are nearly the same for Cases 1 and 2, but the circuit group sizes of routes 1-3, 2-3, 1-4, and 2-4 differ remarkably according to the routing pattern. Routes which are operated as final circuit groups in time periods 22 and 23 have large circuit group sizes. In both Cases the sum of route 1-3 and 2-3 circuit group sizes is nearly equal to that of routes 1-4 and 2-4, so that the total circuit group sizes of the two Cases nearly equal each other. For this sample network, consequently, no significant difference is found from a network efficiency point of view between final circuit group selections among routes which have equivalent idle capacity.

Next. the dimensioning results for the sample network are compared with the results for time fixed routing (Case 3) and direct routing (Case 4). Routing pattern b is selected for time fixed routing since the peak hour traffic demand to/from Node 4 is the largest. The dimensioning results of Cases 3 and 4 are shown in Table 9, and the circuit group capacity for each Node is shown in Table 10. A comparison of these data allows the features of

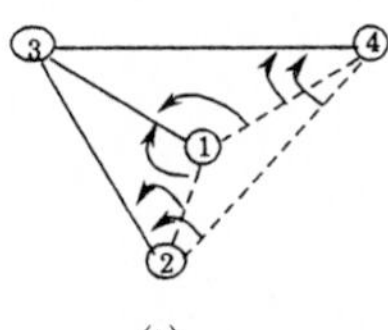
(a)

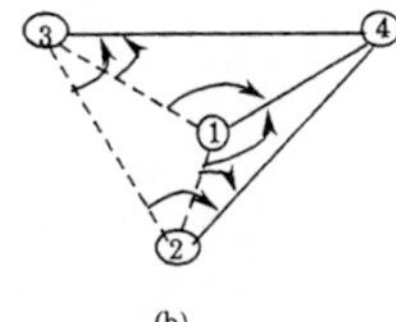
(b)

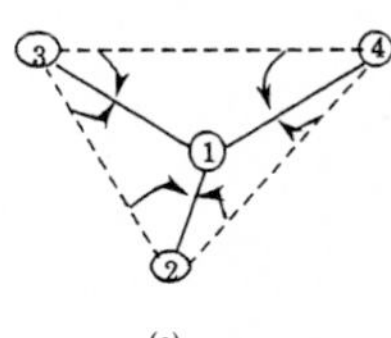
(c)

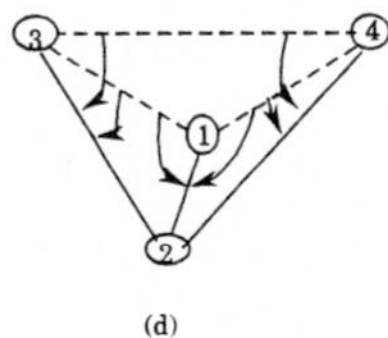
(d)

———— final circuit group
-------- high usage circuit group

Fig. 5 Routing patterns for sample network

Table 7 Hourly routing pattern

Time period	9	10	11	17	22	23
Case 1	a	a	a	b	c	c
Case 2	a	a	a	b	d	d

Table 8 Dimensioning results for dynamic routing scheme

Case 1

Route	1-2	1-3	1-4	2-3	2-4	3-4	Total
Initial Solution	42	71	95	88	88	157	541
1	42	71	95	78	88	157	531
2	42	71	95	68	88	157	521
3	42	71	95	58	88	157	511
4	42	71	95	58	88	150	504
5	42	71	95	58	86	150	502
6	42	71	95	57	86	150	501
7	39	71	95	57	86	150	498
Final result	29	71	95	57	86	150	488

Case 2

Route	1-2	1-3	1-4	2-3	2-4	3-4	Total
Initial Solution	42	88	109	75	106	152	572
1	42	78	109	75	106	152	562
2	42	68	109	75	106	152	552
3	42	58	109	75	106	152	542
4	42	58	109	75	106	150	540
5	42	58	103	75	106	150	534
6	42	58	93	75	106	150	524
7	42	58	83	75	106	150	514
8	42	58	77	75	106	150	508
9	42	58	76	75	106	150	507
10	42	54	76	75	106	150	503
11	39	54	76	75	106	150	500
Final result	35	54	76	75	106	150	496

Table 9 Dimensioning results for time fixed routing scheme and direct routing scheme

Route	1-2	1-3	1-4	2-3	2-4	3-4	Total
Case 3	48	52	117	57	101	170	545
Case 4	64	96	117	96	96	170	639

Table 10 Circuit group capacity per node

Node	1	2	3	4
Case 1	195 (0.704)	172 (0.672)	278 (0.768)	331 (0.864)
Case 2	165 (0.596)	216 (0.844)	279 (0.771)	332 (0.867)
Case 3	217 (0.783)	206 (0.805)	279 (0.771)	388 (1.013)
Case 4	277 (1.000)	256 (1.000)	362 (1.000)	383 (1.000)

The number in parentheses is the circuit group size compared to Case 4 as unity.

dynamic routing in international networks to be summarized as follows:

(1) The utilization of a dynamic routing scheme in the international telephone network can result in savings in total required circuit group sizes of 25% over direct routing, and 10% over time fixed routing schemes.

(2) Dynamic routing allows the circuit group size decreases which are roughly equal for all Nodes, while time fixed routing decreases circuit group size unequally.

4. CONCLUSIONS

A design method for networks with dynamic routing is proposed, and design results for a model international telephone network indicate the potential for significant cost savings. The results of this paper are summarized as follows:

(1) The multiple hour network dimensioning method proposed in this paper treats multiple hour traffic demands simultaneously, and a near optimum solution is obtained regardless of differences in busy hours.

(2) Dynamic routing can realize remarkable savings in the construction cost of international networks.

(3) In networks with dynamic routing, circuit group size reductions are nearly equal for all networked nations.

From the above results, it is clear that the dynamic routing scheme is suited to the international network.

ACKNOWLEDGMENT

The authors wish to acknowledge the continued guidance of Drs. H. Kaji and K. Nosaka of KDD Research and Development Laboratories, and also wish to express their gratitude to the members of the switching system laboratory for helpful discussions on all aspects of this paper.

REFERENCES

[1] G. R. Ash, R. H. Cardwell and R. P. Murry, "Design and optimization of networks with dynamic routing," Bell Syst. Tech. J., Vol.60, pp.1787-1820, 1981.

[2] T. Ohta, "Network efficiency and network planning considering telecommunication traffic influenced by time difference," Proceedings of 7th ITC, Stockholm Sweden, 425/1-425/8, 1973.

[3] R. W. Horn, "A simple approach to dimensioning a telecommunication networks for many hours of traffic demands," Proceedings of ICC, Denver U.S.A., 67.2.1-67.2.5, 1981.

[4] C. J. Truitt, "Traffic engineering techniques for determining trunk requirements in alternate routed networks," Bell Syst. Tech. J., Vol.33, pp.277-302, 1954

[5] C. W. Pratt, "The concept of marginal overflow in alternate routing," Proceedings of 5th ITC, pp.14-20, New York U.S.A., 1967.

TELETRAFFIC ISSUES in an Advanced Information Society
ITC-11
Minoru Akiyama (Editor)
Elsevier Science Publishers B.V. (North-Holland)

LOAD SHARING DYNAMIC ROUTING IN A TELEPHONE NETWORK

Santiago AVILA and Antonio GUERRERO

Centro de Investigación de Standard Eléctrica, S.A.
Madrid, Spain

ABSTRACT

The paper describes the work done on "Load Sharing Dynamic Routing" applied to telephone networks. A model for the traffic in a n-stage telephone network operating under the load sharing policy is developed. An algorithm is described where the optimal load sharing percentages are calculated in order to minimize the global loss of the network assuming a constant mean offered traffic. A centralised control algorithm is described and it is used to dynamically regulate the network by tracking the optimal carried traffic calculated at the previous step. A simulation program has been used to validate the algorithms. Finally some of the conclusions obtained are presented.

1. INTRODUCTION

The Dynamic Routing concept, as opposed to the hierarchical routing one, appeared in the Traffic literature during last years as a consequence of the introduction of modern digital exchanges and CCS techniques.

Szybicki, Bean, etc. [1], developed algorithms to dynamically route the calls progressing through a metropolitan network considering the exchanges as decision points where calls should be diverted according to the expected number of free trunks.

Karstad and Stordahl [2] developed ARIMA models to forecast the network traffic and applied the same routing philosophy as mentioned above.

Ash, Kafker and Khrishnan [3] have applied DNHR to intercity networks. Cameron et al. [4] also report an extension of Szybicki's method to the intercity Canadian network. Field [5] reports results from its application to the Chicago metropolitan network.

Results of the DNHR algorithms performing under overload conditions have being given by Akimpelu [6].

The approach used here is the Load Sharing Dynamic Routing (LSDR) which means the "sharing" of traffic load from any origin to any destination through all allowed routes in the network.

The load sharing policy has been taken from J. Bernussou and A. Titli [7] which they have applied to the overflow traffic routing problem of a telephone network with just one intermediate stage between the origin and destination nodes.

The aim of this paper is to apply Load Sharing Dynamic Routing to all type of traffic flows in a n-stage telephone network in order to keep it working in an environment where the global traffic loss in the long run is minimized. A new traffic model is developed here applying to a far more general network, i.e. a network where routing is allowed throught any number of intermediate nodes.

Looking at the characteristics of the traffic offered to a network two aspects can be distinguished:

- the average busy hour traffic which varies slowly, and
- the fluctuations (random) around the average traffic.

Based on these traffic characteristics the idea is firstly to find the optimal sharing traffic percentages which make the network work with a minimum global traffic loss assuming a constant mean offered traffic, and secondly to design a control procedure to adjust those percentages in order to keep the global loss traffic as low as possible correcting the effect produced by the random traffic fluctuations.

Having distinguished these two basic parts a vertical decomposition of the problem is done taking into account the complexity of the control function.

The two control levels considered in a bottom-up order are:

- Regulation of the network, i.e. direct control over the network once the optimum load sharing parameters have been defined for a given offered traffic.
- Dynamic optimization, i.e. determination of the inputs to the regulation process. Each time the mean offered traffic or trunk matrices substantially change the optimization module calculates the new set points to be tracked by the regulation process.

The paper is then split into two parts, the first one deals with the traffic model of the network and the solution of the optimization problem, the second one is related with the design of a centralized control mechanism.

2. NETWORK DESCRIPTION AND NOTATION

The network is formed up of a set of nodes of different types: end, transit or end-transit nodes.

$$N = \{ i \mid 1 \leqslant i \leqslant n \}$$

The nodes are denoted as type E for end nodes, T for tandem and ET for those nodes having both characteristics.

These nodes are interconnected by links, being the set L defined as:

$$L = \{ (ij) \mid i,j \in N,\ l_{ij} > 0 \}$$

where l_{ij} is the number of circuits between i and j.

The end or end-transit nodes combines between them to form up the set of origin-destination pairs, being this set defined as:

$$OD = \{ (o,d) \mid o,d \in N;\ o,d \notin T \}$$

Associated to the set OD is the offered traffic matrix A, their elements being the mean offered traffic between each origin and destination pair.

Each pair (o,d) has a set of routes, which is defined as the set of all possible routes from o to d, this means direct traffic (if possible) and traffic through any allowed number of stages (intermediate tandems). The set is defined as:

$$R_{od} = \{ r^i_{od} \mid 1 \leqslant i \leqslant n_{od} \}$$

n_{od} = number of allowed routes

r^i_{od} = i-th route from o to d

A route is defined as the series of nodes a call passes through on its path from its origin to its destination. Associated with these sets of routes there are sets of sharing percentages defined by:

$$S_{od} = \{ s^i_{od} \mid 1 \leqslant i \leqslant n_{od} \}$$

s^i_{od}= i-th percentage of traffic flow from o to d through the route $r^i_{od} \in R_{od}$.

The s' must satisfy:

$$\sum_{i=1}^{n_{od}} s^i_{od} = 1 \text{ and } \forall\ i:\ s^i_{od} \geqslant 0$$

3. TRAFFIC MODEL

The following assumptions are considered:

- Offered traffic follows a Poissonian law.
- Holding times of calls are equally distributed according to the negative exponential law with mean taken as the unity for the sake of convenience.
- Traffic process can be modeled by using only the first moment approach.

The apparent simplicity of the resultant network model in which only the first moment of the traffic flows are considered could be surprising when comparing with other classical models such as the Equivalent Random Model, the Interrupted Poisson Process, etc., that were developed to cope with the overflow traffic problem. Nevertheless it should be considered that in the proposed Load Sharing Routing policy there is no traffic overflow and consequently all flows keep the Poissonian characteristic of the original offered traffic. Therefore the first moment approach here seems to be consistent.

The model which is to be used is derived from the well known Markovian formulation of traffic processes, in which the carried traffic variations in a trunk are related with the offered traffic and the loss probabilities.

For any link ij there is a mean carried traffic c_{ij} being

$$\dot{c}_{ij} = - c_{ij} + \sum_{(o,d)\in OD} A_{od} \sum_{\substack{(ij)\subset r,\\ r\in R_{od}}} s_r (1 - p_r) \tag{1}$$

where

$$\sum_{r\in R_{od}} s_r = 1 \tag{2}$$

and

$$s_r \geqslant 0 \tag{3}$$

A_{od} is the mean offered traffic from o to d, p_r is the loss probability in route r and s_r is its associated load sharing percentage.

The calculation of the loss probabilities across a route is a serious problem since they depends on the loss probabilities on all the links forming the route which are not independent. Nevertheless this dependence can be disregarded in large networks since of the many flows passing through a link only one is going to use the next links in the route at the same time.

In such situations the independence of those probabilities can be assumed and then it can be written

$$1 - p_r = \prod_{(ij)\subset r} (1 - p_{ij}) \tag{4}$$

The calculation of p_{ij} is done by means of the following approximation:

$$p_{ij} = E (T_{ij}, l_{ij}) \tag{5}$$

where T_{ij} is a fictitious offered traffic which would give in the stationary situation a carried traffic c_{ij}, i.e. [7]

$$c_{ij} = T_{ij} [1 - E (T_{ij}, l_{ij})] \tag{6}$$

4. OPTIMIZATION

4.1 Optimization Problem

The optimizing criterion chosen here is the global loss in the network (number of calls lost over the whole network).

To calculate the load sharing percentages that minimizes the global loss in the network it is assumed that in the stationary case the offered traffic matrix A remains constant. Under this assumption the criterion can be expressed as

$$\min J(s) = \sum_{(o,d)\in OD} A_{od} \, [1 - \sum_{r\in R_{od}} s_r (1 - p_r)] \qquad (7)$$

subject to the traffic equations in the stationary case, i.e. when $\dot{c}_{ij} = 0$:

$$c_{ij} = \sum_{(o,d)\in OD} A_{od} \sum_{\substack{(ij)\subset r,\\ r\in R_{od}}} s_r \, (1 - p_r) \qquad (8)$$

and (2), (3), (4), (5) and (6).

This is a multivariable non linear optimization problem with equality and non equality constraints which is solved applying a conventional step-size procedure along a direction of descent.

4.2 Some Results

Let consider the network shown in Fig. 1. The network has three local and two tandem exchanges. (The alternate routing pattern is for later use).

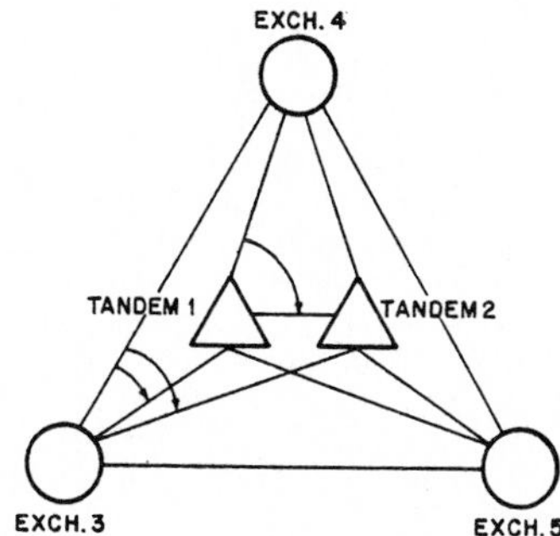

Figure 1

Input data is given in Table 1.

INPUT DATA

TRUNK MATRIX

FROM \ TO	TANDEM1	TANDEM2	EXCH.3	EXCH.4	EXCH.5
TANDEM1	0	30	60	60	60
TANDEM2	0	0	60	60	60
EXCH.3	70	70	0	80	80
EXCH.4	70	70	80	0	80
EXCH.5	70	70	80	80	0

TRAFFIC MATRIX

FROM \ TO	EXCH.3	EXCH.4	EXCH.5
EXCH.3	0.00	150.00	150.00
EXCH.4	110.00	0.00	110.00
EXCH.5	110.00	110.00	0.00

TOTAL OFFERED TRAFFIC 740.00 ERLANGS

Table 1

Table 2 gives the set of routes.

LIST OF COMPLETE ROUTES

ROUTE	
ROUT1 :	EXCH.3 EXCH.4
ROUT2 :	EXCH.3 TANDEM1 EXCH.4
ROUT3 :	EXCH.3 TANDEM1 TANDEM2 EXCH.4
ROUT4 :	EXCH.3 TANDEM2 EXCH.4
ROUT5 :	EXCH.3 EXCH.5
ROUT6 :	EXCH.3 TANDEM1 EXCH.5
ROUT7 :	EXCH.3 TANDEM1 TANDEM2 EXCH.5
ROUT8 :	EXCH.3 TANDEM2 EXCH.5
ROUT9 :	EXCH.4 EXCH.3
ROUT10:	EXCH.4 TANDEM1 EXCH.3
ROUT11:	EXCH.4 TANDEM1 TANDEM2 EXCH.3
ROUT12:	EXCH.4 TANDEM2 EXCH.3
ROUT13:	EXCH.4 EXCH.5
ROUT14:	EXCH.4 TANDEM1 EXCH.5
ROUT15:	EXCH.4 TANDEM1 TANDEM2 EXCH.5
ROUT16:	EXCH.4 TANDEM2 EXCH.5
ROUT17:	EXCH.5 EXCH.3
ROUT18:	EXCH.5 TANDEM1 EXCH.3
ROUT19:	EXCH.5 TANDEM1 TANDEM2 EXCH.3
ROUT20:	EXCH.5 TANDEM2 EXCH.3
ROUT21:	EXCH.5 EXCH.4
ROUT22:	EXCH.5 TANDEM1 EXCH.4
ROUT23:	EXCH.5 TANDEM1 TANDEM2 EXCH.4
ROUT24:	EXCH.5 TANDEM2 EXCH.4

Table 2

The theoretical results showing the optimal carried traffic per link and the optimal load sharing percentages for each route are in Tables 3 and 4.

OPTIMIZING RESULTS - LSDR : LINKS

FROM	TO	CARRIED TRAF.	LOSS PROB.
TANDEM1	TANDEM2	19.101	0.0055
TANDEM1	EXCH.3	44.799	0.0055
TANDEM1	EXCH.4	49.453	0.0254
TANDEM1	EXCH.5	49.453	0.0254
TANDEM2	EXCH.3	46.192	0.0091
TANDEM2	EXCH.4	50.112	0.0308
TANDEM2	EXCH.5	50.112	0.0308
EXCH.3	TANDEM1	62.940	0.0732
EXCH.3	TANDEM2	62.627	0.0676
EXCH.3	EXCH.4	74.059	0.1019
EXCH.3	EXCH.5	74.059	0.1019
EXCH.4	TANDEM1	51.400	0.0024
EXCH.4	TANDEM2	34.148	0.0000
EXCH.4	EXCH.3	63.621	0.0073
EXCH.4	EXCH.5	68.395	0.0262
EXCH.5	TANDEM1	51.400	0.0024
EXCH.5	TANDEM2	34.148	0.0000
EXCH.5	EXCH.3	63.621	0.0073
EXCH.5	EXCH.4	68.395	0.0262

Table 3

OPTIMIZING RESULTS - LSDR : SHARING PERCENTAGES

NO.	VALUE	ON ROUTE	CARRIED TRAF.	LOSS PROB.
*	0.5497	ROUT1	74.059	0.1019
1	0.1996	ROUT2	27.041	0.0968
2	0.0268	ROUT3	3.590	0.1067
3	0.2239	ROUT4	30.349	0.0963
*	0.5497	ROUT5	74.059	0.1019
4	0.1996	ROUT6	27.041	0.0968
5	0.0268	ROUT7	3.590	0.1067
6	0.2239	ROUT8	30.349	0.0963
*	0.5826	ROUT9	63.621	0.0073
7	0.2053	ROUT10	22.400	0.0080
8	0.0268	ROUT11	2.897	0.0170
9	0.1853	ROUT12	20.199	0.0091
*	0.6385	ROUT13	68.395	0.0262
10	0.2096	ROUT14	22.413	0.0278
11	0.0268	ROUT15	2.833	0.0385
12	0.1251	ROUT16	13.340	0.0308
*	0.5826	ROUT17	63.621	0.0073
13	0.2053	ROUT18	22.400	0.0080
14	0.0268	ROUT19	2.897	0.0170
15	0.1853	ROUT20	20.199	0.0091
*	0.6385	ROUT21	68.395	0.0262
16	0.2096	ROUT22	22.413	0.0278
17	0.0268	ROUT23	2.833	0.0385
18	0.1251	ROUT24	13.340	0.0308

TOTAL CARRIED TRAFFIC 702.270 ERLANGS

(THE * MARKS THE NON-BASIC VARIABLES FOR CONTROL)

Table 4

A simulation program has been written to simulate a network operating under the LSDR policy.

Tables 5 and 6 show the simulation results of the studied example. The simulation length is 3 hours and 50,000 calls have been generated in that time, routing has been done with the fixed load sharing percentages calculated theoretically.

All simulation results have been obtained with a 95% confidence and confidence intervals are in the range of 10%.

SIMULATION RESULTS - LSDR

CARRIED TRAFFIC IN TRUNK GROUPS

FROM	TO	CARRIED TRAFFIC
TANDEM1	TANDEM2	17.3453
TANDEM1	EXCH.3	43.7285
TANDEM1	EXCH.4	49.6130
TANDEM1	EXCH.5	49.2658
TANDEM2	EXCH.3	43.1662
TANDEM2	EXCH.4	49.9065
TANDEM2	EXCH.5	51.5676
EXCH.3	TANDEM1	61.3936
EXCH.3	TANDEM2	62.0471
EXCH.3	EXCH.4	73.9595
EXCH.3	EXCH.5	74.0386
EXCH.4	TANDEM1	50.1512
EXCH.4	TANDEM2	33.0724
EXCH.4	EXCH.3	64.2328
EXCH.4	EXCH.5	68.1054
EXCH.5	TANDEM1	48.4078
EXCH.5	TANDEM2	32.1755
EXCH.5	EXCH.3	63.6354
EXCH.5	EXCH.4	69.9184

Table 5

SIMULATION RESULTS - LSDR

END-TO-END OFFERED AND CARRIED TRAFFIC

FROM	TO	OFFERED	CARRIED
EXCH.3	EXCH.4	148.8752	135.8667
EXCH.3	EXCH.5	149.0460	135.5721
EXCH.4	EXCH.3	108.7374	108.1565
EXCH.4	EXCH.5	110.8510	107.4053
EXCH.5	EXCH.3	107.2664	106.6064
EXCH.5	EXCH.4	111.9626	107.5307

TOTAL OFFERED TRAFFIC 736.7385

TOTAL CARRIED TRAFFIC 701.1377

Table 6

The shown results as well as many others, for a lot of different networks and situations, validate the network traffic model.

In order to compare the behavior of the network operating under LSDR with the behavior of the classical alternate fixed routing policy, another simulation program for alternate hierarchical routing policy has been run.

Tables 7 and 8 show the simulation results with the alternate routing policy. Fig. 1 showed the alternate routing pattern for the traffic going from EXCH.3 to EXCH.4 which has been used in this example. Similar patterns apply to the other origin destination pairs, being omitted in the figure for simplicity.

SIMULATION RESULTS - A.R.

CARRIED TRAFFIC IN TRUNK GROUPS

FROM	TO	CARRIED TRAFFIC
TANDEM1	TANDEM2	22.10
TANDEM1	EXCH.3	52.04
TANDEM1	EXCH.4	54.66
TANDEM1	EXCH.5	54.94
TANDEM2	EXCH.3	8.80
TANDEM2	EXCH.4	41.08
TANDEM2	EXCH.5	39.83
EXCH.3	TANDEM1	68.61
EXCH.3	TANDEM2	58.52
EXCH.3	EXCH.4	80.06
EXCH.3	EXCH.5	79.57
EXCH.4	TANDEM1	56.87
EXCH.4	TANDEM2	4.77
EXCH.4	EXCH.3	78.45
EXCH.4	EXCH.5	78.20
EXCH.5	TANDEM1	58.26
EXCH.5	TANDEM2	4.31
EXCH.5	EXCH.3	79.02
EXCH.5	EXCH.4	76.64

Table 7

SIMULATION RESULTS - A.R.

END-TO-END OFFERED AND CARRIED TRAFFIC

FROM	TO	OFFERED	CARRIED
EXCH.3	EXCH.4	148.88	142.58
EXCH.3	EXCH.5	149.05	141.58
EXCH.4	EXCH.3	108.74	107.90
EXCH.4	EXCH.5	110.85	109.11
EXCH.5	EXCH.3	107.27	106.64
EXCH.5	EXCH.4	111.96	109.65

TOTAL OFFERED TRAFFIC 736.74

TOTAL CARRIED TRAFFIC 717.46

Table 8

It is interesting to notice the difference between the carried traffic per link distribution in both cases. The total carried traffic in the alternate routing case (AR) is greater than the LSDR case, but this difference is not significative since it lies inside the confidence interval of the simulation. These results are as it can be expected since the network is well dimensioned and the alternate routing policy behaves optimally in these cases.

The question now is, what happens when the network is not well dimensioned or even more if a link or exchange break down.

Let suppose now that the trunk matrix of the network becomes for any reason the one shown in Table 9, where a partial breakdown of links going from TANDEM.1 to TANDEM.2, EXCH.4 and EXCH.5 is assumed.

INPUT DATA

TRUNK MATRIX

FROM \ TO	TANDEM1	TANDEM2	EXCH.3	EXCH.4	EXCH.5
TANDEM1	0	10	60	20	20
TANDEM2	0	0	60	60	60
EXCH.3	70	70	0	80	80
EXCH.4	70	70	80	0	80
EXCH.5	70	70	80	80	0

Table 9

The LSDR algorithm detects this fact and finds a new set of optimal load sharing percentages to cope with this new situation producing the results shown in Tables 10 and 11. The total carried traffic value in the AR case is the one shown in Table 12.

OPTIMIZING RESULTS - LSDR: LINKS

FROM	TO	CARRIED TRAF.	LOSS PROB.
TANDEM1	TANDEM2	6.818	0.1045
TANDEM1	EXCH.3	46.002	0.0085
TANDEM1	EXCH.4	18.039	0.2849
TANDEM1	EXCH.5	18.039	0.2849
TANDEM2	EXCH.3	45.528	0.0072
TANDEM2	EXCH.4	54.248	0.0969
TANDEM2	EXCH.5	54.248	0.0969
EXCH.3	TANDEM1	49.624	0.0012
EXCH.3	TANDEM2	62.454	0.0647
EXCH.3	EXCH.4	76.058	0.1712
EXCH.3	EXCH.5	76.058	0.1712
EXCH.4	TANDEM1	27.419	0.0000
EXCH.4	TANDEM2	48.361	0.0007
EXCH.4	EXCH.3	63.335	0.0067
EXCH.4	EXCH.5	73.416	0.0871
EXCH.5	TANDEM1	27.419	0.0000
EXCH.5	TANDEM2	48.361	0.0007
EXCH.5	EXCH.3	63.335	0.0067
EXCH.5	EXCH.4	73.416	0.0871

Table 10

OPTIMIZING RESULTS - LSDR: SHARING PERCENTAGES

NO.	VALUE	ON ROUTE	CARRIED TRAF.	LOSS PROB.
*	0.6118	ROUT1	76.058	0.1712
1	0.1536	ROUT2	16.454	0.2858
2	0.0120	ROUT3	1.457	0.1922
3	0.2226	ROUT4	28.201	0.1554
*	0.6118	ROUT5	76.058	0.1712
4	0.1536	ROUT6	16.454	0.2858
5	0.0120	ROUT7	1.457	0.1922
6	0.2226	ROUT8	28.201	0.1554
*	0.5797	ROUT9	63.335	0.0067
7	0.2109	ROUT10	23.001	0.0085
8	0.0082	ROUT11	0.798	0.1109
9	0.2013	ROUT12	21.966	0.0079
*	0.7311	ROUT13	73.416	0.0871
10	0.0201	ROUT14	1.585	0.2849
11	0.0101	ROUT15	0.896	0.1913
12	0.2387	ROUT16	23.693	0.0975
*	0.5797	ROUT17	63.335	0.0067
13	0.2109	ROUT18	23.001	0.0085
14	0.0082	ROUT19	0.798	0.1109
15	0.2013	ROUT20	21.966	0.0079
*	0.7311	ROUT21	73.416	0.0871
16	0.0201	ROUT22	1.585	0.2849
17	0.0101	ROUT23	0.896	0.1913
18	0.2387	ROUT24	23.693	0.0975

TOTAL CARRIED TRAFFIC 661.720 ERLANGS

(THE * MARKS THE NON-BASIC VARIABLES FOR CONTROL)

Table 11

SIMULATION RESULTS - A.R.

END-TO-END OFFERED AND CARRIED TRAFFIC

FROM	TO	OFFERED	CARRIED
EXCH.3	EXCH.4	148.88	94.78
EXCH.3	EXCH.5	149.05	95.57
EXCH.4	EXCH.3	108.74	104.52
EXCH.4	EXCH.5	110.85	85.48
EXCH.5	EXCH.3	107.27	102.84
EXCH.5	EXCH.4	111.96	86.38

TOTAL OFFERED TRAFFIC 736.74

TOTAL CARRIED TRAFFIC 569.56

Table 12

This adaptation capability is the main advantage of LSDR. When bigger networks are being operated it cannot be easy to change the alternate routing pattern to cope with the new situation.

5. CENTRALIZED CONTROL

5.1 Introduction

The optimization model assumes that the traffic offered to the network remains constant disregarding the fluctuations of this traffic around its mean value.

In fact the perturbations induced by the random process associated to the offered traffic are not negligible and the solution obtained with the optimization model would not be applicable in a long term run if significant deviations from the average conditions are produced.

Therefore a continuous surveillance of the state of the network (the mean carried traffic per link) and the updating of the load sharing percentages are necessary in order to smooth the effect of the transient perturbations keeping the network operating near the optimal situation.

To accomplish this objective a control mechanism of the regulation type is proposed. It acts measuring the deviations of the current states of the links from the target ones (the optimal ones calculated before) and modifies the load sharing percentages in order to cope with the actual situation of the network. (See Fig. 2).

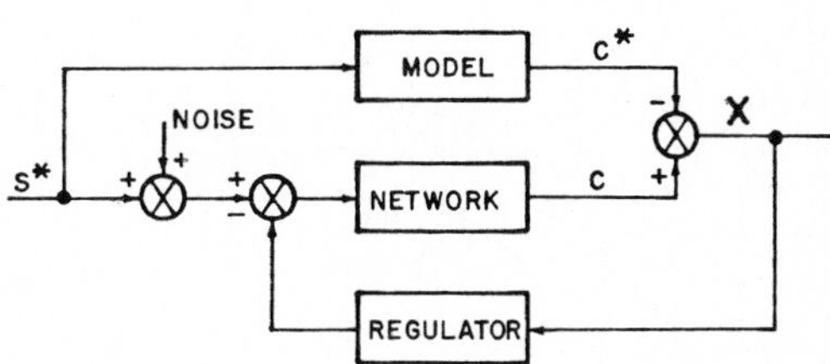

Figure 2

5.2 Control Model

The model of the network we are dealing with is formed up of a set of non-linear differential equations which very schematically can be expressed as:

$$c = f\,(c,s)$$

This model is very difficult to handle and has to be linearized. The first order Taylor expansion of f (c,s) around the optimal point (c^*,s^*) calculated in the optimization step is used and after linearization takes the form:

$$\dot{X} = A\,X + B\,u \qquad (9)$$

where $X = c - c^*$ and $u = s - s^*$. Here A and B are the matrices of the derivatives of f with respect to the c's and s' at the optimal point.

This approach has the drawback of obtaining a traffic model which will only be valid under operating conditions near the optimal values, anyway this is what it is expected to have in a regulation process.

The control model can now be stated as:

$$\min V = \int_{t_o}^{\infty} 1/2\ (X'\ Q\ X + u'\ R\ u)\ dt \quad (10)$$

subject to (9), (2) and (3).

$R > 0$ and $Q \geqslant 0$ are weighting matrices. (' means transpose).

The expression of the cost (10) is a measurement of the states deviations and the control used when the system passes from an initial to a final state in an infinite horizon. The objective is then to find that optimal control vector u which carries the system from the initial state to the final state in a way such that the cost at the final state is minimum.

All the above holds for the stationary case and for those u's which satisfy the equations (2) and (3).

The solution taken for this model is of linear form and is given by

$$u = -K\ X$$

with

$$K = R^{-1}\ B'\ P$$

being P the solution of the steady state matrix Riccati equation

$$A'\ P + P\ A - P\ B\ R^{-1}\ B'\ P + Q = 0$$

The solution fulfills automatically (2) since we have previously expressed one of the s' in every origin destination group as

$$s_{n_{od}} = 1 - \sum_{i=1}^{n_{od}-1} s_i$$

To satisfy contraints (3) it is necessary to be very careful with the selection of the matrix R which weights the control values used in the minimization process.

5.3 Control Results

The same network described in (4.2) is used to implement the regulator. The regulator behavior has been tested with a simulation program and the results are shown in next section.

It is necessary before discussing the results to explain briefly some parameters of capital importance in normal regulation applications, these are the sampling period and the matrix R.

5.3.1 Sampling Period

Sampling period gives the frequency with which the control actuates. Usually results are better as smaller is the sampling period as far as control is concerned. This has otherwise the drawback of increasing measurements, data transmission and processing costs considerably.

Another problem related with this is measurement. It has been said that the model modifies the load sharing percentages depending on the mean carried traffic observed at each link. This observed traffic needs a measurement interval large enough to get consistent measurement values.

A solution for the measurement problem is to consider a moving average process in which a measurement interval large enough (15 minutes) is taken. A new measurement is incorporated at each sampling period (3 minutes), taking away the oldest measurement of the averaging set.

5.3.2 Matrix R

It has been said that matrix R weights the controls in the cost function.

The values of the load sharing percentages are constrained $(0 \leqslant s \leqslant 1)$ and special care must be taken when the optimal values of the s' are closed to the interval limits. In these cases the variations of the values of these load sharing percentages must be very limited in order to prevent them of lying out of their range of validity.

This can be achieved by taking their corresponding values in the main diagonal of R very large, restricting in this way their range of variability.

5.4 Example

The regulator has been implemented in the studied network and its performance is discussed in this section through the results obtained with simulation.

The measurement interval has been taken as 15 minutes which is large enough to obtain confidence measures of the carried traffic at each link.

The sampling period is 3 minutes, i.e. the measurement interval values are updated and the control actions are taken every 3 minutes.

The selected Q matrix has been the identity matrix I and the R matrix is also a diagonal matrix whose elements are either 10,000 or 100,000 depending on their corresponding values of the optimal load sharing parameters, being the greater values corresponding to those cases in which the s' values are less than 0.1 .

Fig. 3 shows the offered traffic from EXCH.4 to EXCH.5 and the corresponding dynamic variations of the load sharing percentages along a simulation period of 3 hours. It can be observed that the s which is responsible of the traffic sent through TANDEM.1 and TANDEM.2 has very small variations which prevent it of going out of its allowed validity range, this has been achieved with the form of the selected R matrix.

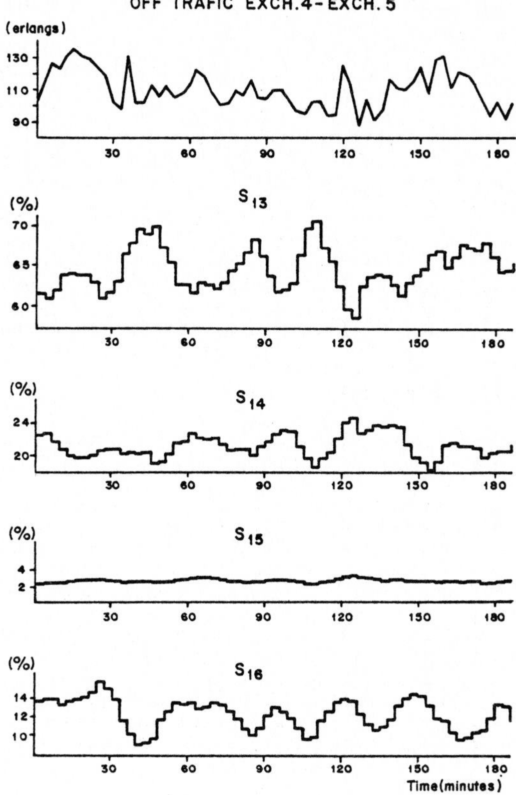

Figure 3

As a measurement of the smoothing effect produced by the regulation, Fig. 4 shows the total accumulated losses of all the links of the network along the total simulation period, being defined the accumulated losses in a link as

$$\sum_{i=1}^{n} x^2_i$$

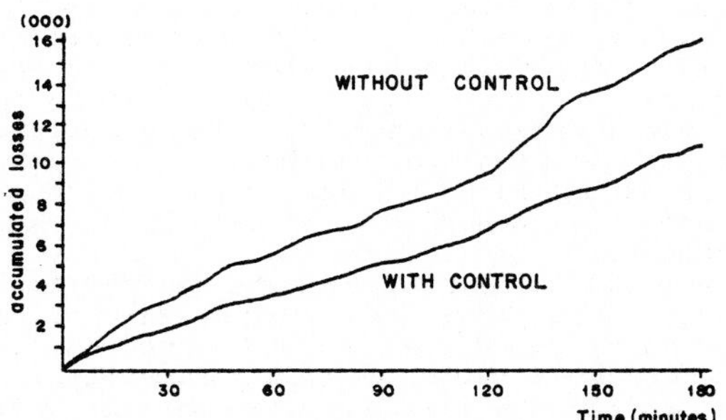

Figure 4

6. CONCLUSIONS

The first moment approach used here for the derivation of the traffic model has been proved to be valid when comparing the results of the theoretical model with those of simulation.

LSDR gives similar results as AR, as far as the total carried traffic is concerned, when a network is properly dimensioned so as to cope with the AR policy. LSDR gives the optimal mean carried traffic in a network under any conditions, being its efficiency greater as worse are the results with AR.

LSDR is adaptive, in the sense that it reacts at the first or second level, against such unpredictable events as traffic fluctuations or links or exchanges breakdowns. This characteristic is what makes the method more attractive.

The regulation does not make any significative improvement in the carried traffic of the network under LSDR policy. This can be surprising at first sight, but this is not so, since the regulator is designed to minimize the effect of the observed traffic fluctuations which are responsible of no few problems.

Finally to say that further studies are being followed up to try to integrate this and other methodologies (Learning Automata theory) in order to consider a mixed decentralized control method.

REFERENCES

[1] E. Szybicki, A.E. Bean, "Advance Traffic Routing in Local Telephone Networks: Performance of Proposed Call Routing Algorithms", Proc. 9th Int. Teletraffic Congress, Torremolinos, 1979.

[2] T. Karstad, K. Stordahl, "Centralized Routing Based on Forecasts of the Telephone Traffic", Proc. 10th Int. Teletraffic Congress, Montreal, 1983.

[3] G.R. Ash, A.H. Kafker and K.R. Krishnan, "Intercity Dynamic Routing Architecture and Feasibility", Proc. 10th Int. Teletraffic Congress, Montreal, 1983.

[4] W.H. Cameron, J. Regnier, P. Galloy and A.M. Savoy, "Dynamic Routing for Intercity Telephone Networks", Proc. 10th Int. Teletraffic Congress, Montreal, 1983.

[5] F.A. Field, "The Benefits of Dynamic Nonhierarchical Routing in Metropolitan Traffic Networks", Proc. 10th Int. Teletraffic Congress, Montreal, 1983.

[6] J.M. Akimpelu, "The Overload Performance of Engineered Networks with Nonhierarchical and Hierarchical Routing", Proc. 10thInt. Teletraffic Congress, Montreal, 1983.

[7] J. Bernussou, A. Titli, "Interconnected Dynamical Systems: Stability, Decomposition and Decentralization", North Holland Systems and Control Series, Vol.5, 1982.

TELETRAFFIC ISSUES in an Advanced Information Society
ITC-11
Minoru Akiyama (Editor)
Elsevier Science Publishers B.V. (North-Holland)

MULTIHOUR DIMENSIONING FOR A DYNAMICALLY ROUTED NETWORK

Ronald HUBERMAN, Serge HURTUBISE, Alexandra LE NIR & Tadeusz DRWIEGA

BNR
Montreal, Canada

ABSTRACT

This paper describes a heuristic algorithm for dimensioning a telephone network to be controlled by BNR's Dynamic Routing (DR) algorithm. DR uses a centralized processor to provide intelligent switches with near-real time alternate route recommendations. The dimensioner incorporates multihour engineering principles to exploit the network's traffic non coincidence. Validation is performed analizing a subset of the Canadian intercity network. The multihour approach extends the trunk saving potential of DR while still maintaining DR's network survivability benefits. Also noteworhty are improvements in network uniformity of grade of service.

1.0 INTRODUCTION

BNR Dynamic Routing (DR) algorithm has been advanced as an effective and viable means of improving network performance ([SZ79], [CA81], [CA83]). The primary benefits cited are increased trunk efficiency (hence reduced trunking costs), improved handling of traffic fluctuations, and improved survivability in case of switch or transmission failure. Full exploitation of these benefits requires new methods of network planning, most notably in the realm of trunk dimensioning. It is asserted here that a sound approach to multihour engineering plays a key role in the realization of the potential benefits.

The trunk dimensioning process is an integral part of the network forecasting function. Based on a point-to-point traffic forecast, this process generates a proposed trunk facility allocation to best meet future demand. The traffic input to this process is an estimate of the customer demand for service and is quite independent of the routing strategy and the current trunking architecture. It is primarily a function of switch deployment and customer behavior. The sources of these data are typically traffic measurements which have been projected to estimate the expected demand in future years.

The trunks obtained by this process are the "logical" circuit requirements between switches in the network. Actual provisioning will require the mapping of these trunk requirements onto physical transmission facilities. Provisioning is not affected by DR, since call routing is irrelevant to provisionning. The only part, then, of the downstream support system directly affected by the routing algorithm is the determination of what logical trunking is required to meet the subscriber traffic demand, i.e., the trunk dimensioner.

In simple terms, the objective of the trunk dimensioning process is to minimize the network trunking costs while satisfying grade of service (GOS) objectives. Within this context, GOS refers to maintaining the probability of loss due to congestion within acceptable limits, and does not bear any relation to hardware service constraints such as transmission quality, dial-tone delay, etc.

The fact that DR changes the rules by which traffic is routed gives rise to different traffic flows than those that would exist under conventional Fixed Hierarchical Routing (FHR). This in turn changes the optimal trunk deployment strategies. Developing the new methodologies is necessary to ensure efficient utilization of network resources. In fact, the efficiency exhibited by an DR dimensioned and controlled network is substantially greater than its FHR counterpart.

Traffic noncoincidence refers to the fact that not all demands for service occur at the same instants in time. Indeed, the entire telephone network is built on the premise that it is not required to provide service to all users at the same time. For dimensioning purposes, traffic noncoincidence often refers to the fact that not all switches in a network reach their peak loads at the same time. The greatest effects of this phenomenon are related to the time of day. In the intercity network, such hourly noncoincidence is primarily due to the effects of different time zones. Metropolitan networks see considerably different hourly behavior in business versus residential areas.

Considering traffic matrices representing the peak loads of several periods of time is important under any routing algorithm to ensure that grade of service is met for all switches, even those which may peak at a time different than an overall network busy hour. However, multihour engineering offers the prospect of reducing trunking cost by exploiting the network idle capacity resulting from traffic noncoincidence. This potential takes on high prominence in a DR environment due to the flexible nature of the routing algorithm. Under conventional FHR routing, the fact that alternate routes are static and generally few limits the amount of noncoincidence which can be exploited.

The wide choice of alternate routes combined with the intelligence of the DR routing processor gives DR a far greater ability to seek out and use capacity idle due to traffic noncoincidence. In short, the same characteristics which allow DR to adapt to unpredictable traffic variations can be depended upon to make use of the noncoincidence predicted by traffic forecasts.

Let us take a moment to put into perspective the effects of traffic noncoincidence, traffic routing, and trunk dimensioning on reducing network trunking cost. Traffic noncoincidence is a (desirable) network phenomenon dependent primarily on customer behavior. In a sense, it provides an upper bound on the possible trunk savings via multiload dimensioning techniques. The routing algorithm determines how much of a network's noncoincidence has the potential to be usable for trunk savings. Flexibility of alternate routes plays a key role here. How much of this "potentially usable noncoincidence" actually gets translated into the realization of trunk savings is dependent on the effectiveness of the multiload dimensioning algorithm.

2.0 THE ROUTING ALGORITHM

A DR network consists of a set of stored program control (SPC) switches, called "participating switches", which have been modified to communicate with a centralized computer, called a routing processor (RP). The basic network architecture consists of switches of two types:

I node — Intelligent nontandem node. A switch which communicates with the central processor and therefore is able to route overflow calls intelligently, but which may not be used as tandem for via traffic.

N node — HPR tandem node. Like an I node, overflow calls are routed intelligently but the use of these switches as tandems for via traffic is permitted.

The sets of I and N nodes may be considered as forming a two-level hierarchy.

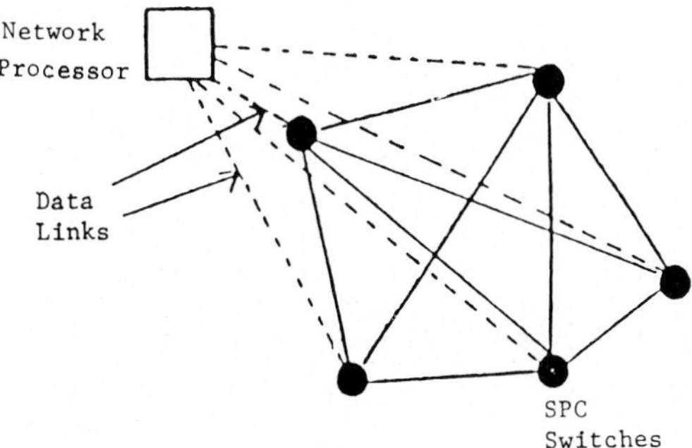

Figure 1. DR Network Architecture.

At time intervals called update cycles, each participating switch sends the RP information on its trunk group occupancies and the state of its own call processing load. The RP performs a calculation called a path selection algorithm which produces a tandem recommendation for each origin-destination pair. Should an alternate route be required within the next cycle this recommendation will be used. The highlights of the path selection algorithm, described in more detail in [CA83], are:

- rapid update cycles, typically 10-15 seconds
- weighted random tandem selection based on the occupancy of the entire two-link path
- direct traffic protection
- protection of overloaded tandem switches

3.0 UNDERLYING CONCEPTS

Before giving the steps of the dimensioning algorithm let us discuss some important concepts underlying it.

3.1 MULTIHOUR ENGINEERING

Rather than consider a single, "busy season", traffic matrix as input to the dimensioning process, multihour engineering assumes the existence of several traffic matrices. These matrices need not reflect actual hourly averages. The only assumption made by the dimensioner is that they represent noncoincident activity, that is, that the matrices do not overlap in time. Combining time periods into traffic "hours" assumes a false coincidence but is necessary to keep the number of input matrices manageable [SW85, BE82].

We may think of multihour engineering as having two objectives; first to ensure an acceptable GOS for all switches, even for those which may reach their peak loads at a time different from the overall "network busy hour", second to reduce the network trunking cost by exploiting the idle capacity resulting from noncoincidence. The first objective may be thought of as a conservative one in that it aims to configure a network which will provide acceptable service under a variety of predictable traffic situations. It is, therefore, foremost a service objective. The second objective may be thought of as improving the efficiency of network trunking resources.

Cameron [CA83] sketched the workings of an algorithm for single-hour dimensioning of a DR network. The algorithm uses two-moment traffic theory and make use of DR concepts first introduced by Lavigne [LA79]. Also presented in [CA83] was a multihour dimensioning method whereby traffic matrices of several hours were combined using the Kruithof procedure [BE76] to produce a single matrix where the total traffic for each switch was adjusted to equal its own peak load over all hours. The success of this method highlighted the significance of providing a switch with enough capacity to handle its own peak load. However, there is a loss of information inherent in the Kruithof merge process. Namely, while the total peak loads of each switch is known, a false coincidence is assumed since it cannot be determined which peaks occur at different times. Also lost are the actual traffic flows, as the different traffic hours may take on varied "shapes". It became apparent that further advancement in multihour dimensioning required enhancing the dimensioning algorithm itself to consider these effects.

The new approach allows access to all the original traffic matrices. The algorithm is still able to cater to a switch's total peak load but now it is able to determine what is happening in the rest of the network at that time. This leads to the "multihour guiding principle" that once enough trunks are provided to "get out of" the switch, additional capacity should only be built when the possibility of using links idle due to non-coincidence has been exhausted. To be able to accomplish this, all traffic flow calculations (overflow, link blocking, trunk efficiency) must be carefully preserved on a per hour basis.

The basic idea, then, of this method is to allocate trunks on the individual links to satisfy the following constraints:

1. The total number of trunks on each bundle must be able to handle the maximum incoming and outgoing traffic taken over all given loads to meet the desired GOS at any time.

2. The trunk allocation must be done in such a way that if, for some load, one part of the network is very busy while the demand on the other switches is below the maximum load for which they were sized, as much as possible of the idle trunks should be used to retandem some of the "busy" traffic before adding new trunks to meet GOS.

3. The network cost must be minimized keeping in mind the survivability aspect. Basically, overflow should be distributed over several possible routes, not totally on the cheapest one in order to maintain a satisfactory diversity.

3.3.2 ECONOMIC CRITERION

The underlying concept of the DR sizer is to keep the traffic on the direct route as long as the cost per unit of traffic carried is lower than the weighted mean cost for carrying the same unit of traffic on alternate routes.

This rule is very similar to the ECCS (Economic CCS) [TR54] principle used to size FHR networks. The only difference is that in FHR networks, since the choice of alternate routes is sequential, only one alternate route need be considered to compute a given ECCS. In DR networks, because many overflow routes must be considered simultaneously, the ECCS is based on a weighted average of the alternatre route costs.

3.3.3 THE CONCEPT OF A BUNDLE

The concept of a bundle is central to the DR sizing algorithm. For a given node, its outgoing trunk group bundle is defined as the set of trunk groups outgoing to all DR tandems. Similarly, the incoming bundle is the set of trunk groups incoming to the node from all DR tandems.

Independent of the particular trunk group chosen, all the calls from a switch must be routed on the trunk group bundle. This means that a bundle must support all the first offered traffic to and from the tandems plus all the overflow traffic intended for nontandem nodes.

When a trunk group is sized, two cases are distinguished:

1. If the trunk group supports only its direct first offered traffic, it is sized using only the economic criterion.

2. If, in addition to its direct traffic, the trunk group also carries overflow traffic, the trunk group is not directly sized. The bundle is sized according to a given grade of service criterion.

3.3.4 MODULE ALLOCATION

When the number of modules in the bundle is determined, the next step is to allocate modules among links in that bundle. The procedure which determines the number of modules for each link should do so in an economical way. Consequently a link should be incremented by adding modules to it as long as it is justified by the cost per carried unit of traffic in comparison to the corresponding cost in the other links in the bundle. To achieve that, the modified ECCS rule, described earlier, is applied. For each link in a bundle the efficiency of the remaining part of the bundle is represented by the weighted average of the other link efficiencies.

3.3.5 TRAFFIC FLOW AND OVERFLOW ALLOCATION

Whereas alternate routing under FHR obeys static tables, overlow in a DR network follows a near-real time recommendation which is proportional to the number of available two-link paths. To distribute overflow while dimensioning, it is therefore necessary to estimate the link occupancies so that the appropriate traffic spread among feasible alternate routes may be determined.

Given a two-link route (i-t and t-j) with known carried traffic, the question becomes one of estimating the average number of two-link paths which will be available. For a single link the average number of available trunks is the number of trunks in the group (N) minus the average number of links in use. This last term is simply the mean of the carried traffic (CM) on the link. If IDLE(x,y) is a random variable representing the number of idle trunks of x-y link we have

$$E(IDLE(x,y)) = N(x,y) - CM(x,y)$$

where E represents the expected value function.

Let us denote A = CEILING [E(IDLE(i,t))] and B = CEILING [E(IDLE(t,j))]. Without loss of generality, assume A <= B. Due to Regnier's [RE82a] approximation let us consider the number of idle trunks on the individual links as independent, discrete random variables, say a and b, taking values 0,1,...,2A and 0,1,...,2B with uniform probabilities 1/(2A+1) and 1/(2B+1) respectively. With this approximation the distribution of the number of idle two-link paths z=min(a,b) can be derived as

$$P(z=k) = P((a=k \text{ and } b=>k) \text{ or } (a>k \text{ and } b=k))$$

$$= \frac{1}{2A+1} * \frac{2B-k+1}{2B+1} + \frac{2A-k}{2A+1} * \frac{1}{2B+1} = \frac{2A+2B+1-2k}{(2A+1)(2B+1)}$$

for k=0,1,...,A. The expectation of this distribution is equal to

$$E(Z) = \frac{k(2A+2B+1-2k)}{(2A+1)(2B+1)} = A - \frac{A^2}{3B} - \frac{A}{2B}$$

Since DR recommendations are proportional to the number of idle routes on a two-link path, the distribution of overflow among a set of feasible tandem routes can be approximated as being proportional to the expected idleness value of each path as computed by the above formula.

Spreading overflow among alternate routes allows the dimensioner to map the routing algorithm and to send larger amounts of overflow to those routes which have more idle capacity. Moreover this flow model also has properties favorable to the multihour concept. Trunks idle due to noncoincidence have a smaller carried traffic than those operating at their engineered load. Distributing overflow based on expected idleness favors the idle alternate routes. The fact that better approximating the DR routing algorithm also capitalizes on noncoincidence is merely a reflection of the fact that the routing algorithm itself sends traffic where there is capacity.

While the need to consider expected idleness seems clear, using the concept alone has its drawbacks. While it is desirable to make use of long-haul alternate routes when there is idle capacity, it would not generally be pragmatic to add new trunks to these routes to cater to such alternate routing. When the network capacity has not yet been "built up" to meet the demand, one still wants to favor construction on the cheapest routes. In short, while the pure expected idleness concept produces a realizable flow, it does not take advantage of the dimensioner's ability to influence the eventual flow through its allocation of capacity.

With this in mind, a hybrid approach was developed. The idea is to still take into account the various alternate route costs, but to ensure the flows assumed by the dimensioner are realizable by the routing algorithm. The approach used was to distribute overflow based on the quotient of expected idleness and cost. The uniformity of service benefits was still strong compared to inverse cost alone, but a better service per cost was found with the hybrid approach. The trade-off here could be of interest as one controlled parametrically to adjust to dimensioning objectives. Increased emphasis on expected idleness tends to favor survivability, whereas the cost aspect improves cost. With this in mind, the algorithm assumes the overflow be distributed proportional to:

$$EI^{beta\_ei} \qquad C^{beta\_c}$$

where beta_ei and beta_c are parameters. Note that

beta_ei = 0 ==> inverse cost rule
beta_c = 0 ==> expected idleness

A sensible range for beta_ei would be [0,1], and for beta_c, [0,10]. Our validation procedures made use of beta_ei = 1 and beta_c = 1, but this could be a place for algorithm fine tuning. This parameter could also be influenced by the application: military networks, for example, lean much more towards expected idleness than cost.

4.0 THE ALGORITHM

In this section the algorithm for dimensioning the DR network is described. It is built upon four main blocks. The motivation of such structure of the algorithm is implied by the DR network architecture.

- All the links between I nodes which carry only direct first-offered traffic from the I-I subnetwork.
- The overflows from I-I links are sent to alternate routes which consist of links connecting I nodes and N nodes. All such links form the I-N subnetwork. This subnetwork carries its own traffic plus the overflow from the I-I subnetwork.
- The overflow from the I-N subnetwork is sent to the N-N subnetwork consisting of all the links between N nodes. The overflows from N to N links are sent to the other links of this subnetwork.

The algorithm can work with one-way or two-way trunks. For two way trunking, we just have to add the traffic from each direction (fold the traffic) together before running the dimensioner. This property comes from the symmetrical nature of the routing algorithm.

4.1 THE MAIN PROCEDURE

The sizing algorithm has the form of consecutive iterations which gradually approximate the final solution, i.e., the trunk requirements. The direction in which the overflows are sent, i.e., from I-I to I-N and from I-N to N-N subnetwork,

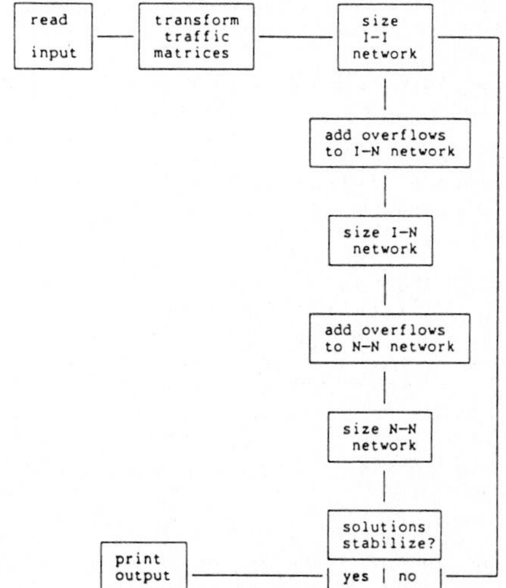

Figure 2. Sizing the DR Network: Basic flow for the main procedure

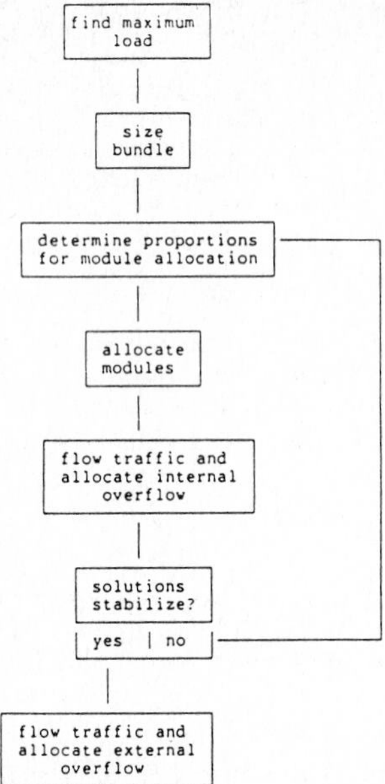

Figure 3. Sizing the I-N subnetwork

determines the order of processing within a given iteration.

Dimensioning the I-N subnetwork affects efficiencies of alternate routes of the previously dimensioned I to I links and a similar relationship exists between the I-N and N-N subnetworks. Therefore, after sizing the N-N subnetwork the whole network is resized until the solution stabilizes.

The global iteration is stopped when the N - N overflow from each link is within desired limit

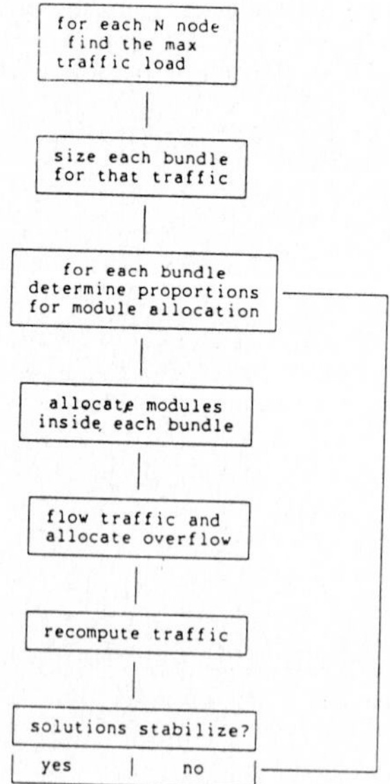

Figure 4. Sizing the N-N subnetwork

from the overflow computed in the previous iteration.

4.2 THE I TO I SUBNETWORK

The I-I subnetwork includes all the links originating and terminating in I nodes. Those trunk groups support only direct first offered traffic (including FHR overflow) similar to the Primary High Usage groups in FHR networks. They are sized using the economic criterion similarly to the ECCS rule, the main difference being that the alternate route cost and efficiency must be computed as weighted averages for all the corresponding I-N links

4.3 THE I-N SUBNETWORK

In this subnetwork, the trunk groups carry their direct traffic and additionnal overflow from I-I and the other I-N links. The bundle is similar to a distributed final and it is sized according to a given grade of service criterion.

The individual links in this bundle play the role of intermediate high usage groups and are sized according to an economic criterion.

4.4 THE N TO N SUBNETWORK

The N to N subnetwork is on the top of the two-level hierarchy of the DR network. An overflow from any N-N link is sent to some other N-N links. Thus N-N links carry overflows from the I-N subnetwork and from the N-N one. It has to be sized after completion of the I-N sizing. The mutual overflow between different N-N links implies that this subnetwork must be sized in a different way than the I-N one.

The procedure of distributing modules inside a bundle is now more complicated than that used in I-N network. The first difference is that bundles can no longer be dealt with one at a time. Rather, bundle sizes for the whole network must first be determined, with associated proportions, after which modules are allocated one at a time to the network as a whole.

5.0 VALIDATION

The validation method for the new algorithm is based on comparison of several networks dimensioned with DR and FHR sizing algorithms. The number of trunks, trunk miles and network cost are analyzed for networks with comparable grade of service. Note that in the following discussion the grade of service indicates blocking probability rather than transmission quality. By comparable grade of service we mean the equivalent average grade of service for traffic loads for which the dimensioning was done.

The grade of service was obtained by simulation. All simulated traffic loads were subjected to medium day-to-day variations (this was the parameter used in sizing which can be varied according to requirements), and to overload conditions. In the latter case the traffic loads were inflated and perturbed several times and the arithmetic mean of blocking probabilities for all perturbations was computed and compared.

The sizing algorithms compared are:

1. GPTM6 - General Purpose Trunking Model, release 6, sizing up mode, a dimensioning algorithm for networks with Fixed Hierarchical Routing developed by BNR. For the multiload case the 'sizing up' procedure was used, starting with the matrix having the greatest total traffic, then considering the one having second greatest total and so on.

2. DIMALG - Dimensioning Algorithm (developed by BNR) - the sizing algorithm for networks in a Dynamic Routing environment. This algorithm is designed for one traffic load, usually the ABSBH, or one traffic matrix obtained by Kruithof merge of several loads. Since it was found that the 'Kruithof' method generally gives better results, we shall use this method for comparison with the new DR sizing algorithm.

3. MDIMALG - Multiload Dimensioning Algorithm - the sizing algorithm for DR networks described in this paper.

5.1 SIZING INPUTS
In order to obtain as realistic a comparison as possible, we have used 'real' inputs to the sizers whenever feasible. The sizing inputs were the following:

1. The Network
   The test network is based on fundamental plans on 1990 traffic distribution for the 38 major switches in Canada. For test purposes, the 38 DR nodes were classified as tandems (N nodes).

2. The Traffic
   The point-to-point traffic loads necessary for dimensioning the network described above were obtained from a Canadian intercity network database. Six different traffic loads were designed to preserve most of the existing traffic noncoincidence. They were obtained from a larger number of point-to-point traffic matrices in the following way.

   From any pair of matrices a third one is constructed containing maxima of corresponding elements. If the sum of the elements of the third matrix does depart greatly from those of the two original it reflects noncoincidence. Therefore these two should be kept apart in different load sets. If it does not depart then the smallest one can be dropped because it does not contain valuable information on noncoincidence.

   This procedure allows for a reduction in the number of traffic matrices and keep the information on the traffic noncoincidence.

3. Modularity
   In all cases modularity of 24 was used. This corresponds to the North American standard for DS-1 transmission rate.

4. Point-to-point unit costs
   The same point-to-point unit cost matrix was used in all computations. It is derived from a cost model taking into account termination cost, a weighted distance factor, an echo suppressor when over a fix distance.

5. Grade of service requirement
   To compare the number of trunks, trunk mileages and network costs obtained by dimensioning with different algorithms, we have dimensioned using the inputs as described in 1-4 of this section. Blocking constraints were specified in the input parameter files so that the resulting grade of service was comparable.

5.2 NUMERICAL RESULTS
In this section we shall compare the numerical results obtained by sizing with different algorithms.

The summary of traffic loads used in dimensioning is:

LOAD NAME	TOTAL TRAFFIC (ERLANGS)
Load 1	10181.75
Load 2	10455.75
Load 3	8958.99
Load 4	10396.14
Load 5	9274.52
Load 6	8720.52
Load 7 (Kruithof merge of loads 1 to 6)	11596.27

NUMERICAL RESULTS

ALGORITHM	NUMBER OF TRUNKS	NETWORK COST	TRUNK MILES
1.FHR	16 560	10.379E7	5.72E6
SIZING UP	(100%)	(100%)	(100%)
2.DR	15 120	9.696E7	5.26E6
KRUITHOF	(91%)	(93%)	(92%)
3.MDIMALG	14 736	9.091E7	4.87E6
IDLE/COST	(89%)	(88%)	(85%)

In the table above we can see that the most economical network for DR is the one obtained by MDIMALG, the DR sizer which allocates the overflow in proportion to expected idleness and in inverse proportion to the cost of alternate route. The trunk miles and network cost are lower than the other networks and the difference in the number of trunks for that network compared to the number of trunks in the results from run 4 is negligible.

5.3 GOS
Producing a cheaper network is a desirable feature of any dimensioner. Another important feature of a sized network is its robustness.

For the GOS validation, we used a BNR traffic simulator. We simulated the network obtained with MDIMALG, DIMALG and GPTM6 with the six engineered traffic loads and four hourly traffic loads. The September traffics were chosen so that the grade of service of the dimensioned networks could be evaluated with realistic traffic. The loads are Sep9, Sep10, Sep13, Sep20, where the numbers indicate the hour of the day (central time). Their total uninflated traffic is:

LOAD NAME	TOTAL TRAFFIC (ERLANGS)
SEP9	9 911.79
SEP10	10 319.37
SEP13	9 753.13
SEP20	8 617.64

Four different overload situations have been considered. We have performed four perturbations and expansions on each traffic matrix (p1, p2, p3, p4) which correspond to estimates of forecasting deviations in traffic loads. Simulation of the networks was also done for traffic loads expanded to account for the medium day-to-day variations (p0). We have found that all simulations resulted in approximately equivalent GOS.

In the following three tables the simulation results with GPTM6, DIMALG and MDIMALG networks show the traffic blocking for the perturbed September hourly traffic loads:

1.GPTM6 % blocking

	SEP9	SEP10	SEP13	SEP20
p0	.41	.65	.41	.39
p1	2.06	2.41	.91	1.42
p2	2.51	1.46	1.55	1.52
p3	1.63	1.49	.65	.49
p4	1.32	.97	.57	1.05
average	1.58	1.39	.82	.97

2.DIMALG % blocking

	SEP9	SEP10	SEP13	SEP20
p0	.86	.91	.49	.27
p1	1.56	2.33	.83	.56
p2	2.56	1.92	1.35	.49
p3	.99	1.92	.92	.38
p4	1.57	1.49	.69	.43
average	1.51	1.71	.85	.43

3.MDIMALG % blocking

	SEP9	SEP10	SEP13	SEP20
p0	.76	.89	.46	.60
p1	1.39	2.14	.66	1.17
p2	2.18	1.55	1.21	1.18
p3	1.17	1.59	.68	.73
p4	1.49	1.23	.63	1.14
average	1.40	1.48	.73	.96

When networks obtained from DIMALG and MDIMALG, both designed to size DR network are compared, the GOS of MDIMALG exhibit more uniform behavior. This is because the new algorithm takes into account traffic flow for each load without any distortion to the loads contrary to the Kruithof method. Also, since with MDIMALG overflow is allocated in a manner which resembles the true overflow allocation in DR, more idle capacity gets utilized and fewer calls get blocked.

6.0 CONCLUSION

All the tests performed, some of which we have included here, provided consistent results. The MDIMALG gives a cheaper network with equivalent grade of service. That confirms the theoretical analysis that the Multiload DIMALG is better in exploiting the spare capacity available in the network in off-peak loads than the DIMALG algorithm, and also more accurately matches the actual traffic flow in DR.

As previous studies indicated, the approximate cost savings of DIMALG with respect to GPTM6 should be about 4 to 9%, depending on traffic forecast and network architecture. With the 38 node network, mapped on an 1988 architecture, we obtained similar results - cost savings of 7% and trunk miles savings of 8%. As the tests indicate, the new DR dimensioning algorithm MDIMALG provided additional cost and trunk mile savings of about 5% for the studied network. Of course these results could vary depending on traffic non-coincidence for a given application.

REFERENCES

[BE76] D. Bear, "Principles of Telecommunication - Traffic Engineering", Institution of Electrical Engineers, London 1976.

[BE83] M.E. Beshai, L.A. Pound, and R. W. Horn, "Traffic Data Reduction for Multiple-Hour Network Dimensioning", Network Planning Symposium Brighton, England, 1983.

[CA81] W.H.Cameron, "Simulation of Dynamic Routing: Critical Path Selection Features for service and Economy", International Conference on Communications, Denver, 1981.

[CA83] W.H. Cameron, J. Regnier, "Dynamic Routing for Intercity Telephone Networks", Tenth International Teletraffic Congress, Montreal, Canada, 1983.

[LA79] M.E. Lavigne, J.R. Barry, "Administrative Concepts in an Advanced Routing Network", Ninth International Teletraffic Congress, Torremolinos, Spain, 1979.

[WS85] W.S.Swain, J.A.Post, E.K.Flindall, "Selection of Forecasting Multihour Traffic Matrices in ITFS", Eleventh International Teletraffic Congress, Kyoto, Japan, 1985.

[SZ79] E. Szybicki, M.E. Lavigne, "The Introduction of an Advance Routing System into Local Digital Networks and its Impact on the Networks Economy, Reliability and Grade of Service", International Switching Symposium, Paris, 1979.

[TR54] C.J.Truitt, "Traffic Engineering Techniques for Determining Trunk Requirements in Alternate Routing Trunk Networks", Bell System Technical Journal, vol. 35, 1956, p.421-514.

TELETRAFFIC ISSUES in an Advanced Information Society
ITC-11
Minoru Akiyama (Editor)
Elsevier Science Publishers B.V. (North-Holland)

ANALYTICAL METHODS TO DETERMINE THE CPU CAPACITY OF AN SPC EXCHANGE WITH RESPECT TO CERTAIN GOS DEMANDS

Karl LINDBERGER and Sven-Erik TIDBLOM

Swedish Telecommunications Administration
Farsta, Sweden

ABSTRACT

In the engineering of telecommunication networks it has become essential to be able to determine the capacity of the SPC exchanges involved. The capacity is dependent on the system structure including a possible overload control but also on GOS demands and the traffic pattern in the particular cases. In this paper the necessary steps to take care of all these conditions mathematically, in a resonably simple way, will be suggested. Thus in the analysis the study of e.g. particular feedback queueing models and variations of the call intensities over the year will be included.

1. INTRODUCTION

The determination of the CPU capacity of an SPC exchange is an important tool in the practical engineering of the network. The way we look at this procedure it must be divided into several steps. The first step which may have its equivalence in other books or papers e.g. [2], [3] and [4], is to analyse the queueing model in a system, where the CPU-queues are not transformed by any overload control. The model is related to the classical feedback model by Takács, [1], but with certain essential modifications. One such modification is that the test process, which represents a particular type of service demand, may have different structure than the background process representing all the types of service demands offered to the CPU.

Thus we have studied in section 2 the total time spent in the system (=system time) for a tagged customer which needs a fixed number of CPU-visits in the feedback system where the background process is of the type in [1].

Another modification is that a feedback call may be delayed in an RP before joining the CPU-queue again. To handle these complications we shall use approximations which however are sufficiently accurate for interesting cases and give results in a simple form.

GOS demands are of the type:

The probability of a certain type of service delay beeing more than α seconds should be less than p. For e.g. dial tone delay we have $\alpha = 1.0$ and $p = 0.03$. In addition we must decide for how many hours a year we can accept this condition not to hold. To do that we must also analyse the variation of the call intensity over the year. This last problem will be taken care of in section 5.

The GOS demands could now be handled with the methods in section 2 if there is no overload control in the system. However in most systems there is.

Thus the second step which we shall discuss in section 3 is how such an overload control may influence on the engineering of the SPC station.

To remind the reader of this problem we shall study the particular overload control we have in our case which roughly behaves like an M/D/1-queueing system and essentially will influence the dial tone delay.

Once we, with or without overload control, have found a capacity with respect to the GOS in terms of the highest load, ρ, a third step is to find the corresponding call intensity. Such a relation can be given by a linear expression in terms of the call mixture as we shall show in section 4.

Finally the fourth step is to find out where in the distribution of call intensities of the year the value obtained in section 4 can be identified so that the extended GOS demand is actually fulfilled. This study of the distribution of call intensities will as we said before be given in section 5.

2. THE QUEUEING MODEL OF THE SPC STATION

Let us first consider an SPC station without overload control. The work of the CPU in the SPC station could be described by the following simplified model.

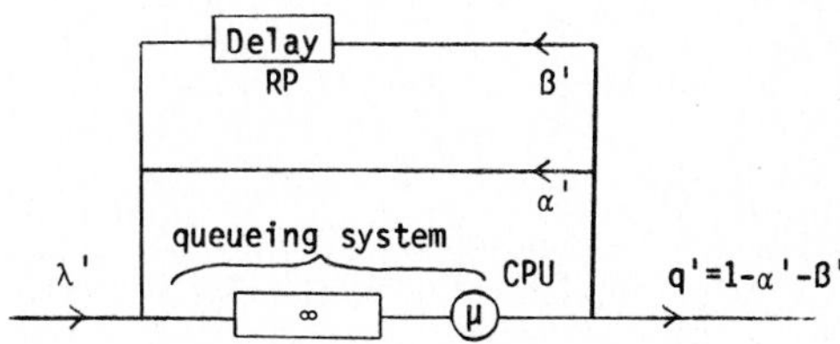

Fig 1 Model for the work of the CPU, where the overload control has been excluded.

Incoming calls arrive according to a Poisson process with intensity λ'. The customer joins the FIFO-queue with an infinite waiting room. The service time in the CPU is exponential with mean μ^{-1}. After service the customer either leaves the system with probability $q'(=1-\alpha'-\beta')$ or is sent through the undelayed feedback loop with probability α' and is placed in the tail of the CPU-queue or is sent to a regional processor (RP) with probability β'. Having been served by the RP the customer joins the CPU-queue again. Thus to pass the system above the customer has to circulate a random number of times through the undelayed and the delayed feedback loop respectively.

The customers circulating in the system of fig 1 generate our background process, which describes the stationary situation of the queueing system. It is not hard to imagine that by a suitable choice of parameters in a model with just an undelayed feedback loop one can create a situation in the CPU-queue which, under stationary conditions, is equivalent to the one in fig 1 at least with respect to the situation for the CPU. In fig 2 we have removed the delayed feedback loop from fig 1 and replaced (λ',q') by (λ,q).

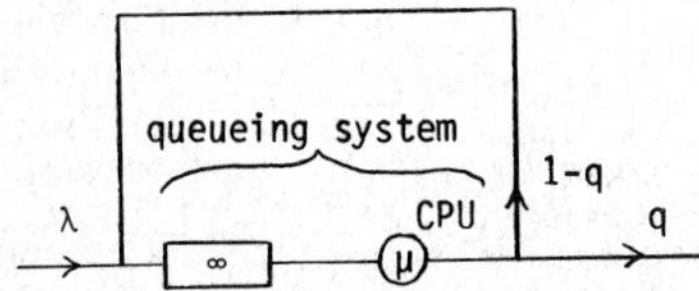

Fig 2 Queueing system with undelayed feedback giving an equivalent description of the stationary situation of the queueing system in fig 1.

The CPU-load is $\lambda'/\mu q'$ in fig 1 and $\lambda/\mu q$ in fig 2. For the queueing system in fig 2 to be equivalent in the stationary situation with the one in fig 1 we must require the CPU-load of the two systems to be same i.e.

$$\lambda/q = \lambda'/q' \quad . \tag{1}$$

We note that the system in fig 2 is created only for the purpose of giving an equivalent description of the stationary situation of the queueing system in fig 1. Later on we shall treat the system time of a tagged customer which, under stationary conditions for the system in fig 1, needs to go k_1 times through the delayed and k_2 times through the undelayed feedback loop. In this treatment we shall use the fact that the background process generating the CPU-queue in fig 1 can be described as the one in the simple system of fig 2 with respect to that problem even though the tagged customer later will be allowed to make RP visits.

Now, let a tagged customer arrive to the system in fig 2. The tagged customer, in contradiction to the customers of the background process, has to pass through the CPU exactly k times (i.e. to pass through the feedback loop exactly k-1 times).

Let $\tilde{\Theta}_k$ = The system time of a tagged customer which has to pass the server in fig 2 exactly k times, $k = 1,2,\ldots$. (2)

Our purpose is now to derive the mean and the variance of the variable (2). From Takács [1] it follows that

$$E\{\exp(-s\tilde{\Theta}_k)\} = U_k(s,1),\ k = 1,2,\ldots\ , \tag{3}$$

where $U_k(s,1)$ is obtained by setting $z = 1$ in

$$U_k(s,z) = \Psi(s+\lambda-\lambda z)U_{k-1}(s,(q+pz)\Psi(s+\lambda-\lambda z)), \tag{4}$$

where $p = q-1$ and U_0 is given by

$$U_0(s,z) = (\mu q-\lambda)/(\mu q-\lambda z) \tag{5}$$

and $\Psi(s)$ is the L-P-transform of the density of the CPU-service time, i.e.

$$\Psi(s) = \mu/(\mu+s)\ . \tag{6}$$

Let

$$_{ij}U_k(s,z) = \frac{\delta^{i+j}U_k(s,z)}{\delta s^i \delta z^j}\ . \tag{7}$$

From (3) and (7) we get that

$$E(\tilde{\Theta}_k^r) = (-1)^r\ {}_{r0}U_k(0,1),\ r = 0,1,\ldots\ . \tag{8}$$

To derive the mean of (2) we set $z = 1$ in (4) and differentiate the so obtained expression with respect to s. This gives

$$_{10}U_k(s,1) = -U_{k-1}(s,\Psi(s))\mu(\mu+s)^{-2} + \\ + \mu(\mu+s)^{-1}\{{}_{10}U_{k-1}(s,\Psi(s)) - \\ - {}_{01}U_{k-1}(s,\Psi(s))\mu(\mu+s)^{-2}\}. \tag{9}$$

By setting $s = 0$ in (9) we get that

$$_{10}U_k(0,1) = -\mu^{-1}+{}_{10}U_{k-1}(0,1) - \\ - {}_{01}U_{k-1}(0,1)\mu^{-1}\ . \tag{10}$$

To be able to solve for $_{10}U_k(0,1)$ in (10) we need an expression for $_{01}U_{k-1}(0,1)$. By setting $s = 0$ in (4) and by differentiating the obtained expression with respect to z we get a difference equation, which when solved yields

$$_{01}U_k(0,1) = \lambda/(\mu q-\lambda),\quad k = 0,1,\ldots\ . \tag{11}$$

For the details in the derivation of (11) we refer to [6]. Inserting (11) into (10) and solving for $_{10}U_k(0,1)$ gives

$$_{10}U_k(0,1) = kq/(\lambda-\mu q),\quad k = 0,1,\ldots\ . \tag{12}$$

By combining (8) and (12) we finally get the mean

$$E(\tilde{\Theta}_k) = kq/(\mu q-\lambda)\ :=\ M_k,\ k = 1,2,\ldots\ . \tag{13}$$

To derive the variance of $\tilde{\Theta}_k$ we start by deriving its second moment according to (8). Differentiation of (9) with respect to s gives an expression for ${}_{20}U_k(s,1)$. By setting s = 0 in this expression and by simplifying we have that

$$ {}_{20}U_k(0,1) = \frac{2(\mu qk+\lambda)}{\mu^2(\mu q-\lambda)} + {}_{20}U_{k-1}(0,1) - $$

$$ - 2\mu^{-1}{}_{11}U_{k-1}(0,1) + \mu^{-2}{}_{02}U_{k-1}(0,1). \quad (14) $$

For the details we again refer to [6]. To be able to solve for ${}_{20}U_k(0,1)$ in (14) we need expressions for ${}_{11}U_{k-1}(0,1)$ and ${}_{02}U_{k-1}(0,1)$. By setting s = 0 in (4) and by differentiating twice with respect to z we get a difference equation, which when solved yields

$$ {}_{02}U_k(0,1) = 2\lambda^2/(\mu q-\lambda)^2, \ k = 0,1,\dots . \quad (15) $$

Differentiation of (4) with respect to each of the two arguments (s,z) gives a difference equation for ${}_{11}U_k(s,z)$. By setting (s,z) = (0,1) and by solving we get

$$ {}_{11}U_k(0,1) = $$

$$ = \lambda\mu q(\mu q-\lambda)^{-3}\{(p+\lambda/\mu)^k + k(\lambda/\mu-q)-1\} . \quad (16) $$

Insertion of (15) and (16) into (14) gives a difference equation, which when solved yields

$$ {}_{20}U_k(0,1) = \frac{kq\{q(k+1)(\mu q-\lambda)+2\lambda\}}{(\mu q-\lambda)^3} - $$

$$ - \frac{2\lambda\mu q\{1-(p+\lambda/\mu)^k\}}{(\mu q-\lambda)^4} . \quad (17) $$

Combination of (8), (13) and (17) gives that

$$ V(\tilde{\Theta}_k) = \frac{kq^2(\mu q-\lambda)^2+2\lambda kq(\mu q-\lambda)-2\lambda\mu q\{1-(1-q+\lambda/\mu)^k\}}{(\mu q-\lambda)^4}, $$

$$ k = 1,2,\dots . \quad (18) $$

The exact variance in formula (18) is a bit complicated. We shall now derive an approximation of it which is quite good for the applications we are interested in. By defining the CPU-load

$$ \rho = \lambda/\mu q , \quad (19) $$

the expression for the mean in (13) becomes

$$ E(\tilde{\Theta}_k) = M_k = k/\mu(1-\rho) , \ k = 1,2,\dots . \quad (20) $$

Let

$$ \beta = q(1-\rho) . \quad (21) $$

By using the definitions (19) - (21) the formula (18) can be written

$$ V(\tilde{\Theta}_k) = M_k^2\{k^{-1}+2k^{-2}\rho\beta^{-2}((1-\beta)^k-1+\beta k)\} . \quad (22) $$

In the applications we shall study, the probability q (see fig 2) is much smaller than 1 and the CPU-load is close to 1, i.e.

$$ \beta \ll 1 . \quad (23) $$

Under (23) the following approximation is quite good

$$ (1-\beta)^k \approx 1-\beta k+\beta^2k(k-1)/2 . \quad (24) $$

Combination of (22) and (24) yields

$$ V(\tilde{\Theta}_k) \approx M_k^2\{\rho+(1-\rho)k^{-1}\} . \quad (25) $$

With the knowledge that k (see (2)) in our applications is rather large (k^{-1} is of the same magnitude as q) we can take one further step in the approximation and finally get

$$ V(\tilde{\Theta}_k) \approx M_k^2 \rho . \quad (26) $$

Now the system time (2) can be written

$$ \tilde{\Theta}_k = \sum_{i=1}^{k} X_i , \quad (27) $$

where

X_i = The time it takes for the tagged customer to pass the queueing system in fig 2 at the i:th passage, i = 1,...,k . (28)

In the sequel we shall need the covariances between the variables (28). We have that

$$ V(\tilde{\Theta}_k) = kV(X_1)+2\sum_{i=1}^{k-1}(k-i)\mathrm{Cov}(X_1,X_{1+i}) . \quad (29) $$

By equating (22) and (29) we find after some calculations that

$$ \mathrm{Cov}(X_1,X_{1+i})=\rho M_1^2(1-\beta)^{i-1}, \ i=1,\dots,k-1. \quad (30) $$

Note that under (23) the covariance (30) does not depend much on i.

Let a tagged customer arrive to the original system in fig 1. The tagged customer needs exactly k CPU visits and k_1 RP visits, i.e. his system time Θ_k^* can be written

$$ \Theta_k^* = \sum_{j=1}^{k_1} D_j + \sum_{i=1}^{k} Z_i , \quad (31) $$

where

D_j = The length of the j:th RP delay, j = 1,...,k_1 , (32)

Z_i = The time it takes for the tagged customer to pass the queueing system in fig 1 at the i:th passage, i = 1,...,k . (33)

Assume that

$$ D_j = d = \text{constant}, \ j = 1,\dots, k_1 . \quad (34) $$

As a matter of fact all the D_i:s are known but their values variate somewhat in which case we let d be their mean. In accordance with the equivalence between the systems in fig 1 and fig 2 we get

$$Z_i \overset{d}{=} X_1 , \quad i = 1,\dots,k . \tag{35}$$

From (31) and (34) - (35) it follows that

$$E(\Theta_k^*) = k_1 d + M_k . \tag{36}$$

For the variance of Θ_k^* we know that

$$V(\Theta_k^*)=kV(X_1)+2\sum_{i=1}^{k-1}(k-i)\mathrm{Cov}(Z_1,Z_{1+i}) , \tag{37}$$

where we have assumed that $\mathrm{Cov}(Z_\ell,Z_{\ell+i})$ does not depend on ℓ. We shall now derive an approximation for the covariance in (37). Let

$$W_i = \sum_{j=1}^{i} Y_j , \tag{38}$$

where

Y_j = The number of passages of the queueing system in fig 1 an undelayed customer could have made between our tagged customer's j:th and j+1:st passage. (39)

We have that the real valued random variable $Y_j \geq 1$.

Note that W_i tells "how many X-variables (see (28)) there is room for on a time axis between the occurence of Z_1 and Z_{1+i}" (see (33)). We get

$$\mathrm{Cov}(Z_1,Z_{1+i}) = E\{(Z_1-M_1)(Z_{1+i}-M_1)\} =$$

$$= EE\{(Z_1-M_1)(Z_{1+i}-M_1) \mid W_i\} =$$

$$= EE\{(X_1-M_1)(X_{1+W_i}-M_1)\} , \tag{40}$$

where we have extended the indices for the variables (28) to noninteger values. Combination of (30) and (40) yields approximately

$$\mathrm{Cov}(Z_1,Z_{1+i}) = \rho M_1^2(1-\beta)^{-1}E(1-\beta)^{W_i} . \tag{41}$$

For the mean

$$f = E(Y_j) \tag{42}$$

we may get the approximation formula

$$f \approx 1 + k_1 d/(k-1)M_1 . \tag{43}$$

In our applications $f \lesssim 10$. In (23) assuming that $\beta < 10^{-2}$, we have that

$$f\beta \ll 1 . \tag{44}$$

The expected value in (41) can be calculated in the following approximate way

$$E(1-\beta)^{W_i} \approx (1-\beta)^{if} \approx (1-f\beta)^i , \tag{45}$$

where the last step follows from (44).

By combining (37), (41) and (45) and by simplifying we get

$$V(\Theta_k^*) \approx$$

$$\approx M_k^2\{k^{-1}+2(1-f\beta)(1-\beta)^{-1}\rho k^{-2}(f\beta)^{-2}((1-f\beta)^k-1+f\beta k)\}. \tag{46}$$

Under (44) the following approximation is quite good

$$(1-f\beta)^k \approx 1-f\beta k+(f\beta)^2 k(k-1)/2 . \tag{47}$$

Insertion of (47) into (46) yields

$$V(\Theta_k^*) \approx M_k^2\rho\{\eta + \frac{(1-\eta\rho)}{k\rho}\} , \tag{48}$$

where

$$\eta = (1-f\beta)(1-\beta)^{-1} . \tag{49}$$

From (23) and (44) it follows that the factor η is close to 1. Moreover, since ρ is close to 1 and k is rather large ($k \sim 10$) we get that the whole bracket in (48) is close to 1, i.e.

$$V(\Theta_k^*) \approx M_k^2 \rho . \tag{50}$$

Note that this variance expression is the same as the one we got in (26), which was valid for a customer who did never visit an RP. This somewhat surprising result relies mainly on the assumption (44). If (44) is not fulfilled we cannot use the variance approximation (50).

As before we think of a customer arriving to the system in fig 1 who needs k CPU-visits and k_1 RP-visits. The GOS for a particular type of service (customer) can now be written in terms of Θ_k^*

$$P(\Theta_k^* > \alpha) \leq p , \tag{51}$$

where α and p are given constants. Primarily we look for the maximal CPU-load which satisfies (51). To be able to do so we need the distribution of Θ_k^*. Under the assumption that the RP-delays are constant (=d) we have shown that

$$\left.\begin{aligned} E(\Theta_k^*) &= k_1 d + M_k \\ V(\Theta_k^*) &\approx M_k^2\rho \end{aligned}\right\} , \tag{52}$$

where ρ and M_k are given by (19) and (20). The most appropriate distribution to adapt to the variable $\Theta_k^*-k_1 d$ would be a gamma distribution.

However, since the value of ρ in our case is believed to be above 0.95 we have that

$$D(\Theta_k^*) \approx M_k , \tag{53}$$

i.e. the mean and the standard deviation of the variable $\Theta_k^*-k_1 d$ are approximately the same ($=M_k$). We then make the approximation that $(\Theta_k^*-k_1 d)M_k^{-1} \sim \mathrm{Exp}(1)$.

By doing so we get that (51) implies the requirement

$$(\alpha-k_1d)M_k^{-1} \geq 3.5 \ , \qquad (54)$$

if we in (51) use the value

$$p = 0.03 \ . \qquad (55)$$

By inserting the expression for M_k (see (20)) into (54) and by solving for ρ we get the requirement

$$\rho \leq 1 - 3.5k\mu^{-1}(\alpha-k_1d)^{-1} \ . \qquad (56)$$

The maximal CPU-load which under (55) fulfills (51) is thus given by the right side of (56). For all types of delay e.g. dial tone delay (i.e. pre-dial tone delay) and post-dial tone delay there are standards of the type (51). What one needs to know to be able to find the maximal CPU-load which fulfills such a standard are (see (56)): the number of required CPU-visits (=k), the number of RP-visits (=k_1), the meanlength of a CPU-visit (=μ^{-1}) and of an RP-visit (=d).

One of our applications, dial tone delay, consists of 14 CPU-visits and 6 RP-visits. The time it takes to satisfy this need of services is denoted by Θ_{14}^*. For the system time Θ_{14}^* we have the requirement

$$P(\Theta_{14}^* \geq 1 \text{ sec}) \leq 0.03 \ . \qquad (57)$$

For the mean of the CPU-visits and RP-visits we use

$$\left.\begin{array}{l} \mu^{-1} = 0.4 \text{ ms} \\ d \ \ = 34 \text{ ms} \end{array}\right\} . \qquad (58)$$

Insertion of the above figures into (56) gives

$$\rho \leq 1-3.5\cdot 14\cdot 0.4(10^3-6\cdot 34)^{-1} \approx 0.975 \ . \qquad (59)$$

Regarding the requirement for the dial tone delay we might thus load the CPU up to 97.5 %. A corresponding investigation has to be done for every type of service which has a delay requirement of the type above. The lowest ρ-value in the set of thus obtained ρ-values is the one which becomes dimensioning.

3. THE OVERLOAD CONTROL

In section 2 we have studied the behaviour of the feedback type of queueing station which was meant to represent our system in the case without overload control. However, many systems have different types of overload control with the purpose to protect the CPU or rather the calls it has already started to take care of from beeing delayed. The protections are constructed for systematic overload situations caused by e.g. some error in the network near this SPC-node. Such an error situation should of course not be included in the normal GOS demands (but perhaps in a more general QOS demand). However, the overload control (OLC) may also be put in action by a normal overload caused by the random character of the call process. In that sense the OLC may have influence on the engineering with respect to the normal GOS demands.

We shall here roughly describe how it works in our specific case.

The OLC is working with a certain load value e.g. ρ = 0.9 as a base. The load is measured every 10 seconds (say) and if that measure indicates a load above 0.9, then only a limited number of new calls will be allowed to join the CPU-queue in the next period and the rest has to wait in an external queue. If now the measured load in this period (10 s) is below 0.9 the limited number of new calls in the following period can be allowed to increase and if this trend should go on we can eventually empty the external queue again.

Now it appears that with a theoretical CPU-load, ρ, close to 0.9 the OLC will be on most of the time and it will approximately accept a limited number of calls corresponding to ρ = 0.9 each period (10 s). The situation in the external queue is thus very similar to an M/D/1-queueing system with service factor

$$\tilde{\rho} = \lambda/\mu_{OLC} \ , \qquad (60)$$

From now on λ is the intensity of new calls and not as in section 2 the total intensity for all the services to all of the calls.

Furthermore μ_{OLC} is the average output intensity for the external queue during busy periods when the OLC is on. On these conditions we have approximately that

$$\tilde{\rho} = \rho/0.9 \ , \qquad (61)$$

where ρ is the theoretical CPU-load and 0.9 was the OLC-base level in our case. Now if we have that the external system under these circumstances can be said to behave approximately like an M/D/1 one with waiting time W we have approximately that

$$E(W) = \tilde{\rho}^2/\{(1-\tilde{\rho})2\lambda\} \qquad (62)$$

and

$$V(W) = E^2(W)\{1+4(1-\tilde{\rho})/3\tilde{\rho}\} \ . \qquad (63)$$

With the base level for the OLC as low as 0.9 (i.e. essentially below 0.975 from section 2) it appears that the "dial tone delay" is the only critical point in the GOS-demands since all the other services will not be extra delayed by the OLC and their critical level is essentially above 0.9. Our important GOS demand is then

$$P(W+\Theta_k^* \geq 1000 \text{ ms}) \leq 0.03 \ . \qquad (64)$$

With $\rho < 0.9$ we can note that the variance part of Θ_k^* in this case is unimportant i.e. the demand can be simplified to

$$P(W+\frac{k}{\mu(1-\rho)} + k_1d \geq 1000 \text{ ms}) \leq 0.03 \ . \qquad (65)$$

Now with (65) and the same values of k, k_1, d and μ as in (59) we can obtain an approximate value of ρ.

A very simple approximation of (65) would be to write it

$$P(W+30+150 \geq 1000) = P(W \geq 820) \leq 0.03 \qquad (66)$$

and regard W as exponentially distributed since from (62) and (63) it follows that $E(W) \approx D(W)$

in this case. With that approach we get $\rho = 0.867$. Simulations show that this approach is accurate enough for the interesting cases near $\rho = 0.9$, e.g. in the case above we get $\rho = 0.865$. The big difference between the ρ-values obtained with and without OLC depends partly on the low OLC-level 0.9 we had to face in the first version. With the OLC-level at 0.94 (say), which seems possible in later SPC-versions, our ρ would be just above 0.90 instead.

What we have described here is just one type of OLC. Another type based on the number of customers in the external queue is e.g. studied in [5].

4. CAPACITY IN TERMS OF CALL INTENSITY

Once we have found the value of the highest acceptable load with respect to our particular GOS demands i.e. in our case $\rho = 0.867$ we must find a correspondence to that value in terms of the call intensity λ. Of course the λ-value connected with a certain ρ is also dependent on the call mixture. We have found that a linear expression can sufficiently well describe that relation i.e. it can be written

$$\rho = \rho_0 + \lambda \sum_{i=1}^{4} b_i p_i \ , \qquad (67)$$

where p_i is the proportion of calls which are internal, outgoing, incoming and of transit-type and ρ_0 is the CPU-load without any traffic. In our case, for the original version of our type of SPC station we had a sufficiently accurate simulation program to help us decide the unknown coefficients by studying a number of call mixtures. Numerically we could have

$$100(\rho-\rho_0) = \lambda\{1.24p_1+1.48p_2+0.93p_3+0.95p_4\} . (68)$$

However it happens that new versions of this type of SPC station appears and it is not always possible to update the simulation program completely each time, but since the systems are similar it can be said that the essential changes are described by new values of ρ_0 and b_i, $i = 1...4$. To determine those values we can use special measurements from e.g. the same station but at 5 different hours of the week with 5 different call mixtures.

It should be told that from the simulations we have drawn the conclusions leading to a linear model including 5 (and not essentially more) coefficients. If however the system should include a number of more complicated extra types of services e.g. call diversion or completion of call to busy subscriber, which are used by an essential proportion of customers, then we have to add a number of terms in (68). With all these methods we should be able to predict the capacity in terms of λ also for a planned new SPC station of the same version as the one we have measured.

5. GOS WITH RESPECT TO THE VARIATIONS IN CALL INTENSITY

Now that we have determined this λ-value we must go back to the GOS-demands and ask how many hours of the year we can accept a higher λ-value with all its consequences e.g. essentially longer dial tone delay.

Suppose that we have the following traffic distribution over the hours of the year: We can find 20 hours a week (including 5 TCBH) for each station in a way that we can say that their offered traffic values could be a r.v. X which approximately belongs to a normal distribution with mean A_0 and standard deviation kA_0. It is understood that all hours with very high call intensity are included in this set. Here A_0 is the value used for the ordinary engineering of the station with respect to e.g. trunks.

The value of k should be based on measurements from a number of different SPC stations but is supposed to be general in our model. When we test the model we should be particularly interested in checking the positive tail of the distribution i.e.

$$P(X/A_0 - 1 > kx) \approx 1 - \Phi(x) \qquad (69)$$

e.g. at x=1.64 and thus $1-\Phi(x)=0.1$. Even though the call intensity λ can be said to be constant for a certain hour, we must regard it, when belonging to an arbitrary hour among the 20 a week we have chosen, as a random variable

$$\Lambda = X/Z = XY \ . \qquad (70)$$

Here Z is the random variable of the mean holding time for an arbitrary hour and $Y = 1/Z$.

Measurements show that the variation of Λ is higher than that of X or Y(Z) and that it doesn't seem too unrealistic to assume that X and Y are independent. Furthermore we can assume that also Y is normally distributed i.e.

$$Y \varepsilon N(1/s_0, c/s_0) \ .$$

Now c can be determined from the inverse of the measurements of mean holding times from different stations. Again we have assumed that we could find a general c. Thus with

$$X' = X/A_0 \varepsilon N(1,k),$$

$$Y' = s_0/Z = s_0 Y \varepsilon N(1,c)$$

and since X' and Y' are both close to 1 it follows that

$$X'Y'-1 \overset{d}{\approx} \log(X'Y') \overset{d}{=} \log X' + \log Y' \approx$$

$$\overset{d}{\approx} X'-1+Y'-1 \ \varepsilon N(0,\sqrt{c^2+k^2}) \ . \qquad (71)$$

Now suppose that we decide to estimate e.g. the 98.5 percentile of each distribution separately in order to determine the highest acceptable Λ-value, λ, according to a certain GOS, then we would have the following approximate relation.

$$P(\Lambda > \lambda = A_0(1+2.18k)s_0^{-1}(1+2.18c)) \approx$$

$$\approx 1-\Phi(2.18(k+c)/(k^2+c^2)^{\frac{1}{2}}) \ . \qquad (72)$$

If k and c are of the same magnitude which our measurements indicate, we have

$$1.3 \leqslant (k+c)/(k^2+c^2)^{\frac{1}{2}} \leqslant 1.4 \qquad (73)$$

i.e.

$$P(\Lambda > \lambda) \approx 0.001 \ . \qquad (74)$$

If our GOS-demand now says that at the most for one hour a year we can accept a call intensity higher than a certain value which is the one that implies the probability of a dial tone delay above 1.0 second to be less than 0.03, then this call intensity could correspond to the λ in (72) and (74).

What we have gained with these rather rough arguments is that if we can not find in each case the tail of the distribution of Λ, we could use our knowledge of the distributions of X and Z separately. If we can not measure the 98.5th percentil directly we can use A_0, s_0 and the general values of k and c to determine

$$\lambda = (1+2.18k)(1+2.18c)A_0/s_0 \ . \qquad (75)$$

For practical purpose we prefer to choose a general factor f such as

$$\lambda = f\lambda_0 = fA_0/s_0 \qquad (76)$$

i.e.

$$A_0 = \lambda s_0/f \ . \qquad (77)$$

Measurements show that in our cases we have

$$0.06 \leqslant k, \ c \leqslant 0.08 \ .$$

Thus with k = c = 0.08 we obtain

$$f = (1+0.08 \cdot 2.18)^2 \approx 1.4 \ . \qquad (78)$$

This factor f = 1.4 is what we use in practice to take care of the variation of the call intensity over the year, and the arguments above can be seen as a sort of explanation to that. The values of A_0 and s_0 are of course available from standard measurements. Other measurements we have referred to here are more special ones.

6. A NUMERICAL EXAMPLE

If we look at a particular SPC station we could find from section 2 that with our GOS for dial tone delay we could accept a load $\rho = 0.975$ in a system without overload control. The GOS for some other service e.g. post dial tone delay could possibly (but not likely) decrease that value somewhat. With the OLC in action and with the base level for it at 0.9 we obtain from section 3 that we can only accept $\rho = 0.867$.

Now suppose that our particular SPC station has the call mixture $p_1 = 0.08$, $p_2 = 0.27$, $p_3 = 0.25$ and $p_4 = 0.4$. Then if it had been of the originally studied version (68), $\rho_0 = 0.13$ and $\rho = 0.867$ would imply an acceptable call intensity of $\lambda = 66$ calls/s. However if it is not of that originally studied version but of a newer version (but with the old OLC-level) then special measurements from another SPC station of this new version could indirectly give us new values of ρ_0 and $b_1 - b_4$ e.g. $\rho_0 = 0.20$, $b_1 = 0.0120$, $b_2 = 0.0140$, $b_3 = 0.0090$ and $b_4 = 0.0090$ which would imply that $\lambda = 63$ calls/s. With such a λ and with a mean holding time for the traffic of the particular station $s_0 = 150$ s we find from (77) in section 5 that the highest acceptable standard measurement traffic value is

$$A_0 = \lambda s_0/1.4 = 6750 \text{ erlang} \ , \qquad (79)$$

where 1.4 is caused by the hour to hour variations of the call intensities (see section 5).

If we finally shall determine the acceptable number of subscribers in the SPC station, N, we have the relation

$$Na = A_0(1-p_4+p_1) \ , \qquad (80)$$

where a is the traffic per subscriber and p_1 and p_4 was the internal and transit call (traffic) proportions respectively. (Here we have assumed that the call and the traffic mixture are the same which is not always true). Thus with e.g. $a = 0.09$ we obtain N = 51000 subscribers. Perhaps a figure of that type is of more interest to certain network planning people than just a capacity value of ρ.

7. FINAL COMMENTS

Of course we are fully aware of the fact that other types of SPC stations with other types of OLC in countries with different GOS demands and traffic profiles must cause changes in almost every step we have discussed here. However the main purpose of this paper is to show that a correspondence to each of these steps may be necessary for the practical engineering of an SPC station. We also wanted to suggest that all the steps can be handled with sufficient accuracy using rather simple expressions which are convenient for practical applications. Thus we still think that our experiences in this field may be of some interest to others.

REFERENCES

[1] L. Takács, "A Single Server Queue with Feedback," Bell Syst. Tech. J., vol.42, pp.506-519, 1963.

[2] L. Kleinrock, "Queueing Systems, vol.II Computer Applications," Wiley, 1976.

[3] D. Rapp, "Some Queueing Models of AXE-10," Proc. 9th Int. Teletraffic Congress, Torremolinos, 1979.

[4] D. Rapp, "Some Delay Models Concerning SPC systems," Techn. Report (TETS-7026) Lund Inst. of Technology, Lund Sweden, 1979.

[5] B. Wallström, "A Feedback Queue with Overload Control," Proc. 10th Int. Teletraffic Congress, Montreal, 1983.

[6] S.E. Tidblom, "Some interesting results for feedback queueing systems," (In Swedish) Report Nst83 002, Televerket Sweden.

TELETRAFFIC ISSUES in an Advanced Information Society
ITC-11
Minoru Akiyama (Editor)
Elsevier Science Publishers B.V. (North-Holland)

THROUGHPUT BEHAVIOR OF SWITCHING SYSTEMS UNDER HEAVY LOAD CONDITIONS

C.H. YIM* and H.L. HARTMANN**

* Korea Elec. Teleco. Res. Ins. (KETRI), Chung-Nam, Korea
** Technical University of Braunschweig, Braunschweig, FRG

ABSTRACT

This paper presents a model for determining throughput of SPC switching systems under heavy load conditions. The model includes three important components: subscriber behavior, switching network and control unit. The model is developed by analyzing the call processing procedure. Mathematical relationships between various characteristic traffic values are derived. The equations form a system of nonlinear equations. These nonlinear equations can be solved iteratively by the method of substitution. As an application of the model proposed we consider a microprocessor-controlled decentralized PABX unit. The results of the numerical method are compared with those of a time-true simulation.

1. INTRODUCTION

The overall system throughput of a SPC exchange is influenced not only by the control structure of the system, but also by the traffic performance of the switching network and the subscriber behavior. The usual approach is to assume that the switching network and the subscriber behavior have no influence on the overall system throughput. Most of the studies on SPC switching systems are carried out under these assumptions (1)-(2). The models developed under these assumptions can not be applied successfully to the overloaded SPC switching systems, as the system behavior under heavy load conditions is not properly considered in the model. For example, the operation response time of a control unit may differ depending on whether or not we take the blocking in the switching network into account, especially when the blocking probability is large under heavy load conditions.

Several papers are published, which deal with system throughput, considering both the switching network and the control unit (3)-(4). For a successful study of system's capability under overload conditions it is necessary to describe the real system processing as realistically as possible and to consider the entire exchange. Szybicki initially proposed a model for an overloaded local switching system which considers the switching network and the control unit as well as the subscriber behavior (5). In the model, the control units are approximately modelled by using an M/G/1-FIFO queueing system while the actual control units have usually a number of tasks of different priority classes.

The purpose of this paper is to develop an analytical model of a SPC switching system which displays the major steady-state behavior of the system under heavy load conditions and which is computationally tractable. In the model, the switching network, the control unit and the subscriber behavior are taken into consideration. As speech connections are usually set up on a loss basis, internal and external blocking probabilities are considered. The control unit operates on a queueing basis and is modelled by an M/G/1 queueing system with nonpreemptive priority discipline. For the subscriber behavior, retrial of call attempts and giving up probabilities during call processing are considered. With this model we can calculate approximately the overall maximum throughput of a switching system and gain insight into measures for overload control.

This paper consists of 5 sections. Following the introduction in Section 1, a description of the model and basic definitions are given in Section 2. The model is developed by analyzing the call processing procedure. In Section 3, mathematical relationships between various characteristic traffic values, including overall system throughput, are derived. The derived equations form a system of nonlinear equations. These nonlinear equations can be solved iteratively by the method of substitution. In Section 4, a microprocessor-controlled decentralized PABX unit is considered as an application example of the proposed model. To test the numerical results, a time-true simulation was run. The results of numerical method and simulation are compared and discussed. The conclusion is given in Section 5.

2. A MODEL OF OVERLOADED SPC SWITCHING SYSTEMS

For a successful study of SPC switching system's call handling ability under overload conditions, it is necessary to describe call processing in the real system as truly as possible and consider the entire exchange. The components considered in this paper are:

a) subscriber behavior
   - call type (internal, transit, incoming, outgoing)
   - call mix (completeness of call attempts)
   - holding time (dialing time, ringing time, conversation time)
   - call repetition ratio

b) switching network
   - internal blocking probability
   - external blocking probability

c) control unit
   - rejection probability due to time-out of dial-tone delay
   - rejection probability due to time-out of ringing-tone delay.

If one analyzes call processing procedure

under consideration of the above mentioned components, one can obtain the model of an overloaded SPC switching system shown in Fig. 1. To explain easily a call processing procedure, an intra-office call connection is considered as an example among four different call types. A calling subscriber lifts his handset off the hook at time t_0. The action of the calling subscriber is detected by scanning of peripheral equipments and delivered to the control unit as an input information to process. The control unit tries to find a free dial reception circuit (register). If the control unit finds a free register, the register is marked at time t_1 and the control unit proceeds to carry out path selection between the calling subscriber and the register. In the case of a successful selection of a path, the calling subscriber receives dial-tone at time t_2. The register is ready to receive dialed digits. Up to this time, there are three possibilities of blocking:

a) there is no free register,

b) at least there is a free register, but no free path,

c) at least there is a free register and a free path, but the dial-tone delay exceeds some time limit.

The blocking probabilities of a) and b) are decided by the traffic behavior of the switching network and denoted by B_1. c) describes the probability B_2 due to time-out of dial-tone delay and is determined by the behavior of the control unit. In these three cases the call attempt is not successful.

Dialing begins at time t_3 and ends at time t_4. The calling subscriber can give up dialing at the beginning and at the end of the dial-tone with probabilities B_3 and B_4 respectively. If the calling subscriber selects all the necessary digits, the control unit evaluates the selected digits and checks the state of a called subscriber. If the called subscriber is free, the control unit tries to find a free junctor for connection of an intra-office call. If the control unit finds a free junctor, it is marked at time t_5 and the control unit proceeds to carry out path selection between the junctor and the called subscriber. In the case of a successful selection of a free path the calling subscriber receives a ringing-tone. Up to this time there are further four possibilities of blocking:

a) the called subscriber is not free,

b) the called subscriber is free, but there is no free junctor,

c) the called subscriber is free and at least there is a free junctor, but there is no free path,

d) the called subscriber is free, at least there are a free junctor and a free path, but the ringing-tone delay exceeds some time limit.

The blocking probabilities a)-c) are decided by the traffic behavior of the switching network and denoted by B_5. The term d) describes the probability B_6 due to time-out of the ringing-tone delay and is determined by the behavior of the control unit.

If the called subscriber lifts his handset off the hook at time t_7, then conversation begins at time t_8. The probability that the called subscriber may not respond to the call is denoted by B_7. The calling subscriber can give up conversation with the probability B_8. If the calling subscriber has dialed a wrong number, the conversation goes to end immediately and the call at-

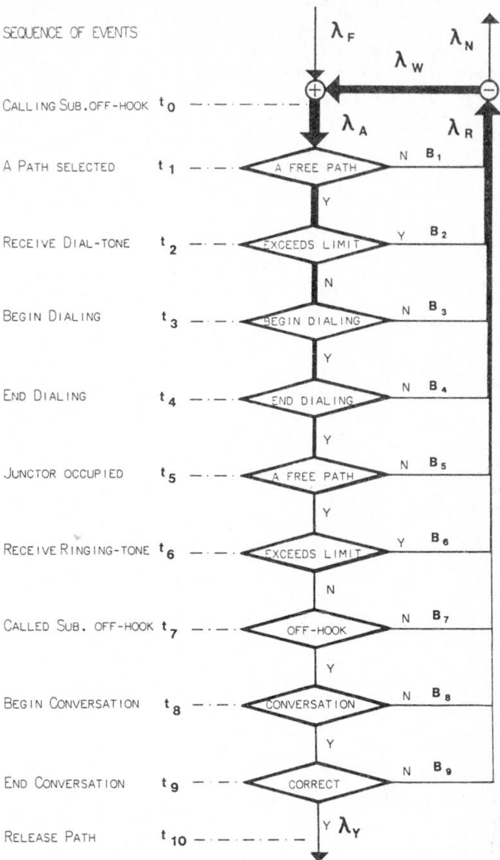

Fig. 1. A model for study of overloaded SPC switching systems.

tempt is not successful with the probability B_9. The conversation goes to end at time t_9 and the control unit carries out disconnection process at time t_{10}. Then the occupied speech path is ready for the further use.

In the model the rate of offered calls λ_A is determined by the rate of fresh calls λ_F and that of the repeated calls λ_W, and the rate of blocked calls λ_R is divided by λ_W and the rate of abandoned calls λ_N. The throughput is denoted by λ_Y. The unsuccessful call attempt makes retrials before it is given up. The retrial probability P_W can be obtained by the quotient of λ_W and λ_R.

The dial-tone delay is characterized by t_2-t_1 and the ringing-tone delay by t_6-t_4. The total holding time of a register t_5-t_1 can be devided by the useful holding time t_4-t_3 which is the duration for digit reception and two useless holding times, t_3-t_1 and t_5-t_4, which are produced by the reaction time of the control unit. The useless holding time which is usually known as blind holding time plays a very important role to the throughput degradation of overloaded SPC switching system. The detailed behavior in relation to this will be discussed in section 4. The total holding time of a junctor $t_{10}-t_5$ can be also devided by the useful holding time t_9-t_8 and two useless holding times, t_8-t_5 and $t_{10}-t_9$.

3. THROUGHPUT ANALYSIS

From the model in the previous section, the basic relationship between various call flow rates can be obtained by

$$\lambda_W = P_W \cdot \lambda_R \qquad (1)$$

$$\lambda_N = (1-P_W) \cdot \lambda_R \qquad (2)$$

$$\lambda_A = \lambda_F + \lambda_W \qquad (3)$$

$$\lambda_R = \lambda_A \sum_{j=1}^{9} B_j \sum_{i=1}^{j} (1-B_{i-1}) \text{ with } B_0=0 \qquad (4)$$

$$\lambda_Y = \lambda_A - \lambda_R = \lambda_F - \lambda_N \qquad (5)$$

The blocking probability of the system may be given by

$$B = \prod_{i=1}^{9} (1-B_i) \qquad (6)$$

The probabilities B_3, B_4, B_7, B_8, B_9 and P_W describe the subscriber behavior. As those values have no direct relationship with traffic offered, they can be determined by the call mix which is specified by measurement or experience. B_1 and B_5 are determined by traffic performance of the switching network, and B_2 and B_6 by the control unit. As those probabilities have close relationship with traffic offered, let us go into details in the following section.

3.1. Blocking due to switching network

The blocking probability of a switching network with n junctors consists of internal and external blocking probabilities, and the total blocking probability B_{tot} is given by

$$B_{tot} = B_{ext} + (1-B_{ext})B_{int} \quad , \qquad (7)$$

where B_{ext} denotes the external probability and B_{int} the internal. For the pure chance traffic of first type, the Erlang's loss formula can be approximately applied to the calculation of B_{ext}

$$B_{ext} = E_{1,n}(A) = \frac{A^n/n!}{\sum_{j=0}^{n} A^j/j!} \qquad (8)$$

where A denotes traffic offered and n the number of junctors. For the pure chance traffic second type, the Engset's loss formula may be used.

$$B_{ext} = B_{Engset} = \frac{\binom{m-1}{n}\left(\frac{\alpha}{\mu}\right)^n}{\sum_{j=0}^{n}\binom{m-1}{j}\left(\frac{\alpha}{\mu}\right)^j} \qquad (9)$$

where m denotes the number of traffic sources, α is the average call attempt rate of a traffic source and μ represents the average holding time.

The calculation of internal blocking probability of a switching network is very complicated. It depends on the traffic offered, network structure and operating strategy

$$B_{int} = f(\text{traffic,structure,strategy}). \qquad (10)$$

There are many methods for calculation of internal blocking probability (6). The ways of various calculating methods are not reproduced in this paper, because the description of them are too extensive.

3.2. Time-out due to control unit

This section deals with the calculation of the probabilities of B_2 and B_6 and some characteristic values of the model which is described in Section 2.

Generally, the programs in SPC switching systems are arranged by levels in such a way that each level contains several programs. In order to efficiently carry out the various program tasks of the levels, each level is assigned a certain priority so that one level may be interrupted by another with higher priority. The real control unit is a very complicated queueing system with different priorities and feedback, and can hardly be solved analytically.

For analytically tractable modelling of the control unit, it is assumed that tasks arrive according to a Poisson process, the number of waiting places for tasks is not limited and the tasks are processed by the discipline of non-preemptive priority. Under these assumptions the control unit can be modelled by an M/G/1 queueing system with non-preemptive priority discipline.

A usual way for dividing the work of a processor consists of two parts: load-independent (ρ_u) and load-dependent (ρ_v) work time. Load-independent work time is the time spent by the central processor executing tasks which are not dependent on the amount of traffic carried by the office. Load-dependent work time is the time spent in performing all the tasks which are generated by the call traffic load in the office. The number of load-dependent tasks which are required for the processing of a call attempt is shown in Tab. 1. The number of events is denoted by y and the kinds of tasks by x. Matrix element Z_{ji} describes the frequency of i-th kind of task in the event j. If we consider the blocking probabilities between events, the arrival rate of i-th load-dependent task is obtained by

$$\lambda_{v,i} = \lambda_A \sum_{j=1}^{y} Z_{ji} \cdot Q_j \qquad (11)$$

with

$$Q_j = (1-B_1)(1-B_2)\dots(1-B_{j-1}) \quad , \; B_0=0$$

where λ_A denotes the call arrival rate. The total arrival rate of load-dependent tasks can then be given by

$$\lambda_{v,t} = \sum_{i=1}^{x} \lambda_{v,i} = \lambda_A \sum_{i=1}^{x} \sum_{j=1}^{y} Z_{ji} \cdot Q_j \qquad (12)$$

The total arrival rate of load-independent tasks is represented by

$$\lambda_{u,t} = \sum_{i=1}^{z} \lambda_{u,i} \qquad (13)$$

where z is the kind of load-independent tasks.

Tab. 1. Number of load-dependent tasks

Sequence of Events \ Tasks	1	2	3	...	x
1	Z_{11}	Z_{12}	Z_{13}	...	Z_{1x}
2	Z_{21}	Z_{22}	Z_{23}	...	Z_{2x}
3	Z_{31}	Z_{32}	Z_{33}	...	Z_{3x}
⋮	⋮	⋮	⋮	⋮	⋮
y	Z_{y1}	Z_{y2}	Z_{y3}	...	Z_{yx}

With the aid of equations (12) and (13) we obtain the total arrival rate of tasks as

$$\lambda_t=\lambda_{v,t}+\lambda_{u,t}=\lambda_A\sum_{i=1}^{x}\sum_{j=1}^{y}Z_{ji}\cdot Q_j+\sum_{i=1}^{z}\lambda_{u,i} \tag{14}$$

This is the total arrival rate of tasks for a call type, e.g. an intra-office call. If we consider the four different types of call, we can get the total arrival rate of tasks

$$\lambda_{t,four}=\lambda_{v,t,io}+\lambda_{v,t,ou}+\lambda_{v,t,in}+\lambda_{v,t,tr}+\lambda_{u,t} \tag{15}$$

where

io : intra-office call,
ou : outgoing call,
in : incoming call,
tr : transit call.

Equation (15) describes the generalized total arrival rate of tasks in a SPC switching system. For further study, however, one type of call is considered in order to represent equations simply.

Fig. 2 shows an M/G/1 queueing system with non-preemptive priority discipline as a model of the control unit. In the model, each task is classified to a certain priority p with $1\le p\le P$. The priority of p is higher than that of p-1. In the same priority class the tasks are processed by the FIFO discipline. For the total arrival rate of tasks we can obtain the following equation:

$$\lambda_t = \sum_{p=1}^{P}\lambda_p \quad , \tag{16}$$

where

$$\lambda_p = \sum_{i\in K_p}(\lambda_{u,i}+\lambda_{v,i}) \tag{17}$$

and K_p is the collection of the priority class p. The value λ_t shows the total arrival rate of tasks of all priority classes.

The average program execution time for all tasks may be given by

$$\bar{h} = \sum_{p=1}^{P}\lambda_p\cdot\bar{h}_p/\lambda_t \tag{18}$$

where $\bar{h}_p$ is the average program execution time of tasks of priority class p and obtained by

$$\bar{h}_p = \frac{1}{\lambda_p}\cdot\sum_{i\in K_p}(\lambda_{u,i}\cdot h_{u,i}+\lambda_{v,i}\cdot h_{v,i}) \ . \tag{19}$$

The occupancy of the control unit for the tasks of priority p can be represented by

$$\rho_p = \lambda_p\cdot\bar{h}_p \tag{20}$$

The total occupancy of the control unit can then be derived as

$$\rho_t = \sum_{p=1}^{P}\rho_p = \sum_{p=1}^{P}\lambda_p\cdot\bar{h}_p \tag{21}$$

From (7) we may obtain the average waiting time w_p for tasks of the priority class p

$$w_p = \frac{w_0}{(1-\beta_p)(1-\beta_{p+1})} \quad ; \quad p=1,2,\cdots,P \ . \tag{22}$$

where

$$w_0 = \sum_{i=1}^{P}\lambda_i\cdot\overline{h_i^2}/2 \tag{23}$$

and

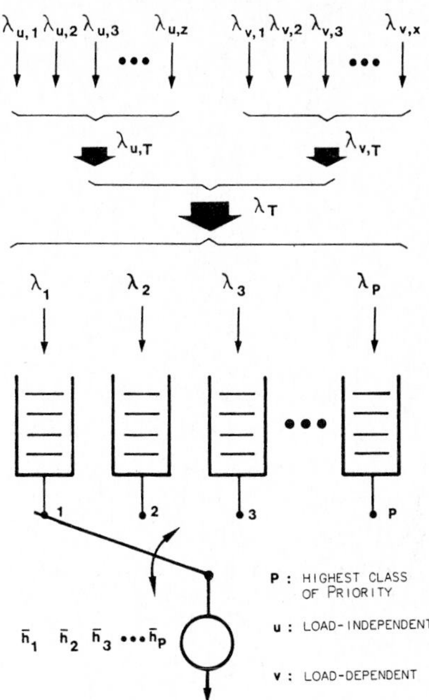

Fig. 2. An M/G/1 queueing system with nonpreemptive priority discipline as a model of control unit.

$$\beta_p = \sum_{i=p}^{P}\rho_i \quad . \tag{24}$$

$\overline{h_i^2}$ is the second moment of the program execution time of a priority class i and may be given by

$$\overline{h_i^2} = \frac{1}{\lambda_i}\cdot\sum_{j\in K_i}(\lambda_{u,j}\cdot\overline{h_{u,j}^2} + \lambda_{v,j}\cdot\overline{h_{v,j}^2}) \ . \tag{25}$$

If we consider a chain of k tasks which have different priority classes and are processed successively, then we obtain the total average waiting time of the chain as

$$t_{w,k} = \sum_{p=1}^{P}r_p\cdot w_p \quad , \tag{26}$$

where r_p is the frequency of the tasks of priority class p in the chain of k tasks. Herein the following relation is valid

$$k = \sum_{p=1}^{P}r_p \quad . \tag{27}$$

If we assume that the total waiting time of k tasks which are processed successively has an exponential density function, then we can obtain the probability that the total running-through time $T_{d,k}$ of k tasks exceeds a time limit t_L as

$$P\{T_{d,k}>t_L\} = \exp\{-\frac{t_L-(t_{h,k}+t_{c,t})}{t_{w,k}}\} \tag{28}$$

where $t_{h,k}$ is the average program execution time of a chain of k tasks and $t_{c,t}$ the average task-independent time of the chain, e.g. counting of time. With the equation (28) we can obtain B_2 and B_6, the probability due to time-out of dial-tone delay and that due to ringing-tone delay, as

$$B_2 = \exp\{-\frac{t_{L,DT}-(t_{h,k_{DT}}+t_{c,DT})}{t_w \cdot k_{DT}}\} \tag{29}$$

$$B_6 = \exp\{-\frac{t_{L,RT}-(t_{h,k_{RT}}+t_{c,RT})}{t_w \cdot k_{RT}}\} \tag{30}$$

where

- $t_{L,DT}$: time limit of dial-tone delay,
- $t_{L,RT}$: time limit of ringing-tone delay,
- $t_{h,k_{DT}}$: total program execution time from off-hook to switching-on of dial-tone,
- $t_{h,k_{RT}}$: total program execution time from end of dialing to switching-on of ringing-tone,
- $t_{c,DT}$: total task-independent time from off-hook to switching-on of dial-tone,
- $t_{c,RT}$: total task-independent time from end of dialing to switching-on of ringing-tone,
- $t_{w,k_{WT}}$: total average waiting time of tasks from off-hook to switching-on of dial-tone,
- $t_{w,k_{RT}}$: total average waiting time of tasks from end of dialing to switching-on of ringing-tone.

3.3. Solution of the equations

In the last two sections, the model was given as a set of equations, (1) through (30). If we examine the equations exactly, it leads to the nonlinear equations

$$\begin{aligned} B_1 &= f_1(B_1,B_2,B_5,B_6) \\ B_2 &= f_2(B_1,B_2,B_5,B_6) \\ B_5 &= f_5(B_1,B_2,B_5,B_6) \\ B_6 &= f_6(B_1,B_2,B_5,B_6). \end{aligned} \tag{31}$$

The equations can be solved by an iterative way. Given a set of B, a new set of B are calculated from the given parameters. If the new set and the old set are the same, a solution has been found.

The first technique is to use a simple substitution method to solve the equation (31). The advantage is simplicity, at the expense of being unable to prove convergence because of the complexity of the problem.

Another solution technique for a set of nonlinear equations is Newton's method. The advantage here is guaranteed convergence. The complexity, however, is far greater than for the method of substitution, since Newton's method requires the computation of the Jacobian matrix at each iteration, which must be evaluated by numerical differentiation. The method which we have used is the substitution method.

4. APPLICATION

In this section, the overload behavior of a periphery control unit with peripheral equipments is analyzed by using the above method. The control unit considered here is the essential part of hierarchically structured control unit of EMS-PABX which is developed by the company Siemens.

4.1. Description of the system considered

The family of EMS (Elektronisch, Mikroprozessorgesteuert, Speicherprogrammiert) is hierarchically structured by microprocessor control units whose tasks are shared by load and function. The family covers various size of system which can be built up from 10 to 12000 subscribers or more. The system considered is a part of the whole system and is shown in Fig. 3. The switching network consists of two stages A and B. In the stage A, 100 subscriber circuits are connected. To guarantee the processing of time critical tasks, the tasks are classified by various priority classes. One can obtain the detailed information about the organization of the control unit in (8).

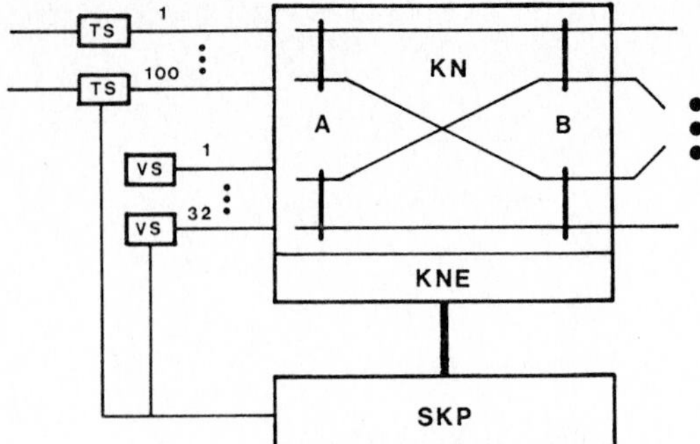

KN : Switching Network TS : Subscriber Circuit
VS : Peripheral SKP : Control Unit
KNE : Switching Network Controller

Fig. 3. A microprocessor controlled PABX system unit.

4.2. Numerical calculation

For the study of the overload characteristic of the system, it is assumed that the system handles 100 % intra-office calls. It is further assumed that a subscriber number of four digits is selected by a rotary dialing. The number of junctors is supposed to be 38 and that of registers to be 26. The internal blocking probability of the switching network is assumed to be zero. The parameters chosen are shown in tables 2, 3 and 4.

The calculation is carried out numerically according to the iterative substitution method. The number of iterations depends upon the system parameters, e.g. the rate of call attempts and the limit of the relative error.

Tab. 2. Selected system parameters for calculation of traffic behavior of the system.

System Parameters Concerning Subscriber Behavior
$m=100$ $P_W=0.8$ $t_c=5s$ $t_{DT}=2s$ $t_{RT}=10s$ $B_3=B_4=0$ $B_7=B_8=B_9=0$
System Parameters Concerning Switching Network
$n_r=26$ $n_j=38$ $B_{int}=0$
System Parameters Concerning Control Unit
$x=16$ $y=10$ Z_{ji}(see Tab.3) $P=14$ K_p(see Tab.4) $h_{v,i}, h_{u,i}, \lambda_{u,i}$(see Tab.4) $t_{L,DT}=3s$ $t_{L,RT}=3s$ $k_{DT}=51$ $K_{RT}=59$ $t_{h,DT}=41.001ms$ $t_{h,RT}=45.032ms$ $t_{c,DT}=140ms$ $t_{c,RT}=260ms$

Tab. 3. Frequency of tasks in each event. (*: time counting)

Sequence of Events \ Program	1 KFEN	2 BSR	3 NTR	4 MEIN	5 ZAV	6 SAV	7 EVA	8 TZGL	9 MAUS	10 AMV	11 VZGL1	12 VZGL2	13 VZGL3	14 VZGL4	15 VSC	16 TSC	Sum
1	-	-	-	-	-	-	-	1	1	-	1	-	-	-	-	1	4
2	5	5	2	4	2	1	10	-	2	6	5	-	3	2 (140)*	-	-	47
3	-	-	5	-	6	-	5	-	-	-	6	-	-	6 (2000)*	-	-	28
4	-	-	1	1	3	44	1	-	1	5	5	42	2	3 (300)*	-	-	108
5	-	-	-	-	1	-	-	-	1	1	1	2	-	1 (100)*	-	-	7
6	5	7	2	5	3	-	10	-	2	7	4	-	4	3 (160)*	-	-	52
7	-	-	11	-	6	-	9	-	-	-	6	-	-	6 (10000)*	-	-	38
8	-	2	3	-	-	1	4	-	-	1	-	-	1	-	-	-	12
9	-	-	-	-	-	1	-	-	-	-	-	-	-	-	-	-	1
10	10	4	11	6	8	1	14	-	6	8	11	-	8	8 (2560)*	-	-	95
Sum	20	18	35	16	29	48	53	1	13	28	39	44	18	29 (15260)*	-	-	392

Tab. 4. Program execution time for load-dependent and load-independent tasks, arrival rate of load-independent tasks and priority classes.

Tasks	Priority Class	$h_{v,i}$(ms)	$h_{u,i}$(ms)	$\lambda_{u,i}$(1/ms)
1	14	1.0	0.1625	0.2
2	12	0.278	0.019	0.1
3	8	0.218	0.034	0.1
4	13	0.8	0.019	0.2
5	7	0.3	0.039	0.1
6	6	1.35	0.039	0.1
7	5	1.18	0.1	0.1
8	4	0.7	0.16	0.1
9	3	0.8	0.021	0.1
10	2	2.0	–	0.1
11	10	0.1	0.083	0.1
12	10	0.125	0.083	–
13	10	0.275	0.083	–
14	10	0.1	0.083	–
15	11	–	1.3	0.1
16	9	0.5	0.5	0.1

4.3. Discussion of the results

To test the numerical results, time-true simulation of the system was run. A detailed description of this simulation is given in (9). The results of simulation are plotted with the statistical confidence of 95 %.

Fig. 4a shows throughput versus call arrival rate of the system considered. For low rate of call attempts the results of numerical method coincide well with that of simulation. From some level of call arrival rate, however, one may observe a little difference of the two results. With the increasing arrival rate of call attempts, the throughput may decrease from a point of call arrival rate. This phenomina is called throughput degradation. The maximum throughput of the system amounts to 1.27 calls per second at 1.4 call attempts per second. The bocking proba-

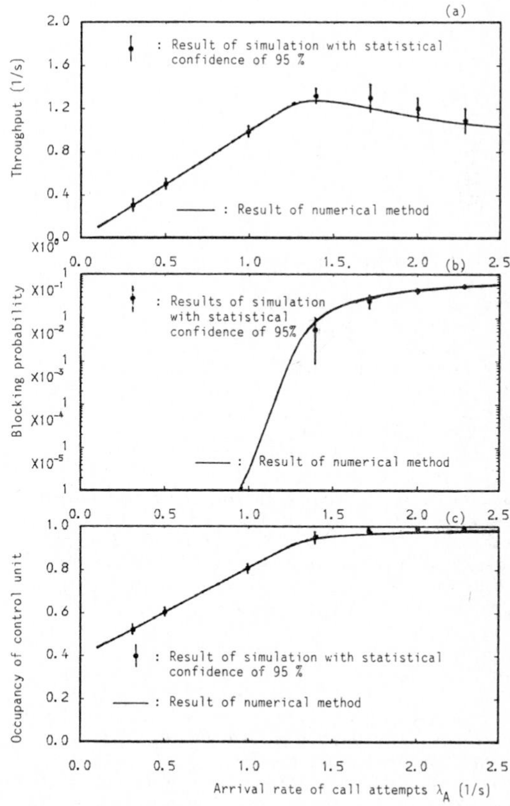

Fig. 4. Characteristical traffic values versus call arrival rate: a) throughput, b) blocking probability, c)average occupancy of control unit.

bility is shown in Fig. 4b. As the case of throughput, there is a little difference between the results of the numerical method and those of the simulation. Fig. 4c shows average occupancy of the control unit versus call attempt rate. From some level of call arrival rate, one can observe a little difference of the results of the numerical method and simulation. With the increasing arrival rate of call attempts, the average occupancy of the control unit increases to the level of 1 Erlang.

Fig. 5 shows the various traffic flows versus the call arrival rate. The results are obtained by the numerical method with the aid of equations (1)-(5). For the calculation, the probability of retrials P_W is assumed to be 0.8. With the results we can catch the relationship between the various traffic flows.

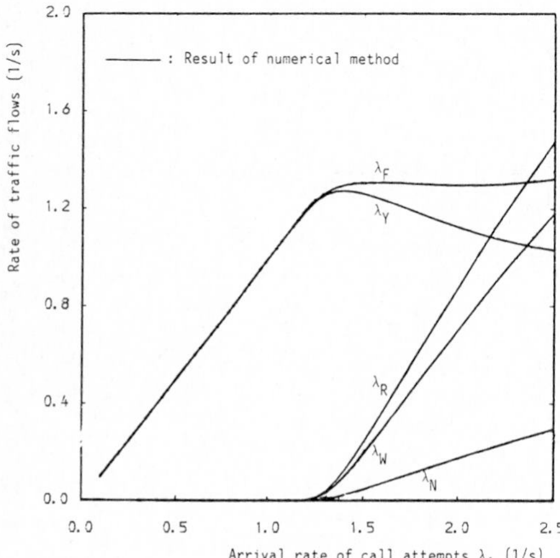

Fig. 5. Various traffic flows:
λ_F : flow rate of fresh calls,
λ_Y : flow rate of successful calls
λ_R : flow rate of blocked calls,
λ_W : flow rate of retrial calls,
λ_N : flow rate of abandoned calls,
λ_A : arrival rate of call attempts.

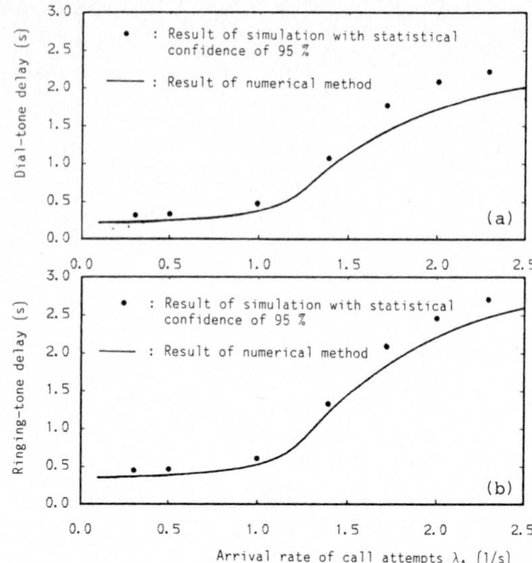

Fig. 6. Average delay time versus call arrival rate: a) dial-tone, b) ringing-tone.

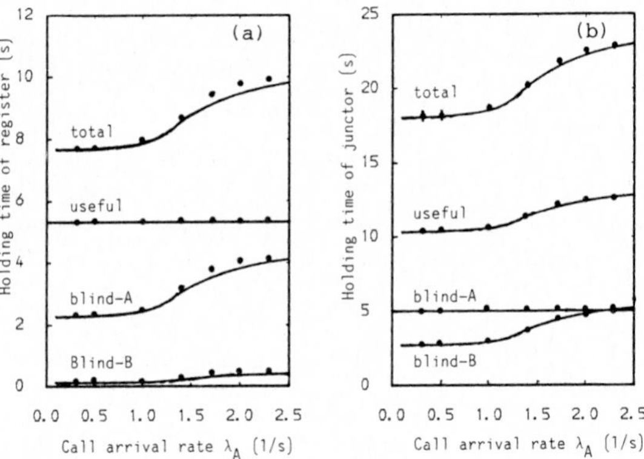

Fig. 7. Average holding time versus call arrival rate: a) register, b) junctor.

Fig. 6a shows the average dial-tone delay versus the call arrival rate. In this case we can observe a little more difference between the results of numerical method and those of simulation. With the increasing arrival rate of call attempts, the average dial-tone delay increases, too. The reason for this is that the waiting time of a task increases with the increasing call arrival rate. The average ringing-tone delay is shown in Fig. 6b. We can observe the similar results as the case of dial-tone delay.

Fig. 7a shows the average total holding time, the average useful holding time, and the average useless holding time of a register before and after dialing versus the arrival rate of call attempts. The useless holding time A (blind A) is the occupied duration of a register before dialing, while the useless holding time B (blind-B) describes the occupied duration after dialing. There is a little difference between the results of numerical method and those of simulation. The average useful holding time is independent of the call arrival rate. The total average holding time increases with the increasing of the call arrival rate. It amounts to 7.7 seconds at 0.5 call attempts per second, while it amounts to 10 seconds at 2 call attempts per second. The control unit causes the enhancement of the useless time which plays an important role for throughput degradation. Fig. 7b shows the average total holding time, the average useful holding time and the average useless holding time of a junctor before (blind-A) and after (blind-B) conversation versus the arrival rate of call attempts. We can observe the similar results as the case of register.

5. CONCLUSION

The analysis has shown that the throughput behavior and some characteristic traffic values can be numerically calculated with the proposed model with a small error. Considering the cost of developing of the simulation programs, we can accept this negligible error. With the proposed model we can observe the conditions when throughput degradation of an overloaded SPC switching system appears. Besides being used to predict steady-state system operation in the presence of overload, such a model should help to gain insight into SPC switching system operation. In addition, the availability of an analytical model allows automatic load control mechanisms to be brought to bear on various overload problems in SPC switching system management.

6. REFERENCES

(1) N. Farber, "A model for estimating the real-time capacity of certain classes of central processors", Inter. Teletra. Cong. (ITC-6), 1970, pp.426/1-7.

(2) J.E. Brand, J.C. Warner, "Processor call carrying capacity estimation for stored program control switching systems", Proc. IEEE, Vol. 65, No. 9, September, 1977, pp.1342-1349.

(3) E. Arthurs, B.W. Stuck, "A theoretical performance analysis of a Markovian switching node", IEEE Trans. Commu., Vol. COM-26, No. 11, November, 1978, pp.1779-1784.

(4) H.L. Hartmann, C.H. Yim, "Approximate description of SPC switching systems by Erlang's and Pollaczek-Khintchine's Formulas", (to be published).

(5) E. Szybicki, "Approximate method for determination of overload ability in local telephone systems", Inter. Teletra. Cong. (ITC-7), 1973, pp.423/1-8.

(6) K. Kümmerle, "Berechnungsverfahren für mehrstufige Koppelanordnungen mit konjugierter Durchschaltung", Dissertation, T.U. Stuttgart, 1969.

(7) L. Kleinrock, "Queueing systems volume 2: Computer applications", John Wiley and Sons, New York. Chichester. Brisbane. Toronto, 1976.

(8) Wallner, "Über das Verkehrsverhalten des Kombinationssteuerwerkes PSt bei EMS OB", Interner Bericht, Siemens A.G., 1977, pp.1-35.

(9) C.H. Yim, "Durchsatzrückgang in überlasteten Wählvermittlungen", Dissertation, T.U. Braunschweig, 1984.

TELETRAFFIC ISSUES in an Advanced Information Society
ITC-11
Minoru Akiyama (Editor)
Elsevier Science Publishers B.V. (North-Holland)

BASIC DEPENDENCES CHARACTERIZING A MODEL OF AN IDEALISED (STANDARD) SUBSCRIBER AUTOMATIC TELEPHONE EXCHANGE

Peter TODOROV and Stoyan PORYAZOV

Telecommunication Research Institute
Bulgarian Academy of Sciences, Institute of Mathematics
Sofia, Bulgaria

ABSTRACT

An idealised Automatic Telephone Exchange (ATE) is considered. A comparison is made between a model without repeated calls (with and without absence with uniform and ununiform activity of the subscribers) and a model with repeated calls. For some cases analytical dependences are given. Mainly a Poisson flow is considered but also Engset flow and a negative binomial case are touched upon. Considered are only losses due to mistakes of the A-subscriber, the absence and occupation of the subscribers, depending on the activity and duration of the occupation of the installations, i.e. only the natural factors limiting the quality of functioning of a switching telephone network. The results obtained are significant for a better estimation of the real systems.

1. INTRODUCTION

There have been made a great number of attempts of simulation of telephone systems with repeated calls (see [1]). The essential thing that distinguishes our approach is that we consider a system of N subscribers connected by means of an idealised ATE and the subscribers' behaviour is described comparatively in a more detailed manner.

Our idealised ATE has: 1) maximum decentralized control without delay which resembles the control of the old step exexchanges but it is digital. The type and duration of the accoustic signals are maximum near to those of the step exchanges. 2) The exchange is with two levels - subscriber and switching subsystems with a full availability. 3) The technical losses are small and can be ignored.

We assume that in the considered interval of time the durations of all the phases of serving that are considered as independent random variables are known and do not change their mean values depending on the intensity of the input flow in calls per second (in contrast to the mean occupation time of a subscriber). Under these conditions we look for a connection between the mean intensity of the input flow on the one hand and the subscribers traffic and other important characteristics of the telephone system on the other hand.

2. MAIN INPUT PARAMETERS

There are considerable differences in the experimental data received under different conditions by different authors because of which the chosen by us values and distributions of the input parameters shown on Figure 1 are illustrative. We have used for example [1,2,3] as well as our observations. When the value is given by an interval (for example 12±3) this means that a uniform distribution in the interval ([12-3, 12+3]) is used. If the value is denoted by "exp" (for example 180 exp) one must understand that the variable has the shown mean value (180) and has an exponential distribution. All the times are given in seconds.

The notations are:

N = number of all the subscribers in the system

λ = mean intensity of the whole primary offered traffic in calls per second

λ' = mean intensity of the primary offered traffic of a free source

<u>Mean holding times:</u>

T_d = dialling performed by the calling A-subscriber (6-7 digits)

T_{md} = dialling when the subscriber thinks he has misdialled the number

T_{ans} = ringing tone when the called B-subscriber answers

T_{abs} = ringing tone when the B-subscriber is absent

T_{busy} = busy tone

T_c = conversation

T_{mc} = conversation that finds misdialling

<u>Times from the hanging up the receiver up to its picking up before the next new attempt, after:</u>

W_{mc} = conversation finding misdialling

W_{busy} = busy B-subscriber

W_{abs} = no answer

W_{md} = misdialling

W_{free} = waiting for the A-subscriber to be free, if he was busy in the interval between two attempts

<u>Probabilities for directioning of the calls to:</u>

P_{abs} = absent subscriber

P_{md} = misdialling

P_{mc} = conversation finding misdialling

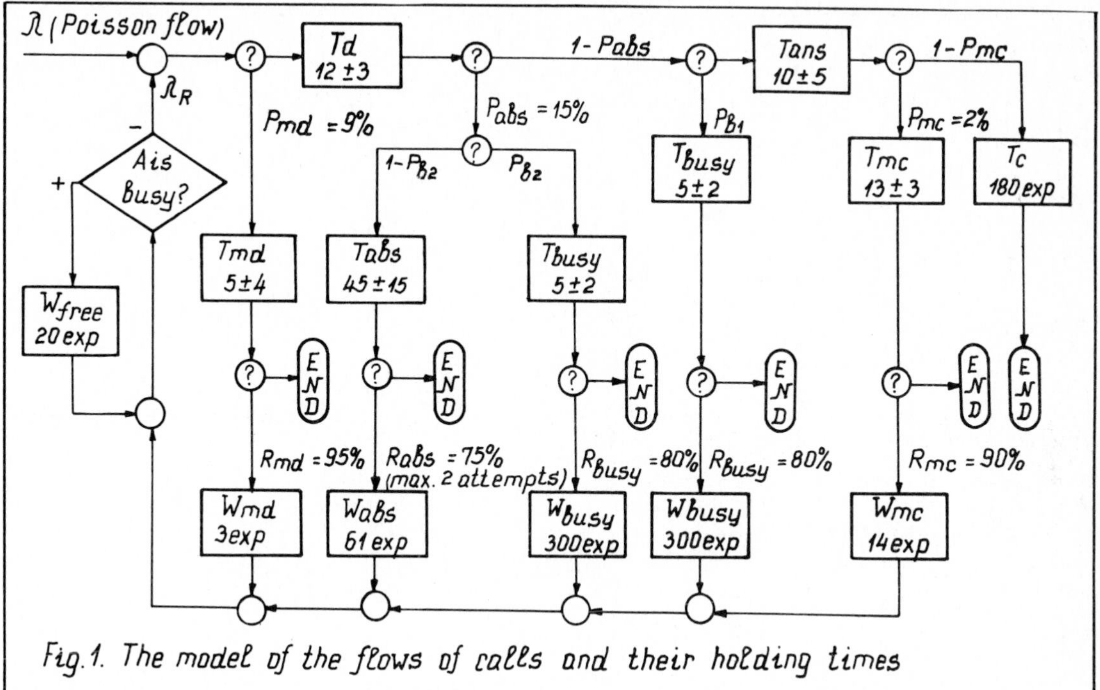

Fig. 1. The model of the flows of calls and their holding times

R_{busy} = busy B-subscriber
R_{abs} = no answer
R_{md} = misdialling

3. MAIN OUTPUT PARAMETERS

P_{busy} = probability of finding the B-subscriber busy
P_{B1} = probability of finding a busy not absent subscriber among all not absent
P_{B2} = probability of finding a busy absent subscriber among all absent
λ_R = average intensity of the whole flow of repeated calls

$$\beta = \frac{\lambda + \lambda_R}{\lambda}$$

A = traffic of the A-subscribers
B = traffic of the B-subscribers
E = efficiency of calls
F = $\dfrac{\text{number of calls ended in a conversation}}{\text{number of all calls}}$
T_A = average occupation time of the A-subscriber (sec)
T_B = average occupation time of the B-subscriber (sec)
T_{AB} = average occupation time of a subscriber
E_t = efficiency of traffic

$$E_t = \frac{\sum T_A \text{ ended in a conversation}}{\sum T_A \text{ for all occupations of the A-subscriber}}$$

$$P'_{abs} = \frac{\text{number of calls directed to an absent subscriber}}{\text{number of all calls after successful dialling}}$$

$P'_{abs} \neq P_{abs}$ if the system includes repeated calls

4. SIMULATION MODEL

is a development of the model [4]. A group of 2000 subscribers is modelled. Its essential particularity is that the subscribers are discernable, i.e. every subscriber has a unique number and other individual characteristics, for example whether he is absent or not. Each call is represented as a transaction in the GPSS-language. The flow of calls corresponds to the shown in Fig. 1. The original flow of calls is directed with equal probabilities only to free subscribers who are not noted as absent. The number of the B-subscriber is chosen with equal probability among all subscribers. The numbers of the A-subscribers and the B-subscribers are written in the parameters of the transaction and without changes are used with the repeated calls.

Since in the interval between two attempts the original A-subscriber can be seized as a B-subscriber or can generate another call as A-subscriber, it is necessary to check before the beginning of each repeated attempt whether the A-subscriber is free. The experiments show that at β = 1.854, for one seizure of an A-subscriber the check cycle including (see Fig. 1) is fulfilled 1.245 times, i. e. it can be assumed that the interval between two calls is influenced directly by the loading of the telephone system (especially at great β).

After terminating of an unsuccessful attempt, in this case with repeated attempts, the call remains in the system with probability R_{md}, R_{abs}, R_{busy} or R_{mc} , depending on the cause for

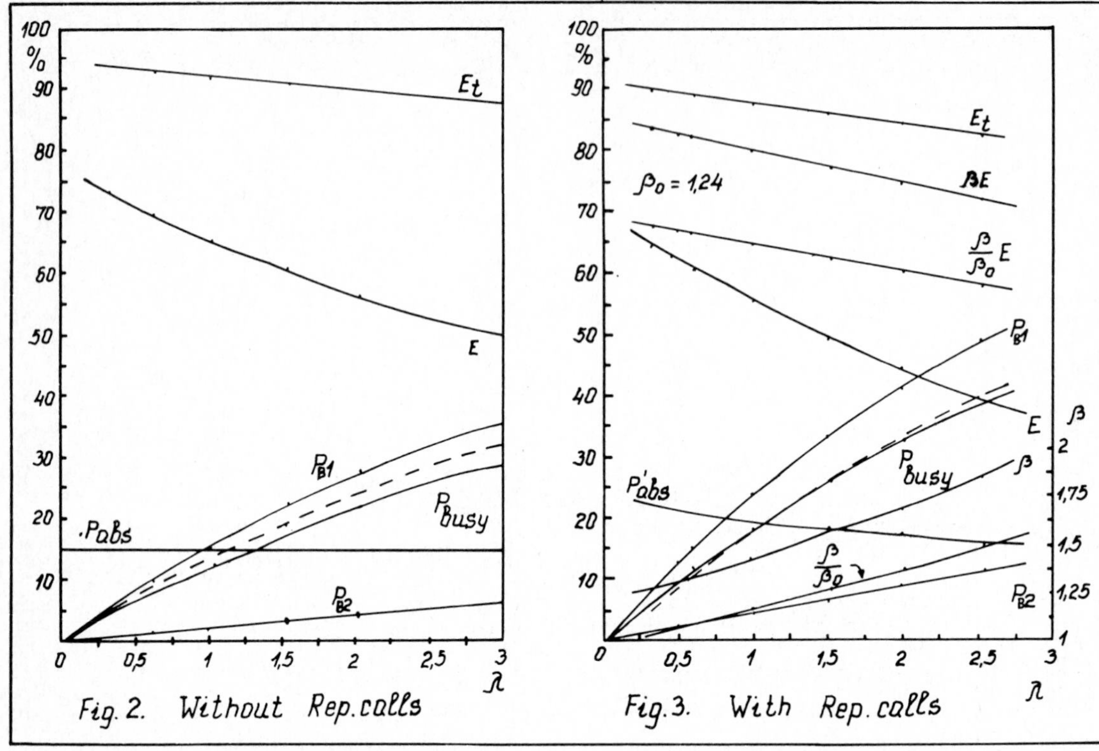

failure. These probabilities are equal for every attempt with two exeptions: 1) Means for registration of up to 10 attempts are provided and for this reason the transactions for the 11-th attempt are rejected by the system. The experiments show that up to λ = 1.526 calls/sec, P_{busy} = 25.781% there is no 10-th attempt and at this value they are 0.013% of all attempts. At the point with the highest simulated value of the input flow - λ = 2.522 calls/sec (P_{busy} = 39.059 %), 0.07% of all calls have made the 10-th attempt. These data show that the registration of more than 10 attempts is indefensible.
2) The other exception is the cases with repeated attempts due to absent subscriber. The known data and our observations show that when the B-subscriber does not answer the A-subscriber makes after a short interval one more (as an exception 2) attempt, mainly to be sure he has not misdialled. We assume that the attempts made after more than 60 minutes are new attempts and not repeated, that is why we do not admit more than 2 attempts ending with "no answer". Of course, as before, so between the two attempts there can be losses due to any of the other described reasons, but it is not possible for the B-subscriber to answer, because he is noted as absent. Naturally the telephone of an absent B-subscriber can be occupied when it rings (average T_{abs} sec). A considerable part of the telephones of the absent subscribers can be occupied due to this reason (see §6.2). The data received from the simulation show that the second call ended in "no answer" under the described conditions can occur even at the 8-th attempt.

The model includes about 1000 punch cards in the GPSS-language, uses about 480 K bytes storage of a computer ES 1040 and needs about 0.07 sec for processing of a call, including input/output operations. For a time unit is chosen a santisecond (0.01s). The shown results are received on the basis of 10000 transactions that left the system through the blocks "end" in Fig.1 after the processes have reached statistical equilibrium. Since statistics are gathered after the end of every attempt, for a system with repeated calls this means that at least 10000 β measurements have been made for every value of λ .

5. MODEL WITHOUT REPEATED CALLS

5.1. Without Absent Subscribers

Let in Fig.1 $P_{abs} = R_{md} = R_{abs} = R_{busy} = R_{me} = 0$. In this case obviously $P_{b1} = P_{busy}$. We shall follow the approach offered in [5] . We consider every phase of service like a multichannel device with unlimited number of channels in which at every moment can be served independently from each other unlimited number of calls. We consider as logically different holding times even such which are received in the same way, if this increases the clearness and convenience at examination and does not change the logic of functioning of the system. For example

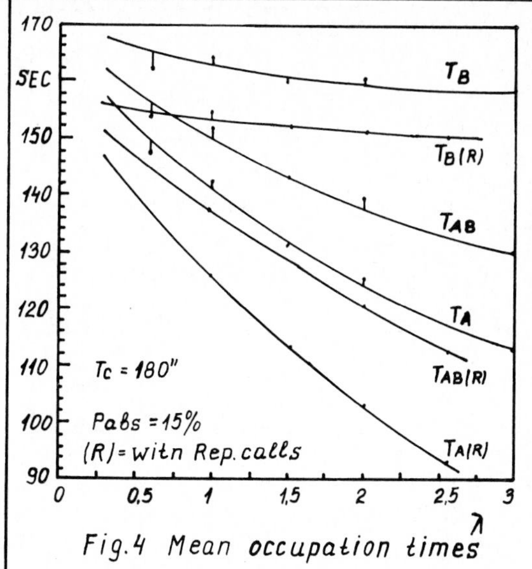

Fig.4 Mean occupation times

in Fig.1 we have two different devices representing the dialling, two representing the conversation an so on.

For λ , λ' and all times of the phases of service we shall consider fulfilled the conditions required for the theorem of Little [6] :

$$\lambda = \lim_{t\to\infty} \lambda_t \quad ; \quad T = \lim_{t\to\infty} T_t \quad , \text{ where} \qquad (1)$$

λ_t and T_t are the average values in the interval $(0, t)$, t is the current time.

From Fig.1, the formula of Little and the saving laws we derive at a stationarity:

$$A + B = \lambda \left[C_1 - P_{busy} C_2 \right] , \qquad (2)$$

where C_1 and C_2 are random variables

$$C_1 = P_{md} T_{md} + (1 - P_{md})(2T_b + T_d) ;$$

$$C_2 = (1 - P_{md})(2T_b - T_{busy}) ;$$

$$T_b = T_{ans} + P_{mc} T_{mc} + (1 - P_{mc}) T_c \quad . \qquad (3)$$

The coefficient 2 before T_b is required, since while a B-subscriber is busy, an A-subscriber is busy too. In this case all subscribers are equal and the calls are directed uniformly to all the subscribers, therefore we can consider:

$$P_{busy} = \frac{A + B}{N} , \qquad (4)$$

From (2) and (4) we receive:

$$P_{busy} = \frac{\lambda}{N} \left[C_1 - P_{busy} C_2 \right] \qquad (5)$$

In the case of a Poisson input $\lambda = \lambda' N$, from where:

$$P_{busy} = \frac{\lambda' C_1}{1 + \lambda' C_2} \qquad (6)$$

By analogy for Engset input and input of negative binomial type we receive correspondingly (7) and (8):

$$P_{busy} = \lambda'(1 - P_{busy})(C_1 - P_{busy} C_2) \qquad (7)$$

$$P_{busy} = \lambda'(1 + P_{busy})(C_1 - P_{busy} C_2) \qquad (8)$$

The expression (5) makes it possible to consider other types of input flows. The simulation data confirm fully the received results [7,8,9,10]. In [11] is considered the case of an Engset input flow, as in case (7), so when the intensity of the input flow decreases proportionally to the absent subscribers. The results show that in Engset input the absence of the A-subscribers as a rule must not be taken into consideration, since a subscriber who has left his telephone can remain a source of calls, creating in this way additional loading of other telephones. The model with Engset input and decrease of flow proportionally to the absence of subscribers gives lower losses in comparison with our observations.

In the mentioned works, in contrast to this work, is considered the case of the subscriber's absence with equal probability. In [11] is given a comparison between the cases with and without absence of the subscribers at Engset and Poisson flows.

As it is seen in Fig.1 the knowledge of P_{busy} gives us the opportunity to calculate all important characteristics of the telephone system in stationarity, since the values of the other input parameters are relatively independent of λ. For example, the conversational traffic (A_c) of the A-subscribers (simultaneously occupied A-subscribers, whose occupation will end with conversation) will be (see Fig.1 with assumptions made):

$$A_c = \lambda(1 - P_{md})(1 - P_{busy}) \left[T_d + T_{ans} + P_{mc} T_{mc} + (1 - P_{mc}) T_c \right]$$

and the specific conversational traffic will be A_c/N . After determining λ/N from (5), we receive:

$$\frac{A_c}{N} = \frac{P_{busy}(1-P_{md})(1-P_{busy})\left[T_d + T_{ans} + P_{mc} T_{mc} + (1-P_{mc}) T_c\right]}{C_1 - P_{busy} C_2}$$

A_c/N is 0 for $P_{busy} = 0$ and 1 and has only one maximum, for example when $P_{busy} = 0.8$. It is important to note that (9) does not depend on the type of input flow (Poisson, Engset or negative binomial case) if it is stationary. Other similar results are shown in [12] . All given results are confirmed with a relative error at 6000 measurements not greater than 2% for all simulated values of λ .

5.1.1. Model when the activity of the subscribers is not uniform

Let us consider the subscriber's traffic with the assumptions made, but when all N subscribers are divided into K categories. The ununiform activity is expressed with regard to the intensity of the offered flow in calls per sec, its directioning and the duration of occupation of the installations. For each category remain in force the already described input parameters and one or two indexes are added to the notations, the first showing the category of the A-subscriber, the second - the category of the B-subscriber. For example, $T_d(I,J)$ is the duration of dialling the number of the B-subscriber from category J, by the A-subscriber from category I. If there is only one index, it refers to the category of the A-subscriber, for example, $T_{busy}(I)$ depends only on the behaviour of the A-subscriber.

The required additional parameters are:

$P(I,J)$ = matrix of the traffic interest. It shows the probability of directioning of a call originating from category I to category J

$L(I)=\lambda(I)/\lambda$, where $\lambda'(I)$ is the offered traffic (calls/sec) by one subscriber of category I and λ is the whole offered traffic by the N subscribers.

$Q(m)=N(m)/N$, where $N(m)$ is the number of the subscribers from category m.

In [5] is shown that at a Poisson flow the k unknowns $P_{busy}(I)$ can be determined as a solution of the following system of k linear equations:

$$\sum_{J=1} A(I)P_{busy}(J)+P_{busy}(I)\left(D(I)+\frac{Q(I)}{\lambda}\right)= B(I)\ , \qquad (10)$$

where

$$A(I)=L(I)Q(I)\sum_{m=1}^{k} P(I,m)\left[T_B(I,m)-T_{busy}(I)\right],$$

$$D(I)=\sum_{m=1}^{k} L(m)\,Q(m)\,P(m,I)\,T_B(m,I)\ ,$$

$$B(I)=L(I)\,Q(I)\sum_{m=1}^{k} P(I,m)\left[T_d(I,m)+T_B(I,m)\right]+$$
$$+\sum_{m=1}^{k} P(m,I)\,L(m)\,Q(m)\,T_B(m,I)\ .$$

After knowing all $P_{busy}(I)$, we can determine all important characteristics describing the modelled system and for each of the categories. In [13] is used this result for 3 categories of subscribers and is shown that not taking into account the ununiform activity of the subscribers when dimensioning PABX, can lead to considerable errors. The data received as a result of the simulation confirm (10) with a relative difference under 2% in the whole work interval of λ.

Another important conclusion for the PABX is that the losses due to "busy" must be accounted not according to the occupation of the connective lines, but according to the occupation of the subscribers from the PABX.

In order to apply the results to local exchanges we need much more and more suitable data in comparison with the available data characterizing the ununiform activity and in particular the mutual interest among the subscribers from different categories. Similar data are given in [14] and are approximately valid for Bulgaria, where the difference between the home, business and administrative categories of subscribers is too small.

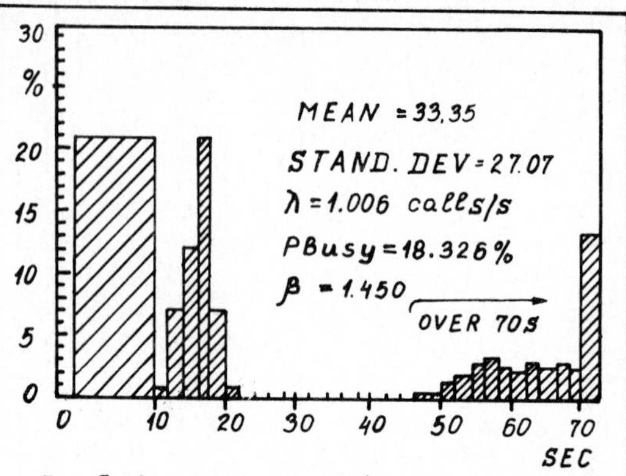

Fig.5 Lost occupation of the A-subscr

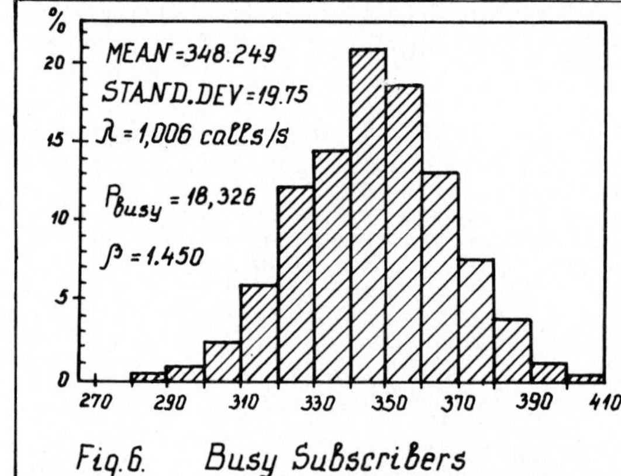

Fig.6. Busy Subscribers

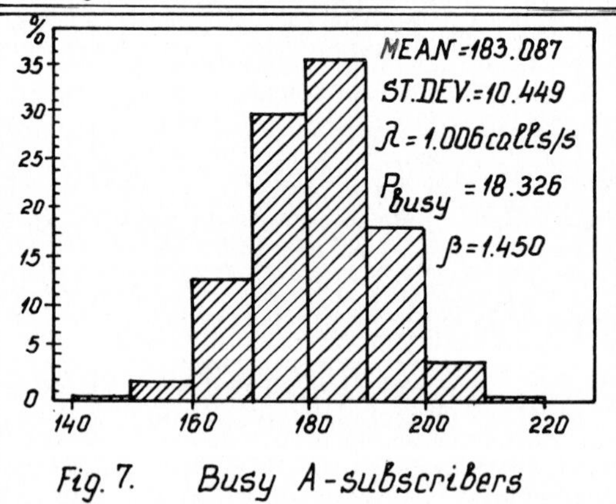

Fig. 7. Busy A-subscribers

5.2. Model with Absence of Subscribers

Let us consider the model in Fig.1 under the condition $R_{md}=R_{abs}=R_{busy}=R_{mc}=0$. The results from § 5.1 are not applicable directly, since here the subscribers are not equal. But we can consider the subscribers consisting of two homogeneous categories: 1) present and 2) absent (with $\lambda'=0$) and use the results from § 5.1.1. In Fig.1 we can easily see that:

$$P_{busy}=(1-P_{md})\left[P_{abs}P_{b2}+(1-P_{abs})P_{b1}\right] \qquad (11)$$

On the other hand, as in each group the subscribers are homogeneous:

$$\frac{A+B}{N} = P_{abs} P_{b2} + (1 - P_{abs}) P_{b1} \qquad (12)$$

In Fig. 2 and 3 are shown the results from the simulation. With a dotted line are plotted the values of $(A+B)/N$ in %. The equation (11) is valid for the measured values up to the first 5 digits and the equation (12) shows a relative difference of 1.9% for $\lambda = 0.326$, that decreases with augmentation of λ (for λ = 0.606 it is 1.16%) and reaches 0.69% for λ = 5.133 calls/sec. The losses from "no answer" decrease with augmentation of λ, since they increase as the common losses from "busy subscriber", so the probability to dial an occupied telephone of an absent subscriber (P_2).

In Fig.2 it is seen that the probability of falling upon a busy subscriber among the present subscribers (P_1) in the work area of λ can exceed 1.26 times the average probability $P_{busy} = 12.23\%$ while $P_2 = 2.15\%$ is 5.68 times lower than P_{busy} and causes 0.28% from all losses. It is seen from (11) that the importance of P_{b2} increases with the increase of P_{abs}. On the other hand, P_{abs} can exceed in some cases even 60%.

E_t is practically a straight line and decreases more slowly than E. E_t depends strongly on the duration of the conversation [8]. In Fig.4 is seen the decreasing of the mean occupation times separately for A,B and all the subscribers. The decrease of T_A is the greatest. This decrease is due to the increase of P_{b1} and P_{b2} (Fig. 1,2), at constant mean values of the holding times.

6. MODEL WITH REPEATED CALLS

Let us consider the full model in Fig.1. The results from the simulation are shown in Fig. 3,4,5,6,7. In Fig.3 is seen considerably more steep increase of P_1, P_2 and P_{busy} in comparison with the case without repeated calls (Fig.2). The values of (A+B)/N (dotted line) approach P_{busy}. This is due to the uneven directing of the flows of the repeated attempts. The dependence (11) is fulfilled again with the same precision, if P_{abs} is replaced by P'_{abs} (plotted on Fig.3). It is seen that only one repeated call due to "no answer" can considerably increase the percentage of calls directed towards absent subscribers. This increase is registered more strongly for the low (near the work area) values of λ.

The equation (12), as it can be expected, is not valid in the case of repeated attempts - the measured relative error is 20%. The observed values of β are a little bit lower than the publicated experimental data [3,15] but this can be explained by the lack of losses in our model, due to technical losses and shortage of installations.

E and E_t are lower and decrease more rapidly than in the case without repeated calls. The shown efficiency (E) is lower in comparison with some publicated data [3,15] but well agrees with our observations upon a system with direct control, without noticeable technical losses.

In [3,14] is indicated that the multiplication βE gives an impression about the traffic which is proportional to the charged traffic. We think that the charged traffic is represented better through the multiplication $\frac{\beta}{\beta_0} E$, where β_0 = 1.24 is the minimal measured by us value of β, when $P_{busy} <$ 3%.

The variables E_t, βE and $\frac{\beta}{\beta_0} E$ are practically straight lines. Round the work point, for which is accepted the value of λ, creating P_{busy} = 12% ÷ 15% P_1, P_{busy} and P_2 are in ratio 1.31/1/0.23.

Fig. 4 shows considerably lower values of the average holding times in comparison with the case without repeated calls.

The ratio A/(A+B) increases slightly faster than in the case without repeated attempts. In both cases, in the whole depicted on Fig.2 and 3 interval of λ, the ratio A/(A+B) varies from 51.92% to 53.89%. For λ = 1 calls/sec without repeated attempts A/(A+B) is 52.197% and with repeated attempts - 52.573%.

In Fig.5 is shown the distribution of the occupation time of the A-subscribers in the case of an unsuccessful attempt. The cases of 1 to 9 sec are due to dialling, considered wrong, from 9 to 22 sec are caused by busy B-subscriber and over 40 sec by not answering B-subscriber. The losses due to conversation, finding wrong dialling are not included, because they pertain to the charged traffic. It is evident that it is very difficult to approximate with only one analytical distribution the results in Fig.5.

In this respect the results shown on Fig. 6 and 7 look better. In Fig. 6 is depicted the distribution of the number of all simultaneously occupied subscribers and in Fig.7 of the number of simultaneously occupied A-subscribers. The distributions in Fig. 5,6 and 7 change as their mean value, so their type, depending on λ. The ratio A_{max}/A_{mean} decreases from 1.45 at P_{busy} = 6.67% to 1.11 at P_{busy} = 39.06%. When P_{busy} = 11.94% it is 1.32.

7. CONCLUSION

A simulation model of a telephone system is worked out, including idealised Automatic Telephone Exchange (ATE). Comparatively in detail is described the behaviour of the subscribers in the case without repeated calls: considered are the subcases with and without absence of the subscribers as well as their ununiform activity. Offered are analytical dependences describing the model without repeated attempts, confirmed with great precision by means of the data received from the simulation models. These dependences are in force for the mean values of the participating in them variables, at a statistical equilibrium of the processes in the modelled system and are independent of the distribution of the values of the participating in them input parameters in the same degree in which the formula of Little is independent.

It is shown that not taking into consideration the ununiform activity of the subscribers can lead to considerable mistakes when dimensioning PABX. The available data from measurement of the uniform activity in real local exchanges are not suitable and not sufficient for the full application of the model worked out. In order to consider the case of permanently absent (in the considered time interval) subscribers it is necessary to use the model with ununiform activity considering the absent subscribers as a separate category. In this case the hypothesis that $P_{busy} \approx$ (A+B)/N [16] is not confirmed.

Given are results from the simulation in the case with repeated attempts and they are compared with data received for the same input parameters, but without repeated attempts. Shown is the considerable deterioration of the considered indices as losses due to busy and absent subscribers, efficiency and average occupation time. Makes an impression a considerable increase of the calls directed towards absent subscribers, for example 23% against 15% in the case without repeated attempts. This is important, having in mind that in today's dynamic world the absence of the subscribers is considerable and shows the necessity of decreasing the cases "the subscriber is busy" and "the subscriber is absent"[17, 18] . Shown are the distributions of the time of the lost occupations and the traffic of the A-subscribers, as well as the traffic of all the subscribers.

In the chosen by us approach, the received dependences have the character of natural laws, since they do not depend on shortage of installations and technical losses. This gives an opportunity to dimension the real ATE reaching optimum. For example, there is a direct relation between the traffic of the A-subscribers and the number of the necessary internal connective lines in ATE. Because of this the found mean values, maximum values and the distribution of the A-subscribers' traffic for each λ have a great practical significance. The worked out model is foreseen to reflect the interval structure of a real ATE with different types of control and in this way it will become a convenient instrument in aid of the telephone administration.

REFERENCES

[1] G.L.Jonin. The Systems with Repeated Calls: Models, Measurements, Results", Proceedings of the Third International Seminar on Teletraffic Theory, Moscow, USSR, p.p. 197-208, 1984.

[2] Evers R. Measurement of Subscriber Reaction to Unsuccessive Call Attempts and the Influence of Reasons to Failure. ITC-7, Stockholm, 1973.

[3] J.-P. Guérineau, G.Pellieux. Nouveaux résultats concernant le comportement de l'abonné du réseau téléphonique de Paris. J. "Commutation et Electronique", Socotel, Paris, No.45, Avril 1974, pp 5p-71.

[4] S.A.Poryazov. A Generalized Simulation Model of a Switching System. Proceedings of the International Conference on Computer-Based Information Servicing (ICCBIS), Varna, Bulgaria, 1983, (in Russian).

[5] S.A.Poryazov. A Method for Analysis of the Telephone Subscribers Traffic. Proceedings of the Fourteenth Spring Conference of the Union of Bulgarian Mathematicians, Sunny Beach, April 6-9, 1985, pp.7. (in Bulgarian).

[6] J.D.C.Little. A Proof of the Queueing Formula $L=\lambda W$. Operations Research, 9, 1961, pp. 383-387.

[7] P.M.Todorov, S.A.Poryazov. Simulation Model of an Isolated Telephone System, Including ATE. Proc. of ICCBIC, Varna, Bulgaria, 1983, pp. 10 (in Russian).

[8] P.M.Todorov, S.A.Poryazov. Simulation Model of an ATE. Proc. of the 14-th Conference of ISAGA, Sofia, Bulgaria, 1983, pp.12.

[9] P.M.Todorov, R.G.Dotcheva, S.A.Poryazov. Basic Characteristics Received from Modelling of PABX with Direct Control. Proc. of ICCBIS, Varna, Bulgaria, 1983, pp. 6.

[10] S.A.Poryazov, P.M.Todorov, R.G.Dotcheva. Research of Some Traffic Characteristics in ATE through Simulation Modelling. J. "Communications", No.4, Sofia, 1983, pp. 20-22.

[11] P.M.Todorov, S.A.Poryazov. Simulation Model of a Subscriber Automatic Telephone Exchange with Poisson and Engset Input Flows. Proc. of the 13-th Spring Conference of the Union of Bulgarian Mathematicians, Sunny Beach, 1984, pp. 457-463.

[12] S.A.Poryazov, P.M.Todorov. Characterizing Idealised Subscriber Automatic Telephone Exchange. Proc. of the Third International Seminar on Teletraffic Theory. Moscow, USSR, 1984, pp. 351-354.

[13] S.A.Poryazov, P.M.Todorov. Influence of the Data Transmission at Dimensioning PABX's. International Conference on Computer-Based Scientific Research, Plovdiv, Bulgaria, 1984, pp. 8.

[14] Lars Engvall. A Socio-economic Study on the Usage of Telephone, Television and Radio Services. Rapport TRIRA-TIDS-8402, Inst. för Teletrafiksystem, Kungl Tekniska Högskolan, Stockholm, Sweden, May 1984, pp. 100.

[15] P.Le Gall. Repeated Calls and Traffic Engineering. J. "Commutation & Electronique". Socotel, Paris, No.56, 1977, pp. 11.

[16] G.Gostony, R. Ranko, R.G.Hapius. La qualité d'écoulement du trafic dans le réseau téléphonique mondial. J. des télécomm., v.46, 1979.

[17] P.M.Todorov. Method of Decreasing the Cases "Called Subscriber Busy" and "Called Subscriber Absent". J. Communications, v.10, Sofia, 1982.

[18] P.M.Todorov. Initial Norms for Dimensioning and Estimation of the Operation of Local Networks and ATE. J. Communications, v.12, Sofia, 1982.

TELETRAFFIC ISSUES in an Advanced Information Society
ITC-11
Minoru Akiyama (Editor)
Elsevier Science Publishers B.V. (North-Holland)

A CLASS OF SCHEDULING POLICIES FOR REAL-TIME PROCESSORS WITH SWITCHING SYSTEM APPLICATIONS

Yonatan LEVY

AT&T Bell Laboratories
Holmdel, New Jersey, USA

ABSTRACT

Consider a queueing system where a single server has to process a queue of primary jobs and also perform background tasks. In this paper, we analyze a class of scheduling policies that allot an uninterrupted segment to background work whenever no primary jobs are present and also when a limit on primary job processing is exceeded. The following performance measures are derived: the delay experienced by primary jobs; the degree of protection given to background work; and the overhead incurred. The results are then used to determine the scheduling parameters that satisfy several performance objectives.

1. INTRODUCTION

One of the options for a switching system is to use real-time processors with a general-purpose operating system. Such an operating system would typically be priority driven, where priorities are assigned to processes based on their required response time. Being highly time critical, call processing has to be implemented on a relatively high priority level, thus having preemptive priority over processes in lower levels, some of which (e.g., administrative and maintenance tasks) can be essential. Under heavy load, the strict priority structure can result in call processing work dominating the processor and thus in excessive delays for lower priority tasks. Therefore, the generic priority discipline has to be complemented by a scheduling mechanism that can override the strict priority rules when necessary. The objective of this scheduling function is to maintain acceptable call-processing delays up to, and above capacity load level, and at the same time provide some protection to lower priority work.

In this paper, the operating system with a dominant process is modeled as a single server with a primary queue and background work. We analyze a class oı scheduling policies that allot an uninterrupted segment of real time to background processes whenever there are no primary jobs to serve and also whenever a limit on continuous primary job execution is exceeded. In addition to inhibiting primary service during those segments, the generic operating system determines which background process is executing. This type of scheduling, besides being easy to implement, enables performance analysis and tuning of the primary jobs without knowing the workload characteristics of the background work. The analysis results in the following performance measures as functions of the scheduling parameters: (i) the delays experienced by primary jobs; (ii) the degree of protection given to background work; and (iii) the real-time overhead. The results are then applied to the scheduling problem by defining three performance objectives and determining the scheduling parameters that meet these objectives.

The remainder of the paper is organized as follows. In Section 2, the queueing model and the scheduling policies are introduced, followed by the analysis that yields the performance measures. Section 3 shows how to set the scheduling parameters by simultaneous evaluation of all the objectives. Finally, in Section 4 we briefly discuss the interaction of the scheduling with overload control.

2. SCHEDULING POLICIES and PERFORMANCE MEASURES

Consider a queueing system where a single server has to process a queue of primary jobs and also perform background (BG) tasks (Figure 1). The primary jobs arrive according to a Poisson process at a rate of λ and require service time of X with mean $\bar{x}$, and let $\rho = \lambda \bar{x}$. Background tasks may arrive from an external source or are generated in the system, and there is always some background work to be done. Primary jobs are served FIFO and ,in general, have higher priority than background work. However, over the long run, the server has to devote at least ρ_l (fraction) of its time to background work. Since background tasks require much longer service times, background work is done in segments, and there is an internal scheduler as part of the background that decides which task to execute.

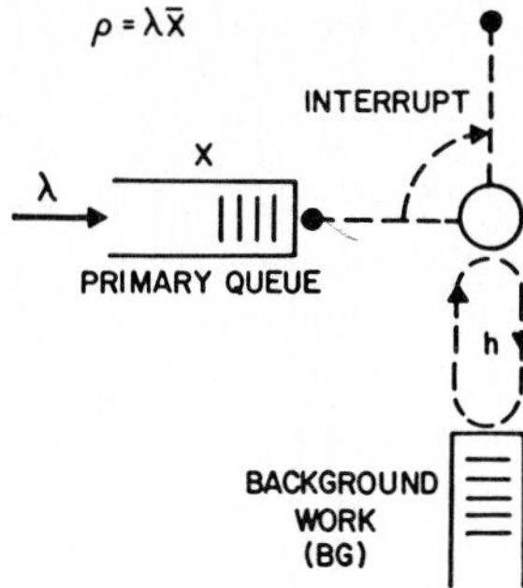

FIGURE 1 THE MODEL

For any random variable X, we denote its distribution function by F_X, $F_X(t) = P\{X \leq t\}$, its Laplace-Stieltjes Transform (LST) by $\tilde{X}(s) = \int_0^\infty e^{-st}\, dF_X(t)$, and its mean by $\bar{x} = E[X]$.

The application of this queueing system to the operating system environment is straightforward. The primary queue consists of call processing jobs, and the service time X is the processing time of the jobs adjusted to account for interrupts by processes with higher, preemptive priority. That is, if the processing time of a job is T, and ρ_h is the total occupancy of processes with higher priority than call processing, then the mean service time of a primary job is

$$\bar{x} = \frac{\bar{t}}{1-\rho_h} . \tag{1}$$

Assuming that interrupts are of length B, and that they are separated by exponential intervals with mean $1/\nu$, then the LST of the adjusted service time X is [1]

$$\tilde{X}(s) = \int_0^\infty e^{-[s+\nu-\nu\tilde{B}(s)]t}\, dF_T(t) .$$

The background work, on the other hand, represents all the tasks with priority lower than call processing. Since these tasks can also be preempted by the high priority processes, all the background requirements and measures have to be scaled by $1-\rho_h$.

A scheduling policy for this queueing system is one that specifies

(i) when the server switches to background work; and

(ii) how long to serve background tasks.

In this paper we focus on a class of policies that allot an uninterrupted real-time segment to background tasks. Specifically, we study the performance of three policies:

1. *U policy* : Switch to background work only when there are no primary jobs in the system and stay for a segment of length U. This policy is also referred to in the queueing literature as server vacations [2,4].

2. *(K,U) policy:* Switch to background work when there are no primary jobs or after continuously serving K primary jobs; segment length is U.

3. *(K,U,V) policy:* Switch to background work for a segment of length V when there are no primary jobs; switch for a segment of length U after continuously serving K primary jobs. If both conditions are met, either U or V can be used; we assume it is V.

The additional control variables increase the complexity of a policy but provide a better control on the performance at various load levels. In fact, the U policy is a limiting case of the (K,U) policy as $K\to\infty$, and the (K,U) policy obviously is a special case of the (K,U,V) policy with $V=U$.

The allotted BG segment (U or V) can be either fixed or a random variable. In most operating systems, even if the required segment length is fixed, due to discrete timing the actual length will be uniformly distributed over an interval equal to the clock period. There is also a fixed real-time overhead, h, associated with each switch to background work. In what follows, h is included in the segment length, but it is a trivial change to include only a portion of h in the BG segment.

Although the server can not serve primary jobs during the BG segment, the U policy is a *strict priority* discipline in the sense that the primary occupancy ρ can approach 1 as the load increases. The other two policies, which may be referred to as *limited priority*, use a limit K on the number of primary jobs served consecutively. They represent a tradeoff between the delay of primary jobs and the degree of protection given to background work. Under both policies, the system is stable (i.e., delay of the primary jobs is finite) as long as

$$\rho < \rho_{\max} = \frac{K\bar{x}}{K\bar{x}+\bar{u}} . \tag{2}$$

In the remainder of this section, the following performance measures are derived for all three policies.

1. The delay of primary jobs. In switching applications this delay is strongly related to important service criteria.

2. The degree of protection for background work. This means the fraction of real time that is reserved for background tasks during each short interval of time. This is different from the fraction

of real time ρ_l required for background work over a long interval, which is usually used to determine capacity. Although we do not discuss response time performance for background jobs, by conservation principles, longer primary job delays for a fixed load imply better performance for background work.

3. Real-time overhead spent on the scheduling function. This accounts for the overhead in switching between primary jobs and background work.

2.1 Primary Job Delays

The delay for all three policies is stochastically longer than the delay for primary jobs with preemptive priority over the background work (standard M/G/1). The U policy is analyzed in Levy and Yechiali [4], and the delay, W_1, for this policy is composed of two independent components: the delay under preemptive priority, W_p, and the residual life, U_r, of the BG segment. A nice intuitive proof of this decomposition can be found in Fuhrmann [2]. Thus the mean delay for the U policy is

$$\overline{W}_1(U) = \overline{W}_p + \overline{u}_r = \frac{\lambda E[X^2]}{2(1-\lambda\overline{x})} + \frac{E[U^2]}{2\overline{u}} . \quad (3)$$

The additional delay due to the BG segments depends only on their length U, and not on the load of primary jobs, which is a nice performance characteristic of this policy.

The (K,U) and (K,U,V) policies are analyzed by studying the process $[i(t),j(t)]$, where $i(t)$ is the number of primary jobs at time t, and $j(t)$ is the number of service completions since the last background segment. Embedding this process at service and BG segment completions results in a Markov chain with a transition matrix of the 'generalized M/G/1' type (see Neuts [5] and references there). This means that algorithmic procedures can be used to compute the queue length distribution and the mean delay, and in the appendix we show that the special structure of this model makes the computations more tractable. Unfortunately, there is no explicit formula for the mean delay. For the (K,U) policy, we have

$$\overline{W}_2(K,U) = \overline{W}_1(U) + f(K,U)$$
$$= \overline{W}_p + \overline{u}_r + f(K,U) .$$

Hence, the delay under the (K,U) policy is composed of three factors: $\overline{W}_p$ depends only on the primary jobs characteristics; $\overline{u}_r$ depends solely on U; and $f(K,U)$ increases with ρ and grows infinitely as $\rho \uparrow \rho_{\max}$. Recently, Fuhrmann and Cooper [3] have shown that the queue length for this model is actually composed of three independent random variables. As illustrated in Figure 2, at low loads the dominant factor is $\overline{u}_r$, while at high loads $f(K,U)$ is the largest component.

The additional parameter V in the (K,U,V) policy provides better control on performance by changing the segment length based on whether the system is empty or not. On first thought, once U is determined, one might choose V longer than U, since in the absence of primary jobs one might as well spend more time doing background work. However, the delays of primary jobs would then become longer. It thus turns out that V smaller than U is more effective. The effect of the parameter V is illustrated in Figure 3, and it is evident that while the delays increase with V for any value of ρ, the effect of V is significant only for ρ values well below $\rho_{\max}$.

The algorithmic procedure described above requires extensive computations. An explicit approximation for the limited priority policies can be obtained by considering a system where the segment U is added to each service time with probability p. A segment V is still taken when the system is empty. The waiting time for this model is identical to that of an M/G/1 queue with service time $Y=X+ZU$ and vacation V (Z is 1 with probability p and is 0 otherwise). So the mean delay is obtained from (3) as

$$\overline{W}(K,U,V) \approx \frac{\lambda E[Y^2]}{2(1-\lambda\overline{y})} + \frac{E[V^2]}{2\overline{v}} , \quad (4)$$

where

$$\overline{y} = \overline{x} + p\overline{u}$$
$$E[Y^2] = E[X^2] + 2p\overline{x}\overline{u} + pE[U^2] .$$

A good upper bound is now obtained by approximating p as a linear interpolation between $p=0$ at $\rho=0$ and $p=K^{-1}$ at $\rho=\rho_{\max}$, which yields

$$p \leqslant \frac{1}{K}\frac{\rho}{\rho_{\max}} = \frac{\lambda(\overline{u}+K\overline{x})}{K^2} .$$

2.2 Protection for Background Work

The U policy does not guarantee any real-time reserve to background work. The real-time fraction, ρ_s, that is reserved for background work by the limited priority policies is given by

$$\rho_s(K,\overline{u}) = \frac{\overline{u}-h}{K\overline{x}+\overline{u}} . \quad (5)$$

Thus, if ρ_s is a requirement, then setting one of the control parameters K or $\overline{u}$ determines the other.

2.3 Overhead

If the overhead per background segment h is 0, then $\rho_s = 1-\rho_{\max}$. For $h>0$, however, the real-time cost $C(\rho)$ is in general a decreasing function of ρ and at $\rho=\rho_{\max}$ it is

$$C(\rho_{\max}) = 1-\rho_{\max}-\rho_s = \frac{h}{\overline{u}}(1-\rho_{\max}) . \quad (6)$$

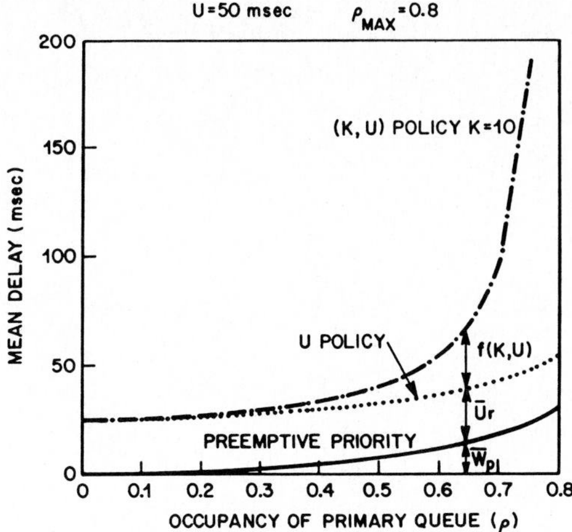

FIGURE 2 DELAY OF PRIMARY JOBS FOR VARIOUS POLICIES

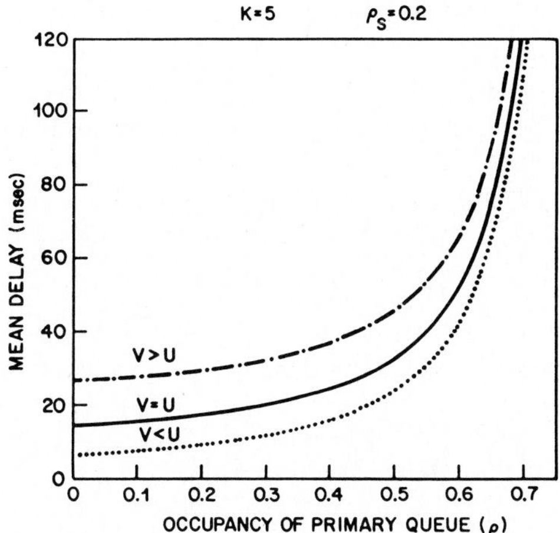

FIGURE 3 (K,U,V) POLICY- EFFECT OF V

For the U and (K,U) policies, where all the BG segments are of length U, the real-time cost for any $\rho \leqslant \rho_{\max}$ is

$$C(\rho) = \frac{h}{\overline{u}}(1-\rho) . \tag{7}$$

Both policies have the same real-time overhead since, over the long run, they have the same expected number of BG segments (as long as $\rho < \rho_{\max}$). Note, however, that the variance of the number of segments per unit time is lower for the (K,U) policy, and hence the response time for background work is better.

For the (K,U,V) policy, the real-time cost is a function of the probability that a BG segment is taken when the primary queue is not empty, and if $\overline{v} < \overline{u}$, then for $\rho \leqslant \rho_{\max}$

$$\frac{h}{\overline{u}}(1-\rho) \leqslant C(\rho) \leqslant \frac{h}{\overline{v}}(1-\rho) .$$

The relationship between $\rho_{\max}$, ρ_s, and $C(\rho)$ for all three policies is illustrated in Figure 4.

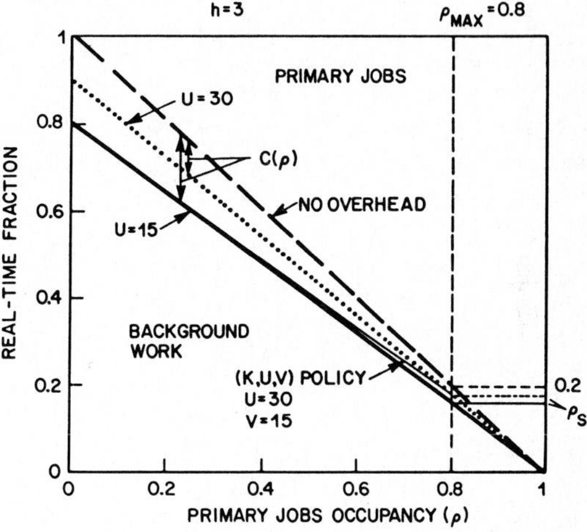

FIGURE 4 REAL-TIME ALLOCATION AND OVERHEAD

3. PERFORMANCE OBJECTIVES and SCHEDULING PARAMETERS

To apply the results of the previous section to the scheduling of real-time processors, we define the following objectives:

(i) Satisfy primary job delay requirements. Here, we use a mean delay objective of the form $\overline{W} < W_o$ at capacity.

(ii) Background work should be protected with minimal capacity loss. We recall that a long-term real-time requirement, ρ_l, is specified*. Hence, if capacity is defined as the occupancy of primary jobs, it can not exceed $1-\rho_l$. The protection level ρ_s, where $0 \leqslant \rho_s \leqslant \rho_l$, may be specified or not.

(iii) Real-time cost of the scheduling should be within specified bounds.

These type of objectives are usually specified for switching systems, with objective (i) related to service criteria and the other objectives to real-time allocation requirements.

* In fact $\rho_l = \rho_o/(1-\rho_h)$, where ρ_o is the actual real-time requirement and ρ_h is the interrupts occupancy (see Equation (1)).

If ρ_s is not specified, objective (iii) and Eq. (7) are first used to determine what should be the magnitude of U. Typically, U will be such that $\overline{W}_1(U) < W_o$ at $\rho = 1-\rho_l$, meaning that if the U policy is used, objective (i) is met at capacity. If ρ_s is not given, one can then determine the maximum ρ_s that meets objective (i) by using the (K,U) policy. As illustrated in Figure 5, the delay at $\rho = 1-\rho_l$ increases with ρ_s, growing infinitely as $\rho_s \rightarrow \rho_l$. Since at $\rho_s = 0, \overline{W} = \overline{W}_1(U) < W_o$, there is some point $\rho_s^* > 0$ where the mean delay equals W_o. Figure 5 also exhibits the tradeoff between the capacity and the protection level, and ρ_s^* is clearly the optimal choice in the sense of providing maximum protection without reducing capacity.

On the other hand, if ρ_s is given, then K is determined from (5) by

$$\hat{K} = \frac{\bar{u}(1-\rho_s)-h}{\rho_s \bar{x}} . \tag{8}$$

Since K has to be an integer, to achieve exactly ρ_s one can vary K randomly between $[\hat{K}]$ and $[\hat{K}]+1$. If the delay requirement is not satisfied, then either the capacity will be lower, or objective (iii) can be relaxed and a smaller U be used. If objective (iii) is not present, then (8) is a relation between $\bar{u}$ and K, and Figure 6 exhibits that for each value of h there is an optimal K^*. However, K^* increases with ρ, and it should be set for high ρ.

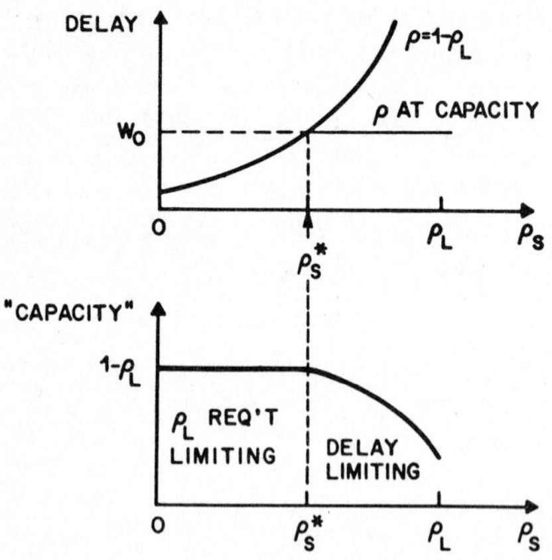

FIGURE 5 DELAY AND CAPACITY AS A FUNCTION OF LEVEL OF PROTECTION

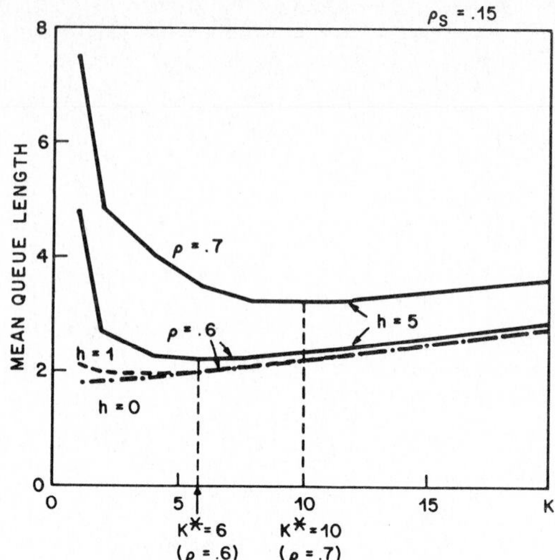

FIGURE 6 EFFECT OF OVERHEAD

The (K,U,V) policy, with $\bar{v} < \bar{u}$, can be used to give a better performance for primary jobs at low loads, and also to yield a little higher ρ_s. However, the real-time cost will be higher, and the lower delays of primary jobs will result in higher response times for lower priority processes. Once V is used, the delay for primary jobs, given ρ_s, will decrease as K (*and* U) increase. In this case, one should be careful not to set U too long, since many background processes are I/O intensive and several shorter segments may be more effective. Hence, the use of the parameter V depends on the particular system and on performance objectives at low loads.

4. INTERACTION WITH OVERLOAD CONTROL

This scheduling approach interacts well with overload controls. It provides short-term protection under overload until the controls affect the call processing load, and also enables the processor to sustain short peaks without activating overload controls. Tasks with priority lower than call processing include some jobs with response time requirements (usually on the order of seconds) and other jobs that can be executed over much longer intervals (order of minutes). The protection provided by the scheduling is meant to guarantee some real time to the first type of jobs. Many of these jobs can be scheduled to minimize contention so that satisfactory response time can be achieved. Overload detectors which are monitored periodically will eventually reduce the call processing load so that all the background tasks can be processed.

APPENDIX

This is a brief description of the computational algorithms for the M/G/1 queue with limited priority. Recall that the model consists of a single server attending a queue of primary jobs and also performing background tasks. Background work is done in uninterrupted segments when:

(i) There are no primary jobs, in which case the segment length is V; or

(ii) K jobs have been served, and then the segment length is U.

The arrival of primary jobs is Poisson with rate λ, and the service time is X. Also, let $A(X)$ be the number of arrivals during X, then

$$a(X)_j = P\{A(X)=j\} = \int_0^\infty e^{-\lambda t}\frac{(\lambda t)^j}{j!}\,dF_X(t) .$$

At any time t, the state of the system is described by $S(t)=[i(t),j(t)]\in\{0,1,2,...\}\times\{0,1,...,K\}$

$i(t)$ = number of (primary) jobs present at time t;
$j(t)$ = number of service completions since the last BG segment.

If the process is observed at epochs of service or BG segment completion, the resulting embedded process is a Markov chain. Given the current state (i,j), the next state is

$$(i',j') = \begin{cases} (A(V),0) & i=0 \\ (i+A(U),0) & i>0,\ j=K \\ (i+A(X)-1,j+1) & i>0,\ j<K \end{cases}$$

Let $\boldsymbol{i} = [(i,0),(i,1),...,(i,K)]$ be the vector of states whose first component is i, then the transition matrix P is

$$P = \begin{bmatrix} B_0 & B_1 & B_2 & B_3 & \cdot & \cdot \\ A_0 & A_1 & A_2 & A_3 & \cdot & \cdot \\ 0 & A_0 & A_1 & A_2 & \cdot & \cdot \\ 0 & 0 & A_0 & A_1 & \cdot & \cdot \\ \cdot & \cdot & \cdot & \cdot & \cdot & \cdot \\ \cdot & \cdot & \cdot & \cdot & \cdot & \cdot \end{bmatrix},$$

where the $(\boldsymbol{i},\boldsymbol{i}')$ block is a $(K+1)\times(K+1)$ matrix, and

$$B_i = \begin{bmatrix} b_i & 0 & \cdot\cdot & 0 \\ \cdot & & & \cdot \\ \cdot & & & \cdot \\ b_i & 0 & \cdot\cdot & 0 \end{bmatrix}$$

$$A_i = \begin{bmatrix} 0 & a_i & 0 & \cdot \\ \cdot & & & \cdot \\ \cdot & & & \cdot \\ 0 & 0 & \cdot\cdot & a_i \\ c_i & 0 & \cdot\cdot & 0 \end{bmatrix} = \begin{bmatrix} \mathbf{0} & a_i I \\ c_i & \mathbf{0} \end{bmatrix}$$

$$a_i = a(X)_i,\quad b_i = a(V)_i,\quad c_i = a(U)_{i-1} .$$

The transition matrix P shows that this model is of the "M/G/1 type" [5] for which there are algorithmic procedures to find the stationary distribution of the queue length. This distribution can be expressed in terms of the first passage time matrices G and L. G is a matrix whose (j,j') entry is the probability that, starting at state $(i+1,j)$, the process will reach stage $\boldsymbol{i}$, and that the first passage to $\boldsymbol{i}$ is into state (i,j').

In general, G satisfies:

$$G = \sum_{n=0}^{\infty} A_n G^n = A_0 + \sum_{n=1}^{\infty} A_n G^n ,$$

and it is computed by successive substitutions, starting with $G_0=(I-A_1)^{-1}A_0$. However, due to the structure of $\{A_i\}$ and $\{B_i\}$ in this problem, the first column of G is zero, so

$$G = \begin{bmatrix} \mathbf{0} & g_0 \\ 0 & G_- \end{bmatrix},$$

and the above equation can be expressed as

$$\begin{bmatrix} g_0 \\ G_- \end{bmatrix} = \begin{bmatrix} \sum_{n=0}^{\infty} a_n G_-^n \\ g_0 \sum_{n=0}^{\infty} c_n G_-^n \end{bmatrix}$$

Hence, instead of matrix multiplications, scalar operations are performed.

The matrix L corresponds to the first passage into stage $\mathbf{0}$; that is, its (j,j') entry is the probability that, starting at state $(0,j)$, the process will return to $\mathbf{0}$, and the first passage back to $\mathbf{0}$ is into $(0,j')$. Since the transition probabilities from $(0,j)$ are independent of j (B_i has equal rows), L in our model has equal rows, so

$$L = \mathbf{1}\boldsymbol{l} ,$$

$$\boldsymbol{l} = \left[b_0,\ g_0 \sum_{n=1}^{\infty} b_n G_-^{n-1} \right]$$

Once L and G are computed, the stationary distributions are readily calculated.

REFERENCES

[1] R. W. Conway, W. L. Maxwell, and L. W. Miller, "Theory of Scheduling," Addison-Wesley, 1967.

[2] S. W. Fuhrmann, "A Note on the M/G/1 Queue with Server Vacations," Operations Research, Vol. 32, pp. 1368-1373, 1984.

[3] S. W. Fuhrmann and R. B. Cooper, "Stochastic Decompositions in the M/G/1 Queue with Generalized Vacations," Operations Research (to appear).

[4] Y. Levy and U. Yechiali, "Utilization of Idle Time in an M/G/1 Queueing System," Management Science, Vol. 22, pp. 202-211, 1975.

[5] M. F. Neuts, "Matrix-Analytic Methods in Queueing Theory," Euro. J. of Op. Res., Vol. 15, pp. 2-12, 1984.

TELETRAFFIC ISSUES in an Advanced Information Society
ITC-11
Minoru Akiyama (Editor)
Elsevier Science Publishers B.V. (North-Holland)

An analysis of processor load control in SPC systems

Jila SERAJ

Aktiebolaget LM Ericsson
126 25 Stockholm, Sweden

ABSTRACT

This paper treats a processor load control method and its properties in detail. Besides it discusses which factors affect the behaviour of this method, and by computer simulations shows that the load profile of calls arriving to the controlled system has a significant influence on the behaviour of this and other similar load control methods.

1. INTRODUCTION

In the evolving ISDN different traffic types are going to be carried by the same network and switching equipment.

Data communication utilizes a wide range of data speeds and, according to previous studies of data traffic characteristics, has different traffic profiles. At the same time users have very different demands on service quality for their applications. The ISDN switching equipment should thus handle a wide range of telematic services, giving satisfactory service to them all. This implies that SPC systems used for ISDN applications will have to meet harder requirements on system characteristics, such as overload protection.

An ideal load control method should have the following properties

a) It should not disturb the normal function of the SPC system as long as the offered load is below the maximum SPC capacity

b) It should protect the system against overload, filled buffers and long delays

c) If a call is accepted it should receive a satisfactory service

d) It must be easy to implement

e) It must be able to handle the sudden jumps in the offered traffic, i.e. must not reject the calls if the offered traffic suddenly decreases and must not accept too many calls if the offered traffic suddenly increases

Two previous studies; ref. 1 and a not yet published study done by I. Anderssen and B. Nilsson at Ericsson; of two different processor load control methods; namely the LOAS method used in AXE telephone exchanges and the "ticket" method used in AXB data switches gave the following result. The ticket method satisfies all the properties listed above when it is applied on the AXB systems while the LOAS method is more suitable for AXE systems. These two studies have been the background of this work. With experiences from these two studies the author proposed that the goodness of a processor load control method is dependent on some factors among them the load profile of the calls handled by the system.

In the following we give a short description of the ticket method, describe a telephone call's and a circuit switched data call's typical load profile, describe a simplified model of the ticket method and present the results of a computer simulation of this model. In the last chapter conclusions from the simulation results are drawn.

2. THE TICKET METHOD

The princple of the ticket method is that every call arriving to the system must buy an entrance ticket. The idea is that the SPC system shall sell as many tickets as it can handle.

The central processor unit of a SPC system usually have several priority levels, aimed for processing jobs with different priorities, like traffic handling, operator commands, I/O, operation and maintenance routines and so on.

Speaking in AX systems terminology, traffic handling jobs are executed in level B, I/O and operator commands in level C, operation and maintenance in level D. Levels above level B are reserved for the operative systems. Figure 1 gives a functional picture of AX systems which interests us in this paper (for more detailed information about AX systems see ref. 3).

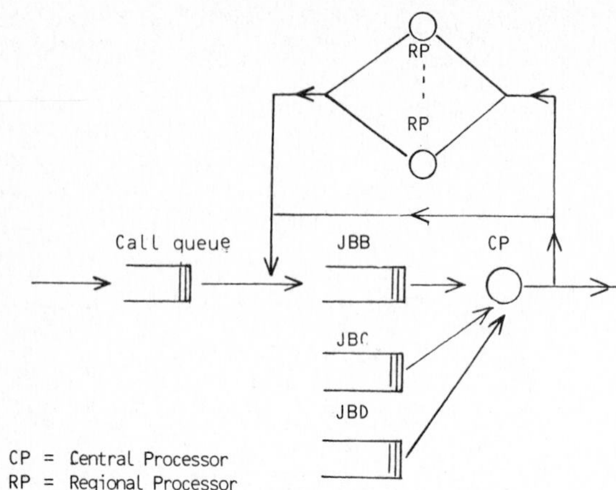

Figure 1.

Arrival of a job to JBB interrupts execution of jobs in level C and D.

To understand the ticket method we have to make some definitions.

- Call queue, a queue containing incoming calls waiting for a ticket.
- Ticket queue, a queue containing available tickets for arriving calls.
- Ticket-coupon, CPU marks used tickets by putting a coupon on them.
- Ticket-coupon queue, a queue for used tickets.
- JBC-basket, a waste basket for coupons torn from ticket-coupons.
- JBC-basket queue, a queue containing empty JBC-baskets.
- JBC-basket-coupon, CPU marks full JBC-baskets by putting a coupon on them.
- JBC-basket-coupon queue, a queue containing full JBC-basket.
- JBD-basket, a waste basket for coupons torn from JBC-basket-coupons.
- JBD-basket queue, a queue containing empty JBD-baskets.

An arriving call queues in the call queue if there are no tickets available in the ticket queue, otherwise it takes a ticket and continues to JBB. The call waits in the JBB for being served by the CPU. After the first visit in the CPU it leaves its ticket; this call is now free to continue through the system. The CPU puts a coupon on this used ticket and put it in the ticket-coupon queue.

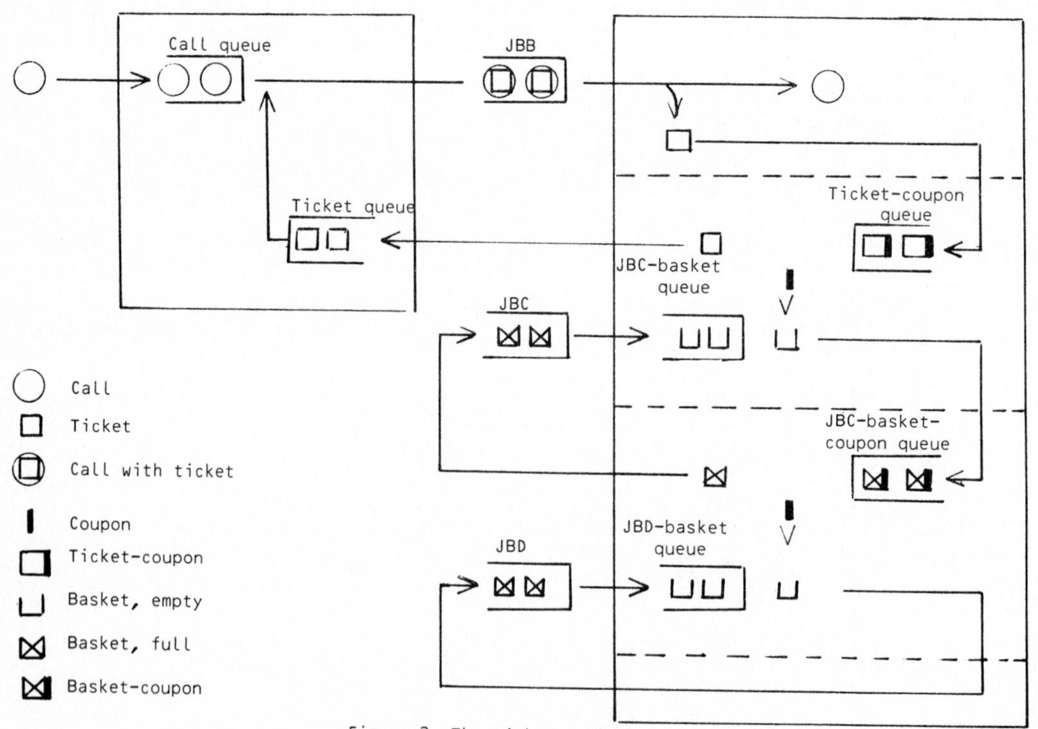

Figure 2. The ticket method

If there is a JBC-basket available, the CPU tears off the coupon from the ticket-coupon, puts the coupon in the JBC-basket and sends back the ticket to the ticket queue.

When a JBC-basket becomes full, the CPU puts a coupon on it and puts it in the JBC-basket-coupon queue.

If there is any JBD-basket available, the CPU tears off the coupon from the JBC-basket-coupon, sends the basket to JBC and puts the coupon in the JBD-basket. When a JBD-basket becomes full, the CPU sends it to JBD.

A full JBC-basket in JBC becomes available when the CPU has executed all the jobs in front of that basket in the JBC. In that case CPU empties the JBC-basket and puts it back in the JBC-basket queue. It works exactly the same way for a full JBD-basket in JBD.

3. A TELEPHONE CALL'S LOAD PROFILE

A telephone call initiates when a subscriber lifts his handset. The SPC recognizes the call attempt and sends a dial tone to the subscriber, which causes a rather short job. After hearing the dial tone the subscriber dials the digits. Arrival of each digit to the SPC causes a short job to receive the digit and store it in some register. After receiving all (or may be 3 or 4 digits) the SPC analyses the received number to choose a path through the group switch and sends ring signal to the called subscriber, and ring tone to the calling subscriber.

Analysing the called subscriber's number, choosing a path and sending of tones and signals causes a rather long job. After sending the tones nothing happens until the called subscriber answers the phone. When the SPC system recognizes that the called subscriber has lifted his handset it connects the two subscribers and stops the sending of tones, which means a rather short job.

When one of the subscribers puts his handset back, the SPC system gets a lot to do, like releasing of the lines, registers, path and so on (ref. 4 and 6). Diagram 1 shows an example of the history of the load offered by a local call.

The times given on the time axis are mean values of time elapsed since the call attempt has been recognized in the SPC system.

4. A CIRCUIT SWITCHED DATA CALL'S LOAD PROFILE

A typical call usually begins with a "call request" signal sent by a data terminal. When the SPC recognizes "call request" it reserves registers for storing the incoming information and sends a "proceed to select" signal to the calling data terminal, which responds by sending the necessary information for establishing the call. Now the SPC system has to analyse all the information at once and to choose a path through the group switch and send a signal to the called data terminal. The called terminal responds the SPC by sending an acceptance signal to it. At this point the SPC connects the two subscribers and they begin to transfer data. During data transfer the SPC system only supervises the call to find out whether any of the subscribers sends a "clear down" request, which causes a negligible load for the SPC.

After receiving the clear down request, the SPC has to disconnect the subscribers, release the path, registers and so on. Diagram 2 shows an example of the history of the load offered by a call in 12 kb/s class (ref. 5).

The times given on the time axis are mean values of the time elapsed since the call request signal has been recognized in the SPC.

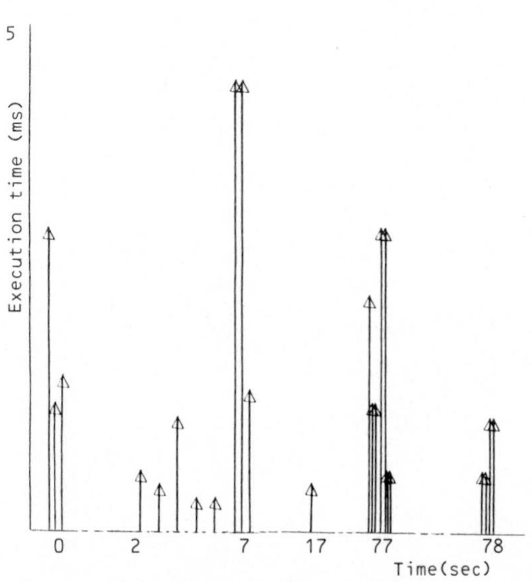

Diagram 1. A telephone call's load profile

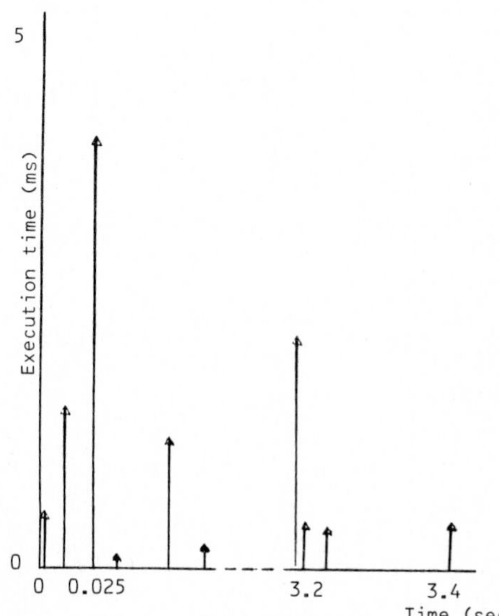

Diagram 2. A circuit switched data call's load profile

5. PROPERTIES OF THE TICKET METHOD

Studies of the ticket method (both very detailed simulations and measurements) have shown that the ticket method is satisfying all of the properties mentioned in chapter 1 when it is applied on the AXB systems (switching nodes in the circuit switched data networks). The traffic studied in chapter 4 is an example of the traffic types handled by AXB systems. Later studies have shown that this load control method does not satisfy all of the above mentioned properties when it is applied to the AXE systems (telephone exchanges). This amazing fact is worth a deaper study to find out which factor/factors affect the performance of the ticket method.

5.1 Proposal

Studying differences between an AXE and an AXB exchange one finds that the central processor structure of AXE and AXB exchanges are quite similar, so that the main factor can not be the exchanges themselves. In ref. 1 the author makes the proposal that the determining factor in performance of the ticket method is the load profile of the calls. To explain the idea more clearly let us look at diagram 1.

Assume a telephone call generates x time units of processor load. When a telephone call arrives to an exchange it generates a processor load less than 20% of x. This leads to dial tone sending to the subscriber. The subscriber reacts to the dial tone and dials the first digit (or pushes the first button), which can take about 2 seconds, a time which is rather long for a SPC exchange. Receiving of each digit causes a load less than 2% of x.

After receiving all the digits which can take about 6 or 7 seconds the exchange analyses the dialled number, chooses the path, sends ring tone and so on. This causes a load about 35% of x. When the called subscriber answers, the SPC has a rather short job for through-connection. While subscribers are connected SPC only supervises the connection and awaits a disconnection request. A disconnection request causes a great deal of job for clearing of the path, releasing the registers and so on. This last job is also about 35% of x. Thus the load of the CPU of the SPC system caused by a telephone call is spread over a very long time.

Now, if we look at the diagram 2 we will find that after less than 0.6 seconds passed from call request recognition, more than 70% of the total load caused by this call have already burdened the CPU of the SPC system. And, after less than 3.5 seconds this call has dissappeared from the whole system.

In other words if the load of calls are concentrated at the arrival time, the ticket method works satisfactorily. But, if the load is spread in time this load profile will affect the ticket method's performance in a negative sence.

To check whether this idea is correct the following simulation model has been considered.

5.2 Simulation model

A central processor with two priority levels B and C has been considered. Level B has strict priority (preemptive resume) over level C.

Calls arrive to this system according to a Poisson process. Service times at each visit to the CPU are independent and exponentially distributed. The ticket method with only two levels is implemented in the model (see figure 3). The capacity of the call queue is 60 calls.

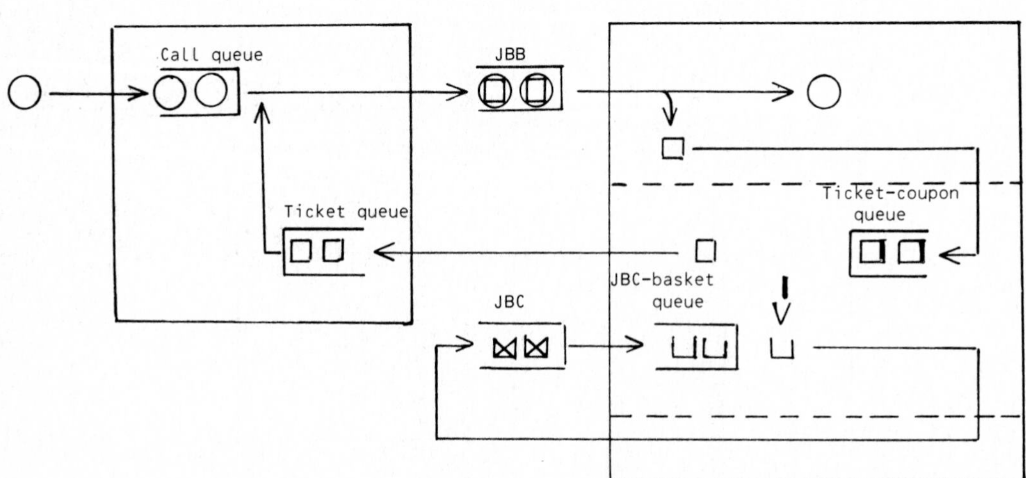

Figure 3. Simulation model

There are always jobs moving around on the C level and also other jobs arrive to the C level at a rather low rate.

To check the hypothesis that the load profile is the determining factor for the performance of the ticket method two options of this model are considered.

Option 1. After having received service in CPU calls will return to JBB with probability α and leave the system with probability $(1-\alpha)$.

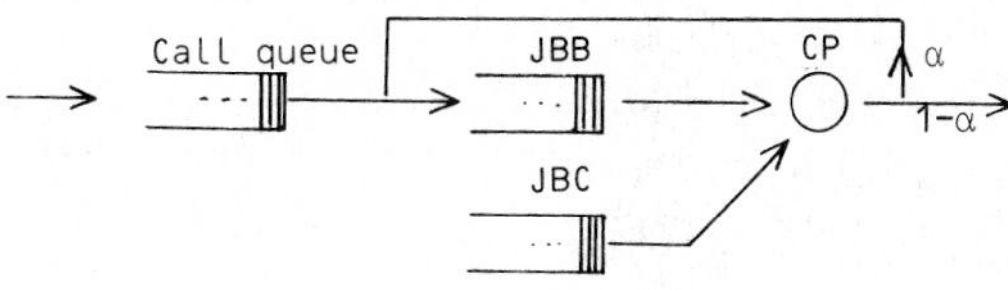

Figure 4. Option 1

Option 2. Calls return to JBB after having received service in CPU with probability β or join an outer delay process with probability γ, or leave the system with probability $1-\alpha$. $\alpha = \beta + \gamma$. The outer delays are independent and exponentially distributed. Calls return to JBB after having been served in the outer delay process.

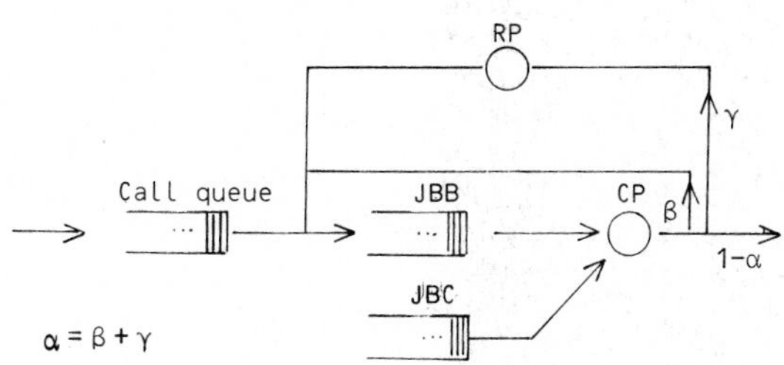

Figure 5. Option 2

The only difference between these two options is the load profile. In option 2 a call's load is spread over time (outer delays), while in option 1 a call's load arrives with the call itself.

5.3 Results

In the simulation the following values have been used. Mean number of feedbacks is fixed to 50 in option 1. In option 2 the mean number of feedbacks via outer delays is fixed to 10 and mean number of direct feedbacks to 40. Mean outer delay time is 2 seconds. Mean service time in the CPU is 0.4 ms.

3 jobs move around in JBC, service times being independent and exponentially distributed with mean 0.1 ms. Operator commands arrive according to a Poisson process with a rate of 10 commands/second. Each operator command job visits the CPU in the mean 3 times before it leaves the system. Service times at each visit are independent and exponentially distributed with mean 0.1 ms.

The offered calling rate increases from 10 calls/second by 10 calls/second every 10 seconds up to 100 calls/second, and then decreases by 10 calls/second every 10 second down to 10 calls/second.

The choice of this method is motivated by the properties a and e in section 1.

The goal of these simulations has been to understand the behaviour of the ticket method rather than to find mean values for performance parameters of the system with good confidence intervals (which is the case in many simulation studies).

Thus statistics has been taken at the end of every 10 seconds interval. In the following diagrams statistics are shown over these intervals to demonstrate effect of changes in call intensity.

Each option has been run several times. In each run of an option certain random sequences are generated. Identically the same sequences are then used for a run of the other option. Therefore the measured values are directly comparable.

The following diagrams are examples of two comparable runs. For the sake of space one can not present the resulting diagrams from all runs. But it is important to point out that the results from other runs are very similar to those presented here.

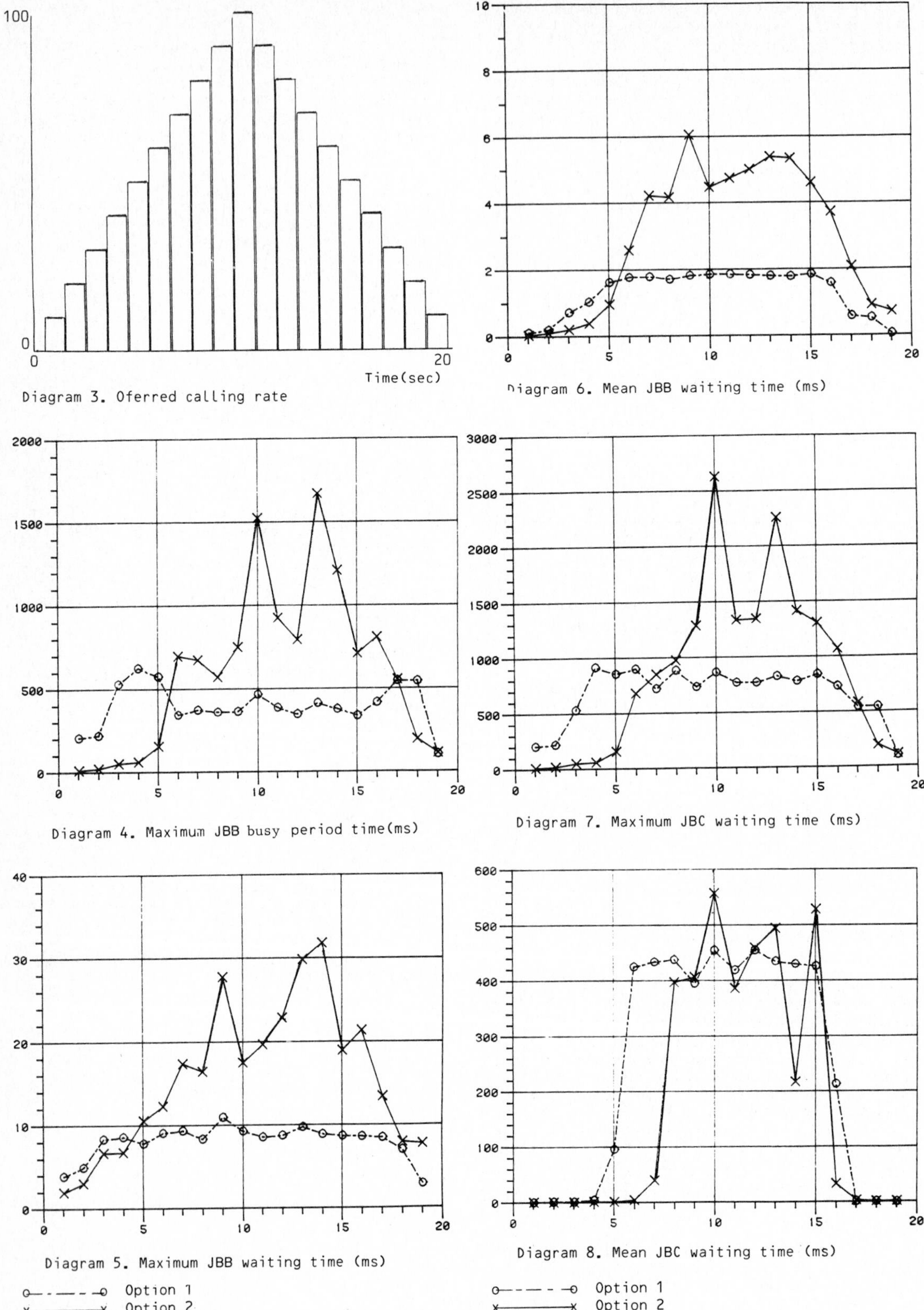

Diagram 3. Oferred calling rate

Diagram 6. Mean JBB waiting time (ms)

Diagram 4. Maximum JBB busy period time(ms)

Diagram 7. Maximum JBC waiting time (ms)

Diagram 5. Maximum JBB waiting time (ms)

Diagram 8. Mean JBC waiting time (ms)

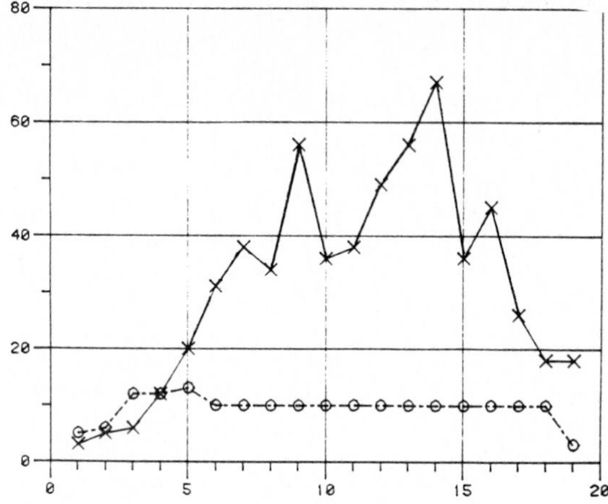

Diagram 9. Maximum JBB queue length

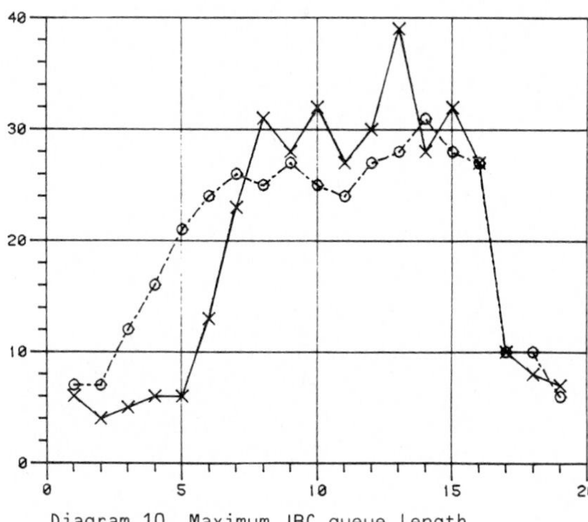

Diagram 10. Maximum JBC queue length

o- - - - -o Option 1
x———x Option2

The above diagrams strongly support the idea that the load profile of the calls has a very strong influence on the behaviour of the system. It should be mentioned here that simulations have been run for other options than those presented here. For example number of tickets, number of baskets, capacity of each basket, mean service time,etc., have been varied within reasonable limits to see their effect on the performance of the studied system. Results show that these variations do not affect the behaviour of the system dramatically.

6. CONCLUSION

We have shown that a load control method's goodness can be highly dependent on the load profile of the offered traffic. The particular method we have studied here sells tickets for the calls it think it can handle. But when the real load of the call does not arrive until late after the acceptance of the call, then this method is not able to adjust its call acceptance to the offered load. In other words the ticket method controls the call acceptance and therefore the load instantaneously. This means that this method is not appropriate for traffic types which do not generate load upon the arrival.

REFERENCES

1. Seraj J. Comparison of two processor load control methods in AXE, Ericsson report XF/SY 84 028, 1984.

2. Wildling K. and Karlstedt T. Call handling and control of processor load in a system. A simulation study. ITC 9, 1976.

3. AXE 10, System survey. Ericsson publication: X/YG 118 806 Ue.

4. Seraj J. Load profile for a local call in AXE. Ericsson report XF/SY 84 035, 1984.

5. Seraj J. Load profile for a data connection in 12 kb/s speed calls with use of concentrators in both directions. Ericsson report XF/SY 84 047, 1984.

6. Vrana C. Studies of traffic control in AXE (in Swedish) Lund Institute of Technology, 1977.

TELETRAFFIC ISSUES in an Advanced Information Society
ITC-11
Minoru Akiyama (Editor)
Elsevier Science Publishers B.V. (North-Holland)

A LEARNING CONTROL APPROACH TO DYNAMIC NETWORK RESOURCE ASSIGNMENTS

Hisao UOSE and Yoshihiro NIITSU

Musashino Electrical Communications Laboratory, NTT
Tokyo, Japan

ABSTRACT

This paper proposes a learning control approach to dynamic network resource assignment control. The proposed method uses a learning approach to approximate the characteristics of the system to be controlled. Once the function is represented, optimum control inputs can be determined by using the prespecified control criteria. The potential function method which is well-known in the area of pattern recognition is used to approximate the function under an actual operating conditions. This control method is applied to two communications system models, a multi-speed circuit switched model and a circuit/packet integrated switched model. Simulation studies are conducted in order to evaluate the method. As a result it is revealed that the method is useful for controlling network systems in which an analytical solution cannot be obtained because of the complexity of the system and the traffic characteristics.

1. INTRODUCTION

In the near future, many communications services are expected to be incorporated in one integrated communications system thanks to the digitization of communications networks. In order to utilize communications resources efficiently in light of this development, it is desirable to establish a dynamic resource assignment control strategy.

However, future communications services will probably have very different traffic characteristics. For example, the introduction of video communications will necessitate wideband communications channels, while teletex service requires a narrower bandwidth than existing telephone services. The holding time of calls will change. The average holding time of conference communications services will be rather long and that of message services short. In addition the distribution of the holding time will differ from that of the conventional telephone network.

Furthermore, service grade specifications for communications services are expected to diversify into many classes. For example, in the case of packet-switched services, services will be classified according to average delay time. For circuit-switched networks, different loss probability values will be the determining factor. All the foregoing factors must be considered if a practical control method is to be found.

Another problem may entail the selection of control criteria for operating the integrated network. These criteria will differ from those used for conventional networks in the complexity of loss probabilities or delays to be controlled, and will probably combine several different evaluation measures.

One effective way to approach this problem is to apply control theory to communications networks. Control theory analytically comes up with the optimum control inputs using the state equation which represent the dynamic characteristics of the system and an objective function which represent the control criteria. However, when communications networks are seen as control objects, many problems appear. These include the nonlinearity, the saturation characteristics and limitation of the control inputs of the system.

Segall, et.al. attempted to apply control theory to communications networks [1]. They approximated the amount of traffic which is retained in each node using a continuous state variable and taking the capacities of each link connected to the specified node as control variables. They showed optimum control inputs can be obtained analytically in a feedback form for the criterion of shortest possible delay time. This approach is desirable because the feedback control is robust to system variations. However, they had to make unrealistic assumptions concerning communications networks to solve the problem. These included assuming unlimited buffer capacities, no call origination during the control due to limitations on input traffic conditions, etc.. These assumptions made in the study must be rectified before actual application can be considered.

A less problematic approach is to apply learning control. In fact, this is the only approach which can be practically applied to communications network control at the moment. Narendra, et al. attempted learning control of telephone network routing [2]. They used learning automata to determine the order of routing link selection at the network nodes. A learning automaton is a kind of probabilistic automaton whose transition probability changes according to the penalty which it gets after each trial. They showed that lower loss probability could be obtained by using the learning control method than with the conventional routing scheme (far-to-near rotation routing) in unbalanced load conditions by simulation study.

In this paper, a learning control approach to communications system is studied from the different viewpoint of Narendra's approach. The problems considered here are concerning to dynamic network resource assignment expected in the future communications systems. Section 2 describes the expected problems in network control arises from the characteristics of the communications networks. Then a learning resource assignment control method is proposed to overcome these problems. Section 3 explains the proposed method using typical communications system models expected in the future, a multi-speed circuit switched model and a circuit/packet integrated switched model. Section 4 examines the performance of the method through computer simulation.

2. COMMUNICATIONS NETWORK CONTROL

This section examines the problems specific to communications network control and proposes a learning control approach to network control.

2.1 Communications network model

(1) Problems specific to communications networks

Communications networks present the following problems when they are seen as control objects.

a. Nonlinearity coming from the nature of the system itself, such as a limited channel number and buffer capacity.

b. Complexity resulting from the large number of states of the system.

c. Limited control inputs due to topological limitations.

These factors must therefore be integrated smoothly into the model.

(2) Network modeling

The following items must be decided upon when determining the modeling procedure.

a. Modeling technique

There are two modeling approaches to choose from, one in which the model is constructed in keeping with structural knowledge of the control object and the other in which it does not incorporate such knowledge but rather it identifies the system using its inputs and outputs.

b. Selection of state variables

The number of simultaneous calls, waiting calls, loss probabilities waiting time, etc., must be determined.

c. Selection of control inputs

The method of assigning resources restricting input calls, routing, etc., must be decided upon.

2.2 Learning approach to network control

Let us consider problem shared common resources in communications networks present to resource assignment control. The dynamic characteristics of the system are difficult to assess because of the complexity of the system. Static characteristics such as the relation between offered traffic and loss probabilities can be assessed for simple systems, for example, those for which Poissonian call origination and the negative exponential distribution of holding time can be assumed. However, in systems where these assumptions can not be made safely even static characteristics become difficult to assess easily.

In order to overcome these problems, a learning control approach is proposed. The method involves using the potential function method to approximate the functions in a learning procedure. This method consists of the two following algorithmic steps;

(i) Approximating the system function which is to represent the system characteristics using the potential function method with state variables, applied control and the observed system outputs by learning.

(ii) Determining the control input to be applied to the system using the approximated function and the objective function to represent the control criteria.

The learning control method does not require detailed knowledge of the system because it only uses input and output pairs of the system, regarding the system as a black box. Using the potential function method to approximate functions has the advantage of being effectively applicable to systems under normal operating conditions without the use of much data.

3. LEARNING RESOURCE ASSIGNMENT CONTROL

This section explains the proposed network control method using sample communications system models.

3.1 Network model

We will consider the resource assignment problem of a multi-speed circuit-switched traffic system with two different call speeds sharing one common resource (a group of channels). Low-speed calls use a single channel of the common channel group and high-speed calls use multiple channels to form a high-speed transmission channel. This kind of system is expected to be used in the multi-service environment of the future. For example, the system has the potential to handle a wide range from high-speed telephone service traffic to low-speed teletex service traffic.

Let us consider the problem of operating this system according to prespecified control criteria. The following criteria, for example, will probably be used in such systems.

<u>a.</u> Minimizing the loss probability of one class of services while keeping the loss probability of another service class to a certain level.

<u>b.</u> Maintaining two loss probabilities of services to a same value.

The former criterion, for example, can be used in the environment in which human-human and machine-machine communications services are mixed in a system. It is in general desirable to give a priority to the services in which human is involved. If the system can be operated according to the criterion, the quality of the human-human communications system is kept to a constant level by worsening the quality of machine-machine communications service.

In the case of multi-speed traffic systems, the loss probabilities of high-speed calls are larger than those of low-speed calls due to the odd number line effects. The trunk reservation method can be used to balance the loss probabilities. However this method can only be used in static, non-varying traffic conditions. Our objective here is to obtain an effective control law even in conditions where traffic varies dynamically.

The model examined here is shown in Fig. 1. In the figure, call 1 is a high-speed call and call 2 is a low-speed call. A learning control mechanism consists of a random switch which restricts the low-speed traffic inputs and a control system which calculates the suitable control inputs from the value of state variables. The measurement system measures state variables such as the amount of offered traffic and the loss probabilities of both calls.

The control procedure can be explained as follows;

(i) The learning part of the control system approximates the function $f(\mathbf{x})$ that represents the characteristics (the relation between the offered traffic and the loss probability) of the system using state variables $\mathbf{x}$ and $\mathbf{y}$, observed value of $f(\mathbf{x})$.

(ii) The control system determines the suitable control input β (the passing rate of the random switch) to satisfy the control criterion based on the approximated function and the object function.

Let us take a function $\mathbf{f}(\mathbf{x})$, which represents the relation among the loss probabilities, the offered traffic and the passing rate of the random switch, as an approximated function here.

$$f^j(\mathbf{x})=bj, \qquad j=1,2 \tag{1}$$

where $\mathbf{x}=\{x1,x2,x3\}$;

x1 : offered traffic 1 (high-speed call)

x2 : offered traffic 2 (low-speed call)

x3 : passing rate of the random switch (β)

bj : loss probability of call j.

The control input β can be determined using the function $\mathbf{f}(\mathbf{x})$, measured variables x1,x2 and the control criteria H, in the following equation;

$$\beta=g(\mathbf{f}(\mathbf{x}),x1,x2,H) \tag{2}$$

where $\mathbf{f}(\mathbf{x})=\{f^1(\mathbf{x}),\ f^2(\mathbf{x})\}$.

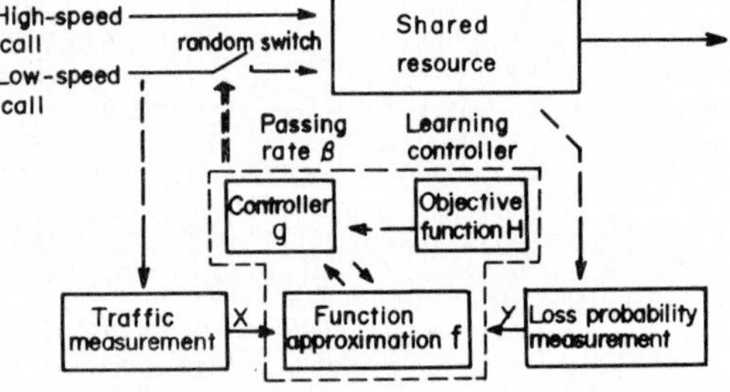

Fig. 1 Multi-speed circuit-switched traffic model

3.2 Usage of of the potential function method to approximate functions

The potential function method was first proposed to solve the learning problem of the pattern recognition machine. However, this method can also be used to approximate unknown multivariable functions using a sample set of the function's input variables and output variables. This method is briefly explained in this section [3].

Suppose $f^*(\mathbf{x})$ is the unknown function which is going to be approximated. Where $\mathbf{x}$ is a m-dimensional variable vector,

$$\mathbf{x}=\{x1,x2,x3,\ldots,xm\}. \tag{3}$$

Let us approximate $f^*(\mathbf{x})$ with the finite number of vector $\mathbf{x}$ and y_* the observed value of $f^*(\mathbf{x})$. If we assume that $f^*(\mathbf{x})$ can be expanded into a sequence of functions $\psi(\mathbf{x})$,

$$f^*(\mathbf{x})=\sum_i \psi_i(\mathbf{x}). \tag{4}$$

Let us then introduce the potential function which is expressed as follows,

$$K(\mathbf{u},\mathbf{v})=\sum_i \psi_i(\mathbf{u})\psi_i(\mathbf{v}). \tag{5}$$

With this function and the values of input and output variables the function $f(\mathbf{x})$ can be approximated using the following repetition algorithm.

$$f_{n+1}(\mathbf{x})=f_n(\mathbf{x}) + \gamma_n s(f_n(\mathbf{x}_{n+1}),f^*(\mathbf{x}_{n+1}))K(\mathbf{x}_{n+1},\mathbf{x}), \tag{6}$$

where $f_n(\mathbf{x})$ is the n-th approximation of the function, $\mathbf{x}_n$ is the realized value of n-th occurrence, γ_n is a convergence coefficient, $s(f^*,f)$ is a function evaluating the degree of approximation and $K(u,v)$ is a potential function. The approximation can be obtained with this algorithm using an initial approximation such as $f_0(\mathbf{x})=0$.

There are some conditions which assure the convergence of the algorithm (6). Especially, we must consider cases in which the function to be approximated is probabilistic as these will inevitably occur when this method is applied to communications networks. The observed value y_n is expressed by true value of the function $f^*(\mathbf{x}_n)$ and noise term ξ_n as follows;

$$y_n=f^*(\mathbf{x}_n)+\xi_n \tag{7}$$

Let us assume that ξ_n satisfies the condition listed below,

(1) random variable ξ_n is independent of the observation time n,

(2) conditional probability distribution of ξ_n under condition $\mathbf{x}_n$ is independent of the observation time n,

(3) the conditional expected value of the random variable is $M\{\xi|\mathbf{x}\}$ and variance $M\{\xi^2|\mathbf{x}\}$ is finite for all $\mathbf{x}$.

It can be readily shown that these conditions are satisfied for the function described in Section 3.1. The function $f_n(\mathbf{x})$ converges with $f^*(\mathbf{x})$ when these conditions are satisfied and function $s(f_n,f^*)$ is the form,

$$s(f_n,f^*)=f_n(\mathbf{x})-f^*(\mathbf{x}) \tag{8}$$

and γ_n satisfies the condition below.

$$\sum_{n=1}^{\infty}\gamma_n=\infty \qquad \sum_{n=1}^{\infty}\gamma_n^2<\infty \tag{9}$$

One of the advantages of this method is that it can accurately approximate most types of functions even when the function can not be expanded by the sequence of functions composing the potential function [3]. The other advantage of the method is that the data used for the algorithm need not be regularly organized but can be distributed at random. Therefore this method is suitable for approximation problems where systematic experiments to obtain the statistical characteristics are impossible. This method is ideal for communications network control because it can use data obtained during the normal operation process of the communications system.

4. SIMULATION STUDY

This section examines the method's convergence and the effects of parameters, traffic variation and control period on its accuracy and speed.

4.1 Simulation study conditions

(1) Communications network model

The validity of the method is verified using the multi-speed circuit-switched traffic model described in the preceding section.

The following assumptions are made.

(i) The distributions of call origination conform to the interrupted Poisson process (IPP) for both calls.
(ii) The offered traffic varies sinusoidary with the T1 and T2 periods and the ratio of v.
(iii) The holding times of the calls conform to the same negative exponential distribution with the mean value of 1/h.
(iv) The amount of traffic is measured accurately.
(v) The random switch passes the call with the probability equal to passing rate on a call-by-call basis.
(vi) Measurement and control is carried out in each control cycle t.
(vii) No trunk reservation is done.

The following numerical conditions are used.

<u>Numerical conditions</u>

The number of circuits	:	100 channels
Input traffic		
average traffic	a1,a2:	50erl
the variation ratio	v:	0~0.4
the variation period		
class 1 call	T1:	3.1 hours
class 2 call	T2:	2.2 hours
Holding time of the calls	h:	100 seconds
Period of the control cycle	t:	7.5 minutes
Speed ratio of calls	u1/u2:	4

(2) Control criteria

The following two control criteria are considered;

<u>criterion 1</u>

maintaining the loss probability of call 1 to 0.05,

<u>criterion 2</u>

maintaining the loss probabilities of call 1 and call 2 to the same level. (b1/b2=1.0)

(3) Selection of potential function

Potential function is expressed as a infinite sum of the sequence of functions, as described in the preceding section. However, a function should be used which can be written in a closed form for a practical usage. In addition, the function which can be expressed as a function of distance between the points **u** and **v** are suitable for implementing the algorithm from the viewpoint of the calculation needed [3]. The function below satisfies all these conditions.

$$K(\mathbf{u},\mathbf{v})=e^{-\alpha^2\rho^2(\mathbf{u},\mathbf{v})}$$

$$=e^{-\alpha^2\sum_{k=1}^{m}(u_k-v_k)} \tag{10}$$

where α is a constant coefficient.

(4) Selection of s and γ

Equation (5) is used as the error evaluation function $s(f^*,f)$. The following coefficient γ is used to satisfy the convergence condition (9).

$$\gamma_n=1/n^k, \qquad k=0.6 \tag{11}$$

4.2 Results

(1) Convergence evaluation of the method

The convergence is quantitatively analysed. As an evaluation value, the square norm of the approximated function and true function, J, is used.

$$J^2= \int_{c1}\int_{c2}(f^*(x_1,x_2)-f(x_1,x_2))^2dx_1dx_2 \tag{12}$$

where $f^*(x)$ is the true function, $f(x)$ is the approximated function and c_1 and c_2 are the function's domain. In order to obtain the true function, a Poisson origination call is applied to the model described in the preceding section in stead of the IPP call [4]. The values obtained for J after 1000 control cycles are shown in Table 1. The numerical conditions used for this evaluation are listed below.

<u>Numerical conditions</u>

The number of circuits		100 channels
Input traffic		
average		50 erlang
variation ratio	v:	0.8
Speed ratio of calls	u1/u2:	4
Range of integration	c1:	30~70 erlang
Range of integration	c2:	30~70 erlang

Table 1. Discrepancies with actual blocking rates using the approximation method

	High-speed calls	Low-speed calls
Actual figures	0.2070	0.0547
Average discrepancy	0.0094	0.0042

Fig. 2 shows the relation between residual error rate and control cycle. These results shows that the residual error rate is under 1 % as an

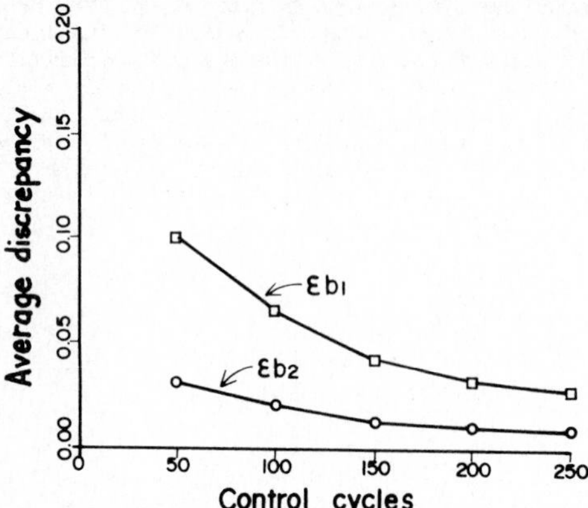

Fig. 2 Average discrepancy vs. number of control cycles

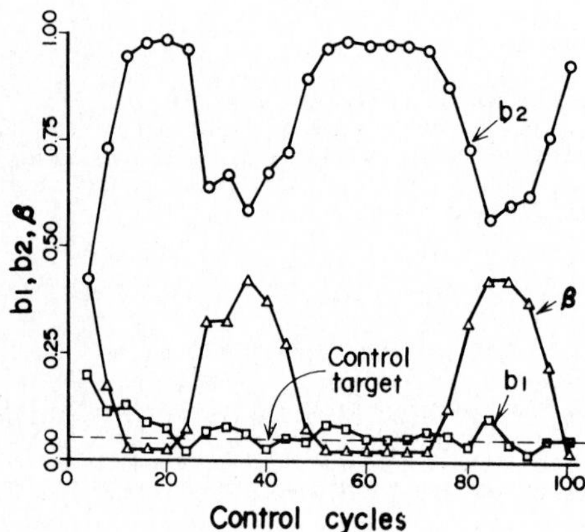

Fig. 3 Control results (Criterion 1)

absolute value. Relative errors in the low traffic region are large compared to the high traffic region. However this does not become a serious problem when the method is actually applied since it is the high traffic region which is the main control object.

(2) Control results

Fig. 3 and Fig. 4 show the results obtained from applying criterion 1 criterion 2. The horizontal axis shows the number of control cycles and the vertical axis shows the control results b1, b2 (loss probabilities of call 1 and call 2), b1/b2 (the ratio of loss probabilities) and β (passing rate of the random switch). The traffic variation ratio is 0.3. The broken lines represent the control target. In the case of criterion 1, the loss probability of call 1 (high-speed calls) is found to maintain a prespecified level (0.05) irrespective of the traffic variation. The loss probability of call 2 (low-speed calls) varies widely because of the restriction placed on the call in order to keep the loss probability of call 1 to a specified value in the high traffic region. In the case of criterion 2, the ratio of loss probabilities of call 1 and call 2 maintained a prespecified level (1.0). In both

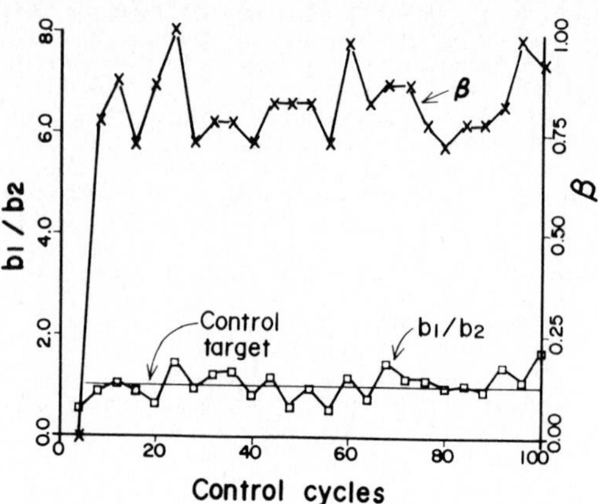

Fig. 4 Control results (Criterion 2)

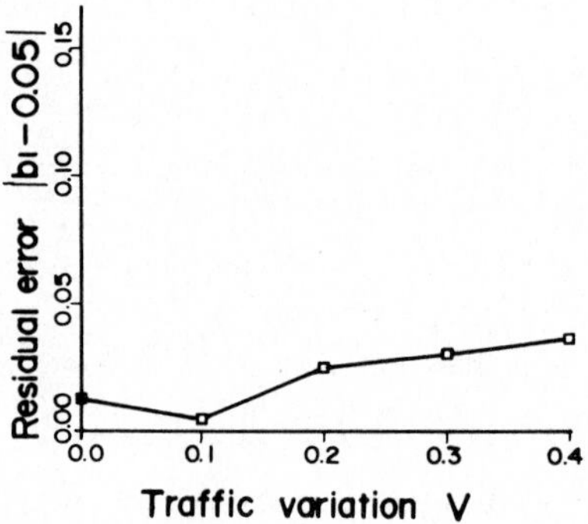

Fig. 5 Traffic variation vs. residual error rate (Criterion 1)

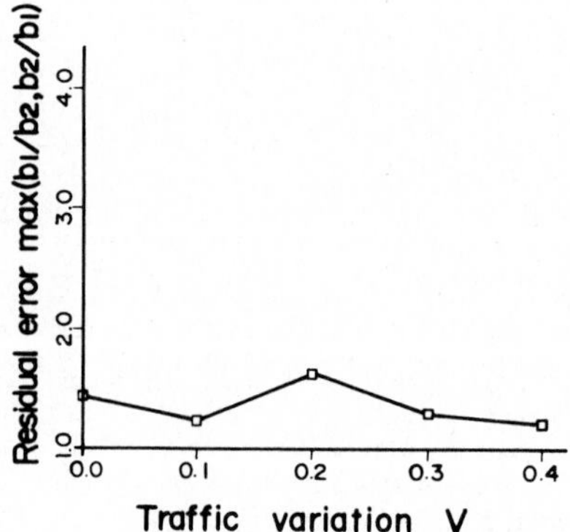

Fig. 6 Traffic variation vs. residual error rate (Criterion 2)

cases, it can be concluded that control was carried out quite effectively according to the control criteria.

(3) Traffic variation and residual error

Fig. 5 and Fig. 6 show the effect of the control cycle on convergence for criteria 1 and 2. The values given were obtained after 120 control cycles. It was expected that the convergence speed would drop when traffic variation is large since the probability density

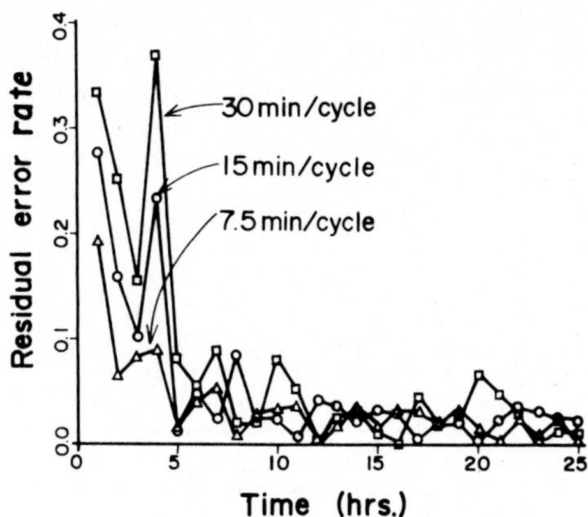

Fig. 7 Control cycle vs. residual error rate (Criterion 1)

of originating data decreases for each point in the function domain. However, the results show that residual errors do not depend on traffic variation. This can be explained by assuming that the number of control cycles applied was sufficient to come up with the function. The residual error rate for the criterion 1 is 3 % and that for criterion 2, 0.3. It can accordingly be concluded that traffic variation does not affect convergence nor the control performance.

(4) Control period and residual errors

The optimum control period depends on the variance of approximated function. Fig. 7 shows the relation between control periods and the residual error rate in the case of criterion 1. The horizontal axis gives time in hours. The plotted lines graph data for control periods of 7.5, 15 and 30 minutes. The figure shows that if the control period can be shortened to include a greater number of control cycles the convergence speed can be improved under the conditions assumed here. This can be interpreted to mean that the number of the control cycles has a greater impact than the loss probability variance. However when the control cycle is too short, it can be expected that convergence speed will be slow due to the variance in measured values.

The above results show that this control method takes a few hours to become effective after operation commences. The residual error rate is small enough to satisfy the control criteria irrespective of the traffic variation.

4.3 Circuit/Packet switched model

To investigate the general applicability of this learning control scheme, we applied it to another type of communications system. In this discussion, learning control method has been applied to the channel allocation control model of the integrated circuit/packet switched system. This model is shown in Fig.8. As a control criterion, "keep the loss probability of circuit-switched call to constant (0.1)" is used. The numerical conditions used for computer simulation are listed below.

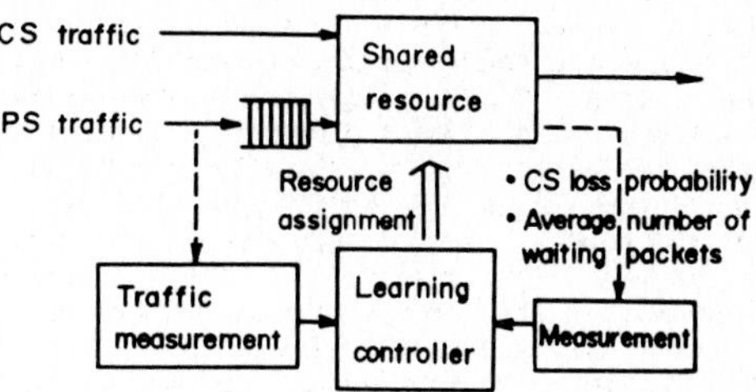

Fig. 8 Circuit/Packet integrated switched traffic model

Numerical conditions

The number of circuits N: 5 channels
Input traffic
CS (circuit-switched) a_{cs}:
traffic 0.4sin{(m-0.5) /5}+1.2 erlang
PS (packet-switched) a_{ps}: 3 erlang
traffic
Holding time of CS call h_{cs}: 10 seconds
(negative exponential distribution)
Holding time of PS call h_{ps}: 0.1 seconds
(unit distribution)
Maximum number of packets
in the system M: 100 packets
Period of control cycle t: 30 minutes

Simulation results from 30 control cycles are shown in Fig.9 (a) and (b) for the loss probability of circuit-switched call and average number of waiting packets of packet-switched call, respectively. In Fig.9(a), the loss probability changes according to traffic fluctuations in the circuit-switched call in the case (i) of fixed channel allocation (with no learning control). But in the case (ii) of learning control, the loss probability converges with the target value of 0.1 after about 10 control cycles.

In Fig. 9 (b), the average number of packets in the case (ii) of learning control increases by at least over 70 (value over 100 are overflow because of the system's restriction on maximum number of calls) This is due to the fact that the loss probability for case (ii) is better than case (i). For both cases, the Fixed Boundary scheme (FB) (neither packet nor circuit traffic can use each other's empty channels) is applied. Increase in the average number of waiting packets can be kept below 13 packets by applying the Movable Boundary scheme (MB) [5] (packet traffic can use empty channels for circuit traffic) instead of FB. These results are shown by (i)' and (ii)' in Fig. 9 (b).

The foregoing discussion demonstrates the effectiveness of this learning control scheme in the integrated switching system which handles a mixture of circuit and packet traffic with different characteristics and required quality, as well as in the multi-speed circuit-switched system.

5. CONCLUSIONS

A learning control approach to communications networks has been studied. A control method suitable for resource assignment problems expected in future communications networks has been proposed. The concept of the control method was explained using the multi-speed circuit-switched traffic system model. Then this method was evaluated through the simulation studies. Results revealed the convergence of the method, the effects of traffic variation and length of control period on the residual error rate and the convergence speed,etc. Furthermore, its general applicability was demonstrated by its successful use in a circuit/packet integrated switched traffic system. In conclusion, the method showed good control performance for models where control laws cannot be obtained analytically. It can be assumed to be effective not only for the systems studied here, but for all communications systems whose complexity make it difficult to determine control laws. The performance of the method can no doubt be further improved in the future by incorporating dynamic characteristics into the model.

ACKNOWLEDGMENT

The authors wish to thank Dr. Keiji Okada and Mr. Masaichi Kajiwara for their invaluable advise.

REFERENCES

[1] Narendra and Mars, "A study of telephone traffic routing using learning algorithms," ICC'81, 1981.
[2] Segall and Moss, "Application of optimal control theory to dynamic routing for computer networks," Proc. JACC, 1976.
[3] Айзерман, Браверман and Розоноэр, "Метод потенциальных ФЧнкций в теории оъучения машин," Издателъство НАУКА
[4] Gimpelson, "Analysis of mixture of wide and narrow band traffic," IEEE Trans. COM13, No.3, 1965.
[5] Fischer and Harris, "A model for evaluating the performance of an integrated circuit- and packet- switched multiplex structure," IEEE Trans. COM-24, No.2, 1976.

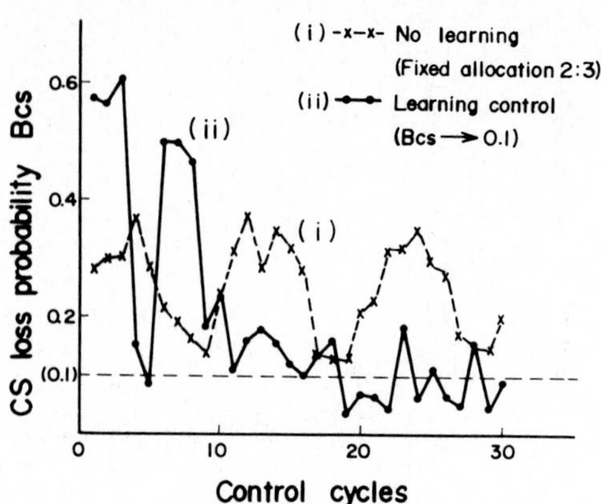

Fig. 9 (a) Loss probability of circuit-switched call

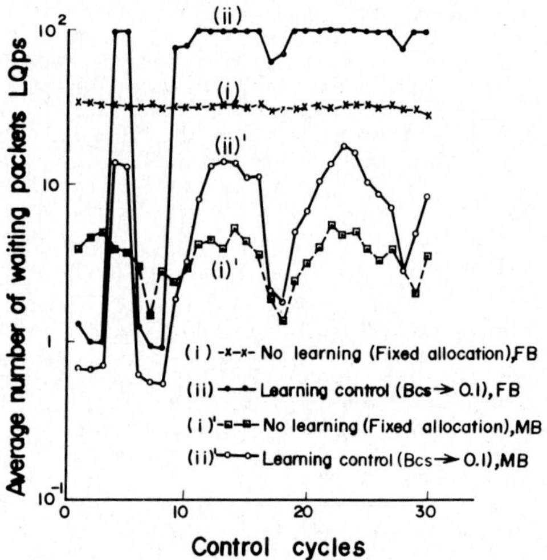

Fig. 9 (b) Average number of waiting packets

TELETRAFFIC ISSUES in an Advanced Information Society
ITC-11
Minoru Akiyama (Editor)
Elsevier Science Publishers B.V. (North-Holland)

PERFORMANCE EVALUATION OF A NEW LEARNING AUTOMATA BASED ROUTING ALGORITHM FOR CALLS IN TELEPHONE NETWORKS

M. T. EL-HADIDI* H. M. EL-SAYED* A. Y. BILAL*,**

* Electronics and Communications Dept. Cairo University, Giza, Egypt
** National Telecomm. Institute Cairo, Egypt

ABSTRACT

With the advent of modern technologies such as SPC switching, CCIS signalling, and computerized traffic measurements, efforts have intensified in search of efficient routing algorithms capable of combating overload conditions in the telephone network. One such algorithm is presented in this paper, with its structure conforming with the recent control methodologies based on learning automata. In addition to its near optimal performance, the proposed algorithm has the important feature of being implementable in a decentralized form. Extensive computer simulations have been carried out which clearly demonstrate the capabilities of the new algorithm.

1. INTRODUCTION

In dealing with overload conditions in a telephone network, two approaches have been generally considered. The first approach is known as "protective control" and uses methods such as Directional Reservation Equipment and Dynamic Overload Controls to protect the network against congestion, by limiting the traffic offered to it [1],[2]. In the second approach, known as "expansive control", the network's capability is expanded by rerouting blocked calls over a sequence of trunks that are not typically selected under normal conditions [3]-[5]. Dynamic rerouting can be accomplished either on a probabilistic basis (load-sharing algorithms), or on a deterministic basis (alternative-routing algorithms). Such expansive controls have become feasible recently with the advent of modern technologies such as SPC switching, CCIS signalling, and computerized traffic measurements. In the present paper, a new algorithm for expansive control is proposed. It has the capability of routing telephone calls throughout the network in a near optimal manner and, in addition, can be implemented in a decentralized form.

2. THE CALL ROUTING PROBLEM

We consider a general telephone network consisting of s source nodes (or exchanges), t transit nodes, and s sink nodes. Within this framework, a call is allowed to originate from any of the source nodes and terminate at any of the sink nodes. Denoting by A_{ij} the offered traffic between the i th source node and the j th sink node, the matrix of offered traffic for the network is given by

$$A = \{A_{ij}\}; \quad i,j = 1, \ldots , s \tag{1}$$

Meanwhile, the available routes for a call between a source and a sink node are determined by the network topology. Let us suppose there are r such routes. Then, depending on the traffic conditions in the network and the capacity of links between nodes, some of these routes would be blocked. We denote by α_{ij}^{ℓ} the probability of using the ℓth route (or more generally the ℓ th routing action), and designate the probability of a call being blocked on such a route by P_{ij}^{ℓ}. The problem of call routing consists of assigning values for α_{ij}^{ℓ}, $i,j = 1,\ldots,s$; $\ell = 1,\ldots,r$; such that the network performance is optimized. Clearly, if each α_{ij}^{ℓ} changes with time, the resulting routing actions would be dynamic. On the other hand, for α_{ij}^{ℓ} equal either to 0 or 1 the actions are deterministic and correspond to the familiar alternative routing type, whereas for $0 < \alpha_{ij}^{\ell} < 1$ they are probabilistic and correspond to the load-sharing type.

In what follows, the network performance is measured in terms of the average number of calls lost over the entire network. Mathematically, this can be expressed as follows :

$$J(\underline{\alpha}) = \sum_{i,j=1}^{s} \sum_{\ell=1}^{r} \alpha_{ij}^{\ell} A_{ij} P_{ij}^{\ell}(\underline{\alpha}) \tag{2}$$

The call routing problem may now be stated as :

Minimize $J(\underline{\alpha})$ with respect to $\underline{\alpha}$

such that $\underline{\alpha} \equiv \{ \alpha_{ij}^{\ell},\ i,j=1,\ldots,s;\ell=1,\ldots,r\}$ satisfies :

$$0 \leqslant \alpha_{ij}^{\ell} \leqslant 1 \ , \quad \sum_{\ell=1}^{r} \alpha_{ij}^{\ell} = 1 \tag{3}$$

3. SOLUTION OF THE CALL ROUTING PROBLEM

To minimize Eq.(2) w.r.t. $\underline{\alpha}$ subject to constraint (3), one needs to evaluate P_{ij}^{ℓ} first. The most accurate analysis for the determination of P_{ij}^{ℓ} is based on a queuing model for the network. In such model, call arrivals are assumed to follow the Poisson distribution, call durations are assumed to be exponential, and the number of calls in the network is assumed to change by at most one call during an infinitismal time interval [6]. A transition probability vector P is then constructed whose components correspond to all combinations of calls on the various routes of the network. The dimension of such vector is

$$\prod_{i,j=1}^{s} \prod_{\ell=1}^{r} (N_{ij}^{\ell} + 1) \ ,$$

where N_{ij}^{ℓ} denotes the capacity of route ℓ between nodes i and j. A differntial equation of the form $\dot{P} = QP$ has to be subsequently solved in order to get the $(s \times s \times r)$ unknown blocking probabilities. Besides the problem of large dimension, it is very difficult to derive the indicated differential equation in the case of a whole network (as opposed to a single trunk group). This is because one should consider the interaction of several routes through a common trunk group as well as the complex dependency of the transition probabilities on $\underline{\alpha}$.

To circumvent the above difficulties, simplified models have been proposed. Among these is the so-called network flow-model, which uses the average network traffics at steady state for modelling the system [7]. Specifically, in a network satisfying the assumptions of : (i) call are assigned single-choice routes on a load-sharing basis, and (ii) "downstream" trunk blocking is independent of "upstream" trunk blocking, one can arrive at the following set of equations :

$$P_{ij}^{\ell} = P_{i\ell} + P_{\ell j} - P_{i\ell} \cdot P_{\ell j} \tag{4}$$

$$Y_{i\ell} = \sum_{j=1}^{s} \alpha_{ij}^{\ell} A_{ij} (1 - P_{\ell j}) \tag{5-a}$$

$$Y_{\ell j} = \sum_{i=1}^{s} \alpha_{ij}^{\ell} A_{ij} (1 - P_{i\ell}) \tag{5-b}$$

$$P_{i\ell} = \frac{Y_{i\ell}^{N_{i\ell}}}{N_{i\ell}!} \Big/ \sum_{m=0}^{N_{i\ell}} \frac{Y_{i\ell}^{m}}{m!} \tag{5-c}$$

$$P_{\ell j} = \frac{Y_{\ell j}^{N_{\ell j}}}{N_{\ell j}!} \Big/ \sum_{m=0}^{N_{\ell j}} \frac{Y_{\ell j}^{m}}{m!} \tag{5-d}$$

where

$P_{ab} \triangleq$ prob. of a call being blocked on trunk ab

$Y_{ab} \triangleq$ "fictitious" offered traffic on trunk ab

$N_{ab} \triangleq$ number of circuits on trunk ab

It soon becomes clear that though the dimension of the equations determining P_{ij}^{ℓ} has been greatly reduced, one should nevertheless solve a set of nonlinear algebraic equations in the α_{ij}^{ℓ}'s. (This fact remains true for the more general network in which assumptions (i) and (ii) are relaxed). The call routing problem (2) and (3), in conjunction with a flow-model description is therefore one of nonlinear optimization. To solve it, one may use a search technique such as the Frank-Wolfe's conditional gradient [8], or the Rosen's gradient projection [9]. In both cases, the gradient $\nabla J(\underline{\alpha})$ has to be evaluated in order to iteratively update the action probability according to :

$$\underline{\alpha}(n+1) = \underline{\alpha}(n) - \mu_n \nabla J(\underline{\alpha}(n)) \tag{6}$$

(In Eq. (6), μ_n denotes the updating step at the n th iteration, and $\nabla J(\underline{\alpha}) = \{\partial J/\partial \alpha_{ij}^{\ell}\}_{i,j,\ell}$).

To appreciate the computational effort involved in calculating (6), we once more consider the network satisfying assumptions (i) and (ii). Here,

$$\frac{\partial J}{\partial \alpha_{ij}^{\ell}} = A_{ij} P_{ij}^{\ell} + \sum_{u,\nu} \sum_{w} \alpha_{u\nu}^{w} A_{u\nu} \frac{\partial P_{u\nu}^{w}}{\partial \alpha_{ij}^{\ell}} \tag{7}$$

where

$$\frac{\partial P_{u\nu}^{w}}{\partial \alpha_{ij}^{\ell}} = \frac{\partial P_{u\ell}}{\partial Y_{u\ell}} \frac{\partial Y_{u\ell}}{\partial \alpha_{ij}^{\ell}} + \frac{\partial P_{\ell\nu}}{\partial Y_{\ell\nu}} \frac{\partial Y_{\ell\nu}}{\partial \alpha_{ij}^{\ell}} - P_{\ell\nu} \frac{\partial P_{u\ell}}{\partial Y_{u\ell}} \frac{\partial Y_{u\ell}}{\partial \alpha_{ij}^{\ell}} - P_{u\ell} \frac{\partial P_{\ell\nu}}{\partial Y_{\ell\nu}} \frac{\partial Y_{\ell\nu}}{\partial \alpha_{ij}^{\ell}}, \quad w = \ell$$
$$= 0 \qquad \text{otherwise} \tag{8}$$

The $P_{u\ell}$'s and the $P_{\ell\nu}$'s are to be evaluated by solving Eqs. (5-a)-(5-d), using successive approximations say, whereas the derivative terms can be obtained by approproaite algebraic manipulations of the same equations. (See [11] for details).

We thus conclude that using a simplified network model does not - alone - circumvent the difficulty of solving the nonlinear optimization problem in a real-time fashion. Besides the computational difficulties, there is also the requirement that all traffic data at every network node (i.e. exchange) be known at some traffic controller, which is responsible for computing the new α_{ij}^{ℓ}'s. Such centralization of the optimal solution can mean tremendous costs that would be associated with establishing data links to all exchanges, and these by themselves can result in reduced system reliability.

4. A NEW LEARNING AUTOMATA BASED ROUTING ALGORITHM

The proposed routing algorithm is a modification of the Frank-Wolfe conditional gradient technique in such a way that it both simplifies the computational effort as well as makes possible decentralized routing decisions [11],[12]. More specifically, in the Frank-Wolfe technique the constraint (3) is satisfied by expressing the recursion (6) as

$$\underline{\alpha}(n+1) = \underline{\alpha}(n) - a\{\underline{\alpha}(n) - [\underline{\alpha}(n) - \frac{\mu_n}{a} \nabla J(\underline{\alpha}(n))]\} \tag{9}$$

and then choosing $\frac{\mu_n}{a}$ so as to bring $\underline{\alpha}(n) - \frac{\mu_n}{a} \nabla J(\underline{\alpha}(n))$ to the boundary of the feasible set. In the new algorithm, the computation of $\nabla J(\underline{\alpha})$ is facilitated by linearizing $P_{ij}^{\ell}(\underline{\alpha})$ about some operating point, which then yields :

$$J = \sum_{i,j} \sum_{\ell} \alpha_{ij}^{\ell} A_{ij} P_{ij}^{\ell}$$

$$\simeq \sum_{i,j} \sum_{\ell} \alpha_{ij}^{\ell} A_{ij} [K_0^{ij} + \sum_{u,\nu,w} K_{u\nu w}^{ij\ell} \alpha_{u\nu}^{w}]$$

Consequently,

$$\frac{\partial J}{\partial \alpha_{u\nu}^{w}} = A_{u\nu} P_{u\nu}^{w} + \alpha_{u\nu}^{w} A_{u\nu} K_{u\nu w}^{u\nu w} + \sum_{i,j} \sum_{\ell} \alpha_{ij}^{\ell} A_{ij} K_{u\nu w}^{ij\ell} \tag{10}$$

Thus, to evaluate $\partial J/\partial\alpha^{w}_{u\nu}$ one needs only to identify the $K^{ij\ell}_{u\nu w}$'s at the operating point. Recalling that these coefficients actually represent the slopes of P^{ℓ}_{ij}'s with respect to $\alpha^{w}_{u\nu}$'s, one may justifiably use the following linear recursive estimator

$$K^{ij\ell}_{u\nu w}(n) = K^{ij\ell}_{u\nu w}(n-1).(1-\gamma_2) + \left[\frac{P^{\ell}_{ij}(n) - P^{\ell}_{ij}(n-1)}{\alpha^{w}_{u\nu}(n) - \alpha^{w}_{u\nu}(n-1)}\right].\gamma_2 \quad (11)$$

together with

$$\begin{aligned} P(m+1) &= P(m).(1-\gamma_1) && \text{if the } (m+1) \text{ st call succeeds} \\ &= P(m).(1-\gamma_1)+\gamma_1 && \text{if the } (m+1) \text{ st call is blocked} \end{aligned} \quad (12)$$

In the above, n is an index for the updating instant, whereas m is a call dependent index. Moreover, γ_1 and γ_2 are constants whose inverses correspond to the estimators' depth of memory. meanwhile, P denotes the blocking probability of a trunk group belonging tothe route under consideration.

Next, to tackle the problem of centralization, Eq. (9) is approximated as follows :

For all $i,j \in \{1,\ldots,s\}$

$$\begin{aligned} \alpha^{\ell}_{ij}(n+1) &= \alpha^{\ell}_{ij}(n) - a\{\alpha^{\ell}_{ij}(n)-1\} && \text{if } \frac{\partial J}{\partial\alpha^{\ell}_{ij}} = \min_{\nu} \frac{\partial J}{\partial\alpha^{\nu}_{ij}} \\ &= \alpha^{\ell}_{ij}(n) - a\{\alpha^{\ell}_{ij}(n)-0\} && \text{otherwise} \end{aligned} \quad (13)$$

(The above approximation says that we replace the boundary point $\underline{\alpha}_{ij}(n) - \frac{\mu_n}{a}\,\partial J/\partial\underline{\alpha}_{ij}(n)$ by the vertix nearest to it in the feasible set.)

Equation (13) represents a linear automaton of the reward-inaction type with a learning step "a" [10]. It is exclusively concerned with the routing of calls between nodes i and j. Moreover, by making the realistic assumption that $K^{ij\ell}_{u\nu w} \simeq 0$ for those nodes which are physically distant from i and j, the automaton needs only to receive traffic data frome those nodes in its immediate vicinity. (See Eq.(10)). Consequently, decentralization of routing actions can be achieved.

5. PERFORMANCE EVALUATION VIA COMPUTER SIMULATION

In order to evaluate the performance of the proposed routing algorithm, a special interactive computer program has been developed in PASCAL [11]. It has the following features : a) the call interarrival times are exponentially distributed in accordance with the offered traffic matrix $\{A_{ij}\}$; b) the call durations are exponentially distributed with an average value T which is the same for all routes; c) routing strategy is updated periodically every "d" seconds according to Eq. (13), with a facility for having routing actions of either the single- or multi-choice type; and d) network conditions are continuously monitored such that its data is available for outputting as desired.

The network used for the simulation results reported below, has 3 source nodes, 4 transit nodes, and 3 sink nodes. The trunk group capacities and the load pattern are as given in Tables 1 and 2, respectively. The average call duration is T = 100 sec and the permitted routing decisions are of the single-choice load-sharing type. After normal operation for a period of 1150 sec, the failure of the transit exchange corresponding to k = 1 is simulated. This condition is maintained for 850 sec during which calls could be routed through transit nodes 2 and 3 only.

A number of simulation experiments were then carried out using the proposed routing algorithm. In one experiment, the algorithm performance was compared to that of the optimal solution detailed in Section 3. For this purpose, several initial trials were conducted in order to determine suitable values for the new algorithm's parameters. This yielded : γ_1= .03, γ_2= .0012, a = .16, and d = 50 sec. The average number of lost calls over 100 replications of simulation runs (each having 2000 sec of simulation time), was then obtained and the results are shown in Fig. 1. Also shown in the figure are the results of the off-line optimal solution for the routing problem, as indicated with the dotted lines. It is clear from Fig. 1 that the automata were able to rapidly "learn" the routing strategy that should be followed under each network condition, and that these strategies are very close to the optimal ones.

In another simulation experiment, the effect of varying the values of γ_1, γ_2, a, and d was studied. For this purpose, the following set of reference parameter values was chosen : γ_1= .07, γ_2= .001, a = 0.1, and d = 50 sec. The results obtained by separately varying each of these parameters are shown in Figs. 2 - 6. Based on these figures one may draw the followin conclusions :

i) Increasing the depth of memory - which is given by $1/\gamma_1$ and $1/\gamma_2$ - leads to a reduction in the responsiveness of the algorithm to random fluctuations, thereby reducing the variance of losses at steady state;

ii) Increasing the learning step "a" causes a more rapid convergence in the event of large amplitude perturbations; however, this is accompanied by an increase in the variance of losses at steady state;

iii) Decreasing the updating period "d" has an influence similar to that of increasing "a".

It is therefore clear that when choosing the parameters of the proposed routing algorithm, a trade-off has to be made between the speed of convergence and the steady state performance. In addition, the choice of the parameter "d" should take into account the permissible rate of data transmission and processing, since this sets a lower limit to "d".

6. CONCLUDING REMARKS

In this paper we have presented a new routing algorithm which offers a number of significant advantages over previously suggested algorithms. On the one hand, it overcomes the prohibitive computational burden and the inescapable centralization requirements, associated with the optimal solution of the routing problem using gradient techniques. On the other hand, it avoids the need for processing traffic data at the impractically high "call-by-call" rate demanded by standard learning automata techniques [3],[10]. This is accomplished while maintaining a near optimal performance that cannot be guaranteed using standard learning automata techniques, especially under heavy traffic conditions.

TABLE 1 Trunk group capacities (# of circuits)

Source Exchange	Transit Exchange 1	2	3	Sink Exchange
1	20/15	12/21	9/30	1
2	23/19	14/21	19/20	2
3	15/15	32/9	11/13	3
4	12/21	16/23	42/18	4

TABLE 2 Normal load pattern (in Erlangs)

Source Exchange	Destination Exchange 1	2	3	4
1	—	6	8	12
2	14	—	7	15
3	5	15	—	17
4	25	19	6	—

REFERENCES

[1] P. J. Burke, "Automatic overload controls in a circuit-switched communications network", in Proc. Nat. Electron. Conf., pp. 667-672, December 1968.

[2] L. A. Gimpelson, "Network management : design and control of communications networks", Electrical Communications, vol. 49, No. 1, pp. 4-22, Jan. 1974.

[3] K. S. Narendra, E. A. Wright, and L. G. Mason, "Application of learning automata to telephone traffic routing and control", IEEE Trans. on Sys., Man, and Cyb., vol SMC-7, No. 11, pp. 785-792, November 1977.

[4] E. Szybicki and A. Bean, "Advanced traffic routing in local telephone networks; performance of proposed call routing algorithms", in Proc. 9 th Int. Teletraffic Congr., Torremolinos, Spain, paper 612, October 1979.

[5] D. G. Haenschke, D. A. Kettler, and E. Oberer, "Network management and congestion in the U.S. telecommunications network", IEEE Trans. on Comm., vol COM-29, No. 4, pp. 376-385, April 1981.

[6] L. Kleinrock, Queuing Systems, vol. 1 : Theory, New York: John Wiley, 1975.

[7] J. Bernussou, J. M. Garcia, I. S. Bonatti, and F. LeGall, "Modelling and control of large scale telecommunication network", IFAC 8 th Triennial World Congress, vol. x, Kyoto, Japan, August 1981.

[8] W. A. Gruver and E. Sachs, Algorithmic Methods in Optimal Control, London: Pitman Pub. Ltd., 1980.

[9] A. V. Fiacco and G. P. McCormick, Nonlinear Programming, New York: John Wiley, 1968.

[10] K. S. Narendra and M. A. L. Thathachar, "Learning automata - a survey", IEEE Trans. on Sys., Man, and Cyb., vol SMC-4, No. 4, pp. 323-334, July 1974.

[11] H. M. El-Sayed, Adaptive and Learning Control of Traffic in Telephone Networks, M. Sc. Thesis, Cairo University, Egypt, Jan. 1984.

[12] H. M. El-Sayed, M. T. El-Hadidi, and A. Y. Bilal, "A learning scheme for routing in telephone networks", in Proc. EUROCON 84 - Computers in Communications and Control, Brighton, Britain, September 1984.

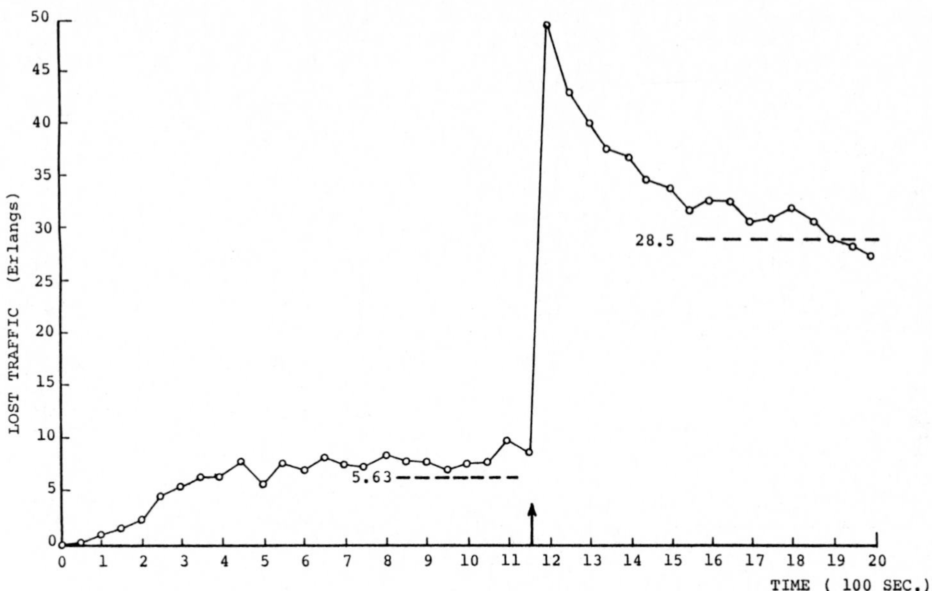

Fig. 1 Behaviour of the new routing algorithm including effect of breakdown

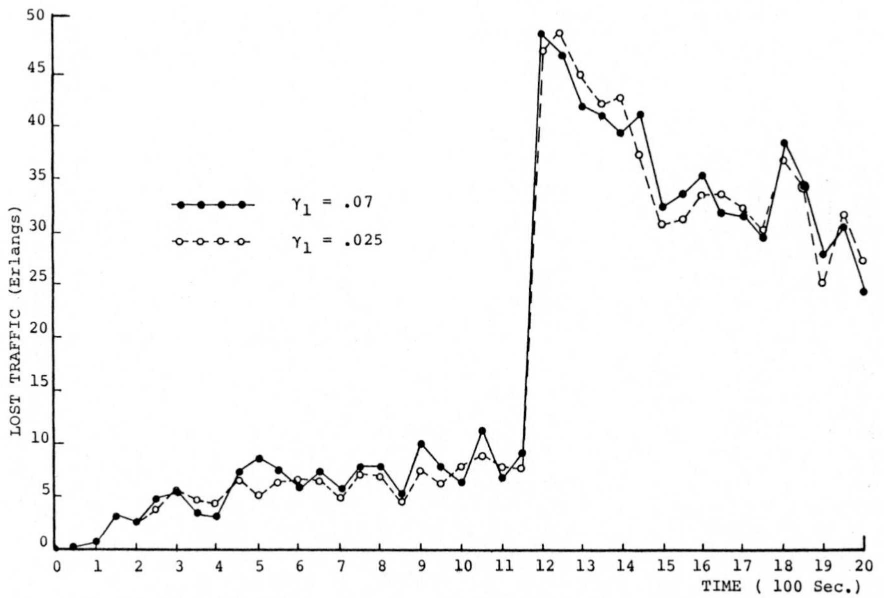

Fig. 2 The effect of parameter γ_1 on the performance of the new routing algorithm

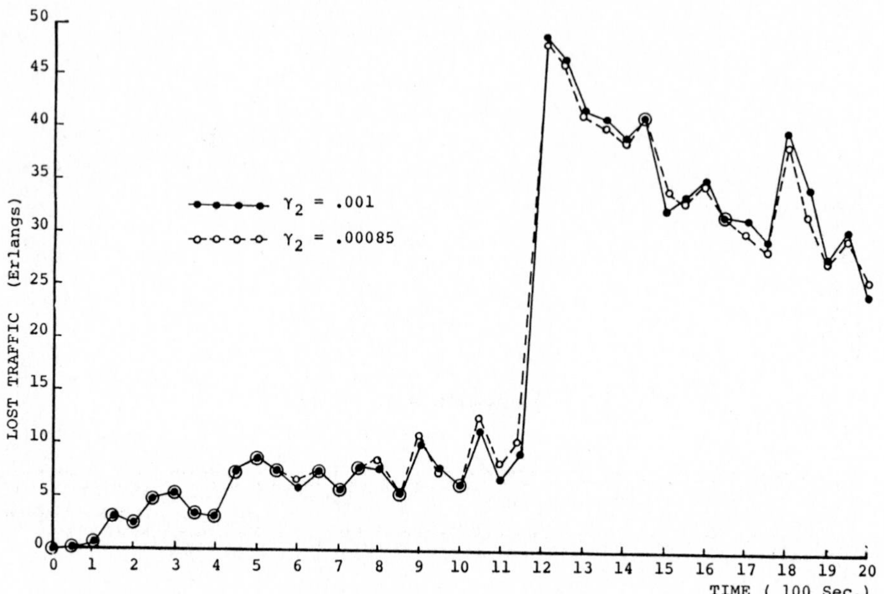

Fig. 3 The effect of parameter γ_2 on the performance of the new routing algorithm

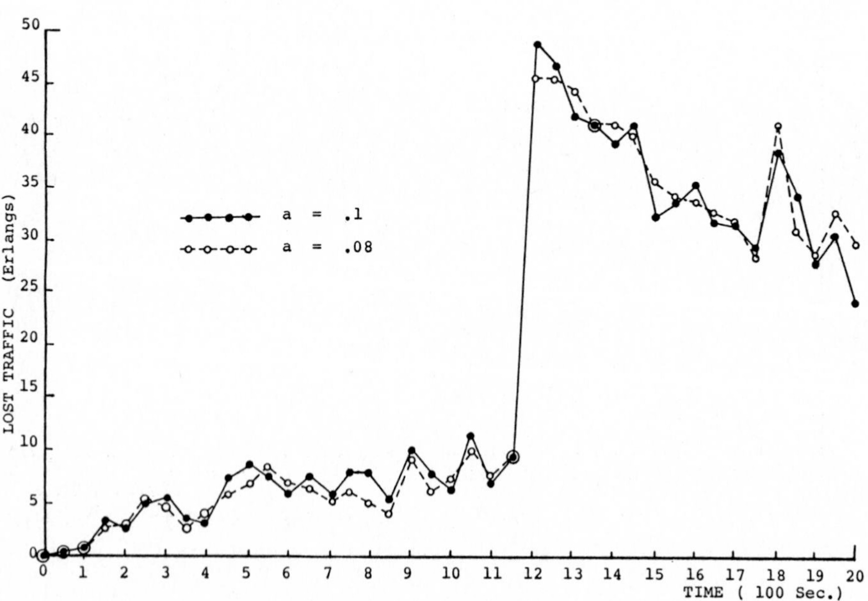

Fig. 4 The effect of decreasing a on the new routing algoritm behaviour

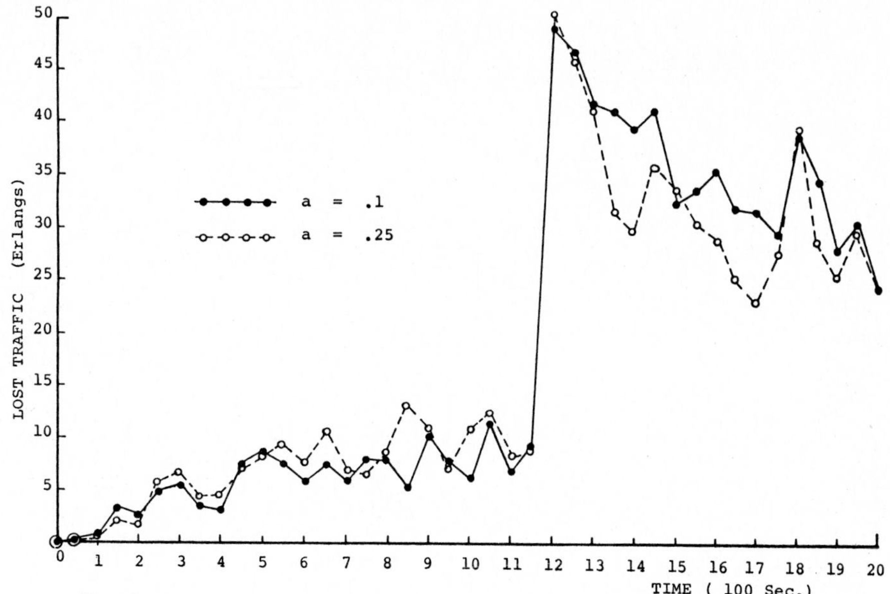

Fig. 5 The effect of increasing a on the new routing algorithm behaviour

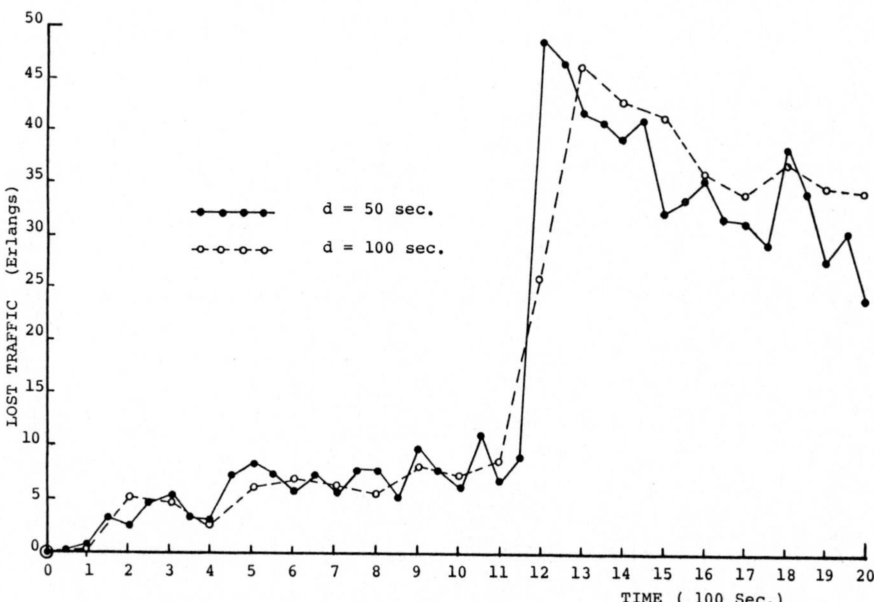

Fig. 6 The effect of slow updating on the new routing algorithm behaviour

TELETRAFFIC ISSUES in an Advanced Information Society
ITC-11
Minoru Akiyama (Editor)
Elsevier Science Publishers B.V. (North-Holland)

MULTIHOUR OPTIMIZATION OF NON-HIERARCHICAL CIRCUIT SWITCHED COMMUNICATION NETWORKS WITH SEQUENTIAL ROUTING

Michal PIORO * and Bengt WALLSTRÖM

Institute of Telecommunications
Technical University of Warsaw
Warsaw, Poland

Department of Communication Systems
Lund Institute of Technology
Lund, Sweden

ABSTRACT

Methods for multihour optimization of link dimensions and routing patterns in circuit switched communication networks employing sequential non-hierarchical call routing strategies are presented. The accuracy of suggested traffic models is evaluated. Time and space requirements are discussed. Examples are shown.

1. INTRODUCTION

Simulation studies and field experiments demonstrate that the traffic handling efficiency of a circuit switched communication network can be increased considerably if a non-hierarchical call routing strategy is used rather than the hierarchical one. Thus cheaper networks can be designed for the prescribed grade of service levels if non-hierarchical routing is utilized.

The rapidly growing number of operating SPC exchanges interconnected by high-speed signalling links makes it possible to apply non-hierarchical routing in existing intercity and metropolitan public telephone networks. In recent years two such strategies were proposed: a sequential strategy called the dynamic non-hierarchical routing (DNHR) [1] and an adaptive one called the advanced routing [2].

Serious attempts are undertaken to implement these techniques in practical networks. In fact DNHR is already operating in a subnetwork of the AT&T intercity telephone network since July 1984, and it is announced in [3] that the entire network of 92 4ESS switches will utilize the DNHR strategy by 1987.

The optimal dimensioning problem for non-hierarchical sequential routing, i.e. the problem of finding the cheapest configuration of link dimensions satisfying the prescibed node-to-node grades of service is harder than the analogous one for hierarchical networks. The main reason is that in order to find the optimal network configuration, not only optimal links dimensions must be found, but also the optimal routing patterns. This is the essential difference as compared with hierarchical networks, in which routing patterns are given in advance.
A non-hierarchical routing pattern optimal for one traffic matrix is not necessarily optimal for another one. Indeed, considerable savings may be achieved if the non-coincidence of busy hours is taken into account by the optimization procedure and optimal routing patterns are found for different hours.

* The paper was prepared during the author's stay at the Lund Institute of Technology.

In this paper we shall deal with the multihour optimization problem for networks employing non-hierarchical sequential routing, i.e. with the problem of finding optimal link dimensions and optimal routing patterns for each traffic hour subject to the node-to-node grade of service constraints. We shall present certain new methods for solving the problem.

2. NETWORK DESCRIPTION

The structure of a circuit switched communication network is represented by a simple graph $G=(V,E)$ with the list of nodes $V=(v_1,v_2,\dots,v_M)$ and the list of links $E=(e_1,e_2,\dots,e_N)$. Each node corresponds to an exchange and each link corresponds to a trunk group interconnecting two exchanges.
If link e_l interconnects nodes v_i and v_j then we write $e_l=(v_i,v_j)$. The number of circuits on link e_l, i.e. the link dimension, will be denoted by n_l. The vector of link dimensions is denoted by $\underline{n}=(n_1,n_2,\dots,n_N)$. The cost of network links, $C(\underline{n})$, is assumed to be a differentiable function of $\underline{n}$ defined for nonnegative real vectors. The function

$$C(\underline{n}) = \sum_{l=1}^{N} c_l n_l \qquad (2.1)$$

will be used as a cost function, where c_l:s are link marginal costs.

An ordered pair of nodes $r=(v_i,v_j)$ is referred to as an origin - destination pair (OD pair). OD pairs will be labelled from 1 to K and denoted by $r_1,r_2,\dots,r_K$.

In order to characterize the traffic offered to a network we use the well known multihour approach [4]. We choose a "typical" period of network operation (e.g. a day). The chosen period is divided into subperiods, called traffic hours, and it is assumed that:

A1. During each traffic hour the stream of calls offered to each OD pair is poissonian with a fixed mean interarrival time. These streams are independent of each other.

A2. Call holding times are independent negative - exponentially distributed random variables of equal mean.

It follows that the traffic offered to each OD pair in

any traffic hour is poissonian. The hours will be labelled from 1 to H. The mean offered traffic to the k:th OD pair in hour h will be denoted by A_k^h. The vector

$$\underline{A}^h = (A_1^h, A_2^h, \ldots, A_K^h) \qquad (2.2)$$

will be referred to as the offered traffic pattern in hour h.

Having established the traffic patterns for a network a call set-up procedure should be defined for each OD pair in each hour. In this paper we shall consider for simplicity a class of procedures with 2-link alternative paths, in which paths are hunted in a predetermined order (other possibilities are mentioned in section 8). A call set-up procedure for OD pair $r=(v_i,v_j)$ in hour h is defined by a routing sequence $S=(t(1),\ldots,t(m))$ of tandem node numbers in the following way. When a call arrives, the direct link (v_i,v_j) is tried first (provided it exists). If it is blocked (or does not exist), the first tandem node number, say t(s), from S, such that the path $(v_i,v_{t(s)},v_j)$ is not blocked, is determined, and the call is set up along this path. If all paths corresponding to S are blocked the call is rejected.

The routing sequence for the k:th OD pair in hour h will be denoted by S_k^h.

Further it will be assumed that:

A3. Call set-up times and call disconnection times are negligible.

With assumptions A1-A3 a network can be treated as an irreducible finite-state Markov chain, and we finally assume that:

A4. In each traffic hour the network is in statistical equilibrium.

For optimization purposes we shall also consider a probabilistic extention [1] of the fixed call set-up procedure described above. Accordingly, to each OD pair k we associate a set of routing sequences $\underline{S}_k=(S_{k1},\ldots,S_{ks(k)})$ and a set of corresponding routing probability distributions $\underline{p}_k^h=(p_{k1}^h,\ldots,p_{ks(k)}^h)$, one for each traffic hour h. For a call of OD pair k in hour h, the direct link is tried first. If it is blocked (or does not exist) a sequence from $\underline{S}_k$, say S_{kq}, is chosen with probability p_{kq}^h and the paths determined by S_{kq} are then hunted in order. If all these paths are blocked the call is rejected.

In the balance of the paper it will be assumed that the following data for any considered network are given:

(i) network's graph $G=(V,E)$

(ii) list of OD pairs $R=(r_1,r_2,\ldots,r_K)$

(iii) traffic patterns $\underline{A}^h=(A_1^h,A_2^h,\ldots,A_K^h)$, $h=1,2,\ldots,H$

(iv) routing sequences S_k^h (or sets $\underline{S}_k$ in the probabilistic case)

and that assumptions A1-A4 are fulfilled.

Let B_k^h denote the equilibrium probability of rejecting a call of OD pair k in hour h. Such a probability will be referred to as the node-to-node grade of service (NNGOS). The set of NNGOS values B_k^h ($h=1,2,\ldots,H$, $k=1,2,\ldots,K$) will be used as the performance measure of the network.

The average grade of service (GOS) in hour h is defined as

$$B^h = \Big(\sum_{k=1}^{K} A_k^h B_k^h\Big) \Big/ \Big(\sum_{k=1}^{K} A_k^h\Big). \qquad (2.3)$$

3. OPTIMIZATION PROBLEMS

The most general optimization problem dealt with in this paper is as follows:

Multihour Optimization Problem (MOP)

Find

* a vector $\underline{n}$ of link dimensions
* a routing sequence S_k^h for each OD pair in each design hour, ($k=1,2,\ldots,K$, $h=1,2,\ldots,H$)

minimizing the cost function $C(\underline{n})$

subject to the grade of service constraints

$$B_k^h \le \beta_k^h \;, \; k=1,\ldots,K, \; h=1,\ldots,H \qquad (3.1)$$

(where β_k^h:s are given blocking limits).

In order to solve MOP we shall consider the following subproblems:

Multihour Optimal Dimensioning Problem (MODP)

For given routing sequences S_k^h ($k=1,\ldots,K$, $h=1,\ldots,H$) find $\underline{n}$ minimizing $C(\underline{n})$ subject to constraints (3.1).

One-hour Optimization Problem (OOP)

This problem is equivalent to MOP with H=1.

One-hour Optimal Dimensioning Problem (OODP)

This problem is equivalent to MODP with H=1.

Routing Optimization Problem (ROP)

For a given vector $\underline{n}$ of link dimensions and a fixed traffic hour h find

* routing sequences S_k^h , $k=1,2,\ldots,K$

minimizing GOS (2.3) in hour h.

Methods for solving the above problems will be presented in the next sections.

4. THE ONE-PARAMETER TRAFFIC MODEL

In this section we describe a conventional one-parameter model [5,7] for the class of networks discussed in section 2. In addition to A1-A4 we make the following assumptions:

A5. The events "link e_l is blocked" (l=1,2,...,N) are independent.

A6. Traffic offered to any link is poissonian, i.e. characterized by one parameter - its mean.

The mean offered traffic to link $e_l=(v_i,v_j)$ in hour h is determined by the sum

$$a_l^h = \sum_k a_{ij}^{kh} \tag{4.1}$$

where a_{ij}^{kh} denotes the mean of the traffic parcel offered to the considered link in hour h and originated at OD pair k.
To see how these parcels can be determined let us consider an OD pair $r_k=(v_1,v_2)$ depicted in Fig.1.

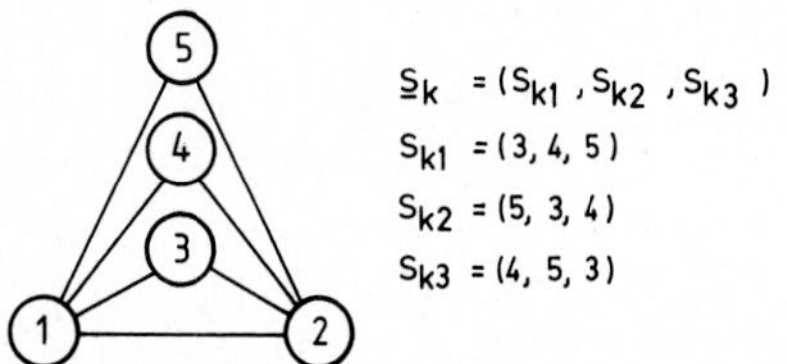

5
4
3
1
2
$\underline{s}_k = (s_{k1}, s_{k2}, s_{k3})$
$s_{k1} = (3, 4, 5)$
$s_{k2} = (5, 3, 4)$
$s_{k3} = (4, 5, 3)$

Fig.1. An OD pair and its alternative paths.

Let b_l^h (or b_{ij}^h) denote the time congestion of link $e_l=(v_i,v_j)$ in traffic hour h, and let $x_l^h=1-b_l^h$ be the corresponding linking probability. Using A1-A6, the mean of the traffic parcel offered in hour h to each link accessible to r_k can be expressed in terms of the linking probability vector $\underline{x}^h=(x_1^h,\ldots,x_N^h)$ and the routing probability vector $\underline{p}_k^h = (p_{k1}^h,p_{k2}^h,p_{k3}^h)$.
For instance $a_{12}^{kh} = A_k^h$ and

$$a_{13}^{kh} = A_k^h(1-x_{12}^h)x_{32}^h(p_{k1}^h+p_{k2}^h(1-x_{15}^h x_{52}^h) + p_{k3}^h(1-x_{14}^h x_{42}^h)(1-x_{15}^h x_{52}^h)). \tag{4.2}$$

Thus each a_l^h is an explicit function of $\underline{x}^h$ and the routing vector $\underline{p}^h=(\underline{p}_1^h,\ldots,\underline{p}_K^h)$, so we may write $a_l^h=a_l^h(\underline{x}^h,\underline{p}^h)$.

Using (4.2) and A6 a system, called the network system of equations NSE1, can be formulated for hour h:

$$x_l^h=1-E(n_l,a_l^h(\underline{x}^h,\underline{p}^h)),\ l=1,\ldots,N \tag{4.3}$$

where E(.,.) denotes Erlang's loss formula. Introducing an appropriate vector function $\underline{F}$, NSE1 can be stated in the vector form:

$$\underline{x} = \underline{F}(\underline{n},\underline{x}^h,\underline{p}^h). \tag{4.4}$$

NSE1 defines a relationship between vectors $\underline{n}$, $\underline{x}^h$, $\underline{p}^h$ for each hour h.
In the sequel we shall frequently omit the superscript h when the hour h is fixed.

Let hour h be fixed. Given $\underline{x}$ and $\underline{p}$, vector $\underline{n}$ can be determined by inverting the equations of NSE1:

$$n_l=D(1-x_l,a_l(\underline{x},\underline{p})),\ l=1,\ldots,N \tag{4.5}$$

where n=D(b,a) is the inversion of b=E(n,a) extended to nonnegative real n.
Thus $\underline{n}=\underline{n}(\underline{x},\underline{p})$ is a function that can be computed by applying (4.1), the generalization of (4.2) and (4.5). Also the partial derivatives of $\underline{n}(\underline{x},\underline{p})$, i.e. $\partial n_l/\partial x_t$ (l,t=1,...,N) and $\partial n_l/\partial p_{kq}$ (l=1,...,N, k=1,...,K, q=1,...,s(k)) can be computed relatively easily using partial derivatives of $a_l(\underline{x},\underline{p})$.

Let us notice that the function $\underline{n}(\underline{x},\underline{p})$ with fixed $\underline{p}$ is not necessarily "one-to-one", although it frequently is. Network examples can be given (cf. section 6) indicating the existence of at least two different $\underline{x}$ vectors which solve NSE1 for the same $\underline{n}$ ($\underline{p}$ fixed). Nevertheless we shall use a function of the form $\underline{x}=\underline{x}(\underline{n},\underline{p})$ for which the vectors $\underline{n}$, $\underline{p}$ and $\underline{x}(\underline{n},\underline{p})$ satisfy NSE1. This function is computed by an iterative procedure for solving NSE1:

$$\underline{x}_0 \text{ fixed (to } \underline{0.9} \text{ say)}$$
$$\underline{x}_1=\underline{F}(\underline{n},\underline{x}_0,\underline{p}),\underline{x}_2=\underline{F}(\underline{n},\underline{x}_1,\underline{p}),\ldots \tag{4.6}$$

The partial derivatives of the function $\underline{x}(\underline{n},\underline{p})$ can be computed using the equations obtained by differentiating NSE1:

$$\frac{\partial x_l}{\partial n_t} = \frac{\partial E(n_l,a_l)}{\partial n_l}\frac{\partial n_l}{\partial n_t} + \frac{\partial E(n_l,a_l)}{\partial a_l}\sum_{j=1}^{N}\frac{\partial a_l}{\partial x_j}\frac{\partial x_j}{\partial n_t},\quad l=1,\ldots,N,\ t=1,\ldots,N. \tag{4.7}$$

$$\frac{\partial x_l}{\partial p_{kq}} = \frac{\partial E(n_l,a_l)}{\partial a_l}\left(\frac{\partial a_l}{\partial p_{kq}} + \sum_{j=1}^{N}\frac{\partial a_l}{\partial x_j}\frac{\partial x_j}{\partial p_{kq}}\right),\quad l=1,\ldots,N,\ k=1,\ldots,K,\ q=1,\ldots,s(k). \tag{4.8}$$

This can be accomplished in two ways. First, after solving NSE1 by means of (4.6), the derivatives $\partial x_l/\partial n_t$ can be obtained by forming a set of N simultaneous systems of linear equations. Each system has N unknowns: $\partial x_l/\partial n_t$ (l=1,2,...,N, t fixed). Coefficients of the equations are obtained from $\partial E(n_l,a_l)/\partial n_l$, $\partial E(n_l,a_l)/\partial a_l$ and $\partial a_l/\partial x_j$ (l, j=1,...,N) computed at the point $\underline{x}(\underline{n},\underline{p})$ resulting from (4.6) (cf. [6,7]).
Then the systems are solved by standard methods.
The derivatives $\partial x_l/\partial p_{kq}$ can be computed in a similar way.

Since for large N this approach may cause numerical difficulties, we suggest another one: to solve (4.7) (or (4.8)) iteratively, along with solving NSE1.

Using formulae analogous to (4.2) NNGOS values can be computed (cf.[7]). Thus NNGOS:s and GOS for hour h are explicit functions of $\underline{x}^h$ and $\underline{p}^h$. They can also be treated as functions of $\underline{n}$ and $\underline{p}^h$, since $\underline{x}^h=\underline{x}^h(\underline{n},\underline{p}^h)$.

5. OPTIMIZATION METHODS BASED ON THE ONE-PARAMETER MODEL

Below we formulate mathematical programs (MP) for solving the optimization problems of section 3.
For this purpose we shall use the functions $\underline{n}=\underline{n}(\underline{x},\underline{p})$ and $\underline{x}=\underline{x}(\underline{n},\underline{p})$ defined in the previous section for a fixed traffic hour h. Also, in order to take the constraints (3.1) into account, we introduce the penalty function

$$P(B,\beta) = \begin{cases} (B-\beta)^2 & \text{if } B > \beta \\ 0 & \text{otherwise .} \end{cases} \quad (5.1)$$

First we shall present programs for solving OODP and OOP for probabilistic call set-up procedures. The vectors $\underline{x}$ and $\underline{p}$ will serve as optimization variables.

MP for solving OODP

For a fixed $\underline{p}$ vector find a vector $\underline{x}$ minimizing

$$C1(\underline{x}) = C(\underline{n}(\underline{x},\underline{p})) + \sum_{k=1}^{K} W_k P(B_k(\underline{x},\underline{p}),\beta_k) \quad (5.2)$$

$$\text{subject to} \quad \underline{0} \le \underline{x} < \underline{1}. \quad (5.3)$$

(W_k:s are penalty factors).

The gradient $\nabla C1(\underline{x})$ of the modified cost function can be determined relatively easily and a simple version of Rosen's gradient projection technique [8] is suggested for solving (5.2).

MP for solving OOP

Find vectors $\underline{x}$ and $\underline{p}$ minimizing

$$C2(\underline{x},\underline{p}) = C(\underline{n}(\underline{x},\underline{p})) + \sum_{k=1}^{K} W_k P(B_k(\underline{x},\underline{p}),\beta_k) \quad (5.4)$$

subject to (5.3) and

$$p_{k1}+\ldots+p_{ks(k)} = 1, \; k=1,\ldots,K$$
$$p_{kq} \ge 0 \;, \; k=1,\ldots,K \;, \; q=1,\ldots,s(k) . \quad (5.5)$$

Although it is possible to solve (5.4) by simultaneous optimization of $\underline{x}$ and $\underline{p}$, we propose a decomposed optimization scheme based on alternating solutions of the following subprograms:

MP1 For a fixed $\underline{x}$ vector find a vector $\underline{p}$ minimizing $C2(\underline{x},\underline{p})$ subject to (5.5).

MP2 For a fixed $\underline{p}$ vector find a vector $\underline{x}$ minimizing $C2(\underline{x},\underline{p})$ subject to (5.3).

For solving MP1 we use a modification of the reduced gradient method described in [9].

The next two programs are based on the function $\underline{x}(\underline{n},\underline{p})$.

MP for solving ROP

For a fixed $\underline{n}$ vector find a vector $\underline{p}$ minimizing

$$C3(\underline{p}) = \sum_{k=1}^{K} A_k B_k(\underline{x}(\underline{n},\underline{p}),\underline{p}) \quad (5.6)$$

subject to (5.5).

The minimization of C3 is equivalent to the minimization of the network's GOS. Again, the reduced gradient method is used for minimizing (5.6).

MP for solving MODP

For given $\underline{p}^h$ vectors (h=1,2,...,H) find $\underline{n}$ minimizing

$$C4(\underline{n}) = C(\underline{n}) + \sum_{h=1}^{H} \sum_{k=1}^{K} W_k^h P(B_k^h(\underline{x}^h(\underline{n},\underline{p}^h),\underline{p}^h),\beta_k^h) \quad (5.7)$$

subject to $\underline{n} \ge \underline{0}$.

For this program Rosen's technique is used.

Finally, we propose the following scheme for solving MOP.

Multihour Optimization Scheme

Initial step

* Solve OOP for the "cluster busy hour" offered traffic pattern $\underline{A}^c=(A_1^c,\ldots,A_K^c)$, where $A_k^c=\max(A_k^h)$ to obtain the initial $\underline{n}_0$ vector.
* Solve OOP for each hour h=1,2,...,H to obtain initial routing probability vectors $\underline{p}_0^1,\ldots,\underline{p}_0^H$.

Multihour step (repeated)

* Solve MODP for fixed $\underline{p}_0^1,\ldots,\underline{p}_0^H$ starting from $\underline{n}_0$ vector. New $\underline{n}$ vector is obtained.
* Solve ROP for fixed $\underline{n}$ for each hour. New $\underline{p}^h$ (h=1,2,...,H) vectors are obtained.
* Update $\underline{n}_0$, $\underline{p}_0^h$ (h=1,2,...,H) and repeat the multihour step until the process converges.

Now let us turn to the original optimization problems for the fixed routing stated in section 3. To solve any of these problems we use the following approach.
First the corresponding problem for probabilistic routing is solved. Then, after obtaining a solution, the resulting distributions $\underline{p}_k=(p_{k1},\ldots,p_{ks(k)})$ are forced to take the fixed routing form:

$p_{kq} = 1$ for some value of q

$p_{ku} = 0$ for $u \ne q$.

This is achieved by introducing penalty functions

$$Q_k(\underline{p}_k) = -(p_{k1}^2+\ldots+p_{ks(k)}^2) \;, \; k=1,\ldots,K \quad (5.8)$$

and solving an additional MP involving these functions. For example, to solve OOP of section 3, we first solve (5.4). Then, for the resulting $\underline{x}$ vector we solve the following MP:

Find $\underline{p}$ minimizing

$$C5(\underline{p}) = C2(\underline{x},\underline{p}) + W \sum_{k=1}^{K} Q_k(\underline{p}_k) \quad (5.9)$$

subject to (5.5),
starting from the vector $\underline{p}$ being a solution to (5.4).

The application of this approach to the remaining problems should be obvious. The approach was found to be efficient, since it can produce a fixed routing without any significant increase of the original cost function (C2 in the above case).

6. A TWO-PARAMETER TRAFFIC MODEL AND RELATED OPTIMIZATION METHODS

The one-parameter traffic model, although simple, is not sufficiently accurate for many network configurations, especially in low and medium loss ranges. Figure 2 illustrates the accuracy of the one-parameter model, and a two-parameter model described in this section.

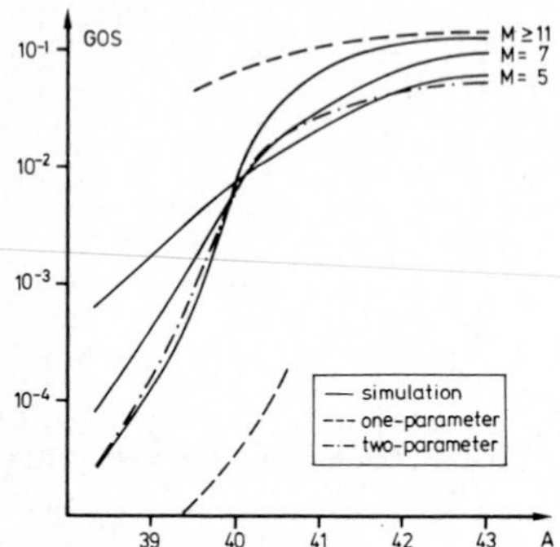

Fig.2. Accuracy of traffic models.

A set of fully connected, one-way link networks with equal link dimensions and symmetric offered traffics is considered. Each OD pair has access to 3 alternative 2-link paths and the routing sequences constitute a symmetric routing (cf. [10]). GOS is plotted vs A. For such symmetric networks the one-parameter model gives the same GOS for any number of nodes M as long as the number of alternative paths is fixed, and so does the two-parameter model. The one-parameter model shows poor results for A $\leq$ 40.6 Erl. and produces double solutions in the range 39.4 $\leq$ A $\leq$ 40.6 Erl. Contrarily, the two-parameter model shows rather accurate GOS values and no double solutions.

We investigated many other symmetric configurations, too [11]. In all cases the two-parameter model gave quite accurate results for the engineering loss levels (0.001 $\leq$ GOS $\leq$ 0.05 say). Double solutions were rarely found and only in cases with many alternative choices (e.g. 9 choices, M=11, A=40, n=51).

The two-parameter model will be formulated for fixed call set-up procedures only. In addition to A1-A4 we shall use the following assumptions:

A5'. All peaked traffic parcels offered to one and the same link encounter the same congestion.

A6'. The traffic offered to any link is well caracterized by its first two moments - mean and variance.

Let t_l^h (or t_{ij}^h) denote the time congestion of link $e_l=(v_i,v_j)$ in hour h and let $y_l^h=1-c_l^h$ (or $y_{ij}^h=1-c_{ij}^h$), where c_l^h denotes the common call congestion for peaked parcels offered to e_l in hour h. Then the formula analogous to (4.2) takes the form:

$$a_{1q}^{kh} = A_k^h \, t_{12}^h \, y_{q2}^h \prod_{u=3}^{q-1} (1-y_{1u}^h \, y_{u2}^h), q=3,4,5 \qquad (6.1)$$

(assuming the fixed routing sequence S_k^h = (3,4,5) in this case).
The traffic offered to a link in a fixed hour h is computed according to (4.1), and $a_l = a_l(\underline{t},\underline{y})$, i.e. a_l is an explicit function of $\underline{t}$ and $\underline{y}$ vectors.

It is also possible to express z_l, the peakedness of the traffic offered to link e_l, as an explicit function of $\underline{t}$ and $\underline{y}$: $z_l=z_l(\underline{t},\underline{y})$. For a way of doing it the reader is referred to [12].

Thus we can formulate a system of equations NSE2, analogous to NSE1:

$$t_l=T(n_l,a_l(\underline{t},\underline{y}),\ z_l(\underline{t},\underline{y}))\ ,\ l=1,\dots,N \qquad (6.2)$$

$$y_l=Y(n_l,a_l(\underline{t},\underline{y}),\ z_l(\underline{t},\underline{y}))\ ,\ l=1,\dots,N \qquad (6.3)$$

where T and Y are appropriate "congestion" functions. In our programs we calculate T from the formula (16) of [13] and then obtain Y from the traffic conservation law.

To determine the functions $\underline{t}=\underline{t}(\underline{n})$ and $\underline{y}=\underline{y}(\underline{n})$ we use an iterative procedure analogous to (4.6). The derivatives $\partial t_l/\partial n_t$ and $\partial y_l/\partial n_t$ (l,t=1,2,...,N) can be computed iteratively as well.

Our next goal is to calculate the vector $\underline{n}$ for a given vector $\underline{x}=(x_1,\dots,x_N)$, where $x_l=1-b_l$ and b_l is the overall call congestion of link e_l. This problem is more difficult than the analogous one of computing $\underline{x} = \underline{x}(\underline{n})$ from NSE1.

Let n=W(b,a,z) be the inversion of the function b=G(n,a,z) resulting from the ERT. The function W is obtained from D and Rapp's formulae. Then a new system of equations can be formulated:

$$t_l=T(W(1-x_l,a_l,z_l),a_l,z_l)\ ,\ l=1,\dots,N \qquad (6.4)$$

$$y_l=Y(W(1-x_l,a_l,z_l),a_l,z_l)\ ,\ l=1,\dots,N\,. \qquad (6.5)$$

Again, this system can be solved iteratively to obtain $\underline{t}$ and $\underline{y}$ vectors for a given $\underline{x}$. The solution yields the $\underline{n}$ vector

$$n_l(\underline{x})=W(1-x_l,a_l(\underline{t},\underline{y}),z_l(\underline{t},\underline{y}))\ ,\ l=1,\dots,N \quad (6.6)$$

since $\underline{t}=\underline{t}(\underline{x})$ and $\underline{y}=\underline{y}(\underline{x})$.
It was found that using $\underline{t}=\underline{1}-\underline{x}$ and $\underline{y}=\underline{x}$ as starting values, the iterative procedure converges rapidly (typically after a couple of iterations).
The partial derivatives $\partial n_l/\partial x_t$ can be determined iteratively as well.

Finally let us note that B_k^h:s, too, are explicit functions of $\underline{t}$ and $\underline{y}$. Thus using the functions $\underline{t}(\underline{n})$ and $\underline{y}(\underline{n})$ or $\underline{t}(\underline{x})$ and $\underline{y}(\underline{x})$, we can treat each B_k^h either as a function of $\underline{n}$ or as a function of $\underline{x}$.

Using the above defined two pairs of functions $(\underline{t},\underline{y})(\underline{x})$ and $(\underline{t},\underline{y})(\underline{n})$, mathematical programs for solving $\overline{\text{OODP}}$ and $\overline{\text{MODP}}$ respectively are formulated in a way analogous to that of section 5. These two-parameter methods are much more accurate, though more complicated and time consuming.

7.TIME AND MEMORY CONSIDERATIONS.EXAMPLES

The optimization methods described in sections 5 and 6 were implemented in FORTRAN on a NORD-100 computer. Below we discuss problems concerning computation times and memory requirements and present some optimization results.

To solve a mathematical program a number of one-dimensional minimization steps must be performed. One gradient computation and a number of objective function computations are performed in each step.

To estimate the time t1 for the objective function computation and the time t2 for the gradient computation it is most convenient to consider the class of symmetric fully connected networks (N=K) with s alternative choices for each OD pair. In the sequel we shall keep s fixed and discuss the dependence of t1 and t2 on N.

The fastest methods among those introduced above are the one-parameter methods based on optimization variables $\underline{x}$ and $\underline{p}$ (cf. (5.2) and (5.4)). The time t1 required to compute the value of C1($\underline{x}$) is proportional to N, and so is the time t2 for computing the gradient. It was found that on NORD-100 t1=0.02N sec. and t2=0.04N sec. for s fixed to 3. Thus assuming a certain average number J (say 20) of C1($\underline{x}$) computations per optimization step we can roughly estimate the time for one step to be t2+Jt1=0.45N sec. This is certainly an acceptable figure. It makes the one-parameter method for solving OODP applicable to large networks as far as the computation time is concerned. Also the memory requirements are low in this case. Approximately 5N real variables are needed for the computations, and these can be kept in the main memory.

Even lower times are achieved for the function C2($\underline{x}$,$\underline{p}$) for fixed $\underline{x}$, since the dependence of C2 on $\underline{p}$ is simpler in nature than its dependence on $\underline{x}$.

One-parameter methods based on $\underline{n}$, i.e. methods for solving ROP and MODP are more time consuming. The time for computing the objective function is I and I·H times greater than t1 for ROP and MODP respectively, where I is the average number of iterations in (4.6), but still the time is proportional to N. Unfortunately the time for a gradient computation is proportional to N^2. Therefore these methods are time consuming when applied to large networks. Also the memory requirements are proportional to N^2 and so it may be necessary to use external memory. These difficulties may be overcome, however, by introducing certain decomposition methods for gradient computation. Such methods will be described in a future paper.

The above estimations are valid for the $\underline{n}$ vector based two-parameter method for solving MODP as well.

Using the $\underline{x}$ vector based two-parameter method for solving OODP, t1 and t2 are proportional to IN and IN^2 respectively, where I is the number of iterations required to solve (6.5). In this case I is small (typically I=4). Moreover, a reasonable approximation may be used for the gradient computation in this case, since we can assume $\partial t_l / \partial x_t = 0$ and $\partial y_l / \partial x_t = 0$ for $l \neq t$. This assumption makes the time t_2 proportional to N.

In table 1 optimization results for a 5 node network with N=K=20 are shown. Each OD pair has access to s=3 alternative paths. In each of H=4 traffic hours 200 Erlangs are offered to the network in total, but these 200 Erlangs are distributed differently in different hours. The cluster busy hour traffic is 284 Erl. The marginal costs per link range from 1 to 6. The design loss level β=0.01 was used for each OD pair in all hours.

	cost	%	GOS	time/step
cluster	728	100	.011	7s.
h=1	561	77	.012	6s.
h=2	517	71	.009	6s.
h=3	561	77	.009	6s.
h=4	610	84	.012	6s.
multihour	636	87	.003,.015,.010,.014	50s.

Tab.1. Optimization results for a 5-node network.

It is seen that the application of the multihour approach gives considerable cost savings as compared with the cluster busy hour approach (13% out of 16% possible).

Similar results for a 7-node network (N=K=42,s=3) are shown in table 2. In this case 624 Erlangs are offered to the network in each of H=2 hours. The cluster busy hour traffic is 744 Erl. The marginal costs are equal to 1 or 2, and β=0.01.

	cost	%	GOS	time/step
cluster	1480	100%	.012	53s.
h=1	1334	90%	.013	50s.
h=2	1283	87%	.012	50s.
multihour	1392	94%	.015,.010	200s.

Tab.2. Optimization results for a 7-node network.

To obtain the above results an initial solution of OOP was first found using the one-parameter approach in each case. Then the appropriate two-parameter methods were applied. The times shown in the last columns of tables 1 and 2 concern the two-parameter methods.

Finally we mention an example of solving ROP. This time we used our method to find the optimal number of alternative choices in a 5-node fully symmetric network with n=48, A=40 Erl. and 0,1,2 or 3 alternative paths available for each OD pair. Using the method we found that in this case a symmetric routing with one alternative path for each OD pair gives minimal GOS. This result was confirmed by simulations.

8. DISCUSSION AND CONCLUSIONS

In sections 5 and 6 we presented a number of optimization methods based on one- and two-parameter traffic models. Each method was obtained by applying a gradient minimization technique to a non-linear mathematical program. The methods are applicable for the optimization problems stated in section 3.

The idea of applying the one-parameter traffic model for engineering non-hierarchical networks with sequential routing is not new and can be found in the early paper [14], where a heuristic dimensioning method is described. A one-parameter method for solving OODP based on the vector of link blockings was introduced in [15]. In fact MP1 of section 5 is a modification of a program presented there. The alternative one-parameter approach based on the vector $\underline{n}$ of link dimensions was suggested in [16]. For OODP however, the $\underline{x}$-based method is much faster. In [16] also two methods for solving ROP were suggested. Our method (cf. section 5 and [7]) is an alternative, non-heuristic approach.

The idea of using the probabilistic routing for optimization purposes was introduced in [1], where a one-parameter method for solving MOP was presented. The multihour optimization scheme described in section 5 is an alternative, more straightforward approach.

To the best of our knowledge two-parameter models were not used to any greater extent in methods for solving optimization problems in non-hierarchical networks with sequential routing, except in [12]. The two-parameter model presented in section 6 is a refined version of the model from [12], however both optimal dimensioning methods of section 6 are new.

All optimization methods presented in this paper were implemented on a computer and run for many network configurations of small and medium size (up to 50 links). The following conclusions can be drawn from the computations.

1. Optimal dimensioning methods based on the one-parameter model frequently produce under-dimensioned networks. Additional difficulties are encountered because of the existence of multiple solutions of NSE1.

2. It has appeared, however, that in most cases, the routing optimization problem can be treated successfully by the one-parameter approach, since for this problem the absolute GOS level is less important.

3. Optimization methods based on our two-parameter model seem to be accurate enough for engineering purposes.

4. Since our mathematical programs are generally not convex there is a risk of ecountering local minima. In order to avoid this and save computation time we propose the following procedure:
   (i) Use the faster one-parameter methods (especially those based on the $\underline{x}$-vector) for different starting points.
   (ii) Reoptimize the cheapest solution obtained by use of the appropriate two-parameter method.

The presented methods can be easily extended to cover networks employing sequential routing with "longer" paths (cf. [12]). In particular the two-parameter optimal dimensioning methods of section 6 can be applied effectively (after certain modification of the penalty function, cf. [12]) for the optimization of hierarchical networks on the NNGOS performance measure basis.

ACKNOWLEDGEMENTS

This paper was prepared under the project "Intelligent Routing" sponsored by LM Ericsson, Stockholm. The authors are further indebted to L.Reneby, Lund Institute of Technology for many enlightening discussions.

REFERENCES

[1] G.R.Ash, R.H.Cardwell and R.P.Murray, "Design and optimization of networks with dynamic routing", Bell Syst. Tech. J., vol 60, no.8, pp. 1787-1820, 1981.

[2] E.Szybicki and M.E.Lavigne, "The introduction of an advanced routing system into local digital networks ...", Proc. Int. Switching Symposium, Paris, 1979.

[3] G.R.Ash and V.S.Mummert, "AT&T carves new routes in its nationwide network", AT&T Bell Labs Record, pp. 18-22, August 1984.

[4] Y.Rapp, "Planning of junction network with non-coincident busy hours", Proc. 6th Int. Teletraffic Congress, Munich, Paper 132, 1970.

[5] P.M.Lin, B.J.Leon and C.R.Stewart, "Analysis of circuit switched networks ...", IEEE Trans. on Comm., vol. com-31, no.6, pp. 754-765, 1978.

[6] M.Pioro, "A concept of a new circuit switched networks optimization method", Referaty Instytutu Telekomunikacji (Technical University of Warsaw), no.110, 1983.

[7] M.Pioro, "Optimal dimensioning problems in non-hierarchical circuit switched networks", Proc.5th Nord. Teletraf. Sem., Trondheim, 1984.

[8] J.B.Rosen, "The gradient projection method for nonlinear programming. Part 1", J. Soc. Indust. Appl. Math., vol. 8, no. 1, pp. 181-217, 1960.

[9] R.J.Harris, "The modified reduced gradient method for optimally dimensioning telephone networks", Australian Telecom. Res., vol. 10, no. 1, pp. 30-35, 1976.

[10] B.Wallström, "On symmetrical two-link routing in fully connected circuit switched networks ...", Proc.5th Nord. Teletraf. Sem., Trondheim, 1984.

[11] M.Pioro, L.Reneby and B.Wallström, "Routing principles in non-hierarchical networks", Technical Report, Lund Institute of Technology, 1984.

[12] M.Pioro, "A uniform approach to the analysis and optimization of circuit switched communication networks", Proc.10th Int. Teletraffic Congress, Montreal, paper 4.3A2, 1983.

[13] A.A.Fredericks, "Approximating parcel blocking via state dependent birth rates", Proc. 10th Int. Teletraf. Congress, Montreal, paper 5.3.2, 1983.

[14] M.Segal, "Traffic engineering of communication networks with a general class of routing schemes", Proc. 4th Int. Teletraffic Congress, London, 1964.

[15] M.Pioro, "A method for optimizing non-hierarchical circuit switched communication networks ..." Referaty Instytutu Telekomunikacji (Technical University of Warsaw), no.95, 1981.

[16] A.Girard, Y.Cote and Y.Ouimet, "A comparative study of non-hierarchical routing", Proc. 2nd Int. Network Planning Symp., Brighton, pp. 70-74, 1983.

TELETRAFFIC ISSUES in an Advanced Information Society
ITC-11
Minoru Akiyama (Editor)
Elsevier Science Publishers B.V. (North-Holland)

USE OF A TRUNK STATUS MAP FOR REAL-TIME DNHR

Gerald R. Ash

AT&T Bell Laboratories
Holmdel, New Jersey

ABSTRACT

Dynamic nonhierarchical routing (DNHR) is a new routing technique that is currently being implemented in the AT&T Communications network. DNHR incorporates a preplanned time-of-day routing strategy that enables a significant increase in the efficiency of the AT&T-C network. Trunk status map routing (TSMR) is an extension of the DNHR concept to a centralized trunk status map (TSM) that provides real-time routing decisions in the DNHR network. Various implementations of TSMR are investigated, and the comparative advantages of several real-time dynamic routing algorithms are illustrated. A simple implementation of TSMR which is a hybrid of time variable routing and fully dynamic routing is shown to yield benefits comparable to more complicated schemes.

1. *INTRODUCTION AND SUMMARY*

Earlier work has described the design, optimization, and control of networks with dynamic routing [1-3]. The term dynamic routing frequently suggests a real-time search for the optimal routing patterns based on current network loads. In its initial implementation, dynamic nonhierarchical routing (DNHR) uses a preplanned timed-variable routing strategy with a limited amount of real-time control. Real-time, traffic sensitive routing is indeed the limiting case of preplanned time-variable routing, and in this paper we investigate the feasibility and benefits of such "true dynamic routing." Here we investigate a central control capability called trunk status map routing (TSMR), which greatly extends the real-time routing capability of DNHR. TSMR uses a real-time network status map to implement a dynamic routing strategy, and we examine the interaction of TSMR with network operation. The trunk status map (TSM) concept involves having an update of the number of idle trunks in each DNHR trunk group sent to a network data base every T seconds. These updates are sent by each stored program control (SPC) switch only when the trunk group status has changed. In return, the TSM, which is located within the common channel signaling network, periodically sends to the SPC switches ordered routing sequences to be used until the next update in T seconds. These routing sequences are determined by the TSM in real time using the TSMR dynamic routing strategy. As T approaches zero seconds, these updates effectively are sent for every call arrival.

In this paper we investigate through simulation techniques a number of approaches to TSMR. These TSMR strategies range from schemes that simply determine the single most idle routing path (which is a one or two-link connection) between each node pair in the network to more complex schemes that periodically reoptimize the entire network routing pattern by solving a large mathematical program. The scheme that appears to be the most attractive operates as follows. The first choice path determined by the design algorithm (the unified algorithm or UA [1], as modified for TSMR [4]) is used if a circuit is available. This first choice path is updated once each routing interval or load-set period (LSP), which is typically one hour to several hours in duration. If the first path is busy, the second path is selected from the list of other paths determined by the UA on the basis of having the greatest number of idle circuits at the time. Hence the TSMR approach is a hybrid of time variable routing and fully dynamic routing. A detailed implementation strategy is described that stabilizes the routing patterns through trunk reservation techniques, and also allows automatic routing controls to augment the TSMR strategy during network overloads and failures.

TSMR provides uniformly better blocking performance than the current implementation of DNHR and a significantly reduced number of crankback messages; both network blocking and crankback messages decrease as the status and routing updates interval, T, decreases. Crankback messages are signaling messages used to inform the originating SPC switch that a DNHR call has been blocked on the second link of a two-link path. An initial evaluation of TSM processing cost and crankback cost as a function of the update interval T places the optimum value of T in the range of about two to eight seconds. The network congestion control strategy developed for TSMR provides performance under overloads and failures that is significantly better than the performance of the current implementation of DNHR.

2. *REVIEW OF DNHR CONCEPTS*

DNHR brings three principal changes to the network plan. First, there is a new network configuration. With DNHR, the AT&T Communications (AT&T-C) network will evolve from its present multiple level structure to a structure consisting of DNHR switches at the upper level with subtending hierarchical switches at the lower level. There is also a new routing technique. DNHR allows the choice of traffic paths to change with time of day and is not constrained by a hierarchical ranking of the switches. This approach greatly expands the flexibility of network routing and thereby permits a large increase in the efficient use of the network. That efficiency reduces the need to construct transmission facilities in the future. Finally, there is a new way of operating and designing the network. DNHR network operations are more centralized and more automated than they are today, and that increases operating efficiency and improves network performance. Several technical feasibility studies have shown that these changes brought about by DNHR are technically feasible, that network performance is comparable to that of the hierarchy, and, with a proper level of automation, that operation of a DNHR network is quite tractable [3].

Figure 1 illustrates the network structure which incorporates DNHR into the AT&T-C intercity network. The DNHR network has only one class of switching system, and end-offices home on DNHR tandem offices by way of exchange access networks. The DNHR portion of the AT&T-C intercity network consists of 4ESS™ switches (and possibly successor

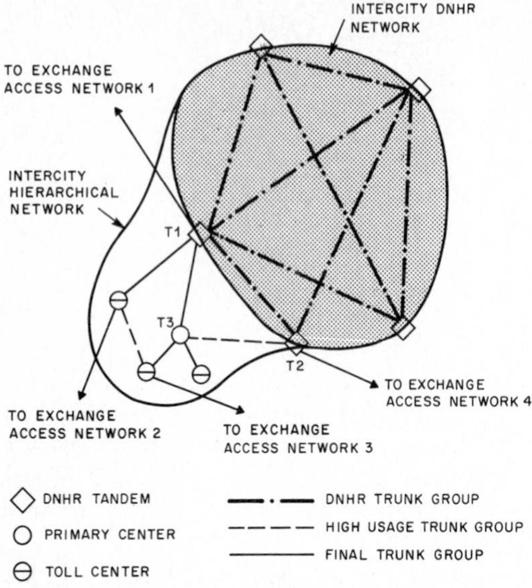

FIGURE 1
DNHR NETWORK CONFIGURATION

ESS™ switches) interconnected by the common channel signaling (CCS) network. Dynamic routing rules are used between pairs of DNHR switches, and conventional hierarchical routing rules are used between all other pairs of AT&T-C switches. These hierarchical switching systems will home directly or indirectly on DNHR switches. Traffic is concentrated in the hierarchical network through the use of conventional homing configurations, as illustrated by tandem T3 in Figure 1. To the subtending hierarchical switches, the DNHR switches appear as a large network of regional centers.

The dynamic routing method illustrated in Figure 2, called "two-link dynamic routing with crankback," capitalizes on two factors: selection of minimum cost paths, as given by the UA design, between originating and terminating switches, and design of optimal, time-varying routing patterns to achieve minimum cost trunking by capitalizing on noncoincident network busy periods. We achieve the dynamic, or time varying, nature of the routing scheme by varying the route choices with time. The routes, which consist of different sequences of paths, are designed to satisfy a node-to-node blocking requirement. Each path consists of one or, at most, two links or trunk groups in tandem. Paths used for routes in different time periods need not be the same. In Figure 2, the originating switch at San Diego (SNDG) retains control over a dynamically routed call until it is either completed to its destination at White Plains (WHPL) or blocked. The control of a call overflowing the second leg of a two-link connection (for instance, the ALBY-WHPL link of the SNDG-ALBY-WHPL path in routing sequence #1) is returned to the originating switch (SNDG) for possible further alternate routing. Control is returned when the via switch (ALBY in the example) sends a CCS crankback signal to the originating switch.

Since many of the intercity traffic demands change with time in a reasonably predictable manner, the routing also changes with time to achieve maximum trunk utilization and minimum network cost. Initially, ten DNHR time periods are being used to divide up the hours of an average business day into contiguous routing intervals called load-set periods (LSPs).

This real-time, traffic-sensitive component of DNHR uses "real-time" paths for possible completion of calls that overflow the "engineered" paths (see Figure 2). The engineered paths are designed to provide the objective blocking performance. The real-time paths, which are also determined by the central forecasting system, can be used only if the number of idle trunks in a group is greater than a specified number of trunks--the reservation level--before the connection is made. This prevents calls that normally use a trunk group from being swamped by real-time routed calls.

The third principal change brought about by DNHR is the way the network is designed and operated. Several operations systems provide centralized functions such as switch planning, trunk forecasting, trunk servicing, routing administration, and network management. An overview of the operations systems used to support DNHR is given in Reference 3. Embedded within the forecasting and servicing systems is the UA, which simultaneously determines the trunking and routing for the entire DNHR network [1].

3. *TRUNK STATUS MAP ROUTING METHOD*

The real-time routing method discussed in connection with the present implementation of DNHR is used to improve network service. Service improvement with real-time routing is significant even with relatively simple procedures. As an alternative to using real-time routing to improve service, we could hold the node-to-node blocking (service) level fixed and use the real-time routing scheme to reduce the number of trunks required to provide that level of service. This alternative approach can produce additional network savings. Furthermore, real-time dynamic routing can improve network performance in the event of network failures, especially when some amount of reserve capacity is available for redirecting traffic flows from their usual patterns. Hence an increased level of real-time decision making might be warranted in the DNHR network.

Trunk status map routing (TSMR) is an extension of the DNHR concept that uses a centralized trunk status map (TSM) to provide real-time routing decisions in the DNHR network. As described in the introduction, the TSM receives, every T seconds from each SPC switch, updates of the number of idle trunks in each DNHR trunk group. These updates are sent only when the number of idle trunks in the trunk group has changed. In return, the TSM periodically sends to the SPC switches ordered routing sequences to be used until the next update in T seconds. These routing sequences are determined by the TSM in real time using the TSM dynamic routing strategy (TSMR). TSMR therefore represents a much more dynamic routing method than the current implementation of DNHR.

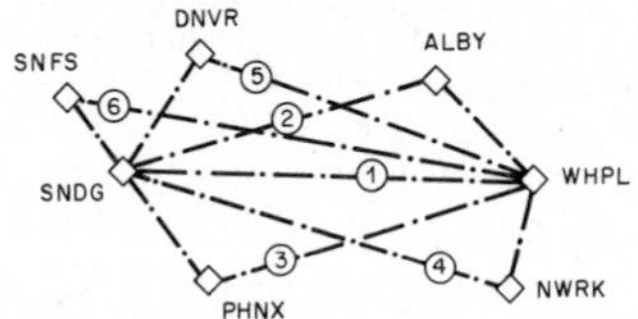

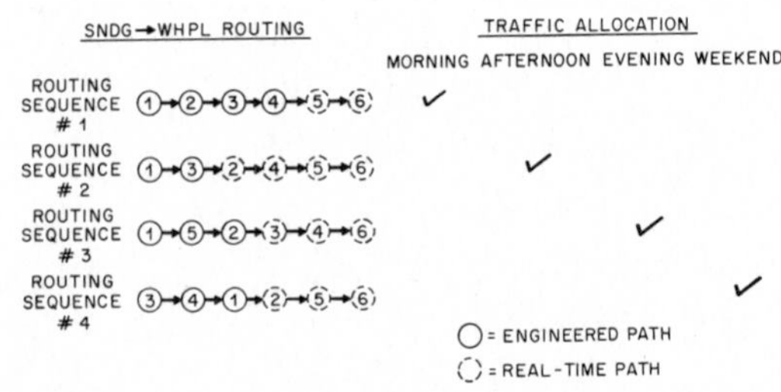

FIGURE 2
TWO-LINK DYNAMIC ROUTING WITH CRANKBACK

In this section we investigate alternative approaches to TSMR as well as the appropriate value of the status and routing update interval (T). We also study an automatic network congestion control strategy that could be implemented by the TSM.

A call-by-call simulation model is used to measure the performance of an engineered network under various routing strategies: DNHR, various alternative TSMR strategies, and TSMR combined with automatic congestion control strategies. A 25-node DNHR network model projected for 1986 is used for the simulation studies. The 25-node model is designed by the DNHR and TSMR design algorithms [1,4] for 16 hours throughout the day (from 8 a.m. through 11 p.m.). The behavior of the network is investigated over a typical two-week period consisting of 10 average business days. Low daily variations are applied to the load for each node-to-node pair in the network simulation. In addition, a systematic daily variation of the total network load is superimposed on the random load variation according to typical load patterns over an average business week.

3.1 Alternative Approaches to TSMR

Various techniques are investigated to determine the most efficient TSMR method. These methods include the following:

1) route each call on the least loaded path (the path having the greatest number of idle circuits) among all candidate paths;

2) route each call first on the direct path, if it exists and is available, or else select the least loaded path;

3) route each call on the first path assigned by the UA, if available, or else select the least loaded path;

4) compute the UA routing sequences that maximize carried traffic for a short-term estimate of the network loads and then apply Method 3 using the new routing sequences.

Methods 1 and 2 do not provide adequate performance for the following reasons. Method 1 favors the path having the largest available capacity, not necessarily the first path, and usually a two-link path. This tendency to favor two-link paths results in a significant redistribution of flows and relatively poor network performance. Method 2 performs considerably better than Method 1 but falls short of the performance of Method 3, which we describe shortly. Method 2 has a problem similar to that of Method 1: since it always selects the direct path as the first choice, and since the UA does not always design the direct path to be the first choice, Method 2 makes the path order too different from that designed by the UA. Hence Method 1 and Method 2 performance is degraded because the actual realization of network flows deviates too far from the network flow patterns designed by the UA.

Method 3, however, performs quite well. It reflects sufficiently accurately the design of the UA by assigning first path traffic to the design first path. Flow on the first choice path accounts for about 80 to 90 percent of the total network traffic flow and hence the routing of this flow must correspond well to the placement of trunks for the network to behave properly. Flows on the second and higher-numbered paths are better assigned by a least-loaded selection method than by a preplanned sequential selection method. This statement is supported by the simulation results shown in Figures 3 and 4.

Figure 3 shows the 10-day average hourly blocking for TSMR Method 3 (for status and routing update interval T equal to five seconds) and also for the current implementation of DNHR; Figure 4 shows the 99th percentile hourly node pair blocking for TSMR and DNHR. These results were obtained with a network designed for DNHR and clearly demonstrate the benefits of TSMR in comparison to the current implementation of DNHR. The details of the TSMR strategy used in these simulations will be described in Section 3.2, 3.3, and 3.4.

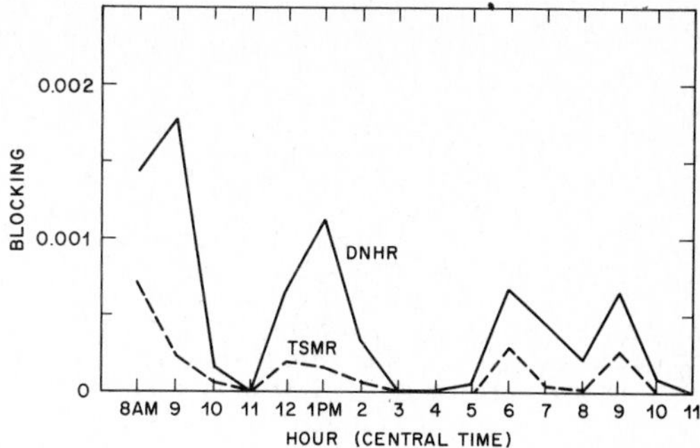

FIGURE 3
AVERAGE NETWORK BLOCKING FOR DNHR AND TSMR
(AVERAGE BUSINESS DAY LOADS)

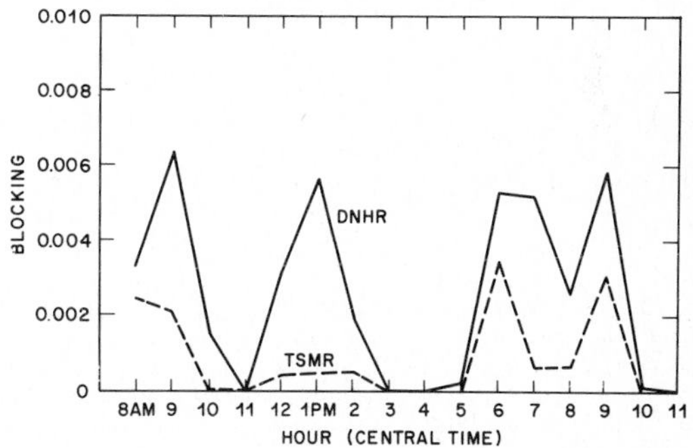

FIGURE 4
99TH PERCENTILE NODE-TO-NODE BLOCKING FOR DNHR AND TSMR
(AVERAGE BUSINESS DAY LOADS)

A simple intuitive explanation of why TSMR Method 3 might complete more calls than the current implementation of DNHR is illustrated by the four-node example shown in Figure 5. The current number of idle circuits in each trunk group is shown. If the DNHR routing sequence for A-D calls is A-D → A-C-D → A-B-D, then the next A-D call arrival will block link AC and link CD, and either an A-C or C-D call arrival will then be blocked. However, the TSMR routing sequence for A-D calls in this network state is A-D → A-B-D → A-C-D. Therefore the next A-D call arrival under the TSMR strategy will not block any additional links. As illustrated by this simple example, TSMR tends to leave capacity on links throughout the network, distributed as uniformly as possible. Because of this property of TSMR, calls arriving in various parts of the network will have a greater chance of being completed than under the hypothesized DNHR routing sequence. Bulfer [5] has established the optimality of the least loaded routing strategy for a class of two stage concentrators.

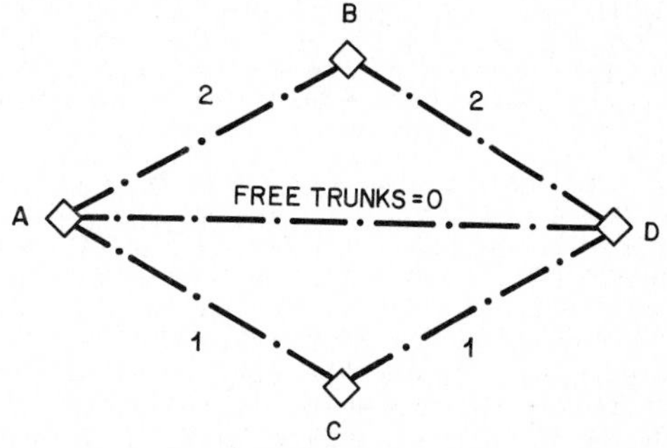

FIGURE 5 ILLUSTRATION OF DNHR vs TSMR

TSMR Method 4 has greater adaptivity to load shifts than Method 3 but is more complex to implement and as such must have better performance to be justified. The details of Method 4 are now discussed. To calculate optimum routing sequences with the UA, as required by Method 4, we assume that these routing patterns are computed in advance of their actual use and that the average load in the future LSP is known precisely. This second assumption of course is an idealization of reality and represents an upper limit on the possible performance of Method 4. We used two methods based on the UA to determine the routing sequences that maximize network flow in the existing network. For each method we find the minimum incremental network capacity required to carry the future LSP load given the existing trunks as available network capacity [2]. That is we minimize

$$\sum_{i=1}^{L} \Delta a_i \, ,$$

where Δa_i are the augmentations above the existing link capacities a_i, and L is the number of links in the network. In this formulation of the UA, we set all incremental link costs to one in order to transform the objective function from incremental network cost to incremental network capacity.

In the first method (which we will call Method 4A), the maximum node-to-node blocking grade of service (GOS) is held at one percent, and the routing and capacity augmentations Δa_i are determined which minimize the objective function. We implement the routing solution, but the capacity augmentations Δa_i produced by the optimization cannot actually be added to the network. Hence the traffic that would have been carried on these augmentations will actually be blocked. But since we have minimized these hypothetical capacity augmentations, this routing solution approximates the minimization of total blocked traffic. In the second method (Method 4B) the node-to-node blocking GOS is raised until there are zero capacity augmentations Δa_i produced by the optimization. This second solution approximates the minimization of the maximum node-to-node blocking.

A comparison of the average network blocking performance of Method 3, Method 4A, and Method 4B is shown in Figure 6. For these results the update interval T is five seconds, and the network is designed for TSMR, as described in Reference 6. As can be seen there is no apparent improvement gained from the more complex methods over Method 3, which suggests that Method 3 may achieve nearly the maximum flow performance. TSMR Method 3 was therefore selected for further study to determine the best switching control logic and TSM logic to implement TSMR.

3.2 *Implementation Strategy for TSMR*

Extensive simulation studies of TSMR Method 3 were used to determine the implementation strategy discussed in this section. We describe the TSMR strategy found to perform the best from these simulation studies.

The SPC switch maintains a TSMR route sequence for each destination. The route sequence stored in the SPC switch is made up of two functional parts. The first part consists of a single path, called the "first-choice path," which is updated by the TSM once each LSP and is made to correspond to the UA selected first-choice path. The second and subsequent paths of the route sequence consist of the "least loaded routing (LLR) paths." These LLR paths are updated every T seconds according to a least loaded criterion applied to the current network status. For each route sequence that needs to be updated, the TSM determines the paths in the current route sequence that need to be changed and transmits these changes to the SPC switch. If the least loaded path differs from the second-choice path currently in the SPC switch, the TSM sends

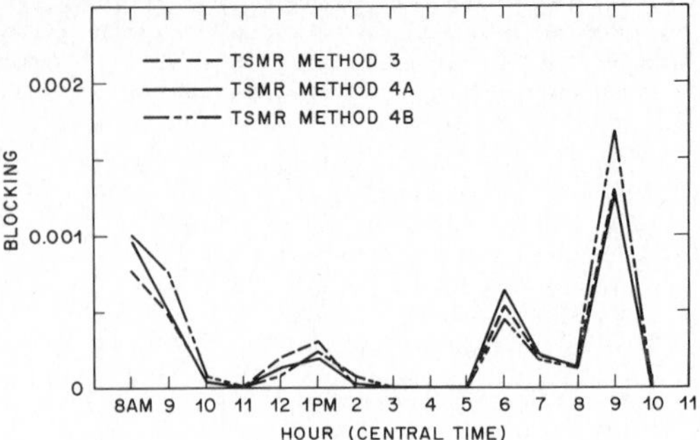

FIGURE 6
AVERAGE NETWORK BLOCKING FOR TSMR METHODS 3, 4A AND 4B

an appropriate message to the SPC switch that changes the contents of the LLR paths to reflect the new least loaded path and also changes the position of the other paths in the route sequence. We found from the simulations that it is sufficient for small values of T to have the TSM compute only the least loaded path and put it second in a route rather than sort the entire route in least loaded order.

For each node pair for which the direct trunk group exists and is the first choice, the TSM applies a thresholding scheme to determine whether a routing update is needed, as follows. Consider the typical case in which the direct path is the first choice and a two-link path is the second choice (case 1 in Figure 7). If the total number of idle circuits on these two choices is greater than a threshold number which is sufficient to

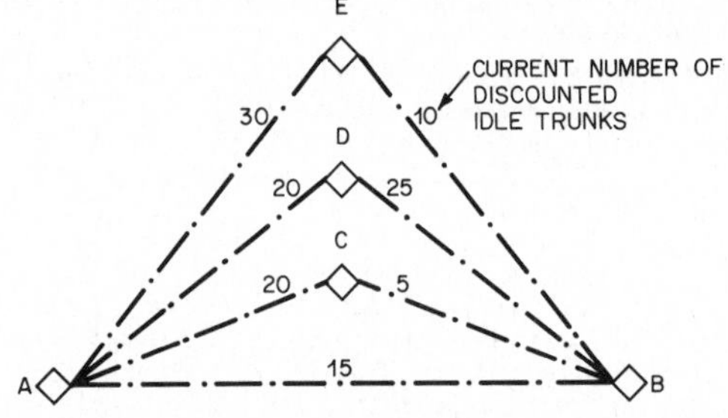

	CASE 1	CASE 2	
UA ROUTING FOR A-B TRAFFIC IN CURRENT LSP	A-B A-C-B A-D-B A-E-B	A-E-B A-C-B A-B A-D-B	
LAST ROUTING FOR A-B TRAFFIC SENT FROM TSM TO SPC SWITCH	A-B A-E-B A-D-B A-C-B	A-E-B A-C-B A-D-B A-B	
CURRENT NUMBER OF IDLE CIRCUITS	A-B=15 A-C-B= 5 A-D-B=20 A-E-B=10	A-B=15 A-C-B= 5 A-D-B=20 A-E-B=10	
NEW ROUTING FOR A-B TRAFFIC TO BE SENT FROM TSM TO SPC SWITCH	A-B	A-E-B	FIRST-CHOICE PATH
	A-D-B A-E-B A-C-B	A-D-B A-C-B A-B	LLR PATHS

CASE 1: FIRST-CHOICE PATH IS THE DIRECT PATH

CASE 2: FIRST-CHOICE PATH IS A TWO-LINK PATH

FIGURE 7
EXAMPLE CALCULATION OF TSM ROUTING

permit completion of all calls likely to arrive over the T-second update interval, then there is no need to reorder the route sequence since the calls can be completed without generating any crankbacks. A threshold that works reasonably well is N/8, where N is the number of trunks on the direct trunk group to the terminating switch. (If the direct group does not exist or is not the first choice, the thresholding scheme is not used). For case 1 in Figure 7, this means that if the number of trunks on the direct group is less than 200, then no routing update message will be sent to the SPC switch. For case 2 in Figure 7, the thresholding scheme is not used. Use of this logic concentrates the TSM processing capacity on those node pairs for which a routing update to the SPC switch is most beneficial. This logic also limits the number of node pairs requiring updated LLR paths.

If there is not a sufficient number of idle circuits according to the thresholding scheme discussed above, and a routing update is potentially required, the next step for the TSM is to determine the current least loaded path. It does this by first retrieving the LLR path choice candidates, for this node pair, from the routing data base memory. These candidates are typically a subset of the one- and two-link paths between the two nodes. In the simulations we find that discounting the number of idle circuits by a small fraction of the trunk group size protects large groups from being selected by a disproportionately large number of node pairs as a via-path candidate when these large groups have temporarily idle trunks. Large groups tend to have short periods with a relatively large number of idle trunks, and overselection of those groups for via traffic during these short periods can be detrimental to the direct traffic routed on the large groups. The simulation studies indicate that discounting the number of idle trunks by N/36 trunks, where N is the trunk group size, works reasonably well in protecting the direct traffic on the large groups. (Groups with fewer than 36 trunks are not affected). Using the discounted idle trunk values, the TSM then determines the number of idle circuits on each path choice and selects the path having the greatest number of idle circuits (which we call the least loaded path). This discounting procedure is applied only in the TSM path ordering logic and is not applied in the SPC switch path selection logic.

If the new least loaded path is the same as the old path or if there are no idle circuits on any candidate path, then a routing update is not sent to the SPC switch because there is no advantage in doing so. Simulations of this logic in combination with the idle circuit thresholds described above predict that only about 10-15 percent of node pairs need routing updates returned to the SPC switch, in the busy hour, for an update interval T of five seconds. This TSM logic therefore tends to hold the use of CCS and SPC switch resources down to reasonable levels.

When the least loaded path does change, then a message is sent to the SPC switch changing the contents of the routing sequence in the SPC switch. As an example, consider case 2 in Figure 7. In the figure the idle trunk quantities are the quantities discounted by N/36 trunks. The current route to switch B has a maximum of four path choices, of which three are LLR path candidates - A-C-B, A-D-B, and A-B - in addition to the two-link, first-choice path A-E-B determined by the UA. The TSM first determines that the second LLR path candidate, A-D-B, has become the least-loaded path, and then the TSM sends LLR path choices one and two to the SPC switch in the order A-D-B and A-C-B. Path choice A-B is not sent to the SPC switch since it remains the third LLR path choice. We call this a "push-down" logic since the new least loaded path replaces the old least loaded path, and the remaining path choices are pushed down one slot. The SPC switch replaces its current LLR path choices one and two with the new paths and leaves the current third LLR path choice unchanged. Hence this logic places the least loaded path in the second-choice position but does not necessarily leave the remaining paths in the order of their available capacity. According to the simulation results, for an update interval T of five seconds, the TSM transmits an average of about four entries in the SPC path sequence with each routing update.

With this implementation of TSMR, the SPC switch does not require any time-varying routing capability of its own, as now implemented for DNHR [6]. All such dynamic routing capabilities are controlled by the TSM, which changes the first-choice path every LSP and also controls the LLR paths in a fully dynamic manner. The SPC switch, however, is the only place where individual trunks are selected and assigned to a particular call. The SPC switch sets up all calls over the selected paths using the present DNHR two-link routing procedure with crankback. The number of crankbacks, however, is greatly reduced by the use of TSMR.

3.3 Status and Routing Update Interval T

Four 10-day simulations were made with values of the status and routing update interval, T, equal to 2, 5, 10, and 30 seconds. For these four values of T, Figure 8 compares the TSMR results on the basis of three performance measures. The first measure is average network blocking, which is the 10-day total number of blocked calls divided by the 10-day total number of originating attempts; the second measure is average number of crankbacks per originating attempt, which is the 10-day total number of simulated crankbacks divided by the 10-day total number of originating attempts; and the third measure is the TSM workload, which is the daily total number of least loaded path searches averaged over the 10 days. We can see from the results that the status and routing update interval should be as short as possible within the limitations of SPC switch and TSM processing if we wish to maximize completions and minimize crankbacks.

There is approximately a two percent penalty in SPC switch capacity due to the real time needed to process crankback messages associated with the implementation of DNHR in the 4ESS switch [3]. If we apply this penalty to the current forecast of switches in the mid-1990's, and if we assume that the switch cost penalty for the TSMR strategy can be allocated in direct proportion to the total number of crankbacks generated, then we obtain the crankback cost as a function of T, which is shown in Figure 9. Also shown in Figure 9 is the total cost of processors needed to support the TSM processing level shown in Figure 8, under the assumption that two processors are needed to support a five-second update interval for a 180-node TSMR network. Here we have assumed that each processor costs $100,000. As shown in Figure 9, no fewer than four processors are required since the TSM is duplicated, because of reliability considerations, as are signal transfer points in the CCS network, and at least two processors are required at each TSM, also for reliability purposes. The results shown in Figure 9 suggest that an update interval T in the range of two to eight seconds will minimize

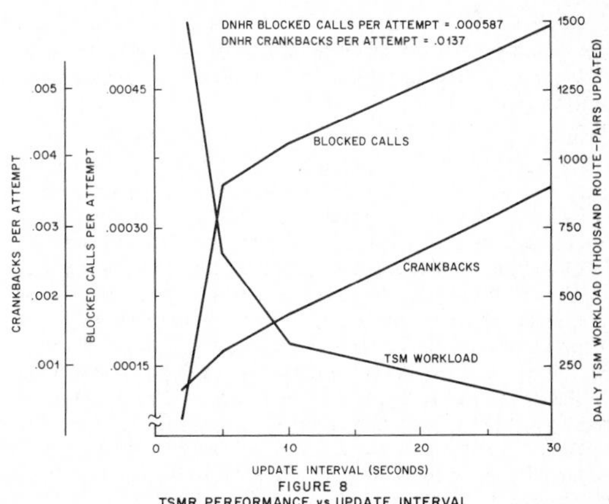

FIGURE 8
TSMR PERFORMANCE vs UPDATE INTERVAL

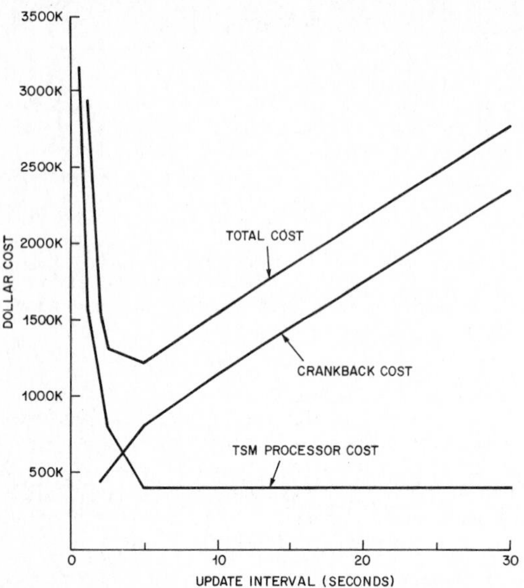

FIGURE 9 PROCESSING & CRANKBACK COST vs UPDATE INTERVAL

the total cost. The tradeoff analysis pictured in Figure 9 assumes that the SPC switch real-time processing load required in addition to the crankback load is not a strong function of the update interval T. If future investigations show that this is not the case, then the optimum value of T will need to be reevaluated.

3.4 Network Congestion Control Strategy for TSMR

In the face of network overloads, failures, or other causes of network congestion, the TSM needs to adjust its routing strategy to best accommodate such conditions. Here we investigate the use of automatic trunk reservation (ATR) controls, busy path removal (BPR) controls, and extended routing logic (ERL) controls that could be implemented by the TSM to help alleviate network congestion. ATR is used to protect the direct trunk group of overloaded node pairs from excess overflow from other node pairs. For traffic subjected to trunk reservation, access to trunks on the direct trunk group is allowed only if the number of idle trunks in the group is greater than a specified number called the reservation level. BPR controls eliminate paths through congested parts of the network when heavy overload conditions exist. ERL allows overloaded node pairs to search out excess capacity available in the network to complete the excess calls. Our simulation studies have investigated a range of ATR, BPR, and ERL strategies; we present here the strategies that we found to perform best.

ATR is triggered automatically for a node-to-node pair if the average node-to-node blocking over a five minute interval is greater than five percent and at least two calls are blocked during the five-minute interval. If ATR is triggered, then ATR is applied to that node pair for the next five minutes. Once triggered, ATR operates in the following manner. Traffic attempting to alternate route over the direct trunk group of a triggered node pair is subjected to trunk reservation. (We used a reservation level of approximately five percent of the number of trunks in each trunk group). This action protects traffic on the direct trunk group of the triggered node pair (if it exists) from interference from traffic on other node pairs. Reservation is applied at both the TSM and the SPC switch. Reservation at the TSM is implemented by having the TSM subtract the number of trunks reserved from the number idle in the process of determining the least loaded path. In no case is trunk reservation applied at the TSM or SPC switch for a one-link (direct) path.

BPR controls are triggered on total network blocking, and, once triggered, BPR removes all busy two-link paths from all route sequences. That is, when the average network blocking over a five minute period exceeds three percent, those two-link paths (excluding the first path) that have zero free circuits are removed from the SPC switch route patterns for the next five-second interval. This strategy results in a uniformly applied restrictive control on the network; all node pairs are triggered at once and are prevented from routing calls through congested parts of the network. BPR controls are also used when an SPC switch signals the TSM that it is in congestion or has failed. In the case of SPC switch congestion, the TSM removes the node as a via point in all routing patterns, and in the case of SPC switch failure the TSM removes both direct and via routing to the failed node.

With the ERL logic, extended searches for idle capacity are triggered for a node-to-node pair whenever the blocking for the pair exceeds one percent over a five-minute period. When the total network blocking threshold used for BPR is triggered, however, then the ERL logic is used only when the node-to-node blocking over a five-minute interval exceeds five percent, with a minimum of two blocked calls. When ERL is triggered, the search for the least loaded path for the triggered node pair is extended to include a larger set of candidate paths.

We simulated the ATR, BPR, and ERL controls over the three hour morning busy period for an average daily load, as well as for 10, 20, and 30 percent general overloads in which each node-to-node traffic load was increased by the overload percentage. We also considered a focused overload on the White Plains switch in which each load to White Plains was increased by a factor of three. Finally, we considered two failure situations which involved a Dallas-Wayne (DLLS-WAYN) trunk group failure and an Anaheim-White Plains (ANHM-WHPL) trunk group failure. The results are given in Table 1, in which the measures used reflect averages over the

TABLE 1

PERFORMANCE COMPARISON OF TSMR AND DNHR

ROUTING	*% OVERLOAD OR FAILURE*	*AVG BLKG*	*99th PERCENTILE BLKG*	*CRKBS PER CALL*
DNHR	0	.00044	.00215	.01131
TSMR	0	.00002	.00010	.00257
DNHR	10	.01382	.03330	.03462
TSMR	10	.00479	.01230	.00931
DNHR	20	.05753	.11558	.08038
TSMR	20	.03016	.13793	.01855
DNHR	30	.10745	.18454	.12002
TSMR	30	.06524	.22876	.05332
DNHR	FOCUS	.01860	.20361	.04878
TSMR	ON WHPL	.00808	.12265	.02865
DNHR	DLLS-WAYN	.00101	.00806	.01556
TSMR	FAILURE	.00004	.00055	.00314
DNHR	ANHM-WHPL	.00187	.01651	.03012
TSMR	FAILURE	.00001	.00000	.00167

three hours of the simulation: the average network blocking is the total number of blocked attempts in three hours divided by the total number of originating attempts; the 99th percentile blocking is the 99th percentile node pair blocking, where each node pair blocking is averaged over the three simulation hours; and the average number of crankbacks per originating attempt is the total simulated number of crankbacks over the three simulation hours divided by the total number of originating attempts. For comparison, we have included results for the current implementation of DNHR. The results indicate that TSMR (with automatic congestion controls) provides improved average blocking performance and comparable or better 99th percentile blocking performance for all the cases studied. The number of crankback messages is also significantly lower with TSMR.

As another indication of peak load performance, Figure 10 shows the hourly blocking performance for TSMR and the current implementation of DNHR under an average Monday load pattern (these loads are normally the highest loads of the week). These results also demonstrate the ability of TSMR congestion controls to increase network flow in comparison to DNHR. Both DNHR and TSMR performance would be improved further by automatic network management controls present in the SPC switch and in the network management support system.

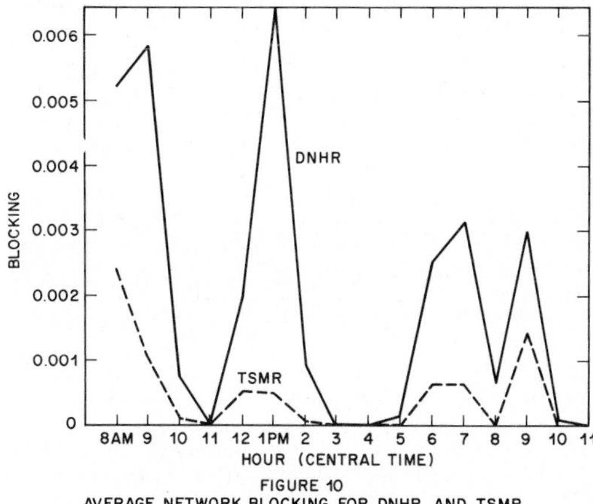

FIGURE 10
AVERAGE NETWORK BLOCKING FOR DNHR AND TSMR (AVERAGE MONDAY LOADS)

4. CONCLUSION

Our conclusions from these studies are that 1) TSMR provides uniformly better blocking performance than the current implementation of DNHR and a significantly reduced number of crankback messages; both network blocking and the number of crankback messages decrease as the status and routing update interval, T, decreases, 2) an initial evaluation of TSM processing cost and crankback cost as a function of the update interval T places the optimum value of T in the range of two to eight seconds; and 3) initial studies show that the TSM can implement effective automatic congestion control strategies.

5. ACKNOWLEDGEMENTS

B. B. Oliver of AT&T Communications first suggested the use of a trunk status map for DNHR. I also benefited greatly from the insights of A. F. Bulfer of AT&T Communications who first suggested investigation of LLR strategies for TSMR. A. H. Kafker of AT&T Bell Laboratories was instrumental in the design of the TSM implementation strategy described in Section 3.2.

REFERENCES

1. Ash, G. R., Cardwell, R. H., Murray, R. P., "Design and Optimization of Networks with Dynamic Routing," BSTJ, Vol. 60, No. 8, October, 1981.

2. Ash, G. R. Kafker, A. H., Krishnan, K. R., "Servicing and Real-Time Control of Networks with Dynamic Routing," BSTJ, Vol. 60, No. 8, October, 1981.

3. Ash, G. R., Kafker, A. H., Krishnan, K. R., "Intercity Dynamic Routing Architecture and Feasibility," Proceedings of the Tenth International Teletraffic Congress, Montreal, Canada, June, 1983.

4. Ash, G. R., forthcoming BLTJ paper on TSMR.

5. Bulfer, A. F., "Blocking and Routing in Two-Stage Concentrators," National Telecommunications Conference, December 1-3, 1975, New Orleans, La.

6. Carroll, J. J., DiCarlo-Cottone, M. J., Kafker, A. H., "4ESS™ Switch Implementation of Dynamic Nonhierarchical Routing," GLOBECOM 1983, November 28 - December 1, 1983, San Diego, Ca.

TELETRAFFIC ISSUES in an Advanced Information Society
ITC-11
Minoru Akiyama (Editor)
Elsevier Science Publishers B.V. (North-Holland)

SOLVING BLOCKING PROBLEMS IN HYBRID NETWORKS

Arik Kashper

AT&T Bell Laboratories,
Holmdel, U.S.A.

Abstract

The recent deployment of a new sophisticated approach to call routing, known as Dynamic Nonhierarchical Routing (DNHR), drastically changed the traditional hierarchical structure of AT&T's toll network and created the need for new traffic network planning methodologies. In this paper, we focus on major aspects of trunk servicing methodology for the new hybrid network environment. In particular, we describe a two-stage optimization procedure that solves trunk group and point-to-point blocking problems in hierarchical and DNHR parts of the network, respectively, in an economical manner.

1. *Introduction*

1.1 *New Network Structure*

With the deployment of the Dynamic Nonhierarchical Routing (DNHR) network, the architecture of the AT&T Communications toll network has changed from its traditional multilevel hierarchy to a hybrid network environment that contains DNHR switching nodes replacing upper levels of the hierarchy and hierarchical switching nodes at the lower levels[1]. The work by Ash, et al, [2] demonstrated that DNHR represents a more flexible and efficient routing scheme that takes full advantage of the expanding Common Channel Interoffice Signaling (CCIS) network and the increased intelligence of the 4ESS™ electronic switching systems. The initial DNHR cutover, in July 1984, included all 10 regional centers, four sectional centers, and two primary centers. By 1987, all 93 of the existing 4ESS switches in the AT&T Communications network will be converted to DNHR switching nodes[1].

At the 10th International Teletraffic Congress held in Montreal in 1983, papers by Ash, et al, [2], Field, [6], Haenschke, et al, [7], and David, et al, [5], described different aspects of the DNHR feasibility, operations, and economic benefits. It was shown that the DNHR method - two-link dynamic routing with crankback - leads to economically attractive, high performance intercity or metropolitan networks and necessitates the development of new more centralized and automated operations support systems.

1.2 *Trunk Network Administration*

Trunk network administration in hierarchical and DNHR networks is composed of two major functions: trunk forecasting and trunk servicing. Trunk forecasting at AT&T Communications network is a biannual process that projects future demands and converts them into economical network designs for several future years. These network designs specify the multiyear schedule of trunk augments/disconnects and corresponding routing changes.

Because of the probabilistic nature of the demand projection, the forecasted network may not provide the desired level of blocking performance for actual traffic loads. Accordingly, a servicing system complements the forecasting process to restore service degradation due to forecast errors [8].

Although trunk forecasting and servicing functions in the hierarchical and DNHR portions of the network are identical, the definitions of acceptable service and the methods to achieve it are quite different. The hierarchical network is forecasted and serviced to guarantee a certain average blocking on final trunk groups, while the DNHR network is forecasted and serviced to achieve a certain level of point-to-point blocking between DNHR switches.

1.3 *Overview*

For the past several years there has been a large effort to improve the methodology of the trunk network administration process. Most of the papers dealt with the trunk forecasting techniques and assumed that all the switching nodes in the network are either hierarchical or DNHR [3,4,9,15]. In contrast, in this paper, we concentrate on several aspects of trunk servicing methodology for the new, hybrid network environment.

More specifically, we will outline a two-stage optimization procedure that combines trunk servicing measures in both parts of the network to restore a desired level of blocking on final trunk groups in the hierarchy and a desired point-to-point blocking level in the DNHR network. We will start by examining the impact of servicing actions in one part of the network on traffic loads offered to the other part. Then, we will formulate a new trunk servicing proposal focusing on several improvements in the hierarchical servicing methodology and the use of trunk reservation in servicing the DNHR part of the network.

2. *Trunk Servicing and Network Interactions*

Servicing systems collect and analyze network measurements on a regular basis to detect blocking problems and, when they exist, develop corrective actions to achieve the desired blocking level in a timely and cost-effective manner. In the hierarchical part of the network, servicing actions are limited to trunk group augmentations [15]. In the DNHR part of the network, however, we can take advantage of the new capabilities of the 4ESS switch and attempt to solve point-to-point blocking problems by changing routing and introducing trunk reservation for problem parcels.

As is illustrated in Figure 1, because the DNHR point-to-point offered loads include overflows from the subtending hierarchy, trunk augmentations in the hierarchy may affect the offered loads and, consequently, the blocking performance of the DNHR part of the network. Similarly, the DNHR servicing measures will change the DNHR carried loads and, therefore, will affect loads offered to the subtending hierarchical switching offices.

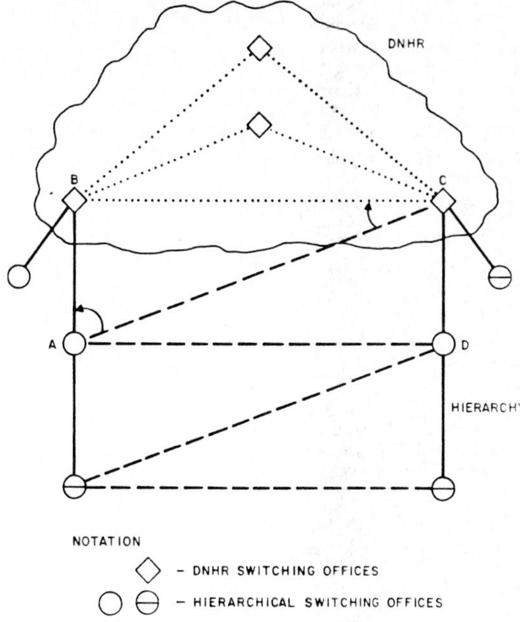

FIGURE 1 HYBRID NETWORK

Our analysis revealed, however, that in the range of engineering interest the impact of the DNHR servicing on the hierarchical part of the network is not significant and may be ignored. Intuitively, this is explained by the fact that a small increase in the carried load between the two DNHR switches generally translates into an increase in the load offered to *all* subtending final groups and, therefore, does not cause a severe service degradation for a particular final group.

Analogously, servicing final groups in the hierarchy may increase point-to-point loads offered to the DNHR part of the network, but, in most cases, will not result in a significant blocking problem for an individual DNHR point-to-point pair. However, if we augment the high-usage group A-C, for example, we will reduce the offered load to the point-to-point pair B-C only. Thus, by augmenting high usage group A-C we may be also solving the DNHR B-C blocking problem.

Consequently, to avoid overtrunking, our servicing procedure will identify blocking problems in the hierarchical part of the network, first, solve them, and then calculate the corresponding change in overflow to the DNHR network. During the second stage of our servicing procedure, we will reevaluate and solve remaining point-to-point blocking problems in the DNHR part of the network.

3. *Hierarchical Trunk Servicing*

3.1 *Background*

In the hierarchical network environment the demand servicing process is based on trunk group measurements collected by the trunk servicing system. To monitor network blocking performance, the statistic $\bar{B} = \frac{1}{n} \sum_{i=1} B_i$ is computed for the busy hour, where B_i represents the fraction of calls blocked in the i-th day of a study period that normally contains 4 consecutive business weeks.

Because of the statistical nature of the demand and relatively small size of the blocking sample, the statistic $\bar{B}$ computed on final groups may deviate from the engineered blocking level, even if the network is correctly sized. Thus, to identify blocking problems one needs to establish servicing thresholds that will account for the volatility of blocking estimates. These thresholds must represent a reasonable compromise between two conflicting objectives: maintaining good service and avoiding unnecessary trunk augmentations.

This problem was addressed in the work by Neal [12] and Szelag [15]. In [12], the distribution for the $\bar{B}$ statistic was obtained. In [15], the false alarm and miss probability concepts were used to convert the $\bar{B}$ distribution into the servicing thresholds. Consequently, in this paper we will concentrate on the solution of blocking problems, assuming that the problems have been identified.

3.2 *Cluster Optimization*

When a statistically significant service problem on the final group has been detected, the group's high blocking can be decreased by adding capacity to the final group and/or reducing the overflow by augmenting subtending high usage groups. In [12], a heuristic servicing algorithm was proposed that achieves the desired blocking level by finding undersized high-usage trunk groups, augmenting them, and, if the problem still exists, resizing the final group.

The algorithm in [12] was developed prior to the introduction of new minimum-cost multiyear trunk forecasting process, which we described at the Tenth ITC [9]. Consequently, in this section we will develop a mathematical model that reflects the new trunk implementation realities and present an algorithm that minimizes the immediate and future costs of solving the final group blocking problem. Note that the current servicing action may affect the cost of planned trunk provisioning activities in the future.

3.2.1 *Notation*

First, we will focus on all the high-usage groups in the network cluster. Note that a network cluster consists of all high-usage groups which originate at a common switching office and overflow to a common final.

To formulate our optimization model for augmenting high-usage groups we need to introduce the following notation:

$T_j(k)$ - number of trunks in service on the subtending group j at forecast period k

$u_j(k)$ - number of planned trunk augments/disconnects on the subtending group j at forecast period k

$d_j(k)$ - the demand in trunks on the subtending group j at forecast period k as specified by the trunk forecasting process [9]

β_j - additional number of trunks on the subtending group j that compensates for one final trunk

δ_j - maximum reduction in the final trunk requirement that can be obtained by augmenting trunk group j

$z(x)$ - deficit in final trunks that we will cover by servicing the subtending high-usage groups

z_j - portion of the deficit (number of final trunks) covered by servicing subtending high-usage group j

M - total number of subtending high-usage

groups in the network cluster

N - number of years in the planning horizon

ρ - discount factor that measures the worth of the next year's dollars in terms of present dollars.

$c_1^{kj}(u_j(k))$ - capital cost defined on Figure 2. On Figure 2, a_1^k and b_1^k represent the per-trunk capital cost and salvage value at forecast period k, respectively.

$c_2^{kj}(u_j(k))$ - labor cost defined on Figure 2. On Figure 2, a_2^k and b_2^k represent the labor costs of connecting and disconnecting one trunk at forecast period k.

$c_3^{kj}(T_j(k)+u_j(k))$ - maintenance cost defined on Figure 2. On Figure 2, a_3^k represents the per-trunk maintenance cost.

Note that $T_j(0)$ represents the number of trunks currently in service, and $u_j(0)$ represents the servicing action to be determined.

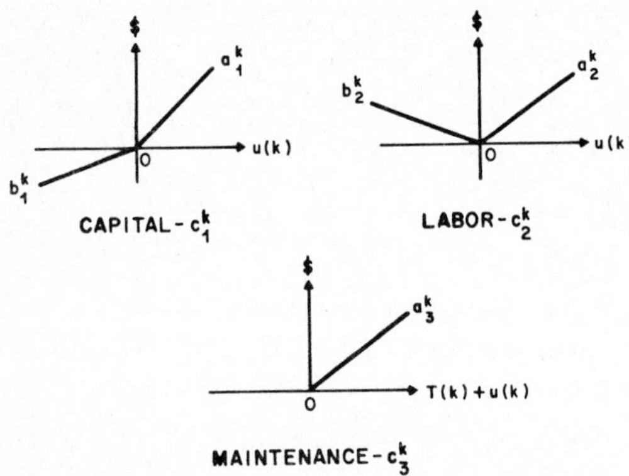

FIGURE 2 TRUNK PROVISIONING COSTS

3.2.2 Mathematical Model

Our goal is to find a servicing policy for the network cluster that will minimize the cost of all trunk rearrangements over the planning horizon

$$\min_x J(x) = F(x) + L(x),$$

where $x(x{=}0,12,24,\ldots)$ is the number of trunks added to the final, $F(x)$ and $L(x)$ are the present worth of all trunk provisioning costs over the planning horizon for the final group and for all subtending high-usage groups respectively.

We would like to note that the method for computing $F(x)$ is provided in [10]. Thus, we can devote out attention to calculating $L(x)$. In particular, we need to find how to cover the remaining deficit in final trunks, $z(x)$, most economically by augmenting the subtending high-usage groups

$$\min L(x) = \min L(z(x)) =$$

$$\min \sum_{k=0}^{N} \sum_{j=1}^{M} \rho^k [c_1^{kj}(u_j(k)) + c_2^{kj}(u_j(k)) + c_3^{kj}(T_j(k)+u_j(k))] \,. \tag{1}$$

We need to minimize $L(x)$ with respect to all present and future rearrangements $u_j(k)$ subject to the following constraints:

1. The total trunk requirement must be sufficient to solve the final blocking problem

$$u_j(0) \geq \beta_j z_j \text{ , and}$$

$$\sum_{j=1}^{M} z_j = z(x) \,. \tag{2}$$

2. The number of trunks in service on the high-usage group j at forecast period k must be greater or equal to the originally forecasted trunk requirement

$$T_j(k) + u_j(k) \geq d_j(k), \qquad k=1,2,..N, \tag{3}$$

where $T_j(k)$ is defined by the recursion $T_j(k+1) = T_j(k) + u_j(k)$.

3. The unknown variables z_j must satisfy feasibility constraints

$$0 \leq z_j \leq \delta_j \,. \tag{4}$$

3.2.3 Servicing Solution

The solution to the optimization problem (1)-(4) exploits the multiyear forecasting algorithm for high-usage groups which we derived in [10]. Indeed, when z_j are fixed, the objective function (1) can be decomposed as follows:

$$\min L(x) = \sum_{j=1}^{M} \min_{u_j} L_j(z_j), \tag{5}$$

where $L_j(z_j)$ represents the present worth of trunk provisioning costs at the subtending group j. Thus, we arrive at a single high-usage group capacity expansion problem: minimize $L_j(z_j)$ subject to constraints

$$T_j(k) + u_j(k) \geq d_j(k), \qquad k=0,1,\ldots N, \tag{6}$$

where $d_j(0) = T_j(0) + \beta_j z_j$.

Therefore, the original trunk servicing problem described by (1)-(4) can be reformulated as follows: find

$$\min_{z_j} \sum_{j=1}^{M} L_j^*(z_j) \tag{7}$$

while solving the final blocking problem

$$\sum_{j=1}^{M} z_j = z(x), \tag{8}$$

where z_j satisfy feasibility constraints, and $L_j^*(z_j) = \min_{u_j} L_j(z_j)$ is computed by a simple, efficient algorithm which is based on the following result.

Under quite general assumptions on the cost parameters, we have shown [10] that the optimal policy, $u_j^*(k)$, has the form

$$u_j^*(k) = \begin{cases} d_j(k) - T_j(k), & \text{if } d_j(k) \geq T_j(k) \\ \min\left[\max_{0 \leq i \leq m}[d_j(k+i) - T_j(k)], 0\right], & \text{if } d_j(k) < T_j(k), \end{cases}$$

where m is the largest integer such that

$$b_1^{k,j} - b_2^{k,j} + \sum_{i=0}^{m-1} \rho^i a_3^{k+i,j} < \rho^m (a_1^{k+m,j} + a_2^{k+m,j}).$$

Now, we can use a standard dynamic programming approach for finding the minimum of (1), i.e.,

$$f_i(y_i) = \min_{z_i}[L_i^*(z_i) + f_{i-1}(y_i - z_i)], \quad i=1,\ldots,M\,, \tag{9}$$

where $f_o(\cdot) = 0$, $0 \leq z_i \leq y_i$, *and* $y_M = z(x)$.

The solution to the dynamic program (9) provides the desired value of L(x) and, consequently, leads to the determination of the optimal (in the sense of minimizing J(x)) trunk servicing policy for the network cluster. We would like to note that an efficient computer implementation of this procedure can be obtained by embedding modularity constraints into the problem formulation.

3.3 Summary of Servicing

To summarize our discussion, in this section we formulate the major steps of the proposed trunk servicing algorithm.

Identify hierarchical blocking problems

The blocking performance of hierarchical networks is determined by the blocking on final trunk groups. Thus, we start by analyzing trunk group data to detect final trunk groups on which the engineered blocking level is violated. If the violation is statistically significant [15], the final group becomes a candidate for servicing.

Define minimum cost cluster augmentations

Next, we use the approach described in Section 3.2.3 to solve the cluster optimization problem (for the candidate final) and determine how many trunks should be added to the final trunk group and how many to its subtending high-usage groups.

Calculate overflow impact of the hierarchical servicing

As we discussed in Section 3.1, when all the hierarchical blocking problems are solved, we then determine the reduction in overflow contributions of the serviced hierarchical part of the network to the point-to-point loads offered to the DNHR part.

4. DNHR Servicing Process

4.1 Background

4.1.1 DNHR Overview

As we have shown in Figure 1, calls overflowing the hierarchical part of the network will enter the DNHR part at a unique entry point and will leave the DNHR part at a unique exit point. Within the DNHR network a call is routed according to a specified sequence of paths. A path may contain either one or two links. If the second link in a two-link connection is busy, the via switch sends a "crankback" CCIS message to the DNHR entry node and then the next path in the sequence is tried. The dynamic nature of DNHR is accomplished by changing the preplanned sequence of paths up to ten times during the day. The ability to adjust the routing allows us to benefit from the noncoincidence of traffic loads and leads to a more efficient network design [3].

4.1.2 Servicing Function

To monitor point-to-point blocking in the DNHR environment, 4ESS switching offices collect point-to-point traffic data. The point-to-point blocking performance is analyzed weekly by the DNHR servicing system to determine which parcels are not receiving adequate service. Similarly to the hierarchical environment, the need for demand servicing is identified if the service violation is statistically significant. If servicing is needed, the DNHR servicing system suggests a combination of trunk augments and routing changes to restore the desired level of point-to-point blocking. In [4], the initial DNHR servicing algorithm was proposed. The algorithm is based on "flow optimization," "engineering," and "blocking correction" routines [3] and results in a servicing solution that includes both trunk augments and routing changes.

It is important to note that in practice trunk augmentations on the demand servicing basis are often expensive and can be implemented only with a minimum of 2 weeks delay. Moreover, there is always a reluctance to add trunks at the end of the busy season. The delay in servicing may result in significant revenue loss as well as customer aggravation. In the DNHR environment, however, we can improve network blocking performance in a timely manner by avoiding trunk augments to the extent possible.

There are two major ways to achieve this goal: *change traffic routings* to maximize the use of existing capacities and *introduce trunk reservation* to provide better service for the high blocking point-to-point pairs. The problem of minimizing the maximum point-to-point blocking in the network by routing changes only was considered by Nassar, [11]. She showed that the initial servicing algorithm [4] does not lead to a routing-only solution of blocking problems and developed a new, heuristic procedure that adjusts the engineered network routing for the difference in forecasted and actual loads. Our task in the section 4.2 will be the investigation of the use of trunk reservation in DNHR servicing.

4.2 The Use of Trunk Reservation

4.2.1 Trunk Reservation Background

Consider a trunk group that handles high-priority and low-priority traffic streams. A high-priority call will be cleared if on arrival there are no idle trunks in the group, whereas a low-priority call will be cleared if on arrival there are less than R+1 idle trunks in the group. This service discipline is called Trunk Reservation (TR) and R is called the reservation level.

Trunk reservation is known as an effective network management technique that preserves the performance of either the hierarchical or DNHR networks under overload conditions [7,14]. We will show that the TR method can be also used to solve some of the DNHR point-to-point blocking problems.

4.2.2 Parcel Blocking with TR

To apply the TR method in servicing, we need to know how to compute the blockings for high and low-priority traffic streams when TR is employed. Under the reasonable assumption that these streams can be modeled by the Interrupted Poisson Processes, the blockings were derived analytically by Songhurst [14]. To simplify the computations, we present less accurate but simpler approximation for the blockings that exploits the decomposition principle introduced in [13].

We assume that the high-priority traffic stream is described by the pair (a_1,z_1), where a_1 is the offered load and z_1 is the peakedness factor $(z_1>1)$; the low-priority traffic stream is random (Poisson) with intensity a_2 (see Figure 3). Applying the decomposition principle, we split the peaked traffic stream into a Poisson component a_1z_1 and a zero-variance component $(a_1z_1 - a_1, 0)$. This is equivalent to the substitution

$$(a_1, v_1) \sim (v_1, v_1) - (v_1 - a_1, 0) \quad ,$$

where $v_1 = a_1z_1$.

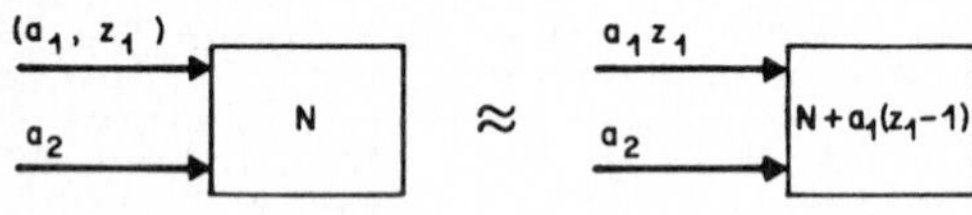

(a_1, z_1) - HIGH PRIORITY PEAKED TRAFFIC

a_2 - LOW PRIORITY POISSON TRAFFIC

a_1z_1 - REMAINING HIGH PRIORITY POISSON TRAFFIC OFFERED TO AN "EQUIVALENT" TRUNK GROUP

FIGURE 3 DECOMPOSITION PRINCIPLE

Dedicating $a_1(z_1 - 1)$ trunks to carry the zero-variance stream component, we arrive at the congestion model of Figure 3. Now, we have two random streams with parameters a_1z_1 and a_2 offered to a trunk group of $S = N + a_1(z_1 - 1)$ trunks. In this case, the probability of blocking for the high and low-priority streams is given by

$$b_1 = \frac{(a_1z_1 + a_2)^{S-R} \, (a_1z_1)^R}{S!} P_o, \quad \text{and}$$

$$b_2 = P_o \, (a_1z_1 + a_2)^{S-R} \sum_{j=S-R}^{S} \frac{(a_1z_1)^{j-S+R}}{j!} \quad ,$$

where P_o is defined by

$$P_0 = \left[\sum_{j=0}^{S-R} \frac{(a_1z_1 + a_2)^j}{j!} + (a_1z_1 + a_2)^{S-R} \sum_{j=S-R+1}^{S} \frac{(a_1z_1)^{j-S+R}}{j!} \right]^{-1} .$$

Our numerical experience shows that for small values of R these formulas provide a satisfactory approximation for parcel blockings. Thus, we can now estimate point-to-point blocking changes when TR is used on one of the trunk groups. Note that the DNHR point-to-point blocking probability is computed by the product of individual path blockings.

4.2.3 Solving Blocking Problems

Now we will outline an heuristic procedure that attempts to minimize the maximum point-to-point blocking in the DNHR network by introducing TR for problem pairs. To preserve high utilization of all DNHR trunk groups we will consider only three trunk reservation levels R=1, 2, 3. These levels proved to be sufficient to significantly change the blocking for problem parcels.

To specify how to use TR in servicing, we use the following iterative procedure:

(i) Assign (upgrade, if necessary) the level of acceptable point-to-point blocking.

(ii) Compute point-to-point blockings for all node pairs and identify the most severe servicing problem.

(iii) Identify trunk groups on which TR should be applied. Determine the trunk group and the reservation level R to obtain the maximum reduction in the point-to-point blocking for the chosen node pair subject to the constraint: point-to-point blocking probabilities for all other pairs should remain less than the level of acceptable blocking defined in (i).

In practical situations, usually, there are many point-to-point pairs with virtually zero blocking probability and few high blocking pairs. According to our numerical experience, in these cases, the TR method can be effectively utilized to solve the DNHR point-to-point blocking problems. In contrast to network management applications, our iterative procedure frequently results in introducing TR to protect overflow rather than direct traffic. On the last choice via-path, for example, where the contribution of the problem parcel is small relative to the other traffic, the application of TR tends to be feasible and beneficial.

4.2.4 DNHR Servicing Overview

To summarize our discussion, in this section we formulate the major steps in the DNHR servicing procedure.

Identify DNHR servicing problems.

We use switch data to detect point-to-point service violations in the DNHR part of the network. We adjust the violations for the effect of the hierarchical servicing. If the resulting blocking problems are statistically significant, the corresponding point-to-point pairs become candidates for servicing.

Relieve blocking problems by routing changes

To reduce the number of blocking problems, we use the procedure described in [11] to change the traffic routing. To minimize the cost of servicing and to avoid the delay, trunk augments are not allowed at this step of the servicing procedure.

Improve network blocking performance by trunk reservation.

For a problem pair, we determine on which trunk groups TR control should be imposed. These trunk groups are selected so as to reduce the blocking for the problem pair with no significant service degradation for other traffic.

Solve blocking problems by trunk augmentation

Finally, as the last resort, trunk group additions [3,4] can be used to bring the network blocking performance to the desired level.

REFERENCES

[1] G. R. Ash and V. S. Mummert, "AT&T Carves New Routes in Its Nationwide Network," Record, AT&T Bell Laboratories, August, 1984, pp. 18-22.

[2] G. R. Ash, A. H. Kafker, and K. R. Krishnan, "Intercity Dynamic Routing Architecture and Feasibility," Proc. 10th Intr. Teletraffic Congress, Montreal, 1983.

[3] G. R. Ash, R. H. Cardwell, and R. P. Murray, "Design and Optimization of Networks with Dynamic Routing," Bell Syst. Tech. J., Vol. 60, October, 1981, pp. 1787-1820.

[4] G. R. Ash, A. H. Kafker, and K. R. Krishnan, "Servicing and Real-Time Control of Networks with Dynamic Routing," Bell Syst. Tech. J., Vol. 60, October, 1981, pp. 1821-1845.

[5] A. J. David and N. Farber, "The Switch Planning System for the Dynamic Nonhierarchical Routing Network," Proc. 10th Int. Teletraffic Congress, Montreal, 1983.

[6] F. A. Field, "The Benefits of Dynamic Nonhierarchical Routing in Metropolitan Traffic Networks," Proc. 10th Int. Teletraffic Congress, Montreal, 1983.

[7] D. G. Haenschke, D. A. Kettler, and E. Oberer, "DNHR: A New SPC/CCIS Network Management Challenge," Proc. 10th Int. Teletraffic Congress, Montreal, 1983.

[8] W. S. Hayward and J. P. Moreland, "Theoretical and Engineering Foundations," Bell Syst. Tech. J., Vol. 62, No. 7, September, 1983, pp. 2183-2207.

[9] A. Kashper, C. D. Pack, and G. C. Varvaloucas, "Minimum-Cost Multiyear Trunk Provisioning," Proc. 10th Int. Teletraffic Congress, Montreal, 1983.

[10] A. N. Kashper and G. C. Varvaloucas, "Trunk Implementation Plan for Hierarchical Networks," AT&T Bell Lab. Tech. J., Vol. 63, No. 1, January, 1984, pp. 57-88.

[11] M. Nassar, Work to be presented at TIMS/ORSA National Meeting, 1985.

[12] S. R. Neal, "Blocking Distributions for Trunk Network Administration," Bell Syst. Tech. J., Vol. 59, No. 6, July, 1980, pp. 829-844.

[13] B. Sanders, W. H. Haemers, and R. Wilcke, "Simple Approximation Techniques for Congestion Functions for Smooth and Peaked Traffic," Proc. 10th Int. Teletraffic Congress Montreal, 1983.

[14] D. J. Songhurst, Paper on Trunk Reservation at the Networks Planning Symposium, Paris, 1980, pp. 214-220.

[15] C. R. Szelag, "Trunk Demand Servicing in the Presence of Measurement Uncertainty," Bell Syst. Tech. J., Vol. 59, No. 6, July, 1980, pp. 845-860.

TELETRAFFIC ISSUES in an Advanced Information Society
ITC-11
Minoru Akiyama (Editor)
Elsevier Science Publishers B.V. (North-Holland)

APPROXIMATE ANALYSIS OF COMMON MEMORY ACCESS CONTENTION IN A MULTIPROCESSOR CONTROLLED SWITCHING SYSTEM

Shuichi SUMITA

Musashino Electrical Communication Laboratory, NTT
Tokyo, Japan

ABSTRACT

This paper presents a queueing network model for common memory access contention and access contention to common data that cannot be referenced by more than one processor in a multiprocessor controlled switching system. The model takes into account the inter-dependence between these types of access contention, which is important for evaluating the performance of a single-bus multiprocessor system. Although this inter-dependence is incorporated into the queueing network model, an exact analysis is difficult. Therefore, an approximate analysis method is proposed based on decomposing the model into two submodels. Approximate analysis results and simulation results are compared and it is shown that the approximation method provides good approximate values. This queueing network model is useful for estimating the call processing capacity of a multiprocessor controlled switching system.

1. INTRODUCTION

Access contention for shared resources presents a major problem in estimating the call processing capacity of a multiprocessor controlled switching system (or simply multiprocessor system), because call processing capacity is reduced due to access contention for shared resources. There are two types of shared resources: hardware resources, such as the common memory, and software resources, such as the common data that cannot be referenced or modified simultaneously by more than one processor. This paper considers access contention to common memory (from now on, referred to as common memory access contention or simply CMC) as well as access contention to common data that cannot be referenced or modified simultaneously by more than one processor (from now on, referred to as common data access contention or simply CDC) in a multiprocessor controlled switching system.

For a multiprocessor system connected by a single common bus (see Fig.1), CMC and CDC cannot be treated independently, because the access rate from a processor to the common memory is affected by CDC and the time during which common data is locked is also affected by CMC. To accurately evaluate the call processing capacity of a multiprocessor system, it is necessary to investigate this inter-dependence between CMC and CDC. A number of papers have presented and analyzed mathematical models for CMC and CDC. However, the inter-dependence between CMC and CDC has been largely neglected. This paper presents a simple queueing network model taking into account the inter-dependence between CMC and CDC.

Several authors have developed discrete-time Markov chain models in evaluating CMC in multiprocessor systems that have a crossbar switch connecting processors and common memory modules [1,2,3,4,5,6]. In this model, as the number of processors or common memory modules increases, the state space of the Markov chain grows larger and more computational time is required to obtain numerical values of the stationary probabilities of the Markov chain. Hence, several methods of approximate analysis have also been proposed [3],[7]. However, when this modeling technique is applied to a system in which processors and common memory modules operate asynchronously or to a system in which each processor has its own local memory and accesses common memory modules at a rate less than that of local memory access, the resulting Markov chain becomes too complex to analyze. To facilitate analysis of these systems, a continuous-time Markov model or a queueing network model has been proposed [8]. This modeling technique has also been used for evaluating CMC in bus-connected multiprocessor systems [9,10,11]. Marsan and Gerla [9] have analyzed CMC in multiple-bus multiprocessor systems. Marsan et al. [10] have developed Markov models for analyzing performance and for comparing several single-bus multiprocessor architectures. Marsan et al. [11] have also investigated CMC in a single-bus multiprocessor system using different modeling approaches (continuous-time Markov chains, queueing networks and stochastic Petri nets). In addition to the above reports, numerous papers related to CMC have already been published [12,13,14].

In addition to CMC, access contention to software resources in computer systems, such as serially reusable programs or common data, have been investigated by several authors. Queueing delay caused by software lockout has been evaluated by a finite source queueing model [8], [15],[16]. More recently, Gilbert [17] and Hoffman and Schmutz [18] have presented queueing network models for evaluating spin locks and suspend locks, respectively. King et al. [19] have evaluated the effects of serial programs on the performance of computer systems. Smith and Browne [20] and Thomasian [21] have emphasized a queueing network modeling for access contention to software resources.

Section 2 describes the multiprocessor system considered in this paper and presents a queueing network model taking into account the inter-dependence between CMC and CDC. Section 3 presents an approximation method for the queueing network model described in section 2. Section 4 gives numerical examples and validates the accuracy of approximate analysis method as compared with simulation results.

2. MULTIPROCESSOR SYSTEM AND QUEUEING NETWORK MODEL

The multiprocessor system considered in this paper is shown in Fig.1. Processors $P_1 \sim P_N$ and a common memory are connected by a single common bus. Each processor has its own individual memory, which can be accessed only by the processor. Programs that are frequently used are stored in the individual memories.

Common memory access contention (CMC) occurs when more than one processor attempt to access the common memory. When the common memory is accessed, the bus is occupied. If a processor attempts to access the common memory when the bus is occupied, that processor will be delayed. A bus-control unit (omitted in Fig.1), which arbitrates CMC, serves delayed requests for accessing the common memory on a first-come first-served basis (FCFS). Thus, instruction execution time in processors becomes longer than when no CMC occurs.

In addition, common data access contention (CDC) occurs. In Fig.1, processors perform call processing functions, such as detection of call origination, translation and task selection, using common data in the common memory. There are several types of common data that must not be referenced or modified simultaneously by more than one processor: the speech path map, call state data etc. If a processor is allowed to use common data while another processor performs call processing using the same common data, the common data may be modified or updated simultaneously by more than one processor. This will cause the normal call processing failure. To ensure normal call processing, it is necessary to serialize requests for accessing the same common data. A Test and Set flag (T&S flag) that indicates whether the data is being accessed or not is given each common data. A Test and Set instruction (T&S instruction) is usually used to ask whether the data is being accessed or not. If the common data is being accessed, the processor executes the T&S instruction repeatedly until it succeeds in accessing the common data. This mechanism is called the spin lock. Thus, the time during which the T&S instruction is executed repeatedly is wasted.

Here, note that there is the inter-dependence between CMC and CDC by the following reasons:

(A) the common memory access rate for each processor cannot be determined independently of CDC, because the number of unsuccessful T&S instructions executed for accessing common data results in an increase in common memory access rate;

(B) the time during which common data is used by a processor (from now on, referred to as common data service time), that is, the time between a set operation and a release operation for a T&S flag, cannot be determined independently of CMC, because total execution time for the instructions executed between the set and release operation for the T&S flag increases as a result of CMC.

This paper analyzes this inter-dependence between CMC and CDC.

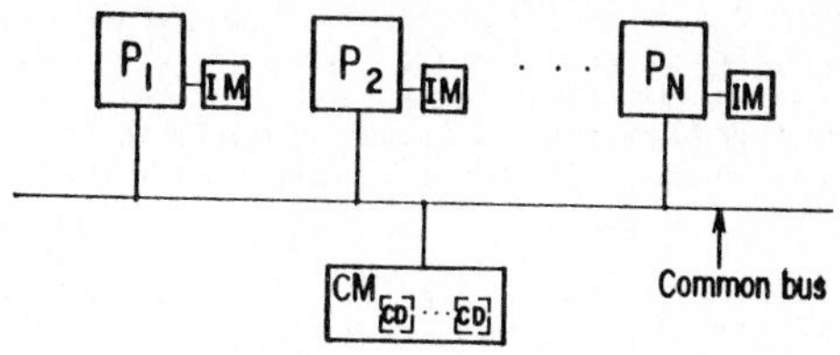

Fig.1. Multiprocessor system.

The following assumptions are made for processors, a bus and common data.

(1) Each processor usually decodes and executes instructions in its individual memory and sometimes accesses the common memory.

(2) When a processor attempts to access the common memory and at the same time the bus is free, the processor holds the bus to access the common memory by some constant time h_b. In other words, the bus holding time is always h_b. Upon the completion of common memory accessing, the bus is immediately released.

(3) The number of common memory accesses per unit time varies with the number of calls offered to a processor per unit time. It is assumed that when the effect of CDC is excluded from consideration, the mean number of common memory accesses per unit time, ν, can be calculated by the following equation of the number of calls offered to a processor per unit time, x (see [22],[23]):

$$\nu = (D_0 + D_1 x)\alpha, \quad (1)$$

where

D_0: the number of instructions executed per unit time for tasks other than call processing;

D_1: the number of instructions executed for completing call processing from the origination to the termination of one call.

α: the mean number of common memory accesses during one instruction execution.

When the effect of CDC is considered, the mean number of common memory accesses per unit time is calculated by:

$$\nu = (D_0 + D_1 x)\alpha + \Delta\nu , \quad (2)$$

where

$\Delta\nu$: the increase in the number of common memory accesses per unit time caused by CDC.

A method for calculating $\Delta\nu$ will be given in section 3.

(4) The intervals between an instant of common memory access completion and an instant of the next common memory access are independent and exponentially distributed random variables with a mean of $1/\lambda_b$ for each processor. Note that the quantity λ_b is defined so as to include the increase in the number of common memory accesses per unit time caused by CDC. The quantity λ_b is referred to as common memory access rate and is calculated using ν . A method for calculating λ_b will be discussed in section 3.

(5) There are K types of common data that cannot be referenced or modified simultaneously by more than one processor. It is assumed that service times of common data i are independent and exponentially distributed random variables

with a mean of $1/\mu_i$. The mean service time of common data i is assumed to be calculated by the mean instruction execution time of a processor, τ, multiplied by the mean number of instructions executed between a set operation and a release operation for T&S flag of common data i, d_i:

$$1/\mu_i = \tau d_i . \qquad (3)$$

τ is defined so as to include the time increase caused by CMC. τ is given by:

$$\tau = \tau_0 + \Delta\tau , \qquad (4)$$

where

τ_0: mean instruction execution time when the effect of CMC is excluded from consideration;

$\Delta\tau$: the increase in mean instruction execution time caused by CMC.

A method for calculating $\Delta\tau$ will be given in section 3.

The number of times that common data i is accessed for completing call processing from the origination to the termination of one call is ℓ_i.

(6) The intervals between an instant of common data access completion and an instant of the next common data access are independent and exponentially distributed random variables with a mean of $1/\lambda_d$. λ_d is referred to as common data access rate. When an access to common data occurs, the common data i is selected with probability q_i. Using ℓ_k defined in (5), q_i is given by:

$$q_i = \ell_i / \sum_{k=1}^{k=K} \ell_k . \qquad (5)$$

The queueing network model taking into account the inter-dependence between CMC and CDC is shown in Fig.2.

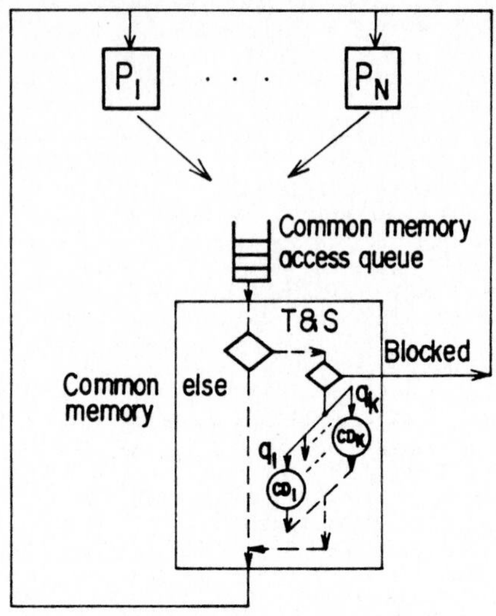

Fig.2. Queueing network model taking into account the inter-dependence between CMC and CDC.

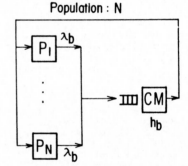

3. APPROXIMATE ANALYSIS

Although the inter-dependence between CMC and CDC is incorporated into the queueing network model in Fig.2, an exact analysis is difficult, because common memory access rate λ_b and common data access rate λ_d cannot be determined independently. Therefore, an approximate analysis method is presented in this section.

The queueing network model in Fig.2 is decomposed into two submodels: the common memory access contention model (CMC model) shown in Fig.3 and the common data access contention model (CDC model) shown in Fig.4. After decomposing the model in Fig.2 into two submodels, each submodel can be analyzed independently. However, to analyze one submodel, an analytic result of the other submodel is required. Thus, an iteration method is used.

3.1 Common memory access contention model (CMC model)

From assumptions concerning the bus described in Section 2, the CMC model is a finite source queueing model M/D/1//N (shown in Fig.3), where N is the number of processors. Using the results of the finite source queueing model M/G/1//N in Takács [24], a mean waiting time for accessing the common memory is given by

$$W_b = (N-1-(1-\pi_0)/(\lambda_b h_b))h_b , \qquad (6)$$

where

$$\pi_0^{-1} = \sum_{i=0}^{i=N-1} \binom{N-1}{i} (1/C_i), \quad C_0 = 1,$$

Population : N

P1 λb
.
.
.
CM
hb
PN λb

Pi : Processor CM : Common memory
λb : Common memory access rate
hb : bus holding time

Fig.3. Common memory access contention model.

$$C_i = \prod_{j=1}^{i} \exp(-j\lambda_b h_b)/(1-\exp(-j\lambda_b h_b)), \quad i=1,2,\ldots,(N-1).$$

To calculate W_b using Eq.(6), it is necessary to determine λ_b according to the number of calls offered to a processor per unit time, x. To obtain λ_b when the number of calls offered to a processor per unit time is x, the following relation can be used:

$$1/\lambda_b = 1/\nu - (h_b + W_b). \qquad (7)$$

Equation (7) is derived as follows: the number of accesses to the common memory per unit time is given by ν and the mean time between successive common memory accesses is given by $(1/\lambda_b+W_b+h_b)$; the fact that the product of both terms is equal to 1 leads to Eq.(7). Here, note that the number of common memory accesses per unit time, ν, is given by: $(D_0+D_1x)\alpha+\Delta\nu$. A variant of Eq.(7) was first derived by Khintchine. This was pointed out in Takács [24].

Thus, W_b and λ_b must be calculated so as to satisfy Eqs.(6) and (7). To do this, the following iteration algorithm is used:

Step 1 Set initial value $W_b^{(0)}=0$ and $k=0$;

Step 2 Compute $\lambda_b^{(k+1)}$, using Eq.(7) in which W_b is replaced by $W_b^{(k)}$;

Step 3 Compute $W_b^{(k+1)}$, using Eq.(6);

Step 4 If $|W_b^{(k+1)} - W_b^{(k)}| < \varepsilon$, where ε has the pre-specified and sufficiently small positive value, then terminate the algorithm. Otherwise, return to Step 2.

In the above algorithm, it is assumed that the increase in the number of common memory accesses per unit time, $\Delta\nu$, is known and fixed. To determine $\Delta\nu$, the analytic results of a CDC model described in the next subsection are used.

Using the mean waiting time W_b, the increase in mean instruction execution time caused by CMC, $\Delta\tau$, is given by

$$\Delta\tau = \alpha W_b, \tag{8}$$

and τ is given by

$$\tau = \tau_0 + \alpha W_b. \tag{9}$$

3.2 Common data access contention model (CDC model)

From assumptions concerning common data in section 2, the CDC model is a closed queueing network model with a single customer class having N customers and (K+1) nodes, where node 0 is an infinite server node and node 1~K are FCFS nodes.

Let $P(m_0,m_1,\ldots,m_K)$ be the state probability that there are m_i customers in node i, i.e, there are m_i processors accessing common data i or waiting for access to common data i and m_0 customers in node 0, i.e. m_0 processors issuing no requests for common data access. This queueing network model has a product form solution [25] and $P(m_0,m_1,\ldots,m_K)$ is given by

$$P(m_0,m_1,\ldots,m_K)=(1/G)(1/m_0!)(1/\lambda_d)^{m_0}\prod_{i=1}^{K}(q_i/\mu_i)^{m_i}, \tag{10}$$

where G is the normalizing constant. To compute the normalizing constant G, the following recursive relation developed by Buzen [26] can be used;

$$\begin{aligned} &G(0,i+1)=1, \quad i=0,\ldots,K; \\ &G(N,1)= (1/\lambda_d)^N/N!; \\ &G(n,i+1)=G(n,i)+(q_i/\mu_i)G(n-1,i+1), \\ &\qquad n=1,..,N,\ i=1,..,K. \end{aligned} \tag{11}$$

where G(n,i+1) is defined as the normalizing constant when the number of processors is n and the number of common data is i. G(N,K+1) is the normalizing constant G to be calculated. Using G(n,i), the mean queue length of requests for accessing common data i, which is defined so as to include the processor currently accessing common data i, L_i, is given by:

$$\begin{aligned} L_i &= \sum_{j=1}^{j=N} P(m_i \geq j) \\ &= \sum_{j=1}^{j=N} (q_i/\mu_i)^j G(N-j,K+1)/G(N,K+1), \end{aligned} \tag{12}$$

and the throughput of requests for accessing common data i, θ_i, is given by:

$$\theta_i = q_i G(N-1,K+1)/G(N,K+1). \tag{13}$$

Appling Little's formula, the mean waiting time for accessing common data i, $W_{d,i}$ is given by:

$$W_{d,i} = L_i/\theta_i - 1/\mu_i. \tag{14}$$

We have presented the algorithm for computing $W_{d,i}$. However, to use the above algorithm, it is necessary to determine λ_d according to the number of calls offered to a processor per unit time, x. To obtain λ_d when the number of calls offered to a processor per unit time is x, the following relation is used:

$$1/\lambda_d = 1/(x\sum_{i=1}^{i=K} \ell_i) - \sum_{i=1}^{i=K} q_i(1/\mu_i + W_{d,i}). \tag{15}$$

The above relation can be derived in the same way as the one for Eq.(7). The number of accesses to common data per unit time is $x\sum_{i=1}^{i=K}\ell_i$ when x calls are offered to a processor per unit time. Since the probability of access to common data i is q_i, the mean time between successive common data accesses is given by $(1/\lambda_d+\sum_{i=1}^{i=K} q_i(1/\mu_i+W_{d,i}))$; the fact that the product of both terms is equal to 1 leads to Eq.(15).

Thus $W_{d,i}$ and λ_d must be calculated so as

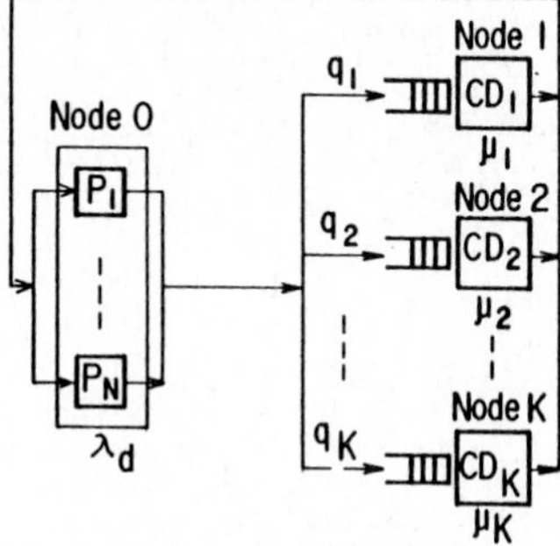

Fig.4. Common data access contention model.

to satisfy Eqs.(14) and (15). To do this, an iteration method similar to the one for computing W_b in the CMC model can be used. The description of the algorithm is omitted here. At this point, note that the mean service time of common data i, $1/\mu_i$, must be modified so as to incorporate an increase in the mean instruction execution time caused by CMC:

$$1/\mu_i = d_i(\tau_0 + \alpha W_b). \tag{16}$$

Using $W_{d,i}$, we can calculate the increase in the number of common memory accesses per unit time caused by CDC, $\Delta\nu$ ($\Delta\nu$ has been used in the analysis of the CMC model discussed in the previous subsection). $\Delta\nu$ can be approximated by:

$$\Delta\nu = x\sum_{i=1}^{i=K} \ell_i W_{d,i} /(\tau_0 + \alpha W_b). \tag{17}$$

3.3 Algorithm

In this subsection, combining the analytic results of the CMC and CDC models, an algorithm is presented for computing W_b and $W_{d,i}$.

The algorithm is as follows:

Step 1

Set $\Delta\nu^{(0)}=0$, $\nu^{(0)}=(D_0+D_1x)\alpha$ and I=0. Compute $W_b^{(0)}$ using the results of the CMC model in 3.1, where ν is replaced by $\nu^{(0)}$ in Eq.(7).

Step 2

Set $\tau^{(I)}=\tau_0+\alpha W_b^{(I)}$. Set $\mu_{d,i}^{-1}=d_i\tau^{(I)}$. Compute $W_{d,i}^{(I)}$ using the results of the CDC model analyzed in 3.2.

Step 3

Compute the increase in the number of accesses to the common memory per unit time, $\Delta\tau^{(I+1)}$, using the following equation:

$$\Delta\nu^{(I+1)} = x\sum_{i=1}^{i=K} \ell_i W_{d,i}^{(I)}/(\tau_0 + \alpha W_b^{(I)}).$$

Set $\nu^{(I+1)}=(D_0+D_1x)\alpha+\Delta\nu^{(I+1)}$ and compute $W_b^{(I+1)}$ using the results of the CMC model analyzed in 3.1.

Step 4

If $|W_b^{(I+1)} - W_b^{(I)}|<\varepsilon$, where ε has the pre-specified and sufficiently small positive value, then set $W_b=W_b^{(I)}$ and $W_{d,i}=W_{d,i}^{(I)}$, and terminate the algorithm. Otherwise $I \leftarrow I+1$ and return to Step 2.

3.4 Calculation of processor occupancy

In this subsection an application to calculating processor occupancy is discussed. As discussed in the previous reports (for example, see [22],[23]), for a single processor controlled switching system, processor occupancy is approximated by a linear equation of the number of calls offered to a processor per unit time. On the other side, this linear approximation cannot be applied to a multiprocessor system, because the effect of CMC and CDC must be considered. Let $f_b(x)$ be the increase in processor occupancy of each processor caused by CMC and $f_d(x)$ be the increase in processor occupancy of each processor caused by CDC when the number of calls offered to a processor per unit time is x. Using Eq.(8), $f_b(x)$ is given by:

$$\begin{aligned} f_b(x) &= (D_0+D_1x)\Delta\tau \\ &= (D_0+D_1x)\alpha W_b. \end{aligned} \tag{18}$$

Using $W_{d,i}$ and ℓ_i, $f_d(x)$ is given by:

$$f_d(x) = x\sum_{i=1}^{i=K} \ell_i W_{d,i}. \tag{19}$$

Taking into account the increase in processor occupancy caused by CMC and CDC, processor occupancy for each processor in a multiprocessor system, f(x), can be approximated by:

$$f(x) = (D_0+D_1x)\tau_0 + f_b(x) + f_d(x). \tag{20}$$

Using Eq.(20) and the algorithm described in 3.3, we can compute processor occupancy as a function of the number of calls offered to a processor per unit time.

4. NUMERICAL EXAMPLES

In this section numerical results are given under the following conditions:

$h_B=1$, $\tau_0=2.2$, $\alpha=0.16$ or 0.32,
$K=5$, $d_1=30$, $\ell_1=30$, $d_2=25$, $\ell_2=15$, $d_3=140$,
$\ell_3=12$, $d_4=100$, $\ell_4=5$, $d_5=140$, $\ell_5=5$,
$D_0=15$ (step) /10000 unit time, and
$D_1=20000$ (step) /1 call .

First, analytic results are compared with simulation results. Figure 5 shows a comparison between the approximate analysis results and simulation results for the mean waiting time for accessing the common memory when N=5 (solid line: analytic results, square: simulation results) and N=8 (dashed line: analytic results, triangle: simulation results). Figures 6 and 7 show the approximate analysis results and simulation results for the mean waiting time for accessing common data under the same conditions as in Fig.5. In either figure, the horizontal axis shows the number of calls offered to a processor per unit time, x. As mentioned in section 2, when a processor attempts to access common data and at the same time the data is being accessed by another processor, the processor trying to access the data executes a T&S instruction repeatedly until it succeeds in accessing the data. This spin lock operation is modeled as a system with repeated calls. However, to facilitate analysis of spin lock, we used a model in which requests for accessing common data form a queue (see Fig.4), assuming that repeated requests for accessing the same common data are independent attempts from each other. we can see that the analytic results and the simulation results are in good agreement. In the present comparison, simulation required more computational time than analytic method. In spite of the approximation, the analytic method proposed in section 3 is useful for performance analysis in the design phase of multiprocessor systems.

Next, we evaluate the effects of the number of common memory accesses during one instruction execution, α, on system performance. Figure 8 shows the increase in processor occupancy caused by CMC, $f_b(x)$, and the increase in processor occupancy caused by CDC, $f_d(x)$, for $\alpha=0.16$ and

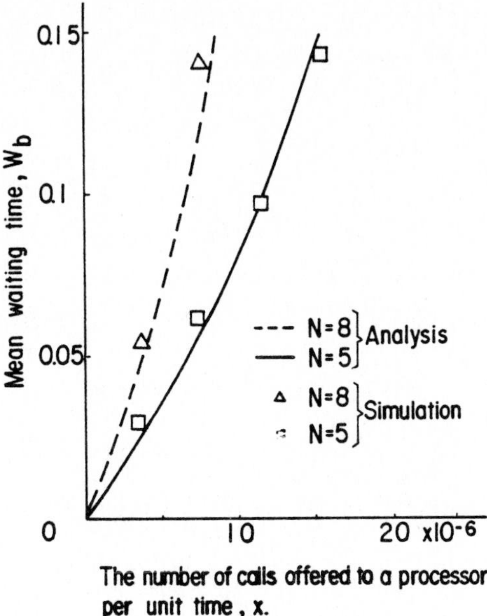

Fig.5. Mean waiting time for accessing the common memory.

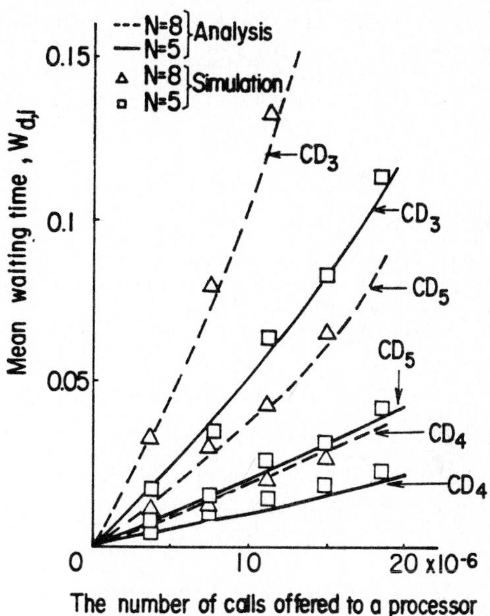

Fig.7. Mean waiting time for accessing common data (common data 3, common data 4, common data 5).

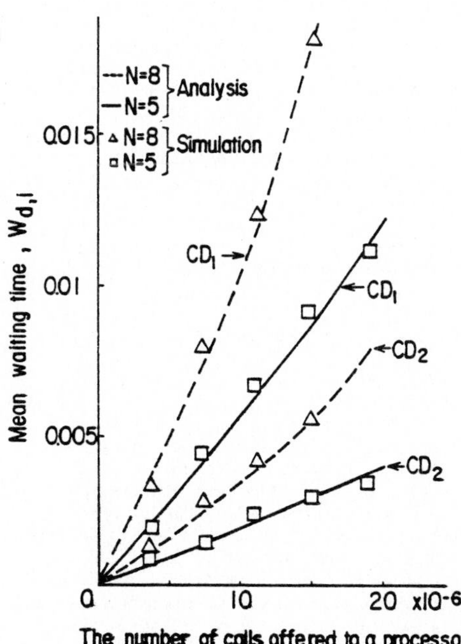

Fig.6. Mean waiting time for accessing common data (common data 1, common data 2).

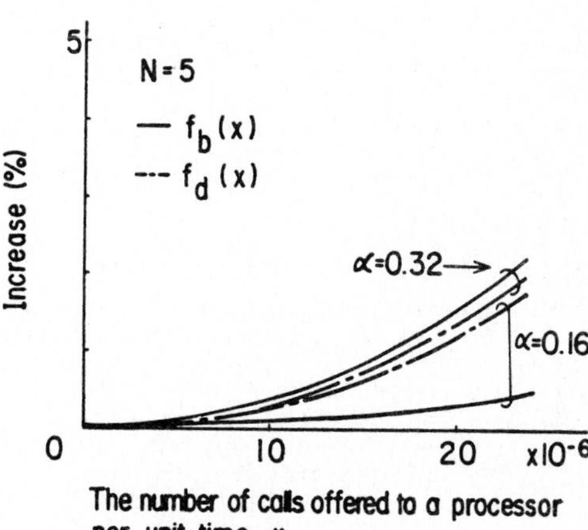

Fig.8. Increase in processor occupancy caused by CMC and CDC.

α=0.32 by the approximate analysis method. The conditions for the other system parameters, such as D_0,D_1,h_b etc., are the same as those described earlier. From Fig.8, we see that while $f_b(x) < f_d(x)$ for α=0.16, $f_b(x) > f_d(x)$ for α=0.32. In other words, as α grows larger, CMC affects the system performance to a greater extent than CDC. From this figure, it can be seen that system performance is largely improved by reducing the influence of CMC.

5. CONCLUSION

This paper has emphasized the significance of the inter-dependence between the common memory access contention and access contention to common data that cannot be referenced or modified simultaneously in multiprocessor controlled switching systems. A queueing network model taking into account this inter-dependence has been developed. An approximate analysis technique based on decomposing the queueing network model has been proposed. When compared with simulation results, it has been shown that the proposed technique provides good approximate values.

Calculation of processor occupancy is essential for estimating the call processing capacity of electrical switching systems. The queueing network model proposed in this paper makes it possible to accurately calculate processor occupancy for each processor in a multiprocessor controlled switching system.

ACKNOWLEDGMENT

The author would like to thank Mr.Kunio Kodaira, Director of the Teletraffic Section, and Dr.Tsuyoshi Katayama and Mr.Konosuke Kawashima, Staff Engineerers in the Teletraffic Section for their thoughtful discussions and helpful comments during the course of this work.

REFERENCES

[1] C.E. Skinner and J.R. Asher, "Effects of storage contention on system performance," IBM Syst. J., vol.18, no.4, pp.319-333, 1969.

[2] D.P. Bhandarkar, "Analysis of memory interference in multiprocessors," IEEE Trans. Comput., vol.C-24, no.9, pp.897-908, 1975.

[3] F. Baskett and A.J. Smith, "Interference in multiprocessor computer systems with interleaved memory," Commun. ACM, vol.19, no.6, pp.327-334, 1976.

[4] C.H. Hoogendoorn, "A general model for memory interference in multiprocessors," IEEE Trans. Comput., vol.C-26, no.10, pp.998-1005, 1977.

[5] K.V. Sastry and R.Y. Kain, "On the performance of certain multiprocessor computer organization," IEEE Trans. Comput., vol.C-24, no.11, pp.1066-1074, 1975.

[6] A.S. Sethi and N. Deo, "Interference in multiprocessor systems with localized memory access probabilities," IEEE Trans. Comput., vol.C-28, no.2, pp.157-163, 1979.

[7] B.R. Rau, "Interleaved memory bandwidth in a model of a multiprocessor computer system," IEEE Trans. Comput., C-28, no.9, pp.678-681, 1979.

[8] J.W. McCredie, "Analytic models as aids in multiprocessor design," Proc. 7th Annu. Princeton Conf. Information Science and Systems, pp.186-191, 1973.

[9] M.A. Marsan and M. Gerla, "Markov models for multiple bus multiprocessor systems," IEEE Trans. Comput., vol.C-31, no.3, pp.239-248, 1982.

[10] M.A. Marsan, G. Balbo and G. Conte, "Comparative performance analysis of single bus multiprocessor architectures," IEEE Trans. Comput., vol.C-31, no.12, pp.1179-1191, 1982.

[11] M.A. Marsan, G. Balbo, G. Conte and F. Gregoretti, "Modeling bus contention and memory interference in a multiprocessor system," IEEE Trans. Comput., vol.C-32, no.1, pp.60-72, 1983.

[12] K. Fung and H.C. Torng, "On the analysis of memory conflicts and bus contentions in a multiple-microprocessor system," IEEE Trans. Comput., vol.C-27, no.1, pp.28-37, 1979.

[13] P.C.C. Yeh, J.H. Patel and E.S. Davidson, "Shared cache for multiple-stream computer system," IEEE Trans. Comput., vol.C-32, no.1, pp.38-47, 1983.

[14] F.A. Briggs and M. Dubois, "Effectiveness of private caches in multiprocessor systems with parallel-pipelined memories," IEEE Trans. Comput., vol.C-32, no.1, pp.48-59, 1983.

[15] S.E. Madnick, "Multi-processor software lockout," Proc. ACM 23rd National Conf., pp.19-24, 1968.

[16] W.A. Wulf and C.G. Bell, "C.mmp-- A multi-mini-processor," Proc. AFIPS Conf., vol.41, part.II, pp.765-777, 1972.

[17] D.C. Gilbert, "Modeling spin locks with queueing networks," Operating Systems Review, vol.12, no.1, pp.29-42, 1978.

[18] J. Hofman and H. Schmutz, "Performance analysis of suspend locks in operating systems," IBM J. Res. Develop., vol.26, no.2, pp.242-259, 1982.

[19] W.F. King,III, S.E. Smith and I. Wladawsky, "Effects of serial programs in multiprocessing systems," IBM J. Res. Develop., vol.18, no.6, pp.303-309, 1974.

[20] C. Smith and J.C. Browne, "Aspects of software design analysis: Concurrency and blocking," Proc. Performance 80, pp.245-253, 1980.

[21] A. Thomasian, "Queueing network models to estimate serialization delays in computer systems," Proc. Performance 83, pp.61-81, 1983.

[22] T. Suzuki, Y. Nunotani and O. Kaneda, "Traffic design and engineering of central processing systems in an electronic switching system," Proc. 6th ITC, no.424, 1970.

[23] J.E. Villar, "Traffic calculations in SPC systems," Proc. 8th ITC, no.611, 1976.

[24] L. Takács, "On a stochastic process concerning some waiting time problems," Theory of probability and its Applications, vol.2, no.1, pp.92-105, 1957.

[25] F. Baskett, K.M. Chandy, R.R. Mnutz, and F.G. Palacios, "Open, closed and mixed networks of queues with different classes of customers," J. ACM, vol.22, no.2, pp.248-260, 1975.

[26] J.P.Buzen, "Computational algorithms for closed queueing networks with exponential servers," Commun. ACM, vol.16, no.9, pp.527-531, 1973.

TELETRAFFIC ISSUES in an Advanced Information Society
ITC-11
Minoru Akiyama (Editor)
Elsevier Science Publishers B.V. (North-Holland)

CONSIDERATIONS ON LOSS PROBABILITY OF MULTI-SLOT CONNECTIONS

J. CONRADT and A. BUCHHEISTER

STANDARD ELEKTRIK LORENZ AG
Stuttgart, Federal Republic of Germany

ABSTRACT

Present telephone connections through pcm networks are set up by switching single-slots with a bit rate of 64 kbit/s. Future ISDN networks demand for switching of multi-slot connections with n x 64 kbit/s. This contribution evaluates loss probabilities of multi-slot calls offered to a trunk group and to a switching network.

For the case of a trunk group with full availability an algorithm is presented which greatly simplifies numerical analysis.

For the case of trunk group with restricted availability and several pcm links the effect of a practical requirement to place all slots of a multi-slot call on one pcm link is studied. It is shown that the effect of such a restriction is not very significant if an adequate hunting strategy is applied. Four strategies are studied which provide almost equivalent grade-of-service.

Probability of internal loss is evaluated for a 2-stage switching network. Three strategies to hunt for internal slots are considered. They diverge widely in their effect. Internal loss of switching network depends also on the strategy to hunt for outgoing trunks.

1. INTRODUCTION

In telephone networks with pcm switching and pcm transmission connections are switched on the basis of 64 kbit/s-channels ("slots"). Future networks like the wideband ISDN or satellite networks demand for "multi-slot connections" comprising n times 64 kbit/s-channels where $n \leq 30$ is usually assumed. Thus bandwidths of up to 2 Mbit/s are obtained. Examples of potential applications are: High quality audio transmission, data file transfer, facsimile transmission, slow scan video conferences.

Performance evaluation of networks with a mix of multi-slot and single-slot calls is complicated since the various calls differ in many traffic characteristics. Furthermore there is a large variation of possible traffic mixes and performance requirements.

Trunk group performance considering mixes of traffic of different bandwidths (i.e. with n x 64 kbit/s-channels or with n x 600 bit/s-channels according to CCITT multiplexing scheme X.50) has been investigated by many authors [1] to [5].

Concern of this contribution is directed to:

- a simple algorithm for fully available trunk groups carrying multi-slot calls
- performance of a trunk group comprising several pcm links with 30 channels each and having the restriction that all slots of a multi-slot call have to be placed on one pcm link
- performance of a switching network carrying multi-slot calls.

General assumptions are:

- independent traffic classes of single-slot and multi-slot calls
- pure chance traffic with same mean holding times for all traffic classes
- lost calls cleared model

Performance is expressed in terms of loss probability. Delay problems may be encountered when multi-slot connections are set-up by a series of independent single-slot-connections using arbitrary pathes through a network. A special application is discussed in [6].

2. TRUNK GROUP PERFORMANCE

2.1 Fully Available Trunk Group

For the case where all slots are available for any class of traffic analytic solutions derived from state equations are given in [1] to [5].

State probability can be described by two different levels of detail: Probability of microscopic state is

$$\text{(1a)} \quad P(x_1, x_2, \ldots x_z) = \frac{T(x_1, x_2, \ldots x_z)}{\sum\limits_{\forall x_i;\ f \leq N} T(x_1, x_2, \ldots x_z)}$$

$$T(x_1, x_2, \ldots x_z) = \frac{A_1^{x_1}}{x_1!} \frac{A_2^{x_2}}{x_2!} \cdots \frac{A_z^{x_z}}{x_z!}$$

$$\bullet\ f = x_1 + 2x_2 + \ldots + zx_z$$

Macroscopic state probability is

$$\text{(1b)} \quad P(x) = \sum_{\forall x_i;\ f=x} P(x_1, x_2, \ldots x_z)$$

x_i number of calls with i slots each, $1 \leq i \leq z$
A_i traffic offered demanding i slots per call
N number of slots of trunk group
z maximum number of slots per call
m number of slots per call; $1 \leq m \leq z$
x number of busy slots; $x = x_1 + 2x_2 + \ldots zx_z$

Probability of loss for a call demanding m slots is

$$\text{(2)} \quad B_m = \sum_{x=N-m+1}^{N} P(x)$$

Equations (1) and (2) comprise terms which may lead to substantial numerical effort. The number of terms to be calculated and to be summed increases rapidly with N due to increasing number of combinations.

To define loss probability a simple recursive algorithm has been developped.

For $\quad U_v = \sum\limits_{f=v} T(x_1, x_2, \ldots x_z)$

the following relation holds

$$U_v = \frac{1}{v} \sum_{i=1}^{z} i\, A_i\, U_{v-i} \quad \text{with } 0 \leq v \leq N,\ U_0 = 1,$$

$$U_s = 0 \text{ for } s < 0$$

Background and proof is given in [7]. The algorithm is as follows:

$$\text{(3)} \quad B_m = B_1(1 + a_N^{-1}(1 + a_{N-1}^{-1}(1 + \ldots a_{N-m+2}^{-1})\ldots)$$

where

$2 \leq m \leq z$

$B_1 = b_N^{-1}$

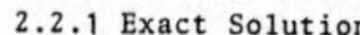

$$b_j = 1 + a_j^{-1} b_{j-1};\ b_0 = 1$$

$$a_j = \frac{A_1}{j} + a_{j-1}^{-1}\left(\frac{2A_2}{j} + a_{j-2}^{-1}\left(\frac{3A_3}{j} + \ldots\ a_{j-z+1}^{-1}\ \frac{zA_z}{j}\right)\ldots\right)$$

$$1 \le j \le N$$

$$A_1, A_z > 0;\ A_2, A_3\ \ldots\ A_{z-1} \ge 0;\ a_s^{-1} = 0 \text{ for } s \le 0$$

The a-values are recursively calculated using (z-1) values gained a step before starting with j = 1 and ending with j = N. The b-values are gained recursively, too. Only z memory cells have to be provided for storage of significant values. To calculate all loss values only about z(z/2 + N) additions and multiplications and zN divisions have to be executed. This algorithm can run on a programmable pocket calculator (e. g. on TI 59 for z ≤14).

Macroscopic state probability P(x) can also be calculated by means of a simple algorithm without summing state probabilites according to equation (1b):

$$P(x) = \left[b_x + a_{x+1}\ (1 + a_{x+2}(1 + \ldots\ a_N)\ \ldots)\right]^{-1} \qquad (4)$$

$$0 \le x \le N$$

$$P(N) = B_1$$

State probability distribution is calculated using all a- and b-values which have to be stored in about 2N memory cells. Run time is determined by execution of about N(N/2 +z) simple additions, multiplications and divisions.

2.2 Trunk Group With Restricted Availability

Cases with various restrictions with regard to slot usage have been investigated in [3,4]. The m slots of a call had to be allocated equally spaced or consecutively or in predetermined groups. Artificial limitations in order to balance loss performance have been additionally studied in [4].

Restriction considered here results from a requirement that all m slots of a call have to be allocated to one pcm link but with full availability as regards the channels on the pcm link.

(Remark: If slots of a multi-slot call were spread over several independently synchronized pcm links a mechanism would be required to align byte sequence. Such a mechanism introduces delay (see also [6]).To avoid this the described restriction has to be observed.)

2.2.1 Exact Solution

The impact of such a requirement has been studied by solving state equations. This allows to calculate the expected small performance differences of various hunting strategies.

Since no general solutions could be found numerical results have been calculated for a limited number of models with the following characteristics:

- r = 3 pcm links
- n = 30 or n = 6
- mix of single slot and one type of n-slot calls

Four hunting strategies have been investigated:

a) Sequential hunting where single-slot calls starting from one side and n-slot calls starting from other side.

b) Sequential hunting from one side for both call types.

c) As strategy a) but single-slot calls hunting for an already partially occupied pcm link by skipping empty pcm links.

d) As strategy a) but single-slot calls hunting for the most heavily loaded pcm link.

These hunting strategies are understood to be superior to others. Reassignment of slots has been considered to be not practical.

Microscopic state is described by variables i,j,k corresponding to the number of calls on each of the 3 pcm links. Values 0 to n indicate number of single-slot calls, value "n+1" seizure by an n-slot call. There are $(n+2)^3$ different states. State probability P(i,j,k) is calculated. The equations for the state probabilities in equilibrium state are given in Appendix to 2.2.1.

Loss probability of single-slot and n-slot calls respectively is:

$$B_1 = \sum_{i,j,k\, =\, n \text{ or } n+1} P(i,j,k) \qquad (5)$$

$$B_n = \sum_{i,j,k\, \ne\, 0} P(i,j,k) \qquad (6)$$

Numerical results are given in figures 1 and 2 illustrating development of loss with n-slot traffic proportion p. It holds

$$p = \frac{nA_n}{A_1 + nA_n} \quad ; \qquad A_1 + nA_n = \text{const.}$$

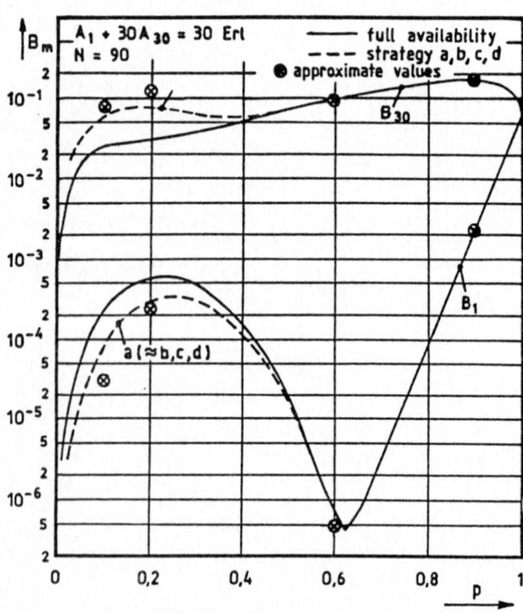

Fig. 1 Loss probability B_1 and B_{30} of trunk group versus n-slot traffic proportion p

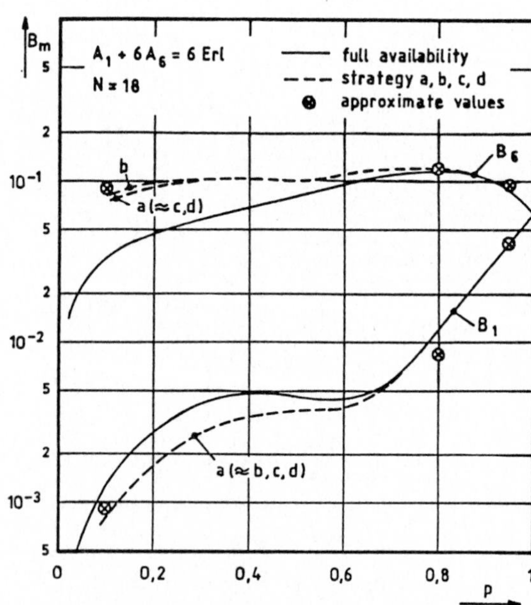

Fig. 2 Loss probability B_1 and B_6 of trunk group versus n-slot traffic proportion p

The following can be read from these figures:

- Course of curves and $B_1 << B_n$ is typical for multi-slot traffic mixes. Minima and maxima have been discussed in [2] for the case of full availability. Reasoning holds approximately for the restricted case considered here.
- The 4 hunting strategies considered cause no significant loss differences.
- B_1 (restricted availability) < B_1 (full availability)
  B_n (restricted availability) > B_n (full availability)
- For p = 100 % it holds $B_n = B_1 = E_{1,3}(A_n)$ where $E_{1,N}(A)$ is Erlang's loss formula
  For p = 0 % it holds $B_1 = E_{1,3n}(A_1)$; $B_n \neq B_1$

2.2.2 Approximate Solution

An approximate calculation without solving the set of state equations has been presented in [4] considering equally spaced slots and three traffic classes.

Requirement followed here is to place all m slots of a m-slot call on one pcm link. The effect of such a requirement is identical to the one considered in the reference cited if only two traffic classes A_1 and A_n are assumed where the multi-slot call occupies a pcm link fully. This approximate method has correspondingly been modified.

Sequential hunting from constant start point is assumed for single slot calls. Hunting strategy for channels of n-slot calls is irrelevant for the approximate calculation.

Loss probabilities of n-slot calls and single-slot calls are calculated consecutively:

$$B_n = \sum_{x=0}^{r} Q_1(x)\, E_{1,r-x}(A_n) \tag{7}$$

$Q_1(x)$ is the probability that x pcm links are partially or fully occupied if there were single-slot traffic only. $Q_1(x)$ can be found from probabilities P_i that the i-th channel relative to start point is occupied:

$$P_i = A_1\,(E_{1,i-1}(A_1) - E_{1,i}(A_1)) \text{ where}$$

$$1 \leq i \leq N;\ N = nr$$

Probability V_j that j-th pcm link is occupied with one or more single-slot calls is

$$V_j = 1 - \left(\prod_{i=(j-1)n+1}^{jn} (1-P_i)\right) \quad \text{where } 1 \leq j \leq r$$

$Q_1(x)$ will be calculated assuming independence of V_j and is based on combinatorial considerations (see [4]). For instance: $Q_1(0) = (1-V_1)(1-V_2)\ldots(1-V_r)$.

$$B_1 = \sum_{x=0}^{r} Q_2(x)\, E_{1,n(r-x)}(A_1) \tag{8}$$

$Q_2(x)$ is the probability that x pcm links are fully occupied by single-slot and n-slot traffic. It is found by weighing the corresponding state probabilities.

$$Q_2(x) = \sum_{u=0}^{r-x} Q_1(u)\cdot P_u(x) \quad \text{where } P_u(x) = \frac{A_n^x/x!}{\sum_{i=0}^{r-u} \frac{A_n^i}{i!}}$$

Note: $Q_1(0)$ can also be found directly

$$Q_1(0) = \frac{1}{\sum_{j=0}^{nr} \frac{A_1^j}{j!}}$$

In actual numerical examples this correct value $Q_1(0)$ can be used to improve the approximate values of $Q_1(x)$ particularly when $Q_1(x) \approx 0$ for $x \geq 2$.

Numerical results found with this approximate calculation are shown in figure 1 and 2. Compared to the exact figures of all 4 hunting strategies it can be said that the approximate method gives sufficiently good results. Accuracy increases with increasing proportion p.

2.2.3 Conclusions Concerning Trunk Group With Restricted Availability

From the exercise on a limited number of models the following conclusions are drawn:

- The four hunting strategies considered are practically equivalent particularly for the normal case of n = 30. Hunting strategy a) or b) should be selected which are considered to be the simplest to implement.
- Effect of the restriction is generally not very significant related to the fully available group if one of the above hunting strategies is applied. Differences of loss probabilities are of minor importance in applications where p > 0,5. In case of unknown p the dimensioning of the trunk group has to take into account the worst B_n. For a given value of B_n it has been calculated that in case of full availability the traffic offered ($A_1 + nA_n$) could be up to 20 % higher than in case of restricted availability.

Single-slot loss profits from the worse performance of the n-slot traffic.

3. SWITCHING NETWORK

A simple one-sided 2-stage network has been investigated. Configuration and graph is shown in fig. 3 and 4.

3.1 Path Search Strategy

Single-slot calls are set up by conditional path search or by step-by-step assuming exhaustive number of trials. Multi-slot calls are split into single-slot requests set up consecutively like single-slot requests (Note: The restriction to place all single-slot requests on one pcm link as followed in section 2 is considered not to be expedient in multi-stage modular switching networks.)

Hunting for a free slot on interstage links depends on strategy chosen. Three strategies have been evaluated:

a) Random hunting
b) Sequential hunting from fixed starting point where single-slot calls start search from one side and single-slot requests of a multi-slot call start from other side.
c) Hunting controlled by a toggle switch directing each single-slot call or single-slot request alternately to one of the two pcm links.

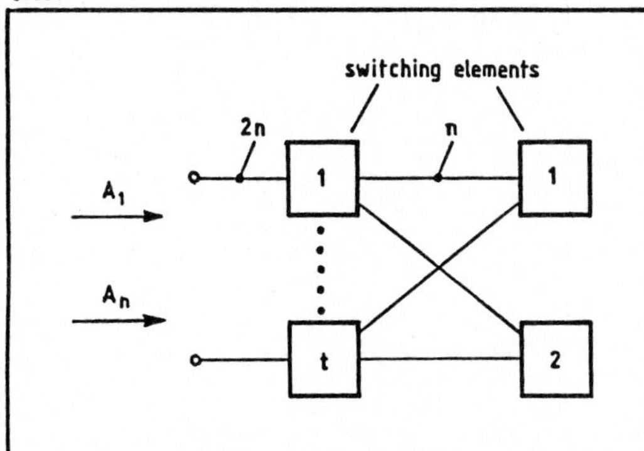

Fig. 3 Switching network considered

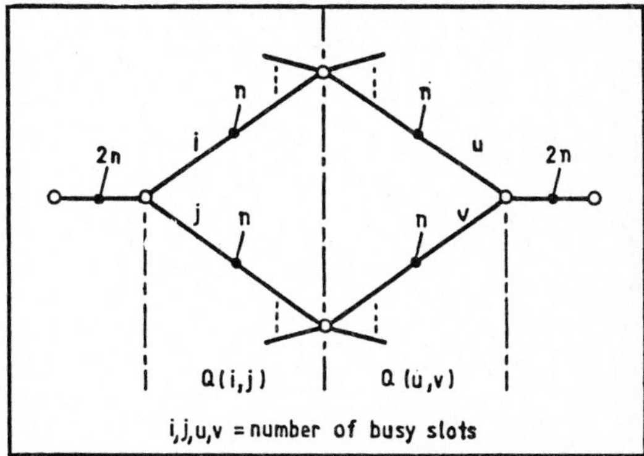

Fig. 4 Graph of switching network considered

3.2 Internal Loss Probability

Internal loss probability B_m is the conditional probability that an m-slot call is blocked in the switching network under the condition that at least m slots are free on the originating and terminating switching element. A mismatch situation of an m-slot call is described by the following relation (see fig. 4):

$$g = \min(n-i, n-u) + \min(n-j, n-v) < m$$

Assuming independence of states between the stages of the switching network and keeping the general prerequisites of section 1 it holds

$$(9) \quad B_m = \frac{\sum_{g<m;\ i+j,u+v\le 2n-m} Q(i,j)Q(u,v)}{\sum_{i+j,u+v\le 2n-m} Q(i,j)Q(u,v)}$$

Q(i,j), Q(u,v) are probabilities of state describing the number of busy slots on the two pcm links of the two interstage link groups i. e. the two halves of the graph (see fig. 4). The state probabilities are calculated for the three hunting strategies defined.

3.2.1 State Probability in Case of Random Hunting (a)

State probability distribution P(x) of a fully available trunk group is given by equation (4) where the variable x describes the total number of busy slots.

In the link group considered here the busy slots are partitioned into i slots on the "upper" and j on the lower pcm link. Since all patterns of i, j busy slots are equiprobable the following relation holds:

$$(10) \quad Q(i,j) = \frac{\binom{n}{i}\binom{n}{j}}{\binom{2n}{i+j}} P(i+j); \quad 0 \le i,j \le n$$

Equation (10) holds for each half of the graph.

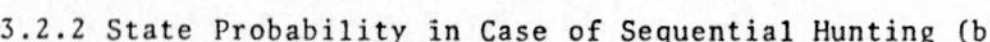

3.2.2 State Probability in Case of Sequential Hunting (b)

State probability P(i,j,k) is calculated where i and j defines the number of busy slots in the upper and lower pcm link and k indicates the number and distribution of n-slot calls.

State equations in equilibrium state and transition rates are given in appendix to 3.2.2. It follows:

$$(11) \quad Q(i,j) = \sum_k P(i,j,k)$$

3.2.3 State Probability in Case of Toggle Switch (c)

State probability P(i,j,k,s) is introduced. Meaning of i,j,k as in sections 3.2.2. Variable s = 0 or 1 corresponds to position of toggle switch. The state equations are given in appendix to 3.2.3. It follows:

$$(12) \quad Q(i,j) = \sum_{k,s} P(i,j,k,s)$$

3.2.4 Numerical Results

Loss probabilities have been calculated assuming equal hunting strategies on both halves of the graph. Figures 5 and 6 show loss curves for a constant traffic offered versus n-slot traffic proportion p. From the loss curves one can read:

- An oscillating course as for trunk groups but more pronounced and with totally different evolution. There are ranges of p where B_{30} becomes smaller than B_1.
- Diverging loss curves for the three hunting strategies. With regard to B_1 figure 5 shows that strategy b is worst while strategies a and c are about equivalent.

With regard to B_{30} figure 6 reveals that in the range of p < 0,5 all strategies are almost equivalent with a slight advantage for strategy b. In the range of p > 0,5 strategy c is much better. Approaching the point p → 1 strategy b is worst, followed by strategy a. For exactly p = 1 it is contrary.

In case of strategy c loss B_1 as well as B_{30} are approaching the value zero.

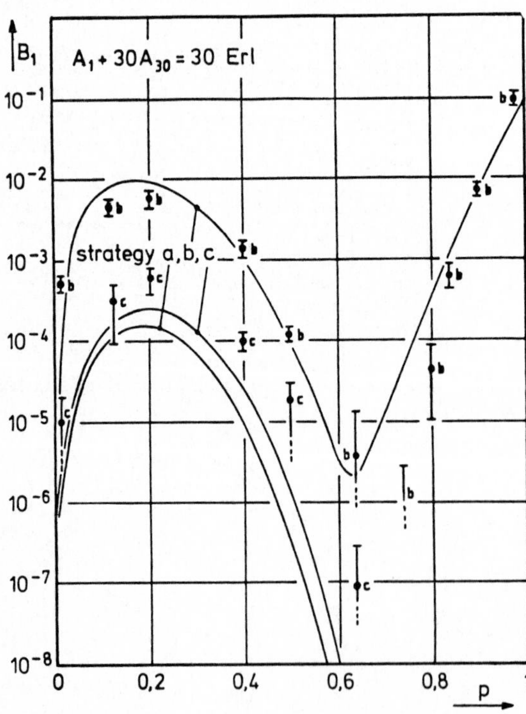

Fig. 5 Internal loss probability B_1 versus n-slot traffic proportion p

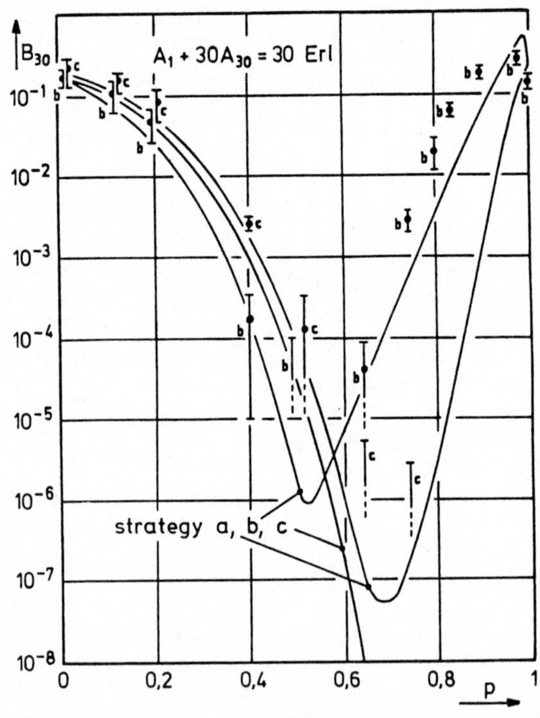

Fig. 6 Internal loss probability B_{30} versus n-slot traffic proportion p

Single-slot calls are blocked if opposite links of the two halves of the graph are fully busy. This condition may be more frequent in case of strategy b than in case of strategy a and c due to "filling" of links. Thus a worse behavior with regard to B_1 could be expected applying strategy b.

Multi-slot calls are blocked under many mismatch conditions (see 3.2). From the cases considered a generally valid explanation cannot be given.

Results of strategy c approaching zero loss are explained by the fact that only states (i,j) = (0,0) or (15,15) can exist where no mismatch is possible.

3.2.5 Simulation Results

Internal loss probability calculated is checked by roulette-type simulation.

An m-slot call selects randomly a first stage switching element out of all which have at least m free slots. The terminating element is selected in the same manner. Thus external loss contributions are excluded. Additionally the traffic has been forced to pass the 2 stages of the network by choosing a relatively large number of first stage elements (t = 100). In practical applications short path connections might take place.

Results have been gained for the cases of strategy b and c and are indicated with 95 %-confidence intervals and mean values (if any) on figures 5 and 6.

In general the simulation results confirm the analytic calculations with deviations to lower and higher loss. Deviations are explained by the fact that the dependencies between the two halves of the graph are neglected in case of analytic calculations and that a lower loss of single-slot calls will be compensated by a higher loss of multi-slot-calls or vice versa.

3.2.6 Impact of External Trunk Hunting Strategy on Internal Loss

So far, internal loss probability has been considered assuming random selection of the two switching elements involved in setting up a call. By random selection of a terminating element the traffic is distributed equally over outgoing trunks. This is a valid approach for fully available trunks.

In case of restricted availability a well adapted hunting strategy is usually applied which may result in traffic imbalances.

The impact of such a trunk hunting strategy on internal loss probability is studied by simulation. A single outgoing trunk group is assumed where single-slot calls and n-slot calls are hunting for slots from different sides (according to strategy a of section 2.2.1). Simulation results are shown in fig. 7. A drastic increase of loss can be seen.

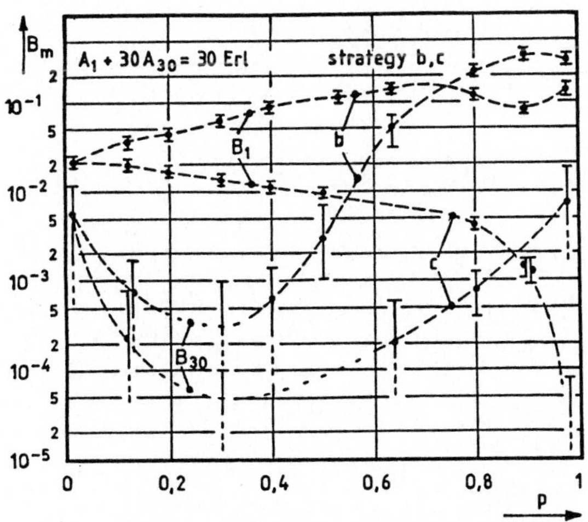

Fig.7 Internal loss B_1 and B_{30} in case of hunting for external trunks from fixed starting point versus n-slot traffic proportion p

3.3 Conclusion on Internal Loss Probability

From the exercise on a simple switching network the following conclusions are drawn:

- There is no general optimum strategy to hunt for internal slots. Loss develops very differently, and drastically with proportion p of n-slot traffic. Selection of a strategy depends on loss objectives to meet, whether B_1 or B_{30} should be minimum or whether equalized losses are required.
- Internal loss is highly sensitive to traffic imbalance which is produced when trunk groups with restricted availability are applied. Loss calculations or simulations should therefore take into account the actual trunk hunting strategy.

4. CONCLUSION

Loss in case of trunk groups with full and restricted availability has been studied by many authors. The investigations in this paper extend the results with regard to trunk group and consider internal loss of switching networks.

For the case of a trunk group with full availability an algorithm is presented which greatly simplifies numerical analysis.

For the case of a trunk group with restricted availability and several pcm links the effect of a practical requirement to place all slots of a multi-slot call on one pcm link is studied. It is shown that the effect of such a restriction is not very significant if an adequate hunting strategy is applied. The four strategies compared provide equivalent grade-of-service.

Probability of internal loss is evaluated for a 2-stage switching network. Three strategies to hunt for internal slots are considered. They differ widely in their effect. Internal loss is also very sensitive to traffic imbalance produced by strategy to hunt for outgoing trunks.

This contribution is based on investigations of simple models derived from practical applications. Investigations of more complex models and with other traffic characteristics are ongoing.

APPENDIX to 2.2.1

State probabilities of trunk group with restricted availability considering hunting strategies a,b,c,d are

$$\begin{aligned} f_0 \cdot P(i,j,k) = {} & f_1 \; P(i-1,j,k) + f_2 \; P(0,j,k) \\ & + f_3 \; P(i,j-1,k) + f_4 \; P(i,0,k) \\ & + f_5 \; P(i,j,k-1) + f_6 \; P(i,j,0) \\ & + f_7 \; P(i+1,j,k) + f_8 \; P(n+1,j,k) \\ & + f_9 \; P(i,j+1,k) + f_{10} \; P(1,n+1,k) \\ & + f_{11} \; P(i,j,k+1) + f_{12} \; P(i,j,n+1) \end{aligned}$$

with $0 \le i,\ j,\ k \le n+1$

Transition rates and validity range are (out of range $f_s = 0$ for $0 \le s \le 12$):

$$\begin{aligned} f_0 &= f_{01} + f_{02} + f_{03} \\ f_{01} &= A_1;\ i < n \text{ or } j < n \text{ or } k < n \\ f_{02} &= A_n;\ i = 0 \text{ or } j = 0 \text{ or } k = 0 \\ f_{03} &= i + j + k;\ i,j,k \ne n+1 \\ & \quad i + j + 1;\ i,j \ne n+1;\ k = n+1 \\ & \quad i + 1 + k;\ i,k \ne n+1;\ j = n+1 \\ & \quad 1 + j + k;\ j,k \ne n+1;\ i = n+1 \\ & \quad i + 2;\ i \ne n+1;\ j,k = n+1 \\ & \quad j + 2;\ j \ne n+1;\ k = n+1 \\ & \quad k + 2;\ k \ne n+1;\ k,j = n+1 \\ & \quad 3\ ;\ i,j,k = n+1 \end{aligned}$$

$$\begin{aligned} f_7 &= i + 1;\ 0 \le i \le n-1 \qquad & f_8 &= 1;\ i = 0 \\ f_9 &= j + 1;\ 0 \le j \le n-1 \qquad & f_{10} &= 1;\ j = 0 \\ f_{11} &= k + 1;\ 0 \le k \le n-1 \qquad & f_{12} &= 1;\ k = 0 \end{aligned}$$

The other rates are given according to hunting strategy (a), (b), (c), (d)

$f_1 = A_1$ for

$1 \le i \le n;$ (a,b)

$1 < i \le n$ or ($i=1$, ($j=0$ or $j \ge n$),($k=0$ or $k \ge n$)); (c)

$1 \le i \le n$, (($j<n,k<n,i>j,i>k$) or ($j<n,k \ge n,i>j$) or ($j \ge n$, $k<n$, $i>k$) or ($j \ge n$, $k \ge n$)); (d)

$f_2 = A_n$ for

$i = n + 1,\ j > 0,\ k > 0$; (a,b,d)

$i = n + 1$; (b)

$f_3 = A_1$ for

$i \ge n,\ 1 \le j \le n$; (a,b)

$(i \ge n,\ (2 \le j \le n$ or $(j=1, (k=0$ or $k \ge n))))$
or $(i=0,\ 2 \le j \le n)$; (c)

$1 \le j \le n$, $((i<n,\ k<n,\ j-1>i,\ j>k)$
or $(i<n,\ k \ge n,\ j-1>i)$
or $(i \ge n,\ k<n,\ j>k)$ or $(i \ge n,\ k \ge n))$; (d)

$f_4 = A_n$ for

$j = n + 1,\ k > 0$; (a,c,d)

$j = n + 1,\ i > 0$; (b)

$f_5 = A_1$ for

$i \ge n,\ j \ge n,\ 1 \le k \le n$; (a,b)

$(i \ge n,\ j \ge n,\ 1 \le k \le n)$ or $(i \ge n,\ j=0,\ 2 \le k \le n)$
or $(i=0, j \ge n, 2 \le k \le n)$ or $(i=0, j=0, 2 \le k \le n)$; (c)

$1 \le k \le n$, $((i<n, j<n, k-1>i, k-1>j)$ or
$(i<n,\ j \ge n, k-1>i)$ or
$(i \ge n, j<n,\ k-1>j)$ or $(i \ge n, j \ge n))$; (d)

$f_6 = A_n$ for

$k = n + 1$; (a,c,d)

$k = n + 1,\ i > 0,\ j > 0$; (b)

The validity of the state equations have been tested according to the relations, that the summation over all f_s, $0 \le s \le 12$ connected to $P(i,j,k)$ with fixed i,j,k must be zero for all possible values i,j,k out of the definition range.

The equations have been solved by applying the "Power" method [8]:

A new value $P_{new}(i,j,k)$ is calculated from the old value $P_{old}(i,j,k)$ by the following procedure:

$$P_{new}(i,j,k) = P_{old}(i,j,k) + \Delta t \cdot d(i,k,k) \text{with}$$

$$\Delta t = \frac{0{,}99}{\max\limits_{i,j,k} |f_0|}$$

$$\text{and } d(i,j,k) = -f_0\, P_{old}(i,j,k) + \sum_{l=1}^{12} f_l P_{old}(\cdot,\cdot,\cdot)$$

The old value $P_{old}(i,j,k)$ is replaced in one step by the new value $P_{new}(i,j,k)$.

The iteration stops when

$$\sum_{i,j,k} |d(i,j,k)| < \varepsilon.$$

APPENDIX to 3.2.2

State probability P(i,j,k), strategy b

The k-value indicates number of n-slot calls and distribution of slots to two pcm links. The k-value defines four areas of validity:

area 1: $k = 0$; no n-slot call

area 2: $1 \le k \le n$; one n-slot call seizing (n-k) slots in the 1st and k slots in the 2nd pcm link

area 3: $k = 2n + 1$; one n-slot call in the 1st link

area 4: $n + 1 \le k \le 2n$; two n-slot calls, 1st n-slot call seizing (k-n) slots on 1st link and (2n-k) slots on the 2nd link, 2nd n-slot call seizing (2n-k) slots on the 1st link and (k-n) slots on the 2nd link.

The state equations read for the four areas of validity defined by the k-value:

area 1:

$k=0$, $0 \le i$, $j \le n$ no n-slot calls

$$\begin{aligned}(A_1\ (1-\delta_{i+j}^{2n}) + A_n\ \delta_{i+j}^{\le n} + i + j)\ P(i,j,0) = {} & A_1\ (1-\delta_i^0)\ P(i-1,j,0) \\ & + A_1 \delta_i^n\ (1-\delta_j^0)\ P(i,j-1,0) \\ & + (i+1)(1-\delta_i^n)\ P(i+1,j,0) \\ & + (j+1)(1-\delta_j^n)\ P(i,j+1,0) \\ & + \delta_i^0\ P(n,j,2n+1) \\ & + \delta_{i+j}^{\le n} \sum_{x=\max(1,i)}^{n-j} P(i+n-x,j+x,x)\end{aligned}$$

(δ = Kronecker symbol)

$$\delta_x^y = \begin{cases} 1 & x = y \\ 0 & \text{otherwise}\end{cases}$$

$$\delta_x^{\le y} = \begin{cases} 1 & x \le y \\ 0 & \text{otherwise}\end{cases}$$

area 2: $1 \le k \le n$; $n - k \le i \le n$; $k \le j \le n$

$$\begin{aligned}(A_1\ (1-\delta_{i+j}^{2n}) + A_n\ \delta_{i+j}^n + i + j - n + 1)\ P(i,j,k) = {} & A_1\ (1-\delta_i^{n-k})\ P(i-1,i,k) \\ & + A_1 \delta_i^n\ (1-\delta_j^k)\ P(i,j-1,k) \\ & + A_n \delta_j^n\ P(i-n+k,j-k,0) \\ & + (i+1-n+k)\ (1-\delta_i^n)\ P(i+1,j,k) \\ & + (j+1-k)\ (1-\delta_j^n)\ P(i,j+1,k) \\ & + \sum_{x=n+1}^{2n} (\delta_i^{2n-x}\ \delta_j^{x-n}\ \delta_k^{x-n} + \delta_i^{x-n}\ \delta_j^{2n-x}\ \delta_k^{2n-x}) \times \\ & \times P(n,n,x)\end{aligned}$$

area 3: $k = 2n + 1$; $0 \le j \le n$

$$\begin{aligned}(A_1\ (1-\delta_j^n) + A_n\ \delta_j^0 + j + 1)\ P(n,j,2n+1) = {} & A_1\ (1-\delta_j^0)\ P(n,j-1,2n+1) \\ & + A_n \delta_j^n\ P(0,j,0) \\ & + (j+1)\ (1-\delta_j^n)\ P(n,j+1,2n+1) \\ & + \delta_j^0\ P(n,n,2n)\end{aligned}$$

area 4: $n + 1 \le k \le 2n$;

$$\begin{aligned}2\ P(n,n,k) = {} & A_n\ (1-\delta_k^{2n})\ P(k-n,2n-k,2n-k) \\ & + A_n\ \delta_n^{2n}\ P(n,0,2n+1) \\ & + A_n\ \delta_k^{2n}\ P(0,n,n)\end{aligned}$$

The methods used to test and solve the state equations are similar to those of Appendix to 2.2.1.

APPENDIX to 3.2.3

State probability P(i,j,k,s), strategy c

It holds: $0 \le i,\ j \le n$; $0 \le k \le 2n + 1$; n even $s = 0$ or 1

The k-value defines 4 areas of validity (as in 3.2.2). The state equations read (δ = Kronecker symbol, s + 1 means s + 1 mod 2)

area 1: $k = 0$; $0 \le i,\ j \le n$; $s = 0,1$

$$\begin{aligned}(A_1\ (1-\delta_{i+j}^{2n}) + A_n\ \delta_{i+j}^{\le n} + i + j)\ P(i,j,0,s) = {} & (\delta_s^0 + \delta_s^1\ \delta_i^n) \cdot A_1 (1-\delta_j^0)\ P(i,j-1,0,s+1) \\ & + (\delta_s^1 + \delta_s^0\ \delta_j^n)\ A_1 (1-\delta_i^0)\ P(i-1,j,0,s+1) \\ & + (i+1)(1-\delta_i^n)\ P(i+1,j,0,s) \\ & + (j+1)(1-\delta_j^n)\ P(i,j+1,0,s) \\ & + \delta_i^0\ P(n,j,2n+1,s) \\ & + \delta_{i+j}^{\le n} \sum_{x=\max(1,i)}^{n-j} P(i+n-x,j+x,x,s)\end{aligned}$$

area 2: $1 \leq k \leq n;\ n - k \leq i \leq n;\ k \leq j \leq n;$
$s = 0,1$

$$(A_1(1-\delta_{i+j}^{2n}) + A_n\delta_{i+j}^{n} + i + j - n + 1)\ P(i,j,k,s)$$
$$= (\delta_s^0+\delta_s^1\delta_i^n)\ A_1(1-\delta_j^k)\ P(i,j-1,k,s+1)$$
$$+ (\delta_s^1+\delta_s^0\delta_j^n)\ A_1(1-\delta_i^{n-k})\ P(i-1,k,s+1)$$
$$+ A_n\ (\delta_i^{\geq n/2}\ \delta_j^{\geq n/2}\ \delta_k^{n/2} + \delta_i^n\ (1-\delta_j^n)\ \delta_k^{>n/2}\ +$$
$$+ \delta_j^n\ (1-\delta_i^n)\ \delta_k^{<n/2} + \delta_i^n\ \delta_j^n\ (1-\delta_k^{n/2}))\ x$$
$$x\ P(i-n+k,j-k,0,s+1)$$
$$+ (i+1-n+k)\ (1-\delta_i^n)\ P(i+1,j,k,s)$$
$$+ (j+1-k)\ (1-\delta_j^n)\ P(i,j+1,k,s)$$
$$+ \sum_{x=n+1}^{2n} (\delta_i^{2n-x}\ \delta_j^{x-n}\ \delta_k^{x-n} + \delta_i^{x-n}\ \delta_j^{2n-x}\ \delta_k^{2n-x})\ x$$
$$x\ P(n,n,x,s)$$

area 3: $k = 2n + 1;\ 0 \leq j \leq n;\ s = 0,1$

$$(A_1\ (1-\delta_j^n) + A_n\ \delta_j^0 + j + 1)\ P(n,j,2n+1,s)$$
$$= A_1\ (1-\delta_j^0)\ P(n,j-1,2n+1,s+1)$$
$$+ A_n\delta_j^n\ P(0,j,0,s+1)$$
$$+ (j+1)(1-\delta_j^n)\ P(n,j+1,2n+1,s)$$
$$+ \delta_j^0\ P(n,n,2n,s)$$

area 4: $n + 1 \leq k \leq 2n$

$$2\ P(n,n,k,s)$$
$$= A_n\ (1-\delta_k^{2n})\ P(k-n,2n-k,2n-k,s+1)$$
$$+ A_n\ \delta_k^{2n}\ P(n,0,2n+1,s+1)$$
$$+ A_n\ \delta_k^{2n}\ P(0,n,n,s+1)$$

ACKNOWLEDGEMENT

The authors wish to express their thanks to W. Hofmann who wrote the programs for generating and solving the state equations as well as the simulation program.

REFERENCES

[1] Fortet, R.M., Grandjean C.H.: Study of Congestion in a Loss System", ITC 4 London, 1964.
[2] Gimpelson, L.A.: Analysis of Mixtures of Wide- and Narrow-Band Traffic, IEEE Trans. on Comm. Technology, vol. 13, no. 3, 1965
[3] Enomoto, O., Miyamoto, H.: An Analysis of Mixtures of Multiple Band-Width Traffic on Time Division Switching Networks, ITC 7 Stockholm, 1973.
[4] Katzschner, L., Scheller R.: Probability of Loss of Data Traffics with Different Bit Rates Hunting One Common PCM Channel, ITC 8 Melbourne, 1976.
[5] Akimaru, H., Tsuneizumi, T., Takahashi, H.: An Analysis of Two-Dimension Traffic Model, June 1983
[6] Nigge, K., Rothenhöfer, K.; Wöhr, P.: Switching for the German Satellite System, Electrical Comm., vol 59, no. 1/2, 1985
[7] Conradt, J.: Ein Algorithmus zur Lösung der Multi-slot Formel,to be published
[8] Stewart, W.J.: A Comparison of Numerical Techniques in Markov Modeling, Comm. of the ACM, vol. 21, no. 2, 1973

TELETRAFFIC ISSUES in an Advanced Information Society
ITC-11
Minoru Akiyama (Editor)
Elsevier Science Publishers B.V. (North-Holland)

EFFICIENCY OF SCHEDULING ALGORITHMS FOR SWITCHING SYSTEMS WITH DISTRIBUTED CONTROL

Wolfgang DENZEL

Institute of Communications Switching and Data Technics
University of Stuttgart, Fed. Rep. of Germany

ABSTRACT

The paper deals with problems on the intermodule communication in switching systems with distributed and modularized control. Two basic scheduling aspects are regarded in the path between an output queue of a processor and the input queue of another processor, namely intelligent polling mechanisms for output queues as well as intelligent task dispatching strategies for load sharing processors. In both cases, performance results are obtained by analytic means for generic queueing models of state dependent dynamic strategies. Comparisons between different simple and intelligent strategies referring to, e.g., the blocking probability, mean waiting time, control overhead, and implementation aspects allow the choice of the suitable strategy and the dimensioning of its parameters under given conditions.

1. INTRODUCTION

The control of digital switching systems becomes more and more distributed by using a large number of microprocessors. Flexibility, extensibility and availability can be obtained by this modularization in the control area. On the other hand, the problems of intermodule communication and the distribution of functions become more difficult.

A generalized model of such a control structure (see Fig. 1) consists of several functional groups of processors, e.g. special peripheral processors, signalling processors or universal processors for more centralized functions. By this, the principle of function sharing is implemented, but within such functional groups of processors the principle of load sharing can be used to reach flexibility and extensibility. The interprocessor communication usually is realized by sending messages into an output queue of the origin processor, from where they are transferred to the input queue of the destination processor by a communication system.

Assuming a centralized communication system, e.g. a bus, an effective access protocol is necessary. An often implemented method is some kind of polling. In the model this is represented by the lower switch. In consequence of the principle of function sharing, but also in the case of momentary overload, intelligent scheduling strategies for the outgoing messages can improve the performance, e.g. a dynamic state dependent polling mechanism. Such strategies are shortly considered in chapter three.

On the other side, the principle of load sharing implies a second scheduling problem, namely an intelligent message dispatching mechanism, e.g. a dynamic state dependent load balancing strategy. This is represented in the model by the upper switch. Different strategies are investigated in detail in chapter two.

Both mentioned intelligent scheduling strategies can commonly be implemented, e.g., in a centralized communication system control, as the bus control in the regarded model. In [1] an integrated switching system with a control structure according to Fig. 1 has been presented. The performance of the whole system has been investigated there by simulation. Since a detailed simulation technique of the whole system is not the suitable method to investigate many different scheduling strategies, this is done in this paper by analytic means for generic models of the two considered scheduling problems. The results allow the choice of the optimal strategy and the dimensioning of its parameters for both cases. So, the whole system can be investigated with the chosen strategy either by simulation or probably by analytic approximations, which eventually can be applied, because simplifications are allowed, if optimal scheduling strategies are used. This could be, e.g., the use of an equivalent single queue model as a replacement for a multiqueue submodel.

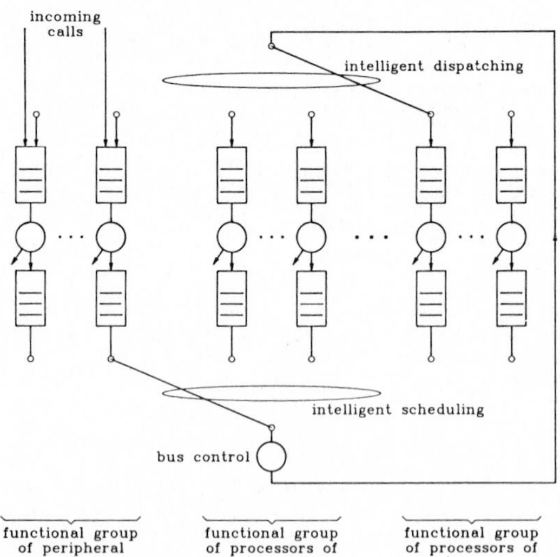

Fig. 1 Generalized queueing model of a modularized switching control

2. PERFORMANCE OF DISPATCHING STRATEGIES

2.1 Dispatching Strategies

Dispatching strategies are taken into consideration, if there exist several processors for a set of functions according to the principle of load sharing. Simple dispatching strategies are regarded here for comparisons with intelligent strategies. In the last case, the dispatcher uses system state informations for the decision, where a message has to be assigned to. Service time oriented strategies are neglected here, as in general the service times cannot be derived from the type of message. Dispatching priorities can also be neglected, because load sharing processors usually are of the same type.

The following dispatching mechanisms are considered:

Strategy 1: Random dispatching, according to assignment probabilities (e.g. equally distributed).

Strategy 2: Ordinary cyclic dispatching.

Strategy 3: Fully dynamic (state dependent) dispatching, that means "join the shortest queue" strategy. In case of equality of shortest queue lengths the dispatching follows, e.g.,
a) strategy 1, equally distributed, or
b) strategy 1, but deterministically to, e.g., the first of the shortest queues, or
c) strategy 2, cyclically, e.g. relative to the last assignment.

Strategy 4: Partially dynamic (state dependent) dispatching, that means according to, e.g., strategy 1, equally distributed among those queues, whose queue length is below a given threshold, and if all queue lengths exceed the threshold the dispatching follows, e.g.,
a) strategy 1, equally distributed, or
b) strategy 3a.

Strategy 5: Dispatching with second attempt, that means according to, e.g., strategy 1, equally distributed. But if the selected queue is full, one second attempt is possible, e.g., according to
a) strategy 1, equally distributed among all other queues, or
b) strategy 1, equally distributed among those queues which are not full, or
c) strategy 3a.

Strategy 1 is regarded, because its analysis is very simple, but in practice it is not so easy to implement as strategy 2. The intelligent strategy 3 is the optimal strategy referring to blocking and waiting time, but its implementation is difficult and the control overhead is high, because the dispatcher has to know the whole system state at any time. Therefore, strategy 4 and strategy 5 are considered, where the dispatcher not so often needs so much of the system state as in strategy 3. The implementation is easier than for strategy 3, especially for strategy 5.

2.2 Basic Queueing Model

Fig. 2 shows the generic queueing model for the investigations of dispatching strategies. A message arriving with rate λ is assigned to one of g queues according to the special dispatching strategy. It gets blocked with probability B, if it cannot be assigned to the queue or the queues, which were defined by the dispatching strategy. A given queue i has s_i waiting places and the appropriate server has the serving rate ε_i. The arrival and service process must be of Markovian type (M), that a state space analysis for the intelligent strategies can be applied. This restriction is indeed not too bad for applications in switching controls. This has been shown in the detailed simulation study of a switching system in [1]. Measurements taken at various points in the model have shown that traffic streams can be described by coefficients of variation for the interevent time between 0.9 and 1.1 in most cases.

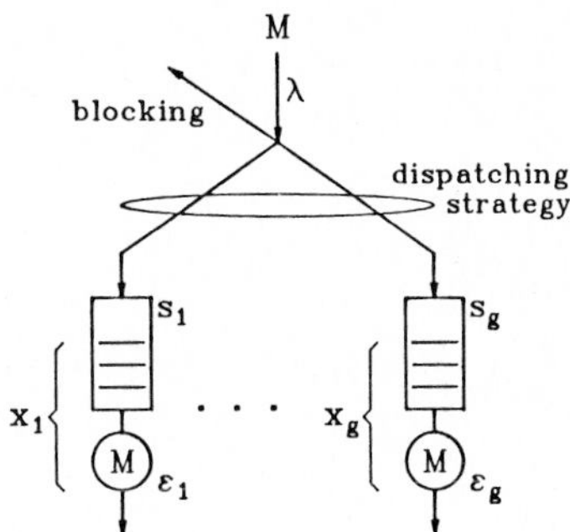

Fig. 2 Basic model for dispatching strategies

2.3 Solution Method

2.3.1 The State Space

The solution of strategy 1 can be reduced to the solution of a M/M/1-s delay-loss system with arrival rate $\lambda_i = p_i \cdot \lambda$, where p_i is the dispatching probability, e.g. $p_i = 1/g$ for equally distributed random dispatching. The characteristic values can be obtained, e.g., from tables [2].

The investigation of strategy 2 can be reduced to the solution of an E_g/M/1-s delay-loss system with Erlang-g arrival process and rate $\lambda_i = \lambda/g$ as above. The characteristic values can easily be obtained by a recursive solution of an appropriate two-dimensional state space.

For all other intelligent strategies state spaces have to be set up with the appropriate state transition rates. For all these cases a state $\underline{X}$ must be at least a g-dimensional vector, where $\underline{X} = (x_1, x_2, \ldots, x_i, \ldots, x_g)$; x_i represents the momentary queue length of queue i, including the server. Strategies with cyclic components must have states of one more dimension for the cycle number.

Fig. 3 shows a simple example of the state space for two queues, each with three waiting places and for the dynamic strategy 3. Each state is represented by the vector $\underline{X} = (x_1, x_2)$. In case of equality of the queue lengths, messages are assigned to queue 1 with the probability p_1 and to queue 2 with $p_2 = 1-p_1$. For strategy 3a p_1 and p_2 have a value of 0.5 and for strategy 3b p_1 is 1.0 and p_2 is 0.

In practical applications there are normally more than two queues with some more waiting places and therefore some hundred or thousand of states. Furthermore, the multi-dimensional state spaces can no more be described in a graphical manner. Thus, a special software tool has been implemented, which allows on the one side a simplified interactive input of states and transitions for smaller but complicated state spaces or, on the other side, the definition of great state spaces by an algorithmic description of the transitions. The program produces the conditional state transition probabilities for arrivals as well as directly the system of linear state equations for the statistical balance. Then the state probabilities can be solved by an iterative Gauss-Seidel algorithm. Out of these state probabilities and the state transition probabilities, characteristic mean values of the dispatching strategies can be calculated as follows.

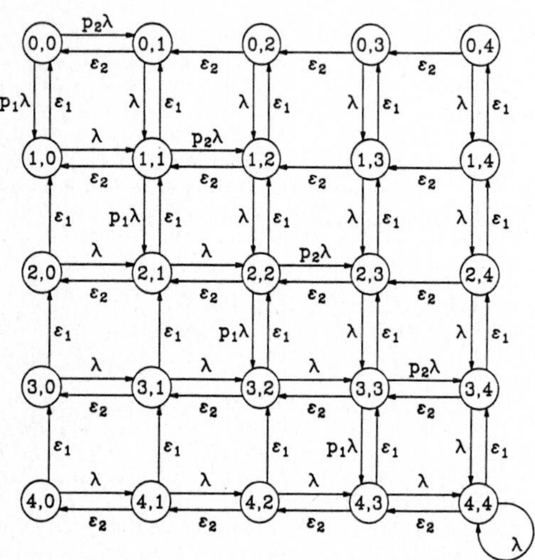

Fig. 3 Example of a state space for dispatching strategy 3

2.3.2 The Characteristic Values

The equations in this paragraph can be applied for state dependent dispatching strategies, as the strategies 3, 4 and 5. Thereby the following assumptions are made:

- g number of queues,
- n_i number of waiting places of queue i, including the server ($n_i = s_i+1$),
- $\underline{X}$ state $(x_1,x_2,\dots,x_g)$,
- $p(\underline{X})$ state probability of the state $\underline{X}$,
- $p_i(\underline{X})$ conditional state transition probability for the transition from state $\underline{X}$ to state $(x_1,x_2,\dots,x_i+1,\dots,x_g)$.

The following characteristic values can be calculated for every subsystem i and some of them also as global values for the whole system:

- Probability, that an arriving customer is assigned to queue i:

$$q_i = \sum_{\underline{X}} p(\underline{X})\cdot p_i(\underline{X}) \tag{1}$$

- Arrival rate at queue i:

$$\lambda_i = q_i\cdot\lambda \tag{2}$$

- Blocking probability at queue i conditioned that a call is assigned to queue i and total blocking probability, i.e. the probability that an arriving call is rejected:

$$B_i = 1/q_i\cdot\sum_{\underline{X}|x_i=n_i} p(\underline{X})\cdot p_i(\underline{X}) \tag{3a}$$

$$B = \sum_{i=1}^{g} q_i\cdot B_i \tag{3b}$$

- Mean load of server i and total mean load of the whole system:

$$Y_i = 1 - \sum_{\underline{X}|x_i=0} p(\underline{X}) \tag{4a}$$

$$Y = \sum_{i=1}^{g} Y_i \tag{4b}$$

- Mean queue length of queue i and total mean number of waiting customers in the whole system:

$$\Omega_i = \sum_{\underline{X}|x_i>0} (x_i-1)\cdot p(\underline{X}) \tag{5a}$$

$$\Omega = \sum_{i=1}^{g} \Omega_i \tag{5b}$$

- Mean waiting time (with respect to all arrivals in queue i) and global mean waiting time for an arriving call at the system:

$$w_{1\,i} = \Omega_i / \lambda_i \tag{6a}$$

$$w_1 = \Omega / \lambda \tag{6b}$$

- Waiting probability in queue i and global waiting probability for an arriving customer at the system:

$$W_i = 1/q_i\cdot\sum_{\underline{X}|0<x_i<n_i} p(\underline{X})\cdot p_i(\underline{X}) \tag{7a}$$

$$W = \sum_{i=1}^{g} q_i\cdot W_i \tag{7b}$$

- Mean waiting time (with respect to the waiting arrivals in queue i) and global mean waiting time for a waiting customer in the whole system:

$$t_{w\,i} = w_{1\,i} / W_i \tag{8a}$$

$$t_w = w_1 / W \tag{8b}$$

- Probability, that queue i is found at threshold $x_i = z$ and probability, that any queue is found at threshold $x_i = z$:

$$p_{th\,i}[z] = \sum_{\underline{X}|x_i=z} p(\underline{X}) \tag{9a}$$

$$p_{th}[z] = \sum_{i=1}^{g} q_i\cdot p_{th\,i}[z] \tag{9b}$$

2.4 Implementation Aspects

Regarding the state dependent strategies and assuming a centralized dispatcher, a certain amount of system state information has to go back from the system to the dispatcher, dependent on the strategy. Fig. 4 shows a model, which describes this fact. Messages arriving with rate λ, have to be assigned to one of the queues, that means, they have to pass a dispatching service phase with service time t_d. Informations of the system state or its changes or blocking informations are sent back from the system or, respectively, from the ports of the queues through the communication system, e.g. a bus, to the dispatcher with rate λ_u, as far as they are needed for the dispatching strategy. These informations are assumed to pass an update service phase with service time t_u in the dispatcher. So, the necessary mean utilization of the dispatcher for a given strategy can be estimated by $t_u \cdot \lambda_u + t_d \cdot \lambda$, and the additional overhead load for the communication system is determined by $\lambda_u \cdot h$, where h is the transfer time for a message of the communication system. As λ_u is a function of the used strategy and its implementation, in the following the minimal necessary rate λ_u is used as a comparison measure for the control overhead of a dispatching strategy. λ_u is assumed to include informations on blocking, because losses usually are not allowed in switching controls.

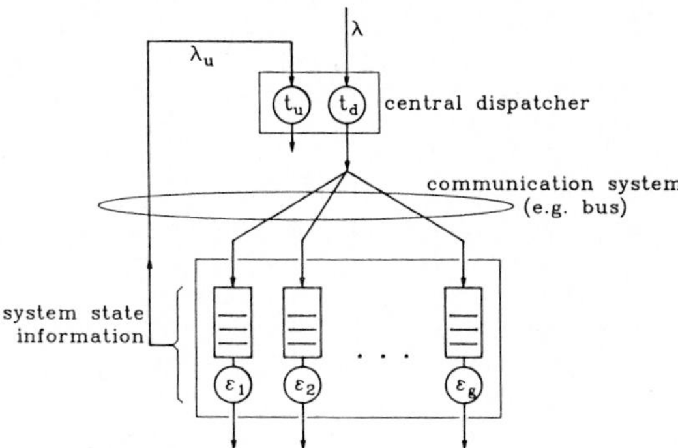

Fig. 4 Model for state dependent dispatching control

In the following some implementation methods and the appropriate rates λ_u are given:

Strategy 1 and 2: Normally no system state information is necessary, but to avoid losses, for each blocked message one back message contributes to λ_u. Hence, it follows:

$$\lambda_u = \lambda \cdot B \tag{10}$$

Strategy 3: As the arriving messages directly reach the dispatcher, only one information is necessary for each termination of a service phase in the system, to allow the dispatcher to reconstruct the system state. Therefore, the minimal necessary rate λ_u is:

$$\lambda_u = Y \cdot \sum_{i=1}^{g} \varepsilon_i = \lambda \cdot (1 - B) \tag{11}$$

Strategy 4: For each arrival, one information is necessary, whether the threshold th is reached, and for each termination of a service phase in the system, one information is necessary, whether the queue length falls below the threshold th. Hence, it follows:

$$\lambda_u = \lambda \cdot p_{th}[th-1] + \lambda \cdot (1 - B) \cdot p_{th}[th] \tag{12}$$

Strategy 5: Each arrival at a full queue causes at the first attempt one blocking message. A queue is found full with the probability $p_{full} = p_{th}[z=n_i]$. For the second attempt, the strategies 5a, b and c must be distinguished:

a) If the second attempt is blocked, one more blocking message is necessary. Hence, it follows:

$$\lambda_u = \lambda \cdot (p_{full} + B) \tag{13a}$$

b) After the first attempt, from each queue, which is not full, one information is necessary. Therefore, we have:

$$\lambda_u = \lambda \cdot p_{full} \cdot (1 + g \cdot (1 - p_{full})) \tag{13b}$$

But the impementation method of strategy 4 can also be applied here with th = n_i.

c) After the first attempt, from each other queue one information is necessary, that means g messages all over. Thus, it follows:

$$\lambda_u = \lambda \cdot g \cdot p_{full} \tag{13c}$$

2.5 Results

All considered strategies have been investigated in numerous parameter studies. Some typical examples of the obtained results are presented in the following.

First, the differences between the strategies should be illustrated by the global blocking probability B, the global mean waiting time t_w, and the waiting probability W versus the offered load, defined by $\rho = \lambda / \Sigma \varepsilon_i$. In the second example, the influence of the number of queues on the blocking probability B will be shown. Finally, the control overhead rate λ_u, as defined in section 2.4., is demonstrated for different strategies.

Fig. 5 shows the blocking probability B for a system with g = 4 queues, each with s_i = 6 waiting places. The strategies 1,2,3, 5a and 5b are considered. The dynamic strategy 3 is the best, while the random strategy 1 is the worst case. Thereby, strategy 3a and 3b have exactly the same results for the global characteristic values. In all investigated parameter studies, the blocking probability of the cyclic strategy 2 lies approximately in the middle between strategies 1 and 3 in logarithmic scale. Furthermore, two examples for strategy 5, namely 5a and 5b, are also shown in the diagram. The curves for different thresholds of the queue length of strategy 4, which are not included in the diagram, are located in the range between strategies 1 and 3, where every blocking level can be achieved through a proper choice of the thresholds.

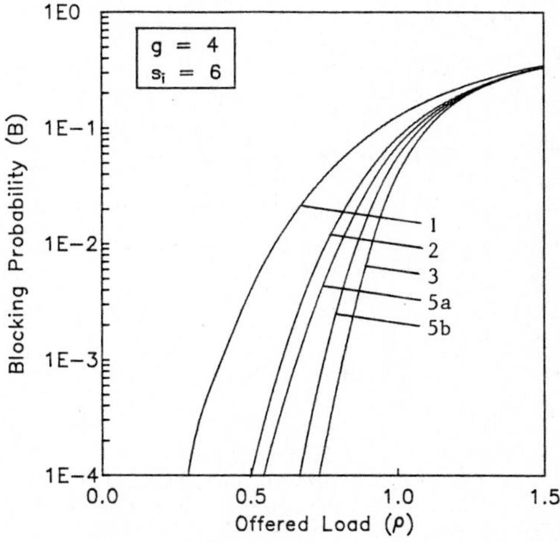

Fig. 5 Blocking probability versus offered load for dispatching strategies

Regarding the mean waiting time t_w of the waiting customers for the same system, Fig. 6, the areas of normal load and overload must be separated. In the normal load range, strategy 3 is already the best, strategy 1 is the worst case, and strategy 2 lies between them. In the overload situation, the waiting time for strategy 3 is longer, because the throughput is higher in this case. Referring to the waiting time, the strategies 4 and 5 do not work very well, because intelligent dispatching is only active in the high load range. The examples of strategies 5a and 5b demonstrate this fact.

To get an idea of the mean waiting time of all arriving customers, the waiting probability W is shown in Fig. 7 for the same system. Due to the higher blocking probability in the overload case, the waiting probability decreases in this range.

The effect, that the cyclic strategy 2 is quite better than the random strategy 1 was already shown in other context in [7].

Summarizing all obtained results, the following statements can be given. The tendencies are approximately independent of the number of waiting places, that means, that the differences between the strategies remain relatively constant with increasing number of waiting places, although, of course, the blocking probability decreases and the waiting time increases. On the other hand, the difference between the strategies increases with the number of queues, but constant offered load per queue. This effect is the greater the more intelligent the strategies are, since the probability that an arriving customer finds a free place is the greater the more queues there are disposable for the dispatching strategy. This fact is demonstrated for the blocking probability B of the strategies 1, 2 and 3 in Fig. 8 for a system with $s_i = 2$ waiting places for each queue and constant offered load of $\rho = 0.5$. Strategy 1 naturally is independent of the number of queues, strategy 2 gets better and for strategy 3 the blocking probability decreases almost exponentially.

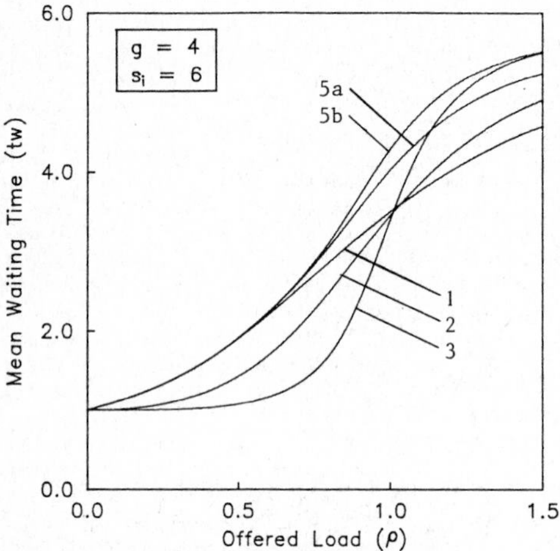

Fig. 6 Waiting time versus offered load for dispatching strategies

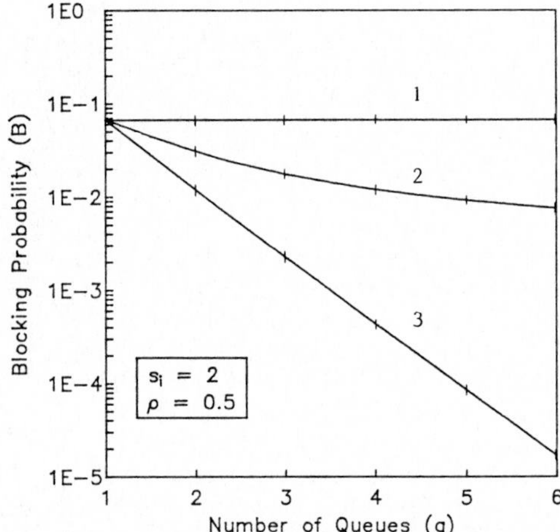

Fig. 8 Blocking probability versus number of queues for dispatching strategies

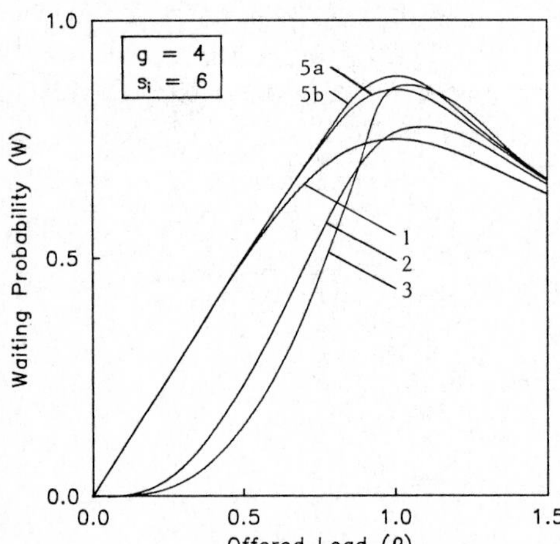

Fig. 7 Waiting probability versus offered load for dispatching strategies

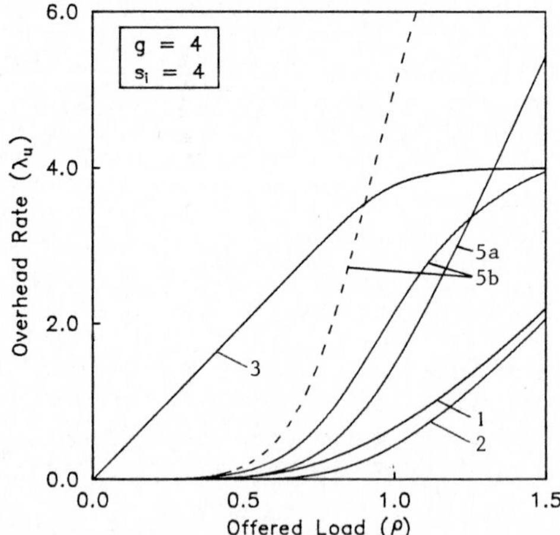

Fig. 9 Overhead rate versus offered load for dispatching strategies

Finally, the control overhead rate λ_u will be discussed, which is shown in Fig. 9. In the normal load range, strategy 3 is the worst case, but λ_u remains saturated at a relatively small level in the overload case, whereas the other strategies increase rapidly in the overload range. The non-intelligent strategies 1 and 2 naturally are the best strategies under the overhead aspect. For strategy 5b two curves are shown. The dashed curve stands for the application according to equation (13b), while the bold curve stands for the application according to equation (12), which also can profitably be implemented here.

3. PERFORMANCE OF POLLING STRATEGIES

3.1 Polling Strategies

As in chapter 2 for dispatching strategies, some corresponding polling mechanisms can be considered. Here, also many subcases could be regarded, but this is ommitted in the following list of examples, since the results do not differ significantly:

Strategy 1: Random polling, according to scheduling probabilities (e.g. equally distributed).

Strategy 2: Ordinary cyclic polling.

Strategy 3: Fully dynamic (state dependent) polling, that means, that the longest queue is the next queue being served.

Strategy 4: Partially dynamic (state dependent) polling, that means according to, e.g., strategy 1, if the queue lengths are below a given threshold. But if a queue exceeds this threshold, the appropriate queue is the next queue being served.

Strategy 5: Like strategy 4, but with an hysteresis range of the queue length instead of one fixed threshold.

3.2 Basic Queueing Model

Fig. 10 shows the generic queueing model for the investigations of polling strategies. Thereby, g queues are polled by the polling server with the service rate ε corresponding to the scheduling strategy. A queue i has s_i waiting places and an arrival rate λ_i. The arrival and service processes are supposed to be of Markovian type, because of the same reasons as in the model for dispatching mechanisms.

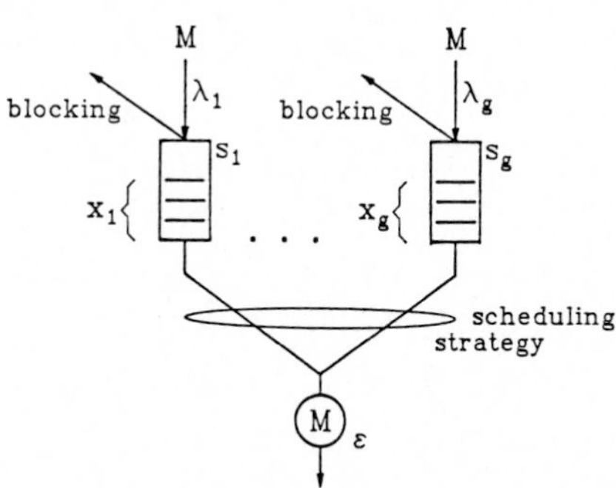

Fig. 10 Basic model for polling strategies

3.3 Solution Method

As the solution is very similar to that for dispatching strategies, a detailed description can be ommitted here.

The state variables are the numbers x_i of occupied places in each queue. One additional state for the case of a free server must be used. For strategy 2 one more state variable is necessary for the cycle number. For strategy 5 one more state variable is necessary for each queue to indicate between the hysteresis thresholds, whether at last the upper limit has been exceeded or the lower limit has been crossed.

3.4 Results

Although polling mechanisms have also been investigated in numerous parameter studies, only one example will be shown here, because the differences between the strategies are very small.

Fig. 11 shows the blocking probability B and Fig. 12 the mean waiting time t_w of the waiting customers for a symmetrical system with g = 4

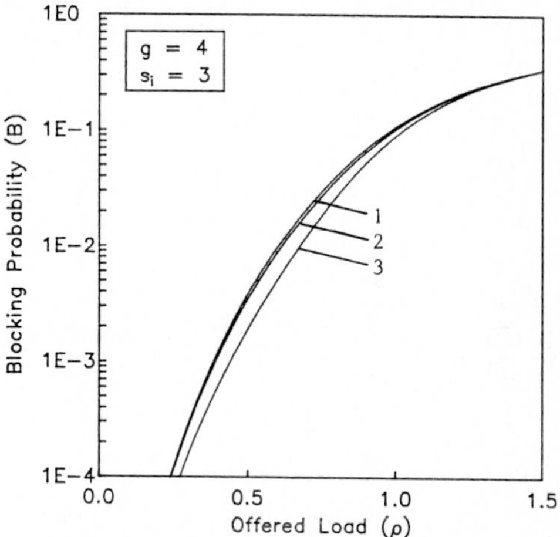

Fig. 11 Blocking probability versus offered load for polling strategies

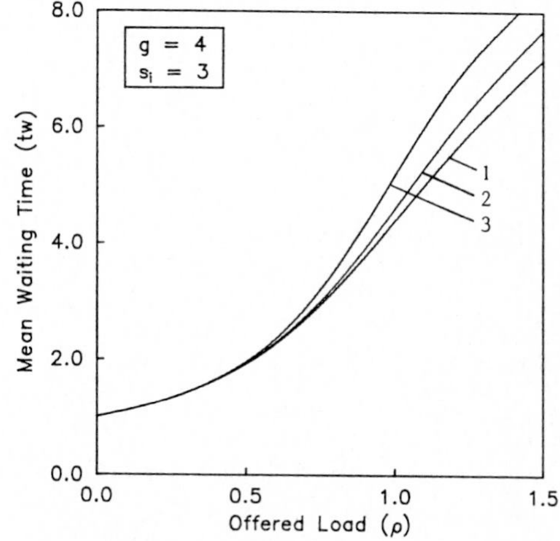

Fig. 12 Waiting time versus offered load for polling strategies

queues, each with $s_i = 3$ waiting places, versus the offered load $\rho = \Sigma\lambda_i/\varepsilon$. The curves show that the cyclic strategy 2 has almost the same blocking probability as the random strategy 1, while the dynamic strategy 3 is insignificantly better. The waiting time is almost the same for all strategies in the normal load range, but in the overload range the intelligent strategies are worse.

All cases for different thresholds of strategies 4 and 5 lie between strategies 1 and 3, and a comparison between strategy 4 with a threshold th and strategy 5 with an hysteresis range around th renders almost the same results.

So, it seems to be not very usefull to implement state dependent polling strategies, but the advantage must be seen in the behaviour for unsymmetrical load and momentary overload, since a state dependent strategy automatically includes dynamic priorities and therefore adapts itself to the queue-individual loads. Investigations on these effects are in work at present.

4. CONCLUSIONS

The implementation of intelligent state dependent load dispatching mechanisms in systems with modularized control according to the principle of load sharing improves the performance. The blocking probability and the waiting time can be evidentually reduced by the mentioned strategies 3, 4, or 5, compared to simple mechanisms, as the strategies 1 or 2. The effectiveness of the intelligent strategies, especially strategy 3, is the greater the more servers are sharing the load, but the performance improvements must be payed for by a higher amount of control overhead. Therefore, the strategies 4 and 5 can be a compromise, especially strategy 5b. The following table gives the designer of such control systems a general summary of the results for the mentioned strategies:

Disp. Strategy	Blocking Probab.	Waiting Time Normal Load	Waiting Time Overload	Control Overhead Normal Load	Control Overhead Overload
1	high	high	low	low	low
2	medium	medium	medium	low	low
3	low	low	high	high	medium
4(a) *)	high ... low	high ... low	low ... high	low ... high	low ... high
5(b)	low	high	high	low	medium

*) Parameter: Threshold th of the queue length

If the principles of these dispatching mechanisms are also applied for intelligent state dependent polling strategies, the effectiveness of intelligent strategies, as the mentioned polling strategies 3, 4, or 5, do not improve the performance very much for symmetrical load. The blocking probability can be lowered insignificantly, but the waiting time and the amount of control are in principle higher. Nevertheless, intelligent polling mechanisms can improve the performance for unsymmetrical and dynamically changing load, since these strategies are adaptive. Under dynamic and overload aspects, the application can although be profitable, but these effects are not investigated in this paper.

For the realization of intelligent scheduling strategies, a centralized control unit, e.g., a bus control unit, is suitable. The overhead of the control information which has to be transferred from the single modules to this scheduling control unit, has been estimated in this paper for the various dispatching mechanisms. But a decentralization of the scheduling control is also imaginable, e.g., being located in a microprocessor control of the communication system access ports of each unit. In this case, the scheduling control information messages with a fixed destination must be replaced by broadcast messages.

Furthermore, analytic performance investigations of the whole system can be facilitated by the implementation of the best scheduling strategies, because then certain approximations can be applied. So, e.g., the blocking probability for the dispatching strategy 3 can be approximated quite well by the blocking probability of a single queue M/M/g-s delay-loss system with $s = \Sigma s_i$ waiting places.

ACKNOWLEDGEMENT

The author would like to express his gratitude to Prof. P. Kuehn, Head of the Institute of Communications Switching and Data Technics, University of Stuttgart, for supporting this work.

REFERENCES

[1] W. Denzel, W. Weiss, "Modelling and Performance of a New Integrated Switching System for Voice and Data with Distributed Control", Proc. 10th Int. Teletraffic Congress, Montreal, paper 1.1-3, 1983.

[2] P. Kuehn, "Tables on Delay Systems", Inst. of Communications Switching and Data Technics, University of Stuttgart, 1976.

[3] L. Kleinrock, "Queueing Systems", Vol. I/II, Wiley + Sons, New York, 1975/76.

[4] R. Schassberger, "Ein Wartesystem mit zwei parallelen Warteschlangen", Computing 3, pp. 110-124, 1968.

[5] U. Herzog, P. Kuehn, "Comparison of some Multiqueue Models with Overflow and Load-Sharing Strategies for Data Transmission and Computer Systems", Symposium on Computer-Communications Networks and Teletraffic, Polytechnic Institute of Brooklyn, pp. 449-472, 1972.

[6] G. Császár, R. Konkoly, T. Szádeczky-Kardoss, "Optimal Allocation of Calls to Private Automatic Branch Exchange Operators", Budavox Telecommunication Review, Budapest, Vol. 4, pp. 1-6, 1982.

[7] M. Buchner, S. Neal, "Inherent load balancing in step by step switching systems", BSTJ 50, pp. 135-357, 1971.

[8] H. Takagi, L. Kleinrock, "Analysis of Polling Systems", Japan Science Institute Research Report, 1985.

[9] P. Tran-Gia, T. Raith, "Multiqueue Systems with Finite Capacity and Nonexhaustive Cyclic Servive", Inst. of Communaications Switching and Data Technics, University of Stuttgart, to be published.

TELETRAFFIC ISSUES in an Advanced Information Society
ITC-11
Minoru Akiyama (Editor)
Elsevier Science Publishers B.V. (North-Holland)

AN APPROACH FOR ANALYSIS OF A CLASS OF DISTRIBUTED SPC SYSTEMS

Wojciech BURAKOWSKI

Institute of Telecommunication,
Technical University of Warsaw, Poland

Dariusz BURSZTYNOWSKI

Industrial Institute of Automation
and Measurement, Warsaw, Poland

ABSTRACT

The paper presents approximate methods for evaluation of mean waiting time in a priority tandem-queue system fed by a single task sequence according to a Poissonian process. A task sequence consists of several tasks executed one after the other in a prescribed order. Each task is characterized by: the position in sequence, the number of server the task is executed by, a non-preemptive priority assigned to it and service time (constant). Within each priority the FIFO discipline is assumed.
Exact methods for calculating mean waiting time of tasks executed in the first server of the considered system are known. In this paper approximate evaluation of mean waiting time of tasks executed in the second server is proposed.

1. INTRODUCTION

This work is motivated by the need of analytical methods to analyse control systems in a class of distributed SPC switching systems. These systems are designed to serve different processes activated by external events. More specifically, the service of an event requires execution of a task sequence.
These systems are generally modelled as open queueing networks with the number of servers corresponding to the number of processors. The input process is determined by the number of different types of external events and their arrival process. The type of an event corresponds to this task sequence which is to be executed in the system as a consequence of an arrival.
A task sequence consists of several tasks served one after the other in prescribed order. Each task is characterized by the following set of parameters:

- position in the sequence;
- number of server the task is executed by;
- a non-preemptive or preemptive priority;
- the service time distribution.

Tasks from the sequence are executed consecutively according to their positions in the sequence. A task is said to be activated when the previous task has been served (the first task is activated at the moment of an external event arrival).
Taking the above said into account, we arrive at an open queueing network with priorities (non-preemptive and preemptive) and internal (within one server) as well as external (between servers) feedbacks.

State of art

In the case of a single-processor control system we are able to calculate exact values of mean waiting time for all tasks executed in the system. This issue was studied with success by several authors [1], [2], [3], [4] (in [2] the author considers one task sequence case only but the method may be generalized to the multi - task sequence case). The proposed methods are based on a set of linear equations in which the mean waiting time for particular tasks are the unknows.
Attempts of calculating the distribution of response time for tasks were made by Fontana [8]. Unfortunately, the proposed model is constrained to two priority system case.

Generally, the analysis of queueing networks with priorities and feedbacks is very complex. At present no satisfactory solution has been proposed. The existing approximate methods of analysis of general queueing networks (e.g. [9], [10]) do not consider priorities. On the other hand, approximations which take priorities into account(e.g. [5], [7], [11]) are suitable only for BCMP network representations and fail in the case of internal feedback. Few realistic approximations were proposed to cope with control systems in SPC switching systems. For example, Villen [6] considers the impact of internal and external feedbacks on waiting time for particular tasks but without priorities (processors independence also assumed).

Given the present state of art, the most realistic approach to the analysis of a multi-processor system with priorities and feedbacks seems to be assuming independence between processors and analysing them separately using one of the methods presented in [1], [2], [3], [4].
It is also worth noting that in most practical SPC systems the service time is approximately constant [12]. From analytical viewpoint this is a serious difficulty since constant service times introduce strong correlation between processors. An attempt to study this correlation is one of the goals of the paper.
The system considered in this paper is simplified with respect to the general system we have described earlier. It consists of two servers and one type of task sequence. External feedback is not considered. Approximate methods to evaluate mean waiting time for tasks executed in the second server are proposed.
In section 2. the considered model is presented in detail. Approximate methods for particular

subsystems are discussed in section 3.

2. THE MODEL

The considered model consists of two servers in tandem (Fig. 1) such that:

- one type of task sequence is executed;
- the external events arrive accoring to a Poisson process with parameter λ (i.e. the first task of a sequence is activated at the moment of its arrival);
- each task sequence consists of several tasks, T, executed one after the other in a prescribed order; a subsequent task is activated at the moment in which its preceding task has been served;
- task i (i=1,...,T) is characterized by the following set of parameters:
  - position in the sequence i;
  - number of server the task is executed by (j=1,2);
  - non-preemptive priority - p; (p=1,2; 1 is the high, 2 is the low);
- for each priority the queue discipline is FIFO;
- there is no feedback between servers.

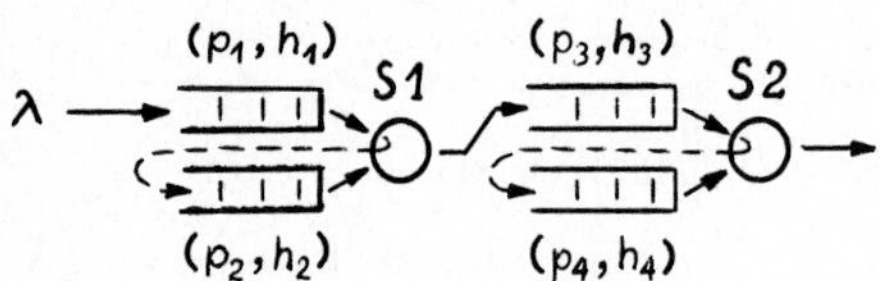

Fig. 1. The general model

Using one of the methods presented in [1], [2], [3] or [4] we can calculate exact values of mean waiting time for tasks executed by server no. 1. Approximate methods of evaluating mean waiting time in second server for several subsystems of the model from Fig.1. are proposed in the next section.

3. ANALYSIS

In this section we describe methods that are used in the analysis of several subsystems of our system from Fig. 1. First of all we analyse a simple queue-tandem without priorities and feedbacks. We show simple closed formula for mean waiting time of tasks in the second server. This can be extended to the case of many servers in series. Next, we consider two cases of three-task sequences such that: i) two first tasks with different priorities are executed in the first server; ii) two last tasks with different priorities are executed in the second server.

3.1. Subsystem no.1.

The considered subsystem is depicted in Fig.2.

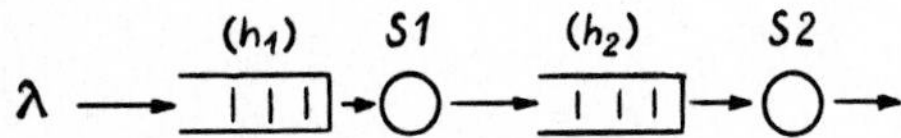

Fig.2. The subsystem no.1

Let:

h_1, h_2 - constant service times of task 1 and 2;
λ - intensity of Poisson stream;
W_1,W_2 - mean waiting time of tasks 1 and 2.

From the M/D/1 system solution the expresion for W_1 is:

$$W_1=(h_1/2)\,\rho_1/(1-\rho_1), \tag{1}$$

where $\rho_1=\lambda\cdot h_1$.

It appears that we may calculate the value of W_2 from the simple expression:

$$W_2=\begin{cases} W_2^*-W_1, & \text{for } h_2>h_1 \\ 0, & \text{for } h_2\leqslant h_1 \end{cases} \tag{2}$$

where:

$W_2^*=(h_2/2)\,\rho_2/(1-\rho_2)$ (see expression (1)).

Formula (2) may be extended to the case with n servers in series. Let W_i denote mean waiting time of tasks executed by the i-th server (i=2,...,n):

$$W_i=\begin{cases} W_i^*-\sum_{j=1}^{i-1} W_j, & \text{for } h_i>\max_j h_j \\ 0 & \text{otherwise} \end{cases} \tag{3}$$

where $W_i^*=(h_i/2)\cdot\rho_i/(1-\rho_i)$ (see expressions (1) and (2)).

Formulas (2) and (3) had been verified by simulation for a wide range of i, λ, h_i values. Excellent accordance of analytical and simulation results was observed in all cases. Unfortunately, the authors failed to prove these formulas.

In Appendix the general expression for the mean waiting time in the second server in a tandem-queue with constant service times is derived. It converges to the expression (2) in some special cases. The above said suggests that (2) is generally true.

3.2. Subsystem no.2

The system is depicted in Fig. 3

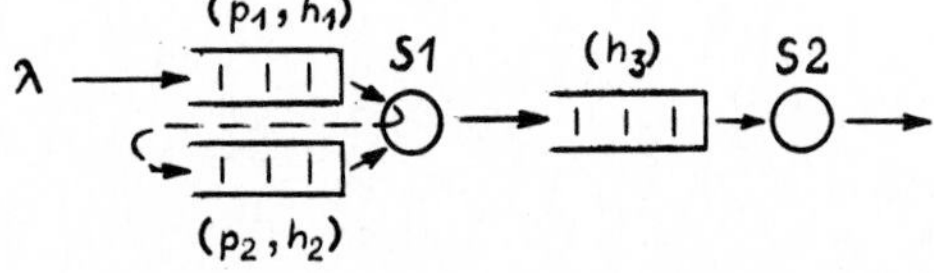

Fig.3. Subsystem no.2

The task sequence consists of three tasks. Two first tasks have different priorities and are executed in the first server.

Clearly, for $h_3\leqslant h_2$, W_3 equals 0. Thus $h_3>h_2$ is the interesting case.

We distinguish two cases:

A: $p_1=2$, $p_2=1$;

B: $p_1=1$, $p_2=2$.

3.2.1. Case A

From the viewpoint of analysis, we can reduce this system to simple tandem system from section 3.1 and use the expression (2) with (h_1+h_2) being the total service time in the first server. This results from the fact that $W_2=0$.

3.2.2. Case B

In this case, we can not calculate the value of W_3 in simple way. Thus, we use an approximation. Consider a "test-event" occuring at random. The test-event is said to activate the

"test-sequence" containing "test-tasks".
The approximation is based on the following:

i) At the moment of test-event arrival (the moment of test-task 1 activation) the average number of tasks present in the system (and in each queue separately) follows from the Little's theorem;

ii) The value of the test-task 3 mean waiting time, W_3, depends on the following factors:
   1° the number of tasks present in server 1 at the activation moment of test-task 1 (excluding the task with priority 2 being in service at this moment).
   2° the service time of all tasks present in server 2 (queue and service) at the moment when first task from among those defined in 1° leaves server 1 (we denote this moment as t_1).

Let W_3' and W_3'' denote complements of W_3 resulting from factors 1° and 2°, respectively. Then:

$$W_3 = W_3' + W_3'' \tag{4}$$

The approximations for W_3' and W_3'' are shown below.

$\underline{W_3'}$

The mean number of the tasks from p.1°, say L_2, is given by the following expression:

$$L_2 = W_1\lambda + \rho_1 + W_2\lambda \tag{5}$$

All these tasks at the activation moment of test-task 2 have priority 2 and they will leave server 1 before the activation moment of test-task 3. We assume that the distribution of the number of these tasks is approximated by the geometrical distribution with parameter:

$$\hat{\rho} = L_2/(1+L_2) \tag{6}$$

Now:

$$W_3' = \sum_{n=1}^{\infty} p_n \cdot W_3'(n), \tag{7}$$

where:
- $p_n = (1-\hat{\rho})\hat{\rho}^n$
- $W_3'(n)$ - conditional mean waiting time of test-task 3 resulting from the presence of n tasks defined in p.1°

In order to evaluate $W_3(n)$, we must know the distribution function of random variable, say X, describing the time interval between consecutive output moments from server 1 of tasks described in p. 1°.
The X takes values from a numerable set, say H, $H=\{h_2+k\,h_1;\ k=0,1,\ldots\}$ with probabilities:

$$Pr(X=h_2+k\cdot h_1) = F_k(\rho_2,\rho_1)\cdot \exp[-(\rho_2+k\rho_1)] \tag{8}$$

where:

$F_0(a,b)=1$

$F_1(a,b)=a$

$F_2(a,b)=(a^2/2!)+a\,F_1(b,b)$

$F_3(a,b)=(a^3/3!)+(a^2/2!)\,F_1(2b,b)+a\cdot F_2(b,b)$

$F_4(a,b)=(a^4/4!)+(a^3/3!)\,F_1(3b,b)+(a^2/2!)\,F_2(2b,b)+$
$+a\,F_3(b,b)$

....
e.t.c

We introduce a new random variable X_1 which takes values from the set, say H_1, $H_1=\{h_3-h_2-k\cdot h_1;\ k=0,1,\ldots\}$ with probabilities:

$$Pr(X_1=h_3-h_2-k\cdot h_1) = Pr(X=h_2+k\cdot h_1) \tag{9}$$

X_1 describes the influence on the waiting time of test-task 3 of a single task served in server 1 just before test-task 2.
Next, we assume server 1 is empty at moment t_1 (in reality, it may be occupied by tasks from p. 2°).
We calculate $W_3'(1)$ from the expression:

$$W_3'(1) = \sum_{x:x>0} x\cdot Pr(X=x) \tag{10}$$

In order to evaluate $W_3'(2)$, we make two following steps:

<u>step 1:</u>
Define random variable X_2^1 which takes only non-negative values from set H_1 with probabilities:

$$\forall x>0 \quad P_r(X_2^1=x)=Pr(X_1=x) \tag{11}$$

$$Pr(X_2^1=0)=\sum_{x:x\leq 0} Pr(X_1=x)$$

<u>step 2:</u>
Introduce random variable $X_2^2=X_2^1+X_1$ (X_1 and X_2^1 are independent). The distribution of X_2^2 can be calculated as the convolution of (9) and (11), and thus:

$$W_3'(2)=\sum_{x:x>0} x\,Pr(X_2^2=x) \tag{12}$$

Let H_2 denote the set of values of X_2^2. For calculating $W_3'(3)$ we use the above algorithm in the following way: in step 1., we put new random variable X_3^1 and H_2 instead of X_2^1 and H_1, respectively; in step 2. we introduce random variable X_3^2, $X_3^2=X_3^1+X_1$; and so on.

$\underline{W_3''}$

At the beginning we define a factor, say D, which express the decrease of influence of tasks from p. 2° on the waitng time of test-task 3.
The value of D is obtained similary as W_3':

$$D=\sum_{n=1}^{\infty} p_n\cdot D(n), \tag{13}$$

where n is defined as previously.
To evaluate D(n), we calculate the n-fold convolution of the distribution function (9). Let X_n denote the sum of n random variables each distributed identically as X_1. Thus:

$$D(n)=\sum_{x:x<0} x\cdot Pr(X_n=x) \tag{14}$$

Finally, we obtain W_3'' using the following algorithm:

i/ At the beginning we calculate $\hat{W}_3$ which is defined as the total service time of tasks present in server 2 at moment t_1:

$$\hat{W}_3 = \int_0^B (B-y)\, f_1(y)\, dy \,/(1-\rho_3)\quad , \tag{15}$$

where:

- $B=W_3' \cdot \rho_3+\rho_3(h_3/2) - \rho_2(h_2/2)$
- $A=(W_1+h_1)/(1-\rho_1)-\rho_2(h_2/2)+D\cdot\rho_3$
- $f_1(y)=(1/A)\cdot \exp(-y/A)$

ii/ We calculate W_3'' using $\hat{W}_3$ and D:

$$W_3''=\int_0^{\hat{W}_3}(\hat{W}_3-y)\, f_2(y)\, dy \tag{16}$$

where:

$$f_2(y)=(1/D)\cdot\exp(-y/D)$$

<u>Checking by simulation</u>

In table 1. we compare sample values of W_3 calculated from formulae (4)÷(16) – W_3 approx, with simulation results – W_3 sim. Simulations results are given with .05 confidence coefficient. In all cases h3=5, so evaluation of W_3 assuming independence between servers leads to the value 2.5.

Table 1.
$\lambda=.1$

h_1	h_2	h_3	W_3'	D	W_3sim	W_3approx
2			1.4	0	2.9-3.2	2.48
4	0		4.45	0	6.2-6.8	5.64
6			13.15	0	16 - 18	15.3
2			1.07	.02	1.9-2.2	1.77
4	2	5	3.15	.58	3.75-4.25	3.5
6			9.0	5.76	9 - 10	9.1
2			.45	.38	.46 - .48	.47
4	4		1.09	5.96	1.02-1.09	1.09
5.95			1.8	6.72	1.73-1.87	1.81

Table 1 shows that: i) The max. value of relative error is about 15%; ii) The value of W_3' is always smaller than W_3 and, iii) for large values of the factor D and, as a consequence, for small W_3'', the approximate results are very close to simulation.

3.3 Subsystem no.3

The considered subsystem is depicted in Fig.4.

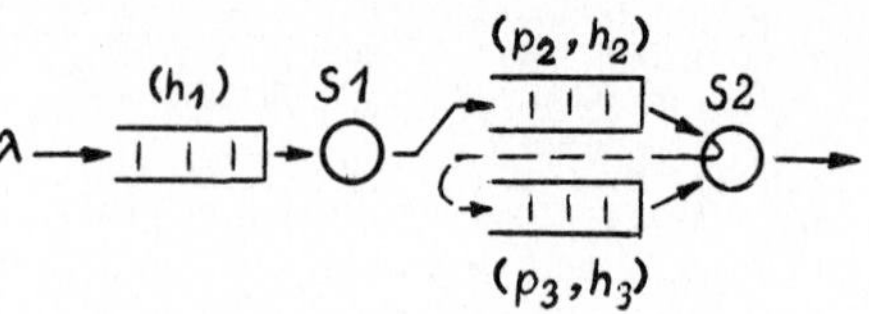

Fig.4. The subsystem no.3

It is obvious, that $W_2=W_3=0$ for $(h_2+h_3)\leqslant h_1$. We distinsuish two cases for $(h_2+h_3)> h_1$:

A: $h_2 \geqslant h_1$

B: $h_2 < h_1$

3.3.1. Case A

In this case, the expression for W_2 and W_3 take the following form:

$$W_2=W_2^* - W_1\ ,\qquad W_3=W_3^*\ , \tag{17}$$

where W_2 and W_3 denote mean waiting time (for tasks 2 and 3, respectively) for tasks executed by server 2 given Poissonian input:

$$W_2^* = [\rho_2(h_2/2) + \rho_3(h_3/2)]/(1-\rho_2)\ ,$$

$$W_3^* = (W_2^* + h_2 + W_3^*)(\rho_2+\rho_3)\ .$$

Expression (17) is a generalization of formula (2). As before, the obtained evaluations are very close to simulation results.

3.3.2. Case B

Unfortunately, in this case we are not able to calculate W_2 and W_3 separately. So, we use an approximation to evaluate the sum (W_2+W_3):

$$W_2+W_3 \cong [W^* - W_1 + (W_1+h_1+W_2)\cdot\rho_2\cdot A]/(1-\rho_2) \tag{18}$$

where:

- $W^*=[(h_2+h_3)/2]\cdot(\rho_2+\rho_3)/(1-\rho_2-\rho_3)$
- $A=1-h_1/(h_2+h_3)$

<u>Checking by simulation</u>

In Tables 2. and 3. we compare the values of (W_2+W_3) calculated from (17)-(18) – (W_2+W_3) approx, assuming independence between processors – (W_2+W_3) ind, and simulation results – (W_2+W_3) sim. Simulation results are given with .05 confidence coefficient.

Table 2.

λ=.1

h_1	h_2	h_3	(W_2+W_3) sim	(W_2+W_3) approx	(W_2+W_3) ind
		2	4.7-5.42	4.61	10.2
	4	3	10.0-13.2	11.63	16.3
		4	21.8-25.8	25.37	29.3
		5	53.0-67.0	66.74	70.2
		3	3.1-3.6	3.6	7.7
	3	4	8.0-9.0	9.8	13.0
		5	29.3-22.8	20.97	24.2
		6	45.5-57.5	56.25	59.1
5		4	2.5-2.81	2.9	7.1
	2	5	7.0-8.3	7.76	11.7
		6	16.0-18.6	17.76	20.5
		7	46.0-52.0	48.55	50.1
		5	2.0-2.2	2.38	5.1
	1	6	5.9-6.7	6.56	9.2
		7	14.0-16.0	15.43	18.9
		8	39.0-44.0	42.93	45.1

Table 3.

λ=.1

h_1	h_2	h_3	(W_2+W_3) sim	(W_2+W_3) approx	(W_2+W_3) ind
2			19.8-22.21	21.05	
4	5	2	18.5-20.7	19.96	21.3
6			10.7-12.0	9.53	
2			15.1-16.7	16.03	
4	3	4	14.0-15.7	14.96	16.28
6			7.0-8.0	7.46	
2			12.2-13.5	12.75	
4	3	4	10.0-11.7	11.28	13.0
6			6.0-6.8	6.05	
2			10.0-11.0	10.46	
4	2	5	8.3-9.2	9.3	10.7
6			4.3-5.1	5.02	

Tables 2., 3. show that: i) The proposed approximations (17,18) give results close to simulation in all cases; ii) Max. relative error by using formula (18) is less than 10%; iii) Formula (17) is very correct; iv) Corelation between servers is very strong, so the approximation by assuming independence between servers is only acceptable for (W_2+W_3)ind $\gg W_1$.

4. CONCLUSIONS

We have proposed approximations for mean waiting time of tasks executed by the second server in some tandem-queue systems given non-preemptive priorities, internal feedback and constant service times. According to the tables from section 3., the results obtained with the aid of proposed methods are in good accordance with simulation results. As follows from the tables, even small changes of service times often lead to large variations in mean waiting times. Though these variations can be predicted in some cases by an intuition still they are difficult to evaluate by means of reasonably simple analysis assuming independence of processors. In fact, knowing the performance measures of independly analysed one-processor systems we can say almost nothing (e.g. see Table 1. or 3.) about the behaviour of the whole system.
The authors did not succeed to create a uniform analytical approach to all considered cases (systems). Further authors efforts are focused in this direction.

REFERENCES

1. B.Simon, "Priority queues with feedback", Journal of the ACM, no.1, 1984
2. J.N. Daigle, C.E.Houstis, "Analysis of a task oriented multipriority queueing system", IEEE Trans. on Com., no.11, 1981
3. M.Villen, "Average response times in an M/G/1 queue with general feedback and priorities", Proc. 10th Int. Teletraffic Congress, Montreal, 1982
4. W.Burakowski, D.Bursztynowski, "An approach to the analysis of M/G/1 queue with general feedback and non-preemptive priorities", submitted for Intrn. Conf. on Modelling Techniques and tools for performance Analysis, France, 1985
5. J.T.Morris, "Priority queueing networks", B.S.T.J., no.8, 1981
6. M.Villen, G.Morales, "Traffic analysis of a class of distributed SPC systems", Proc. 9th Int. Teletraffic Congress, Torremolinos, 1979
7. J.S. Kaufman, "Approximation methods for networks of queues with priorities", Performance Evaluation, no.4, 1984
8. B.Fontana, "Queue with two priorities and feedback: Join queue-length distribution and response time distributions for specific sequences", Proc. 10th Int. Teletraffic Congress, Montreal, 1982
9. W.Whitt, "The queueing network analyzer", B.S.T.J., no.9, 1983
10. P.Kuhn, "Approximate analysis of general queueing networks by decomposition", IEEE Trans. on COM, no.1, 1979
11. W.Smitt, "Approximate analysis of marcovian queueing networks with priorities", Proc. 10th Int. Teletraffic Congress, Montreal, 1982

12.M. Dąbrowski, J.Jakubicki, M.Jarociński, "Aspecte der CHILL - implementierung fur eine Vermittlungstelle", Fachkolloqium Informetionstechnic, Dresden, 1985, p. A22K.

Appendix

The considered model is depicted in Fig. 2. We assume that $h_2 \geqslant h_1$.
We may express the mean waiting time in the second server, W_2, as follows:

$$W_2=\sum_{i=0}^{\infty} \sum_{j=0}^{\infty} p(i,j) \cdot W_2(i,j) , \qquad (1A)$$

where:

- $p(i,j)$ $(i,j=0,1,...)$ is the probability that upon an external arrival there are i and j tasks (including service) in the first and second server, respectively;
- $W_2(i,j)$ is the conditional mean waiting time of the second task of a sequence (in the second server) given that upon activation of the first task of this sequence the system is in state (i,j).

It is easy to show that:

$$W_2(i,j)=\begin{cases} i(h_2-h_1) & i\geqslant 0,\ j=0 \\ (h_2-h_1)\left[h_2(2_i+1)-h_1\right]/\ 2h_2 & i\geqslant 1,\ j=1 \\ ih_2+(j-1)h_2+h_2/2+h_1 & i=0,\ j\geqslant 2 \\ ih_2+(j-1)h_2+h_2/2+h_1/2+ & \\ +ih_1 & i\geqslant 1,\ j\geqslant 2 \end{cases} \qquad (2A)$$

Next, putting (2A) into (1A), we obtain, after some algebra, the following general expression for W_2:

$$W_2=W_2\varrho_2+\varrho_2 h_2/2+W_1\varrho_2+h_1\varrho_2-\varrho_1 h_1/2-W, \qquad (3A)$$

where:

$$W=p_2(1)h_1^2/2h_2+h_1\cdot\sum_{j=2}^{\infty} p(0,j) \quad +$$

$$+\ h_1/2\cdot\sum_{i=1}^{\infty}\left[p(i,0)+p(i,1)\right] ;$$

$$p_2(1)=\sum_{i=0}^{\infty} p(i,1) \quad .$$

Using expression (1A) and denoting $W_2^*=\varrho_2 h_2/2(1-\varrho_2)$, we can rewrite (3A) as follows:

$$W_2=W_2^*-W_1+(h_1\varrho_2-W)/(1-\varrho_1) . \qquad (4A)$$

It may be shown that in two important limiting cases (i.e for $h_1 \to h_2$ and $h_1 \to 0$, h_2-constant), W tends to $\varrho_1 h_2$ yielding expression (2).

TELETRAFFIC ISSUES in an Advanced Information Society
ITC-11
Minoru Akiyama (Editor)
Elsevier Science Publishers B.V. (North-Holland)

AN OVERLOAD CONTROL STRATEGY FOR DISTRIBUTED CONTROL SYSTEMS

M. VILLEN-ALTAMIRANO, G. MORALES-ANDRES and L. BERMEJO-SAEZ

Centro de Investigacion de Standard Electrica, S.A.
Madrid, Spain

ABSTRACT

An overload control strategy for fully distributed control systems is presented. The overload control has to deal with overload situations in a network of processors, in which different overload patterns can affect to few, many or all the processors.

The proposed strategy has a distributed structure which matches with the architecture of these systems: Each processor has its own overload detection mechanism, and control actions are taken to protect each processor against overload. The paper discusses how to coordinate the actions taken to control the overload of each one of the processors to obtain a good performance of the system as a whole during overload.

The guidelines of the strategy proposed in this paper have been followed in the design of the overload control strategy of System 12 (*), where simulation tests have proved its effectiveness: A high throughput with a good grade of service for accepted calls is ensured under any overload situation.

1. INTRODUCTION

Overload control in previous stored program control systems had to cope with overload in a centralized configuration; see [1-4] among others. Paper [5] studies overloads in a distributed configuration, under an overload level which only affects one processor.

In this paper, fully distributed control systems are studied, in which the overload control has to deal with overloads in a network of processors. Different overload levels and patterns, focussed or general overloads, affecting to few, many or all the processors, are considered.

This paper is based on the work done for designing an overload control for System 12 [6]. The distributed structure of the system presented new features of the problem of controlling the overload, such as how to cope with overload situations only affecting a part of the processors, how to avoid the propagation of the overload effects, or when to take actions to control the overload of processors which treat calls previously treated by other processors; in few words, how to coordinate the actions to be taken to control the overload of each one of the processors in a way which, in turn, matches with the distributed structure of the system. The specific solution adopted for System 12, as well as an evaluation by simulation of the performance obtained, are described in [7]. In this paper, we present those conclusions which are not specific to System 12. The characteristics of the distributed systems for which these conclusions could be applied are:

- The system is based on a network of processors which communicate among themselves through a virtually non-blocking medium.
- Different pools of processors are distinguished according to the function or group of functions which they perform.
- Within each pool of processors, each processor can be dedicated to a group of lines, trunks or types of calls, or, conversely, all of them can work as a dynamic load sharing pool for all the calls of the exchange. Intermediate situations are also covered.

The development of the overload control strategy has been based both on simplified analytical models and on a full scale multinode simulation:

Several analytical tools have been developed to model, in a simplify manner, the different aspects of the general problem and to observe trends. They have been the basis for establishing the guidelines of the strategy.

A simulation model has been developed to check the validity of the conclusions obtained with the analytical models, and to study the integration of all the aspects in the solution adopted for System 12. It has also been used for refining the strategy, tuning its parameters and evaluating its performance. The simulation model is not described here, since a similar model is described in paper [8], also presented to this Congress.

2. OVERVIEW OF THE STRATEGY

The proposed strategy can be summarized as follows: Each processor has an overload detection mechanism to detect its own overload situation. Control actions are taken in different points along the treatment of the calls. Each processor has associated one or several control points where actions are taken based on the overload status of this processor. Each control action is always based on the overload status of an individual processor, even in the case in which a control point is associated to several processors. In this case, the action taken when more than one processor is overloaded is the logical OR of the actions corresponding to each one of the overloaded processors.

Each processor requires its own overload detection mechanism, since different overload situations can affect to different processors.

The proposal of control actions based on the overload status of individual processors, instead of being based on a combined information, gives a distributed structure to the strategy. The coordination of the actions taken by the diverse processors is given by the appropriate choice of the control points and by the selfregulation of the detection/action mechanism of each processor; this mechanism graduates the intensity of the actions according to the level of overload offered to the processor, which, in turn, does not only depend on the traffic demand, but also on the actions taken by other processors. This coordination has proved to be enough to obtain an efficient performance of the proposed strategy, whose distributed structure matches with the distributed structure of these systems.

3. OVERLOAD DETECTION

As said before, each processor requires its own

(*) System 12 is an ITT trademark.

overload detection mechanism. The same mechanism is proposed for all of them. Although a better solution could be obtained for each processor if the mechanism is tailored to the functions which it performs, a common mechanism fits better to the usual flexibility of these systems, in which the distribution of the functions among the processors can change with the applications.

Two parameters which are common for all the processors regardless the functions they perform, will be analyzed as overload indicators. One of them is a parameter related to the spare call handling capacity of the processor in a certain period. The other one is a parameter related to the call handling work waiting for execution in each moment. The discussion here will be done with two parameters easily measured in the processors of System 12: the load, related to the spare capacity, and the queue length, related to the pending work.

In a first step, two simple detection mechanisms will be presented. From the discussion of the pros and cons of each one, the need of a more sophisticated one will be shown. Finally, a combined proposal, which takes the pros of each of the previous ones, is given.

Three criteria will be applied to evaluate the goodness of each detection mechanism:

- Reliability to avoid the triggering of the overload control actions when there is not overload.
- Capability to ensure a specified throughput and response time (within a compromise between them) during a maintained overload situation, regardless of the overload level.
- Capability to protect the system against peaks or abrupt changes of the offered traffic.

The detection mechanisms will be analyzed at the light of the following simplified traffic model: the processor works as a M/M/1 model, in which each arrival is a call and the control action is assumed to be rejection of arriving calls when the detection mechanism indicates overload. The trends observed with this simplified model have been confirmed in the full scale simulation.

3.1 First simple detection mechanism

This mechanism is based on the observation of the queue and declares overload situation when the queue length, Q, reaches a certain threshold, T.

According to the criterion of reliability, the threshold, T, has to be greater than a certain value in order to have a negligible probability of call rejection under normal traffic conditions. The probability of call rejection is equal to the probability of reaching the threshold and is given by:

$$\Pr\,[Q=T] = \frac{1-a}{1-a^{T+2}}\, a^{T+1} \qquad (1)$$

where a is the load in normal conditions. For numerical comparison purposes, the following values could be considered:

$$a = 0.6 \text{ Erl.} \;;\quad \Pr\,[Q=T] \leqslant 10^{-4}$$

For these values, this mechanism is reliable if a threshold equal or higher than 16 is chosen.

During a maintained overload, the throughput C (accepted call rate) and the average response time, RT, are given by:

$$C = \lambda\cdot\frac{(\lambda h)^{T+1}-1}{(\lambda h)^{T+2}-1} \qquad (2)$$

$$RT = \left[\frac{1}{1-\lambda h} + (T+1)\cdot\frac{(\lambda h)^{T+1}}{(\lambda h)^{T+1}-1}\right]\cdot h \qquad (3)$$

where: λ is the offered call rate.
h is the average processing time of a call.

Fig.1 plots response time against throughput being the offered call rate a parameter, for several values of the threshold. It can be inferred from Fig. 1 that both the throughput and the response time depend on the offered call rate and, for any fixed value of T, a reasonable compromise between them, valid in the full range of possible overload levels, cannot be found.

Thus, according to the second criterion, this mechanism is not at all efficient.

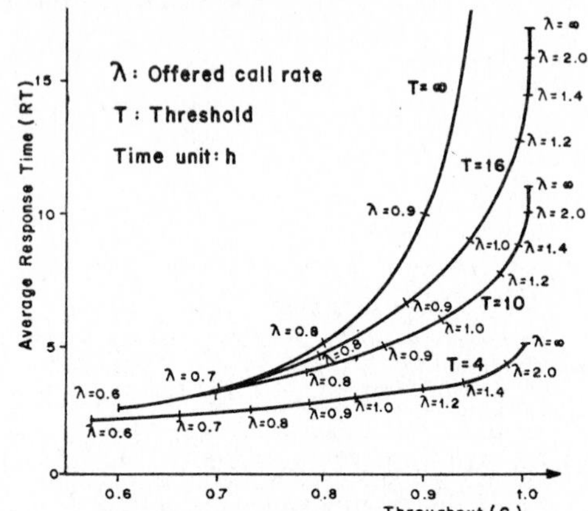

Fig. 1 Response time versus throughput for the first simple detection mechanism.

As far as to the capability to protect the system against peaks or abrupt changes of the overload levels is concerned, this mechanism is quite efficient. The mechanism ensures a limit in the response time, T·h, and in the required memory for any situation. This limit (a response time of 17·h, 6.8 times higher than in normal conditions if T = 16) looks reasonable for short peak conditions or abrupt changes.

3.2 Second simple detection mechanism

This mechanism consists of the observation of the load during periods of time, P, and declares overload situation if the load during the period reaches a certain threshold T.

This detection mechanism has to be complemented with a mechanism which regulates the appropriate percentage of calls to be rejected within each period. It could be as follows: If during a given period the load is greater than T the percentage of rejected calls is increased in the next period; in the opposite case the percentage is decreased.

In order to evaluate the probability of rejecting calls in normal conditions, the following formula has been derived, using the event counting method [9]. It calculates the probability of exceeding a load T during a period P, when the average offered load is a, in an M/M/1 queue.

$$\Pr[\geqslant T] = e^{-(1+a)P/h}\cdot\sum_{i=0}^{\infty}\frac{[(1+a)P/h]^{i}}{i!}\cdot\sum_{r=0}^{i}\binom{i}{r}\cdot T^{i-r}\cdot(1-T)^{r}\cdot\left[(1-A)\cdot\sum_{k=0}^{i-r}a^{k}\cdot L(i,r,k) + a^{i-r+1}\right] \qquad (4)$$

where:

$L(0,r,k)=1 \quad \text{if } r+k \geqslant 1 \quad ; \quad L(i,0,0)=0$

$$L(i,r,0)=\frac{a}{1+a}\cdot L(i-1,r-1,1) + \frac{1}{1+a}\cdot L(i-1,r-1,0) \qquad \text{if } i \geqslant 1,\ r \geqslant 1$$

$$L(i,r,k)=\frac{a}{1+a}\cdot L(i-1,r,k+1) + \frac{1}{1+a}\cdot L(i-1,r,k-1) \qquad \text{if } i \geqslant 1,\ k \geqslant 1$$

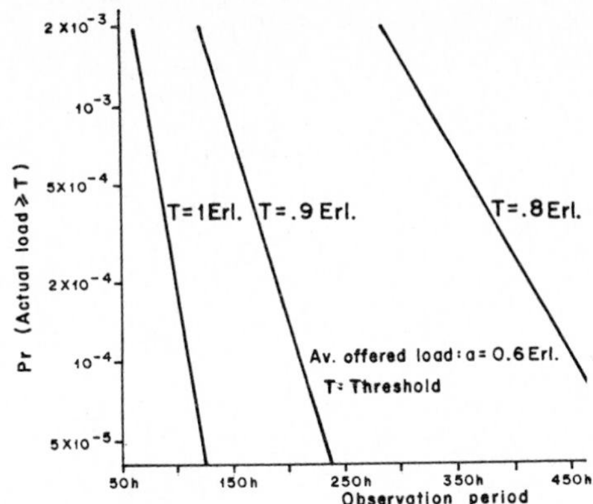

Fig. 2 Probability of the load in a period exceeding a threshold in a M/M/1 model.

Results from this formula are given in Fig. 2.

If the probability of call rejection in normal conditions has to be smaller than 10^{-4}, as in the first mechanism, the probability of exceeding T during a period must be smaller than $10^{-4}/\pi$, being π the proportion of calls which are rejected in the first step once the overload has been detected.

If we assume $\pi = 0.1$, the required values of the period P are 313·h, 141·h and 71·h if the thresholds are 0.8, 0.9 and 1.0, respectively.

During a maintained overload, this mechanism is able to obtain an average throughput C approximately equal to T/h (except if T is close to 1), regardless the overload level. The response time will be in the same order of magnitude as in a M/M/1 model with offered load T, i.e.,

$$RT = h/(1-T) \qquad (5)$$

Reasonable requirements of throughput and response time could be satisfied if the thershold is 0.8 Erl. A higher threshold would produce an excessive response time.

On the other hand, this mechanism is not efficient to protect the system against peaks or abrupt changes of the overload level. During a time t in which λ calls per second are offered, the queue is increased, if $\lambda h > 1$ and no actions are taken, in the following value:

$$\Delta Q = t \cdot (\lambda - 1/h) \qquad (6)$$

A time in the order of the observation period (313·h) is required to detect overload when it arises. Thus, if an overload with $\lambda = 1.8/h$ (200% overload) abruptly arises, the queue length is about 250 when the first control actions start. Thus, the system is without an efficient protection during transitions.

3.3 Proposed mechanism

From the previous discussion, we see that a more more sophisticated mechanism is required. The solution proposed here has proved to be very efficient under the three established criteria. The proposal is as follows:

The overload status is determined by comparing the queue length with a dynamical threshold. This threshold is periodically updated by comparing the actual processor load during the previous period with a threshold load, called "target load". The values of the dynamical threshold of the queue are ordered in a table as follows:

T1	T1	...	T1	T2	T3	...	Tm

(first n positions: n times) with: T1 > T2 > T3 > ... > Tm

A cursor points a position of the table in each period. During normal traffic conditions, the cursor will point the leftmost position. If the load during a period is greater than the target load, the cursor moves one step to the right; otherwise, it moves one step to the left.

The mechanism can be tuned to have a negligible probability of rejecting calls by means of an appropriate choice of the values of T1 , n and P (observation period). T1 will have a similar value to T in the first simple mechanism, and n·P will have a value in the order of P in the second simple mechanism. Factor n allows a value of P smaller than in the second simple mechanism.

During a maintained overload, the performance is good, as it was in the second simple mechanism. The throughput corresponding to the target load is obtained under any overload condition, since the appropriate rejection percentage is regulated by the dynamical adaptation of the threshold value. The regulation can be more efficient than in the second simple mechanism, since a smaller value of P allows a quicker adaptation during overload. The response time is even better than that corresponding to an offered load equal to the target load and no control actions (see Fig.1 for the effect of the queue limitation on the response time for a same throughput).

During normal load condition, this mechanism is so good as the first simple mechanism to protect the system against peaks or abrupt increases of the traffic levels, and even better during overload, since the threshold of the queue is smaller.

This mechanism can be extended to allow the treatment of calls according to priorities by defining more than one threshold in each cursor position, each threshold being associated to the rejection of calls of a priority. The proposed table for the case of two priorities is as follows:

(first n positions: n times)

T1	T1	...	T1	T1	T1	...	T1	T2	T3	...	Tm
T1'	T1'	...	T1'	T2'	T3'	...	0	0	0	...	0

being: T1> T1' ; T1> T2>...> Tm ; T1'> T2'>...> 0

This table ensures that, except few calls during abrupt transitions, higher priority calls are only rejected if the overload level is such that the rejection of all the lower priority calls is not enough to control the overload.

The efficiency of this mechanism has been proved by simulation in a real life system, as well as its dynamical adaptation to very different traffic situations: different call types, different overload levels and different impact of the actions taken by other processors.

4. OVERLOAD CONTROL ACTIONS

When a processor is overloaded, some calls which would use this processor cannot be accepted. The non accepted calls can inmediately be rejected or delayed in order to try to be accepted later; even in this case, there are calls which will have to be finally rejected. At least in the case in which a call has to be rejected in an advanced phase, the inmediate rejection is recommended. In any case, the delay before rejection intends to prevent rejection of calls due to transitory peaks during normal traffic conditions, and the previously proposed detection mechanism reaches this objective without requiring this additional delay.

As said in section 2, the main problem is to know which are the most appropriate points to reject calls when each processor is overloaded. The discussion of the appropriate control points, or rejection points, is independent from the fact that the rejection is inmediate or after a delay.

Studies on single processor systems [1-4] have shown that the shorter the time spent by the processor with a rejected call, the better the throughput obtained in overload situation. In a distributed system, this principle applies to each particular processor for calls rejected due to its own overload.

The validity of this principle for calls rejected due to overload of other processors is discussed in the next sections, as a part of the general problem of determining which are the most appropriate rejection points when the dependence among several processors is considered.

4.1 Study of basic configurations

Fig.3 shows different relations between processors of two stages, with respect to the traffic handled by them. The arrows mean traffic flows and the order in which an arrow crosses the processors is the order of the beginning of the treatment of calls of this flow by the processors. The rectangular boxes mean rejection points to be investigated. The letter besides a rejection point indicates the ratio of calls arriving to this point which are not rejected there.

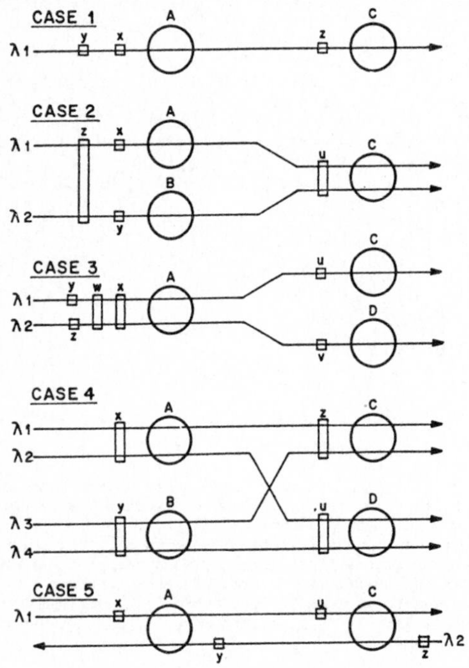

Fig. 3 Basic configurations of processors of two stages.

These models will be analytically studied assuming that the detection/rejection mechanisms fully reach their objectives, i.e.:

- The load carried by a processor is just its target load if calls are rejected due to its overload.
- First priority calls are rejected due to overload of a processor only if all the second priority calls treated by this processor are being rejected.

Rejection of calls in points such as x of case 1 can be done after a small treatment by processor A. This small treatment is neglected in the analytical study.

The simulation has proved that the results of this analytical study are quite accurate and that the conclusions apply to an actual system.

Let us study each one of the five cases shown in Fig.3. Common notation is as follows:

λi is the call rate of flow i.

λin is the value of λi in normal conditions.

hp is the average processing time in processor P of an accepted call.

$h'p$ is the average processing time in processor P of a call rejected after (at least partial) treatment by processor P; e.g. $h'a$ is the processing time in A of a call rejected in z in case 1.

kp is the target load minus the fixed overhead of processor P, i.e., the available load for call handling of processor P during overload.

C is the total throughput, i.e., the total number of accepted calls from all the traffic flows considered in each case.

Case 1:

Two alternatives of rejection will be studied:

Alt. a: Rejection of calls is done before the treattreatment by any processor.

Alt. b: Rejection due to overload of processor C is done after the calls have been at least partially treated by processor A.

a) Rejection in x if A is overloaded and in y if C is overloaded. The values of x and y are selfregulated to satisfy the following conditions:

$$\text{If } \lambda 1\cdot ha\cdot y \leqslant ka \implies x = 1$$

$$\text{If } \lambda 1\cdot ha\cdot y \geqslant ka \implies x = \frac{ka}{\lambda 1\cdot ha\cdot y}$$

$$\text{If } \lambda 1\cdot hc\cdot x \leqslant kc \implies y = 1$$

$$\text{If } \lambda 1\cdot hc\cdot x \geqslant kc \implies y = \frac{kc}{\lambda 1\cdot hc\cdot x} \qquad (7)$$

With the values of x and y obtained from equations above, the throughput is given by:

$$C = \lambda 1\cdot x\cdot y = \text{Min}\left[\lambda 1, \frac{ka}{ha}, \frac{kc}{hc}\right] \qquad (8)$$

A usual criterion considered in the dimensioning rules is that each processor has a reserve capacity for accepting moderate overloads without requiring control actions. This criterion can be expressed in this way:

$$\lambda 1n\cdot ha \leqslant (1/r)\cdot ka \;;\; \lambda 1n\cdot hc \leqslant (1/r)\cdot kc \qquad (9)$$

where r is a number higher than 1, which will be called here "Overload Reserve Factor" (ORF).

If this criterion has been considered, formula (8) becomes:

$$C \geqslant \text{Min}\,(\lambda 1, r\cdot\lambda 1n) \qquad (10)$$

b) Rejection in x if A is overloaded and in z if C is overloaded. The values of x and z are selfregulated to satisfy:

$$\text{If } \lambda 1\cdot ha\cdot z + \lambda 1\cdot h'a(1-z) \leqslant ka \implies x = 1$$

$$\text{If } \lambda 1\cdot ha\cdot z + \lambda 1\cdot h'a(1-z) \geqslant ka \implies x = \frac{ka}{\lambda 1\cdot ha\cdot z + \lambda 1\cdot h'a(1-z)}$$

$$\text{If } \lambda 1\cdot hc\cdot x \leqslant kc \implies z = 1$$

$$\text{If } \lambda 1\cdot hc\cdot x \geqslant kc \implies z = \frac{kc}{\lambda 1\cdot hc\cdot x} \qquad (11)$$

The throughput obtained from these equations is the same as in alternative a.

Therefore, since the optimum throughput is obtained in both cases, it is not necessary to take any precaution of rejecting calls due to overload of a processor before being treated by the other processor.

Case 2:

Two alternatives equivalent to those in case 1, have been studied: In both alternatives the rejection due to overload of A and B is done in x and y respectively, but the rejection due to overload of C is done in z in one alternative and in u in the other one.

The application of the method of case 1 gives an optimum throughput in both alternatives, which is:

$$C = \text{Min}\left[\lambda 1+ \lambda 2, \lambda 1+ \frac{kb}{hb}, \frac{ka}{ha} + \lambda 2, \frac{ka}{ha} + \frac{kb}{hb}, \frac{kc}{hc}\right] \qquad (12)$$

If the criterion of the ORF has been followed when dimensioning, the throughput becomes:

$$C \geqslant \text{Min}\ (\lambda 1, r \cdot \lambda 1n) + \text{Min}\ (\lambda 2, r \cdot \lambda 2n) \qquad (13)$$

Thus the same conclusions of case 1 are extended to this case.

Case 3:

Three alternatives have been studied in this case. In all three alternatives the rejection due to overload of A is done in x. The rejection due to overload of C or D is done in this way:

Alt. a: Selective rejection before treatment by A: Rejection in y due to C and rejection in z due to D.

Alt. b: Selective rejection after (at least partial) treatment by A: Rejection in u due to C and rejection in v due to D.

Alt. c: Indiscriminate rejection before treatment by A: Rejection in w due to C or D.

The study of the three alternatives shows that the optimum throughput is only obtained in alternative a, while there is a loss of throughput in alternatives b and c.

For reason of extension, the throughput is only given when the criterion of the ORF has been followed when dimensioning. The values obtained are:

$$C \geqslant \text{Min}(\alpha 1 \cdot \lambda 1, r \cdot \lambda 1n) + \text{Min}(\alpha 2 \cdot \lambda 2, r \cdot \lambda 2n) \qquad (14)$$

being:

In alt. a: $\alpha 1 = 1 \quad , \quad \alpha 2 = 1$

In alt. b: $\alpha 1 = \text{Min}\left[1, \dfrac{r \cdot \lambda 1n + r \cdot \lambda 2n \cdot (h'a/ha)}{\lambda 1 + \lambda 2 \cdot (h'a/ha)}\right]$

$$\alpha 2 = \text{Min}\left[1, \frac{r \cdot \lambda 1n \cdot (h'a/ha) + r \cdot \lambda 2n}{\lambda 1 \cdot (h'a/ha) + \lambda 2}\right]$$

In alt. c: $\alpha 1 = \text{Min}\left[1, \dfrac{r \cdot \lambda 2n}{\lambda 2}\right] \quad , \quad \alpha 2 = \text{Min}\left[1, \dfrac{r \cdot \lambda 1n}{\lambda 1}\right]$

Alternative a cannot be implemented in many practical cases, due to the fact that the usage of a processor by a call depends on decissions taken after its (at least partial) treatment by the previous processor. Thus, alternatives b and c are often the practical options.

The analysis of formula (14) shows that the throughput with alternative b is always equal to or better than the throughput with alternative c. If alternative b is used, the smaller the value of h'a, the better the throughput obtained; if h'a were nil the throughput would be equal to that of alternative a, and if h'a tends to infinite, the limit of the throughput is that of alternative c. In practice, $h'a \leqslant ha$.

The following is a quantification of the loss of throughput of alternatives b and c, with respect to the optimum, when the load of the processors is just the load limit given by the application of the criterion of the ORF:

From (14) it can be derived that loss of throughput can only occur when there is focussed overload of one of the traffic flows $\lambda 1$ or $\lambda 2$, but it will not occur with general overload ($\lambda 1/\lambda 1n = \lambda 2/\lambda 2n$).

Fig.4 shows the throughput for the three alternatives for the case of focussed overload of flow $\lambda 1$. Flow $\lambda 2$ is kept constant at the level of normal traffic conditions ($\lambda 2n$). The ORF is assumed to be r = 1.45, which is the value usualy adopted in the dimensioning of System 12. Three cases of the ratio between $\lambda 1n$ and $\lambda 2n$ are considered: 3, 1 and 1/3.

The figure shows that, while an important loss of throughput can occur with alternative c, the throughput of alternative b is not far from the optimum one, being always greater than the throughput in normal traffic conditions for reasonable overload levels.

Consequently, selective rejection is always recommended, even when it implies (alternative b) to reject calls after the treatment by other processors. The shorter the time spent in this treatment, h'a, the better the throughput obtained.

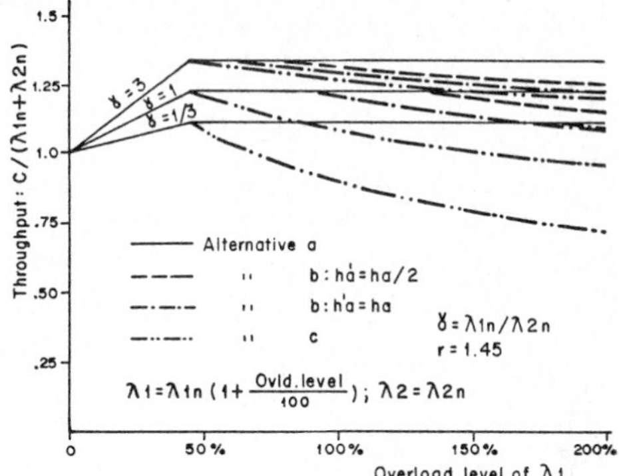

Fig. 4 Throughput in case of focussed overload in the three alternatives of case 3.

Case 4:

Alternatives of case 3 have been studied. Results obtained are similar: optimum throughput is only obtained in alternative a, and the loss of throughput with respect to this optimum is much smaller with alternative b than that with c. For reasons of extension, the only throughput expressed here is that of alternative b when the criterium of the ORF is followed in the dimensioning, and for h'a = ha (worst practical case).

$$C \geqslant \text{Min}(\alpha \cdot \lambda 1 + \beta \cdot \lambda 3, r \cdot \lambda 1n + r \cdot \lambda 3n) +$$
$$+ \text{Min}(\alpha \cdot \lambda 2 + \beta \cdot \lambda 4, r \cdot \lambda 2n + r \cdot \lambda 4n) \qquad (15)$$

being:

$$\alpha = \text{Min}\left[1, \frac{r \cdot \lambda 1n + r \cdot \lambda 2n}{\lambda 1 + \lambda 2}\right] \quad , \quad \beta = \text{Min}\left[1, \frac{r \cdot \lambda 3n + r \cdot \lambda 4n}{\lambda 3 + \lambda 4}\right]$$

The analysis of this formula shows that a loss of throughput, with respect to the optimum, equivalent to that of case 3 is obtained with alternative b when there is an overload focussed towards either processors C or D (traffic flows $\lambda 1 + \lambda 3$ versus $\lambda 2 + \lambda 4$); but a throughput equal to the optimum of alternative a is obtained with a general overload or an overload focussed towards either processors A or B if imbalance between processors C and D is not produced.

Case 5:

A situation of two processors with crossed traffic is studied in this case, in order to see if conclusions of case 1 can be extended. The proposal is rejection in x and y when A is overloaded and rejection in u and z when B is overloaded. The problem is to know if the optimum throughput is obtained when both traffic flows have the same priority in both processors (x = y, u = z) or if priority between flows has to be established. Thus, there are three alternatives for each processor:

Alt. a: The flow coming from the other processor has higher priority than the flow starting in this processor. In processor A, it means: $x > 0$ implies $y = 1$.

Alt. b: Both flows have the same priority. In processor A, it means: $x = y$.

Alt. c: Opposite to alternative a, that is: $y > 0$ implies $x = 1$.

Considering all the combinations with the two processors, 9 alternatives are obtained which are called aa, ab, ac, ba, etc.

Equations to calculate x, y, z and u have been established under the same methodology as in previous cases. For example, x and y are obtained for alternative a in processor A from the following equations:

$$\text{If } \lambda 1\cdot ha\cdot u + \lambda 1\cdot h'a\cdot(1-u) + \lambda 2\cdot ha\cdot z \leqslant ka \implies$$

$$\implies x=1 \ , \ y=1$$

$$\text{If } \left\{ \begin{array}{l} \lambda 1\cdot ha\cdot u + \lambda 1\cdot h'a\cdot(1-u) + \lambda 2\cdot ha\cdot z \geqslant ka \\ \lambda 2\cdot ha\cdot z \leqslant ka \end{array} \right\} \implies$$

$$\implies x=\frac{ka - \lambda 2\cdot ha\cdot z}{\lambda 1\cdot ha\cdot u + \lambda 1\cdot h'a\cdot(1-u)} \ , \ y=1$$

$$\text{If } \lambda 2\cdot ha\cdot z \geqslant ka \implies x=0 \ , \ y=ka/\lambda 2\cdot ha\cdot z \qquad (16)$$

The result of the study is that the optimum throughput is only obtained with alternatives aa, ab, ac, ba, and ca. It is given by:

$$C \geqslant \mathrm{Min}(\lambda 1+ \lambda 2, r\cdot\lambda 1n + r\cdot\lambda 2n) \qquad (17)$$

if the criterion of the ORF has been followed when dimensioning.

Thus, the conclusion is that the optimum throughput is obtained if the assignment of priorities in at least one of the two processors is done in the following way: The flow coming from the other processor has a higher priority than the flow starting in this processor. No condition has to be satisfied by the priorities in the other processor.

This study has assumed that the processing time required in one processor by the calls of both flows is the same, and based on this assumption, alternatives aa, ab, ac, ba and ca, produce the same throughput. If this assumption is not valid, the throughput differs from one to another of the 5 alternatives. In addition, the global priority given to flow $\lambda 1$ with respect to $\lambda 2$ differs from one to other alternative. Both points of view have to be considered in the choice of one of the alternatives.

Case 5 could be extended to more complex configurations in a similar manner as case 1 has been extended to cases 2, 3 and 4. Global conclusions would be similar to those already obtained.

4.2 Extension to complex configurations

The study done in section 4.1 can easily be extended to the case of two stages of processors with several processors in each stage. In the case of several processors working as a dynamical load sharing pool, all of them will have the same overload level, and the percentage of calls rejected by each one will be the same. If it occurs in each of the two stages, case 1 of section 4.1 will represent this situation. Values of ka and kc in this extension will be the total available load for call handling of all the processors of the corresponding stage during overload. Case 2 and case 3 are representative of the situation in which the processors of one of the two stages work as a dynamical load sharing pool.

If each processor of a pool or stage is dedicated to a group of lines, trunks or types of calls, the overload level of each processor can be different. In the case of general overload, if a balanced dimensioning a balanced dimensioning of the processors has been done according to the criterion of the ORF, the performance can be studied by means of case 1. In the case of focussed overload, case 4 is representative of the situation if the processors of each stage can be classified into two groups in the following way: All processors of a group have a certain overload level, the same for all of them, and all processors of the other group have other overload level. In the first stage, processor A can represent one of the groups, with ka equal to the total available load for call handling of the processors of this group, and processor B can represent the other group. The same can be done with processors C and D for the second stage. Any other focussed overload situation can be considered as an intermediate case between a general overload and the type of focussed overload represented by case 4.

The extension of the above conclusions to a real life configuration with more than two stages has been checked by simulation. They can be summarized as follows:

- The rejection of calls due to overload of a processor used by all the calls, or belonging to a dynamical load sharing pool used by all the calls, can be done in any moment before the treatment of the calls by this processor, regardless whether they have been treated or not by other processors.
- The rejection of calls due to overload of a processor used by part of the calls must be selective. In this case, it is desirable that the treatment of the rejected calls by other processors be as small as possible (if these processors also treat calls which are not treated by the former). When these two objectives conflict each other, the criterion of selective rejection must prevail.
- The loss of throughput, in case of focussed overload, due to the above mentioned conflict has been evaluated in cases 3 and 4 for a two stage configuration. In case of several stages, a focussed overload only affecting to processors of one stage has the same impact as in the two stage configuration. In case of focussed overload affecting to processors of several stages, the simulation has shown that the additional impact of an additional stage, also affected by focussed overload, quickly decreases as the number of stages increases. Thus, the throughput obtained for any case of reasonable focussed overload is better than the throughput in normal conditions regardless the number of stages, if a reasonable value for the ORF has been chosen when dimensioning.
- In case of several stages of processors with crossed traffic, the rules given in case 5 must be satisfied for any pair of stages of the whole configuration.

4.3 Treatment according to priorities

One of the main objectives of an overload control strategy is to handle calls according to specified priorities. Specifically, the coordination with the network management strategy requires to assign higher priority to the incoming calls than to the originating calls.

The proposed overload detection mechanism allows that each processor treats calls according to priorities: Rejection of high priority calls due to overload of a processor is only produced if the rejection of all the low priority calls using this processor is not enough to control the overload.

The effective treatment according to priorities in an exchange as a whole does not only require the treatment according to priorities in each processor, but also that the distribution of functions among the processors and the dimensioning of these allow this treatment:

If the high priority calls use different processors than the low priority calls, the treatment of calls according to priorities can only be based on a dimensioning of the processors treating high priority calls with a ORF higher than that of the processors treating low priority calls.

On the opposite side, if all the processors treat (in normal conditions) the same proportion of high and low priority calls, the treatment according to priorities in each processor will be fully effective to produce a global treatment according to priorities: The throughput of high priority calls in case of overload is the same as if only these calls would exist in the exchange and the dimensioning of the processors would have been done with an ORF equal to $r\cdot\lambda T/\lambda H$, being:

r : ORF (for all the calls) used for dimensioning, as defined in 4.1.

$\lambda T/\lambda H$: ratio of total rate of calls to rate of high priority calls in the whole exchange in normal conditions.

The total throughput is the same as if there would not exist priorities.

In the general case, in which the proportion of high and low priority calls can differ from one to other processor, an effective treatment according to priorities requires that the dimensioning rules consider, for each processor, apart from the ORF for all the calls, r, an overload reserve factor for high priority calls, R, being R > r. It means that the load due to high priority calls of each processor must not exceed 1/R times the available load for call handling during overload.

If $R = r \cdot \lambda T/\lambda H$, the same throughput of high priority calls and of total calls as in the previous case is guaranteed. It implies an absolute preference of the high priority calls, in the sense that the handling of the maximun number of them can imply not to handle any low priority call.

If a more moderate preference is desired, a value of R smaller than $r \cdot \lambda T/\lambda H$ should be chosen. In this case, it should be avoided that a stage of processors different from the first one used by the high priority calls works as bottleneck for these calls. Otherwise, low priority calls can be rejected in the first stages in order to treat high priority calls which, in turn, are rejected in the stage which works as bottleneck for them. Thus although the throughput of high priority calls corresponding to an ORF equal to R would be guaranteed, the total throughput corresponding to an ORF equal to r would not be guaranteed. To avoid it, the dimensioning load due to high priority calls in the processors of the first stage and the value of R must be accordingly chosen, in such a way that R has the maximum value allowed by that load or vice versa.

4.4 Other aspects to be considered

Apart from the aspects previously considered, the following ones must be taken into account in the design of the control actions:

- Up to now, it has been considered that the combined detection/rejection mechanism ensures a load in each processor equal to its target load and that the throughput and response time corresponding to this load are obtained. This objective is achieved if the actions taken to avoid overload of a processor have immediate impact on the load of the processor. If the time between the moment in which an action is taken and that in which it is effective is large, oscillations of the load of the processor are produced and, consequently, worst response times are obtained and even the target load is not ensured. Therefore, the rejection of calls due to overload of a processor must not be done, from this point of view, much time before the instant in which the calls would have used the processor.
- The points of rejection due to overload of a processor can be placed in the own processor or in other ones. The rejection in the own processor should only be adopted if the time spent by the processor in the treatment of a rejected call can be limited to a small percentage of the time spent in the treatment of an accepted call. (A limit of 5% for this percentage has been considered in System 12). If the rejection point is located in another processor, a communication mechanism of the overload status has to be implemented. On one hand, the communication has to ensure a quick updating of the information in the processor where the rejection point is placed; otherwise, problems of load oscillation would appear, with the effects commented in the previous paragraph. On the other hand, the processing time spent in the communication must be minimized. The achievement of these two objectives can condition the choice of the rejection points.
- Apart from the above general considerations, the implementation of the control actions have to consider the specific characteristics of the system. In the case of System 12, its modular structure requires a modular overload control. Moreover, the system has an "architecture for change" [6], that is, a flexible architecture which allows a large variety of applications and an evolution in technology and services.
Thus, the overload control strategy should not consider the particularities of each application, but it has been designed [7] to be sufficiently general to allow, by changing its parameters, to be successfully applied to the variety of applications and to the future evolution.

A compromise solution must be searched for the implementation of the proposed strategy in any particular system, since some of the given considerations conflict each other. The full scale simulation is the best tool to reach this objective.

5. CONCLUSIONS

A proposal of overload control strategy for fully distributed control systems has been presented. The strategy has a distributed structure which matches with the architecture of these systems:

Each processor has its own overload detection mechanism, and control actions are taken to protect each processor against overload. A sufficiently general detection mechanism has been designed which can be applied to any processor, regardless of the functions which it performs. Control actions which protect each processor have been designed to be consistent with those which protect the other ones. The required coordination among processors is provided in this way without need of a centralized coordinator.

The general strategy proposed here has proved to be efficient when it has been implemented in System 12. Simulation results [7] have shown that a throughput a 40% higher than that corresponding to normal load conditions is practically maintained in most cases of severe overload situations. In some cases of focussed overload the throughput may be lower, but always significantly higher than that corresponding to normal load conditions. Under severe overload -either general or focussed- the grade of service for accepted calls is always kept equal or better than that corresponding to a 40% general overload.

REFERENCES

[1] J.A.G. Higuera and C. D. Berzosa, "Progressive Method of Overload Control for Processors", 9th ITC, Torremolinos, 1979.

[2] P. Somoza and A. Guerrero, "Dynamic Processor Overload Control and its Implementation in Certain Single-Processor and Multiprocessor SPC Systems", 9th ITC, Torremolinos, 1979.

[3] A. Briccoli, "Comparison of Regulation Methods for Traffic Overloads in SPC Systems", 9th ITC, Torremolinos, 1979.

[4] F.C. Schoute, "Adaptive Overload Control for an SPC Exchange", 10th ITC, Montreal, 1983.

[5] B.T. Doshi and H. Heffes, "Analysis of Overload Control Schemes for a Class of Distributed Switching Machines", 10th ITC, Montreal, 1983.

[6] "System 12 Digital Exchange", Electrical Communication, vol. 59, no. 1/2, 1985.

[7] G. Morales-Andres and M. Villen-Altamirano, "System 12 Traffic Overload Control", Electrical Communication, vol. 59, no. 1/2, pp. 74-79, 1985.

[8] R.N. Andries et al., "Simulation of Distributed Microprocessor Control in Digital Switching Systems", 11th ITC, Kyoto, 1985.

[9] W.M. Jolley, "Analysis of Delays by Counting Events", 9th ITC, Torremolinos, 1979.

TELETRAFFIC ISSUES in an Advanced Information Society
ITC-11
Minoru Akiyama (Editor)
Elsevier Science Publishers B.V. (North-Holland)

A NEW ROUTING ALGORITHM FOR MULTIPLE SERVER GROUPS

James C. FRAUENTHAL, Brad A. MAKRUCKI and David J. HOUCK, Jr.

AT&T Bell Laboratories,
Holmdel, New Jersey, USA

ABSTRACT

This paper considers the design and analysis of a multi-queue, multi-server group, queueing system that is intended to provide a specified grade of service (e.g., specified mean waiting times) to different classes of customers. The system routes customers from queues to server groups (which may have different service time distributions for a given customer type) in order to obtain mean waiting times below the objectives while also requiring the smallest number of servers to do so. Analytic results are described for an example of the general system.

1. INTRODUCTION

This paper considers the queueing system depicted in Figure 1. There are Q queues each having an associated stream of arriving customers. Furthermore there are G groups of servers each of which contains s_g servers of similar type. Different server groups may have different service characteristics. The connectivity of queues to server groups capable of providing service is given by a relation R where qRs if and only if server group s can serve type q customers. In addition there is a routing matrix which describes the sequence of server groups scanned when a customer arrives and an idle server must be sought. With each queue is associated a mean waiting time objective, o_q. The system is to provide actual mean waiting times as close to the objectives from below as possible. To obtain mean waiting times near their objectives, group sizes and the customer service priority rules must be determined.

Analytical and simulation results are presented for a simplified system where each queue has an associated primary service group and there is one overflow server group (containing s_o servers) that is capable of serving customers from any queue. It is desired to provide mean waiting times less than or equal to their objectives with the smallest number of servers required to do so. That is, the following problem is to be solved:

$$\min \sum_g s_g$$
$$\text{subject to: } \overline{W}_i \le o_i \qquad i = 1, \cdots, Q$$

where $\overline{W}_i$ is the mean waiting time for customers entering queue i. The minimization should be performed over all algorithms that prioritize customers for service. Considered here are only algorithms which make decisions at departure epochs of customers from overflow servers, preemptive service is not considered in this application. Furthermore only algorithms that make choices which depend on the present state of the system will be considered in the system analysis. The algorithm and analysis described constitute the first phase of problem study required in solving the complete optimization problem.

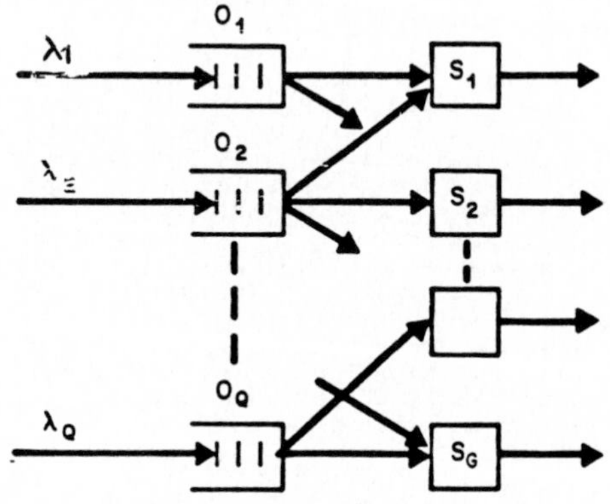

FIGURE 1

Applications of the minimization problem and customer routing include load balancing situations in computer systems [1]. For example, in multiprocessor systems where different processors are specialized to execute certain program constructs efficiently it may be advantageous to channel arriving jobs into appropriate job queues for execution on specialized hardware. Alternatively, during times of arrival rate imbalance it may be desirable to send jobs to slower processors in an effort to meet job queue waiting time objectives.

This paper is organized as follows: section 2 describes the queueing system and its operation. Section 3 describes simulation results obtained for larger systems. The results from simulations are presented because of present analysis size and complexity limitations. Section 4 describes an analysis that provides approximations for determining system behavior; it includes the ability to change the customer service algorithm for the overflow servers. Section 5 is the conclusion.

2. SYSTEM OPERATION

Although the analysis shown in section 4 generalizes to many queues and servers groups, we restrict attention to the two queue/three server group system shown in Figure 2.

Whenever an arrival occurs, the new customer is placed in the queue corresponding to its type. If server able to serve the customer is available, the customer proceeds directly to the server. Routing rules which employ thresholds to hold customers from using slower servers are not considered here. Qualified server groups are in general checked for availability in order of increasing average service time.

When a customer completes service, a check is made of all queues that can be handled by the idle server. Primary servers accept customers only from their primary queues; overflow servers choose their next customer from all queues that are not empty. This decision is based on the value of a random variable[1] called the service affinity, $\tilde{V}_q$. An overflow server chooses to serve queue l when

$$\tilde{V}_l \geq \tilde{V}_q \quad \textit{for all } q \in \{j \neq l : \tilde{Q}_j(t_d^-) > 0\}$$

where t_d^- is the time just before the departure and $\tilde{Q}_j(t_d^-)$ is the length of queue j at t_d^-. If $\tilde{V}_l = \tilde{V}_q$ then the server chooses uniformly to break service affinity ties.

A version of delay ratio queueing [2], which is a time dependent priority scheme, is used to define service affinities. An overflow server, upon becoming free, chooses to serve the queue which is providing the worst service to its arriving customers, relative to its mean waiting time objective. An analytically tractable definition of $\tilde{V}_q$ using queue length objectives rather than waiting times is:

$$\tilde{V}_q = \frac{\tilde{Q}_q(t_d^-)}{\lambda_q o_q}.$$

Where $\lambda_q o_q$ corresponds to the objective mean queue length, and $\tilde{Q}_q(t_d^-)$ is the instantaneous queue length.

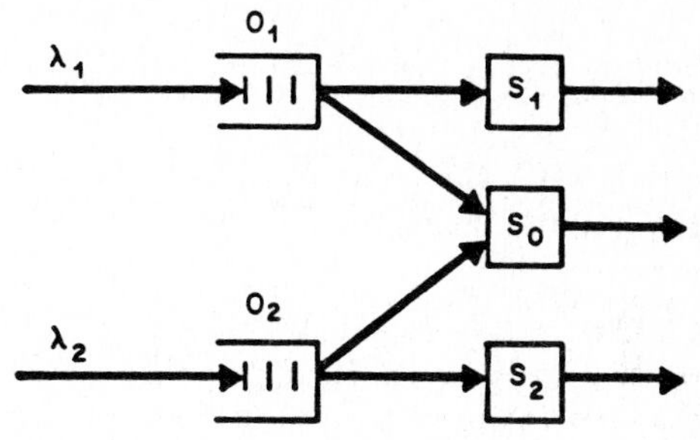

FIGURE 2

3. THE SIMULATION MODEL

The aim of the simulation program is to compare different configurations of server groups. This is accomplished efficiently by pairwise comparison of results obtained with different server configurations but identical streams of traffic.

3.1 SIMULATION RESULTS

Assume that customers can be distinguished by type at time of arrival. In addition to the two dedicated (primary) server groups, each of which handles only one type of traffic, there is an overflow group which is able to handle both kinds of customers. Whenever a member of the overflow group handles a call of either type, the service is slower than it would have been had a primary server provided it. This extra service time, which is attributable to adapting to varying demands on the servers, is central to the model. If there is no extra service time associated with mixed customer demands, it is always best to have just a single server group. Thus, the performance of the overflow system is intimately associated with the magnitude of the extra service time.

This example problem has been studied because it appears to represent a particularly hard situation to manage. The difficulty is based in part on the fact that the overflow server group does not have its own queue, so all customers handled by a member of the group incur a service time penalty. This in turn implies that it is undesirable to have the ratio of the

1. A $(\tilde{\ })$ is used to denote a random variable, and a $(\bar{\ })$ is used to denote its mean value.

number of overflow to primary servers too large as this leads to a large service time for customers that use the overflow servers. On the other hand, if the ratio is too small, the load balancing effects of customer routing will be lost.

The basic properties of the system whose operation is simulated are listed in the table below.

Customer Type	1	2
Objective Delay	6	2
Average Service Time	25	30
Customer Fraction (%)	55	45

These values have been chosen so that the first offered loads for both types of traffic are about equal. This does not make the delay ratio (the expected delay divided by the delay objective) equal for both call types. The small imbalance is included intentionally as it tends to illustrate quite dramatically the ability of the customer routing scheme to equalize the delay ratios. It will be noted that, for all examples run without an overflow server group, the type 1 customers receive service better than their objective, while the type 2 customers receive service worse than their objective since the independent groups have not necessarily been sized so as to provide optimum service.

Results are generated for a variety of different numbers of servers. Each simulation experiment consists of fixing the total number of servers and then generating the identical arriving stream of traffic to differently configured server groups.

The first offered load is chosen for each configuration so that the delay ratios are about unity when there is no customer routing. The range of values employed for the occupancy ran from a low of 0.75 for a group of 20 servers to a high of 0.945 for a group of 240.

Figures 3a-3c employ the same basic layout. The total number of servers is listed at the top, along with the service time penalty (STP). This quantity states, as a percentage, the extra time required by a server in the overflow group to handle a customer. It is assumed that the service time penalty is the same percentage for both call types. The horizontal axis lists the number of overflow servers deployed. To this information must be added the knowledge that there are equal numbers of primary servers of the two types, and that the total sums to the number at the top of the figure. The purpose is not to indicate proper group sizes required but to show the effects of customer routing on system performance.

Figures 3a-3c plot the delay ratio against the number of overflow servers for 20, 80 and 240 servers, respectively. Notice that the scales on both axes are the same on all three figures. There are several other things to notice in these figures. First, as the number of overflow servers increases, the lines representing the delay ratio for the two call types tend to come together. This means that one effect of customer routing is to make each type of traffic receive more nearly equivalent service relative to its prestated goal. This is particularly evident in Figure 3a, which shows the type 1 traffic to be receiving better than objective service and the type 2 traffic to be receiving worse than objective service when no overflow servers are available. As the number of overflow servers increases, the delay ratios become nearly equal. Second, notice that there is an initial decrease in the delay ratios with increasing number of overflow servers, followed by a deterioration of

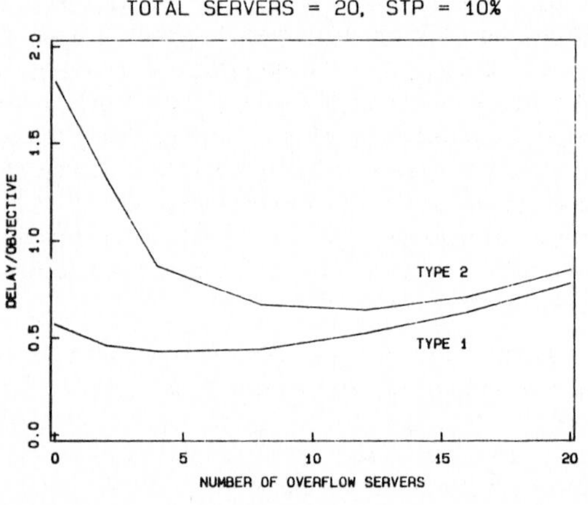

FIGURE 3a.

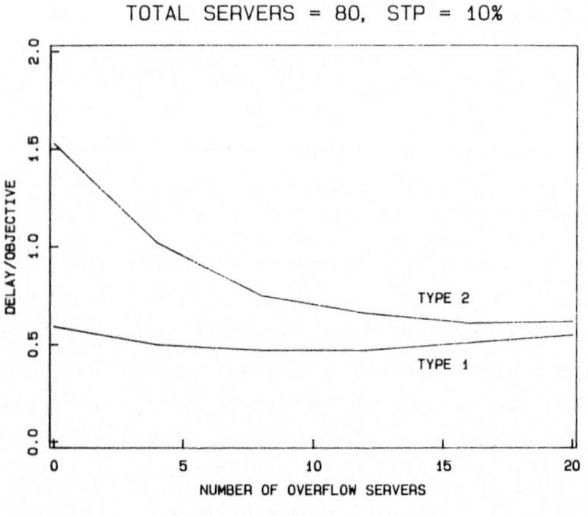

FIGURE 3b.

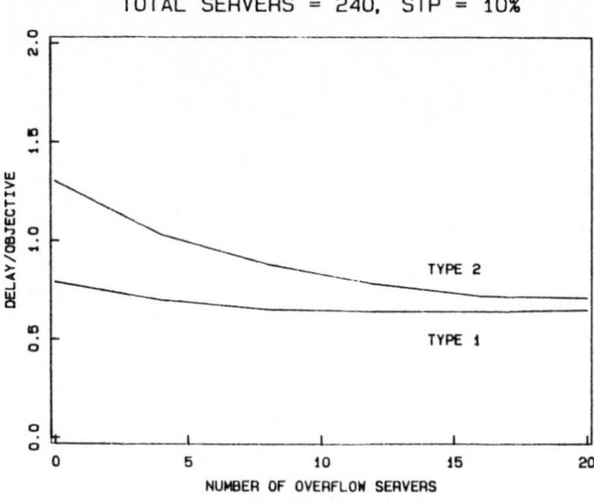

FIGURE 3c.

service with further increase in the number of servers. The initial improvement is due to the benefits associated with customer routing; the subsequent decline is the consequence of the extra service time incurred by the overflow servers. In effect, once there are too many overflow servers, all of whom operate 10% slower than the primary servers, the benefits of customer routing are lost to increased service time. Third, and perhaps most unexpected, observe that the optimal number of overflow servers changes quite slowly with the total number of servers.

4. *QUEUEING ANALYSIS*

In the analysis exponential and independent interarrival and service times are assumed. Service rates are defined as:

$$\bar{X}_j = \frac{1}{\mu_j} = \textit{mean service time}$$

for type j customers in their primary servers

$$\bar{X}_{oj} = \frac{1}{\mu_{oj}} = \textit{mean service time}$$

for type j customers in the overflow servers

Quantities to be calculated include server utilizations, mean waiting times, and rates of customer flow from queues to server groups.

Let ρ_i be the utilization of type i primary servers and ρ_o be the utilization of overflow servers. Let λ_i^o be the rate at which customers flow from queue i to overflow servers. Obviously $\rho_i < 1, \rho_o < 1$ is sufficient to ensure positive recurrent queue lengths for all queues. Let ρ_{oi} be the utilization of an overflow server due to type i customers, $\rho_o = \sum_i \rho_{oi}$.

An exact calculation may be derived for the two queue, single overflow group as follows: the system state may be denoted by a 4-tuple process $\{(\tilde{N}_1(t), \tilde{N}_2(t), \tilde{G}_1(t), \tilde{G}_2(t)); t \geq 0\}$; where $\tilde{N}_i(t)$ is the number of customers in queue and primary group i; and $\tilde{G}_i(t)$ is the number of type i customers in the overflow group. This approach requires a detailed description of state transitions that depends on the particular service affinity function chosen. The state space is large and would lead to intractable solution complexity for large systems. To alleviate these two drawbacks approximate processes have been devised.

The approach to developing approximations has been to decouple the above process into two separate processes $\{(\tilde{N}_i(t), \tilde{G}_i(t)); t \geq 0\}$ while accounting for coupling effects in transition rates in the decoupled processes. Furthermore consider steady-state results and define for $q = 1, 2$:

$$g_{ij}(q) = \lim_{t \to \infty} Pr(\tilde{N}_q(t) = i, \tilde{G}_q(t) = j).$$

Since exponential assumptions are used, there are no lattice distribution complications, and by assumption, the processes are positive recurrent. Equations for utilizations, queue length probabilities, and flow rates may now be written:

$$\rho_i = \frac{\lambda_i - \lambda_i^o}{s_i} \bar{X}_i$$

$$\begin{aligned} \rho_o &= \rho_{o1} + \rho_{o2} \\ &= \frac{\bar{G}_1}{s_o} + \frac{\bar{G}_2}{s_o} \\ &= \frac{1}{s_o}\left[\sum_{k=1}^{s_o} k\left(\sum_{l=0}^{\infty} g_{lk}(1)\right) + \sum_{k=1}^{s_o} k\left(\sum_{l=0}^{\infty} g_{lk}(2)\right)\right] \end{aligned}$$

$$\lambda_i^o = \bar{G}_i \mu_{oi}$$

and

$$\lim_{n \to \infty} Pr(n^{th} \textit{ customer from queue } i \textit{ is served by an overflow server}) = \frac{\lambda_i^o}{\lambda_i}.$$

Queue length probabilities may be written as:

$$Q_i(l) = \lim_{t \to \infty} Pr(\tilde{Q}_i(t) = l) = \begin{cases} \sum_{j=0}^{s_i} \sum_{k=0}^{s_o} g_{jk}(i) & l = 0 \\ \sum_{k=0}^{s_o} g_{l+s_i,k}(i) & l > 0 \end{cases}$$

The process state diagram for queue 1 is shown in Figure 4. Note that the width of the state diagram for process i is constant at $s_i + 1$ and the structure of the arcs is independent of the particular service affinity function considered, provided the customers to be served by overflow servers are chosen based only on the discrete state of the system at departure epochs.

Levels 0 through s_i-1 rates are derived in a straightforward manner. Define:

$$\delta_j(i) = Pr(\hat{G}_{\bar{i}}(t_a^-) < s_o - j \mid \hat{N}_i(t_a^-) = s_i, \hat{G}_i(t_a^-) = j)$$

which is the probability that an arriving customer at queue i that finds all s_i servers busy may go into overflow service immediately given the state of subsystem i. $\bar{i}$ is the other queue in the two queue system, t_a is the arrival time.

Next define

$$\beta_{ij}(q) = Pr(\text{customer type } q \text{ is chosen over customer type } \bar{q} \text{ at an overflow server departure epoch} \mid \hat{N}_q(t_d^-) = i, \hat{G}_q(t_d^-) = j)$$

where t_d is the departure time for a customer from an overflow server (regardless of customer type).

In level s_i the branching of customers into queue or overflow service is achieved by multiplying arrival rates by $\delta_j(i)$.

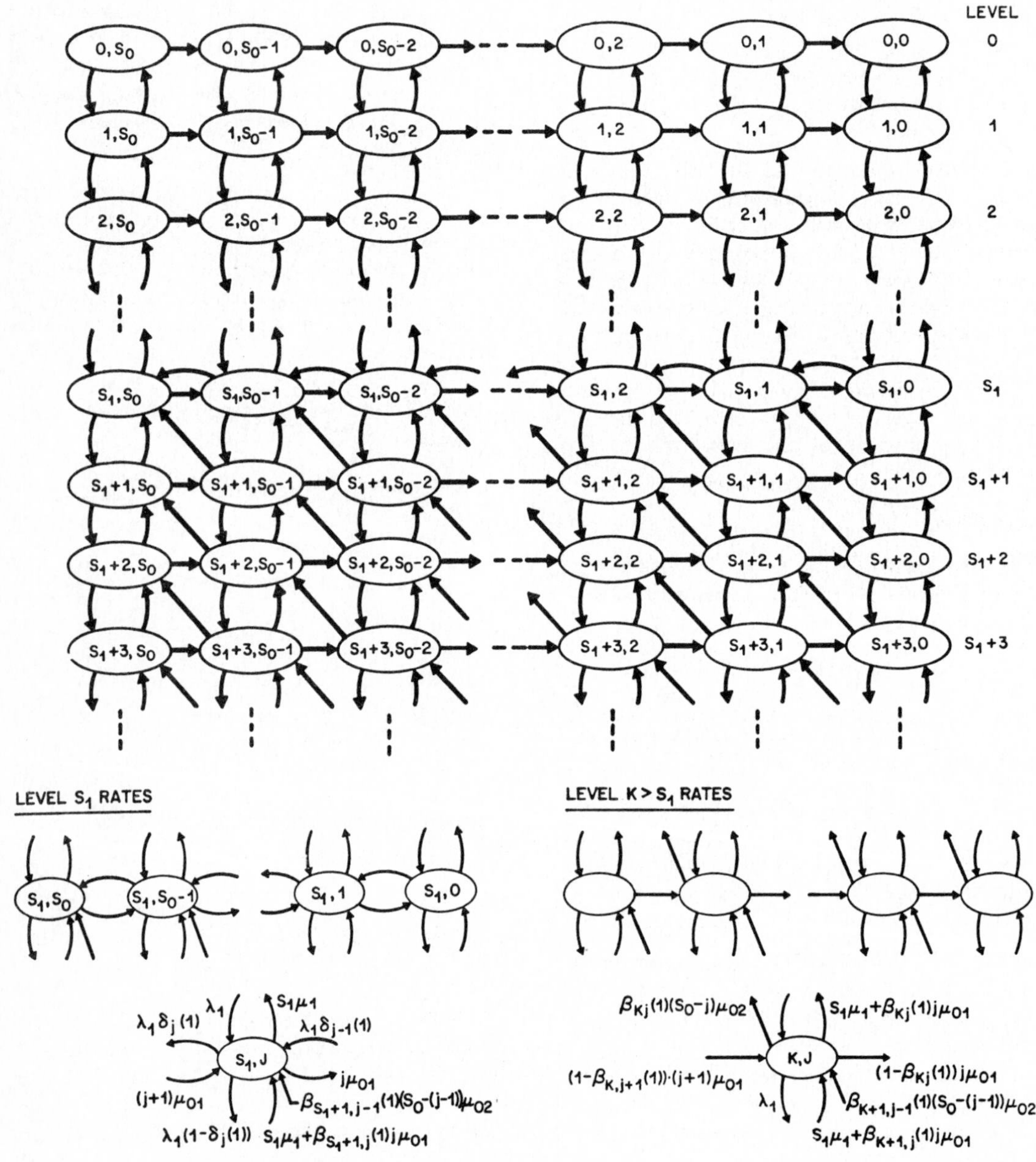

FIGURE 4.

In states in levels $k > s_i$ (for which there is a customer in queue) a branching probability implied by $\beta_{ij}(q)$ is present. Implicit in the state diagram are self loops which are taken in process i when an overflow server completes a type $\bar{i}$ customer service and again chooses to serve a type $\bar{i}$ customer. That is, at the completion epoch, process i leaves its present state and re-enters it if it does not obtain control of another overflow server. This approximates subsequent departures from overflow servers as a Bernoulli process concerning the choice of customer type to be served.

This time invariant approximation allows a conditional subsystem to be used in estimating $\delta_j(q)$ and $\beta_{ij}(q)$. Given that $\bar{G}_i(t) = j$ there is a subsystem consisting of queue $\bar{i}$ and $s_o - j$ overflow servers. The subsystem is depicted in Figure 5. The solution for the subsystem's steady-state queue length distribution is straightforward and is amenable to solution using matrix geometric techniques. Let primed quantities be those associated with subsystem behavior. Denote the subsystem steady-state queue length probabilities as:

$$Q'_{k,n}(m) = \lim_{t\to\infty} Pr(\bar{Q}'_m(t) = k \quad \text{with } n \text{ overflow servers in the subsystem})$$

$$g'_{l,j,n}(m) = \lim_{t\to\infty} Pr(\bar{N}'_m(t) = l,\ \bar{G}'_m(t) = j \quad \text{with } n \text{ overflow servers in the subsystem})$$

Approximating $\beta_{ij}(q)$ as if the subsystem reaches steady-state during its existence we get:

$$\beta_{ij}(q) = Pr(\bar{V}_q(t_d^-) > \bar{V}_{\bar{q}}(t_d^-) \mid \bar{N}_q(t_d^-) = i, \bar{G}_q(t_d^-) = j) + \frac{1}{2} Pr(\bar{V}_q(t_d^-) = \bar{V}_{\bar{q}}(t_d^-) \mid \bar{N}_q(t_d^-) = i, \bar{G}_q(t_d^-) = j)$$

$$\approx \hat{\beta}_{ij}(q) = Pr\left(\frac{\bar{Q}'_{\bar{q}}(t_d^-)}{\lambda_{\bar{q}} o_{\bar{q}}} < \frac{\bar{Q}_q(t_d^-)}{\lambda_q o_q} \,\Big|\, \bar{N}_q(t_d^-) = i, \bar{G}_q(t_d^-) = j\right) + \frac{1}{2} Pr\left(\frac{\bar{Q}'_{\bar{q}}(t_d^-)}{\lambda_{\bar{q}} o_{\bar{q}}} = \frac{\bar{Q}_q(t_d^-)}{\lambda_q o_q} \,\Big|\, \bar{N}_q(t_d^-) = i, \bar{G}_q(t_d^-) = j\right) \quad i > s_q$$

$\hat{\beta}_{ij}(q)$ is an estimator of $\beta_{ij}(q)$. The decomposition approach ignores dependency of the state of $Q'_{\bar{q}}(t_d^-)$ on the queue length of queue q at t_d^- (it exploits only the dependence on $\bar{G}_q(t_d^-)$). Approximating departure point distributions (from the actual system) with arrival point (or general time) distributions allows t_d^- to be ignored, observing that $\bar{N}_q(t_d^-) = i$ implies $\bar{Q}_q(t_d^-) = i - s_q$ and defining $\gamma = \lambda_{\bar{q}} o_{\bar{q}} / \lambda_q o_q$ we get

$$\hat{\beta}_{ij}(q) = \begin{cases} \sum_{k=0}^{\gamma(i-s_q)-1} Q'_{k,s_o-j}(\bar{q}) + \frac{1}{2} Q'_{\gamma(i-s_q),s_o-j}(\bar{q}) & \gamma = \text{integer} \\ \sum_{k=0}^{\lfloor \gamma(i-s_q) \rfloor} Q'_{k,s_o-j}(\bar{q}) & \text{otherwise} \end{cases}$$

If $\{\bar{Q}'_{\bar{q},s_o-j}(t); t \ge 0\}$ is not positive recurrent because

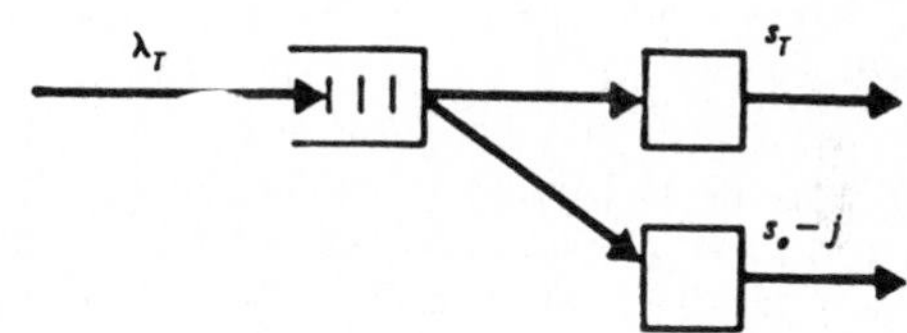

FIGURE 5.

ANALYSIS VS. SIMULATION RESULTS										
					SIM	ANALYSIS	SIM	ANALYSIS		
s_1, s_2, s_o	λ_1	λ_2	o_1	o_2	$\bar{W}_1$	$\bar{W}_1$	$\bar{W}_2$	$\bar{W}_2$	% $error_1$	% $error_2$
2, 2, 2	0.4	0.6	1.0	1.0	.183	.176	.373	.360	-4	-1
	0.65	0.65	1.0	1.0	.816	.864			6	
	0.5	0.5	1.0	1.0	.264	.252			-5	
8, 8, 2	2.23	2.73	0.2	0.6	.124	.109	1.187	1.126	-12 *	-4
	2.295	2.295	1.0	1.0	.385	.374			-3	
8, 8, 4	2.49	3.05	0.2	0.6	.112	.113	.709	.729	1 *	3
	3.42	2.28	0.2	0.6	.358	.373	.227	.215	4	-5 *
	3.42	2.28	0.2	2.0	.337	.334	.264	[illegible]	-1	-11 *
0, 0, 4	0.267	0.401	1.0	1.0	.242	.172	.284	.294	-29 *	4 *
	0.234	0.434	1.0	3.0	.168	.160	.321	.310	-5	-3

Entries denoted with a (*) required extrapolations tobe done.

TABLE I. Analysis and simulation results.

$\lambda_{\bar{q}} \geq s_{\bar{q}}\mu_{\bar{q}} + (s_o - j)\mu_{o\bar{q}}$ for some j then the above calculations logically give $\hat{\beta}_{ij}(q) = 0$ (we take all the queue length probability mass to be concentrated at ∞). This is inaccurate in that the decomposed system does not stay in existence long enough for a steady-state to be reached (because the number of overflow servers serving type q customers changes at some point in time with high probability, unless one queue is in saturation while the other is not). Therefore approximations for $\hat{\beta}_{ij}(q)$ will be made for $j \geq$ smallest J such that $\lambda_{\bar{q}} \geq s_{\bar{q}}\mu_{\bar{q}} + (s_o - J)\mu_{o\bar{q}}$. A simple scheme has been used which is based on extrapolating from $\beta_{ij}(q)$ for $j < J$ to j's $\geq J$.

5. CONCLUSIONS

This paper investigates a model for a multi-queue, multi-server group queueing system. The particular example problem studied has two customer types and three server groups, one of which is dedicated to each customer type and the third is is able to server all customers.

Simulation results demonstrate that the distribution algorithm works efficiently. With a fixed total number of servers, as the number of overflow servers is increased, each type of traffic receives better service. Further, there is an optimal number of overflow servers which depends upon the magnitude of the service time penalty paid for having the overflow servers present. In addition, as the total number of servers increases sharply, the optimal number of overflow servers increases rather slowly.

An analytic model for the same problem is discussed, and various approximations are discussed to make the problem computationally manageable. Results from the analytic model are compared with simulation results for several different numbers of servers. It is found that the mean waiting times compare favorably, with errors smaller than 10%.

Work underway will extend the analytic model so it includes non-exponential service times, larger numbers of customer types and better methods of approximation.

REFERENCES

[1] T.P. Yum, and H. Lin, "Adaptive Load Balancing for Parallel Queues with Traffic Constraints," *IEEE Trans. on Communications*, Vol. COM-32, No. 12, Dec. 1984, pp. 1339-1342.

[2] M. Segal, *On the Delay-Ratio Queue Discipline*, 9th International Teletraffic Congress, Torremolinos, Spain, session 41.

TELETRAFFIC ISSUES in an Advanced Information Society
ITC-11
Minoru Akiyama (Editor)
Elsevier Science Publishers B.V. (North-Holland)

THE ANALYSIS OF TRANSIENT PERFORMANCE FOR NODAL FAILURE IN AN ALTERNATIVE ROUTING NETWORK

John BONSER

British Telecom Research Laboratories
Felixstowe, UK
IP11 8XB

ABSTRACT

A new integrated digital trunk network is currently being implemented by British Telecommunications plc (BT). This network and the switching systems in it have been extensively studied by the BT Teletraffic Division. This paper describes one area of the new network and a transient traffic profile that can occur under network failure conditions. The switching system processor, its load control scheme and a simulation model of the processor are outlined in the paper; the model has been used to predict processor performance under transient traffic conditions. The results of the simulation study and an approximate mathematical model describing them and suitable for use in network studies are presented.

1 INTRODUCTION

The plan for an integrated digital network in the United Kingdom, now being implemented by British Telecommunications plc (BT), envisages a completely new trunk network structure. This new network, and the switching units to be employed in it have been extensively studied by the BT Teletraffic Division. This paper describes one of the studies conducted on the switching units and [1] describes a study of the network itself.

The switching units have been studied under a range of traffic conditions including normal loads, steady state overloads, transient overloads and partial failure conditions. A study of the switching unit processor performance under a transient traffic loading that could occur in the new network has been conducted using a simulation model of the processor. The processor itself, the load control scheme employed, and the simulation model are all discussed in the paper along with results obtained from the simulation of the transient overload.

A full list of abbreviations used in the text is contained in Table 1.

2 PSEUDO DIRECTIONAL NETWORK

In the new trunk network structure some 400 existing analogue trunk switching units are being replaced by about 60 Digital Main Switching Units (DMSUs). Most DMSUs will be singly located, fully interconnected and will switch traffic in both directions. In some large cities, however, the volume of trunk traffic to be switched will warrant two or more DMSUs.

A particular case arises where two DMSUs are located in a large city and are required to be mutually supportive in the event of a system failure. To meet this need, a Pseudo-Directional network has been designed, as shown in Figure 1. Normally, one of the DMSUs carries all traffic incoming to the city and the other all traffic outgoing from the city. In the event of an imbalance between outgoing and incoming traffics, Automatic Alternative Routing (AAR) is used to overflow traffic from the heavily loaded unit to the lightly loaded unit, which then carries traffic in both directions. Each DMSU secures the other and in the event of one DMSU failing, all calls proper to that unit would be offered to the other. This situation has been judged to represent the worst possible case of DMSU overload likely to be encountered in the UK network. It was therefore selected for a teletraffic investigation of DMSU processor performance under extreme transient overload.

Table 1 Abbreviations

The following abbreviations are used in the text:

AAR	Automatic Alternative Routing
AP	Applications Program
BT	British Telecommunications plc
CCP	Call Control Process
CPU	Central Processor Unit
DMSU	Digital Main Switching Unit
DSP	Digital Switching Process
PA	Process Allocator
PLC	Processor Load Control
SP	Signalling Process
UK	United Kingdom

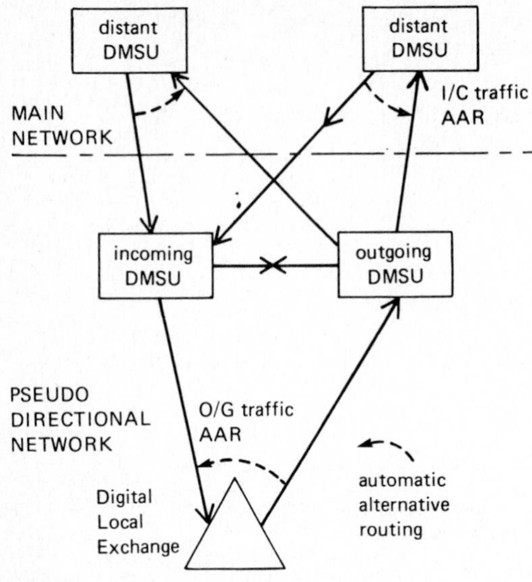

Figure 1

Pseudo Directional Network

3 TRANSIENT CALL ATTEMPT PROFILE

The transient call attempt profile offered to a system under system failure conditions clearly depends on a very wide range of factors including network configuration, repeat attempts and network management. It was not necessary to consider all these factors in detail for the Pseudo Directional Network study since the main purpose of the study was to demonstrate the satisfactory operation of Processor Load Control under the severe transient call attempt overload conditions that could arise with the failure of one DMSU. Thus a 'worst case' transient was constructed, after consulting [2].

Many of the parameters defining the transient profile including:

traffic type
peak calling rate
transient duration

were arbitrarily selected to give the worst possible representation of the transient profile likely to be offered to the processor. Variations of this profile have been investigated to allow the effect of variation in the parameters to be determined.

The transient call attempt profile for which results are given in this paper is shown in Figure 2. It corresponds to the entire traffic carried on the failed DMSU in Figure 1, being offered to the remaining DMSU in an interval of 24 seconds. This profile was added to steady state traffic levels of normal load and 50% overload to obtain a total call attempt profile. This total call attempt profile, of course represents a gross overload of the DMSU processor.

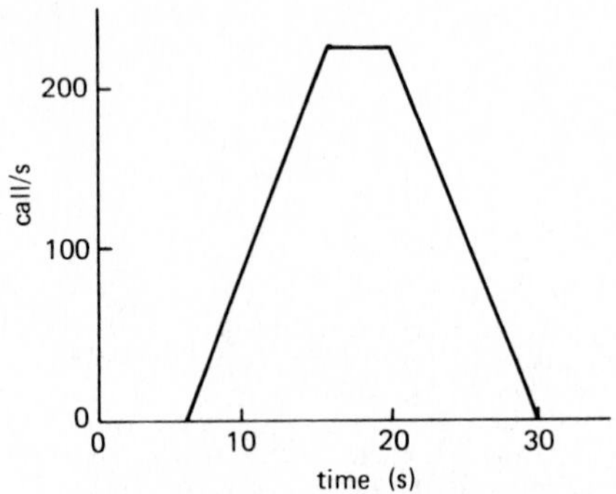

Fig. 2 Transient call attempt profile

4 DMSU PROCESSOR

The control processor which is used in each DMSU in the Pseudo Directional Network and the teletraffic simulation model of the processor are described in [3].

The processor is outlined in Figure 3. This shows that it consists of up to four CPU's and their associated hardware collectively termed a cluster. Clusters can be interconnected to provide additional processor power if required.

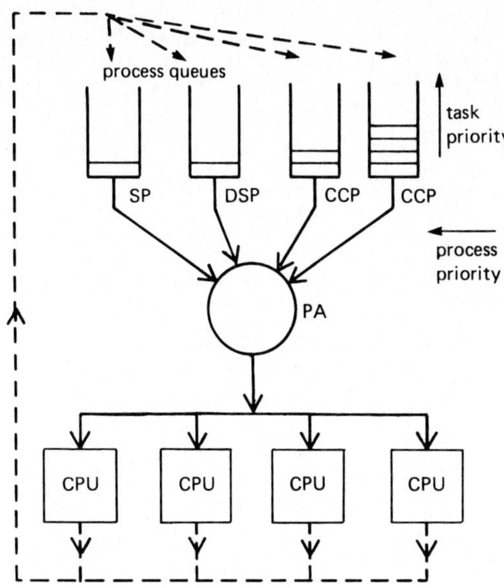

SP = Signalling process
DSP = Digital switching process
CCP = Call control process
PA = Process allocator

Fig. 3 DMSU processor

The software on the processor is organised into a number of Application Programs (APs) performing such functions as signalling, switching, and call control. These APs are served in priority order. Each AP has its own process queue and the aggregate of all process queues is termed the System Queue. The scheduling of the APs and the communication between them is controlled by the Process Allocator (PA).

Thus the processor is a multiserver priority queueing system and it has been simulated as such when determining the system performance under transient traffic conditions.

5 PROCESSOR LOAD CONTROL

The purpose of processor load control (PLC) is to regulate the work performed by the multiprocessor so that, under call-attempt overload or failure conditions, the number of tasks within the process queues does not build up to excessive levels. Should the total number of tasks exceed the maximum allowed in the System Queue, the consequences would be dire because the multiprocessor would cease handling any work at all and would need to be restarted.

The multiprocessor uses a dual scheme to perform load control as shown in Figure 4. First, workload limits are used to control the acceptance or rejection of new calls (represented by sequences of the APs) such that the maximum number of calls in the setup phase at any instant does not exceed a given workload limit. A monitoring period of approximately 5 seconds is specified (the periodicity is adjustable) and in each period information is gathered on the CPU occupancies, the calls accepted, the calls rejected and the maximum number of calls in the setup phase. Based on this information, new workload limits are calculated for the next monitoring period to ensure that the average CPU occupancy does not exceed a specified value, say 0.9.

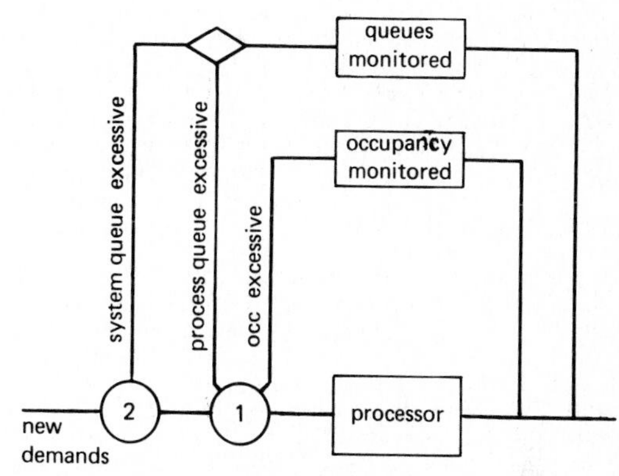

1 new calls rejected and routed to tone

2 all inputs inhibited

Fig. 4 Processor load control

The second part of the scheme involves thresholds on the process queues and the system queue. If the number of tasks within a process queue, or the system queue exceeds a preset, upper threshold, an overload is indicated. End of overload occurs when the number of tasks is reduced below another preset, lower threshold. There exists a degree of flexibility as to the action invoked when an overload is in progress, but, generally, a process queue overload causes all new calls to be rejected for its duration and a system queue overload causes the input of all tasks from the peripheral units to be inhibited.

The mode of operation of the dual load control scheme is such that, under call-attempt overload, regulation of the CPU occupancy enables the acceptance or rejection of new calls to be controlled in a stable and smooth manner. In more severe circumstances, such as a large transient overload, the thresholds provide direct protection against an excessive number of tasks in a process queue or the system queue.

6 PROCESSOR SIMULATION MODEL

The performance of the processor has been observed using a simulation model. The model has been designed to simulate the interaction between PA , PLC and the AP's for a single cluster with the added facility of being able to represent intercluster communication for a multicluster configuration. It has been progressively developed over a period of years from the early processor design concepts to the current multiprocessor, multicluster, configuration. This development of the model has been facilitated because, firstly, the process structure of the software has been carried forward almost unchanged through the major design changes of the multiprocessor hardware, and secondly, the decision was taken at an early stage that most of the multiprocessor design details would be defined through input data to the model. New multiprocessor designs have therefore been relatively easy to model by changing this data.

The model has been programmed in PL/1 and uses the Telesim event-by-event simulation package which has been specifically developed by the BT Teletraffic Division for teletraffic performance analysis. The Telesim package which runs on an IBM 3081 computer, undertakes the scheduling of events and provides for the handling of histograms and confidence interval routines. The model thus consists of a set of event-by-event action blocks, and Telesim schedules the action blocks to run in the order in which they occur in real time.

The model provides the following main facilities:

i a variable number of CPUs
ii a variable number of APs
iii a variable number of process sequences, representing call progression through the APs
iv timing values for AP run times and PA run times
v results giving details of:

CPU occupancies
AP occupancies
AP queueing delays
AP queue lengths
processor grade of service
behaviour of PLC parameters
history of the above during a simulation run

The model can generate a number of call arrival patterns including random (Poissonian) arrivals. The mean interarrival time of these random arrivals can itself be a function of time and this facility enables the model to be used to simulate the transient traffic conditions occurring in the Pseudo-Directional Network.

7 RESULTS

The processor performance observed using the simulation model is shown in Figure 5, 6 and 7. Each Figure includes two curves corresponding to the observed performances of processors during the application of the transient call attempt profile from an initial steady state load of

- normal load (full line)
- 50% call attempt overload (broken line)

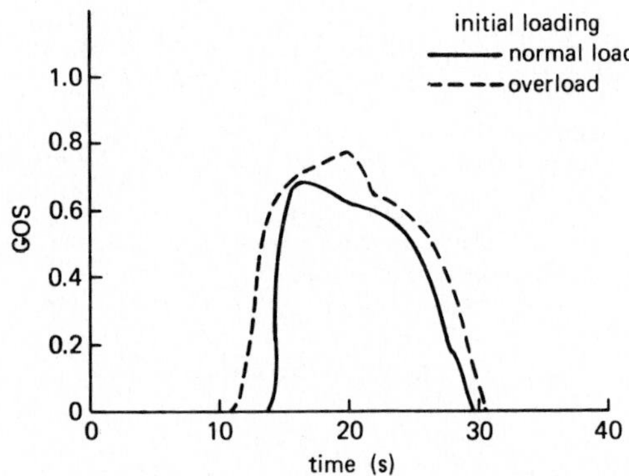

Fig. 5 Processor grade of service

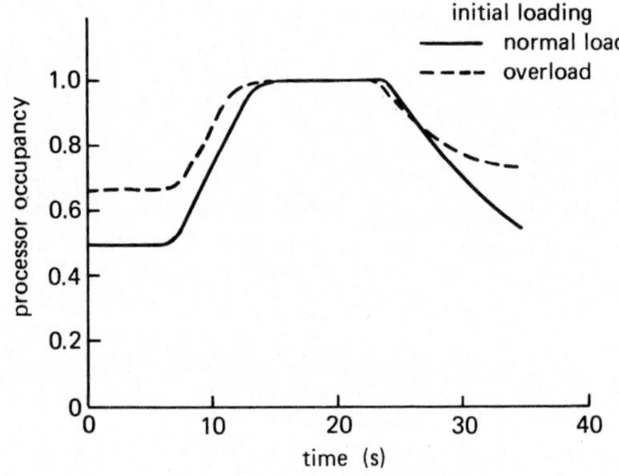

Fig. 6 Processor occupancy

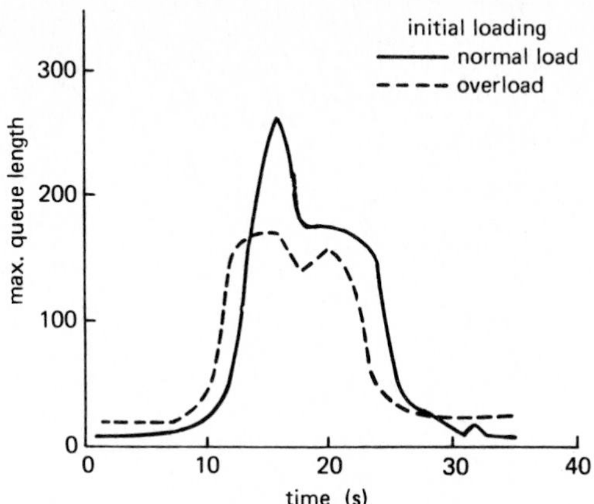

Fig. 7 Processor queue length

Figure 5 shows that the processor continues to process some calls successfully during the transient period albeit at a poor grade of service, and recovers rapidly when the transient is removed.

The processor occupancy observed under these transient conditions is illustrated in Figure 6. This shows that the occupancy rapidly rises to almost 1.0. At this occupancy the processor is accepting as many calls as possible and rejecting the remainder.

Figure 7 demonstrates two interesting features about the processor queue lengths observed:

1 System queue length rises higher when the processor is initially carrying normal load than when it is initially overloaded. This reflects the fact that the procesor load control parameters are set to accept a smaller increase in call attempt rate at overload than at normal load

2 System queue length remains below its threshold, set for the purpose of this study at 300, and well below its limit of 600 throughout the transient. This means that, referring to Figure 4, all new calls are either carried or rejected and no new demands from the signalling systems are inhibited. Hence all calls that have been accepted and are in setup continue to completion and all cleardowns are sucessfully processed.

Thus the results show that under severe transient overloads the processor will continue to accept some call attempts. The remainder will be rejected and routed to tone in a controlled manner, with normal operation quickly being restored once the transient overload condition is removed.

Further results showed the effect of varying the somewhat arbitrarily chosen transient parameters:

traffic type -

the major effect of the traffic type is to determine the loading offered to the processor

peak calling rate -

the peak calling rate primarily determines processor occupancy and if that becomes excessive determines the grade of service occuring during the transient

transient duration -

the results show that the processor recovers rapidly from transient overloads and thus the transient duration has little effect on processor performance other than determining the time at which the processor returns to normal.

8 PERFORMANCE MODEL

In studies of networks containing switching systems, such as [4], it is desirable to have a simple model of system performance. In studies of the steady state performance of these networks the grade of service at a system is often obtained from the equations:

$$a + r = T$$
$$a.A + r.R = K$$

where

a = accepted calling rate
r = rejected calling rate
T = total (offered) calling rate

A = processing effort to accept a call
R = processing effort to reject a call
K = total processing effort available

These can, of course, readily be solved to give a grade of service for the system at any given calling rate.

The equations do not give a good description of the performance observed during the transient conditons. This is primarily because they assume equal arrival rates for call set-ups and cleardowns which is clearly not valid during transient conditions. Under these conditions the performance is better described by

$$a(1) + r(1) = T$$
$$a(1).A(1) + a(2).A(2) + r(1).R(1) + r(2).R(2) = K$$

where

a(1) = rate at which new calls are accepted
a(2) = rate at which accepted calls cleardown

and so on for other terms

The system grade of service can be evaluated from these equations in a similar manner to that used earlier, provided that the cleardown rates, a(2) and r(2), are available.

Thus the performance of a switching system under transient traffic conditions, can be obtained from equations similar to those used to obtain the performance under steady state conditions. It is interesting to observe, however, that the result is non-Markovian as the current state of the system is dependent upon previous states, in as much as these determine the call cleardown rates. The implications of this observation have not been investigated.

9 CONCLUSION

This paper has described a network with AAR and the transient traffic conditions that can occur in such a network. It has also described the processor provided at the switching nodes, the load control scheme employed in the processor, and a computer simulation of the processor.

The simulation model was used to evaluate the performance of the processor under transient traffic conditions and the results of this analysis are presented. An approximate mathematical model of the observed performance, suitable for inclusion in network studies, is also presented.

10 ACKNOWLEDGEMENTS

The author wishes to thank his colleagues, including particularly Dr R H Thompson, Mr K J Miller and Mr J L C Grimbly, for their contributions to this paper. Acknowledgement is made to the Director of System Evolution and Standards Department of British Telecom for permission to publish this paper.

11 REFERENCES

[1] D J Songhurst, C R Becque, M Lebourges

Analysis and Dimensioning of Non-Hierarchical Telephone Networks
Paper submitted to:
11th International Teletraffic Congress
Kyoto, Japan, 1985

[2] Conference of European Postal and Telecommunications Administrations (CEPT)
Performance Requirements for Digital Transient Exchanges
Recommendation T/CS 68-04

[3] Dr R H Thompson
Modelling a multiprocessor designed for Telecommunication Systems control
ICL Technical Journal pp119-130
November 1984

[4] Minoru Akiyama
Extraordinary Traffic Handling in the Telecommunication Network
Electronics and Communications in Japan
Vol 65-B, No 12, 1982

TELETRAFFIC ISSUES in an Advanced Information Society
ITC-11
Minoru Akiyama (Editor)
Elsevier Science Publishers B.V. (North-Holland)

ON SYMMETRICAL TWO-LINK ROUTING IN CIRCUIT SWITCHED NETWORKS

Ulf KÖRNER and Bengt WALLSTRÖM

Department of Communication Systems
Lund Institute of Technology
Lund, Sweden

ABSTRACT

For fully connected, symmetrical circuit switched networks we study the problem how to arrange two-link routing in accordance with certain symmetry conditions. Design rules are derived for a certain class of symmetrical schemes that are successfully applied in cases where the number of nodes is prime. Other cases, including networks with two-way links, are considered more briefly.

1. INTRODUCTION

Most telephone networks today have, and have always had, a hierarchical topology. Though alternative routing is frequently used and meshed structures are found in the upper layers, the routing principles used are nevertheless to be characterised hierarchical.

However, recent investigations [1] [2] [3] [4] have indicated that great advantages are to be found when using a non-hierarchical network structure in, at least, parts of the network. The advantages lie in a more homogeneous distribution of traffic load over the network. Busy-hour traffics in different time-zones as well as those generated in city- and suburb-areas could be spread over non busy parts of the network and thus communication resources would be better shared.

We shall focus mainly on networks with one-way links, and discuss only briefly solutions for networks with two-way links.

The basic conditions for meshed networks using non-hierarchical routing-strategies are now being fulfilled as we get more and more SPC-systems and exploit more intelligent signalling principles, like in number seven.

The symmetrical, fully connected network structure forms a natural starting point for investigations of non-hierarchical routing concepts for circuit switched telephone plant. This paper is focused on the question of how to route calls in such a network in accordance with a proposed symmetry condition.

2. THE PROBLEM IN ESSENCE

Let us consider the five node fully connected network graph of fig. 1. Each arc between two nodes of the graph is to represent two circuit groups, one for each switching direction. The similar network with M nodes has M(M-1) links and OD-pairs (Origination - Destination node pairs). Calls are routed by first choice on direct links and by subsequent alternative choices on two-link paths.

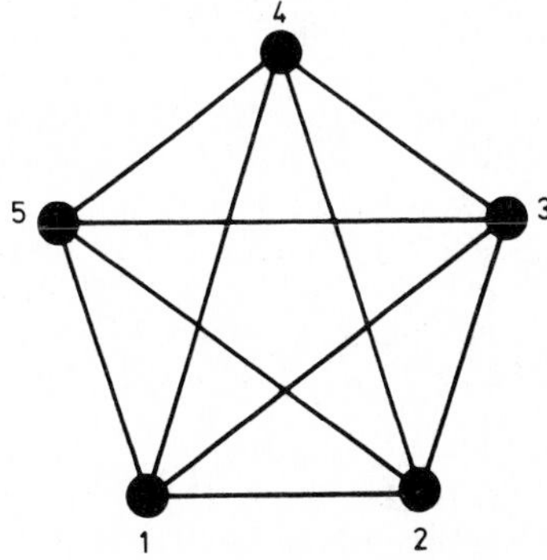

Fig. 1. Fully connected network (M=5).

Various schemes may be used for the selection of a free path among the M-2 two-link paths accessible to an OD-pair. With sequential hunting a link appearing in early choice two-link paths of several OD-pairs will clearly be more heavily loaded than a link accessed mainly by late choices. The routing schemes to be considered here have the property of producing equal loading on all links for any method of selection among free two-link paths. This is achieved, firstly by partitioning the M(M-1)(M-2) two-link paths into M-2 disjunct sets, secondly by associating to each such set a routing matrix having the following properties:

(i) An element of a routing matrix, $r_{ij}^{k} \in (1,2,...,M)$; $1 \leq i,j \leq M$; $i \neq j$; $1 \leq k \leq M-2$; is the number of a tandem node connecting the OD-pair (i,j)

Thus a routing matrix provides precisely one two-link path to each of the M(M-1) OD-pairs. On average each link must appear twice per routing matrix. As a definition of symmetric routing we propose that

(ii) each link of the network must appear precisely twice in each routing matrix, once as the first link and once as the second link of a path.

Routing (and loading) symmetry is thus provided by each single routing matrix and is consequently not dependent on the path selection procedure. Moreover, we may reduce the number of alternative choices by deleting arbitrary routing matrices and still retain routing symmetry.

DESTINATION NODE

ORIGINATING NODE	1	2	3	4	5
1	-	3	4	5	2
2	4	-	5	3	1
3	2	1	-	5	4
4	3	5	2	-	3
5	3	4	1	2	-

Fig. 2. Routing matrix.

A routing matrix for a five node network is shown in fig. 2. It does not fulfill the symmetry condition (ii), since e.g., link 31 appears three times while link 14 appears once only. A systematic search for symmetric solutions is somewhat simplified if we put $r_{ii} = i$, $i=1,2,\ldots,M$, and require that each of the numbers 1,2,..,M appear one time in each row and column.

We wish to find M-2 such matrices with $r_{ij}^{k} \neq r_{ij}^{l}$ for $k \neq l$, $i \neq j$.

For lower values of M solutions can be obtained by a search procedure based on the rule given above (fig. 3), but the method quickly becomes impracticable when M is increased. Indeed, a network with 6 nodes required several hours of computer time, and no more than two simultaneous symmetric routing matrices were found.

M = 3

1	3	2
3	2	1
2	1	3

M = 4

1	3	4	2		1	4	2	3
4	2	1	3		3	2	4	1
2	4	3	1		4	1	3	2
3	1	2	4		2	3	1	4

Fig. 3. Symmetric solutions for small networks.

Luckily, however, we have found a faster procedure, which seems to produce complete solutions (M-2 matrices) whenever M is a prime number.

3. A CLASS OF CYCLICAL ROUTING SCHEMES

Let us consider the special array of two-link paths shown in fig. 4. It contains (M-1)(M-2) sets of M two-link paths each connecting a particular set of M OD-pairs. The set of paths in position (i,j) will be denoted P[i,j]. The first and second links of P[i,j] are arranged in a simple cyclical manner defined by (i,j). This, of course, is just one of many possible ways of grouping links and paths. Possibly we may obtain M-2 symmetric routing matrices by cleaver partitioning of the P-sets into M-2 sub-sets of size M-1.

Let us look into what kind of restrictions such a choice must obey. Firstly, any routing matrix must have just one node number r_{ij} in each position.

SECOND LINKS

FIRST LINKS	12 23 34 45 51	13 24 35 41 52	14 25 31 42 53	15 21 32 43 54	i
12 23 34 45 51	2	3	4	-	1
13 24 35 41 52	3	4	-	1	2
14 25 31 42 53	4	-	1	2	3
15 21 32 43 54	-	1	2	3	4
j	1	2	3	4	

Fig. 4. A Cyclic Array of Two-Link Paths (M=5).

Examining fig. 4 it is seen that sets P[i,j], such that i+j=constant, connect the same sets of OD-pairs. A closer examination indicates that the same is true more generally for sets P[i,j] such that (i+j)mod M = constant, where

$$(p)\bmod q = \begin{cases} r \text{ if } p=sq+r \\ q \text{ if } p=sq \end{cases}$$

$$p = 1,2,\ldots$$
$$q = 1,2,\ldots$$
$$r = 1,2,\ldots,q-1$$
$$s = 1,2,\ldots$$

(This differs from the standard modulous definition where (sq)mod q=0).

In fig. 4 the sub-sets defined by (i+j)mod 5 = 1,2,3,4 are marked.

Thus, to obtain a proper routing matrix we must choose precisely one square from each of the sub-sets P[i,j] such that (i+j)mod M = l, l=1,2,...,M-1.

Secondly, it is seen that if the M-1 chosen P-sets belong to different rows and columns, then our symmetry condition will be fulfilled, as well. Note, that for M odd the sub-set P[i,i] i=1,2,...,M-1 defines a symmetric routing matrix. Thus we have obtained simple heuristic rules for the design of symmetric routing schemes composed of particular sets of M links. A formal proof of the rules is given in Appendix.

4. PRIME NUMBER OF NODES

The class of routing schemes defined above has proved very useful when M is a prime number. Indeed we have found a simple algorithm, that has not so far failed to produce complete symmetric solutions in such cases.

To explain the procedure let us consider the solution for M=11 represented by a matrix (s_{ij}), fig. 5, where

1	2	6	7	5	3	9	8	4	0
3	1	9	2	8	6	4	7	0	5
7	8	1	5	4	2	3	0	6	9
6	3	4	1	7	9	0	2	5	8
4	9	5	6	1	0	8	3	7	2
2	7	3	8	0	1	6	5	9	4
8	5	2	0	9	7	1	4	3	6
9	6	0	3	2	4	5	1	8	7
5	0	7	4	6	8	2	9	1	3
0	4	8	9	3	5	7	6	2	1

Fig. 5.

we have written $s_{ij}=k$, $1\leq i,j\leq M-1$, $i+j\neq M$, $k=1,2,\ldots,M-2$, to indicate that P[i,j] is used for the k:th routing matrix and $s_{i,M-i}=0$, $i=1,2,\ldots,M-1$.

For k=1 we use the main diagonal, i.e. $s_{ii}=1$; $i=1,\ldots,M-1$. The elements s_{ij} chosen for k=2,3,4 and 5 form simple patterns that are defined by the equations

$$j=(id)\bmod M; \qquad i=1,2,\ldots,M-1 \qquad (1)$$

$$i=(jd)\bmod M; \qquad j=1,2,\ldots,M-1 \qquad (2)$$

$$j=[(M-i)d]\bmod M; \qquad i=1,2,\ldots,M-1 \qquad (3)$$

$$i=[(M-j)d]\bmod M; \qquad j=1,2,\ldots,M-1 \qquad (4)$$

respectively, with d=2.

For k=6,7,8 and 9 the same equations apply with d=3.

The (general ?) algorithm reads:

1. $s_{ii}=1$; $i=1,2,\ldots,M-1$

   $s_{ij}=0$; $i\neq j$

2. Stop if M-2 subsets have been found. d=min(j) such that $s_{1j}=0$. Assign two new values of k to s_{ij} according to the patterns given by eqs (1) and (2).

   If the first element, s_{1j}, determined by eq. (3) is not zero, then return to 2.

3. Assign two new values of k to elements s_{ij} determined by eqs (3) and (4). Return to 2.

d	k								
2	2,3	2-5	2-5	2-5	2-5	2-5	2-5	2-5	2-5
3			6-9	6-9	6-9	6-9	6-9	6-9	6-9
4					10,11	10-13	10-13	10-13	10-13
5				10,11	12-15		14-17	14-17	14-17
7						14-17	18-21		18-21
8								18-21	
9								22-25	
11									22-25
12								26,27	26-29
M->	5	7	11	13	17	19	23	29	31

Table 1. Values of d and k for some solutions obtained by the algorithm.

Having determined the matrix $\{s_{ij}\}$, the elements of the k:th routing matrix are obtained from the formula (cf. Appendix):

$$r^{k}_{n,(n+i+j)\bmod M} = (n+i)\bmod M \quad n=1,2,\ldots,M$$

where the M-1 values of (i,j) are obtained from one of the equations (1) - (4) determined by k.

In general, when M is a prime, it seems that many different symmetrical schemes may exist, and the algorithm developed above produces just one of them. Thus for M=5 we found five additional solutions not belonging to the class considered here.

5. NON-PRIME NUMBER OF NODES

The class of routing schemes defined in section 3 seems to be strictly applicable only when M is a prime number. When M is an odd number, however, we may design "fairly symmetric" solutions in a simple way. The general idea should be clear from the following $\{s_{ij}\}$-matrix for M=9.

1	2	3	4	5	6	7	0
7	1	2	3	4	5	0	7
6	7	1	2	3	0	5	6
5	6	7	1	0	3	4	5
4	5	6	0	1	2	3	4
3	4	0	6	7	1	2	3
2	0	4	5	6	7	1	2
0	2	3	4	5	6	7	1

Fig. 6. "Fairly symmetric" solution for M=9.

Except for k=1, we have somewhat relaxed the symmetry condition that each k must appear precisely once in each column and row.

For M even we have not yet obtained strictly symmetric solutions except for the case M=4 (fig. 4). Neither have we found any simple "fairly symmetrical" solution like the one shown in fig. 6. Networks of practical sizes, however, seldom use more than a few of the M-2 possible alternative choices and a small number of symmetric routing matrices can generally be determined rather quickly by a computer program based on the rules given in section 2.

Moreover, there is a possibility of using a somewhat weaker symmetry condition than the one proposed in section 2. To assure equal loading of links in the symmetric network considered here, it seems unnessessary to stipulate that each link must appear once as first link and once as second link in each routing matrix. A wider class of useful schemes will be at hand if we just maintain that each link must appear twice in each routing matrix.

6. TWO-WAY LINKS

It is obvious, that the routing matrices obtained for networks with one-way links (case I) are applicable to networks with two-way links (case II) as well. The assumptions stipulated for case I are sufficient but not nessessary for case II. Thus, the methods, presented in the previous sections, could be used, when we treat two-way links. This implies, that solutions for case II are of interest only if there does not exist a solution for case I, or if the solution of the former has qualities (not treated here), that outshine those of the latter. From now, when we talk about solutions for case II, we mean solutions, that do not fulfill case I.

Though the constraints under case II are lighter than those under case I, we have not found a general method to produce routing matrices for networks with two-way links.

The presented methods for case I assume each link to be used twice, once as the first link in a two-link path and once as the second. If we transfer this to the two-way link network, we must claim each link to be used four times, twice as the first link in a two-link path and twice as the second. This in turn leads to that our M-2 routing matrices must obey the following rules:

1. In row i, at most two elements may be equal to k, $k \neq i$;
2. If m elements in row i are equal to k, than 2-m elements in row k must be equal to i.
3. In column j, at most two elements may be equal to l, $l \neq j$;
4. If n elements in column j are equal to l, than 2-n elements in column l must be equal to j.

Take for instance the five node network, where the following three matrices do obey the rules above

$$\begin{array}{ccccc} 1 & 4 & 2 & 3 & 3 \\ 4 & 2 & 4 & 5 & 1 \\ 2 & 4 & 3 & 5 & 2 \\ 3 & 5 & 5 & 4 & 1 \\ 3 & 1 & 2 & 1 & 5 \end{array} \qquad \begin{array}{ccccc} 1 & 5 & 5 & 2 & 4 \\ 5 & 2 & 1 & 3 & 3 \\ 5 & 1 & 3 & 1 & 4 \\ 2 & 3 & 1 & 4 & 2 \\ 4 & 3 & 4 & 2 & 5 \end{array}$$

$$\begin{array}{ccccc} 1 & 3 & 4 & 5 & 2 \\ 3 & 2 & 5 & 1 & 4 \\ 4 & 5 & 3 & 2 & 1 \\ 5 & 1 & 2 & 4 & 3 \\ 2 & 4 & 1 & 3 & 5 \end{array}$$

Fig. 7. The five node network.

However, similarly as for the one-way link network, we may relax the constraints. For the two-way link network, we may require just every link to be used four times. Then, the following matrices for a six-node network form a solution.

$$\begin{array}{cccccc} 1 & 5 & 2 & 2 & 2 & 2 \\ 3 & 2 & 5 & 6 & 3 & 3 \\ 4 & 4 & 3 & 1 & 4 & 4 \\ 5 & 5 & 5 & 4 & 1 & 2 \\ 6 & 6 & 6 & 6 & 5 & 3 \\ 3 & 4 & 1 & 1 & 1 & 6 \end{array} \qquad \begin{array}{cccccc} 1 & 3 & 4 & 5 & 6 & 3 \\ 5 & 2 & 4 & 5 & 6 & 4 \\ 2 & 5 & 3 & 5 & 6 & 1 \\ 2 & 6 & 1 & 4 & 6 & 1 \\ 2 & 3 & 4 & 1 & 5 & 1 \\ 2 & 3 & 4 & 2 & 3 & 6 \end{array}$$

$$\begin{array}{cccccc} 1 & 6 & 6 & 6 & 3 & 4 \\ 4 & 2 & 1 & 1 & 1 & 1 \\ 5 & 6 & 3 & 2 & 2 & 2 \\ 3 & 3 & 6 & 4 & 3 & 3 \\ 4 & 4 & 1 & 2 & 5 & 4 \\ 5 & 5 & 5 & 5 & 2 & 6 \end{array} \qquad \begin{array}{cccccc} 1 & 4 & 5 & 3 & 4 & 5 \\ 6 & 2 & 6 & 3 & 4 & 5 \\ 6 & 1 & 3 & 6 & 1 & 5 \\ 6 & 1 & 2 & 4 & 2 & 5 \\ 3 & 1 & 2 & 3 & 5 & 2 \\ 4 & 1 & 2 & 3 & 4 & 6 \end{array}$$

Fig. 8. The six node network.

In this case we state

1. In row i, at most four elements may be equal to k, $k \neq i$;
2. If m elements in row i are equal to k, than at most 4-m elements in row k may be equal to i.
3. If n elements in row k are equal to i, than at most 4-m-n elements in column i may be equal to k.
4. If p elements in column i are equal to k, than 4-m-n-p elements in column k must be equal to i.

CONCLUSIONS

We have studied the problem of attaining symmetric loading and loss performance in symmetric, fully connected networks employing two-link alternative routing. First a natural symmetric condition based on the routing matrix concept was defined for an M node network. It does garantee the required symmetry properties irrespective of the number of alternative choises performed (from 1 to M-2) and of the hunting procedure used on the two-link paths. A particular class of cyclical routing schemes was then introduced and a set of rules for its application was derived. These schemes appeared to be particularly useful for networks with a prime number of nodes, yielding complete solutions in all cases considered so far. A fast solution algorithm was found but it remains to have it proved.

Moreover, the above mentioned class of routing schemes, could be used to design "fairly symmetric" solutions in a simple way, whenever the number of nodes is odd. For even number of nodes, however, the class appeared rather unsuitable. As an alternative we suggested a somewhat simpler, but still sufficient, symmetry condition, that will increase the number of useful solutions. It remains to investigate these possibilities.

Finally we studied briefly the symmetric routing problem for networks using only two-way links.

Symmetry conditions analogous to those used for one-way link networks were stated and examples were shown. In general the two-way conditions are less demanding, but we have not so far found any general solution method, other than the algorithm given in section 4 for prime numbers of nodes. There is one, possibly important, case, however, for which these solutions do not apply, viz when it is required that each routing matrix must provide just one two-link path for both traffic directions between two nodes. This case is not considered in this paper.

Obviously there are many unsolved mathematical problems connected to symmetrical routing. For instance we have proved nothing about the existence of solutions (just observed that M=6 is a stubborn number). The discoveries for prime number of nodes were made to a great deal by good luck and we have not proved the solution.

In addition there are several interesting extensions of the problems that call for further study, i.e..

(i) Two- and three-link routing, etc.

(ii) Partly connected networks.

(iii) Sensitivity to non-symmetric traffic loading.

(iv) Optimality in view of traffic intensity variations.

APPENDIX - PROOF OF DESIGN RULES GIVEN IN SECTION 3.

In this Appendix all additions, subtractions and equalities are taken mod M, according to the definition given in section 3.

The following definitions will be used.

D1. L(n,m) is the (unidirectional) link from node n to node m. The length of L(n,m) is m-n. Referring to fig. 1 it is seen that the length of a link is simply equal to the number of by-passed nodes plus 1.

D2. The set of two-link paths using first links of length i and second links of length j is

$$P[i,j] = \{L(n,n+i),L(n+i,n+i+j) \mid n=1,2,...,M\}$$

$$1 \leq i,j \leq M-1, \quad i+j \neq M$$

D3. The set of OD-pairs connected by P[i,j] is

$$OD[i,j] = \{(n,n+i+j) \mid n=1,2,...,M\}$$

The entities defined above have the following notable properties:

P1. There are M-1 possible link lengths: 1,2,...,M-1. There are M links of length l, l=1,2,...,M-1.

P2. Since $(i,j) \neq (i',j') \Rightarrow P[i,j] \cap P[i',j'] = \Phi$ and i,j = 1,2,...,M-1, $i+j \neq M$, there are (M-1)(M-2) disjunct P-sets. Each of them contains M paths. Together they constitute the set of all M(M-1)(M-2) paths.

P3. Paths of the set P[i,j] contain all links of length i as first links and all links of length j as second links.

P4. The set OD(i,j) contains M OD-pairs. The M paths of the set P(i,j) connect M different OD-pairs.

P5. $i+j = i'+j' \Rightarrow OD[i,j] = OD[i',j']$
$i+j \neq i'+j' \Rightarrow OD[i,j] \cap OD[i',j'] = \Phi$
P(i,j) sets belonging to a subgroup with i+j = constant connect the same OD-pairs. Otherwise they connect different OD-pairs.

Theorem. Let $C_1,C_2,...,C_{M-2}$ be a partition of the set $\{P[i,j] \mid 1 \leq i,j \leq M-1, i+j \neq M\}$ with the properties

(i) $|C_k| = M-1$, k=1,2,...,M-2

(ii) for each k,

$$P[i,j] \in C_k \;\&\; P[i',j'] \in C_k \;\&\; (i,j) \neq (i',j')$$

$$\Rightarrow i \neq i' \;\&\; j \neq j' \;\&\; i+j \neq i'+j'$$

If all paths from P-sets belonging to C_k are used for the choice number k, then a symmetric routing is obtained.

Proof. By P4 and P5 all OD-pairs are connected by C_k. If $C_k = \{P[i_1,j_1],P[i_2,j_2],...,P[i_{M-1},j_{M-1}]\}$ then both $(i_1,i_2,...,i_{M-1})$ and $(j_1,j_2,...,j_{M-1})$ are permutations of the set (1,2,...,M-1) and by P3 every link appears exactly once as first link and once as second link among the k:th choice paths.

Remark 1. For M odd $\{P[i,i] \mid i=1,2,...,M-1\}$ is a subset satisfying conditions (i) and (ii) of the theorem. (Note that (2i) mod M = (2i') mod M in this case.)

Remark 2. Any set C of M-1 P-sets with the property $P[i,j] \in C \;\&\; P[i',j'] \in C \;\&\; (i,j) \neq (i',j') \Rightarrow i+j \neq i'+j'$ defines a proper routing in the sense that all OD-pairs are connected by C.

Remark 3. Any matrix $\{s_{ij}\}$, i,j=1,2,...,M-1, such that

(i) $s_{i,M-i} = 0$, i=1,2,...,M-1

(ii) $s_{ij} \in (1,2,...,M-2)$, $i+j \neq M$,

(iii) each number $k \in (1,2,...,M-2)$ appears exactly once in each row, in each column and in each of the subsets

$$\{s_{ij} \mid i+j = l;\; l=1,2,...,M-1\},$$

defines a symmetric routing:

$$P[i,j] \in C_k \Leftrightarrow s_{ij} = k$$

Remark 3 above gives the main rule for the design of a symmetric routing pattern. By remark 1, when M is odd, we may use the set $\{P[i,i] \mid i=1,2,...,M-1\}$ as one of the M-2 subsets of P, say C_1. Accordingly we put $s_{ii}=1$, i=1,2,...,M-1.

ACKNOWLEGEMENTS

This report was prepared within the project "Intelligent Routing" sponsered by Ericsson, Stockholm. The authors also wish to thank Michal Pioro, Technical University of Warsaw, and Lars Reneby, Lund Institute of Technology for valuable advise and enlightening discussions. A particular thank goes to Marianne Greiff and Lena Svensson for skilful typing and drawing.

REFERENCES

[1] G.R. Ash, R.H. Cardwell and R.P. Murray, "Design and optimization of networks with dynamic routing", Bell Syst. Tech. J., vol 60, no. 8, 1981.

[2] M. Pioro and B. Wallstöm, "Multihour optimization of non-hierarchical circuit switched communication networks with sequential routing", Proc. 11th Int. Teletraffic Congress, Kyoto, 1985.

[3] C. Grandjean, "Call routing strategies in telecommunication networks", Proc. 5th Int. Teletraffic Congress, New York, 1967.

[4] E. Szybicki and A.E. Bean, "Advanced traffic routing in local telephone networks; Performance on proposed call routing algorithms", Proc. 9th Int. Teletraffic Congress, Torremolinos, 1979.

TELETRAFFIC ISSUES in an Advanced Information Society
ITC-11
Minoru Akiyama (Editor)
Elsevier Science Publishers B.V. (North-Holland)

COMPARATIVE EVALUATION OF CENTRALIZED/DISTRIBUTED TRAFFIC ROUTING POLICIES IN TELEPHONE NETWORKS

J.M. GARCIA*, F. LE GALL*, C. CASTEL**, P. CHEMOUIL‡, P. GAUTHIER‡, G. LECHERMEIER‡‡

LAAS-CNRS*, DERA-CERT**, CNET‡, DTIF-DGT‡‡
France

ABSTRACT

This paper deals with adaptive traffic routing in telephone networks. At present dimensioning and traffic routing are determined once a year in order to meet some predefined grade of service requirements. For technological reasons, i.e. electromechanical switching centres, call routing is now defined according to some fixed static procedure. In order to provide protection against important disturbances, the network has then to be overdimensioned at the planning stage. The emergence of stored program control networks consisting of electronic switching centres interconnected by common channel signalling links is making possible a new strategy - adaptive traffic routing -, which allows real-time reaction to changes of the network state due to overloads or failures. Several adaptive traffic routing algorithms are therefore described and compared on a testbed network in order to improve network performances.

INTRODUCTION

At present the traffic routing used in most telephone networks is the fixed alternate routing. At each switching centre, calls are routed according to destination using only local informations. The incoming call is routed on a first choice trunk group or overflows over second choice trunk group when the first one is blocked. This routing pattern is purely decentralized. A telephone network planned for this routing policy gives satisfying performances under normal conditions. However in case of significant disturbances, this policy does not make efficiently use of the network resources.

Recent advances in telecommunications such as the introduction of stored program control networks consisting of electronic switching centres interconnected by common channel signalling links allows the elaboration of more sophisticated routing policies. Theses methods are intended to take into account real-time conditions in order to provide better network performances. Two strategies have been considered to realize network control : one approach considers the problem at the planning stage and is based on forecasted traffic data. Then it consists in planning reserve routes which are used under overload conditions and updated according to dayly variations [1]. This method, referred to as dynamic routing, is an implementation of feedforward control.
A second approach is to consider real-time measurements to design traffic routing algorithms which take into account the real-time behaviour of the telephone networks [2], [3], [4], [5], [6]. This strategy refers to as adaptive routing and acts as a closed loop-type control.
In this paper we are concerned with the second approach and we present four adaptive routing algorithms. These methods have been investigated according to items such as time scale analysis, information structure and routing scheme. In the first method the routing algorithm works on a two time scale basis and leads to a hierarchical control structure : In a centralized way a coordinator determines the optimal static routing over a mid-term period according to aggregated information over the whole network. At the local level, the optimal routing is implemented according to local observations.
In the second method, a decentralized algorithm is realized using the learning automata approach [2]. The approach has been enhanced in order to allow overflow when a route is blocked [6].
Last two distributed algorithms are presented : The first one, based on a deterministic rule, is an application of the residual capacity approach [3]. The second algorithm takes into account the stochastic nature of the system to update the oveflow routes.
The main purpose of this paper is to evaluate the performances of the proposed adaptive traffic routing policies and to quantify the improvements with the real-time control. A study of the behaviour of the algorithms is presented considering a testbed network, subset of Paris network.
Attention is then focused on the comparative evaluation of the methods under different operating conditions.
We first describe briefly the routing algorithms that have been developed. We then present the testbed network and the typical disturbances which may affect a network. Next we present the performances obtained with each algorithm, using event by event simulations. Finally a comparison of these methods with the present fixed alternate routing scheme shows the improvements of the network performances when implementing adaptive traffic routing.

* Laboratoire d'Automatique et d'Analyse des Systemes, Toulouse, France
** Département d'Etudes et de Recherche en Automatique, Toulouse, France
‡ Centre National d'Etudes des Télécommunications, Issy-les-Moulineaux, France
‡‡ Direction des Télécommunications d'Ile de France, Paris, France

DESCRIPTION OF THE ROUTING POLICIES

This section is devoted to the description of the routing policies. We only point out the main characteristics of each algorithm. Therefore the methods are briefly presented. A complete development of the traffic routing algorithms can be found in [6].

The four proposed routing policies are state-dependent : the routing of a call may change according to the network state and the input rates (measured). The call by call routing for all policies is decentralized and sequential i.e. at each call arrival the switching centre sets up a route to the call without exchanging information with others centres. The route that is chosen of course depends on the parameters of the routing policies.

Some important differences between the four routing policies exist at several levels :

* The local routing may use either multi-choice load sharing or alternative routing tables.
* The algorithm that changes the parameters of the local controllers can be :
  - decentralized : each node updates its own parameters, independently of the others, with a partial information on the network.
  - distributed : a set of nodes, but not all, exchanges their state information and routing parameters.
  - centralized : The state information of the whole network is used by a coordinator that changes the whole set of routing parameters.
* Measurements operated on the network.
* Time scale of the routing policy i.e. the frequency with which the routing parameters are updated.

For each algorithm the offered traffic is assumed to be unknown. Hence the offered traffic matrix is estimated according to measurements such as carried traffic over trunk groups or call counting.

The networks under study are one-level transit networks with one way trunks (see Fig.1). Moreover we consider that:

- transit switches are dedicated to the transit function; no traffic originates or terminates at transit centres.
- only transit switches may be considered as transit points.

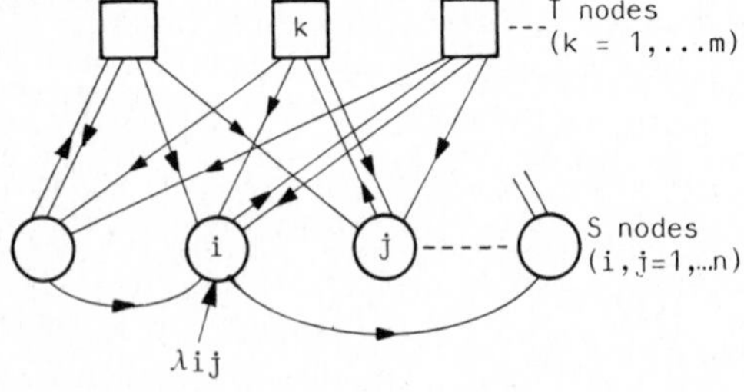

Figure 1 - A One Level Transit Network -

When direct links exist between two nodes, they are used in first choice to route the concerned calls. When a direct link is blocked, the calls overflow to the transit level.

The selected transit centre will depend on the routing policy. In the experiments, inter-transit links will not be considered but the routing policies presented here extend to that case without any problem.

The notations of variables and indices associated to this system are :

N_{ij} : Capacity of trunk group ij
λ_{ij} : Rate of traffic flow (i,j) Poisson distributed
τ : Average holding time, exponentially distributed and equal for all traffic flows.
$A_{ij} = \lambda_{ij}\tau$: element of the traffic matrix $[A_{ij}]$
X_{ij} : carried traffic over trunk group ij
π_{ij} : blocking probability of trunk group ij
α_{ikj} : load sharing parameter i.e. percentage of traffic flow ij routed via transit k

• A centralized policy : Feedforward load-sharing

This centralized approach is based on the quasi-static operation of the network : inputs are assumed to be stationary during an interval of time long enough to get statistical measurements of the system. A coordinator computes periodically (i.e. every 15mn) the optimal load sharing parameters of the routing policy according to the estimation of the offered traffic matrix during the last period. These parameters are used in each switching centre to route the calls during the next time interval.

The optimization model is a non linear multi-commodity flow problem.

Let $z_{ij}(n)$ be the number of calls from flow ij measured during the time interval n. The offered traffic matrix $[A_{ij}]$ can be estimated for the period n+1 with the recursive algorithm [3] :

$$[A_{ij}]^{(n+1)} = F([A_{ij}]^{(n)}, [z_{ij}]^{(n)})$$

The optimal load sharing parameters $\alpha^*_{ikj}(n+1)$ to apply during period n+1 are the solution of the non linear optimization problem [7] :

$$J = \min_{\alpha_{ikj}} \sum_{i,j} (A_{ij} - X_{ij}) \qquad (1)$$

with $\sum_k \alpha_{ikj} = 1 \quad \forall\ i, j$ and $\alpha_{ikj} \geq 0$

If $\alpha_{ikj} < 1$, flow control is achieved (the policy may refuse a call even if an idle path exists). The carried traffic X_{ij} are the steady-state solutions of the implicit system [8] :

$$\dot{X}_{ik} = -X_{ik}/\tau + \sum_j \alpha_{ikj}\lambda_{ij}(1-\pi_{ik})(1-\pi_{kj})$$
$$\dot{X}_{kj} = -X_{kj}/\tau + \sum_i \alpha_{ikj}\lambda_{ij}(1-\pi_{ik})(1-\pi_{kj}) \qquad (2)$$

where X_{ik} and X_{kj} are determined as follows :

$$X_{ik} = A_{ik}(1-\pi_{ik}) \text{ and } \pi_{ik} = E[A_{ik},N_{ik}]$$
$$X_{kj} = A_{kj}(1-\pi_{kj}) \text{ and } \pi_{kj} = E[A_{kj},N_{kj}] \qquad (3)$$

A_{ik} and A_{kj} are fictitious offered traffic and $E[.,.]$ is the Erlang-B loss function.

The problem (1), (2), (3) can be solved using relaxation techniques combined with feasible direction optimization models.

In fact using a single choice policy is not optimal for resources utilization; then the use of overflow policy has been added :
if the first choice trunk group ik is blocked, the call towards node j to be routed over this trunk group overflows over a second choice trunk group with a probability β_{iklj}, $(\sum_{l\neq k}\beta_{iklj} \leq 1)$.

β can be computed by solving an optimization problem like the preceding one or with an heuristic solution. For example a simple solution could be:

$$\beta_{iklj} = \alpha_{ilj}/(1-\alpha_{ikj}) \qquad (4)$$

- A Decentralized policy : The Learning Automata

The learning automata update their action probability (choice of a route) according to the completion or the rejection of the call and improve their own performance call by call. The method results in a load sharing policy implemented at each origin node. The factor that affects the performance is the reinforcement scheme for the updating of the action probabilities. In general a reinforcement scheme can be represented as :

$$\alpha(n+1) = T[P(n), \alpha(n), x(n)] \qquad (5)$$

where T is an operator, $\alpha(n)$ and $x(n)$ are respectively the action and the input of the automaton at time n, P(n) is the state probability vector governing the choice of the route at stage n.

The reinforcement schemes can be classified according to the operator T. Linear schemes have been studied for application to telephone network management [2] :

* The L_{RP} automaton (linear reward-penalty) updates its actions at each call with a reward if the call is successful and a penalty if it is blocked.

* The L_{RI} automaton (linear reward-inaction) updates its actions only in the case of success. It is shown [10] that this automaton converges to the solution $\pi_i = \pi_j$ where π_i is the blocking probability associated to action $\alpha(i)$. Moreover in the case where the π_i are linear functions of P(i), this solution is optimal in the sense that the overall losses are minimized. This is the reason why we focus our attention on L_{RI} schemes. In the case of L_{RI} automata, the reinforcement scheme has been enhanced by allowing overflow when the first trunk group is blocked. Then action probabilities are updated after the first and the second (when necessary) attempt.
The two stage automaton is therefore:
If action $\alpha(i)$ has been chosen at step n :

- if $\alpha(i)$ is successful :

$$\begin{aligned} P_{n+1}(i) &= P_n(i) + a(1 - P_n(i)) \\ P_{n+1}(j) &= P_n(j) - aP_n(j) \qquad i\neq j \end{aligned} \qquad (6)$$

- if $\alpha(i)$ is unsuccessful, the action probability $P_{n+1}(i)$ remains unchanged and the call reattempts. If action j is chosen the automaton is updated as follows :

- if action $\alpha(j)$ is successful then:

$$\begin{aligned} P_{n+1}(j) &= P_n(j) + a(1 - P_n(j)) \\ P_{n+1}(k) &= P_n(k) - aP_n(k) \qquad k\neq j \end{aligned} \qquad (7)$$

- if action j is unsuccessful, the action probability $P_{n+1}(j)$ is not updated.

Other improvements of the learning automata can be made such as :

- the adaptation of the reinforcement parameter a according to the importance of each traffic flow.
- the design of periodic reinforcement schemes. In fact the call by call automaton is unrealistic due to the size of telecommunication network and due to the number of traffic flows. Therefore automata can be updated periodically according to observations over a time interval or aperiodically (i.e. every N calls).

- Distributed algorithms

The routed schemes presented in this section consist in a periodic modification (i.e. every 60 seconds) of the overflow paths which are used only in case of saturation of the first choice trunk group. The performance of this control structure depends on the reconfiguration period and the quality of the congestion index. Then the choice of overflow paths is done on the basis of the least congestion rule which aims at reducing the number of calls lost during the next period.

In this approach, each centre updates its alternative routing tables considering all possible paths according to measurements. These are the state of outgoing trunk groups and informations provided by adjacent centres concerning the congestion index of their own outgoing trunk groups.

Two congestion indices have been considered :

* Residual capacity

This congestion index is based on trunk group idleness and is very easy to compute. The residual capacity of a trunk group is defined as :

$$C_{ij} = N_{ij} - S_{ij} \qquad (8)$$

where N_{ij} is the capacity of the trunk group and S_{ij} the number of busy trunks. This index indicates the possibility of routing C_{ij} calls. The extent to a path ikj is made by taking :

$$C_{ikj} = \mathrm{Min}(C_{ik}, C_{kj}) \qquad (9)$$

Then, the second choice route (transit center k^*) is selected as follows :

$$k^*(i,j) = \mathrm{Arg}^t\mathrm{Max}\{C_{ikj}\} \qquad (10)$$

This index is independent of the offered traffic $\lambda_{ij}\tau$. But, for a given value of C, it would be obviously different to affect a traffic flow with a high rate λ or a low one. To prevent this fact, some new indices, such as C_{ij}/λ_{ij}, have been tested. The simulation results have showed that this sophistication did not improve the performances of the testbed network.

To take into account the probabilistic nature of the process, a second congestion index have been investigated.

* First time probability to overflow

The overflow paths being determined periodically, the new index gives an information on the evolution of the network during one period. This index is the probability that no call overflows before the end of the time of the reconfiguration $\pi(t_0>T)$, where t_0 is the first time when a call overflows over a trunk group and T is the reconfiguration time period.
The greatest this probability is, the lower the losses are. The calculation uses the Markov chain model with a supplementary sink state (Fig. 2).

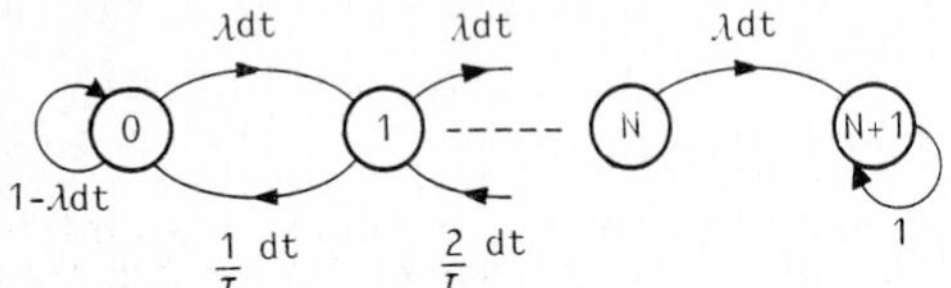

Figure 2 - Markovian Transition Diagram -

The dynamical equations are the following :

$$\begin{aligned}
\dot{\pi}_0 &= -\lambda\pi_0 + \tfrac{1}{\tau}\pi_1 \\
\dot{\pi}_1 &= \lambda\pi_0 - (\lambda + \tfrac{1}{\tau})\pi_1 + \tfrac{2}{\tau}\pi_2 \\
&\vdots \\
\dot{\pi}_N &= \lambda\pi_{N-1} - (\lambda + \tfrac{N}{\tau})\pi_N \\
\dot{\pi}_{N+1} &= \lambda\pi_N
\end{aligned} \tag{11}$$

with π_i the probability to be in state S_i at time t. The interesting probability is given by :

$$\pi(t_0<T) = \pi_{N+1}(T) \tag{12}$$

The solution of the differential system can be easily obtained with approximative methods. The calculation of this index for a route ikj supposes the independency between two tandem trunk groups :

$$\pi_{ikj}(t_0>T) = \pi_{ik}(t_0>T)\pi_{kj}(t_0>T) \tag{13}$$

To determine the best overflow paths with this index it is necessary to estimate the offered traffic and to enumerate the whole possible affectations. The traffic estimation is previously mentioned. In order to reduce the huge number of affectations to be tested a step by step method has been used. Let us consider that the overflow paths for the destination $j_1, j_2 \ldots j_n$ have already been fixed, then the next overflow route at node i is found by :

$$(k,j) = \mathrm{Arg}^t[\min_{j \neq j_1, j_2 \ldots j_n} \max_k (\pi_{ikj})] \tag{14}$$

DESCRIPTION OF THE EXPERIMENTS

The comparison of the described algorithms first requires the choice of a representative network and the consideration of significant disturbances affecting the network.

- Testbed network

The node network is a subset of eight electronic switches (S) and three transit switches (T) from the Paris network. It was chosen in order to have a model network close to a real possible trial network that could be used in a few years. The traffic matrix was taken out from the projected matrix for 1985 (see Table 1). The total traffic is 407.5 erlangs.

	S1	S2	S3	S4	S5	S6	S7	S8
S1	0.0	1.9	2.9	2.0	2.7	5.2	25.5	6.5
S2	2.0	0.0	5.4	7.4	24.7	6.9	2.3	4.5
S3	2.3	6.9	0.0	5.7	5.2	11.7	3.1	6.4
S4	2.9	9.5	5.0	0.0	13.2	8.8	4.0	7.5
S5	2.1	21.5	4.4	9.9	0.0	6.6	2.0	4.6
S6	3.4	5.3	9.9	5.8	5.4	0.0	4.0	19.2
S7	29.3	2.2	3.8	3.9	2.3	6.1	0.0	4.7
S8	4.1	5.0	5.9	6.8	6.0	27.1	4.3	0.0

Table 1 - Offered Traffic Matrix (erlangs) -

We first sized the links with usual rules of fixed alternate routing :
- Direct trunk groups exist when the traffic exceeds a creation threshold of 5 erlangs. Each source node is linked to the hierarchical transit node only at the transit level.
- Each transit switching centre is tied to each destination switching centre.

This network is in fact useless for adaptive traffic routing since there is no possibility to update the routing tables. Therefore the structure of the network was modified in order to cope with adaptive routing and all the origin switching centres were connected to the three transit centres. This network was resized according to real time routing schemes and we obtained 846 trunks (i.e. 2% less than in the hierarchical network). Two networks were in fact obtained considering the two different routing schemes used in the algorithms (load sharing and overflow). For each network the grade of service requirements were met under nominal load.

- Operating conditions

In developing adaptive traffic routing algorithms we are mainly interested in the network behaviour under abnormal conditions. We then considered typical disturbed conditions such as traffic overload, trunk group failure or transit centre breakdown.

* Traffic overload D1

A mass calling was considered resulting in 50% overload traffic towards node S1. The overload was implemented as shown in Fig.3 to take into account the transient of the system.

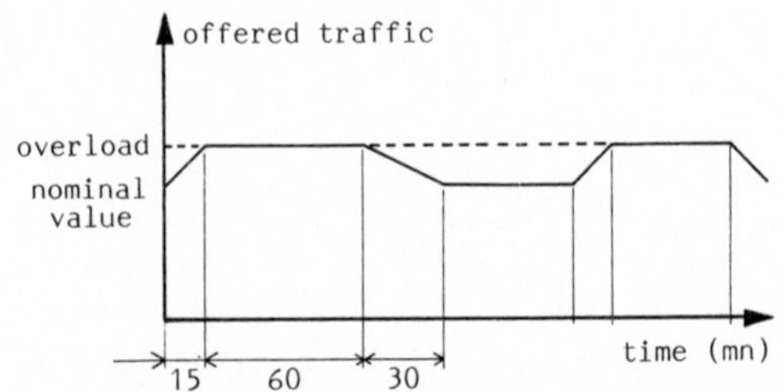

Figure 3 - Offered Traffic Variations -

* Trunk group failure D2-D3

Transmission failures often affect telephone networks. Then in order to provide some transmission security to Paris network, trunk groups are physically birouted. A transmission failure would thus affect half of a trunk group. In our experiments we considered strong disturbances such as multiple failures :

- first a double correlated failure D2 has been simulated which resulted in the capacity reduction of the two trunk groups S8-S6 (15 trunks instead of 30) and S8-T3 (8 trunks instead of 16). Such a failure may exist since these two trunk groups can have the same circuit routing.
- second a double uncorrelated failure D3 concerned the links S4-T3 (9 trunks instead of 19) and T1-S5 (6 trunks instead of 12).

* Terminal equipment failure D4-D5

For these two disturbances we considered the case when terminal equipments are out of order and result in the capacity reduction of half the trunks towards transit centre (case D4) or from the transit node (case D5).

* Transit centre breakdown D6

The last disturbance concerns the breakdown of the transit centre T2 which is therefore isolated from the network.

PERFORMANCE EVALUATION

• Simulation conditions

As the subnetwork we study could not be isolated from the real network, extensive use of event by event simulations was made to analyze the network behaviour under abnormal conditions.

For each structural disturbance the simulation time was equal to 5 hours and in the case of overload traffic it was equal to 10 hours.
The reconfiguration time was chosen to 15mn for the centralized algorithm and 60s for the distributed algorithms. Learning automata were implemented on a call by call basis.
The global performance index we primarily consider is the total traffic loss in the network. This result is shown in Table 2 for each algorithm and each disturbance. The performance using the present network with fixed routing rule is given in order to evaluate the benefit of adaptive traffic routing.

• Performance analysis

From simulation results, we point out the following conclusions :

* First we highlight the improvement of the performances in case of structural disturbances. Table 2 shows the adequacy of all the policies described in the paper. The total losses are reduced from 20% to 60% compared to the fixed routing.

* Second in case of strong overload D1, adaptive routing does not provide better performances than the fixed routing. One reason is that the network is rather loaded and resources are not available; the routing cannot perform. However in case of traffic disymmetry, i.e. traffic overload in a part of the network and underload in another part, improvement of performance is expected as adaptive routing aims at maximizing the capacity allocation to calls. Another reason is that the network with adaptive routing has less capacity (846 trunks) than the network with fixed rule (866 trunks).

* Third Table 2 shows that adaptive routing cannot really be compared together. Depending on the disturbance case, methods are better or worse than the others.

- Concerning the feedforward load sharing policy, the routing is fixed during 15 mn for a mean behaviour and is not updated according to real time traffic variations or to information on the network (blocking) during the time interval. Though performance are good, they can be improved by taking into account more frequent measurements at the local level.
- The learning automata works on a real-time basis but the information used to update the control is rather "poor".In fact the measurements (completion or rejection of calls) give a state information on a path, as the residual capacity approach does. However this measurement does not

Table 2 - Total Rating Loss (%) -

Routing rule / Disturbance	Fixed rule	Centralized policy: Feedforward loadsharing	Decentralized policy: Learning automata	Distributed policy: Residual capacity	Distributed policy: Overflow time
Traffic overload D1 towards S1	1.64	2.14	2.32	1.61	1.55
Trunk group failure D2 S8-S6 S8-T3	3.34	2.58	3.14	2.54	2.56
Trunk group failure D3 S4-T3 T1-S5	1.74	1.11	1.35	1.04	1.00
Equipment failure D4 from T2	2.91	2.16	1.37	1.04	1.40
Equipment failure D5 to T2	2.84	1.92	2.57	1.63	1.60
Transit node breakdown D6	10.65	3.67	4.40	4.69	4.80

indicate any information about traffic intensity along the path and, as no other information is available concerning the rest of the network, results are not uniform.

- The performance of the distributed algorithms is due mostly to the information pattern used in the control design. The reconfiguration time is short enough to allow rerouting in case of real time traffic variations. Moreover the information is very useful : the periodic observation gives a good measurement of the state of the routes and the aperiodic information on the blocking avoids the routing on saturated paths. Algorithms taking into account the probabilistic nature of the traffic (overflow time) do not perform better than the simple deterministic approach (residual capacity). However in case of high rate traffic, the first approach would give better results as the traffic estimation would be more reliable.

All the above conclusions concern the global performances of the proposed algorithms. Another item of major importance in the grade of service requirements lies in the distribution of losses. We found that all the methods we designed smooth the losses throughout the network and provide therefore fairness amongst the traffic flows. As an example the maximum rating loss among all the traffic flows is given in Table 3 for the distributed approach and the fixed routing. It can be seen that in all cases the maximum loss is highly reduced.

Routing rule / Disturbance	Fixed rule	Residual capacity	Overflow time
normal conditions	4.8	2	1.9
D1	9.4	7.7	7.7
D2	34.4	24.8	26
D3	31	9	8
D4	38	5	9
D5	37	23.3	22
D6	100	59	58

Table 3 - Maximum Rating Loss (%) -

Further results were obtained in the analysis of the algorithms. In the case of the distributed approach, better performances were obtained when decreasing the reconfiguration time period to 30 seconds. Furthermore, as sophistication of the algorithms did not appear to improve the performances, it seemed interesting to test some heuristic methods in the case of the centralized approach. It was found that they did perform quite as well as the optimal solution, and then it would be more advisable to use the simplest methods in real operating conditions.

CONCLUSION

In this paper, a performance evaluation of several adaptive traffic routing policies was made, considering a testbed network. It was shown that adaptive traffic routing can improve the network performances in case of disturbance. All the methods provide robustness to the network and tend to equalize the losses throughout the network.

However the performance depends on the measurements that can be made on the network. Furthermore, the time period is of major importance and performance improvement can be obtained with a short reconfiguration time. Finally it was found that different dimensioning rules should be associated with each control. These rules may result in capacity reduction, decreasing then network costs.

Studies are currently undertaken which concern networks dimensioning with adaptive routing as well as a sensitivity analysis of the algorithms in order to determine the main parameters of the control design in the performance.

These studies are the preliminary steps to the implementation of adaptive traffic routing policies in a network that could be designed in the near future for a field trial in France.

This research has been supported by the CNET under grants n° 80 35 124 and n° 82 1B 188.

REFERENCES

[1] G.R. ASH, R.H. KAFKER,J.R. KRISHNAN : "Servicing and Real-time Control of Networks with Dynamic Routing". Bell. Syst. Tech. Journal vol.50 n°8 (1981)

[2] K.S. NARENDRA, E.A. WRIGHT, L.G. MASON : "Application of Learning Automata to Telephone Traffic Routing and Control". IEEE vol. SMC 7 n°11 (1977)

[3] E. SZIBICKI, A.E. BEAN : "Advanced Traffic Routing in Local Telephone Networks : Performance of Proposed Call Routing Algorithms". Conf. ITC 9 Torremolinos (1979)

[4] B. HENNION : "Le Partage de Charge dans la Téléphonie Adaptative". L'Echo des Recherches n°2 (1978)

[5] L.G. MASON, A. GIRARD : "Control Techniques and Performance Models for Circuit Switched Networks". Conf. CDC 82 Orlando (1982)

[6] G. BEL, P. CHEMOUIL, J. BERNUSSOU, J.M. GARCIA, F. LE GALL : "Adaptive Traffic Routing in Telephone Networks" Large Scale Systems Journal vol.6 n°3 (1985)

[7] J. BERNUSSOU, J.M. GARCIA, I.S. BONATTI, F. LE GALL : "Modelling and Control of Large Scale Telecommunication Networks" Proc. 8th IFAC World Congress Kyoto (1981)

[8] F. LE GALL : "One Moment Model for Telephone Traffic" Applied Math. Modelling vol.6 (1980)

[9] F. LE GALL, J.M. GARCIA, J. BERNUSSOU : "A One Moment Model for Telephone Traffic Application to Blocking Estimation and Resource Allocation" Proc. of 2nd Int. Symp. on Performance of Computer-Communication Systems, Zûrich (1984)

[10] P.R SRIKANTAKUMAR, K.S. NARENDRA : "A Learning Model for Routing in Telephone Networks" SIAM Journal of Control and Optimisation vol.20 n°1 (1982)

TELETRAFFIC ISSUES in an Advanced Information Society
ITC-11
Minoru Akiyama (Editor)
Elsevier Science Publishers B.V. (North-Holland)

STATE DEPENDENT ROUTING OF TELEPHONE TRAFFIC AND THE USE OF SEPARABLE ROUTING SCHEMES

T. J. OTT and K. R. KRISHNAN

Bell Communications Research
Morristown, New Jersey 07960 U.S.A.

ABSTRACT

In a modern stored-program-controlled network, a considerable amount of information on the state of the network can be used for traffic routing. We investigate the problem of optimum state-dependent routing as a Markov decision process, and introduce a class of simple, easily-implementable routing schemes, the so-called "separable" schemes. Numerical results, for very simple networks, obtained with a member of this class called the DRS (Direct-Routing-derived-Separable) routing scheme indicate an almost-optimum performance.

1. INTRODUCTION

In this paper, we study the problem of optimal routing of telephone traffic when information about the "state" of the network is available for routing decisions.

Traffic routing is evolving from time-invariant, hierarchical schemes to time-dependent, non-hierarchical schemes such as Dynamic Non-Hierarchical Routing (DNHR) introduced by AT&T Communications [1]. With the use of network state information, as in the Trunk Status Map [2], more general state-dependent routing schemes become feasible. In these state-dependent schemes, for every individual call attempt, an individual "optimal" routing decision is made on the basis of a considerable amount of information on the state of the network (numbers of busy and idle trunks in the various trunk groups, etc.) at the time of the call attempt.

The problem of devising efficient routing schemes is, and will remain at least in the near future, of considerable interest because of its impact on network provisioning cost [1]. In another one or two decades this may very well change. If, indeed, optical fibers are going to make transmission an order of magnitude cheaper than switching, there will no longer be much of a reason to spend switching resources to save on transmission. Until that time, research into efficient routing schemes remains of great interest, possibly the more so because occasionally efficient routing may delay the point where capacity expansion is necessary. Thus it may make it possible to delay committing to a specific technology, possibly until a better technology becomes available.

At any point in time the state of the network can be described by the state of all current connections (routes and elapsed times since start). Thus, optimal state dependent routing is a problem of optimal control of a semi Markov process on a huge state space. With standard assumptions, this problem can be reformulated as a semi-Markov decision process (still with a state space of astronomical size) or as a (gigantic) non-linear mathematical programming problem [3].

In section 2 we state the standard simplifying assumptions which allow us to formulate the routing problem as a Markov Decision problem. In section 3 we give a brief overview of some aspects of the theory of Markov Decision processes, specialized for the problem at hand. We will see that even in the simplified model the truly optimal scheme is both very hard to find and, even if found, probably unimplementable anyhow. The two factors which make these problems so hard are the size of the state space (the state space can easily have more then 10^{100} states) and the fact that, in order to compare the relative desirability of two different routes, the state of trunk groups not in either of these routes may be relevant.

At the end of section 3 we will describe a class of routing schemes which we call separable schemes. These separable schemes have the properties that in order to compare the relative desirability of two routes, only information on trunk groups in those routes is used, and, moreover, that the comparison is simple enough to make the schemes implementable. Also, at least some of these separable schemes can be very good.

In section 4 we discuss the class of "direct routing schemes" and also introduce the related, larger class of "non-alternate routing schemes".

In section 5 we obtain from these schemes other, separable schemes for which an intuitive argument is given which makes plausible that they are very good schemes, quite possibly almost indistinguishable in performance from the theoretically optimal (but unknown and probably unimplementable) scheme. These are the DRS scheme (direct-routing-derived separable) and the NARS scheme (non-alternate-routing-derived separable). These schemes, which are separable and therefore simple enough to be implemented, and which are also expected to have very good performance, are the real raison d'être of this paper.

One of the reasons that these DRS and NARS schemes are likely to be very effective is that while older methods mostly take a defensive stance towards the statistical fluctuations inherent in telephone traffic, these DRS and NARS schemes can be said (with some justification, see section 5), to positively attempt to use these fluctuations to increase throughput and thus decrease blocking.

The actual performance of separable schemes can be directly computed only for very simple networks. In section 6 we present, for a few very simple networks, numerical results on the performance of DRS schemes, and compare them with the performance of the theoretically optimal scheme.

Work is under way to extend this comparison to more realistic networks, and also to obtain a comparison of DRS and NARS routings with DNH routing. Together with A. Federgruen of Columbia University we have developed a method which may make it possible to obtain bounds for the difference in performance between the DRS and NARS schemes on the one hand and the theoretically optimal scheme on the other, even if the latter scheme is not known. Implementing this method still is a quite nontrivial job.

2. NETWORK DESCRIPTION AND ASSUMPTIONS

The network under consideration has N nodes (telephone switches). There are $K = \frac{1}{2} N(N-1)$ trunkgroups. Trunkgroup k ($1 \leqslant k \leqslant K$) contains s_k trunks, with $s_k \geqslant 0$. The network need not be fully connected ($s_k = 0$ is allowed).

We assume that for each node pair (i,j) there is a Poisson process of intensity $\lambda_{i,j}$ of call attempts (end to end between i and j). Each such call is either blocked (all routes between i and j which are allowed contain at least one trunkgroup k where all s_k trunks are busy) or rejected (at least one allowed route from i to j has at least one free trunk in all its trunk groups, but yet the call attempt is rejected to protect future call attempts), or routed. In the future we do not always distinguish between blocked and rejected calls. We assume that blocked and rejected calls are cleared.

For each node pair (i,j) there may be a large number of allowed routes, say over trunkgroup $k_1 \equiv$ node pair (i,j) (the direct route), and over n two-link routes,

$$\text{trunkgroups } (k_{2\mu}, k_{2\mu+1}) \equiv \text{nodes } (i, l_\mu, j), (\mu = 1, 2, \ldots, n) \quad (2.1)$$

and possibly over multi-link routes of the type

$$\text{trunkgroups } (k_1, k_2, \ldots, k_n) \equiv \text{nodes } (i, l_1, l_2, \ldots l_{n-1}, j) \quad (2.2)$$

Usually, only direct routes and two-link routes are considered. We do not need to make that restriction in this section.

Throughout this paper we make two assumptions about the stochastic behavior of this network.

The first is that all calls have holding times which are exponentially distributed, "independent of everything else," and have expected value 1. (The average length of a call is used as unit of time).

The second is that as soon as a call has been routed on an n-hop route as in (2.2) it becomes n independent calls on the n trunkgroups involved, each with an independent holding time as above.

As a result, the <u>state of the network</u> at any time is described by the <u>states</u> of the K trunkgroups, i.e., the numbers of busy trunks in the K trunkgroups.

Let $x_k(t)$ ($1 \leqslant k \leqslant K$) be the number of busy trunks in trunkgroup k at time t. The state of the network at time t is then described by the vector.

$$\underline{x}(t) = (x_1(t), \ldots, x_K(t)) \quad (0 \leqslant x_k(t) \leqslant s_k). \quad (2.3)$$

Let X denote the state space of this network. Clearly, X contains

$$|X| = \prod_{k=1}^{K} (s_k + 1) \quad (2.4)$$

different states. For example, with $N = 20$, $K = 1/2 \times 20 \times 19 = 190$ and $s_k = 20$ for all k, this gives

$$|X| = 21^{190}, \quad (2.5)$$

which is considerably more than 10^{80}, the estimated number of elementary particles in the universe!

The state of the network can change only at call terminations and at epochs of call attempts. At call terminations no decision needs to be made. At call attempt epochs a <u>routing decision</u> needs to be made: either block (reject) or accept, and if the decision is to accept, a route must be chosen.

The decision depends on the node pair (i,j) and on the state $\underline{x}$ of the network at the time of the call attempt. The routing policy $\mathscr{P}$ thus is a map which assigns a routing decision to every ($\underline{x}$, (i,j)).

The effect of a decision to route a call attempt (i,j) arriving in state $\underline{x} = (x_1, \ldots, x_K)$ over trunkgroups $(k_1, \ldots, k_n) \equiv$ *nodes* $(i, l_1, \ldots, l_{n-1}, j)$ is a transition

$$\underline{x} \rightarrow \underline{x} + \sum_{\nu=1}^{n} \underline{e}_{k_\nu} \quad (2.6)$$

where $\underline{e}_k$ is the k - th unit vector: the vector with a 1 at location k and zeros at all other locations.

The effect of a call completion on trunkgroup k is a transition

$$\underline{x} \rightarrow \underline{x} - \underline{e}_k, \quad (2.7)$$

and because of the assumption that an n-hop call behaves like n independent calls, there are no simultaneous call completions at different links.

The effect of a call attempt which is blocked or rejected is no state transition, but in this case there is a lost call.

The rate of an (i,j) call attempt in state $\underline{x}$ is $\lambda_{i,j}$ independent of $\underline{x}$. The rate of a call termination in trunkgroup k, while in state $\underline{x}$, is x_k (dependent on $\underline{x}$).

The objective is to find a policy $\mathscr{P}$ which minimizes the loss rate, i.e., which minimizes the average number of lost (blocked or rejected) calls per unit of time.

With the assumptions made in the beginning of this section, finding the optimal policy $\mathscr{P}$ is a continuous time Markov decision process on the finite state space X.

There are several ways to find the optimal policy $\mathscr{P}$. by Howard's value determination - policy iteration method [4], by dynamic programming [5], or by solving a linear programming problem (with about $|X|$ variables and more than $|X|$ constraints, see [6]).

3. MARKOV DECISION PROCESSES

In this section we give a quick overview of some aspects of the theory of Markov Decision processes, specialized for the problem at hand.

If $\mathscr{P}$ is any policy (a map which assigns a routing decision to every ($\underline{x}$, (i,j))), Howard's value determination step finds the loss rate g corresponding with $\mathscr{P}$ (the average number of lost calls per unit of time if $\mathscr{P}$ is used) and values $v(\underline{x})$ $(\underline{x} \epsilon X)$ by solving a system of | X | linear equations with | X | unknowns.

The value $v(\underline{x})$ is the "relative value" of starting in state $\underline{x}$, in the sense that there exists some constant c such, that

$$E\,[\textit{number of calls lost in } [0, t] \mid \underline{x}(0) = \underline{x}, \textit{policy } \mathscr{P}] = g\,t + v\,(\underline{x}) + c + o(1)\ (t \rightarrow \infty)\ . \tag{3.1}$$

If a set of (relative) values $w(\underline{x})$ $(\underline{x} \in X)$ is given (this set may or may not be derived from a policy $\mathscr{P}$) it gives rise, in its turn, to a policy $\mathscr{P}$:

Assign the following costs:

$$\text{Cost of blocking or rejecting} = 1\ , \tag{3.2}$$

and the cost of routing over a route as in (2.2), (2.6) is: cost of routing over trunkgroups $(k_1, k_2, \ldots, k_n)$

$$= w\,(\underline{x} + \sum_{\nu=1}^{n} \underline{e}_{k_\nu}) - w\,(\underline{x})\ . \tag{3.3}$$

Now find the decision which minimizes the cost over (3.2), (3.3). i.e., if no feasible allowed routes exist, block (cost = 1). If at least one feasible, allowed route exists: minimize (3.3) over all feasible, allowed routes. If the minimum is <1: use the route which minimizes (3.3). Otherwise, reject the call (cost = 1).

The result is a routing decision for every ($\underline{x}$, (i,j)), i.e., a policy $\mathscr{P}$. We now have:

Theorem 3.1 Let $w(\underline{x})$ $(x \epsilon X)$ be given and construct $\mathscr{P}$ as above. Define S ($\underline{x}$) as the set of node pairs for which, if an (i,j) call attempt is made while the network is in state $\underline{x}$, the call is blocked or rejected. Define $\underline{y}(\underline{x}, (i,j))$ as the state the network moves to if an (i,j) call attempt is made while the network is in state $\underline{x}$, so that

$$\underline{y}\,(\underline{x}, (i,j)\,) = \underline{x} \textit{ iff } (i,j) \epsilon S\,(\underline{x})\ . \tag{3.4}$$

Define g ($\underline{x}$) as

$$g\,(\underline{x}) = \sum_{(i,j)\,\epsilon\, S\,(\underline{x})} \lambda_{i,j} + \sum_{(i,j)} \lambda_{i,j}\,(w(\underline{y}\,(\underline{x},(i,j))) - w\,(\underline{x})) \tag{3.5}$$

$$+ \sum_{k:x_k \geqslant 1} x_k\,(\,w\,(\underline{x} - \underline{e}_k) - w\,(\underline{x})),$$

define g_{opt} as the optimal (minimal) loss rate, and define g^* as the loss rate corresponding to $\mathscr{P}$. Then:

$$\min_{\underline{x}\,\epsilon\, X} g\,(\underline{x}) \leqslant g_{opt} \leqslant g^* \leqslant \max_{\underline{x}\,\epsilon\, X} g\,(\underline{x}). \tag{3.6}$$

Moreover, $\mathscr{P}$ is an optimal policy if and only if there is equality throughout (3.6).

Proof: See [5]. Howard's value determination - policy iteration method [4] chooses an arbitrary policy $\mathscr{P}_0$ and then constructs a sequence of better and better (in the sense of (3.7) below) policies $\mathscr{P}_0, \mathscr{P}_1$,...as follows:

From policy $\mathscr{P}_\ell$, determine the loss rate g_ℓ and (relative) values $v_\ell\,(\underline{x})$ by the value determination method. Use these values $v_\ell\,(\underline{x})$ as in theorem 3.1 to construct policy $\mathscr{P}_{\ell+1}$. (This is the policy iteration step).

We now not only have that

$$g_0 \geqslant g_1 \geqslant \ldots\ldots \geqslant g_l \geqslant g_{l+1} \geqslant \ldots \tag{3.7}$$

but also that after a finite number of iterations we reach the optimal policy $\mathscr{P}_L$ (for which equality holds throughout (3.6)).

Remark 3.1. Because of the size of the state space X, any policy which is given in the form of a table of routing decisions is unimplementable. Theorem 3.1 showed us that many policies can be given in the form of a value function w(·). When there exists a mechanism to compute w ($\underline{x}$) quickly for any $\underline{x}$, we in fact have an implementable policy, since for any $(\underline{x},(i,j))$ only a few routes need to be compared.

If we can find a value function w (·) which not only leads to an implementable policy but for which, also, the difference between the RHS and the LHS in (3.6) is small, we clearly have found a very good policy. We have not yet succeeded in doing this.

A special class of value functions which lead to implementable policies are the separable value functions:

Definition: A value function w (·) is called separable if it has the form.

$$w\,(\underline{x}) = \sum_{k=1}^{K} w_{x_k\,,\,k}\ . \tag{3.8}$$

A policy $\mathscr{P}$ is called separable if it is generated by a separable value function.

Remark 3.2 The value function generated by a separable policy need not be separable!

Remark 3.3 It is clear that a separable policy is implementable. In fact, (3.3) reduces to cost of routing over trunkgroups $(k_1, k_2, \ldots, k_n)$

$$= \sum_{\nu=1}^{n} (w_{x_{k_\nu}+1, k_\nu} - w_{x_{k_\nu}\,,\,k_\nu}) \tag{3.9}$$

In addition, it is useful to note that such separable schemes use essentially the same information as the DNHR scheme.

Remark 3.4 In section 5 we will present a separable value function v for which, even though the RHS and the LHS in (3.6) are not close together, an intuitive argument can be made which indicates that for the resulting separable policy $\mathscr{P}$ "probably" g^* and g_{opt} are very close together. Depending on whether the separable value function v was derived from a direct routing scheme or from a non-alternate routing scheme we call $\mathscr{P}$ a DRS (direct-routing-derived separable) or a NARS (non-alternate-routing-derived separable) scheme.

4. DIRECT ROUTING AND NON ALTERNATE ROUTING

In this section we describe a class of routing schemes which, while not necessarily optimal or even good, make it easier to introduce, in the next section, the DRS and NARS routing schemes.

Definition In a non-alternate routing scheme there is for every node pair (i,j) a set of $m(i,j)$ permitted routes. Route $m(1 \leqslant m \leqslant m(i,j))$ in this set has form

$$\text{route } m \equiv \text{trunkgroups} \quad (k_1^{(m)}, \ldots, k_{n_m}^{(m)})\ . \tag{4.1}$$

If an (i,j) call attempt is made, the system attempts, with probability $p_m\,(i,j)$, to route this call over route m. If this route happens to be available the call is so routed, otherwise the call is blocked: no alternate route is tried.

A special case of non-alternate routing is direct routing: In this scheme only direct, single hop routing is allowed.

For direct routing, it is clear that the different trunkgroups are independent $M|M|s_k|\ s_k$ blocking systems. It is likely that for many other non-alternate routing schemes this approximately still is true.

In section 5 we will derive the DRS and the NARS schemes, respectively, from the direct routing scheme and a (still to be chosen) non alterate routing scheme. This is done (exactly for DRS and approximately for NARS) by obtaining the (Markov Decision process) value function of the original scheme and using it (as in section 3) to obtain the separable scheme.

This means (see (3.7)) that the DRS scheme is at least as good as the direct routing scheme, while (almost certainly) the NARS is at least as good as the original non-alternate routing scheme.

An intuitive argument will be given which indicates that the differences in fact are sizeable.

It seems intuitively clear that a very good or optimal non-alternate routing scheme will produce a very good NARS scheme. The problem of finding the optimal non-alternate routing scheme is under investigation.

5. SEPARABLE ROUTING

Suppose we have a non-alternate routing scheme $\mathscr{P}_0$ which, although clearly not the best state - dependent routing scheme, at least is very good in the class of non-alternate routing schemes. Suppose this routing scheme $\mathscr{P}_0$ has the property that, practically speaking, the K trunkgroups are independent $M|M|s_k|s_k$ blocking systems.

Let trunkgroup k have arrival rate of call attempts λ_k. Note that if trunkgroup k corresponds with the node pair (i,j), then, in general

$$\lambda_{i,j} \neq \lambda_k \tag{5.1}$$

since, in the first place, not all call attempts (i,j) need be assigned to the group k and, in the second place, some other traffic may be assigned to k. For direct routing, of course, equality holds in (5.1).

Let us now do the following thought experiment: at time zero, let the system be in state $\underline{x}$ and let a call attempt (i,j) be made. For this one call we can make "any" routing decision, but from that call on, the non alternate routing policy $\mathscr{P}_0$ assumed in the beginning of this section will be used. What is now the optimal way to route this one call?

For each trunkgroup k, for each $0 \leqslant j \leqslant s_k - 1$, compute the probability.

$$v_{j+1,k} - v_{j,k} \tag{5.2}$$

that if the trunkgroup is in state j at time zero (has j occupied trunks) and a special tagged, customer is added at time zero, then (at least) one future call will be blocked on trunkgroup k during the lifetime of the tagged call (assuming a Poisson arrival stream of intensity λ_k).

It is easily seen that then

$$v_{j+1,k} - v_{j,k} = \frac{B(s_k,\lambda_k)}{B(j,\lambda_k)}, \tag{5.3}$$

where

$$B(j,\lambda) = \frac{\frac{\lambda^j}{j!}}{\sum_{i=0}^{j} \frac{\lambda^i}{i!}} \tag{5.4}$$

is the Erlang-B function.

Let us now evaluate the effect of routing the tagged call over route $(k_1,k_2,\ldots,k_n)$. On trunkgroup k_ν the probability that this causes future blocking of a call, and therefore the expected value of the increase in the number of future calls blocked, is

$$v_{x_{k_\nu}+1,k_\nu} - v_{x_{k_\nu},k_\nu}. \tag{5.5}$$

If the probability that the tagged call will block a future call on two trunkgroups simultaneously is zero (this is true for direct routing and "probably almost true" for non alternate routing) then routing the tagged call over trunkgroups $(k_1,k_2,\ldots,k_n)$ increases the expected number of future lost calls by

$$\sum_{\nu=1}^{n} (v_{x_{k_\nu}+1,k_\nu} - v_{x_{k_\nu},k_\nu}). \tag{5.6}$$

Hence, we can use (as in theorem 3.1) the separable value function

$$v(\underline{x}) = \sum_{k=1}^{K} v_{x_k,k}. \tag{5.7}$$

Since every individual routing decison is based on a probably fairly good guess of the expected value of the increase in the number of future call blockings it causes, these schemes must be expected to perform quite well, possibly almost as well as the truly optimal scheme.

6. NUMERICAL RESULTS

A comparison was made between the policy obtained by exact solution of Howard's equations and the DRS policy, derived from the separable v functions corresponding to direct routing. On account of the large sizes of state-space that are encountered, the calculations were limited to small networks but, it should be emphasized that the actual implementation of the 'separable' method is not limited by the size of the state-space; its requirements are much more modest, since they involve, for each node-pair, calculations only for the trunkgroups on the admissible paths.

Figure 1 compares network blocking under the truly optimal routing scheme with that under the DRS scheme, at various load intensities, on several fully connected symmetrical networks; in each network, all trunkgroups have the same number of trunks and all node-pair loads are equal. The admissible routes for each node-pair consist of the direct link and every possible two-link path. The blocking under direct routing is included only to make the point that, at high enough loads, direct routing tends to be optimal. The results suggest that at both low and high loads, the DRS scheme is practically as good as the truly optimal scheme, with a rather modest range of load intensities where the difference is noticeable.

7. CONCLUSION

This paper has presented an analysis of the problem of optimal state-dependent routing of telephone traffic as a Markov decision process. A class of simple, easily implementable routing schemes (the separable schemes) is introduced. An intuitive argument is given which indicates that some of these separable schemes (the DRS scheme and some NARS schemes) give very good performance. Numerical results given for the DRS scheme, for very simple networks, suggest that the performance of this scheme is quite close to the optimum.

REFERENCES

[1] G. R. Ash, A. H. Kafker, and K. R. Krishnan, "Intercity Dynamic Routing Architecture and Feasibility," Proceedings of the 10th International Teletraffic Congress, Paper 3.2.2, Montreal, June, 1983.

[2] G. R. Ash and A. H. Kafker, "Use of Trunk-Status Map for Real-Time DNHR: Routing Method and Network Design," ORSA/TIMS 1983, Paper TA5.3, Orlando, Florida, November, 1983.

[3] V. E. Benes, "Programming and Control Problems Arising from Optimal Routing in Telephone Networks," Bell System Technical Journal, Vol. 45, pp. 1373-1438, November, 1966.

[4] R. A. Howard, "Dynamic Programming and Markov Processes," the M.I.T. Press, Cambridge, Massachusettes, 1960.

[5] A. Federgruen and P. J. Schweitzer, "A Survey of Asymptotic Value-Iteration for Undiscounted Markovian Decision Processes," in "Recent Developments in Markov Decision Processes," edited by R. Hartley, L. C. Thomas, and D. J. White, Academic Press, New York, 1980, pp 73-110.

[6] D. P. Heyman and M. J. Sobel, "Stochastic Models in Operations Research," Vol. II, McGraw-Hill, New York, 1984.

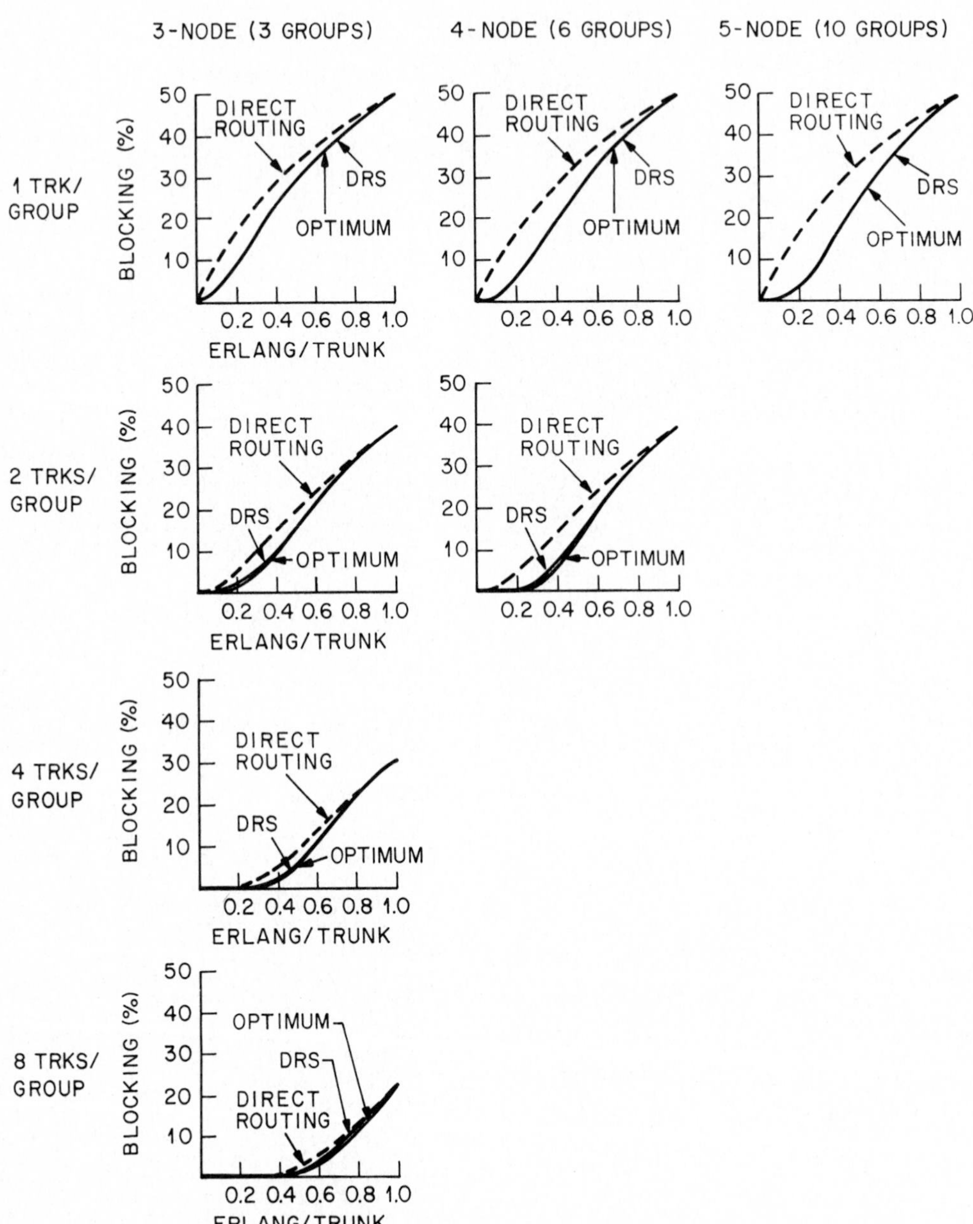

FIGURE 1 NETWORK BLOCKING IN FULLY-CONNECTED SYMMETRICAL NETWORKS.

TELETRAFFIC ISSUES in an Advanced Information Society
ITC-11
Minoru Akiyama (Editor)
Elsevier Science Publishers B.V. (North-Holland)

SIMULATION OF DISTRIBUTED MICROPROCESSOR CONTROL IN DIGITAL SWITCHING SYSTEMS

R.N. ANDRIES, M. GRUSZECKI, J. MASSANT, G.H. PETIT, P. VAN ESBROECK

Bell Telephone Manufacturing Company
Antwerp, Belgium

ABSTRACT.

Throughput and performance evaluation/prediction in digital switching systems with distributed microprocessor control required the development of new traffic simulation tools. These tools (simulators) are highly modular and can handle a wide range of system configurations, facilities and traffic environments without significant changes in the programs. A large number of simulations supported the system design and validated the performance objectives. The present paper shows the structure of these simulators, gives examples of results and briefly discusses the main traffic characteristics of S 12 distributed microprocessor control. It also compares the simulation results with those obtained with analytical tools.

Keywords : multinode simulation, performance and throughput evaluation, large queuing systems, distributed microprocessor control and digital switching.

1. INTRODUCTION.

Digital switching systems with distributed microprocessor control constitute a distinct and unique class in digital switching. ITT System 12, built on advanced VLSI components, programmed in high level CCITT CHILL language and featuring a variety of integrated voice and data switching services [1], is the first fully distributed digital switch and one of the largest microprocessor systems used at present.

For instance, a system of this type connecting some 20.000 telephone and data lines, may use several hundred microprocessors interconnected via a **digital switching network (DSN)** to control its operation (Fig. 1).

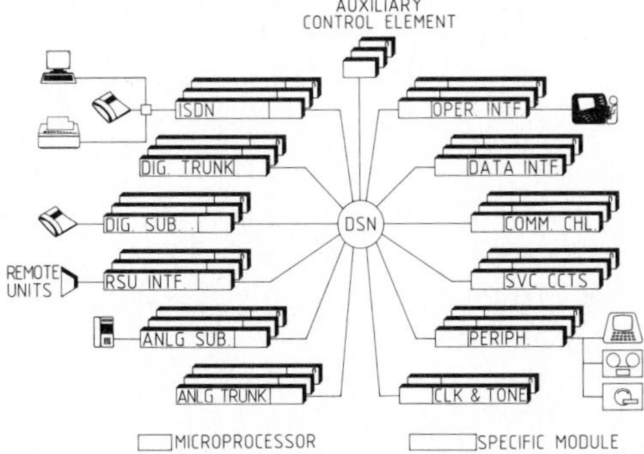

Fig. 1 Distributed microprocessor control communicating via the DSN.

S 12 distributed control architecture is very different from previously known SPC control structures :

— Switching and path search is accomplished autonomously by each of the DSN switching elements (custom-made VLSI's) without the intervention of the processors. The same DSN is used to switch the control messages, as well as the circuit and packet traffic. To send a control message from one processor to another a «**virtual path**» through the DSN is set up and broken down immediately after transmission.

— Although each individual control element is in fact a standard microprocessor, the distributed control as a whole is not a sequential (Von Neumann) machine as the processing of calls and messages is mostly done in parallel by the different processors and DSN elements.

— Each call is handled by a few microprocessors only, independently of all others. A total system failure is therefore very unlikely. Should any of the individual microprocessors fail, a spare one will take over its function.

A possible microprocessor reload concerns only a limited number of lines or trunks (60 or 128 lines or 30 trunks), active calls are not effected.

— The same basic architecture supports various types of switching functions, like voice and data traffic switching, ISDN, advanced switching capabilities, etc.

Compared to other control architectures, especially those with central SPC, the traffic characteristics of the fully distributed S 12 microprocessor control are also radically different :

— The call and message handling capacity viz. the number of microprocessors, increases in a linear way (Cf. 3.3.) when the exchange size or number of call attempts or messages grows. Hence, the maximum call handling capacity is not limited; the same basic control architecture covers a wide range of exchange sizes from very small to very large offices, stepping up in a smooth way.

— As a consequence of the fully distributed control architecture, replication of functions and autonomous switching of interprocessor messages by the DSN elements, the queuing characteristics of distributed control in small and large exchanges are basically the same (Cf. 3.3.).

— The DSN through which the microprocessors communicate is virtually non-blocking [2] (see also 3.3.).

Traffic studies of the above described system are mainly concerned with the behaviour, throughput and grade of service analysis in multinode queuing networks (distributed control) in normal, high and overload situations. This includes the behaviour under various load types created by voice/data, **Integrated Services Digital Network (ISDN)** and other advanced switching services.

These studies also allow the simulation of real operating conditions and the performance prediction before the actual exchange installation.

They require appropriate tools capable of keeping pace with the rapid evolution of digital switching technology. This implies a significant innovation effort as concerns the design of traffic simulators, especially the development of a powerful **multinode simulator (MNS)**. This simulator uses the principle of flexible modelling [3], it is highly modular and the modules used are reusable for different environments and system functions. From the above basic version, various specific versions have been derived.

The simulators are supported by additional modules which perform the automatic preparation of the input. They also format and analyse the output and link the simulators with the general systems support tools (e.g. system software production).

They range from rather simple single processor models (or few processors used in a specific sub-system) to full models representing the complete distributed control network.

All these simulators use the time-true discrete-event sequencing method and are usually of the full call model type [4].

The paper is structured as follows :

— Section 2 gives a brief description of the multinode simulator, the microprocessor's model and functions. It also outlines the organisation of the **Operating System**. The same section describes the support modules and their linking to system support tools.

— Section 3 briefly discusses the main traffic characteristics of S 12 distributed control and includes examples of multinode simulation results.

— Section 4 compares the results of MNS runs with analytical formulae.

The simulation and support programs are written in GPSS-V (7,8) FORTRAN and PL1 and run on IBM3033, 3081-k, 3084-Q and Amdahl 470/V8 machines. The full detail multinode model of a medium size exchange e.g. 200000 BHCA requires 3072 kbyte memory and several hours of computing time for the larger simulations.

2. MULTINODE SIMULATOR.

The S 12 distributed microprocessor control has been modelled in the multinode simulator in a simple and straightforward way as a collection of sequential/parallel machines and processes. The simulation model consists of two fundamental parts :

— System model.

— Model of traffic environment.

System model.

The basic building block of the system model is a symbolic representation of one microprocessor node taking into account its accurate internal organization, all its essential functions, queues, feedbacks, resume/non-resume interrupt priority handling, etc. (Cf. 2.1.). This very detailed representation of the actual **Operating System (OS)** is necessary, since essentially the same OS supports many system variants and switching functions. This fundamental module is repeated in each node of the multiprocessor system.

The interprocessor communication is accomplished in each node via a specific sub-module representing the **Network Handler (NH)** with its message reception and transmission task priorities. The (virtual non-blocking) [2] DSN, via which the different processors communicate, is not simulated but is represented in the model by the appropriate transmission delays.

The repetitive concept of the simulator and an algorithm that allocates the processor functions during the MNS run, result in a flexible and sophisticated mechanism that controls a large number of nodes and allows all the required modes of operation (e.g. processors working in load, function or device sharing modes, switching-in of a preloaded spare processor, hot stand-by, etc.).

Traffic environment model. Basic call (or message) scenario concept.

A basic **call (or message) scenario** is a detailed list of all tasks and events to be handled in a parallel/sequential order by the different processors. The scenarios also indicate the type of processor to be used by each task group, the task priority, execution time and their linkage. A full pattern consists of several basic call (or message) scenarios so that all call or message types are represented in the pattern. The task execution times and system parameters are taken from measurements on each processor and function type in the laboratory model.

Traffic generation modules create the required types of processes and initiate tasks or control the scenarios. The generation of calls (or messages) is achieved by repeating the basic scenarios and by modifying the required parameters. The above mechanism for creating calls (and/or non-voice messages) and modelling traffic is very flexible and can be used to realize a variety of traffic conditions. In the example of a combined local and transit telephone exchange, all types of calls will use :

— Gamma distributed process for call arrivals (with Poisson process as a specific case).

— Exponentially distributed service times for all call phases.

— Distribution law for signalling phase will be choosen depending on the type of signalling used.

The basic call (or message) scenario concept can be used to simulate many other interesting features :

— Short term traffic peaks can be created by introducing additional call or message types starting and terminating at specific times in the simulation run. If required, the peaks can be repeated following an assumed distribution law.

— Overflow traffic can be simulated using e.g. an interrupted Poisson process.

The fidelity of representation of the real call details and possibility of using many call types increases the accuracy of the traffic environment simulation.

2.1. Internal structure of one node.

The processor functions are partitioned into priority task groups that interact via system kernel interfaces. Task scheduling and dispatching are done by a non-interruptible operating system nucleus (backbone scheduler of the simulation) which acts on a flexible priority scheme within the different priority groups (Fig. 2) :

— Clock driven procedures are used to scan for asynchronous external events.

— An interface scan procedure used to retrieve information on interface events related to the digital switching network.

— Both clock driven and fast scan procedures are non-interruptible and defer the time consuming tasks to a medium priority task group of **Event Handlers (EH)** via an event handler priority queue.

Event handlers run with interrupts enabled to detect new events.

— Processes perform the bulk of transaction processing and are message driven via Message Ready Queues (abbrev. **Message Queues MQ**).

Messages are entered to the queues by event handlers (e.g. incoming messages through the network via the network handler to the message queue) or by processes (e.g. internal messages for another process in the same processor) according to the task priority within the low-priority groups (FIFO discipline per priority task).

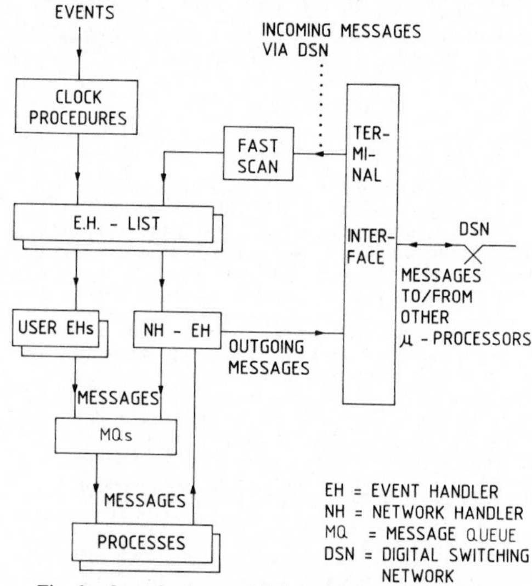

Fig. 2 Organisation and linkage of tasks within one node

The backbone scheduler of the simulator (Fig. 3) can handle :

— 60 different clock interrupt service procedures.

— An interface task.

— 25 event handlers, of which the highest task in the medium-priority group is the network handler.

— 8 process levels of message queues, numbered from 0 through 7, in the low-priority groups.

For clock interrupts, occurring at fixed periods, a detailed interrupt mechanism, taking into account the saving and restoring of the environment, has been implemented.

— Event handlers are interruptible but must run to completion after the execution of the interrupt service procedures.

— Processes are also interruptible and placed on an **Interrupt Control Block (ICB)**, also numbered from 0 through 7, to allow handling of higher priority tasks.

2.2. Input/output support.

This section outlines the support software for the automatic preparation of inputs and the analysis of simulation outputs.

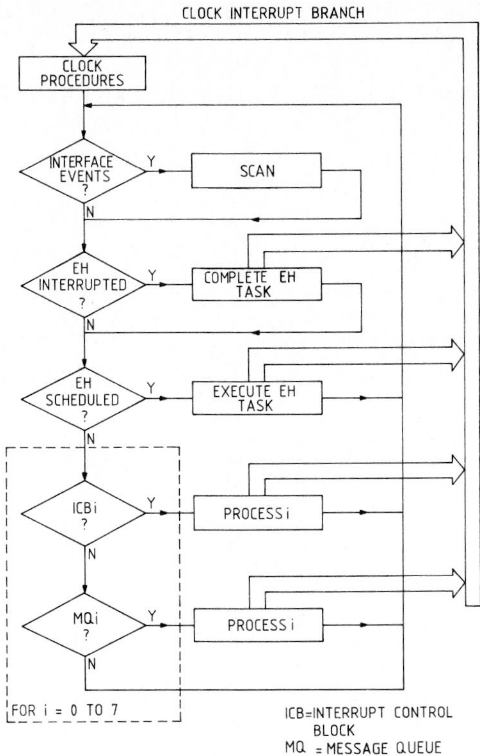

Fig. 3 Priority and interrupt handling in one node.

Input composer.

The simulator requires a typical input pattern consisting of linked transactions representing not only the external (outside plant) events but also defining all tasks to handle those events, as specified in the call scenarios.

The translation of the original call scenarios into the simulator input format is done by a software composer system :

— In a first step, the call scenarios are translated into a manageable format in which all interprocessor paths and message types are identified.

— In a second step, all execution times for processes, data base actions and procedure relations are introduced.

— In a third step, a final translation transforms the different message types into a corresponding number of GPSS transactions, thereby maintaining the sequential and/or the parallel order of the message flow. In order to perform this in the most efficient way, predefined GPSS modules are developed for each type of message sending or receiving operations.

During the execution of this input composer program, statistics are collected concerning the execution times per type of scenario and this per microprocessor and per specific task priority.

Output composer.

The output composer formats the GPSS simulation results and prints the required statistics :

— The average value and distribution, per priority and per sample, for each task load, queue waiting time and queue contents.

— Average and distribution for all response times to be evaluated (**Response time** is the time required to perform a predefined sequence of linked tasks, e.g. connection of the dial tone delay).

— Confidence intervals, based on the t-Student test, for each mean value of interest.

3. MNS STUDIES AND RESULTS.

A large number of MNS simulations has been performed to support system design, validate the performance objectives and to investigate the general traffic characteristics of the S 12 distributed microprocessor control. The examples given in the present section have been selected to illustrate two aspects of these studies :

— Sub-section 3.2. discusses in some detail the performance characteristics of distributed control in a small S 12 telephone exchange giving both the individual processor and the distributed control behaviour as a whole. These simulations use a full detail model (Cf. section 2), practically without any simplification of the actual operating system, event sequences and their linking, appropriate hardware representation, etc. Sub-section 3.1. gives the traffic environment used in these simulations.

— Sub-section 3.3. selects a few simulation results to show some general characteristics of S 12 distributed control. These simulations have been performed with a model using a number of simplifications to reduce the run times. These simplifications have been further carefully checked by a large number of separate simulations to confirm that no significant loss of accuracy will occur in the final and overall results. Due to lack of space the study of model simplifications will be discussed in a future publication.

Note also that further simulation studies (to investigate the overload prevention and control mechanisms) have been performed elsewhere for the same type of distributed control [12].

3.1. Exchange configuration and traffic environment.

Multinode configuration (Fig. 4).

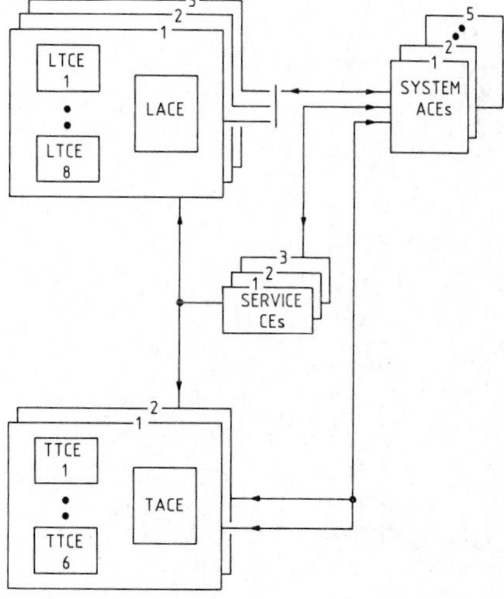

Fig. 4 Simulation model of a small exchange.

The control functions of the nodes are distributed over two major levels :

— The first level handles external events and issues commands to set up paths through the DSN. It consists of three processor types : **Line and Trunk Terminal Control Elements (LTCE and TTCE)**, which supervise the subscriber resp. trunk lines and the Service TCEs, which supervise groups of signalling circuits.

— The second and higher level is represented by **Auxiliary Control Elements (ACE)**. These perform functions such as call control (LACE and TACE) and Resource Management, Administration and Maintenance. These system functions reside in the System ACEs.

Section 3.2. describes the simulation results of an exchange with 49 processors, 39 in the first level and 10 in the second level, controlling 1440 subscriber lines and 360 trunks.

— Model of traffic environment.

Basic call scenarios : a list of all call types used in the simulation, including the fully linked sequences of events used is termed the basic call scenario. A typical call pattern consists of several basic call scenarios so that all types of calls used in the simulation are represented with an appropriate call mix. Originating and incoming call attempts use an exponentially distributed (Poisson) process for the determination of the call interarrival times, service times of each call phase and between the call phases. The distribution laws for the signalling phase is chosen in accordance with the type of signalling. Once a call attempt is generated, it is offered at random to one of the TCEs; for terminating and outgoing call attempts the same principle of random selection is used. If required, the selection can be directed towards specific control elements resulting in imbalance and allowing observation of system reaction to focussed load situations. In order to accomplish a correct linkage between different call phases (e.g. interdigit times) additional interarrival time distributions are used. The traffic model used is therefore the full call model [4].

	SIMULATION 1	SIMULATION 2
Call mix :		
— Internal	0.1	0.1
— Originating outgoing	0.5	0.5
— Incoming terminating	0.4	0.4
— Transit	0.0	0.0
— Load per line	0.20 Erl.	0.25 Erl.
— Load per trunk	0.67 Erl.	0.84 Erl.

Table 1 Traffic characteristics of two MNS runs.

3.2. Simulation Results.

Total processor loads (including overhead), obtained after a simulation of 1 transitory period (Cf. Appendix) and 10 sample periods of 40 sec. real time each, are given in table 2.

CONTROL ELEMENT	SIMULATION 1 TOTAL LOAD		SIMULATION 2 TOTAL LOAD	
	MEAN	CONF.	MEAN	CONF.
LTCE	0.294	± 0.002	0.302	± 0.001
TTCE	0.285	± 0.002	0.295	± 0.003
SVCE	0.462	± 0.027	0.568	± 0.028
LACE	0.519	± 0.020	0.626	± 0.025
TACE	0.542	± 0.024	0.638	± 0.020
System ACE 1	0.426	± 0.021	0.639	± 0.018
System ACE 2	0.508	± 0.012	0.632	± 0.017
System ACE 3	0.523	± 0.005	0.634	± 0.010
System ACE 4	0.519	± 0.013	0.633	± 0.009
System ACE 5	0.528	± 0.014	0.641	± 0.012

Table 2 Total load per processor in Erlang. The confidence interval is obtained from a two-tailed t-Student test.

Simulation results as presented, are further subdivided into two categories :

— Results concerning the internal behaviour of a single processor.

— Results which depend on the multi-processor system behaviour. These are mainly response times as discussed below.

— **Category 1 : Results per Single Processor.**

The processor load has only a small impact on the waiting time and the queue contents of the high-priority queues (Cf. table 3).

(Note : in the tables and drawings the average value for all processors of the same type is used where applicable, i.e. for LTCE, TTCE, LACE and TACE).

CE-LOAD (Erl.)	CLOCK	FAST SCAN	NH
LTCE 0.294	0.003 ± 0.0	0.771 ± 0.048	1.302 ± 0.082
0.302	0.004 ± 0.0	0.759 ± 0.035	1.305 ± 0.062
LACE 0.519	0.052 ± 0.003	1.014 ± 0.077	1.300 ± 0.074
0.626	0.071 ± 0.004	1.039 ± 0.073	1.286 ± 0.064

Table 3 Average waiting time in ms and confidence interval for the high priority queues in LTCE and LACE.

The bulk of transaction processing is situated in the low-priority task group which is driven via messages placed on the Message Queue (i = 4 on fig. 3). This task group is more strongly influenced by an increase in the processing load than the high-priority tasks, which feed the lower levels (Cf. Table 4).

CE TYPE	SIMULATION	LOAD (Erl.)	QUEUE WAITING TIME		QUEUE CONTENTS
			MEAN	95 %	MEAN
LTCE	1	0.294	3.01	—	0.007
	2	0.302	3.02	—	0.010
TTCE	1	0.285	1.82	—	0.006
	2	0.295	1.76	—	0.007
SVCE	1	0.462	4.59	20.7	0.130
	2	0.568	8.26	36.4	0.318
LACE	1	0.519	8.75	34.0	0.317
	2	0.626	13.81	49.6	0.670
TACE	1	0.542	7.60	37.5	0.276
	2	0.638	12.09	57.0	0.558
System ACE 1	1	0.426	7.52	40.0	0.061
	2	0.639	22.45	90.7	0.342
System ACE 2	1	0.508	19.06	89.5	0.159
	2	0.632	37.75	162.0	0.421
System ACE 3	1	0.523	12.43	52.7	0.176
	2	0.634	22.12	88.1	0.408
System ACE 4	1	0.519	9.14	37.8	0.178
	2	0.633	16.25	61.5	0.414
System ACE 5	1	0.528	12.20	52.3	0.198
	2	0.641	19.81	74.5	0.419

Table 4 Low-priority message queue contents and waiting times (in ms) for simulations 1 and 2.

An example of the waiting time distribution for the LACE processors is given in fig. 5.

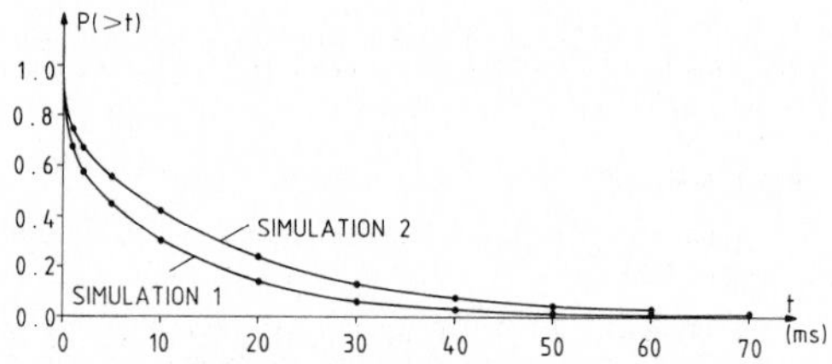

Fig. 5 Waiting time distribution of the low priority message queue in the LACE.

Fig. 6 presents the influence of the average processor load on the average queue contents for the LACE.

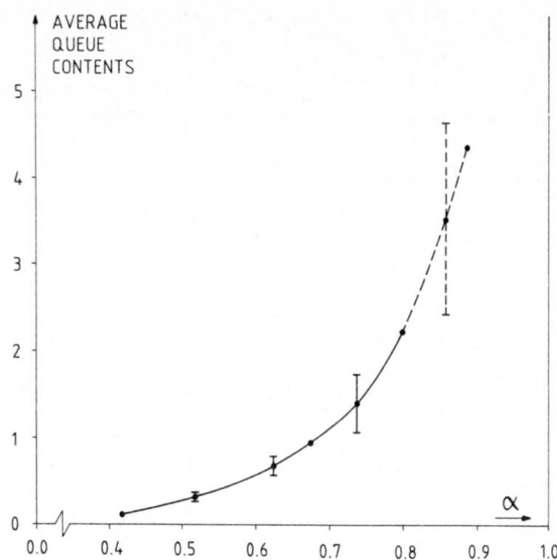

Fig. 6 Average queue contents for the low priority MQ as a function of the microprocessor load α.

— **Category 2 : Multi-processor Results.**

The present sub-section is concerned with systems behaviour under normal and high-load conditions with reference to response times as a function of the processor loads in table 2.

Fig. 7 and 8 represent 2 typical response time distributions, respectively for subscriber preselection and incoming trunk preselection. There is always a minimum processing and transmission time for linked tasks within and between the different processors. This is reflected in the horizontal line for P(RT>t) = 1.0.

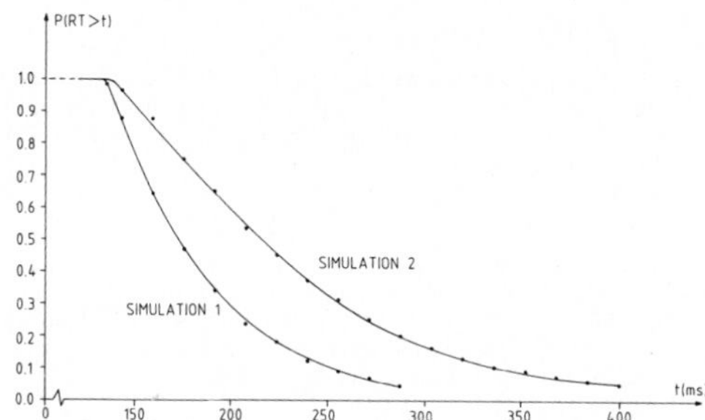

Fig. 7 Dial tone delay distribution.

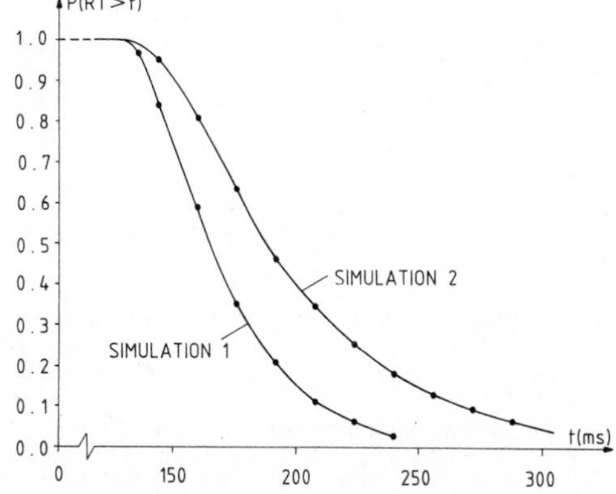

Fig. 8 Incoming preselection delay distribution.

Fig. 9 sets out the average subscriber preselection response time as a function of the load. The average load value of the LACEs is put in abscissa to be able to represent the load increase.

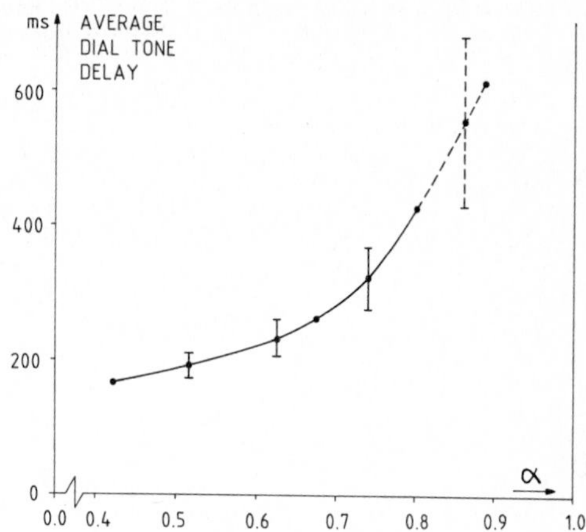

Fig. 9 Average dial tone delay as a function of the LACE processor load.

The response time performance is in accordance with CCITT requirements for normal and high-load. For overload the overload control mechanism is activated. This assures a good response time for the accepted calls [12]. Fig. 9, however, shows results of simulations without the implementation of the overload control mechanism; this is emphasized by the dotted line for the highest traffic values.

3.3. General characteristics of S 12 distributed control.

— **Exchange size :**

To demonstrate the performance of S 12 distributed control for exchanges of different sizes, two simulation examples are shown :

— (A) - exchange with 2.880 lines.

— (B) - exchange with 14.400 lines.

The above exchanges are simulated under normal load (NL) and high-load (HL : 1.5 times the normal load).

In exchange A several system functions are combined in a few System ACEs, whereas in exchange B each system function is assigned to a dedicated System ACE.

Exchange configuration, input parameters and traffic environment are also slightly different from those described in subsection 3.1. The average values of 3 response times for a local call, i.e. dial tone delay, ringing tone sending and through connection delay are compared in table 5.

EXCHANGE	DIAL TONE	RING TONE	THROUGH CONNECTION
A NL	175 (± 7)	200 (± 7)	34.3 (± 0.3)
B NL	178 (± 3)	201 (± 4)	34.4 (± 0.4)
A HL	264 (± 16)	292 (± 18)	36.2 (± 0.7)
B HL	268 (± 10)	295 (± 10)	35.9 (± 0.4)

Table 5 Comparison of local call response times (in ms) for two exchanges of a different size and under different load conditions. The two-tailed t-Student test confidence intervals (95 %) are given in parentheses.

The results shown in table 5 are averages for approximately 2.400 and 12.000 local calls for exchange types A and B respectively. They are very similar in both cases. The delay time distributions obtained from the same simulations are also very similar for exchanges A and B.

The good agreement between the average response times and delay distribution results for the two exchanges as well as other studies not shown here, lead to the following observation :

— **The S 12 distributed control queuing behaviour is independent of the exchange sizes for the same type of traffic and load.** The above property implies that studies of larger distributed control configurations do not provide new information about the queuing behaviour.

— **The interprocessor message traffic.**

The same simulation also shows that the interprocessor message traffic constitutes a small fraction of the total switched traffic in the DSN. Fig. 10 demonstrates that **the interprocessor traffic increases in a linear way when the switching office size grows** (for a constant traffic per line).

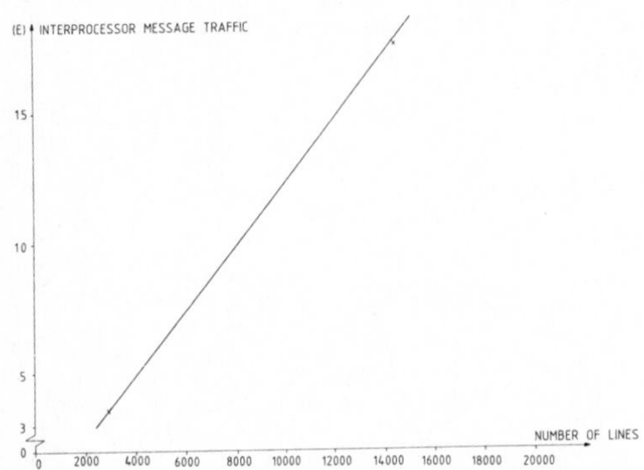

Fig. 10 Interprocessor traffic as a function of the exchange size.

— **Virtual non-blocking in the DSN.**

One of the basic conditions for the successful handling of interprocessor messages is **a very low blocking in the digital switching network** via which the processors communicate. This is demonstrated in fig. 11, which shows the blocking for interprocessor messages and for comparison also the DSN blocking for an outgoing call as a function of the load per line in the largest S 12 configuration. The outgoing call congestion comprises the following connections : subscriber line to receiver, line to outgoing trunk and selected trunk to sender. The calculations were done according to the congestion formulae given in ref. [2].

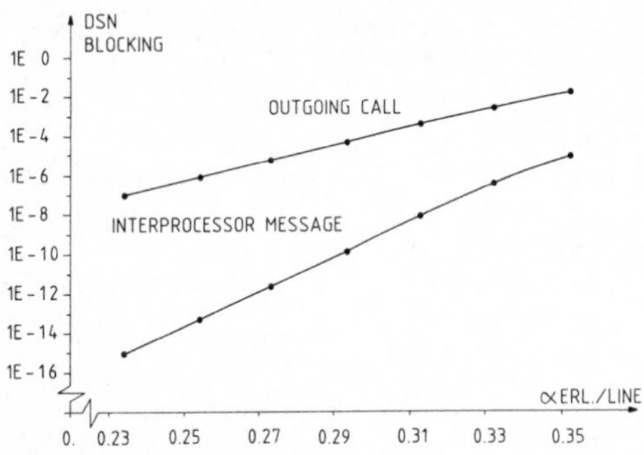

Fig. 11 Interprocessor message and outgoing call blocking in the DSN as a function of load per line.

4. COMPARISON WITH AN ANALYTICAL TOOL.

The analytical performance evaluation tool used is based on the work of M. Villen [5,6]. It allows the calculation of the Average Response Times in an M/G/1 Queue with General Multiple Feedback and any Arrangement of Preemptive/Non-Preemptive Priorities. For the evaluation of the overall response times in a multiprocessor environment, the sum of the average response times in each of the individual processors participating in the specific task sequence is taken.

The inputs require the definition of the task sequences, their service time distributions, priority and feedback assignment etc. as well as the arrival rates for each of them.

Table 6 gives an example comparing the simulated and calculated response times for a normal load situation in a distributed microprocessor control of a System 12 telephone exchange.

		ANALYTICAL TOOL	MNS SIMULATION
Local call	Dial tone delay	189	192
	Through connection delay	27	29
	Call release delay	180	181
Outgoing call	Dial tone delay	189	192
	Through connection delay	130	122
	Call release delay	183	170
Incoming call	Incoming response delay	179	172
	Through connection delay	47	40
	Call release delay	178	174

Table 6 Comparison of calculated and simulated average response times in ms.

As it can be seen from the above table, the agreement between the simulated and the calculated results is very good. This has been confirmed by many runs in the whole range of interest. The calculated results lie within the confidence intervals of the simulated results.

Note that the simulation results, compared with those obtained with analytical formulae, were derived from the full detail model (Cf. section 2 and 3). The good agreement between the simulation and analytical formulae results has been found to be mainly due to precise priority, feedback and task linking representation in the formulae as well as the inclusion of exact overheads and transmission times.

5. CONCLUDING REMARKS.

A full detailed Multinode Simulator written in GPSS and PL1 has been developed and employed for the study of large distributed microprocessor control systems. The experience obtained with this simulator shows that it is an adequate tool to verify and prove the performance for a wide range of exchange sizes and types including performance prediction before the actual installation.

— Simulations of a large range of configurations and projects confirmed the excellent quality, performance, throughput and queuing behaviour of the System 12 distributed control architecture. Under normal load conditions the processors are loaded up to approx. 0.6 Erl., incl. the fixed overhead.

— Under high-load, i.e. 1.4 to 1.5 times the normal load, all calls are accepted and correctly handled, satisfying the grade of service specified for high-load situations (Cf. 3).

— Under all severe overload situations, the high-load throughput is maintained and a good grade of service is ensured for the accepted calls. The selective rejection of calls in excess of system capacity, allows calls to be handled according to priority [12].

— S 12 with its distributed control architecture, virtual non-blocking digital switching network and mostly parallel and independent call or message handling by processors, makes the system architecture ready for present and future traffic environments among ISDN features, wideband switching and other advanced switching facilities.

Analysis of results obtained with the above tool also allows the following general observations concerning the queuing behaviour of S 12 distributed control :

— The study of very large configurations does not bring new information because the system functions become replicated and characteristics repetitive (Cf. 3.3.). The delay and response time characteristics are practically independent of the office size for the same type of traffic (Cf. table 5).

— The interprocessor traffic constitutes a very small fraction of the total switched traffic in the DSN, independently of the office size. When the switching office grows, the interprocessor traffic increases in a linear way, proportionally to the number of messages or calls to be switched.

— The digital switching network through which the microprocessors communicate is virtually non-blocking. Simulation of the control structure (for a reasonable duration of the runs) cannot consider the DSN itself. This approximation has an entirely negligible effect on the results.

— The analytical formulae available at present allow the calculation of average delays and response times, load of the processors and many other interesting characteristics for this kind of queuing network (Cf. 4). Overall response time distributions are still not available by analytical means but several approximations have been tried out and compared with simulation results.

— A full scale simulation is still recommended as a reliable means of comprehensive performance check (distributions, transitory/burst situations, control mechanisms and many other characteristics which will be very difficult to obtain by analytical means alone).

The cost of simulations will drastically decrease as a result of using new simulation systems, compilers etc. e.g. the GPSS-H compiler [9] which reduces the run times at least 4 times.

6. REFERENCES.

[1] «Electrical Communication», vol. 59, no. 1/2, (Issue devoted to digital switching with distributed microprocessor control S 12), 1985.

[2] J.R. de los Mozos and A. Buchheister, «ITT 1240 Digital Exchange : Traffic Handling Capacity», Electrical Communication, vol. 56, no. 2/3, pp. 207-217, 1981.

[3] O.G. Soto et al. : «An approach to Flexible Modelling and Simulation for Control Processor Analysis». Proceedings of the 9th International Teletraffic Congress, Torremolinos, 1979.

[4] M. Gruszecki, «Throughput Evaluation in Digital Switching Systems with Distributed Microprocessor Control», Microprocessing and Microprogramming, vol. 11, no. 3/4, pp. 207-215 (Published by NorthHolland), 1983.

[5] M. Villén Altamirano, «Average Response Times in an M/G/1 Queue with General Feedback and Priorities», Proceeding of the 10th International Teletraffic Congress, Montreal 1983, paper 5.2.5.

[6] M. Villén Altamirano, «Average Response Times in an M/G/1 Queue with General multiple Feedback and any Arrangement of Preemptive/non-Preemptive Priorities», (to be published), 1985.

[7] T.J. Schriber, «Simulation Using GPSS», John Wiley & Sons, NY, 1974.

[8] «General Purpose Simulation System V», User's Manual, IBM Publication SH20-0851.

[9] J.O. Henriksen, R.C. Crain, «GPSS/H», User's Manual, Wolverine Software Corporation, 1983.

[10] L. Kleinrock, «Queuing Systems», Volume II : Computer Applications, John Wiley & Sons, NY, 1976.

[11] T.J. Schriber, R.W. Andrews, «Interactive Analysis of Simulation Output by the Method of Batch Means», Simulation Conference ACM, NY, pp. 512-524, 1979.

[12] G. Morales Andres and M. Villén Altamirano : «System 12 Traffic Overload Control». Electrical Communication, vol. 59, no. 1/2, 1985.

APPENDIX : TRANSITORY PERIOD AND CONFIDENCE INTERVAL ESTIMATION.

The output of the initialisation run (in which the length of the transition period has to be determined) was first batched into n exclusive adjacent groups of equal time sizes. The test for initialisation bias was then based on the comparison of the sample means of x consecutive batches (b1, b2, ..., bx) of 4 sec. with those of the next x batches (b2, b3, ..., bx, bx+1) and this replicated n-x+1 times for various values of x. In practice, the truncation threshold was chosen when the sample means for $x=4$ flatten out. This simple heuristic approach showed its validity as the complete simulation results became available. Figure 12 and 13 give two typical examples of the initialisation bias of the load, average queue content, average waiting time of one particular processor and an average total response time respectively. The comparison of these figures with others (not shown here) revealed the following facts :

— Truncation points differ from processor to processor and the selection of these points should take into account the joint behaviour of all output series.

— The total response times have the largest transition period while the load of the processors has the lowest one. It is evident that the highest truncation point fixes the transient period of the simulation runs e.g. minimum 40 sec. real time for a typical load situation. This does not mean that 40 sec. is sufficient for other load situations.

Confidence intervals for the mean simulation output variables were gathered using the batched means method since it required only a single simulation run. As concerns the batch size, it should be noted that all observed mean values in each batch should have very little correlation with adjacent batch means. This leads to confidence intervals which are superior in terms of well defined measures of effectiveness [11]. Elementary statistical methods can then be used to construct a confidence interval on the process means. Consequently, 10 batch sizes of a time length equal to the transition period for a particular simulation run were used to perform a t-Student statistical analysis. A 95 % confidence level has been imposed.

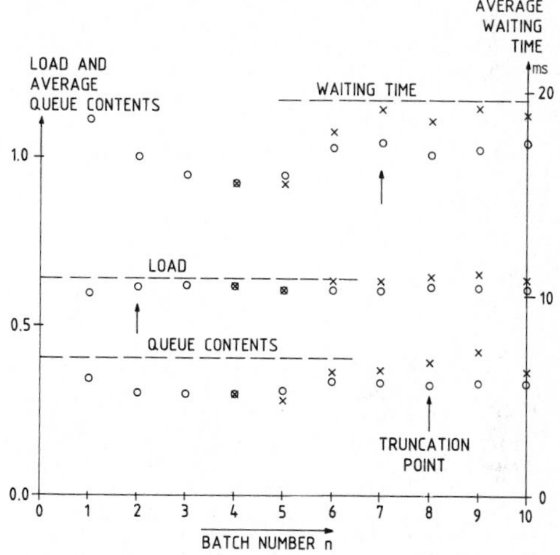

o Cumulative

x For n = 10 and x = 4

——— Steady state characteristic after a complete simulation has been performed.

Fig. 12 Example of a transition period study for a particular processor of the MNS. The average waiting times are expressed in units of time.

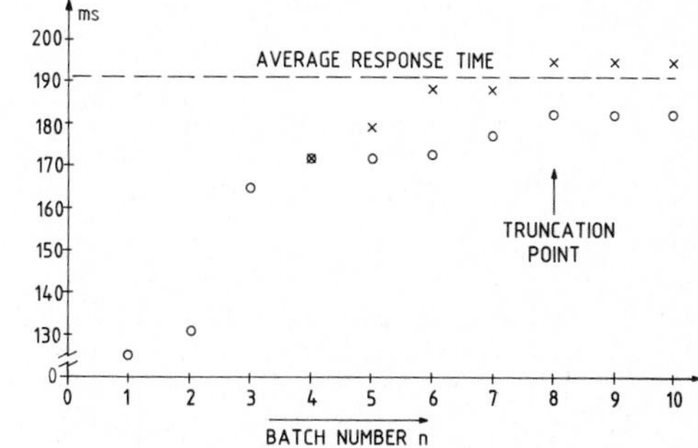

o Cumulative

x For n = 10 and x = 4

——— Converged and averaged value reached at the end of the complete simulation run, which is ten times longer than the transition period.

Fig. 13 Initialisation bias for a response time measurement, for which the truncation threshold has been taken to be 8 transitory batches.

TELETRAFFIC ISSUES in an Advanced Information Society
ITC-11
Minoru Akiyama (Editor)
Elsevier Science Publishers B.V. (North-Holland)

SIMULATION OF TELETRAFFIC SYSTEMS: SPECIFICATION LANGUAGES, SIMULATION LANGUAGES, SOFTWARE

Gerard JONIN, Janis SEDOL, Victor SUPE

Computing Center of Latvian State University,
Riga, USSR

ABSTRACT

Some extensions and further development are offered for the specification language SDL, which is recommended by CCITT for the functional description of the telephone exchanges with stored program control.

Simulation system SPALM85 and SDL/PL, developed in Computing Center of the Latvian State University, are based on the extended SDL. The software is implemented on EC computers, compatible with IBM/360.

Basic terms, tools, characteristics and facilities are described for both simulation systems, as well as their application field during the project stage and investigations of communication systems.

1. INTRODUCTION

With the increase of complexity of telecommunication systems, the design of telephone exchanges with stored program control, the developing computer technique and engaging of computers in the structure of communication systems and networks, the simulation method becomes more important and frequently it is the only way of the investigation of the characteristics of many systems. Simulation languages and applied program packages are developed to process the simulation. To the current moment more than 500 simulation systems are designed throughout the world and new systems are developing [1-3]. The development is caused by new application fields and the introduction of the tools and technologies, which are giving new facilities and are decreasing the expenses. This may be related both to the design and investigations of communication systems, and to the design, debugging and maintenance of simulation programs.

The efficiency of investigations is considerably dependent on the choice of simulation language and it's software. The modern requirements for program tools suppose the presence of the specification language, too, for simplification of the project development, the functional description of the system to be simulated and the better understanding between the programmers and system investigators. The tradicional simulation systems and languages either do not use such tools for specification, or the tools are too close to the syntax of the simulation language (as, for example, in GPSS). As a result, the difficulties arise in communication between the designers, the investigators of the real systems and the programmers. This is the main source of errors and the lack of correspondence between the real object and its model.

The proposed simulation systems SPALM85 and SDL/PL are based on the specification language SDL, which is recommended by CCITT [4]. At the same time the simulation languages and their software have some features, which determine their conveniences and possibilities of their use, as well as their application fields.

The system SPALM85 for the first time provides the common way for both the simulation model design and the algebraic calculations by solving state equations using iteration method. The building of the simulation program in the form of the single module gives a possibility to obtain very high technical and economical showings (the resources of main storage of computer, the processor time for experiment). The system is easy to learn and simple in the usage.

The system SDL/PL is provided for simulation of complex systems, where the components of the model may be developed separately, using independent translation, data input and output of results, as well as separate debugging for each component. The libraries of model components may be stipulated. The complexity of models are limited only by the resources of the available computer. The software for SDL/PL is implemented taking into account special debugging tools for simulation programs. The system is applied to the simulation of communication systems and for the design of debugging and testing tools for the software of quasielectronic and electronic telephone exchanges. The system has the facilities for the exchange of signals and data with the programs,designed in other programming systems, as it is stipulated by PL/1 programming language.

In addition, both simulation languages have different coding style

(SPALM85-laconic, consise style, SDL/PL - similarity to the natural language), which may satisfy the different tastes of users.

2. SPECIFICATION LANGUAGES

CCITT has recommended Specification and Description Language (SDL) [4] for the functional description of telephone exchanges with stored program control. The symbols, used in this language, are presented in Fig.1.

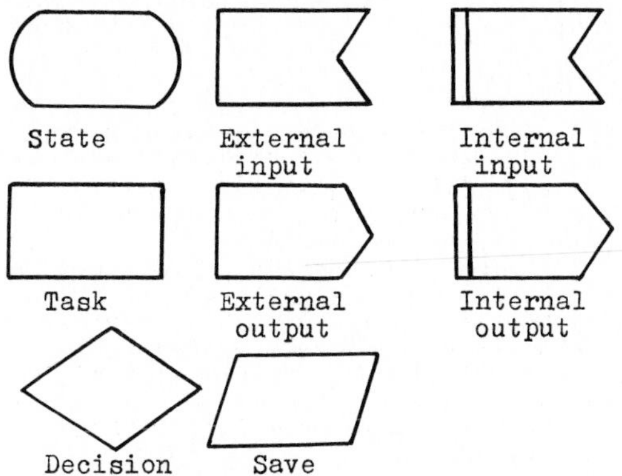

Fig.1 SDL Symbols.

The tools of SDL give the possibility to describe the actions in systems in convenient and visual way. However, the time characteristics are not represented. It is the cause, why the language is insufficient for the description of systems to be simulated. Let us introduce additional tools for the description of systems to be simulated [5]: the delay of the signal, the delay of the process, the absorbing state, and additional types of signals - those, which get lost, waiting, as well as generating and cancelling signals. These tools are represented in Fig.2.

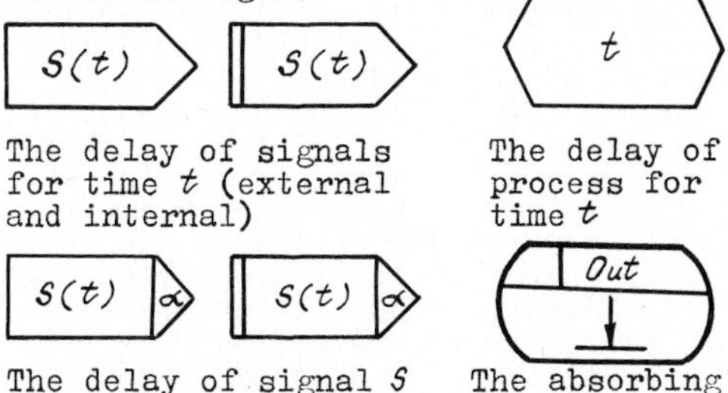

Fig. 2 Additional tools for SDL.

The delay of process instance is the abbreviation of the group of actions, which contains the delay of internal signal, and this group is shown in Fig.3. For the type of signal α following designations are used: G - for generating signal, W - for waiting signal, C - for cancelling signal, L - for the signal, which gets lost. The generating signal creates new process instance, which proceeds according to the associated transition string. The handling of the waiting signals and the signals, which get lost, is following: if the process instance exists in the associated states, then the signals of both types transfer the process instance into transition. In the case of absence of process instance, the waiting signal moves to the queue of signals for the given state and is handled, when the process instance arrives. The signal of type L in this case is got lost and no actions upon the state are processed. The cancelling signal has an influence only upon the queue of signals. This signal activates the waiting signal with the same name and then both signals get lost. The absorbing state is invented, too. If the process instance enters such a state, then it leaves the system.

The addition of such tools makes the interaction of process instances more exact and widens the application field of SDL. The observed extended SDL is directly used in system SDL/PL.

To develop the language SDL further, the original and the generalized states are introduced. The original state (Fig.4) is invented for the creation of the very first process instance, which begins its actions at the initial time moment. This state needs not the associated inputs. The generalized state combines the functions of state and delay. The realization of a generalized state is shown in Fig.5a, and the proposed abbreviation - in fig.5b.

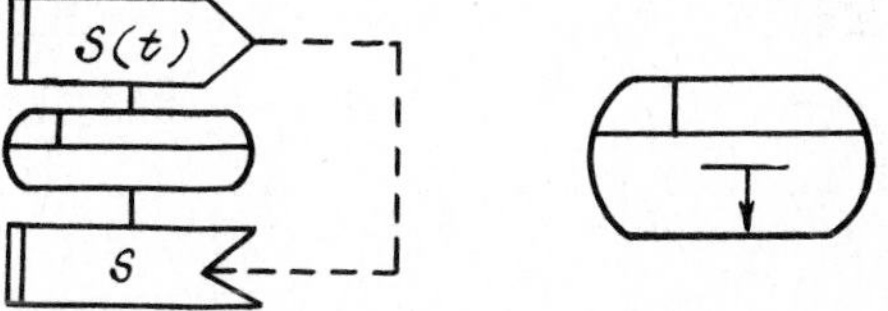

Fig. 3 The broadened form of the delay of process instance.

Fig.4 The original state.

When using such a figure the process instance may proceed to the transition also in case, when no signal is accepted, but the delay time has run out. In this case the process instance leaves the state along the line or arrow, which has not associated input symbol. With the introduction of the generalized state there is no need for the special delay symbols. The delay then is only particular case of a state, when there are not associated inputs.

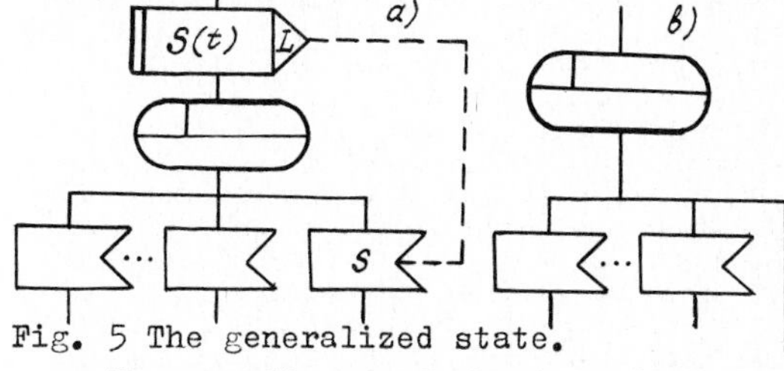

Fig. 5 The generalized state.

The expedience of the introduction

of both the original and the generalized states is corroborated by the experience of the design and application of the system SPAIM85, which is the further development of the SPAIM system [6,7], based on SDL.

According to the established traditions, the system SPAIM85 offers the graphical symbols for the generalized state, the original state and the absorbing state, as well as for the decision, and they are different from those of SDL (Fig.6).

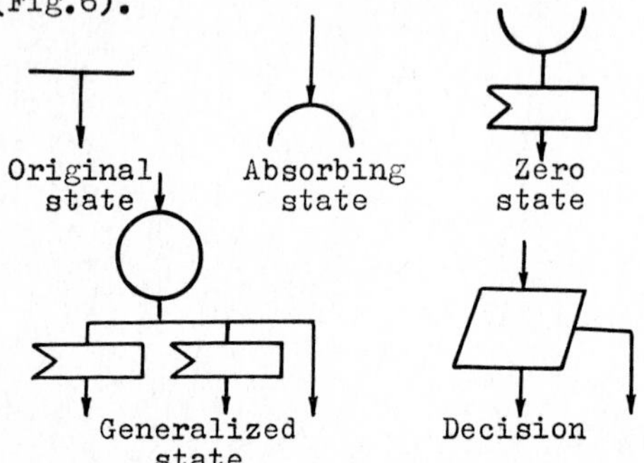

Fig.6 The graphical symbols of SPAIM85.

Moreover, SPAIM85 does not distinguish external and internal signals (input, output), uses only two types of signals (G,L). "Save" operation is not used,and the delay is not distinguished from generalized state.

There is zero state introduced in SPAIM85. It serves as the source of process instances for generating signals.

For frequently used diagram elements there are new designations (derivative actions). Fig.7 shows the source of process instances in its full (Fig. 7a) and abbreviated form (Fig.7b).

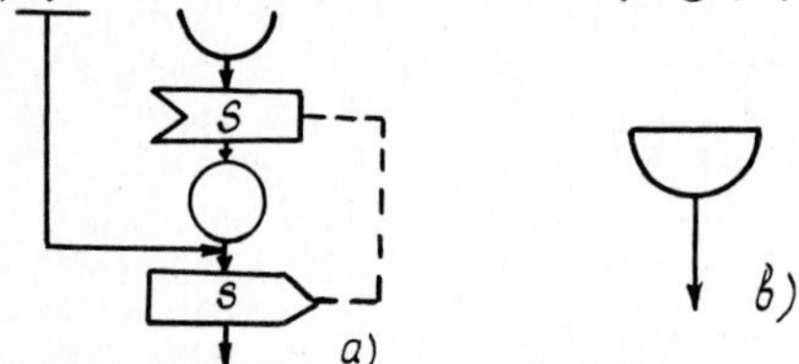

Fig.7 The Source in SPAIM85.

3. SIMULATION SYSTEM SPAIM85

3.1 The Structure

Simulation System SPAIM85 contains the specification language SITA, the programming language PAL85 and the simulation languages SITA-ITER (for algebraic calculations by iteration method) and SITA-SIM (for Monte-Carlo simulation). A model in language SITA is represented in the form of a diagram, built from the actions of the extended SDL (the symbols shown in Fig.6 and Fig.7). The simulation languages SITA-ITER and SITA-SIM are designed in a common way (SITA-ITER is the subset of SITA-SIM), and the program in SITA-ITER may be used both for the algebraic calculations and simulation. The further discussion concerns the language SITA-SIM only. This language is the expansion of the programming language PAL85 by the procedures for simulation elements.

3.2 The Principles of PAL85

The program in PAL85 consists of statements, which are separated by blanks. Identifiers are used to designate arrays, blocks, procedures and other objects. Identifiers consist of letters only. Arithmetical and logical statements are designated by standard operators: "+,-,*,/,**" for arithmetic operations and "=,<,>,<=,>=,<>" - for logic operations. The operator "#" designates the transfer of control, as well as the address to a block or a procedure. Statements, containing ":" symbol, are called "tittles" and are used to separate the logical parts of the program. They are also used as labels.

Let us consider the basic types of statements. The declaration part begins with the statement "::" and contains the declarations of all the variables and arrays, used in the program. The declaration statement begins with the identifier and contains the information on the type and dimensions of the array (in particular case it is single variable). The type REAL is designated by the operator "*", the INTEGER type - by the absence of "*". The declaration of the initial values consists of the symbol "=" and the number record. The initial values are declared for the array, declared by the preceeding declaration statement.

The block head statement contains the block identifier and the symbol ":". The label statement consists of the number, which is called "label number",and of the character ":". The assignment statement has the form "variable=expression". There is abbreviated form for the assignment statement, where the left side is omitted. This gives the opportunity to write "A+B" instead of "A=A+B". The value of the expression is assigned to the first variable in the expression. The variable from the array is designated by the identifier of the array and the number of the variable in this array. For the indexed variables the indeces are given in the square brackets. The statement for control transfer to label has the form "#label-number".The conditional transfer statement has the form "the-correlation-of-two-expressions # label-number". The transfer is executed when the correlation is satisfied. The ommition of the label number means the exit from the current block (in case of both types of transfer statements). The address to block has the form "# block-identifier". The address to procedure has the form "#procedure-identifier (parameter-list)". Parameters are separated by commas. Numbers, variables, blocks and labels may be used as parameters. In case of parameter-block or parameter-label this parameter begins with the character "#".

3.3 Simulation Procedures

The procedures for simulation operations are given in Table 1.

Table 1 Simulation procedures in the language SITA-SIM

Name	Identifier	Parameters
Generator (source)	GEN	Flow intensity
State	ST	State number, intensity of exits
Output of signal	OUT	Label of input statement
Input of signal	IN	State number
Transfer by probability	PR	Probability, label for transfer

3.4 Model Examples

Let us consider two model examples, built in SITA-SIM.

Example 1. Fully available schema with limited waiting. The call flow with intensity L arrives in the schema with V lines. There are M places for waiting. The waiting call leaves the queue with intensity N and gets lost. The call, which arrives when all the places for waiting are full, gets lost, too. The average service time is equal to 1.

The diagram for this model in SITA language is shown in Fig.8.

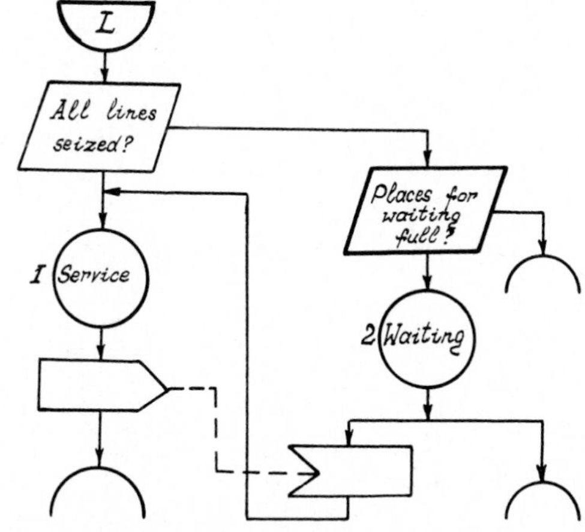

Fig.8 The diagram for fully available schema with limited waiting.

Simulation program uses the following standard designations: Q - the number of states, X - the employment vector for states, Y - the vector of state capacities, SET - the preparation block, ALG - the block for simulation algorithm. The program, including initial data V=3, L=2,M=4,N=1/2, is following:

```
  :: V=3   L*=2.  M=4    N*=.5
     Q=2   X%2   Y%2
SET: Y1=V    Y2=M
ALG: #GEN(L)  X1=V#2
  1: #ST(1,1.) #OUT(#3)  #
  2: X2=M#    #ST(2,N)  #
  3: #IN(2)   #1
```

Example 2. The system with internal link, repeated calls and preliminary service. There are N subscribers. The free subscriber sends the call with intensity L/N to any other subscriber. There are lines. The calling subscriner seizes the line for the preliminary service with V average time 1/A. If there is no free line available, or if the called subscriber is busy after preliminary service, then the call goes to the state, from which the repeated call is made with intensity M or the call gets lost with intensity S. The repeated call is handled in the same way as the new one. If the called subscriber is available, then the conversation takes place between the two subscribers after the preliminary service. The average conversation time is equal to 1/B. After this the subscribers and seized lines are released. Fig.9 shows the diagram for the discussed model.

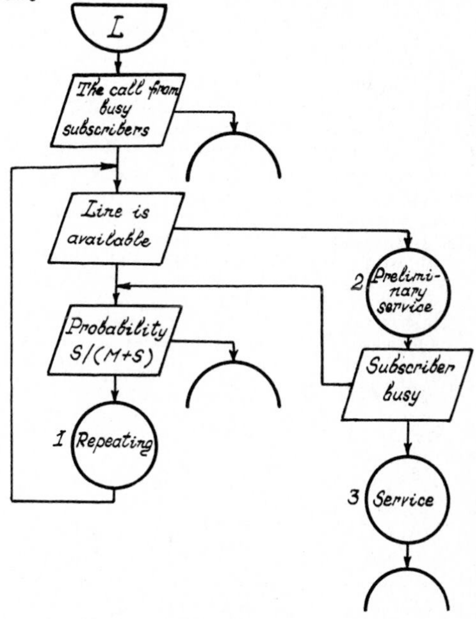

Fig.9 The diagram for the model with repeated calls.

The simulation program including initial data N=4,V=2,L=1/2,A=2,B=1,M=4, S=1/10, is the following:

```
  :: N=4  V=2  L*=.5   A*=2.
     B*=1.  M*=4.  S*=.1
     Q=3  X%3  Y%3  E*  R*
SET: Y1=N Y2=N  Y3=N  R=S/(M+S)
ALG: #GEN(L)  E=(X1+X2+2*X3)/N
     #PR(E,#4)  1: X2+X3<V#3
```

```
2:  #PR(R,#4) #ST(1,M) #1
3:  #ST(2,A) E=(X1+X2+2*X3)/(N-1)
    #PR(E,#2)  #ST(3,B) 4: #
```

The programs of both examples may be used both for the calculations with iteration method (at low values of parameters) and for simulation (in general case).

The program of the example 2 has been used for the calculations at several values of L. The results show the decreasing of the paying load at high values of L (Fig.10).

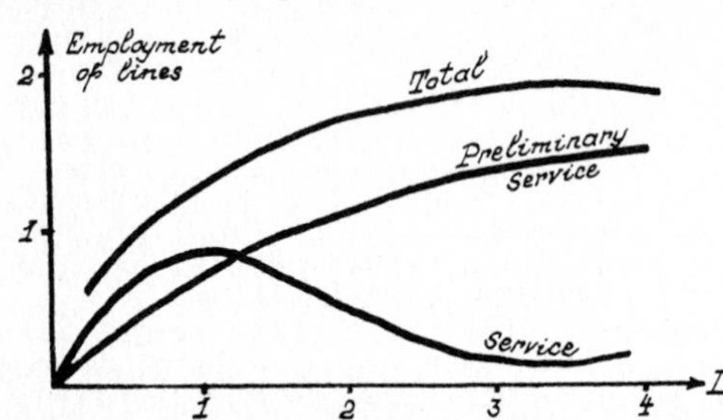

Fig.10 The dependence of the average number of seized lines on the L value.

4. SIMULATION SYSTEM SDL/PL

4.1 General Information

The simulation system SDL/PL is the tool for the discrete event simulation of the complex systems, described or designed, using specification language SDL. This system is provided for the debugging of real time control programs by the method of simulation of the controlled equipment.

The system consists of the simulation language SDL/PL and its software. The language SDL/PL is built on the basis of the specification language SDL with supplements and the programming language PL/1. The same principles of the model design may be implemented as well as on the basis of other programming language (for example, SIMULA-67 [8]).

4.2 The Structure of Simulation Program

Unlike the representation of the model in the form of a single module in system SPAIM85, the simulation program in SDL/PL contains the collection of the model components - process descriptions, which is supplemented by the model description. Moreover, the model may interact with the programs in other programming languages.

4.2.1 Process Description

The process description (PD) is the description of a component of the model. The PD consists of the head of the description, the declaration part, the functional part, the blocks for the initialization and preparation of statistical data, as well as for results of simulation, the block for debugging actions and the end part of the description. Several process instances of the same kind may interact in the model according to one PD. Each of those process instances may have the individual characteristics - the passport of process instance. The head of the PD provides the name for the description and the set of variables for the passports of process instances. The declaration part consists of the DECLARE statements of PL/1 for all the variables in common use for all the process instances in this description.

The functional part corresponds to the SDL diagram for this PD, and describes the functional algorithm, common to all the process instances in this PD. Special statements are used for the representation of SDL actions, the syntaxis for those statements is close to the language SDL/PR [4] - the program form of SDL. Some actions, such as tasks and decisions, require flexible algorithmical facilities, and they are represented by PL/1 statements. Other SDL symbols are represented by Special Statements (STATE, INPUT, OUTPUT, DELAY, etc.). Taking into account such a way of model design, when the components of the model -process descriptions are developed by different programmers, PD provides special blocks for independent actions for the data input, the processing of Statistical operations, the output of the simulation results.

To decrease the number of errors in the debugging process of the simulation program and its components, and to standartize these operations, the language SDL/PL provides special facilities for debugging. The principal difference from the other programming languages is the possibility to specify the debugging operations outside the functional algorithm - in the special DEBUG-block in PD. The sequence of the processing of such actions is controlled by the software according to the specifications stipulated by the user. When the debugging tools are not necessary, they are deleted automatically by the software.

4.2.2 Model Description

The model description serves for the consolidation of separate PD into united model. Special operations are provided for the description of model structure, data input/output and preparation sequence, the sequence of the starting of PD-s by signals, the description of the initial state of model, the conditions and orders for applying the debugging tools, as well as the sequence of the experiment (the duration of time for it). Moreover, actions may be provided for various algorithmic purposes (preparation of data, calculation of the experiment conditions) in terms of PL/1 statements, including the calling of external procedures in other programming languages.

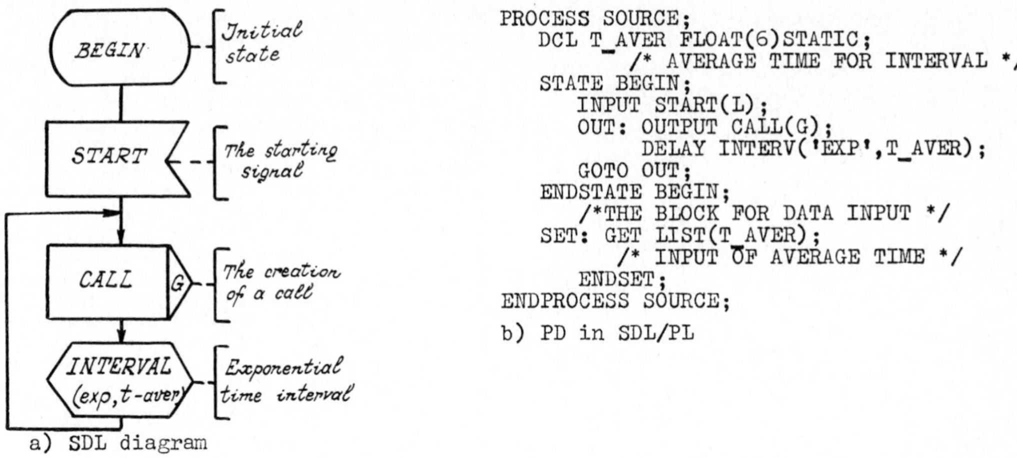

a) SDL diagram

```
PROCESS SOURCE;
   DCL T_AVER FLOAT(6)STATIC;
          /* AVERAGE TIME FOR INTERVAL */
   STATE BEGIN;
      INPUT START(L);
      OUT: OUTPUT CALL(G);
           DELAY INTERV('EXP',T_AVER);
      GOTO OUT;
   ENDSTATE BEGIN;
      /*THE BLOCK FOR DATA INPUT */
   SET: GET LIST(T_AVER);
         /* INPUT OF AVERAGE TIME */
      ENDSET;
ENDPROCESS SOURCE;
```

b) PD in SDL/PL

Fig.11 The description of process-source.

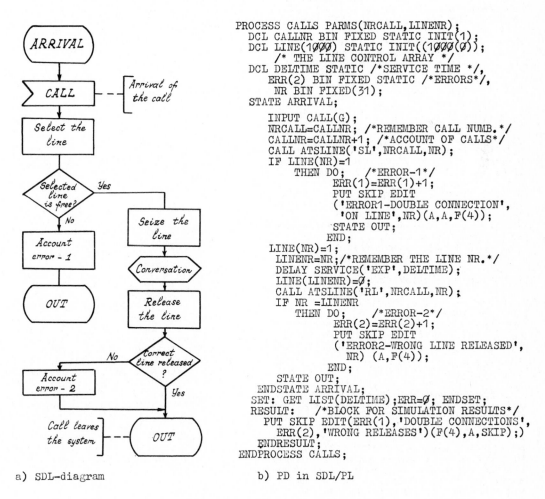

```
PROCESS CALLS PARMS(NRCALL,LINENR);
  DCL CALLNR BIN FIXED STATIC INIT(1);
  DCL LINE(1ØØØ) STATIC INIT((1ØØØ(Ø));
      /* THE LINE CONTROL ARRAY */
  DCL DELTIME STATIC /*SERVICE TIME */,
     ERR(2) BIN FIXED STATIC /*ERRORS*/,
      NR BIN FIXED(31);
  STATE ARRIVAL;

     INPUT CALL(G);
     NRCALL=CALLNR; /*REMEMBER CALL NUMB.*/
     CALLNR=CALLNR+1; /*ACCOUNT OF CALLS*/
     CALL ATSLINE('SL',NRCALL,NR);
     IF LINE(NR)=1
         THEN DO;   /*ERROR-1*/
               ERR(1)=ERR(1)+1;
               PUT SKIP EDIT
               ('ERROR1-DOUBLE CONNECTION',
                'ON LINE',NR)(A,A,F(4));
               STATE OUT;
              END;
     LINE(NR)=1;
      LINENR=NR;/*REMEMBER THE LINE NR.*/
      DELAY SERVICE('EXP',DELTIME);
      LINE(LINENR)=Ø;
      CALL ATSLINE('RL',NRCALL,NR);
      IF NR =LINENR
         THEN DO;     /*ERROR-2*/
               ERR(2)=ERR(2)+1;
               PUT SKIP EDIT
               ('ERROR2-WRONG LINE RELEASED',
                 NR) (A,F(4));
              END;
      STATE OUT;
   ENDSTATE ARRIVAL;
  SET: GET LIST(DELTIME);ERR=Ø; ENDSET;
  RESULT:   /*BLOCK FOR SIMULATION RESULTS*/
    PUT SKIP EDIT(ERR(1),'DOUBLE CONNECTIONS',
      ERR(2),'WRONG RELEASES')(F(4),A,SKIP);)
   ENDRESULT;
ENDPROCESS CALLS;
```

a) SDL-diagram

b) PD in SDL/PL

Fig.12 The description of process-call.

4.3 The example: the debugging of the program for line selection

Let us consider the following example of the application of the system SDL/PL. Let us debug the program (from the control software of the telephone exchange), provided for the selection of line for the connection through the telephone exchange. Let us develop the model for the "environment" of such program. In the model, the exponentially distributed flow of incoming calls arrives at the exchange, and in the moments of the establishment of the connection and of the ringing off the model calls the proper program in the control software (let us name this program ATSLINE). Three parameters are given to this program - the function ('SL' - seize line, 'RL' - release line), the number of the call and the parameter to return the number of the line to be seized or released. The model will consist of two process descriptions - the source of calls (Fig.11) and the calls (Fig.12). The second PD describes the actions of several process instances simultaneously. Each of process instances will remember in its passport the individual characteristics - the number of the call (NRCALL) and the number of the line, chosen for this call (LINENR).

The PD of call will execute the diagnostic check for the program ATSLINE (the printout of messages and account of errors). Such a model (the model description is shown in Fig.13) may be used for the debugging of various algorithms of data maintenance and selection of the line for the call. The implementation of the program ATSLINE may be in any programming language, which is accessible from PL/1.

```
MODEL DBGATS;
  INCLUDE SOURCE,CALLS;/*THE STRUCTURE*/
   MAKE SOURCE(BEGIN);/*CREATE SOURCE*/
   SET;         /*DATA INPUT*/
   OUTPUT START(L);/*START THE SOURCE*/
   SIMULATE 5ØØ;/*EXPERIMENT LENGTH*/
   RESULT;     /*OUTPUT OF RESULTS*/
 ENDMODEL DBGATS;
```

Fig.13 The model description for the debugging of the line selection program.

4.4 The Software: Structure And Application

The software of SDL/PL is implemented in operation system OC for computers EC (the analogue of IBM/360). It contains: the preprocessor SDL/PL, which generates the PL/1 text for the model; the program for the analysis of PL/1 compilers output to bring it to accord to the listing of SDL/PL preprocessor; the programs for model control; the programs for debugging regime support; service programs.

The preprocessor of SDL/PL generates the program in PL/1 for each PD or model description (either in standard or debugging regime). Then the compiler of PL/1 works on the generated text. The further steps are the work of the program of analysis and link editing by the means of operation system. As a result, we obtain the load module either for process description, or model description, or the collection of them.

According to the principles of SDL/PL, the components of the model may be separately translated and debugged. The goal model is formed by the linkage editor of the operation system. At this stage the external procedures and the programs from the SDL/PL run-time software are added to the model.

5. CONCLUSIONS

The experience has shown, that the described in the paper supplemented SDL provides the convenient means for the description of the communication systems to be simulated. The simulation systems SPALM85 and SDL/PL are developed on the basis of the supplemented SDL.

SPALM85 system is applied to the investigations of teletraffic systems and their probability characteristics in cases, when no program is involved (in other programming languages). The system SDL/PL is oriented for the simulation of complex systems and is used for complex debugging of the software of telephone exchanges with stored program control.

REFERENCES

[1] E.Kindler, "Simulation Programming Languages", SNTL, Prague, 1980. /In Czech/

[2] W.Kreutzer, "Patterns of modelling: towards a conceptual basis for discrete event simulation", Simuletter, vol.11, no.3 and 4, pp.7-23, 1980.

[3] Tuncer I.Ören, "A personal view on the future of simulation languages", Proc.UKSC conf.comput.simulat., Chester, pp.294-304, 1978.

[4] "Report on the meeting held in Geneva from 5 to 16 December 1983, Part III - SDL Recommendations", CCITT Document COMXI-R, Geneva, January 1984.

[5] G.Jonin, "The description of systems to be simulated using SDL", Automatica y vytchislitelnaya technika, no.1, pp.30-33, 1982. /In Russian/

[6] G.Jonin, J.Sedol, "Simulation of teletraffic systems", Radio y Svyaz, Moscow, 1982. /In Russian/.

[7] G.Jonin, J.Sedol, "Simulation language SPALM", Proc.9th Int.Teletraffic Congress, Torremolinos-Spain, pp.524/1-524/4, 1979.

[8] E.Kindler, V.Supe, "Simulation of Telecommunication Networks", Proc.12th SIMULA User's Conf., Budapest, pp.111-120, 1984.

TELETRAFFIC ISSUES in an Advanced Information Society
ITC-11
Minoru Akiyama (Editor)
Elsevier Science Publishers B.V. (North-Holland)

A Universal Environment Simulator for SPC Switching System Testing

Wolfram LEMPPENAU
Phuoc TRAN-GIA

Institute of Communications Switching and Data Technics
University of Stuttgart, Fed. Rep. of Germany

ABSTRACT

Besides of queueing system analysis and system simulations, the environment simulation provides a realistic test technique for communication systems. In this paper the concept of a universal environment simulator is presented, whereby realistic customer behaviour as well as subscriber-system interaction are considered. Implementation and performance aspects are discussed. The simulator is implemented by means of a multiprocessor structure operating in a function sharing mode according to a distributed control strategy. The Universal Environment Simulator UNES provides a tool to investigate switching system performance, e.g. call handling capacity under designed load as well as under overload , or the effectivity of overload control strategies. The performance of the environment simulator itself is investigated using a queueing network model, where system characteristics like message transfer or message circulation delay caused by the simulator are discussed.

1. A UNIVERSAL ENVIRONMENT SIMULATOR CONCEPT

During the past decade, a number of digital, stored program controlled (SPC) switching systems have been developed and introduced, which replace the electromechanical system generation. This tendency can also be recognized in current developments of private automatic branch exchanges (PABX's). In accordance with advances in hardware and software methodologies, the architectural complexity of this system generation increases rapidly. As a consequence, more powerful performance investigation methods are required in order to support the system design and to ensure a proper system performance.

Fig.1 illustrates a systematical overview of various established performance evaluation methods, where the importance of system simulations can be clearly recognized. Opposite to the two model oriented and queueing system oriented simulation levels with different degrees of abstraction and complexity, the environment simulation provides the most realistic test technique for telecommunication systems.

In the literature a number of environment simulators have been presented. Most of them are designed for specific systems to be tested [5-10]. Thus, they are system-dependent and can only be applied to the dedicated systems. Other approaches [1-4] deal with more system-independent concepts; they are designed for use in simulations of a relatively small number of subscribers. Most of known environment simulators do not take into account the dependency of the subscriber behaviour on system reactions (feedback effects, e.g. repeated attempts, subscriber impatience, etc.) as well as subscriber-system interactions.

In this paper the concept of the universal environment simulator UNES will be presented. The simulator is designed for telephone switching systems for up to one thousand connected subscribers. The interconnection and the communication to an arbitrary switching system is realized by an interface which is independent of the target system. The random subscriber behaviour is described in terms of arbitrarily chosen distribution functions and includes also real-time system reactions. Therefore, realistic system loads, which model stationary as well as nonstationary overload conditions, can be generated and offered to the system for test purposes. In order to characterize and to simulate overload traffic streams, e.g. for performance investigations of overload control mechanisms, the offered traffic can be realized by means of short-term, nonstationary load patterns (c.f. [12]), whereby realistic effects like repeated attempt phenomena can be taken into account.

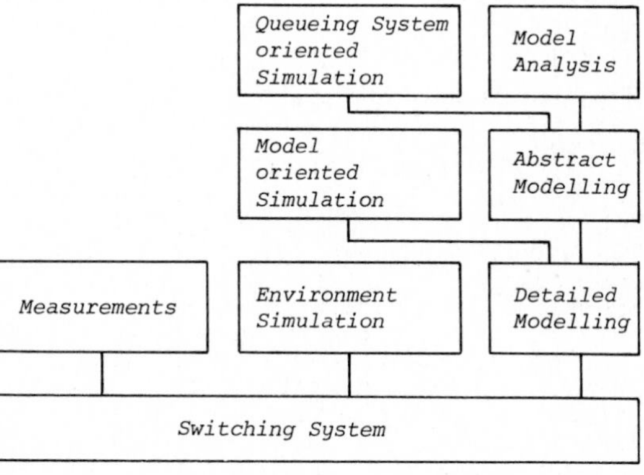

Fig. 1 Performance Investigation Methods for Switching Systems

The main features of the developed universal environment simulator will be briefly outlined. The subscriber behaviour is modelled and specified in the form of a SDL diagram (SDL: CCITT Functional Specification and Description Language) and is embedded in a multiprocessor structure environment. Thus, the subscriber behaviour model is programmable depending on the desired application. The interface to the target system, i.e. the switching system to be tested, is designed in a system-independent manner, represented by a set of telephonic events in conjunction with a messaging system.

In order to generate the random subscriber reactions, e.g. to simulate effects like dialling before dial tone conditions, incompleted dialling, call abandonments, subscriber impatience, etc., a large number of programmable distribution function types by means of a hardware random number generator are provided.

2. SYSTEM DESCRIPTION

2.1 System Overview

The functional structure and the software hierarchy of the environment simulator UNES are illustrated in Figs. 2 and 3, respectively. The simulator consists of three functional modules :

- System Control Module (SCM)
- Subscriber Behaviour Module (SBM)
- Target System Interface (TSI).

The simulator modules are implemented by means of microprocessor-based control units. Intermodule communication is done by message interchanging via the system bus.

2.1.1 System Control Module (SCM)

The System Control Module supervises the whole activities of the environment simulator. It will be distinguished between the configuration phase and the simulation phase. During the configuration phase, the environment simulator can be accessed and programmed by the user on the SCM. The configuration data to be stored can be either entered interactively through the man machine interface or loaded from mass storage or host computers. These data are subdivided into program parts for the modules TSI and SBM and tables, which are dedicated for use in the Random Number Generator (RNG), Message Transformer (MTR) and Subscriber Finite State Machines (FSM). During the simulation phase, the Simulation Control (SIC) monitors the system activities, starts and stops simulation runs by activating/deactivating the SBM and accepts messages from the SBM for measurement and statistic purposes. Messages from the SBM can be stored in a trace buffer or saved simultaneously in mass storage and host computer.

2.1.2 Subscriber Behaviour Module (SBM)

The main function of the Subscriber Behaviour Module is to model the designed number of subscribers and trunks connected to the system. It generates telephonic events (e.g., off-hook, on-hook, digits, etc.) for the simulated subscriber groups according to subscriber behaviour models and the reactions of the system to be tested.

The module controls also the signalling activities to the TSI during the simulation phase. Messages generated by a simulated subscriber are transmitted to the TSI and messages from the Target System (TAS), received via the TSI, are directed to the addressed subscriber,

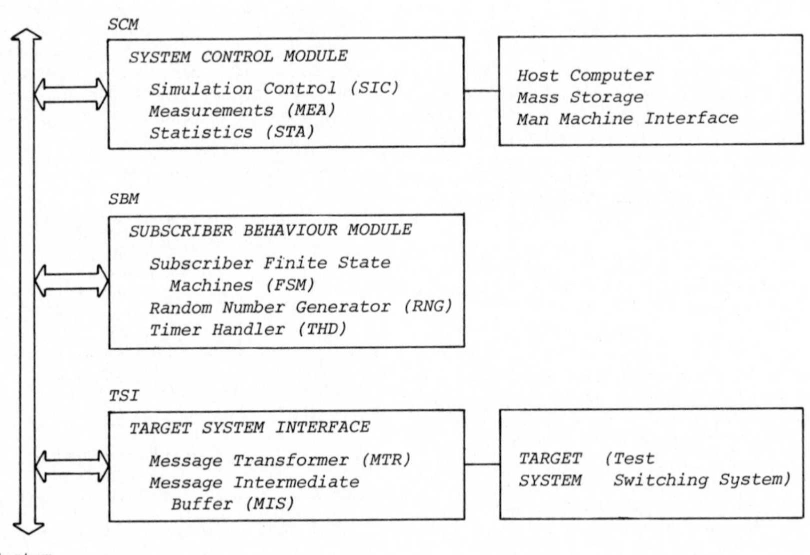

Fig. 2 *Functional Structure of the Universal Environment Simulator UNES*

e.g. to the appropriate FSM. The FSM accepts the message and stops the current subscriber-individual software timer by sending out a stop-timer message to the Timer Handler (THD). In order to model a subscriber in conjunction with his individual behavioural timing properties (impatience interval, wait for reattempt time, interdigit intervals, etc.), a software-timer is allocated to each simulated subscriber process, controlled by the THD. The subscriber process, which defines the behaviour of the subscriber in detail, is described by means of finite state machines in the form of SDL diagrams containing random branching probabilities and subscriber-oriented timer length distribution functions.

The Random Number Generator (RNG) is implemented by means of a multiplicative congruential random number generator providing uniformly distributed random variates in the interval (0,1] (c.f. [13]) in conjunction with a table-driven distribution function transformer. The transformer, which contains the desired and programmable distribution function types, inverts the uniformly distributed random numbers into required random numbers according to predefined distribution functions. Based on random numbers obtained from the random number generator, the FSM determines the action to be done. This action can be either an immediate or a delayed reaction from the subscriber to the test switching system. In the case of an immediate reaction, the corresponding message to the TSI will be sent. Otherwise, the delay is realized by starting an appropriate software-timer according to the corresponding random variable with a given distribution function. At the timeout epoch, the THD will inform the FSM by messaging. Based on the current state and on the obtained random numbers, the FSM sends a message to the TAS and plans the next subscriber action by starting the software timer. These facts will lead to decomposition and branching of messages, which will be modelled in detail in section 3.

Inside of the FSM, each subscriber or trunk is represented by an individual random-driven process. The actual state of a process is located in the individual data area of the simulated subscriber, where references to the specific data for a subscriber-type (e.g., behaviour-oriented time periods, probabilities for actions/reactions, facilities, etc.) are also stored.

2.1.3 Target System Interface (TSI)

The Target System Interface controls the communication with the Target System (TAS) connected to the environment simulator. Functionally, it consists of a Message Transformer (MTR), which transforms the message alphabets (e.g., coding and numbering of subscribers, digits, etc.) used internally in the simulator and in the test system. On the other hand, the TSI has to perform the flow control function. On the physical level the transmission of messages on the bidirectional data link is protected by parity- and timeout-mechanisms. In the case of a transmission error, the simulation is stopped and the actual state of the entire configuration is frozen. In order to enable more accurate measurements of message delay and system reaction times, the TSI is allowed to report the sending and receiving instants of messages to the system control module through the subscriber behaviour module.

2.2 Software Hierarchy

Based on the functional structure described above, the software structure of the environment simulator is built up in three levels (Fig.3). Level 1, represented by the SCM, includes the software for the overall system control and statistic evaluations. By entering the configuration phase, level 1 takes control over the whole system and initiates directly the data and program areas of levels 2 and 3, which stand for

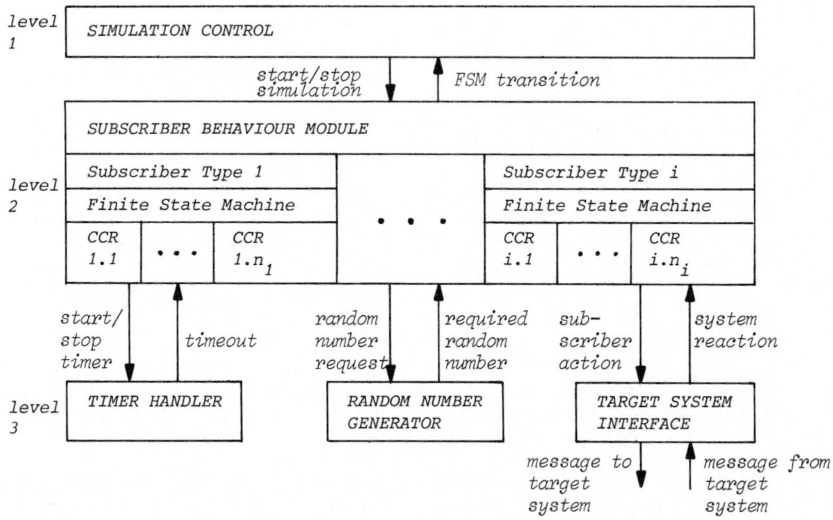

Fig. 3 Software Structure of the Universal Environment Simulator UNES (CCR : Call Control Record)

the SBM and TSI, respectively. During the configuration phase, the operating control units of SBM and TSI are deactivated. This enables level 1 to address the local program/data storage of levels 2 and 3 directly as data area.

During the simulation phase, there is no direct access to the local storage areas of levels 1, 2 and 3. By starting a simulation run, levels 2 and 3 will be activated and the interlevel communication has to be executed via message interchanging. Level 1 deactivates its overall system control, activates its statistic evaluation task and transfers the control over the system and messaging processes to level 2. Additionally, level 2 supervises the individual subscriber processes realized as Finite State Machines (FSM) in accordance with their appropriate data and program areas. The simulated subscribers are divided into groups of subscriber type (numbered from 1 to i, c.f. Fig. 3); each subscriber type represents a particular behavioural environment (telephone lines, trunk groups, subscribers with extended facilities, etc.) with their different attributes. The actual state of a subscriber and his attributes are stored in a Call Control Record (CCR). The states, their transitions and their attributes are linked together to form a universal software interface up to level 1, which is provided to simulator users.

Thus, a desired subscriber type behaviour can be programmed in terms of state transition diagrams, which consist of behaviour-dependent branching probabilities, random variables for timer periods in conjunction with programmable distribution functions. The simulation of the time-dependent and random-driven subscriber behaviour gets assistance of level 3. This level includes the Timer Handler (THD), the Random Number Generator (RNG) and the Target System Interface (TSI). For the definition of a subscriber reaction, the FSM requests and receives random numbers from the RNG. Actions initiated by subscriber processes (in terms of timeouts) are predefined by the FSM by starting the dedicated software timer. Since there is always a subscriber action planned for the future, the subscriber-individual software timers are always active. The determination of a timeout event is based on the simulation time supervised by the THD. The whole message flow between the three software levels during a simulation run is controlled by the FSM.

In order to estimate the performance of the described environment simulator, the functional modules and the messaging structure will be considered and mapped into a queueing model. The detailed model and its investigation are the subject of the next section.

3. PERFORMANCE OF THE ENVIRONMENT SIMULATOR

In order to estimate the performance and the maximum call throughput capacity of the environment simulator UNES, a detailed queueing model is developed, which will be investigated by means of computer simulations. In the model, both traffic levels, the call level and the message level (subcalls, telephonic events [11, 12]), are taken into account.

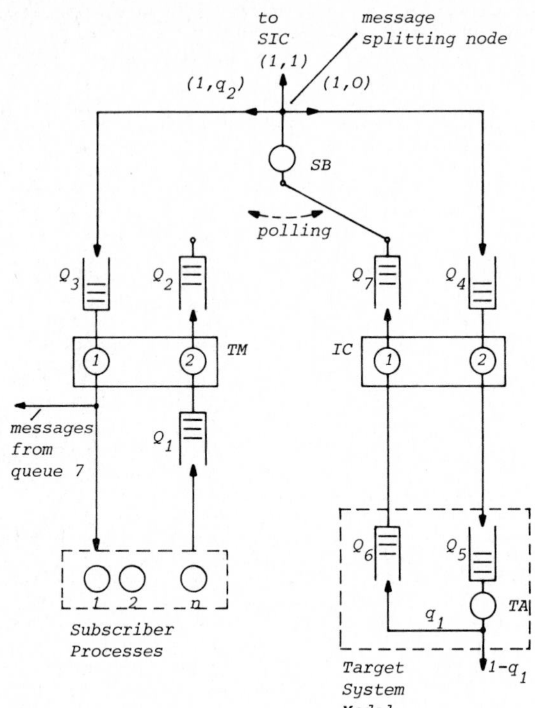

Fig. 4 *Queueing Model for the Universal Environment Simulator UNES*

3.1 Model Description

The detailed model of UNES is depicted in Fig. 4 in the form of a queueing network. In the following, the model components will be briefly described.

3.1.1 Server Stations

The server stations correspond to the functional units presented in section 2. The timer handler control unit is modelled by the server station TM (Timer Manager), which is connected to the multi-server group of subscriber processes. This server group consisting of n servers stands for the subscriber-individual software timers as described in section 2.1.2. The departure process of this server group forms the timeout message traffic, which initiates subscriber activities.

The server station SB represents the message handling activities corresponding to the state transitions of the subscriber behaviour in the FSM.

The server IC models the control unit of the TSI module, which supervises the transfer protocol of telephonic messages to the target system as well as the intermodule communication towards the SBM.

The reaction time of the test system is approximately described by means of the server TA.

3.1.2 Traffic Sources

Since the queueing system is a closed queueing network, there exist no external traffic sources. However, due to the splitting of messages after being served in the subscriber behaviour module, message generation in the sense of a branching process is taken into account.

At the message splitting node (c.f. Fig. 4), depending on the origination i (i=1,2 for queues 2,7 respectively) and the destination j (j=1,2,3 for server stations TM, SIC and IC), a message coming from i will generate a group of messages of size g_{ij} routed to the destination j with the probability q_{ij} according to the group size matrix

$$\underline{G} = \{g_{ij}\} = \begin{pmatrix} 1 & 1 & 1 \\ 2 & 2 & 0 \end{pmatrix} \qquad (3.1)$$

and the message branching matrix

$$\underline{Q} = \{q_{ij}\} = \begin{pmatrix} 1 & 1 & 1 \\ q_2 & 1 & 0 \end{pmatrix}. \qquad (3.2)$$

Considering the message traffic generated by subscribers during the set-up phase, the conversation phase and the releasing phase of a call, the message traffic generated by a particular subscriber is modelled by means of an interrupted Poisson process (IPP, c.f. [14]); the "on"-phase of the process corresponds to the call set-up phase, the "off"-phase stands for the conversation and the idle phases of the considered subscriber. Thus, the intermessage intervals of a subscriber, which are approximated by the timer period lengths, is described in the model by the random variable T_{SUB}. In general, T_{SUB} is assumed to be distributed in accordance with the interarrival time distribution of a generalized interrupted Poisson process [14]. By the choice of Markovian "on"- and "off"-phases for the IPP process, T_{SUB} follows a 2nd-order hyperexponential distribution function.

3.1.3 Traffic Flows

Timeout messages, which represent subscriber-oriented telephonic events to be handled with, are offered to queue 1 and subsequently served by the timer manager. After service in the TM station, messages will wait in queue 2 for the next polling instant initiated by the subscriber behaviour control unit SB. Reaching the message splitting node, the observed message originating from queue 2 will branch into three messages, one for each direction towards the TM, SIC and IC servers, as described with the matrix notation above.

The subsequent message offered to queue 3 corresponds to another start-timer message, while the message representing the telephonic event to be transmitted to the target system will wait for transmission in queue 4. Since every activity inside level 2 has to be reported to the SIC, a subsequent message must be generated and sent to the module SCM.

A message received from the target system will be processed in the server station IC (Interface Control Unit) and subsequently waits to be polled in queue 7. Based on the current state of the addressed subscriber process, the branching procedure will take place after service in SB, according to eqns. (3.1) and (3.2).

Messages with origination queue 7 and destination server TM will leave the system after service in TM. Messages sent to the simulation control module will not further affect the model.

3.1.4 Server Scheduling

The server stations operate according to the following schedules :

i) Server TM
While queue 1 is served exhaustively in a first-in, first-out order, a nonexhaustive service is implemented for queue 3.

ii) Server SB
A cyclic nonexhaustive service in conjunction with a polling mechanism is applied.

iii) Server IC
Nonexhaustive servive is implemented for incoming messages (queue 6) and exhaustive service is designed for the outgoing direction (queue 4).

3.1.5 Target System Modelling

The target system, i.e. the switching system to be tested is modelled by means of a single server station with the service time T_{TAR}, which can be thought of as the reaction time of the test system. After being processed in the switching system, a system reaction upon a message will be created and sent back to the simulator with the probability q_1.

The Target System Model can be changed to adequately model specific target system structures, e.g., using an infinite server model, or a more complex queueing network.

3.2 Results and Discussion

Simulation is provided to determine the delays of messages whereby the following delays are considered (c.f. Fig. 4)

T_{IN} Input delay :
Receiving and recognizing delay of input messages, i.e. system reactions, caused by the message processing structure in UNES. This delay is accounted from the receive instant at IC until the departure time in server SB.

T_{OUT} Output delay :
Sending delay for messages to be transmitted to the target system. This delay is accounted from the departure instant in the subscriber process until the completion instant of the transmission at server IC.

T_{CIR} Circulation delay :
Time interval from the timeout occurance of a subscriber process until the next timer of the considered process is activated. This turn-around time is measured by observing a message from departure instant in the subscriber process server group (timeout), via the following model components: queue 1, server TM, queue 2, server SB, queue 3, server TM.

3.2.1 Simulation Parameters

In order to estimate the performance limitations of UNES, the parameters of the model shown in Fig. 4 will be chosen in the following for worst-case considerations. Thus, all service phases, which are likely to have coefficient of variations less than unity (i.e. rather have deterministic or hypoexponential distribution), are assumed to be of Markovian type with means of one millisecond. In the simulative investigations presented below, the probabilities are chosen as q_1=0.5 and q_2=0.75. Due to observations done in test systems, the target system reaction time is modelled to be negative exponentially distributed with mean 50msec.

The random variable T_{SUB} for server processes follows a hyperexponential distribution of 2nd order. In conjunction with the standard balancing equation for this type of distribution function, T_{SUB} is determined by the mean $E[T_{SUB}]$ and the coefficient of variation $c[T_{SUB}]$.

Simulation results will be depicted in the following with their 95% confidence intervals.

3.2.2 Results

In Figs. 5 and 6 the mean values of input, output and circulation delays are depicted as functions of the number n of simulated subscribers and the subscriber traffic intensity $\alpha_{SUB} = 1/E[T_{SUB}]$, respectively. In Fig. 5 it can be clearly seen that for low subscriber traffic intensities, say $\alpha_{SUB} < .2$/sec, no queueing delays have been registered. The expected values of delays, especially the circulation delay, increase rapidly at higher subscriber traffic range, dependent on the number of simulated subscribers. The rapid increase of the circulation delay is caused by the waiting time in queue 3 (c.f. Fig. 4) in conjunction with the scheduling of server TM and the branching of additionally superposed messages from the target system. It should be noted here that the characteristics shown in Fig.5 correspond to worst-case parameters; the subscriber traffic intensities to be simulated are normally given at α_{SUB}= .01/sec ($E[T_{SUB}]$ = 100sec).

The dependency of average delays on the number of simulated subscribers is illustrated in Fig. 6, where the assumption of an extremely high subscriber traffic intensity is made ($E[T_{SUB}]$ = 4sec, i.e. one message per 4 seconds from each subscriber on average, including his conversation and idle times). It can be recognized in this diagram, that even with this assumption, the delay characteristics of the environment simulator are in a reasonable range, which has no strong influence on the simulation accuracy, for up to one thousand simulated subscribers.

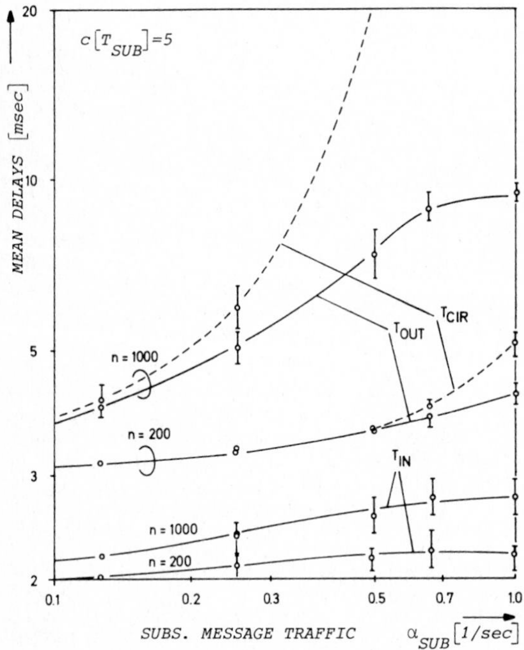

Fig. 5 *Mean Delays vs Subscriber Message Traffic Intensity*

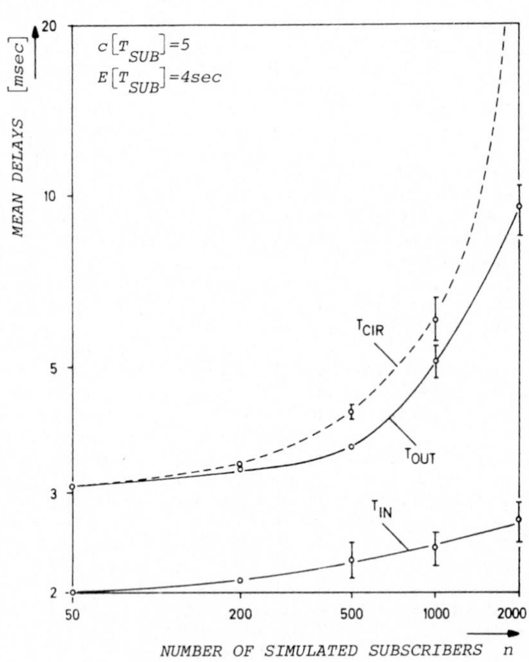

Fig. 6 *Mean Delays vs Number of Simulated Subscribers*

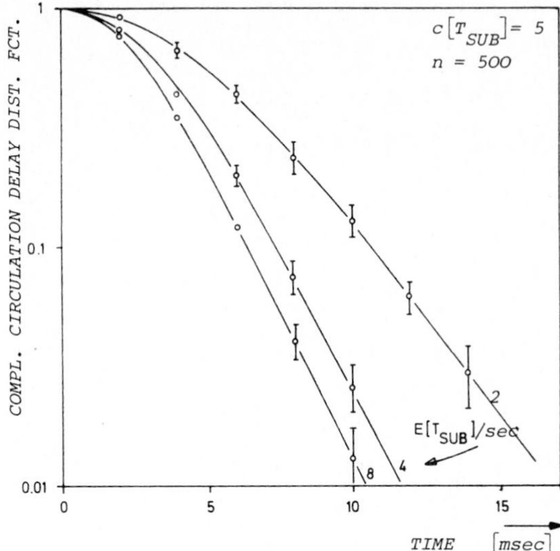

Fig. 7 *Complementary Circulation Delay Distribution Function*

Fig. 7 depicts the complementary distribution function of the message circulation time for different values of $E[T_{SUB}]$, where the influence of the subscriber traffic intensity on the delay characteristics is illustrated.

4. CONCLUSIONS AND OUTLOOK

In this paper, the concept as well as implementation and performance aspects of the universal environment simulator UNES have been presented, which is developed for telephone switching systems. The simulator provides a universal performance evaluation tool for switching systems, where realistic phenomena, which strongly affect the switching system performance, like subscriber impatience or repeated attempts can be considered and investigated.

The environment simulator allows us to investigate switching system performance under designed load and overload conditions. Furthermore, it can be applied to evaluate the effectivity of overload control strategies in switching systems, whereby detailed modelling of subscriber behaviour including considerations of the system feedback is arbitrarily programmable and adaptable. The performance of the simulator is investigated by means of a queueing network model. By means of the investigation, performance measures like the limiting load generation capability under a given message traffic characteristic and message delays are obtained. The concept can be extended to model subscribers operating with new services, which will be provided in current and future system developments, e.g., in ISDN-featured systems (ISDN: integrated services digital network).

ACKNOWLEDGEMENT

The authors would like to express their thanks to Prof.P.J. Kuehn for his interest in this project, to J. Saegebarth (University of Dortmund) for helpful discussions on random number generation, to R. Lehnert (TeKaDe - Philips Kommunikations Industrie, Nuremberg) for stimulating discussions on test requirements for switching systems and to T. Raith for helps with simulations.

REFERENCES

[1] Johner W., "LCS - ein Anrufsimulator für Ortsvermittlungsstellen", Elektrisches Nachrichtenwesen 55(1980)3, 217-220.

[2] Metzger R.M., Staber E., "Anrufsimulator UCS für die Prüfung von Fernleitungsausrüstungen",Elektrisches Nachrichtenwesen 55(1980)3, 210-216.

[3] Dael G., Amarger D., Marrois C., "AROMAT Telephone Call Monitoring and Recording Equipment", Comm. & Transm. 1983, pp. 39-52.

[4] Jansson B.,Johansson K.,Ostlund T.,"LPB 110, A System for Controlled Test Traffic", Ericsson Review No.1 1984.

[5] Haugk G., Tsiang S.H., Zimmerman L., "System Testing of the No. 1 Electronic Switching System",Bell Syst.Tech.J. (1964)9,2575-2593.

[6] Gruszecki M., Cornelis F., "Anwendung des Umweltsimulators zur Bestimmung der Leistungsfähigkeit von rechnergesteuerten Vermittlungssystemen",Elektrisches Nachrichtenwesen 51(1976)2, 119-123.

[7] Foucault C.C., Cerny D.J., Ponce de Leon L., "TCS-Umweltsimulation", Elektrisches Nachrichtenwesen 48(1973)4, 461-465.

[8] Del Coso Lambreabe M.A., "Umweltsimulation für Echtzeit-Software", Elektrisches Nachrichtenwesen 54(1979)2, 132-138.

[9] Grantges R.F., Sinowitz N.R., "NEASIM: A General-Purpose Computer Simulation Program for Load-Loss Analysis of Multistage Central Office Switching Networks",Bell Syst.Tech.J. (1964)5, 965-1005.

[10] Gruszecki M., "ENTRASIM: Real Time Traffic Environment Simulator for SPC Switching Systems", Proc. 8th ITC, Melbourne 1976.

[11] Dietrich G., Salade R., "Subcall-Type Controlled Simulation of SPC Switching Systems" Proc. 8th ITC, Melbourne 1976.

[12] Tran-Gia P., "Subcall-oriented Modelling of Overload Control in SPC Switching Systems", Proc. 10th ITC, Montreal 1983, pp.5.1.1.

[13] Ide H.D., Sagebarth J., "On Properties of Random Number Generators and their Influence on Traffic Simulation",Proc. 10th ITC, Montreal 1983, pp.2.4.6.

[14] Tran-Gia P.,"A Renewal Approximation for the Generalized Switched Poisson Process", Proc. Int. Workshop on Applied Math. and Perf/Rel. Models of Comp./Comm. Systems, Pisa, Italy 1983 (North-Holland 1984) 167-179.

TELETRAFFIC ISSUES in an Advanced Information Society
ITC-11
Minoru Akiyama (Editor)
Elsevier Science Publishers B.V. (North-Holland)

OVERLOAD CONTROL IN A HIERARCHICAL SWITCHING SYSTEM

David MANFIELD, Bob DENIS, Kalyan BASU, Guy ROULEAU

BNR
Ottawa, Canada

ABSTRACT

Large digital switching systems have distributed call processing implemented in a hierarchical architecture with high-level call processing handled by a central processing facility, and low-level functions handled by a large number of peripheral processors. In a large digital switch without overload controls the loss of throughput when switch real-time attempt capacity is exceeded is catastrophic, due to the extremely steep load-service relations at the point of switch capacity. In this paper the requirements for overload controls in a distributed switch are investigated through the use of a novel, distributed, call-based simulator which is modeled on the actual switch architecture. The performance of a successful set of overload controls is demonstrated, and rationalized in terms of the differing functional requirements for overload controls at each level of the call processing hierarchy.

1. INTRODUCTION

Most modern digital switching systems have distributed processing of telephone calls implemented through a hierarchical switch architecture. High-level call processing functions requiring centralized resources are handled in a central processor (e.g., network path selection, service circuit allocation, translation, routing). Low-level call processing functions are handled by a large number of peripheral units, each one in control of a number of lines or trunks. The low-level call control functions include idle line scan, DP digit collection, line supervision, tones and ringing. The two levels of the control hierarchy communicate by means of messages to synchronize the concurrent real-time call processing. Both the peripheral controller (PC) and the central control (CC) maintain state machines for a particular call.

The switch real-time processing capacity determines the maximum attempt load and throughput that the switch can handle. The determination of this capacity is a complex problem in itself, and we do not address it here. Suffice it to say that the real-time capacity of all the switch processing elements needs to be considered. In the work reported here it is assumed that it is the CC real-time which is the limiting system resource, because all calls are funneled through the CC from all the PC.

From the engineering point of view, the dial tone delay characteristic of the CC is a very steep function of the attempt load when the switch is operating near the point of capacity. This steep load-service relation is almost purely a result of the large switch size and not of the software architecture. The large switch size makes it particularly susceptible to overloads, due to the destructive interaction between the subscriber abandon behavior and the sudden transition to high dial tone delay which occurs just beyond switch capacity. Rapid degradation in throughput characterized by low call completion rates results. This effect has already been described in the literature, e.g., [1]. What distinguishes large switches is the speed at which throughput drops beyond the switch capacity limit.

In comparison with the overload controls developed for analog SPC systems [1-5], the controls we discuss here in relation to distributed digital SPC switches are significantly different. Analog SPC switches historically have a monolithic, single-processor control. This processor has direct control over the peripheral line scanning, and therefore overload controls can be implemented by switch-wide counting mechanisms to regulate the rate of accepted new originations, based on feedback of some direct real-time indicator. In contrast to this, a distributed digital switch does not have direct control over the periphery from the central control. The challenge is to construct controls for all the call processing elements in the control hierarchy which effectively cooperate to produce globally optimal overload response.

In this paper we address the requirements and problems related to the design of a distributed overload control mechanism for a large digital local (Class 5) office. The contribution of this paper is the definition of a successful set of controls, and their performance evaluation through the use of a novel quasi-distributed overload simulator which is modeled on the actual switch architecture. In section 2 the general requirements for a distributed overload control are discussed. In Section 3 some details of the switch operation are given and in Section 4 the design details of the distributed simulator are given. The performance results and discussion are contained in Section 5.

2. OVERLOAD CONTROL REQUIREMENTS

While there may be instances where it is possible to exploit certain features of a particular system architecture to implement overload controls, there are nonetheless implementation principles which remain invariant. We focus on the principles as they pertain to distributed switching systems.

The primary requirement is for separate overload controls at both the PC and CC levels of call control. Although the nature of the controls at each level is different, they must work together to protect the global switch performance. This goal of cooperativeness is not necessarily easy to achieve, given the speed of response required, and the possibility of only some peripherals being overloaded.

The CC controls must protect the CC real-time, which is the primary central resource. This must be done by limiting the rate at which new work enters the CC and ultimately leads to the discarding of calls. The key requirements for the CC controls is that they consume minimum real time to execute. Comparison of some alternatives is done in Section 6. Certain switch architectures [6, 7] have an inherent advantage in the priority queue structure supported by the CC scheduler, in that origination work is kept separate and hence directly controllable under overload. In general, however, there is an implicit requirement that once an origination event is processed by the CC, the call should be guaranteed for completion, excepting, of course, in cases of speech path congestion in the network, peripherals or outgoing trunks where treatment is required.

The role of the peripheral overload controls is quite different, and must be complementary to the CC controls. The PC is not a global resource, and for the purposes of our work here we assume that PC real time is not an issue. The requirement for the PC is to dynamically throttle its load to the CC while at the same time preserving the integrity of its lines which may have had calls subject to CC overload control action (discard, or long delay). If a call is discarded or suffers long delay in the CC, the line must not be prevented from making subsequent attempts i.e., the line should not be left "high and dry". The grade of service offered by the switch under overload will not be good, but this is unavoidable regardless of architecture. There is however a key requirement that the subscriber who goes off-hook and waits long enough should eventually receive dial tone.

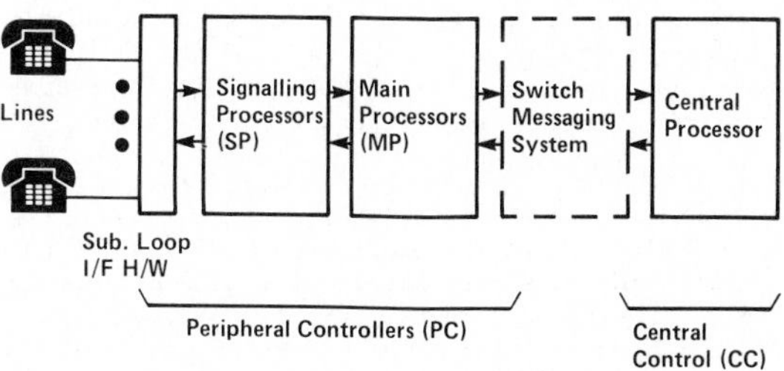

Figure 3.1 Functional Processing Elements

Looking at the switch as a whole, the requirements for the global controls are to maintain switch throughput during short-term load peaks by smoothing the load without loss of calls (though it is possible that some calls will experience delay), and to maintain continuity of service during sustained overloads when normal grades of service cannot be met. In a distributed switch there is always the possibility of only parts of the system being overloaded, in which case the controls should not be active unless the integrity of central resources is threatened. This means that spare resources should be automatically directed at the overloaded parts of the system up to the point where the global response begins to deteriorate.

3. SYSTEM DESCRIPTION

3.1 Overview

The hardware and software details of a particular switch architecture are very complex, and many of the details are not relevant to the study of overload performance. Therefore we give here only a partial representation to illustrate the principles of hierarchical control and to allow us to concentrate at the system level on the operational characteristics of the components whose performance have a direct impact on the overload behavior. The system overview concentrating on the real-time functional blocks is depicted in Figure 3.1.

The subscriber loop hardware interface presents the true state of the line. The PC is made up of a group of separate functional processors which communicate via messages. Line state changes and DP signaling information are detected by the loop interface circuitry which is scanned by the Signalling Processor (SP), whose functions are line signaling and supervision. Signalling information and call events are passed to the Main Processor (MP) which maintains the individual call state machines and implements the peripheral call processing logic under overall control from the CC. The CC maintains the central call state machines and is responsible for all the high-level call processing such as receiver allocation, digit translation and routing. Both the PC(MP) and the CC maintain a set of priority scheduling queues for the execution of call-processing and non-call-processing tasks.

The key elements of the overload controls are in the MP in the periphery and the CC, corresponding to the major elements in the call processing hierarchy. The queuing structure for processing line call originations is depicted in Figure 3.2. The SP periodically scans all idle lines at a nominal rate of once every 200 ms and reports switch-hook state changes to the MP via messages placed in scan report queue buffers in common memory. The MP checks the report queue periodically for incoming messages and when it is able, processes these messages according to the local state machine (seizing whatever resources are necessary) and sends off a message to the CC via the switch messaging system which is not explicitly depicted in Figure 3.2. Messaging delays between the PC and CC are very short

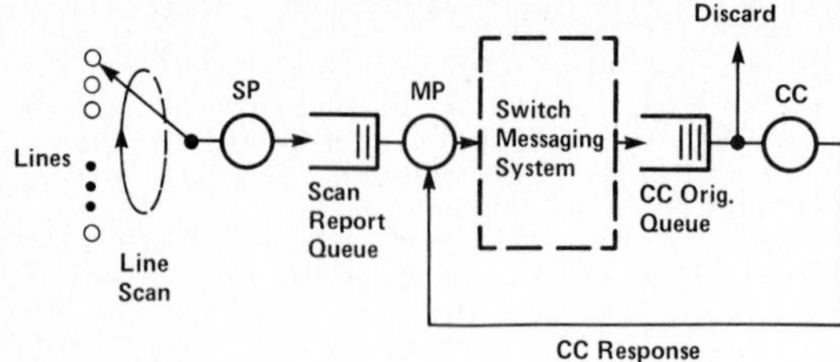

Figure 3.2 Origination Queueing and Processing

compared to other delays in the system and are hence not a concern for overload control. For every message sent by the PC, the CC (normally) provides a response which is used to drive the peripheral call state machine.

3.2 Central Control

The central control real-time is the primary limiting system resource governing the attempt capacity. The important features of the CC call processing with respect to overload performance are the following.

Firstly, call progress messages are processed with higher priority than call origination messages. This ensures that calls in set-up and disconnect phases are protected because this work is done in preference to the processing of origination events which represent new work in the system [6, 7]. This has been demonstrated both by simulation and by field studies to be an inherently stable arrangement for call processing.

Secondly, to avoid long CC origination queue delays which would otherwise result under overload conditions, there is a CC mechanism for discarding originations corresponding to abandoned calls. This can be done before any execution of the call processing code, based on a delay threshold in the origination queue, at virtually zero real-time cost when the origination is taken off the head of the origination queue. The indirect effect of this control is to limit the length and hence the delays in the CC origination queue. The design goal is to prevent wasting real-time on ineffective subscriber attempts.

3.3 Peripheral Controls

The SP scans each line periodically and off-hook events result in an origination message being written into a buffer on the scan report queue. If no buffer is available, then the state change is ignored and will be picked up again the next time that the line is scanned (if the line is still off-hook). If the line abandons before it is scanned then the attempt is never seen by the system. The SP then waits for the MP response to the origination.

The scan report queue is checked at closely-spaced intervals by the MP for incoming messages. When an origination message is processed, the line state information is updated and a message prepared for sending to the CC. At the same time a timer is started to signal a timeout if no CC response to this message is received. While the timer is active, the MP sends no more origination phase messages to the CC, representing a one-at-a-time flow control of events from lines in the origination phase. The CC response is recognized by the terminal number contained in each call processing message, and if the response is for the terminal which last sent a message to the CC, then the MP timer is reset to allow the MP to send further messages to the CC if there are any.

The peripheral controls are characterized by two parameters, namely the number of scan report buffers, and the timeout associated with the one-at-a-time flow control for origination phase messaging.

There is one more critical component of the peripheral controls which is related to the maintenance of line integrity. Knowing that it is possible for the CC to discard calls under the action of its own overload controls, it is necessary for the PC to recover lines that have had calls discarded so that further attempts can be made by them. In normal operation, the PC depends upon receiving a CC response for every message that it sends and, under overload, such a response may not be forthcoming, intentionally or not. Further, it is necessary to guarantee that the patient subscriber eventually receives dial tone regardless of the degree of overload. To accomplish these dual purposes, the PC has the facility whereby when a timeout for the CC response to an origination phase message occurs, the line state is idled to allow a system-generated reorigination. This aspect of the implementation is critical for preserving line integrity between the PC and CC and is very important in the robustness of the global overload controls.

4. SIMULATOR DESCRIPTION

4.1 Overview

The principal challenge for an overload control simulator is one of time scale. It is necessary to both model the CC scheduling operations which take place on a time scale of microseconds, and it is also necessary to model the subscriber call traffic process which is on the order of an average call holding time (100-200 seconds). For overload performance investigation a call-based simulator is required, because the relationship between successive subcalls of a single call [8] needs to be preserved in order to model the process of call abandon and reattempts which depends on the instantaneous switch response, and which is at the heart of the system overload performance. Sub-call simulation is not possible because the causal link between subscriber events and switch delays is broken.

The switch simulator is modular and has a hierarchical structure based directly on the architecture of the switching system. The structure is depicted in Figure 4.1 and has three major modular components for simulating respectively the central control, the line controllers and the trunk controllers. Although the objective is to study line traffic overloads, the trunk traffic must be simulated in order to provide a realistic traffic environment.

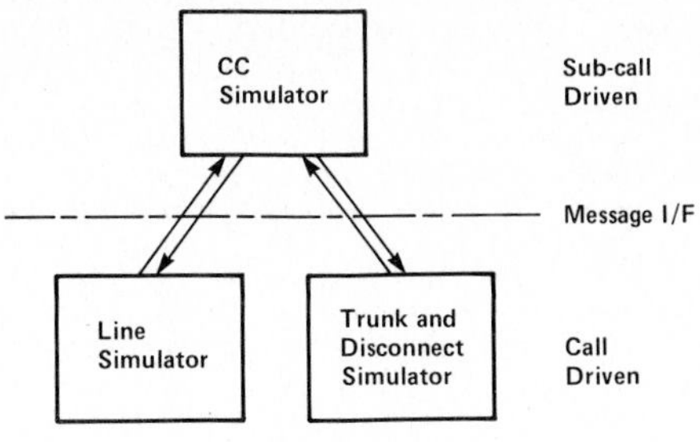

Figure 4.1 Simulation Structure

4.2 CC Simulator

The CC simulator itself resembles a subcall simulator in that it is driven by messages from the peripheral simulators, representing the various call events. The peripheral simulators for line and trunk peripherals are call-based (at least for the set-up phases). The peripheral simulators generate subscriber events, do the peripheral event processing, and process the CC responses. The subscriber behavior is integrated into the peripheral simulators. In a more clean, but not necessarily more efficient, design it may be desirable to have the subscriber models separate.

The call-based simulator preserves the timing information between subcalls (events) of the same call. The subcall sequence for a line call is depicted in Figure 4.2. Purely from the point of view of simulation run time and storage requirements it is not possible to explicitly simulate the entire length of each call. To do so would involve keeping track of the order of at least twenty thousand simultaneously active calls, most of which would be in the conversation phase. It is much simpler to simulate the disconnect subcalls as a stream of independent events, governed only by a current count of the number of calls in the conversation phase. This reduces the simulation warm-up time and storage requirements by an order of magnitude. Subcall sequences are only preserved during the call set-up phases.

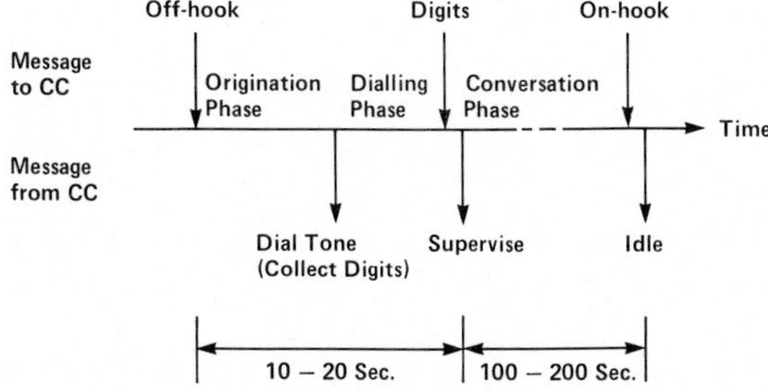

Figure 4.2 Completed Call Phases (Line Originator)

4.3 Peripheral Simulators

The major simulation problem, however, remains the difference between the CC and PC time scales, and the event-list management problems which result. The solution used here was to implement a hierarchy of simulation clocks, all driven from the central CC simulator. The range of values is depicted in Table 4.1.

MODULE	CLOCK PERIOD
CC Simulator	50-100 μs
Line Simulator	0.1-0.2 s
Trunk and Disc Simulator	0.5-1.0 s

Table 4.1: Distributed Clock Periods

The coarseness of each of these clocks represents the relative accuracy required to acceptably model the traffic processes in each of the simulation modules. The CC subcall simulator therefore needs to have only one more event type added to its event list, namely a "peripheral processing" event which results in the line simulator being invoked. Periodically, this event also results in the trunk and disconnect simulator being invoked. This approach limits the simulation horizon of the CC simulator, and the number of future peripheral events to one, excluding incoming call processing messages. Huge amounts of processing time are saved this way, because it avoids the need to sort long event lists repeatedly.

This finite simulation horizon method has some implications for the peripheral call processing. It is not acceptable to have peripheral events generated in batches at the peripheral clock epochs, resulting in a periodic batch input to the CC simulator. Rather, at a peripheral clock epoch, the events generated are spread over the succeeding clock interval according to a Poisson process. This is easy to do because the order statistics of a unformly-distributed random variable are exponentially distributed.

Each call in the set-up phase is characterized by a unique sequence number, and a list of call attributes which define the state of the call. The principal attributes are:

- call Type (line-to-line, etc.)
- call disposition (completed, uncompleted, treatment, etc.)
- DP/DT (lines) or DP/MF (trunks)
- phase of call (origination, dialing, etc.)

In addition the peripheral simulator maintains, via the call attribute vectors, three state machines for each call while in the origination phase. These are respectively the actual line state (on-hook, off-hook), the state of the origination event, and the state of the abandon event. Since the call processing is distributed, each of the CC and PC processors may have a different view of the line state, and the actual line state may be different again. Each active call is checked at the epochs of the peripheral clock to test for call progressions (e.g., abandon), depending on its current state. If the state is the change, the peripheral state machines are updated and (possibly) a message is generated to be sent to the CC and is allocated a time in the next clock interval using the method described previously.

4.4 Subscriber Behaviour Model

Subscriber response to switch delay is characterized by abandoned calls and successive reattempts. When the dial tone delay is protracted there is a marked rise in the number of calls which abandon before dial tone, and calls where dialing commences before dial tone, resulting in a partial-dial abandon [1]. From the CC point of view, a partial-dial abandon due to dial tone delay looks the same as an abandon before dial tone, since in each case it receives the same messages, namely an origination followed by an abandon. (A digits message only goes to the CC when a full set has been collected.) The only difference between the two types of abandon is that the partial-dial abandon occurs a little later than the false-start abandon.

The subscriber behavior model for line calls was based on the following parameters:

- Mean time to abandon due to DTD — 5 s (Const.+NED)
- Probability of reattempt after abandon — 0.8 (Geometric)
- Mean wait before reattempt — 30 s (NED)

The characterization for incoming (trunk) calls was based on the following parameters:

- Mean wait for Start-to-Dial signal — 5 s (Constant)
- Probability of reattempt — 1.0 First reattempt; 0.0 Subsequent reattempts

These numbers are based in part on results reported in [7, 9, 10].

When considering subscriber behavior, recall that there is a residual level of abandoned calls even when the switch dial tone delay is so small as to be unnoticeable. These abandons are due to subscriber action independent of switch response and are modeled differently. In the framework of the peripheral simulator presented here, this residual level of abandoned calls was modeled by a separate class of call disposition (uncompleted) in the call attribute vectors.

5. RESULTS AND DISCUSSION

The results shown in this stion are some of the final results coming from the investigation of a number of different alternatives for the switch overload controls. The objective here is not only to demonstrate the effectiveness of the final design, via the simulation vehicle outlined in Section 4, but also to look at the reasons for its success as compared to other design alternatives.

The overload performance for a large switch without overload control is qualitatively presented in [1] and we do not reproduce it here, except to note the almost vertical drop in throughput when the point of switch capacity is exceeded. This behavior is characteristic of any large switch.

Turning now to the introduction of overload controls, the major results are depicted in Figure 5.1, where normalized switch throughput is plotted as a function of the normalized switch offered load (nominal). The normalized offered load of 1.0 corresponds to the switch CC real-time attempt capacity, which depends primarily on the call mix. (For the overload studies discussed here a typical Class 5 office call mix was used, but the results are in no way peculiar to the call mix used). The parameter for the family of throughput curves given in Fig. 5.1 is the number of message buffers in the PC scan report queue (recall Fig. 3.2). When the repeat queue is allowed to be long, say 15, then the

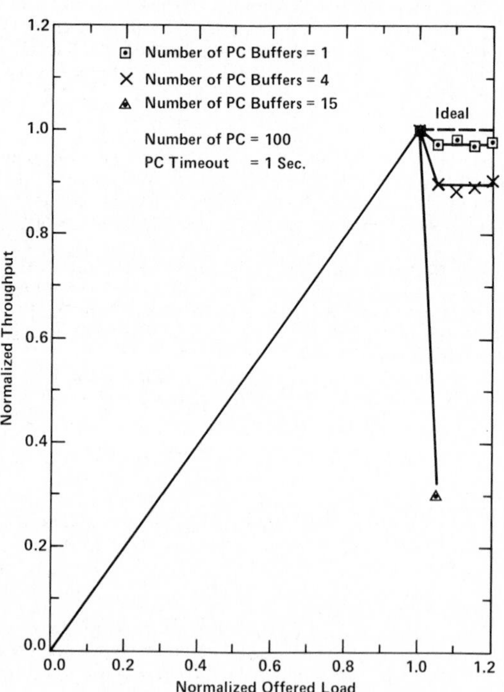

Figure 5.1 Throughput vs Offered Load

overload performance is very poor and is indistinguishable from the switch performance without any overload controls at all. As the number of report buffers is reduced, so the overload performance improves, until with a single buffer we reach near-optimal throughput performance which is stable and sustainable regardless of the degree of overload.

The scan report queue functions solely as a buffer between the SP idle-line scanning and the MP peripheral call processing. If delays in this queue are large then the probability of call abandon before the origination can be processed by the MP is very high. This means that either an abandon message soon follows, or the origination is processed by the CC only to find the line state on-hook when the CC response reaches the MP. This results in much wasted time in the MP and the CC. Under overload the report queue is almost always full, and is emptied FIFO by the MP at a rate which is largely determined by the MP timeout, since the CC response will be slow. For the sake of illustration, if the report queue is 15 long, and the timeout is 1 sond, an origination has a wait of the order of 15 sonds to be processed and sent to the CC, and in this time the call has almost surely been abandoned. By choosing the report queue to be the minimal size of one, the PC delays are kept short so that when an origination is processed it has a good chance of getting dial tone before abandon.

This does not mean that subscribers get short dial tone delay, since the effect of the control is to keep the excess call load outside the system at the line scan level. Many attempts will not get dial tone before abandon, but this is unavoidable. Although dial tone delay is long with overload controls, it is still shorter than the delay without controls, as can be seen in Figure 5.2.

The size of the report queue alone is not sufficient to give rise to the system stability under overload. The stability results from the peripheral reorigination facility. When the CC slows down under a CC overload, the MPs begin to time out waiting for responses to origination phase messages, leading to the generation of system reoriginations. These reoriginations have the effect of tying up the MP-CC messaging via the MP timeout, and throttling the load to the CC. In other words, if an origination ties up the MP for one sond, a single reorigination means that the MP is tied up a total of two sonds for the same call, thus reducing the effective MP attempt rate. Just enough reoriginations are generated to balance the CC capacity, constituting an automatic negative feedback control between CC and PCs triggered by the CC origination delay. In this way the switch throughput is maintained independently of the magnitude of the overload.

The secondary PC/MP parameter is the value of the timeout for CC response to an origination phase message. The impact of the timeout value is depicted in Figure 5.3, where the switch throughput is plotted against MP timeout, for a fixed nominal normalized offered load of 1.15 (15% overload). Throughput is plotted both with and without overload controls. Without controls there is a point at which the timeout alone is sufficient to throttle the load enough to deload

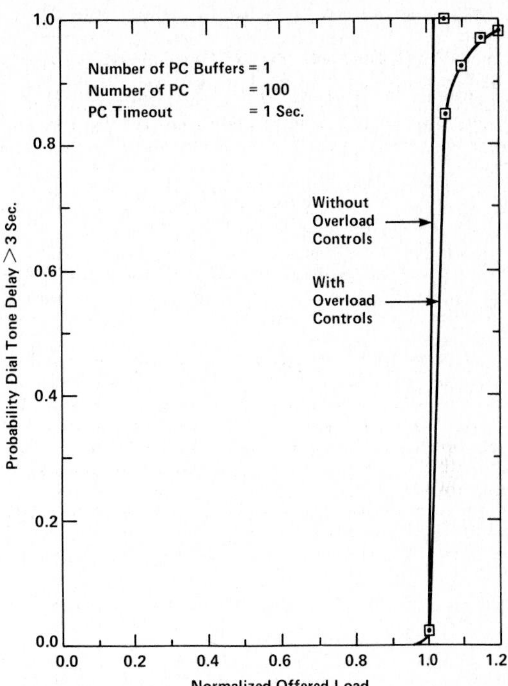

Figure 5.2 Probability Dial Tone Delay > 3 Sec. vs Offered Load

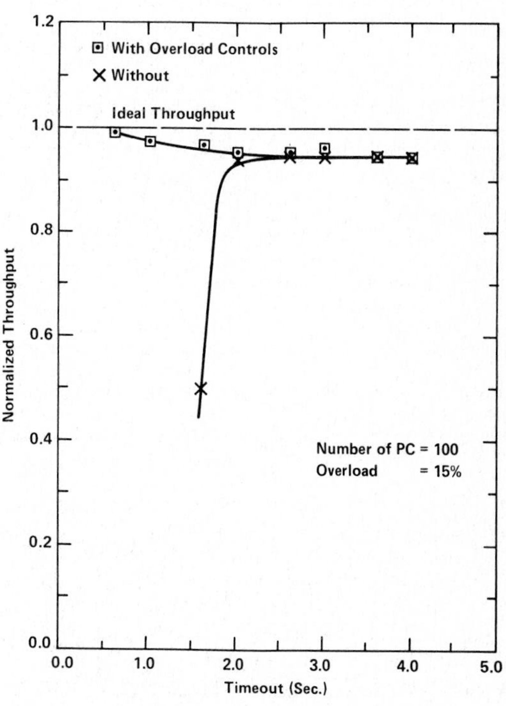

Figure 5.3 Throughput vs PC Timeout

the CC and restore throughput. The location of this point depends on the switch configuration and call mix. Below the critical timeout value, CC delays are very high and the call completion rate very low. With the overload controls in place the picture changes dramatically, and we see that the throughput does not vary much with timeout value. This is because the negative-feedback overload control is self-tuning and settles to a near-ideal operating point. In short, when the timeout is chosen to be short, more system reoriginations are generated to throttle the MP. When the timeout is long, fewer reoriginations are generated to achieve virtually the same operating point.

There is an important conclusion to be drawn from the above arguments, and it is that regardless of the number of PCs terminated on the switch and the traffic offered by each PC, regardless of the call mix, regardless of the office real-time capacity, the CC throughput is stable under overload. This is true because the controls are inherently adaptive. It means that there are no overload control parameters that need to be "tuned" for each switch individually, which can be seen not to be the case for overload control mechanisms in some other switching systems.

A side benefit of the controls is that by protecting the CC from line overloads, the switch response times for incoming trunk calls can be maintained, given that the incoming load itself is not presenting an overload to the switch. Trunk overload controls are a separate issue and we do not discuss them further here.

There has been recent work reported on the use of LIFO service strategies for peripheral origination queues as an overload control measure [1, 11]. The principle behind this idea is that the most recent origination should be reported first because it has the best chance of success. This strategy was tried in the overload simulator for the PC report queue, but was not successful. While the throughput response was significantly better than that without controls, the throughput did not remain stable under overload and dropped off with increasing load. In any case, for the successfully implemented controls with a single report buffer, the queue service discipline (FIFO or LIFO) becomes irrelevant.

6. CONCLUSIONS

The system features and simulation testing of the overload performance for a distributed, hierarchical switching system have been discussed. The key features which lead to good overload performance have been identified as:

- Priority processing of call progress work in the Central Control
- The strict limitation on scan report queue length in the line peripherals
- The automatic system reorigination mechanism

These factors together lead to a set of cooperative overload controls whose stability is independent of office configuration and call mix. The overload controls have already been introduced into offices in the field, and although no office has yet been subject to a full-scale overload, the controls have demonstrated positive field behaviour under some limited abnormal traffic conditions.

The switch overload simulator which was used as the testing and validation tool was based on a novel quasi-distributed architecture for maximum run-time efficiency. The major features of the simulator were:

- Separate CC and PC clocks reflecting the different time scales required for central and peripheral processing
- The finite-horizon simulation used for the peripheral controllers to limit the CC simulation event list to a small size.

The overload simulator is being evolved to study a wider class of system performance problems. In particular it is being extended to look at trunk traffic overloads, the overload performance of new peripherals being designed for the switch, and the switch response to abnormal maintenance activity in the face of carrier failure.

REFERENCES

[1] L. Forys, "Performance Analysis of a New Overload Strategy". Proc. ITC 10, Montreal, 1983.

[2] K. Basu et al, "Real Time Simulator for Performance Evaluation and Overload Control Design of an SPC System". Proc. ITC 10, Montreal, 1983.

[3] F. Schoute, "Optimal Control and Call Acceptance in an SPC Exchange". Proc. ITC 9, Spain, 1979.

[4] F. Schoute, "Adaptive Overload Control for an SPC Exchange". Proc. ITC 10, Montreal, 1983.

[5] M. Somoza and A. Guerrero, "Dynamic Processor Overload Control and its Implementation in Certain Single Processor and Multiprocessor SPC Systems". Proc. ITC 9, Spain, 1979.

[6] B. Penney and J. Williams, "The Software Architecture for a Large Telephone Switch". IEEE Trans. on Comm., 30, 1369-1378 (1982).

[7] J. DenOtter and R. Pfeffer, "DMS-100 Proves Performance in the Field". Telesis, 10, 20-24 (No. 2, 1983).

[8] G. Dietrich and R. Salade, "Subcall Type Control Simulation of SPC Switching Systems". Proc. ITC 8, Melbourne, 1976.

[9] J. Roberts, "Recent Observations of Subscriber Behaviour". Proc. ITC 9, Spain, 1979.

[10] L. Burkard et al, "Customer Behaviour and Unexpected Dial Tone Delay". Proc. ITC 10, Montreal, 1983.

[11] B. Doshi and H. Heffes, "Analysis of Overload Control Schemes for a Class of Distributed Switching Machines". Proc. ITC 10, Montreal, 1983.

TELETRAFFIC ISSUES in an Advanced Information Society
ITC-11
Minoru Akiyama (Editor)
Elsevier Science Publishers B.V. (North-Holland)

Liquid models for a type of information buffer problems

by L. Kosten

formerly: Delft University of Technology,
The Netherlands

Summary

Results of earlier papers by Anick *et al*. and by the present author are generalised. A central processor, working at a uniform speed, receives messages to be processed from a number of sources. A buffer stores information that cannot be handled directly. The problem is that of buffer overflow. The generalisation consists in the possibility of more general laws than the exponential one for message durations.

INTRODUCTION

1. Introduction; basic model. In a paper by Anick, Mitra and Sondhi [1], further referred to as AMS, the following model has been thoroughly dealt with. A data handling switch receives messages from N identical, mutually independent sources, which each can be "on" or "off". The changes off $\rightarrow$ on and on $\rightarrow$ off take place according to Poisson processes with densities λ and 1, respectively. A source that is "on", fills a central buffer at a rate of 1 unit of information per unit of time. When r sources are "on", the total buffer filling rate is r. The buffer is drained by an output channel (e.g. a distribution switch) at a maximum rate c, the depletion rate. Whenever $r>c$, the level of the buffer increases at rate $r-c$; when $r<c$, this level decreases at rate $c-r$, until the buffer is empty. For the total system stochastic stationarity is assumed. The actual problem is that of determining the rate of overflow of the buffer, given a buffer of a certain size. It is extremely complicated. We shall replace it by the following problem. The size of the buffer is supposed to be infinite. Instead of an overflow-rate we ask for G(u), the probability of the buffer-level exceeding a value u. The system cannot be stationary unless the depletion-rate c exceeds the overall incoming-rate $N\lambda/(1+\lambda) \triangleq \beta$.

The model described has been completely analysed by AMS. It is advised to read this paper -which is easily accessible- before turning to the present paper. In possible applications any overflow of the buffer results in serious technical troubles. Hence, overflow -or the correlated situation of some high buffer level- should be a very rare event, i.e. G(u) should be very small. Consequently, the proof in AMS of the existence of an asymptotic approximation:

$$G(u) \sim C\, e^{-\alpha u} \quad (u \uparrow \infty) \tag{1.1}$$

together with easy ways to compute the exponential coefficient α and the amplitude C, is very important!

We shall call the group of sources in the basic model described a Homogeneous Exponential Group (HEG). "Homogeneous" refers to the sources being identical. The fields of application at once suggest the following variants of the model to be important:

(i) the message-durations (i.e. the "on"-times) need not be distributed exponentially; as variants we shall consider phase- (or Erlang-) distribution and hyperexponential distribution (cf. [5], [3]), with constant duration as a limiting case of an infinite number of phases; see sections 8 and 7;

(ii) sources are different; for the exponential type e.g. they have their own parameters $\lambda, \mu\ \gamma$; λ is the ending rate of "off"- periods, μ of "on"-periods and γ the filling-rate (for that source); this leads to the conceptions of the Non-homogeneous Exponential Group (NEG), dealt with in sec. 6; normally the NEG will consist of a number of homogeneous subgroups; this case has already been treated by Kosten [4].

In all variants considered the "off"-times are supposed to be exponential. Various models for ∞ sources have been studied in Kosten [3]. Cf. the "note at correction" at the end of section 10.

Before turning to the variant indicated, sections 2/5 will be devoted to some general theory.

For all the variants to be considered it will turn out that there exists an asymptotic approximation of type (1.1), where the exponential coefficient α can be determined analytically. The determination of the amplitude C, however, resists all analytic attack for those cases (i.e. apart from sheer formal solutions)! It should be found either by simulation (expensive) or by some heuristic approach, to be discussed in sec. 9.

2. General Markov-driven source. All models indicated above, to the exception of the constant "on"-time case, can be described by the following general model. The momentary filling-rate of the buffer depends on the momentary state of an m-state Markov process with continuous time-parameter (cf.fig.1). Let $\bar{M}$ be its (non-degenerate) infinitesimal generator. (cf.Cohen [2]).
We define:

$$\underline{r} \triangleq \text{ the stochastic state}$$

$$f_r(t) \triangleq P\{\underline{r} = r, \text{ at time } t\}$$

$$\bar{f}(t) \triangleq \left(f_1(t),\ldots, f_m(t)\right)$$

The vector $\bar{f}(t)$ satisfies:

$$\frac{d\bar{f}}{dt} = \bar{M}\,\bar{f}(t) \tag{2.1}$$

When the state is r, the filling-rate is defined to be γ_r. The buffer depletion-rate is c. Furthermore, we define:

$\underline{u} \triangleq$ the stochastic buffer-content (or level).

$F_r(u,t) \triangleq P\{\underline{r} = r \wedge \underline{u} \leq u$, at time $t\}$

$\bar{F}(u,t) \triangleq (F_1(u,t),\ldots, F_m(u,t))$

The rate of increase of $P\{\underline{r} = r \wedge \underline{u} \leq u\}$ by $\underline{r}$-changes is $[\bar{M}\,\bar{F}(u,t)]_r$.

The rate of level-decrease is $c-\gamma_r$ (i.e. the reverse of level-increase, when negative). Hence, the rate of increase of said probability by change of level is $(c-\gamma_r)\cdot \partial F_r(u,t)/\partial u$. The sum of those rates is:

$$\frac{\partial F_r}{\partial t} = [\bar{M}\bar{F}(u,t)]_r - (\gamma_r-c)\frac{\partial F_r}{\partial u} \tag{2.3}$$

Now, let $\bar{I}$ be the identity-matrix and $\bar{\Gamma}$ the diagonal-matrix:

$\bar{\Gamma} \triangleq \text{diag}\,\{\gamma_1,\ldots, \gamma_m\}$. Then:

$$\frac{\partial \bar{F}}{\partial t} + (\bar{\Gamma} - c\bar{I})\frac{\partial \bar{F}}{\partial u} = \bar{M}\bar{F}(u,t) \tag{2.4}$$

Assuming stationarity of the Markov process and dropping the argument t we have:

$$\bar{M}\bar{f} = \bar{0}\ ;\ \bar{1}'\,\bar{f} = 1 \tag{2.5}$$

$\bar{1}$ being the unit-vector: $\bar{1} \triangleq (1,\ldots,1)$. When the average filling-rate is less than the depletion-rate:

$$\bar{1}'\,\bar{\Gamma}\,\bar{f} < c \tag{2.6}$$

we may assume stationarity for the total system.
Again dropping argument t:

$$(\bar{\Gamma}-c\bar{I})\frac{d\bar{F}}{du} = \bar{M}\bar{F}(u) \tag{2.7}$$

Now, assume that none of the filling-rates γ_r equals the depletion-rate. Then the matrix $(\bar{\Gamma}-c\bar{I})^{-1}\bar{M}$ has m eigenvalues $z^i (i=1,\ldots,m)$. Let $\bar{\phi}^i$ be the right eigenvector corresponding to z^i:

$$\{\bar{M} - z^i(\bar{\Gamma}-c\bar{I})\}\bar{\phi}^i = \bar{0} \quad (i = 1,\ldots,m) \tag{2.8}$$

The determinantal equation

$$\det\,\{\bar{M}-z(\bar{\Gamma}-c\bar{I})\} = 0 \tag{2.9}$$

is of degree m, <u>both in z and in c</u>.

No confluence is supposed to apply. Then the general solution of (2.7) is:

$$\bar{F}(u) = \sum_{i=1}^{m} a_i\bar{\phi}^i\, e^{z^i u} \tag{2.10}$$

the a_i being constants, as yet unknown.
As $\bar{F}(\infty)$ must be finite, terms pertaining to eigenvalues with positive real part must be absent. The column-sums of $\bar{M}$ being zero, there is one zero eigenvalue, the contribution of which in (2.10) must be $\bar{F}(\infty)$. Now:

$F_r(\infty) = P\{\underline{r} = r \wedge \underline{u} \leq \infty\} = P\{\underline{r} = r\} = f_r$. Hence, said contribution must equal $\bar{f}$, which can be obtained from (2.5).

Consequently, (2.10) reduces to:

$$\bar{F}(u) = \bar{f} + \sum_{(\forall i | \mathrm{Re}\, z^i < 0)} a_i \bar{\phi}^i e^{z^i u} \tag{2.11}$$

For the determination of the remaining coefficients a_i we use the obvious fact that an empty buffer is inconsistent a filling rate γ_r exceeding the depletion-rate c:

$$F_r(0) = 0 \qquad (\forall r | \gamma_r - c > 0) \tag{2.12}$$

In connection with (2.11) this yields the following set of equations in the unknown coefficients a_i:

$$\sum_{(\forall i | \mathrm{Re}\, z^i < 0)} a_i (\bar{\phi}^i)_r = -f_r \qquad (\forall r | \gamma_r - c > 0) \tag{2.13}$$

It is to be expected that the number of equations equals the number of unknowns. This leads to the following

LEMMA: the number of eigenvalues z^i of $(\bar{\Gamma} - c\bar{I})^{-1}\bar{M}$ with negative real part equals the number of elements γ_r exceeding c.

This lemma has recently been proved by P. Sonneveld [6]

Once the eigenvalues with negative real part have been determined, the eigenvectors follow from (2.8) and the coefficients a_i from (2.13). Then the final solution is formally known from (2.10):

$$G(u) = 1 - \bar{1}'\, \bar{F}(u) =$$

$$= \sum_{(\forall i | \mathrm{Re} z^i < 0)} a_i (\bar{1}' \bar{\phi}^i) e^{z^i u} \tag{2.14}$$

When z^* is the rightmost eigenvalue in the negative half-plane (supposed to be unique), the asymptotic approximation of type (1.1) now reads:

$$G(u) \sim -a_0 (\bar{1}' \bar{\phi}^*) e^{z^* u} \tag{2.15}$$

The determination of z^* mostly is relatively easy. There are, however, mostly unsurmountable difficulties in determining the amplitude. We shall call $z^* = -\alpha$ the dominant transient's exponent, or simply the <u>dominant</u>.

3. The Decomposition Theorem

The source part of the system in fig. 1 may be specified by the matrices $\bar{M}$ and $\bar{\Gamma}$. Together with the depletion rate c the system is completely described. Hence, we shall speak about "the" system $[\bar{M}, \bar{\Gamma}; c]$. The eigenvalues z^i and eigenvectors $\bar{\phi}^i (i=1,\ldots,m)$, which originally are connected with the matrix $(\bar{\Gamma} - c\bar{I})^{-1}\bar{M}$, will also be considered as attributes of the system $[\bar{M}, \bar{\Gamma}; c]$.

In most cases the source part consists of a number n of sources or of sub-groups, not necessarily identical (cf. fig.2). Each of the n components can be described by its own matrixpair $\bar{M}_j, \bar{\Gamma}_j$ of dimensionality $m_j (j=1,\ldots,n)$. Hence, the total system may be described by

$$[\bar{M}, \bar{\Gamma}; c] = [\bar{M}_1, \bar{\Gamma}_1; \ldots; \bar{M}_n, \bar{\Gamma}_n; c] \tag{3.1}$$

Let $\underline{r}_1, \ldots, \underline{r}_{m_j}$ be the stochastic states of the n Markov processes induced by the $\bar{M}_j$. Then we can define the following variables (again under assumption of stationarity):

$$F_{r_1 \ldots r_n}(u) \triangleq P\{\underline{r}_1 = r_1 \wedge \ldots \underline{r}_n = r_n \wedge \underline{u} \leqslant u\} \tag{3.2}$$

They form an n-dimensional array of size $[1:m_1, 1:m_2, \ldots, 1:m_n]$ (in Fortran notation). If this be preferred, they also can be ordered (e.g. lexicographically) in one vector of size $m = m_1 \cdot \ldots m_n$. The array mentioned will be denoted by $\hat{F}(u)$.

A special type of array is the Cartesian products of n vectors:

$$\hat{V} = \bar{v}_1 \times \bar{v}_2 \times \ldots \times \bar{v}_n \tag{3.3}$$

defined by:

$$[\hat{V}]_{r_1, \ldots r_n} = [\bar{v}_1]_{r_1} \times [\bar{v}_2]_{r_2} \times \ldots \times [\bar{v}_n]_{r_n} \tag{3.4}$$

It is possible to construct the matrices $\bar{M}$ and $\bar{\Gamma}$ from the component matrices. This is, however, an awkward procedure, not being instructive at all. It is best to choose another way of representing the total system. We shall denote by $\hat{M}_j\ \hat{F}(u)$ the array, obtained by letting the $m_j * m_j$ matrix $\bar{M}_j$ act on each of the m/m_j vectors of size m_j contained in the array $\hat{F}(u)$. I.e. for <u>one</u> element of the resultant array all m indices of $\hat{F}(u)$ are kept fixed but for the j-th. In an analogous way we define $\bar{\Gamma}_j\ \hat{F}(u)$.

The probability $P\{\underline{r}_1 = r_1 \wedge \ldots \wedge \underline{r}_n = r_n \wedge \underline{u} \leqslant u\}$ increases in the following ways and at the following rates:

(i) by change of sub-state $\underline{r}_j$ at the rate $[\bar{M}_j\ \hat{F}(u)]_{r_1 \ldots r_n}$

(ii) by buffer-filling from source j: the filling-rate is $[\bar{\Gamma}_j]_{r_j r_j}$; hence, the forming-rate is

$$-[\bar{\Gamma}_j]_{r_j r_j}\, d\hat{F}_{r_1 \ldots r_n}/du = -\frac{d}{du}[\bar{\Gamma}_j \hat{F}]_{r_1 \ldots r_n} \quad (j=1,\ldots,n);$$

(iii) by buffer-depletion at rate $c\, d\hat{F}_{r_1 \ldots r_n}/du$.

Owing to stationarity the sum-total is zero:

$$\sum_{j=1}^{n} [\bar{M}_j\ \hat{F}(u)]_{r_1 \ldots r_n} - \frac{d}{du} \sum_{j=1}^{n} [\bar{\Gamma}_j\ \hat{F}(u)]_{r_1 \ldots r_n} + c \frac{d}{du} \hat{F}_{r_1 \ldots r_n}(u) = 0,$$

or, in matrix-array notation:

$$\sum_{j=1}^{n} (\bar{\Gamma}_j \frac{d}{du} - \bar{M}_j\ \hat{F}(u)) = c \frac{d}{du} \hat{F}(u) \tag{3.5}$$

This is the formal equivalent of (2.7) for the total system. Accordingly, there are $m=m_1 \ldots m_n$ eigenvalues. Let z be a generic eigenvalue and $\hat{\phi}$ the associated eigenarray:

$$\sum_{j=1}^{n} (z\bar{\Gamma}_j - \bar{M}_j)\ \hat{\phi} = c\ z\ \hat{\phi} \tag{3.6}$$

Assume that z is a generic eigenvalue of all the systems $[\bar{M}_j,\ \bar{\Gamma}_j; c_j]$ ($j = 1,\ldots, m$); let $\bar{\phi}_j$ be the associated eigenvectors (cf. 2.8):

$$(z\ \bar{\Gamma}_j - \bar{M}_j)\ \bar{\phi}_j = z\ c_j\ \bar{\phi}_j \tag{3.7}$$

If:

$$\sum_{j=1}^{n} c_j = c \tag{3.8}$$

and:

$$\hat{\phi} = \bar{\phi}_i \times\ldots\times \bar{\phi}_n \tag{3.9}$$

z is also an eigenvalue of the total system $[\bar{M},\bar{\Gamma};c] = [\bar{M}_1,\bar{\Gamma}_1;\ldots;\ \bar{M}_n,\bar{\Gamma}_n;\ c]$ and $\hat{\phi}$ the associated eigenarray. The proof is straightforward by substitution in (3.6). But also the reverse is true. Suppose that z is an eigenvalue of the total system $[\bar{M},\ \bar{\Gamma};c]$. Solve the n equations (cf. 2.9):

$$\det\{\bar{M}_j - z(\bar{\Gamma}_j - c_j\bar{I}_j)\} = 0 \qquad (j=1,\ldots,n) \tag{3.10}$$

for the c_j. This yields m_j values for c_j ($j=1,\ldots,n$). Taking each of the

$m_1\ldots m_n$ values $c^{\xi} = \sum_1^n c_j$, this yields $m_1\ldots m_n$ systems $[\bar{M}_1,\bar{\Gamma}_1;\ldots;\ \bar{M}_n,\bar{\Gamma}_n;c\ ]$

for which z is an eigenvalues. Otherwise, there would be $m_1\ldots m_n+1$ solutions of the $m_1\ldots m_n$ degree equation (2.9) in c. Hence, we have proved the following.

DECOMPOSITION THEOREM.

The quantity z is an eigenvalue of the total system $[\bar{M},\bar{\Gamma};c] \triangleq [\bar{M}_1,\bar{\Gamma}_1;\ldots;$ $M_n,\bar{\Gamma}_n;c]$ if and only if there exists a partitioning (3.8) such that z is an eigenvalue for each of the subsystems $[\bar{M}_j,\bar{\Gamma}_j;c_j]$ ($j=1,\ldots,n$). The eigenarray of the total system associated with z is the Cartesian product (3.9) of the subsystems eigenvectors.

As the decomposition is purely fictitious, the components c_j of the partitioning need not be positive, nor even real.

4. Homogeneity of sources; confluence.

Let the n sources be identical with source data $\bar{M}',\bar{\Gamma}'$. Then the total system is $[\bar{M},\bar{\Gamma}']$.Then the total system is $[\bar{M},\bar{\Gamma};c] = [\bar{M}',\bar{\Gamma}',\ldots,\bar{M}',\bar{\Gamma}';c]$. The number of states, essential to describe the total filling system shrinks from $m_1\ldots m_n$ to $\binom{n+m-1}{n}$, where m is the number of possible states of each of the n sources. The Decomposition Theorem remains valid. However, the partitioning (3.8) should be replaced by

$$\sum_{i=1}^{m} s_i c_i = c \quad (s_i = 0,\ldots,n;\ \sum_1^n s_i = n) \tag{4.1}$$

For a given value z, equation (2.9) has m solutions $c_i(j = 1,\ldots,m)$. This yields $\binom{m+n-1}{n}$ values c^{ξ} for the left-hand side of (4.1). For z to be an eigenvalue of $[\bar{M},\bar{\Gamma};c]$,c must be one of those values c^{ξ}.

5. Use of Decomposition Theorem.

For (homogeneous and non-homogeneous) groups of exponential and hyperexponential sources it is possible to detect the partitionings and hence the eigenvalues, in particular the dominant. Let us now consider a general homogeneous system $[\bar{M},\bar{\Gamma};c] = [(\bar{M}',\bar{\Gamma}')^n;c]$. Consider the subsystem $[\bar{M}', \bar{\Gamma}'; c/n]$. It has m eigenvalues. One of those is z=0, another one the dominant $z^* = -\alpha$. From the Decomposition Theorem it follows that those eigenvalues are also eigenvalues of the total system (equipartition of c over the n subsystems). The value z=0 clearly is connected with the total system's stationary behaviour. We adopt as a working hypothesis that $z^* = -\alpha$ is not only just an eigenvalue but the dominant one. It has been proved for homogeneous groups of exponential and hyperexponential sources. It has not been refuted in other cases.

6. Non-homogeneous group of exponential sources (NEG)

First the case of a single exponential source is treated. The source data are:

$$\bar{M} = \begin{pmatrix} -\lambda & \mu \\ \lambda & -\mu \end{pmatrix} \quad , \quad \bar{\Gamma} = \begin{pmatrix} 0 & 0 \\ 0 & \gamma \end{pmatrix} \tag{6.1}$$

The solution of (2.9) as an equation in z is:

$$z = \begin{cases} 0 \\ \frac{\lambda}{c} - \frac{\mu}{\gamma - c} \end{cases} \tag{6.2}$$

Those z-values are real. The solution of (2.9) as an equation in c is the two-valued function:

$$c = c(z) \triangleq \left[(\lambda+\mu+\gamma z) \pm \sqrt{(\lambda+\mu+\gamma z)^2 - 4\lambda\gamma z}\right] / 2z \tag{6.3}$$

Both branches are decreasing functions of z. When $z<0$, the +sign yields a negative value, the −sign a positive one. Cf. fig. 3. For $z\uparrow 0$ the latter value is $\beta \triangleq \lambda\gamma/(\lambda+\mu)$, the average filling-rate.
Now, consider the case of n non-identical exponential sources (NEG). The source data are λ_i, μ_i, γ_i $(i=1,\ldots,n)$. Also the relevant quantities β and c(z) will be indexed by i. Given some value z, there are 2^n real values $c^{\$} = \sum_{i=1}^{n} c_i(z)$ by taking either the + or the −sign in each of the (indexed) expressions (6.3) for $c_i(z)$. If and only if c equals one of those values $c^{\$}$, z is an eigenvalue of the total system, according to the Decomposition Theorem. When $z<0$, the largest c −value, say $c^*(z)$, is obtained by taking −signs only (cf. fg. 4 for the case n=2). It is clear that $c^*(z)\in(B,C)$ with $B \triangleq \sum_1^n \beta_i$, $C \triangleq \sum_1^n \gamma_i$, the average and maximal total filling rate. We observe that $c^*(0^-) = B$, $C^*(-\infty) = C$.

Now, take the inversion of $c = c^*(z)$, to be called $z = z^*(c)$. For $c\in(B,C)$ it is the largest negative eigenvalue, i.e. the dominant of the total system. Hence, given some total system depletion rate c, the dominant $z=z^* \triangleq -\alpha$ is implicitely given by:

$$c = \frac{1}{2z} \sum_{i=1}^{n} \left[(\lambda_i+\mu_i+\gamma_i z) - \sqrt{\{(+\lambda_i+\mu_i+\gamma_i z)^2 - 4\lambda_i\gamma_i z\}}\right] \tag{6.4}$$

It can easily be obtained by inverse interpolation. When the non-homogeneous group consists of some homogeneous sub-groups, the terms in (6.4) per sub-group are equal. This case has already been treated in [4].

Though we easily obtain the dominant for the NEG case, contrary to the HEG case there is no realistic analytic way of obtaining the amplitude of the dominant transient.

7. Hyperexponential source group.

Though it is possible to consider more complicated cases, we immediately turn to the simple hyperexponential source case, where "on"-times belong to one of two exponential classes with different averages; at the beginning an "on-time is picked from those classes with a certain probability. The diagram of one source is (cf. [5], [3]):

$$\textcircled{1} \underset{2\sigma\mu}{\overset{\sigma\lambda}{\leftrightarrows}} 0 \underset{2\tau\mu}{\overset{\tau\lambda}{\rightleftarrows}} \textcircled{2} \qquad \gamma\downarrow \quad \downarrow\gamma \qquad (\sigma+\tau = 1)$$

and hence:

$$\bar{M}' = \begin{pmatrix} -\lambda & 2\sigma\mu & 2\tau\mu \\ \sigma\lambda & -2\sigma\mu & 0 \\ \tau\lambda & 0 & -2\tau\mu \end{pmatrix}; \ \bar{\Gamma}' = \text{diag}\ \{0,\gamma,\gamma\} \tag{7.1}$$

The determinantal equation (2.9) yields one value z=0 and further 2 values, satisfying:

$$(y-\sigma A)(y-\tau A) = \sigma\tau B^2$$

with

$$y \triangleq c(c-\gamma)z \ ; \ B \triangleq \lambda(\gamma-c); \ A \triangleq 2\mu\ c\ -B \tag{7.2}$$

Some college analytical geometry results in fig. 5., showing the relation between the two non-zero values of z and c. The z-values are real and distinct. For negative values of z there are three values c(z): $c'(z) > c''(z) > c'''(z)$, which can be obtained from the tertiary equation (7.2) (c=0 is a false root). Obviously, for the region of feasibility $c \in (\beta,\gamma)$, the dominant is given by the inversion $z = z'(c)$ of $c=c'(z)$.

Now, consider a non-homogeneous group of n hyperexponential sources. A lower index i (i=1,...,n) refers to all quantities for the separate sources: λ_i, μ_i, γ_i, etc. By a line of argument analogous to that of sect. 6, we obtain the dominant z* for the total group by the following procedure. For any z we compute $c_i'(z)$ (i=1,...,n) as the largest root c of (7.2) with $\lambda=\lambda_i$, $\mu=\mu_i$, $\gamma=\gamma_i$, $\tau=\tau_i$.

Then:

$$c(z) \triangleq \sum_1^n c_i'(z) \tag{7.3}$$

As well as the components, c(z) is a decreasing function. By inverse interpolation we obtain the dominant z* as the value z, for which c(z) equals the given value c. The region of feasibility is $c \in (B,C) \triangleq (\sum_1^n \beta_i, \sum_1^n \gamma_i)$.

When this procedure is applied to a homogeneous group, it is obvious that all components of (7.3) will be constantly equal. This proves that the working hypothesis of sect. 5 is also valid for homogeneous groups of hyperexponentical sources.

8. Homogeneous phase source group

The "on"-periods consist of m phases. Each phase is exponential with termination rate μ_i for phase i. When a source is in phase i it fills the buffer at a rate $\gamma_i (i=1,\dots,m)$. The $\bar{M}'$ and $\bar{\Gamma}'$ matrices (per source) are $(m+1) \times (m+1)$ (cf. [5], [3]):

$$\bar{M}' = \begin{bmatrix} -\lambda & - & - & - & \mu_m \\ \lambda & -\mu_1 & & O & \vdots \\ & & \ddots & -\mu_{m-1} & \vdots \\ & O & & \mu_{m-1} & -\mu_m \end{bmatrix} \quad ; \quad \bar{\Gamma}' = \text{diag}\,\{0, \gamma_1; \dots, \gamma_m\} \qquad (8.1)$$

The determinantal equation (2.9) in z now reads:

$$(\lambda - cz)\{\mu_1 + (\gamma_1 - c)z\}\dots\{\mu_m + (\gamma_m - c)z\} = \lambda_1 \mu_1 \dots \mu_m \qquad (8.2)$$

One root is zero. The dominant is easily found by numerical means. Using the working hypothesis of sect. 5 the dominant for an n source is obtained by calculating the dominant from (8.2) after changing c into c/n.

Mostly, we stick to the simple phase model with identical phases: $\gamma_i = 1$, $\mu_i = m$ $(i=1,\dots,n)$. The average "on"-time now is 1. We then have:

$$(1 - \frac{c}{\lambda} z)(1 + \frac{1-c}{m} z)^m = 1 \qquad (8.3)$$

For $m \to \infty$ this yields the case of a constant "on"-time:

$$(1 - \frac{c}{\lambda} z)\, e^{(1-c)z} = 1 \qquad (8.4)$$

9. The method of the "Equivalent HEG".

In the sections 6/8 various models have been investigated, where determination of the dominant was possible by analytical means. Analytical calculation of the amplitude C is only possible in the case of the HEG (homogeneous exponential group). Now, the dominant is the exponent of asymptotic approximations (1.1) and for rough guesses as to asymptotic behaviour it is far more important than the amplitude C. Nevertheless, it would be worth while to dispose of a method to guess C, albeit in some heuristic way.

In the following we propose to substitute for some source group under consideration by an "Equivalent HEG", i.e. a HEG that on intuitionistic grounds is supposed to yield a fair representation of the group considered (in any case as far as the behaviour for large u-values in involved). The HEG may be characterised by 4 structural parameters, viz. λ, μ, γ and n (the number of sources). We denote it by HEG (n,λ,μ,γ), n being the number of identical sources. We then shall try to adjust those structural parameters in such a way, that both the HEG and the studied general group have equal values for 4 other parameters, to be called behavioral parameters. Those parameters must be chosen in such a way that they have a meaning for more general source groups and can be determined for them in a not too complicated way.

One of those parameters is pretty obvious. We may assume that the value of the dominant (α) is known for the studied group (in connection with a depletion rate c). It then is not but natural to demand that the equivalent HEG shows the same dominant value α in connection with a depletion rate c. In [1] the value α for HEG $(n,\lambda,1,1)$ is given as:

$$\alpha = (1+\lambda-n\lambda/c) \; / \; (1-c/n) \tag{9.1}$$

By dimensional arguments it can be generalised for HEG (n,λ,μ,γ) to

$$\alpha = (\mu+\lambda-n\lambda\gamma/c) \; / \; (\gamma-c/n) \tag{9.2}$$

This parameter has a special bearing on the dynamic behaviour of the source group. Now, let $\underline{s}$ be the stochastic variable that represents the filling rate of a source or group. Then, we adopt for 3 more behavioral parameters the first 3 <u>cumulants</u> of $\underline{s}$:

$$\begin{aligned} A &\triangleq K_1(\underline{s}) && \text{(average)} \\ V &\triangleq K_2(\underline{s}) && \text{(variance)} \\ S &\triangleq K_3(s) && \text{(characteristic of skewness)} \end{aligned} \tag{9.3}$$

For a large class of sources those cumulants are very easy to compute, viz. those sources that are of the "off/on" type. This means that the Markov process has two types of states: in the "off"-states the filling rate is 0, in the "on"-states it has a unique value γ. This happens to be the case for exponential sources and simple phase and hyperexponential sources, as well as "constant". When p and q are the probabilities of the Markov process being in an "on" or "off"-state $(p+q=1)$, we eventually find:

$$A = p\gamma \;;\quad V = pq\gamma^2 \;;\quad S = pq(q-p)\gamma^3 \tag{9.4}$$

For the simple cases meant we have:

$$p = \lambda/(\lambda+\mu) \;;\; q = \mu/(\lambda+\mu) \tag{9.5}$$

Let the general source group consist of n sources, having stochastic filling rates $\underline{s}_1,\ldots,\underline{s}_n$. We suppose the cumulants of $\underline{s}_1,\ldots,\underline{s}_n$ are known: $A_1,\ldots, A_n$; $V_1,\ldots, V_n$; $S_1,\ldots, S_n$. As the sources supposedly are mutually independent, so are $\underline{s}_1,\ldots,\underline{s}_n$, and the cumulants are additive:

$$A = \sum_1^n A_i \;;\; V = \sum_1^n V_i \;;\; S = \sum_1^n S_i \tag{9.6}$$

the indices i referring to the separate sources.

Once, α,A,V and S are known for the general group they are also known for the equivalent HEG. Let us call the latter's structural parameters $n^*, \lambda^*, \mu^*, \gamma^*$.

From (9.2), (9.4) and (9.5) the structural parameters of the HEG can be solved:

$$\left.\begin{aligned} n^* &= A^2V/(V^2 - AS) \\ \gamma^* &= (2V^2 - AS) \;/\; AV \\ \lambda^* &= \alpha\,(\gamma^*-c/n^*) \;/\; n^*\gamma^*(1/A-1/c) \\ \mu^* &= (n^*\gamma^*/A - 1)\,\lambda^* \end{aligned}\right] \tag{9.7}$$

In the AMS model μ and γ are 1. Hence, the calculated quantities, as well as the depletion rate c, need scaling:

$$n' = n^* \; ; \; \lambda' = \lambda^*/\mu^* ; \; c' = \mu^* c/\gamma^* \qquad (9.8)$$

The amplitude C, being dimensionless, is not affected by the scaling. Now, C is computed for the AMS model with the values n', λ',c' according to the prescription found in AMS. It will be taken as the value for the amplitude in (1.1) for the actual model too.

When the actual group is a homogeneous group of "off/on" sources (like the cases of sect. 7 and 8), the situation as well as g are equal for all sources. It then follows from (9.5)-(9.8) that the parameters n', λ' and c' are the same for all those models. Hence, they all have the same amplitude C in the equivalent HEG approximation.

10. Experimental verification and conclusions

In a number of cases we have checked the asymptotic expression (1.1), as obtained according to the aforegoing sections, by the results of simulation. The outcome is to be found in figs. 6 and 7, where also the 3σ-confidence limits of the simulation are shown.

In the first place a number of cases have been taken, all with a homogeneous group of n=10 sources with λ=0,4. The average "on"-period duration is $1/\mu$=1. For this duration we considered phase-distribution with k=1,2,∞ (i.e. constant) and hyperexponential distribution with σ=0,5; 0,3 and 0,2. For k=1 and σ=0,5 they are identical to exponential distribution.

For the exponential case α and C are calculated analytically by the AMS method. For the other cases α has been obtained according to sections 7 and 8. Those results are correct on the proviso that the working hypothesis of sect. 5 be right. Now, C has been obtained by the heuristic equivalent HEG method of sect 9. This means that for those cases C has been taken equal to that of the exponential case. The results (cf. fig. 6) are:

phase	hyp.exp.	C	α
k = ∞	(const)	0.156	1.878
k = 2		0.156	1.072
k = 1	σ = 0,5 (exp)	0.156	0.772
	σ = 0,3	0.156	0.585
	σ = 0,2	0.156	0.408

As another example we took the case of two homogeneous exponential subgroups (already dealt with in [4]):

$$\begin{array}{ll} n_1 = 25 & n_2 = 50 \\ \lambda_1 = 0.4 & \lambda_2 = 0.6 \\ \mu_1 = 1.5 & \mu_2 = 0.75 \\ \gamma_2 = 2 & \gamma_2 = 1 \end{array}$$

$$c=38$$

According to sect. 6 we first calculate α by inverse interpolation of (6.4), here reading:

$$c = \sum_1^2 n_i \left[(\lambda_i+\mu_i+\gamma_i z) - \sqrt{\{(\lambda_i+\mu_i+\gamma_i z)^2 - 4\lambda_i\gamma_i z\}}\right]/2z \qquad (10.1)$$

yielding: $\alpha = -z = 0{,}270$.

From (9.4) we obtain:

$$A = A_1+A_2 = \frac{n_1\lambda_1\gamma_1}{\lambda_1+\mu_1} + \frac{n_2\lambda_2\gamma_2}{\lambda_2+\mu_2} = 32{,}75$$

and similary V = 29,0 and S = 20,62.

Then from (9.7) and (9.8) with a slight rounding to integer n'-value:

$n' = 190$; $\lambda' = 0.195$; & $c' = 35.94$.

Employing those values in the AMS model yield the amplitude for the equivalent HEG: C = 0.126. Hence, the equivalent HEG asymptotic approximation is:

$$G(u) \sim 0.126\, e^{-0.270\, u} \qquad (10.2)$$

This has been shown in fig. 7, together with simulations results ($\pm 3\sigma$).

From the diagram it is clear that the exponents α from the previous section do no shown serious discrepancies from situation results. This indicates that the working hypothesis of sect.5 is probably right for the phase and hyperexponential groups considered (for the non-homogeneous exponential and hyperexponential case it is correct).

The heuristic values for the amplitude C are none too good. In the absence of a better method they may probably do as a substitute for a thorough simulation, which is very expensive.

Note at correction (10/5/85)

It can be proved that the dominant of a general Markov-driven source system is non-multiple and hence real (cf [6]). Consequently, the method used in sec. 6 is applicable to general cases too.

REFERENCES

[1] D. Anick, D. Mitra and M.M. Sondhi, Stochastic theory of a data-handling system with multiple sources, BSTJ 61(1982) 1871-1894.

[2] J.W. Cohen, (1969), The Single Server Queue, North-Holland, Amsterdam.

[3] L. Kosten, Stochastic theory of a multi-entry buffer,
Delft Progress Report;
part 1 : 1 (1974) 10-18
part 2 : 1 (1974) 44-50
part 3 (together with O.J. Vrieze) : 1 (1975) 103-115.

[4] L. Kosten, Stochastic theory of data-handling systems with groups of multiple sources, Int. Symp. on Perform. of Comp.-Comm. Syst., Zürich, March 1984.

[5] P.M. Morse (1958), Queues, Inventories and Maintenance, John Wiley & Sons, Inc., New York.

[6] P. Sonneveld, Private communication, (to be published).

Acknowledgement

The author wishes to express his thanks to Mr. T.C.A. Mensch and Mrs. Lieneke Lekx for their assistance and patience (with me) in doing simulations and preparing the copy.

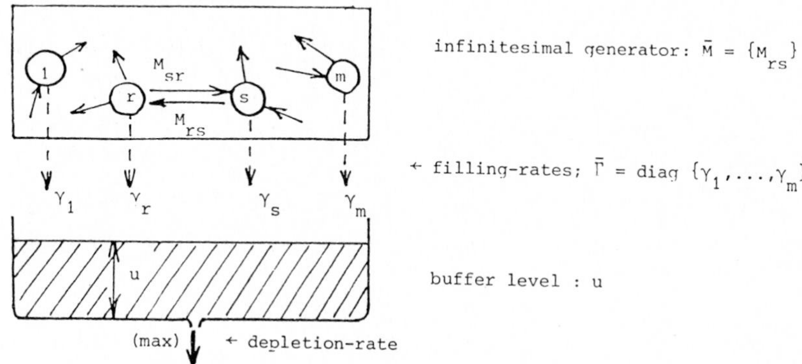

fig. 1. : Liquid model of Markov-process controlled buffer system

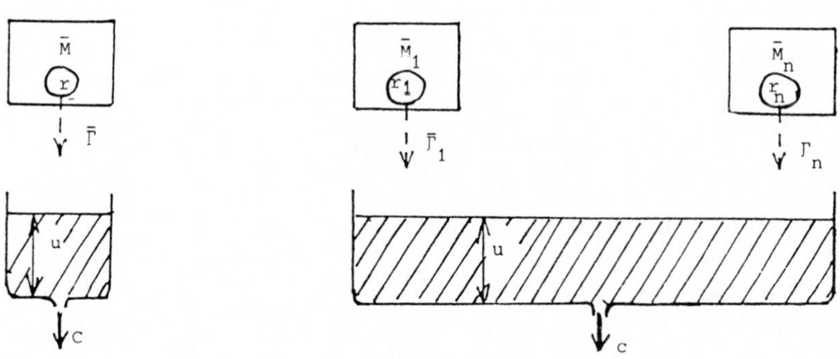

fig. 2: Decomposition of buffer system

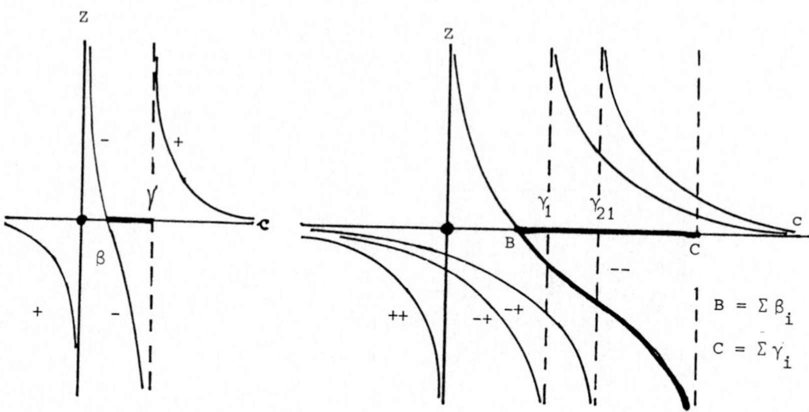

fig. 3

fig. 4

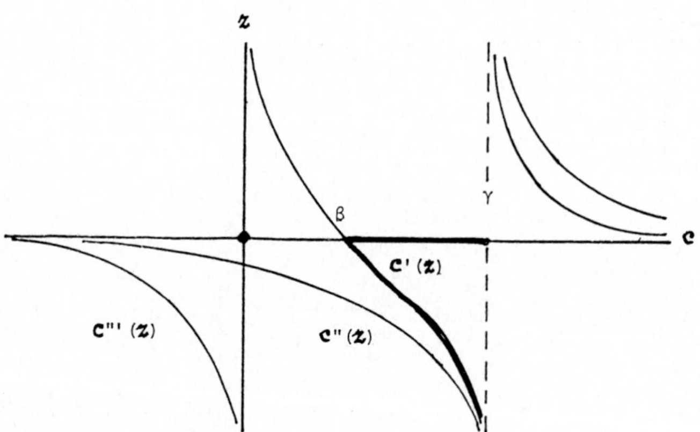

fig. 5: eigenvalues of single hyperexponential source

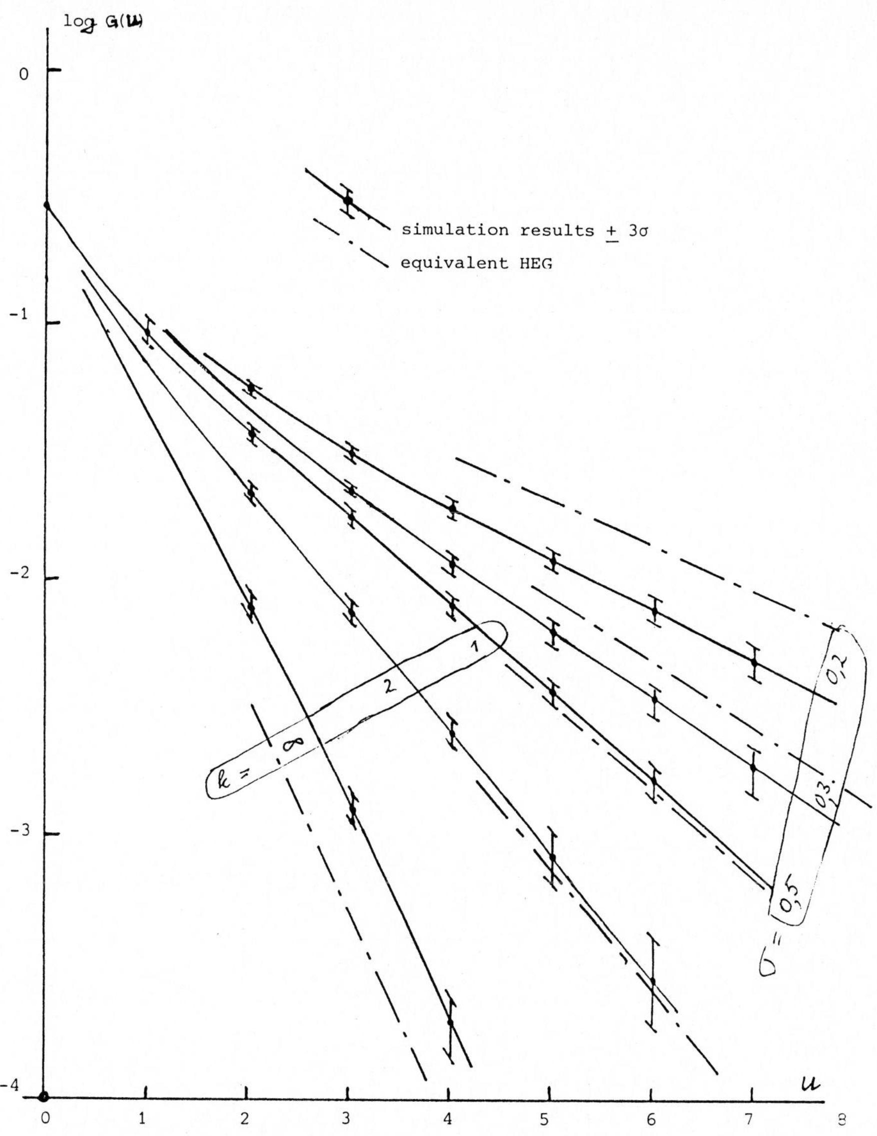

Fig. 6: simulation results and equiv. HEG asymptotes for phase-k and hyperexp. sources

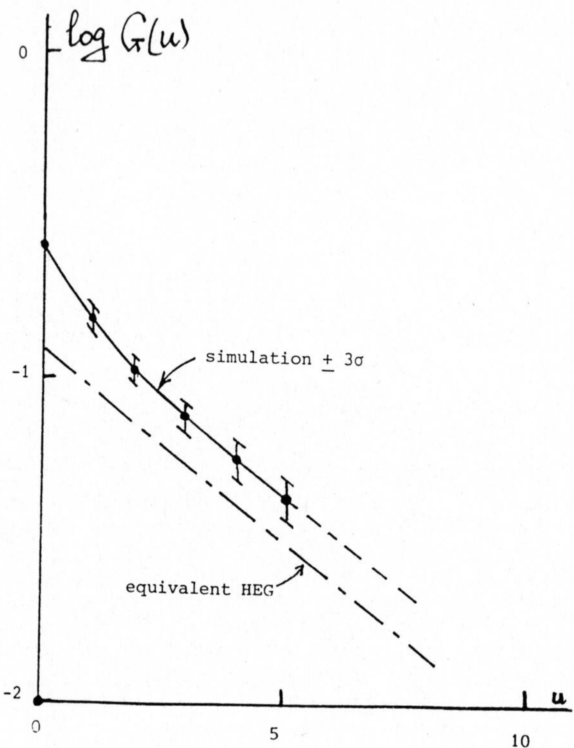

fig. 7: simulation results and equiv. HEG asymptote for NEG example

TELETRAFFIC ISSUES in an Advanced Information Society
ITC-11
Minoru Akiyama (Editor)
Elsevier Science Publishers B.V. (North-Holland)

DISCRETE-TIME PRIORITY QUEUES WITH PARTIAL INTERFERENCE[1]

Moshe SIDI

Department of Electrical Engineering
Technion - Israel Institute of Technology
Haifa 32000, Israel

ABSTRACT

A class of discrete time priority queueing systems with partial interference is considered. Packet-radio communication networks that use a certain mode of operation fall into this class. In these systems N nodes share a common channel to transmit their packets. One node uses a random access scheme while other nodes access the channel according to their preassigned priorities. Packet arrivals are modeled as discrete-time batch processes, and packets are forwarded through the network according to fixed prescribed probabilities.

Steady-state analysis of the class of systems under consideration is provided. In particular, we present a recursive method for the derivation of the joint generating function of the queue lengths distribution at the nodes in steady-state. The condition for steady-state is also derived. A simple example demonstrates the general analysis and provides some insights into the behavior of systems with partial interference.

1. INTRODUCTION

The survey paper by Kobayashi and Konheim [1] discusses many models of discrete-time queueing systems. Such systems have been receiving increased attention in recent years [2-4] due to their usefulness in modeling and analyzing various types of communication systems. Packet-switched communication networks with point-to-point links between the nodes, where data packets are of a fixed length, motivated most of these models. The models in [2-4] are of tandem nature since in point-to-point networks the transport of a packet from its source to its destination involves the transmission of the packet over a succession of links. The fixed packet length assumption induces the discrete-time nature of the models.

In this paper we consider a class of discrete-time priority queueing systems with partial interference. Consideration of these systems have been primarily motivated by the class of packet-switched communication networks called the multi-access/broadcast networks, or packet-radio networks. In these communication networks all nodes share a common channel through which they transmit their packets and from which they extract packets destined to them, hence the multi-access nature of these networks. In addition, when a node transmits a packet through the shared channel, all nodes that are within its transmission range hear this transmission, thus inducing the broadcast nature of the system.

[1]This research was carried out at the Massachusetts Institute of Technology, Laboratory for Information and Decision Systems with partial support provided by the National Science Foundation under grant NSF-ECS-831098.

We assume that the channel time axis is slotted into intervals of size equal to the transmission time of a packet. All packets are assumed to be of fixed and equal size. The nodes are synchronized so that they may start transmission of a packet only at the beginning of a slot, hence the discrete-time nature of the system. All nodes are assumed to have *infinite* buffers.

One of the most crucial issues in multi-access networks is the protocol required to transmit packets on a shared channel in a distributed environment. For a survey of multi-access protocols the reader is referred to [5]. The design and analysis of multi-access protocols is not trivial. This is due to the following two facts that hold for packet-radio networks: (i) If two or more nodes transmit packets during the same slot to the same node, then the overlap in transmission destroys all packets involved in the transmission; (ii) A transmitting node is unable to receive packets transmitted by other nodes of the system. These two facts together with the broadcast nature of the network give rise to statistical dependence between the queues at the nodes of the network. In most cases this dependence is rather complicated and therefore, there is little hope to obtain analytical results for general multi-access protocols and for general network configurations. The purpose of this paper is to analyze a rather general network configuration with a specific mode of operation.

One mode of operation that can be accomplished in multi-access networks is a conflict-free mode. This can be achieved if every node knows perfectly which are the nodes that have packets ready for transmission at the beginning of each slot. This is possible in systems that have a central scheduler that schedules the transmissions according to information it receives from the nodes, or in systems where the nodes exchange this information between themselves [6]. Generally, any order of transmission can be used, in particular, priority can be easily implemented. Yet, if there are some nodes that cannot exchange information with the scheduler or with other nodes, on which nodes have packets ready for transmission, then their transmissions cannot be accommodated in a conflict-free mode of operation and they should use some random access scheme [5].

The class of discrete-time queueing systems that we consider in this paper consists of systems having $N-1$ nodes that access the channel in a conflict-free mode according to fixed priorities that are preassigned to them. No two nodes have the same priority and a given node is allowed to use the channel in a given slot only if it has a packet ready for transmission and all nodes with higher priority have empty queues. In addition, there is an extra node in the system that cannot be accommodated in the conflict-free mode of operation and therefore is allowed to use the channel in any slot on a random basis. If the node uses the channel along with any other node then their packets are destroyed and must be retransmitted, hence the interfering feature of the systems under consideration.

To enhance the network structure of the problem we attach to each node a given probability distribution that indicates the probabilities that a packet transmitted by the node is forwarded to one of the other nodes or to the outside of the system.

Outside sources feed the nodes of the system with new packets. An important feature of this paper is that these sources are allowed to depend on each other. Thus we are able to characterize a rather general class of batch arrival processes.

Several discrete-time queueing systems that have been previously investigated [7-9] are related to the class of systems considered in this paper. In [7] a "loop system" in which nodes transmit packets only to the outside of the system, the arrival processes are independent and there is no interference, has been considered. In [4] and [8] two-node systems have been analyzed and in [9] no interference is allowed.

The paper is organized as follows: In Section 2 we describe the model along with the assumptions and several definitions and notations that we use throughout the paper. In Section 3 we present the steady-state analysis of the class of systems under consideration. In particular we develop a method for deriving the joint generating function of the queue lengths at the nodes and we give the ergodicity condition for the system. Moments of the queue lengths at the nodes can be derived from the generating function and average time delays can be obtained by using Little's law [10]. Finally, in Section 4 we give an example that demonstrates the general analysis and provides some insight into the behavior of systems with partial interference.

2. MODEL DESCRIPTION

We consider a discrete-time queueing system in which the time axis is divided into intervals of equal size referred to as slots. The slots correspond to the transmission time of a packet and all packets are assumed to be of the same fixed size. The system consists of N nodes and packets arrive randomly to the nodes from N sources that in general, may be correlated. Let $A_i(t)$, $i=1,2,...,N$, $t=0,1,2,...$ be the number of packets entering node i from its corresponding source during the time interval $(t,t+1)$. The input process $\{A_i(t)\}_{i=1}^{N}$ $t=0,1,2,...$ is assumed to be a sequence of independent and identically distributed random vectors with integer-valued elements. Let the corresponding probability distribution and generating function of the input processes be:

$$a(i_1,i_2,\cdots,i_N)=Prob\{A_1(t)=i_1,A_2(t)=i_2,\cdots,A_N(t)=i_N\}$$

$$i_j=0,1,2,\cdots \quad 1\le j\le N \tag{1a}$$

$$F(\underline{z}) = E\left\{\prod_{i=1}^{N} z_i^{A_i(t)}\right\} \tag{1b}$$

where we use the notation $\underline{z} = (z_1,z_2,\cdots,z_N)$.

All nodes share a common channel for transmission of their packets, and transmissions are started only at the beginning of a slot. No more than one packet may be transmitted in any given time slot by a single node. Using some conflict-free protocol the channel is made available to nodes $i=1,2,\cdots,N-1$ according to a fixed priority. Specifically, node i $(1\le i\le N-1)$ transmits the packet at the head of its queue whenever the queues at nodes $1,2,\cdots,i-1$ are empty and the one at node i is nonempty. Node N is a special node that cannot participate in the conflict-free protocol and therefore apply a random access protocol. At the beginning of each slot for which the queue at node N is nonempty, a coin with probability of success p is tossed. In case of a success node N transmits the packet at the head of its queue; otherwise it remains silent. Whenever node N transmits while another node i $(1\le i\le N-1)$ is also transmitting, then both transmissions are unsuccessful and the two nodes must retransmit the packets at the head of their queues according to the protocols described above.

In any case, when a node i $(1\le i\le N)$ transmits a packet successfully, then the packet joins node j $(1\le j\le N, j\ne i)$ with probability $\vartheta_i(j)$ or leaves the system with probability $\vartheta_i(0)$. We assume here that $\vartheta_i(i)=0$. All packets received by a node from an outside source or from other nodes, are buffered in a common outgoing queue and transmitted on a first-come first-served basis. It is assumed that packets indeed arrive at every node of the system, so that there is no node that is empty with probability 1 (in other words, nodes that are alwasys empty are ignored). Finally we assume that the buffers at the nodes are *infinite*. A schematic figure of a node i in the system is depicted in Fig. 1.

3. STEADY-STATE ANALYSIS

To describe the evolution of the queue contents at the nodes, we need several definitions. Let $L_i(t)$ $1\le i\le N$, $t=0,1,2,...$ be the number of packets at node i at time t and let $U(L_i(t))$ $(1\le i\le N, t=0,1,2,...)$ be a binary-valued random variable that takes value 1 if $L_i(t)>0$ and 0 otherwise. Let V be a binary-valued random variable that takes values 1 and 0 with probabilities p and $\bar{p}=1-p$ respectively. Also let $D_i^j(t)$, $1\le i\le N$, $0\le j\le N$, $t=0,1,2,...$ be a binary-valued random variable that takes value 1 if a packet is successfully transmitted from node i to node j at time t, where $j=0$ corresponds to the case that the packet leaves the system.

Using these definitions it is easy to see that the system under consideration evolves for $t=0,1,2,\cdots$ as follows:
For $1\le i\le N$,

$$L_i(t+1) = L_i(t)+A_i(t)+\sum_{m=1}^{N} D_m^i(t)$$

$$-V_i(t)U(L_i(t))\prod_{m=1}^{i-1}[1-U(L_m(t))] \tag{2a}$$

where

$$V_i(t) = \begin{cases} 1-VU(L_N(t)) & 1\le i\le N-1 \\ V & i=N \end{cases} \tag{2b}$$

Notice that $V_i(t)$ is a binary valued random variable and for $1\le i\le N-1$ it can be interpreted as the interference indicator at time t, i.e. it indicates whether or not node N interferes with the transmission of node i at time t. Clearly, $\{L_i(t)\}_{i=1}^{N}$ is a vector Markov chain. Assume that this Markov chain is ergodic (we shall derive the condition for this later), let us consider the steady-state joint generating function of the queue lengths distribution,

$$G(\underline{z}) = \lim_{t\to\infty} E\left\{\prod_{i=1}^{N} z_i^{L_i(t)}\right\} \tag{3}$$

For notational convenience, let us define the following boundary generating functions:

$$G_0(\underline{z}) = G(\underline{z}) \tag{4a}$$

$$G_i(\underline{z}) = G(\underline{z})\Big|_{z_1=z_2=\cdots=z_i=0} \quad 1\le i\le N \tag{4b}$$

$$\hat{G}_i(\underline{z}) = G_i(\underline{z})\Big|_{z_N=0} \quad 0\le i\le N-1 \tag{4c}$$

Notice that by our definition $G_N(\underline{z})=\hat{G}_{N-1}(\underline{z})$ is a constant representing the steady-state probability that the system will be empty. Finally let us define the following polynoms:

$$Q_i(\underline{z}) = \vartheta_1(0) + \sum_{m=1}^{N} \vartheta_i(m)z_m \quad 1\le i\le N \tag{5}$$

Theorem 1

With the above notations the following holds:

$$G(\underline{z})=F(\underline{z})\{G_N(\underline{z})+[G_{N-1}(\underline{z})-G_N(\underline{z})][\bar{p}+pz_N^{-1}Q_N(\underline{z})]+$$
$$+\sum_{i=1}^{N-1}[\hat{G}_{i-1}(\underline{z})-\hat{G}_i(\underline{z})]z_i^{-1}Q_i(\underline{z})+ \tag{6}$$
$$+\sum_{i=1}^{N-1}[G_{i-1}(\underline{z})-G_i(\underline{z})-\hat{G}_{i-1}(\underline{z})+\hat{G}_i(\underline{z})][p+\bar{p}\,z_i^{-1}Q_i(\underline{z})]\}$$

The formal proof of Theorem 1 is straightforward. Let us give here an intuitive explanation for (6). The right-hand side of (6) is a multiplication of the generating function of the joint arrival process, that by our assumptions is independent of the state of the system, and an expression that indicates, for the various states that the system may be at, which node is transmitting and how packets are moved within the network. Specifically, $G_N(\underline{z})$ corresponds to the case that the queues at all nodes are empty. $G_{N-1}(\underline{z})-G_N(\underline{z})$ corresponds to the situation that all nodes except node N are empty, therefore with probability p a packet leaves node N and joins another node or leaves the system according to the probabilities $\vartheta_N(j)$ $0\le j\le N$. $\hat{G}_{i-1}(\underline{z})-\hat{G}_i(\underline{z})$ for $1\le i\le N-1$ corresponds to the situation that node N is empty as well as nodes $1,2,\cdots,i-1$ and node i has a packet for transmission. Then, a packet leaves node i and joins another node or leaves the system according to the probabilities $\vartheta_i(j)$ $0\le j\le N$. Finally, the term $G_{i-1}(\underline{z})-G_i(\underline{z})-\hat{G}_{i-1}(\underline{z})+\hat{G}_i(\underline{z})$ for $1\le i\le N-1$ corresponds to the case that nodes $1,2,\cdots,i-1$ are empty and nodes i and N have both packets for transmission. In this case with probability p the two nodes interfere and no packet is moved, while if node N remains silent (this happens with probability $\bar{p}=1-p$) then a packet leaves node i and joins another node or leaves the system as before. Rearranging (6) we obtain:

$$G(\underline{z}) = F(\underline{z})\frac{\sum_{i=1}^{N}H_i(\underline{z})G_i(\underline{z}) + \sum_{i=0}^{N-1}\hat{H}_i(\underline{z})\hat{G}_i(\underline{z})}{1-F(\underline{z})[p+\bar{p}\,z_1^{-1}Q_1(\underline{z})]} \tag{7a}$$

where

$$H_i(\underline{z})=\begin{cases}\bar{p}\,[z_{i+1}^{-1}Q_{i+1}(\underline{z})-z_i^{-1}Q_i(\underline{z})] & 1\le i\le N-2\\ 1-2p+pz_N^{-1}Q_N(\underline{z})-\bar{p}\,z_{N-1}^{-1}Q_{N-1}(\underline{z}) & i=N-1\\ p[1-z_N^{-1}Q_N(\underline{z})] & i=N\end{cases} \tag{7b}$$

and

$$\hat{H}_i(\underline{z})=\begin{cases}p[1-z_1^{-1}Q_1(\underline{z})] & i=0\\ p[z_{i+1}^{-1}Q_{i+1}(\underline{z})-z_i^{-1}Q_i(\underline{z})] & 1\le i\le N-2\\ p[1-z_{N-1}^{-1}Q_{N-1}(\underline{z})] & i=N-1\end{cases} \tag{7c}$$

In (7) we encounter a common phenomena in interfering queues, namely that the generating function $G(\underline{z})$ is expressed in terms of several boundary functions. In order to uniquely determine $G(\underline{z})$ in (7) we will have to determine $2N-1$ boundary functions,[2] $G_i(\underline{z})$ $1\le i\le N$ and $\hat{G}_i(\underline{z})$ $0\le i\le N-2$. In what follows, we develop the method for obtaining these boundary functions. The basic idea is to first express $\hat{G}_i(\underline{z})$ $i=0,1,\cdots,N-2$ (in this order) in terms of $\hat{G}_j(\underline{z})$ $i+1\le j\le N-1$. Then $G_i(\underline{z})$ $i=1,2,\cdots,N-1$ is expressed in terms of $\hat{G}_j(\underline{z})$ $0\le j\le N-1$ and $G_j(\underline{z})$ $i+1\le j\le N$. Finally the constant $G_N(\underline{z})$ is determined from the normalization condition and using backward substitutions all the boundary functions are determined. Along the above process we mainly use the analytic properties of the generating function $G(\underline{z})$ in the polydisc $|z_i|<1$ $1\le i\le N$.

[2]Notice that in a general system where each node can interfere with any other node we might have up to 2^N-1 boundary functions to determine. An example for such a system is a network that all nodes use a random access policy.

In order to proceed we shall need the following Lemma:

Lemma 1: Let $F(\underline{z})$ be the generating function of the joint arrival process (1b), $Q_1(\underline{z})$ the function defined in (5) and $0\le p\le 1$. Then for given $|z_i|<1$ $2\le i\le N$, the following equation in z_1,

$$F(\underline{z})[pz_1 + (1-p)Q_1(\underline{z})] = z_1 \tag{8}$$

has a unique solution $z_1=z_1(z_2,z_3,\cdots,z_N)$ in the unit circle $|z_1|<1$.

Proof: Let $|z_1|=1$ and $|z_i|<1$, $2\le i\le N$. We distinguish between two cases: The first is the case that packets do arrive to some node l, $2\le l\le N$, from its corresponding source. The second is the case that no packets arrive to nodes $2\le l\le n$ from their corresponding sources. Our assumption that packets do arrive to all nodes implies that in the latter case, packets do arrive at node 1 from its corresponding source and it routes some of them to at least one of the nodes l, $2\le l\le N$.

Case 1. There exists some node l $(2\le l\le N)$ for which the probability that a packet will arrive to it from its corresponding source is strictly positive, i.e., there exists $a(i_1,i_2,\cdots,i_N)>0$ for some i_1 and some $i_l>0$ $(2\le l\le N)$. Therefore,

$$|F(\underline{z})[pz_1+(1-p)Q_1(\underline{z})]|\le|F(\underline{z})| =$$
$$= \left|\sum_{i_1=0}^{\infty}\sum_{i_2=0}^{\infty}\cdots\sum_{i_N=0}^{\infty}a(i_1,i_2,\cdots,i_N)\prod_{j=1}^{N}z_j^{i_j}\right|$$
$$\le \sum_{i_1=0}^{\infty}\sum_{i_2=0}^{\infty}\cdots\sum_{i_N=0}^{\infty}a(i_1,i_2,\cdots,i_N)|z_l^{i_l}| \tag{9}$$
$$< \sum_{i_1=0}^{\infty}\sum_{i_2=0}^{\infty}\cdots\sum_{i_N=0}^{\infty}a(i_1,i_2,\cdots,i_N)=1=|z_1|.$$

Hence, applying Rouche's theorem [11] the claim is proved in this case.

Case 2. Packets arrive at node 1 and it routes some of them to at least one of the nodes l $(2\le l\le N)$, i.e., there exists $\vartheta_1(l)>0$ for some $2\le l\le N$. Therefore,

$$\left|F(\underline{z})[pz_1+(1-p)Q_1(\underline{z})]\right|\le\left|p+(1-p)Q_1(\underline{z})\right| = \tag{10}$$
$$= \left|p+(1-p)[\vartheta_1(0)+\sum_{i=2}^{N}\vartheta_1(i)z_i]\right|<p+(1-p)=1=|z_1|$$

Hence, applying Rouche's theorem the proof is completed.

∎

Let $\sigma_1(z_2,z_3,\cdots,z_N)$ (for simplicity σ_1) denote the unique solution of (8). Let $\underline{z}^{(1)}$ denote the vector $\underline{z}$ with its first component z_1 replaced by σ_1. Using a similar proof as for lemma 1 we can show that for $|z_i|<1$, $3\le i\le N$, the following equation in z_2,

$$F(\underline{z}^{(1)})[pz_2+(1-p)Q_2(\underline{z}^{(1)})] = z_2 \tag{11}$$

has a unique solution in the unit circle $|z_2|<1$. Let $\sigma_2(z_3,z_4,\cdots,z_N)$ denote this solution and $\underline{z}^{(2)}$ denote the vector $\underline{z}$ with its first component z_1 replaced by $\sigma_1(\sigma_2(z_3,z_4,\cdots,z_N),z_3,\cdots,z_N)$ and its second component z_2 replaced by $\sigma_2(z_3,z_4,\cdots,z_N)$. Continuing this procedure we have the following lemma that recursively determines the unique functions $\sigma_i(z_{i+1},z_{i+2},\cdots,z_N)$ $2\le i\le N-1$ as follows:

Lemma 2: With the above notations and for $2\le i\le N-1$, the following equation in z_i,

$$F(\underline{z}^{(i-1)})[pz_i+(1-p)Q_i(\underline{z}^{(i-1)})] = z_i \tag{12}$$

has a unique solution in the unit circle $|z_i|<1$ for $|z_j|<1$, $i+1\le j\le N$. Here $\underline{z}^{(i-1)}$ denotes the vector $\underline{z}$

with the variables z_j replaced by σ_j for $1 \le j \le i-1$. This unique solution is denoted by $\sigma_i(z_{i+1}, z_{i+2}, \cdots, z_N)$.
The proof of this Lemma is similar to the proof of Lemma 1.

If we let $p=0$ and $z_N=0$ in Lemma 1 and 2 and we use the recursions defined by (8) and (12) for this case, then the unique functions $\hat{\sigma}_i(z_{i+1}, z_{i+2}, \cdots, z_{N-1})$ $1 \le i \le N-2$ are defined, i.e., $\hat{\sigma}_1$ is the unique solution in the unit circle $|z_1|<1$ of the equation $F(\hat{\underline{z}})Q_1(\hat{\underline{z}})=z_1$ where $\hat{\underline{z}}=(z_1, z_2, \cdots, z_{N-1}, 0)$ given that $|z_i|<1$ $2 \le i \le N-1$. $\hat{\sigma}_i$ $2 \le i \le N-2$ is the unique solution in the unit circle $|z_i|<1$ of the equation $F(\hat{\underline{z}}^{(i-1)})Q(\hat{\underline{z}}^{(i-1)})=z_i$ where $\hat{\underline{z}}^{(i-1)}$ is the vector $\hat{\underline{z}}$ with $z_1=\hat{\sigma}_1$, $z_2=\hat{\sigma}_2, \cdots, z_{i-1}=\hat{\sigma}_{i-1}$ given that $|z_j|<1$ $i+1 \le j \le N-2$. We are now armed enough to attack the problem of determining the $2N-1$ boundary functions.

Determination of the boundary functions $\hat{G}_i(\underline{z})$ $0 \le i \le N-2$

Letting $z_N \to 0$ in (6) we obtain:

$$\hat{G}_0(\underline{z}) = F(\hat{\underline{z}})\{\hat{G}_{N-1}(\underline{z})+pQ_N(\hat{\underline{z}})G'_{N-1}(\underline{z}) + \tag{13a}$$

$$+\sum_{i=1}^{N-1}[\hat{G}_{i-1}(\underline{z})-\hat{G}_i(\underline{z})]z_i^{-1}Q_i(\hat{\underline{z}})\}$$

where $\hat{\underline{z}}=(z_1, z_2, \cdots, z_{N-1}, 0)$ and

$$G'_{N-1}(\underline{z}) = \frac{dG_{N-1}(\underline{z})}{dz_N}\Big|_{z_N=0} \tag{13b}$$

Notice that $G'_{N-1}(\underline{z})$ is a constant.
Rearranging (13a) and noticing that by definition $\hat{G}_{N-1}(\underline{z})=G_N(z)$ we obtain:

$$\hat{G}_0(\underline{z})=F(\hat{\underline{z}})\frac{E(\hat{\underline{z}})+\sum_{i=1}^{N-2}D_i(\hat{\underline{z}})\hat{G}_i(\underline{z})}{1-F(\hat{\underline{z}})z_1^{-1}Q_1(\hat{\underline{z}})} \tag{14a}$$

where,

$$E(\hat{\underline{z}})=[1-z_{N-1}^{-1}Q_{N-1}(\hat{\underline{z}})]\hat{G}_{N-1}(\underline{z})+pQ_N(\hat{\underline{z}})G'_{N-1}(\underline{z}) \tag{14b}$$

$$D_i(\hat{\underline{z}})=z_{i+1}^{-1}Q_{i+1}(\hat{\underline{z}})-z_i^{-1}Q_i(\hat{\underline{z}}) \tag{14c}$$

Notice that in (14) the boundary function $\hat{G}_0(\underline{z})$ is expressed in terms of the boundary functions $\hat{G}_i(\underline{z})$ $1 \le i \le N-1$ and the constant $G'_{N-1}(\underline{z})$. Now using the analytic property of $G_0(\underline{z})$ we immediately obtain the following result:

Theorem 2: Let $\hat{\sigma}_1$ and $\hat{\underline{z}}^{(1)}$ be as defined before. Then,

$$\hat{G}_1(\underline{z})=F(\hat{\underline{z}}^{(1)})\frac{E(\hat{\underline{z}}^{(1)})+\sum_{i=2}^{N-2}D_i(\hat{\underline{z}}^{(1)})\hat{G}_i(\underline{z})}{1-F(\hat{\underline{z}}^{(1)})z_2^{-1}Q_2(\hat{\underline{z}}^{(1)})} \tag{15}$$

This is true since $\hat{G}_0(\underline{z})$ is an analytic function in the polydisk $|z_i|<1$ $1 \le i \le N-1$. Then in this polydisk whenever the denominator of $\hat{G}_0(\underline{z})$ vanishes, the numerator must also vanish. Since the denominator of $\hat{G}_0(\underline{z})$ vanishes at $\hat{\sigma}_1$, we have from (14) that:

$$E(\hat{\underline{z}}^{(1)}) + \sum_{i=2}^{N-2}[z_{i+1}^{-1}Q_{i+1}(\hat{\underline{z}}^{(1)})-z_i^{-1}Q_i(\hat{\underline{z}}^{(1)})]\hat{G}_i(\underline{z}) = \tag{16}$$

$$= [\hat{\sigma}_1^{-1}Q_1(\hat{\underline{z}}^{(1)})-z_2^{-1}Q_2(\hat{\underline{z}}^{(1)})]\hat{G}_1(\underline{z})$$

which together with the fact that $F(\hat{\underline{z}}^{(1)})\sigma_1^{-1}Q_1(\hat{\underline{z}}^{(1)})=1$ imply (15).

Now, exploiting the similarity between (14) and (15) and repeating the above procedure for $i=2,3, \cdots, N-2$ we obtain the following result:

Theorem 3: Let $\hat{\sigma}_i$ and $\hat{\underline{z}}^{(i)}$ $2 \le i \le N-2$ be as defined before. Then for $2 \le i \le N-2$,

$$\hat{G}_i(\underline{z})=F(\hat{\underline{z}}^{(i)})\frac{E(\hat{\underline{z}}^{(i)})+\sum_{j=i+1}^{N-2}D_j(\hat{\underline{z}}^{(i)})\hat{G}_j(\underline{z})}{1-F(\hat{\underline{z}}^{(i)})z_{i+1}^{-1}Q_{i+1}(\hat{\underline{z}}^{(i)})} \tag{17}$$

The proof of (17) is similar to that of (15).

Now, using (17) for $i=N-2$ we have,

$$\hat{G}_{N-2}(\underline{z})=F(\hat{\underline{z}}^{(N-2)})\frac{E(\hat{\underline{z}}^{(N-2)})}{1-F(\hat{\underline{z}}^{(N-2)})z_{N-1}^{-1}Q_{N-1}(\hat{\underline{z}}^{(N-2)})} \tag{18}$$

and since $\hat{G}_{N-2}(\underline{z})$ is an analytic function for $|z_{N-1}|<1$ we obtain from (18) and (14b) that

$$pG'_{N-1}(\underline{z}) = \hat{G}_{N-1}(\underline{z})\frac{\sigma_{N-1}^{-1}Q_{N-1}(\hat{\underline{z}}^{(N-1)}) - 1}{Q_N(\hat{\underline{z}}^{(N-1)})} \tag{19}$$

Substituting (19) in (18) we get $\hat{G}_{N-2}(\underline{z})$ expressed in terms of the constant $\hat{G}_{N-1}(\underline{z})$. Using (17) for $i=N-3, N-4, \cdots, 2$, and then (15) and (14) we obtain all the functions $\hat{G}_i(\underline{z})$ $0 \le i \le N-2$ expressed in terms of the constant $\hat{G}_{N-1}(\underline{z})=G_N(\underline{z})$. Specifically, as we shall need it later let us define the function $k(\hat{\underline{z}})$ as follows:

$$k(\hat{\underline{z}}) = \hat{G}_0(\underline{z}) / G_N(\underline{z}) \tag{20}$$

Determination of the boundary functions $G_i(\underline{z})$ $1 \le i \le N-2$

To obtain the boundary functions $G_i(\underline{z})$ $1 \le i \le N-2$ we use a similar procedure as for $\hat{G}_i(\underline{z})$ $0 \le i \le N-2$. Let us first rewrite (7a) as follows:

$$G(\underline{z})=F(\underline{z})\frac{H(\underline{z}) + \sum_{i=1}^{N}H_i(\underline{z})G_i(\underline{z})}{1-F(\underline{z})[p+\bar{p}\,z_1^{-1}Q_1(\underline{z})]} \tag{21a}$$

where $H_i(z)$ $1 \le i \le N$ are defined in (7b) and $H(\underline{z})$ is a known function up to the constant $G_N(\underline{z})$. $H(\underline{z})$ is given by:

$$H(\underline{z})=\sum_{i=0}^{N-1}\hat{H}_i(\underline{z})\hat{G}_i(\underline{z}) \tag{21b}$$

$\hat{H}_i(\underline{z})$ are defined in (7b).

Using Lemmas 1 and 2 we immediately obtain the following result:

Theorem 4: Let σ_i, $\underline{z}^{(i)}$ $1 \le i \le N-1$ be as defined in Lemmas 1 and 2. Then for $1 \le i \le N-2$ we have:

$$G_i(\underline{z})=F(\underline{z}^{(i)})\frac{H(\underline{z}^{(i)}) + \sum_{j=i+1}^{N}H_j(\underline{z}^{(i)})G_j(\underline{z})}{1-F(\underline{z}^{(i)})[p+\bar{p}\,z_{i+1}^{-1}Q_{i+1}(\underline{z}^{(i)})]} \tag{22a}$$

and

$$G_{N-1}(\underline{z})= -\frac{H(\underline{z}^{(N-1)}) + H_N(\underline{z}^{(N-1)})G_N(\underline{z})}{H_{N-1}(\underline{z}^{(N-1)})} \tag{22b}$$

We will demonstrate how (22a) is proved for $i=1$. Then by induction one can easily obtain (22a) and (22b). Since $G(\underline{z})$ is an analytic function for $|z_i|<1$ $1 \le i \le N$ and since the denominator of $G(\underline{z})$ vanishes at σ_1, we have from (21a) that:

$$H(\underline{z}^{(1)}) + \sum_{i=2}^{N}H_i(\underline{z}^{(1)})G_i(\underline{z}) + H_1(\underline{z}^{(1)})G_1(\underline{z}) = 0 \tag{23}$$

Using the definition of $H_1(\underline{z}^{(1)})$ from (7b), i.e. $H_1(\underline{z}^{(1)})=\bar{p}\,[z_2^{-1}Q_2(\underline{z}^{(1)})-\sigma_1^{-1}Q_1(\underline{z}^{(1)})]$ and the fact that $F(\underline{z}^{(1)})[p+\bar{p}\,\sigma_1^{-1}Q_1(\underline{z}^{(1)})]=1$ we get immediately (22a) for $i=1$.

Now in (22b) $G_{N-1}(\underline{z})$ is expressed in terms of the constant $G_N(\underline{z})$. using (22a) for $i=N-2, N-3, \cdots, 1$ we finally have all the boundary functions $G_i(\underline{z})$ $1 \le i \le N-1$ expressed in terms of the constant $G_N(\underline{z})$.

Now that we have already determined $\hat{G}_i(\underline{z})$ $0 \le i \le N-2$ and $G_i(\underline{z})$ $1 \le i \le N-2$ in terms of the constant $G_N(\underline{z})$, the problem is reduced to that of determining this constant.

Determination of the constant $G_N(\underline{z})$

To determine the constant $G_N(\underline{z})$ let us first prove the following:

Theorem 5: For $1 \le l \le N$ let,

$$r_l = \frac{\partial F(\underline{z})}{\partial z_l}\Big|_{z_1=z_2=\cdots=z_N=1} \tag{24a}$$

and

$$\lambda_l = r_l + \sum_{j=1}^{N} \lambda_j \vartheta_j(l) \tag{24b}$$

Then the following holds:

$$\lambda_i = \bar{p}\,[G_{i-1}(\underline{1}) - G_i(\underline{1})] + p\,[\hat{G}_{i-1}(\underline{1}) - \hat{G}_i(\underline{1})] \quad 1 \le i \le N-1 \tag{25a}$$

$$\lambda_N = p\,[G_{N-1}(\underline{1}) - G_N(\underline{1})] \tag{25b}$$

where,

$$G_i(\underline{1}) = G_i(\underline{z})\Big|_{z_{i+1}=z_{i+2}=\cdots=z_N=1} \quad 0 \le i \le N-1 \tag{26a}$$

$$\hat{G}_i(\underline{1}) = \hat{G}_i(\underline{z})\Big|_{z_{i+1}=z_{i+2}=\cdots=z_{N-1}=1} \quad 0 \le i \le N-2 \tag{26b}$$

and $G_N(\underline{1}) = \hat{G}_{N-1}(\underline{1})$ is just the constant we are looking for.

Proof: For $1 \le i \le N$, let us derive both sides of eq. (6) with respect to z_i and substitute $z_1 = z_2 = \cdots = z_N = 1$. Then for $1 \le i \le N-1$ we obtain

$$0 = r_i + [G_{N-1}(\underline{1}) - G_N(\underline{1})]p\,\vartheta_N(i) + \tag{27a}$$

$$+ \sum_{\substack{j=1\\ j\ne 1}}^{N-1} [\hat{G}_{j-1}(\underline{1}) - \hat{G}_j(\underline{1})]\vartheta_j(i) - [\hat{G}_{i-1}(\underline{1}) - \hat{G}_i(\underline{1})] +$$

$$+ \sum_{\substack{j=1\\ j\ne 1}}^{N-1} [G_{j-1}(\underline{1}) - G_j(\underline{1}) - \hat{G}_{j-1}(\underline{1}) + \hat{G}_j(\underline{1})]\bar{p}\,\vartheta_j(i)$$

$$- \bar{p}\,[G_{i-1}(\underline{1}) - G_i(\underline{1}) - \hat{G}_{i-1}(\underline{1}) + \hat{G}_i(\underline{1})]$$

and

$$0 = r_N - p\,[G_{N-1}(\underline{1}) - G_N(\underline{1})] \tag{27b}$$

$$+ \sum_{i=1}^{N-1} [\hat{G}_{i-1}(\underline{1}) - \hat{G}_i(\underline{1})]\vartheta_i(N)$$

$$+ \sum_{i=1}^{N-1} [G_{i-1}(\underline{1}) - G_i(\underline{1}) - \hat{G}_{i-1}(\underline{1}) + G_i(\underline{1})]\bar{p}\,\vartheta_i(N)$$

where in (27) we used the fact that $G(\underline{1}) = G_0(\underline{1}) = 1$. Rearranging (27) we get for $1 \le i \le N-1$:

$$0 = r_i + [G_{N-1}(\underline{1}) - G_N(\underline{1})]p\,\vartheta_N(i) + \tag{28a}$$

$$+ \sum_{\substack{j=1\\ j\ne i}}^{N-1} \{\bar{p}\,[G_{j-1}(\underline{1}) - G_j(\underline{1})] + p\,[\hat{G}_{j-1}(\underline{1}) - \hat{G}_j(\underline{1})]\}\vartheta_j(i)$$

$$- \{\bar{p}\,[G_{i-1}(\underline{1}) - G_i(\underline{1})] + p\,[\hat{G}_{i-1}(\underline{1}) - G_i(\underline{1})]\}$$

and

$$0 = r_N - p\,[G_{N-1}(\underline{1}) - G_N(\underline{1})] \tag{28b}$$

$$+ \sum_{i=1}^{N-1} \{\bar{p}\,[G_{i-1}(\underline{1}) - G_i(\underline{1})] + p\,[\hat{G}_{i-1}(\underline{1}) - \hat{G}_i(\underline{1})]\}\vartheta_i(N)$$

In (28) we have N linear equations with N unknowns $\bar{p}\,[G_{i-1}(\underline{1}) - G_i(\underline{1})] + p\,[\hat{G}_{i-1}(\underline{1}) - \hat{G}_i(\underline{1})]$ for $1 \le i \le N-1$ and $p\,[G_{N-1}(\underline{1}) - G_N(\underline{1})]$. Apparently (25) solves these equations.

From (25) we obtain:

$$\sum_{i=1}^{N-1} \lambda_i = \bar{p}\,[1 - G_{N-1}(\underline{1})] + p\,[\hat{G}_0(\underline{1}) - \hat{G}_{N-1}(\underline{1})] = \tag{29}$$

$$= \bar{p}\,[1 - \lambda_N/p - G_N(\underline{1})] + p\,[\hat{G}_0(\underline{1}) - G_N(\underline{1})]$$

Therefore,

$$G_N(\underline{1}) - p\hat{G}_0(\underline{1}) = \bar{p}\,[1 - \lambda_N/p] - \sum_{i=1}^{N-1} \lambda_i \tag{30}$$

Recalling that $\hat{G}_0(\underline{z}) = k(\hat{\underline{z}})G_N(\underline{z})$ we finally have that:

$$G_N(\underline{1}) = \frac{\bar{p}\,[1 - \lambda_N/p] - \sum_{i=1}^{N-1} \lambda_i}{1 - pk(\hat{\underline{1}})} \tag{31}$$

where $k(\hat{\underline{1}}) = k(\hat{\underline{z}})\big|_{z_1=z_2=\cdots=z_{N-1}=1}$. (31) implies that the condition for steady-state is:

$$\sum_{i=1}^{N-1} \lambda_i < \bar{p}\,(1 - \lambda_N/p) \tag{32}$$

Rewriting (32) as:

$$\lambda_N < p\,(1 - \sum_{i=1}^{N-1} \lambda_i/\bar{p}) \tag{33}$$

we can explain the steady-state condition intuitively as follows: Clearly, node N is the bottleneck of the system. If it is heavily loaded, then the fraction of time that the channel is used by the other $N-1$ nodes is $\sum_{i=1}^{N-1} \lambda_i/\bar{p}$, so the fraction of time that the channel is available for node N for successful transmissions is $1 - \sum_{i=1}^{N-1} \lambda_i/\bar{p}$. As node N transmits with probability p when nonempty the rate of its successful transmissions is $p(1 - \sum_{i=1}^{N-1} \lambda_i/\bar{p})$ which for stability must be greater than the total arrival rate to the node. Consequently, (33) should hold.

Having obtained the joint generating function $G(\underline{z})$ we can derive, at least in principle, any moment of the queue lengths at the nodes. Specifically, if we denote by $\bar{L}_i$ the average queue length at node i in steady-state, then

$$L_i = \frac{\partial G(\underline{z})}{\partial z_i}\Big|_{z_1=z_2=\cdots=z_N=1} \tag{34}$$

Assuming that packets arrive at the nodes only at the end of a slot, then using Little's law [10] we may also obtain the average time delays at node i denoted by T_i as follows:

$$T_i = L_i/\lambda_i \tag{35}$$

where λ_i is the total arrival rate at node i as defined in (24b). The total average time delay in the system is obtained by applying Little's law to the whole system and it is given by:

$$T = \sum_{i=1}^{N} L_i / \sum_{i=1}^{N} r_i \tag{36}$$

where r_i is the arrival rate at node i from its corresponding source as defined in (24a). The total average delay T is clearly a function of the transmission probability p. Obviously, as p decreases, the total average delay increases since node N transmits rather rarely. Also when p increases the total average delay also increases since there are many conflicts in the transmissions. Consequently, there is some intermediate value of p (that depends on the arrival processes to the nodes) that minimizes the total average delay in the system. This will be demonstrated in the example given in Section 4.

4. EXAMPLE

In this section we will use a simple example in order to show some details of the general solution method developed in the previous section. The example consists of the network of Fig. 2, where packets arrive to nodes 1 , 2 , 3 and node 2 forwards its packets to node 1. Consequently, $Q_1(\underline{z})=Q_3(\underline{z})=1$; $Q_2(\underline{z})=z_1$ (here $\underline{z}=(z_1,z_2,z_3)$). We shall also assume that:

$$F(\underline{z}) = r_1 z_1 + r z_2 z_3 + 1 - r_1 - r$$

i.e., during each slot a packet arrives to node 1 with probability r_1, with probability r a packet arrives to both nodes 2 and 3 and with probability $1-r_1-r$ no packet arrives to the system. Then using (8), (12) for $z_3=0$, $p=0$, we obtain:

$$\hat{\sigma}_1 = 1-\frac{r}{1-r_1}$$

$$\hat{\sigma}_2 = (1-\frac{r}{1-r_1})^2$$

Using (19), (18) and (14) we have:

$$pG'_2(0,0,0) = \frac{r}{1-r_1-r}G(0,0,0)$$

$$\hat{G}(0,z_2,0) = G(0,0,0)$$

$$\hat{G}(z_1,z_2,0) = G(0,0,0)[1 + \frac{r_1}{1-r_1-r}z_1]$$

Using (31) we have that:

$$G(0,0,0) = \frac{\bar{p}(1-r/p)-(r_1+2r)}{1-p(1-r)/(1-r_1-r)}$$

and the condition for steady-state is:

$$\bar{p}(1-r/p) - (r_1+2r) > 0$$

From (8) and (11) we obtain:

$$\sigma_1(z_2,z_3) = (1-f(z_2,z_3) - r_1\bar{p} - \sqrt{\Delta})/2r_1\bar{p}$$

where,

$$f(z_1,z_3) = p(rz_2z_3 + 1 - r_1 - r)$$

$$\Delta = (1-f(z_2,z_3) - r_1\bar{p})^2 - 4r_1\bar{p}f(z_2,z_3)$$

and $\sigma_2(z_3)$ is the solution of

$$\sigma_2(z_3) = \sigma_1^2(\sigma_2(z_3),z_3)$$

in the unit circle $|\sigma_2|<1$.

From (15) and (17) we obtain:

$$G(0,0,z_3)=G(0,0,0)\frac{p(z_3^{-1}-1)+\frac{r_1p}{1-r_1-r}(1-\sigma_1(\sigma_2(z_3),z_3))}{1-2p+pz_3^{-1}-\bar{p}\,\sigma_1^{-1}(\sigma_2(z_3),z_3)}$$

and

$$G(0,z_2,z_3) = \{G(0,0,0)[p(1-z_3^{-1})+\frac{r_1p}{1-r_1-r}(1-\sigma_1(z_2,z_3))$$

$$+ G(0,0,z_3)(1-2p+pz_3^{-1}-\bar{p}\,z_2^{-1}\sigma_1(z_2,z_3)\}/$$

$$\{\bar{p}\,(\sigma_1^{-1}(z_2,z_3)-z_2\sigma_1(z_2,z_3)]\}$$

Finally we have that:

$$G(z_1,z_2,z_3) = F(z_1,z_2,z_3)\{G(0,0,0)[p(1-z_3^{-1})+$$

$$+\frac{r_1p}{1-r_1-r}(1-z_1)]++G(0,0,z_3)(1-2p+pz_3^{-1}-\bar{p}\,z_2^{-1}z_1)+$$

$$+G(0,z_2,z_3)\bar{p}\,(z_2^{-1}z_1-z_1^{-1})\}/[1-F(z_1,z_2,z_3)(p+\bar{p}\,z_1^{-1})]$$

The explicit expressions for the average delays in the system are too complicated to be given here. To give some insight into the behavior of this network we plotted these quantities in Fig. 3-5. In Fig. 3 T_1, T_2, T_3 and T are plotted as a function of $r=r_1$ for $p=0.4$. In Fig. 4, these quantities are plotted as a function of p for $r_1=r=0.05$. As we can see, for small values of p, the queue is built up only at node 3 (since it is rarely transmitting) while for large values of p, queues are built up at all the nodes and this is due to the interference.

As we see, there is an optimal transmission probability p^* that minimizes the total delay in the system. In Fig. 5 T_{min} -- the minimal total delay in the system is plotted as a function of $r=r_1$. It is intersting to mention that $p^* \approx 0.34$ and it is almost insensitive to the value of $r=r_1$. Also T_{min} is not very sensitive to small variations in p^*.

Acknowledgement

I would like to thank the Rothschild Foundation for their generous support and Professor R.G. Gallager for inviting me to spend the year at M.I.T.

REFERENCES

[1] H. Kobayashi and A.G. Konheim, "Queueing Models for Computer Communications System Analysis", *IEEE Transactions on Communications*, Vol. COM-25, pp. 2-28, Jan. 1977.

[2] J. Hsu and P.J. Burke, "Behavior of Tandem Buffers with Geometric Input and Markovian Output", *IEEE Transactions on Communications*, Vol. COM-24, pp. 358-361, March 1976.

[3] A.G. Konheim and M. Reiser, "Delay Analysis for Tandem Networks", *ICC*, pp. 265-269, June 1977.

[4] J.A. Morrison, "Two Discrete-Time Queues in Tandem", *IEEE Transactions on Communications*, Vol. COM-27, pp. 563-573, March 1979.

[5] F. Tobagi, "Multiaccess Protocols in Packet Communication Systems", *IEEE Transactions on Communications*, Vol. COM-28, pp. 468-488, April 1980.

[6] M. Scholl, "Multiplexing Techniques for Data Transmission Over Packet-Switched Radio Systems", Ph.D. Thesis, Computer Science Department, UCLA, 1976.

[7] A.G. Konheim and B. Meister, "Service in a Loop System", *J. Assoc. Mach.*, Vol. 19, pp. 92-108, 1972.

[8] M. Sidi and A. Segall, "Two Interfering Queues in Packet-Radio Networks", *IEEE Transactions on Communications*, Vol. COM-31, pp. 123-129, Jan. 1983.

[9] M. Sidi and A. Segall, "Structured Priority Queueing Systems with Applications to Packet-Radio Networks", *Performance Evaluation*, pp. 264-275, 1983.

[10] J.D.C. Little, *Operations Research*, Vol. 9, pp. 383-387, 1961.

[11] E.T. Copson, "Theory of Functions of a Complex Variable", Oxford University Press, London, 1948.

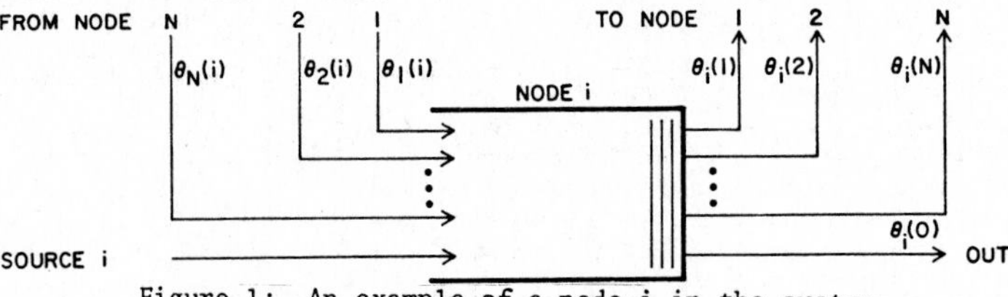

Figure 1: An example of a node i in the system

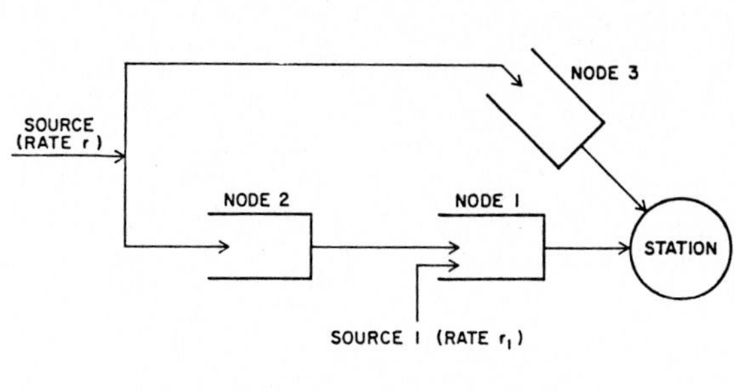

Figure 2: Example Network.

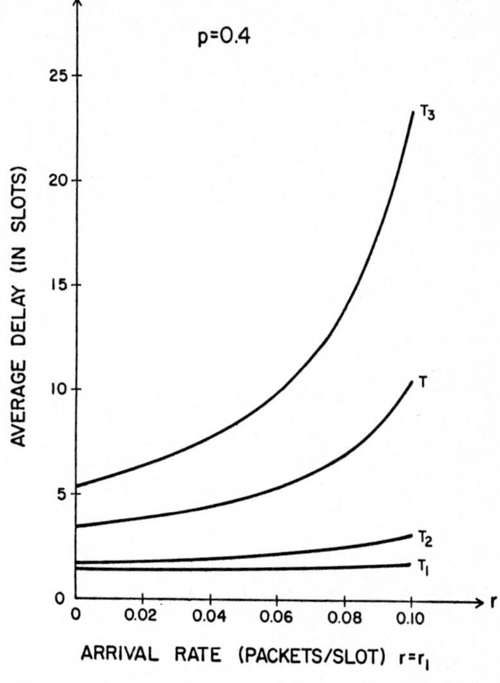

Figure 3: Average delays versus arrival rate.

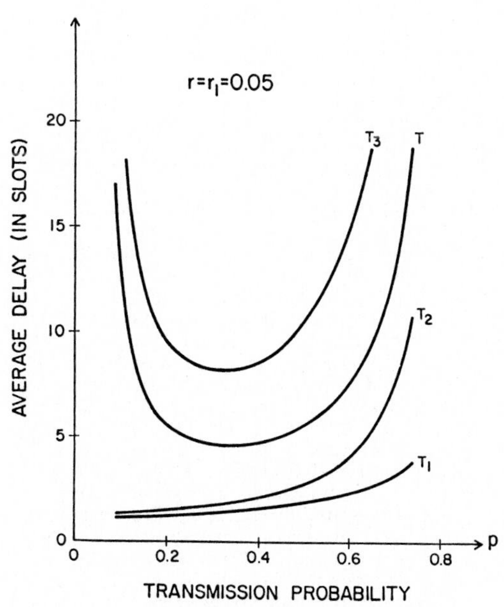

Figure 4: Average delays versus transmission probability.

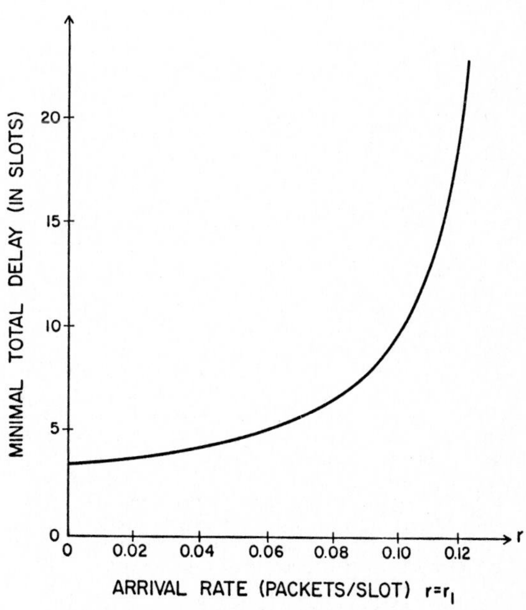

Figure 5: Minimal total delay versus the arrival rate.

TELETRAFFIC ISSUES in an Advanced Information Society
ITC-11
Minoru Akiyama (Editor)
Elsevier Science Publishers B.V. (North-Holland)

ANALYSIS ON A SATELLITE CHANNEL WITH TWO TYPES OF DOWNLINK ERROR

*Kunio GOTO, **Katsuhiko NISHITO, and *Toshiharu HASEGAWA

*Dept. of Appl. Math. and Physics, Fac. of Engineering, Kyoto University
Kyoto, Japan
**Integrated Switching Development Division, NEC, Abiko
Chiba, Japan

Abstract

In this paper, we consider a point-to-multipoint data communication system using satellite channel which is suffering from transmission error caused by local bad weather around terrestrial stations. Then we propose and analyse a retransmission scheme for broadcast connection, in which retransmission times are reduced by abandoning almost destinations under bad condition. To analyse the performance of the scheme, a mathematical model of the transmission system that includes separate error processes for uplink and downlink, and that has two types of downlink error, is used. Upper and lower bounds of mean transmission delay are obtained, and mean system delay is approximately calculated by them. Through numerical results, the system proposed here is found to be of good performance.

1. Introduction

With the recent advent of computer application such as document distribution, updating of local database, and geographically dispersed storage of multiple backup copies of indispensable records, it is likely that the volume of point-to-multipoint multi-packet data traffic will become significant because of broadcast property of the communication satellite. Calo and Easton [1] proposed a broadcast protocol for file transfers to multiple sites. Fujiwara [3] analysed the performance of a point-to-multipoint file transfer system using satellite with uniform transmission error. But satellite communication system suffers intensely from local bad weather, [2]. Actually, it is probable that it is clear in some region on the earth, and storming in some others. And in point-to-multipoint data communication systems, a packet is retransmitted until all the receivers have correctly caught it. So, receivers under bad weather prevent ones under fine weather from receiving new data packets following the one which has not been correctly received by some of receivers under bad weather. In other words, receivers under fine weather are obliged to receive the same packet over and over again. As a result of it, local bad condition reduces the performance of the conventional point-to-multipoint satellite communication system.

Then we propose a new retransmission scheme in which the station transmitting a multi-packet message to multiple destination can ignore the target stations sending back negative acknowledgement too many times. The ignored target stations give up receiving the job, and request it using another point-to-point file access channel.

2. Model

The data communication system considered in this paper is fundamentally based on the one in [4]. There are many terrestrial stations on the earth. Each station can communicate with each others using several satellite channels. One of them is broadcasting channel for file transfer, and some of them is the file access channel like one in [6]. We consider only the former, not the latter and the other channels.

Channel time of satellite is assumed to be divided into slots and synchronized among all stations. Each slot consists of two parts. One is called S_a part, and another is S_d part. S_a part is used for target station to send back an negative acknowledgement(NACK) and has a few bits' transmission time. S_d part has enough length for one packet's transmission.

All stations are assumed to be homogeneous and the number of them is finite N. Each station has some local users and receives newly originated data files from them. This data file is called a job. The arrival process of jobs to a station is as follows. At each station, jobs are assumed to arrive according to a Poisson process with parameter λ (jobs/slot). It is also assumed that this arrival process is independent among stations. Each job consists of multi-packet, the number of which is assumed to obey geometric distribution with parameter ℓ. That is, the probability that a job has i packets is $\ell(1-\ell)^{i-1}$. As a mater of course, each station has finite capacity to store and process jobs, and it can process only one job at a time. Namely, if jobs arrive at a station while processing, they will be lost. A job arriving at a station must be transmitted to several target stations using a satellite channel. There are two types of target stations. n_1 target stations have relatively high downlink error rate, and n_2 target stations have low downlink error rate.

The satellite is used as a transponder, that is, it receives packets from a ground station on one frequency band, and then sends back them to all ground stations on another frequency band. One round trip propagation delay is denoted by R(slots).

2.1 Access Scheme

Channel access scheme is basically R-ALOHA[4]. So, each station having a job to be transmitted sends a packet including the information of the job with probability q at S_d part of the next unreserved slot. The packet is called declaration packet and assumed to be error free by effective error correcting code in order to announce the transmission of the job to all the destinations(target stations). After one round trip propagation delay R, all station will know by sensing the broadcast channel whether the declaration packets have collided or not, that is, more than one station have sent them or not. Without collision, every S-th slot (S>R) until completion of the job's transmission will be implicitly reserved for the station that transmitted a declaration packet. So, access will have succeeded. Otherwise, same process should be repeated in the future unreserved slots.

2.2 Acknowledgement and Retransmission

The station which has succeeded in access, called sender, can continue to transmit packets in sequence in the every S-th reserved slot. The (n_1+n_2) target stations, called receivers, inspect whether the received packet is without transmission error or not. Transmission error may occur in uplink with probability e_u per packet, and in downlink to n_1 receivers with e_{d1}, and to n_2 receivers with e_{d2} ($>e_{d1}$). Also assumed that uplink and downlink error are independent because each station can monitor its own transmission and control transmission power.

Scheme 1

The receivers which detect transmission error send back to the sender NACKs in the S_a part of next reserved slot. NACKs have sufficient redundancy and can be correctly received by the sender after R-slots. By receiving NACKs, the sender knows more than one receivers couldn't receive the packet sent correctly, and retransmits it in the slot 2S after it's previous transmission. End of the job's transmission can be detected by receiving an end flag attached in the header of the last data packet.

In Scheme 1, retransmission times are not limited. So, in the (n_1+n_2) receivers, the n_2 receivers suffering intensely from downlink error send back NACK on the troubled packet to the sender too many times, and the sender must retransmit the same packets over and over again. Consequently, transmission delay grows so large that performance of the system drops remarkably. In order to avoid such a bad situation, retransmission times should be better to limited. Then we propose Scheme 2 as below.

Scheme 2

Basic retransmission rule is the same as scheme 1. Retransmission times are limited as follows.

1. Sender does not transmit the same packet over M times (include first time transmission).
2. Receiver does not send back NACK on the same packet over M times.
3. Once a receiver has sent back NACK on the same packet M times, it must not send back NACK for successive packets and gives up receiving the job. The receiver is called dead receiver. The sender only knows whether dead receivers exist or not, and does not know how many they are.

In addition, dead receivers will request the sender to transmit the job on alternative point-to-point file access channel.

Fig.3.1 is an example to show the transmission of a job by Scheme 2. The job consists of length is four packets, and limit of retransmission M is 2. At the beginning of transmission, the sender transmits the declaration packet correctly. Next, it sends the first data packet P_1, and P_1 results in failure at some of the receivers. At time t_2, the sender has not known whether P_1 was received correctly or not. Then NACK on P_1 and next data packet P_2 are transmitted at S_a and S_d part in the next reserved slot, respectively. One round trip delay R after t_2, the sender knows P_1 failed, and retransmit it. Supposing that transmission error occurrs again in retransmission of P_1, the receivers which have sent NACK on P_1 send back the last NACK on P_1 at time t_3 and give up receiving the job, because M=2. At time t_3, the sender has not received any NACKs in the packet P_2, so it ensures that P_2 is received correctly by all the receivers, and transmits P_3 in the next reserved slot. At time t_4, nevertheless the sender has received the NACK on P_1 again, it transmits P_4 because P_1 had transmitted M(=2) times. At time t_5, the last packet P_4 is transmitted again, because the sender knows transmission of P_1,P_2, and P_3 has been completed, but does not know about P_4 yet. The packet P_4 transmitted at t_5 is called dummy packet. At time t_6, the sender knows completion of the job. And the next reserved slot should be empty to announce to all the station that the reservation is released. So, the job's completion is at time t_8. The packet P_4 transmitted at t_5 is called dummy packet.

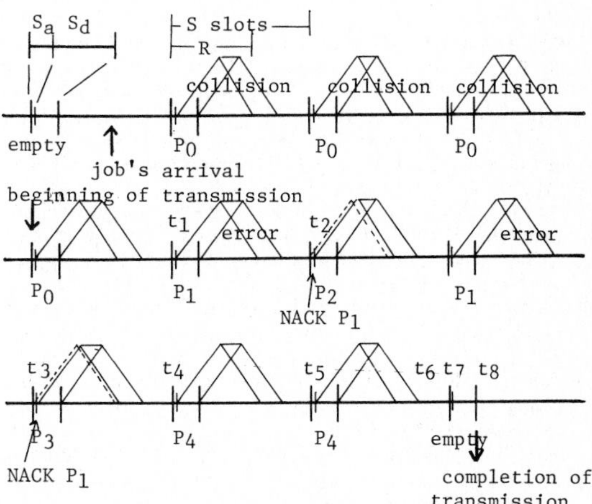

Fig.2.1. Acknowledgement and Retransmission.

3. Analysis

Mean system delay D(slots) is defined here as the time interval from the arrival epoch of a job at a station until the job's completion. D is divided into two parts, access delay D_A and transmission delay D_T. D_A is the time interval from a job's arrival at a station until the declaration packet is transmitted without collision (at the time reservation for succeeding data packets is made). D_T is the rest of D, that is the duration from a successful access until the job's completion.

If the retransmission times has no restriction (Scheme 1), D_T may be very large suffering from local bad weather. We are most interested in how much mean transmission delay is reduced by adopting the restriction of retransmission times (Scheme 2). As a matter of course, as D_T is smaller, D_A is smaller.

In this section, upper and lower bounds of D_T are analyzed at first. Then using them, approximate analysis of D_A is discussed.

Following notations are used throughout this paper.

- N: number of all stations,
- R: one round trip propagation delay (slots),
- S: cycle of reservation (slots),
- λ: Poisson arrival rate of job at a station per one slot,
- ℓ: parameter of geometric distribution of job's length,
- B_u: bit error rate in uplink,
- B_{d1}: lower bit error rate in downlink,
- B_{d2}: higher bit error rate in downlink,
- B: packet length (bits),
- M: maximum allowable transmission times of one packet,
- n_1: number of target stations with lower downlink error rate,
- n_2: number of target stations with higher downlink error rate.

Note that n_1+n_2 is constant in the analysis below.

3.1 Analysis of Mean Transmission Delay

Fig.3.1 shows an example of transmission of a job according to the retransmission scheme 2 in previous section, provided that M=2. Number of packets contained in the job is 3, and according to the occurred errors eight patterns are in transmission order. It is very difficult to obtain the exact solution of D_T even in the case of such a small job, because of the complexity of transmission order. Then let us consider the following virtual system in order to derive lower bound of D_T.

In the virtual system, it is assumed that the NACK returns instantaneously to the sender. Under this assumption, the sender can transmit packets according to increasing order. Fig.3.2 shows the order of transmission in the virtual system on the same job in Fig.3.1. Duration in each pattern necessary for completing the sample job is same as in Fig.3.1. Difference is the probability of each pattern. In Fig.3.2, at time t_1 some of target stations (probably ones with higher downlink error rate) may give up receiving the job, while t_2 in Fig.3.1. Thus the probability of pattern 3 and 4 in Fig.3.2 is slight greater than in Fig.3.1. To the contrary, the probability of pattern 1 and 2 is less than in Fig.3.1 for the same reason. Considering that the time interval of transmission of the job in pattern 3 and 4 is shorter than in pattern 1 and 2, mean delay in Fig.3.2 is slightly smaller than mean delay in Fig.3.1. In general, it is evident that mean transmission delay in the virtual system 1, $\underline{D}_T$, is slightly smaller than D_T.

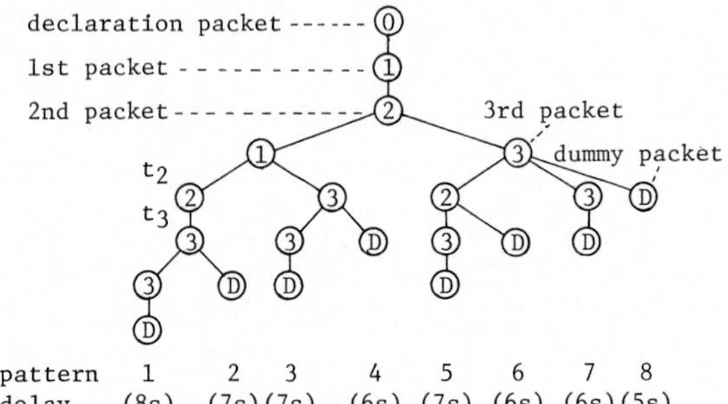

Fig.3.1. Transmission order in real system.

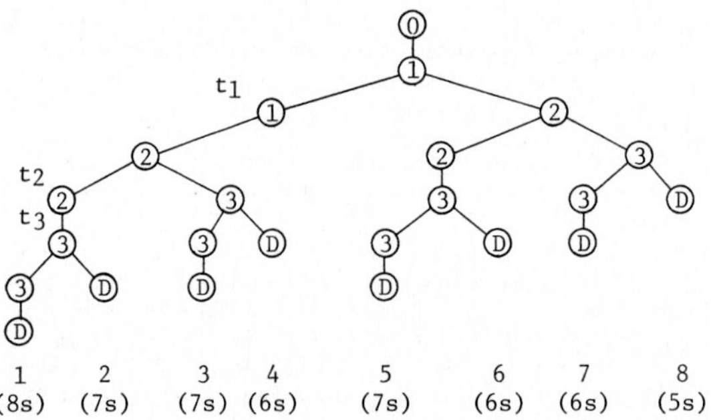

Fig.3.2. Transmission order in virtual system 1.

Next, we consider upper bound of D_T, $\tilde{D}_T$. The following virtual system 2, is introduced. Assuming that transmission of packets with odd sequence number and that of even are independent, transmission order is simplified. An example is shown in Fig.3.3, 3.4. Difference between real system and virtual system is the time when some of target stations give up receiving the job, that is, t_2 or t_3 in Fig.3.4 while t_2 in Fig.3.1. So, mean transmission delay in virtual system 2, $\tilde{D}_T$, is a little greater than D_T.

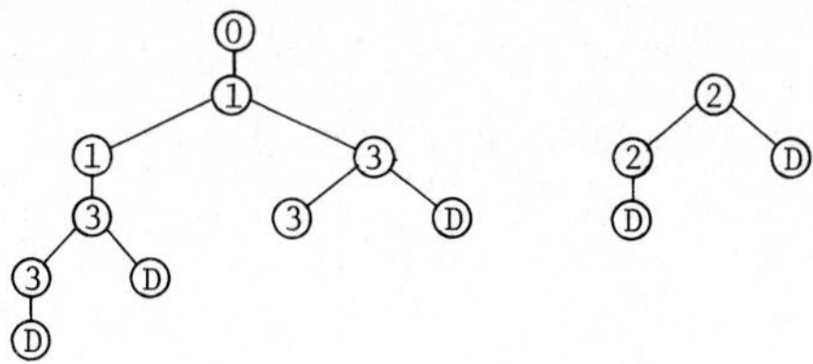

Fig.3.3. Transmission orders with odd and even packets.

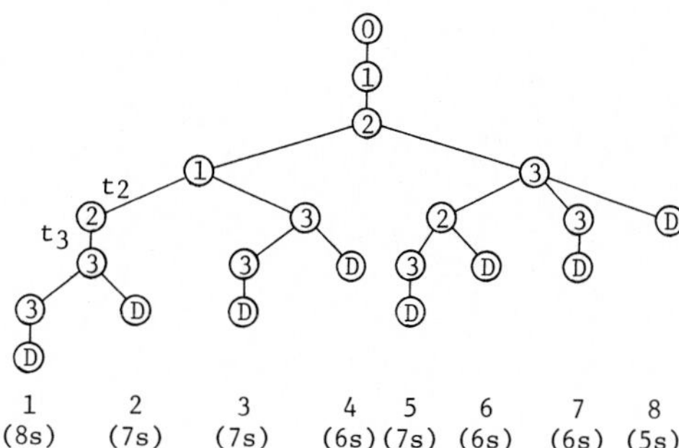

Fig.3.4. Transmission order in virtual system 2

Through the above discussions, we have

$\underline{D}_T < D_T < \tilde{D}_T$.

3.2 Lower Bound of Mean Transmission Delay

Let $P_m(i,j_1,j_2|k_1,k_2)$ be the probability that j_s $(s=1,2)$ receivers with downlink error rate per packet, e_{ds}, have received the m-th data packet after the i-th $(1 \leq i \leq M)$ transmission of it, given that (m-1)st data packet was correctly received by k_s receivers with e_{ds} $(s=1,2)$ respectively. And e_u denotes uplink error rate per packet. It should be noted that for m=1, $k_s=n_s$ $(s=1,2)$ from the assumption that the declaration packet was received correctly. Then $P_m(i,j_1,j_2|k_1,k_2)$ is given as,

$$P_m(i,0,0|0,0)=\begin{cases} 1 & (i=1) \\ 0 & (2 \leq i \leq M) \end{cases}, \quad (m \geq 2)$$

$$P_m(i,j_1,j_2|k_1,k_2)$$
$$=\begin{cases} \sum_{v_1=0}^{j_1} \sum_{v_2=0}^{j_2} P_m(i-1,v_1,v_2|k_1,k_2) \cdot A^{(k_1,k_2)}_{(v_1,v_2),(j_1,j_2)} \\ \quad (0 \leq j_s \leq k_s,\ s=1,2, \text{ and } (j_1,j_2) \neq (k_1,k_2)) \\ \sum_{v_1=0}^{j_1} \sum_{v_2=0}^{j_2} P_m(i-1,v_1,v_2|k_1,k_2) \cdot A^{(k_1,k_2)}_{(v_1,v_2),(j_1,j_2)} \\ \quad - P_m(i-1,k_1,k_2|k_1,k_2) \\ \quad ((j_1,j_2)=(k_1,k_2)). \end{cases}$$

Where,

$$e_u=1-(1-B_u)^B,$$
$$e_{d1}=1-(1-B_{d1})^B,$$
$$e_{d2}=1-(1-B_{d2})^B,$$

and

$$A^{(k_1,k_2)}_{(v_1,v_2),(j_1,j_2)}$$
$$=\begin{cases} (1-e_u)\binom{k_1-v_1}{j_1-v_1}(1-e_{d1})^{j_1-v_1} \cdot e_{d1}^{k_1-j_1} \cdot \binom{k_2-v_2}{j_2-v_2} \\ \quad \cdot (1-e_{d2})^{j_2-v_2} \cdot e_{d2} \\ \quad (0 \leq v_s \leq j_s \leq k_s,\ s=1,2, \text{ and } (v_1,v_2) \neq (j_1,j_2)) \\ e_u+(1-e_u)e_{d1}^{k_1-v_1} \cdot e_{d2}^{k_2-v_2} \\ \quad (0 \leq v_s=j_s \leq k_s, \text{ except } v_s=j_s=k_s, s=1,2) \\ 1 \quad (v_s=j_s=k_s). \end{cases}$$

$A^{(k_1,k_2)}_{(v_1,v_2),(j_1,j_2)}$ is the probability that j_s $(s=1,2)$ receivers with e_{ds} have received correctly after the last transmission of it, given that v_s receivers with e_{ds} had received the current packet correctly before the last transmission of it, and that k_s receivers with e_{ds} had received all the previous packets correctly.

From the above equations, $P_m(i,k_1,k_2|k_1,k_2)$ $(1 \leq i \leq M-1)$ and $P_m(M,j_1,j_2|k_1,k_2)$ $(0 \leq j_s \leq k_s,\ s=1,2)$ are obtained, and satisfy the normalizing condition.

$$\sum_{i=1}^{M-1} P_m(i,k_1,k_2|k_1,k_2)+\sum_{j_1=0}^{k_1}\sum_{j_2=0}^{k_2} P_m(M,j_1,j_2|k_1,k_2) = 1$$

(for $m \geq 2$, $0 \leq k_2$, for m=1, $k_s=n_s$, s=1,2).

Finally, the relation between (m-1)th and m-th data packet is described. Probability that a sender transmits the m-th data packet i times, $Q_m(i)$, and probability that j_s receivers with e_{ds}, s=1,2, have received the m-th packet correctly within M times, $R_m(j_1,j_2)$, are obtained by following equations,

for m=1

$$Q_1(i)=\begin{cases} P_1(i,n_1,n_2|n_1,n_2) & (1 \leq i \leq M-1) \\ \sum_{j_1=0}^{n_1}\sum_{j_2=0}^{n_2} P_1(M,j_1,j_2|n_1,n_2) . & (i=M) \end{cases}$$

for $m \geq 2$

$$Q_m(i)$$
$$=\begin{cases} \sum_{k_1=0}^{n_1}\sum_{k_2=0}^{n_2} P_m(i,k_1,k_2|k_1,k_2) \cdot R_{m-1}(k_1,k_2) \\ \quad (1 \leq i \leq M-1) \\ \sum_{k_1=0}^{n_1}\sum_{k_2=0}^{n_2}\sum_{j_1=0}^{k_1}\sum_{j_2=0}^{k_2} P_m(i,j_1,j_2|k_1,k_2) \cdot R_{m-1}(k_1,k_2) \\ \quad (i=M) \end{cases}$$

$$R_1(j_1,j_2)=\begin{cases} P_1(M,j_1,j_2|n_1,n_2) \\ \quad (0 \leq j_s \leq n_s, s=1,2, \text{ and } (j_1,j_2) \neq (n_1,n_2)) \\ \sum_{i=1}^{M} P_1(i,n_1,n_2|n_1,n_2) \quad (j_s=n_s, s=1,2) \end{cases}$$

for $m \geq 2$

$$R_m(j_1,j_2) = \begin{cases} R_m(0,0)+\sum_{k_1=0}^{n_1}\sum_{k_2=0}^{n_2} P_m(M,0,0|k_1,k_2)\cdot R_{m-1}(k_1,k_2) & (j_1=j_2=0) \\ \sum_{i=1}^{M-1} P_m(i,j_1,j_2|j_1,j_2)\cdot R_{m-1}(j_1,j_2) + \sum_{k_1=j_1}^{n_1}\sum_{k_2=j_2}^{n_2} P_m(M,j_1,j_2|k_1,k_2)\cdot R_{m-1}(k_1,k_2) . & \end{cases}$$

$(0 \leq j_s \leq n_s, s=1,2, \text{ and } (j_1,j_2)\neq(0,0))$

From the above results, lower bound of mean transmission delay, $\underline{D}_T$, is given as

$$\underline{D}_T = \sum_{j=1}^{\infty} S\{\sum_{m=1}^{j}\sum_{i=1}^{M} i\cdot Q_m(i)+2\}\ell(1-\ell)^{j-1} .$$

The second term in parentheses, 2, correspond to the time interval necessary for both the transmission of the declaration packet and dummy state.

Let r_s be mean number of target stations with e_{ds}, s=1,2, which survive until the completion of a job, respectively.

r_s are given as,

$$r_s = \sum_{m=1}^{\infty}\ell(1-\ell)^{m-1}\sum_{j_1=0}^{n_1}\sum_{j_2=0}^{n_2} j_s\cdot R_m(j_1,j_2) \quad (s=1,2).$$

3.3 Upper Bound of Mean Transmission Delay

$\tilde{D}_T$ is simply obtained by using the results of preceding section. Let d(j) be defined as,

$$d(j) = \begin{cases} \sum_{i=1}^{M} i\cdot Q_1(i) & (j=1) \\ 2\sum_{m=1}^{j/2}\sum_{i=1}^{M} i\cdot Q_m(i) & (j=2,4,6,8\cdots) \\ 2\sum_{m=1}^{(j-1)/2}\sum_{i=1}^{M} i\cdot Q_m(i)+\sum_{i=1}^{M} i\cdot Q_{(j+1)/2}(i) . & (j=3,5,7,9,\cdots) \end{cases}$$

Then we get,

$$\tilde{D}_T = \sum_{j=1}^{\infty} S\{d(j)+2\}\ell(1-\ell)^{j-1} .$$

3.4 Mean Transmission Delay of Scheme 1

We obtain mean transmission delay of scheme 1, D_T^*, in order to compare with that of scheme 2. In the case using scheme 1, exact numerical solution of D_T^* is also simply obtained from the results of preceding section.

Let k denotes the mean transmission times necessary for a single packet to be received by (n_1+n_2) receivers. k is given as

$$k = \sum_{i=1}^{\infty} i\cdot P_1(i,n_1,n_2|n_1,n_2) .$$

Let $L=1/\ell$ be the mean number of packets consisting in a single job. Then D_T^* is given as

$$D_T^* = S(Lk+2) .$$

$\underline{D}_T$, $\tilde{D}_T$, and D_T^* can be calculated with the desirable accuracy by accomplishing the summation necessary times.

3.5 Discussion about Mean Access Delay

Various methods to obtain mean access delay of R-ALOHA have been proposed [3], [4]. We obtained upper and lower bounds of mean transmission delay. They are adopted as approximate values of mean transmission delay. The approximate analysis on mean transmission delay in [3] is adopted, because it gives mean access delay when mean transmission delay is obtained.

In the paper, reduced system state X is defined as

$$X=(N_T,N_A) ,$$

where

N_T: the number of stations transmitting its job.

N_A: the number of stations waiting for successful access.

Then transition probability from state x to x' is approximately defined including terms of λ, S, R, q, and D_T. Equilibrium equations are solved numerically, and expectations $E[N_T]$, and $E[N_A]$ are obtained. Finally system delay D and D_A are obtained as

$$D=1/\lambda'$$

$$\lambda'=E[N_T]/N\cdot D_T$$

$$D_A=D-D_T .$$

4. Numerical Results

First, we compare the values of $\underline{D}_T$ and $\tilde{D}_T$ with the values obtained from the simulation model [5]. Fig.4.1 shows $\underline{D}_T$ and $\tilde{D}_T$ as a function of maximum allowable retransmission times, M. Simulation results are plotted by symbols while analytical results are shown by solid lines. $\tilde{D}_T$ is close to $\underline{D}_T$. As proposed in 3.1, they are considered to give good approximate values of D_T and can be utilized to calculate approximate values of D_A. Especially, $\underline{D}_T$ is a better approximate value of D_T, because its analysis takes into consideration that packets with odd number influence ones with even. Fig.4.1. shows that it is effective to restrict retransmission times when higher bit error rate in downlink, B_{d2}, is large.

Figs. from 4.2 to 4.4 show D_T, r_1,r_2 respectively as a function of B_{d2}. In Fig.4.2, the case of M= corresponds to the case of using scheme 1. It is shown that our transmission scheme gives good performance when $B_{d2}>1\times10^{-4}$ and M=1,2,3. In order to determine the optimal value of M, we must investigate r_1 and r_2. Fig.3.3 indicates that receivers with lower downlink error rate do not suffer heavily from transmission error. If $M\geq2$, almost all receivers with B_{d1} will survive, that is they will receive jobs completely. Fig.4.4 shows that almost receivers with B_{d2} survive when $B_{d2}\leq1\times10^{-4}$ and $M\geq3$. Fig.4.5 shows D_T as a function of n_2, where $n_1+n_2=25$. It indicates D_T is not much influenced by the increase of n_2 when M=1,2,3. Through Figs.4.2,4.4,4.5, we propose that the best value of M is 3 in the case. Concerning system delay, Fig.4.6 shows D versus M in the case of relatively long packets. It indicates additional delay(each value of D minus minimum value) can be considerably reduced by setting M=2 or 3.

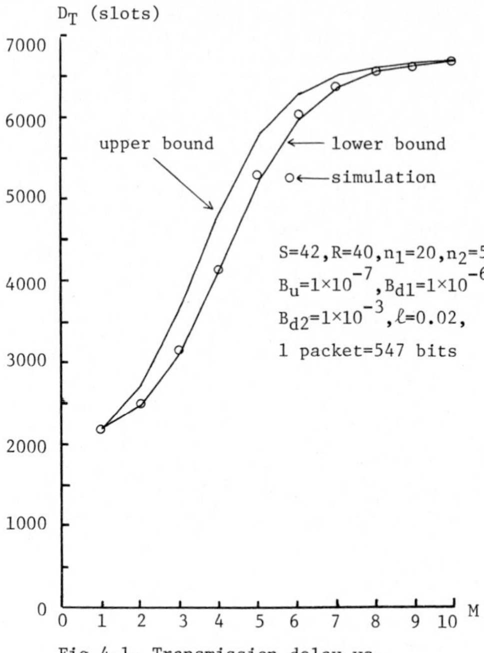

Fig.4.1. Transmission delay vs. maximum allowable retransmission times.

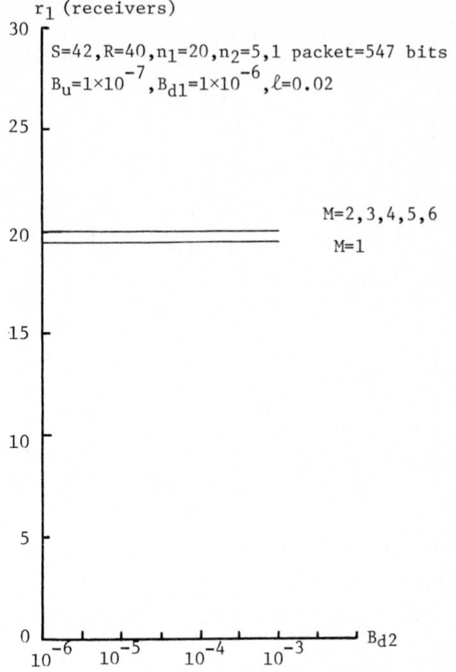

Fig.4.3. Mean number of survived receivers with lower downlink error rate vs. lower bit error rate in downlink.

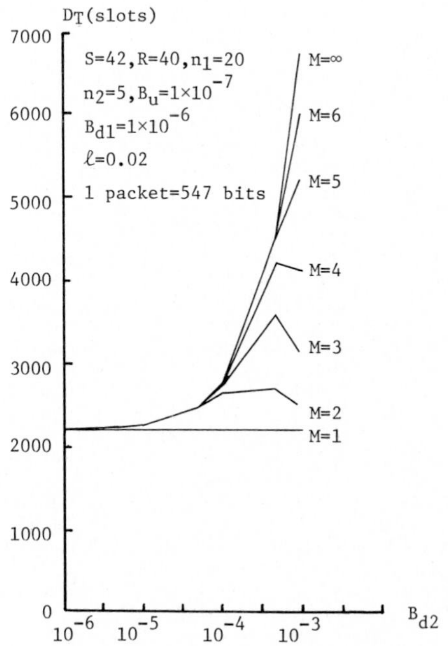

Fig.4.2. Transmission delay vs. higher bit error rate in downlink.

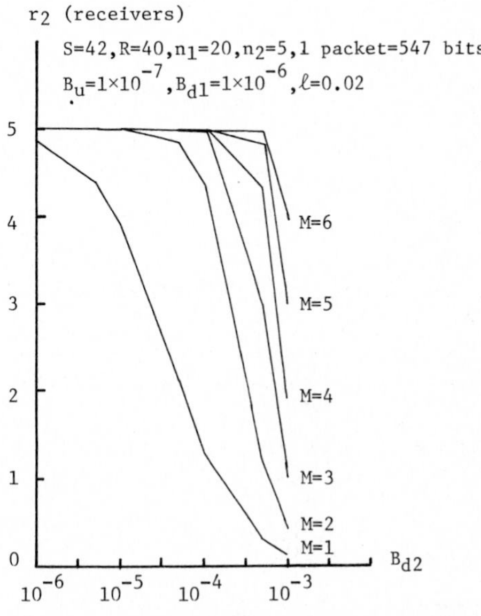

Fig.4.4. Mean number of survived receivers with higher downlink error rate vs. higher bit error rate in downlink.

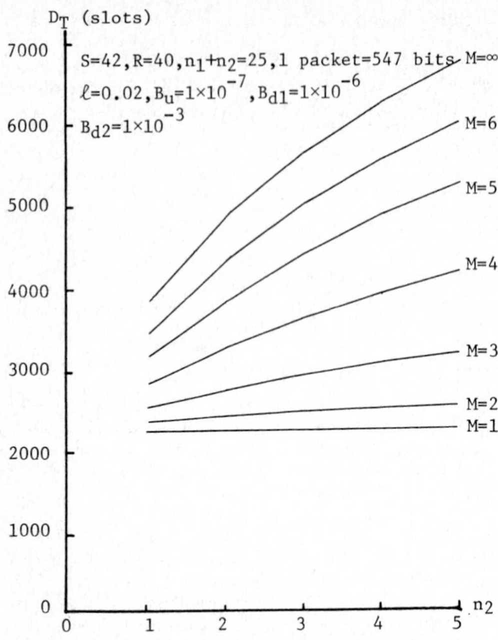

Fig.4.5. Transmission delay vs. number of receivers with higher downlink error rate.

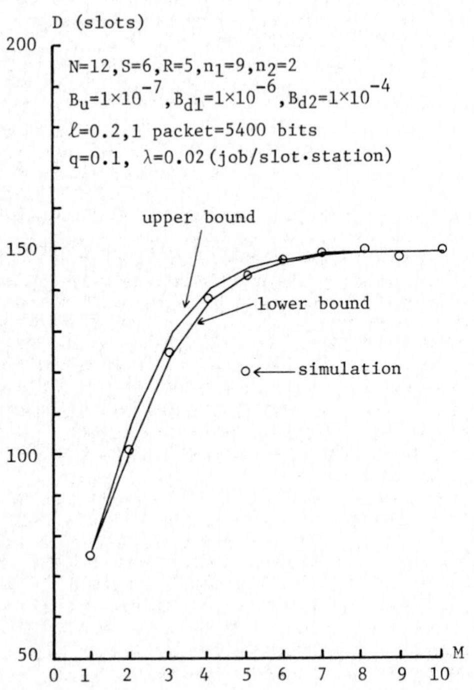

Fig.4.6. System delay vs. maximum allowable retransmission times.

5. Conclusion

In this paper, we considered point-to-multipoint multi-packet data communication system using satellite channel, which suffers from transmission error caused especially by local bad weather around terrestrial stations. It is that on uplink, bit error rate is one constant while on downlink there are two types of bit error rate. This assumption makes the model more realistic. Then we proposed a retransmission scheme for broadcast connection, in which retransmission times are reduced by abandoning almost receivers under bad condition. That is, a sender does not transmit the same packet over M times, and a receiver does not send back NACKs on the same packet over M times. Once a receiver has sent NACKs on the same troubled packet, it must not return a NACK any more, and must gives up receiving the multi-packet data. The retransmission scheme can be adopted in fixed assigned scheme and reservation scheme such as TDMA, R-ALOHA.

Here, performance characteristics of the system based on R-ALOHA which adopts the retransmission scheme was investigated. Upper and lower bounds of mean transmission delay were obtained, and mean access delay was calculated using them. Through the numerical results, interesting properties of the system are shown and the system proposed here is found to be of good performance. Analytical results are well verified by simulation results. Especially compared with the case that times of retransmission is not limited, mean system delay is extremely reduced. By setting M=2, additional delay is halved for the victim of almost unreachable destinations.

In this model, it is not taken into account that traffic on the other channels increases by abandoned destinations. And it is necessary to consider the influence of abandoned receivers. Satellite was used here as transponder. If the satellite is equipped with large buffer and processing capacity, it is possible for the satellite to change M dynamically by observing NACKs and control the access probability according to the fluctuation of arrival rate. These problems are left for future works.

References

[1] S.B. Calo and M.C. Easton, "A Broadcast Protocol for File Transfers to Multiple sites," IEEE Trans. on Commun., VOL.COM-29, No.11, Nov., pp.1701-1707(1981).

[2] A.B. Carleial and J. Kono, "Link Availability in Satellite Telecommunication Subject Correlated Rain Attenuation at the Ground Terminals," ICC'82, pp.1B.2.1-1B.2.5.

[3] S. Fujiwara, "Analysis for Satellite Communication System with Transmission Error," Master Thesis, Kyoto University, 1983.

[4] S.S. Lam, "An Analysis of the Reservation-ALOHA Protocol for Satellite Packet Switching," Proc. of ICC, pp.27.3.1-27.3.5(1978).

[5] T. Takine, "Simulation Study on a Satellite Communication Channel with Two Types of Downlink Error (in Japanese)," Graduate Thesis, Kyoto University, 1984.

[6] A.S. Tanenbaum, "Computer Networks," Prentice-Hall, INC, 1981.

TELETRAFFIC ISSUES in an Advanced Information Society
ITC-11
Minoru Akiyama (Editor)
Elsevier Science Publishers B.V. (North-Holland)

IMPROVING MEAN DELAY IN DATA COMMUNICATION NETWORKS BY NEW COMBINED STRATEGIES BASED ON THE SRPT - PRINCIPLE

Carmelita GOERG   Xuan Huy PHAM

Technical University Aachen
Aachen, Federal Republic of Germany

ABSTRACT

Considering queueing systems in data communication networks one can identify a special feature of the communication system: the service time, which is proportional to message length, is generally known in advance. This is a precondition for applying the queueing strategy SRPT (Shortest Remaining Processing Time first), which offers the shortest mean delay time among all conceivable strategies. This paper analyzes new combined strategies based on the SRPT principle. First, a preemptive priority system combined with the SRPT strategy within each priority class is presented. Second, a combination of Round Robin and SRPT including overhead is described. Both strategies show a considerable reduction of the mean delay time in comparison with the same strategies combined with FIFO instead of SRPT.

1. INTRODUCTION

In communication networks transmission scheduling is one of the most important factors influencing the performance of the system. It is well known that the application of the service time dependent strategy SRPT (Shortest Remaining Processing Time first) offers the minimum mean number of jobs or messages in the system and the minimum mean delay time. This is achieved by always selecting the job with the shortest remaining processing time and by preempting the job in service if necessary. A theoretical proof of the optimality of SRPT with respect to mean delay time was given by Schrage in [8]. In spite of this advantage the strategy SRPT has not been used for job scheduling in computers because of the requirement that the service times of jobs must be known in advance. But this condition is generally satisfied in data communication networks due to the following feature: the service time of each job is proportional to the message length and is therefore known in advance. This feature allows the application of the pure SRPT strategy [7] as well as strategies which are built on the SRPT principle in order to improve the performance of communication systems, especially for service time distributions with a large coefficient of variation.

In this paper two new combined strategies based on the SRPT principle are presented and analyzed. First, a preemptive priority system is considered in section 2, which uses the pure SRPT strategy within each priority class. The main aim of this investigation is to demonstrate the optimality character of SRPT in priority queues. A further interesting aspect of such a priority queue with SRPT is the quantitative performance of the mean delay in lower priority classes (not preferred classes) in comparison with the corresponding priority queue with the classical FIFO strategy (First In First Out). For the model $M_i/G_i/1$ with a Poisson input process M_i and a general service time distribution G_i for the i-th priority class $(i=1,2,..,n)$ the mean delay time is determined for each class individually and evaluated for typical application examples.

Second, a combination of the RR strategy (Round Robin) and SRPT including overhead is considered. The idea for this new combined strategy originated from the fact that the preemption mechanism of the SRPT strategy is rather difficult to implement, especially if overhead has to be considered [1]. For SRPT it is necessary to evaluate the selection upon every arrival. This disadvantage can be avoided by introducing the RR preemption mechanism giving each job a maximum time slice. In section 3 the mean delay for the model M/G/1 with the combined SRPT/RR strategy including constant overhead time is evaluated and compared to the FIFO, RR, and SRPT strategies. In data communication networks the RR strategy corresponds to packet switching, i.e. a time slice corresponds to a packet. The SRPT/RR strategy can be utilized for packet switching in a similar way. Basically this strategy selects the next packet of the message with the least number of remaining packets. In this way the mean number of messages and the mean delay are reduced.

The results for both models are presented as formulas and diagrams for typical examples, which show the potential improvement of mean delay time in comparison with the commonly used strategies. Throughout the paper the terms job and message are used synonymously. A list of symbols at the end of the paper contains the notations used in the formulas.

2. PREEMPTIVE PRIORITIES WITH SRPT

Priority queueing is widely used in many real-time computer and data communication networks. Presently preferred scheduling strategies within the queue of each priority class are FIFO, RANDOM etc., which are independent of the service time and thus have the same mean delay time

$\overline{TD}_i$ [4]. In data communication networks the service time of a message is known in advance so that the service time dependent strategy SRPT may be used within each class in order to reduce the mean delay time $\overline{TD}_i$ to its lowest possible value compared to any other strategy. Improving mean delay in each class also improves the overall mean delay of the system.

2.1 Preemptive Priority Model with SRPT

The considered preemptive priority queueing model, which is shown in fig. 2.1, is of the type $M_i/G_i/1$ with Poisson input intensity λ_i, mean service time $\overline{TB}_i$ and the coefficient of variation c_i for each priority class i. The basic load of each class is $\rho_i = \lambda_i \overline{TB}_i$. The priorities are defined externally and are considered to be static. The class with the lowest index (i=1) has the highest priority. Each message entering the queue of class i has a known service time TB_i. Within each class the strategy SRPT is applied.

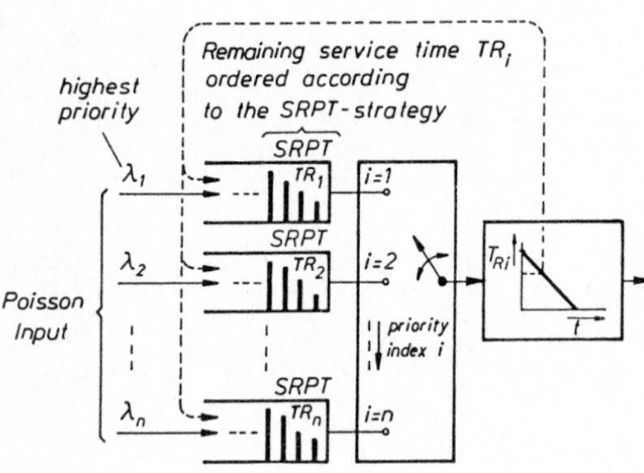

Fig. 2.1: Preemptive priority model with SRPT.

For this reason the model has the classification $M_i/G_i/1$-PRE-SRPT which is characterized by two preemption mechanisms, that is first the preemptions due to the processing of higher priority classes and second the preemptions by SRPT within each class. Other strategies e.g. the nonpreemptive priority strategy with the nonpreemptive strategy SPT (Shortest Processing Time first) are treated in [5].

2.2 Analysis of the Mean Delay $\overline{TD}_i$

For the considered model the mean delay time $\overline{TD}_i$ for each priority class i will be determined. $\overline{TD}_i$ is the mean delay of a message of class i between arrival and departure. The calculation of $\overline{TD}_i$ is done in three steps [5].

a) Approximate Model for each SRPT Queue

Due to the fact that the SRPT strategy is based on a strict preference of the message with the shortest remaining service time, the processing within each SRPT queue may be interpreted as a new preemptive priority control with a priority arrangement according to the value of the remaining service time. An approximate priority model is constructed by dividing the time axis presenting the service time into m equidistant domains with the time slot T. Thus a message whose remaining service time TR_i satisfies the relation $(j-1)T \leq TR_i < jT$ belongs to subclass j of the new approximate priority model. For a new message the service time TB_i corresponds to the remaining service time TR_i. If a message starts out in subclass j it passes through all subclasses j,j-1,...,1 as its remaining service time is reduced until it leaves the system. The exact solution is obtained as the limit of infinitely many classes and time slot $T \to 0$.

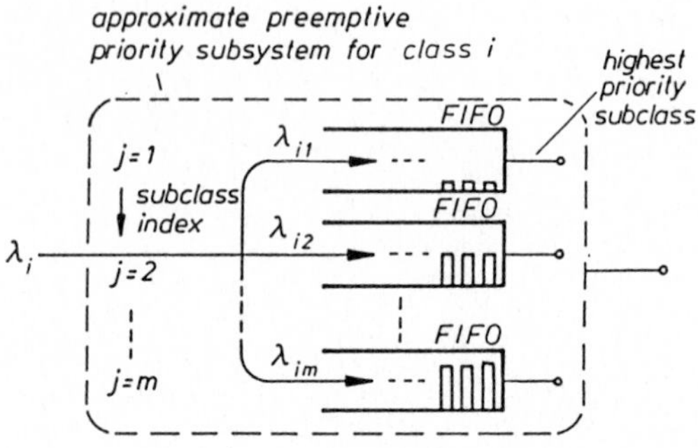

Fig. 2.2: Approximation of the SRPT queue of priority class i by a new preemptive priority subsystem.

The new preemptive priority queue, which is an approximation for the SRPT queue, is independent of the initial priority system. Therefore the approximate model is called priority subsystem. Messages belonging to the domain $0 < TB_i < T$ have the highest priority subclass and the messages of the domain $(m-1)T \leq TB_i < \infty$ the lowest (m-subclass) priority subclass. The approximate model for the SRPT queue is shown in fig. 2.2. Within each priority subclass ij (index i shows the original priority class and j the subclass belonging to the value of the service time) the strategy is FIFO. Through this approximation the complete priority system with SRPT is transformed into a new priority system with internal priority subsystems containing only FIFO queues.

For each priority subclass within each approximate priority subsystem i the input intensity λ_{ij} has the following form:

$$\lambda_{ij} = \lambda_i q_{ij} \tag{2.1}$$

$$q_{ij} = \int_{(j-1)T}^{jT} f_{Bi}(t)dt \tag{2.2}$$

with the new distribution for each subclass:

$$f_{Bij}(t) = \begin{cases} f_{Bi}(t)/q_{ij} & \text{for } (j-1)T \leq t < jT \\ 0 & \text{otherwise} \end{cases} \tag{2.3}$$

The corresponding service time for each subclass is now called TB_{ij}. The basic load for each subclass is:

$$\rho_{ij} = \lambda_{ij}\overline{TB}_{ij} \qquad (2.4).$$

b) Mean Delay for the Approximate Model
For the approximate model the mean delay time is determined by applying the classic priority approach [4], [3]. The mean delay time for a message of the class ij is given by:

$$\overline{TD}_{ij} = \overline{TW}_{ij} + \overline{TC}_{ij} \qquad (2.5)$$

with the mean initial waiting time $\overline{TW}_{ij}$ as the mean waiting time of a message from entry to the first service and the mean completion time $\overline{TC}_{ij}$ as the mean time from the first service to the departure of the message. Due to the fact that for the mean initial waiting time $\overline{TW}_{ij}$ preemptions in higher priority classes are not essential, $\overline{TW}_{ij}$ may be calculated exactly as for the case with no preemptions. $\overline{TW}_{ij}$ may be expressed by the following sum:

$$\overline{TW}_{ij} = \overline{TE}_{ij} + \overline{TG}_{<i} + \overline{TN}_{<i} + \overline{TG}_{ij} + \overline{TN}_{ij} \qquad (2.6)$$

with the mean remaining service time $\overline{TE}_{ij}$ of a message which belongs to a priority class higher than i or to a subclass higher than j. A detailed calculation of $\overline{TE}_{ij}$ may be found in [5].

$$\overline{TE}_{ij} = \frac{1}{2}\sum_{r=1}^{i-1}\lambda_r M_2\{TB_r\} + \frac{\lambda_i}{2}\int_o^{jT} t^2 f_{Bi}(t)dt + \lambda_i\frac{(jT)^2}{2}\int_{jT}^{\infty} f_{Bi}(t)dt \qquad (2.7).$$

$\overline{TG}_{<i}$ is the total mean service time of all messages found in the system upon arrival, which belong to the higher priority classes:

$$\overline{TG}_{<i} = \sum_{r=1}^{i-1}\sum_{v=1}^{m}\rho_{rv}\overline{TW}_{rv} \qquad (2.8).$$

$\overline{TN}_{<i}$ is the total mean service time of higher priority messages which enter the queue during the mean waiting time $\overline{TW}_{ij}$:

$$\overline{TN}_{<i} = \overline{TW}_{ij}\sum_{r=1}^{i-1}\rho_r \qquad (2.9).$$

$\overline{TG}_{ij}$ is the total mean service time of messages found in the system upon arrival belonging to the observed class i, which have a shorter service time:

$$\overline{TG}_{ij} = \sum_{v=1}^{j}\rho_{iv}\overline{TW}_{iv} \qquad (2.10).$$

$\overline{TN}_{ij}$ is the total mean service time of messages with shorter service time of the observed class i, which enter class i during the mean waiting time $\overline{TW}_{ij}$:

$$\overline{TN}_{ij} = \overline{TW}_{ij}\sum_{v=1}^{j-1}\rho_{iv} \qquad (2.11).$$

The development of eq. (2.6) by using eqs. (2.8) to (2.11) and by eliminating the recursive form gives:

$$\overline{TW}_{ij} = \overline{TE}_{ij}/(\theta_{i-1\ j}\cdot\theta_{i-1\ j-1}) \qquad (2.12)$$

$$\theta_{ij} = 1 - \sum_{r=1}^{i}\rho_r - \sum_{v=1}^{j}\rho_{iv} \qquad (2.13).$$

For the completion time the following expression is derived [5]:

$$\overline{TC}_{ij} = \sum_{v=1}^{i-1} T/\theta_{i-1\ v-1} + \left(b_{ij} - (j-1)T\right)/\theta_{i-1\ j-1} \qquad (2.14).$$

Adding eqs. (2.12) and (2.14) results in the formula for the mean delay $\overline{TD}_{ij}$ defined in eq. (2.5) of the approximate model in fig. 2.2.

c) Exact Formula for the Mean Delay
The exact formula for the mean delay $\overline{TD}_i$ is obtained by letting the number of FIFO classes within each priority class grow to infinity:

$$\lim_{\substack{m\to\infty\\ T\to 0\\ jT\to b_i}} \overline{TD}_{ij} = \overline{TD}_i(b_i) \qquad (2.15).$$

The mean delay $\overline{TD}_i(b_i)$ for a message of length b_i is then given by:

$$\overline{TD}_i(b_i) = \frac{1}{2\left(1-\sum_{r=1}^{i-1}\rho_r-\rho_i P_1(b_i)\right)^2}\Big(\sum_{r=1}^{i-1}\lambda_r M_2\{TB_r\} + \lambda_i M_2\{TB_i\}P_2(b_i) + b_i^2\lambda_i(1-P_o(b_i))\Big) + \int_o^{b_i}\frac{dt}{1-\sum_{r=1}^{i-1}\rho_r-\rho_i P_1(t)} \qquad (2.16).$$

In eq. (2.16), $P_o(b_i)$, $P_1(b_i)$, $P_2(b_i)$ are the abbreviated terms of the formula:

$$P_h(b_i) = \frac{1}{M_h\{TB_i\}}\int_o^{b_i} t^h f_{Bi}(t)dt \qquad (2.17)$$

denoting the distribution function of order h of the service time TB_i. The final mean delay $\overline{TD}_i$ is obtained by

$$\overline{TD}_i = \int_o^{\infty}\overline{TD}_i(t)f_{Bi}(t)dt \qquad (2.18).$$

A comparison of eq. (2.16) with the corresponding formula for a preemptive priority queue with FIFO within each class shows a similar structure [3].

2.3 Numerical Results
In Fig. 2.3 the mean delay time for the model $M_i/M_i/1$-PRE-SRPT related to the mean service

time is shown for the case that the considered 5 priority classes have the same input intensity $\lambda_i = \lambda/5$ and the same mean service time. The corresponding FIFO curves are presented in dashed lines for each priority class to allow a comparison. For the examples in this section the following equations are required:

$$\lambda = \sum_{i=1}^{n} \lambda_i \;; \qquad \overline{TB} = \sum_{i=1}^{n} \frac{\lambda_i}{\lambda} \overline{TB}_i \qquad (2.19).$$

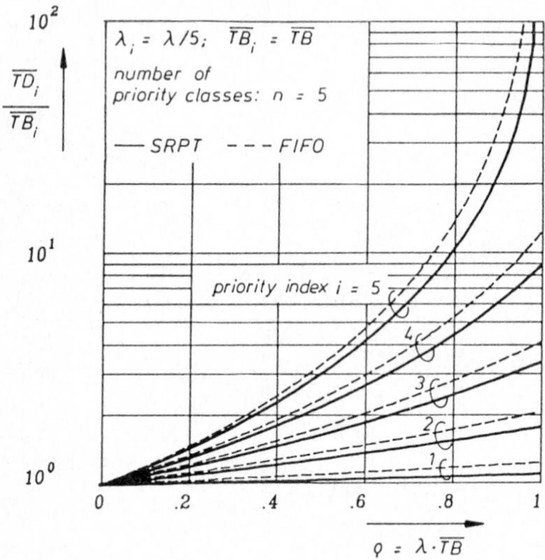

Fig. 2.3: Modell $M_i/M_i/1$-PRE-SRPT
Comparison of the mean delay for the case "all messages have the same input intensity and the same mean service time".

The mean delay time for the queue with SRPT is generally better than for the queue with FIFO. The improvement increases with the offered traffic ρ and with the priority index i. The strict preference of higher priority classes by preemptive control causes a relatively higher population (offered traffic) of waiting messages in lower classes, which means that the SRPT selection mechanism can be applied to more messages thus being more effective for lower priorities.

In fig. 2.4 the mean delay time for the model $M_i/H_{2i}/1$-PRE-SRPT with 5 priority classes is shown for the case that the input intensity λ_i and mean service time $\overline{TB}_i$ increase proportionally to i. The service time distributions are hyperexponential with the same coefficient of variation $c_i = 2.0$. The improvement of SRPT in comparison with FIFO is in this case more distinct than in fig. 2.3, especially for high values of ρ.

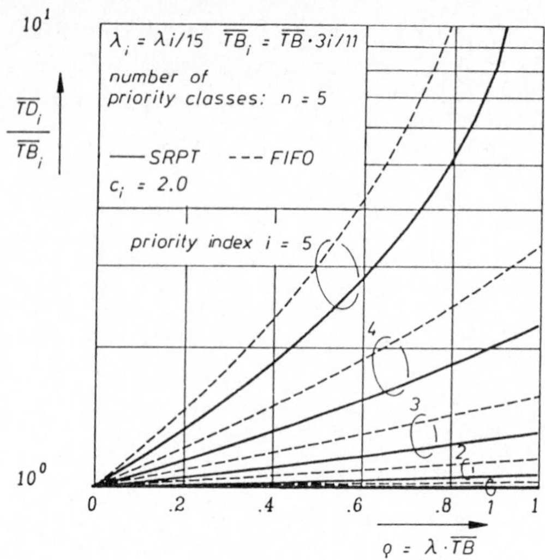

Fig. 2.4: Model $M_i/H_{2i}/1$-PRE-SRPT
Comparison of the mean delay for the case "input intensity and mean service time proportional to priority index i".

3. ROUND ROBIN WITH SRPT: SRPT/RR

One of the main disadvantages of the pure SRPT strategy is the necessity to preempt a job depending on the service time of a new arrival, which means that preemptions can occur at any time. Restricting preemptions and giving a job a maximum time slice before it is preempted leads to a new strategy: a combination of the SRPT and RR (Round Robin) strategy called: SRPT/RR.

3.1 The SRPT/RR - Model with Overhead

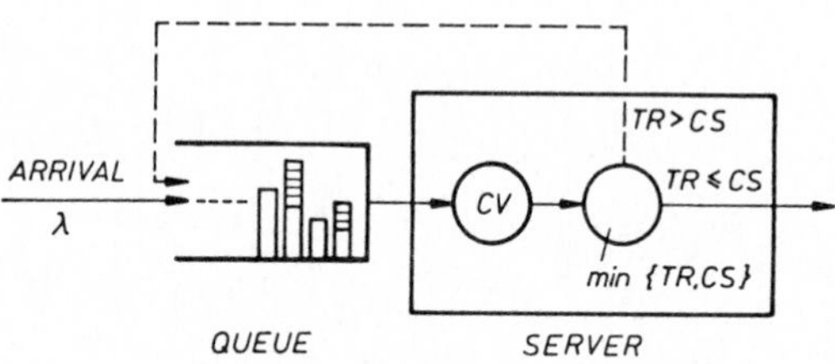

Fig. 3.1: M/G/1-SRPT/RR model with overhead time CV, maximum time slice CS, and remaining service time TR.

Fig. 3.1 shows the M/G/1 model for the SRPT/RR strategy. Jobs arrive with the Poisson arrival rate λ in a single server system, where service times are independent and identically distributed according to a general distribution. If the server is idle the next job with the shortest remaining service time is selected.

The model includes overhead as a constant setup time CV, which is needed every time a job enters service. A constant takedown time can also be modelled by including it in CV, because the order is not of importance in this model. A job is then served for a maximum time slice CS. If the job requires more processing after it has received its maximum time slice, it returns to the queue, otherwise it leaves the system. The queue is sorted according to the remaining service time of the jobs.

There are two ways of interpreting this system. First, one can say it is Round Robin with a queue sorted according to the remaining service time. Second, one can view it as SRPT with preemptions restricted to the end of a maximum time slice.

3.2 Method of Analysis

Fortunately the analysis of the SRPT/RR strategy is less complex than the analysis of the individual strategies with overhead. The methods used for those strategies can be combined without the need for the more complex parts of those methods. The integrals are not as deeply nested as for the SRPT strategy with overhead [1] and the matrix operations of the RR strategy [6],[2] can be eliminated.

The analysis for the mean delay with overhead $\overline{TD}^*$ can be divided into two parts: the mean initial waiting time $\overline{TW}(b)$ and the mean completion time $\overline{TC}(b)$ for the observed job B with service time b.

The total mean delay is then obtained by a weighted integration with the probability density function $f_B(t)$ of the service time random variable TB:

$$\overline{TD}^* = \int_0^\infty \{\overline{TW}(t) + \overline{TC}(t)\} f_B(t)dt \qquad (3.1).$$

For the derivation of $\overline{TW}(b)$ the same approach as for the SRPT strategy with overhead [1] can be used:

$$\overline{TW}(b) = \overline{TE}(b)/([1-D(b^-)][1-D(b)]) \qquad (3.2)$$

$$D(x) = \lambda \int_0^x \overline{TB}^*(t) f_B(t)dt \qquad (3.3),$$

where $\overline{TE}(b)$ is the mean initial waiting time due to the R-job in service when the B-job with service time b arrives. D(x) is the total load due to jobs with service time $TB \leq x$ and $\overline{TB}^*(b)$ is the service time with overhead for a job with service time b, which consists of the service time b and the overhead for every time slice:

$$\overline{TB}^*(b) = b + N(b)CV \qquad (3.4)$$

$$N(b) = [b/CS]^+ \qquad (3.5)$$

$[x]^+$: smallest integer value greater than or equal to x (3.6).

The derivation of $\overline{TE}(b)$ is similar to [1]. Two cases have to be distinguished. If the remaining service time of the R-job at the end of the present time slice is smaller than b then the R-job is completed before the B-job, otherwise only the present time slice is completed.

The formula for $\overline{TE}(b)$ contains the length R(t) of the last time slice of a job with service time t and the index N(t-b) of the time slice of a job with service time t after which the remaining service time is less than b:

$$\overline{TE}(b) = \lambda \cdot CV^2/2 + \lambda \int_0^\infty R(t)(CV + R(t)/2)f_B(t)dt$$

$$+ \lambda \cdot (CV+CS)^2/2 \int_0^b \left(N(t)-1\right)^2 f_B(t)dt$$

$$+ \lambda \cdot (CV+CS) \int_{b^+}^\infty M(t,b)\, f_B(t)dt \qquad (3.7)$$

$$R(t) = t - \left(N(t)-1\right)CS \qquad (3.8)$$

$$M(t,b) = (CV+CS)\left\{\left(N(t)-N(t-b)\right)^2 + N(t-b)-1\right\}/2$$

$$+ \left(N(t)-N(t-b)\right)(CV+R(t)) \qquad (3.9).$$

The mean completion time $\overline{TC}(b)$ for a job with service time b consists of the mean completion times $\overline{TC}_j$, one for every time slice $j=1...N(b)$:

$$\overline{TC}(b) = \sum_{j=1}^{N(b)} \overline{TC}_j \qquad (3.10).$$

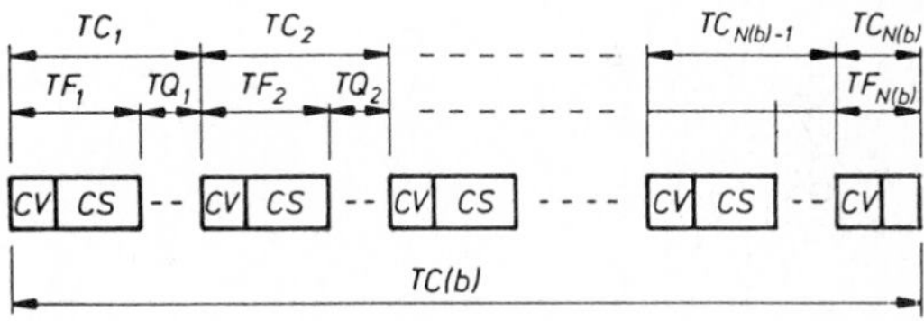

Fig. 3.2: Interval partition for completion time TC(b).

The mean completion time $\overline{TC}_j$ for the j-th interval consists of the time slice and overhead $\overline{TF}_j$ and the mean waiting time $\overline{TQ}_j$:

$$\overline{TC}_j = \overline{TF}_j + \overline{TQ}_j \qquad j = 1...N(b)-1 \qquad (3.11)$$

$$\overline{TF}_j = CV + CS \qquad j = 1...N(b)-1 \qquad (3.12)$$

$$\overline{TC}_{N(b)} = \overline{TF}_{N(b)} = CV + R(b) \qquad (3.13).$$

Note that $TF_j = \overline{TF}_j$ and $TB^*(b) = \overline{TB}^*(b)$ because the number of time slices for the B-job is deterministic.

The waiting time TQ_j consists of the sum of service times with overhead for all jobs with service time $TB < b-jCS$ that arrive during TF_j and TQ_j, because these jobs are served before the next time slice of the B-job is assigned. These jobs arrive with the Poisson rate $\lambda F_B(b-jCS)$ and have a mean service time with overhead $\overline{TB}^*(TB < b-jCS)$:

$$\overline{TQ}_j = \lambda F_B(b-jCS)\{\overline{TF}_j + \overline{TQ}_j\}\overline{TB}^*(TB < b-jCS) \qquad (3.14)$$

$$\overline{TB}^*(TB < b-jCS)F_B(b-jCS) = \int_0^{b-jCS} \overline{TB}^*(t)f_B(t)dt \qquad (3.15).$$

Solving the recursive equation results in:

$$\overline{TQ}_j = \overline{TF}_j\ D(b-jCS)/[1-D(b-jCS)] \qquad (3.16)$$

$$\overline{TC}_j = \overline{TF}_j/[1-D(b-jCS)] \qquad (3.17)$$

$$\overline{TC}(b) = CV + R(b) + \sum_{j=1}^{N(b)-1} \frac{CV + CS}{1-D(b-jCS)} \qquad (3.18).$$

Eqs. (3.2) and (3.18) complete the formula for the mean delay $\overline{TD}^*$ eq. (3.1).

3.3 Evaluation of SRPT/RR

For the computational evaluation of SRPT/RR integrals have to be evaluated that are generally not given as closed formulas. The numerical method used is the same as in [1] for the SRPT strategy: the continuous distribution function is approximated by a discrete distribution and the integrals are then evaluated as sums. In the following diagrams all time dependent variables are normalized with respect to the mean service time, i.e. $\overline{TD}^*$/overtextTB, μCV, and μCS are used.

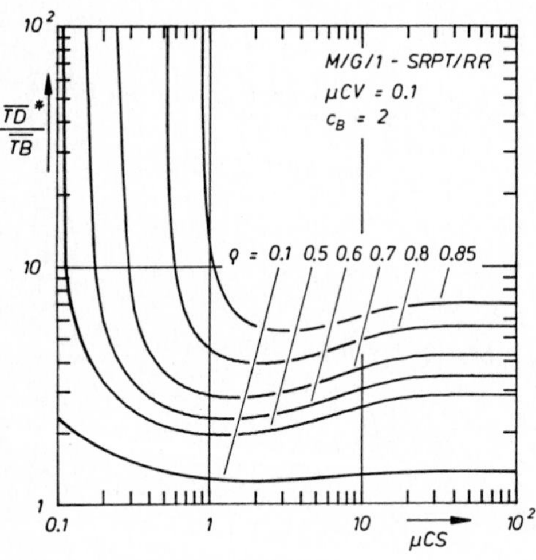

Fig. 3.3: Influence of maximum time slice CS.

For a given arrival rate, service time distribution, and overhead time the remaining parameter which influences system performance is the maximum time slice CS. Fig. 3.3 shows the influence of the maximum time slice CS. For small values of CS additional overhead time increases mean delay, for large values of CS the strategy tends to the non-preemptive SPT (Shortest Processing Time first) strategy. Inbetween these two extremes there is an optimal maximum time slice CS_{opt}.

In view of fig. 3.3 and similar results for different service time distributions the maximum time slice should be chosen somewhat larger than the mean service service time assuming overhead in the range of 10-20 % of mean service time.

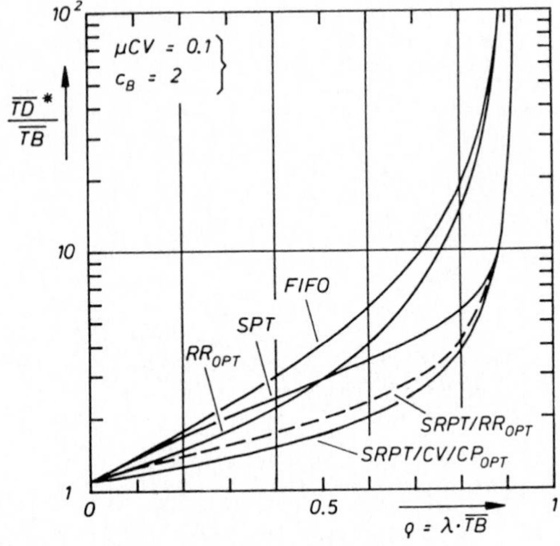

Fig. 3.4: Comparison of strategies.

In fig. 3.4 SRPT/RR is compared to FIFO, RR, SPT, and SRPT with overhead CV and preemption gap CP [1]. To assure a fair comparison the optimal maximum time slice or optimal preemption gap was chosen for each strategy. FIFO and RR show a similar behavior with respect to mean delay, but with preemptions improving the performance of the RR strategy. The SPT strategy displays the possible performance improvement of mean delay especially for higher loads if the service time is known in advance. But the strategy coming closest to SRPT/CV/CP is the SRPT/RR strategy, which combines the advantages of the SRPT and RR strategy provided that service times are known in advance.

A potential application area for this strategy in data communication networks is packet switching. The packet length then corresponds to the maximum time slice and overhead corresponds mainly to the packet frame. In networks with virtual or static routing this strategy selects the next packet of the message with the least number of remaining packets. For an equal number of remaining packets the message with the shortest last packet is chosen. In addition the FIFO rule is used to resolve ambiguity. To implement this strategy packets are numbered in reverse order for each message. Together with a message number and the length of the last packet the specific order within the queue for this strategy can be maintained. An additional advantage of this strategy is the preferential treatment of control messages, which are usually the shortest type of message. A simulation study is planned to investigate the behavior of this strategy in networks.

4. CONCLUSION

The analysis of the SRPT strategy combined with priorities and the Round Robin strategy has shown that mean delay can be reduced considerably especially for the hyperexponential case, if the service time is known in advance. Introducing external priorities allows overriding the internal priority mechanism based on the remaining service time, which is necessary for practical applications. The combination with the Round Robin strategy on the other hand simplifies the preemption mechanism.

Future research should investigate the possibilities offered by further combining these strategies as preemptive priorities with SRPT/RR in each priority class. Similar to [9] several preemption mechanisms have to be studied for this combination.

ACKNOWLEDGEMENTS
The authors would like to thank Prof. Dr.-Ing. F. Schreiber for his continous support. Special thanks is also extended to the many students that helped to produce the results of this paper.

REFERENCES

[1] Goerg, C.: Queueing System M/G/1: The Optimal Strategy SRPT in Comparison with the Round Robin Strategy Including Overhead. Doctoral Thesis, Technical University Aachen, 1983 (in German).

[2] Heck, J.: Queueing System M/G/1: Analytical Evaluations for the Round Robin Strategy Including Overhead. Master Thesis, Technical University Aachen, February 1984.

[3] Kesten, H.; Runnenberg, J.Th.: Priority in Waiting-Line Problems. I. and II. Nederlandse Akademie van Wetenschappen, Amsterdam, Proc., Series A, 60 (1957), 312-336.

[4] Kleinrock, L.: Queueing Systems, Volume I: Theory, Volume II: Computer Applications. John Wiley and Sons, New York 1975, 1976.

[5] Pham, X.H.: A Computer Aided System for the Evaluation of Analytical Results in Queueing Theory. Doctoral Thesis, Technical University Aachen, 1984 (in German).

[6] Sakata, M.; Noguchi, S.; Oizumi, J.: An Analysis of the M/G/1 Queue under Round Robin Scheduling. OP. RES. 19 (1971), 370-385.

[7] Schrage, L.; Miller, L.W.: The Queue M/G/1 with the Shortest Remaining Processing Time Discipline. OP. RES. 14 (1966), 670-684.

[8] Schrage, L.: A Proof of the Optimality of the Shortest Remaining Processing Time Discipline. OP. RES. 16 (1968), 687-690.

[9] Wang, Y.T.: An Analysis of a Round-Robin Schedule with Preemptive Priorities. Performance 1981, North - Holland Publishing Company, 1981.

LIST OF SYMBOLS

$\overline{TX}$	mean of random variable TX.
$M_2\{TX\}$	second moment of random variable TX.
λ λ_{ij} λ_i	Poisson input intensity for the complete system, for priority class ij, and for priority class i.
TB TB_i TB_{ij}	service (processing) time random variable for the complete system and for priority class i and subclass ij.
$F_B(t)$ $f_B(t)$ $f_{Bi}(t)$ $f_{Bij}(t)$	probability distribution function of TB, and probability density function of TB, TB_i, and TB_{ij}.
$\mu=1/\overline{TB}$ c_B c_i	mean service rate and coefficient of variation for TB and TB_i.
TB^* $TB^*(b)$	actual service time: service time with overhead, and actual service time for a job with service time b.
ρ ρ_i ρ_{ij} ρ^*	basic load: load without overhead for the complete system and for priority class i and subclass ij, actual load: load with overhead.
D(x)	total load due to jobs with TB $\leqq$ x.
CV CS	overhead (setup) time, and maximum time slice for Round Robin (RR).
N(b)	number of overhead times for a job with service time b.
TD TD_i TD_{ij}	total delay for the complete system, for a job of priority class i, and subclass ij.
TD^* $TD^*(b)$	total delay with overhead for the complete system and for a job with service time b.
TR TR_i	remaining service time for the complete system, and for priority class i.
B-job	observed job with service time b.
R-job	job in service with remaining service time r.
TW TW(b) TW_i TW_{ij}	initial waiting time: from entry to first service, for a job with service time b, for priority class i, and subclass ij.
TE(b) TE_i	initial waiting time for the B-job and for a job of priority class i due to the R-job in service.
TC TC(b) TC_{ij}	completion time: from first service to exit, for a job with service time b, and for a job of priority class ij.
TC_j TF_j TQ_j	completion time, actual service time, and waiting time for the j-th time slice.

TELETRAFFIC ISSUES in an Advanced Information Society
ITC-11
Minoru Akiyama (Editor)
Elsevier Science Publishers B.V. (North-Holland)

A RELIABILITY ORIENTED APPROACH TO THE TOPOLOGICAL DESIGN OF PACKET SWITCHED NETWORKS

F. Bernabei*, A. Leccese**, A. Pattavina*, A. Roveri*

* INFOCOM Department, University "La Sapienza" of Rome (Italy).
** Telespazio S.p.a., Rome (Italy).

ABSTRACT

The topological design of a backbone packet switched network is here faced. The synthesis problem is solved by means of two procedures: the first one is applied to a logical context, while the second refers to a physical environment. This approach allows to adopt different cost-functions for the switching and the transmission networks. Moreover performance criteria can be used that take into account the joint evolution of both the traffic and the failure - and - repair processes. An application example shows the feasibility of the proposed design method.

1. INTRODUCTION

About the topological design of communication networks, present research activities are addressed to the case of network models more adequate tha the ones adopted in the classical approaches. The objective is a design, in which logical, i.e. traffic-related, and physical, i.e. failure-related, phenomena are jointly and suitably taken into account.

According to this trend, one of our recent papers (Ref. [1]) introduced a new approach for solving the topological design problem concerning the backbone section of a hierarchical packet switched network (PSN); this paper reports on the further results obtained following the guidelines sketched in the above-mentioned work.

Our design methodology is based on two distinct graph models of the switching/transmission resources and adopts performance criteria that take into account the joint evolution of both the traffic process (T-process) and the failure-and-repair process (F/R-process). The two graph models, i.e. the logical and the physical one, are separately the contexts of these evolving processes: in particular the logical graph (LG), which is the model of the switching network, is associated to the T-process, while the physical graph (PG), which represents the transmission network, refers to the F/R-process.

Moreover, about the performance criteria used to qualify the resulting design, our approach aims to verify the network capability to carry out a given offered traffic, while considering the joint evolution of the T- and F/R-process.

In the following, Section 2 is devoted to describe our design objectives, by outlining the general key points of our methodology as composed of two procedures - the LG-procedure and the PG-procedure - that mutually interact for performing the synthesis of the switching and the transmission network, respectively. Details on these two procedures and on their respective contexts are given in Sections 3 and 4, respectively. Then, network performance criteria are introduced in Section 5. Finally an application example of the proposed design method is illustrated in Section 6.

2. DESIGN OBJECTIVES

As well known, the optimum topological design of communication networks starts from the identification of two design functions, both depending on network resources: they are here called the p-function (performance f.) and the c-function (cost f.). The p-function describes a significant performance to be provided by the network depending not only on the amount of network resources, but also on the way in which they are used. The c-function is related to the amount of network resources necessary to the providing of the required performances. In general the optimum design consists in determining the network resources by minimizing any one of these functions, the other acting as a constraint. Our approach relies upon the minimization of the c-function under a given constraint on the p-function.

Under these basic statements, our design method can be seen as composed of two procedures: the LG and the PG ones.

The LG procedure receives the mean offered traffic matrix and the switching network topology (LG topology) as input data. It consists in determining the node/branch capacities and in performing the traffic routing through the network. This determination is obtained by means of the above mentioned constrained minimization technique. The proper LG c-function is identified with the weighted amount of processing and transferring components; the LG p-function refers to the T-process and in particular to the total mean packet delay in the network.

The PG procedure, assuming an initial topology for the transmission network (initial PG topology), starts from the capacity assignment obtained in the LG procedure. It consists in performing the circuit routing of the transmission resources. Also in this case a constrained minimization has to be applied, involving proper c- and p-functions. The PG c-function is related to the amount of transmission components; the PG p-function is expressed by a quantity related to the F/R-process and in particular to reliability aspects.

As regards the LG problem, its relevant generalization could consist in starting from only partial data about the LG topology (e.g. from the switching node locations only). This more general problem will not be here considered, even if various literature contributions are available as possible hints of solution.

3. LOGICAL NETWOK SYNTHESIS

The starting points to this synthesis are given by: (1) the model of the switching network (Par. 3.1); (2) a suitable choice of the p- and c-functions (Par. 3.2). The LG design objectives are the synthesis of a minimum cost switching network satisfying the prescribed congestion target. The corresponding procedure is largely based upon the classical approaches to the PSN topological design, suitably modified to take into account the different modelling of the logical network (Par. 3.3).

3.1. LG design modelling

The adopted model of the switching network takes into account only those parameters that turn to be essential to show the feasibility of the proposed design procedure. So, aspects concerning, for example, stand-by facilities, are intentionally disregarded.

The LG context is described by the following items:

- N_L and B_L are the sets including the switching nodes and the logical branches, respectively; $|N_L|=n_L$, $|B_L|=b_L$, in which $|X|$ denotes the cardinality of the set X;
- a function $t_L:B_L \rightarrow N_L x N_L$ is given in order to specify the switching network topology;
- **S** is a b_L-dimensional vector, whose k-th component ($k \in B_L$) represents the shortest path distance between the ending nodes of the k-th logical branch;
- **R** is a $n_L x n_L$ matrix, in which r_{ij} ($i,j \in N_L$) denotes the mean offered traffic in the busy hour of the i-j relation [(i-j)-traffic]; a balanced offered traffic is assumed, i.e. $r_{ij}=r_{ji}$, and $r = \sum_{i,j} r_{ij}$ ($i,j \in N_L$; $i>j$) indicates the mean value of half of the total input traffic; r_{ij} and r are assumed measured in packet/s; qr_{ij} and qr indicate the same quantities expressed in bit/s, q denoting the mean packet length;
- **C** and **F** are two b_L-dimensional vectors, in which c_k and f_k ($k \in B_L$) indicate the normalized values of the unidirectional capacity (capacity module amount) and the mean traffic flow, respectively, supported by the k-th logical branch; the normalizing value is the basic capacity module $\hat{c}_0$ of the transmission systems (e.g. $\hat{c}_0$=64 kbit/s); the relation $f_k < c_k$ must be always satisfied for the ergodicity of the T-process; moreover the total mean traffic flow is denoted by $f = \sum_k f_k$ ($k \in B_L$);
- **E** and **G** are two n_L-dimensional vectors, in which e_k and g_k ($k \in N_L$) indicate the normalized values of the processing capacity and the relevant mean throughput, respectively, supported by the k-th node; e_k and g_k are assumed normalized in the same way as c_k and f_k; the condition $g_k < e_k$ holds; the total mean throughput of nodes in the network is denoted by $g = \sum_k g_k$ ($k \in N_L$); finally n_e denotes an integer such that $n_e\hat{c}_0$ is the minimum available module of processing capacity.

The traffic routing policy consists in assigning each (i-j)-traffic to one (or more than one) end-to-end logical path; on each path a parcel of the r_{ij} traffic flows. As a result of such a routing strategy, each end-to-end logical path is composed by some logical branches having a total mean number equal to $f\hat{c}_0/rq$.

About the representation of functions performed by each node and its inlets/outlets, we refer to a particular couple of inlet/outlet. The corresponding functions are here modelled by a cascade of two single server queues: the first queue (input queue) models the processing unit of the node; the second queue (output queue) models the communication facilities offered by the considered outgoing data link.

Two remarks must be added to such a modelling. At first, only one input queue is available for all the inlets, while one output queue exists for each outgoing link. Secondly, by referring to the OSI layered architecture, the output and the input queues perform basically the 1-st and 3-rd layer functions, respectively. The data link functions (error recovery, time-outs, window mechanisms, etc.) are disregarded for the sake of simplicity, even if some literature contributions would suggest how to approximately take into account the 2-nd layer functionalities (see, e.g. Ref [2]).

According to Kleinrock's independence assumption, the following hypotheses hold: external Poisson input at each node, exponential packet length distribution, infinite storage at nodes inlets and outlets, fixed routing, error-free transmissions, independence between interarrival and service times at each queue. Thus each input and output queue is of the $M/M/1/\infty/\infty$ type and the overall PSN can be modelled by a Jackson's queueing network.

On the basis of these simplifying hypotheses, the T-process evolution can be analyzed straigthforwardly, in statistical equilibrium conditions, by means of well known methods. That holds also with reference to the determination of the probability distribution functions of the end-to-end packet delays (see e.g. Ref [3]).

3.2. LG design figures

As mentioned in Section 2 the adopted LG c-function D(C,E) describes the weighted amount of processing and transferring resources. In particular we assume:

$$D(\mathbf{C},\mathbf{E}) = 2\sum_{i=1}^{b_L} d_i c_i + w \sum_{k=1}^{n_L} e_k \qquad (1)$$

in which d_i is a coefficient related to the shortest path distance s_i between the ending nodes of the i-th logical branch; w is the cost factor of a node, which is assumed independent of the node processing capacity.

About the LG p-function, we refer to a single packet and to the total average source-destination mean packet delay T. This delay accounts for the time elapsing from the packet arrival at the source node of the network to the end of its processing at the destination node.

The total mean packet delay T can be determined on the basis of the above model of switching network and can be expressed, as a function of the vectors **C**, **F**, **E**, **G**, in the form

$$T = \frac{1}{r}\sum_{i=1}^{b_L}\left[\frac{f_i}{c_i-f_i} + \frac{\hat{c}_0}{q} f_i T_i\right] + \frac{1}{2r}\sum_{k=1}^{n_L}\frac{g_k}{e_k-g_k} \qquad (2)$$

in which T_i denotes the channel propagation time on the i-th logical branch.

3.3. LG design methodology

With reference to the model introduced in Par. 3.1, the function t_L, the matrix **R** and the vector **S** are assumed as input data, together with other parameters involved in Eqs. (1) and (2) (e.g. the coefficients d_i, w and the channel propagation time). By denoting with T_{obj} the LG p-function target, the LG design problem consists in finding a set of integer capacities (vectors **C** and **E**) for nodes and branches, and a traffic flow distribution (vectors **F** and **G**), such that the LG c-function [Eq. (1)] is minimized under the delay constraint $T \leq T_{obj}$ [see Eq. (2)]. This is a classical CFA problem (e.g. see Ref. [4]) for which a procedure getting the global minimal solution is not known.

The adopted procedure consists in two steps: (1) solving a CFA problem with capacities expressed by real numbers; (2) finding a set of integer capacities in such a way that the corresponding LG c-function has, under the delay constraint, a value as near as possible to that computed in step (1).

As regards the step (1), a flow assignment and a capacity assignments are iteratively carried out, as long as the network cost D(**C**,**E**) decreases. In particular the modified Flow Deviation (FD) algorithm described in Ref. [4] is adopted for the traffic routing; it basically consists in a shortest path routing operated with branch lengths $z_i = dD/df_i$ given by

$$z_i = \frac{2d_i c_i}{f_i} + \sum_{k \in N_i} \frac{we_k}{g_k} + T_i \frac{2\sum_{k=1}^{b_L} d_k(c_k - f_k) + \sum_{k=1}^{n_L} w(e_k - g_k)}{\frac{qr}{\hat{c}_0}\left(T_{obj} - \frac{1}{r}\sum_{k=1}^{b_L} \frac{\hat{c}_0}{q} f_k T_k\right)} \tag{3}$$

in which N_i is the set of the two nodes on which the i-th branch terminates. It is outlined that this procedure has the characteristic of determining only one logical path for each traffic relation.

The capacity values in Eq (3) are computed by means of the method of Lagrange multipliers (Ref. [5]), that is

$$c_i = f_i + Q\sqrt{\frac{f_i}{d_i}}\,; \qquad e_k = g_k + Q\sqrt{\frac{g_k}{w}}$$

in which

$$Q = \frac{\sum_{i=1}^{b_L}\sqrt{d_i f_i} + \frac{1}{2}\sum_{k=1}^{n_L}\sqrt{wg_k}}{r\left(T_{obj} - \frac{1}{r}\sum_{i=1}^{b_L}\frac{\hat{c}_0}{q} f_i T_i\right)} .$$

It is pointed out that this iterative CFA procedure, starting from a random flow distribution, converges towards a solution which is only a local minimum of the c-function. Thus, by repeating the procedure for different starting flows, different local minima are found, the best of which is selected.

The step 2 consists in finding a new network with integer capacities as required by physical considerations. That is accomplished by approximating each c_i ($i \in B_L$) and e_k/n_e ($k \in N_L$) value to its first non-inferior integer. For the obtained new network the flow distribution **F** does not represent the best traffic routing. Thus a flow assignment procedure, i.e. the classical Flow Deviation procedure (Ref. [6]) is performed, in which the branch lengths $z_i = dT/df_i$ for the shortest path routing are now given by

$$z_i = \frac{1}{r}\frac{c_i}{(c_i - f_i)^2} + \frac{\hat{c}_0}{qr} T_i + \frac{1}{2r}\sum_{k \in N_i}\frac{e_k}{(e_k - g_k)^2} .$$

This algorithm provides the global minimum for the total mean packet delay.

The above integer approximation determines a more expensive network, whose delay performance T is definitely better than the target T_{obj}. It is outlined that in many cases, depending on the selected $\hat{c}_0$, n_e and on the obtained utilization factors f_i/c_i and g_k/e_k, the relative performance improvement $(T_{obj} - T)/T_{obj}$ is not negligible. Thus a module reduction procedure can be carried out to reduce the c-function values, while satisfying the delay constraint.

The basic idea of this procedure is to reduce that capacity, which determines the minimum worsening of delay performances per unit of c-function decreasing. In particular, if the integer capacities network is called the "old network" and T_{old} and D_{old} denote the values assumed by the p-and c-functions, respectively, the reduction procedure acts as follows:

(a) starting from the actual network, verify which is the effect of removing either only one module from a branch capacity, or only n_e modules from a node capacity; that corresponds to compute the relevant delay T_{new}, resulting from a new FD procedure, and the new cost D_{new};

(b) physically reduce the capacity of that network element which has showed the minimum ratio $(T_{new} - T_{old})/(D_{old} - D_{new})$ in the tests of step (a) and assume T_{new}, D_{new} and the new network as T_{old}, D_{old} and the old network, respectively; then go to the step (a).

Steps (a) and (b) are iterated as long as all the tests operated in step (a) provide at least one T_{new} value smaller than T_{obj}.

4. PHYSICAL NETWORK SYNTHESIS

In this context the starting points are represented by: (1) the model of the transmission network (Par. 4.1); (2) the set **C** of logical capacities determined by the LG synthesis; (3) a suitable choice of the p- and c-functions (Par. 4.2). The objective is to operate a minimum cost circuit routing compatibly with the assigned reliability target.

The synthesis procedure (Par. 4.3) is iterative; it starts from a single path routing of each capacity and afterwards, if required by the reliability constraints, it performs a multipath routing. Each step of the procedure identifies a particular allocation of logical capacities on the transmission resources according to a minimum cost criterion. The iteration is stopped as soon as the reliability constraints are satisfied.

4.1. PG design modelling

The transmssion network is modelled by the physical graph (PG), whose nodes and branches sets will be denoted by N_P and B_P, respectively; $|N_P| = n_P$, $|B_P| = b_P$. The branches of PG represent the transmission circuits, while the nodes are representative of the circuit terminals. The PG topology is described by the function t_P: $B_P \rightarrow N_P \times N_P$.

The B_P set is weighted by: (1) the branch length vector **L**, whose h-th component ($h \in B_P$) indicates the length of the h-th transmission circuit; (2) the branch capacity vector **M**, whose h-th component ($h \in B_P$) indicates the capacity module amount allocated on the h-th transmission circuit.

In order to complete this modelling two main aspects must be considered: the transmission technique to be used and the characterization of the F/R process.

About the transmission technique, a complete digital solution is assumed, with a basic channel rate of 64 kbit/s and by applying time division multiplexing. Therefore each PG branch together with its ending nodes consists of circuit terminals (such as digital multiplexes and line interfaces at each end-point), and of line devices (such as transmission medium and, where required, intermediate digital regenerators suitably spaced).

With reference to the F/R process, which results from the failure proneness of the transmission network and from the adopted overall maintenance policy, we assume that: (1) failures on physical components are statistically independent of each other; (2) negative exponential distributions characterize both interfailure and repair times for each transmission component and are input data to the problem.

Let S denote the set of the relevant states assumed by the transmission network during the evolution of the F/R-process. The cardinality of this set is determined by all the relevant failure configurations involving down-condition of single physical branches (including their ending nodes). In the $x \in S$ state the set B_{Px} includes all the up-branches, while $\bar{B}_{Px} = B_P - B_{Px}$. The state x=0 corresponds to $B_{Px}=B_P$.

If $\mathbf{V}(p)$ is a b_P-dimensional vector, whose h-th component $v_h(p)$ indicates the availability of p capacity modules in the h-th physical branch ($h \in B_P$), then the limiting probability p_x of the state x ($x \in S$) is given by

$$p_x = \prod_{h \in B_{Px}} v_h(m_h) \prod_{h \in \bar{B}_{Px}} [1-v_h(m_h)] \tag{4}$$

by taking into account that the total equipped capacity in the h-th physical branch is equal to m_h. In the following $\sum_x p_x$ ($x \in S$) is denoted by P_S. The identification of the set S can be suggested by the purpose of limiting the computational burden, while maintaining the P_S value enough near to the unity.

The availability $v_h(p)$ can be computed by means of $v_h(p)=[v_h^{(1)}(p)]^2 v_h^{(2)}$, by denoting with $v_h^{(1)}(p)$ and $v_h^{(2)}$ the corresponding availabilities of the circuit terminals and the line devices, respectively. While the latter availability is straightforward to be computed, the former is more difficult to be evaluated. Based on a markovian approach, the analysis of a multiplex unit has been carried out in Ref. [7].

4.2. PG design figures

As regards the cost aspect, a staircase function of the equipped transmission capacity is considered. Such a c-function, referred to the h-th transmission circuit, is assumed to depend on two factors: the equipped capacity m_h and the length l_h of the transmission circuit. So in a given network configuration a new capacity to be routed determines a cost increase only if a change in the kind or number of equipped multiplex systems is needed. Thus the shortest path between two switching nodes does not always represent the minimum cost routing for the required logical capacity between them.

In the PG context the p-function is reliability-oriented and expressed in terms of the first-order statistics of the F/R-process; in particular an availability function is introduced with reference to each logical capacity. By considering the k-th logical capacity c_k, the availability function $a_k(p)$ is defined as the probability that a share of p modules ($0<p\leq c_k$) of the c_k ones is working.

In order to introduce the evaluation method for $a_k(p)$, let i and j ($i,j \in N_L$) be the switching nodes at the ends of the k-th logical branch. Since, in general, multiple paths support the c_k capacity, a given share of $p<c_k$ modules between i and j can be obtained by means of different combinations of module amounts on each path supporting the c_k capacity. Obviously the sum of the modules considered on each path equals p. If $G_{kh}(p)$ is the event of p modules available out of the c_k ones within the h-th combination, then

$$a_k(p) = \Pr\{ \bigcup_h G_{kh}(p)\} \tag{5}$$

in which the union U is extended to all the feasible combinations providing p modules from i to j. It is to be noted that $a_k(p)$ is a monotonically non-increasing function of the p available modules.

The definition of such a p-function allows the pursuing of effective reliability goals for the transmission network synthesis. Good objectives seem to be: (1) the providing, with a very high probability, the availability of limited shares, i.e. $< 20\%$ for example, of the total installed logical capacity; (2) the assuming that the unavailability of great capacity shares, i.e. $> 80\%$ for example, does not overcome a prescribed value. These goals are consistent with the effect of the multipath routing procedure; in fact availability improvements at small capacity shares must cause availability degradations at great capacity shares.

4.3. PG design methodology

The problem of minimum cost circuit routing without constraints has been extensively investigated in literature; in particular, as staircase functions seem to be the most effective way to model the costs of transmission circuits, the λ_1-optimum algorithm (Ref. [8]) is adopted. It is remarked that this algorithm guarantees the convergence of the procedure towards a sub-optimal solution, if referred to the most general cost minimization problem.

The synthesis relies upon the iteration of the basic λ_1-optimal algorithm: it consists in finding the minimum cost single-path routing for all the logical capacities; one at a time is processed, the others being unchanged. That is accomplished by removing from the actual network the c_k capacity, as resulting from the circuit routing in the preceding iteration, and by finding its new best routing. If the c-function value at the i-th iteration is less than the value at the (i-1)-iteration, the basic λ_1-optimal algorithm is performed again, otherwise the λ_1-optimal solution has been found. The iterative performing of the basic λ_1-optimal algorithm will be referred to as λ_1-procedure.

The basic idea for improving reliability performances is the adoption of a multipath circuit routing, meaning that more than one path is selected to support a given capacity in accordance with suitable capacity shares on each path. Between the expected reliability improvements (which are as greater as the path number increases) and the related computational burden of the synthesis procedure, a reasonable trade-off suggests the adoption of a two-path routing.

In general a two path routing of a capacity is more expensive than a single path routing; then the constrained synthesis consists in iteratively applying the λ_1-procedure for decreasing capacity module amounts on the second path, as long as the reliability constraint is not satisfied by the $a_k(p)$ function.

The adopted approach, starting from the minimum cost single path circuit routing, basically consists in gradually and iteratively applying a 2-path routing on one or more c_k's properly selected in order to satisfy the reliability constraints. As in general the extra-cost due to the 2-path routing is proportional to the percentage of circuits diverted on the second path, the rule is followed to proceed iteratively at increasing step sizes of that percentage.

5. NETWORK PERFORMANCE EVALUATION

In Sections 3 and 4 the network synthesis has been carried out separately in the LG and PG contexts under performance constraints referring, on the one hand, to only traffic phenomena, and, on the other hand, to only failure/repair phenomena. Now an attempt is made in order to define performance figures - grades of service (GOS) - aiming to characterize the behaviour of the resulting network in terms of joint evolution of congestion and reliability performances.

For this purpose. congestion phenomena are here supposed to reach statistical equilibrium conditions in each F/R-process state assumed by the network during its steady state evolution. In particular the significant PG failure configurations must be mapped onto the LG structure, in such a way that the T-process evolution can be evaluated with reference to the actual availability of logical resources as resulting from real failures in the physical network.

A significant network GOS figure can be identified with reference to the probability distribution function of the packet end-to-end delay; the network GOS figure can be assumed as the value of such a distribution function computed for a specific value t_{ref} of its argument. Thus this GOS figure (ranging from 0 to the unity) represents the probability that the considered packet end-to-end delay is not greater than t_{ref}. Consequently, for a given t_{ref}, higher GOS values correspond to better network performances.

In general the above GOS figure is dependent on: (1) the specific F/R network state, which has been considered within the set S; (2) the specific traffic relation assumed in the network analysis; (3) the specific traffic level which is supposed to load the network. About this last item, let $b_{ref}R$ denote the reference offered traffic matrix assumed for the network GOS evaluation; the multiplicative coefficient b_{ref} ($b_{ref}>0$) will be referred to as traffic level.

As a basic GOS figure $H_{ij}^{(x)}(t_{ref},b_{ref})$ denotes the GOS value characterizing the network loaded by the traffic level b_{ref}, considered in the state $x \in S$ and with reference to the specific (i-j)-traffic relation; therefore $H_{ij}^{(x)}(t_{ref},b_{ref})$ is the probability that the end-to-end delay of packets belonging to the (i-j)-traffic is not greater than t_{ref} in the network state $x \in S$ and under the traffic level b_{ref}. It is understood that, if increasing values of b_{ref} are considered starting from 0, the function $H_{ij}^{(x)}(t_{ref},b_{ref})$ is non-increasing; moreover if the (i-j)-relation is disconnected in the state x, then $H_{ij}^{(x)}(t_{ref},b_{ref})=0$, independently of the b_{ref} and t_{ref} values.

Starting from the basic $H_{ij}^{(x)}(t_{ref},b_{ref})$ definition, other more general GOS figures can straightforwardly defined by performing, either separately or jointly, averaging operations in the F/R state set S, or in the set of the traffic relations, or in the range of the feasible traffic levels. As far as the traffic level averaging is concerned, we outline that the LG synthesis has been carried out on the basis of the specific traffic level $b_{ref}=1$. In this condition the mean value T has been constrained to be not greater than T_{obj}. Neverthless, in actual network operation the traffic level b_{ref} is a random variable whose distribution must be specified in order to delimit the range of b_{ref} values to be investigated in the network performance evaluation. A reasonable choice seems to be the assuming a probability density function $K_{b_{ref}}(z)$ for b_{ref} such that $T \leq T_{obj}$ (or equivalently $z \leq 1$) with a prescribed probability P_{ref} (e.g. equal to 0.95).

The average operated within the set of the traffic relations, the other parameters being fixed, provides this other GOS figure

$$H^{(x)}(t_{ref},b_{ref}) = \frac{1}{r} \sum_{i,j \in N_L} r_{ij} H_{ij}^{(x)}(t_{ref},b_{ref}) \quad (6)$$

Moreover, if the F/R state set or the set of feasible traffic levels are considered separately to average the preceding GOS figure, two other figures can be obtained as follows

$$H(t_{ref},b_{ref}) = \frac{1}{P_S} \sum_{x \in S} p_x H^{(x)}(t_{ref},b_{ref}) \quad (7)$$

$$H^{(x)}(t_{ref}) = \int_0^{\infty} H^{(x)}(t_{ref},z)\, K_{b_{ref}}(z)\, dz \quad (8)$$

Finally, by jointly operating the last two averages, we obtain the following global GOS figure

$$H(t_{ref}) = \frac{1}{P_S} \sum_{x \in S} p_x H^{(x)}(t_{ref}) \quad (9)$$

In conclusion we remark that, even if other GOS figures derived from the basic one could have been defined, those merit figures here introduced seem the most significant to characterize network performances.

6. APPLICATION EXAMPLE

A case study is here considered as an application example of the proposed procedure for a PSN topological design. Figure 1 shows the LG topology, in which $n_L=12$, $b_L=21$; the mean offered traffic matrix **R** is given in Table 1 and a mean packet length q=1200 bit has been assumed; furthermore $\hat{c}_0$=64 kbit/s and n_e=8. About the LG c-function, as given in Eq (1), the position $d_i=A^{(L)}+B^{(L)}s_i$ has been made, in which $A^{(L)}$=12, $B^{(L)}$=2; moreover w=200. The delay target T_{obj} equals 80 ms and the propagation delay has been fixed equal to $5\cdot10^{-6}$ s/km.

A PG initial topology coincident with the LG one has been assumed, i.e. $N_P=N_L$, $B_P=B_L$ and $t_P=t_L$. The PG c-function, for the h-th transmission circuit, is given by $n(A^{(P)}+B^{(P)}l_h)$, in which the vector **L** is provided in Table 2, $A^{(P)}$=12, $B^{(P)}$=2 and n=31, compatibly with the 1-st order European PCM multiplex. Only this kind of multiplex is supposed to be adopted in this application example.

As regards reliability aspects, with reference to the h-th transmission circuit, the unavailability $1-v_h^{(2)}$ of line devices is contained in Table 2, while the unavailability $1-v_h^{(1)}(p)$ of circuit terminal can be expressed by the approximate law $(60+12p)10^{-6}$. The unavailability target for each logical branch is given in Fig. 2, showing also the resulting $1-a_k(p)$ function, k corresponding to the logical branch 9-11, as obtained by means of Eq. (5) at the end of the circuit routing procedure.

The results of the LG synthesis, i.e. the vectors **C**, **F**, **E**, **G** (expressed in the normalized form as defined in Par. 3.1), and of the PG synthesis, i.e. the vector **M**, are summarized in Table 2. It is observed that some of the initial physical branches have not been used by the circuit routing procedure. Moreover, by referring to the PG c-function, we can report that the cost increase due to the two-path circuit routing (as imposed by the constraints described in Fig. 2) has resulted of the order of 30%, with respect to the synthesis with a single path routing.

About the performance analysis of the synthesized network, failures of only one physical branch at a time have been considered. Moreover, in each x network state and for each considered traffic level b_{ref}, the function $H^{(x)}(t_{ref},b_{ref})$ has been computed on the basis of the traffic routing resulting from a FD procedure.

By examining now some samples of the obtained GOS figures, Figs. 3 and 4 show the $H^{(x)}(t_{ref},b_{ref})$ behaviour [as obtained from Eq. (6)] in the state x=0 and (as a sample for failure states) in the other state corresponding to the failure of the 5-9 physical branch. In these Figures the four curves refer to different traffic levels: in particular the b_{ref} values adopted for x≠0 (Fig. 4) are half of those adopted for x=0 (Fig. 3). It is to be reamarked that a maximum traffic level b_{refM} can be identified for each network state different from a disconnected one: for each traffic level $b_{ref}\geq b_{refM}$, the network GOS is equal to 0 independent of the selected t_{ref}. In the above considered states the values b_{refM}=1,02 and 0.53 have been obtained, the former value referring to x=0. Figures 3 and 4 show how GOS figures degrade, for a given t_{ref}, as the traffic level increases.

In the Fig. 5, the GOS figure $H(t_{ref},b_{ref})$ [as obtained from Eq. (7)] is plotted, for four t_{ref} values, as a function of the traffci level b_{ref} in the range (0-1.02). By considering t_{ref} values near to T_{obj} (t_{ref} = 80 ms), the GOS value shows to be better than 0.9 for a wide range of b_{ref} (b_{ref}<0.85), while it becomes about 0.5 for a traffic level equal to the design one (b_{ref}=1).

As regards the distribution density function $K_{b_{ref}}(Z)$, a 5-th order erlangian function has been assumed, by imposing P_{ref}=0.95. The resulting $H^{(x)}(t_{ref})$ function [as obtained from Eq. (8)], for the same states considered in Figs. 3 and 4, is plotted in Fig. 6. It shows that the GOS figure for the state x=0 is as much better than the GOS figure for the state x≠0 as much b_{ref} increases: that is apparently due to the different b_{refM} values characterizing the two considered states.

Finally Fig. 7 plots the $H(t_{ref})$ function representing a global merit figure according to Eq. (9). It is outlined that the curve A of Fig. 6 is practically the same as the one in Fig. 7, since the probabilistic weight of the state x=0 is largely predominant over that of all the other states $x\in S$.

7. CONCLUSIONS

The design method here introduced has showed to be feasible and effective in the assumed contexts and under the choosen design functions. However some topics appear to be attractive for further study. We refer in particular to: (1) the definition of perturbation algorithms to be applied for modifying the starting LG topology in case of unsatisfied delay constraint; (2) the identification of a more significant model for representing the LG context with specific reference to protocol implications.

REFERENCES

[1] A. Pattavina, A. Roveri, A. Leccese: "Minimum cost communication network design with reliability and congestion constraints", Proc. ICCC84, Sydney, pp. 434-439, 1984.

[2] P.J. Kuehn: "Modelling and analysis of computer networks - decomposition techniques, transient analyisis and protocol implications", Proc. ICC84, Amsterdam, paper 41.6, 1984.

[3] V. Bonaventura, S. Cacopardi, M. Decina, A. Roveri: "Service availability of communication networks", Proc. NTC80, Houston, pp. 15.2.1-15.2.6, 1980.

[4] M. Gerla, L. Kleinrock: "Topological design of distributed computer networks", IEEE Trans. on Comm., Vol. COM-25, N. 1, pp. 48-60, 1977.

[5] L. Kleinrock: "Communication nets, stochastic message flow and delay", McGraw Hill, New York, 1964.

[6] L. Fratta, M. Gerla, L. Kleinrock: "The flow deviation method: an approach to store and forward computer communication network design", Networks, Vol. 3, pp. 97-133, 1973.

[7] A. Leccese, A. Pattavina, A. Roveri: "Reliability analysis of a non-redundant repairable multiplexing unit", submitted for publication to IEEE Trans on Rel., 1983.

[8] M. La Cava, F. Mazzarella, U. Mocci: "Istradamento ottimo di circuiti con funzioni costo del tipo a gradinata", F.U.B. Internal Report, Rome, 1973.

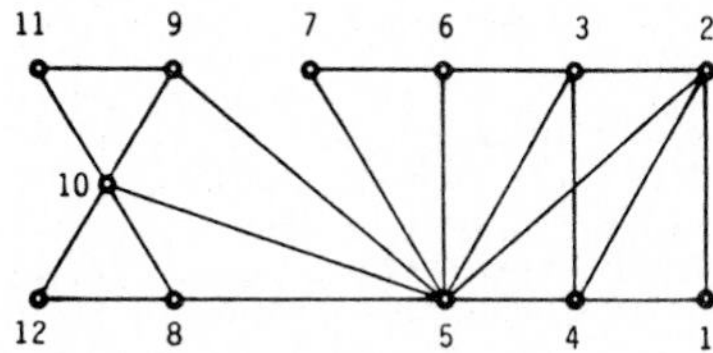

Fig. 1 - Network topology

Table 1: Mean offered traffic matrix **R** (packet/s)

0	64.0	53.3	21.3	21.3	21.3	53.3	53.3	64.0	53.3	106.7	106.7
64.0	0	21.3	53.3	21.3	42.7	53.3	64.0	64.0	64.0	106.7	106.7
53.3	21.3	0	42.7	21.3	53.3	21.3	42.7	64.0	53.3	42.7	53.3
21.3	53.3	42.7	0	21.3	42.7	64.0	53.3	42.7	64.0	64.0	42.7
21.3	21.3	21.3	21.3	0	21.3	21.3	21.3	21.3	21.3	21.3	64.0
21.3	42.7	53.3	42.7	21.3	0	21.3	64.0	53.3	53.3	64.0	21.3
53.3	53.3	21.3	64.0	21.3	21.3	0	53.3	21.3	21.3	64.0	42.7
53.3	64.0	42.7	53.3	21.3	64.0	53.3	0	42.7	64.0	21.3	53.3
64.0	64.0	64.0	42.7	21.3	53.3	21.3	42.7	0	21.3	64.0	64.0
53.3	64.0	53.3	64.0	21.3	53.3	21.3	64.0	21.3	0	53.3	42.7
106.7	106.7	42.7	64.0	21.3	64.0	64.0	21.3	64.0	53.3	0	42.7
106.7	106.7	53.3	42.7	64.0	21.3	42.7	53.3	64.0	42.7	42.7	0

Table 2: Design input data and output results

LG context						PG context			
data link	**C**	F	switch node	E	G	transm. circuit	**L** (Km)	$1-V^{(2)}$ $(\times 10^{-3})$	**M**
1 - 2	2	1.69	1	24	23.20	1 - 2	80	1.32	16
1 - 4	11	9.91	2	32	25.77	1 - 4	50	.82	27
2 - 3	2	1.69	3	24	20.04	2 - 3	70	1.13	15
2 - 4	1	0.80	4	40	38.23	2 - 4	100	1.63	0
2 - 5	10	9.19	5	88	85.67	2 - 5	200	3.29	0
3 - 4	2	1.66	6	24	20.06	3 - 4	90	1.47	0
3 - 5	6	5.30	7	24	16.40	3 - 5	100	1.63	0
3 - 6	3	2.58	8	40	35.44	3 - 6	60	.97	18
4 - 5	18	16.25	9	24	20.38	4 - 5	80	1.32	25
5 - 6	8	7.20	10	48	44.05	5 - 6	70	1.13	26
5 - 7	7	6.53	11	32	24.67	5 - 7	150	2.44	6
5 - 8	15	14.26	12	32	24.00	5 - 8	280	4.60	19
5 - 9	7	6.52				5 - 9	300	4.91	19
5 - 10	16	15.22				5 - 10	350	5.72	0
6 - 7	2	1.67				6 - 7	80	1.32	5
8 - 10	3	2.48				8 - 10	40	.66	30
8 - 12	10	8.70				8 - 12	60	.97	1
9 - 10	3	2.52				9 - 10	60	.97	22
9 - 11	2	1.54				9 - 11	100	1.63	1
10 - 11	12	10.92				10 - 11	60	.97	13
10 - 12	4	3.30				10 - 12	50	.82	13

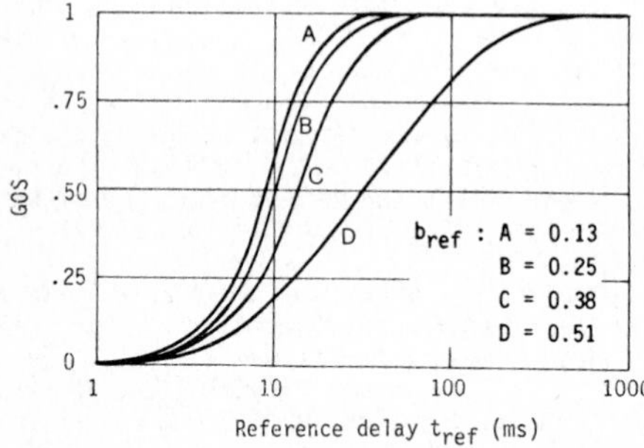

Fig. 4 - Figure $H^{(x)}(t_{ref},b_{ref})$ for the 5-9 failure

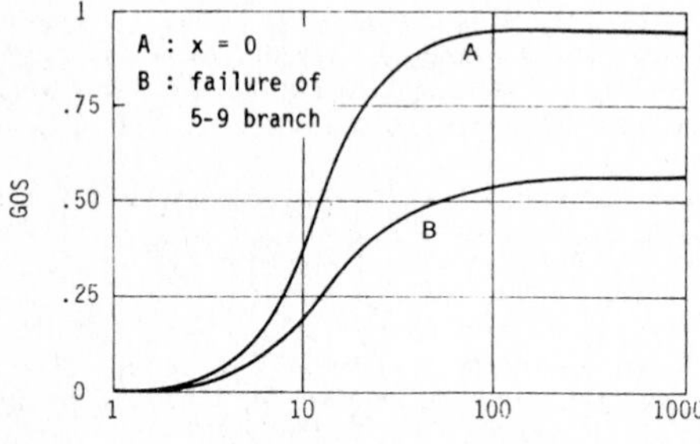

Fig. 6 - Figure $H^{(x)}(t_{ref})$ for two different states

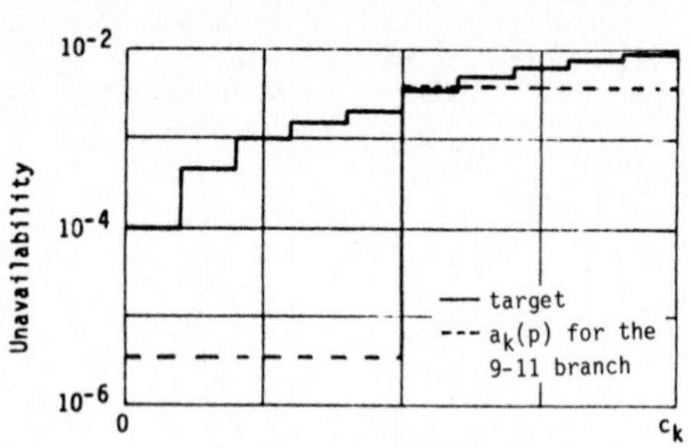

Fig. 2 - Reliability target

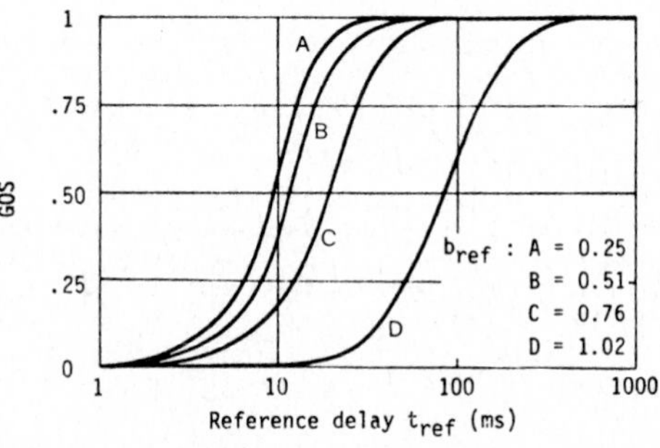

Fig. 3 - Figure $H^{(x)}(t_{ref},b_{ref})$ for the state x=0

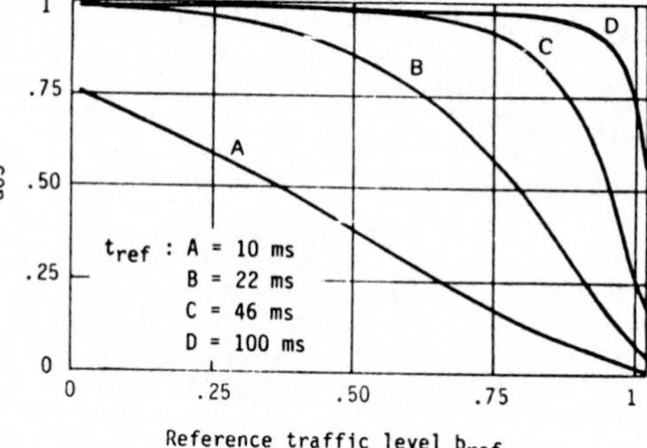

Fig. 5 - Figure $H(t_{ref},b_{ref})$

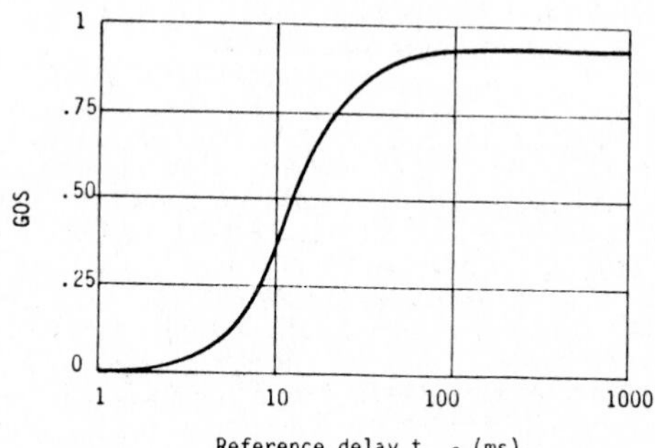

Fig. 7 - Global merit figure $H(t_{ref})$

TELETRAFFIC ISSUES in an Advanced Information Society
ITC-11
Minoru Akiyama (Editor)
Elsevier Science Publishers B.V. (North-Holland)

MIXING OF TRAFFIC ON A TRUNK GROUP: CALCULATION OF BLOCKING PROBABILITIES USING AN EXTENSION OF THE ERT METHOD

J. LABETOULLE

CENTRE NATIONAL D'ETUDES DES TELECOMMUNICATIONS
CNET/PAA/ATR, 38 Rue du Général Leclerc
92131 Issy les Moulineaux, FRANCE

ABSTRACT

We deal in this paper with a problem arising when several kinds of traffic are offered to a circuit switched network. It can be considered that telephone networks may be used for both telephone calls and data transfers. The mean holding times for these two kinds of traffic will be very different. It is thus necessary to develop a method for disigning such networks. We propose here a method which allows firstly to calculate the mean and variance of overflowing traffics when Poisson traffics are offered, and secondly to determine the blocking probability of a last choice trunk group when several peaky traffics, for each class of users, are offered. Finally, numerical results are given to validate the method, and dimensioning considerations are discussed.

1- INTRODUCTION

This study is focused on the evaluation of a circuit switched network to which are offered several kinds of traffics. It is not strictly speaking a study concerning integration of services but we are concerned with the utilization of a telephone network in which several classes of customers are introduced. In fact, it is more and more usual to utilize a telephone network to carry telematic communications: terminal to terminal or terminal to computer connections. These kinds of communication have duration characteristics very different from those of classical telephone calls. This is especially true in France with the wide scale introduction of small terminals at the subscriber's home (the "Minitel" terminals used for the Electronic Directory). Local communications due to the utilization of the "Minitel" for the electronic directory may be quite short (1 or 2 minutes) compared with the holding time of a local call. Conversly, a terminal to computer connection in the toll network may be very long (half an hour or more) compared with the inter-city calls. It is thus natural to study the grade of service of a network when these kinds of communications are introduced. A related problem is the dimensioning of such networks.

The study will be conducted in the following way:

1- the study of the overflowing process when Poisson traffics are offered. The steady state system of equations which can be derived leads to a linear system which can only be solved numerically by matrix inversion. An approximate solution is then presented, to evaluate the mean and variance of the overflowing traffic for each class of traffic. The results are compared with the exact results mentioned above and show a maximum relative error of about 10% for the second moment.

2- when a trunk group receives several overflow streams, it is necessary to calculate its blocking probability. The first part of the study has given the mean and variance of each flow, for each customer class. We can therefore establish the global characteristics of the flow offered to the last choice trunk group. From this information, we calculate, as in the ERT method, an "equivalent trunk group" which allows us to calculate the blocking probability. The result is obtained by inversion of the system of equations of the approximate method developed in part 1. This results in very simple calculations and numerical values are compared with simulations.

3- The last part of the paper validates the method and gives some considerations on the dimensioning problem, when the proportion of each traffic varies.

The assumptions of the study are the following: two classes of customer are considered, each one being characterized by the mean call holding time, whose values are $1/\mu_1$ and $1/\mu_2$. For each class, the duration of the calls are supposed to be exponentially distributed. The values of μ_1 and μ_2 must be considered as different.

To model the network, the method will be copied on classical techniques used for telephone networks: it is supposed that trunk groups can be ordered, so that trunk group i can be calculated as soon as all trunk groups j, $j<i$ have been evaluated. The problem is then to calculate the mean and variance of the overflowing traffic for each trunk group when the offered traffic is Poisson. When the traffic is peaky (in the case of an overflow), it is necessry to evaluate the blocking probability of the trunk group. These are classical assumptions used to calculate telephone networks, at least for hierarchical networks when end to end blocking probability is not the criterion. This classical theory will be extented here for the case of a mixture of traffics.

2- CALCULATION OF OVERFLOWING TRAFFICS

We suppose now that the two flows are offered to a trunk group of size N. We will use the following notations:

- μ_1 and μ_2 are the rates of service for the two classes of customers,

- λ_1 and λ_2 are the arrival rates, which are supposed Poisson,

- m_{i1}, m_{i2}, v_i are mean, second moment and variance of overflow traffic i,

- A_i is the offered traffic for class i, equal to λ_i/μ_i,

- A is the total traffic, equal to A_1+A_2.

The following results can be derived from the classical theory:

$$P(N) = E(A,N) = \frac{A^N/N!}{\sum_{i=0}^{i=N} A^i/i!} \qquad (2\text{-}1)$$

$$m_{i1} = A_i\ P(N) \quad ,\ i=1,2 \qquad (2\text{-}2)$$

$$P(i,j) = P(0)(A_1^i/i!)(A_2^j/j!) \qquad (2\text{-}3)$$

P(i,j) is the steady state probability of having i customers of class 1 and j customers of class 2 on the first choice trunk group. P(i) is the steady state probability of having a total of i customers.

We also have:

$$P(i) = P(0)\ A^i/i!$$

The only real problem is thus to calculate the variance or the second moment of each overflowing traffic.

2-1 EXACT SOLUTION

Define P(i,j,k) to be the steady state probability of having:
- i customers of class 1 on the first choice trunk group,
- j customers of class 2 on the first choice trunk group,
- k customers of class 1 overflowing.

i and j can vary from 0 to N, and k varies from 0 to infinity. The following system (equations 2-4) can be derived from the Chapman-Kolmogorov theory:

$$(\lambda_1+\lambda_2)P(0,0,0) = \mu_1P(1,0,0)+\mu_2P(0,1,0)+\mu_1P(0,0,1)$$

$$(\lambda_1+\lambda_2+i\mu_1+j\mu_2)P(i,j,0) = \lambda_1P(i-1,j,0)+\lambda_2P(i,j-1,0)+(i+1)\mu_1P(i+1,j,0) +(j+1)\mu_2P(i,j+1,0)+\mu_1P(i,j,1) \qquad i+j<N$$

$$(\lambda_1+i\mu_1+j\mu_2)P(i,j,0) = \lambda_1P(i-1,j,0)+\lambda_2P(i,j-1,0)+\mu_1P(i,j,1) \qquad i+j=N$$

$$(\lambda_1+\lambda_2+i\mu_1+j\mu_2+k\mu_1)P(i,j,k) = \lambda_1P(i-1,j,k)+\lambda_2P(i,j-1,k)+(i+1)\mu_1P(i+1,j,k) +(j+1)\mu_2P(i,j+1,k)+(k+1)\mu_1P(i,j,k+1) \qquad i+j<N$$

$$(\lambda_1+i\mu_1+j\mu_2+k\mu_1)P(i,j,k) = \lambda_1P(i-1,j,k)+\lambda_2P(i,j-1,k)+\lambda_1P(i,j,k-1) +(k+1)\mu_1P(i,j,k+1) \qquad i+j=N$$

Define $f(i,j) = \sum_{k=0}^{\infty} kP(i,j,k)$

We multiply each equation of the previous system by k, and sum over k to find the following system (2-5):

$$(\lambda_1+\lambda_2+\mu_1)f(0,0)=\mu_1f(1,0)+\mu_2f(0,1)$$

$$(\lambda_1+\lambda_2+i\mu_1+j\mu_2+\mu_1)f(i,j) = \lambda_1f(i-1,j)+\lambda_2f(i,j-1) +(i+1)\mu_1f(i+1,j)+(j+1)\mu_2f(i,j+1) \qquad i+j<N$$

$$(i\mu_1+j\mu_2+\mu_1)f(i,j)=\lambda_1f(i-1,j)+\lambda_2f(i,j-1)+\lambda_1P(i,j) \qquad i+j=N$$

By this method, we obtain a system of (N+1)(N+2)/2 equations with the same number of unknown variables f(i,j). A numerical resolution by inversion of the linear system is then possible. The summation of all equations of system 2-5 leads to:

$$m_{11} = \sum_{i,j} f(i,j) = \frac{\lambda_1}{\mu_1} P(N)$$

This relation may be used to verify the accuracy of the variables f(i,j). To calculate the second moment of overflow traffic 1, it is necessary to multiply each equation of system 2-4 by k^2 and to sum all these equations over i and j. The terms of the form $k^3P(i,j,k)$ will disappear after summation, and the following result is finally obtained after some simplifications:

$$m_{12}=\sum_{i,j,k} k^2P(i,j,k)=\frac{\lambda_1}{\mu_1}\ (P(N)+\sum_{i+j=N} f(i,j))$$

The previously calculated coefficents f(i,j) are thus sufficent to calculate the first two moments.

$$m_{11} = \sum_{i,j} f(i,j) \qquad (2\text{-}6)$$

$$m_{12} = m_{11} + \sum_{i+j=N} f(i,j) \qquad (2\text{-}7)$$

2-2 APPROXIMATE CALCULATION.

The exact method presented in the previous section is certainly not satisfactory from a practical point of view. It is thus necessary to imagine an approximate method, which should be fast enough and whose accuracy will be evaluated by comparison with the exact analysis.

The basic idea which will be devoped now is to model the first choice trunk group by a switch, which will be remain closed during a negative exponentially distributed time, with a rate α, and will remain closed during an exponential time with rate β. This method is derived from the classical simplified IPP (Interrupted Poisson Process) of Kuczura [1]. Figure 1 explains the model.

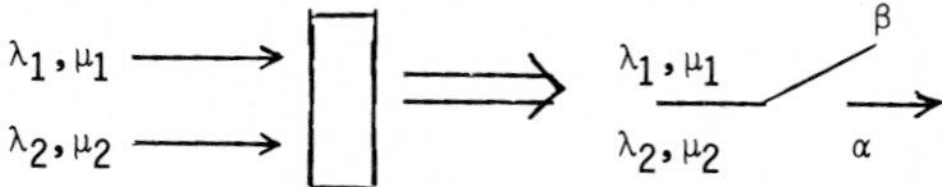

figure 1: The simplified IPP model

The full IPP calculates three parameters (λ,α,β) to match the three moments of the overflowing traffic. Here, as in the simplified IPP, λ is kept to its real value, α and β are calculated to match the two first moments. However in our problem, λ is not defined since there are two traffics. We have therefore to evaluate the parameters of the IPP using an "equivalent flow", whose arrival rate is $\lambda = \lambda_1+\lambda_2$ and whose mean service time μ is such that $A = \lambda/\mu$:this is the exact flow when $\mu_1 = \mu_2$. The switch is in fact defined as if both traffics have the same holding time. This is justified by the fact that the steady state probabilities of the trunk group are independant of the service rates for each class.

The equivalent flow generates an overflow traffic whose mean and variance are known from the classical formulas:

$$m = AP(N)$$

$$v = m\left(1 - m + \frac{A}{N+1+m-A}\right) \qquad (2\text{-}8)$$

The IPP for this equivalent flow gives a renewal process whose Laplace transform is given by:

$$g(s) = \frac{\lambda\,(\beta + s)}{s^2 + s\,(\lambda + \alpha + \beta)} \qquad (2\text{-}9)$$

From this last formula, it is known that:

$$P(N) = \frac{\beta}{\alpha + \beta}$$

$$m_2 - m = v + m^2 - m = m\,\frac{g(\mu)}{1 - g(\mu)} \qquad (2\text{-}10)$$

This gives:

$$g(\mu) = \frac{m_2 - m_1}{m_2}$$

So $g(\mu)$ can be calculated directly from (2-8) and (2-10). Then, in (2-9), $\alpha+\beta$ can be replaced by $\beta/P(N)$ or by $\beta A/m$ to obtain a linear equation in β.

Finally, the solution for α and β is :

$$\beta = \frac{\lambda\mu - g(\mu)\mu^2 - \lambda\mu g(\mu)}{\lambda g(\mu) + \lambda g(\mu)/m - \lambda}$$

$$\alpha = (A\beta - \beta m)/m$$

We apply this simplified model to each individual flow. The IPP defined by α and β is applied to each individual traffic 1 and 2 to obtain:

$$m_{i2} = m_{i1} + m_{i1}\,\frac{g_i(\mu_i)}{1 - g_i(\mu_i)}$$

with:

$$g_i(\mu_i) = \frac{\lambda_i\,(\beta + \mu_i)}{\mu_i^2 + \mu_i(\lambda_i + \alpha + \beta) + \lambda_i\beta}$$

Thus, we obtain an approximation for the second moment and variance of each individual overflowing traffic. Of course, this method does not consider the covariance between overflowing traffics, but it will be shown in the following sections that this approximation is reasonable.

2-3 EVALUATION OF THE APPROXIMATE SOLUTION

The first remark is that the proposed model gives exact results in the special case where $\mu_1=\mu_2$. This property can easily be verified, but the proof will not be given here. It consists in verifying that the known formulae are found, that is (cf [2]):

$$E(n_i^2) = p_i^2\,E(n^2) - p_i(1-p_i)E(n)$$

with $p_i = A_i/A$.

For the general case where μ_1 and μ_2 are not equal, we give a comparison of results in table 1. Several cases are studied, with A = 10 and N = 10. First, we examine the case where $A_1 = A_2$, then the case where the traffics are not equal. For each case, the ratio μ_1/μ_2 varies from 1 to 20. We have only shown the results concerning the second moment, since the first is given exactly.

The results are certainly not perfect, but give only a 10% maximum relative error. The error is less than that in most cases, however it is increasing with the ratio of the holding times. We will show below that the results are sufficient to solve the problem of cluster networks, when several trunk groups overflow on a common trunk group.

A_1	A_2	λ_1	λ_2	μ_1	μ_2	flow 1: proposed method	flow 1: exact method	flow 2: proposed method	flow 2: exact method
N = 10									
5	5	5	5	1	1	2.7782	2.7782	2.7782	2.7782
		10	5	2	1	2.9319	2.9371	2.6104	2.6442
		15	5	3	1	3.0040	3.0290	2.5206	2.5836
		20	5	4	1	3.0459	3.0892	2.4647	2.5507
		25	5	5	1	3.0732	3.1318	2.4265	2.5289
		50	5	10	1	3.1337	3.2372	2.3370	2.4813
		100	5	20	1	3.1673	3.3037	2.2840	2.4552
8	2	8	2	1	1	6.0822	6.0822	0.7020	0.7020
		16	2	2	1	6.2171	6.2335	0.6657	0.6698
		24	2	3	1	6.2680	6.3118	0.6505	0.6566
		32	2	4	1	6.2947	6.3598	0.6421	0.6494
		40	2	5	1	6.3112	6.3922	0.6368	0.6448
		80	2	10	1	6.3451	6.4676	0.6256	0.6352
		100	2	20	1	6.3626	6.5117	0.6196	0.6302
2	8	2	8	1	1	0.7020	0.7020	6.0822	6.0822
		4	8	2	1	0.7492	0.7461	5.8723	5.9354
		6	8	3	1	0.7785	0.7755	5.7165	5.8625
		8	8	4	1	0.7985	0.7967	5.5962	5.8186
		10	8	5	1	0.8130	0.8128	5.5006	5.7891
		20	8	10	1	0.8500	0.8578	5.2170	5.7211
		40	8	20	1	0.8742	0.8914	4.9935	5.6811

Table 1 : Comparison of second moments: exact and approximate solutions

3- MIXTURE OF OVERFLOW TRAFFICS.

3-1 THE MODEL.

For network dimensioning, it is necessary to evaluate classical clusters. We now examine the case of several trunk groups overflowing onto a final choice trunk group. The following figure 2 also gives the notation used:

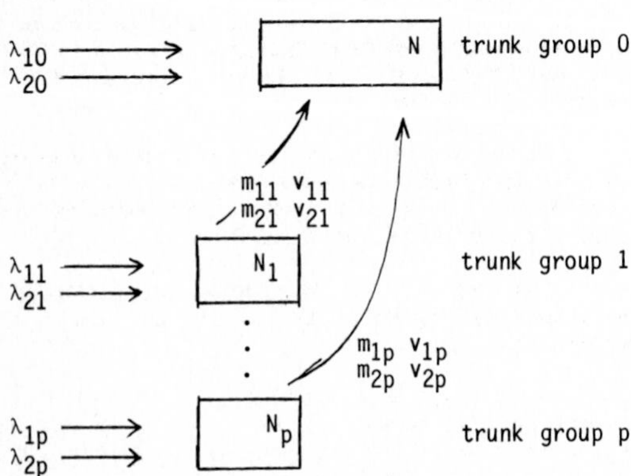

figure 2 : Case of multiple overflow

The previous section has shown how to calculate the two first moments of each individual overflow traffic. Using the assumption of independence of traffics, the following expressions can be derived:

$$M_1 = \lambda_{10} + \Sigma\, m_{1i} \quad ; \quad V_1 = \lambda_{10} + \Sigma\, v_{1i}$$

$$M_2 = \lambda_{20} + \sum_i m_{2i} \quad ; \quad V_2 = \lambda_{20} + \sum_i v_{2i} \tag{3-1}$$

The problem is to evaluate the lost traffic on the last choice trunk group of size N. In a first step , we will calculate the parameters of a switch (α and β) and equivalent input rates λ_1 and λ_2, by a method which is the exact converse of that presented in section 2-2. The four unknown variables α, β, λ_1, λ_2 must be derived from the equations:

$$M_1 = \frac{\lambda_1}{\mu_1}\,\frac{\beta}{\alpha+\beta} \tag{3-2}$$

$$M_2 = \frac{\lambda_2}{\mu_2}\,\frac{\beta}{\alpha+\beta} \tag{3-3}$$

$$\frac{V_1 + M_1^2 - M_1}{M_1} = \frac{g_1(\mu_1)}{1 - g_1(\mu_1)} \tag{3-4}$$

$$\frac{V_2 + M_2^2 - M_2}{M_2} = \frac{g_2(\mu_2)}{1 - g_2(\mu_2)} \tag{3-5}$$

with:

$$g_1(\mu_1) = \frac{\lambda_1\,(\beta + \mu_1)}{\mu_1^2 + \mu_1(\lambda_1 + \alpha + \beta) + \lambda_1\beta} \tag{3-6}$$

$$g_2(\mu_2) = \frac{\lambda_2\,(\beta + \mu_2)}{\mu_2^2 + \mu_2(\lambda_2 + \alpha + \beta) + \lambda_2\beta} \tag{3-7}$$

It is in fact quite easy to solve this system. First, it can be noted that $g_1(\mu_1)$ and $g_2(\mu_2)$ are known from (3-4) and (3-5). Define X_1 by the following equation which can be obtained from (3-6):

$$X_1 = M_1\left(\frac{1}{g_1(\mu_1)} - 1\right) = M_1\frac{\mu_1(\mu_1+\alpha+\beta)}{\lambda_1(\beta+\mu_1)} = \frac{\beta}{\alpha+\beta}\,\frac{\mu_1+\alpha+\beta}{\mu_1+\beta}$$

We find:

$$1 - X_1 = \frac{\alpha}{\alpha+\beta}\,\frac{\mu_1}{\mu_1+\beta} \tag{3-8}$$

and in a similar way:

$$1 - X_2 = \frac{\alpha}{\alpha+\beta}\,\frac{\mu_2}{\mu_2+\beta} \tag{3-9}$$

As X_1 and X_2 are known, β can be obtained from:

$$\frac{1 - X_1}{1 - X_2} = \frac{\mu_1}{\mu_2} \frac{\mu_2 + \beta}{\mu_1 + \beta} \quad (3\text{-}10)$$

All other variables are then derived directly from (3-8) (to get α) and (3-2) and (3-3) to obtain λ_1 and λ_2.

The problem is now to calculate the blocking probability of the final trunk group. The first idea, which seems natural, is to calculate it directly from the IPP. Define $\lambda=\lambda_1+\lambda_2$ and $\mu=\lambda/A$. These data will be the input of the model described in figure 3:

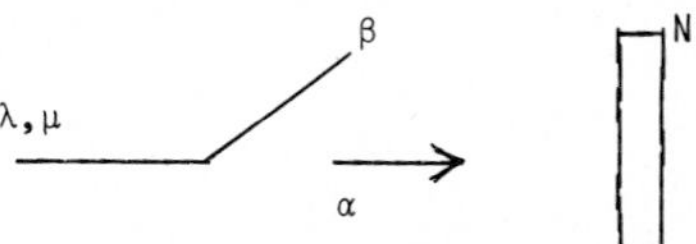

Figure 3: The model to calculate the blocking probability.

We define $P_1(i)$ to be the probability that the trunk group is in state i when the switch is closed, and $P_2(i)$ the corresponding probability when it is open. The following system of equations (3-11) can be established:

$$(\lambda+\alpha)P_1(0) = \beta P_2(0) + \mu P_1(1)$$

$$\beta P_2(0) = \beta P_1(0) + \mu P2(1)$$

.....

$$(\lambda+i\mu+\alpha)P_1(i) = \beta P_2(i) + \lambda P_1(i-1) + (i+1)\mu P_1(i+1)$$

$$(i\mu+\beta)P_2(i) = \alpha P_1(i) + (i+1)\mu P_2(i+1)$$

.....

$$(N\mu+\alpha)P_1(N) = \beta P_2(N) + \lambda P_1(N-1)$$

$$(N\mu+\beta)P_2(N) = \alpha P_1(N)$$

The matrix of this system is triangular, so that values of $P_1(i)$ and $P_2(i)$ can be obtained with a single loop, with a complexity no greater than the calculation of the Erlang formula. The blocking probability is then equal to $P_1(N)*(\alpha+\beta)/\beta$.

This solution will be referred as solution I in the following section which analyses the results. It will be shown that in some cases this solution, although it is simple, gives erroneous values.

It is also possible to imagine a solution consisting in calculating an "equivalent trunk group", (following the idea of the ERT method), which corresponds to our IPP. Two methods are then possible:

a) The total traffic offered to the equivalent trunk group is known: $A=\lambda_1/\mu_1+\lambda_2/\mu_2$. It is also known that the blocking probability of the equivalent trunk group $P(N^*)$ is equal to $\beta/(\alpha+\beta)$. N^* can thus be obtained by inversion of the Erlang formula $P(N^*)=E(A,N^*)$.

b) We define $\lambda=\lambda_1+\lambda_2$ and $\mu=\lambda/A$. It is then possible to calculate the variance of the "equivalent overflowing traffic" using:

$$g(\mu) = \frac{\lambda(\beta + \mu)}{\mu^2 + \mu(\lambda+\alpha+\beta) + \lambda\beta}$$

The usual formula for the variance,

$$V = M\left(1 - M + \frac{A}{N^* + 1 + M - A}\right)$$

allows us to calculate N^* in a linear way without the use of the Erlang formula.

In fact, in the general case, the two values obtained for N^* may be slightly different, leading to different values for the blocking probability. Numerical calculations, compared with simulations, show that Erlang inversion gives better results for the blocking probability. But the conclusion is that the equivalent trunk group and the IPP are not in exact correspondance. To match the two models, it may be necessary to modify the value of the ratio μ_1/μ_2.

It seems in fact unnatural to modify the values of μ_1 and μ_2. The following argument shows that it may be necessary to match the model of section 2: Consider a cluster with two overflowing trunk groups where $\lambda_{12}=0$ and $\lambda_{21}=0$. In this case, μ_2 has no significance in trunk group 1 and similarly μ_1 for trunk group 2. Only the ratios λ_{11}/μ_1 and λ_{22}/μ_2 affect the caracteristics of the overall overflowing traffic. It is then necessary to calculate the ratio μ_1/μ_2 to obtain a realistic model. The problem is to calculate this ratio, knowing that the system of equations of the system is not linear. The equation $\beta/(\alpha+\beta)=E(A,N^*)$ must be added to the system of equations (3-2) to (3-7), μ_2 being an unknown variable. The way to solve the system is the following:

1- choose $\mu_1=1$, μ_2 large enough,

2- increase μ_1 by a step (1 for example),

3- At each step, calculate α, β, λ_1, λ_2 as in the beginning of this section, and N^* as in solution b,

4- when the sign of the difference between $\beta/\alpha+\beta$ and $E(A,N^*)$ changes, modify the step of evolution of μ_1 (divide by 10 for example) and modify the sense of variation of μ_1,

5- iterate steps 2 to 4 until the difference between $\beta/\alpha+\beta$ and $E(A,N^*)$ is as small as desired.

The resulting model is then such that it corresponds exactly to the case of section 2. The blocking probability of the last choice trunk group is equal to $E(A,N+N^*)(\alpha+\beta)/\beta$.

3-2 EVALUATION OF THE METHOD

Some results and comparisons with simulation runs are summurized in table 2. The relative proportion of the holding times for each type of traffic seems to be the critical parameter of the problem. In the first part of table 2 (examples 1 to 5), x is the same on each trunk group and the results (theoretical and simulation) show that the blocking probabilities are constant and therefore independent of x. This is probably a quite important result and the following assertion can be formulated:

"If the proportion x of the traffic of each class is the same on all trunk groups in a cluster, then the blocking probability is independent of x, and therefore the performance can be evaluated using the equivalent random theory."

It is certainly not possible to prove this assertion with the model presented here since the values for the blocking probabilities may vary with x, but with a very small relative variation. I have tried to prove this result directly but without success.

The second part of table 2 (examples 6 to 11) shows the results when the coefficients x are not equal for each trunk group. It can be observed that now the blocking probabilities vary quite widely: from 0.082 to 0.123 with the model, from 0.083 to 0.097 with the simulation. The conclusion is that the mixture of traffics may have a significant effect on blocking probabilities and therefore on network dimensioning.

First trunk group : Offered traffic 8 Erlang; Size 10 trunks
Second trunk group : Offered traffic 6 Erlang; Size 7 trunks

Last choice trunk group : Size 6 trunks

example nb.		1	2	3	4	5	6	7	8	9	10	11
DATA	A_{11}	0.8	2	4	6	7.2	2	6	4	0.8	7.2	4
	A_{12}	7.2	6	4	2	0.8	6	2	4	7.2	0.8	4
	x_1	10%	25%	50%	75%	90%	25%	75%	50%	10%	90%	50%
	A_{21}	0.6	1.5	3	4.5	5.4	4.5	1.5	5.4	5.4	0.6	0.06
	A_{22}	5.4	4.5	3	1.5	0.6	1.5	4.5	0.6	0.6	5.4	5.94
	x_2	10%	25%	50%	75%	90%	75%	25%	90%	90%	10%	1%
Method I	P_1	0.011	0.011	0.011	0.011	0.011	0.012	0.010	0.015	0.014	0.010	0.008
	P_2	0.017	0.017	0.017	0.017	0.017	0.018	0.015	0.023	0.022	0.015	0.012
	P	0.092	0.092	0.092	0.092	0.092	0.101	0.082	0.123	0.117	0.084	0.065
Method II	P_1	0.011	0.011	0.011	0.011	0.011	0.013	0.011	0.013	0.016	0.014	0.011
	P_2	0.016	0.016	0.016	0.016	0.016	0.019	0.017	0.020	0.024	0.022	0.017
	P	0.088	0.088	0.088	0.087	0.087	0.103	0.093	0.113	0.130	0.119	0.091
Simulation	P_1	0.012	0.012	0.011	0.012	0.012	0.012	0.012	0.013	0.012	0.010	0.012
	P_2	0.015	0.016	0.015	0.014	0.014	0.014	0.015	0.015	0.014	0.015	0.014
	P	0.088	0.090	0.087	0.087	0.087	0.091	0.083	0.097	0.095	0.084	0.084
P by ERT		0.089										

P1: proportion of lost traffic for traffics offered to the first trunk group
P2: " " " " second trunk group

P : Blocking probability of the last choice trunk group

Table 2: Validation of the method

The method I (using the initial values for μ_1 and μ_2) seems to be more stable when the parameter x varies. However the validity of this model can be questioned when the coefficient x for one stream takes a very small value compared to the other stream: this is the case of example 11 with x_1 = 50% and x_2 = 1%.

Model II gives perfect results when all the x's are equal, but is unstable in the converse case. We have found no rational explanation for this phenomenon. The reason is probably that the covariance between overflow traffics has not be considered. Further research in that direction is necessary.

4 CONSIDERATIONS ON DIMENSIONING AND CONCLUSIONS.

Still with the same examples as in section 3-2, we have tried to dimension the network such that the blocking probability of the last choice trunk group is equal to 1%. The same set of 11 examples has been considered, in two cases: firstly with exactly the same traffics as those presented in table 2 (data part); secondly with a Poisson additional traffic of 3 Erlangs offered directly to the last choice trunk group such that x=50%.

The results for the dimensioning of the last choice trunk group are summerized in table 3. The first line corresponds to the first case without additional traffic, and the second line to the case where the three Erlang traffic is added. We have used model I to solve this problem. It can be considered that example 11 is suspicious since the results of table 2 were not in agreement with the simulation.

These two sets of examples show that the dimensioning of the network may vary (in these examples by about 20%) with the variation of the proportion of each class of traffic.

The whole study can easily be extended to the case where more than two classes of customers are considered. To get the mean and variance of overflow traffics, the same model as in section 2-2 can be used without difficluties. The converse problem (for the blocking probabilities and for dimensionning) imply the recalculation of the ratios of the holding times.

This study can be considered as a first approach to the problem of mixture of traffics. It may probably be fruitfull to investigate the effect of covariance of flows, which was not considered here. It seems very difficult to obtain a better model than the one presented without adding a lot of complexity. It can be also concluded that mixture of traffics has no influence on grade of service as soon as the proportions of each class of traffic remain in a narrow space for each stream.

BIBLIOGRAPHY

[1] A. Kuczura and D. Bajaj, "A method of moments for the analysis of a switched communication network's performance", I.E.E.E Trans. on Com., Vol 25 n°2, pp. 185-193, 1981

[2] R.B.Cooper "Introduction to queueing theory", North Holland, 1977

[3] R.I.Wilkinson "Theories for the toll traffis engineering in the U.S.A.", B.S.T.J., Vol 35, pp.421-514, 1956.

First trunk group : Offered traffic 8 Erlang; Size 10 trunks Second trunk group : Offered traffic 6 Erlang; Size 7 trunks											
example nb.	1	2	3	4	5	6	7	8	9	10	11
case 1	10	10	10	10	10	10	9	11	10	9	9
case 2	11	11	13	12	12	13	12	12	12	12	14

Table 3 : Size of the last choice trunk group for 1% blocking probability.

TELETRAFFIC ISSUES in an Advanced Information Society
ITC-11
Minoru Akiyama (Editor)
Elsevier Science Publishers B.V. (North-Holland)

A STATE DEPENDANT ONE MOMENT MODEL FOR GRADE OF SERVICE AND TRAFFIC EVALUATION IN CIRCUIT SWITCHED NETWORKS

Françoise LE GALL, Jacques BERNUSSOU and Jean-Marie GARCIA*

Laboratoire d'Automatique et d'Analyse des Systèmes du C.N.R.S.
7, avenue du Colonel Roche - 31077 Toulouse Cédex, France

ABSTRACT

Handling relatively simple and accurate performance evaluation models constitutes a must for the decision makers in the communication networks management problem. This paper is adressed to the particular problem of circuit switched networks : traffic modelling and performance evaluation. A single moment model is derived which tries to overcome some of the limitations due to the relative roughness of the first moment models.

First, the model presented enables to tackle in some way the different natures of telephone traffic : poisson, peaky and under-variant. Moreover, being an overall modelization i.e., a model dealing with the entire network and with all the flows it naturally gives the point to point grade of service which is a very important index to test the quality of a routing scheme. Not all the details of the presented model are rigorously mathematically given but most of them are derived exactly for simple cells and then validated by means of numerical simulations.

1. INTRODUCTION

In the present time the communication networks are one of the most important element for the social and even profisional life. The paper we are giving is adressed to the circuit switched networks and more precisely to the problem of modelling the traffic and estimating the blocking probabilities. A very important measure for a circuit switched network is the point to point blocking which defines the grade of service for the network. This index fixes the probability of being able to establish a communication between any two points in the network. Of course, the point to point blocking depends on a number of different factors: first of all the load i.e. the node to node traffic demands, the link capacities and then the routing used.

Evaluating the performance of a circuit switched network can be performed using event by event simulation programs. Such programs are generally huge programs highly time consuming when dealing with realistic networks. Having in hand some analytical means or at least aggregated models for performance evaluations is welcome. Different approaches have been developped up to now and they can be classified according to the number of moments for the stochastic process which is the telephone traffic. Among the three moment methods the most well known is the Interrupted Poisson Process method IPP [7 - 8]. Let us also mention some attempt based on the use of Cox function [5].

Numerous are the two moment methods and of course we must first mention the pioneering work by Wilkinson and his equivalent trunk method [14]. After him, a lot of workers have proposed some refinements especially in the case of networks, in order to propose the evaluation of individual losses for the different flows in the network [3 - 4 - 6 - 11].

Finally, one moment models have also been derived [2 - 9 - 10 - 12 - 13]. One of the main interest in using such models stands in the fact that the point to point blocking estimation is relatively easy to perform. Of course, they can be attacked on the fact that they use a relatively poor information concerning the true telephone traffic. It is to the author's opinion a criticism that cannot be done so easily to the model presented here. In fact single moment methods are usually developped under the following assumptions.

- A1 - Call arrival is a poisson process
- A2 - Call holding time has a negative exponential distribution
- A3 - The network is in statistical equilibrium
- A4 - Nodes are non blocking
- A5 - Call set up time is negligible
- A6 - Blocking calls are cleared and do not return
- A7 - Link blocking are statistically independant
- A8 - Call arrival on any link in the network is a Poisson process.

The two last assumptions are the one that prevent the first moment model to furnish excellent results. In the paper we drop A7 and we manage to take into account the real nature of the traffic crossing the links (under or over variant). Another interest of the method is that it provides a model which deals with the overall network and not in a link by link manner like many others methods.

The paper is divided into four parts. In the second introductory section we develop a model concerning a network managed with a single choice policy at each commutation node (load sharing). This simple case enables us to fix the main ideas and to justify some approximations made. In the third part we consider a two choice routing policy, and in a same way develop a model taking into account joint blocking probabilities. In the fourth section the most general model is presented and remarks are made to help the realization of a general numerical algorithm. The last section pre-

* This work has been done under grant DAII-CNET 82-1B-188.

sents some numerical experiments for different types of routing policies on the same network.

2. MODELLING A SINGLE CHOICE ROUTING POLICY

At each commutation node, the routing is performed according to a load sharing policy

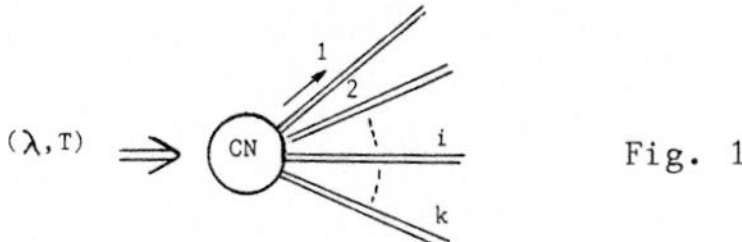

Fig. 1

A call traffic with parameters λ (arrival rate) and T (mean call duration) enters the commutation node C.N.

It is divided between the outgoing links proportionnally to load sharing parameters α_i with :

$$\sum_{i=1}^{k} \alpha_i = 1 \qquad (1)$$

With such a routing procedure the offered traffic to each link ($\alpha_i \lambda$ T in erlangs) remains poisson. For the simple cell of figure 1 and for each of the links a markov chain type model can be defined by assuming that over an infinitely small time interval only one event can occur (either a call arrival or a call end).

For link i, capacity N_i one has, $P_j(t)$ being the probability of j calls present on the link at time t.

$$\begin{cases} \dot{P}_j(t)=\alpha_i\lambda P_{j-1}-(\alpha_i\lambda+\frac{j}{T})P_j+\frac{j+1}{T}P_{j+1} & \begin{cases} 0 \leqslant j < N \\ P_k=0 \text{ if } k<0 \end{cases} \\ \dot{P}_N(t)=\alpha_i \lambda P_{N-1} - \frac{N}{T} P_N \end{cases} \qquad (2)$$

which in the steady state furnishes the well known solution

$$P_j = \frac{A_i}{j} P_{j-1}; \qquad A_j = \alpha_i \lambda T$$

and . the Erlang-B function, the probability of link i being blocked

$$P_{N_i} = E(A_i,N_i) = \frac{A_i^{N_i} / N_i!}{\sum_{j=0}^{N_i} \frac{A_i^j}{j!}}$$

. the losses $A_i\ E(A_i,N_i)$

. the carried traffic $A_i\ (1 - E(A_i,N_i))$

For the mean state of link i in the transient state a straightforward calculation leads to :

$$\dot{X}_i = - X_i/_T + \alpha_i \lambda\ (1 - P_{N_i}(t)) \qquad (3)$$

which shows that the mean state obeys a fluid type equation

$$\text{variation} = \text{inflow} \quad (\tfrac{A_i}{T}(1-P_{N_i}(t)) - \text{outflow} \quad (X_i/_T)$$

Based on the fluid analogy and then equation like (3) it is easy to end up with a model for networks in the whole and not link by link models.
The problem for (3) is the presence in it of the $P_{N_i}(t)$ term, the transient blocking probability which has been approximated (see [1]) in the following way :

$$\begin{cases} X_i(t) = Y_i(t)\ (1 - E(Y_i(t),\ N_i)) \\ P_{N_i}(t) = E\ (Y_i(t),\ N_i) \end{cases} \qquad (4)$$

This approximation becomes the true value in the statistical equilibrium.
Let us now consider a more complex network shown in fig. 2.

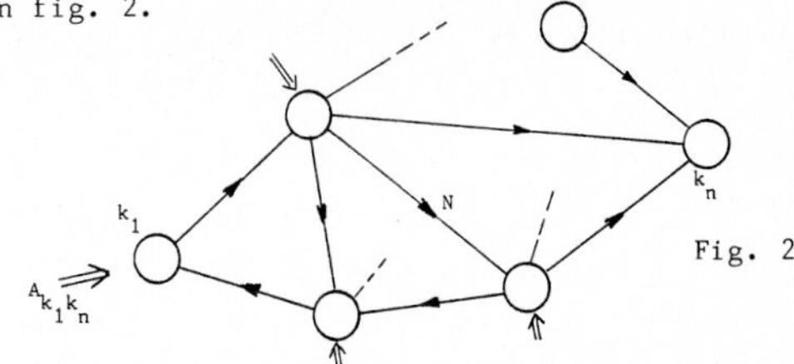

Fig. 2

Each link $k_r k_s$ is assumed unidirectional of given capacity say $N_{k_rk_s}$ and between two nodes k_1k_n is offered a traffic $A_{k_1k_n}$. Each node routes the calls according to the load sharing technique.
Given the routing parameters, one is able to find all the different pathes joining two distinct nodes, i.e., all the chains that can be used to establish a call (ex : $k_1\ k_2,\ldots,k_{n-1}\ k_n$). For the part of the traffic $A_{k_1k_n}$ processed through this path, a straightforward extension of equation (3) furnishes :

$$\dot{X}_{k_1k_2..k_n} = -\frac{X_{k_1k_2...k_n}}{T} + \alpha_{k_1k_2..k_n}\lambda_{k_1k_n}(1-P_{k_1k_2..k_n}) \qquad (5)$$

where $X_{k_1k_2...k_n}$ is the mean state of a route concerning the given flow $k_1...k_n$

$\lambda_{k_1k_n}$ is the arrival rate between nodes k_1 and k_n ($A_{k_1k_n} = \lambda_{k_1k_n}$ T)

$\alpha_{k_1k_n..k_n} = \alpha_{k_1k_2}\cdot\alpha_{k_2k_3}\cdots\alpha_{k_{n-1}k_n}$; $\alpha_{k_ik_j}$ is the load sharing parameter between node i and node j.

$P_{k_1k_2..k_n}$ is the blocking probability of the path $k_1k_2...k_n$.

In the fourth section, the calculation of $P_{k_1..k_n}$ will be developped.

3. MODELLING A TWO-CHOICE ROUTING POLICY

Let us consider the cell of Fig. 3.

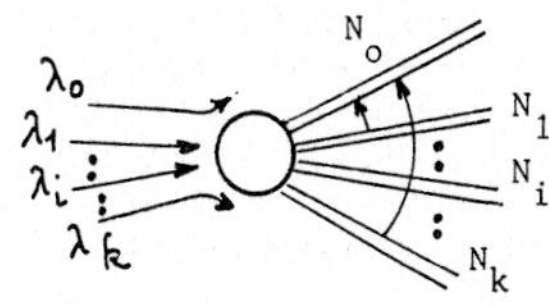

Fig. 3

Traffics with λ_o and λ_i as respective arrival rates are offered in first choice to links of capacity N_o, N_i i=1,...k. If link (N_i,i ⩾ 1) is blocked, the call λ_i is offered to link N_o (overflowing procedure).
A continuous time Markov system is written for the probability $P(e_j)$ of state e_j where :

- . e_j is the state $(j_o..j_i..j_k)$, with j_i calls on link N_i i=0...k
- . δ_i is the nul vector except element i equals 1: $\delta_i = (0_o...1_i...0_k)$.

$$\dot{P}(e_j)=\sum_{i\neq o}\lambda_i P(e_j-\delta_i)+(\lambda_o+\sum_{i\neq o}\lambda_i^*)P(e_j-\delta_o) \quad (6)$$

$$-\sum_i(\lambda_i^+ + \frac{j_i}{T})P(e_j)+\sum_i\frac{(j_i+1)}{T}P(e_j+\delta_i)$$

with $P(e) = 0$ if $\exists\, j_i > N_i$ or $\exists\, j_i < 0$

$\lambda_i^* = 0$ if $j_i < N_i$

$= \lambda_i$ if $j_i = N_i$

$\lambda_i^+ = 0$ if $j_i = N_i$

Let X_i be the average traffic on link N_i :

$$X_i = \sum_{j_o=0}^{N_o} \cdots\cdots \sum_{j_k=0}^{N_k} j_i\, P(e_j)$$

then (see[9]) :

$$\begin{cases}\dot{X}_o=-\frac{X_o}{T}+\lambda_o(1-P_{N_o}(t))+\sum_{i=1}^{k}\lambda_i(P_{N_i}-P_{N_oN_i})\\ \dot{X}_i=-\frac{X_i}{T}+\lambda_i(1-P_{N_i}),\quad i=1,2,\ldots,k\end{cases} \quad (7)$$

and

$$\begin{cases}X_i=Y_i(1-E(Y_i,N_i)),\ P_{N_i}=E(Y_i,N_i)\\ X_i+X_o=Y_{io}(1-E(Y_{io},N_i+N_o));P_{N_oN_i}=E(Y_{io},N_i+N_o)\end{cases} \quad (8)$$

Numerical experiments proved that (7-8) give an accurate solution in the stationnary case $\dot{X}=0$ (see for example [9]).

It is easy to see that such an approach enables to take into account the peaky nature of the overflowing traffic :

$$Y_o > \lambda_o T$$

and

$$Y_{oi} > (\lambda_o+\lambda_i)T$$

In (7) it is interesting to splite the equation concerning X_o into as much equations as there are many different traffics using link N_o.

$$\begin{cases}\dot{\overline{X}}_o = -\overline{X}_{o/T}+\lambda_o(1-P_{N_o}(t))\\ \dot{\overline{X}}_i = -\overline{X}_{i/T}+\lambda_i(P_{N_i}-P_{N_oN_i}),\ i=1,2,\ldots,k\end{cases} \quad (9)$$

$X_o = \sum_{i=o}^{k}\overline{X}_i$. This writing points out how easy is the individual evaluation of the losses for each traffic. It introduces the basic idea for the following : on the N_o link are merged two sorts of traffic:the first choice traffic (λ_o), and the overflowed traffics (λ_i, $i \geqslant 1$). The idea to be pursued is to develop a fluid type equation of type (9) for the part of each traffic passing through a given path. The next section is devoted to the general problem.

4. THE GENERAL MODEL

Given a network with routing policies at each commutation nodes (we will restrict ourselves to the two choice policies), the first step, before modelling, is to list all the distinct pathes that can be followed by a traffic corresponding to each couple origin-destination. This task depends on the network topology and on the routing strategy. In the case of circuit switched networks, the complexity in the enumeration of the pathes is reduced by the fact that they are restricted to have a low number of trunks.

In the network of Fig. 2 we denote by $K_1,K_2,\ldots,K_l$ all the different pathes between the nodes k_1 and k_n and we first suppose that we are dealing with OOC (originated office control) : we define $\alpha_1,\ldots,\alpha_l$ the load sharing parameters corresponding to the different possible pathes and the overflowing tables.

By aggregation performed on the markovian model one is able to write down a dynamic model for the two choice policy of the form :

$$\dot{X}_{K_i}=-X_{k_i}/T+\alpha_{K_i}\lambda_{k_1k_n}(1-P_{K_i})+\sum_{j\in J_i}\alpha_{K_j}\lambda_{k_1k_n}(P_{K_j}-P_{K_jK_i}) \quad (10)$$

where J_i defines the set of routes K_j overflowing on route K_i

P_{K_i} is the blocking probability of route K_i

$P_{K_iK_j}$ the joint blocking probability of routes K_i and K_j

$$P_{K_i}=P_{k_1k_2\ldots k_i\ldots k_n} = \text{Prob}\left[k_1k_2 \text{ blocked or } \ldots \text{ or } k_{n-1}k_n \text{ blocked}\right]$$

Developping the calculations it is shown that :

$$P_{K_i}=P_1^1 - P_2^1 + P_3^1 + \ldots (-1)^{max-1}P_{max}^1 \quad (11)$$

where P_l^1 is the summation of the joint blocking probability of the combination of l trunks belonging to the same path (here K_i)

In a similar way $P_{K_iK_j}$ is written as

$$P_{K_iK_j} = P_2^2 - P_3^2 + P_4^2 \ldots + (-1)^{max-2}P_{max}^2 \quad (12)$$

where P_l^2 is the summation of the joint blocking probability of the combination of l trunks with at least one trunk belonging to path K_i and one trunk belonging to path K_j.

Now some care must be taken in computing the different terms P_i^n n=1,2. It is to be noticed that X_{K_i} represents the average number of calls from k_1 to k_n on the path K_i.

X_{K_i} is splited into as much flows $\overline{X}_{K_i}$ as there are many different traffics using path K_i (see for example equation 9).

To have the state Z_{ab} of a link a-b, which is necessary to compute the different blocking probabilities, simply asks for a summation over all the pathes (X_{K_1}) going through the link a-b. If there is only one path K_1 (sections 2 and 3) then $Z_{ab} = X_{K_1}$. For the computation of terms P_1^1 (blocking probability of one trunk group), we refer to equation (4). For the joint blocking probability of two or more trunk groups we consider two cases.

<u>First case</u> : the trunk-groups belong to different pathes (calculation of the term P_i^1 $i \geqslant 2$). The joint blocking probability of two trunk groups lm and nq is computed according to :

$$Z_{lm} + Z_{nq} = Y_{lm\,nq}(1-E(Y_{lm\,nq},\ N_{lm}+N_{nq}))$$
$$\text{prob (lm and nq)} = E(Y_{lm\,nq},\ N_{lm}+N_{nq}) \quad (13)$$

which is the direct extention of (8).

<u>Second case</u> : some of the trunk groups belong to the same path (calculation of terms P_i^n $i > n$). In that case one has to tackle with conditional probabilities. For instance in the calculation of a term of P_3^2, if link lm belongs to path K_i and links nq and rs belong to path K_j one has :

prob(lm and nq and rs) = prob(lm and rs/nq).prob (nq).
where prob(lm and rs/nq) is given by :

$$Z_{lm}+Z'_{rs} = Y'_{lm,rs}\ (1-E(Y'_{lm,rs},N_{lm}+N_{rs}))$$

$$\text{prob(lm and rs/nq)} = E(Y'_{lm,rs},\ N_{lm}+N_{rs}) \qquad (14)$$

where Z_{lm} is the mean occupancy state of trunk group lm and Z'_{rs} the mean occupancy state of trunk group rs taking into account the fact that link nq is blocked :

$$Z'_{rs} = Z_{rs} + \sum_{E_{nq}} X_1\ (\frac{N_{nq}}{X_{nq}} - 1)$$

X_1 : origin-destination flow using trunk group rs

E_{nq} : set of flows X_1 using also trunk group nq.

More generally a term of P_i^n, $i > n$ is computed as

$$\text{prob}(E_i) = \text{prob}\ (e_n/E_{n-i})\ \text{prob}\ (e_{n_2}/E_{n-i-n_2})\ldots$$

$$\text{prob}\ (e_{n_f})$$

where E_r is a set of r links with s links ($s < r$) belonging to different pathes

e_s is a set of s links belonging to different pathes.

The generalization of system (10) which has been written for a OOC two choice control can be done, for other types of control (multi-choice, SOC) and networks (bidirectional links).
For the sake of simplicity we consider now a reduced network for which the fundamental equations are written for some different types of control.

5. A REDUCED NETWORK - NUMERICAL RESULTS

Let us deal with the two level network of Fig. 4 which is constituted by 3 transit centers (CT), four origine-destination centers and unidirectional links. The indexes i,k,j will be attached to the origin, transit, and destination nodes respectively.

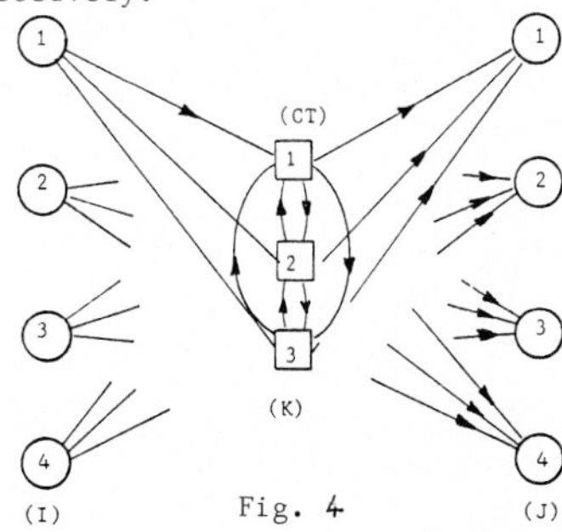

Fig. 4

Three types of experiments are performed; for each of them the equations are given below.

* The first is a SOC type control : for each (i,j) flow the call first attempts a first choice route $i-k_1-j$; if the outgoing link (ik_1) is blocked, them it attempts a second choice route $i-k_2-j$.
The equations are given by :

$$\dot{X}_{ik_1j} = -X_{ijk_1}/T + \lambda_{ij}(1-P_{ik_1j}) \qquad (15)$$

$$\dot{X}_{ik_2j} = -X_{ik_2j}/T + \lambda_{ij}(P_{ik_1} - P_{ik_1\ \text{and}\ ik_2j}) \qquad (16)$$

where P_{ikj} stands for $P_{ik\ \text{or}\ kj}$.

* The second is a OOC type control : for each (i,j) flow a first choice route is defined say ik_1j; if ik_1 or k_1j is blocked, then the call attempts a second choice route ik_2j.
For the ik_1j first choice route, equation (15) holds;
For the second choice route :

$$\dot{X}_{ik_2j} = -X_{ik_2j}/T + \lambda_{ij}(P_{ik_1j} - P_{ik_1j\ \text{and}\ ik_2j}) \qquad (17)$$

* The third one is a SOC type control where the transit nodes are also commutation nodes. The description begins as for the first control and is completed as follows : at node k_1 (first choice) the call can overflow on link (k_1k_2) if the link k_1j is blocked.
Here one has to consider three different pathes for the routing of each flow (i-j) with :
equation 15 for the ik_1j first choice route
equation 16 for the ik_2j route
and for the ik_1k_2j route

$$\dot{X}_{ik_1k_2j} = -X_{ik_1k_2j}/T + \lambda_{ij}\Big[P_{\overline{ik_1}\ \text{and}\ k_1j} - P_{\overline{ik_1}\ \text{and}\ k_1j\ \text{and}\ k_1k_2j}\Big] \qquad (18)$$

where $P_{\overline{ik_1}\ \text{and}\ ik_2}$ means Prob$[ik_1$ non blocked and ik_2 blocked$]$

We give below some example for the calculations of the blocking probabilities involved in these equations

. $P_{ik_1j} = P_{ik_1} + P_{k_1j} - P_{k_1j/ik_1} \cdot P_{ik_1}$

with

$$\begin{cases} P_{k_1j/ik_1} = P(Y_{k_1j/ik_1},\ N_{k_1j}) \\ Z'_{k_1j} = Y_{k_1j/ik_1}\Big[1-E(Y_{k_1j/ik_1}, N_{k_1j})\Big] \end{cases} \qquad (19)$$

$$Z'_{k_1j} = Z_{k_1j} + X_{ik_1j}\left[\frac{N_{ik_1}}{Z_{ik_1}} - 1\right]$$

. $P_{ik_1\ \text{and}\ ik_2j} = P_2^2 - P_3^2$

$$= P_{ik_1\text{and}\ ik_2} + P_{ik_1\text{and}\ k_2j} - P_{ik_1\text{and}\ k_2j/ik_2} \cdot P_{ik_2} \qquad (20)$$

where * $P_{ik_1\ \text{and}\ ik_2}$ and $P_{ik_1\ \text{and}\ k_2j}$ are computed according to (13)

* $P_{ik_1\ \text{and}\ k_2j/ik_2} = E\ (Y.,\ N_{ik_1} + N_{k_2j})$

$$Z_{ik_1} + Z'_{k_2j} = Y.\ [1-E(Y.,\ N_{ik_1}+N_{k_2j})] \qquad (21)$$

Z'_{k_2j} is computed similarly as Z'_{k_1j} in (19)

. $P_{\overline{ik_1}\ \text{and}\ k_1j} = P_{k_1j} - P_{ik_1\text{and}\ k_1j} = P_{k_1j} - P_{k_1j/ik_1} \cdot P_{ik_1}$

. Finally :

$$P_{\overline{ik_1}\ \text{and}\ k_1j\ \text{and}\ k_1k_2j} = P_{k_1j\ \text{and}\ k_1k_2j} - P_{ik_1\ \text{and}\ k_1j\ \text{and}\ k_1k_2j}$$

where $P_{ik_1 \text{ and } k_1j \text{ and } k_1k_2j} =$

$$= P_{ik_1} \cdot \left[P_{k_1j \text{ and } k_1k_2/ik_1} + P_{k_2j \text{ and } k_1j/ik_1} - P_{k_1j \text{ and } k_2j/ik_1 \text{ and } k_1k_2} \cdot P_{k_1k_2/ik_1} \right]$$

and

$$\begin{cases} P_{k_1j \text{ and } k_1k_2/ik_1} = E(Y, N_{k_1j} + N_{k_1k_2}) \\ Z'_{k_1j} + Z'_{k_1k_2} = Y \left[1 - E(Y, N_{k_1j} + N_{k_1k_2}) \right] \\ Z'_{k_1j} \text{ is given in (19)} \\ Z'_{k_1k_2} = Z_{k_1k_2} + \sum_n X_{ik_1k_2n} \left(\dfrac{N_{ik_1}}{Z_{ik_1}} - 1 \right) \end{cases} \tag{22}$$

and so and

Z_{ab} in the different equations is the summation over all X going through link a-b.

For the numerical experiments, the offered traffic in Erlangs is given in table 1, the links capacities in tables 2, 3 and 4 and the routing table (first and second choice nodes) in table 5.

i \ j	1	2	3	4
1	-	13.	19.	25.
2	12.	-	10.	17.
3	15.	25.	-	16.

Table 1

i \ k	1	2	3
1	20	25	30
2	25	20	15
3	20	20	35

Table 2 : N_{ik}

k \ j	1	2	3	4
1	11	19	15	24
2	19	9	10	19
3	5	22	14	28

Table 3 : N_{kj}

k_1 \ k_2	1	2	3
1	-	5	-
2	-	-	5
3	5	-	-

Table 4 : $N_{k_1k_2}$

i-j	1-2	1-3	1-4	2-1	2-3	2-4	3-1	3-2	3-4
1st choice : k_1	1	2	3	2	3	1	1	3	2
2nd choice : k_2	2	3	1	3	1	2	2	1	3

Table 5 : routing table

The mean occupancies of the trunk-groups obtained by the model (first line) and by the event by event simulation (second line) are given in tables 6.1 and 6.2 for the first experiment, in tables 7.1 and 7.2 for the second experiment and in tables 8.1, 8.2 and 8.3 for the last one. For the event by event simulation, the trust interval is computed for a trust level of 0.9. The efficiencies are given in table 6.3 for the first experiment, in table 7.3 for the second and in table 8.4 for the third.

Table 9 gives the overall carried traffic for the three policies.

i \ k	1	2	3
1	13.05 12.90±0.41	9.46 9.18±0.06	22.47 22.77±0.42
2	16.65 16.58±0.48	11.88 11.99±0.45	9.41 9.43±0.30
3	9.63 9.64±0.11	14.33 14.71±0.29	19.91 19.60±0.19

Table 6.1

k \ j	1	2	3	4
1	9.62 9.63±0.11	12.51 12.90±0.41	0.28 0.0	16.92 16.58±0.48
2	11.75 11.99±0.45	0.28 0.0	9.18 9.17±0.06	14.46 14.71±0.29
3	0.09 0.0	19.47 19.60±0.19	9.31 9.42±0.30	22.92 22.77±0.42

Table 6.2

i \ j	1	2	3	4	
1		0.982 0.969	0.483 0.481	0.920 0.915	model simulation
2	0.986 0.984		0.959 0.955	0.971 0.978	
3	0.641 0.636	0.779 0.765		0.923 0.918	

Table 6.3

i \ k	1	2	3
1	13.91 14.35±0.40	10.80 11.53±0.34	25.40 25.77±0.30
2	18.02 18.20±0.41	11.95 12.42±0.42	8.07 7.92±0.27
3	13.07 14.03±0.36	16.29 16.69±0.27	21.49 22.10±0.36

Table 7.1

k \ j	1	2	3	4
1	9.57 9.62±0.11	14.83 15.30±0.38	2.42 2.82±0.30	18.18 18.83±0.38
2	14.21 14.75±0.41	1.62 2.38±0.33	9.18 9.15±0.06	14.04 14.37±0.31
3	0.76 1.03±0.17	19.46 19.52±0.19	12.01 12.14±0.17	22.73 23.11±0.36

Table 7.2

i \ j	1	2	3	4	
1		0.996 0.997	0.731 0.765	0.931 0.962	model simulation
2	0.984 0.990		0.973 0.989	0.971 0.988	
3	0.849 0.891	0.918 0.943		0.948 0.976	

Table 7.3

i \ k	1	2	3
1	13.18 13.15±0.39	12.96 12.77±0.22	23.19 23.10±0.48
2	16.64 16.67±0.50	11.55 11.72±0.40	9.04 8.82±0.33
3	12.37 12.32±0.27	14.67 14.81±0.36	21.78 22.19±0.39

Table 8.1

k \ j	1	2	3	4
1	9.61 9.58±0.12	13.98 14.48±0.40	1.16 0.98±0.16	17.64 17.40±0.46
2	14.03 14.12±0.42	0.64 0.66±0.16	9.18 9.17±0.07	14.71 14.83±0.32
3	0.27 0.32±0.09	19.47 19.50±0.20	11.49 11.36±0.25	23.19 23.15±0.41

Table 8.2

$k_1 \rightarrow k_2$	
1 → 2	3.21 3.51±0,15
2 → 3	3.83 4.03±0.10
3 → 1	3.42 3.81±0.15

Table 8.3

i \ j	1	2	3	4	
1		0.984 0.973	0.666 0.673	0.955 0.943	model simulation
2	0.958 0.957		0.918 0.898	0.973 0.979	
3	0.827 0.828	0.851 0.848		0.945 0.959	

Table 8.4

	model	simulation
offered traffic	152 Erlangs	
carried traffic case 1	126,79 Erl.	126,80 Erl.
case 2	138,99 Erl.	143,06 Erl.
case 3	135,38 Erl.	135,56 Erl.

Table 9

CONCLUSION

We have derived a first moment model for telephone traffic and the point to point blocking estimation. Due to space limitation, we have restricted the numerical experiments. However the ones given show the concordance in the flow efficiencies between a monte carlo simulation and the first moment model. This is certainly due to the consideration of assumption A7 and A8 of the introduction. Furthermore, we would like to insist on the fact that it is a global model that enables to give at once the point to point blocking. As for the numerical resolution of the model it can be simply done using standard numerical integration technique to reach the steady state point or to use by means of relaxation techniques the non linear algebraic system corresponding to the steady state.

REFERENCES

[1] J. BERNUSSOU, J.M. GARCIA, I.S. BONATTI, F. LE GALL, Modelling and control of a large scale telecommunication network, Proc. 8th triennal World Congress IFAC, Kyoto, pp. 52-58, vol. X, 1981.

[2] W.S. CHAN, Recursive algorithms for computing end-to-end blocking in a network with arbitrary routing plan, IEEE Trans. Commun. vol. COM-28, n°2, pp. 153-164, 1980.

[3] L.E.N. DELBROUCK, A unified approximate evaluation of congestion functions for smooth and peaky trafic, IEEE Trans. Commun. vol. COM-29, n° 2, pp. 85-91, 1981.

[4] P.J. DESCHAMPS, Analytic approximation of blocking probabilities in circuit switched communication networks, IEEE Trans. Commun. vol. COM-27, n° 3, pp. 603-606, 1979.

[5] J.P. GUERINEAU, J. LABETOULLE and M. LEBOURGES A three moments method for traffic in a network based on a coxian input process, Proc. 10th Int. Teletraffic Cong., Montreal, 1983

[6] S. KATZ, Statistical performance analysis of a switched communication network, Proc. 5th Int. Teletraffic Cong., New-York, 1967.

[7] A. KUCZURA and D. BAJAJ, A method of moments for the analysis of a switched communication network's performance, IEEE Trans. Commun. Vol. COM-25, n° 2, pp. 185-193, 1977.

[8] A. KUCZURA, The interrupted Poisson process as an overflow process, Bell Syst. Tech. J., vol. 52, n° 3, pp. 437-448, 1973.

[9] F. LE GALL and J. BERNUSSOU, An analytical formulation for grade of service determination in telephone networks, IEEE Trans. Commun. vol. COM-31, n° 3, pp. 420-424, 1983.

[10] P.M. LIN, B.P. LEON and C.R. STEWART, Analysis of circuit switched networks employing originating office control with spill forward, IEEE Trans. Commun. vol. COM-26, n° 6, pp. 754-765, 1978.

[11] D.R. MANFIELD and T. DOWNS, Decomposition of traffic in loss systems with renewal input, IEEE Trans. Commun. vol. COM-27, n° 1, pp. 44-58, 1979.

[12] D.R. MANFIELD and T. DOWNS, On the one moment analysis of telephone traffic networks, IEEE Trans. Commun. Vol. COM-27, n° 8, pp. 1169-1174, 1979.

[13] K.S. SCHNEIDER and D. MINOLI, An algorithm for computing average loss probability in a circuit-switched communication network, IEEE Trans. Commun., vol. COM-28, n° 1, pp. 27-32, 1980 .

[14] R.I . WILKINSON, Theories for toll traffic engineering in the USA, Bell Syst. Tech. J., pp. 421-515, March 1956.

TELETRAFFIC ISSUES in an Advanced Information Society
ITC-11
Minoru Akiyama (Editor)
Elsevier Science Publishers B.V. (North-Holland)

ON SOME NUMERICAL METHODS FOR THE COMPUTATION OF ERLANG AND ENGSET FUNCTIONS

Jerzy KUBASIK

Institute of Electronics, Technical University of Poznań
Poznań, Poland

ABSTRACT

In the paper simple, efficient and accurate numerical methods for the evaluation of Erlang and Engset loss formulas are presented. The paper contains a review of practical formulas enabled to construction of convenient programs that rapidly evaluate some most frequently used traffic problems. The following problems are discussed: determination of probability of loss, determination of permissible offered traffic, determination of number of servers using both Erlang and Engset functions. Furthermore, approximations of Erlang B function for noninteger number of trunks are mentioned. Several of these methods are based on new improved procedures that are first presented in this paper. The described methods were tested on a computer over a wide range of arguments.

1. INTRODUCTION

The Erlang and Engset functions for lost call systems are the basic formulas in telecommunications traffic theory. They are the relation between four quantities: loss probability call congestion - B, offered traffic intensity - - A, number of traffic sources - S, and number of servers - N. Erlang B formula is valid for the infinite value of S.

The importance of these functions in current traffic engineering is well known. Many tables and charts have been published, e.g. [2, 4, 5, 9, 13, 17], that allow one more readily to apply these formulas in practice. However, many current traffic analysis and design problems can be solved only with the aid of computers. From the analysis point of view, "table or chart look ups", in a computer can be time and space consuming inaccurate, and inconvenient. It is therefore advantageous to have methods or subroutines available, which can easily be incorporated into the main program for routine maximum likelihood estimation.

The methods presented below have these necessary capabilities and thereby eliminate the need for "table look ups" or interpolation in existing tables.

This paper does not fully develop the theories of all the formulas used. Instead it, gives enough description and explanation to make the development of the computational formulas understand able. For theoretical accuracy one should consult the references.

2. ERLANG LOSS FUNCTION

2.1 Determination of Loss Probability

The Erlang B function arises in the study of the M/M/n/n queue. The probability of loss, i.e. probability that an arriving call is rejected because no trunk is available, is expressed by Erlang loss formula, when a Poisson stream of calls offering A Erl to a fully available group of N trunks.

The Erlang probability function is usually defined as

$$E(N,A)=(A^N/N!)\ /\ \sum_{i=0}^{N} (A^i/i!) \quad . \qquad (1)$$

Values of E(N,A) can be conveniently calculated using the recurrence relation which has been known since the days of Erlang

$$E(N,A)= AE(N-1,A)/[AE(N-1,A)+N] \quad , \qquad (2)$$

where the initial value $E(0,A)=1$.

Dill and Gordon [4] have proposed other method, in which E(N,A) is exactly expressed as sum of terms

$$E(N,A)^{-1} = \sum_{j=0}^{N} T_j \quad , \qquad (3)$$

where the terms T_j are obtained using the following recurrence

$$T_j = T_{j-1}\cdot(N-j+1)/A \quad , \qquad (4)$$

with $T_0=1$.

The calculations can be terminated before j=N because, during the succesive computation, the individual term eventually becomes very small and has no significant effect to the series sum. In this way, instead use N times the formula (2), one may add up only a several terms (4) to obtain the required value B with the satisfactory accuracy. Equation (3) may be used for efficient calculation of the blocking probability for any size trunk group.

2.2 Case of Noninteger Number of Servers

The Erlang loss formula can only have any physical significance when N is a positive integer. The case of N not being an integer has, however, practical significance in many problems concerning the telecommunication network dimensioning or determination of blocking probability in the multistage switching networks.

For this reason it is also desirable to extend the definition for continuous number of devices. There are known numerous equivalent forms of the expansion for Erlang loss formula. The most frequently used expressions are the following

$$E(X,A)^{-1} = \int_0^{\infty}(1+t/A)^X \cdot e^{-t}dt \quad , \qquad (5)$$

$$E(X,A)^{-1} = A\int_0^{\infty}(1+t)^X \cdot e^{-At}dt \quad , \qquad (6)$$

$$E(X,A)^{-1} = e^A A^{-X} \int_A^{\infty} t^X e^{-t}dt \quad , \qquad (7)$$

$$E(X,A)^{-1} = e^A A^{-X}\,\Gamma(X+1,A) \quad , \qquad (8)$$

where

$$\Gamma(X,A) = \int_A^{\infty} t^{X-1} \cdot e^{-t}dt \quad , \qquad (9)$$

is the Incomplete Gamma function of the second kind. In these formulas X denotes the noninteger number of trunks.

Because all of these expressions contain the integrals or special functions, they are not convenient for numerical evaluation and for practical use. Therefore, many authors have developed the approximations of Erlang function for X.

It should be noted that the definition (1) as well as the continuous expansions (5), (6), (7), and (8) satisfy the recurrence formula (2), which is valid for all nonnegative X. Furthermore, Rapp [15] has shown that this relation will not propagate errors in the initial value $E(X_o,A)$. Therefore, X_o can be chosen to be the fractional part of X, i.e. $0 \leqslant X_o < 1$, and if $E(X_o,A)$ can be easily computed, then applying (2) $X-X_o$ times will yield $E(X,A)$.

Rapp [15] has proposed the approximation of $E(X_o,A)$ based on quadratic interpolation, namely

$$E(X_o,A) \cong C_o + C_1X_o + C_2X_o^2 \quad , \qquad (10)$$

where

$$C_o = 1 \quad ,$$

$$C_1 = -(2+A)/(1+3A+A^2) \quad , \qquad (11)$$

$$C_2 = 1/(1+4A+4A^2+A^3) \quad .$$

Another expression for computing $E(X_o,A)$ has been given by Szybicki [18]

$$E(X_o,A) \cong \left[(2-X_o)\cdot A+A^2\right]/(X_o+2A+A^2) \quad . \qquad (12)$$

Next approximation developed from Newton's interpolation formula has been described by Jagerman [7] as follows

$$E(X,A) \cong B^{1-X_o} B_1^{X_o} \left[B_1^2/(BB_2)\right]^{X_o(1-X_o)/2} \qquad (13)$$

In this formula B, B_1, and B_2 denote E(N,A), E(N+1,A), and E(N+2,A), respectively, where N is the integral part of X. Using this formula one obtain value of loss probability directly for X and not for X_o.

Farmer and Kaufman [6] have proposed to calculate the $E(X_o,A)$ by means of the Laguerre quadrature, i.e. the Erlang function given by definition (5) is approximated by the sum

$$E(X_o,A)^{-1} \cong \sum_{i=1}^{n} w_i(1+t_i/A)^{X_o} \quad , \qquad (14)$$

where w_i and t_i are the weight factors and abscissas respectively and they are given in [1]. Naturally, the better accuracy is achieved by increasing n.

Another method, in which the series expansion of Gamma function is applied, is described by Urmoneit [19]. He express the Erlang function in the following way

$$E(X_o,A)^{-1} = \Gamma(X_o+1)\sum_{n=0}^{\infty}(C_n-D_n) \quad , \qquad (15)$$

where

$$C_n = A^{n-X_o}/n! \quad ,$$

$$D_n = A^{n+1}/\Gamma(X_o+n+2) \quad . \qquad (16)$$

The above coefficients can be computed recurrently

$$C_n = C_{n-1}\cdot A/n \quad ,$$

$$D_n = D_{n-1}\cdot A/(X_o+n) \quad , \qquad (17)$$

with the starting values

$$C_o = A^{-X_o} \text{ , and } D_o = A/\Gamma(X_o+2) . \qquad (18)$$

The process of calculation of the C_n and D_n is stopped when given accuracy indicated by $|C_n-D_n|$ is obtained. The Gamma function in the formulas (15) and (18) can be computed by the series expansion

$$\Gamma(Y) = \sum_{k=1}^{26} c_k\cdot Y^k \quad . \qquad (19)$$

The factors c_k are given in [1, 19].

One should remember that for Gamma function the following recursive relation is valid

$$\Gamma(Y+1) = Y\cdot\Gamma(Y) \quad . \qquad (20)$$

Therefore, to save the computation time, the procedure (19) can be used only one times. The $\Gamma(X_o+1)$ in (15) must be calculated from (19), while for X_o+2 in (18) it may be obtained from the recurrence (20).

The next method described here is the Lévy-Soussan procedure [12], which uses the continued-fraction expansion of the Incomplete Gamma function. The Erlang B function can be calculated as the continued-fraction expansion of some function $F(A,-X) = E(X,A)^{-1}$. The value of $F(A,-X)$ lays in the interval between P_n and I_n, which are the even and odd approximants of F respectively, defined as

$$P_n(A,-X) = \frac{A|}{|A-X} + \frac{X|}{|A-X+2} + \frac{2(X-1)|}{|A-X+4} + \ldots + \frac{n(X-n+1)|}{|A+2n-X} , \qquad (21)$$

$$I_n(A,-X) = 1 + \frac{X|}{|A-X+1} + \frac{X-1|}{|A-X+3} + \ldots + \frac{(n-1)(X-n+1)|}{|A-X+2n-1} . \qquad (22)$$

The notation

$$F(X) = b_o + \frac{a_1|}{|b_1} + \frac{a_2|}{|b_2} + \frac{a_3|}{|b_3} + \ldots \qquad (23)$$

can be replaced by a more legible form of continued-fraction

$$F(X) = b_o + \cfrac{a_1}{b_1 + \cfrac{a_2}{b_2 + \cfrac{a_3}{\ldots}}} \qquad (24)$$

Simultaneous computation of P_n and I_n is particulary convenient. The computation sequence is stopped when the first of two conditions is met: limit of n was exceeded or when $|(I_n-P_n)/I_n| < \varepsilon$ is obtained, where ε is the given accuracy. The first condition means that required accuracy is unobtainable by using this method.

The last method proposed in this section uses the basic method of numerical evaluation of Incomplete Gamma function $\gamma(X,A)$. The γ is defined as

$$\gamma(X,A) = \int_0^A t^{X-1}e^{-t}dt . \qquad (25)$$

Since

$$\Gamma(X,A) = \Gamma(X) - \gamma(X,A) , \qquad (26)$$

the function $\gamma(X,A)$ can be used for determination of blocking probability from Erlang formula. Substituting $\Gamma(X+1,A)$ by (26) into (8) yields the following expression for $E(X,A)$

$$E(X,A)^{-1} = A^{-X}e^{A}\left[\Gamma(X+1) - \gamma(X+1,A)\right] \qquad (27)$$

Furthermore, it can be shown that

$$\lim_{A\to\infty} \gamma(X,A) = \Gamma(X) . \qquad (28)$$

Hence, the function γ may be used for computation of the Gamma function by choosing relatively large value of A.

The Incomplete Gamma function [1] may be written as

$$\gamma(X,A) = X^{-1}A^{X}e^{-A}\, M(1,1+X,A) , \qquad (29)$$

where

$$M(1,1+X,A) = 1 + \sum_{j=1}^{\infty} A^j / \prod_{k=1}^{j}(X+k) , \qquad (30)$$

is the Confluent Hypergeometric function or Kummer's function, which can be represented by

$$M(1,1+X,A) = \sum_{j=0}^{\infty} T_j , \qquad (31)$$

and the terms T_j are obtained using the following recurrence formula

$$T_j = T_{j-1} \cdot A/(X+j) \qquad (32)$$

with $T_o=1$.

One should mention the asymptotic expansion of Erlang loss function proposed by Akimaru and Takahashi [3]. They have found the general representation of $E(X,A)$ in the following form

$$E(X,A)^{-1} = \sum_{j=0}^{\infty} a_j X^{(1-j)/2} . \qquad (33)$$

Because the method for determination of the factors a_j is complicated, formula (33) cannot be easily applied for practical computations of $E(X,A)$.

There exist some other methods, e.g. the described in [6], but they are very complicated and can be used only in special situations.

2.3 Determination of Traffic Intensity

Determination of the permissible offered traffic for a given number of servers and given value of loss probability is frequently required in the practical design of communication systems. Erlang formula (1) can be solved for A directly, only for

$$N=1 \qquad A=B/(1-B) , \text{ and} \qquad (34)$$

$$N=2 \qquad A=\left[B+\sqrt{B(2-B)}\right]/(1-B) . \qquad (35)$$

Although this equation for larger values of N cannot be easily inverted and calculation of A by an explicit formula is unobtainable, numerical methods can be used. This problem has been previously dealt with by several authors.

Rappaport [16] presented scheme of determination of carried traffic C and offered traffic A using method of false position /regula falsi/. He proposed to solve for C the equation

$$F(C) = E(C) - B = 0 , \qquad (36)$$

where $E(C)=E\left[N,C/(1-B)\right]$.

The solution must lie in the interval from 0 to N. The described method contain three parts. To reduce the number of iterations a simple method of finding roots, e.g. method of bisections, was used first. Initial search was applied to isolate the solution within an interval N/100. In the first part of iterations, beginning with C_{i-1} and C_i as the leftmost and rightmost points of interval, a new value C_{i+1} is calculated using formula

$$C_{i+1}=\frac{C_{i-1}E(C_i)-C_iE(C_{i-1})+B\cdot(C_i-C_{i-1})}{E(C_i)-E(C_{i-1})} \tag{37}$$

This new value replaces that value of C_{i-1} or C_i, for which the function (36) has the same sign. The process is repeated until the denominator of (37) becomes smaller than 0.01B. Second part of iterations is described by the formula

$$C_{i+1}=C_i-P\cdot F(C_i) \quad , \tag{38}$$

where the constant P is calculated after first part of iterations by

$$P=(C_i-C_{i-1})/[E(C_i)-E(C_{i-1})] \quad . \tag{39}$$

After each step a new value replaces the former value C. Iterative process is stopped when given accuracy is obtained. Then offered traffic can be determined as $A=C/(1-B)$.

Szybicki [18] described special numerical method for offered traffic calculation from Erlang function. His method is given by the formula

$$A_{i+1}=A_i-\frac{A_i[E(A_i)-B]}{E(A_i)[N+1-A_i+A_iE(A_i)]} \quad , \tag{40}$$

where $E(A)=E(N,A)$ and suitable starting value

$$A_o = N/(1-B) \quad . \tag{41}$$

Both presented methods are the first order convergence procedures. Because the first derivative of Erlang loss formula with respect to A is well known

$$E'(A)= E(A)[N/A-1+E(A)] \quad , \tag{42}$$

some faster numerical methods can be applied to determine A. There is Newton's iterative method for calculation of a simple zero of a nonlinear equation

$$F(A) = 0 \quad . \tag{43}$$

In this method iterations are defined by

$$A_{i+1} = A_i-F(A_i)/F'(A_i) \quad . \tag{44}$$

This is the second order convergence method. It can be used to solve the discussed problem, if an appropriate function, which have characteristic needed in this procedure is chosen. For example it may be the function

$$F(A) = A[E(A)-B] \quad . \tag{45}$$

First derivative of this function is given by

$$F'(A) = E(A)-B+A\cdot E'(A) \quad . \tag{46}$$

The other authors prefer to use the function

$$F(A) = E(A)-B \tag{47}$$

as the left side of equation (43). The derivative of (47) is identical as (42). The proposed starting values A_o are also different. In [7] we have

$$A_o = N/(1-B)-1 \quad , \tag{48}$$

whereas, authors of [6] suggest to assume

$$A_o = N \quad , \text{ and}$$

$$A_1 = N - [E(N)-B]/E(N)^2 \quad , \tag{49}$$

where E(N) can be approximated by

$$E(N) = E(N,N) \cong \sqrt{2/(\pi N)} \quad . \tag{50}$$

In [4] Newton's method is applied with the function

$$F(A) = 1/E(A) - 1/B \quad . \tag{51}$$

There is proposed a new, third order convergence method, based on Halley's formula [14]

$$A_{i+1}=A_i-\frac{F(A_i)}{F'(A_i)}-\frac{F(A_i)^2\,F''(A_i)}{2[F'(A_i)]^3} \tag{52}$$

As function F the expression (45) can be used. Second derivative of this function with respect to A is given by

$$F''(A) = E'(A)\cdot(N-A+2)+E(A)N/A \quad . \tag{53}$$

This procedure is more complicated than Newton's method, but it needs less iterations to obtain the result with the same accuracy. Number of iterations needed to obtain one element with the accuracy to six signifficant figures and processing time per element are compared in [10]. The proposed method makes it possible to calculate Erlang loss function significantly faster than known procedures. This method is especially attractive for small values of loss probability.

One should be confirm that all presented in this section methods can be applied for computation both with integer as with noninteger numbers of servers in Erlang formula.

2.4 Determination of Number of Trunks

The problem of determination of minimum number of devices or trunks, which are able to serve the offered traffic with given grade of service, is also frequently studied. Since the expression for the derivative of Erlang loss function with respect to X is rather complicated, three approaches are possible for computation of the required value.

First, which yields the solution in the integer numbers, is based on the repetitions of calculations of E(N,A) for a given A and for increasing N=1, 2, 3,..., using the recurrence formula (2) The process is continued until the obtained, in each step, E(N,A) becomes smaller than the assumed probability B. The value N, for which the last computation is performed, is the required number of servers [19].

The second possible approach is the application of Newton's or other iterative methods with approximate values of the derivative. Jagerman [7] has proposed the following approximation of the derivative of Erlang function with respect to X

$$E'(X) = E(X)\cdot\frac{1/(2A\alpha) - \ln\alpha}{1-E(X)/(2\alpha)} , \qquad (54)$$

where

$$\alpha = E(X) + (1+X)/A , \qquad (55)$$

and $E(X)=E(X,A)$.

It is possible to apply the another approximation of $E'(X)$ given in [18]

$$E'(X) = \left[E(X+y) - E(X-y)\right] / (2y) , \qquad (56)$$

where $y \ll 1$, is the free chosen small real number.

Knowing $E'(X)$ we can solve for X the nonlinear equation

$$F(X) = 0 , \qquad (57)$$

where

$$F(X) = E(X) - B , \qquad (58)$$

using the Newton's method, in which each iteration is given by

$$X_{i+1} = X_i - F(X_i)/F'(X_i) . \qquad (59)$$

The derivative $F'(X_i)$ is simply equal to $E'(X_i)$, and the suitable starting value is $X_o=(A+1)\cdot(1-B)$ [7].

Third possible method employs the iterative method for finding a simple zero of the equation (57) other than Newton's method, e.g. regula falsi [6]. In this method, beginning from X_{i-1} and X_i as leftmost and rightmost points of interval, in which the solution must lie, a new value X_{i+1} is calculated using the formula

$$X_{i+1} = \frac{X_{i-1}E(X_i) - X_iE(X_{i-1}) + B(X_i - X_{i-1})}{E(X_i) - E(X_{i-1})} \qquad (60)$$

This new value replaces that value of X_{i-1} or X_i for which the function (58) has the same sign. As the starting values X_o and X_1 two numbers must be chosen, for which $F(X_o)$ and $F(X_1)$ have the opposite signs, e.g. $X_o=0$ and $X_1=A$. Hence, $E(0)=1$ and from (50)

$$E(A) = E(A,A) \cong \sqrt{2/(\pi A)} ,$$

the next X, i.e. X_2 is obtained immediately

$$X_2 \cong A\ 1-B \Big/ \left[1-\sqrt{2/(\pi A)}\right] , \qquad (61)$$

whereafter, if it is necessary, one continue the iteration using (60).

For the calculation of $E(X)$ in all cases any method described in section 2.2 may be used.

Naturally, the other approaches are possible for solving this problem more accurate. The search for a simple, fast, and accurate method for evaluate the first and the next derivatives of the Erlang loss function with respect to the number of devices seems to be especially attractive. Such a method should be applied in higher order convergence methods for fast determination of X.

3. ENGSET FUNCTION

3.1 Determination of Loss Probability

The Erlang loss formula is valid with the assumption that the Poisson distribution is used as the traffic source model, which implies infinite number of traffic generating sources. There are many situations, however, where the traffic is originated by a small number of subscribers, e.g. in PBXs, small rural exchanges, line concentrators. In such cases the Engset /truncated Binomial/ distribution is a better traffic model than Poisson distribution, but it is also more difficult to manipulate numerically.

The well-known Engset function is an implicit function. When traffic A Erl generated by S independent sources with equal calling intensities is offered to a full availability group of N trunks, the probability of call congestion B is a function of S, N, A, and also B itself. It is usually stated in the form

$$B=B(S,N,A) = \frac{\binom{S-1}{N}\left[\dfrac{A/S}{1-(1-B)A/S}\right]^N}{\sum_{i=0}^{N}\binom{S-1}{i}\left[\dfrac{A/S}{1-(1-B)A/S}\right]^i} \qquad (62)$$

For convenience in writing, the following abbreviations are used

$$\alpha = a/\left[1-a(1-B)\right] , \qquad (63)$$

and

$$a = A/S . \qquad (64)$$

The "a" is the average traffic generated per source in Erl, whereas, "α" means the average traffic originated by one source when it is free. Thus, from (62) we obtain

$$B(S,N,\alpha) = \binom{S-1}{N}\alpha^N \Big/ \sum_{i=0}^{N}\binom{S-1}{i}\alpha^i \qquad (65)$$

The blocking probability /time congestion/ in the Engset systems is defined by

$$E(S,N,\alpha) = \binom{S}{N}\alpha^N \Big/ \sum_{i=0}^{N}\binom{S}{i}\alpha^i . \qquad (66)$$

with α given by (63). There exist the interrelation formulas between B and E

$$B(S,N,\alpha) = E(S-1,N,\alpha) \quad , \tag{67}$$

$$E = \frac{S}{S-N} \cdot \frac{B}{1+\alpha(1-B)} \quad , \tag{68}$$

$$B = \frac{(S-N)\cdot E\cdot(1+\alpha)}{S+(S-N)\cdot E\cdot\alpha} \quad . \tag{69}$$

In this section problem of determination B with respect to S, N, and A will be discussed. The case of integer number of sources and servers will only be developed.

Beforehand, the methods for computation of B when S, N, and α are given. First of them is a well-known recurrence relation

$$B(N) = \frac{(S-N)\cdot\alpha\cdot B(N-1)}{N+(S-N)\cdot\alpha\cdot B(N-1)} \quad , \tag{70}$$

where B(N) is the abbreviation from $B(S,N,\alpha)$. The initial value B(0) is evidently equal to unity. The other form of this formula has been given by Joys [8] for the inverse of B

$$I(N) = 1+I(N-1)\,N/[\alpha\cdot(S-N)] \quad , \tag{71}$$

where

$$I(N) = 1/B(S,N,\alpha) \quad , \tag{72}$$

and I(0)=1. This method has been used, e.g. by Rubas for preparing the tables [17]. Analogous to the (3) for the Erlang function we can write (65) as a sum of terms

$$I(N) = \sum_{j=0}^{N} T_j \quad . \tag{73}$$

The terms T_j are succesively computed using the recurrence formula

$$T_j = T_{j-1}(N-j+1)\,/[\alpha\cdot(S-N+j-1)] \quad , \tag{74}$$

with $T_0=1$. The calculations can be terminated before j=N because the individual terms for increasing indexes j become very small and have no significant effect on the series sum.

This new method may be used for fast calculation of loss probability from Engset function especially for large trunk groups.

Now we return to the problem of computation of B when the value A, not α, is given. However, (65) is not an explicit equation, since α is not a constant, but a function of B (63). Therefore, an iterative procedure is necessary. For this purpose an auxiliary function

$$D = B - F(B) \quad , \tag{75}$$

is used, where F(B) is the new estimate of B computed from (70), (71), or (73) with α obtained from (63) using original B. For computational efficiency equation (71) has been simplified to the following form

$$1/\alpha = C + B \quad , \tag{76}$$

where C=S/A-1, and hence, C is a constant for a given combination of traffic and the number of sources.

The above described method, accompanied with the Joys recurrence (71), was employed in [2, 17].

3.2 Determination of Traffic Intensity

The determination of the permissible offered traffic for a given other quantities from Engset distribution is also the frequently required problem. Since from (63) and (64) we have

$$A = S/(1/\alpha+1-B) \quad , \tag{77}$$

the equation

$$B(S,N,\alpha) = B \quad , \tag{78}$$

where B is the given probability of loss, may be solved for α and the result used for the calculation of A.

Simple solution of (78) can be found directly only for

$$N=1 \quad \alpha=B/[(1-B)\cdot(S-1)] \quad , \text{ and} \tag{79}$$

$$N=2 \quad \alpha=\frac{B}{(1-B)\cdot(S-2)}\left[1+\sqrt{\frac{1+2(S-2)}{(S-1)\,B/(1-B)}}\right] \tag{80}$$

For N greater than two (78) cannot be solved directly for α, especially it is impossible to find the general explicit formula for all values of N. The iterative procedure must be used.

Driksna and Wormald [5] have used the simplest, but very slow, method of bisections. Here, the method, which employes the Newton's iterative procedure, will be proposed. The iterations are given by

$$\alpha_{i+1} = \alpha_i - F(\alpha_i)/\,F'(\alpha_i) \quad . \tag{81}$$

The function $F(\alpha)$ can be chosen as

$$F(\alpha) = 1/B - 1/B(S,N,\alpha) \quad . \tag{82}$$

According to (73) the above equation may be written as

$$F(\alpha) = 1/B - \sum_{j=0}^{N} T_j \quad , \tag{83}$$

with the terms defined reccurently by (74) or directly performed as function of α

$$T_j = \frac{N!}{(N-j+1)!}\cdot\frac{(S-N-1+j)!}{(S-N)!}\,\alpha^{-j} \tag{84}$$

Hence, the first derivative of $F(\alpha)$ with respect to α has the following form

$$F'(\alpha) = \alpha^{-1} \sum_{j=0}^{N} j\cdot T_j \quad . \tag{85}$$

Since the similar sum of series appears in the formulas for computation of $F(\alpha)$ (83) and its derivative (85), they may be calculated simultaneously. The suitable starting value α_0 is any number belonging to the interval $(0,1\rangle$.

In the same way the second and the next derivatives of Engset function can be calculated to apply in the numerical procedures with higher degree of convergence, e.g. in the Halley's method [14].

3.3 Determination of Number of Devices and Number of Sources

Methods for determination of number of devices when A, B, and S are given or determination of number of sources generating the traffic A Erl, which is carried by N trunks with given grade of service are also frequently required in many teletraffic problems.

Since the simple formulas for the partial derivative of Engset loss function with respect to N or S are not known, the method of succesive calculation of probability of loss can be applied. From (63) and (64) with given A, B, and S, α is obtained and whereafter, by repetitions of calculations of $B(S,N,\alpha)$ for increasing $N = 1, 2,..$ using the recurrence formulas (70) or (71). The process is stopped when the obtained in N-th step value $B(S,N,\alpha)$ becomes smaller than assumed grade of service B.

The recurrence formula for increasing S has been also given by Joys [8]

$$I(S) = \frac{S+N+1}{S+1}\left[(1+\alpha)I(S-1)-\alpha\right], \qquad (86)$$

where

$$I(S) = 1/B(S,N,\alpha) \qquad (87)$$

and the starting value $I(N+1)=[\alpha/(1+\alpha)]^N$. The method for determination of S is a little more complicated, because the value α is not a constant, but a function of S (63). Hence the iterative procedure should be employed, analogous to this, which is used for determination of B with respect to A.

One should be noted that the above methods yield the integer results only.

4. CONCLUSIONS

All methods described in this paper were tested on ODRA 1305 computer /compatible with ICL 1900/. The programs written in FORTRAN have been used to calculate tables over a wide range of parameters.

Number of iterations needed to obtain one element with the assumed accuracy and the processing time have been compared for some algorithms in [10].

Subroutines, which illustrate all described above methods appears in [11].

REFERENCES

[1] M. Abramowitz, I.A. Stegun, "Handbook of mathematical functions", Dover Publ., New York, 1965.

[2] M. Agosthazi, G. Gosztony, E. Uxa, "Engset charts", BUDAVOX Telecom. Rev., Special Ed., 1973.

[3] H. Akimaru, H. Takahashi, "Asymptotic expansion for Erlang loss function and its derivative", IEEE Trans. Commun., vol. COM-29, no. 9, pp. 1257-1260, 1981.

[4] G.D. Dill, G.D. Gordon, "Efficient computation of Erlang loss functions", COMSAT Tech. Rev., vol. 8, no. 2, pp. 353-370, 1978.

[5] V.V Driksna, E.G. Wormald, "Engset traffic tables", Telecommun. J. of Australia, vol. 16, no. 6, pp. 154-156, 1966.

[6] R.F. Farmer, I. Kaufman, "On the numerical evaluation of some basic traffic formulae", Networks, vol. 8, pp. 153-186, 1978.

[7] D.L. Jagerman, "Methods in traffic calculations", Bell Lab. Tech. J., vol. 63, no. 7, pp. 1283-1310, 1984.

[8] L.A. Joys, "Variations of the Erlang, Engset and Jacobaeus formulae", Proc. 5th Int. Teletraffic Congress, New York, pp. 107-111, 1967.

[9] M.M. Jung, "Loss probability charts calculated with the formula of Engset", Philips Telecommun. Rev., vol. 23, no.4, pp. 186-201, 1962.

[10] J. Kubasik, "Fast computation of Erlang B function", Proc. of IEEE Mediterranean Electrotech. Conf. MELECON'83, Athens, vol. 1, paper A4.13, 1983.

[11] J. Kubasik, "Subroutines for the computation of Erlang and Engset functions", Inst. of Electronics, Tech. Univ. of Poznań, Report PR-11/1/1985/A, 1985.

[12] G. Levy-Soussan, "Numerical evaluation of the Erlang function through a continued-fraction algorithm", Electr. Commun., vol. 43, no. 2, pp. 163-168, 1968.

[13] C. Palm, "Table of the Erlang loss formula", Kungl. Telestyrelsen, Stockholm, 1964.

[14] A. Ralston, "A first course in numerical analysis", McGraw Hill Book Comp., New York, 1965.

[15] Y. Rapp, "Planning of junction network in a multi-exchange area", Ericsson Tech., vol. 20, no. 1, pp. 77-130, 1964.

[16] S.S. Rappaport, "Calculation of some functions arising in problems of queueing and communications traffic", IEEE Trans. Commun., vol. COM-27, no. 1, pp. 249-251, 1979.

[17] J. Rubas, "Table of Engset loss formula", Melbourne, 1969.

[18] E. Szybicki, "Some numerical methods used for telephone traffic theory applications", Ericsson Tech., vol. 20, no. 2, pp. 203-229, 1964.

[19] W. Urmoneit, "Die Aufbereitung der verkehrstheoretischen Grundlagen zur Bemessung von Nachrichtennetzen", Der Fernmelde-Ingenieur, vol. 37, no. 4, pp. 1-39, no. 6, pp. 1-32, no. 8, pp. 1-36, 1983.

TELETRAFFIC ISSUES in an Advanced Information Society
ITC-11
Minoru Akiyama (Editor)
Elsevier Science Publishers B.V. (North-Holland)

COMPUTATION OF SECOND DERIVATIVES OF ERLANG'S B AND WILKINSON'S FORMULAE AND ITS APPLICATION ON PLANNING OF JUNCTION NETWORKS

Endre TOTH

Planning and Investing Institute of Hungarian PTT
Budapest, Hungary

ABSTRACT

In the case of junction networks having minimum cost the derivatives of the cost function according to the number of circuits in the high usage routes are zero. These derivatives form a system of non-linear simultaneous equations, suitable for the determination of the number of circuits. For this purpose the use of the Newton-Raphson method is the most advantageous, but it requires the first and second derivatives of the Erlang B formula. These derivatives cannot be expressed simply by a formula, but a polinom of third degree may be found giving a good approximation of the Erlang B formula in the inteval $0 \leq x \leq 1$. For values of $x > 1$ the derivatives may be approximated by recursive formulas. The recursion does not give accurate results if $x < 0$ but here too we may find good approximations

1. INTRODUCTION

It is well known that generally the cost of a junction network is the lowest if it contains high usage routes. No explicit formula exists for the dimensioning of networks of this type but Rapp's method is suitable for this purpose. The essence of this method is that the number of circuits in the high usage routes should be varied until the cost of the whole network becomes the minimum.

A disadvantage of this method is that finding the minimum cost by experimenting with new circuit numbers in the high usage routes takes much time. It would be quite reasonable to calculate the number of circuits in the high usage routes, resulting in a network of minimum cost, from the derivative of the overall cost, according to the number of circuits in the high usage routes.

The overall cost can be calculated by the following equation:

$$C = \sum_j B_j N_j \tag{1}$$

where N_j=number of circuits in route j
B_j=cost of one circuit in route j(for the sake of simplicity let us suppose that $\partial B_j/\partial N_j = 0$)
C=cost of the whole junction network

The derivative of the overall cost according to the number of circuits in route i:

$$\partial C/\partial N_i = B_i + \sum_{i \neq j} \partial N_j/\partial N_i \cdot B_j \tag{2}$$

The overall cost is the minimum where $\partial C/\partial N_i = 0$.

As $N_j(N_i)$ is not an explicit function the derivative can only be determined step-by-step and N_i can be obtained by iteration. It is expedient to use the Newton-Raphson method. For this purpose we should use the derivative of eqation (2):

$$\partial^2 C/\partial N_i^2 = \sum_{i \neq j} B_j \cdot \partial^2 N_j/\partial N_i^2 \tag{3}$$

This calculation method is treated in the present paper. The paper also includes a detailed calculation for dimensioning triangular networks and a generalization of the method for the cases of more complex polycentric networks.

2. TRIANGULAR NETWORKS WITH FULL AVAILABILITY

The structure of such a networks may be seen on Fig.1.

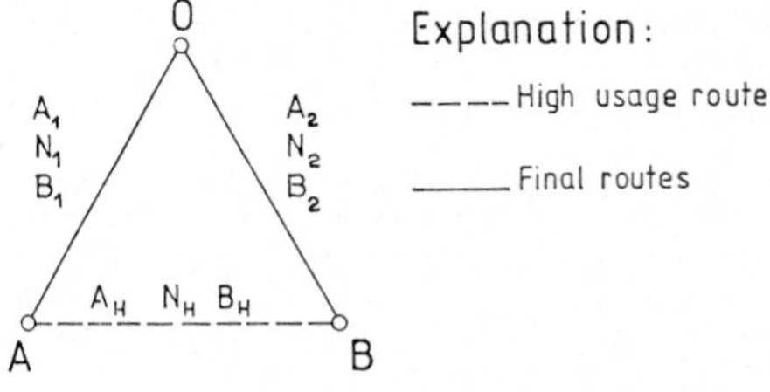

Fig.1 Structure of the triangular networks.

The traffic overflowing from route A-B passes over to routes A-O and O-B.

2.1 Dimensioning final routes

Let us suppose that the value of N_H is given.

The mean value of overflowing traffic (M_H) and its variance (V_H) are the following:

$$M_H=A_HE_H \quad (4)$$

$$V_H=M_H(1-M_H+A_H/t_H) \quad (5)$$

where $E_H = A^{N_H}e^{-A}/\int_A^{\infty} x^{N_H}e^{-x}dx \quad (6)$

and

$$t_H=1+N_H+M_H-A_H \quad (7)$$

By adding the traffic on the final route (A_f) to the mean of the overflowing traffic we obtain the mean of the total traffic of that final route:

$$M_f=M_H+A_f \quad (8)$$

The variance of total traffic on the final route may be obtained similarly:

$$V_f=V_H+A_f \quad (9)$$

Based on Wilkinson's ERT (Equivalent Random Theory) method [11], an equivalent traffic value (A^*) and an equivalent circuit number (N^*) may be found with which the following equations may be written:

$$M_f=A^*E(A^*,N^*) \quad (10)$$

and $V_f=M_f(1-M_f+A^*/t^*) \quad (11)$

where $t^*=1+N^*+M_f-A^* \quad (12)$

Using A^*, N^* and the loss figure for the final route we may calculate the circuit number in the final route. For the calculation of the loss in the final route from the average loss in the whole network the following equation may be considered [9]:

$$E_o\sum A=E_fA^* \quad (13)$$

where E_o=average loss of the junction network

$\sum A$=total traffic of the final routes in pure tandem networks

E_f=loss figure of the final route

E_f can be obtained from equation (13).

On the other hand E_f may also be expressed by another equation:

$$E_f=f(A^*,m+N^*) \quad (14)$$

where m=number of circuits in the final route (on Fig.1. $m=N_1$ or $m=N_2$)

Equation (14) is suitable for the calculation of m.

2.2 First derivative of final routes according to the high usage routes

The expression $\partial m/\partial N_H$ in equation (2) can be calculated in a similar way as the number of circuits in the final route.

We may form dM_H and dV_H differentials:

$$dM_H=\frac{\partial M_H}{\partial N_H}\cdot dN_H \quad (15)$$

$$dV_H=\frac{\partial V_H}{\partial N_H}\cdot dN_H \quad (16)$$

Since during the optimalization only the number of circuits will vary and the traffic will not, the derivatives according to traffic is zero.

Since the complete change in the mean and variance of the overflowing traffic appears on the final route, we may write:

$$dM_f=dM_H \quad (17)$$

and $dV_f=dV_H \quad (18)$

On the other hand the following equations are also valid:

$$dM_f=\frac{\partial M_f}{\partial A^*}\cdot dA^*+\frac{\partial M_f}{\partial N^*}\cdot dN^* \quad (19)$$

$$dV_f=\frac{\partial V_f}{\partial A^*}\cdot dA^*+\frac{\partial V_f}{\partial N^*}\cdot dN^* \quad (20)$$

From equations (19) and (20) dA^* and dN^* can be calculated.

$$dA^*=(\frac{\partial V_f}{\partial N^*}\cdot dM_H-\frac{\partial M_f}{\partial N^*}\cdot dV_H)/\triangle \quad (21)$$

$$dN^*=(\frac{\partial M_f}{\partial A^*}\cdot dV_H-\frac{\partial V_f}{\partial A^*}\cdot dM_H)/\triangle \quad (22)$$

$$\text{where } \triangle=\frac{\partial M_f}{\partial A^*}\cdot\frac{\partial V_f}{\partial N^*}-\frac{\partial V_f}{\partial A^*}\cdot\frac{\partial M_f}{\partial N^*} \quad (23)$$

From equations (21)-(23) $\partial A^*/\partial N_H$ and $\partial N^*/\partial N_H$ derivatives can be calculated.

$$\frac{\partial A^*}{\partial N_H}=(\frac{\partial V_f}{\partial N^*}\cdot\frac{\partial M_H}{\partial N_H}-\frac{\partial M_f}{\partial N^*}\cdot\frac{\partial V_H}{\partial N_H})/\triangle \quad (24)$$

$$\frac{\partial N^*}{\partial N_H}=(\frac{\partial M_f}{\partial A^*}\cdot\frac{\partial V_H}{\partial N_H}-\frac{\partial V_f}{\partial A^*}\cdot\frac{\partial M_H}{\partial N_H})/\triangle \quad (25)$$

Now we may form the differentiate of E_f using equation (13).

$$dE_f=-\frac{E_o\sum A}{(A^*)^2}\cdot dA^* \quad (26)$$

Alternatively we may express the same differentiate from equation (14).

$$dE_f=\frac{\partial E_f}{\partial A^*}\cdot dA^*+\frac{\partial E_f}{\partial(m+N^*)}\cdot(dm+dN^*) \qquad (27)$$

From equations (26) and (27) can be expressed.

$$dm=\frac{-\frac{E_o\Sigma A}{(A^*)^2}+\frac{\partial E_f}{\partial A^*}}{\frac{\partial E_f}{\partial(m+N^*)}}\cdot dA^*-dN^* \qquad (28)$$

The derivative $\partial m/\partial N_H$ using the above equation is the following:

$$\frac{\partial m}{\partial N_H}=-\frac{\frac{E_o\Sigma A}{(A^*)^2}+\frac{\partial E_f}{\partial A^*}}{\frac{\partial E_f}{\partial(m+N^*)}}\cdot\frac{\partial A^*}{\partial N_H}-\frac{\partial N^*}{\partial N_H} \qquad (29)$$

By substituting the expression of the derivative $\partial m/\partial N_H$ in equation (2) we may calculate the number of circuits in the high usage routes, N_H, in the network of minimum cost.

2.3 Second derivative of the final routes according to the high usage routes

Solution of the (2) equation by the Newton-Raphson method requires the calculation of the derivative $\partial^2m/\partial N_H^2$ that is the derivative of $\partial m/\partial N_H$ according to N_H.

$$\frac{\partial^2 m}{\partial N_H^2}=\frac{\frac{2E_o\Sigma A}{(A^*)^3}\cdot\frac{\partial A^*}{\partial N_H}-\frac{\partial^2E_f}{\partial A^*\partial N_H}}{\frac{\partial E_f}{\partial(m+N^*)}}\cdot\frac{\partial A^*}{\partial N_H}+$$

$$+\left(\frac{\partial m}{\partial N_H}+\frac{\partial N^*}{\partial N_H}\right)\left(\frac{\frac{\partial^2A^*}{\partial N_H^2}}{\frac{\partial A^*}{\partial N_H}}-\frac{\frac{\partial^2E_f}{\partial(m+N^*)\partial N_H}}{\frac{\partial E_f}{\partial(m+N^*)}}\right)-$$

$$-\frac{\partial^2N^*}{\partial N_H^2} \qquad (30)$$

where

$$\frac{\partial^2N^*}{\partial N_H^2}=\left(\frac{\partial^2M_f}{\partial A^*\partial N_H}\cdot\frac{\partial V_H}{\partial N_H}+\frac{\partial M_f}{\partial A^*}\cdot\frac{\partial^2V_H}{\partial N_H^2}\right)/\triangle-$$

$$-\left(\frac{\partial^2V_f}{\partial A^*\partial N_H}\cdot\frac{\partial M_H}{\partial N_H}+\frac{\partial V_f}{\partial A^*}\cdot\frac{\partial^2M_H}{\partial N_H^2}\right)/\triangle-$$

$$-\frac{\partial N^*}{\partial N_H}\left(\frac{\partial^2V_f}{\partial N^*\partial N_H}\cdot\frac{\partial M_f}{\partial A^*}+\frac{\partial V_f}{\partial N^*}\cdot\frac{\partial^2M_f}{\partial A^*\partial N_H}\right)/\triangle+$$

$$+\frac{\partial N^*}{\partial N_H}\left(\frac{\partial^2M_f}{\partial N^*\partial N_H}\cdot\frac{\partial V_f}{\partial A^*}+\frac{\partial M_f}{\partial N^*}\cdot\frac{\partial^2V_f}{\partial A^*\partial N_H}\right)/\triangle \qquad (31)$$

and

$$\frac{\partial^2A^*}{\partial N_H^2}=\left(\frac{\partial^2V_f}{\partial N^*\partial N_H}\cdot\frac{\partial M_H}{\partial N_H}+\frac{\partial V_f}{\partial N^*}\cdot\frac{\partial^2M_H}{\partial N_H^2}\right)/\triangle-$$

$$-\left(\frac{\partial^2M_f}{\partial N^*\partial N_H}\cdot\frac{\partial V_H}{\partial N_H}+\frac{\partial M_f}{\partial N^*}\cdot\frac{\partial^2V_H}{\partial N_H^2}\right)/\triangle-$$

$$-\frac{\partial A^*}{\partial N_H}\left(\frac{\partial^2V_f}{\partial N^*\partial N_H}\cdot\frac{\partial M_f}{\partial A^*}+\frac{\partial V_f}{\partial N^*}\cdot\frac{\partial^2M_f}{\partial A^*\partial N_H}\right)/\triangle+$$

$$+\frac{\partial A^*}{\partial N_H}\left(\frac{\partial^2M_f}{\partial N^*\partial N_H}\cdot\frac{\partial V_f}{\partial A^*}+\frac{\partial M_f}{\partial N^*}\cdot\frac{\partial^2V_f}{\partial A^*\partial N_H}\right)/\triangle \qquad (32)$$

For the practical calculation we may use the following approximation:

$$\partial m/\partial N_H\sim[th(aN_H+b)-1]/2 \qquad (33)$$

Substituting this approximation in equation (2) we get

$$\partial C/\partial N_H\sim B_o+B_1[th(a_1N_H+b_1)-1]/2+$$
$$+B_2[th(a_2N_H+b_2)-1]/2 \qquad (34)$$

A further approximation can be made in practice for equation (34):

$$\partial C/\partial N_H\sim B_o+$$
$$+(B_1+B_2)[th(aN_H+b)-1]/2 \qquad (35)$$

The values of a and b can be obtained from the differentiates $\partial C/\partial N_H$ and $\partial^2C/\partial N_H^2$ at a former value of N_H.

$$a_i=\frac{\left[\frac{\partial^2C}{\partial N_H^2}\right]_{i-1}\cdot\frac{B_1+B_2}{B_o-(\partial C/\partial N_H)_{i-1}}}{\frac{(\partial C/\partial N_H)_{i-1}-B_o}{B_1+B_2}+1} \qquad (36)$$

$$b_i=\frac{1}{2}\cdot\ln\frac{1+k}{1-k}-a_i\cdot(N_H)_{i-1} \qquad (37)$$

where

$$k=2[(\partial C/\partial N_H)_{i-1}/B_o-1]+1 \qquad (38)$$

a_i=i-th value of a

b_i=i-th value of b

$(N_H)_{i-1}$=(i-1)-th value of N_H

$(\partial C/\partial N_H)_{i-1}$=(i-1)-th value of $\partial C/\partial N_H$

$(\partial^2C/\partial N_H^2)_{i-1}$=(i-1)-th value of $\partial^2C/\partial N_H^2$

With this method the speed of finding a value which approximates the real N_H with a defined precision can be increased.

3. POLYCENTRIC NETWORKS WITH FULL AVAILABILITY

With certain additions the dimensioning principles we have derived for triangular networks may also be applied in the case of polycentric networks.

For example consider the two-level polycentric network on Fig. 2.

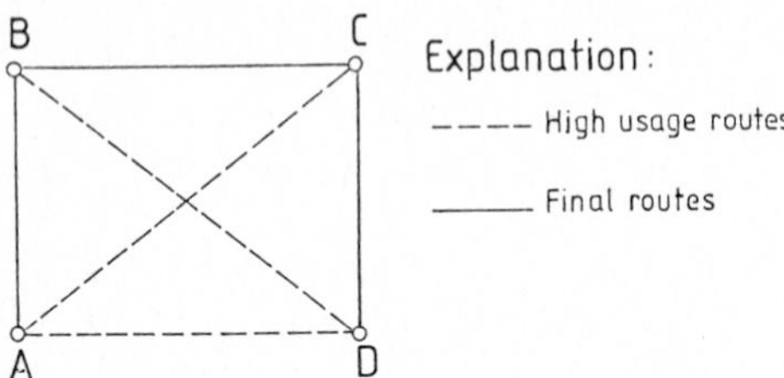

Fig.2 Structure of two-level polycentric network.

When dimensioning such networks the following new conditions and principles should be applied.

1./ Route C-D is loaded with the mean value and variance of the traffic overflowing from route A-D and that carried on route A-C.
We may calculate these values with the well known formulas [10].
For the optimalization the derivatives of these expressions are also needed, did the derivation may be performed with the aid of the formulas presented in this paper.

2./ The resultant traffic on route C-D may also be smoothed and in this case the equivalent circuit number is negative [5] .
Erlang's B formula and its derivative to which negative circuit number belongs can be approximated with formulas introduced later in this paper.

3./ High usage routes have influence on one another, for this reason the method that was suitable for triangular networks, cannot be applied directly.
For example the traffic on route A-C is dependent on the traffic overflowing from route A-D that is on the number of circuits in this route. On the other hand the number of circuits in route A-D is dependent on the portion of its traffic that may overflow to routes A-C and A-B, and this depends on the number of circuits in route A-C.
The effect of interdependence of high usage routes is reduced by the fact that overflowing traffic only from route A-D and not from route A-C is passed on to route B-D, and because of the double overflow only an unsignificant part of the traffic on route A-D is passed on to route B-D even in case of larger loss probabilities in the routes A-D and A-C.
So in practice we may dimension route B-D as the high usage route of the triangular network B-C-D without considering the traffic overflowing from A-D to B-D.
But we cannot calculate routes A-D and A-C independently of each other. At first it is expedient to take 10% loss probabilities for route A-D and then routes A-D and A-C should be optimalized alternatively. In practice after two or three steps of optimalization we reach the optimum of the whole junction network.

Even more complex junction networks can be optimalized in a similar way.

4. OPTIMALIZATION OF NON FULLY AVAILABLE GROUPS

The principles described in the previous points of the paper may be extended to non fully available groups (graded or link system etc.) as well. In such cases instead of Erlang's B formula we should employ formulas expressing the loss probabilities of the non-fully available group under study (for example O'Dell's formula for gradings, the appropriate Jacobaeus formula for link connections). The mean value and variance of the overflowing traffic should be calculated based on formulas or methods valid for non-fully available groups (for example Lotze's formulas [8] for gradings and Herzog's calculations for link systems [7]). The derivatives necessary for the optimalization of the junction network deduced from these values.

5. FORMULAS APPROXIMATING THE DERIVATIVE OF ERLANG'S FORMULA ACCORDING TO THE NUMBER OF CIRCUITS

It is known that the derivatives of Erlang's formula cannot be expressed by explicit functions because of the Γ, ψ and $\partial\psi/\partial x$ functions it contains. For this reason it would be useful to approximate these functions by polynoms in the interval $0 \le x \le 1$ and to calculate them with recursive formulas if $x > 1$.

Rapp's parabola gives a very good approximation of Erlang's loss functions in the interval (0,1). While the first derivative of this parabola is still a good approximation of the first derivative of the Erlang B function, the second derivatives of the two functions differ significantly.

Let us approximate Erlang's formula with a polynom of third degree instead of Rapp's parabola:

$$E \sim C_0 + C_1 x + C_2 x^2 + C_3 x^3 \qquad (39)$$

The coefficients of the polynom may be calculated in the following way:

i.) Form the first and second derivatives of Erlang B formula at $x=0$ and $x=1$.

ii.) Apply the recursive formula approximating Erlang's B function and its first and second derivatives.

We now can calculate coefficient

$$C_o=1 \tag{40}$$

To express C_1 we may write a quadratic equation:

$$aC_1^2+bC_1+c=0 \tag{41}$$

where
$$a=A^2/(1+A)^3 \tag{42}$$

$$b=1+4A/(1+A)^2+A^2/(1+A)^4+ \\ +A(1-A)/(1+A)^3 \tag{43}$$

$$c=(3+6-2A^3)/(1+A)^4 \tag{44}$$

We should consider only the positive discriminant when solving the quadratic equation:

$$C_1=(-b+\sqrt{b^2-4ac})/2a \tag{45}$$

The value of C_2 can be calculated from C_1, and C_3 can be expressed with C_2 and C_1:

$$C_2=A(1+A)^2-3/(1+A)- \\ -C_1\left[2+A/(1+A)^2\right] \tag{46}$$

$$C_3=-1/(1+A)-C_1-C_2 \tag{47}$$

With forming the derivatives of formula (39) we obtain a good approximation of derivatives of Erlang B formula:

$$E'\sim C_1+2C_2x+3C_3x^2 \tag{48}$$

$$E''\sim 2C_2+6C_3x \tag{49}$$

where E' = the first derivative of Erlang B formula according to the circuit number

E''= the second derivative of Erlang B formula according to the circuit number

For values of $x>1$ the Erlang formula and its derivatives may be calculated with the aid of recursive formulas. It could be proven that the error of calculation does not increase if the number of circuits does.

We may obtain the third, fourth or further derivatives of Erlang's B function and may form approximating polynoms of fourth, fifth etc. degree, but in practice these are too complicated increasing computation time, and are unnecessary from the viewpoint of accuracy. (The error of the third degree polynom is less than 10^{-4}, satisfactory for practical purposes.)

For negative circuit numbers the recursive formulas may not be used for the error would be unacceptably large.

But if we extend Farmer-Kaufman's method [6] to negative circuit numbers we may obtain a good approximation of Erlang's formula even in this case.

$$E\sim(A-x)/A/\{1-x/[(A-x)^2+2A]\} \tag{50}$$

By forming the derivatives of (50) we get a good approximation of even the derivatives of Erlang's formula for negative circuit numbers.

$$E'\sim EN_1/[N_2(N_2+x)]-(1+x/N_2)/A \tag{51}$$

$$E''\sim(N_1E'-2Ex)/[N_2(N_2+x)]- \\ -[2x(A-x)+N_2+x]/(AN_2^2)+ \\ +[2EN_1(A-x)]/[N_2^2(N_2+x)]+ \\ +EN_1/[N_2^2(N_2+x)]+ \\ +2EN_1(A-x)/[N_2(N_2+x)^2] \tag{52}$$

where
$$N_1=A^2+2A-x^2 \tag{53}$$

and
$$N_2=(A-x)^2+2A-x \tag{54}$$

6. CONCLUSIONS

In this paper a new method has been proposed for the optimalization of junction networks, making use of the first and second derivatives of the Erlang and Wilkinson formulas.

With this method we may spare at least one step in the optimalization of a simple triangular network as compared to the "classic" method (saving would be 10-20 steps in case of more complicated networks).

The saving can also be expressed in computer time which is about 10-20% depending on traffic volume and the complexity of the network.

Further saving of computer time may be achived if the derivative of the cost function is approximated with a logistic curve instead of the Newton-Raphson iterative method. The curve gives a good approximation of the exact function except for some extreme cases - for example if the cost ratio is nearly 1 or if the traffic on the final route is less than 1 Erlang.

On the other hand the application of the second derivative results in cost savings of 1-2% as compared to the faster but less accurate network optimum approximating methods, in a large network this is quite a sum.

For the calculations derivatives of Erlang's B function should be used. These cannot be expressed by explicit functions, and in this paper an approximating method has been proposed for the calculation of said derivatives which also saves computer time compared to integral approximations or infinite series used earlier.

REFERENCES

1 H.Akimaru, T. Nishimura, "Derivatives of trunk functions for full availability systems,"(Internal Report - Seiko Hokoku No.3550 - of ECL), pp.1-5.

2 H. Akimaru, "Optimum design of switching systems," Review of the Electrical Communication Laboratory, vol.10, no.7-8, July-August, pp.385-401, 1962.

3 H. Akimaru, T. Nishimura, "The derivatives of Erlang's B formula," Review of the Electrical Communication Laboratory, vol.11, no.9-10, September-October, pp.428-445, 1983.

4 H. Akimaru, H. Tokushima, T. Nishimura, "Derivatives of Wilkinson formula and their application to optimum design of alternative routing systems," 9th Int. Teletraffic Congress, Torremolinos, pp.1-6, 1979.

5 G. Bretschneider, "Extension of the equivalent random method to smooth traffics," Proc. 7th Int. Teletraffic Congress, Stockholm, pp.1-9, 1973.

6 R.F. Farmer, I. Kaufman, "On the numerical evaluation of some basic traffic formulae," Networks, vol.8, no.2, pp.153-186, 1978.

7 U. Herzog, "A general variance theory applied to link systems with alternate routing," Proc.5th Int. Teletraffic Conress, New York, pp.398-406, 1967.

8 A.Lotze, "The variance of overflow traffic behind gradings of arbitrary type," Proc. 4th Int. Teletraffic Congress, London, pp.1-39, 1964.

9 Y.Rapp, "Planning of junction network in a multi-exchange area. I.General principles," Ericsson Technics, vol. 20, no.1, pp.77-130, 1964.

10 B. Walström, "Congestion studies in telephone systems with overflow facilities," Ericsson Technics, vol.22,no.3, pp.189-351, 1966.

11 R.I. Wilkinson, "Theories for toll traffic engineering in the U.S.A.," Bell System Technical Journal, vol.35, no.3, pp.421-514, 1956.

APPENDIX

Values of derivatives used in point 2.

$\partial E/\partial A=E(N/A-1+E)$

$\partial E/\partial N=-E\psi$

$\partial M/\partial A=tM/A$

$\partial M/\partial N=A\cdot\partial E/\partial N=-M\psi$

$\partial V/\partial A=V/t-t(M^2-V)/A$

$\partial V/\partial N=AM(M\psi-1)/t^2+(M^2-V)\psi$

$\partial^2 E/\partial N^2=E(\psi^2-\partial\psi/\partial N)$

$\partial^2 E/\partial A^2=E(N/A-1+E)(N/A-1+2E)-EN/A$

$\partial^2 M/\partial N^2=A\cdot\partial^2 E/\partial N^2=M(\psi^2-\partial\psi/\partial N)$

$\partial^2 M/\partial A^2=tM(2M+N-A)/A^2-E$

$\partial^2 V/\partial N^2=AM\psi(1-M\psi)/t^2+$
$+AM[2(1-M\psi)^2/t+M(\partial\psi/\partial N-\psi^2)]/t^2-$
$-(2M^2\psi+\partial V/\partial N)\psi+(M^2-V)\partial\psi/\partial N$

$\partial^2 V/\partial A^2=2V/t^2-VM/(At)+V/A+$
$+t(1-M)(M^2-V)/A^2+$
$+t^2(3M^2-V)/A^2$

$\partial^2 E_f/\partial A^*\partial N_H=$
$=[\partial(A^*+m+N^*)/\partial N_H][(N^*+m)/A^*-1+2E_f]\partial E_f/\partial A^*+$
$+E_f/A^*[\partial(m+N^*)/\partial N_H-\partial A^*/\partial N_H(N^*+m)/A^*]$

$\partial^2 E_f/\partial(N^*+m)\partial N_H=$
$=\partial^2 E_f/\partial(N^*+m)^2(\partial m/\partial N_H+\partial N^*/\partial N_H)+$
$+\partial A^*/\partial N_H\cdot E_f[\partial E_f/\partial(N^*+m)+1/A^*]-$
$-(\partial A^*/\partial N_H)^2\psi$

$\partial^2 M_f/\partial A^*\partial N_H=$
$=(\partial t/\partial N_H-t/A^*\partial A^*/\partial N_H)M_f/A^*+$
$+t/A^*\partial M_H/\partial N_H$

$\partial^2 M_f/\partial N^*\partial N_H=$
$=\partial^2 M_f/(\partial N^*)^2\cdot\partial N^*/\partial N_H+\partial A^*/\partial N_H\cdot E_N^*+$
$+\partial A^*/\partial N_H\cdot E_N^*[(1-2E_N^*)A^*-N^*-1]\psi$

$\partial^2 V_f/\partial A^*\partial N_H=$
$=1/t\cdot\partial V_f/\partial N_H-$
$-[V_f/t^2+(M_f^2-V_f)/A^*]\partial t/\partial N_H-$
$-t\cdot(2M_f\partial M_H/\partial N_H-\partial V_H/\partial N_H)/A^*-$
$-t(V_f-M_f^2)/(A^*)^2\cdot\partial A^*/\partial N_H$

$\partial^2 V_f/\partial N^*\partial N_H=$
$=2A^*M_f(1-M_f\psi)/t^3\cdot\partial t/\partial N_H-$
$-(M_f\cdot\partial A^*/\partial N_H+A^*\cdot\partial M_H/\partial N_H)(1+M_f\psi)/t^2-$
$-A^*M_f/t^2\cdot\partial^2 M_f/\partial N^*\partial N_H-$
$-(M_f-V_f/M_f)(\partial^2 M_f/\partial N^*\partial N_H+\psi\cdot\partial M_H/\partial N_H)+$
$+(2M_f\cdot\partial M_H/\partial N_H-\partial V_H/\partial N_H)\cdot\psi$

where

$$t = N + 1 - A + M$$

$$\partial t/\partial N_H = \partial N^{*}/\partial N_H + \partial M_H/\partial N_H - \partial A^{*}/\partial N_H$$

$$\psi = -\partial \ln E/\partial N$$

$$\partial \psi/\partial N = (1/E \cdot \partial E/\partial N)^2 - 1/E \cdot \partial^2 E/\partial N^2$$

TELETRAFFIC ISSUES in an Advanced Information Society
ITC-11
Minoru Akiyama (Editor)
Elsevier Science Publishers B.V. (North-Holland)

INTEGRATED TOOL FOR COMPUTER NETWORK SYSTEM CONFIGURATION PLANNING

Hiroyuki KITAJIMA and Kazuhiko OHMACHI

Systems Development Laboratory, Hitachi, Ltd.
Kawasaki, Japan

ABSTRACT

One of the most important tasks in developing computer network systems is system configuration planning. This paper presents a man-machine interactive tool named ISCP/NET (integrated tools for system configuration planning / network) which assists system designers in optimizing computer network system configuration from the viewpoints of performance, reliability and cost. ISCP/NET features the following : 1) Communication sub-networks and computer systems are handled simultaneously; 2) A new capacity assignment method is proposed which can assume discrete capacities, as well as produce the cost/effective assignment to satisfy user requirements even in partial failure.

1. INTRODUCTION

One of the most important tasks in developing computer network systems is system configuration planning to obtain a cost/effective network system configuration. This activity is increasing in difficulty as computer networks increase in scale and deal with more varied service requests. There has been much research on network system optimization (1). However, the following considerations should be noted:

1) Although the research has mainly dealt with performance, reliability is another important aspect to consider.

2) Research has been confined to the planning of communication sub-networks. However, in order to evaluate and/or optimize the performance and reliability to end users, it is desired that the tool can deal with communication sub-networks and computer systems at the same time.

3) Concerning the problems of flow assignment and capacity assignment, most previous research has tried to minimize the mean response time in communication subnetworks for all service requests. However, because each service request may have its own response time and throughput requirements, these should be optimized.

4) As for the capacity assignment, most previous research has assumed continuous capacity (2) which is not realistic in most cases.

This paper presents an integrated method called ISCP/NET (integrated tools for system configuration planning / network) which assists system designers in configuration planning of computer network systems. The main features of ISCP/NET are the following :

a) ISCP/NET is able to deal with communication sub-networks and computer systems at the same time, evaluate performance and cost as well as reliability, and optimize flow assignment and capacity assignment.

b) A new method of flow assignment is proposed which consider response time and throughput for all types of service requests.

c) A new capacity assignment method is proposed which assumes not only discrete capacities but partial failure.

d) The above proposed methods are implemented as a man-machine interactive computer aided design tool with a graphic display.

2. OUTLINE OF ISCP/NET

The planning and design procedure using ISCP/NET is as follows (see Fig. 1).

1) Analysis

Users of ISCP/NET first input the initial plan of the system configuration as well as the data on projected system load. In case of system expansion, the current system configuration and system load data are input. Performance, reliability and cost indices for the initial plan are predicted using evaluation models and basic data on performance, reliability and cost, which have been stored in the tool's data table in advance.

2) Synthesis

Given performance and reliability targets, ISCP/NET autmatically generates an improved plan from either flow assignment, capacity assignment or function assignment.

Based on the above results as well as their own judgements, users create the optimal system configuration.

The input and output in terms of ISCP/NET are summarized in Table 1. The ISCP/NET's characteristics from the viewpoints of utility are as follows :

a) Model base

In terms of computer networks' nodes (e.g. computer systems, communication processors) ISCP/NET users can define and store various kinds of performance and reliability models in this tool, as well as reuse and modify them. This enables the common use of nodes' models. Besides, some standard models are already stored in this tool.

b) Components' data base function

In this tool, performance and reliability values of various types of hardware and software components can be stored. Users can define their initial system configuration by first selecting the appropriate model from the model base in a), and then assigning the type and number to each component in the model.

c) Interactive functions

Model building as described in a) and b) is carried out interactively through a character display and a keyboard. The estimation and optimization results are displayed along with a schematic diagram of the system configuration on a graphic terminal. If a graphic terminal is not available, characteristic output is also possible.

3. ANALYTICAL FUNCTIONS

3.1 Performance Evaluation Model

This model estimates the flow pattern in the communication sub-network and the performance with regard to the whole system including computer systems.

1) network performance evaluation model

The key to flow pattern estimation in the communication sub-network is the modeling of routing policies, e.g. adaptive routing, virtual circuit and parmanent circuit. Although most of the practically used routing policies are sophisticated, simulation seems unsuitable because it requires too much computing time for a large network. Therefore, each typical routing policy is formulated as a nonlinear multicomodity network flow problem, and approximately solved through an incremental assignment (IA) method. The IA method is a kind of unconstrained search technique for a nonlinear network flow problem with convex objective function (3). As an example, the IA method is applied to the flow pattern estimation for adaptive routing in the following way :

(Phase 1)

(i) Add a directed dummy link to every pair of nodes in the network, and set its delay time to a very large value.

(ii) Start from the state without any flow in the network.

(iii)For each type of service requirement **t**, assign ($0<\varepsilon<1$) of its required throughput λ_t (transaction/sec) to the minimal path from t's origin to destination. As this results , the flow and, thus, the delay of each node and link in the network are modified.

(iv) Repeat step (iii) $1/\varepsilon$ times.

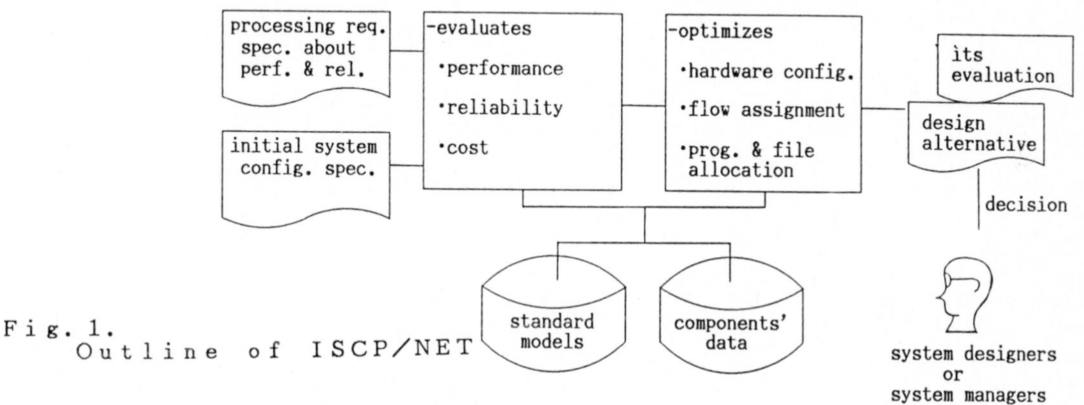

Fig. 1. Outline of ISCP/NET

Table 1 Summary of Input and Output

		Performance	Reliability
INPUT	req. spec.	each service req.'s · required throughput & response time	each service req.'s · allowable unavailability
	first spec. of syst. config.	System config. of · network(s) · computer systems if necessary	
	components' spec.	Hardware · speed (MIPS etc.) · allowable utilization (%) Software · dynamic steps , · # I/O	components' · MTBF · MTTR
OUTPUT	evaluation	· avg. response time · utilization (%)	· unavailability for service req. · performance degradation in failure
	alternatives	· hardware configuration · flow assignment of each service req. in network(s) · prog. & file allocation	

(Phase 2)

(v) For each type of service request t, reduce the amount of t's flow in every link l, f_{lt}, to $(1-\delta)f_{lt}$.

(vi) For each t, assign $\delta \cdot \lambda_t$ to the minimal path from t's origin to destination. As a result of this, the flow and the delay of each node and link in the network are modified.

(vii) Repeat(v)-(vi) until no significant (or greater than the threshold value) change in the flow pattern occurs.

(viii) From the obtained flow assignment, not only throughput f_t and mean response time R_t in the network for each t but also the utilization of each node and link are easily calculated.

(2) computer systems performance evaluation model

The mean response time for each type service request and the utilization of each component in each computer system in the computer network are estimated by an open queueing network model. Here, the system is assumed to be modeled as a network of servers, each of which is expressed in terms of an analytical queueing model such as M/M/s/n, M/G/1 and G/G/1. Although this assumption is not correct, the authors' and others' experiences with this model (4) have convinced them of its usefulness.

By simultaneously using the above network as well as computer systems parts, the response time and the throughput in the whole system for each type of service request can be estimated within a short computing time.

3.2 Availability Evaluation Model

This model estimates the availability of each type of service request. The model expresses an evaluated system with the probability network. Each of the nodes and the links in a network represents a system component such as cpu, disk, communication processor and communication link, and is provided with the availability value, i.e. MTBF/(MTBF+MTTR), of the corresponding component.

For each type of service request t, the set of possible paths from its origin to destination are enumerated (5). Since the availability of each component is generally close to 1.0, the calculation of t's availability based on the path sets is prone to generate a computational error. Therefore, the path sets for t are first transformed into the minimal cut sets, and the inclusion and exclusion formula is applied to this result to produce an estimation of t's availability.

4. SYNTHETIC FUNCTION

4.1 Flow Assignment Model

Except for the networks employing dynamic routing policies where route(s) for each type of service results is(are) dynamically updated in order to balance the load in the network, the routing table must be updated corresponding to the changes in the traffic requirements on the network. As stated in the introduction, most of the previous research has aimed to minimize the mean response time of all service requests, while this model is interested in satisfying the throughput as well as response time requirements for each type of service request. The flow assignment model is formulated as follows :

$$\text{minimize} \sum_t w_t(1-x_t/\lambda_t) \quad (1)$$

$$\text{with respect to } (x_{et}) \quad (2)$$

subject to

$$R_t \leq R_t^+, \qquad t=1,\dots,T \quad (3)$$

$$\sum_t x_{et} \triangleq x_e \leq Q_e, \; e=1,\dots,E \quad (4)$$

$$\sum_{e\in\delta^+(v_k)} x_{et} - \sum_{e\in\delta^-(v_k)} x_{et} = \begin{cases} f_t & (v_k=O_t) \\ 0 & (v_k\neq O_t,D_t) \\ -f_t & (v_k=D_t) \end{cases} \qquad t=1,\dots,T \quad (5)$$

$$0 \leq f_t \leq \lambda_t, \qquad t=1,\dots,T \quad (6)$$

where

- t : the type of service request
- w_t : the expected amount of damage caused when a portion of t's throughput requirement is not attained
- x_t : t's throughput
- λ_t : t's throughput requirement
- R_t : t's mean response time in the network
- R_t^+ : t's response time requirement
- T : the total number of t
- e : the index of element (node or link)
- E : the total number of e
- x_{et} : t's throughput in e
- Q_e : e's capacity
- $\delta^+(v_k)$, $\delta^-(v_k)$: the set of links inputting and outputting from node v_k, respectively
- O_t, D_t : t's origin and destination, respectively.

The above (in)equations represent the constraints of response time (3), link and node capacity (4), flow conservation (5) and throughput (6). The above is a nonlinear multi-commodity network problem, with convex objective function and constraint functions, and to which the IA method is again applied. To apply the IA method, the above problem has to be transformed into unconstrained optimization problem whose objective function is convex in terms of x_{et}. Moreover, if the partial derivative of the objective function in terms of x_{et} is expressed with the weight of the link or node, efficient calculation using a shortest path search technique can be employed. Note the following fact :

$$R_t = \Big(\sum_{e=1,E} x_{et}\, R_{et}(x_e)\Big)/x_t \qquad t=1,\dots,T \quad (7)$$

where $R_{et}(x_e)$ is the mean system time (i.e. the mean service time and the mean wait time) for t in e. For example, if element e behaves like M/M/1, $R_{et}(x_e) = 1/\mu_e(1 - x_e/Q_e)$ where μ_e^{-1} is the mean service time of e. From the above observation, problem (1)-(6) is modified as follows :

$$\text{min. } G(x_{et}) = \sum_{e=E+1,\ E+M} \sum_{t} (w_t x_{et}/\lambda_t) + \sum_{t=1,\ T} \pi_t(p, R_t) + \sum_{e=1,\ E} \psi_e(q, x_e) \quad (8)$$

In (8), (e, e= E+1, E+M) denote directed dummy links between every pair of nodes in the network for dealing with the unattained throughput, and each of whose weight value is w_t for t, t = 1,...,T. $\pi_t(p, R_t)$ is penalty function in terms of the response time constraint for t with q being a positive parameter. $\pi_t(q, R_t)$ is convex in terms of x_e and

$$\lim_{p\to\infty} \pi_t(p, R_t) = \lim_{p\to\infty} \partial\pi_t/\partial x_e \quad \begin{cases} = 0 & R_t \leq R_t^+ \\ = \infty & R_t > R_t^+ \end{cases} \quad (9)$$

$(1/p)\exp(p(R_t-R_t^+))$ is an example of $\pi_t(p, R_t)$. $\psi_e(q, x_e)$ is the penalty function in terms of e's capacity constraint and q is a positive parameter.

$\psi_e(q, x_e)$ is convex in terms of x_e and

$$\lim_{q\to\infty} \psi_e(q, x_e) = \lim_{q\to\infty} \partial\psi_e(q, x_e)/\partial x_e \quad \begin{cases} = 0 & x_e \leq Q_e \\ = \infty & x_e > Q_e \end{cases} \quad (10)$$

An example of $\psi_e(q, x_e)$ is $(1/q)\exp(q(x_e-Q_e))$. The IA method is applied to problem (8)-(10) as follows :

(Phase 1)

(i) Between every pair of nodes in the network, place directed dummy nodes with very large delay time values.

(ii) Start from the state where $f_t = 0$ for every t.

(iii)Calculate the partial derivative of $G(x_{et})$ in (8) for each x_{et}, e = 1,..., E, t = 1,..., T. Then make it the weight value of the respective link and node. In the obtained weighted network, for each t, t=1,...,T assign the traffic, $\varepsilon\cdot\lambda_t$, to its shortest (i.e. minimum weight sum) path, where $0 < \varepsilon < 1$.

(iv) Repeat step (iii) $1/\varepsilon$ times.

(Phase 2)

(v) For each t and e, reduce $\delta\cdot x_{et}$ from t's throughput in element e, where $0 < \delta < 1$.

(vi) Calculate the partial derivative of $G(x_{et})$ for each x_{et} , e = 1,..., E, t = 1,..., T. Next, make it the weight value of the respective link and node. Then, for each t, t=1,...,T, assign the traffic, $\delta\cdot\lambda_t$, to its shortest path.

(vii)Repeat (v)-(vi) until no significant (or greater than the threshold value) change in the flow pattern occurs.

(viii)From thus obtained flow assignment, performance measures such as throughput f_t, t = 1,...,T, mean response time R_t, t = 1,...,T and the utilization of each node and link are easily calculated.

4.2 Capacity Assignment Model

This model aims to obtain the capacity for each component (link and node) in the network such that the total cost of links and nodes is minimized while satisfying performance requirements. As stated in the introduction, the main features of this model are that (1) this model assumes discrete capacity, and (2) this model can calculate the capacity assignment which is able to satisfy performance requirements in any case of specified partial failure modes. First, the model for the case without failure is described, and then the model which can consider specified partial failure.

4.2.1 Capacity assignment model

This model is formulated to :

$$\text{minimize} \quad Z = \sum_i \sum_j y_{ij} C_{ij} \quad (11)$$

with respect to (y_{ij}) (12)

subject to

$$y_{ij} = \begin{cases} 1 & \text{if } j \text{ is assigned to } i \quad (13) \\ 0 & \text{otherwise} \quad (14) \end{cases}$$

$$i = 1,\dots, I,\ j = 1,\dots, J$$

$$\sum_j y_{ij} = 1, \qquad i = 1,\dots, I \quad (15)$$

$$R_t = \sum_{i=1,\ I} (x_{it} R_{it})/\lambda_t \leq R_t^+ \qquad t = 1,\dots, T \quad (16)$$

where

C_{ij} : the cost of jth capacity for element (node or link) i in the network, and is assumed that $Q_{ij} > Q_{ij+1}$ and $C_{ij} \geq C_{ij+1}$ for i=1,...,I, j=1,...,J-1

R_t, R_t^+, λ_t : see (1), (3) in 4.1

x_{it} : t's throughput in i

In the above, each element i must select one of its capacity alternatives. This problem is a nonlinear binary programming problem with multiple choice constraints. An exact solution called a lexicographical enumeration method (6) is applicable to this programming problem if (i) the number of variables is not large and (ii) any of the objective functions and the right- and left- hand sides of the constraints are either non-decreasing or non-increasing for the increase of variable's value (i's capacity in this case).

In (11)-(16), $\sum_i\sum_j y_{ij}C_{ij}$ and $\sum_j y_{ij}$, i =1,...,I satisfies the above condition (ii). This can be also assumed for R_t in (16) in many cases. For example, if every element i in the system is assumed to be an M/M/1 server, then $R_{it} = \mu_{it}^{-1} (1 - \sum_t x_{it}/ Q_i)^{-1}$ where μ_{it}^{-1} is i's mean service time, and Q_i i's capacity. In this case, it is obvious that $\partial R_t/\partial Q_i \leq 0$ and therefore R_t is non-increasing for Q_i's increase. From the above observation, ISCP/NET employs the lexicographical enumeration method for capacity optimization of a small system.

However, this exact solution takes so much computing time for a large system, thus, a heuristic algorithm was newly developed to obtain suboptimal capacity assignments in reasonable computing time even for a large system. This algorithm consists of three phases as follows ;

Heuristic algorithm

(Phase 1) capacity increase for throughput requirements

(i) Apply the flow assignment procedure in 4.1 under the current capacity assignment $C = (c_1, \cdots, c_n)$ where n is the number of elements in the network. If the routing policy of the network is dynamic (i.e. having the capability of loadbalancing by itself), the flow pattern estimation procedure in 3.1 is applied instead.

(ii) Let $T' = (t_1', t_2', \cdots)$ denote the types of service requests whose throughput requirements are not met. If $T' = \emptyset$, go to (vi). Otherwise, for each $t_i' \in T'$, identify the string of elements on t_i' 's path, $E_i' = (e_{i1}', e_{i2}', \cdots)$, whose utilizations are close to their acceptable limits.

(iii)For each string of elements in (ii), E_i', do the following : Increase the capacity of every element by one grade, and apply the flow assignment (or flow pattern estimation if the network is dynamic). If any element of E_i' has no upper grade, the problem is infeasible and stop the procedure. Otherwise, calculate sensitivity measure α_i such that ;

$$\alpha_i \triangleq \sum_{t_i' \in T'} w_i'(\Delta \lambda_i')/\Delta P_i' \tag{17}$$

where $\Delta\lambda_i'$ is t's throughput increase, $w_i'(\Delta\lambda_i')$ is the amount of damage if $\Delta\lambda_i'$ is not attained, and $\Delta P_i'$ is the increase of cost by upgrading E_i's capacity.

Downgrade E_i's capacity to C in (i).

(iv) Identify the maximum α_i in (iii), and increase the capacity of corresponding E_i' by one grade. Set the thus modified capacity assignment to C.

(v) Repeat i)-iv) until the throughput requirement is met for every $t \in T$.

(Phase 2) capacity increase for response time requirements

(vi) Apply the flow assignment (or flow pattern estimation), and let $T'' = (t_1'', t_2'', \cdots)$ be the types of service requests which do not satisfy the response time requirements. If $T'' = \emptyset$, go to phase 3. Otherwise, identify the elements on the paths of T'' whose wait time is large, and let them be $E'' = (e_1'', e_2'', \cdots)$.

(vii) For each $e_i'' \in E''$, increase its capacity by one grade and apply the flow assignment (or flow pattern estimation). If every $e_i'' \in E''$ has no upper grade, stop the procedure because the problem is infeasible. Otherwise, calculate sensitivity measure β_i such that ;

$$\beta_i \triangleq \sum_{t_j'' \in T''} \lambda_j \cdot \Delta R_j''/\Delta P_i \tag{18}$$

where λ_j is the throughput of t_j, $\Delta R_j''$ is the amount of response time decrease for t_j'' and ΔP_j is the increase in cost by upgrading e_i'' 's capacity.

Decrease e_i'' 's capacity by one grade.

(viii) Identify the maximum β_i in (vii), and increase the capacity of corresponding e_i'' by one grade.

(ix) Repeat (vi)-(viii) until every $t \in T$ satisfies its response time requirement.

(Phase 3) capacity decrease satisfying capacity and response time constraints

(x) Apply the flow assignment (or flow pattern estimation) and let $E^* = (e_1^*, e_2^*, \cdots)$ be the elements in the system with small wait time. If $E^* = \emptyset$, then go to (xiv).

(xi) For each $e_i^* \in E^*$, decrease its capacity by one grade and apply the flow assignment (or flow pattern estimation). Calculate sensitivity measure γ_i such that ;

$$\gamma_i \triangleq \sum_{t_j \in T} \lambda_j \cdot \Delta R_j/\Delta P_i \tag{19}$$

Here, λ_j is t_j's throughput, ΔP_i is the decrease of cost by downgrading e_i^*'s capacity, and

$$\Delta R_j \begin{array}{ll} = R_j^+ - R_j & \text{if } R_j \le R_j^+ \\ = \infty & \text{if } R_j > R_j^+ \text{ or } \lambda_j < \lambda_j^+ \end{array} \tag{20}$$

where R_j is t_j's mean response time, and R_j^+ and λ_j^+ are R_j's and λ_j's target values, respectively.

Increase e_i^*'s capacity by one grade.

(xii) Identify the minimal γ_i in (xi), and decrease the capacity of corresponding e_i^+ by one grade.

(xiii) Repeat (x)-(xii).

(xiv) Stop.

It is obvious that the above heuristic algorithm provides us with a suboptimal capacity assignment as well as performance measures such as mean response time and throughput to each type of service request, and utilization of each element, the flow assignment (or flow pattern) and the total cost.

4.2.2 Capacity assignment model (with failure)

One of the reasons why a mesh configuration is preferred to a hierarchical one is its fault tolerance ability. The capacity assignment model was newly proposed in order to obtain a suboptimal capacity assignment that can satisfy the performance requirements in any prespecified failure mode. Here, a failure mode is identified with the set of elements which fail in the failure mode. This model's procedure is as follows :

(i) Apply the capacity assignment model (in 4.2.1) assuming no failure, and calculate the expected amount of damage caused by each of prespecified failure modes such that ;

$$\delta_k \triangleq \lambda_k/\mu_k \sum_{t_i \in T} f_{ik}\, d_i\,, \quad k = 1, \cdots, K \tag{21}$$

where δ_k is the expected amount of damage caused by failure mode k, λ_k is the k'th failure rate, μ_k is the k's repair rate, t_i denotes the ith type of service request, T denotes the set of t_i, f_{ik} is the amount of t_i's flow in the set of elements assumed to fail in failure mode k, and d_i is the amount of damage in case a unit of t_i's throughput requirement is not satisfied.

(ii) The prespecified failure modes (k) are arranged into (k^*) in order of (δ_{k^*}) . Then, in this order, do (iii).

(iii) Generate k^*th failure mode, and apply the phase 1 and 2 of the capacity assignment model in 4.2.1. If any capacity increase occurs as a result of this, modify the system's capacity. Restore the elements' capacity corresponding k^*th failure mode to the state just before generating the k^*th failure mode.

5. APPLICATION

Some of the results obtained by applying ISCP/NET to a hypothetical problem are briefly described as follows. : The hypothetical designers' first system configuration and the users' requirements are shown in Fig. 2 and Table 2, respectively. For simplicity, the system is assumed to consist of only communication sub-network which takes the adaptive routing policy. The performance evaluation results are shown in Fig. 2. The results of capacity assignment assuming no failure as well as those assuming that any one link in the network may fail are shown in Fig. 3 and Fig. 4, respectively.

6. CONCLUSIONS

The tool presented in this paper has been applied to many systems in their initial development and evolution stages, and proved to save many man-hours in system configuration tasks as well as to improve the cost/effectiveness of the system configuration. This tool is expected to play an increasingly important role as computer network systems continue to increase their size and complexity, the presented tool is expected to play a more and more important role.

ACKNOWLEDGMENTS

The authors would like to acknowledge the support of Dr. Jun Kawasaki, general manager of Systems Development Laboratory, Hitachi, Ltd., and Dr. Koichi Yamamoto, Mr. Seiji Kobayashi, Mr. Yasufumi Fujii, Mr. Shoichiro Yamaguchi of Hitachi, Ltd., and Mr. Tohru Ohno of Hitachi Information Networks, Ltd. for their useful comments on this work.

REFERENCES

(1) Karlgaard, D.G., et. al., "ASSET - A Set of Automated Network Design and Performance Management Tools", IEEE Cmputer '82.

(2) Kleinrock, L., "Queueing Systems-Vol.II", John Wiely & Sons, 1976.

(3) Fiacco, A.V. & McCormick, G.P., "Nonlinear Programming: Sequential Unconstrained Minimization Techniques", John Wiely & Sons, 1968.

(4) Samari, N.K., et. al.., "The Analysis of Distributed Computer Networks using M/D/r & M/M/1 Queues, Proceeding of Distributed Computing Conference", pp.143-155, 1979.

(5) Abraham, J.A., "An Improved Algorithm for Network Reliability", IEEE Trans. Reliability, vol. R-28, April 1979.

(6) Sasaki, R., et. al., "A New Technique to Optimize System Reliability", IEEE Reliability, vol. R-32, no.2, June 1983.

(7) Kitajima, H., "A Method for Hierarchically Distributed Computer System Design Minimizing Cost under Response Time Constraints", Journal of Electrical Engineering of Japan, vol.104, no.5, May 1984. (in Japanese)

(8) Kitajima, H., "A method for Optimally Allocating Functions to Distributed Computers Considering Availability Constraints and Independence among Functions", ibid vol.103, Nov. 1983. (in Japanese)

Table 2 Service requests

origin	dest.	throughput req. (trans/sec)	message length (bit)	allowable response (sec)	overflow ratio* (%)
A	G	0.66	3250	7	14
B	G	0.28	4060	5	57
I	G	0.4	3070	5	14
H	G	1.35	3100	4	14
E	H	0.25	3400	9	57
E	G	0.46	4060	9	57
F	H	0.1	3440	7	57
F	G	0.7	3440	7	57
D	G	0.2	3400	6	14
C	G	0.7	3400	6	14
B	C	0.6	3440	5	57
J	K	0.92	3440	3	14

* overflow for initial configuration

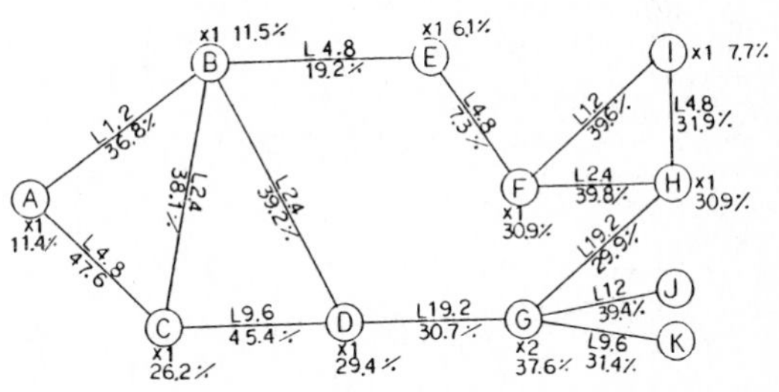

L + number : the speed of the communication line (kbps)
x + number : the assigned number of the component
% : utilization of the component
↑ : change from the initial assignment

Fig.2 Initial configuration & its evaluation (hypothetical system)

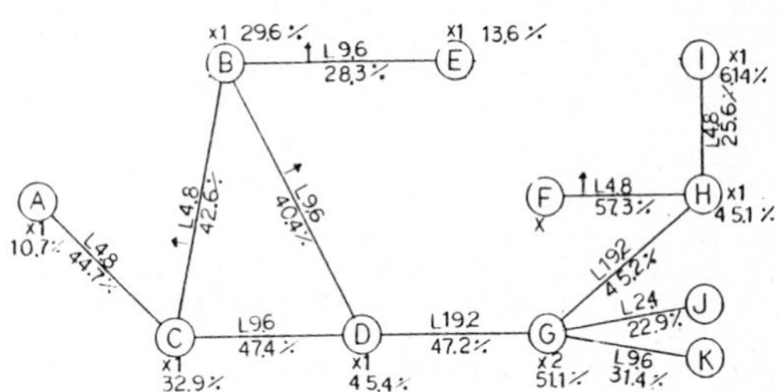

Fig.3 Capacity assignment assuming no failure

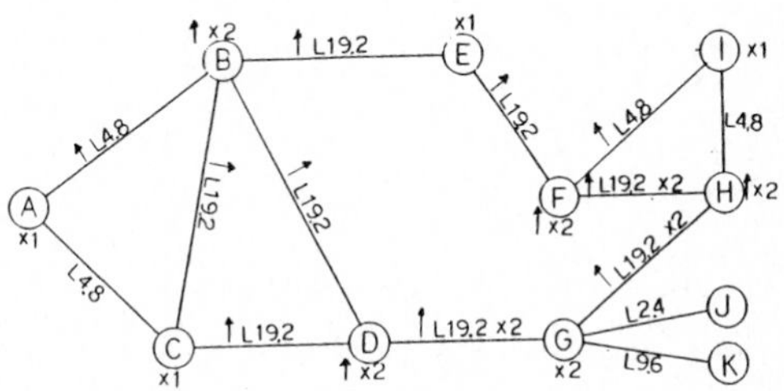

Fig.4 Capacity assignment assuming failure

TELETRAFFIC ISSUES in an Advanced Information Society
ITC-11
Minoru Akiyama (Editor)
Elsevier Science Publishers B.V. (North-Holland)

COMPUTER-AIDED PLANNING OF PUBLIC PACKET-SWITCHED NETWORKS

A.BURATTIN, U.MAZZEI, C.MAZZETTI

SIP - Direzione Generale - Via Flaminia,189 00196 Roma (Italy)

ABSTRACT

The pressure of the expanding data communication market and the emerging competition in providing value added services lead telecommunications operating companies in most countries to deploy nationwide public packet networks; the need then arises for a cost effective planning and design of the relevant network structures.

The computer-aided planning tools presented in this paper enable the planner to find an optimal placement of packet switches and concentrators, and as well to dimension the interswitch trunk network, thus attaining a minimum-cost solution through an iterative procedure. The heuristics adopted in this approach overcome the intrinsic difficulties of the theoretical optimization and match the main practical constraints and requirements.

The flexible use of this planning methodology is stressed by some case study results relevant to the Italian territory for different scenarios of network size and throughput of the packet switches.

1. INTRODUCTION

A great emphasis is presently placed by telecommunication operating companies on the deployment of public packet switched networks, as adequate solutions to comply with most of the expected growing needs for non-voice services and to effectively compete in the expanding data communications market. In addition to data and text transmission, packet networks can also provide a suitable infrastructure to support other emerging services, like message handling and value added services, thus well positioning the operating companies to cope with the private networks competition and to enlarge their presence beyond the role of leased lines providers.

From the network standpoint, the basic advantage of packet switching relies on the efficient use of long-haul transmission facilities through the statistical multiplexing of the data messages. Main advantages of packet switching from the user viewpoint are that it allows subscribers using different protocols and data speeds to transparently communicate with each other, and that the tariffs are primarily based on the volume of transmitted information, rather than on holding times and distances. That makes the packet mode particularly attractive for bursty and interactive data transmission, while circuit switching (especially in a digital environment) will remain competitive for bulk data transfer, like computer-to-computer communication, slow-scan video and fast facsimile.

Packet networks are initially implemented as autonomous from the existing telephone networks, with access provided via both dial-up and dedicated lines; it is worth noting that this arrangement does not contradict the worldwide foreseen evolution toward the ISDN.

In fact, the first stage of commercial ISDN implementations (late '80s) will be in most countries characterized by the deployment of few ISDN local switches allowing a selected stratum of subscribers to access a wide range of services on different network facilities, including packet switching, through digital integrated loop arrangements (144 Kbit/s and 2 Mbit/s). Even in a more mature stage (during '90s), when the extension of ISDN capabilities will be such to gradually attract on it all the packet users, the backbone part of the packet network will remain for many years a specialized resource actually autonomous from the circuit switched facilities, until new integrated packet/circuit techonologies at trunk level will emerge.

Based on these considerations, a long cycle of life for the public packet network infrastructure can be foreseen with a short-medium term rapid growth in terms of user number, traffic amount, territorial extension; consequently, the need arises for a cost-effective planning and dimensioning.

The paper presents a planning metodology (Sect.2) and a computer tool (Sect.3) specifically intended for optimizing the design of nationwide packet networks. The case study results relevant to the Italian territory (Sect.4) enlights the flexible and practical use of this approach.

2. PLANNING METHODOLOGY

The adopted approach reflects the typical environment of a telecommunications operating company, i.e. where:

- a forecast is given of the amount and the territorial distribution of the actual or potential demand, and as well a traffic model and a service quality objective are defined;

- cost figures, modularities and performances of available packet switches and concentrators are given;

- transmission facilities, at both subscriber and trunk levels, are the same currently used in the telephone network, so that their cost figures and topology layouts are known.

The planning problem then basically consists in searching for the minimum-cost structure of the overall network satisfying the subscriber demand and the performances requirements.

2.1. Network structure

The basic architecture of a packet network is made up of two main parts: access network and interswitch network, as in fig.1.

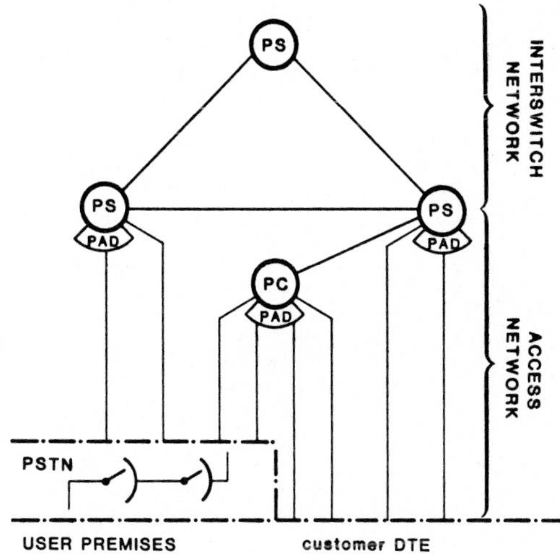

Fig.1 - Packet network architecture

The access network provides the subscriber data terminal equipments (DTE) with a dial-up or a dedicated line access to the packet switches (PS), where switching and routing functions are performed; the large costs relevant to the provisioning of individual subscriber lines can often justify the use of remote packet concentrators (PC), where packets from different users are statistically multiplexed into a single data channel connecting the PC equipment to a parent PS node.

Apart from the different transmission speeds, the DTEs fall into two broad categories: packet DTEs directly interfaced to the packet network by the standard X.25 protocol, and non-packet DTEs (BSC/SDLC and X.28 are the most diffused types), for which a packet-assembly-disassembly (PAD) function is required; the PAD conversion can be suitably provided by the network itself as integrated in the PC and/or PS equipments. The X.28 DTEs have a low speed of transmission in the range 300-1200 bit/s, while packet DTEs have medium-high speed ranging from 2.4 to 64 Kbit/s.

The interswitch network provides the interconnection among access PS nodes by means of dedicated trunk circuits usually operated at 9.6 or 64 Kbit/s. Two basic alternatives for the interswitch network structure can be considered:

a) meshed structure, when at least one trunk circuit connects each pair of access PS nodes;

b) tandem structure, when additional PS nodes provide a transit switching function and each access PS is connected to at least one transit PS; besides, traffic patterns can economically justify direct links between pairs of access PSs, just as currently practiced on the telephone network.

In general, the meshed structure is the most economical solution when only a few access swhitches are placed, while the tandem structure proves in for large size networks.

An intermediate solution may consist in incorporating the transit switching function into the access PS nodes having spare capacity: this arrangement can be seen as an evolutionary step for passing from the meshed to the tandem structure rather than a steady-state network structure.

2.2. Design inputs

The basic input of every network design obviously regards the subscribers to be served, in terms of number of subscribers, their distribution on the territory and relevant traffic characteristics.
As the packet network is accessed by the users (both in the dial-up and in the dedicated mode) through existing loop and local trunk facilities and also considering the forecast uncertainties, it is useful that the subscriber distribution is referred to quite large portions of territory and consistently with the structure of the telephone network. As an example, the results of Sect.4 are carried out starting from the number of subscribers forecasted in each of the 231 District areas of the Italian telephone network.

Moreover, either for the whole territory or for each subarea, the percent distribution of user DTEs must be given between packet and non-packet terminals and between dial-up and dedicated access, and as well their distribution in the different classes of transmission speeds. Beside the dimensioning of equipments providing the PAD function and the dial-up access, these input data sensibly affect the cost of the access lines due to the wide range of cost of the modems at different speeds.

A key point of network dimensioning is, in addition, the definition of the subscriber traffic characteristics, tightly correlated to the DTE types and to the exploited services. Current measurements and experiences on existing packet networks indicate that reference values of peak-hour average traffic in packet per second are of the order of 1 p/s and .1 p/s for packet and non-packet terminals respectively.

As far as network elements are concerned, the main input data are the capacities of the adopted packet switches and concentrators in terms of maximum traffic (throughput in p/s) and maximum number of ports (subscriber accesses and trunks); relevant costs can be espressed as a linear function of the number p of ports in the form C_o+c.p, when C_o is the start-up cost of of the equipment and c the incremental cost.

Transmission costs per circuit can also be expressed as linearly varying vs. the link length d_{km} in the form A+B.d_{km}, where A represents a fixed component (multiplexing, line terminals, etc.) and B the sum of repeater equipments and physical bearers costs per Km. For planning purposes it is appropriate to average those cost figures on the different transmission facilities actually adopted in the telephone network, where different A and B costs are to be used for 9.6 Kbit/s or 64 Kbit/s trunk circuits.

The dimensioning of the trunk circuits consists in defining for earch link the optimal number of circuits as a function of the relevant packet traffic; for moderate traffic flows often a single circuit is sufficient and only the choice among the allowed speeds (e.g. 9.6 or 64 Kbit/s) might be needed. In any case the link dimensioning calls for considering the queue process of the packet at the input of the transmission channel on which they are statistically multiplexed; in fact the time delay involved in this queue process is the measure of the channel performances in transferring data packets, just as the loss probability equivalently measures the performances of a telephone trunk group in handling voice traffic.

Under appropriate assumptions regarding the statistics of the packet length and the arrival process of thé packets, the average packet delay T can be simply computed [1] as

$$T = L/(C-RL)$$

where

C= capacity of the transmission channel (bits/s)
R= packet traffic rate (p/s)
L= average packet length (bits/p)

This formula shows that the average delay T goes to infinity as the channel utilization factor RL/C approaches the unity; a good trade off of transmission cost vs. delay performance is usually achieved by setting this utilization factor around the value .5. Moreover, it must be considered that the traffic rate R also includes the additional packets generated by the network for its own operational needs, like billing, maintenance, etc.; starting from the user traffic R_u, the total traffic R can be roughly computed as $R = K R_u$, where the coefficient K can reasonably vary in the range 1.1 - 1.3.

2.3. Optimization approach

The optimization problem, classically modeled as the search for the minimum-cost structure of the network, basically consists of two interconnected parts: i) placement of switches and concentrators, and homing of the customer lines, ii) design of the trunks and the transit switches of the interswitch network. Several optimization algorithms have been proposed in the literature, mainly based on linear programming techniques for i), and on the application of graph theory and multicommodity flows for ii).

Apart from the inherent complexity of those algorithms especially when applied to large size networks, the main limitation of those theoretical approaches is that they prevent the use of detailed input data and the respect of the many constraints encountered in a practical environment. To overcome these difficulties, a more pragmatic way is here followed, allowing the planner to easily generate and explore different feasible solutions, so that the overall optimality is ultimately attained through an iterative process of cost comparisons. This methodology can flexibly incorporate a large range of practical constraints and it is implemented by the interactive computer procedure described in the following section.

On addressing firstly the problem of placing PCs and PSs on the territory, it is evident that the candidate nodes to be considered are coincident with the telephone network centers corresponding to the elementary subareas (districts) for which the user forecasts are given; in fact, subscriber loops and local transmission facilities of the existing telephone network are starly oriented toward these centers, being natural points of aggregation. These centers lie in turn on the backbone transmission network and it is then clear that the transmission facilities layout constitues the main constraint in determining PC and PS locations among the candidate nodes in order to minimize transmission costs.

With reference to the scheme of fig.2, the concept of serving area can be properly applied by defining:

- a concentration area (CA), as the aggregate of adjacent districts served by one or more PCs located in a single node;

- a switching area (SA), as the aggregate of adjacent CAs served by one or more PSs located in a single node; no PC equipment could be needed in the CA where the switching node is present, unless for PAD function not being provided by PS.

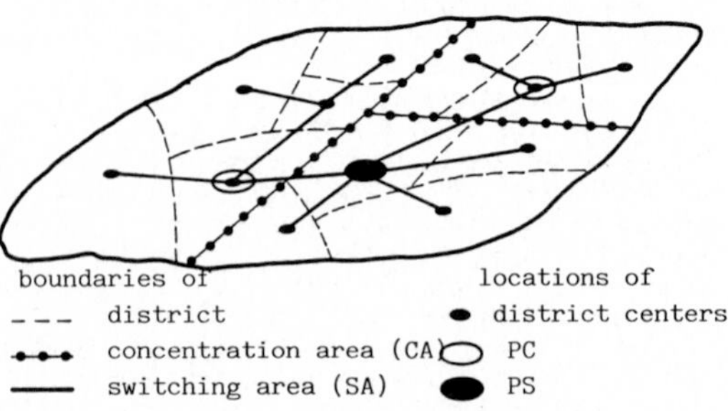

Fig.2 - Network model for placing PCs and PSs

If the backbone transmission layout is given as a tree topology, different patterns of merging the subareas to form CAs and SAs according to this topology can be simply generated by the planner by varying two thresholds TR_{CA} and TR_{SA}, relevant repectively to a fill factor of PC and PS equipments, in terms of number of access ports or of traffic capacity. If, for example, according to the forecasted user traffic the actual capacity of the adopted PC equipment is limited by the available number of ports rather than by its throughput, the threshold TR_{CA} should fix a minimum number or users, and the merging process for constructing each concentration area CA is stopped as soon as the sum of the users of the relevant subareas exceedes TR_{CA}. The same process applies to the construction of SAs by the threshold TR_{SA} and starting from a predetermined solution for the concentration areas. A similar heuristic method has been adopted in [2] .

Note that practical contraints could in certain cases prevent to home to the PC equipment all the users of the relevant AC area (e.g.user having high-traffic or special protocol interfaces only provided by PS); in such cases, special "homing criteria" can be incorporated in the previous procedure.

Once a feasible solution for PC and PS placement has been generated, the design of the interswitch network can be carried out. A basic input is the traffic matrix among the packet switches: when reliable forecasts are lacking, a simple rule is to assume the traffic between two PSs directly proportional to the product of their amounts of generated traffic.

As for the access network, different structures of the interswitch network can be generated under the planner's control; two variable inputs are here used:

- the number TS of tandem switches to be placed in addition to the access switches only for performing the transit switching function (TS=0 in the special case of the meshed structure); the locations of these tandem switches can be simply chosen looking at the most important nodes of the network;

- the average cost per packet C_s of tandem switching; by varying the value C_s, the economy of direct trunks between pairs of access switches can be varied, so that, for a given TS, different trunking configurations can be compared.

A great simplification in the dimensioning of the interswitch network can be gained by externally fixing proper criteria for routing the traffic. As in normal operating condition both the cost and the time delay performances are deteriorated by multiple tandem switching of the traffic, a suitable criterion is to allow at most a single transit for all traffic parcels; a routing matrix can then specify for each pair of nodes which tandem is designated to switch all the relavant traffic or the overflow from the direct trunk.

In building this routing matrix a balanced distribution of traffic load on the transit switches must be attempted; by this way the fill factor of the transit switches is easily controlled only by varying the cost parameter C_s in order to prevent the saturation of their throughput capacities by means of direct trunks.

Note that by this design approach other requirements might be met regarding network performances. Beside the use of the channel utilization factor defined in Sect.2, time delay performances can be also controlled by the parameter C_s, i.e. by forcing more traffic on the direct trunks. Moreover, a path availability requirement can be met by acting on the routing matrix, thus ensuring, for example, that a minimum of two tandems are connected to each access switch: by this way a sufficient interconnection of the tandems greatly enhances the network robustness versus trunk failures.

3. COMPUTER TOOLS

A set of computer programs has been developed to implement the planning methodology described in the previous section. Interactive facilities have been incorporated allowing the designer to explore different solutions and to find an optimal structure of the global network by iterating the procedure, or to perform sensitivity analyses. Fig. 3 is a simplified flow-chart composed of three main blocks:

i) determining the areas served by access switches and concentrators (AREA);
ii) dimensioning the access network (ACCESS);
iii) dimensioning the interswitch network (INTER).

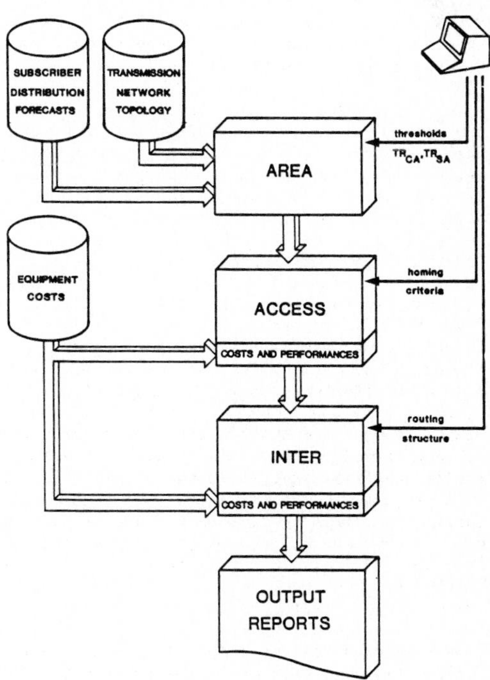

Fig.3 - Flow-chart of the computer tools

Input data to the program AREA are the subscriber distribution on the territory and subdivision into different classes of DTEs, together with the intercity trasmission layout which is supplied to the program in the form of a tree topology.

The branches of this topology are sequentially explored and its nodes are merged starting from the peripheral ones.The process is stopped when the corresponding aggregation of areas collects a threshold number of subscribers TR_{CA} (see sect 2.3): this area becomes a "concentration area". The same process is used to merge these last areas into "switching areas": in this case the threshold parameter TR_{SA} may consist in a larger number of subscribers or in a traffic amount.

Having divided the territory into areas served by the PCs and the PSs, the following program ACCESS carries out the homing of subscriber lines to these equipments, together with their consequent dimensioning. The main characteristics of the equipment used (e.g. maximum throughput and number of ports) may be changed by the designer who has also the option of using two different types of packet switches, according to the size of each single area. Options are left to the network designer regarding the homing criteria, e.g. high-bit rate DTEs in concentration areas not provided with a PS may be homed either to the local PC or to the parent PS according to a distance threshold. Then the program ACCESS makes the detailed economical evaluation of the single network components and of the overall access network. Finally, performance evaluations of time delay are made on this part of the network and fed to the next program, INTER, together with the structure of the access network: number and location of concentrators and switches, as well as their throughputs.

The program INTER carries out the optimization of the interswitch network once given the number and location of transit nodes, together with a routing policy as shown at sect. 2.3. The cost of this part of the packet network and the overall performances of time delay are calculated.

Output reports contain all details of resulting network structure, equipment installations, costs and network performances.

4. APPLICATION EXAMPLE

4.1. Description of the study cases

The computer tool described at sect.3 has been extensively used to perform network optimization and sensitivity analyses. In this section some application examples are shown all referring to the Italian territory where the public packet network (ITAPAC [3]) would reach a maximum size of the order of 50 000 subscribers when the emerging ISDN should "freeze" the dedicated data networks and gradually incorporate them.

As mentioned in sect 2.2., the total amount of subscribers has been subdivided into 231 areas of the Italian territory according to forecasted figures. The different areas have been classified into a number of classes according to their characteristics (e.g. industrial, residential, rural) and for each class a percent partitioning has been given of subscribers located in the main town and those located elsewhere in the area and of different classes of data terminals in terms of bit rate and packet or non-packet mode.

The link between DTE and the packet network port has been considered as implemented on physical circuits through baseband modems in the case of dedicated line access when DTE and port are in the same town and on telephone circuits through voice-band modems according to CCITT V-series Recommendations in the other cases, including the dial-up access.

Packet switches (PS) have been assumed having a maximum capacity of 800 ports (X.25 and X.75) and a maximum throughput of about 700 p/s, including network-service data. Bigger packet switches having a throughput of about 2500 p/s have been also considered.

Packet concentrators (PC) have been considered having up to 32 ports at low bit rate and 8 ports at high bit rate or up to 64 ports at low bit rate only. Packet assembly-disassembly (PAD) function is incorporated in PCs, also.

For these examples, interswitch circuits have been considered implemented on analog facilities, i.e. on telephone circuits for 9,6 Kbit/s links and on Primary FDM Groups for 64 Kbit/s links.

The three application examples described in the following sections are relevant to:

i) the optimisation of the global network for 10000 subscribers (fig.4 and 5);
ii) the optimisation of the interswitch network structure for a number of subscribers ranging from 5000 to 40000 (fig.6);
iii) the impact of the packet switch capacity on the network costs (fig.7).

4.2. Optimization of the global network

The case refers to the cost optimization of the global network for 10 000 subscribers.

Fig.4 shows the costs of the network for different numbers of switching areas setting different thresholds TR_{SA} for subscriber-generated traffic, once found sub-optimal solutions for the establishment of concentration area. Curve t indicates the total cost and shows a minimum at 10-12 switching areas. Curve a represents the cost of the interswitch network, including the transit switches; this portion of the total cost increases when the number of switching areas is increased. Curve c represents the cost of access packet switches, concentrators and of the links between them, and also increases with the number of switching areas, i.e. with the peripherization of the access to the network. Increases of costs of switching are compensated by the decrease of cost of subscriber network, i.e. circuits and modems shown in curve b.

The figure shows also that the cost of subscriber network is comparable with that of concentration and access switching, whilst the cost of interswitch network is considerably lower, though not completely optimized, as only analog transmission facilities have been considered.

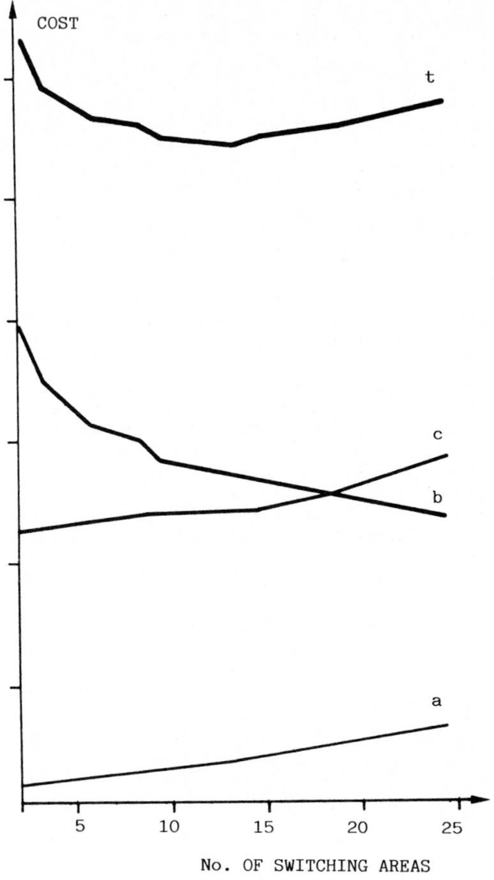

Fig.4 - Cost components vs. the number of switching areas

a) interswitch network
b) subscriber network
c) PS and PC equipments
t) overall network

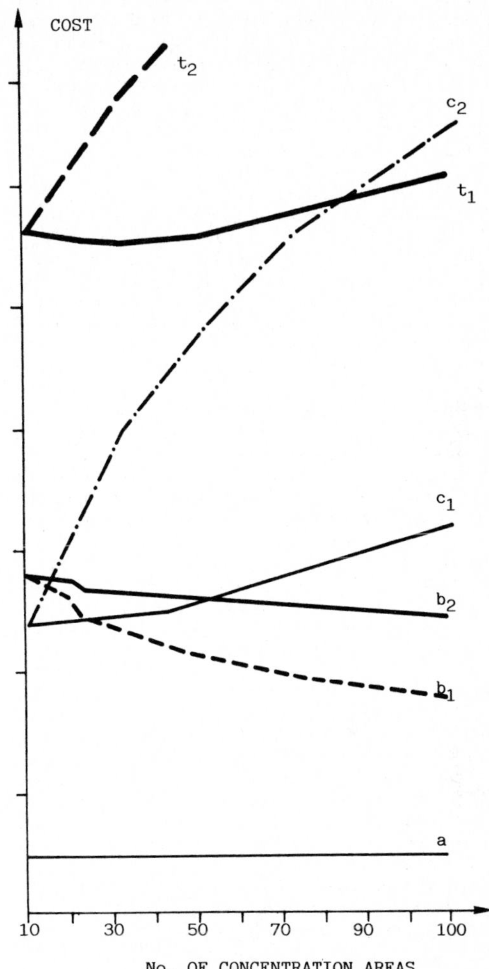

Fig.5 - Same cost components of fig.4 when varying the number of concentration areas for two different criteria of DTE homing (see sect 4.2)

Fig.5 shows the same cost elements for different numbers of concentration areas, given the number of switching areas, equal to 11. Two different cases have been considered for the homing of medium and high speed terminals. The solid lines refer to the case where only low-speed terminals are homed to packet concentrators, while other DTEs are connected to packet switches. It can be seen that the curves relevant to subscriber network and access switching/concentration costs are similar to those of fig.4, whilst the cost of interswitch network is constant having fixed the amount and location of access packet switches.

The dotted lines refer to the case where the medium and high speed DTEs located in the same towns where a PC is installed, are connected to it, instead than to a farther PS. It can be seen that the savings which are obtained in the subscriber network, due mainly to the shorter lines and wider use of cheaper baseband modems, are made useless by the greater increase of packet concentrators costs, so that total costs are considerably higher than for the previous case.

4.3. Optimization of interswitch network

This example refers to the optimization of the interswitch network structure at different sizes of the packet network, i.e. for different number of subscribers. The following structures have been considered: fully meshed (i.e. no transit switch) and tandem with 2, 4 and 6 transit switches.

Fig.6 shows the minimum cost of the interswitch network in the previous four cases, for a number of subscribers ranging from 5000 to 40000. It can be seen that the fully-meshed and the two-tandem structures have nearly the same cost at 5000 subscribers, but as soon as the total traffic increases with the number of subscribers, it is convenient to use an increasing number of transit switches.

4.4. Effect of packet switches capacity

For the above examples packet switches having a maximum throughput of about 700 p/s have been considered. To asses the economical effect of switch capacity, in this example, a switch having a maximum throughput of about 2 500 p/s has been compared with the previous one.

The new switch has been assumed to have a start-up cost of about 70% higher than the former and the same incremental cost per port.

Fig.7 shows the cost per subscriber of the global network, at different amounts of subscribers from 5000 to 40000, implementing the network with 700 p/s switches (dotted line) and with 2500 p/s switches (solid line). It can be seen that savings can be obtained with the bigger switch starting at 20 000 subscribers. It can be seen also that quite important is the scale-effect in the network cost-per-subscriber, as it reduces of about 20% when the subscribers increase from 5000 to 40000 and use is made of the smaller switch. With the bigger one the effect is obviously more remarkable and may be estimated about 25%.

REFERENCES

[1] L.Kleinrock - "Queueing Systems" - John Wiley & Sons

[2] S.W.Johnston, P.L.Shuhmann - "Design and optimization of packet networks" - 2nd Int. Network Planning Symposium, Brighton (UK), march 1983

[3] R.Parodi, F.Cataldi, N.Corsi - "Evolutionary prospects of packet switching in Italy" - ISS'84, Florence (Italy), may 1984

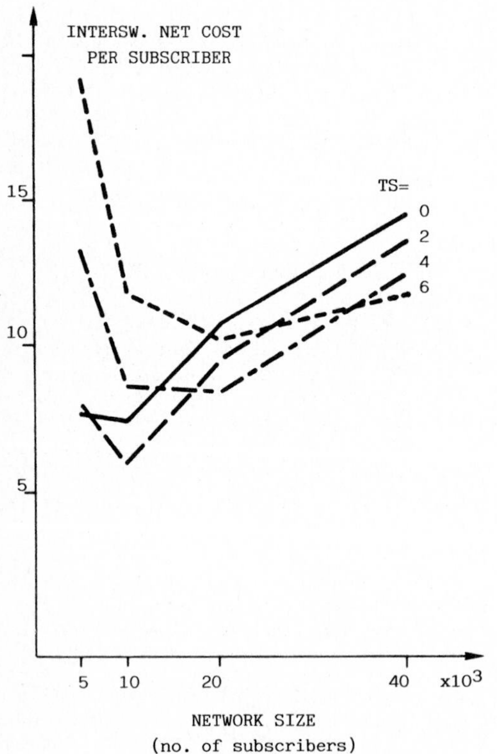

Fig.6 - Effect of the number TS of tandems on the interswitch network cost per subscriber for different network sizes.

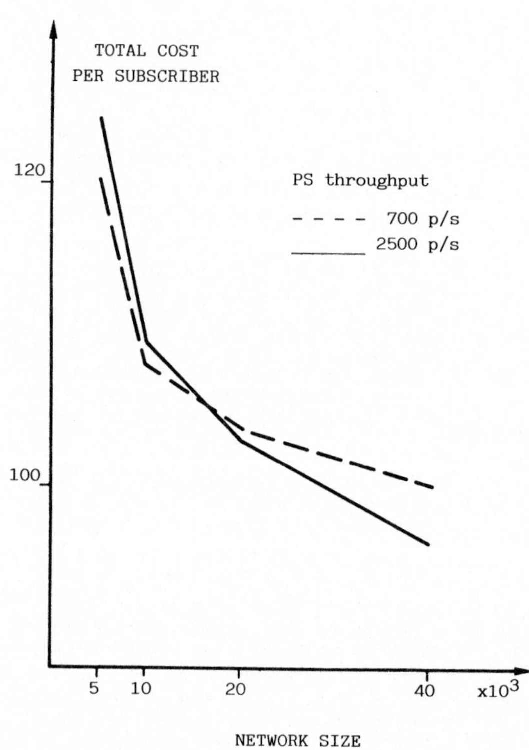

Fig.7 - Effect of the PS throughput on the total cost per subscriber for different network sizes.

TELETRAFFIC ISSUES in an Advanced Information Society
ITC-11
Minoru Akiyama (Editor)
Elsevier Science Publishers B.V. (North-Holland)

CAPACITY ASSIGNMENT IN PACKET-SWITCHING NETWORKS: A HIERARCHICAL PLANNING APPROACH(*)

Paolo M. CAMERINI and Francesco MAFFIOLI

Centro Studi Telecomunicazioni Spaziali of CNR
Dipartimento di Elettronica Politecnico di Milano
Milano, Italy

ABSTRACT

We address the problem of planning the capacity augmentation of a packet-switching network with given routing and topology under uncertainty of future costs and traffic requirements. Recent advances in hierarchical planning and lagrangean relaxation techniques allow to minimize the present cost plus the expected value of future costs, without exceeding a specified threshold for the average delay.

1. INTRODUCTION

The general problem of planning the development of telecommunication networks consists in determining topology, routing and capacities satisfying given constraints on cost, throughput, and grade of service, over a given time horizon.

The formidable task of finding an optimal solution to this general problem becomes even more difficult when, as it is often the case, we have to deal with uncertain future conditions. For this reason, it has been found convenient to split the planning into a number of simpler problems, whose solution methods can be utilized heuristically to tackle the whole.

Among the many examples of this approach, we would like to remember /1,2/ in the field of circuit-switching and /3/ for packet-switching networks.

However, the uncertainty on future cost/traffic conditions is not taken into account in these works and indeed seldom considered in the literature. In order to begin exploring the possibility of implementing more general decision tools, in this paper we focus on the problem of planning the (discrete) capacity augmentation of a packet-switching network with given routing and topology under uncertainty on future costs and traffic requirements.

For this problem we propose a lagrangean relaxation approach, extending the method used in /4/ for similar, simpler problems. The results concerning our method are based on recent advances on multiple-choice knapsack /5,6/ and hierarchical planning problems /7,8/.

Section 2 illustrates how the network design problem can be formulated by a hierarchical planning model; Section 3 thouroughly describes the methods for obtaining lower bounds and approximate solutions. Section 4 presents some encouraging computational results: a rationale for this good behaviour is given in Section 5.

2. THE PROBLEM

The problem of planning the (discrete) capacity augmentation of a packet-switching network with given routing and topology under uncertainty on future costs and traffic requirements, can be formulated as follows.

Given:

- a threshold τ to the average delay for the network;
- a finite set of L possible future traffic/cost conditions, each with probability p_ℓ ($\ell = 1,2,\dots,L$);
- a finite set $C_1, \dots, C_K$ of possible capacity levels to be assigned to each of the J links;
- total (external) traffic γ_ℓ that will be offered under future condition ℓ ($\ell = 1,2,\dots,L$) and total known traffic γ_o presently offered to the network;
- on each link j, the (internal) traffic $f_{j\ell}$ under future condition ℓ ($\ell = 1,2,\dots,L$) and present known traffic f_{jo};
- a relationship (such as $t_{j\ell}(k)=f_{j\ell}/[\gamma_\ell(C_k-f_{j\ell})]$ in /9/) between the previous items and the average delay $t_{j\ell}(k)$ on link j under future ($\ell = 1,2,\dots,L$) or present ($\ell = 0$) condition ℓ;
- the (known) cost $b_j(k)$ of assigning capacity C_k to link j, at present time;
- the cost $a_{j\ell}(k,h)$ of future installation of capacity C_h on link j under condition ℓ, if capacity C_k is installed on the same link at present time.

Find:

$$\bar{z} = \min_{\underset{\sim}{y} \in S} \left\{ \sum_{1}^{J} {}_{j}\, b_j(y_j) + \sum_{1}^{L} {}_{\ell}\, p_\ell\, v_\ell(\underset{\sim}{y}) : \sum_{1}^{J} {}_{j}\, t_{jo}(y_j) \leq \tau \right\}, \quad (1)$$

(*) A preliminary version has been presented at ORSA/TIMS 1984, Dallas, Texas, Nov. 26-28,1984.

where for $\ell = 1,2,\ldots,L$

$$v_\ell(y) = \min_{x \in S} \left\{ \sum_{1}^{J} a_{j\ell}(y_j,x_j) \; : \; \sum_{1}^{J} t_{j\ell}(x_j) \leq \tau \right\} \quad (2)$$

and $S = \{1,2,\ldots,K\}^J$ is the space of the decision vectors $y = (y_1,\ldots,y_J)$ and $x = (x_1,\ldots,x_J)$ of present and (respectively) future capacity assignment.

Such a model is of the hierarchical planning type /7/, as it involves two phases: a first (aggregate) phase aiming at minimizing the cost of present capacity installations, plus the average cost of future optimal installations; and a second (deterministic) phase, where the actual traffic/cost condition is known, and the cost of capacity augmentation is minimized. The values of $a_{j\ell}(k,h)$ and $b_j(k)$ being arbitrary, such a formulation is capable of modeling any type of relationship (e.g. concave non-linear) between capacity updating and cost. Similarly, the average delay on each link may be any specified function of capacity assignment and traffic condition.

The second phase of this capacity planning problem is a multiple-choice knapsack problem /5/, thus implying that the whole problem is *NP-hard*. The theory of computational complexity of combinatorial problems has been applied on various occasions to provide fundamental insights into their inherent difficulty: we refer to /10/ for an informal introduction and to /11/ for a thorough exposition. Suffice it to say that the theory has allowed the identification of a large class of NP-complete problems, with the following two important properties:

(i) no NP-complete problem is known to be easy, i.e., solvable by an algorithm whose running time is bounded by a polynomial function of problem size;
(ii) if any NP-complete problem would turn out to be easy, then they would all be easy.

All these problems are recognition problems, which require a yes/no answer. The optimization problems that correspond to many of them are at least as difficult and are called NP-hard. Many notorius problems such as 0-1 programming, traveling salesman, plant location and knapsack problems are NP-hard. Hence, establishing NP-hardness of a problem yields strong circumstantial evidence against the existence of a polynomial-time algorithm for its solution. This makes it easier to accept the inevitability of enumerative optimization methods or of fast approximation algorithms, where finding tight bounds to the value of the optimum plays a crucial role.

3. LOWER BOUNDS AND APPROXIMATE SOLUTIONS

In this section we propose a lagrangean relaxation method for finding lower and upper bounds to $\bar{z}$, together with corresponding feasible solutions. We assume in the following that the highest capacity level available for each link is large enough to ensure the existence of at least one feasible solution, i.e. a vector y and L vectors x satisfying the average delay requirements; such vectors are called *feasible*.

For each (not necessarily feasible) vector y,

$$v_\ell(y) \geq u_\ell(y,\mu_\ell) = \min_{x \in S}\{-\mu_\ell \tau + \sum_{1}^{J} [a_{j\ell}(y_j,x_j) + \mu_\ell \, t_{j\ell}(x_j)]\}, \quad (3)$$

$\ell = 1,2,\ldots,L,$

where μ_ℓ is any non-negative value of the lagrangean multiplier corresponding to the delay constraint in (2). For any $\mu_o \geq 0$, let

$$u_o(y,\mu_o) = -\mu_o \tau + \sum_{1}^{J} \mu_o \, t_{jo}(y_j). \quad (4)$$

Let

$$\beta(y) = \sum_{1}^{J} b_j(y_j), \quad (5)$$

and

$$\Psi(y,\mu) = \beta(y) + \sum_{0}^{L} p_\ell \, u_\ell(y,\mu_\ell), \quad (6)$$

where $\mu = (\mu_o, \mu_1,\ldots,\mu_L)$ and $p_o = 1$.

Theorem 1. For any function $M : S \rightarrow (\mathbb{R}^+)^{L+1}$,

$$\min_{y \in S} \Psi(y,M(y)) \leq \bar{z}. \quad (7)$$

Proof. Let $\bar{y}$ be any optimal vector in (1) and $\bar{\mu} = M(\bar{y})$. Then

$$\bar{z} = \beta(\bar{y}) + \sum_{1}^{L} p_\ell \, v_\ell(\bar{y})$$

$$\geq \beta(\bar{y}) + \sum_{1}^{L} p_\ell \, v_\ell(\bar{y}) + u_o(\bar{y},\bar{\mu}_o)$$

(since $\bar{y}$ is feasible and $\bar{\mu}_o \geq 0$)

$$\geq \beta(\bar{y}) + \sum_{0}^{L} p_\ell \, u_\ell(\bar{y},\bar{\mu}_\ell) = \Psi(\bar{y},M(\bar{y}))$$

(because of (3)).

Let $S_f \subseteq S$ be the set of all feasible vectors y.

Corollary 1. Over all choices of M in (7),

$$z' = \min_{y \in S_f} \max_{\mu \geq 0} \Psi(y,\mu) \quad (8)$$

yields the highest lower bound to $\bar{z}$.

Proof. For any vector $y \in S$, $u_\ell(y,\mu_\ell)$ is a concave piecewise-linear function of μ_ℓ, $\ell = 1,2,\ldots,L$. Because of the feasibility assumption, $u_\ell(y,\mu_\ell)$ cannot tend to $+\infty$ as μ_ℓ tends to $+\infty$, so that for some $\mu' \geq 0$,

$$u_\ell(y,\mu'_\ell) = \max_{\mu_\ell \geq 0} u_\ell(y,\mu_\ell).$$

For any feasible vector y, $u_o(y,\mu_o)$ is either identically zero, or tends to $-\infty$ as μ_o tends to $+\infty$, so that

$$\Psi(y,\mu') = \max_{\mu \geq 0} \Psi(y,\mu), \quad (9)$$

where $\mu' = (0, \mu'_1,\ldots,\mu'_L)$. Since μ' depends on $\underset{\sim}{y}$, we write $\underset{\sim}{\mu}' = \underset{\sim}{M}'(\underset{\sim}{y})$, where $\underset{\sim}{M}'$ denotes a function ranging over the feasible vectors.

For any unfeasible vector $\hat{\underset{\sim}{y}}$, $u_o(\hat{\underset{\sim}{y}},\mu_o)$ tends to $+\infty$ as μ_o tends to $+\infty$, hence $\Psi(\hat{\underset{\sim}{y}},\underset{\sim}{\mu})$ is unbounded from above. Let then H be a large enough positive constant, such that for all feasible vectors $\underset{\sim}{y}$ and $\underset{\sim}{\mu} = (H,0,\ldots,0)$,

$$\Psi(\underset{\sim}{y},\underset{\sim}{M}'(\underset{\sim}{y})) < \Psi(\hat{\underset{\sim}{y}},\underset{\sim}{\mu}).$$

We can therefore extend the range of $\underset{\sim}{M}'$ to the unfeasible vectors $\hat{\underset{\sim}{y}}$, by defining $\underset{\sim}{M}'(\hat{\underset{\sim}{y}}) = (H,0,\ldots,0)$, so that

$$\min_{\underset{\sim}{y}\in S} \Psi(\underset{\sim}{y},\underset{\sim}{M}'(\underset{\sim}{y})) = \min_{\underset{\sim}{y}\in S_f} \Psi(\underset{\sim}{y},\underset{\sim}{M}'(\underset{\sim}{y})) = z' .$$

Because of theorem 1, z' is a lower bound to $\bar{z}$. Because of (9), for any other function $\underset{\sim}{M} : S \to (\mathbb{R}^+)^{L+1}$,

$$\min_{\underset{\sim}{y}\in S} \Psi(\underset{\sim}{y},\underset{\sim}{M}(\underset{\sim}{y})) \le \min_{\underset{\sim}{y}\in S_f} \Psi(\underset{\sim}{y},\underset{\sim}{M}(\underset{\sim}{y})) \le \min_{\underset{\sim}{y}\in S_f} \Psi(\underset{\sim}{y},\underset{\sim}{M}'(\underset{\sim}{y})),$$

and z' is the highest possible lower bound over all choices of $\underset{\sim}{M}$. □

The lower bound (8) could be evaluated by computing for each feasible $\underset{\sim}{y}$, $\max_{\underset{\sim}{\mu}\ge 0} \Psi(\underset{\sim}{y},\underset{\sim}{\mu})$.

This value, as already pointed out in the proof of Corollary 1, can be obtained through (6), by evaluating for each $\ell = 1,2,\ldots,L$,

$$u_\ell(\underset{\sim}{y},\mu'_\ell) = \max_{\mu_\ell\ge 0} u_\ell(\underset{\sim}{y},\mu_\ell).$$

This in turn is the lower bound proposed in /1,3/ for the multiple-choice knapsack problem (2), and can be computed in polynomial time.

Applying straightforwardly the results of /6/, a corresponding upper bound to v_ℓ is

$$V_\ell(\underset{\sim}{y}) = \sum_{1}^{J}{}_j\; a_{j\ell}(y_j,x_{\ell j}) \tag{10}$$

where $\underset{\sim}{x}_\ell$ is a suitable feasible vector among those solving the minimization of (3) with $\mu_\ell = \mu_\ell'$. Moreover Proposition 1 of /6/ ensures that $V_\ell(\underset{\sim}{y}) - u_\ell(\underset{\sim}{y},\mu'_\ell)$ is not greater than $\varepsilon = \max_{j,h,k,\ell} a_{j\ell}(k,h)$.

Let now $y' \in S_f$ and $\underset{\sim}{\mu}' \ge 0$ be such that

$$z' = \Psi(\underset{\sim}{y}',\underset{\sim}{\mu}') = \min_{\underset{\sim}{y}\in S_f} \max_{\underset{\sim}{\mu}\ge 0} \Psi(\underset{\sim}{y},\underset{\sim}{\mu}).$$

Hence

$$Z' = \beta(\underset{\sim}{y}') + \sum_{1}^{L}{}_\ell\; p_\ell V_\ell(\underset{\sim}{y}') \tag{11}$$

is an upper bound to $\bar{z}$,

$$z' = \beta(\underset{\sim}{y}') + \sum_{1}^{L}{}_\ell\; p_\ell u_\ell(\underset{\sim}{y}',\mu'_\ell)$$

and

$$Z'-z' = \sum_{1}^{L}{}_\ell\; p_\ell \left[V_\ell(\underset{\sim}{y}')-u_\ell(\underset{\sim}{y}',\mu'_\ell)\right] \le \varepsilon . \tag{12}$$

Unfortunately, the evaluation of the lower bound (8) suggested above would require to solve a number of maximization problems growing exponentially with J.

Therefore we propose in the following a different, and possibly worse lower bound, which however can be computed much more efficiently.

In fact Theorem 1 implies that, for any $\underset{\sim}{\mu} \ge 0$

$$w(\underset{\sim}{\mu}) = \min_{\underset{\sim}{y}\in S} \Psi(\underset{\sim}{y},\underset{\sim}{\mu}) \le \bar{z}, \tag{13}$$

so that

$$z'' = \max_{\underset{\sim}{\mu}\ge 0} w(\underset{\sim}{\mu}) = w(\underset{\sim}{\mu}'') \tag{14}$$

is a lower bound to $\bar{z}$, although Corollary 1 indicates that $z'' \le z'$. In order to compute z'', we first observe that $w(\underset{\sim}{\mu})$ is the minimum of K^J functions of $\underset{\sim}{\mu}$, each of them being the sum of L+1 piecewise-linear concave functions of μ_ℓ, $\ell = 0,1,\ldots,L$ (see (6), (4) and (3)). It follows that $w(\underset{\sim}{\mu})$ is also a piecewise-linear concave function of $\underset{\sim}{\mu}$, and any local maximum is a global one. Moreover for each $\underset{\sim}{\mu} \ge 0$, both $w(\underset{\sim}{\mu})$ and a subgradient $\nabla(w(\underset{\sim}{\mu}))$ can be computed in polynomial time. In fact $O(\tilde{K}^2JL)$ steps suffice for computing:

(i) $w(\underset{\sim}{\mu}) = \xi(\underset{\sim}{\mu}) + \sum_{1}^{J}{}_j \min\{\theta_j(\underset{\sim}{\mu},y_j,x_{1j},\ldots,x_{Lj}) : y_j,x_{1j},\ldots,x_{Lj}\in\{1,\ldots,K\}\}$

where

$$\xi(\underset{\sim}{\mu}) = -\tau(\mu_o + \sum_{1}^{L}{}_\ell\; p_\ell \mu_\ell),$$

$$\theta_j(\underset{\sim}{\mu},y_j,x_{1j},\ldots,x_{Lj}) = b_j(y_j)+\mu_o t_{jo}(y_j) + \sum_{1}^{L}{}_\ell\; p_\ell\left[a_{j\ell}(y_j,x_{\ell j})+\mu_\ell t_{j\ell}(x_{\ell j})\right],$$

for $j = 1,\ldots,J$;

(ii) $\nabla(w(\underset{\sim}{\mu})) = (\partial w/\partial\mu_o, \partial w/\partial\mu_1,\ldots,\partial w/\partial\mu_L)$

where

$$\partial w/\partial\mu_o = -\tau + \sum_{1}^{J}{}_j\; t_{jo}(\bar{y}_j),$$

$$\partial x/\partial\mu_\ell = p_\ell\left[-\tau + \sum_{1}^{J}{}_j\; t_{j\ell}(\bar{x}_{\ell j})\right]$$

for $\ell = 1,\ldots,L$,

and $\bar{y}_j$, $\bar{x}_{\ell j}$, $\ell = 1,\ldots,L$ are any values in $\{1,\ldots,K\}$ minimizing θ_j for $j = 1,\ldots,J$. (Note that these values and the corresponding subgradient are not necessarily uniquely determined for each vector $\underset{\sim}{\mu}$.)

In order to maximize $w(\underset{\sim}{\mu})$, many techniques are available. One of them is the modified subgradient method of /12/, taking into account the constraint $\underset{\sim}{\mu} \ge 0$. Because of (13), the final value obtained is a lower bound to $\bar{z}$, even when the optimum $\underset{\sim}{\mu}''$ is reached only approximately.

An upper bound Z'' and feasible vectors corresponding to $\underset{\sim}{\mu}''$ can be obtained as follows. Because of the optimality of $\underset{\sim}{\mu}''$, one of the two following cases must occur.

(a) There exists a set of vectors $\bar{\underset{\sim}{y}}$, $\bar{\underset{\sim}{x}}_\ell$, $\ell = 1, \ldots, L$, minimizing θ_j, $j = 1,\ldots,J$, such that all the components of the corresponding subgradient $\nabla(w(\underset{\sim}{\mu}''))$ are zero.

(b) There exists a set of vectors $\underset{\sim}{y}^f$, $\underset{\sim}{x}^f_\ell$, $\ell = 1,\dots,L$, such that $\underset{\sim}{y}^f$ corresponds to $\partial w(\underset{\sim}{\mu}'')/\partial\mu_0 \leq 0$ and for each $\ell = 1,\dots,L$, $\underset{\sim}{x}^f_\ell$ corresponds to $\partial w(\underset{\sim}{\mu}'')/\partial\mu_\ell \leq 0$.

In the first case $\bar{\underset{\sim}{y}}$, $\bar{\underset{\sim}{x}}_\ell$, $\ell = 1,\dots,L$ are feasible and optimal vectors; in the second case, although optimality is not obtained, we still have a set of feasible vectors $\underset{\sim}{y}^f$, $\underset{\sim}{x}^f_\ell$, and an upper bound

$$Z'' = \beta(\underset{\sim}{y}^f) + \sum_{\ell=1}^{L} p_\ell \sum_{j=1}^{J} a_{j\ell}(y^f_j, x^f_{\ell j}). \qquad (15)$$

The feasible vectors $\underset{\sim}{y}^f$, $\underset{\sim}{x}^f_\ell$ can be evaluated by solving ties in the minimization of θ_j in favour of choices which yield minimum delays. Even when only an approximation to $\underset{\sim}{\mu}''$ is available this policy is likely to give feasible vectors.

4. COMPUTATIONAL RESULTS

A FORTRAN code has been implemented along the lines of the previous sections and an extensive testing is in progress.

The initial phase of this testing aims at collecting evidence on the quality of the bounds and the approximate solutions obtained, by generating a set of random instances of the capacity assignment problem. To this aim, we have more in mind the simplicity of the simulation model, than the need of testing our method on realistic data.

This latter need will be hopefully fulfilled in a second phase of our experimentation.

Even in this first phase, however we have tried to simulate some features of realistic data.

In particular we have considered the two following models for generating the present end-to-end traffic requirements γ^o_{hi}:

(T.1) a randomly generated matrix Γ whose elements γ^o_{hi} are independently and uniformly distributed in $[0, 10]$;

(T.2) a matrix Γ whose elements γ^o_{hi} are proportional to $s_h \cdot s_i$, where $\underset{\sim}{s}$ is a randomly generated vector of the (relative) "sizes" of each communication center; the components s_i are independently and uniformly distributed in $[0.1, 1]$.

The internal traffic f_{jo} on each link j is derived from Γ by routing each external traffic γ^o_{hi} along a path of minimum length: the length of each link is also randomly generated, uniformly and independently in the intervals A = $[1,10]$ or B = $[10,100]$. Of course this particular choice of routing has the only purpose of generating the internal traffics.

For each link j, the present cost $b_j(k)$ is the sum of a part b'(k) depending on the link capacity according to Table 1, and a part b"(j) depending on the link length λ_j either

(C.A) proportional to λ_j if it ranges in interval A, or

(C.B) according to Table 2, if λ_j ranges on interval B.

Table 1

k	C_k	b'(k)
1	0.0	0
2	1.2	452
3	2.4	1020
4	3.6	1950
5	4.8	2910
6	6.0	3950
7	7.2	4500
8	8.4	5000
9	9.6	5486

Table 2

range of λ_j	b"(j)
$0 < \lambda_j \leq 10$	626
$10 < \lambda_j \leq 15$	968
$15 < \lambda_j \leq 30$	1822
$30 < \lambda_j \leq 60$	3188
$60 < \lambda_j \leq 120$	3985

For what concerns the future traffic/cost conditions, nine different scenarios, each with the same probability of occurrence, have been considered by combining three traffic and three cost conditions. In order to simplify input data, the routing has been assumed to remain the same in all scenarios, as well as the pattern of the external traffics, but for a multiplicative factor, which takes the values 1, 1.5, 2 in the three traffic conditions. Each cost of future installation is given by

$$a_{j\ell}(k,h) = \chi_\ell\left[b''(j)+b'(h)-\frac{1}{2}\,b'(k)\right]$$

where χ_ℓ takes the values .8, 1, 1.2 in the three cost conditions.

In order to avoid generating instances for which the average delay requirements are either unsatisfiable or trivially satisfiable with the minimum capacity level, we have normalized all the traffics so that for each instance the most loaded link is given an interval traffic equal to 9.

Table 3 summarizes the computational results obtained so far. The first four rows refer to the different traffic/cost models described above. The three columns correspond to different sizes of the network, in terms of the number of communication centers. Each entry reports three figures obtained by averaging over twenty randomly generated instances. (More specifically, five instances have been generated for each one of

Table 3

no. of centers / traffic/cost models	5	10	15
T.1/C.A	1.60(1.88)	1.04(1.19)	0.79(0.89)
	226	379	392
T.1/C.B	0.67(1.84)	0.52(1.51)	0.37(1.28)
	220	346	282
T.2/C.A	1.61(1.86)	1.13(1.37)	0.78(0.92)
	229	339	311
T.2/C.B	0.59(1.66)	0.38(1.21)	0.26(1.11)
	202	339	344
Average	1.12(1.81)	0.77(1.32)	0.55(1.05)
	219	350	332

four values of τ, selected appropriately in order to give neither too tight, nor too loose requirements.) The first upper figure measures the quality of the upper bound Z" by the relative difference

$$\frac{Z''-z''}{Z''} \cdot 100 \ .$$

The second upper figure in parenthesis reports

$$\frac{Z''-z''}{Y''} \cdot 100 \ ,$$

where Y" is the cost of the approximate solution when only the costs depending on the link capacities are considered. This aims at avoiding the masking effect of the costs depending on link lengths only. The lower figure in each entry measures the computational effort by reporting the number of evaluations of $w(\underset{\sim}{\mu})$.

The average results corresponding to each network size are shown in the last row.

It appears that all relative differences shown in Table 3 are quite small and tend to decrease when the problem size increases. A similar behaviour for the upper bound Z' is theoretically derived in the next section.

5. ASYMPTOTIC PERFORMANCES

Let us measure the quality of the upper bound Z' by the relative difference

$$\eta' = \frac{Z'-z'}{Z'} \ .$$

Similarly as in /6/, we investigate the asymptotic behaviour of the sequence of random variables $\eta' = \eta'(J)$, as J tends to infinity with K, L fixed and assuming a reasonable stochastic model for the cost of the capacities. Specifically, let us assume that:

(i) $a_{j\ell}(k,h) = 0$ for all j,ℓ and all $k \geq h$;

(ii) $b_j(k)$ and $a_{j\ell}(k,h)$ for all j,ℓ and all $k < h$ are realizations of independent r.v.'s, with a common distribution defined over $\mathbb{Z}^+ = \{1,2,\dots\}$, and with finite variance;

(iii) $\sum_{o}^{L}{}_{\ell}\, p_\ell\, n_\ell = \Omega(\sqrt{J})$, where n_ℓ is the number of links j for which $f_{j\ell} > 0$, and hence either $b_j(y_j)$ or $a_{j\ell}(y_j,x_{\ell j})$ or both, must be greater than zero for all feasible vectors $\underset{\sim}{y}$ and $\underset{\sim}{x}_\ell$.

Notice that assumption (iii) is very mild, since it is satisfied when the average number of links with non zero traffic is not smaller than the number of links in a connected network.

Because of (12), $\eta'(J) \leq \varepsilon/Z' \leq \varepsilon/(\sum_{o}^{L}{}_{\ell}\, p_\ell\, n_\ell)$.

Known results /7, 13/ from extreme order statistics imply that

$$\lim_{J\to\infty} \frac{\varepsilon}{\sqrt{J}} = 0 \quad \text{a.e.}$$

From assumption (iii) it follows that

$$\lim_{J\to\infty} \eta'(J) = 0 \text{ a.e.} \qquad (16)$$

For what concerns the quality of the upper bound Z", measured by the relative difference

$$\eta'' = \frac{Z''-z''}{Z''}$$

it is not straightforward to obtain an inequality similar to (12), uniquely in terms of the costs of the capacities. Nevertheless, it appears that an upper bound to Z"-z" could be found assuming a suitable model also for the average delays, thus providing a result similar to (16). In fact it can be proved that among the vectors $\bar{\underset{\sim}{y}}$ corresponding to $\bar{\underset{\sim}{\mu}}$, a procedure similar to REFINE of /3/ yields in polynomial time a feasible and an unfeasible vector which differ only in one component. The question of finding a model (as unrestrictive as possible) implying the asymptotic optimality of Z", is still under investigation.

ACKNOWLEDGMENTS

The authors wish to thank their colleagues F. Borgonovo and C. Vercellis for stimulating discussions on the model and the algorithm, respectively, and are indebted to C. Carli and M. Parini for their careful implementation and testing of the method.

REFERENCES

/ 1/ B. Yaged Jr., "Minimum cost routing for dynamic network models", *Networks* 3 (1973) 193-224.

/ 2/ M. Bonatti, P.M. Camerini, L. Fratta, G. Gallassi, F. Maffioli, "A dynamic planning method for telecommunication networks and its performance evaluation for district trunk networks", *10th International Teletraffic*

Congress (Montreal, 1983), Session 2.1, paper n. 5.

/ 3/ M. Gerla, "The design of store-and-forward networks for computer-communications", Ph. D. dissertation, Dept. of Computer Science, UCLA, 1973.

/ 4/ V.K.M. Whitney, "Lagrangean optimization of stochastic communication system models", in *Proc. of the Symp. on Computer-Communications Networks and Teletraffic* (J. Fox ed.) Polytechnic Press of the Polytechnic Institute of Brooklyn, N.Y. (1972) 385-395.

/ 5/ R.D. Armstrong, D.S. Kung, P. Sinha and A. A. Zoltners, "A computational study of a multiple-choice knapsack algorithm", *ACM Trans. on Math. Software* 9 (1983) 184-198.

/ 6/ P.M. Camerini and C. Vercellis, "The matroidal knapsack: a class of (often) well-solvable problems", *Operations Research Letters* 3 (1984) 157-162.

/ 7/ M.A.H. Dempster, M.L. Fisher, L. Jansen, B.J. Lageweg, J.K. Lenstra and A.H.G. Rinnooy Kan, "Analytical evaluation of hierarchical planning systems", *Operations Research* 29 (1981) 707-715.

/ 8/ J.K. Lenstra, A.H.G. Rinnooy Kan and L. Stougie, "A framework for the probabilistic analysis of hierarchical planning systems", in *Stochastics and Optimization* (F. Archetti and F. Maffioli eds.) *Annals of Operations Research vol. 1*, Baltzer, Basel (1984) 23-42.

/ 9/ M. Schwartz, *Computer-Communication Network Design and Analysis*, Prentice-Hall, Englewood Cliffs, N.J. (1977).

/10/ F. Maffioli, "The complexity of combinatorial optimization algorithms and the challenge of heuristics", in *Combinatorial Optimization* (eds. N. Christofides, A. Mingozzi, P. Toth, C. Sandi), J. Wiley & Sons (1979) 107-129.

/11/ M.R. Garey, and D.S. Johnson, *Computers and Intractability: a Guide to the Theory of NP-completeness*, Freeman, San Francisco, 1979.

/12/ P.M. Camerini, L. Fratta and F. Maffioli, "On improving relaxation methods by modified gradient techniques", *Math. Programm. Study* 3 (1975) 26-34.

/13/ J. Galambos, *The asymptotic Theory of Extreme Order Statistics*, Wiley (1978).

TELETRAFFIC ISSUES in an Advanced Information Society
ITC-11
Minoru Akiyama (Editor)
Elsevier Science Publishers B.V. (North-Holland)

AN ALGORITHM FOR TRAFFIC ROUTING PROBLEM

Xiong Jian LIANG

Beijing Institute of Posts and Telecommunications
Beijing, China

ABSTRACT

In designing multicommodity network, routing is an important subproblem.

We may set up a programming mathmatical model to get an optimal solution, but it is very diffcult to solve it, even though we use the computer.

In this paper we find out an approximate solution, we consider the traffic is carried by a given network through shared "links" of fixed capacity. For each source-destination pair, traffic is assumed to be restricted to a small number of permissible paths, each consisting of a string of links from source to destination.

We use "Shift" and "Diversion" methods to maximize the amount of traffic which can be accommodated by allocating traffic among permissible paths.

PROBLEM

In designing and planning a telecommunication network, we may consider:

i) Selection of nodes or tandem center;

ii) Configuration of the network or selection of links connecting nodes;

iii) Allocation of traffic to links or paths

iv) Determing the capacity of the network;

v) Estimating the quality of service (delays, reliability etc.)

vi) Economical studies'

The routing problem addressed here was formulated as a subproblem to be solved as a component of such an overall "devide-and-conquer" approach to network design. We assume that the network topology is given; We assume that the nodes have sufficient capacity to do their job; We assume that the traffic demand is given, except that we don't know how much of that traffic the network can handle.

Our problem is simply to assign traffic to paths from source to destination to maximize the amount of traffic the network can handle.

Let the nodes, links, and paths be numbered as follows:

Nodes are indexed by n=1,2,------,M.

Links are indexed by i=1,2,------,K.

Paths are indexed by p=1,2,------,Q.

Such as figure 1, in this network involves 6 nodes, n=1,2,----6, 7 links i=1,2,---7, 49 paths.

For each path P, let

L(P)=set of links in path P;

F(P)=fraction of the total traffic demand having the same sources and destination nodes as does P;

X(P)=fraction of F(P) allocated to path P.

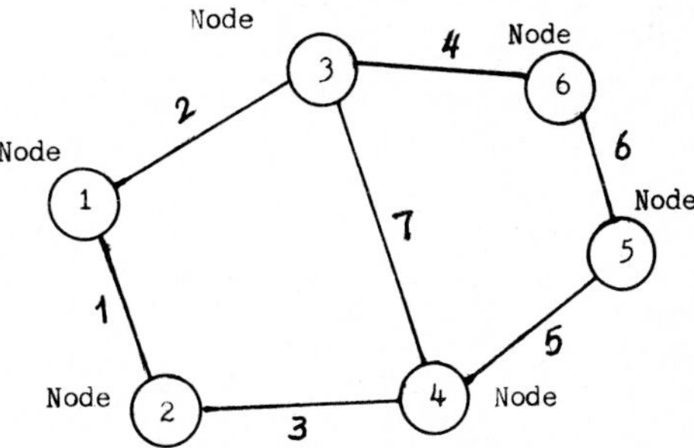

Fig. 1 Test Network

Also, let T be the total traffic demand for all node pairs.

For each link i, let

C(i) = capacity of link i;

P(i) = the set of paths which contain link i;

Y(i) = "load" on link i.

By the load on link i we mean

$$Y(i) = \sum TF(P)X(P)/C(i) \qquad (1)$$

The objective is to maximize T. There are three types of constraints which must be met. First, all of the specified traffic demand must be allocated to permissible paths. That is, for each pair of nodes (m,n) with $1 \leq m < n \leq M$

$$\sum X(P) = 1 \qquad (2)$$

Secondly, all link loads are limited to no more than a factor of 1.0. That is, for all $1 \leq i \leq K$

$$T \sum F(P)\, X(P)/C(i) = Y(i) \leq 1 \qquad (3)$$

And finally, all allocations of traffic to paths must be non-negative. That is,

$$X(P) \geq 0 \text{ for all } p= 1,2,\ldots.Q. \qquad (4)$$

The problem as posed is easily converted into a linear programming (LP) problem.

Note that the size of this LP problem is:

Number of Variables = Q + K + 1

-One for each path, X(P);

-One slack variable for each link, and

-One objective function variable,

Number of constraints = M(M-1)/2 + K

-One for each pair of nodes, and

-One for each link.

So when the number of links and nodes increase, the LP algorithms computation will be duplicated. The resources required are substantial.

We seek a quick and inexpensive way of obtaining approximate solutions to such problems.

APPROXIMATE SOLUTION

Considering the quality (efficiency and delay time), only "minimum hops" paths -containing the fewest links connecting each node pair- were permitted for the algorithm development. Fig. 1, this test network contains 49 non-looping paths, but only 23 permissible paths (minimum hops), and 9 unique traffic paths -the minimum hops paths are unique for some node pairs (1-2,1-3,1-5,2-4, 2-6,3-4,3-5,4-6,5-6)- and 14 separated paths.

The permissible paths of each pair shown in Table 1.

Table 1

Node	Link							Separated
Pair	1	2	3	4	5	6	7	Path
1-2	*							
1-3		*						
2-4			*					
3-5				*				
4-6					*			
5-6						*		
3-4							*	
1-5		*		*				
2-6			*		*			
1-4	*		*					1
		*					*	2
1-6	*		*		*			3
		*		*		*		4
		*		*			*	5
2-3	*	*						6
			*				*	7
2-5	*	*		*				8
			*		*	*		9
			*	*			*	10
3-6				*		*		11
					*		*	12
4-5					*	*		13
				*			*	14

Obviously, node pairs which paths are unique traffic paths only have one choice to allocate their traffic. So the key point is to allocate the traffic into separated paths for the node pairs which have seversl separated paths.

If the test network traffic demand as Table 2

We can compute the network minimum hops link data from Table 1 and Table 2. The unique traffic shown in Table 3.

In Table 3, the unique traffic of link 2 equals the traffic of node pairs 1-3 plus 1-5.

Now, the problem is how to allocate the remaining traffic into separated paths.

First, we allocate the traffic routing based on equal splitting of traffic over the permissible paths shown in Table 4.

Table 2 Traffic Matrix

Node	1	2	3	4	5	6
1	-	4	6	8	7	5
2	-	-	5	8	6	4
3	-	-	-	6	5	4
4	-	-	-	-	9	8
5	-	-	-	-	-	7
6	-	-	-	-	-	-

Table 3 Unique Traffic

Link Number	1	2	3	4	5	6	7
Unique Traffic	4	13	12	12	12	7	6

Table 4 Equal Splitting

Path	Link 1	2	3	4	5	6	7
1	4.00		4.00				
2		4.00					4.00
3	1.67		1.67		1.67		
4		1.67		1.67		1.67	
5		1.67			1.67		1.67
6	2.50	2.50					
7			2.50				2.50
8	2.00	2.00		2.00			
9			2.00		2.00	2.00	
10			2.00	2.00			2.00
11				2.00		2.00	
12					2.00		2.00
13					4.50	4.50	
14				4.50			4.50
Unique	4	13	12	12	12	7	6
Total	14.17	24.83	24.17	24.17	23.84	17.17	22.67

In Table 4. Link 2, it's load is 24.83 unit which is the maximum for link.We call this "Saturated Link". By the term "Saturated Link" we mean a link carrying a load which is the maximun for any link in the network.

In order to maximize T, we should balance the link load and find the way to minimize the "Saturated Link"load.

We use the "Shift" method to shift the traffic from the "most saturated path" to the "least saturated path" over all pairs of nodes. By the term "most saturated path", we mean a path containing a most saturated link; by the term "most saturated link" we mean a link carrying a load which is the maximun any link in the permissible paths of the node pair. Such as node pair 1-4, it has two separated paths: Path 1, it includes two links (1,3); Path2, it includes two links (2,4). The load of most saturated link is 24.83 path 2 is the "most saturated path". In path 1, the maximun link load is 24.17. We called it as "least saturated link" load.

So in this case, we shift (24.83-24.17)/2 = 0.33 from path 2 to path 1.

The amount of traffic shifted from the most saturated path to the least saturated path is limited to the minimun of :

1) The amount which balances the load on most saturated links in the most saturated and least saturated paths;

2) The amount of traffic on the most saturated path.

Table 5 After Shifting

Path	Link 1	2	3	4	5	6	7
1	4.33		4.33				
2		3.67					3.67
3	1.67		1.67		1.67		
4		1.67		1.67		1.67	
5		1.67			1.67		1.67
6	2.50	2.50					
7			2.50				2.50
8	2.00	2.00		2.00			
9			2.00		2.00	2.00	
10			2.00	2.00			2.00
11				1.83		1.83	
12					2.17		2.17
13					4.50	4.50	
14				4.50			4.50
Unique	4	13	12	12	12	7	6
Total	14.50	24.50	24.50	24.00	24.01	17.00	22.50

Now we use the "Shift" method shifting 0.33 units of traffic from path 2 to path 1 balacing the loads on links 2 and 3, and shifting 0.17 units of traffic from path 11 to path 12 balacing the loads on links 4 and 5. Once these two shifts are performed, links 2 and 3 are saturated. Shown in Table 5.

Only using "Shift" method, it may has a deadlock condition. We set up a "Diversion" method, in this "divert" loop, the paths connect-ing a given node pair are examined to determine the number of saturated links in each. Any of the paths which have a positve traffic allocation and which contain more than one saturated link is called "multisaturated". If one or more multisaturated paths exist, a small amount of traffic(currently limited to 10% of the traffic between the node pairs) is "divert" from the most multisaturated path (the one with the most saturated links) and split equally among all other paths connecting the particular node pairs.

The need for this "divert" process is best illustrated by example. Consider the test network (Fig. 1) with the traffic demand of Table 6.

Table 6 Traffic for Divert Case

Node	1	2	3	4	5	6
1	-	2	1	0	0	3
2		-	2	1	0	0
3			-	2	0	0
4				-	0	0
5					-	0
6						-

Assuming that all links are of unit capacity and that only minimum hops paths are permitted, Table 7 lists the initial equal allocation of traffic to all available paths together with the link loads.

Table 7 Initial Equal Allocation

Path Nodes	X(p)	Link Loads 1	2	3	4	5	6	7
1-2	1.0	2	0	0	0	0	0	0
1-3	1.0	0	1	0	0	0	0	0
1-2-4-6	1/3	1	0	1	0	1	0	0
1-3-4-6	1/3	0	1	0	0	1	0	1
1-3-5-6	1/3	0	1	0	1	0	1	0
2-4	1.0	0	0	1	0	0	0	0
2-1-3	1/2	1	1	0	0	0	0	0
2-4-3	1/2	0	0	1	0	0	0	1
3-4	1.0	0	0	0	0	0	0	2
TOTAL LINK LOADS		4	4	3	1	2	1	4

(In Table 7 link loads are expressed in absolute traffic terms rather than as fractions).

Every permissible path connecting nodes 1 and 6,and nodes 2 and 3 contains a saturated link. Therefore, the"Shift" loop cannot find any change which redues the most saturated link load for the paths connecting any node pair. This traffic pattern constitutes a "shift deadlock". It does not however, minimizethe maximum load on any link.

If 0.5 units of traffic is "divert" from the path (through nodes) 2-1-3 to the path 2-4-3 and the traffic on path 1-3-4-6 is "divert" to path 1-3-5-6, we obtain the results shown in Table 8.

With this routing pattern the maximum link load is reduced to 3.5 from 4.0 for the "shift deadlock" pattern of Table 8. The divert features attempts to avoid such "shift deadlock"

The second feature added to enchance convergence was the reodering the pairs of nodes for considering in the mean loops which reallocate traffic,without tnis, reodering convergence was found to be quite slow for some cases. Several different reodering algorithm have been considered. the following principles seem to provide the best scheme:

- Prioritize node pairs according to the numbers of connecting paths and the numbers of saturated links in the paths.
- Top priority should go to node pairs which contain both saturated paths and unsaturated paths.
- Higher priority should go to node pairs having lower numbers of connecting paths.
- Node pairs which have saturated connecting paths should be higher priority than those with only unsaturated paths.
- Possibly, the order of node pairs should be forced to change.

An iterative heuristic algorithm called MAXFLOW was developed to compute practical solutions to restricted, symmetric, multi-commodity routing problems. MAXFLOW shown in Fig. 2.

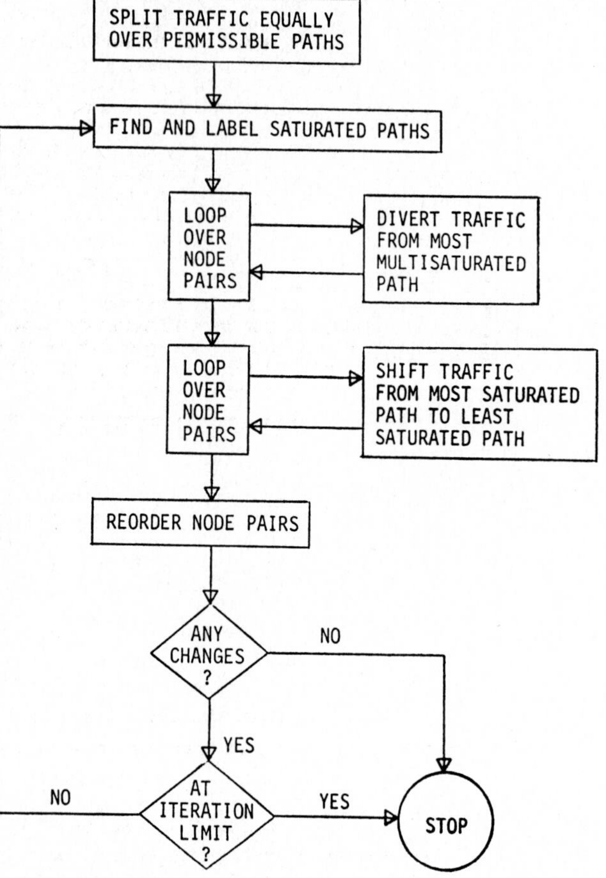

Fig. 2 MAXFLOW Summary

Table 8 After Diverting

Path Nodes	X(p)	Link Loads 1	2	3	4	5	6	7
1-2	1.0	2	0	0	0	0	0	0
1-3	1.0	0	1	0	0	0	0	0
1-2-4-6	1/3	1	0	1	0	1	0	0
1-3-4-6	0	0	0	0	0	0	0	0
1-3-5-6	2/3	0	2	0	2	0	2	0
2-4	1.0	0	0	1	0	0	0	0
2-1-3	1/4	0.5	0.5	0	0	0	0	0
2-4-3	3/4	0	0	1.5	0	0	0	1.5
3-4	1.0	0	0	0	0	0	0	2
TOTAL LINK LOADS		3.5	3.5	3.5	2	1	2	3.5

REFERENCES

1) G. Danzig,"Linear Programming and Extension", Princeton, N.J. 1963
2) H. Frank, M. Gerla, W. Chow, "Issues in the Design of Large Distributed Network", Proceeding of the IEEE National Telecommunications Conference, Atlanta, Ga, Nov. 1973
3) M. Gerla, W. Chow, H. Frank, "Computational Considerations and Routing Problems for Large Computer Communication Networks", Proceedings of the National Conference, Atlanta, Ga, Nov. 1973
4) A. Tanenbaum, "Computer Networks", Englewood Cliffs, N.J.,1981
5) H. Frank, I.T. Frisch, "Communicat ion, Transmission, and Transportation Networks", Reading, Mass.: Addision-Wesley, 1971.
6) L.R. Ford, and D.R. Fulkerson, "Flows in Networks", Princeton, N.J., 1962.
7) Howard Cravis, "Communications Network Analysis", Lexingt-onBooks, D.C. Heath and Company 1981.
8) W.L. Price, "Graphs and Networks, An Introduction.", London: Butterworth, 1971.

TELETRAFFIC ISSUES in an Advanced Information Society
ITC-11
Minoru Akiyama (Editor)
Elsevier Science Publishers B.V. (North-Holland)

CONTROLLING TRANSIENTS AND OVERLOADS IN COMMON CHANNEL SIGNALING NETWORKS

Jacqueline AKINPELU and Ronald SKOOG

AT&T Bell Laboratories
Holmdel, New Jersey 07733

ABSTRACT

With traffic overloads and signaling link failure/recovery events, large queues of messages can build up and cause significant transients to propagate through a signaling network. These transients must be controlled if network delay and message loss, duplication, and mis-sequencing objectives are to be met. In this paper we consider CCITT Signaling System No. 7 Level 3 congestion control and signaling link changeover/changeback procedures. We describe transient queueing models and analysis methods used to study the transients associated with overload controls and link changeover/changeback events, and we develop control strategies and methods for setting parameter values so as to achieve a desired network performance.

1. INTRODUCTION

With traffic overloads or signaling link failure/recovery events, large queues of messages can build up and cause significant transients to occur in a signaling network. These transients must be controlled if network delay and message loss, duplication, and missequencing objectives are to be met. For network overload it is assumed that the network congestion is controlled by using the CCITT Signaling System No.7 (SS7) Level 3 congestion control procedures [1]. For signaling link failure/recovery, it is assumed that the SS7 changeover/changeback procedures are used.

In setting congestion control thresholds, the main objectives are to control the traffic causing overload before significant delays build up in the network, to exercise these controls without having to discard messages in the network, and to exercise the controls only when there is a traffic overload. Regarding the last objective, an important relationship exists between setting congestion control thresholds and the transients induced by signaling link failure/recovery events.

When a signaling link fails, SS7 uses changeover (CO) procedures to divert traffic from the failed link to alternative links. The changeover procedures avoid message loss, duplication, and missequencing by using a handshaking procedure between the two ends of the failed signaling link and a message retrieval procedure to transfer messages from the failed link to the alternative links. When a signaling link recovers from a failure, traffic must be diverted from the alternative signaling links to the recovered link. Changeback (CB) procedures are used to accomplish this, and these procedures use handshaking and timeout methods to control message loss, duplication and missequencing.

An important aspect of these CO/CB procedures is that a significant queue of messages will build up in the transmit buffer of the failed link in the case of CO, and in the alternative link in the case of CB. This is because it takes a certain length of time to perform the handshaking or timeout procedures and, in the case of link failure, time to detect link failure. When this queue of messages is sent to the active link designated to carry the redirected traffic, care must be taken to ensure that the transmit buffers of these active links are not filled to levels exceeding the Level 3 congestion control thresholds, for if this were to happen, congestion controls would be invoked when there was no traffic overload.

In this paper we develop transient queueing system models and analysis methods to characterize the queues in signaling link transmit buffers during CO/CB and congestion control procedures, and we use these results to develop methods for setting parameter values and congestion control thresholds to achieve a desired network performance. In the next section we describe the relevant SS7 procedures, and then in Section 3 we provide the motivation for controlling CO/CB transients and choosing congestion control thresholds. In Sections 4 and 5 we develop the models and analysis for controlling CO/CB transients and choosing congestion control thresholds, respectively. A summary of the results and conclusions are given in Section 6.

2. SUMMARY OF RELEVANT SS7 PROCEDURES

2.1 The Signaling Link Buffers

There are three buffers associated with a SS7 signaling link: the transmit buffer, the retransmit buffer, and the receive buffer. A message arriving at a link for transmission is placed in the transmit buffer associated with the link. When the message is transmitted, it is removed from the transmit buffer, and a copy is stored in the retransmit buffer until a positive acknowledgment is received from the remote end of the link. In this way, the message is available in case it must be retransmitted due to link errors, remote Level 2 congestion, or some other anomaly. When a transmitted message reaches the remote end of the link, it enters a receive buffer and undergoes Level 2 processing. If it passes the error check and is accepted by Level 2, it is sent to Level 3 processing and a positive acknowledgment is sent to the transmitting end. If the message is not accepted, it is discarded, and a negative acknowledgment is sent to the transmitting end. Level 2 then discards all subsequent messages until it receives and accepts a retransmitted copy of the rejected message. When the transmitting end receives a positive acknowledgment, it removes the corresponding message and all those preceding it from the retransmit buffer. When it receives a negative acknowledgment, it retransmits the corresponding message and all those transmitted after it. With this procedure, all messages are maintained in sequence.

2.2 The Changeover/Changeback Procedures

A signal unit error rate monitor at the receiving end of a link is used to determine the rate of signal unit errors and decide if the link should be removed from service. This is the most frequent cause for failing signaling links. A "leaky bucket" overflow scheme is used wherein each errored signal unit is given a plus one count, and the acceptance of 256 signal units is given a minus one count when the existing total count is positive. A link is declared failed when the total count reaches 64.

When the receiving end of a signaling link decides to fail the link, it discards all subsequent received messages and sends a changeover order to the far end of the link. When the far end receives the CO order it stops transmission and sends a CO acknowledgment to the receiving end. It then determines the alternative signaling links on which to route the traffic, and enters a message retrieval procedure. In the message retrieval procedure, the signaling point retrieves the messages from the retransmit buffer that have not been acknowledged by the far end, retrieves all messages from the transmit buffer, and sends all retrieved messages, in their proper sequence, to the alternative links. All messages received during this process are queued and sent to the alternative links in proper sequence.

When a failed link recovers and is ready to be used, a changeback procedure is used to divert traffic back to the recovered link. The procedure begins with one end sending a CB declaration. In response, the far end sends a CB acknowledgment after it has processed all signal units relating to the traffic being diverted. When a signaling point receives the CB acknowledgment message, it then initiates a message retrieval procedure, analogous to that used for CO, to redirect traffic back to the recovered link.

When the two ends of a signaling link do not have a path over which to send the CO or CB messages, alternative procedures are invoked in which timers are used to delay transmission of messages on the new path. These delays are necessary to avoid out-of-sequence messages. The details on these procedures are given in Recommendation Q.704 in [1]. The major point for our purposes is that for all CO and CB procedures, there is a time interval during which transmission of messages is stopped, to ensure against out-of-sequence messages, and this results in a queue buildup that will subsequently be sent to an active link. The length of this time interval will vary depending on the particular implementation, but a value of at least 300 ms appears to be necessary for most implementations. For the examples in this paper, we have assumed a value of 350 ms for this time interval. For CO there is an additional queue buildup during the time before link failure is detected.

2.3 The Level 3 Congestion Control Procedures

Level 3 congestion is caused by an unusually high rate of message arrivals at the signaling link, which results in an extraordinary buildup of messages in the transmit buffer. The purpose of the Level 3 congestion control procedure is to reduce the amount of traffic offered to the link.

Associated with the Level 3 congestion control procedure is a set of transmit buffer thresholds. Threshold T is the congestion onset threshold. When the occupancy of the transmit buffer (in bytes) reaches T, Level 3 congestion is declared. New messages are still accepted for transmission over the link. However, when a new message arrives at the link destined for signaling point (SP) X, a notification is sent to the originating SP of the message informing it to stop sending traffic to destination X. These actions continue until the transmit buffer occupancy drops to the congestion abatement threshold A. At this point, the link returns to normal operation. Whenever the transmit buffer occupancy exceeds the congestion discard threshold D, no new messages are accepted for transmission over the link.

The above description of the transmit buffer thresholds assumes that all messages are treated the same. In fact, the SS7 Level 3 protocol provides for as many as four message priorities as a national option. In this case, every message carries with it an indication of its priority. These priorities provide the network with a means of giving selective treatment to messages during Level 3 congestion. To do this, three sets of transmit buffer thresholds, $\{T_i,A_i,D_i\}$, i=1,2,3, are specified. When threshold T_i is reached, congestion state i is declared. Only those new messages with priority $i-1$ or less result in the sending of congestion notifications to the originating SPs; higher priority messages are accepted as normal. The congestion notifications carry an indication of the congestion state i, as well as the affected destination X, and the originating SPs stop only those messages to X with priority $i-1$ or less. When the transmit buffer occupancy drops to threshold A_i, new messages with priority $i-1$ arriving at the link are queued as normal. Whenever threshold D_i is exceeded, all new messages with priority $i-1$ attempting to enter the transmit buffer are discarded. Priority 3 messages are always accepted as long as there is transmit buffer space available.

3. CONTROLLING TRANSIENTS AND DETERMINING CONGESTION THRESHOLDS: MOTIVATION

In this section we discuss the general nature of the CO/CB transients and the method to be studied for controlling them. We also discuss the criteria to be used in setting congestion thresholds and the nature of the transients that must be analyzed to determine these thresholds. The models and analysis related to these issues are discussed in Sections 4 and 5.

3.1 Controlling CO/CB Transients and Determining T_1

From the above discussion of the CO and CB procedures it is clear that the CO transients will be more severe than those for CB. This is because the message buildup for CO includes the time to detect link failure as well as the time to exchange CO messages. Also, with CO the link to which the traffic is sent is carrying its nominal load in addition to the redirected traffic. Therefore, we will restrict our attention to the CO situation.

We will assume that the traffic from the failed link is sent to one alternative link, and we will assume that both links have the same nominal load consisting of a Poisson message arrival process of rate λ (msg/sec) and a message length distribution with first and second moments $\overline{B}$ and $\overline{B^2}$, respectively. The ratio $\overline{B^2}/\overline{B}^2$ will be denoted by η. The signaling link transmission rate will be denoted by C (bytes/sec), and it is assumed that the retrieval process sends data at some fixed transmission rate C_r (bytes/sec). The signaling link nominal utilization, denoted by ρ, is $\lambda\overline{B}/C$, and we define the utilization of the retrieval process as $\rho_r = \lambda\overline{B}/C_r$.

The transients will be considered from the time (t=0) the message retrieval process begins. The size of the buffer buildup at the failed link will be expressed as $K\overline{B}$ bytes. Here, K is the number of messages in the buffer at time t=0. For our numerical examples we have used $K = 60$. This corresponds to a 56 Kbps link with $\rho = 0.4$, $\lambda = 100$ msg/sec, $\overline{B} = 28$ bytes, a 100 percent message error rate during a failure event, and 350 ms to carry out the CO procedures prior to retrieval.

We can model the transients by the queueing system shown in Figure 1a. Queue 1 (Q1) represents the buffering of messages at the failed link, before and during the retrieval process, and $\mu_r = C_r/\overline{B}$ is the average message service rate from Q1 during retrieval. Q1 is in the system for the length of its busy period, and it serves the load λ normally destined to the failed link. The transmit buffer of the alternative link is represented by Queue 2 (Q2), and it receives its nominal load λ in addition to the load from Q1. The average message service rate is denoted by $\mu = C/\overline{B}$. It is important to note that in Q1 the server is assumed to remove messages by removing bytes at a fixed rate, and not by removing each message as a unit. This assumption is made to simplify the analysis; it should not be viewed as a practical limitation. Also, it is assumed that the number of messages queued in Q2 in steady state will be small compared to the buildup during CO, so we can assume that Q2 is empty at $t = 0$.

From the results in the next section, it can be seen that the transients in Q1 and Q2 are closely approximated by what one would intuitively expect. We will denote the number of bytes in Q1 at time t by M_t and the number in Q2 by N_t. Consider the average number of bytes in these queues. Q1 will be in a busy period while it works off the initial $K\overline{B}$ bytes and the additional load that arrives during its busy period. The net rate at which bytes are removed is $(\mu_r-\lambda)\overline{B}$. Thus, the expected number of bytes in Q1 should decrease linearly in time from $K\overline{B}$ with slope $(\lambda-\mu_r)\overline{B}$, and its busy period has an expected value of $K/(\mu_r-\lambda)$. During the busy period of Q1, Q2 is receiving bytes at rate $(\mu_r+\lambda)\overline{B}$ and removing bytes at rate $\mu\overline{B}$. Thus, its expected number of bvtes should grow linearly in time with slope $(\mu_r+\lambda-\mu)\overline{B}$. After the Q1 busy period, Q2 will be receiving messages at rate 2λ, so its expected number of bytes should decrease in time with slope $(2\lambda-\mu)\overline{B}$. Note also that the expected number of bytes in the total system will start at $K\overline{B}$ and decrease at rate $(\mu-2\lambda)\overline{B}$ during the busy period of Q2.

These relationships are illustrated in Figure 1b. It is clear that the basic means of controlling the level of message buildup in Q2, without discarding messages, is by changing the retrieval service rate μ_r. By decreasing μ_r the maximum value for the expected number of messages in Q2 can be decreased. The criteria for choosing T_1 is that the probability of exceeding T_1 on the alternative link during a changeover event should be very small. For a given level of message buildup, K, the only means for reducing T_1 is to reduce μ_r. However, by decreasing μ_r the delays for the messages in Q1 are increased. Thus, the basic tradeoff in choosing μ_r is between the reduction of T_1, and thereby a reduction in the delays incurred before congestion controls are invoked during Level 3 congestion, and the decrease in delays for some messages during changeover events. This tradeoff is quantified in Section 4.

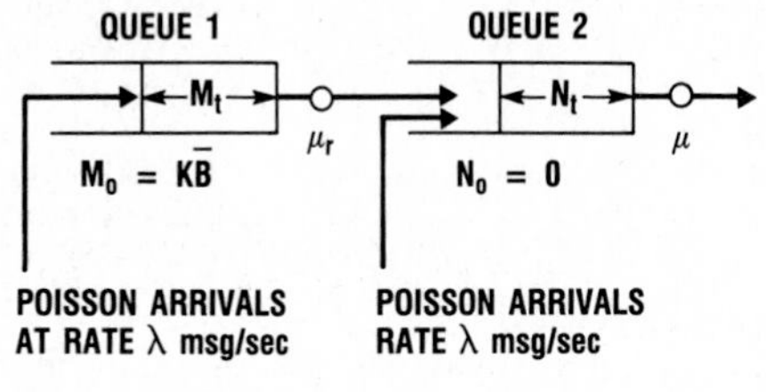

(a)

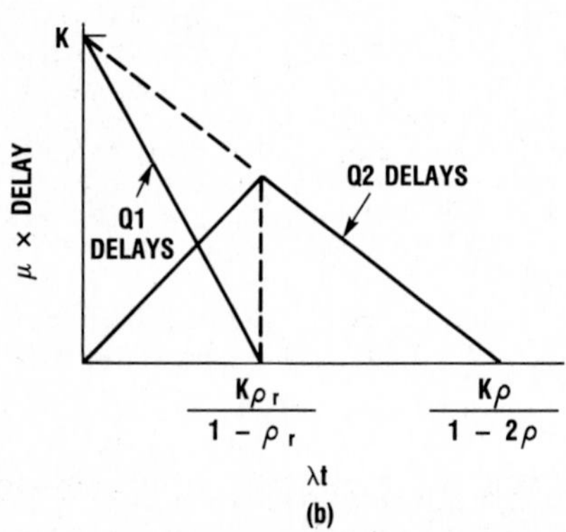

(b)

FIGURE 1 (a) THE CHANGEOVER QUEUEING MODEL
(b) THE EXPECTED QUEUEING DELAYS

3.2 *Determining D_i*

The motivation behind choosing the thresholds D_i is to minimize the discard of messages during Level 3 congestion. Let us assume that at time t=0, the transmit buffer occupancy reaches T_i and the link utilization is greater than one. Subsequently arriving messages with priority $i-1$ or less will trigger congestion notifications to their originations, resulting in an eventual reduction of this traffic. In addition, there is some background traffic (messages with priority i or higher) which is unaffected by the controls. Despite the removal of the controlled traffic from the link, the buffer occupancy may continue to grow, and if the threshold D_i is reached, newly arriving messages with priority $i-1$ will be discarded.

Suppose that ρ_t is the total link utilization at time t, ρ^* is the utilization due to the background traffic, and N_t is the buffer occupancy at time t. We assume that $\rho_0 > 1$. Because of the controls, ρ_t will converge to ρ^*. We consider the case of $\rho^* < 1$. In this case, the buffer occupancy N_t will eventually converge to that associated with a utilization of ρ^*. In general, this means that the buffer occupancy will peak and then gradually decrease. This is illustrated in Figure 2. In this case, we want to choose D_i to be larger than the peak buffer occupancy.

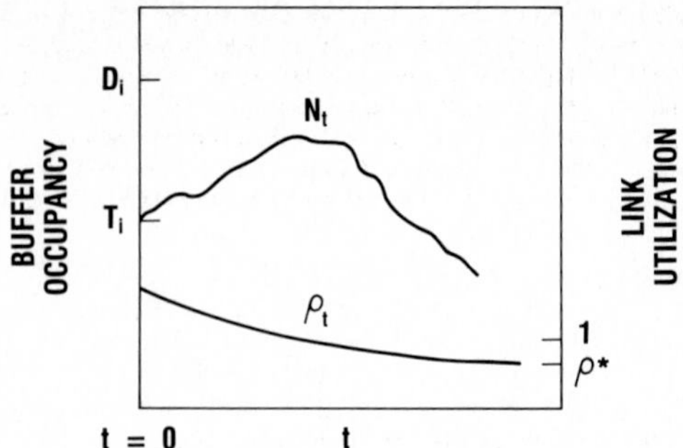

FIGURE 2 DETERMINING D_i

3.3 *Determining T_i, $i>1$, and A_i*

To understand the motivation behind choosing T_i, $i>1$, assume that the threshold D_{i-1} is reached; then all priority $i-2$ messages destined for the link are discarded. If the utilization due to the higher priority messages is less than one, the buffer occupancy should eventually settle to some satisfactory steady-state level without further controls. Hence, T_i should be set high enough above D_{i-1} so that the probability of reaching T_i with a link utilization less than one is very small.

The motivation behind choosing A_i is similar. We assume that the Level 3 congestion controls have reduced the link utilization to less than one, and as a consequence the buffer occupancy is decreasing. When the buffer occupancy reaches A_i, this should imply that the current level of traffic can be sustained without triggering congestion state i again, i.e., without reaching threshold T_i. Consequently, A_i should be chosen far enough below T_i so that with a link utilization less than one the probability of reaching T_i again is very small.

4. *ANALYSIS OF CHANGEOVER TRANSIENTS*

There are two aspects of the changeover transients that we want to investigate. One is the nature of the delays and how these change as the retrieval rate is changed. The other is what the threshold level should be so that the probability of exceeding this threshold during the CO process is acceptably small. To address these questions, we will first obtain results for queues Q1 and Q2, and then apply these results to the specific questions regarding delay and threshold levels.

4.1 *The Model and Analysis of Queues Q1 and Q2*

Our objective is to determine the mean values for M_t and N_t, and to determine the threshold $\Psi(\epsilon,t)$ such that $Pr[N_t > \Psi(\epsilon,t)] < \epsilon$. We are concerned with situations for which the probability that Q2 is in a busy period is equal to or very close to one. When this condition is not met, the delays and queue size in Q2 are sufficiently small that we can consider the transient effects to have passed. Thus, in modeling the behavior of Q2, we can assume it is in a busy period.

The approach we will use to determine $\Psi(\epsilon,t)$ is to determine the mean and variance of N_t, and then find $\Psi(\epsilon,t)$ assuming N_t is normally distributed. The assumption that N_t has a normal distribution is reasonable for the effects we are investigating. We are looking for significant accumulations of messages in Q2 due to random arrivals. Since the accumulations consist of a large number of random events, central limit theorem arguments can be used to justify a normal distribution. If the random events do not lead to a significant increase in queue size, then the queue behavior is largely deterministic, and we can characterize the system easily.

We will obtain expressions for the mean and variance of N_t by determining its characteristic function $\hat{N}(s\ ;t)$. Let τ_r denote the busy period of Q1. If $t < \tau_r$ then $N_t = (\mu_r - \mu)\bar{B}t + r_t$, where r_t is the random variable representing the total bytes arriving over period $(0,t]$ from a Poisson message arrival process of rate λ. Thus, r_t has mean $\lambda t\bar{B}$, variance $\lambda t\overline{B^2}$ and characteristic function $exp[\lambda t(B(s)-1)]$, where $B(s)$ is the characteristic function of the message length distribution [2]. Since r_t is the only random term in N_t, it follows that the characteristic function for N_t, given $\tau_r > t$, is

$$\hat{N}(s\,;t|\tau_r > t) = \exp\left\{\lambda t[B(s) - 1] - (\mu_r - \mu)\bar{B}ts\right\}$$

If $\tau_r \leqslant t$ then $N_t = \mu_r\tau_r\bar{B} - \mu t\bar{B} + r_t$ where r_t is the random variable representing the number of bytes received from a Poisson message arrival process of rate λ over the interval $(0,\tau_r]$ and rate 2λ over the interval $(\tau_r,t]$. It follows that, for a given $\tau_r \leqslant t$, r_t has characteristic function $\exp\left\{(2\lambda t - \lambda\tau_r)[B(s) - 1]\right\}$, and the characteristic function of N_t, given a $\tau_r \leqslant t$, is

$$\hat{N}(s\,;t|\,\tau_r) = \exp\left\{(2\lambda t - \lambda\tau_r)[B(s) - 1] + (\mu t - \mu_r\tau_r)\bar{B}s\right\}$$

The characteristic function for N_t is then given by

$$\hat{N}(s\,;t) = \int_0 \hat{N}(s\,;t|\,\tau_r)p(\tau_r)d\tau_r + \hat{N}(s\,;t|\tau_r > t)\;Pr[\tau_r \geqslant t] \quad (1)$$

where $p(\tau_r)$ is the pdf for τ_r.

To evaluate (1) we need information on the distribution of τ_r. Using arguments along the lines used by Takács [3], we will determine the characteristic function of τ_r, and from that development it will be seen that τ_r is the sum of a large number of independent 1-busy periods (a 1-busy period is the busy period when a queue begins with one message). As a consequence, we can use central limit theorem arguments to establish that τ_r is approximately normal. Thus, what we need to evaluate (1) is the mean and variance of τ_r.

Observe that Q1 will be busy for the time $\tau_m = K\bar{B}/C_r = K/\mu_r$ required to remove the initial $K\bar{B}$ bytes. During this time, there will be a random number, j, of message arrivals, where j is Poisson distributed with mean $\lambda\tau_m$. Each of these j arrivals can be viewed as generating a 1-busy period, and the busy period τ_r is then τ_m plus the sum of these j independent 1-busy periods. Let $\beta(s)$ denote the characteristic function of a 1-busy period. Then $\beta^j(s)$ is the characteristic function of the sum of j independent 1-busy periods. Using this with the Poisson distribution for j, it follows that the characteristic function for τ_r is

$$\hat{\tau}_r(s) = \exp\left\{\lambda\tau_m(\beta(s)) - 1) - \tau_m s\right\}$$

It then follows (see [2] p.230 for details on evaluating derivatives of $\beta(s)$) that

$$\bar{\tau}_r = \frac{K}{\mu_r}\left[\frac{1}{1 - \rho_r}\right]$$

$$\sigma^2_{\tau_r} = K\frac{\lambda\eta}{(\mu_r - \lambda)^3}$$

where $\eta = \overline{B^2}/\bar{B}^2$.

The mean and variance of N_t can be evaluated by taking the first and second derivatives of (1) with respect to s and evaluating the integrals, using the normal approximation for the distribution of τ_r, and setting $s = 0$. Carrying out this procedure gives the following result for $\bar{N}_t$:

$$\frac{\bar{N}_t}{\bar{B}} = (\lambda + \mu_r - \mu)t + [K + \lambda t - \mu_r t]\Phi\left[\frac{t - \bar{\tau}_r}{\sigma_{\tau_r}}\right] - (\mu_r - \lambda)\frac{\sigma_{\tau_r}}{\sqrt{2\pi}}\exp\left[-\frac{1}{2}\left[\frac{t - \bar{\tau}_r}{\sigma_{\tau_r}}\right]^2\right] \quad (2)$$

where $\Phi(x)$ denotes the standard normal distribution function. Recognizing that Q2 is in a busy period during the Q1 busy period, the expected number of bytes in the whole system is $\bar{M}_t + \bar{N}_t = (K + 2\lambda t - \mu t)\bar{B}$. Using this with (2) gives

$$\frac{\bar{M}_t}{\bar{B}} = [K - (\mu_r - \lambda)t]\left[1 - \Phi\left[\frac{t - \bar{\tau}_r}{\sigma_{\tau_r}}\right]\right] + (\mu_r - \lambda)\frac{\sigma_{\tau_r}}{\sqrt{2\pi}}\exp\left[-\frac{1}{2}\left[\frac{t - \bar{\tau}_r}{\sigma_{\tau_r}}\right]^2\right] \quad (3)$$

This result for $\bar{M}_t$ does not depend on Q2, and it is a general result valid for M/G/1 queues. Note, however, we have assumed Q1 does not receive load after its busy period. This result is an improvement over the approximation given in [4] where an M/M/1 queue was assumed. Figure 3 illustrates the results in (2) and (3); it is seen that simple linear approximations for these mean values are very accurate.

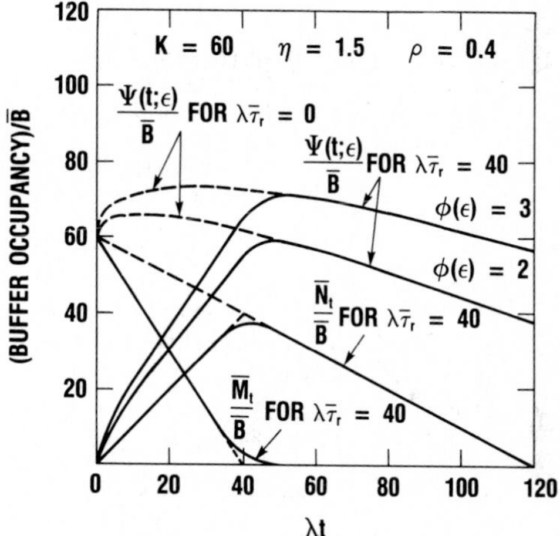

FIGURE 3 MEAN QUEUE LENGTHS FOR Q1 AND Q2 AND THRESHOLDS $\Psi(t;\epsilon)$

Using the assumption, discussed above, that N_t is normal, the expression for $\Psi(\epsilon,t)$ is obtained using (1) to find the variance of N_t. Let $\Phi(x)$ denote the standard normal distribution function, and define $\phi(\epsilon)$ by $\Phi(\phi(\epsilon)) = 1 - \epsilon$. Then the threshold $\Psi(\epsilon,t)$ is given by

$$\Psi(\epsilon,t) = \bar{N}_t + \phi(\epsilon)\sigma_{N_t} \quad (4)$$

We find that for $t \leqslant \bar{\tau}_r - 2\sigma_{\tau_r}$, $\sigma^2_{N_t} \approx \lambda t\overline{B^2}$, and for $t \geqslant \bar{\tau}_r + 2\sigma_{\tau_r}$, $\sigma^2_{N_t} \approx 2\lambda t\overline{B^2}$. Note that for $\bar{\tau}_r = 0$, $\sigma^2_{N_t} = 2\lambda t\overline{B^2}$. This follows from (1), but it can also be deduced by recognizing that in this case all $K\bar{B}$ bytes from Q1 are immediately sent to Q2, and Q2 receives a Poisson message stream of rate 2λ. From these results it is seen that the effects of controlling the rate of flow into Q2 are significant only for $t \leqslant \bar{\tau}_r + 2\sigma_{\tau_r}$. For larger t the system behaves as if there were no control (i.e., as if $\bar{\tau}_r = 0$).

These results are illustrated in Figure 3. In these examples we have chosen $\mu_r = \mu$, $\rho = 0.4$, and $\eta = 1.5$. The thresholds $\Psi(\epsilon,t)$ are shown for $\phi(\epsilon) = 2$ and 3, which correspond to ϵ, the probability of exceeding $\Psi(\epsilon,t)$, of approximately 0.02 and 0.0015, respectively. Also shown are the thresholds $\Psi(\epsilon,t)$ for $\bar{\tau}_r = 0$; this illustrates how the system behavior approaches the case when no control is used on Q1.

4.2 Analysis of Mean Delays

We are interested in the mean delays seen by messages arriving to the system after the start of the message retrieval process. There are two streams of messages to consider: those that first enter Q1 during its busy period, and those that first enter Q2. Both streams have the same delay after the Q1 busy period because Q1 is removed and all messages enter Q2. Messages entering Q2, either from Q1 or directly, have a mean delay in Q2 equal to $N_t/(\mu\overline{B})$, where N_t is given by (2), and t is the time the message enters Q2. The delay in Q1 is $\overline{M_t}/(\mu\overline{B})$ with $\overline{M_t}$ given by (3).

With the above results, it follows that the mean delays in Q1 and Q2 have the simple intuitive characteristics discussed in Section 3.1. To obtain the total delay experienced by messages entering Q1, we must add the Q1 and Q2 delays. This gives an expected total delay for messages entering Q1 at time t of $(K/\mu)(1+\rho_r) - (1-\rho-\rho\rho_r)t$. Figure 4 illustrates the delays for the two streams of traffic.

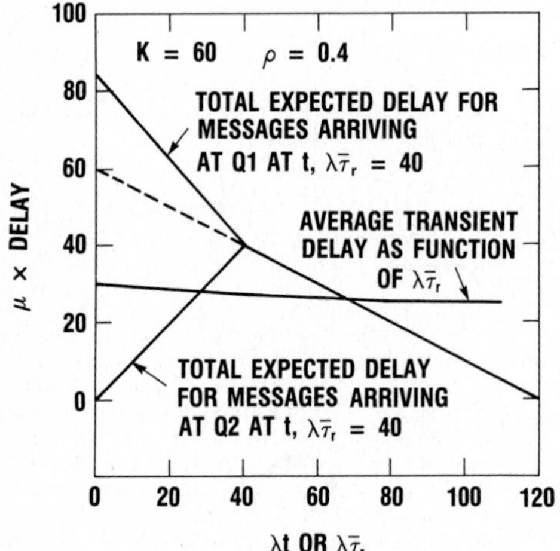

FIGURE 4 ILLUSTRATION OF MEAN TOTAL DELAY FOR MESSAGES ENTERING Q1 AND Q2, AND THE AVERAGE DELAY OVER THE TRANSIENT INTERVAL

Another quantity of interest is the average delay over all messages during the transient. To illustrate this we have averaged the above delays over the expected busy period of Q2. The result is

$$\begin{matrix}\text{Average}\\ \text{Transient}\\ \text{Delay}\end{matrix} = \frac{K}{2\mu}\left[1 - \frac{\lambda\overline{\tau}_r}{\lambda\overline{\tau}_r + K}\left(\frac{1-2\rho}{2\rho}\right)\right] \tag{5}$$

This average transient delay is also illustrated in Figure 4 as a function of $\lambda\overline{\tau}_r$. It is interesting to note that by increasing the expected busy period of Q1, the Q2 delays are decreased, but the average delay, over all messages, decreases very little. Thus, the transient effects in Q2 can be decreased, but only at the expense of the messages passing through Q1.

4.3 Controlling Threshold Levels in Q2

In order to set congestion threshold T_1 so that the probability of invoking congestion control during a CO transient is acceptably small, we need to know the maximum of $\Psi(\epsilon,t)$ over t. Figure 5 shows this maximum, $\Psi_{max}(\lambda\overline{\tau}_r;\epsilon)$, as a function of the Q1 expected busy period. It is interesting that as $\phi(\epsilon)$ is increased, the decrease in $\Psi_{max}(\lambda\overline{\tau}_r;\epsilon)$ for a given increase in $\overline{\tau}_r$ becomes less. It is also seen that for larger values of $\phi(\epsilon)$, the expected busy period of Q1 must be rather large before a significant reduction in $\Psi_{max}(\lambda\overline{\tau}_r;\epsilon)$ is achieved. As a result, one must be willing to increase delays in Q1 substantially to achieve reductions in threshold levels for larger values of $\phi(\epsilon)$. It should also be noted that significant errors would be obtained if the threshold values were determined on the basis of mean values.

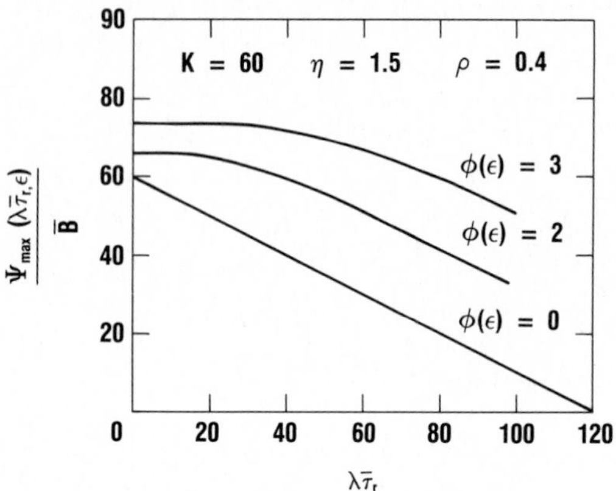

FIGURE 5 THRESHOLD VALUES FOR Q2 AS A FUNCTION OF $\lambda\overline{\tau}_r$

5. ANALYSIS FOR DETERMINING CONGESTION THRESHOLDS

Analysis of the first congestion threshold, T_1, was discussed in Section 4. In this section, we analyze the remaining congestion thresholds. Our approach is an iterative one: given T_i, we determine D_i; given D_{i-1}, we determine T_i; and, given T_i, we determine A_i. The basis of the analysis is identical to that discussed in Section 4. We analyze the occupancy of the transmit buffer, N_t, under certain assumptions. Then, assuming N_t is normally distributed, we determine a function $\Psi(\epsilon,t)$ such that $Pr[N_t > \Psi(\epsilon,t)] < \epsilon$. This function $\Psi(\epsilon,t)$ is given by (4). We then set the threshold under consideration based on $\Psi_{max}(\epsilon) = \max_t \Psi(\epsilon,t)$. In certain cases, this maximum can be computed explicitly. Again, we assume that the probability that the queue is in a busy period is very close to one.

5.1 Determining D_i

Let N_t be the transmit buffer occupancy at time t (in bytes) and assume that $N_0 = T_i$. Also let a 0 subscript refer to controlled traffic and a 1 subscript refer to uncontrolled traffic. Then λ_i will denote the message arrival rate at $t=0$, $\overline{B_i}$ and $\overline{B_i^2}$ will denote the first and second moments of the message length distributions, and τ will be the time between sending a congestion notification and the resultant reduction in traffic at the congested link.

We assume that the controlled traffic and uncontrolled traffic form two independent Poisson streams. Initially, during the interval $(0,\tau]$, before the control begins to take effect, the traffic rate to the congested link is $\lambda_0\overline{B}_0+\lambda_1\overline{B}_1$. After time τ, the traffic rate gradually decreases toward $\lambda_1\overline{B}_1$ as the controlled traffic is removed from the link. We assume that the initial traffic rate is sufficiently large so that the buffer occupancy begins to increase toward infinity (i.e., $\lambda_0\overline{B}_1+\lambda_1\overline{B}_1>C$). We also assume that, once the controlled traffic is removed, the buffer occupancy will decrease to an acceptable level (i.e., $\lambda_1\overline{B}_1<C$). We want to choose D_i so that the probability that the buffer occupancy exceeds D_i at any time $t>0$ is very small. This will be realized if

$$D_i > \Psi_{max}(\epsilon)$$

for suitable choices of λ_0, λ_1, and ϵ.

To determine the mean and variance of N_t, let us assume that at time $t = 0$, the controlled traffic consists of V independent Poisson origination-destination pairs, each with message rate λ_0/V. Each congestion notification affects traffic for one such pair. If we let Z_t denote the number of pairs that have been sent congestion notifications in $(0,t]$, then Z_t is a binomial random variable, and

$$\overline{Z}_t = V\left[1-\exp\left(-\lambda_0 t/V\right)\right]$$

$$\sigma^2_{Z_t} = V\left[1-\exp\left(-\lambda_0 t/V\right)\right]\exp\left(-\lambda_0 t/V\right) .$$

Now, let $\lambda_{0,t}$ be the message arrival rate for the controlled traffic at time t. Then, for $0 \leqslant t \leqslant \tau$, $\lambda_{0,t} = \lambda_0$, and for $t > \tau$, $\lambda_{0,t} = \lambda(V - Z_{t-\tau})/V$. Thus, it follows that

$$E(\lambda_{0,t}) = \begin{cases} \lambda_0 , & 0 \leqslant t \leqslant \tau \\ \lambda_0 \exp\left(-\lambda_0(t-\tau)/V\right) , & t > \tau \end{cases}$$

and

$$var(\lambda_{0,t}) = \begin{cases} 0 , & 0 \leqslant t \leqslant \tau \\ \dfrac{\lambda_0}{V}\left[1-\exp\left(-\lambda_0(t-\tau)/V\right)\right] E(\lambda_{0,t}) , & t > \tau . \end{cases}$$

Now, we divide the interval $(0,t]$ into M slots each of length t/M and define Y_m as the number of bytes of controlled traffic that arrive in slot m. For large values of M,

$$E\left[Y_m | \lambda_{0,s}, \frac{(m-1)t}{M} < s \leqslant \frac{mt}{M}\right] \approx \overline{B}_0 \lambda_{0,mt/M} \frac{t}{M}$$

$$var\left[Y_m | \lambda_{0,s}, \frac{(m-1)t}{M} < s \leqslant \frac{mt}{M}\right] \approx \overline{B^2_0} \lambda_{0,mt/M} \frac{t}{M} .$$

It follows that

$$\overline{Y}_m \approx \overline{B}_0 E\left[\lambda_{0,mt/M}\right] \frac{t}{M}$$

$$\sigma^2_{Y_m} \approx \overline{B^2_0} E\left[\lambda_{0,mt/M}\right] \frac{t}{M} + \left[\overline{B}_0\right]^2 var\left[\lambda_{0,mt/M}\right] \frac{t^2}{M^2} .$$

If R_t is the number of bytes of controlled traffic that arrives during $(0,t]$, then

$$\overline{R}_t = \lim_{M\to\infty} \sum_1^M \overline{Y}_m = \overline{B}_0 \int_0^t E(\lambda_{0,s}) ds$$

$$\sigma^2_{R_t} = \lim_{M\to\infty} \sum_1^M \sigma^2_{Y_m} = \overline{B^2_0} \int_0^t E(\lambda_{0,s}) ds ,$$

since the sum over the second term of $\sigma^2_{Y_m}$ goes to zero. By combining the controlled and uncontrolled traffic at time t, we get the following:

$$\overline{N}_t = T_i + \overline{B}_0 \int_0^t E(\lambda_{0,s}) ds + \overline{B}_1 \lambda_1 t - Ct$$

$$= \begin{cases} T_i + \overline{B}_0 \lambda_0 t + \overline{B}_1 \lambda_1 t - Ct , & 0 < t \leqslant \tau \\ T_i + \overline{B}_0 \left[\lambda_0 \tau + V - V\exp\left(-\lambda_0(t-\tau)/V\right)\right] \\ \quad + \overline{B}_1 \lambda_1 t - Ct , & t > \tau \end{cases}$$

$$\sigma^2_{N_t} = \overline{B^2_0} \int_0^t E(\lambda_{0,s}) ds + \overline{B^2_1} \lambda_1 t$$

$$= \begin{cases} \overline{B^2_0} \lambda_0 t + \overline{B^2_1} \lambda_1 t , & 0 < t \leqslant \tau \\ \overline{B^2_0} \left[\lambda_0 \tau + V - V\exp\left(-\lambda_0(t-\tau)/V\right)\right] + \overline{B^2_1} \lambda_1 t , & t > \tau . \end{cases}$$

The function $\Psi_{max}(\epsilon)$ can be computed numerically. If T_i is sufficiently large, then our assumption that the probability that the queue is in a busy period is close to one will be satisfied over the interval where $\Psi(\epsilon,t)$ achieves its maximum.

5.2 Determining T_i, $i>1$, and A_i

Let us assume that at time $t=0$, the buffer occupancy is D_{i-1} and the arrival rate (corresponding to the uncontrolled traffic) is a constant, λ. The first and second moments of the message length distribution for the incoming traffic are given by $\overline{B}$ and $\overline{B^2}$, and we assume that $\lambda\overline{B} < C$. It is readily shown that $\overline{N}_t = D_{i-1} + (\lambda\overline{B} - C)t$, $\sigma^2_{N_t} = \lambda t \overline{B^2}$, so that

$$\Psi(\epsilon,t) = D_{i-1} + (\lambda\overline{B} - C)t + \phi(\epsilon)\left[\lambda t \overline{B^2}\right]^{1/2} .$$

To determine $\Psi_{max}(\epsilon)$, we differentiate $\Psi(\epsilon,t)$ with respect to t and set the derivative equal to zero. This gives

$$\Psi_{max}(\epsilon) = D_{i-1} + \frac{3\phi(\epsilon)^2 \lambda \overline{B^2}}{4(C - \lambda\overline{B})} .$$

In this case, we choose $T_i > \Psi_{max}(\epsilon)$. Similar arguments can be applied to show that A_i should satisfy

$$A_i < T_i - \frac{3\phi(\epsilon)^2 \lambda \overline{B^2}}{4(C - \lambda\overline{B})} .$$

6. CONCLUSIONS

In this paper we have developed models and approximation techniques that can be used to establish control parameters and congestion thresholds so as to achieve a desired network performance under overload and signaling link failure conditions. We have demonstrated a close relationship between the transient effects that arise under signaling link changeover/changeback events and the setting of congestion control thresholds. The first congestion threshold is set high enough that the probability that it is exceeded during a CO transient is less than some chosen value. The control strategy for reducing changeover/changeback transient effects is to regulate the flow of data from the failed link to the alternative link during the message retrieval process. The first congestion threshold can be reduced by increasing the time to complete the CO retrieval process. We have shown that the amount the threshold value can be reduced, for an increase in the time to complete the retrieval process, can be substantially less than what would be predicted on the basis of mean values.

The effect of controlling the retrieval rate from Q1 on the behavior of Q2 is significant only for times less than the expected Q1 busy period. For times greater than the expected busy period of Q1, the behavior of Q2 rapidly approaches what it would be in the system with no throttling control at Q1.

The basic tradeoff in setting the first congestion threshold is between the delays that will be experienced during a network overload condition and transient delays during changeover/changeback events. The results of this paper allow this tradeoff to be quantified for specific network implementations.

We have also shown in this paper how to set the remaining congestion thresholds so as to keep the probability of lost messages as small as desired. There is a tradeoff here between delay and lost messages, and the results of this paper allow this tradeoff to be quantified.

Acknowledgment

The authors would like to thank D.A. Kettler for supporting this work and participating in helpful discussions.

REFERENCES

[1] CCITT Study Group XI, "Specifications Of Signaling System No.7, Red Book," Volume VI Fascicle VI.7, to be issued in 1985.

[2] R. B. Cooper, "Introduction to Queueing Theory, Second Edition," North Holland, New York, 1981.

[3] L. Takács, "Introduction to the Theory of Queues," Oxford University Press, New York, 1962.

[4] R. A. Skoog, "Transient Considerations in the Performance Analysis of Ring-Based Packet Switches," Proc. of the IFIPWG7.3/TC6 Second International Symposium on the Performance of Computer-Communications Systems, Zurich, Switzerland, March, 1984.

TELETRAFFIC ISSUES in an Advanced Information Society
ITC-11
Minoru Akiyama (Editor)
Elsevier Science Publishers B.V. (North-Holland)

APPLICATION OF BAYESIAN TELETRAFFIC MEASUREMENT TO SYSTEMS WITH QUEUEING OR REPEATED ATTEMPTS

Robert WARFIELD and Gavin FOERS

Telecom Australia Research Laboratories
Melbourne, Australia

ABSTRACT

In this paper Bayesian Statistical Inference is applied to the estimation of traffic parameters for systems with queueing and systems with repeated attempts. Algorithms for computing estimators of the mean and variance of the offered traffic and the mean service time are developed. A method for incorporating constraints on the variance to mean ratio of the offered traffic is also introduced.

1. INTRODUCTION

In References [1] through [3], the authors have described an application of Bayesian Statistical Inference to the problem of teletraffic measurement. The particular cases dealt with are loss systems with or without internal congestion. The theory on which this work is based can be found in References [4] through [7], and related work is described in References [8] through [10]. This method appears to offer some potential advantages over exisitng methods, as discussed in the concluding Section of this paper. However, it has not yet been applied in practice.

The work reported in the present paper extends the application of Bayesian Statistical Inference to two additional areas of practical significance: systems with queueing, and systems with repeated attempts. For the models of the two new applications, the algorithms for computing estimators are derived. The resulting estimation algorithms are somewhat similar to those previously derived for loss systems without repeated attempts.

The method of estimating parameters is extended to allow the variance to mean ratio of the offered traffic to be constrained either to a fixed value or to a range of values. Introducing a constraint on the variance to mean ratio eliminates the possibility of deriving unrealistic or physically impossible estimates of the mean of the offered traffic.

In Section 2 of this paper previous results are briefly summarised. At the end of Section 2, problems that can arise in the application of the basic results are discussed. In Section 3 systems with queueing are considered: unconstrained estimation is dealt with in 3.1, and constrained estimation is introduced in 3.2. Section 4 deals with systems having repeated attempts. A simple model of repeated attempts is proposed in 4.1. Unconstrained and constrained estimation of the parameters of the model are dealt with in 4.2 and 4.3 respectively. Numerical results for a simple example are given in Section 5, and conclusions are given in Section 6.

2. SUMMARY OF PREVIOUS RESULTS

For convenience, a brief summary is given of the relevant results reported in References [1] to [3]. The traffic process being observed is modelled as a Markov process specified by parameters ($\underline{a}$, $\underline{b}$, $\underline{c}$), where

- $\underline{a}$ is an R+1 dimensional vector of arrival coefficients.
- $\underline{b}$ is an R+1 dimensional vector of blocking probabilities (loss factors).
- $\underline{c}$ is an R+1 dimensional vector of departure coefficients.
- r is the state of the system (number of calls in the system),
- R is the maximum possible value of the state - for a queueing system this is the number of servers plus the number of waiting places; for a loss system it is just the number of servers.

Individual components of all vectors are denoted by subscripts, for example components of $\underline{a}$ are:

$$a_r \quad (r = 0,1,\ldots,R) .$$

It is assumed that the group occupancy is observed as a function of time, and that the time of occurrence of every unsuccessful bid is also observed. Therefore, the observations are taken as: the time of occurrence of each event (bid or departure), denoted

$$t(i) \qquad (i = 0,1,\ldots,k);$$

and the state of the system prior to each event,

$$r(i) \qquad (i = 0,1,\ldots,k).$$

It can be shown that the following statistic constitutes an "information state" for the system - that is, a recursively computable, sufficient statistic of fixed, finite dimension:

$$(r(0), r(k), \underline{s}(k), \underline{u}(k), \underline{v}(k))$$

where $\underline{s}(k)$, $\underline{u}(k)$, and $\underline{v}(k)$ are each R+1 dimensional vectors with r'th components defined by:

$s_r(k)$ is the total time spent in state r,

$u_r(k)$ is the total number of unsuccessful bids which occurred while the system was in state r,

$v_r(k)$ is the total number of successful bids which occurred while the system was in state r,

with all statistics being collected over the period from the 0'th event to the k'th event.

Let the R+1 dimensional vector $\underline{w}(k)$ be defined by

$w_r(k)$ is the total number of departures which occurred while the system was in state r.

This statistic can be computed from:

$$w_0(k) = 0 , \quad (1)$$

and, for r = 1,...,R,

$$w_r(k) = v_{r-1}(k) + e(r(0), r(k), r) , \quad (2)$$

where the function e(.,.,.) is defined by

$$e(r(0), r(k), r) = \begin{cases} -1 \text{ if } r(0) < r \leq r(k) \\ +1 \text{ if } r(k) < r \leq r(0) \\ 0 \text{ otherwise .} \end{cases} \quad (3)$$

It is desired to estimate the unknown parameters $(\underline{a}, \underline{b}, \underline{c})$ from the sufficient statistic given above. For this purpose, the following function is introduced:

$p_k(\underline{a}, \underline{b}, \underline{c})$ is the likelihood kernel, at stage k, of the unknown parameters.

The likelihood kernel may be written directly by taking the probability density function of the observations as a function of the unknown parameters $(\underline{a}, \underline{b}, \underline{c})$, and eliminating any factor which does not depend on the unknown parameters. In this case, it is given by the expression below, in which the dependence on k of s_r, u_r, v_r, w_r, and $p(\underline{a}, \underline{b}, \underline{c})$ is not shown for brevity:

$$p(\underline{a}, \underline{b}, \underline{c}) = \prod_{r=0}^{R} \{ (1-b_r)^{v_r} \; b_r^{u_r} \; a_r^{(u_r + v_r)} \; c_r^{w_r} \exp(-s_r(a_r + c_r)) \} \quad (4)$$

The probability density function of $(\underline{a}, \underline{b}, \underline{c})$, conditional on the observations, is equal to the function $p(\underline{a}, \underline{b}, \underline{c})$ multiplied by a normalising constant and by the prior density function of the unknowns. The normalising constant is computed by numerical integration of the function $p(\underline{a}, \underline{b}, \underline{c})$. The prior density function must be assigned, and may incorporate prior information regarding $(\underline{a}, \underline{b}, \underline{c})$.

The results summarized above are based on a general model. Several special cases have also been dealt with in [1] to [3]. In this paper we consider models which involve negligible internal congestion. This is appropriate for modern digital switching systems with reasonable levels of traffic loading. Hence, as explained in [3], the term

$$(1-b_r)^{v_r} \quad b_r^{u_r}$$

is set equal to unity.

The offered traffic is modelled as being either Binomial, Poisson, or Negative Binomial. Hence, the arrival coefficients are

$$a_r = (S-r)\,\alpha \quad (5)$$

where

S is the effective number of sources,

α is the arrival rate per free source.

Note that S and α may be both positive (Binomial), or both negative (Negative Binomial). Poisson traffic corresponds to the limit as S approaches infinity and α approaches zero, with the product $S\alpha$ remaining constant.

Assuming Negative Exponential service times, the departure coefficients are

$$c_r = r\,\mu \quad (6)$$

where

μ is the mean service rate,

h is the mean service time,

and μ is related to h by

$$\mu = 1 / h \quad (7)$$

Estimators of the unknown parameters can be found by computing the values of the parameters which maximise the logarithm of $p(\underline{a}, \underline{b}, \underline{c})$. The resulting estimators may be interpreted as Maximum Likelihood Estimators, or, equivalently, as Maximum Aposteriori Probability Density Estimators for the case of uniform prior density. The logarithm of the function p can be written as:

$$\ln(p) = \sum_{r=0}^{R} \{ n_r \ln(\alpha(S-r)) + w_r \ln(r\mu) - s_r(\alpha(S-r) + r\mu) \} \quad (8)$$

where

$$n_r \stackrel{\Delta}{=} u_r + v_r \quad (9)$$

Necessary conditions to be satisfied by the estimators are found by equating the partial derivatives of ln(p) to zero. Firstly, differentiating with respect to μ and equating to zero yields

$$\sum_{r=0}^{R} (- r\, s_r + w_r / \mu) = 0 \,. \qquad (10)$$

Differentiating with respect to α yields

$$\sum_{r=0}^{R} (-s_r(\hat{S}-r) + n_r / \hat{\alpha}) = 0 \,, \qquad (11)$$

and with respect to S yields

$$\sum_{r=0}^{R} (-\hat{\alpha}\, s_r + n_r / (\hat{S}-r)) = 0 \,. \qquad (12)$$

Defining

$$N \triangleq \sum_{r=0}^{R} n_r \,, \qquad (13)$$

$$W \triangleq \sum_{r=0}^{R} w_r \,, \qquad (14)$$

$$T \triangleq \sum_{r=0}^{R} s_r \,, \qquad (15)$$

and average occupancy

$$\bar{r} \triangleq (1/T) \sum_{r=0}^{R} r\, s_r \,, \qquad (16)$$

it follows that

$$\hat{\mu} = W / (\bar{r}\, T) \,, \qquad (17)$$

$$\hat{\alpha} = N / (T (\hat{S} - \bar{r})) \,, \qquad (18)$$

and $\hat{S}$ satisfies

$$\sum_{r=0}^{R} (n_r / (\hat{S}-r)) = T \hat{\alpha} \qquad (19)$$

$$= N / (\hat{S} - \bar{r}) \,. \qquad (20)$$

The last equation does not have a general closed form solution for $\hat{S}$. However, it can be solved iteratively as described in [3].

The offered traffic is defined to be the traffic that would be carried on an infinite group of servers with the arrival and departure coefficients as above. For cases where the inequality

$$\alpha\, h > -1 \qquad (21)$$

is satisfied, the mean of the offered traffic is given by:

$$m = S\, \alpha\, h / (1 + \alpha\, h) \qquad (22)$$

and the variance is given by

$$v = S\, \alpha\, h / (1 + \alpha\, h)^2 \qquad (23)$$

Hence, estimators of m, v, and h can be written as:

$$\hat{m} = \hat{S}\, \hat{\alpha}\, \hat{h} / (1 + \hat{\alpha}\, \hat{h}) \qquad (24)$$

$$\hat{v} = \hat{S}\, \hat{\alpha}\, \hat{h} / (1 + \hat{\alpha}\, \hat{h})^2 \qquad (25)$$

$$\hat{h} = 1 / \hat{\mu} \qquad (26)$$

For cases where

$$\hat{\alpha}\, \hat{h} < -1 \qquad (27)$$

then the mean of the corresponding offered traffic is undefined. The physical interpretation of this is that the arrival rate, as a function of the state, increases faster than the departure rate. Hence as more servers are made available to handle the traffic, the carried traffic increases without limit. This problem arises from the physically untenable assumption that the arrival rate as a function of the state can be extrapolated linearly to any value of r. Clearly, from physical arguments, the arrival rate cannot continue to exceed the departure rate as the number of servers increases without limit.

Even as the term $\hat{\alpha}\hat{h}$ approaches -1, $\hat{m}$ and $\hat{v}$ become very sensitive to small errors. A practical means of overcoming these problems of sensitivity to errors and undefined offered traffic is to introduce a realistic constraint on the variance to mean ratio of the offered traffic. This is dealt with in following Sections.

3. SYSTEMS WITH QUEUEING

The systems considered are those with a single queue which has a finite number of waiting places, and which provides access to a finite group of servers. Internal congestion is taken to be negligible. The offered traffic is modelled as being either Binomial, Poisson, or Negative Binomial. Hence, the arrival coefficients are as given in the previous Section.

Assuming Negative Exponential service times, the departure coefficients are

$$c_r = x\, \mu \qquad (28)$$

where

x is the number of busy servers, given by

$$x = \begin{cases} r & \text{for } 0 \leq r \leq R-Q \\ R-Q & \text{for } R-Q < r \leq R \end{cases} \qquad (29)$$

where

Q is the number of waiting places,

R-Q is the number of servers.

3.1 Unconstrained Estimation of Parameters

The logarithm of $p(\underline{a},\underline{b},\underline{c})$ can be written as:

$$\ln(p) = \sum_{r=0}^{R} \{ n_r \ln(\alpha(S-r)) + w_r \ln(x\,\mu) - s_r(\alpha(S-r) + x\,\mu) \} \quad (30)$$

Necessary conditions to be satisfied by the estimators are found by equating the partial derivatives of ln(p) to zero. Differentiating with respect to μ and equating to zero yields

$$\sum_{r=0}^{R} (- x\,s_r + w_r / \hat{\mu}) = 0 . \quad (31)$$

Hence, the estimator of μ is

$$\hat{\mu} = W / (T\,\overline{x}) \quad (32)$$

where

$$\overline{x} \triangleq (1/T) \sum_{r=0}^{R} x\,s_r . \quad (33)$$

The estimators for α and S are exactly as given in Section 2, since the derivatives of the logarithm of p with respect to these two variables are unchanged.

3.2 Constrained Estimation of Parameters

As discussed in Section 2, it is possible to produce unrealistic estimates of the mean offered traffic in some cases. To overcome this problem, the constraint

$$\frac{v}{m} = z \quad (34)$$

is introduced, where

z is the variance to mean ratio (or peakedness) of the offered traffic, and is assumed to be known.

The case of z=1 must be dealt with separately later. For other cases, it is equivalent to use the constraint

$$\mu = \frac{z}{1-z}\,\alpha \quad (35)$$

The logarithm of p now becomes

$$\ln(p) = \sum_{r=0}^{R} \{ n_r \ln(\alpha(S-r)) + w_r \ln(x\,\alpha \frac{z}{1-z}) - s_r(\alpha(S-r) + x\,\alpha \frac{z}{1-z}) \} \quad (36)$$

Equating the appropriate partial derivatives to zero yields the following necessary conditions for the estimators:

$$\hat{\alpha} = (N+W) / (T (\hat{S} - \overline{r} + \frac{z}{1-z} \overline{x})). \quad (37)$$

and

$$\sum_{r=0}^{R} \frac{n_r}{\hat{S} - r} = T\,\hat{\alpha} \quad (38)$$

$$= (N+W) / (\hat{S} - \overline{r} + \frac{z}{1-z} \overline{x}) . \quad (39)$$

This last equation can be solved iteratively by adjusting the value of S until

$$S - (N+W) / \sum_{r=0}^{R} \frac{n_r}{S - r} = \overline{r} + \frac{z}{1-z} \overline{x} . \quad (40)$$

For the case of z=1, S must approach infinity and α must approach zero in such a way that

$$S\,\alpha = a_0 . \quad (41)$$

Any non-negative value of μ will satisfy z=1. Hence, the arrival rate is estimated by

$$\hat{a}_0 = N / T \quad (42)$$

and the service rate by

$$\hat{\mu} = W / (T\,\overline{r}) . \quad (43)$$

4. SYSTEMS WITH REPEATED ATTEMPTS

4.1 Repeated Attempt Model

When congestion is high, it is possible that repeated attempts will be a significant proportion of all bids. The objective adopted here is to eliminate the effect of repeat attempts on estimates of the parameters of the offered traffic. The assumption being that if sufficient servers are provided then repeated attempts will become insignificant.

For the limited objective described above, the following model of repeated attempt behaviour is proposed. While there is no congestion the arrival rate is modelled as before. Once an unsuccessful bid occurs the arrival rate jumps to some constant value, presumably higher than normal. The arrival rate remains at this higher value until the next successful bid, at which time it reverts immediately to the normal level depending on the state of the system.

It is emphasized that this approach is proposed as a means of eliminating the effect of repeated attempts on the estimation of offered traffic. It is not intended to provide a good model for predicting repeated attempt behaviour.

To analyse the model of repeated attempt behaviour, a second state variable is introduced, namely:

q is a state variable which takes the value 0 for normal arrival rate, and 1 for higher than normal arrival rate.

ρ is the value of arrival rate for q=1.

Figure 1 shows the state transition diagram for this model.

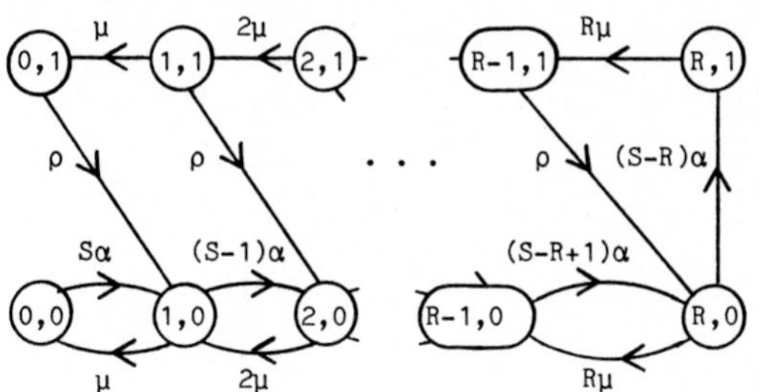

Fig. 1 State Transition Diagram for Repeated Attempt Model.

It is desired to solve for P(r,q) for r=0,...,R and q=0,1, where

P(r,q) is the steady state probability of the state (r,q).

First consider the subset of the state-space defined by

$$\{ (i,1) \mid i < r, \text{ for some } r < R \}.$$

This subset can be described as the set of all states (i,1) such that i is less than some fixed value r, which in turn is less than R. This subset may be entered only if a departure occurs while the system is in state (r,1). Hence the rate of entry into this subset is just $r\mu$. An arrival occurring while in any state within the subset will cause an exit from the subset, hence the rate of exit is given by

$$\rho \sum_{i=0}^{r-1} P(i,1)$$

Equating the rate of entry to the rate of exit (for steady state) gives

$$r\mu\, P(r,1) = \rho \sum_{i=0}^{r-1} P(i,1) \tag{44}$$

Thus, starting with an arbitrary constant for P(0,1), the values of P(r,1) for r=1,...,R-1 can be computed.

Consider the state (R,1). This state will be entered if and only if an unsuccessful bid occurs while the system is in state (R,0). Hence the rate of entry into this state is $(S-R)\alpha\, P(R,0)$. The state will be exited if and only if a departure occurs. Thus the rate of exit is $R\mu\, P(R,1)$. Equating the rates of entry and exit,

$$P(R,0) = \frac{R\mu}{(S-R)\alpha}\, P(R,1) \tag{45}$$

which gives the value of P(R,0) (still in terms of an arbitrary constant for P(0,1)).

To work down from state (R,0) to state (1,0), consider subsets of the form

$$\left\{ (i,q) \;\middle|\; \begin{matrix} q=0,\ i=r,\ldots,R \\ q=1,\ i=r-1,\ldots,R-1 \end{matrix} \right\}$$

The only way of entering this subset is for an arrival to occur while in state (r-1,0). The rate of entry is thus

$$(S-r+1)\alpha\, P(r-1,0).$$

Exit from the subset may be effected by a departure from state (r-1,1) or a departure from state (r,0). Thus the rate of exit is

$$r\mu\, P(r,0) + (r-1)\mu\, P(r-1,1).$$

Equating the rates of entry and exit gives

$$P(r-1,0) = \frac{\mu}{(S-r+1)\alpha} \left(r\, P(r,0) + (r-1)\, P(r-1,1) \right) \tag{46}$$

Finally, by similar argument,

$$P(0,0) = \frac{\mu}{S\,\alpha}\, P(1,0) \tag{47}$$

All the values of P(r,q) are thus computed in terms of an arbitrary constant assigned to P(0,1). The values obtained are then normalised to give the steady state probability distribution. From which time congestion is given by

$$E = P(R,0) + P(R,1) \tag{48}$$

and call congestion is given by

$$B = \frac{\rho\, P(R,1) + (S-r)\,\alpha\, P(R,0)}{\rho \sum_{r=0}^{R} P(r,1) + \sum_{r=0}^{R} (S-r)\alpha\, P(r,0)}\,. \tag{49}$$

In computing the offered traffic, the effect of repeated attempts is ignored, since congestion will never occur with unlimited servers. Hence the expressions for mean and variance of the offered traffic remain unchanged. Similar problems still arise for $(\alpha\, h) \leq -1$.

4.2 Unconstrained Estimation of Parameters

The notation used previously must be extended to cope with the two-dimensional state-space for the model introduced in the previous Section. The statistics to be collected and processed are now denoted by:

$s_{r,q}(k)$ is the total time spent in state (r,q),

$u_{r,q}(k)$ is the total number of unsuccessful bids which occurred while the system was in state (r,q),

$v_{r,q}(k)$ is the total number of successful bids which occurred while the system was in state (r,q),

$n_{r,q}(k)$ is the sum of $u_{r,q}(k)$ and $v_{r,q}(k)$,

$w_{r,q}(k)$ is the total number of departures which occurred while the system was in state (r,q).

with all statistics being collected over the period from the 0'th event to the k'th event.

The likelihood kernel is now given by (with the dependence on k omitted for brevity):

$$p = \prod_{r=0}^{R} \{ ((S-r)\alpha)^{n_{r,0}} (r\mu)^{w_{r,0}} \exp(-s_{r,0}((S-r)\alpha + r\mu))$$

$$\rho^{n_{r,1}} (r\mu)^{w_{r,1}} \exp(-s_{r,1}(\rho + r\mu)) \} \quad (50)$$

As a notational convention, the sum of a variable over all possible values of one subscript is denoted using a period to replace the appropriate subscript. For example,

$$s_{r.} = s_{r,0} + s_{r,1} \quad (51)$$

and

$$s_{.0} = \sum_{r=0}^{R} s_{r,0} \quad (52)$$

Using this notation for s, u, v, w, and n, the logarithm of p can be written as

$$\ln(p) = \sum_{r=0}^{R} \{ w_{r.} \ln(r\mu) - s_{r.} r\mu + n_{r,0} \ln((S-r)\alpha) - s_{r,0} (S-r)\alpha + n_{r,1} \ln(\rho) - \rho s_{r,1} \} \quad (53)$$

Taking the partial derivative with respect to ρ and equating to zero gives

$$\hat{\rho} = \frac{n_{.1}}{s_{.1}} \quad (54)$$

Following a similar procedure for μ, it is also found that

$$\hat{\mu} = (w_{.0} + w_{.1}) / (T \, \bar{r}) \quad (55)$$

where, in this context,

$$\bar{r} \triangleq (1/T) \sum_{r=0}^{R} r \, s_{r.} \quad (56)$$

and

$$T \triangleq s_{.0} + s_{.1} \quad (57)$$

Similarly, it can be shown that

$$\hat{\alpha} = n_{.0} / (s_{.0} (\hat{S} - \bar{r}_0)) \quad (58)$$

where

$$\bar{r}_0 \triangleq (1/s_{.0}) \sum_{r=0}^{R} s_{r,0} \, r \quad (59)$$

By a similar procedure, $\hat{S}$ must satisfy

$$\sum_{r=0}^{R} \frac{n_{r,0}}{(\hat{S} - r)} = n_{.0} / (\hat{S} - \bar{r}) \quad (60)$$

As before, this may be solved iteratively.

4.3 Constrained Estimation of Parameters.

As before, the constraint

$$\mu = \frac{z}{1-z} \alpha \quad (61)$$

is introduced. The logarithm of p is now

$$\ln(p) = \sum_{r=0}^{R} \{ w_{r.} \ln(r \alpha \frac{z}{1-z}) - s_{r.} r \alpha \frac{z}{1-z} + n_{r,0} \ln(\alpha(S-r)) - s_{r,0} \alpha(S-r) + n_{r,1} \ln(\rho) - s_{r,1} \rho \} \quad (62)$$

By the procedure of equating partial derivatives to zero, it is found that the estimator for ρ is the same as in the case of unconstrained estimation, and that, for z not equal to one,

$$\hat{\alpha} = \frac{w_{.0} + w_{.1} + n_{.0}}{s_{.0} (\hat{S} - \bar{r}_0 + \bar{r} \frac{T}{s_{.0}} \frac{z}{1-z})} \quad (63)$$

and finally that $\hat{S}$ must satisfy

$$\sum_{r=0}^{R} \frac{n_{r,0}}{\hat{S} - r} = \frac{w_{.0} + w_{.1} + n_{.0}}{\hat{S} - \bar{r}_0 + \bar{r} \frac{T}{s_{.0}} \frac{z}{1-z}} \quad (64)$$

which may be solved iteratively.

As before, the case of z=1 must be treated separately. The estimators of ρ and μ are the same as for unconstrained estimation, and the estimator of a_0 is

$$\hat{a}_0 = n_{.0} / s_{.0} \quad (65)$$

5. COMPUTATION

A computer program has been written to compute estimates as described above, and has been tested using simulation. The results of a small example problem are presented here by way of illustration. The problem considered is a system with repeated attempts and having the following parameters:

Number of Servers, R = 3

Effective Sources, $S = 6$

Arrival rate per free source, $\alpha = 1.0$

Mean service rate, $\mu = 1.0$

Arrival rate after unsuccessful bid, $\rho = 10.0$

From the values given above, the offered traffic has the following parameters:

Mean, $m = 3.0$

Variance, $v = 1.5$

For this system, twenty batches were run, each representing about twenty mean service times. For conversation traffic with a mean holding time of three minutes this represents an hour of measurement. The results for unconstrained and constrained estimation of the mean and variance of the offered traffic are shown in Table 1, below. The numbers in that table represent the sample mean of twenty estimates, and their sample standard deviation.

Table 1. Results of Simulation

	Constraint		
	None	z = 0.5	z = 1.0
$\hat{m}$	3.122	3.025	3.828
Std. Dev.	0.581	0.312	0.604
$\hat{v}$	1.826	1.513	3.828
Std. Dev.	1.178	0.156	0.604

Constrained estimates may be used in practice when the variance to mean ratio is known (or assumed) to be equal to a specified value, or to lie in a specified range. In the latter case, a search must be performed over the specified range of z values to find the constrained estimator that yields the greatest value of the likelihood function.

These results demonstrate the working of the estimation algorithms. The accuracy of the estimators has not yet been studied quantitatively.

6. CONCLUSIONS

The method described in this paper allows Bayesian Statistical Inference to be applied to the two applications of systems with queueing and systems with repeated attempts. The additional refinement of constraining the estimated variance to mean ratio of the offered traffic eliminates the possibility of producing an unrealistic estimate of offered traffic due to a poor estimate of arrival and departure rates.

The method has been tested on small problems by simulation. Further studies using real data are needed before recommending adoption of the method as a practical alternative to classical methods.

There are several potential advantages offered by the method such as the following. It processes all available information, and therefore may prove more accurate than present methods. The conditional probability density function of the parameters of the offered traffic can be computed. This could be used for dimensioning based on Decision Theory (as described in [3]). Also, it is possible that the method could be extended to allow for time-varying traffics. This could be of use in applications such as Network Management.

ACKNOWLEDGEMENT

The permission of the Chief General Manager, Telecom Australia, to present this paper is gratefully acknowledged

REFERENCES

[1] Warfield R.E., "A Study of Teletraffic Measurement and Other Estimation Problems" Ph.D. Thesis, 1980, University of New South Wales, Sydney, Australia.

[2] Warfield R.E., "Bayesian Analysis of Teletraffic Measurement", Australian Telecommunications Research, Vol. 15, No. 2, 1981, pp. 43-51

[3] Warfield R.E. and Foers G.A., "Application of Bayesian Methods to Teletraffic Measurement and Dimensioning" Tenth International Teletraffic Congress, Montreal, Canada, 1983. Reprinted in Australian Telecommunications Research, Vol. 18, No. 2, 1984, pp. 51-58.

[4] Billingsley, P., "Statistical Methods in Markov Chains", Annals of Mathematical Statistics, Vol. 32, 1961, pp. 12-40.

[5] Cox, D.R., "Some Problems of Statistical Analysis Connected With Congestion", Proceedings of the Symposium on Congestion Theory, University of North Carolina Press, 1964, pp. 289-316.

[6] Reynolds J.F., "On Estimating the Parameters of a Birth and Death Process", Australian Journal of Statistics, Vol. 15, No. 1, 1973, pp. 35-43.

[7] Box G.E.P. and Tiao G.C., "Bayesian Inference in Statistical Analysis", Addison-Wesley, 1973

[8] Benes, V. E., "A Sufficient Set of Statistics for a Simple Telephone Exchange Model", Bell System Technical Journal, July 1957, pp. 939-963.

[9] Songhurst D., "Variance of Observations on Markov Chains" Eighth International Teletraffic Congress, 1976, Melbourne, Australia.

[11] Rubas J. and Warfield R.E., "Determination of Confidence Limits for the Estimates of Congestion" Ninth International Teletraffic Congress, 1979, Torremolinos, Spain. Reprinted in Australian Telecommunications Research Vol. 14, No. 1, 1980, pp. 7-17.

TELETRAFFIC ISSUES in an Advanced Information Society
ITC-11
Minoru Akiyama (Editor)
Elsevier Science Publishers B.V. (North-Holland)

A GENERAL (rHβ) FORMULA OF CALL REPETITION: VALIDITY AND CONSTRAINTS

Géza GOSZTONY

BHG Telecommunications Works
Budapest, Hungary

ABSTRACT

The /rHβ/ formula is a generally valid relationship among the efficiency rate, the average perseverance and the repetition coefficient. Its derivation is based on a theory using parameters which depend on the reason of call attempt failure and on the attempt number and it assumes a hierarchical structure of information. This structure determines the reactions of the user when he faces unsuccessful call attempts and it differs from the structure of the network in which connections are built up. The /rHβ/ formula holds not only for subscriber-to-subscriber connections but also for any Reference Point-to-subscriber connection if parameters valid at the Reference Point are used in the latter case. Comparison with measurements gives satisfactory results. Practical application is simple since the measurement of the efficiency rate can easily be performed.

INTRODUCTION

Call repetition is a major problem in countries with an underdeveloped telephone network. However even in countries with developed networks about 30% of all national call attempts is due to repetition. Earlier and recent statistics have shown that call completion may differ remarkably due to the called country [LIU 79] [ROBE 79] [LEWI 83]. The harmful effects of call repetition together with their financial impact should not be underestimated.

In practice the effect of call repetition should be considered in the following areas [LEWI 83]:

- estimation of offered traffic,
- identification of bottlenecks in case of overload,
- network management strategies,
- evaluation of improvements in call completion /e.g. by removal of a given reason of failure/.

These facts justify the continuous interest manifested by several papers presented to the 10th ITC [SONG 83] [LEWI 83] [LUBA 83] [NIVE 83] which refer directly or implicitly to this problem, by the recent ITC Seminar papers of [JONI 84] [LEGA 84] [SHNE 84] [STAS 84] and [STEP 84] and also by the new CCITT Recommendation E.501 [CCITT 85] about the determination of offered traffic, which is the first result of a study to be continued.

The phenomenon of call repetition can be described by additional parameters as perseverance, repetition interval, repetition coefficient, efficiency rate, etc., the determination of which is a tedious work even with SPC techniques of our days. The main problem is that measurements must be performed on the level of individual failure types and call strings, since average values refer only to a traffic and failure pattern which exists in the time of observation. Problems of the evaluation of call completion improvements can e.g. not be handled by calculations based on average values [GOSZ 76]. Mathematical models taking several types of call attempt failures and attempt numbers explicitly into consideration are fairly complex [STEP 83].

It is therefore very important to find general relationships which themselves are simple, refer to average parameters but can be derived from models using a detailed description of call repetition.

In Section 1 of the paper some intrinsic problems of call repetition are summarized which are perhaps not so well known as the problems of mathematical modelling or of the measurement of characteristic parameters. Section 2 and the Appendices present the outlines of a general theory and the derivation of a basic formula. The practical importance of the formula is pointed out in the short Section 3. /Some knowledge about call repetition is assumed./

The present investigation is a general approach of modelling call repetition in complex structures to show that some practical results can be achieved even in the absence of detailed analytical theories.

1. BASIC PROBLEMS OF REPEATED ATTEMPTS CALCULATIONS

Theoretical and practical considerations referring to repeated attempts are faced with some principal difficulties. To solve these rather general problems seems to be a hard task if possible at all.

1.1 THE INFORMATION STRUCTURE - NETWORK STRUCTURE CONFLICT

The users of point-to point telecommunication networks are normally not familiar with the structure of and technology used in the network concerned. They react according to the information received. Experience, habits, temper, psychological reactions, briefly: human factors have an important influence. This refers also to call repetition.

The reasons of call attempt failures /as congestion, called party busy, no answer, technical failure/ or more exactly the relevant information appears to the user in an ordered form corresponding to the sequence of establishing a connection.

This sequence represents an information hierarchy starting e.g. with the presence or absence of a dial tone and at the end of which information on the state of the called party is received.

The user answers the received information, the sequence may be different for different types of calls, but should be /and is normally/ unambiguous for a given type of call. The structure of the information hierarchy does not necessarily correspond to the structure of the network, only a limited interaction exists. A congestion tone e.g. may be due to a single direct trunk group, but it may also be caused by a complex alternate routing network in an other case. Reaction is the same, but the calculation of failure probabilities is completely different.

Parameters of call repetition as perseverance probabilities and reattempt times are directly related to human factors i.e. users' reactions. Relevant traffic engineering considerations require the probabilities of call attempt failures in a form which corresponds to the information structure, but they can only be calculated taking the actual network structure into account. Present day calculations of call repetition are far from being able to handle the complex structures of existing networks. On the other hand, the information structure covers in many cases world-wide networks.

This infromation structure - network structure conflict is very likely the main obstacle in performing practical calculations with repeated calls.

1.2. FEED-BACK AND OFFERED TRAFFIC

Some estimation of the traffic to be handled by a given exchange or network is required for all traffic engineering considerations. This traffic, generally called offered traffic /see CCITT Rec.E. 600 [CCITT 85] / can not directly be identified with the traffic which would be originated by the users if there were no restriction of handling it. The difference is caused by unsuccessful call attempts /occuring in all practical cases/ which produce feed-back.

If call attempts which have already seized traffic handling resources in a given group of the network encounter failures in establishing a connection then the average holding time of these call attempts might be different from the average holding time of successful call attempts. The average and distribution of the holding time referring to the group in question may change. This effect should not be confused with e.g. the zero holding times in the lost-calls-cleared model, since it refers to call arrempts which have already penetrated into the network. /Example given in [GOSZ 83]./

If, further, call repetition takes place, then additional call attempts appear and the characteristics of the input process may also change.

The feed-back outlined above has thus an important impact on traffic engineering.

1.3. OFFERED TRAFFIC AND CHANGING ENVIRONMENT

In an indirect way, through calling rate and holding time measurements offered traffic can be determined. This offered traffic, however, is valid only for that state of the network which existed in the time of the measurement. Every change in the network due to enlargements, rearrangements, routing, etc. may have an influence on the efficiency rate behind the group considered, and thus may cause changes also in the actual value of offered traffic, even if the traffic demand of the users remains the same.

Considering that offered traffic is mainly used in engineering calculations in a future oriented way, the main difficulty can be summarized as follows: how is it possible to determine the offered traffic in a future situation from measurement results referring to the present.

From the difficulties summarized above one might feel severe reservations as regards the practical feasibility of investigations with repeated attempts. However there is a doubtless need to perform the necessary calculations of everyday work and one can not wait for the progress in teletraffic theory. Therefore the basic formula, given in the next Section, being simple but generally valid seems to have some practical importance.

2. OUTLINES OF A GENERAL THEORY OF CALL REPETITION

2.1. INFORMATION STRUCTURE AND PARAMETERS

The procedure of setting up a connection can be seen in a flowchart form in Fig.2.1 There are N pieces of serially arranged information on call attempt failures which will be called reasons of failures in the following.

The probability that an ith call attempt of a call string which arrives to the jth reason will fail is denoted by F_{ji}. The probability that the same call will repeat if failed is denoted by H_{ji}. The essential feature of the hierarchical

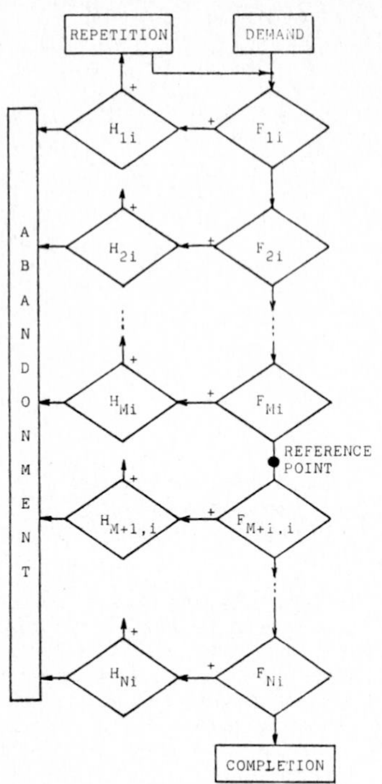

Fig.2.1 The information structure of the establishing of a connection. Flowchart with retrials.

structure is that the jth reason can only be arrived at if the preceding /j-1/ reasons have been successfully passed. The reasons of failure act independently. /A simple reason of failure can be substituted by a group of mutually exclusive reasons without affecting the validity of the considerations below./

Reattempt intervals do not appear in this model, their effect is assumed to be considered in the actual value of the F_{ji}-s. The input process at the jth reason of failure /i.e. at the input of the corresponding group of resources/ is determined by the preceding failure probabilities. It is assumed that this can be taken into account by the F_{ji}-s which are constant values.

The H_{ji} perseverance probabilities represent the condensed experience of users and are assumed to be constant values as well.

The use of constant F_{ji} failure probabilities seems to be straightforward if a steady state is investigated. H_{ji} type probabilities can be measured [ROBE 79], there is however no experience as regards their stability. In any case it can be assumed that they do not change with random traffic variations and are only sensitive to permanent traffic changes.

F_{ji} and H_{ji} may depend on the reason of failure experienced by the /i-1/th attempt of the call string. /About the selection effect involved see [MYSK 76]/ This feature has been neglected i.e. the call string is assumed to have no memory.

2.2 CHARACTERISTICS AND THE AVERAGE PERSEVERANCE FORMULA

Parameters to characterize call repetition can be defined taking different aspects into account. One can consider

- ith call attempts
- call demands /first calls/ and
- call attempts in general.

Relevant parameters are of course interrelated. These paramaters are summarized in Appendix A, together with the derivation of the average perseverance formula which reads:

$$H = \frac{\beta - 1}{/1-r/\beta} \qquad /2.1/$$

where β = average number of call attempts per call demand, repetition coefficient,
H = average perseverance,
/1 - r/= F = average failure probability.

This formula can easily be derived from simple approximate mathematical models using average H and F values. It appears also in some investigations with more general assumptions [MYSK 76] [SONG 83] [GOSZ 76] . Now in Appendix A it has been shown that it holds generally for the hierarchical structure described assuming F_{ji} failure probabilities and H_{ji} perseverance probabilities which depend on the reason of failure and the attempt number. However in all these investigations it was assumed that the model described a subscriber-to-subscriber connection.

2.3. THE LEVEL M REFERENCE POINT

In Fig.2.1 between the Mth and /M+1/th reason of failure one can find the "Reference Point M". Let us assume that measurements are carried out here. /Example: outgoing calls in an international gateway./

Let us consider the ith call attempt of the call string which is the first one being successful at level M. At Reference Point M this will be a call demand. If this attempt fails between M and M+1, the next call attempt successful at level M will be the first repeated attempt at this Point, irrespective of those attempts which fail between levels 1 and M, etc. Thus at this Reference Point one can determine an intensity of call demands, perseverance probabilities, etc., which of course will differ from similar parameters related to the original connection.

2.4. GENERAL VALIDITY OF THE AVERAGE PERSEVERANCE FORMULA

The formula of /2.1/ is valid not only for the subscriber-to-subscriber connection but also for the Reference Point M-to-subscriber connection.

This can be proven in two ways

- indirectly by making use of the fact that no special assumptions were made regarding H_{ji} and F_{ji},

Table 2.1 Comparison of observed and calculated average perseverance values /Measurement data from [LEWI 83]./

Country	First attempts	r	β	H	H_c	ΔH %	% of 5th and greater atts
A	1 184	.147	3.88	.835	.870	4.19	40.3
B	461	.291	2.36	.768	.813	5.86	19.3
C	349	.245	2.09	.670	.691	3.13	18.1
D	798	.380	1.45	.493	.502	1.83	5.5
E	5 314	.456	1.37	.484	.496	2.48	6.4
F	1 805	.593	1.34	.607	.621	2.31	5.6
G	5 730	.272	2.28	.748	.772	3.21	25.4
H	2 414	.306	2.21	.755	.789	4.50	18.8
I	4 299	.308	2.21	.758	.791	4.35	19.9

- directly by deducing M level characteristics using H_{ji} and F_{ji} values of the subscriber-to-subscriber case.

The former approach assumes that the considerations of Appendix 1 were carried out with perseverance probabilities and failure probabilities valid at Reference Point M.

The latter approach, outlined in Appendix B starts from the original subscriber-to-subscriber connection but presents traffic engineering characteristics which are valid at the Reference Point. It takes into account that call attempts may fail and abandon before Reference Point M, therefore

- there are call demands which are present at the originating subscribers level but do not appear at Reference Point M;
- many call attempts can be made before reaching Reference Point M for the first time or between two subsequent M level attempts;
- after an unsuccessful attempt at the jth reason of failure, j=M, M+1,...N, there are two possibilities to abandon: /i/ no subsequent attempt is made, /ii/ subsequent attempts do not reach Reference Point M.

It has been shown that the basic formulae of making at least a given number of retrials are formally equivalent for the subscriber-to-subscriber and the Reference Point M connections if in the latter case conditional probabilities are introduced. The condition is that at least one call attempt of a call string was able to reach Reference Point M. The interpretation is very simple: an observer sitting in Reference Point M and having no knowledge about preceding parts of the information/network structure will see a similar situation and may use the same general mathematical description as for the subscriber-to-subscriber case, of course with H_{ji} and F_{ji} values valid for this situation.

2.5. EXAMPLE

In [LEWI 83] detailed measurement data can be found which refer to calls directed to nine foreign countries and were observed in an international gateway. From these data one can easily calculate β, r and H. Using the measured β and r values an average perseverance, H_c, corresponding to /2.1/ was determined and compared with H. All these appear in Table 2.1 together with $\Delta H = |H-H_c|/H$.

ΔH is in all but one case less than 5%. In the original measurement 5th and subsequent attempts formed a single category. ΔH is about 2.5% if there were relatively few attempts in this category, i.e. if the averaging effect of the measurement was small for long call strings. Without analysing possible effects of other factors /e.g. measurement error/ it can be concluded that the egreement of the two perseverance values is rather good.

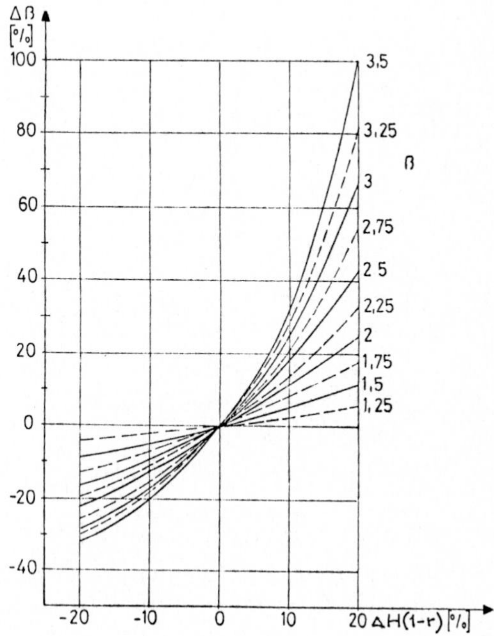

Fig.3.1 The error of β calculation resulting from the error of the H/1-r/ product

3. PRACTICAL IMPORTANCE OF THE /rHβ/ FORMULA

The formula of /2.1/ describes a general relationship between efficiency rate, average perseverance and repetition coefficient and will be referred to in the following as the /rHβ/ formula. Some features of this formula are summarized below.

3.1. The /rHβ/ formula

- includes all possible types of call attempt failures and is therefore not limited to failure probabilities being targets of traffic engineering calculations,
- holds for any network structure, since the basic hierarchical information structure is generally valid,
- justifies repeated attempt measurements at any level of the information structure supposing that the results will be used in considerations referring to the same level.

3.2. The amount of repeated attempts or the relevant inefficient traffic can easily be determined by the /rHβ/ formula. One parameter, the efficiency rate /answer signal ratio/ can be measured in a rather simple way and there is normally some information on possible H values. Therefore in practice it is recommended to use an estimated H value and to calculate β. The resulting error of β is shown in Fig. 3.1.

3.3. An example of how to estimate offered traffic after having enlarged a trunk group in a repeated attempts environment can be found in the Appendix of CCITT Recommendation E.501 on offered traffic [CCITT 85]. A possible way to solve the problem of determining the offered traffic in a changing environment /see 1.3/ is presented there by assuming among others the validity of the /rHβ/ formula.

REFERENCES

[CCIT 85] CCITT Recommendation:"Estimation of traffic offered to international trunk groups"- E.501. Red Book,Vol.III.,Fasc.III.1., ITU Geneva, 1985.

[GOSZ 76] Gosztony,G.:"Comparison of calculated and simulated results for trunk groups with repeated attempts" - 8.ITC,Melbourne,1976.321/1-11.

[GOSZ 83] Gosztony,G.:"CCITT - a chance to unify the basic traffic engineering practice" - 10.ITC, Montreal,1983.231/1-8.

[JONI 84] Jonin,G.L.:" The system with repeated calls: models, measurements, results" - 3.ITC Seminar, Moscow,1984.Preprint Book,197-208.

[LEGA 84] Le Gall,P.:"The repeated call model and the queue with impatience" - 3.ITC Seminar, Moscow, 1984.Preprint Book,278-289.

[LEWI 83] Lewis,A., Leonard, G.:"Measurements of repeat call attempts in the intercontinental telephone service" - 10.ITC, Montreal,1983.242/1-4.

[LIU 79] Liu,K.S.:"Direct distance dialling call completion and customer retrial behaviour" - 9.ITC, Torremolinos,1979.144/1-7., BSTJ.59.1980.3. 295-312.

[LUBA 83] Lubacz,J.:"Circuit switched communication networks analysis - a call delay point of view" - 10.ITC,Montreal,1983.221/1-7.

[MYSK 76] Myskja,A., Aagesen,F.A.:"On the interaction between subscribers and a telephone system" - 8.ITC Melbourne,1976.322/1-8.

[NIVE 83] Nivert,K., Gunnarsson,R., Sjostrom,L.E.:"Optimum distribution of congestion in a national trunk network. An economical evaluation as a base for grade of service standards" - 10.ITC,Montreal, 1983.239/1-5.

[ROBE 79] Roberts,J.W.:" Recent observations on subscriber behaviour" - 9.ITC,Torremolinos,1979.147/1-8.

[SONG 83] Songhurst,D.J.:"Subscriber repeat attempts, congestion and quality of service: a study based on simulation" - 10.ITC,Montreal,1983.115/1-7.

[SHNE 84] Shneps-Shneppe,M.A., Petrov,A.F.:"Optimum distribution of overall grade of service in hierarchical telephone networks with repeated attempts" - 3.ITC Seminar,Moscow,1984.Preprint Book,377-386.

[STAS 84] Stastny,M.:"On the substitution of the basic retrial model for a complex loss model with retrials" - 3.ITC Seminar,Moscow,1984.Preprint Book,395-399.

[STEP 83] Stepanov,S.N.:" Mathematical methods for the dimensioning of systems with repeated attempts" /in Russian/ - Nauka,Moscow,1983.p.230.

[STEP 84] Stepanov,S.N.:"Estimating of characteristics of multilinear systems with repeated calls" - 3.ITC Seminar, Moscow,1984.Preprint Book,400-409.

APPENDIX A

CHARACTERISTICS OF CALL REPETITION

Characteristics are based on the structure of Fig.2.1. H_{ji} perseverance probabilities and F_{ji} failure probabilities as defined in Para.2.1 are used according to the assumptions made ibidem.

AA. Let us consider an ith call attempt of a call demand. From Fig.5.1 one can easily deduce the following:

- probability that the ith call attempt succeeded

$$r_i = \prod_{j=1}^{N} (1 - F_{ji}) \qquad /A.1/$$

- probability that the ith attempt failed

$$(1-r_i) = \sum_{j=1}^{N} f_{ji} = \sum_{j=1}^{N} F_{ji} \prod_{k=0}^{j-1} (1 - F_{ki}) \qquad /A.2/$$

probabilities that the failed ith attempt has been repeated or abandoned are:

$$p_i = \sum_{j=1}^{N} f_{ji} H_{ji} \quad \text{and} \qquad /A.3/$$

$$v_i = \sum_{j=1}^{N} f_{ji} (1 - H_{ji}) \qquad /A.4/$$

respectively. $F_{0i}=0$ per definitionem.

AB. In the considerations above it was assumed that the call demand in question has already made /i-1/ attempts previously. Of course i attempts do not belong to all call demands. The probability of making an ith attempt can be expressed using p_k probabilities defined in /A.3/. p_i is the probability that an /i+1/th attempt will be made if i attempts have already been made. Therefore the

- probability of making an ith attempt /i.e. at least i attempts/ is:

$$P_i = P(\geq i) = \prod_{k=0}^{i-1} p_k \qquad /A.5/$$

with $p_0=1$, since a demand definitely makes a first attempt.

Let β/i/ denote the probability that a call demand makes exactly i attempts. This happens if the ith attempt is successful or if the unsuccessful ith attempt is abandoned. The fact of having made /i-1/ attempts previously is now taken into account.

$$\beta(i) = P_i r_i + P_i v_i = S_i + V_i \qquad /A.6/$$

β/i/ is a discrete probability distribution, therefore:

$$\sum_{i=1}^{\infty} \beta(i) = S + V = 1. \qquad /A.7/$$

Here S_i and V_i are the probabilities that the ith attempt of the call demand is successful and abandons, respectively.

AC. From the call demands' aspect one can define the following parameters:

- probability that a call demand is ultimately successful: S
- probability that a call demand will ultimately abandon: V=1-S
- average number of call attempts made by a call demand - repetition coefficient:

$$\beta = M(\beta(i)) = \sum_{i=1}^{\infty} P(\geq i) = \sum_{i=1}^{\infty} \prod_{k=0}^{i-1} p_k. \qquad /A.8/$$

Here P/$\geq$ i/ as being the complementary distribution function of β/i/ was used. The limits in /A.7/ and /A.8/ exist if either $H_{ji} < 1$ or $F_{ji} < 1$ is fulfilled. It should be noted that β refers to any call demand, irrespective its final success.

AD. The relative weight of call attempt failures is manifested by the proportion of unsuccessful attempts due to a particular cause of failure. This relative weight has an impact also on the average perseverance which is, as already mentioned, situation dependent.

Let d/i/ denote the probability that a call demand encounters exactly i unsuccessful attempts. This happens if repetition is discontinued after the ith unsuccessful attempt or if the /i+1/th attempt is successful. At least i unsuccessful attempts are experienced by a call demand if the ith attempt is unsuccessful. The relevant probability is:

$$D_i = P_i (1 - r_i) \qquad /A.9/$$

The last parameter referring to call demands to be defined is thus the:

- average number of unsuccessful call attempts per call demand:

$$D = M(d(i)) = \sum_{i=1}^{\infty} D_i = \sum_{i=1}^{\infty} (1-r_i) \prod_{k=0}^{i-1} p_k = \beta - S. \qquad /A.10/$$

This last result is obvious.

AE. Considering call attempts the parameters below may be given:

- efficiency rate $\qquad r = S/\beta \qquad$ /A.11/

  which can be deduced e.g. indirectly from β and D by forming the probability that a call attempt is unsuccessful: /1-r/=D/β.

The second parameter in this group is the average perseverance probability which is a weighted average of the H_{ji} values. The weighting factors can be deduced from /A.2/ and /A.9/ as follows. Let us consider an unsuccessful ith attempt. The probability of being unsuccessful for the jth reason of failure is given by the members of the sum of /A.2/ i.e. by f_{ji}. Therefore the average perseverance related to the ith attempt is:

$$H_i = \frac{\sum_{j=1}^{N} f_{ji} H_{ji}}{(1 - r_i)} = \frac{p_i}{(1 - r_i)}. \qquad /A.12/$$

The weighting factors in general can be gained from /A.9/ being:

$$D_{ji} = P_i f_{ji}.$$

Thus the

- average perseverance probability is:

$$H = \frac{\sum_{i=1}^{\infty} \sum_{j=1}^{N} D_{ji} H_{ji}}{\sum_{i=1}^{\infty} \sum_{j=1}^{N} D_{ji}} = \frac{\sum_{i=1}^{\infty} p_i P_i}{\sum_{i=1}^{\infty} D_i} = $$

$$= \frac{\sum_{i=1}^{\infty} (1-r_i) P_i - \sum v_i P_i}{\sum_{i=1}^{\infty} D_i} = \frac{\beta - S - V}{D}. \qquad /A.13/$$

Taking /A.7/ and /A.11/ into account:

$$H = \frac{\beta - 1}{(1-r)\beta} \qquad /A.14/$$

is achieved.

APPENDIX B

CHARACTERISTICS OF CALL REPETITION AT REFERENCE POINT M

The parameters characterizing call repetition at Reference Point M can be defined in a similar way to that included in Appendix A. The same assumptions hold, the same designations were used.

BA. Let us consider that call attempt of a call demand which appears as gth successful attempt at Reference Point M. The following probabilities can be defined:

- probability that the gth M level attempt will succeed

$$r'_g = \sum_{i=g}^{\infty} \prod_{j=M+1}^{N} (1 - F_{ji}) \frac{P(M_g)_i}{P(M_g)} \qquad /B.1/$$

where

$$P(M_g) = \sum_{i=g}^{\infty} P(M_g)_i = r_{M_g} + \sum_{i=g+1} r_{M_i} \prod_{s=g}^{i-1} \sum_{j=1}^{M} f_{js} H_{js}$$

and $\qquad r_{M_i} = \prod_{j=1}^{M} (1 - F_{ji})$

- probability that the gth M level attempt will fail

$$(1 - r'_g) = \sum_{j=M+1}^{N} f'_{jg} =$$

$$=\sum_{i=g}^{\infty}\left\{F_{M+1,i}+\sum_{j=M+2}^{N}F_{ji}\prod_{k=M+1}^{j-1}(1-F_{ki})\right\}\frac{P(M_g)_i}{P(M_g)} \qquad /B.2/$$

- probabilities that the failed gth M level attempt will be repeated or abandoned are:

$$p'_g=P(M_{g+1})\sum_{i=g}^{\infty}\left\{F_{M+1,i}H_{M+1,i}+\sum_{j=M+2}^{N}F_{ji}H_{ji}\prod_{k=M+1}^{j-1}(1-F_{ki})\right\}\frac{P(M_g)_i}{P(M_g)} \qquad /B.3/$$

$$v'_g=\sum_{i=g}^{\infty}\left\{F_{M+1,i}(1-H_{M+1,i})+\sum_{j=M+2}^{N}F_{ji}H_{ji}\prod_{k=M+1}^{j-1}(1-F_{ki})\right\}\frac{P(M_g)_i}{P(M_g)}+\left(1-P(M_{g+1})\right)\frac{p'_g}{P(M_{g+1})}. \qquad /B.4/$$

$P/M_g/_i$ and $P/M_g/$ above are the probabilities that after an unsuccessful /g-1/th attempt level M has been successfully passed by the ith attempt or at all, respectively. Of course $g\leq i$ is required since an ith attempt can not pass level M $g>i$ times.

A gth M level attempt can be originated by any $i\geq g$ call attempt with the probability $P/M_g/_i$. Therefore these probabilities appear as weighting factors in the calculation of r'_g, f'_g, p'_g and v'_g.

p'_g is the probability that a /g+1/th attempt will be made if the gth attempt has failed. However not any new attempt will necessarily be a /g+1/th attempt since the new attempt will appear only at the "entrance" and not at Reference Point M. The fact of reaching Reference Point M is taken into account by $P/M_{g+1}/$.

The first term of v'_g is the probability that that call was abandoned without making a new attempt. The second term takes those attempts into account which have been repeated but in spite of repeated trials no /g+1/th attempt appeared at Reference Point M.

BB. The probability of making gth attempt can be expressed using the p'_g probabilities defined in /B.3/.

- Probability of making a gth M level attempt /i.e. at least g M level attempts/ is:

$$\left.\begin{aligned}(P\geq 1)&=P(M_1)\\(P\geq g)&=P(M_1)\prod_{k=1}^{g-1}p'_k \quad \text{for } g\geq 2\end{aligned}\right\} \qquad /B.5/$$

The probability of making a first attempt equals definitely one at the entrance. However at Reference Point M the first attempt appears only with probability $P/M_1/$.

Let $\beta'/g/$ denote the probability that a call demand makes exactly g attempts at Reference Point M. This happens if the gth attempt is successful or if the /between M+1 and N/ unsuccessful gth attempt is abandoned. The fact having made /g-1/ M level attempts previously is now taken into account by the P'_g probabilities

$$\beta'_g=P'_g r'_g+P'_g v'_g=S'_g+V'_g. \qquad /B.6/$$

Since at Reference Point M the first attempt appears only with probability $P/M_1/$, the $\beta'/g/$ discrete distribution should also be interpreted for g=0. Thus:

$$\sum_{g=0}^{\infty}\beta'(g)=1-P(M_1)+\sum_{g=1}^{\infty}\beta'(g)=$$
$$=1-P(M_1)+S'+V'=1. \qquad /B.7/$$

BC. The /B.5/ probabilities and the /B.6/ distribution are "entrance" oriented. Let the event of having reached Reference Point M for the first time be M_1. One can define the following conditional probabilities:

$$\left.\begin{aligned}P(\geq 1\mid M_1)&=P'_1=1\\P(\geq g\mid M_1)&=P'_g=\prod_{k=1}^{g-1}p'_k \quad \text{for } g\geq 2\end{aligned}\right\} \qquad /B.8/$$

and

$$\left.\begin{aligned}\beta'(g\mid M_1)&=\beta'(g)/(1-P(M_1))\\\sum_{g=1}^{\infty}\beta'(g\mid M_1)&=1 \quad \text{for } g\geq 1\end{aligned}\right\} \qquad /B.9/$$

BD. The P'_g probabilities and the $\beta'(g|M_1)$ distribution are completely equivalent to the P_i probabilities and the $\beta/i/$ distribution of Appendix A. Therefore the average perseverance formula can be derived in the same way and will not be detailed here.

ACKNOWLEDGEMENTS

The author would like to express his sincere thanks to his colleague Ms.M. Ágostházi for many valuable discussions and for the help given during the preparation of the paper.

The permission of BHG Telecommunication Works to publish this work is highly acknowledged.

TELETRAFFIC ISSUES in an Advanced Information Society
ITC-11
Minoru Akiyama (Editor)
Elsevier Science Publishers B.V. (North-Holland)

AN AUTOMATICALLY REPEATED CALL MODEL IN NTT PUBLIC FACSIMILE COMMUNICATION SYSTEMS

*Hisayoshi INAMORI, *Muneo SAWAI, **Takaya ENDO and **Katsuhiro TANABE

*Musashino Electrical Communication Laboratory, NTT
Tokyo, Japan

**Yokosuka Electrical Communication Laboratory, NTT
Yokosuka, Japan

ABSTRACT

A queueing system featuring batch arrival and call repetition at a constant interval is presented as a traffic model of the NTT facsimile network, which is a store-and-forward network laid over the telephone network. This network has a multi-address calling service function and an automatic re-calling function. The multi-address calling service enables a document to be sent to several terminals simultaneously. The automatic re-calling function allows facsimile documents to be resent when calls are rejected for some reason, such as all circuits being busy in the network. An approximation algorithm for analyzing transfer delay and loss probability of the model is proposed for the dimensioning of facsimile networks. The numerical results of the approximation algorithm are compared with simulation results in order to demonstrate its high accuracy.

1. INTRODUCTION

New public facsimile communication network services are now being offered in Japan through the Facsimile Intelligent Communication System (FICS). The FICS is constructed by laying a store-and-forward network over the existing telephone network in order to keep initial investment as low as possible. In the store-and-forward network, facsimile storage and conversion equipment (STOC) provides various new telecommunication services such as confidential delivery and multi-address calling. Using the multi-address calling service, a subscriber can deliver a document to many terminals simultaneously.

In conventional communication, any call which finds all circuits or the terminal busy is lost. However, in the FICS, to make the best use of the characteristics of the store-and-forward service, call intents which encounter congestion are repeated automatically by STOC instead of being lost. It is important for the network design to analyze the characteristics of automatically repeated calls, and to estimate repetition probability and transfer delay.

Up to now, many papers treated repeated calls focused on subscriber behavior in telephone networks [1-6]. In these papers, the authors assumed that calls arrive at circuits between a toll switch and a local switch as single calls. They further assumed that the repetition intervals were random. However, in the FICS, calls arrive at the circuits as groups and a constant repetition interval is adopted to simplify call processing.

This paper provides a good approximation algorithm for analyzing the traffic characteristics of a queueing model that features batch arrival and call repetition at a constant interval. This model is applied to the facsimile network with the multi-address delivery service function and the automatic re-calling function. The numerical results of repetition probability and transfer delay are presented and compared with simulation results to validate the accuracy of the approximation.

2. THE NTT PUBLIC FACSIMILE COMMUNICATION NETWORK SERVICES

NTT started its new public facsimile communication network service in Tokyo and Osaka in September 1981 with the inauguration of the Facsimile Intelligent Communication System (FICS-1). This system is more economical than telephone facsimile and provides various additional services. Following FICS-1, the second version of the system, FICS-2, was developed to expand the service area nationwide at a low cost and to provide communication services featuring media conversion[7,8].

2.1 Network Configuration

The NTT facsimile network configuration is shown in Fig.1. The lower part of the network is shared by the existing telephone network. Local switches (LS) and subscriber loops in the existing telephone network are utilized, so that FICS can keep the initial investment as low as possible. The upper part is only for facsimile service and this part is responsible for terminal control and particular services.

The connection control functions needed in the lower part of the facsimile network such as facsimile communication charging and facsimile subscriber file checking are centralized in special toll switches (TS-FX) equipped with facsimile control functions.

The STOC, which is mainly responsible for handling communication processing services such as confidential delivery, facsimile memory box and multi-address calling, is the most important piece of equipment in this store-and-forward network.

2.2 Basic Operation Outline

The system is operated as follows.

(a) When a calling end TS-FX receives a request from a calling terminal, the TS-FX asks the calling end STOC whether it can store the document or not.

(b) If the STOC has sufficient memory capacity to store the document, the STOC sends the response signal to the terminal through the TS-FX and the LS.

(c) When the terminal receives the signal, the terminal sends the document to the STOC through the LS and the TS-FX.

(d) The calling end STOC sends the document to the called end STOC.

(e) The called end STOC asks the called end TS-FX to make connection to the called terminal.

(f) When the connection is set up, the called end STOC sends the document to the called terminal through the called end TS-FX and the called end LS.

2.3 Service Outline

Various kinds of communication processing services such as basic calling, multi-address calling and confidential delivery services can be offered using STOC and TS-FX. The basic calling service is applied to one-to-one facsimile communication in which a document has a single destination. Using the confidential delivery service, users can keep their privacy. When a document requiring this service arrives at a called end STOC, a confidential delivery notice is delivered to the called terminal. If the correct key number is sent from the terminal to the STOC, the document is then sent to the terminal.

2.3.1 Multi-Address Calling

Using the multi-address calling service, a subscriber can send a document to many terminals simultaneously. Copies of the document are made in called end STOCs and delivered to the called terminals through the TS-FXs and the LSs. When the number of copies of a document requested by a multi-address calling service is larger than the number of idle circuits between the TS-FX and the LS, the idle circuits are filled with calls and the remaining calls wait for automatic re-calling.

The maximum number of destinations able to be requested in a multi-address calling is 100.

2.3.2 Automatic Re-Calling

In some cases facsimile documents, once received by a called end STOC, cannot be delivered immediately to the called terminal. Before the delivery of a document from the called end STOC to the called terminal, a call set-up signal is sent from the STOC to the terminal through the TS-FX and the LS. The signal cannot be sent to the terminal mainly in the following three cases (see Fig.2).

(a) All circuits between the TS-FX and the LS are busy.

(b) The called terminal is busy.

(c) The terminal has run out of recording paper or is turned off.

If a signal is rejected, the signal for the same document is repeated from the called end STOC T minutes later. In the cases (a) and (b), if the signal is repeated 5 times and still encounters rejection, the STOC gives up sending the call set-up signal (see Fig.2 (2)).

For the case (c), only one additional calling attempt is made since there is little chance of delivering the documents successfully.

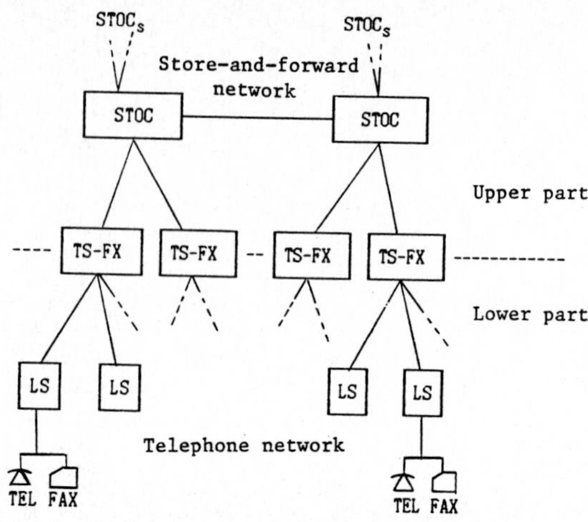

STOC: Facsimile storage and conversion equipment
TS-FX: Toll switch equipped with facsimile service control facility
LS: Local switch of telephone network

Fig.1 NTT facsimile network configuration

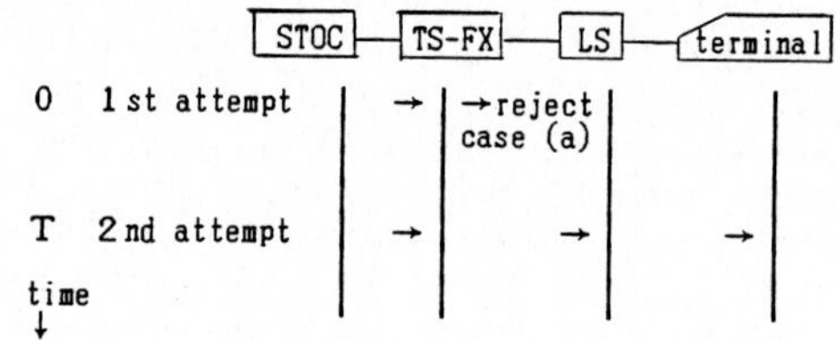

(1) Example of successful access

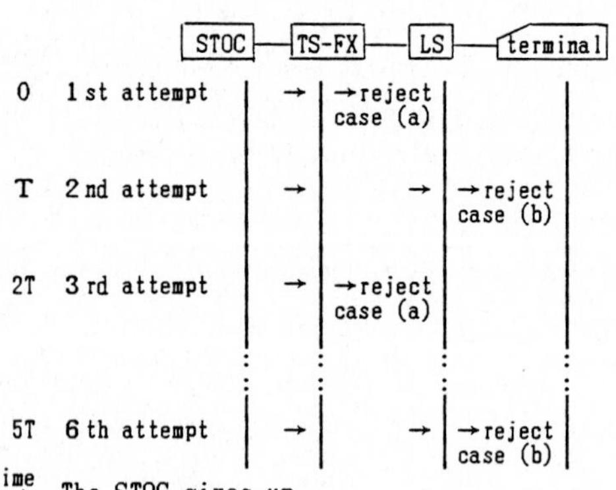

(2) Example of failed access

case (a): All circuits busy
case (b): Terminal busy
→: Call set-up signal
T: Repetition interval

Fig.2 Automatic re-calling function

2.3.3 Grade of Service (GOS)

The GOS for facsimile calls is classified by delay and loss probability[8]. GOS standard values are used for dimensioning resources in the FICS network, e.g., the number of circuits between TS-FX and LS.

Subscribers often compare the GOS in FICS service with that of telephone service since facsimile terminals can also be used as the telephone facsimile terminals. The basic concepts of GOS specification in FICS service are therefore based on those in the telephone facsimile service to avoid dissatisfaction by subscribers. We must evaluate the GOS considering the influence of multi-address calling and automatic re-calling functions which are not offered by telephone facsimile service.

2.4 Traffic Characteristics

Fig.3 shows examples of holding time distribution statistics of the circuits between TS-FX and LS. The holding time is determined by the following factors.

(a) Document size (e.g. ISO A4, A5)

(b) Density of document scan

(c) Document pattern complexity

(d) Facsimile coding scheme (e.g. modified-Huffman coding scheme)

(e) Transmission speed

The mean holding time for the examples is 90 seconds. The repetition interval in FICS is 5 minutes, which is three times as much as the mean holding time.

Fig.4 shows examples of the number of destinations distribution statistics for the multi-address calling. It can be seen that the number of destinations is distributed over a wide range. Therefore, when designing the network, we must take the effect of the group arrivals into account.

Fig.5 shows an example of rejection probability statistics for k-th attempts in a STOC. Many TS-FXs are connected to a STOC, so the probability values in the figure show mean values over all groups of circuits between TS-FXs and LSs. It can be seen that the rejection probabilities for repeated attempts are considerably larger than that of first attempts.

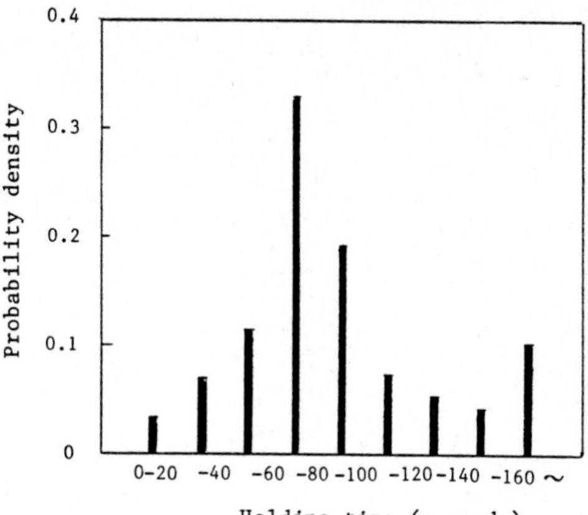

Fig.3 Examples of holding time per document distribution statistics

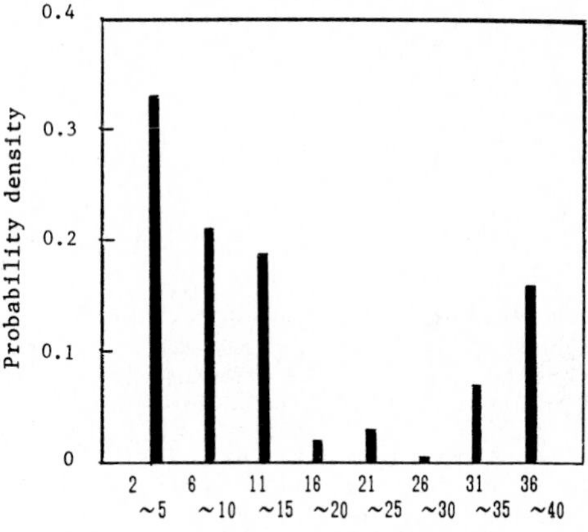

Fig.4 Examples of the number of destinations distribution statistics for multi-address calling

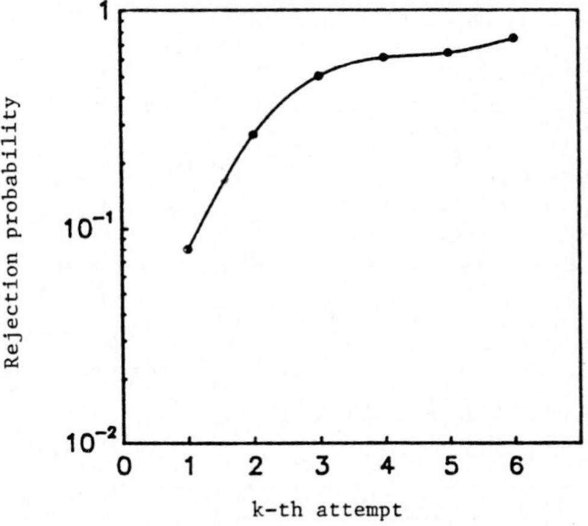

Fig.5 An example of rejection probability statistics for k-th attempt in a STOC

3. QUEUEING MODEL AND APPROXIMATION ALGORITHM

3.1 Queueing Model

The circuits between a called end TS-FX and a called end LS are modeled as a queueing system that accepts batch arrivals and repeated calls. NTT is responsible for the loss in the network, so that case(a), all circuits busy, is only taken into account in the model.

There are two types of calls: those which arrive as single calls for ordinary service, and those which arrive as a group for multi-address service. Let us refer to the former as individual calls and the latter as group calls. This queueing model with mixed inputs is described below more precisely.

(a) The arrival instants of group calls and individual calls for first attempts constitute stationary Poisson processes with an arrival rate λ_g and λ_i, respectively.

(b) There are S circuits and each call can access to any one of the circuits.

(c) The number of calls included in an arriving batch (i.e., batch size) is an independent and identically distributed random variable with an arbitrary distribution $H_g(j)$ for $1 < j \leqq g$. Mean batch size is denoted by m.

(d) The holding times of both types of calls are independent and identically distributed random variables having a negative exponential distribution with a mean $1/\mu$.

(e) If an arriving batch size of group calls is larger than the number of idle circuits, all the idle circuits are filled with calls and the remaining calls are rejected and attached to the re-calling queue. Individual calls are rejected when they find all circuits busy and attached to the re-calling queue. The calls in the re-calling queue then try to access the circuits T (repetition interval) minutes later as repeated calls.

(f) The number of repetitions is limited to M. If a call is repeated M times and still encounters congestion, it leaves the system immediately.

(g) A call which is repeated k times is called a k-th repeated call or a call at (k+1)-th attempt.

3.2 Approximation Algorithm

It is difficult to solve this model exactly because of the constant repetition interval. Therefore, this paper presents an approximation algorithm. In the case of exponentially distributed repetition intervals and single arrivals, the model can be solved exactly [9].

The algorithm is divided into two parts. The first part uses an iteration method in which Poisson arrivals of repeated call groups are assumed. This part offers the rejection probability for first attempts. In the second part, the characteristics of constant repeating calls are taken into account and rejection probability for repeated attempts are offered.

3.2.1 First Part of the Algorithm

When the repetition interval is long enough to ignore the effect of the previous attempt, this model is approximated by the $M^{[x]}/M/S$ loss system (batch arrival according to a Poisson process/exponential holding time/loss system with S servers). In order to approximate the model by the $M^{[x]}/M/S$ loss system in this way, we must make the following assumptions.

(1) Repeated calls belonging to one group arrive at the circuit as a repeated call group and the group size is maintained from the rejected epoch.

(2) Repeated call groups arrive independently according to a Poisson process, so that the offered load is obtained by adding the load of all repeated attempts and first attempts.

The steady state difference equations of $M^{[x]}/M/S$ are [11]

$$\left.\begin{aligned} &\lambda P(0) + \mu P(1) = 0, \\ &(\lambda+i\mu)P(i) + (i+1)\mu P(i+1) \\ &\qquad + \sum_{j=1}^{\min(g,i)} \lambda P(i-j)H(j)=0, \end{aligned}\right\} \quad (1)$$

where, for $0 \leqq i \leqq S$,

$P(i)=Pr$ (arriving batch finds i calls in the system);

for $1 \leqq j \leqq g$,

$H(j) = Pr$ (batch size is j).

λ is a total arrival rate.

These parameters are determined by the following algorithm.

(step 1)

If we ignore the existence of repeated calls, P(i) can be obtained from Eq.(1), substituting the arriving rate and batch size probability for first attempts.

$$\left.\begin{aligned} &\lambda=\lambda_i+\lambda_g, \\ &H(1)=\lambda_i/(\lambda_i+\lambda_g). \\ &\text{For } 2 \leqq j \leqq g, \\ &H(j)=\lambda_g H_g(j)/(\lambda_i+\lambda_g), \end{aligned}\right\} \quad (2)$$

then P(i) can be derived from Eq.(1).

Let

$A(0)=\lambda$,

$G(0,j)=H(j)$, for $1 \leqq j \leqq g$,

as the initial values for next step.

(step 2)

Considering the overflow of (k-1)-th repeated calls becomes k-th repeated calls, arrival rate A(k) and batch size probability $G(k,\ell)$ for the k-th repeated calls having ℓ calls in the batch are obtained recursively from k=1 to k=M+1 as follows:

$$\left.\begin{aligned} &A(k)=\sum_{\ell=1}^{g}\sum_{j=\ell}^{\min(g,\ell+S)} P(S+\ell-j)G(k-1,j)A(k-1), \\ &G(k,\ell)=\sum_{j=\ell}^{\min(g,\ell+S)} P(S+\ell-j)G(k-1,j)A(k-1)/A(K). \end{aligned}\right\} \quad (3)$$

(step 3)

Adding the load of all the repeated calls and first call attempt as an offered load, the following equations are obtained:

$$\left.\begin{aligned} &\lambda=\sum_{k=0}^{M} A(k), \\ &H(j)=\sum_{k=0}^{M} A(k)G(k,j)/\lambda. \end{aligned}\right\} \quad (4)$$

(step 4)

From Eq.(1), we obtain the new state probabilities P(i).

(step 5)

Compare the previous P(i) with the new P(i). If the difference between them is larger than ε (iteration indicator), return to step 2, or else we get a new A(k) and $G(k,\ell)$ from Eq.(3) and the iteration is finished. For practical usage, this iteration indicator must be determined as a sufficiently small value.

The similar type of iteration algorithm was introduced by Lam [10]. He applied the algorithm to a queueing model with finite buffer and re-calling functions.

3.2.2 Second Part of the Algorithm

The characteristics described below can be obtained by observing the sample path of the number of calls in the system (see Fig.6).

(a) There are two types of repeated call groups. The first type (TYPE 1) occurs when the arriving batch size is larger than the number of idle circuits. This batch creates a busy period (all circuits are busy). The second type (TYPE 2) represents the groups which arrive during busy periods.

(b) The interval between successive rejected calls is a constant t, and the same interval t is maintained while these calls are repeated.

On the other hand the mean busy period for the GI/M/S loss system is derived from dividing the mean holding time by the number of circuits. If the number of circuits is large enough, busy periods are considerably shorter than the holding time.

Considering the above facts, we can assume the following:

(a) The steady state probability that TYPE 1 attempts or first attempts finds the number of circuits occupancy being i is P(i) derived by the first part of the algorithm.

(b) TYPE 1 calls always arrive just before the arrival of TYPE 2 calls, if there are.

(c) TYPE 1 calls and first call attempts become TYPE 2 with probability P(S) when they access to the circuits.

(d) Once a batch encounters a busy period, this batch is repeated with another group which has the batch size probability shown below,

$$F(\ell)=\sum_{j=\ell+1}^{\min(\ell+S,g)} H(j)P(S-j+\ell),$$

$$\text{for } 0 \leqq \ell \leqq g-1. \tag{5}$$

Based on the above assumption, we can add the following steps to the first part of the algorithm, and complete the algorithm as a whole.

(step 6)

Let D(i,j,k) and E(i,k) denote the arrival rate for all calls and individual calls respectively, where j is the number of calls in the batch under consideration, i is the number of calls arriving just before the arrival of the batch indicated by j and k is the number of times of repetition. This arrival rate is defined in the regions $0 \leqq i \leqq (g-1)(k-1)$, $1 \leqq j \leqq g$ and $0 \leqq k \leqq M+1$. The batch indicated by j is taken to mean "the batch in question ", and the batch indicated by i to mean "the interfering batch". The batch in question includes both TYPE 1 and TYPE 2 calls, while the interfering batch includes only TYPE 1 calls. For k=0, the rates correspond to the arrival rate for the first attempts. As the batch size of the interfering batch is zero for first attempts, we can obtain the following equations:

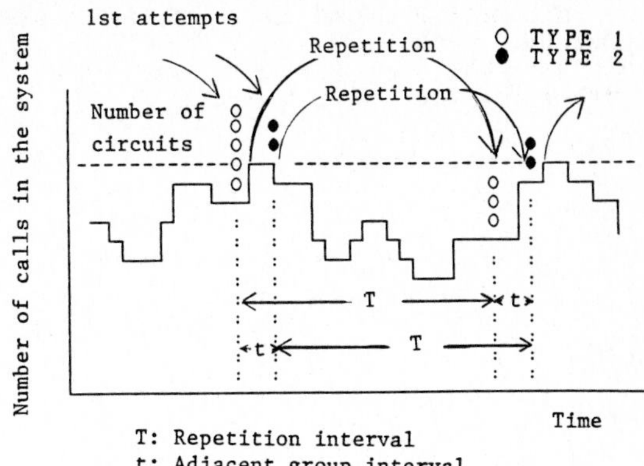

Fig.6 Example of repeated calls

$$\left.\begin{aligned}
&D(0,1,0)=\lambda_i .\\
&\text{For } 2 \leqq j \leqq g,\\
&D(0,j,0)=\lambda_g H_g(j),\\
&E(0,0)=\lambda_i .
\end{aligned}\right\} \tag{6}$$

(step 7)

The arrival rate D(i,j,k) and E(i,k) are determined recursively from k=1 to k=M using Eq.(6) as the initial value.

$$\left.\begin{aligned}
&\text{For } 1 \leqq i \leqq (g-1)(k-1) \text{ and } 1 \leqq j \leqq g,\\
&D(i,j,k) = \sum_{\ell=0}^{\min(g-1,i)} D(i-\ell,j,k-1)F(\ell)P(S)\\
&\quad + \sum_{\ell=1}^{\min((g-1)(k-1)-i,S)} D(i+\ell,j,k-1)P(S-\ell).\\
&\text{For } i=0 \text{ and } 1 \leqq j \leqq g,\\
&D(0,j,k) = D(0,j,k-1)F(0)P(S)\\
&\quad + \sum_{r=0}^{\min((g-1)(k-1),S)} \sum_{\ell=\max(1,r)}^{\min(g-j+r,S)} P(S-\ell)\\
&\quad \cdot D(r,j+\ell-r,k-1).\\
&\text{For } 0 \leqq i \leqq (g-1)(k-1),\\
&E(i,k) = \sum_{\ell=0}^{\min(g-1,i)} E(i-\ell,k-1)F(\ell)P(S)\\
&\quad + \sum_{\ell=1}^{\min((g-1)(k-1),S)} E(i+\ell,k-1)P(S-\ell).
\end{aligned}\right\} \tag{7}$$

In Eqs.(7), the first terms indicate that once calls find all circuits busy, the number of calls in the interfering batch increases by ℓ according to $F(\ell)$. The second terms correspond to the phenomena whereby the number of calls in the interfering batch decreases by the number of idle circuits which the calls find.

3.2.3 Performance Measures

Let us define the offered load $L_\ell(k)$ for the k-th repeated calls which are obtained from the arrival rates, where ℓ=1, ℓ=2 and ℓ=3 indicate all calls, individual calls and group calls, respectively.

From D(i,j,k) and E(i,k), we obtain the load of k-th repeated calls:

$$\left.\begin{aligned} L_1(k) &= \sum_{j=1}^{g}\sum_{i=0}^{(g-1)(k-1)} jD(i,j,k)/\mu, \\ L_2(k) &= \sum_{i=0}^{(g-1)(k-1)} E(i,k)/\mu, \\ L_3(k) &= L_1(k)-L_2(k). \end{aligned}\right\} \quad (8)$$

The performance measures are formulated as follows.

(a) The rejection probability of the k-th attempt

$$R_\ell(k)=L_\ell(k)/L_\ell(k-1). \quad (9)$$

(b) The loss probability

$$B_\ell=L_\ell(M+1)/L_\ell(0). \quad (10)$$

(c) Mean transfer delay

Transfer delay is defined as the time interval from the first attempt until the successful attempt.

$$W_\ell=\frac{\sum_{k=1}^{M} kT(L_\ell(k)-L_\ell(k+1))}{(L_\ell(0)-L_\ell(M+1))} \quad (11)$$

3.2.4 Simplified Algorithm

The offered load $L_\ell(k)$ can also be obtained by the first part of the algorithm. In this case, the offered load is obtained as follows:

$$\left.\begin{aligned} L_1(k) &= \sum_{j=1}^{g} jG(k,j)A(k)/\mu, \\ L_2(0) &= \lambda_i/\mu, \\ L_2(k) &= L_2(k-1)P(S), \\ L_3(k) &= L_1(k)-L_2(k). \end{aligned}\right\} \quad (12)$$

From Eqs. (9), (10), (11) and (12), the performance measures are obtained. Algorithm 1 and algorithm 2 are defined as follows:

algorithm 1: step 1 - step 7,
algorithm 2: step 1 - step 5.

Algorithm 2 can provide the same rejection probability of the first call attempts as algorithm 1. The computation time of algorithm 2 is considerably shorter than that of algorithm 1. Algorithm 2 can be used for rough estimation as a substitution of algorithm 1.

3.2.5 Numerical Results

The rejection probability for k-th repeated calls for $\mu=1$, $\lambda_i=10$, $\lambda_g=0.5$, m=g=20, T=3 and S=30 is shown in Fig.7. The values obtained in simulation, algorithm 1 and algorithm 2 are indicated as dots, triangles and crosses, respectively. The algorithms offer a good approximation for first attempts. For individual calls, simulation shows that the rejection probability for repeated calls is considerably larger than that of calls at the first attempt. Algorithm 2 does not consider this phenomenon, and can only provide constant values for the rejection probability of individual calls. On the other hand, algorithm 1 offers a good approximation for repeated attempts as well as for first attempts.

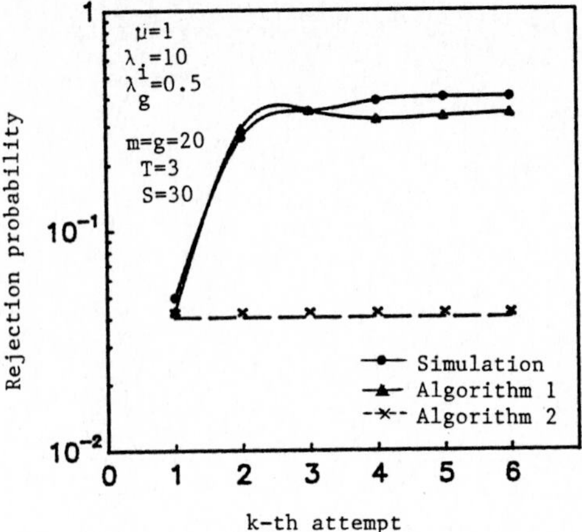

(1) Individual calls

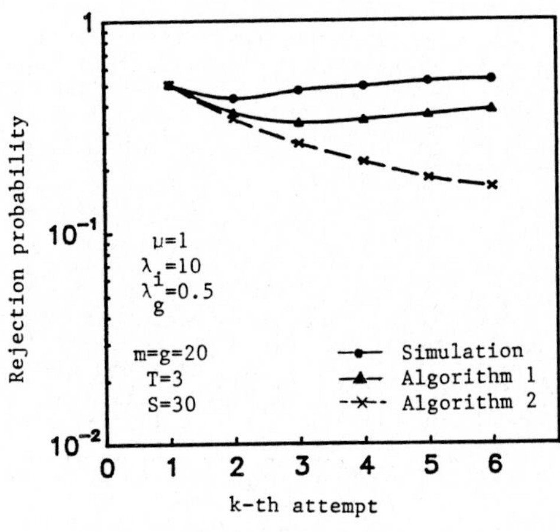

(2) Group calls

Fig.7 Rejection probability for K-th attempt

The relation between mean transfer delay and mean batch size for $\mu=1$, $\lambda_i=10$, $\lambda_g=10/m$, T=3 and S=30 is shown in Fig.8. This figure shows that algorithm 1 is required for individual calls evaluation, though algorithm 2 is enough for group calls.

The relation between mean transfer delay and repetition interval is shown in Fig.9. This figure shows that the effect of previous attempts can be ignored when the repetition interval is longer than three times of the mean holding time.

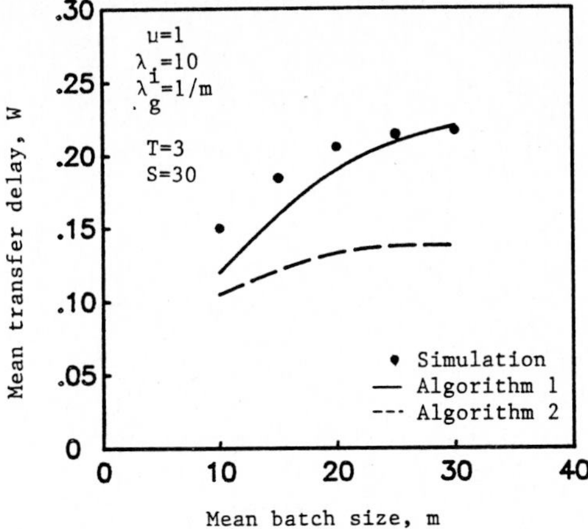

(1) Individual calls

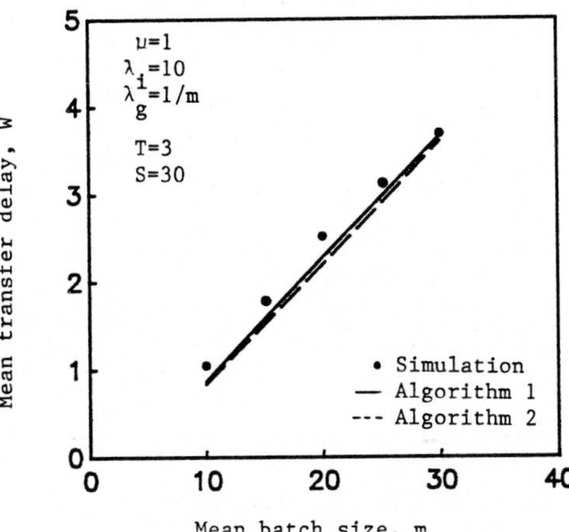

(2) Group calls

Fig.8 Mean transfer delay versus mean batch size

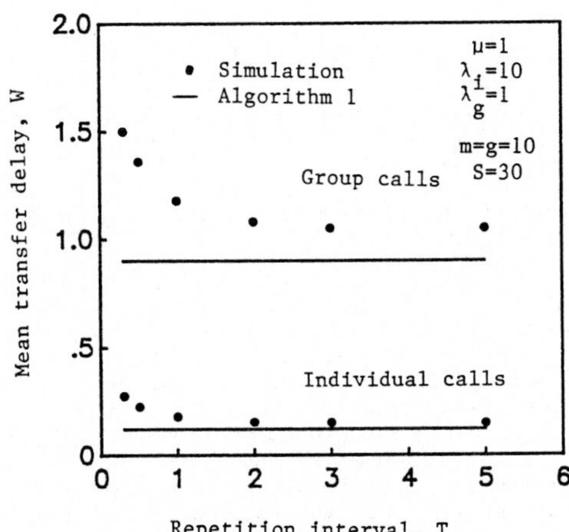

Fig.9 Relation between mean transfer delay and repetition interval

4.CONCLUSION

We have developed an approximation algorithm for a loss system featuring batch arrival and call repetition at a constant interval. This algorithm can be used in designing networks which offer multi-address delivery service and automatic re-calling service.

By comparing numerical and simulation results, the high accuracy of the approximation was demonstrated.

This algorithm can easily be extended to other batch arrival models with repeated calls.

ACKNOWLEDGMENT

The authors wish to express their thanks to Mr. Kunio Kodaira, Chief of the Teletraffic Section of Musashino Electrical Communication Laboratory, NTT, and to Dr. Tsuyoshi Katayama, Staff Engineer of the same section, for their guidance and encouragement.

REFERENCES

[1] K.S. Liu, "Direct-distance-dialling call completion and customer retrial behaviour", ITC9, paper 144(1979).
[2] J.W. Roberts, "Recent observations of subscriber behaviour", ITC9, paper 147(1979).
[3] G. Gosztony and G. Honi, "Some practical problems of the traffic engineering of overloaded telephone networks", ITC8, paper 141(1979).
[4] G. Gosztony, "Comparison of calculated and simulated results for trunk groups with repeated attempts", ITC8, paper 321(1976).
[5] A. Elldin, "Approach to the theoretical description of repeated call attempts", Ericsson Technics, 23, 3, pp.345-407(1967).
[6] D.J. Songhurst, "Subscriber repeat attempts, congestion, and quality of service: a study based on network simulation", ITC10, session 1.1, paper 5(1983).
[7] K. Yuki, F. Kanaya and H. Ishikawa, "Facsimile intelligent communication system (FICS-2) outline", Review of the ECL, NTT, Jpn.,31, 4, pp.467-474(1983).
[8] T. Endo, M. Sawai and F. Adachi, "Traffic Design and Service Quality in FICS-2", Review of the ECL, NTT, Jpn., 33, 1, pp.13-20(1985).
[9] J.W. Cohen, "Basic problems of telephone traffic theory and the influence of repeated calls", Philips Telecommun. Rev., 18, 2, pp.49-100(1967).
[10] S.S. Lam, "Store-and-forward buffer requirements in a packet switching network", IEEE Trans. Commun., COM-24, 4, pp.394-403(1974).
[11] D.R. Manfield and P. Tran-Gia, "Analysis of a finite storage system with batch input arising out of message paketization", IEEE Trans. Commun., COM-30, 3, pp.456-463(1982).

TELETRAFFIC ISSUES in an Advanced Information Society
ITC-11
Minoru Akiyama (Editor)
Elsevier Science Publishers B.V. (North-Holland)

FAULT LINES' INFLUENCE TO THE VALUES OF PROBABILISTIC CHARACTERISTICS AND THEIR MEASUREMENT ACCURACY IN SYSTEM WITH EXPECTATION AND REPEATED CALLS

Naumova E.O., Shkolny E.I.

Institute for Problems of Information Transmission
USSR Academy of Sciences Moscow, USSR

ABSTRACT

The paper deals with statistical properties of service quality characteristics and capacity of full-available group with repeated calls and partly or completely faulty lines. A new method is applied to statistical study of Markovian teletraffic systems. Values of means and variances of characteristics for lost calls' coefficient, blocking probability and average number of busy lines are obtained. Problems of engineer algorithm constructing to detect non-blocking faults in telephone and telegraph circuits are discussed.

1. INTRODUCTION

At last years due to telephone and telegraph circuits' development it becomes necessary to elaborate and use in practice methods of automatic fault detection in lines and equipment. If faulty lines are not detected, the calls continue coming to them not being satisfactory serviced. Such calls cause repeated ones that additionally load the circuit, decrease service quality and leads to untime wear of equipment. Fault detection method based on measurement analysis of service quality characteristics (lost calls coefficient and blocking probability) and capacity (average number of busy lines) is considered to be economic at last years. To provide high reliability of detection it is necessary to obtain enough accurate characteristics' measurements. To make an optimal plan for the measurement accuracy and duration required means and variances of characteristics shoud be studied. The paper presents results of such study for full-available group model with partly or completely faulty lines. Some numerical data are also presented.

2. MODEL OF SYSTEM

Consider a full-available group of V lines and a buffer with $n < \infty$ expectation places (see Fig.1). The group of lines can be conditionally presented as consisting of 3 groups with V_k, $k=\overline{1,3}$, $(V=V_1+V_2+V_3)$ of good, partly and completely faulty lines each correspondingly. Assume that the faulty lines are not detected and receive calls as good ones. Line occupation in the third group never ends in conversation. Lines of the second group are faulty with probability γ.

A Poissonian flow of primary calls of intensity λ comes to the group of lines. The incoming call with probability δ_k, $k=\overline{1,3}$, (see later) occupies any line of the k-th groups for preservice time, exponentially distributed with a parameter α. Preservice time is the mean time from the moment of line occupation up to its liberation due to subscriber non-answer, subscriber line being busy, unsatisfactory service quality. After preservice in the k-th group with probability Q_k, $k=\overline{1,3}$, a call leaves the line and with probability H_3 forms one source of repeated calls (SRC) and with probability $1-H_3$ leaves the system finally. Values Q_k are determined from equations

$$Q_1=Q,\quad Q_2=Q+(1-Q)\cdot\gamma,\quad Q_3=1, \qquad (1)$$

where Q is mean probability of subscriber non-answer, subscriber line being busy, or call blocking in network sections following considered group. After preservice in k-th group with probability $1-Q_k$, $k=\overline{1,3}$, the line does not become free and the main service - conversation starts. Such service duration is exponentially distributed with parameter equal to one.

Calls, coming at times, when V lines are busy, occupy one place each in the buffer. If there are no free places then the primary call with probability H_1 forms one SRC and with probability $1-H_1$ leaves the system finally. The call's waiting time in the buffer is bounded by exponentially distributed random value with parameter ν. Having not received service during this period the call leaves the buffer and with probability H forms one SRC and with probability 1-H leaves the system finally.

Each SRC sends repeated calls in independent exponentially distributed intervals with parameter μ. Service algorithm of repeated calls differs from one of primary calls by the fact that blocking repeated call with probability H_2 shoud be repeated and with probability $1-H_2$ leaves the system finally that decreases the number of SRC by a unit. Occupation of line or of expectation place by a repeated call is accompanied by immediate liquidation of one SRC.

3. SYSTEM CHARACTERISTICS AND OTHER RANDOM VALUES

Consider a stepped continuous in the left part process $J(\tau), \tau \geq 0$, with states $i=(i_k, k=\overline{1,7}) \in G$, where $i_1=\overline{0,i_2}$, $i_3=\overline{0,i_4}$, $i_5=\overline{0,V_3}$ - is the number of lines, occupied by preservice in the first, second and third groups correspondingly; $i_2=\overline{0,V_1}$, $i_4=\overline{0,V_2}$ is the number of lines occupied in the first and second groups correspondingly; $i_6=\overline{0,n}$ - is the number of busy places in the buffer; $i_7=\overline{0,n_1}$ - is the number of SRC. It is not difficult to see that the process $J(\tau)$ is homogeneous transitive conservative Markovian process.

Consider the most typical system characteristics corresponding to measurement interval $[0,t]$. The characteristics have the form

$$\mathcal{I}_u(t)=S_{z_u}(t)/S_{f_u}(t),\ u=\overline{1,3} . \qquad (2)$$

and mean the following. Denote by $G_1 = \{i: i_6=n\}$ a subset of states all places in the buffer being busy, in which the incoming calls are blocked. Characteristic $\mathcal{I}_1(t)$ means lost calls' coefficient in states $i \in G_1$. Values $S_{z_1}(t)$, $S_{f_1}(t)$ in (2) mean corresponding number of blocking (primary and repeated) calls in states $i \in G_1$ and of all the calls incoming during time t. Characteristic $\mathcal{I}_2(t)$ is loss probability (probability of blocking) of the system, and $S_{z_2}(t)$ in (2) is the duration stay of process $J(\tau)$ in states $i \in G_1$ during time t, $S_{f_2}(t)=t$. Characteristic $\mathcal{I}_3(t)$ is the common number of busy lines in a group and $S_{z_3}(t)$ is equal to $\int_0^t (i_2(\tau)+i_4(\tau)+i_5(\tau))\,d\tau$, $S_{f_3}(t)=t$.

Consider a vector of values

$$S(t)=\{S_0(t),\ S_1(t),\dots,\ S_4(t)\} \qquad (3)$$

where

$$S_1(t)=S_{z_1}(t),\ S_2(t)=S_{f_1}(t),\ S_3(t)=S_{z_2}(t),\ S_4(t)=S_{z_3}(t),\ S_0(t)=S_{f_2}(t)=S_{f_3}(t)=t. \qquad (4)$$

Denote by

$$h_{ij} = \{h_{ij}(0),\ h_{ij}(1),\dots,h_{ij}(4)\} \qquad (5)$$

the vector of random values $h_{ij}(r)$, $r=\overline{0,4}$, corresponding to one-step stay of process $J(\tau)$ in state $i \in G$ with condition that the process will transfer into state $j \in G$. Value $h_{ij}(0)=h_i(0)$ is the duration of one-step stay of process $J(\tau)$ in state i, exponentially distributed with a parameter

$$\lambda_i=\lambda+i_7\mu+\alpha(i_1+i_3+i_5)+i_2-i_1+i_4-i_3+i_6\nu \qquad (6)$$

Value $S_r(t)$, $r=\overline{0,4}$, in (3) is equal to the sum of values $h_{ij}(r)$, $r=\overline{0,4}$, at time $[0,t]$.

In (5) values $h_{ij}(r)$, $r=\overline{1,2}$ are equal

$$h_{ij}(1)=\varkappa_1 h_{ij}(2),\quad h_{ij}(2)=\varkappa_{ij}(\lambda,\mu), \qquad (7)$$

where $\varkappa_1$ - is the indicator of subset G_1 blocking states and $\varkappa_{ij}(\lambda,\mu)=1$, if the process $J(\tau)$ transition from state i to state j for one step (later we shall denote it by $i \to j$) takes place due to primary or repeated call incoming, otherwise $\varkappa_{ij}(\lambda,\mu)=0$. Values $h_{ij}(r)$, $r=\overline{3,4}$, are determined from equation

$$h_{ij}(3)=h_i(3)=\varkappa_1 h_i(0),\quad i,j \in G,$$
$$h_{ij}(4)=h_i(4)=(i_2+i_4+i_5)h_i(0),\quad i,j \in G. \qquad (8)$$

From (7), (8) it is seen that values $h_{ij}(r)$, $h_{jk}(f)$, $r,f=\overline{0,4}$, corresponding to two-step transition $i \to j \to k$ of process $J(\tau)$ are independent.

Consider a vector of random values

$$h_i=\{h_i(0),\ h_i(1),\dots,h_i(4)\},\quad i \in G. \qquad (9)$$

Vectors (5) and (9) are connected by stochastic equation

$$h_i=\sum_{j \in G} \varkappa_{ij} h_{ij},\quad i \in G, \qquad (10)$$

where $\varkappa_{ij}$ is the indicator of $i \to j$ transition of process $J(\tau)$. The indicator has distribution

$$p_{ij}=\Pr\{\varkappa_{ij}=1\}=\lambda_{ij}/\lambda_i,$$
$$\Pr\{\varkappa_{ij}=0\}=1-p_{ij},\quad i,j \in G, \qquad (11)$$

where λ_{ij} is intensity of $i \to j$ transition of the process $J(\tau)$. Values $h_i(r)$, $r=0,3,4$, in (9) are determined by equations (6), (8). Values $h_i(1)$, $h_i(2)$ are determined from equations

$$h_i(1)=\varkappa_1 h_i(2),\quad h_i(2)=\varkappa_i(\lambda,\mu),\quad i \in G, \qquad (12)$$

where $\varkappa_i(\lambda,\mu)$ is the indicator of primary or repeated call incoming to the state $i \in G$ of process $J(\tau)$.

Consider the vector of random values

$$S_{ij}=\{S_{ij}(0),\ S_{ij}(1),\dots,S_{ij}(4)\},\quad i,j \in G, \qquad (13)$$

that is obtained by summing of the vectors (5) during the time $S_{ij}(0)$ of one transition of process $J(\tau)$ from state i to state j.

4. MEANS AND VARIANCES OF CHARACTERISTICS

Introduce the notation

$$h_{ij}^{(u)}=h_{ij}(r_u)-\theta_u h_{ij}(f_u),\quad i,j \in G,\quad u=\overline{1,3}, \qquad (14)$$

$h_i^{(u)}=h_i(r_u)-\theta_u h_i(f_u)$, $i\in G$, $u=\overline{1,3}$,

$S_{ij}^{(u)}=S_{ij}(r_u)-\theta_u S_{ij}(f_u)$, $i,j\in G$, $u=\overline{1,3}$,

$L(r)=\sum_{i\in G} P(i)\lambda_i Mh_i(r)$, $r=\overline{0,4}$,

$\theta_u=\lim M\mathcal{I}_u(t)$, $t\to\infty$, $u=\overline{1,3}$,

where $P(i)$, $i\in G$, are the stationary probabilities of Markovian process $J(\tau)$. In [1] a theorem is proved for characteristics of semi-Markovian process $J(\tau)$ with finite set of states $i\in G$. For a particular case of Markovian process $J(\tau)$ the theorem takes the form.

Theorem. If for homogeneous transitive conservative Markovian process $J(\tau)$ with finite set of states $i\in G=\{i=\overline{0,N}\}$ the following conditions hold:

a) values $h_{ij}(r)$, $i,j\in G$, $r=r_u$, f_u, $u=\overline{1,q}$, are non-negative;

b) values $h_{ij}(r)$, $h_{jk}(f)$, $i,j,k\in G$, $r,f,=r_u,f_u$, $u=\overline{1,q}$, corresponding neighbouring steps of process $J(\tau)$ is transition $i\to j\to k$ are independent;

c) means $Mh_i(r)$ are such that $0\le Mh_i(r)<\infty$, $i\in G$, $r=r_u,f_u$, $u=\overline{1,q}$, and for each fixed value $r=r_u,f_u$ such value of $i\in G$ exists that $Mh_i(r)>0$;

d) $M[h_i(r)h_i(f)]<\infty$, $i\in G$, $r,f=r_u$, f_u. Then for $t\to\infty$ with probability equal to unit $\mathcal{I}_u(t)\to\theta_u$, $u=\overline{1,q}$, and the vector of characteristics $\{\mathcal{I}_u(t),\ u=\overline{1,q}\}$ is asymptotically normal

i.e. $\{\sqrt{t}\,[\mathcal{I}_u(t)-\theta_u],\ u=\overline{1,q}\}\Rightarrow$

$$\Rightarrow N(0,\ \|d_{uz}\|,\ u,z=\overline{1,q}), \tag{15}$$

where $\theta_u=L(r_u)/L(f_u)$, $u=\overline{1,q}$,

$$d_{uz}=\sum_{i\in G}P(i)\{\sum_{j\in G}\lambda_{ij}\,[Mh_{ij}^{(u)}MS_{jo}^{(z)}+Mh_{ij}^{(z)}MS_{jo}^{(u)}]+\lambda_i M(h_i^{(u)}h_i^{(z)})\}/(L(f_u)L(f_z)),$$

and $MS_{oo}^{(u)}=0$, $u=\overline{1,q}$, and $MS_{io}^{(u)}$, $i\ne 0$, are determined from the equation systems

$$\lambda_i MS_{io}^{(u)}-\sum_{\substack{j\in G\\ j\ne 0}}\lambda_{ij}MS=\lambda_i Mh_i^{(u)},\quad i\in G, u=\overline{1,q}, \tag{16}$$

having for fixed value of u the only solution.

For the model of system p.2 it is not difficult to make sure that the conditions a), b) of the theorem hold. Find moments for values $h_i(r)$ and control the realization of conditions c), d). From (6)-(8) we can find

$$\begin{aligned} &Mh_i(0)=1/\lambda_i,\quad Mh_i^2(0)=2/\lambda_i^2,\\ &Mh_i(1)=Mh_i^2(1)=\chi_1(\lambda+i_7\mu)/\lambda_i,\\ &Mh_i(2)=Mh_i^2(2)=(\lambda+i_7\mu)/\lambda_i,\\ &Mh_i(3)=\chi_1/\lambda_i,\quad Mh_i^2(3)=2\chi_1/\lambda_i^2,\\ &Mh_i(4)=(i_2+i_4+i_5)/\lambda_i,\\ &Mh_i^2(4)=2(i_2+i_4+i_5)^2/\lambda_i^2. \end{aligned} \tag{17}$$

Formulae for moments $M[h_i(r)h_i(f)]$, $r\ne f$, we omit. From (17) and the formulae omitted follows that the conditions c), d) of the theorem hold.

From (14), (15), (17) we can find means of characteristics for $t\to\infty$

$$\begin{aligned} &\theta_1=(\sum_{i\in G_1}(\lambda+i_7\mu)P(i))/(\lambda+\mu\sum_{i\in G}i_7P(i)),\\ &\theta_2=\sum_{i\in G_1}P(i),\\ &\theta_3=\sum_{i\in G}(i_2+i_4+i_5)P(i). \end{aligned} \tag{18}$$

Values d_{uz} in (15) for $u=z=\overline{1,3}$ are the main members in the decomposition of variances of characteristics

$$D\mathcal{I}_u(t)=d_{uu}/t+o(1/t),\quad t\to\infty \tag{19}$$

Means $Mh_{ij}^{(u)}$, $u=\overline{1,3}$, in (15) we can find from (6)-(8), (14)

$$\begin{aligned} &Mh_{ij}^{(1)}=\chi_{ij}(\lambda,\mu)(\chi_1-\theta_1),\\ &Mh_{ij}^{(2)}=Mh_i^{(2)}=(\chi_1-\theta_2)/\lambda_i,\\ &Mh_{ij}^{(3)}=Mh_i^{(3)}=(i_2+i_4+i_5-\theta_3)/\lambda_i. \end{aligned} \tag{20}$$

Values $L(f_u)$, $u=\overline{1,3}$, in (15) are equal

$$\begin{aligned} &L(f_1)=\lambda+\mu\sum_{i\in G}i_7P(i),\\ &L(f_2)=L(f_3)=1. \end{aligned} \tag{21}$$

Means $Mh_i^{(u)}$, $M(h_i^{(u)})^2$, $u=\overline{1,3}$, in (15), (16) we can find from (14), (17)

$$\begin{aligned} &Mh_i^{(1)}=(\chi_1-\theta_1)(\lambda+i_7\mu)/\lambda_i,\\ &Mh_i^{(2)}=(\chi_1-\theta_2)/\lambda_i,\\ &Mh_i^{(3)}=(i_2+i_4+i_5-\theta_3)/\lambda_i,\\ &M(h_i^{(1)})^2=(\lambda+i_7\mu)[\chi_1(1-2\theta_1)+\theta_1^2]/\lambda_i,\\ &M(h_i^{(2)})^2=2(\chi_1-\theta_2)^2/\lambda_i^2,\\ &M(h_i^{(3)})^2=2(i_2+i_4+i_5-\theta_3)^2/\lambda_i^2. \end{aligned} \tag{22}$$

Denote by χ_k, $k=\overline{2,4}$, the indicators of stay of process $J(\tau)$ in states $i\in G_k$, $k=\overline{2,4}$, where

$G_2=\{i:i_7\ne n_1\}$, $G_3=\{i:i_2+i_4+i_5=V\}$,

$G_4=\{i:i_6=0\}$

and by l_k vector

$l_k=(i_k=1,\ i_l=0,\ l=\overline{1,6},\ l\ne k)$.

Then equation system (16) for $MS_{jo}^{(u)}=x_u(i)$, $u=\overline{1,3}$, has the form

$$\begin{aligned} &[\lambda_i-\chi_1(\lambda(1-\chi_2H_1)+i_7\mu H_2)]x_u(i)-\\ &-\{(1-\chi_3)[\lambda\delta_1 x(i+l_1+l_2)+\lambda\delta_2 x_u(i+\\ &+l_3+l_4)+\lambda\delta_3x_u(i+l_5)+i_7\mu\delta_1x_u(i+l_1+\\ &+l_2-l_7)+i_7\mu\delta_2x_u(i+l_3+l_4-l_7)+i_7\mu\delta_3\times\\ &\times x_u(i+l_5-l_7)]+\chi_4[\alpha i_1Q_1H_3x_u(i-l_1-l_2+\\ &+l_7)+\alpha i_1Q_1(1-\chi_2H_3)x_u(i-l_1-l_2)+\\ &+\alpha i_3Q_2H_3x_u(i-l_3-l_4+l_7)+\alpha i_3Q_2(1-\chi_2\times\\ &\times H_3)x_u(i-l_3-l_4)+\alpha i_5H_3x_u(i-l_5+l_7)+\\ &+\alpha i_5(1-\chi_2H_3)x_u(i-l_5)+(i_2-i_1)x_u(i-l_2)+\\ &+(i_4-i_3)x_u(i-l_4)]+\alpha i_1(1-Q_1)x_u(i-l_1)+\\ &+\alpha i_3(1-Q_2)x_u(i-l_3)+\chi_3[\lambda x_u(i+l_6)+ \end{aligned}$$

$$+i_7\mu x_u(i+l_6-l_7)]+[\alpha(i_1Q_1+i_3Q_2+i_5)(1-\chi_2H_3)+i_6\nu(1-\chi_2H)]x_u(i-l_6)+[\alpha(i_1Q_1+i_3Q_2+i_5)H_3+i_6\nu H]x_u(i-l_6+l_7)+(i_2-i_1)x_u(i+l_1-l_6)+(i_4-i_3)\times x_u(i+l_3-l_6)+\chi_1[\alpha H_1x_u(i+l_7)+i_7\mu(1-H_2)\times x_u(i-l_7)\}=\lambda_i Mh_i^{(u)}, \quad i\in G,\ i\neq 0 . \qquad (23)$$

Members without sense in (23) should be omitted. Stationary probabilities P(i) are determined from the equation system

$$P(i)-\sum_{j\in G}\lambda_{ji}P(j)=0, \quad \sum_{i\in G}P(i)=1 . \qquad (24)$$

Intensities λ_{ji} in the evident form are not difficult to write down, if evident form of intensities λ_{ij} from equation system (23) is used. We omit the full writing of system (24). Equation systems (23), (24) were computed by the method of successive approximate Gauss-Zejdel (it is proved that this method is convergent to systems (16), (24) with approximately the same rate).

5. NUMERICAL RESULTS

Computed numerical results are presented on Fig.2-7. On Fig.2, 4, 6 curves of values of means θ_u, $u=\overline{1,3}$, and on Fig.3, 5, 7 of variances d_{uu}, $u=\overline{1,3}$, of characteristics $\pi_u(t)$, $u=\overline{1,3}$. The first curve on Fig.2-7 corresponds to the following set of parameters of the model:

$V=V_1=4$, $n=0$, $\mu=5$, $\alpha=8$, $H_1=H_2=H_3=0{,}83$, $Q=0{,}5$.

Let us point for curves 2-4 the values of parameters that are not equal to the ones of the curve 1: the second curve -

$n=\nu=3$, $H=0{,}83$;

the third curve -

$V=V_3=4$;

the fourth curve -

$V=V_3=4$, $n=\nu=3$, $H=0{,}83$.

On Fig.2-7 it is seen the great difference between the means' and variances' values in the systems without faulty lines (curves 1, 2) and in the systems with completely faulty lines (curves 3, 4). Such difference can be used for the systems automatic non-blocking fault detection in telephone circuit. For this purpose any of characteristics $\pi_u(t)$, $u=\overline{1,3}$, may be used and faith interval

$$\theta_u \pm U_\alpha\sqrt{d_{uu}/t}=\theta_u \pm \varepsilon_u$$

may be built, where U_α is twosided α-quantil of normal distribution N(0,1) (for example, for $\alpha=0{,}95$ $U_\alpha=1{,}96$). For the given absolute ε_u (or relative $\Delta_u=\varepsilon_u/\theta_u$) width of the faith interval measurement duration T_u, $u=\overline{1,3}$, can be found by formula

$$T_u=d_{uu}U_\alpha^2/\varepsilon_u^2=d_{uu}U_\alpha^2/(\Delta_u\theta_u)^2, \quad u=\overline{1,3}.$$

Measurement duration T_u during which the faults in the group of lines will be detected with good accuracy ε_u is the less the greater number of faulty lines are in the group.

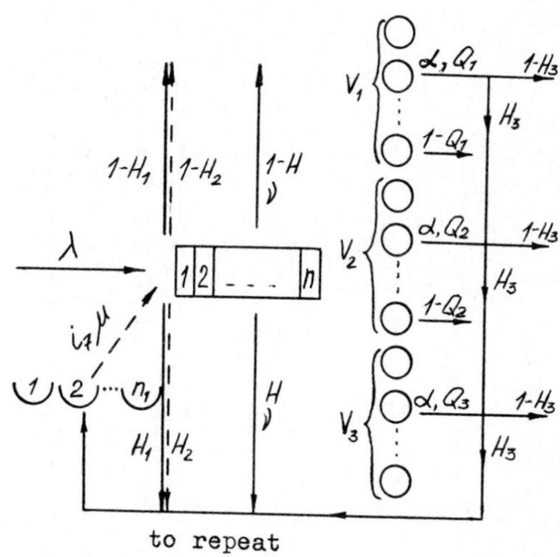

Fig.1

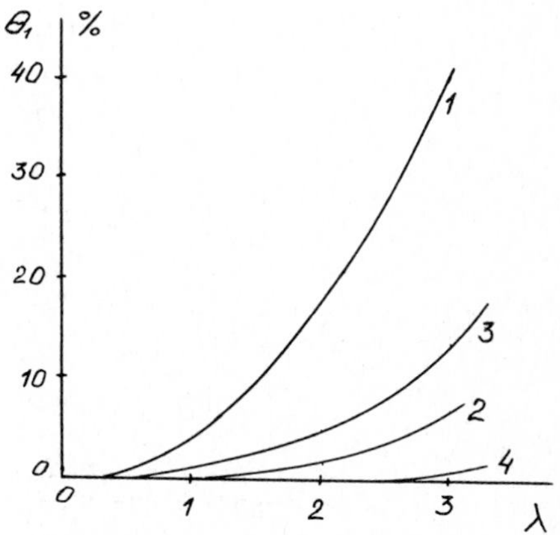

Fig.2

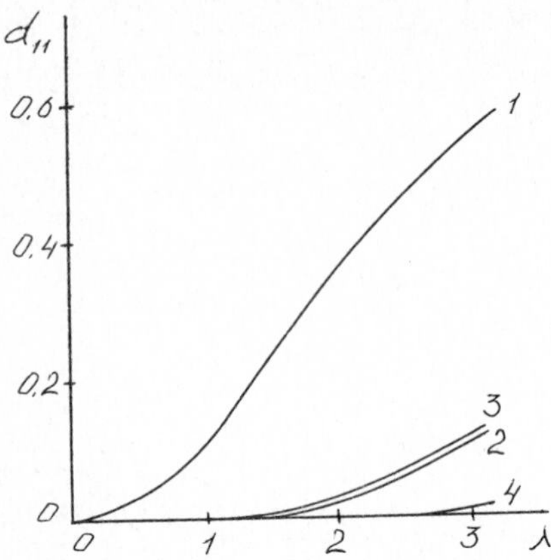

Fig.3

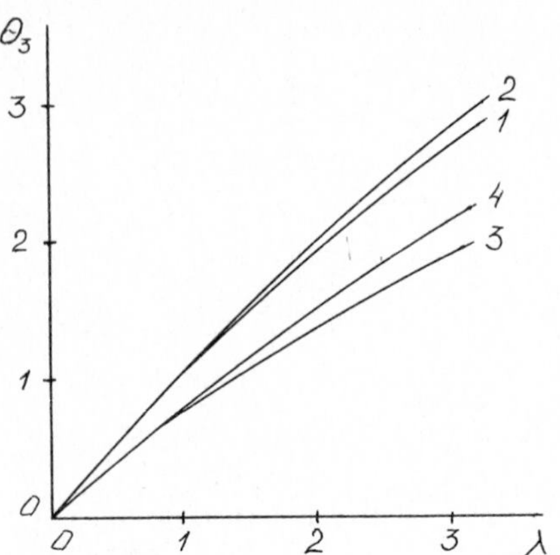

Fig.6

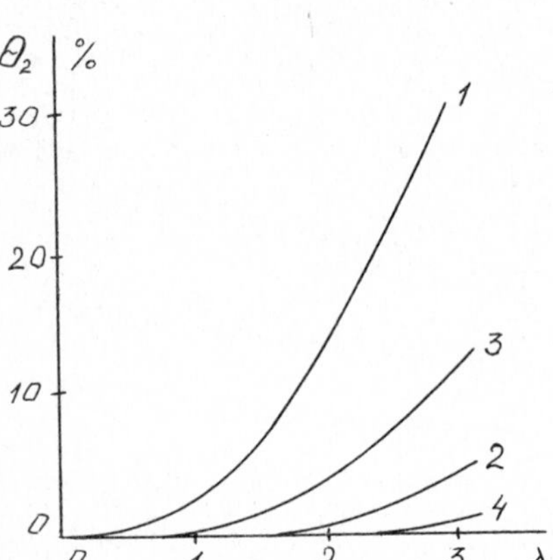

Fig.4

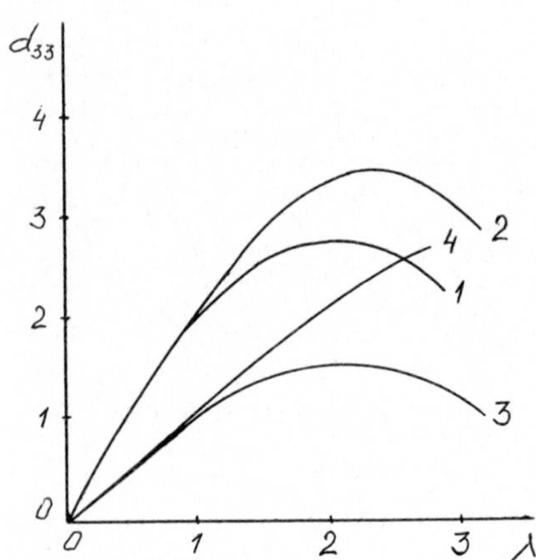

Fig.7

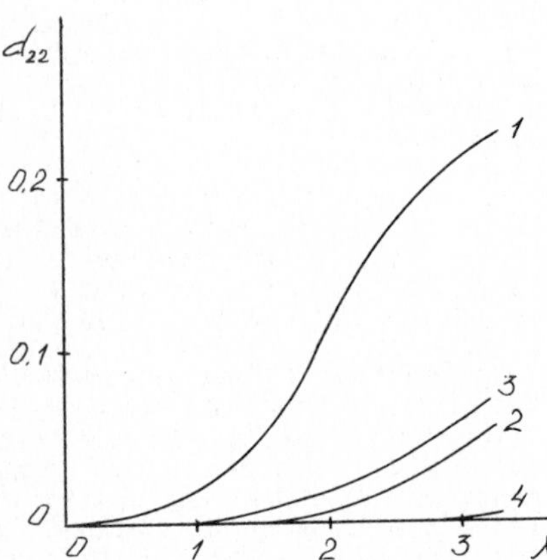

Fig.5

6. CONCLUSION

The authors suppose that the full analysis of numerical results, partly presented on Fig.2-7, makes possible to build a convenient engineering algorithm for non-blocking fault detection in the telephone and telegraph circuits. Using of the method of statistical study of teletraffic Markovian systems, presented above, admits the investigation of the other interesting models of the systems.

REFERENCES

1 E.I. Shkolny, "Variances and covariances of statistical estimates of characteristics of semi-Markovian systems with finite set of states," Problemy peredachi informatsii, vol.XXI, no.1, 1985.

TELETRAFFIC ISSUES in an Advanced Information Society
ITC-11
Minoru Akiyama (Editor)
Elsevier Science Publishers B.V. (North-Holland)

APPROXIMATE ANALYSIS OF SYSTEMS WITH REPEATED CALLS AND MULTIPHASE SERVICE

A.KHARKEVICH, I.ENDALTSEV, E.MELIK-GAIKAZOVA, N.PEVTSOV, S.STEPANOV

Institute for Problems of Information Transmission
USSR Academy of Sciences Moscow, USSR

ABSTRACT

A model of full-available bunch with Poissonian flow of primary calls is considered. Service time consists of three exponentially distributed phases. According to the cause of the refusal there are two types of repeating subscribers in the model: due to all the types of losses that the subscriber explains by blocking and due to the no answer of the called subscriber. For given model a consequence of approximate calculation algorithms is built, based on conservation laws and on changing of the structure of input flow of calls and service schedule. Besides the method description the paper contains analytical and numarical study of the accuracy of proposed method. The paper considers the results of the application of the model to the calculation of characteristics of direct bunch of channels in toll telephone system.

1.INTRODUCTION

The number of repeated calls increases with raise of the level of telephone system automatization. The influence of repeated calls is higher in the toll telephone systems. The necessity of studying of repeated calls' problem leads to the appearance of models, in which different aspects of interaction between subscriber and servicing system are taken into account. Among the great number of papers on the problem it is sufficient to mention only some works published in the proceedings of the last teletraffic congresses, where one can find references to the earlier studies. We name the papers of Le Gall P. [1], Ionin G.L., Sedol I.I. [2], Evers R. [3], Gosztony G. [4], Shneps-Shneppe M.A. [5] and some others. They state the essence of the problem and give the results of studies of some models with repeated calls.

The application of the proposed models is possible in the case when there is accurate or effective enough approximate method of calculation of their characteristics. Due to the large dimension of stochastic processes, describing the functioning of model with repeated calls, we succeeded in obtaining accurate formulae or calculation algorithms only for a small class of models for the simplest cases of subscriber behaviour. As result, the only way to estimate the characteristics is to use approximate methods. Constructing the models with repeated calls different aspects of the interaction between subscriber and service system should be taken into account. It follows that a whole class of models appeared with different assumptions on subscriber behaviour after refusal. For that reason among the approximate methods the most important are those that can be easily formalized and extended from one subscriber behaviour model to another. Some results obtained in constructing of such methods are presented in the book [6] recently published. The English review can be find in [7]. Our purpose is to apply methods given in [6] to calculation of the model of the direct bunch of channels in toll telephone system.

2. MODEL

2.1 Operation Scheme

A Poisson flow of primary calls of intensity A comes to full-available bunch of v lines. The incoming call occupies an arbitrary free line for the time of connection that is exponentially distributed with a parameter equal to w_1. The phase corresponds to the passing time of the call from the calling subscriber to the called subscriber. After the first phase with a probability P_1P_2 the subscriber continues occupying the line for the second and third servicing phases, that are exponentially distributed with parameters $w_{2'}$ and w_3.

With probability $P_1(1-P_2)$ the duration of the second phase is exponentially distributed with parameter equal to $w_{2''}$ and after its completion the subscriber frees the line. Here $(1-P_1)$ is the probability of called subscriber being busy, and $(1-P_2)$ is the probability of non-answer of called subscriber. The values of $1/w_{2'}$ and $1/w_{2''}$ are the mean listening time of ringing tone. It depends on if the conversation takes place or not. In such way the process of occupying the line by the call is described. Now pass to the formation of repeated call sources (RCS).

Primary or repeated call entering the system with probability a gets refusal before the input of the bunch. This event with probability H leads to appearance of RCS of the first type, from which the repeated call comes in a random time, exponentially distributed with parameter equal to s_1. The cause of the appearance of such RCS is thelosses on the input into the toll telephone system. RCS of the first type is also formed in all the cases when subscriber explains the refusal by blocking. The event takes place with probability H if all the lines are busy, after unsuccessful ending of the first phase of service and if the called subscriber is busy. RCS of the second type is formed with probability Z due to non-answer of the called subscriber The repeated call in this case comes in random time, exponentially distributed with parameter s_2. The relation between s_1 and s_2 usually has a form $s_1 > s_2$. Model scheme is given in the Fig.1.

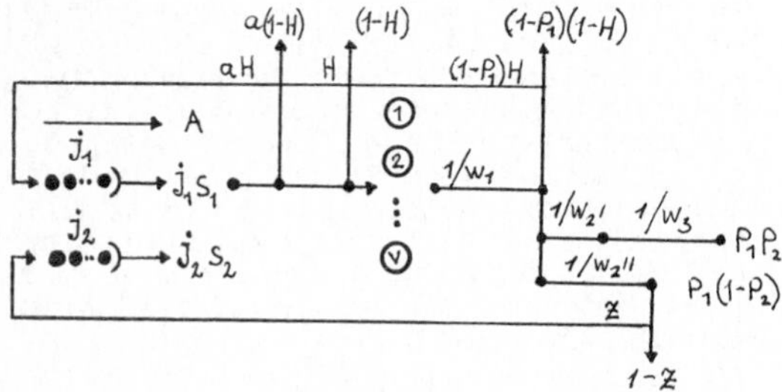

Fig.1

Model operation is described by Markovian process of the form

$$Y(t) = (j_1(t), j_2(t), i_1(t), i_{2'}(t), i_{2''}(t), i_3(t)),$$

where $j_k(t)$ - is the number of RCS of k-th type at time t k=1,2 and $i_1(t)$, $i_{2'}(t), i_{2''}(t), i_3(t)$ are the number of bunch lines, occupied at time t for corresponding service phase.

Denote by $P_{j_1,j_2,i_1,i_{2'},i_{2''},i_3}$ probabilities of stationary states of model that further for the aim of simplicity we shall denote by $P_{j,i}$. Here $j = (j_1, j_2)$, $i = (i_1, i_{2'}, i_{2''}, i_3)$. Using the standart procedure for Y(t) it is possible to write out the system of statistical equilibrium equations that we do not present here for being bulky.

2.2 Probability Characteristics of Model

Let $l = i_1+i_{2'}+i_{2''}+i_3$. Let us define some probabilistic characteristics of the model:

- the probability of blocking for primary calls due to all lines of the bunch busy

$B = \sum_j \sum_{l=v} P_{j,i}$, (here $\sum_j$ means summation with respect to all j_1, j_2 in interval from 0 to ∞);

- the mean number of subscribers of k-th type who repeat their calls

$$J_k = \sum_j \sum_{l \leq v} P_{j,i}\, j_k, \qquad k=1,2;$$

- the mean number of bunch lines occupied for corresponding service phase

$$I_a = \sum_j \sum_{l \leq v} P_{j,i}\, i_a, \quad a = 1, 2', 2'', 3;$$

- the total loss probability for the calls arriving at the input of the bunch

$$B_c = ((AB+J_1's_1+J_2's_2)(1-a) + I_1w_1(1-P_1) + I_{2''}w_{2''}) / ((A+J_1s_1+J_2s_2)(1-a)),$$

(here $J_k' = \sum_j \sum_{l=v} P_{j,i}\, j_k, \quad k=1,2$).

- the mean number of repeated calls to one primary

$$M = (J_1s_1+J_2s_2) / A ;$$

- the mean number of refusals to one connection

$$Q = ((A+J_1s_1+J_2s_2)a + (AB+J_1's_1+J_2's_2)(1-a) + I_1w_1(1-P_1) + I_{2''}w_{2''})) / (I_3w_3) ;$$

By the similar way it is possible to introduce some other characteristics of the given model.

2.3 Conservation Laws

The intensities of coming and servicing by the system flows of calls are connected by conservation laws, that play an important role in the study of the models with repeated calls. Taking account of the above notations we have the following relations:

$$J_1s_1 = (A+J_1s_1+J_2s_2)aH + (AB+J_1's_1+J_2's_2)(1-a)H + I_1w_1(1-P_1)H; \tag{1}$$

$$J_2s_2 = I_{2''}w_{2''}Z;$$

$$I_1w_1P_1P_2 = I_{2'}w_{2'};$$

$$I_1w_1P_1(1-P_2) = I_{2''}w_{2''};$$

$$I_{2'}w_{2'} = I_3w_3;$$

$$(A(1-B)+(J_1-J_1')s_1+(J_2-J_2')s_2)(1-a)=I_1w_1;$$

The method of proof of such relations for systems with repeated calls has been considered in general form in [6].

These relations may be used for solving many problems connected with repeated calls. One of them is finding relationships between different probabilistic characteristics of the model. For example using (1) all I_a, a=1,2',2",3 can be expressed through $I = I_1+I_{2'}+I_{2''}+I_3$ (here I is the mean number of busy lines of the bunch) by the following formulas

$$I_a = If_a,$$

where

$$f = 1 + \frac{w_1P_1P_2}{w_{2'}} + \frac{w_1P_1(1-P_2)}{w_{2''}} + \frac{w_1P_1P_2}{w_3} ;$$

$$f_1 = \frac{1}{f} ; \; f_{2'} = \frac{w_1P_1P_2}{fw_{2'}} ; \; f_{2''} = \frac{w_1P_1(1-P_2)}{fw_{2''}} ;$$

$$f_3 = \frac{w_1P_1P_2}{fw_3} .$$

With the help of conservation laws we can easy find the necessary ergodicity condition on the value of A for the case of absolutely persistent subscriber (H=1). Really, from (1) we have

$$A = Iw_1P_1(1-(1-P_2)Z) \; / \; f. \qquad (2)$$

It is true that $I \leqslant v$. Using this inequality we have

$$A \leqslant vw_1P_1(1-(1-P_2)Z) \; / \; f. \qquad (3)$$

After direct substitution in (1) and system of statistical equilibrium equations it is easy to verify that the realization of the equality in (3) contradicts to the irreducibility of the Markov process Y(t). Thus we have finally the necessary condition

$$A < vw_1P_1(1-(1-P_2)Z \; / \; f. \qquad (4)$$

Now we turn to application of conservation laws in construction of algorithms of approximate analysis models with repeated calls. This problem would be discussed in the following section.

It is clear that to find the characteristics of introduced model it is sufficient to know only the values of B, J_k (k=1,2), $L' = (AB+J_1's_1+J_2's_2)$, I_a. These characteristics we shall further name the maine characteristics and only for them we shall write approximate formulae.

3. APPROXIMATE DESIGN OF MODEL

3.1 The Principle of Construction of Approximate Algorithms

We shall change the operation scheme of the initial model so that the stochastic process, describing the transformed (simplified) model, should have lesser dimension and more convinient for calculation matrix of transition intensities. We shall construct the simplified model by changing the input flow structure and schedule of call service. As estimates of probabilistic characteristics of the studied model we take values of corresponding characteristics of the simplified model. In order not to loose the accuracy we demand for estimates found to hold relations, corresponding in the form to conservation laws (1) of the original model. For this purpose we shall introduce into the simplified model some unknown parameters whose specific values will be determined from conservation laws considered as implicit equations with respect to these parameters.

This method being applied to the introduced model gives a possibility to construct at once a number of approximate algorithms if perfome the following actions:

1.At first we simplify the service process. We do not distinguish line occupation for specific phase, but fix only the total number of busy lines i. Assume that in the stationary state among i busy lines $y_a i$ lines are occupied for a-th service phase a = 1, 2', 2", 3. Otherwise the scheme of operation of the original model does not change.

2.Using the fact that s_1 is considerably greater than s_2 replace the flow of repeated calls, formed due to non answer of called subscriber, by Poissonian with intensity x_2. Otherwise the scheme of operation of the original model does not change.

2a.For the case when a=0 we replace additionally a flow of repeated calls formed due to calling subscriber being busy, by Poissonian with intensity $x_{1,1}$. Otherwise the scheme of operation of the original model does not change.

3.Replace the flow of repeated calls, formed due to blocking, by Poissonian with intensity x_1. Realization of the last two steps means that we replace the whole flow of repeated calls by Poissonian with intensity $x = x_1 + x_2$.

The values of unknowns y_1, $y_{2'}$, $y_{2''}$, y_3, x_1, x_2, $x_{1,1}$ are found by iteration procedure from the solution of implicit equations, obtained from the conservation laws. At every step the dimension of Markovian process, describing the simplified model, decreases. After the first step we have 3-dimensional process, after the second - 2-dimensional, after the second (a) - 2-dimensional but with more convinient for calculations matrix of transition intensities, after the third - 1-dimensional.

Consider the realization of each of the steps of the proposed constructing procedure of approximate algorithms more precisely.

3.2 An Algorithm Based on Replacing the Multi-Phase Service Time to One-Phase

To this algorithm we pass after transformations mentioned in the first step. Given simplified model is described by 3-dimensional Markovian process. Let us denote by $P(j_1,j_2,i)$ stationary probabilities of the model. Here j_1,j_2 is the number of repeating subscribers of the corresponding type, and i - the number of busy lines in the bunch. Using the standart procedure, it is possible to write out the system of statistical equilibrium equations, connecting $P(j_1,j_2,i)$. However, it is bulky to be present. For the estimate of probabilistic characteristics of original model B, I_a, J_1, J_2,

L' we introduce corresponding characteristics of simplified model $B_1, I_{a,1}, J_{1,1}, J_{2,1}, L'_1$, which are defined as follows:

$$B_1 = \sum_j P(j_1,j_2,v) \; ; \qquad (5)$$

$$I_{a,1} = \sum_j \sum_{i=0}^{v} P(j_1,j_2,i) i y_a, \quad a=1,2',2'',3;$$

$$J_{k,1} = \sum_j \sum_{i=0}^{v} P(j_1,j_2,i) j_k, \quad k=1,2 \; ;$$

$$L'_1 = \sum_j P(j_1,j_2,v)(A+j_1 s_1+j_2 s_2).$$

Values of introduced estimates depend on unknown parameters y_a, $a=1,2',2'',3$. To find them let us use the condition of holding relations (1) for estimates (5). It is not difficult to show that these relations should really be held if for y_a the equations hold

$$y_a = f_a, \quad a = 1,\ 2',\ 2'',\ 3. \qquad (6)$$

So to find estimates (5) for probabilistic characteristics of the original model it is necessary to solve the system of statistical equilibrium equations that connects probabilities $P(j_1,j_2,i)$ for $y_a = f_a$, $a = 1,\ 2',\ 2'',\ 3$. As the matrix of given system has not any properties that make easier its calculation, we have to use iteration method to solve it. In the view of it it is necessary to limit the number of unknowns in the system of statistical equilibrium equations. For this purpose it is usually assumed that the number of repeating subscribers is limited and does not exceed an integer N, being chosen large enough. The error that we obtain in computing of characteristics after such reducing of state space of the studied process is investigated in [6].

3.3 Algorithm Based on Transition to One-Phase Service Time and on the Flow of Repeated Calls Partly Replaced by Poissonian

To this algorithm we pass after transformations mentioned in the second step. The simlified model studied here operates as follows. Two Poissonian flows of calls with intensities A and x_2 come to a full available bunch of v lines. If the call is refused service it will arrive next time in a random time that is exponentially distributed with a parameter equal to s_1. Besides, as before only the total number of busy lines i is fixed in the model, and the number of lines, occupied for the a-th phase is determined as iy_a. Such model is described by 2-dimensional Markovian process. Denote by $P(j,i)$ stationary probabilities of the model: here j - is the number of repeating subscribers and i - is the number of busy lines.

Probabilities $P(j,i)$ are connected by a system of statistical equilibrium equations of a form:

$$P(j,i)((A+x_2)(1-a(1-H))+ \qquad (7)$$
$$+js_1(1-aH)+i(y_1 w_1(1-P_1)+y_{2''}w_{2''}+y_3 w_3)) =$$
$$= P(j,i-1)(A+x_2)(1-a)+P(j-1,i)(A+x_2)aH+$$
$$+ P(j+1,i-1)(j+1)s_1(1-a) + P(j+1,i)(j+1)\times$$
$$\times s_1 a(1-H) + P(j-1,i+1)(i+1)y_1 w_1(1-P_1)H +$$
$$+ P(j,i+1)(i+1)(y_1 w_1(1-P_1)(1-H)+y_{2''}w_{2''}+$$
$$+ y_3 w_3), \quad j = 0,1,\ \ldots,\ i = 0,1,\ldots,v-1;$$

$$P(j,v)((A+x_2)H+js_1(1-H)+v(y_1 w_1(1-P_1)+$$
$$+y_{2''}w_{2''}+y_3 w_3)) = P(j,v-1)(A+x_2)(1-a) +$$
$$+ P(j-1,v)(A+x_2)H + P(j+1,v-1)(j+1)s_1(1-a)+$$
$$+ P(j+1,v)(j+1)s_1(1-H), \quad j = 0,1,\ \ldots .$$

For $P(j,i)$ a normalization condition is satisfied. We use as estimates of probabilistic characteristics of original model corresponding characteristics of simplified model B_2, $I_{a,2}$, $J_{1,2}$, $J_{2,2}$, L'_2 which are defined as follows

$$B_2 = \sum_{j=0}^{\infty} P(j,v); \quad I_{a,2} = y_a \sum_{j=0}^{\infty} \sum_{i=0}^{v} P(j,i)i,$$
$$a = 1,\ 2',\ 2'',\ 3 \; ; \qquad (8)$$

$$J_{1,2} = \sum_{j=0}^{\infty} \sum_{i=0}^{v} P(j,i)j; \quad J_{2,2} = x_2/s_2 \; ;$$

$$L'_2 = \sum_{j=0}^{\infty} P(j,v)(A+x_2+js_1).$$

Values of introduced estimates depend on unknown parameters y_a, $a = 1,\ 2',\ 2'',\ 3$; x_2. To find them we shall use the condition that relations (1) hold for estimates (8). It is not difficult to show that these relations should really held if y_a is determined by relations (6) and the value of x_2 is the solution of implicit equation

$$x_2 = I_{2'',2}(x_2) w_{2''} Z. \qquad (9)$$

That can be solved by iteration method.

Thus for estimating the probability characteristics of the original model with help of (8) it is necessary to solve the system of equations (7) for a set of values $x_{2,r}$ that are convergent to x_2. The nonzero elements of the matrix of the system of equations (7) marked with asterisks are listed in Table 1, with $v = 2$, $N = 2$. Here N is the quantity that limits the number of repeating subscribers $j \leq N$. It is easy to see that with the numeration of the unknowns listed in Table 1 the matrix has a tridiagonal block structure. For $a \neq 0$ the determinants of upper and lower diagonal blocks are not equal to the zero. To solve (7) we use the matrix method of Gauss ($a \neq 0$) or iteration method.

Table 1 Nonzero elements of the matrix of the system of equation (7)

j,i	0,0	0,1	0,2	1,0	1,1	1,2	2,0	2,1	2,2
0,0	*	*		*					
0,1	*	*	*	*	*				
0,2		*	*		*	*			
1,0	*			*	*		*		
1,1		*		*	*	*	*	*	
1,2			*		*	*		*	*
2,0				*			*	*	
2,1					*		*	*	*
2,2						*		*	*

For $a=0$ case it is possible to obtain more convenient for calculations matrix of linear equation system if the original model is additionally transformed as in the step 2a. In this case in the simplified model presented at the beginning of the section 3.3 a Poissonian flow of calls of intensity $x_{1,1}$ is added, and it is assumed that after the first phase being failed a call with a probability equal to one leaves the system. Denote by $P(j,i)$ probabilities of stationary states for the simplified model of this type. They are connected by a system of equilibrium equations of a form

$$P(j,i)(A+x_2'+js_1+i(y_1w_1(1-P_1)+y_{2''}w_{2''}+y_3w_3)) = P(j,i-1)(A+x_2') + P(j+1,i-1)(j+1)s_1 + P(j,i+1)(i+1)\times(y_1w_1(1-P_1)+y_{2''}w_{2''}+y_3w_3), \quad j=0,1,\ldots,\quad i=0,1,\ldots,v-1; \tag{10}$$

$$P(j,v)((A+x_2')H+js_1(1-H)+v(y_1w_1(1-P_1)+y_{2''}w_{2''}+y_3w_3) = P(j,v-1)(A+x_2') + P(j+1,v-1)(j+1)s_1 + P(j-1,v)(A+x_2')H + P(j+1,v)(j+1)s_1(1-H), \quad j=0,1,\ldots,.$$

Here $x_2' = x_2 + x_{1,1}$ and for $P(j,i)$ the normalization condition is sutisfied.

By the analogous way the estimates B_3, $I_{a,3}$, $J_{1,3}$, $J_{2,3}$, L_3' of corresponding probabilistic characteristics of the original model are defined:

$$B_3 = \sum_{j=0}^{\infty} P(j,v) \; ; \tag{11}$$

$$I_{a,3} = y_a \sum_{j=0}^{\infty} \sum_{i=0}^{v} P(j,i)i, \quad a=1,2',2'',3;$$

$$J_{1,3} = \sum_{j=0}^{\infty} \sum_{i=0}^{v} P(j,i)j + x_{1,1} / \varepsilon_1 \; ;$$

$$J_{2,3} = x_2 / s_2 \; ; \quad L_3' = \sum_{j=0}^{\infty} P(j,v)(A+x_2+x_{1,1}+js_1).$$

Further with aid of the condition of holding of relations (1) for defined estimates (11) we obtain that

$$y_a = f_a, \quad a = 1,\ 2',\ 2'',\ 3$$

and x_2' is determined as a solution of implicit equation

$$x_2' = I_{2'',3}(x_2')w_{2''}Z + I_{1,3}(x_2')w_1(1-P_1)H. \tag{12}$$

If x_2' is the solution of (12) then x_2, $x_{1,1}$ are determined from

$$x_2 = I_{2'',3}(x_2')w_{2''}Z, \quad x_{1,1} = x_2' - x_2.$$

Equation (12) is solved by the iteration method. Thus for estimating the probability characteristics of the original model with help of (11) it is necessary to solve the system of statistical equations (10) for a set of values $x_{2,r}'$ that are convergent to x_2'.

The nonzero elements of the matrix (10) marked with asterisks are listed in Table 2 with $v = 2$, $N = 2$. It is easily seen in the Table that the matrix in this case also has a tridiagonal block structure, but the structure of low diagonal blocks admits to reduce the solution of the whole system to the solution of (N+1)-th subsystem with (v+1) unknowns in each of them [2].

Table 2 Nonzero elements of the matrix of the system (10)

j,i	0,0	0,1	0,2	1,0	1,1	1,2	2,0	2,1	2,2
0,0	*	*							
0,1	*	*	*	*					
0,2		*	*		*	*			
1,0				*	*				
1,1				*	*	*	*		
1,2			*		*	*		*	*
2,0							*	*	
2,1							*	*	*
2,2						*		*	*

3.4 Algorithm Based on the Application of Erlang's Formula

It is constructed after transformations of the original model mentioned in step 3. All the flow of repeated calls is replaced by Poissonian with intensity x. Denote by $P(i)$ probabilities of stationary states in the simplified model of this type. Here i - is the number of busy lines. Probabilities $P(i)$ are determined by equations

$$P(i) = \frac{d^i/i!}{1+d+d^2/2!+\ldots+d^v/v!}, \quad (13)$$

$$d = \frac{(A+x)(1-a)}{y_1w_1(1-P_1)+y_{2''}w_{2''}+y_3w_3}.$$

For estimating the probabilistic characteristics of the original model we shall introduce the corresponding characteristics B_4, $I_{a,4}$, $J_{1,4}$, $J_{2,4}$, L'_4 of the simplified model that are defined as follows

$$B_4 = P(v);\ I_{a,4} = y_a \sum_{i=0}^{v} P(i)i, \quad (14)$$
$$a=1,2',2'',3;$$

$$J_{1,4} = (x-I_{2'',4}w_{2''}Z) / s_1 ;$$

$$J_{2,4} = I_{2'',4}w_{2''}Z / s_2 ;\quad L'_4 = (A+x)P(v).$$

Further with the help of condition of holding relations (1) for introduced estimates (14) we obtain that $y_a = f_a$, $a = 1, 2', 2'', 3$ and x is determined by solving the implicit equation

$$x = \frac{A(aH+(1-a)(P+B_4(g)(H-P))}{1-aH-(1-a)(P+B_4(g)(H-P))}, \quad (15)$$

where

$$g = (A+x)(1-a)f/w_1,\ P = H-P_1(H-(1-P_2)Z).$$

Equation (15) is solved by iteration method.

Thus, for estimating probabilistic characteristics of the original model with help of (15) it is necessary to use Erlang's formula for a set of values x_r that are convergent to x.

In conclusion of the section we note that realization of natural limitations to input parameters of the original model the equations (9),(12),(15) always has a unique solution (see [6]).

4. APPROXIMATE ALGORITHMS' ACCURACY

4.1 Analytical Results

Denote by B_0, $I_{a,0}$, $J_{1,0}$, $J_{2,0}$, L'_0 values of corresponding probabilistic characteristics of the original model: $B_0 = B$, $I_{a,0} = I_a$ etc. Let us write the asymptotic formulae for B_k, $I_{a,k}$, $J_{1,k}$, $J_{2,k}$, L'_k, $k = 0,1,2,4$ for $A\to 0$ and $A\to\infty$.

1.Intensity of primary calls A tends to 0.

$$I_{a,k} - \frac{A(1-a)ff_a}{w_1(1-aH-(1-a)P)} = o(A), \quad (16)$$
$$a=1,2',2'',3 ;$$

$$J_{1,k} - \frac{AH(1-P_1(1-a))}{s_1(1-aH-(1-a)P)} = o(A) ;$$

$$J_{2,k} - \frac{A(1-a)P_1(1-P_2)Z}{s_2(1-aH-(1-a)P)} = o(A) ;$$

2.Intensity of primary calls $A\to\infty$.

$$B_k - 1 + \frac{vw_1(1-H)}{f(1-a)A} = o(\tfrac{1}{A}) ; \quad (17)$$

$$I_{a,k} - vf_a + \frac{vw_1(1-H)f_a}{f(1-a)A} = o(\tfrac{1}{A}) ;$$
$$a=1,2',2'',3$$

$$J_{1,k} - \frac{AH}{s_1(1-H)} + \frac{vw_1P_1H(1-(1-P_2)Z)}{s_1(1-H)f} - \frac{vw_1^2P_1H(1-(1-P_2)Z)}{s_1f^2(1-a)A} = o(\tfrac{1}{A}).$$

Using (17) and (1) it is easy to write the asymptotic formulae for $J_{2,k}$ and L'_k. If a=0, then (16),(17) hold for k=3 also. The way of finding of asymptotic formulae for the original model (k=0) and for all the kinds of estimates (k=1,2,3,4) can be found in [6].

We proved that the introduced estimates (5),(8),(14) are asymptotically exact for $A\to 0$ and $A\to\infty$. This is a very important result that makes it possible to expect a high accuracy for the obtained approximate algorithms. This conclusion is confirmed by numerical calculations.

4.2 Numerical Study of Approximate Algorithms' Accuracy

In Table 3 we listed the results of an exact and approximate with respect to formulas (5),(8),(14) calculations of probability characteristics B and $J = J_1+J_2$.

The input parameters of the model take the following values: v=2, H=0.5, a=0.1, s_1=20, s_2=5, $w_1=\infty$, $w_{2'}=w_{2''}$=10, w_3=2, P_1=1, P_2=0.5, Z=0.5.

Table 3 Results of an exact and approximate calculation

A	B	B_1	B_2	B_4
0.5	0.01910	0.01910	0.01916	0.01922
2	0.1703	0.1705	0.1676	0.1808
3	0.2742	0.2745	0.2680	0.2974
5	0.4366	0.4371	0.4259	0.4782
10	0.6579	0.6589	0.6478	0.7018
20	0.8204	0.8214	0.8116	0.8444
A	**J**	**J_1**	**J_2**	**J_4**
0.5	0.03260	0.03260	0.03260	0.03266
2	0.1259	0.1259	0.1255	0.1269
3	0.1842	0.1842	0.1836	0.1860
5	0.2950	0.2949	0.2941	0.2975
10	0.5569	0.5570	0.5563	0.5592
20	1.064	1.064	1.064	1.065

An analysis of the numerical data confirms a sufficiently high accuracy of the introduced estimates in the cases of small and large losses, in particular.

5. UTILIZATION OF MODEL IN TOLL TELEPHONE SYSTEM

The introduced model was utilized for calculating of characteristics of operation quality of direct channel bunch in toll telephone system. Calculation results were presented as nomograms of dependencies of main probabilistic characteristics B_c, M, Q, I, J_1, J_2 etc (10 in total) from loss probability B (it ranges in two intervals from 0 to 10% and from 0 to 95%). Due to great design volume the calculations were held commonly according to the simplest algorithm, presented in section 3.4.

Input parameters take the values: $v=12 \div 300$, $H=0.6, 0.75, 0.9$, $a=0.15$, $1/w_1=5$, $1/w_{2'}=15$, $1/w_{2''}=45$, $1/w_3=250$, $P_1=0.65$, $P_2=0.85$, $Z=0.3$, $1/s_1=20$, $1/s_2=5$ (here all values 1/. in sec.).

For the example we show in Fig.2 the depencies B_c from B and J from B for different values of v and H=0.75. Let us mention some problems that can be solved with help of nomograms calculated.

i) Estimate of characteristics of circuit operation quality. If the value of one of 10 characteristics, presented on nomograms is known, it is very easy to estimate the values of others. For example if B_c (the general loss probability) is measured, then using the nomograms that expressed dependence B_c from B we find B and then - using the remaining nomograms - all additional characteristics of operation quality of circuits M, Q, etc.

ii) Estimate of circuit operation quality from the point of view of subscriber. It is given by average number of losses to one connection Q.

iii) Estimate of circuit operation quality from the point of view of administration. It is given by comparison between paid load, corresponding to conversation If_3 and different types of lost load: due to the called subscriber being busy $If_1(1-P_1)$, due to non-answer of called subscriber $If_{2''}w_{2''}(1/w_{2''}+1/w_1)$.

iv) Estimate of circuit control elements operation. It can be found from the number of repeated calls to one primary M.

Besides its obvious practical importance nomograms obtained have a certain theoretical value, that consists in the fact that with their help we succeeded in forecasting a number of interesting relations between different characteristics of model with repeated calls. The following two properties of the (14) are the most interest: 1) dependence of B_c, Q, M from B does not change with the increase of v ; 2) dependence of B_c from B is a linear function. Exact numarical calculations (see Table 3) shows that the approximate method has a high accuracy. From here follows that the properties pointed out should also hold (approximately) for exact values of characteristics. To test the hypothesis we consider some examples of accurate calculation of model of this type.

In Fig.3 it is shown the dependence of B_c and Q correspondingly from B for the following values of input parameters: v= 1, 10, H=0.75, a=0.15, $s_1=10$, $s_2=5$, $w_1=\infty$, $w_{2'}=w_{2''}=10$, $w_3=2$, $P_1=1$, $P_2=0.85$, Z=0.3. The data presented confirms the correctness of the hypothesis.

6. CONCLUSIONS

On an example of complicated enough model of full available bunch with multiphase service and two types of repeated calls' sources we showed the application of methods introduced in [6] of approximate calculation of models with repeated calls, based on consequent simplification of the original model with the help of conservation laws. An advantage of the calculation procedure is its evident formalization, achieved by conservation laws. From here follows that method can be easily extended from one subscriber behaviour model to another.

REFERENCES

1 P.Le Gall, "Sur l'influence des répetitions d'appels dans l'écoulement du trafic téléphoniques", Proc.6th Int.Teletraffic Congress, Munich, N432, p.1-7, 1970.

2 G.L.Ionin, I.I.Sedol, "Telephone systems with repeated calls", Proc.6th Int. Teletraffic Congress, Munich, N435, p.1-5, 1970.

3 R.Evers, "Measurement of subscriber reaction to unsuccesful call attempts and the influence of reasons of failure", Proc.7th Int.Teletraffic Congress, Stockholm, N544, p.1-8, 1973.

4 G.Gosztony, "Comparison of calculated and simulated results for trunk groups with repeated attempts", Proc.8th Int. Teletraffic Congress, Melbourne, N321, p.1-11, 1976.

5 M.Shneps-Shneppe, "The effect of repeated calls on communication system", Proc.6th Int. Teletraffic Congress, Munich, N433, p.1-5, 1970.

6 S.N.Stepanov, "Numarical Methods for Computing Systems with Repeated Calls", Nauka, Moscow, 1983, (in Russian).

7 S.N.Stepanov, "Estimation of characteristics of multilinear systems with repeated calls", Proc.Int. Seminar on Teletraffic Theory, Moscow, p.400-409, 1984.

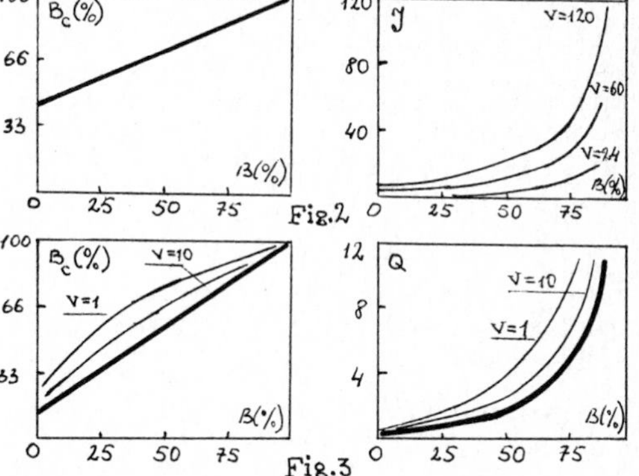

Fig.2

Fig.3

TELETRAFFIC ISSUES in an Advanced Information Society
ITC-11
Minoru Akiyama (Editor)
Elsevier Science Publishers B.V. (North-Holland)

COMPARATIVE EVALUATION OF HIERARCHICAL AND NON-HIERARCHICAL PACKET NETWORKS

Shin-ichi KURIBAYASHI and Toyofumi TAKENAKA

Musashino Electrical Communication Laboratory, N.T.T.
Tokyo, Japan

ABSTRACT

There has been a rapid increase in traffic volume in packet switching networks due to the recent growth in data communications. Consequently, the ease with which networks can be expand to keep pace with this increase has become an important consideration.

This paper discusses a hierarchical network configuration as one possible solution to the network expansion problem, and evaluates it from the cost efficiency and administrative viewpoints. A simulation analysis indicates that the hierarchical network is a promising solution to the network expansion problem from the cost efficiency viewpoint as well as the administrative viewpoint.

1. INTRODUCTION

The packet switching network in Japan, DDX-P, has experienced a rapid growth as well as in many countries [1]. Thus, the ease of network expansion has become a topic of concern when network configuration is considered [2],[3],[4].

This paper examines a hierarchical network configuration as one possible solution to the network expansion problem and evaluates it from the cost efficiency and administrative viewpoints. In order to calculate the network cost, a simulator is developed which uses a rectangular model reflecting the geographical conditions in Japan. Any value may be entered in the simulator for parameters affecting network cost such as traffic volume, the traffic flow matrix, and the transmission-to-switching costs ratio. Network data transmission delay, which is specified in CCITT recommendation X.135, is also taken into account. Network configuration flow and network configuration techniques are given in section 2. The network configuration simulator is explained in section 3. Simulation results are used to compare the cost of hierarchical and non-hierarchical networks in section 4. Hierarchical and non-hierarchical networks are compared from the network administrative viewpoints in section 5.

2. NETWORK CONFIGURATION FLOW AND NETWORK CONFIGURATION

Network configuration flow and network configuration techniques for both hierarchical and non-hierarchical networks are described in this section. A two-level network is adopted in the hierarchical network and a partially-connected network is adopted in the non-hierarchical network.

2.1 Network configuration flow

The packet networks described in this paper are constructed as shown in Fig. 1.

(1)< Installation of local switches and concentrators >

The installation of local switches and concentrators, and the assignment of concentrators to specific local switches are determined by the ADD method (explained in section 2.2).

(2)< Installation of dedicated circuits between local switching nodes >

The decision to install full-duplex dedicated circuits between local switching nodes is made by the DROP method in the case of the non-hierarchical network and by the TRIANGULAR method in the case of the hierarchical network (explained in section 2.3).

(3)< Installation of transit switching nodes >

In the hierarchical network, a transit switching node is installed in the city in each Primary Area (PA) which handles the largest volume of traffic. In this paper, a PA is defined as the area where transit traffic issuing from local switching nodes is concentrated. Within a PA, only a transit switching node can pass transit traffic between local switching nodes. Each local switching node is associated with exactly one PA.

(4)< Installation of transit circuits between a local switching node and the transit switching node, and between transit switching nodes >

In the hierarchical network, the number of transit circuits is calculated according to the transit traffic volume between a local switching node and the transit switching node. Moreover, all transit switching nodes in the network are interconnected.

(5)< Setting of routing table for switching nodes >

Route selection table for packet transmission and circuit selection table in each selected route are set for each switching node according to the routing method proposed by

KANEMAKI et al. [5]. The following conditions are also considered:

(a) Building-block-type switching equipment is used in each switching node [6]. This reduces the initial cost of switching node installation and the expansion cost of the switching function.

(b)In the hierarchical network, the transit switching function is implemented as a part of some specified local switching node, thereby reducing the installation cost of transit switching facilities. Moreover, within the same PA, local switches are connected to all transit switches to enhance network reliability.

2.2 Installation of local switching nodes

The decision to install local switching nodes is made by the following ADD method.

(1)< Two-level hierarchical network >

First, the city which handles the largest volume of traffic in each PA is determined. A local switch is installed in that city. The following procedure is then repeated until a local switch or a concentrator is installed in every city. A decision to install a local switch in a particular city is made by comparing the cost of installing the local switch to the minimum total cost of installing a concentrator and the circuits connecting the concentrator with a target local switch previously installed in another city.

(2)< Non-hierarchical network >

The same procedure is used for the non-hierarchical network as would be used for the hierarchical network with a single large PA.

2.3 Installation of dedicated circuits

The installation of dedicated circuits between local switching nodes is determined by the following two methods:

(1)< Two-level hierarchical network >

The following TRIANGULAR method is used. The decision to install dedicated circuits within a PA or between PAs is made by comparing the cost of installing dedicated circuits between local switching nodes to the total cost of installing transit circuits and transit switching in the transit switching node (Fig. 2).

(2)< Non-hierarchical network >

The following DROP method is used. This method begins with the interconnected network solution and removes dedicated circuits between local switching nodes. The algorithm finds the route of the least available dedicated circuits and considers it as a candidate for removal. If the cost of dedicated circuits in the candidate route is greater than the cost in the transit route, all circuits in the candidate route are removed. If topological constraints for network data transmission delay can not be met, they should not be removed. This procedure is repeated until all routes have been considered.

The DROP method differs from the TRIANGULAR method in the following way. In the DROP method, any local switching node can be used as a transit switching node. On the other hand, in the TRIANGULAR method only specified switching nodes can function as transit switching nodes [Fig. 3].

3. NETWORK CONFIGURATION EVALUATION TOOL

In order to evaluate and design the packet network configuration, Packet Network Configuration Evaluation System (PACOS) has been developed as a simulation tool which can be carried out with a personal computer. The PACOS constructs the packet switching network according to the procedure described in section 2 and calculates the network cost. It can also design the network for any combination of given traffic conditions. The following factors which are assumed to affect network cost are considered ; traffic volume, traffic flow matrix and the transmission-to-switching costs ratio. Moreover, the network data transmission delay is controlled by restricting the maximum number of switches traversed by a packet.

The PACOS consists of three main blocks; the OS block, NETALG block and NETDATA block [Fig. 4]. The NETALG block consists of the hierarchical and non-hierarchical network construction algorithm sub-blocks which implement the network configuration procedures described in section 2 and calculate the network cost. The NETDATA block consists of the man-machine interface and floppy disk control sub-blocks, which initialize and modify such simulation conditions as switching equipment cost, circuit cost and traffic conditions. The output of the NETALG block is the network cost and network topology of hierarchical or non-hierarchical networks. Moreover, the PACOS is provided with an interface for entering the traffic flow matrix and the distance between switching nodes automatically. This is accomplished by entering the traffic pattern (uniform, 1-peak and 2-peak) and the distance between adjacent switching nodes.

4. EVALUATION OF THE HIERARCHICAL NETWORK

4.1 Traffic model

A rectangular model is used which reflects the geographical conditions in Japan. The ellipses in Fig. 5 are candidate cities for installing a switching node. The following three traffic matrixes are evaluated ; uniform, 1-peak and 2-peak patterns [Fig. 6]. The calculation algorithm for the traffic matrix, the distance between switching nodes, and the PA structure are shown in Table 1.

4.2 Evaluation

Simulation results obtained by PACOS are shown in Figures 7, 8 and 9. In the simulations, the maximum number of switches traversed by a packet is constrained to 4. It is clear from the simulation results that the cost of a two-level hierarchical network with several PAs is almost equal to that of a partially-connected non-hierarchical network which would be the most economical of all network topologies. Moreover, this has been shown to be true, even if the traffic pattern changes.

The reasons for this are as follows:

(1) The TRIANGULAR method has almost the same ability as the DROP method for deleting dedicated circuits with a low traffic volume and for using transit circuits effectively (Figures 7 and 8). However, in a hierarchical network with many PAs, transit circuits are not used effectively when the amount of total traffic originating in the network is small, because all transit switching nodes have been interconnected to enhance the network reliability and ensure the network data transmission delay.

(2) The biggest difference between hierarchical and non-hierarchical networks is that while the transit switching is implemented only in specified switching nodes in the former, it can be implemented in every switching node in the latter. Because of this, the non-hierarchical network seems to use the network resources (circuits and switching equipment) more effectively than the hierarchical network. However, it has been shown in (1) that circuits are used effectively in both network topologies. Therefore, the attractiveness of the non-hierarchical network seems to lie in its reduced switching node cost due to the shared use of switching facilities by both the local and transit switching nodes. However, this is not true in the case of the hierarchical network where the transit switching function is implemented as part of some specified local switching node.

5. ADMINISTRATIVE ASPECTS OF THE HIERARCHICAL NETWORK

From the results in section 4, it is clear from a cost efficiency viewpoint that the two-level hierarchical network is not inferior to the non-hierarchical network.

This section discusses the administrative aspects of hierarchical and non-hierarchical networks. In the non-hierarchical network, the optimization of resource allocations is made by the DROP method as described in section 2.3. In this method, the allocation of transit switching facilities and circuits would be changed according to traffic conditions (i.e. the traffic volume and traffic flow matrix). Because of this, whenever traffic conditions change, routing tables for switching nodes must be rewritten and transit circuits must be reinstalled.

On the other hand, in the hierarchical network the reallocation of transit switching facilities and circuits would not be so tightly controlled by changes in traffic condition, because only the specified switching node could take the part of the transit switching facility.

It can be seen from the above discussion that a hierarchical network is superior to a non-hierarchical network from administrative viewpoint.

6. CONCLUSIONS

This paper discussed the hierarchical network configuration as one possible solution to the problem of network expansion, and evaluated it from the cost efficiency and the administrative viewpoints. It was demonstrated from the simulation analysis that the cost of a hierarchical network with several PAs is almost the same as that of a partially-connected non-hierarchical network which would be the most economical of all network configurations. Moreover, a hierarchical network, in which only specified switching nodes can function as transit switching nodes, is more adaptable than a non-hierarchical network to network expansion or to changes in traffic flow condition.

It is concluded that the hierarchical network offers a promising solution to the problem of network expansion from the cost efficiency viewpoint as well as the administrative viewpoint.

ACKNOWLEDGMENTS

The authors wish to express their gratitude to Mr. Ikuo HONGO, chief of the Packet Switching Systems Section, and Mr. Ken-ichi KURODA, staff engineer, of the Musashino Electrical Communication Laboratory for their guidance and encouragement during the course of this work.

REFERENCES

[1] H. ITOH, C. ENDOH, M. NAKAMURA and M. IIKURA, "DDX: The first five years," ICCC'84 (1984).

[2] S. KURIBAYASHI and T. TAKENAKA, "Analyses of packet network configuration Elements," Paper of the technical group on Switching, IECE, JAPAN, SE84-1 (1984)

[3] S.W. JOHNSTON and P.L. SCHUHMAN,"Design and optimization of packet networks," Networks (1983).

[4] S.W. JOHNSTON and R.K. POOLE, "Optimization of packet networks with remote concentrators," GLOBECOM 83.

[5]K. KANEMAKI,et al.,"A routing method on a loosely coupled multiprocessor type packet switching system," Paper of the Technical group on Switching, IECE, JAPAN, SE83-139 (1983).

[6] M. NISHIWAKI,et al.,"Architecture of a distributed packet switching system", Paper of the technical group on Switching, IECE, JAPAN, SE84-121 (1984).

[7] J. DRESSLER, J.A.C. GOMES, R.J.MANTEL, J.M. MEPUIS and E.J. SARA, "COST201:A European Research Project, a flexible procedure for minimizing the costs of a switched network taking into account mixed technologies and end-to-end blocking constraints," ITC 10 (1983).

[8] M. YAMASHITA and M. AKIYAMA, "Hierarchical structure of telecommunication networks," Trans. IECE, JAPAN, Vol.J64-B, No.4 (1981).

[9] S. KURIBAYASHI and T. TAKENAKA, "A study on network configuration for packet switching network shared by long and short size packets," Paper of the technical group on Information Network, IECE, Japan, IN84-61 (1984).

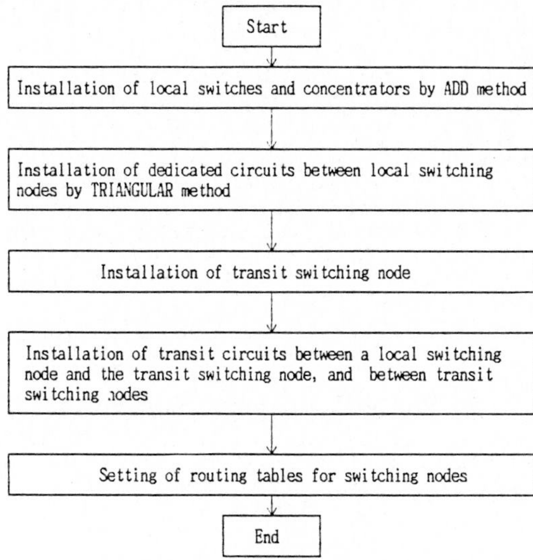

Fig. 1 Network configuration flow of a hierarchical network.

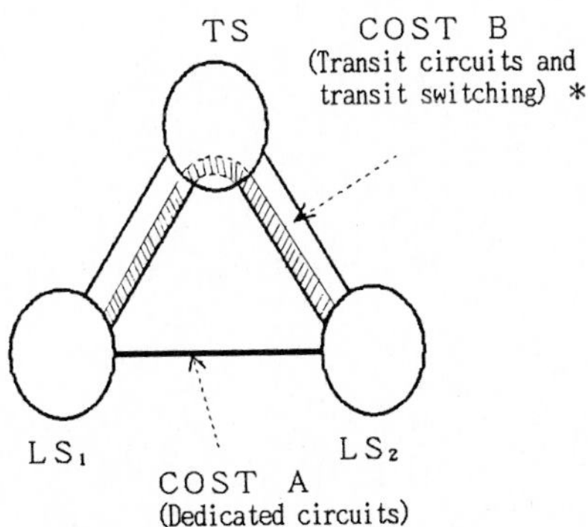

*Other transit traffic may be transmitted through the same transit circuit and transit switch.

If COST A ≦ COST B then install dedicated circuits between LS_1 and LS_2.

<Note>
・LS : Local switch, TS : Transit switch

Fig. 2 TRIANGULAR method.

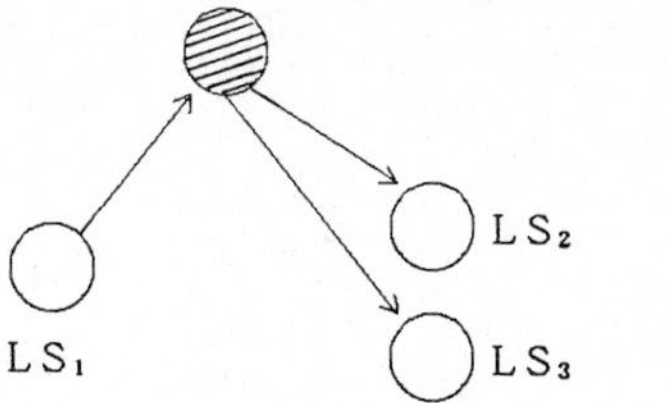

(1) HIERARCHICAL NETWORK

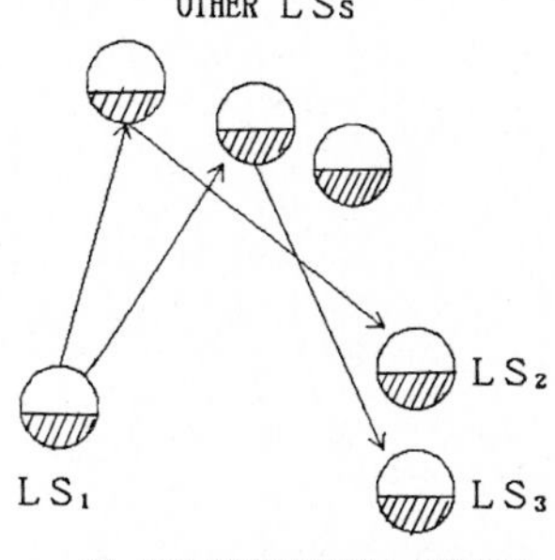

(2) NON-HIERARCHICAL NETWORK

▨ : TRANSIT SWITCHING FUNCTION

Fig. 3 Difference between hierarchical and non-hierarchicalnetworks.

Originating traffic distribution

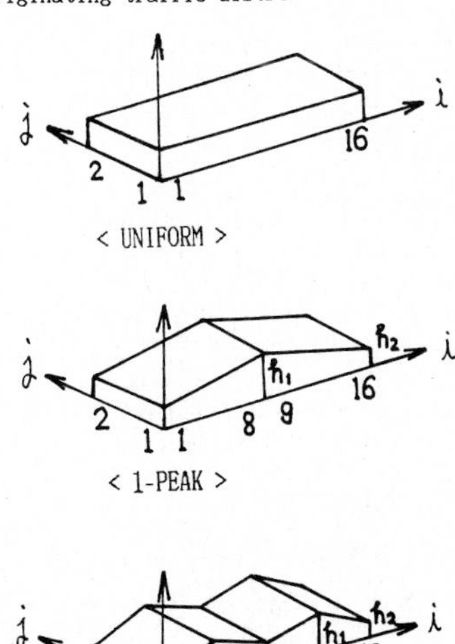

Fig. 6 Originating traffic distribution.

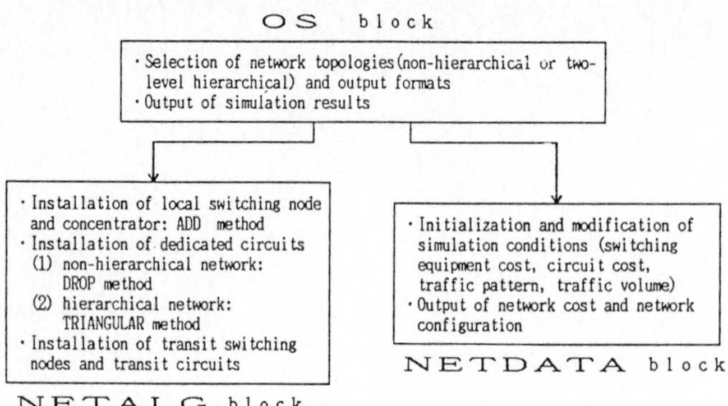

OS block
· Selection of network topologies (non-hierarchical or two-level hierarchical) and output formats
· Output of simulation results
· Installation of local switching node and concentrator: ADD method
· Installation of dedicated circuits
(1) non-hierarchical network: DROP method
(2) hierarchical network: TRIANGULAR method
· Installation of transit switching nodes and transit circuits
NETALG block
· Initialization and modification of simulation conditions (switching equipment cost, circuit cost, traffic pattern, traffic volume)
· Output of network cost and network configuration
NETDATA block

Fig. 4 Packet network configuration evaluation system (PACOS).

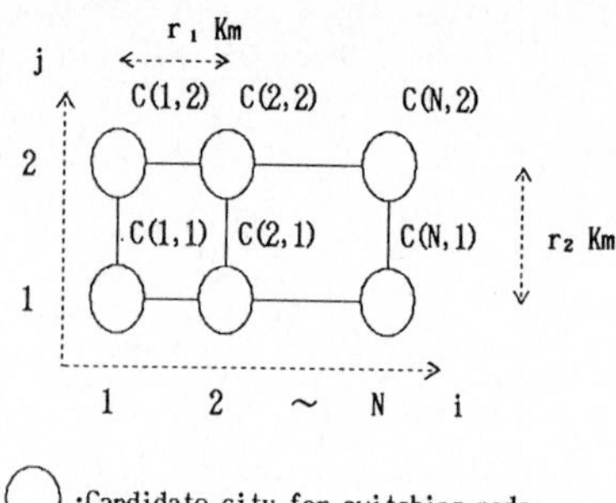

Fig. 5 A rectangular model.

Table 1. Traffic flow matrix in the rectangular model.

(1) The ratio of traffic flow from city C(i,j) to city C(m,n)

UNIFORM	· $Rg/(N_1 \times N_2)$ · $(1-Rg)/N_1$	$(i,j) \neq (m,n)$ $(i,j) = (m,n)$
1-PEAK	· $\dfrac{Rg \times A(m,n) \times A(i,j)}{\sum_{(s,t)\neq(i,j)} A(S,T) \times \sum_{p,q} A(p,q)}$ · $\dfrac{(1-Rg) \times A(i,j)}{\sum_{p,q} A(p,q)}$ * A(i,j) is the number of total packets originating in city C(i,j) and given by $A(i,j) = k \times (1 + (i-1)/(N_3-1) \times (D-1))$ $(1 \leq i \leq N_3)$ $= k \times (1 + (N-i)/(N_3-1) \times (D-1))$ $(N_3+1 \leq i \leq N)$	$(i,j) \neq (m,n)$ $(i,j) = (m,n)$
2-PEAK	· $\dfrac{Rg \times A(m,n) \times A(i,j)}{\sum_{(s,t)\neq(i,j)} A(S,T) \times \sum_{p,q} A(p,q)}$ · $\dfrac{(1-Rg) \times A(i,j)}{\sum_{p,q} A(p,q)}$ * A(i,j) is the number of total packets originating in city C(i,j) and given by $A(i,j) = k \times (1 + (i-1)/N_4 \times (D-1))$ $(1 \leq i \leq N_4+1)$ $= k \times (1 + (2N_4+2-i)/N_4 \times (D-1))$ $(N_4+2 \leq i \leq 2N_4)$ $= k \times (1 + (i-(2N_4-1))/N_4 \times (D-1))$ $(2N_4+1 \leq i \leq 3N_4-1)$ $= k \times (1 + (N-1)/N_4 \times (D-1))$ $(3N_4 \leq i \leq N)$	$(i,j) \neq (m,n)$ $(i,j) = (m,n)$

Note:
- $N_1=2N$, $N_2=2N-1$, $N_3=[N/2]$, $N_4=[N/4]$
  $[S]$ = Maximum integer equal to or less than S
- Rg = the ratio of outgoing traffic to total traffic originating in each local switching node.
- D = the traffic ratio of the highest peak to the lowest peak in the 1-PEAK and 2-PEAK conditions.
- k=constant

(2) The distance between city C(i,j) and city C(m,n)

$$|i-m| \times r_1 + |j-n| \times r_2$$

(3) PA structure in a rectangular model

M: The number of PAs

	Included city C(i,j) (J=1,2)
PA 1	i= 1~ [N/M]
·	·
PA t	i= [N/M] ×(t-1) + 1 ~ [N/M] ×t
·	·
PA M	i= [N/M] ×(M-1) + 1 ~N

In this paper, N=16, $r_1=r_2=150$Km

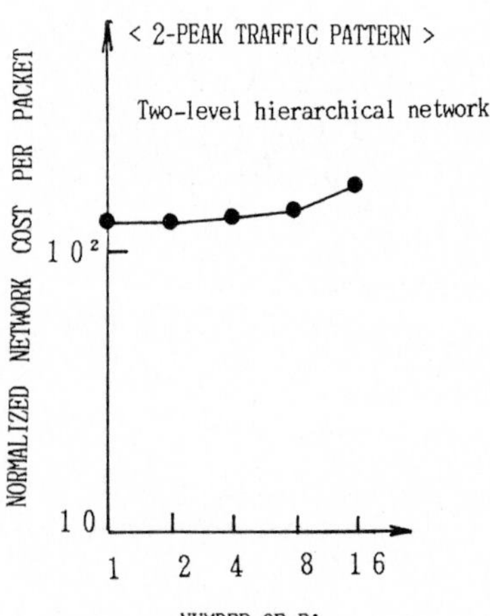

Fig. 9 Simulation results III.

(Rg=0.8, D=10, Total originating traffic=10^4, The transmission-to-switching costs 10^{-2})

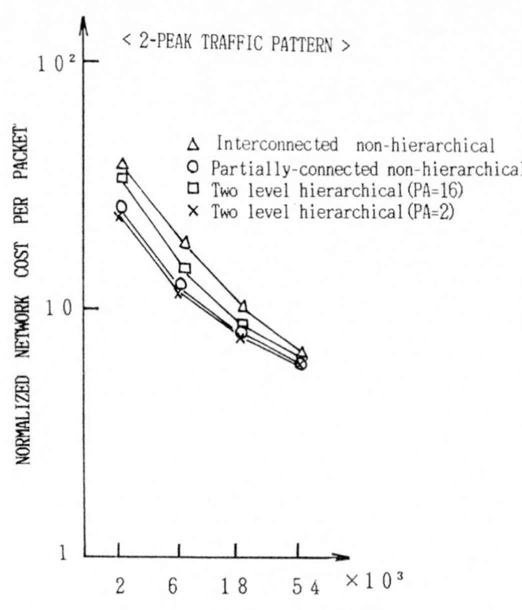

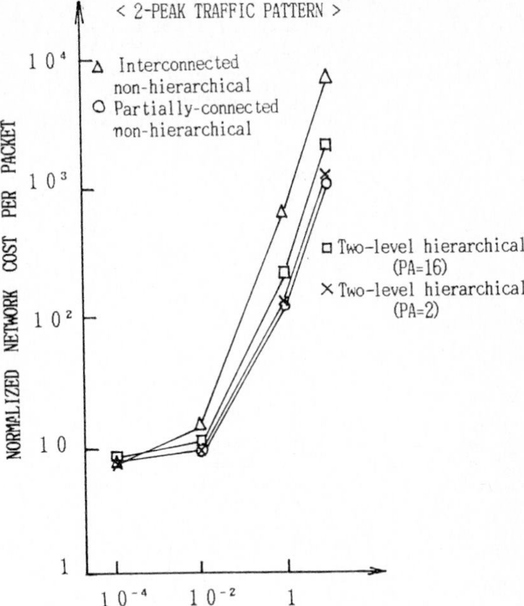

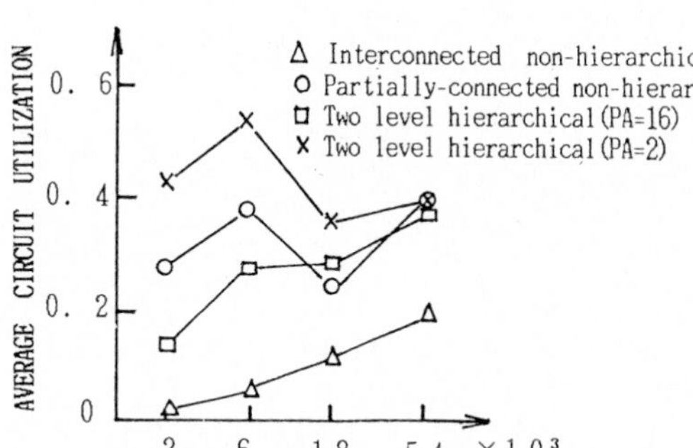

Fig. 7 Simulation results I.
(Rg=0.8, D=10, The transmission-to-switching costs ratio=10^{-2})

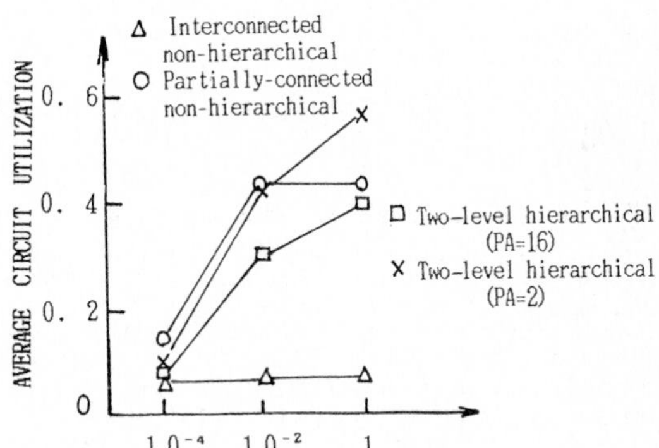

Note: The transmission-to-switching costs ratio is the ratio of initial cost of transit circuit installation to initial cost of switching node installation.

Fig. 8 Simulation results II.
(Rg=0.8, D=10, Total originating traffic=10^4)

TELETRAFFIC ISSUES in an Advanced Information Society
ITC-11
Minoru Akiyama (Editor)
Elsevier Science Publishers B.V. (North-Holland)

METHODS FOR PREDICTING SUBSCRIBER BEHAVIOR IN THE CHOICE OF DIGITAL DATA SERVICES

Mark A. WILLIAMSON and Dennis L. JENNINGS

Bell Communications Research
Livingston, New Jersey, USA

ABSTRACT

Methods for predicting subscriber behavior in the choice of digital data services are described, including novel approaches to market segmentation and subscriber choice modeling. Three-way multidimensional scaling, preference mapping, and complete linkage clustering are used to cluster subscribers into groups which are homogeneous with respect to their perceptions and valuations of the non-price features of various existing and new data services. These clusters form new market segments which cut across traditional industry segments. Discrete choice models are used to describe how subscribers trade-off the price and non-price characteristics of service alternatives when considering replacement of their current data networks. The use of these models in forecasting demand and usage is also discussed.

1. INTRODUCTION

The initial validation of the methods described in this paper is based on a market research sample of 300 business data service subscribers from three Local Access and Transport Areas (LATAs) in the Southeastern United States. The LATAs chosen were known to be major data markets. Six data services were included in the study; three existing services and three new service concepts. The three existing services included analog dial-up and private line services and a digital private line service. The three new service concepts included two packet service concepts and a circuit switched digital service concept. Additional validation of these techniques is planned and will be based on a recently completed survey of over 600 business data service subscribers in another part of the United States.

2. THE DECISION PROCESS

In order to model subscriber choice behavior, it is necessary to develop a set of assumptions that describe the decision process. In the case of data services, we assume that there are two basic steps by which a subscriber makes appropriate choices. These assumptions provide a set of hypotheses that guide the demand analysis work and that can be tested for validity.

First, it is assumed that the subscriber restricts his attention to those services which are appropriate for his business applications. Only services which have certain key features required to satisfy the subscriber's data communications needs are deemed appropriate.

Second, it is assumed that the subscriber chooses among appropriate services in a rational manner; that is, he chooses what he determines to be the best value. It is further hypothesized that the subscriber's value for a particular service can be expressed as the difference of his "Willingness to Pay" (WTP) for the service features and the cost that would be incurred in actually implementing and using the service.

3. MARKET SEGMENTATION

In modeling the first assumption, we considered two related questions. First, what are the salient features that subscribers pay attention to and use to compare services? Data services can vary on many different dimensions, but subscribers may attend to only a few features that are relevant to their business needs. Once the salient dimensions are identified, we can turn to the second question. What is the relative importance of the salient features? Although a subscriber may compare services on several dimensions, one of these dimensions may be most important in determining which service(s) the subscriber purchases.

3.1. Identification of salient features

Three-way multidimensional scaling was the statistical technique used to identify the salient features. To obtain the data necessary for this type of analysis, a sample of analog private line subscribers were first given information about the engineering features of the six data services. They were then asked to rate the similarity of all pairs of these six services. From this information, it was possible to deduce the features that a particular subscribers paid attention to. The technique we used, INDSCAL [1], displays the services as a configuration of points in an n-dimensional space, known as a "group perceptual space". This configuration is constructed such that the distance between any two services is a linear function of the perceived similarity of those two services. The dimensions in this space correspond to perceptual dimensions subscribers use when comparing the services. The location of the points (services) show where subscribers perceive each service as being located on the dimensions.

Although subscribers were given information about nine different engineering features of the data services, only three dimensions appeared to be salient to the group of analog private line subscribers. These three dimensions accounted for 21, 20, and 16% of the variation in subscribers' ratings, respectively. Each additional dimension accounted for much less variance, and was not interpretable. The locations of the six services on each of these dimensions are shown as letters in Figure 1. While the first and second dimensions each corresponded to a specific engineering feature, the third dimension did not. It appeared to correspond to a service capability that results from a combination of aspects of several engineering features.

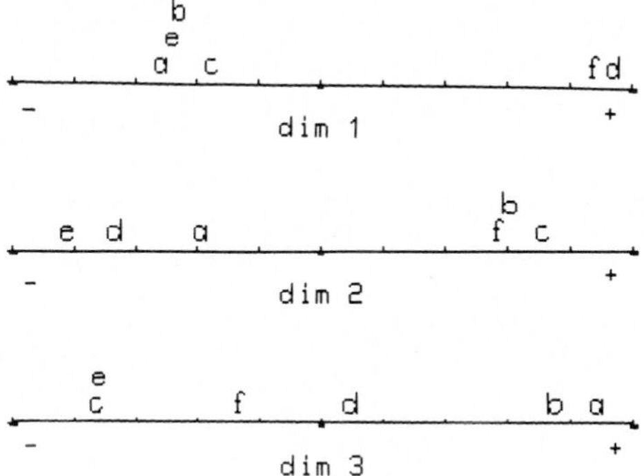

Figure 1. Group Perceptual Space: One Dimensional Plots

3.2. Relative importance of the salient features

The next series of analyses determined the relative importance of these three dimensions. In order to get at this information, we used the technique known as preference mapping. We used a version of this technique called PREFMAP [1]. The data for these analyses were obtained by asking each respondent to rate the appropriateness of each service for his data communication needs currently being served by analog private lines. Answers to this question are assumed to reflect the relative importance of the salient features. For example, a subscriber who values a particular feature would rate services which have the feature as more appropriate than services which do not have the feature. In preference mapping, each subscriber's ratings of appropriateness for the services are regressed onto the coordinates of the services in the 3-dimensional group perceptual space. A vector model of PREFMAP was used that yields for each subscriber a preference vector that can be plotted in the group perceptual space to produce a "joint perceptual-preference map". This vector is selected so that the correlation between the projections of the services onto this vector and the respondent's ratings of appropriateness is maximized. The correlation between a perceptual dimension and this vector indicates how important that dimension was in determining a respondent's ratings of appropriateness.

In Figure 2 the endpoints of the vectors for a subset of subscribers are plotted in a two-dimensional subspace of the joint perceptual-preference map, and are shown as asterisks. In the figure, each vector is normalized to unit length. This 2-dimensional figure suggests that some customers preferred the positive end of dimension 1, while others preferred the positive end of dimension 2.

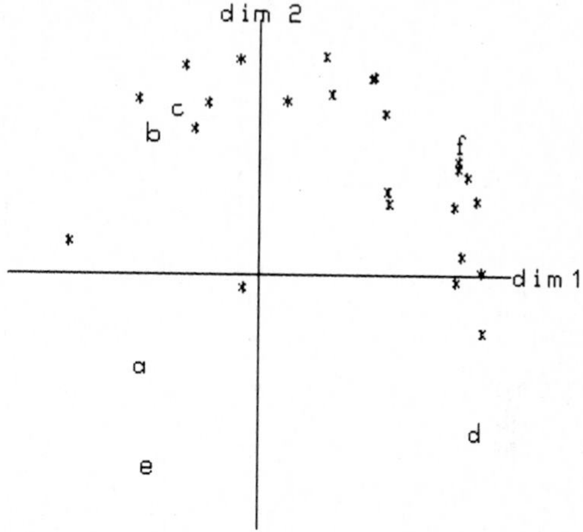

Figure 2. Joint Perceptual-Preference Map: Two Dimensional Projection

3.3. Identification of market segments

In the next step of the analysis, we determined if there were subsets of subscribers who had similar preference vectors. The arc distance between the endpoints of two vectors was used as a measure of dissimilarity. These between vector distances were used as the input to a complete linkage hierarchical clustering analysis. This analysis indicated that analog private line subscribers consisted of two segments of customers who valued the salient features in very different ways, and confirmed the observation made at the end of Section 3.2 The first segment consisted of subscribers who have a very strong preference for services at the positive end of the first dimension. Since new data services are not located at this end of the dimension, this segment tended to rate new services as inappropriate for their needs. The second segment consisted of subscribers who have a very strong preference for services at the positive end of the second dimension. Since some new data services are located at this end of the dimension, this segment tended to rate a subset of the new services as appropriate.

In the final phase of these analyses, we examined the differences between these two segments on a number of variables. We first looked at the industry types (e.g., manufacturing) that made up each segment. These analyses indicated that only in one industry type, banking, was there a difference in the percentage of that type of subscribers in the two segments. Such small differences did not reliably differentiate the two segments.

In contrast to this pattern of results, there were differences in variables reflecting the size, data communication expenditures, use of current services, PBX and CENTREX use, number of locations in a LATA, and amount of intra-LATA traffic of the subscribers in the two segments. Logistic discriminant analysis indicated that these differences did significantly differentiate the two segments. If a company did not have this much information about its subscribers, it would have to obtain customer specific information before it could use these findings to assign a subscriber that was not in the sample surveyed to a segment.

3.4. Discussion

The findings described in this section could be used by a BOC in a number of ways. First, these findings can give a BOC insight into the nature of its data communication market. This information could be used by a BOC in positioning its products and in designing marketing strategies. Second, it could be used in designing new services. Finally, the salient features identified via multidimensional scaling can be used in building discrete choice models for assessing new service demand (Section 4.4).

4. MODELING OF SUBSCRIBER CHOICE

4.1. Gathering choice data

In order to understand how subscribers trade-off features versus price in choosing among available data services, it was necessary to gather data indicating how customers would choose among the alternative services under various pricing scenarios. The basic question to be answered was how data communications customers would restructure their data networks, given the availability of an expanded set of alternative services under various price scenarios. The challenge was to present this question in a way which would be understandable for respondents and would provide the necessary data for building multinomial logit and probit choice models.

It was anticipated that these price response questions would be among the most difficult for respondents to answer. Choosing among data services is a complex process in that the respondent must first determine which services have the features required to meet his data communications needs. Having determined the set of suitable services, he must then weigh both price and non-price characteristics of the various alternatives in making a final choice. This process is further complicated by the fact that the respondent may represent a company with multiple business applications, each with its own set of data communications needs and requirements.

The existence of multiple applications may cause an individual subscriber to exhibit different choice behaviors in meeting the needs of his various applications. While this appears to suggest that choice data be gathered separately by application, this has not proven feasible. Because of the large number of potential applications, and because individual customers have multiple applications, this approach would be costly and require the administration of excessively long, detailed market research questionnaires. Furthermore, evidence from one of our earlier market research efforts suggests that a single named application may give rise to different data communications needs for different customers. Fortunately, there is a relatively simple solution to this dilemma. Since all applications served by a particular data service are at least comparable in that the service features are sufficient to serve their data communications needs, it is natural to group applications according to current service type and model subscriber choice separately for each such group. Respondents were therefore asked to make separate choices for each of their current data services. The results quoted in subsequent sections are based on their choices for their existing analog private lines.

The questions themselves were broken into small components and were structured to give both interviewer and respondent the opportunity to check for consistency. For example, when queried about his analog private lines, a respondent was first reminded of the number of analog private lines used by his company within the study LATA. He was then asked how many he would like to see replaced, given a set of alternatives at specific prices. Finally, he was asked how many of the replacements would - correspond to each alternative. Consistency required that the number of replacements by alternative add to the total number of replacements and that the total number of replacements be less than the total number of analog private lines.

Although some customers may benefit from the consolidation of several current data lines onto a single new service line (e.g., several analog private lines onto a single digital line), the above procedure forced respondents to substitute on a one-for-one basis. This greatly simplified the respondent's task. Furthermore, the resulting data are sufficient to determine how customers trade off the price and non-price attributes of the various services. The potential for consolidation is more appropriately assessed in a pricing and reconfiguration model (Section 5).

4.2. Fitting discrete choice models

It is tempting to assume that responses to the questions described above will approximate a sequence of independent first choices. Under this scenario, the data would be treated as if subscribers make independent decisions about individual lines, and discrete choice models (e.g., multinomial logit and probit models) could be fit using individual lines as the basic sampling unit. Our recent market research has failed to support this assumption. When considering replacement of current data lines, most respondents treated lines of a given service as a unit; that is, they preferred to either keep all such lines or trade them off for lines of a single alternative service. Of those customers who split their lines among two or three alternatives (three was the maximum observed), many represented subscribers with a large number of lines. This may have indicated the implementation of more than one application or may have reflected a desire to "try out" a new service.

Because the segments described in Section 3.3 partition subscribers into groups which value the salient features of data services in fundamentally different ways, it is natural to attempt to fit segment-specific choice models. In doing this, we found that the WTPs for the service alternatives differed by segment and generally reflected segment-specific feature preferences. Further work is required to assess any potential for bias which may result from fitting segment-specific models.

4.3. Multinomial logit models

The following simple model illustrates the application of multinomial logit models. Suppose that a subscriber currently uses a single line of a given data service. His utility for service j is assumed to be of the form

$$U(j) = a(j) - bB(j) + e(j), \quad (1)$$

where j=1 denotes the subscriber's current service, a(j)/b denotes a "representative" subscriber's WTP for service j over his current service, and B(j) denotes the monthly bill the subscriber would incur in utilizing service j. The e(j)'s are assumed to be independently and identically distributed according to the extreme value distribution over the population of subscribers. The model is easily extended to customers with N lines by multiplying the right hand side of (1) by N. Models allowing multiplication by a function of N are also being investigated.

More general logit models, incorporating additional subscriber and service attributes, can also be described. A general framework for such models is given in [2].

4.4. Multinomial probit models

If the e(j)'s in (1) are assumed to be independent and identically distributed normals, the resulting model is known as the independent probit and yields WTP estimates nearly identical to those produced by the logit model. In light of its computational advantages, the logit model is therefore to be preferred. Both models have been criticized, however, due to their "independence of irrelevant alternatives" (i.i.a.) property. Briefly, the i.i.a. property states that the relative odds of one service being chosen over another is independent of the presence or absence of other service alternatives [3]. Clearly, this does not hold for all conceivable choice situations.

Assuming the e(j)'s in (1) to be jointly normal and allowing for nonzero correlations gives rise to a class of correlated probit models. Within this class are models which are not constrained by the i.i.a. property. Hausman and Wise [3] have described methods for constructing correlated probit models which can be used to express a subscriber's WTP for a service alternative as a sum of his WTPs for its salient features. Under such models, WTPs for different features are assumed to have independent normal distributions across the subscriber population. WTPs for sums of these features (i.e., services) are correlated to the extent that their features overlap. The salient features identified by multidimensional scaling (Section 3.1), together with a "feature" indicating whether an alternative is the subscriber's current service or not have been used to construct correlated probit models in the present context. While initial results indicate that such models do not fit the data as well as do logit models when all alternatives are included, additional verification is required. Given their freedom from the i.i.a. assumption, these models may prove useful when some alternatives must be deleted (e.g., when a service is not deployed in certain areas).

5. FORECASTING DEMAND

One of the major benefits of modeling subscriber decision behavior is that it leads to improved methods for forecasting demand. Figure 3 illustrates how customer decision models can be incorporated into a computerized system for forecasting the demand for new data services. This allows the extrapolation from a limited market research sample to an entire data market.

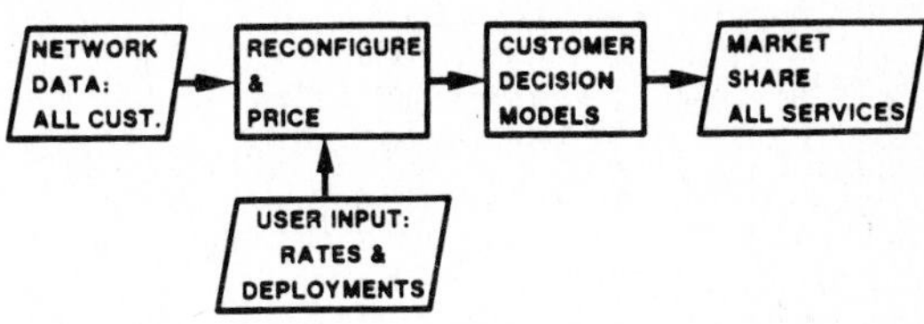

Figure 3. Demand Analysis System

The input database for such a system consists of network information for all a BOC's data communications customers in a particular geographical area. For each subscriber, the database includes the company name and industry code, together with information on the subscriber's data circuits, including circuit identifier, endpoints for circuit segments (identifying the serving wire center), line speed, and service type.

With the detailed picture of a subscriber's data network provided by the input database described above, one can determine reasonable scenarios for reconfiguring the network using alternative data services. Each of these scenarios can be priced out, using rates and deployments together with the likely usage carried on such a network (Section 6).

Having determined appropriate replacement scenarios and their corresponding prices, customer decision models can be applied to determine how a typical subscriber would then trade-off price versus service features for each scenario. The result is a set of probabilities indicating the likelihood with which a typical customer would choose each scenario. Total market shares are determined by aggregating these likelihoods over the entire subscriber database.

Note that rates and deployments are listed as user inputs. By varying these inputs, the user can study how various pricing and deployment scenarios will impact data services demand.

6. THE ROLE OF USAGE

For some services, a subscriber's monthly bill is a function of the usage on his data lines. Figure 4 illustrates how a subscriber's usage can affect the attractiveness of different data service alternatives. Circuit switched digital services are often high speed services for which the total monthly bill increases with total connect time. As a result, this service is most attractive for those applications which require the transmission of large quantities of data over a relatively short period of time (i.e., batch applications). Pricing for packet switched services, on the other hand, is often sensitive to the quantity of data transmitted, and is therefore more attractive for applications which transmit small amounts of data over long periods of time (i.e., interactive applications). Figure 4 exhibits hypothetical boundaries which delineate a "region of attractiveness" for each service. Since any measure of attractiveness must be a function of relative take rates, these regions cannot depend solely on usage, but must account for rates and subscriber WTP as well.

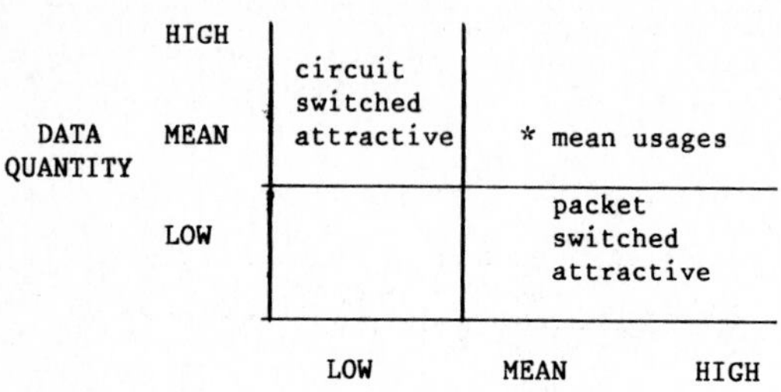

Figure 4. The Role of Usage

Given that demand is to be forecast for varying rates and deployments, the discussion above points to the need for a thorough understanding of subscriber usage patterns. It is clear that knowledge of a few summary measures (e.g., usage means) will not suffice, since, as indicated in Figure 4, regions of attractiveness are not necessarily defined by such summary measures. At the minimum, one needs a knowledge of the bivariate distribution of data quantity and total session time across the subscriber population. Ideally, this knowledge should be based on actual usage measurements covering the entire subscriber population. This information could be called upon as needed in the reconfiguration and pricing model described in Section 5. In practice, such measurements are often impossible to obtain, particularly for private line services. This makes it necessary to depend upon usage distributions constructed from market research surveys. Using such distributions, it is possible to simulate the usage of any given subscriber. This simulated usage can then be fed to the reconfiguration and pricing algorithm for use in usage sensitive pricing schemes.

By obtaining separate distributions for each current service type, it is possible to simulate shifts in usage which will occur as current lines are replaced by alternative services. Recall that for each subscriber in the input database, the customer decision models compute the likelihoods with which such a customer would choose a variety of reconfigured networks. By assigning the usage on the subscriber's current network among the reconfigured networks proportional to these likelihoods, and aggregating over the subscriber base, we can estimate the expected usage shifts which will occur under any given rates and deployments scenario.

REFERENCES

[1] J. D. Carroll, "Individual differences and Multidimensional Scaling," in A. K. Romney, R. N. Shepard, and S. B. Nerlove (eds.), Multidimensional Scaling: Theory and Applications in the Behavioral Sciences, Vol 1, Theory, Seminar Press, New York, pp. 105-175, 1972.

[2] D. McFadden, "Conditional Logit Analysis of Qualitative Choice Behavior," in P. Zarembka (ed.), Frontiers of Econometrics, Academic Press, New York, pp. 105-142, 1973.

[3] J. A. Hausman and D. A. Wise, "A Conditional Probit Model for Qualitative Choice: Discrete Decisions Recognizing Interdependence and Heterogeneous Preferences," Econometrica, vol 46, no. 2, pp. 402-426, 1978.

TELETRAFFIC ISSUES in an Advanced Information Society
ITC-11
Minoru Akiyama (Editor)
Elsevier Science Publishers B.V. (North-Holland)

INVESTIGATIONS ON TRAFFIC CHARACTERISTICS AND NETWORK BEHAVIOUR IN DATEX-P

Raimund TRIERSCHEID and Klaus NÜSSLER

Deutsche Bundespost, Fernmeldetechnisches Zentralamt (FTZ)
Darmstadt, Federal Republic of Germany

ABSTRACT

In this paper the main results of statistical investigations and measurements of the West German packet switching system DATEX-P are presented.
These measurements were made from 1983 to early 1985. Traffic distributions in DATEX-P are shown and subscriber groups HTS, MTS and LTS are introduced.
The paper evaluates the influence of traffic patterns on network performance and deals with the impact of the measurements on network performance and node engineering. The investigations are interpreted and a quality of service indicator, the Performance Factor (PF) is proposed.

1 INTRODUCTION

In August 1980 the Deutsche Bundespost's Packet Switched Digital Network (PSDN) DATEX-P began commercial service with 17 switches at 17 node sites. DATEX-P offers a broad range of service protocols and speeds for packet mode and non-packet mode terminals. For non-packet mode services the PAD (Packet Assemble and Disassemble) function is provided by the network.

1.1 Current Network Status

As of December 1984 7 812 ports were in service. The December 1983 count of ports in service was 4 179. This is an annual growth rate of 87% which correlates well with the forecast data. As this data show a continuous high rate of growth the Deutsche Bundespost expects about 30 000 connections by the end of 1987.

At the end of 1984 the network comprised of 35 nodes and two network control centres (NCC). Nodes are interconnected by trunks at speeds of 64 kbps.

Two switching centres in Düsseldorf and Frankfurt process international traffic. 29 international trunks (9.6 Kbps) according to CCITT-Recommendation X.75 connect DATEX-P directly to foreign packet switched data networks.
The first international trunk at a speed of 64 Kbps to Switzerland will be in service by the end of 1985.

The service mix remained static from 1983 to 1984:

	Portion of total ports		
	X.25	ITI	BSI+DSI+X.75
December 83	77%	19%	6%
December 84	79%	17%	6%

Table 1. Service Mix

The main service protocol is by far X.25. DATEX-P offers several speeds for X.25 from 2.4 kbps to 48 kbps. The speed mix remained static from 1983 to 1984 as:

X.25/HDLC Service				
Speed/kbps	2.4	4.8	9.6	48
December 83	30%	28.0%	41%	1.0%
December 84	31%	28.5%	40%	0.5%

Table 2. Speed Mix

1.2 Reasons for Traffic Measurements

The investigations of traffic characteristics and studies on user behaviour can be divided in two main areas of interest:
a) Improving the cost efficiency of installed or planned system components,
b) offering the best possible quality of service for the user.

The resulting traffic data can be applied to a variety of activities, such as planning, tariff studies, cost optimization and detection of service degradations.
The objective of all traffic investigations is to maximize the utilization of all resources within the constraints of grade of service, in general called node engineering.

2 STRUCTURE OF A DATEX-P NODE

In the following chapters the main concern will be to investigate how traffic flow and distribution influence LP and node engineering. For a basic understanding the physical model used throughout this paper is illustrated in Fig.1.

Fig.1 DATEX-P NODE PHYSICAL MODEL

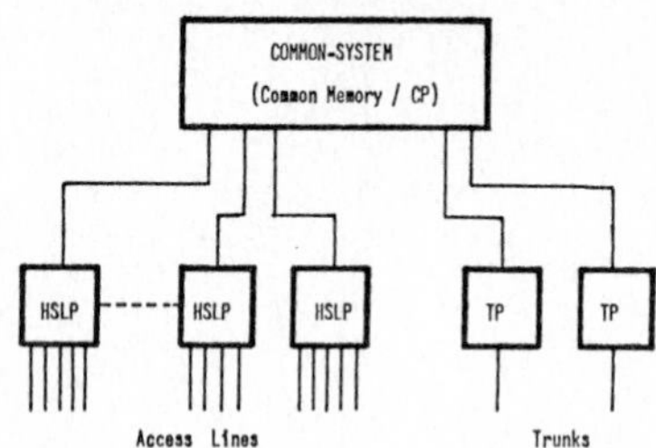

There are three significant resource components of the node for engineering purposes. These are:

a) High Speed Line Processor (HSLP) with
   * processing power - to handle packet processing
   * local memory - to store program code, service data, packet messages and process control blocks
   * physical fanout - to accomodate subscriber lines
b) Trunk Processor (TP) with
   * trunk bandwidth - to handle traffic to and from other nodes
c) Common System with
   * common memory - to store service data, control and message blocks
   * common bus bandwidth- to ensure operation of the multi-processor system
   * control processor (CP) processing power - to handle call set-ups, clears and administrative functions
   * processor fanout - to accomodate desired processor types

The most important component of node engineering is line processor (LP) engineering. This importance is due to the relatively large proportion of total nodal resources which is attributed to line processing and to the frequency with which changes are made that affect the LP.

3 TRAFFIC MEASUREMENTS

3.1 General Aspects

The traffic measurements are based on two sources. The Data Collection Centre (DCC) collects DATEX-P statistics of the main network components at 15 minute intervals. Hence this paper uses this time interval as a basic unit for statistical measurements.

The 15 minute intervall of a day during which the busiest traffic behaviour occurs is defined as the busy quarter hour (BQH). In this paper the BQH is restricted to a nodal environment without operational hardware and software problems.

The second source for traffic measurements are the accounting records. The record e.g. keeps track of call set-up/clear and duration times and is therefore useful for user behaviour investigations.

3.2 General Traffic Measurements

3.2.1 Network Traffic Trends

The subject of growth in a PSDN can be viewed from two perspectives: number of ports and the traffic volume resulting from the connected ports. These two parameters are important, since they directly affect the revenue stream that can be expected from the number of subscribers.
As one can see in Fig.2, a doubling of the number of subscribers may not result in an automatic doubling of equipment. The exact cost impact will depend solely on the effect on the network traffic characteristics. If the traffic per port drops, then the cost of provisioning drops since less common equipment is needed.

Fig.2

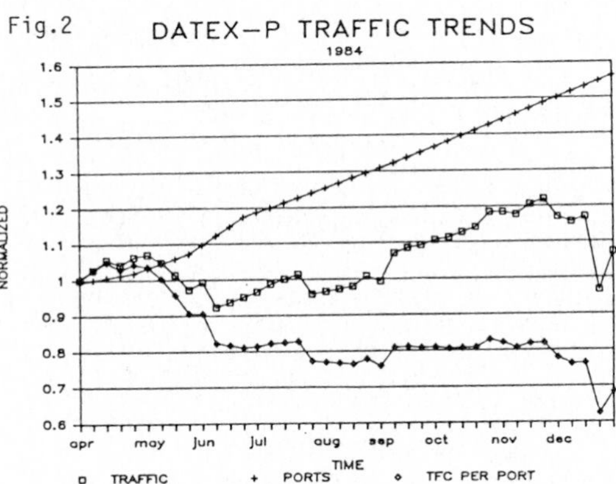

Fig.2 is a good example, how sensitive the subscribers react to economical influences. Traffic per port dropped significantly over the May-June time frame. The per port erosion of traffic cannot be explained by changes in the make up of the ports since the proportion of X.25 ports has not changed. A good explanation might be that the customers reacted with their applications to the announcement of a change (higher rates) in the tariff structure. The second drop in traffic reflects the industrial strike period in Summer 1984. A recovery to the normal trend can be seen. The December drop is due to Christmas holidays, as the recovery in Jan/Feb 1985 showed.

3.2.2 Distribution of Traffic Volume

One of the most costintensive components of a digital network is the line access system. DATEX-P uses currently static multiplexers to reduce the backhaul costs. Concentrators with a more economical use of the bandwidth are far more cost effective but require from the network engineer detailed knowledge of the geographical distribution and traf-fic volume of the connected data terminals.

Hence a long term investigation was made to determine, if and how DATEX-P could use concentrators without losing an adequate quality of service.

Fig.3 shows the distribution of the traffic volume found.

Fig.3 DISTRIBUTION OF TRAFFIC VOLUME

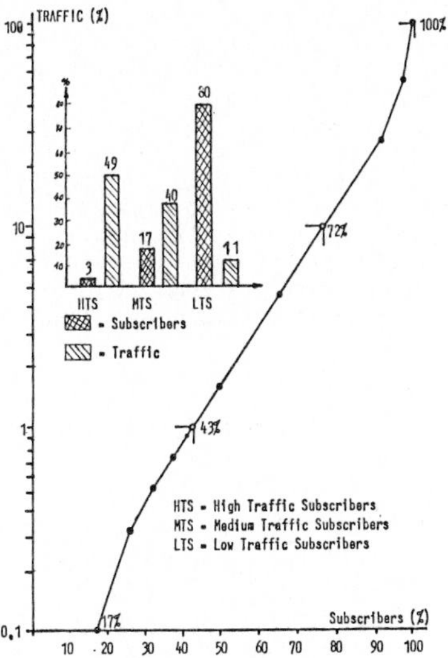

In DATEX-P segments (1 segment=max.64 octets) are used as accounting units. Subscriber categories could be based on the amount of segments they produced. Thus the subscribers were classified into three groups:

Group	Volume/day	Subscribers '83	Subscribers '84	Traffic '83	Traffic '84
LTS	< 5 KSegm.	80%	80%	13%	11%
MTS	5..50KSegm.	17%	17%	37%	40%
HTS	> 50KSegm.	3%	3%	50%	49%

Table 3 Subscriber Traffic Groups

The HTS group consists of about one third subscribers who use SVCs and two third use PVCs. The SVCs produce less than 30% of the HTS traffic. This reflects the situation of the remainder of the network, where the SVCs produce just 30% of the whole traffic. It should be noted, that the logical channel distribution of LTS and MTS is 67% SVCs to 33% PVCs.

Based on the geographical distribution of the data terminals it could be determined that about 60% of the LTS group could economically be connected to concentrators. The movement between the 3 traffic groups stayed far below 10% during a period of 1 year. Hence the costly risk of having to remove traffic intensive MTS or HTS from the concentrator is very low.

3.3 Traffic Profiles

3.3.1 Network Traffic Profiles

The hourly traffic variation is based on LP statistics for the number of frames received and sent. The relationship of LP frames per second to data packets per second (DPPS) is approximately constant. As long as the relationship between frames and DPPS is static, the use of LP frames for analysis purposes is acceptable.

The values on Fig.4 are based on the total sum of all network LP frames for one quarter hour for each hour from 05.00 to 21.00 hours. The totals of each quarter hour for four days in 1984 (Apr 26,Jun 12,Sep 11 and Dec 12) are then averaged for Fig.4.

The chart shows a fast increase in traffic during the morning with slower drop off in the afternoon. There is a noticeable decrease at noon. The business day will be defined as from 07.30 to 18.00 so that postponable maintenance activities should not occur within this interval. One important observation from this chart is that the time of day tariffs (day tariff from 8.00 - 18.00) do not have a major impact on the traffic distribution. In fact the traffic after 18.00 to 21.00 is mainly due to international traffic.

Fig.4 TOTAL LP TRAFFIC

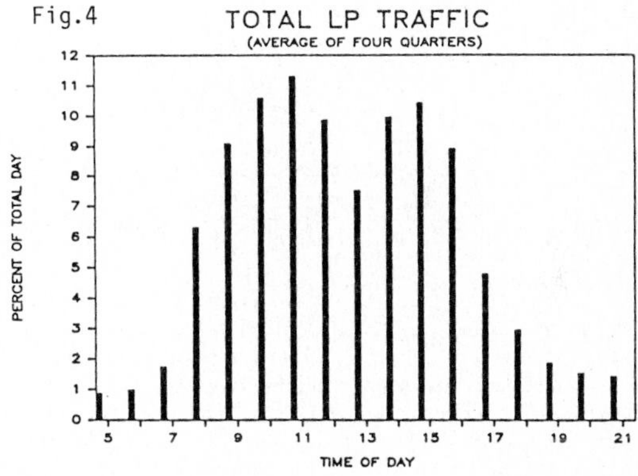

The present night tariff reduction of 45% of the peak tariff rate seems not to be an incentive to shift larger proportions of the traffic to night hours.

Fig.5 DAILY TRAFFIC VARIATIONS

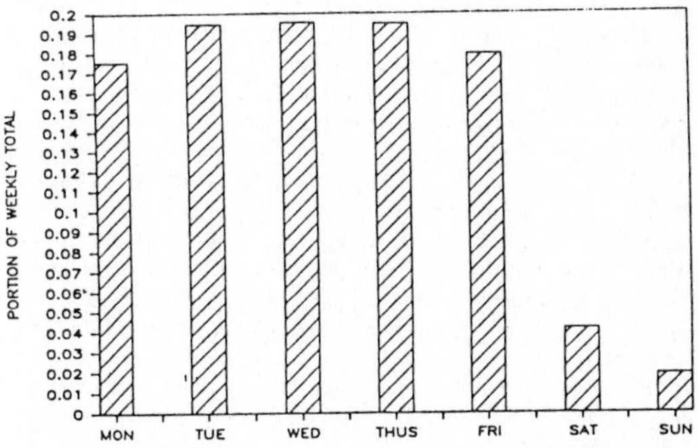

The daily traffic variation is illustrated in Fig.5. The chart is based on a series of measurements during 1984. The results are consistent with Fig.4 in that the network traffic is concentrated in the business days.

3.3.2 Node Traffic Profiles

Though most of the hourly traffic profiles of the single nodes is quite similar to the traffic profile of the network, there were some significant deviations found in the profile of some single nodes.

Fig.6 shows an example of a node with a different hourly traffic profile. Further investigations showed that the node's behaviour was due to one subscriber whose application (international traffic) had an impact of the traffic profile of this node (notice the shifted BQH and traffic distribution until midnight).

Fig.6 NODE TRAFFIC PROFILE

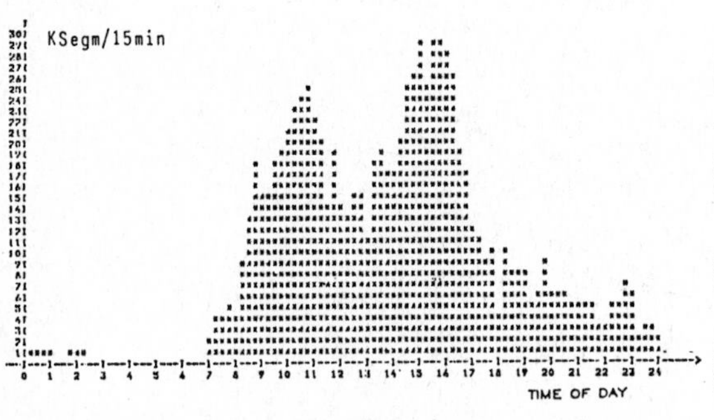

With the knowledge of node traffic profiles maintenance activities can be influenced so that postponable work can be pushed to off-hours.

3.3.3 LP Traffic Profiles

Since October 1984 DATEX-P is using a set of integrated statistic programs for operational and planning purposes. The Operational Measurements and Analysis Tool gives the network operator and engineer an opportunity to analyse the statistics of the most important network components.

The analyst is able to predefine thresholds and automatically run benchmark tests to flag only those network components that show an unexpected or abnormal behaviour. As the LP's behaviour is of major concern for engineering a node, Fig.7 and 8 show the plot of the hourly traffic profile on LPs.

The LP contains local memory, some of which is set aside for the local work queue to transfer messages between processes to and from the common system. To check the node condition the number of free memory blocks is sampled approximately every 4 seconds. A running total of these samples is reported to the node every minute.

The results of this procedure are compared to the current throughput rate to find a correlation of the statistics.

Throughput and utilization increase and decrease at the same time in a normal manner. The memory drops when the throughput increases because of the allocation of message blocks from the initial memory free queue.

The key is to compare simultanously the memory use with the throughput rate and the processor utilization to indicate the node condition. If the processor utilization is high while the throughput is low, a trouble situation is indicated.

Fig.7 illustrates the plot of an abnormal behaviour. Throughput and utilization increase and decrease at the same time, however the utilization is off-set from the throughput. Normally the curves run close together in parallel.

Fig.7 LP TRAFFIC PROFILE

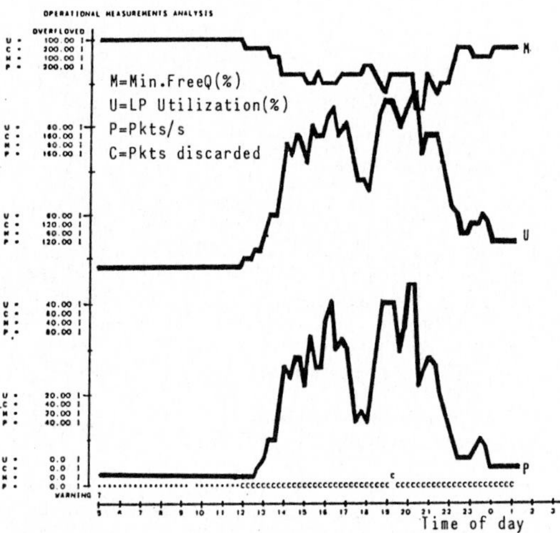

In this case there is a constant off-set between the LP utilization and the packet processing curve. The scales of the U- and P-curve are chosen in such a way that the two curves run close together, when the LP behaves normally.

The off-set means a high background utilization that can be caused by an unstable line or faulty modem, or by having the LP configured incorrectly for the type of service it is serving (e.g. mismatched speeds).

Another example of how valuable LP plots are for the network engineer and operator is shown in Fig.8. In the abnormal range the processor experiences little or no throughput but has a high background utilization.

This may indicate a processor problem, a test condition or a bad line/modem. The significant drop in the utilization curve indicates that the operation group detected the reason for the abnormal behaviour and solved the problem so that the utilization curve behaves as expected.

Fig.8 LP TRAFFIC PROFILE

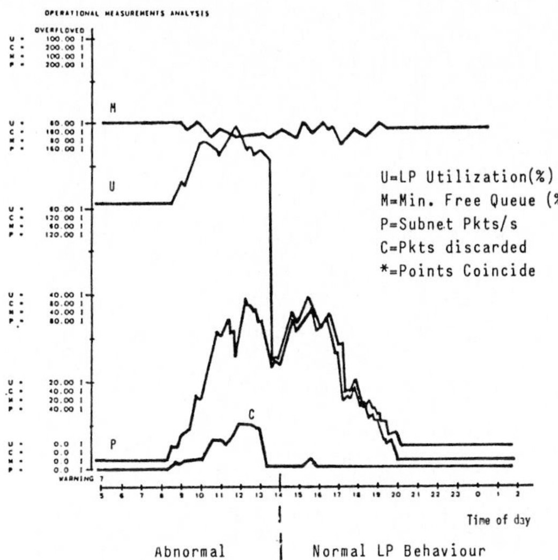

4 LP TRAFFIC ENGINEERING

4.1 Quality of Service Indicator (PF)

Fig.7 and 8 illustrate the problems resulting from inefficient line operation. In a normal operating environment it is not desirable to detect inefficient LP behaviour through this means; consequently the concept of Performance Factor (PF) is introduced.

PF is simply the ratio of LP frame throughput to the LP processor utilization. All things being equal, the LP should be able to process a certain number of frames for each increment in LP utilization. The absolute number of frames processed per % increment is not important. What is important is to flag those few LPs which are not being used to service frames but other interrupts (protocol errors, modem status changes etc.).

By removing the non-productive interrupts, the effective throughput of the network can be improved.

4.2 Calculation of LP Utilization

Node performance optimization requires the careful engineering of the Line Processor. The engineering of a node may be addressed in two different ways. The first approach begins with a standard node configuration of LP's with service protocols. This approach is concerned with how many access lines of different traffic characteristics can be terminated. The second approach begins with a given number of access lines of different traffic characteristics to be terminated on the node or site. The main concern of the second approach is how many LP's and therefore how many nodes of what configuration are required.

The major consideration for the LP is CPU utilization. A target utilization is chosen and engineered towards.

There are 3 phases in the engineering process:

i) Prediction: Estimation of traffic for new lines;
ii) Monitoring: The predicted resource utilization is compared to actual measurements on the operating node;
iii) Adjustment: Addition or removing of lines to meet the target values of LP utilization

The currently recommended target CPU utilization is 60% for the BQH. Values of about 80% indicate an overload situation that is defined as the utilization level at which customers experience degraded service such that customer complaints will occur.

To predict the usage of LP's the network engineer has to calculate the total LP utilization TUT.

TUT is defined as the sum of the utilization fractions UTILj due to each of the lines assigned to the LP and the utilization fractions CUTj due to call setup and call clearing.

As a result of a series of measurements in the live network and theoretical investigations some formulas were defined to be used before the line assignment process.

In equation form the total utilization TUTj of LPj is:

$$TUTj = UTILj + CUTj \qquad (E2.1)$$

The value of UTILj can be calculated once the following has been found:

* throughput of each line, Ri;
* packet processing time for each line type, Ti;
* LP capacity for each line type, Ci.

The LP utilization due to line i is given by:

$$UTILi = 100 * Ri/Ci \qquad (E2.2)$$

Hence $UTILj = \sum_i UTILi = \sum_i 100 * Ri/Ci$ (E2.3)

As an example the value of UTILj for X.25/HDLC lines can be calculated as follows:

$$R = S * Uin/(8*100*(L+Hl)) \text{ DPPS} \qquad (E2.4)$$

where: S = line speed in bps
Uin = percent line utilization into the LP due to data packets only
L = average packet length in octets
Hl = overhead bytes (8 bits) for data packet, packet RR and frame RRs.

Averaged values of DATEX-P: L = 48
Hl = 30

The Hl value assumes low % piggybacking at the frame and packet level that was found as the most dominating behaviour.

The total line processor throughput capacity, at 100% utilization (saturated) is:

$$C = 1000/T \text{ DPPS} \qquad (E2.5)$$

Where: (1 DPPS in + 1 DPPS out) = 2 DPPS;
T = total packet processing time in ms as determined for each service.

The packet processing time T for X.25/HDLC lines is given as:

$$T = q + r * L \qquad (E2.6)$$

where: q = overhead processing time in ms
r = length-dependent packet processing time in ms per octet
L = average packet length in octets

E2.6 was found to be accurate to within 6% for packet lengths of 32 to 128 octets. For the current software release the following values were measured:

q = 15 ms and r = 0.035 ms/octet

Table 4 shows the currently used values for the number of X.25/HDLC lines at different line utilizations for an estimated LP utilization of 10%.

Line Speed/ bps	Line Utilization					
	5%		10%		15%	
	L = 48	128	48	128	48	128
2400	15.6	27	7.8	13.5	5.2	9
4800	7.8	13.5	3.9	6.8	2.6	4.5
9600	3.9	6.8	1.9	3.4	1.3	2.3

Table 4 Estimated nr. of lines for 10% LP UTIL

The fraction of LP utilization due to call set up and call clear operations in the BQH is calculated with:

$$CUTj = Ncc/Kcc * 100\% \qquad (E2.7)$$

Where: Ncc = avg. nr. of combined call setup and clearings per second
Kcc = Ncc at LP saturation = 5.6 as measured in lab environment for X.25 through one LP only

A series of snapshots of the user behaviour showed and traffic analysis confirmed that presently the fraction of processor utilization CUTj due to call activities is negligible in DATEX-P.

The average call duration time for the SVCs is 6033 seconds on an average business day. The bandwidth of call durations is from less than 80 seconds up to more than 19000 seconds per call. Unless the user behaviour changes significantly the utilization is not a function of call activities.
When DATEX-P began commercial service the user behaviour was unknown and thus the prediction of the traffic amount per line had to be conservative to avoid overload situations just at the beginning.

After a period of intensive studies on the network and subscriber behaviour in 1983 and 1984 it could be summarized that the initial predictions of the initial phase were too high. The net result of the original guidelines can be illustrated by Fig.9 which shows the current LP utilization.

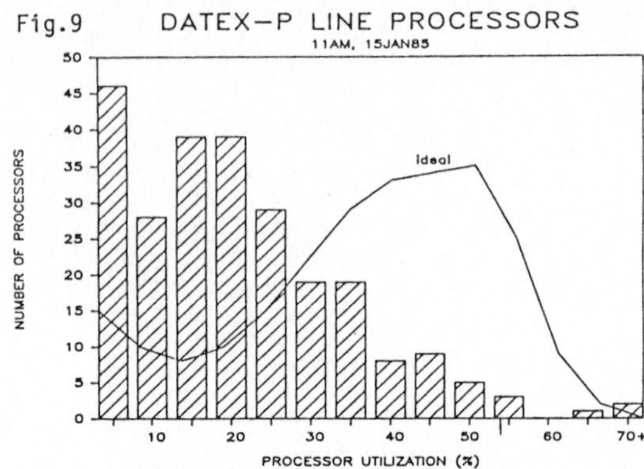

From a resource utilization perspective the ideal shape means the optimal use of the common equipment without degrading the quality of service. Current efforts is to engineer the processors to carry more load thus shifting the LP utilization upward.

5 TRAFFIC PROFILES OF USER GROUPS

5.1 General Aspects

In the assignment process it is important for the network engineer to know as much as possible about the available statistics of the subscriber lines and the node that will have to serve the lines.
To provide the customer with the best quality of service the current node behaviour, especially the LP behaviour should be known. The hourly LP traffic profile should be used to define the LP's BQH and distribution of the traffic amount.

If the user behaviour during the business day is also known the two traffic patterns should be normalized to a common base and then compared. Usually the network provider doesn't know too much about new lines and their future traffic behaviour. In fact even the subscriber's estimation is not very accurate.
To avoid overload situations and to use the provided LP bandwidth efficiently the single lines assigned to the LP should consist of a mixture of lines with different user behaviour.

In the worst case all lines of the same LP produce the traffic only at the same time of day during a short period of time. Though the LP is idle almost all the time, the connected subscribers could experience a degradation of quality of service, if their traffic volume is high enough.
The theoretically ideal model is to combine lines with equally distributed BQHs. Thus the LP's load is distributed over the whole business day.
To approach the ideal model investigations in DATEX-P were made to determine traffic profiles of special user groups with different business oriented traffic distributions.

Once typical shapes for special user groups are found, the prediction of the user's behaviour and estimation of the traffic volume of a single line is more accurate than to take an average of all existing subscriber lines.

5.2 Business Oriented Traffic Profiles

The investigation is based on the 300 busiest subscriber lines in DATEX-P. At the end of 1984 these lines consisted of 74% HTS and 26% MTS producing about 55% of the whole traffic volume. Most of the lines (99%) used a speed of 9.6 Kbps.

The initial research started with three user groups:

* Banking business,
* insurance business and
* department store business (including mail-order firm applications).

Fig.10 TRAFFIC PROFILE/BANKING BUSINESS

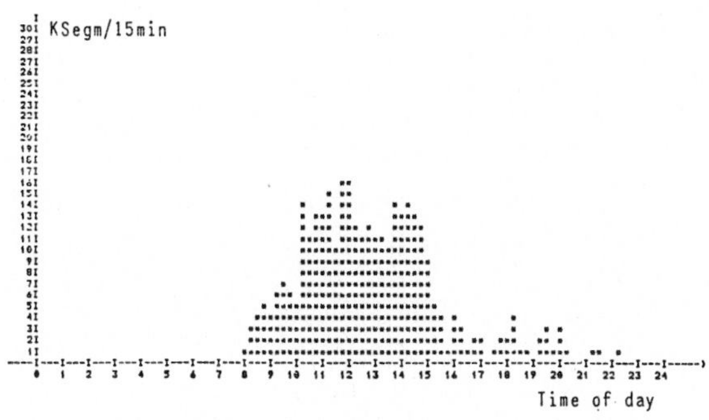

These three business groups represented a total traffic percentage of 19% and included 10% of the 300 busiest lines.

Fig.11 TRAFFIC PROFILE/INSURANCE BUSINESS

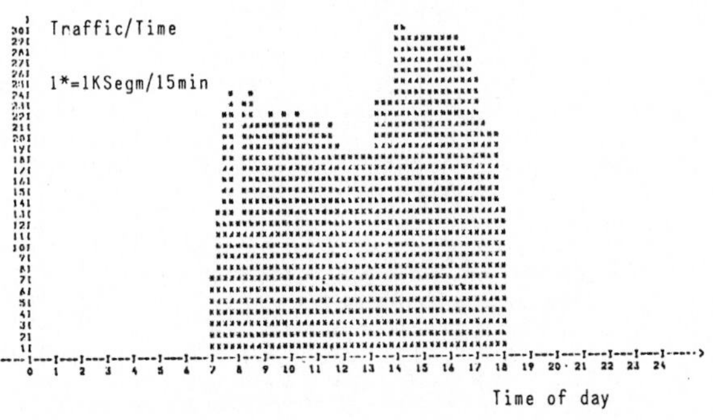

For every business oriented group a FORTRAN program plotted the traffic profile of an average business day. Then 5 samples of single lines, that didn't belong to the 300 busiest lines, but belonged to the investigated user group, were plotted and the shapes of the traffic profile were then compared.

In most of the cases a good correlation could be determined. The main deviations could be seen for department store business. Closer analysis revealed that this was due to the impact of not having separated Host to Terminal from Terminal to Host communications.

Fig.12 TRAFFIC PROFILE/
DEPARTMENT STORE BUSINESS

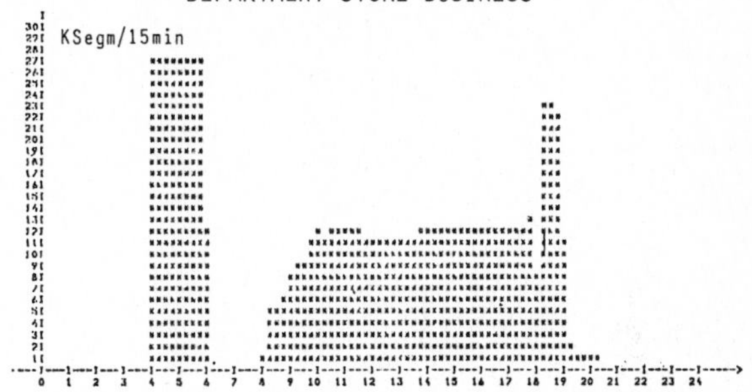

Fig.10 to Fig.12 illustrate clearly the different user behaviour of the business oriented groups. While the banking and insurance business profiles reflect the behaviour of the whole group as well as the behaviour of a single line, the traffic profile in Fig.12 is only accurate for Terminal to Host traffic in the time frame from 08.00 to 19.00.
The night traffic from 04.00 to 06.00 is due to the summarized Host to Terminal communication. The business day traffic is mainly based on interactive dialogue sessions with only short responses from the Host. The night traffic is due to file transfers from Host to Terminal.

Further investigations will have to determine more typical user groups and their traffic profile based on separated Host to Terminal and Terminal to Host communications.

6 CONCLUSIONS AND FURTHER RESEARCH

The paper has analyzed the distribution of the traffic volume in DATEX-P and introduced three main subscriber groups that are classified by the produced traffic amount.
The impact of traffic distribution and user behaviour on network and node engineering has been discussed and the influence on the quality of service has been shown. To have a quick overview of the condition of a node, a quality of service indicator, the Performance Factor (PF) has been proposed. Further researches will be initiated to cover the remaining business oriented groups to provide the network engineers as well as the marketing groups with the findings. With the added information the network utilization, both current and planned, can be optimized.

TELETRAFFIC ISSUES in an Advanced Information Society
ITC-11
Minoru Akiyama (Editor)
Elsevier Science Publishers B.V. (North-Holland)

INTERACTIVE VOICE APPLICATION HANDLING IN WIDE-AREA PACKET-SWITCHED NETWORKS

Davide GRILLO

Fondazione Ugo Bordoni
Rome, Italy

ABSTRACT

The performance of a packet-switched wide-area network for integrated data and voice service is investigated under the assumption of a traffic mix resulting in network resources mainly loaded with data traffic. The analysis focuses on two issues impacting the transport delay, namely probability of allocating network resources to one traffic type in contention cases and reduction of voice redundancy depending on network congestion level. The algorithm used for reducing voice redundancy bases on embedded ADPCM coding and a particular bit assembling within voice packets.

1. INTRODUCTION

As a first step toward services integration in packet-switched networks, voice and data integration is envisaged, (1), (2), (3), (4), (5). In this paper the problem of handling both voice and data is addressed by referring to CCITT X.25 protocol for data application and by adopting a voice-oriented, X-series compatible and simplified protocol for voice, described in (6) and (7), thus exploiting the advantages deriving from protocol commonality.

Among other issues needed to be considered in order to adapt the X.25 protocol for voice handling, flow control plays a significant role. Due to the high interaction between partners in a voice session and the stringent requirements on transfer time, excessive delay in information delivery is to be faced by speeding up voice transport. This can be achieved , i.a. by making use of priorities in allocating network resources, by reducing voice load at the expence of fidelity, or by a combination of the two.

In (8) various schemes for adapting the voice sources encoding rate to the network congestion level, as reported by the sinks, are examined in a voice only environment.

In this paper, which expands under several aspects a preliminary analysis reported in (9), both random allocation of network resources according to different probabilities and load adaptation are considered. While allocation of resources is ruled by servicing dedicated voice and data queues, load adaptation is pursued by keeping the encoding rate at the source fixed and dropping off voice bits as packets travel through the network, depending on actual congestion level. Such an action aims at meeting delay constraints. Prerequisite for its adoption is that it does not corrupt too much the affected speech segments, i.e. does not make objectionable the speech quality at the relevant application level. Bit dropping is operated through an algorithm activated locally to the switching nodes in the long distance network. The reduction is accomplished by adopting embedded ADPCM coding, (10), conveniently assembling single bits of voice samples inside each packet and intervening on less significant of them.

In the analysis emphasis is given to the bursty nature of voice packet arrival process. Although various investigations address characterization of arrival statistics for packet voice, they mainly deal with voice streams flowing through a single network facility, (11), (12), (13).

Due to the complexity of the analysis that considers routing through tandem links and 99.th percentile end-to-end delays, a simulation approach has been adopted.

The paper is organized as follows: Section 2 presents the voice handling protocol and the load reduction algorithm; Section 3 describes the network model used in the analysis; Section 4 presents results referred to various cases ; finally, Section 5 reports on the major findings.

2. VOICE HANDLING ISSUES

2.1 X.25-compatible Protocol Features

As for the voice handling protocol, reference will be made to the mentioned simplified, X.25-compatible version, (6), (7). Accordingly, error control is missing at both layer 2 and 3, and flow control is replaced by dropping off less significant bits comprised in the information field as a means of affecting the packet transport delay. In particular, layer 3 functions are limited to packet addressing and routing according to the virtual circuit technique, and possibly to bit dropping. X.25 layer 2 procedures, besides those performing zero stuffing and destuffing and error detection, are inhibited. Current X.25 flag patterns are retained to ensure bit transpanrency and frame synchronization. In addition, new bit configurations in the address field are used to discriminate between voice and data frames. Information frames do not need to support send and receive sequencing, and error detection is the only action to be taken at layer 2.

2.2 Signal Encoding and Packetizing

Packetizing delay corresponding to 16 ms speech and 32 kb/s ADPCM coding, (10), have been assumed. Using of 4 bits for expressing the difference between adjacent samples, a voice packet consists of 128 samples for a total of 512 bits.

Since according to embedded ADPCM coding, each subsequent bit used for representing the difference between two samples adds resolution to the representation, this technique lends itself for voice load reduction and graceful degradation of speech quality. In order to ease possible dropping of less significant bits, a packet is structured into a certain number of fields, each containing bits of equal significance related to different samples. Fig. 1 shows the structure of a voice packet as assumed in the analysis. The upper part of the figure represents the sequence of 128 voice samples before assembling. The lower part shows how the single bits are moved into the various fields, 16, into which a packet is assumed to be partitioned. In particular, the bits in the first position in each sample are ordered in sequence in the first four fields (each field consisting of 32 bits), the bits in the second position in each sample are ordered in the subsequent four fields, and so on, so that the significance of fields decreases from one packet extreme to the other.

Dropping a certain number of fields, starting from the extreme corresponding to less significant of them, amounts to reducing the redundancy of the affected samples in the same way as if an encoding rate lower than the original one were used.

2.3 Signal Reduction Algorithm

In each node, before transmitting a packet, the state of the relevant output queue is inspected. The queue state is expressed by the nunmber of bits associated with the information fields of the voice packets waiting for transmission, i.e.

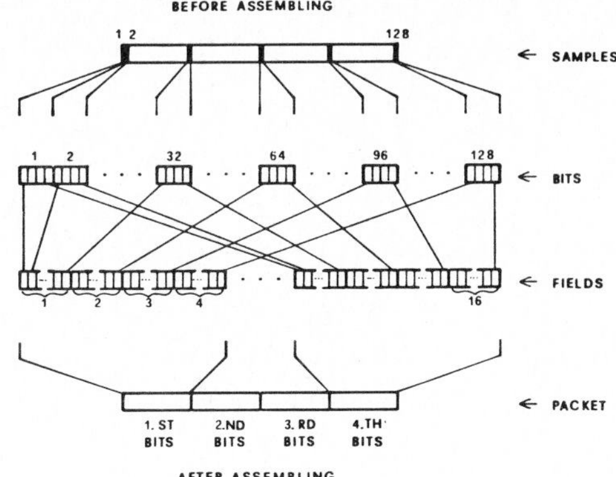

Fig. 1 Bit assembly in a voice packet.

the bits belonging to the packet in the process of being transmitted are not counted. If this number exceeds a stated threshold, the packet to be transmitted is not allowed to contain more than a specified number of fields.

Should the actual number of fields be greater than allowed, then so many fields are dropped (and lost) as necessary to meet the limitation. Packets may be reduced up to a specified extent, i.e. once the maximum reduction has been reached, subsequent reduction is inhibited. A generalization of the algorithm consists in setting several thresholds for describing the state of the queue, with associated number of fields allowed for transmission. In this analysis, threshold values are chosen as integer multiples of a basic threshold, the parameter of the algorithm. A packet may be reduced up to half the original size and allowed fields linearly decrease by one unit passing from one threshold to the upper next threshold.

3. THE MODEL

Elements of the queueing network model, represented in Fig. 2, are switching nodes and transmission channels. Nodes are provided with memories of three types: buffer storage for incoming frames waiting to be processed, memories for storing data packets temporarily inhibited to transmission due to window flow control, memories for storing packets waiting for transmission facilities to become available.

Full-duplex transmission mode is assumed and simultaneous transmission activity on both ways is modelled through distinct and independent servers, one for each direction.

Voice and data frame contending for network resources, both switching and transmission, join separate queues. Inside each queue the servicing algorithm is FCFS, whereas for the inter-queue servicing algorithm random allocation of resources according to specified probabilities is considered.

A data frame enters the network through the input data queue in the origin node and, after being locally processed, it either joins the data queue associated with the transmission channel, or it is held in the node, according to whether the window flow control allows for transmission or not. Upon reaching the downstream node, a data frame joins the input data queue and, after processing, an acknowledgement for the upstream node is issued . Piggybacked acknowledgements on information frames are considered, as well as use of a same frame for collectively acknowledging multiple data frames.

A voice frame enters the network through the voice input queue and after obtaining service it is routed to the voice output queue associated with the transmission channel leading to the subsequent node. Depending on the state of the voice workload waiting for service, a reduction in the information field of the packet in the process of being transmitted may be operated.

3.1 Data Flow Control

Admission into the network of single data packets is controlled by monitoring the packet population, both voice and data, within the long distance network at the instant of packet arrival. Data packets are denied access when the population exceeds a specified threshold. Backpressure is exercised of halted data packets on data sources and the process of data packet arrival is interrupted ; the process is resumed when data packets are allowed again into the network.

In case of backpressure, chek for admission.population level is performed at randomly and exponentially distributed intervals, long on the average half the data packet interarrival time. The rationale for this admission control is that voice applications should not experience delay which hampers natural speech interaction, and that data service may be -to a certain extent- sacrificed to meeting voice service requirements. Link-by-link window flow control within the long distance network is performed at layer 2. Use of extended numbering is assumed and the considered window size is 128. Dedicated acknowledgement packets of 80 bits are employed, when data frames to be used for piggybacking are not timely available.

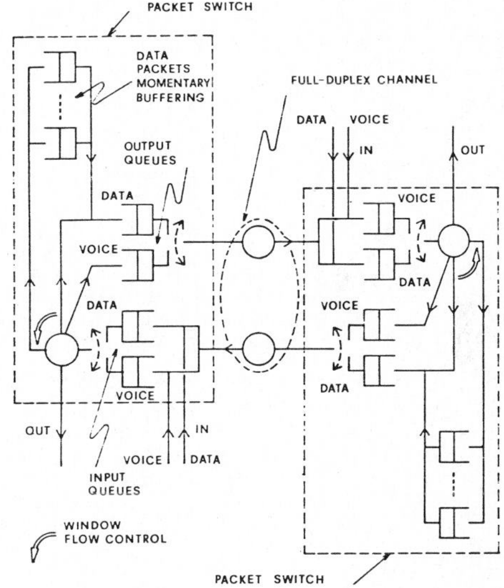

Fig. 2 Queueing network model of long distance network.

3.2. Environment

Ample buffer space and an error-free environment are assumed, thus excluding recovery actions due to out-of-sequence or information contents corruption, respectively.

3.3. Switching Nodes and Transmission Channels

Processing of a frame in a node , independently of the type, takes a constant time, .125 ms, i.e. nodes can handle up to 8000 packets per second . Full-duplex transmission facilities in the long distance network operate at 2.048 Mb/s. Transmission time is proportional to (fixed) frame length.

3.4. Traffic Issues

Bidirectional, balanced exchange of information is assumed between partners involved in a voice call or data session. Traffic mix resulting in 70% data and 30% voice workload on transmission channels, unless otherwise specified, has been taken as reference. Voice signal statistics in (14) are used to model voice traffic. Due to the different dynamics related to the arrival of calls and packets, the number of calls simultaneously in progress has been kept fixed. Voice calls are individually modelled as alternating "activity" and "silence" pha

ses. Activity phases (talkspurts) correspond to the emission of a burst of frames, whose number depends on talkspurt duration, exponentially distributed, and voice encoding rate.
The total length of a voice frame is 592 bits.
As modelling of data traffic is concerned, data frames are individually spaced and data sessions are collectively considered. The length of information and acknowledgement data frames is 1096 and 80 bits, respectively. Voice frames belonging to the same talkspurt are spaced by the packetizing time, whereas an exponential distribution is assumed for the time between submission to the network of a data frame, or last voice packet in a burst, and start packetizing of the next frame.

3.5. Performance Metrics and Targets

The reference end-to-end connection is depicted in Fig. 3.

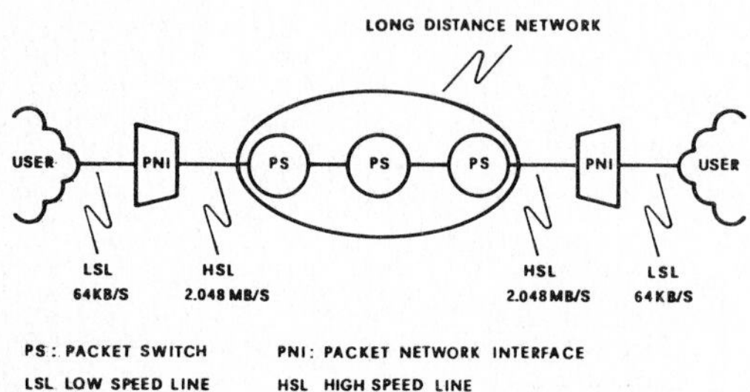

Fig. 3 End-to-end reference connection.

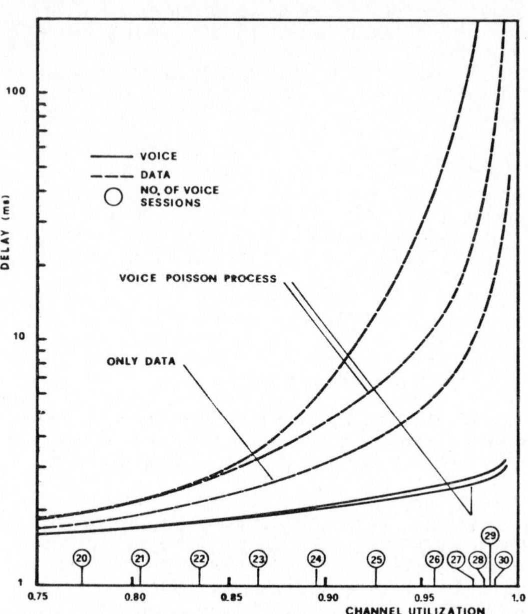

Fig. 4 Average transport delays in the long distance network for .65 voice servicing probability.

While the long distance network is modelled as described above, the access network is modelled by considering that each access side typically comprises a low speed line, 64 Kb/s, connecting the user with a packet handler, and a high speed line, 2.048 Mb/s, from the packet handler to the long distance network. The influence of the access network is accounted for by assuming that high and low speed lines behave as M/M/1 queues. Typical utilization for low and high speed lines ranges between .3 to .5 and .7 to .8 respectively. By combining extreme utilization values, two options for the access network are referred to in what follows, namely "low utilization" i.e. .3 and .7 and "high utilization", i.e. .5 and .8. The design values related to these options are to be intended in case no backpressure on data is exercised and as related to .8 and .9 design utilization of the channels in the long distance network respectively. Different utilization of the channels reflects into different utilization of the access network segments.
Delay in the long distance network is defined as the time elapsed from joining the input queue in the origin node until leaving the destination node. In the case of data traffic, this delay also includes the time spent at the network entry point waiting for admission into the origin node queue.
The end-to-end delay consists of variable components (time spent for accessing transmission and processing facilities) and constant components (packetizing, transmission and switching time). The constant components are assumed to contribute 44 ms to the end-to-end delay, resulting from the following break-down: 32 ms for encoding and decoding, 10 ms due to propagation, 2 ms for processing in the packet network interfaces (1 ms each). As for the variable components, (theoretical) delay distributions in the access network are combined by convolution with the (experimental) distributions for the long distance network.
A value of 250 ms for the 99.th percentile voice end-to-end delay has been taken as service target, (15).

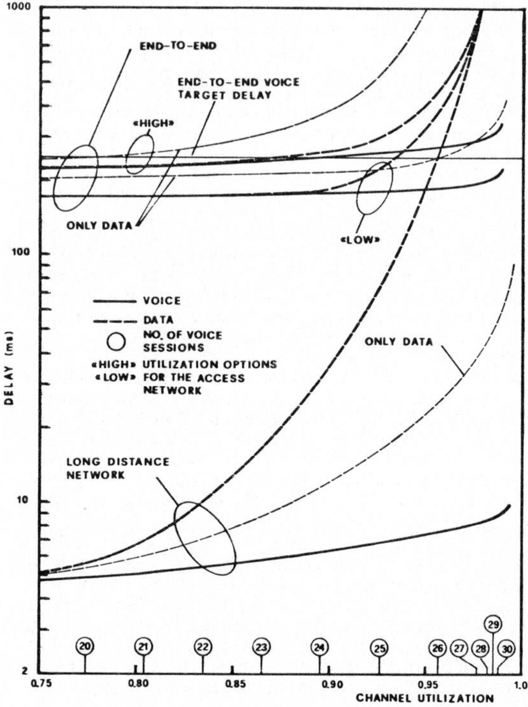

Fig. 5 99.th percentile transport delays in the long distance network and end-to-end for .65 voice servicing probability.

4. RESULTS

The analysis relates to the reference connection of Fig. 3, i.e. the long distance network consists of three nodes connected through two transmission channels. Data traffic is only exchanged between directly connected nodes, whereas voice traffic is exchanged between any node pair and hence it is also routed through one transit node. Symmetry for the offered load, both voice and data, submitted from any node is assumed and destination nodes are addressed equitably. The packet population limiting the admission into the

long distance network is equal to 1000.
Fig. 4 and Fig. 5 show how servicing voice traffic with probability .65, i.e. twice more frequently than data traffic, respectively impacts the average and the 99.th percentile delay. Delays in the figure are shown against the channel utlization as it results by increasing both voice and data offered load while keeping constant their proportion in the traffic mix. Data load increases by increasing the packet arrival rate, whereas voice load increases by increasing the number of simultaneous voice sessions.
On the horizontal axes mapping between channel utilization and number of voice sessions is also shown. Since the length of voice packets is practically half the length of data packets, this could be viewd as a "fair" policy for allocating network resources. As results demonstrate, the two traffic types actually obtain quite different service.
In Fig. 4 the voice average delay in the long distance network increases very slowly by increasing channel utilization, whereas the data delay increases steeply. In the same figure the data delay corresponding to the absence of voice load is also depicted. The difference between the data delays in the two cases increases very rapidly and, for 95% utilization, the ratio between them is almost one order of magnitude in favour of the case in which only data are handled.
This phenomenon is due to a great extent to the bursty nature of voice traffic. As a demonstration, by replacing the voice packet arrival process with a Poisson process, the difference reduces considerably, as it is evidenced in the same figure.
This penomenon is due to a great extent to the bursty nature of voice traffic. As a demonstration, by replacing the voice packet arrival process with a Poisson process the difference reduces considerably, as it is evidenced in the same figure.
This observation is in line with the finding in (13), according to which at high resource utilization a Poisson process does not satisfactorily model the compound arrival process of multiple voice sources .
No reduction effect is detectable for voice traffic whose delay remains practically unaffected, as a demonstration that voice service protection is satisfactorily robust.
Fig. 5 evidences the combined effect of delays in the long distance and in the access network on the 99.th percentile end-to-end delay. The voice end-to-end delay is only very marginally affected by the delay experienced in the long distance network under both low and high utilization options for the access network, and only slightly increases with the load. Anyway, for the high utilization option, voice service requirements are no more met above 90% channel utilization. On the contrary, the 99.th percentile data end-to-end delay increases steeply above 90% utilization, due to the dramatic increase of the related delay in the long distance network.
In the same figure it is also shown both the 99.th percentile long distance network and the 99.th percentile end-to-end delays, for the case in which only data are handled.
As a general observation, the delay contributed by the access network conditions the 99.th percentile end-to-end delay up to channel utilization in the long distance network between 85% and 90%. Beyond that limit the behaviour of the long distance network clearly appears dominant. As the trend exhibited by the voice delay shows , there is practically no room left for improvement of voice handling and hence redundancy reduction would not pay for voice.
In conclusion, voice experiences very low delay and service is well within requirements, except for a slight deviation in the case of high channel utilization in the long distance network combined with the high utilization option for the access network. However this result is obtained at the expense of data traffic which is severely penalized. This situation requires that data traffic be favoured in accessing network resources, for example by adopting less unbalanced servicing probabilities for data and voice.
Figure 6 to 9 refer to a case in which the two traffic types are serviced with the same probability, i.e. .5 .
As for the average delay in the long distance network, Fig. 6, when no voice reduction is operated, results indicate that the difference between voice and data delay reduces to a great extent, with voice delay experiencing a drastic increase relative to the former case , although still below the data delay. Data are also handled somewhat worse for utilization up to

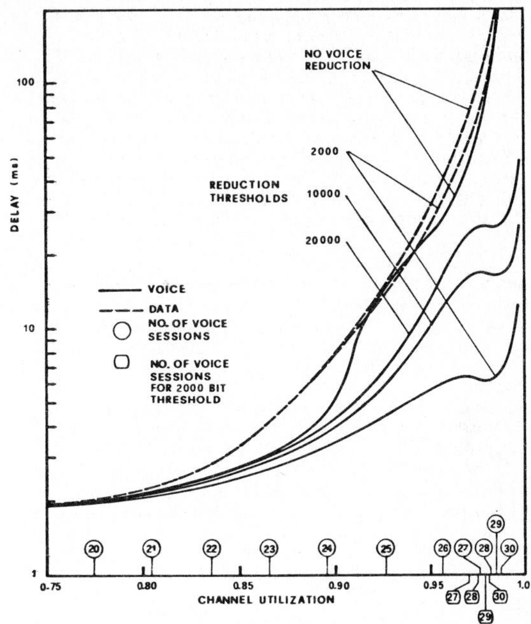

Fig. 6 Average transport delays in the long distance network for .5 voice servicing probability.

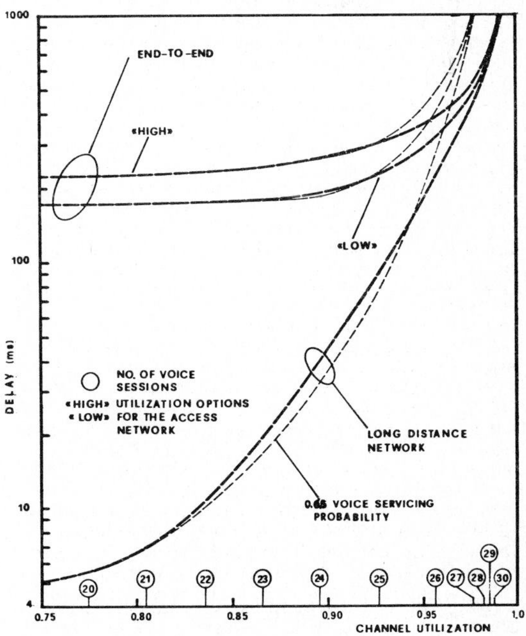

Fig. 7 99.th percentile data transport delays in the long distance network and end-to-end for .5 voice servicing probability.

90% and benefit of some improvement above that value.
The explanation is that parithetically handling voice and data is detrimental to voice because traffic peaks cannot be handled timely enough with the incurred delay increase.

On the other hand, the share of voice traffic is apparently too small to let the data traffic profit of equitable access to network resoureces, at least for moderate utilization.
In this utilization range, data traffic even suffers from voice not leaving the long distance network more quickly. At higher utilization, the load intensity of the two traffic types overrides any other influence and equitable access operates in the intended direction for data. A similar improvement for data traffic is also exhibited by the 99.th percentile end-to-end delay, Fig. 7. In the figure the present case is contrasted with the former one for comparison's sake.
Poor service for voice may be faced in this case by voice redundancy reduction. In Fig. 6 the effect of reducing redundancy as a function of the reduction threshold is shown. As it is apparent, the objective of impacting the voice delay is reached and -according to expectattions- the reduction is more pronounced as the utilization increases and the reduction threshold decreases. For the lower bit reduction threshold (2000), the average delay sensibly narrows the delay shown in the former case.
An indication of the voice redundancy reduction paid for obtaining this result is given in Fig. 8, where the percentage reduction of voice packet depending on reduction threshold is plotted against the channel utilization. In the figure distinction is also made between one-hop and two-hop routed packets. In the worst case, i.e. 2000 bit reduction threshold and two-hop routing, voice packets reduction does not exceed 8%, except for limiting utilization values.

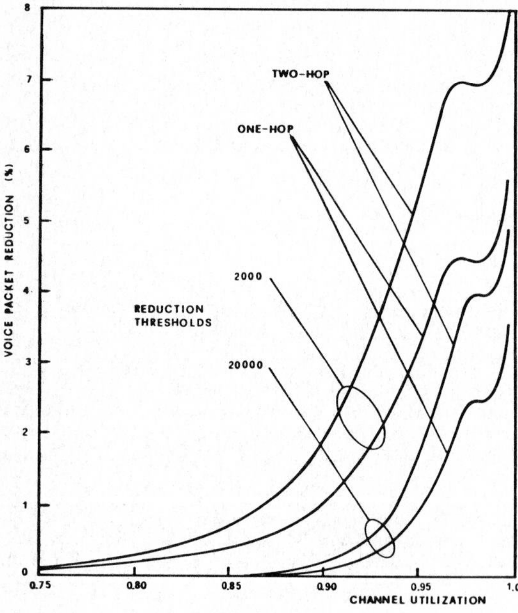

Fig. 8 Percentage voice redundancy reduction for .5 voice servicing probability.

Stated differently, this means that in the worst case about 40 bits per voice packet are lost on the average, which maps into somewhat more than one field containing the less significant bits. It is worthy of noting that, by increasing utilization, two-hop routed packets proportionally lose less redundancy than one-hop routed packets. This suggests that reduction operates almost independently on the first and the second hop for two-hop routed packets at moderate utilization, whereas reduction in the first hop conditions reduction in the second hop at higher utilization. As a consequence , it may be induced that the affectiveness of the algorithm progressively reduces with the number of hops.

Fig. 9 shows the effect of voice reduction on the 99.th percentile delay both in the long distance network and end-to-end. The delay in the long distance network exhbits a trend very similar to that of the average delay. Quite significant appears to be the effect of on the end-to-end delay which drops to values meeting the specified service target up to 95% utilization under the high utilization option for the access network and 2000 bit reduction threshold. Considering the relatively small voice redundancy reduction paid for the improvement obtained for both data and voice traffic, the reduction algorithm seems to perform efficiently.

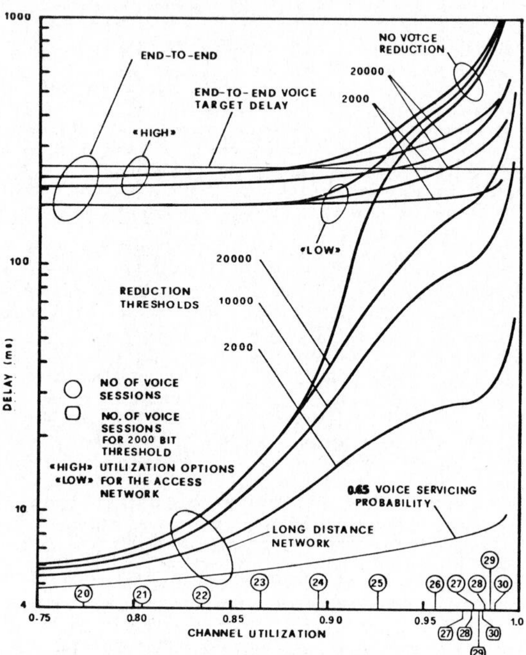

FIG. 9 99.th percentile voice transport delays in the long distance network and end-to-end for .5 voice servicing probability.

Since the positive effects on delays are obtained by combining service probabilities and reduction of redundancy, one could infer that further gain in the end-to-end delay could be achieved by favouring data access to network resources beyond the parithetical situation.
Fig. 10 shows the effect of confining to .3 the probability of service for voice traffic, a proportion which equals the voice share in the traffic mix. As results demonstrate, this setting has disastrous consequences on both voice and data traffic. Two striking effects are evident. Firstly, voice traffic delays increase to such an extent that they overtake data delays, with the latter also increasing considerably with respect to the preceding case . Secondly, due to the vertical increase in the voice delays, the channel utilization is limited to 90%. The results shown relate to 20000 bit reduction threshold. The situation does not improve significantly for lower thresholds.
From this case it can be deduced that bursty voice traffic and unbalance in the voice and data packet sizes prevent data traffic from being strictly favoured in accessing network resources.
Besides delays, another issue is essential in evaluating the effectiveness of combined service priority and voice reduction policy, namely data throughput. Fig. 11 shows the data throughput against the channel utilization for the last two considered cases.

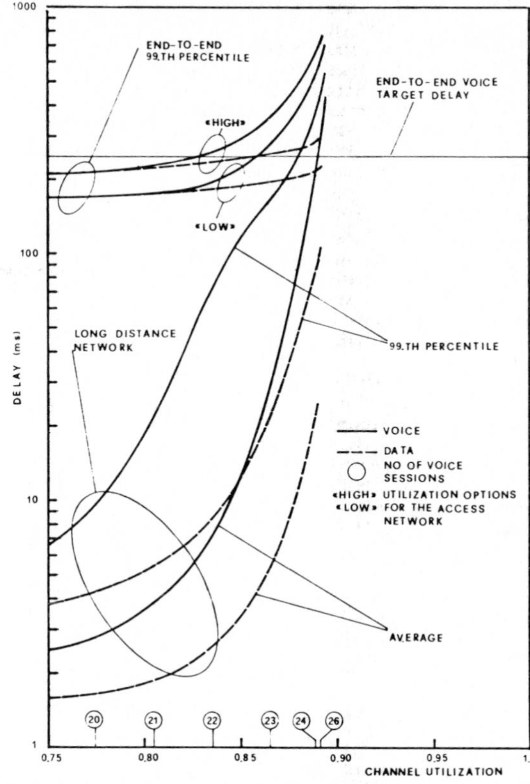

Fig. 10 Average and 99.th percentile transport delays in the long distance network and end-to-end for .3 voice servicing probability.

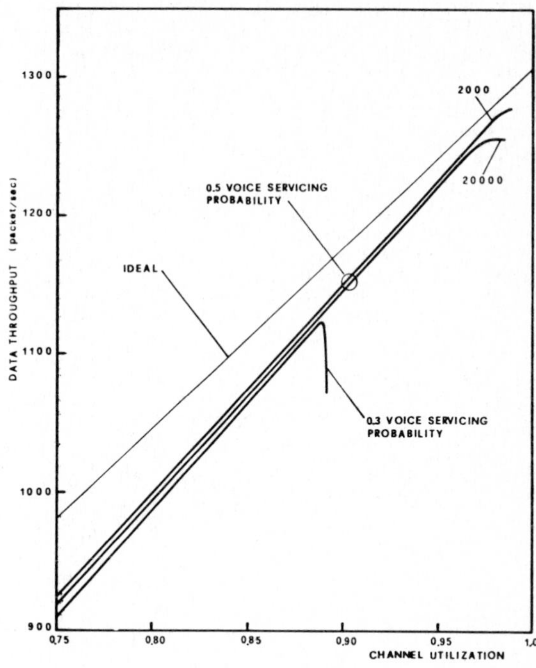

Fig. 11 Data throughput.

Throughput for .5 voice servicing probability linearly increases with utilization, except at almost limiting utilization values, with small impact from backpressure and negligible impact from voice reduction. Quite different is the behaviour for .3 voice servicing probability. In this case, throughput follows linearly with the utilization up to 90% and then drops steeply, due to backpressure. Consequently, strictly favouring data traffic reveals detrimental also with reference to data throughput. In the same figure it is also shown that deviation from ideal throughput characteristics due to bandwidth consumed by axnowledgement traffic is of little entity, independently of the case considered.

5. CONCLUSIONS

A manyfold of factors impact voice and data service in the considered model. The conclusions drawn from the analysis focused on probability of allocating network resources to one traffic in contention cases and reduction of voice redundancy, can be summarized as follows:

- carrying a mixture of voice and data in a packet switched network employing an X.25-derived protocol for voice, requires that delays for the two traffic types be traded off. Voice delay requirements are met if resources are allocated to one traffic according to a probability that is proportional to the packet length for the other traffic . Anyway, data need to be protected against bandwidth capture of voice, due to the bursty nature of related load submission process;
- protection of data traffic can be achieved by increasing the related servicing probability. Positive effects on data traffic obtained by equitably allocating network resources are balanced by service worsening for voice. To face this situation, reduction of voice redundancy through the analyzed algorithm appears effective in meeting specified voice service targets at the expense of a modest bit loss;
- pronounced unbalance in the service probability favouring data traffic are detrimental to both traffic types and reduce bandwidth utilization;
- the effectiveness of the algorithm is due to high responsiveness in levelling traffic peaks, made possible through continuous monitoring of the node output queues and an appropriate setting of the voice redundancy reduction threshold. This suggests that the algorithm is best employed in those situations where the bursty nature of voice traffic is retained;
- the 99.th percentile end-to-end delay appears to be heavily conditioned by the performance of the access network up to sustained channel utilization.

A certain exposure of the algorithm is associated with possible interventions on contiguous packets belonging to the same conversation segment. Related effects on the quality of the signal deserves a separate analysis which is beyond the scope of this paper.

6. REFERENCES

(1) J. W. Forgie, "Network speech implications of packetized speech," Annu. Rep.Defense Commun. Agency, M.I.T. Lincoln Lab., Lexington, MA, DTIC AD-A45455, Sept. 30, 1976

(2) I. Gitman and H. Frank, "Economical analysis of integrated voice and data networks: a case study," Proc. IEEE, vol. 66, no. 11, pp. 1549-1570, 1978

(3) J. S. Turner and L. F. Wyatt, "A packet network architecture for integrated services," Proc. GLOBECOM '83, San Diego, paper no. 2.1, pp. 45-50, 1983

(4) C. J. Weinstein and J. W. Forgie, "Experience with speech communication in packet networks," IEEE Journal on Selected Areas in Communications, vol. SAC - 1, no. 6, pp. 963-980, 1983

(5) W. L. Hoberecht, "A layer network protocol for packet voice and data integration," IEEE Journal on Selected Areas in Communications, vol. SAC - 1, no. 6, pp. 1006-1013, 1983

(6) M. Listanti and F. Villani, "An X.25-compatible protocol for packet voice communication," Computer Communications, vol. 6, no. 1, pp. 23-31, 1983

(7) M. Listanti and F. Villani, "Voice communication handling in X.25 packet switching networks," Proc. GLOBECOM '83, San Diego, paper no. 2.4, pp. 66-70, 1983

(8) T. Bially, B. Gold and S. Seneff, " A technique for adative flow control in integrated packet network," IEEE Trans. Commun., vol. COM - 28, no. 3, pp. 325-333, 1980

(9) D. Grillo, F. Villani, M. Calabrese and R. Pietrojusti, "Impact of low and high bit-rate coding in an integrated packet network," Proc. international Switching Symposium (ISS '84), Firenze, paper no. 42.B.1, 1984

(10) D. J. Goodman, "Embedded DPCM for variable bit rate transmission," IEEE Trans. Commun., vol. COM - 28, no. 7, pp. 1040-1046, 1980

(11) T. E. Stern, "A queueing analysis of packet voice," Proc. GLOBECOM '83, San Diego, paper no. 2.5, pp. 71-76, 1983

(12) B. G. Kim, "Characterization of arrival statistics of multiplexed voice packets," IEEE Journal on Selected Areas in Communications, vol. SAC - 1, no. 6, pp. 1133-1139, 1983

(13) Y. C. Jenq, "Approximations for packetized voice traffic in statistical multiplexer," Proc. IEEE INFOCOM '84

(14) J. P. T. Brady, "A statisical analysis of on-off patterns in 16 conversations," Bell Syst. Tech. J., vol. 47, no. 1, pp. 73-91, 1968

(15) J. G. Gruber and N. H. Le, "Performance requirements for integrated voice/data networks," IEEE Jornal on Selected Areas in Communications, vol. SAC - 1, no. 6, pp. 981-1005, 1983

TELETRAFFIC ISSUES in an Advanced Information Society
ITC-11
Minoru Akiyama (Editor)
Elsevier Science Publishers B.V. (North-Holland)

DESIGN OF SWITCHING NETWORKS COMPOSED OF UNIFORM TIME-SPACE ELEMENTS

Andrzej JAJSZCZYK

Institute of Electronics, Technical University of Poznań
Poznań, Poland

ABSTRACT

In the paper the problem of selecting minimum cost switching networks composed of uniform time-space elements is discussed. The continuous and discrete optimization methods are used. It is shown that nonblocking and rearrangeable switching networks having a regular structure should have as few stages as possible in order to minimize the cost. The dynamic programming technique is adopted for discrete optimization. The relevant formulas are derived and samplings of results of optimization are presented. Different network structures are compared and discussed.

1. INTRODUCTION

The design concepts for time-division switching networks are changing. These changes are due to the introduction of uniform elements which combine time and space switching in the same chip or in the same printed circuit board. In such a uniform element information can be transferred from any channel of any incoming PCM link to any channel of any outgoing PCM link. Owing to physical limitations such as the number of pins of the IC can, chip complexity or power consumption a uniform time-space element can switch a limited number of PCM links, e.g. 4, 8 or 16. In order to obtain networks having a greater capacity than the capacity of a single element we can construct switching networks containing the elements connected in various ways.

The design of multistage networks composed of uniform time-space elements differs in many aspects from the design of space-division switching networks and conventional time and space division systems. The main difference involves limitations of switch size and the measure of cost. It is assumed throughout this paper that a uniform element can switch n incoming and n outgoing PCM links having f channels each. Such an element is denoted by E(n,f). Some links and/or channels may not be used, so n and f represent only the upper limits. Crosspoint minimization does not make any sense when LSI technology is used, so the cost C_N of a network having N incoming and outgoing channels is defined as its total number of time-space switches. A network is called optimal if it has a minimal cost relative to the possible realizable networks having the same number of input/output channels.

We shall restrict ourselves to nonblocking and rearrangeable networks. Such networks are essential for data switching and have been studied [12], or used in practice [4]. The rearrangeable networks described in this paper can be treated as quasinonblocking ones when rearrangement procedures are not employed. Such networks can also offer nonblocking service for some chosen terminals [9].

In Section 2 we present one- and multistage switching networks composed of uniform time-space elements. Both regular and irregular structures are considered. Section 3 is devoted to continuous optimization of some specific networks. In Section 4 the problem of choosing the optimum number of stages is discussed. Relevant theorems are formulated and proven. Section 5 deals with discrete optimization. The optimization method is based on the dynamic programming principle. Some optimal network structures are compared and discussed.

2. NETWORK STRUCTURES

The maximum capacity of a single time-space element is limited by the current level of technology. To obtain larger capacities we can connect more than one element. The simplest method is based on parallel connections between time-space elements, as is shown in Fig.1. Such a network is nonblocking. However, the number of elements needed for networks constructed in such a way grows exponentially as the capacity increases [5].

Using uniform time-space elements we can design multistage networks. A two-stage network can be used in practice. However, such a structure does not make it possible to achieve the nonblocking property [5]. The most popular solution uses a three-stage network as the basic structure [2],[7]. The three-stage network is shown in Fig.2. The structure is similar to that proposed by Clos [3]. The main difference is the multiple linkage connection between pairs of switches in successive stages and that each link can carry more than one channel. It has been shown that such a network is nonblocking in the strict sense if and only if [7]:

$$m \geqslant 2\left\lfloor \frac{u_i f_i - 1}{v f_o} \right\rfloor + 1 ,$$

where the symbol $\lfloor x \rfloor$ denotes the greatest integer less than or equal to x, and rearrangeable if and only if [10]:

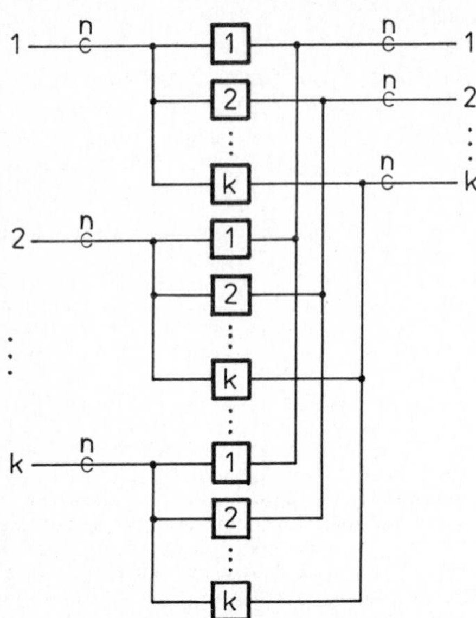

Fig.1. One-stage switching network composed of uniform time-space elements

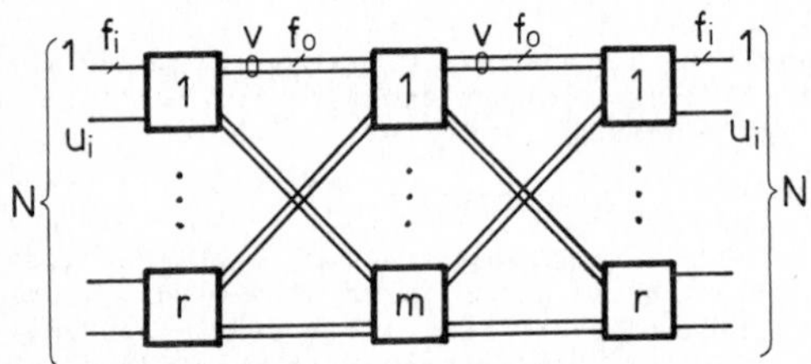

Fig.2. Three-stage switching network. N-number of input/output channels, u-number of PCM links, v-number of PCM links between the same switches, f-number of channels in a PCM link

$$m \geqslant \left\lfloor \frac{u_i f_i - 1}{v f_o} \right\rfloor + 1 \ .$$

Usually, the general s-stage network is obtained by replacing each middle-stage switch in the (s-2)-stage network by a complete three-stage network (see Fig.3). Using this method we can construct five-, seven-, nine-, etc. stage switching networks. We shall refer to these networks as regular networks. The maximum capacity of regular networks is closely related to the number of stages [7], [10]. The relatively large number of stages results in a considerable delay of signals transmitted through the network, as well as requiring complicated path search procedures. To overcome the above-mentioned close relation between the capacity and the number of stages, as well as for savings in

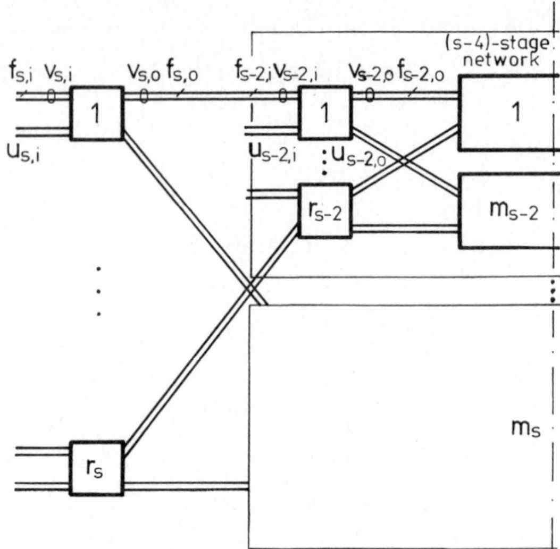

Fig.3. General s-stage network

time-space elements, a different network architecture has been proposed [6]. In this case the basic structure is also formed by the three-stage network. However, in order to obtain larger capacities the switches in one or more stages are replaced by one-stage nonblocking networks. In some cases it is justifiable to replace the switches in networks having more than three stages.

3. CONTINUOUS OPTIMIZATION OF THREE-STAGE NETWORKS

For the three-stage network we have (see Fig.2):

$$C_N = 2r + m \ ,$$

where $r = N/(u_i f_i)$, and m for nonblocking and rearrangeable networks is obtained from formulas of Section 2. It should be noted that the parameters in the formula for C_N are positive integers and that the following inequalities hold: $u_i \leqslant n$, $f_i \leqslant f$, $f_o \leqslant f$, $v \leqslant n/r$, $v \leqslant n/m$.
These constraints are due to the physical limitations of the switching element's size. C_N is an integer function of integer variables. However, this function has been handled as a continuous one [7], [10]. The results may not be entirely correct from a numerical point of view, but do give a good approximation as to the way the various parameters may vary. An example of the cost function for a rearrangeable network with $N = 1024$, $f_i = f_o = f = 32$, and $n = 16$ is presented in Fig.4. It can be observed that the cost reaches a minimum when v is maximal. It is also seen that the products $u_i f_i$ and $v f_o$ can be treated as single variables. When we set the first derivative of the cost function equal to zero for a fixed v, we obtain the optimal number of channels incoming for each first stage switch:

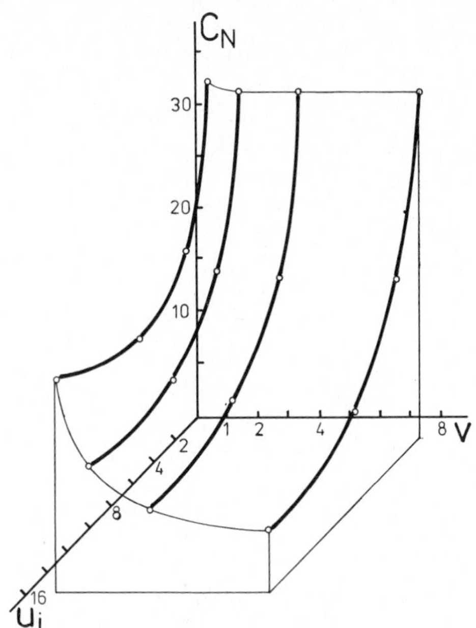

Fig.4. Cost versus u_i and v for a rearrangeable network with N=1024, $f_i=f_o=f=32$, n=16

$$(u_if_i)_{opt}=\begin{cases}\sqrt{Nvf} & \text{for nonblocking networks}\\ \sqrt{2Nvf} & \text{for rearrangeable networks}\end{cases}$$

In the case where $(u_if_i)_{opt}$ is not an integer or is not realizable, the choice of the nearest realizable values has been advised [7]. However, such a procedure may lead to non-optimal results. For example, let us consider the rearrangeable network with N = 1568 composed of elements having n = 8 and f = 32. From our given formula we have:

$$(u_if_i)_{opt} = \sqrt{2Nvf_o} \approx 316.8 ,$$

that is

$$u_{i,opt} \approx 9.9 .$$

So, according to the proposed procedure, the nearest realizable value $u_i = 8$ is chosen. However, the network with $u_i = 8$ contains 22 switches, and the network with $u_i = 7$ contains 21 switches.

Two useful theorems can be formulated.

Theorem 1: Consider a three-stage switching network nonblocking in the strict sense with N incoming and N outgoing channels composed of uniform time-space elements E(n,f). If $n = 2^x$ and $N = 2^y$, where x and y are positive integers, then $(u_if_i)_{opt} \geqslant nf/2$.

Proof: Because $(u_if_i)_{opt}=\sqrt{Nvf}$, we have $\sqrt{Nvf} \geqslant nf/2$. After rearrangement we obtain

$$N \geqslant \frac{n^2f}{4v} .$$

Because the maximum possible value of N is equal to $n^2f/2$, we have $N \geqslant N_{max}/2v$. We shall show that this inequality is always true for $n = 2^x$ and $N = 2^y$. When we set v = 1 we obtain $N \geqslant N_{max}/2$. If for a given network capacity $N < N_{max}/2$ holds, then it is always possible to construct the corresponding network with v = 2; in this particular case, $N \geqslant N_{max}/4$. In the case of $N < N_{max}/4$ we increase v, etc. Theorem 1 is thus proved.

Remark 1. Theorem 1 is also true for $N \neq 2^y$, provided that N has a sufficient number of divisors. So, in the design process we should choose structures having at least nf/2 incoming and outgoing channels per outer stage switch.

Remark 2. Lemma 1 in [7] stating that the maximum capacity of a single first-stage switch in a three-stage nonblocking switching network composed of uniform elements is equal to $\lfloor (n+1)/2 \rfloor f$ is true for v = 1. However, in some cases the actual capacity of a single switch can be greater. For example, for N = 640, n = 16, and f = 32 we have v = 5 and $(u_if_i)_{opt} = 320$.

Theorem 2: Consider a three-stage rearrangeable switching network with N incoming and N outgoing channels composed of uniform time-space elements E(n,f). If $n = 2^x$ and $N = 2^y$, where x and y are positive integers, then $(u_if_i)_{opt} \geqslant nf$.

Proof: Because for a rearrangeable network $(u_if_i)_{opt} = \sqrt{2Nvf}$, we have $\sqrt{2Nvf} \geqslant nf$. After rearrangement we obtain

$$N \geqslant \frac{n^2f}{2v} .$$

Because the maximum possible value of N is equal to n^2f, we have $N \geqslant N_{max}/2v$. This inequality is always true for $n = 2^x$ and $N = 2^y$ by the arguments presented for Theorem 1. Theorem 2 is thus proved.

Remark 3. Theorem 2 is also true for $N \neq 2^y$, provided that N has a sufficient number of divisors. This means that we should construct three-stage networks in which the outer stage switches are used to their full capacity. However, it should be noted that for some values of N, especially those having a small number of divisors, the number of channels per switch is less than nf (see the above example for N = 1568).

Remark 4. Because the cost of a rearrangeable network reaches a minimum for $v = v_{max}$, the optimal network should have r equal to m, if this is possible.

For networks composed of more than three stages a continuous optimization can also be employed, although the mathematical expressions become more complex [7].

4. THE SEARCH FOR THE OPTIMAL NUMBER OF STAGES

It is clear that the cost of a network depends on the number of switching stages. In space-division networks the increase in the number of stages when the network grows beyond a certain point results in crosspoint savings. The situation is quite different for regular networks composed of uniform time-space elements. In such networks the main reason for implementing a given number of stages is the close dependence between the capacity of the network and the number of stages. It was shown elsewhere that for a given N it is impossible to construct a network which has any number of stages [7], [10]. However, only the minimal number of stages is limited.

The symbol $M(N,s)$ is used to denote the class of all regular switching networks having N incoming and N outgoing channels, composed of s stages. $C(\nu)$ denotes the cost of the network ν. $C_{min}(N,s) = \min\{C(\nu): \nu \in M(N,s)\}$. The following theorem has been formulated and proven [7].

Theorem 3: Consider regular networks nonblocking in the strict sense having the same capacity N, and composed of uniform time-space elements. If $s < s'$ then $C_{min}(N,s) < C_{min}(N,s')$.

In other words, Theorem 3 can be formulated as follows. Let us suppose that for a given N it is possible to construct two or more regular nonblocking networks which have a different number of stages s and s' ($s' > s$). These networks are composed of identical time-space elements. Then it is always possible to construct a network with s stages, which contains fewer time-space elements than any network with s' stages.

A similar theorem can be formulated for rearrangeable networks.

Theorem 4: Consider regular rearrangeable networks having the same capacity N, and composed of uniform time-space elements. If $s < s'$ then $C_{min}(N,s) < C_{min}(N,s')$.

Proof: We shall prove the theorem for s = 3 and s' = 5, and then the proof can be generalized for any s and s'.

For a five-stage network we shall denote $v_{5,o} = v'$ and for a three-stage network $v_{3,o} = v$ (see Fig.3). The theorem will first be proved for $v' = v$, and then for $v' \neq v$. The costs of five-stage and three-stage networks are denoted by C' and C, respectively. The letters having (') designate the parameters of a five-stage network. The switch obtained by replacing a single middle-stage switch by the three-stage network will be called the "major" switch.

We shall consider three cases.

Case 1: $r_3 = r_5'$. For a given N and for $r_3 = r_5'$ we have $u_{3,i}f_{3,i} = u'_{5,i}f'_{5,i}$. Hence, $m_3 = m_5'$. Each "major" switch in a five-stage networks contains more than one uniform time-space element. Thus $C' > C$.

Case 2: $r_3 > r_5'$. For this case, it is always possible to construct a three-stage network with $r_3 = r_5'$. Hence, from Case 1, $C' > C$.

Case 3: $r_3 < r_5'$. We shall consider the three-stage network having a maximal capacity N_{max}. For such a network $v = 1$, $N_{max} = n^2f$, and $(u_{3,i}f_{3,i})_{max} = nf$. From Theorem 2 we have

$$(u_{3,i}f_{3,i})_{max} < (u_{3,i}f_{3,i})_{opt} .$$

After rearrangement we obtain: $r_{3,max} > r_{3,opt}$

In Case 3 we have $r_5' > r_{3,max}$. This means that as r_5' increases, the network falls further from optimality. Thus, the five-stage network has more switches (counting only "major" and outer stage switches) than the minimum cost three-stage network having the same capacity.

This result is also true for networks with $N < N_{max}$. This is due to the fact that in the five-stage network r_5' is greater than $r_{3,max} = n$ (if not, the problem can be solved as for Cases 1 or 2). With decreasing N, $r_{3,opt}$ is also decreasing, and, hence, the inequality $r_{3,max} > r_{3,opt}$ is true. This completes the proof of the theorem for $v' = v$.

We shall examine the influence of v' in a five-stage network for the increase of v' in a five-stage network for the cost of this network. It can be observed that the cost of a three-stage network is greater than $2\lceil N/n \rceil$, i.e. the minimal cost of its outer stages ($\lceil x \rceil$ denotes the smallest integer greater than or equal to x). Let us suppose that in the five-stage network v' is increased x times. This results in a decrease of the number of "major" switches by x times, and at the same time it results in an increase of their size by x times, as well. Hence, the following inequality holds for the cost of a single "major" switch:

$$C_3' > 2\lceil xN_3'/n \rceil,$$

where N_3' is the number of incoming channels for a "major" switch before the increase of v'. It can be seen that N_3' is at least equal to n/2. In the opposite case, it would be possible to increase v in the three-stage network. Thus, $C_3' > 2\lceil x/2 \rceil$. Hence, $C_3' > x$.

This means that if the number of "major" switches is decreased x times then the cost of one such switch increases more than x times. So, the cost of a five-stage network is always greater than the cost of an optimal three-stage network having the same number of incoming and outgoing channels. This completes the proof for s = 3 and s' = 5. The theorem is also true for any s because of the iterative structure of multistage networks. Theorem 4 is thus proved.

5. DISCRETE OPTIMIZATION

5.1. Derivation of Dynamic Programming Formulas

As was shown in Section 3, continuous optimization, and the rounding connected with it, can be a source of errors. However, the synthesis problem of selecting a network having the minimum cost can be solved using discrete optimization. An original method, based on a combinatorial argument has been developed for space-division rearrangeable switching networks [1],[8]. However, this method is not applicable for networks composed of uniform time-space elements.

Because of the iterative structure of multistage networks optimization will be treated as a multistage decision problem and will be solved by using the dynamic programming method. Dynamic programming is based on Bellman's optimality principle: an optimal decision policy has the property that whatever the initial state of the system and the initial decision are, the rest of the decisions must form an optimal policy according to the state produced by the first decision.

A simplified s-stage nonblocking network with $f_{k,i} = f_{k,o} = v_{k,i} = v_{k,o} = 1$ $(k = 3,5,\ldots,s)$ will be used to illustrate the application of dynamic programming to optimization of switching networks. In this case the s-stage network is defined by the set of parameters $u_{k,i}$, r_k, m_k; and the condition under which the network is nonblocking is reduced to the following formula:

$$m_k \geqslant 2u_{k,i} - 1, \quad \text{for } k = 3,5,\ldots,s.$$

If we assume that $m_k = 2u_{k,i} - 1$, the optimization problem may be formulated as follows:

$$\min\{2r_s+(2u_{s,i}-1)[2r_{s-2}+(2u_{s-2,i}-1)[\ldots[2r_3+(2u_{3,i}-1)]\ldots]]\};$$

$u_{s,i}r_s \geqslant N$;

$u_{k,i}r_k \geqslant r_{k+2}$, for $k = 3,5,\ldots,s-2$;

$u_{k,i} \leqslant n$, $r_k \leqslant n$, for $k = 3,5,\ldots,s$;

$u_{k,i}$, r_k integer, for $k = 3,5,\ldots,s$.

Using the optimality principle the problem may be reformulated [11]. Let us denote by x_t the number of inlets for a t-stage network. $T(x)$ is the set of decisions (u_i,r) permitted by x, where

$x_t = r_{t+2}$, for $t = 3,5,\ldots,s-2$;

$T(x) = \{(u_i,r) \mid 0<u_i\leqslant n,\ 0<r\leqslant n,\ 2u_i-1\leqslant n,\ u_ir\geqslant x\}$.

The optimization problem is now embedded in a set of problems:

$$\min\{2r_t+(2u_{t,i}-1)[2r_{t-2,i}+(2u_{t-2,i}-1)[\ldots[2r_3+(2u_{3,i}-1)]\ldots]]\},\ \text{for } t = 3,5,\ldots,s;$$

$u_{s,i}r_s \geqslant N$,

$u_{k,i}r_k \geqslant x_k$,

$u_{k,i} \leqslant n$, $r_k \leqslant n$ integer

$x_k = r_{k+2}$,

for $k = 3,5,\ldots,t$.

The optimum of this problem is denoted by: $f_t(x)$, for $0<x\leqslant N$, $t = 3,5,\ldots,s$.

The function $f_3(x)$ can be easily determined for all $0<x\leqslant N$ from the definition:

$$f_3(x) = \min\{(2r+2u_i-1) \mid (u_i,r)\in T(x)\}.$$

To determine the function $f_5(x)$ describing the minimal cost of the five-stage network, the principle of optimality is used. It should be remembered that no matter what the first decision (u_i,r) is, the continuation must be optimal, according to the value of r obtained during the optimization of three-stage networks. The value r denotes the number of inlets for each three-stage network which forms a "major" switch for a five-stage network. The best solution for the five-stage network, therefore, can be found by comparing the total costs resulting from the different decisions (u_i,r) and from their optimal continuations:

$$f_5(x) = \min\{[2r+(2u_i-1)f_3(r)] \mid (u_i,r)\in T(x)\},$$

for $0 < x\leqslant N$. Similarly, after determining $f_{t-2}(x)$, we obtain

$$f_t(x) = \min\{[2r+(2u_i-1)f_{t-2}(r)] \mid (u_i,r)\in T(x)\},$$

for $0 < x \leqslant N$.

5.2. General Optimization Formulas

In Section 5.1 the expressions for optimization of simplified multistage networks were derived. Now, the general formulas will be presented.

For regular nonblocking networks composed of uniform time-space switches we have:

$$f_t(x) = \min\left\{2\left\lceil\frac{x}{u_if_i}\right\rceil+\left(2\left\lceil\frac{u_if_i-1}{vf_o}\right\rceil+1\right)f_{t-2}\left(\left\lceil\frac{x}{u_if_i}\right\rceil vf_o\right)\right\}$$

where $t = 3,5,\ldots,s$,

$u_i\leqslant n$, $f_i\leqslant f$, $f_o\leqslant f$, $v\left(2\left\lceil\frac{u_if_i-1}{vf_o}\right\rceil+1\right)\leqslant n$, and $\left\lceil\frac{x}{u_if_i}\right\rceil vf_o \leqslant N_{t-2,\max}$.

It has been shown that for regular networks nonblocking in the strict sense the following holds [7]:

$$N_{s,\max} = \left(\left\lfloor\frac{n+1}{2}\right\rfloor\right)^{\frac{s-1}{2}} fn\ .$$

It should be noted that by Theorem 3 the lower limit of x is also known.

Similarly, for regular rearrangeable networks we have:

$$f_t(x) = \min\left\{2\left\lceil\frac{x}{u_if_i}\right\rceil+\left(\left\lceil\frac{u_if_i-1}{vf_o}\right\rceil+1\right)f_{t-2}\left(\left\lceil\frac{x}{u_if_i}\right\rceil vf_o\right)\right\},$$

where $t = 3,5,\ldots,s$,

$u_i\leqslant n$, $f_i\leqslant f$, $f_o\leqslant f$, $v\left(\left\lceil\frac{u_if_i-1}{vf_o}\right\rceil+1\right)\leqslant n$, and $\left\lceil\frac{x}{u_if_i}\right\rceil vf_o \leqslant N_{t-2,\max}$.

For regular rearrangeable networks [10]:

$$N_{s,\max} = n^{\frac{s+1}{2}} f\ .$$

The lower limit of x results from Theorem 4.

The optimal solution for a given t is found by using enumeration methods. The essence of these methods is to enumerate all solutions, i.e. all integer vectors in the given range, and to choose according to the objective function the best one from among those satisfying the constraints of the problem. Enumeration can be used for our problem because a high proportion of solutions is implicitly examined, i.e. excluded in large groups without direct checking, as they are shown to violate the set of constraints. It should be noted that we find the parameters of a t-stage network using a table of the optimal solutions for (t-2)-stage networks.

Formulas analogous to those presented above can also be derived for the irregular networks described in Section 2.

5.3. Numerical Examples

The formulas presented in Section 5.2 have formed the basis for a set of computer programs which have been used for the calculation of extensive tables of nonblocking and rearrangeable networks composed of uniform time-space elements of various sizes. The programs have been written in FORTRAN and run on an ODRA 1305 computer (compatible with an ICL 1900). Each table enabling us to choose the optimum structure of a network having a capacity up to 100 000 channels was calculated using a few minutes of processing time.

We shall discuss some results of optimization. Fig.5 and Fig.6 show the dependence between the cost, expressed by the number of time-space elements, and the capacity of optimum regular nonblocking and rearrangeable networks composed of elements of two different sizes. It is seen that it is possible to obtain very large capacities by using a relatively small number of uniform time-space elements. Moreover, for the same capacity, rearrangeable networks require significantly fewer elements than nonblocking ones. It is also seen that the increase of the number of stages causes a significant "jump" of the cost. It should be noted that the continuous lines between the jumps approximate only discrete values of the cost. However, these discrete changes are significantly lower than the jumps presented in the figures.

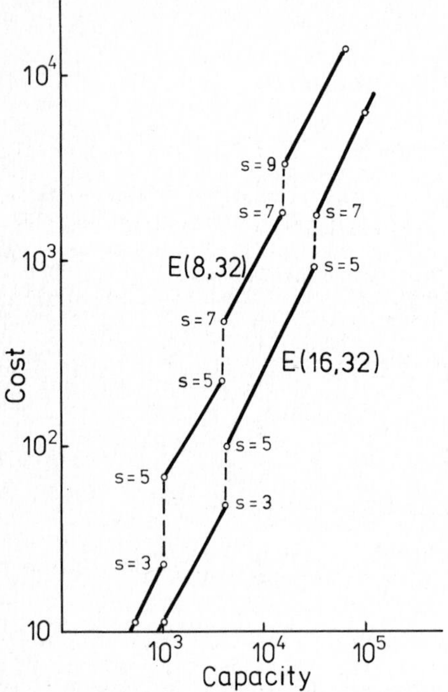

Fig.5. Cost versus capacity for regular nonblocking networks

In Tables 1 and 2 the parameters of chosen optimal regular networks are shown. Table 1 shows the results of optimization of regular nonblocking networks composed of elements E(8,32). It can be seen that for most networks we have $u_{k,i} = n/2$, where $k = 3,5,\ldots s$. In our case $u_{k,i} = 4$. This result is consistent with Theorem 1 presented in Section 3. However, in some cases $u_{k,i} < n/2$; for example, for N = 1056 we have $u_{5,i} = 3$. This effect is due to the relatively small number of divisors of N. It should be noted that for some values of N some of the incoming and outgoing channels remain unused. For example, the network having N = 520 and the network having N = 640 have the same structure, but the first of these networks has 120 unused channels. Except for the presented case, for optimal networks we have $f_{k,i} = f_{k,o} = f = 32$ where $k = 3,5,\ldots,s$.

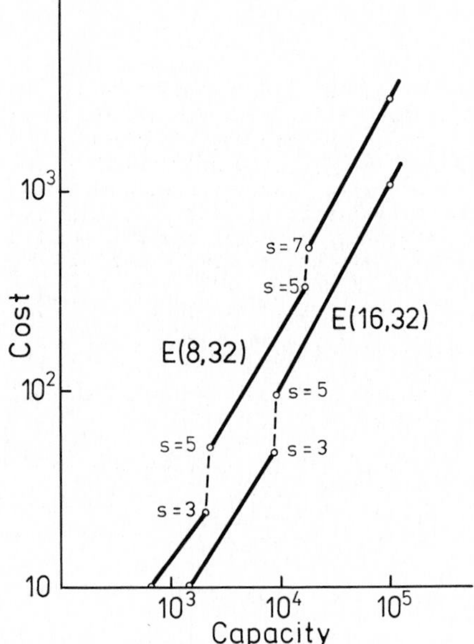

Fig.6. Cost versus capacity for regular rearrangeable networks

It is seen that for the network having

Table 1. Sample results for regular nonblocking networks composed of uniform elements E(8,32)

N	s	$u_{9,i}$	$v_{9,o}$	r_9	$u_{7,i}$	$v_{7,o}$	r_7	$u_{5,i}$	$v_{5,o}$	r_5	$u_{3,i}$	$v_{3,o}$	r_3	C
512	3	-	-	-	-	-	-	-	-	-	4	2	4	11
520	3	-	-	-	-	-	-	-	-	-	4	1	5	17
576	3	-	-	-	-	-	-	-	-	-	4	1	5	17
"	3	-	-	-	-	-	-	-	-	-	3	1	6	17
640	3	-	-	-	-	-	-	-	-	-	4	1	5	17
1024	3	-	-	-	-	-	-	-	-	-	4	1	8	23
1056	5	-	-	-	-	-	-	3	1	11	4	2	3	67
4096	5	-	-	-	-	-	-	4	1	32	4	1	8	225
8192	7	-	-	-	4	2	64	4	1	32	4	1	8	803
49152	9	3	1	512	4	1	128	4	1	32	4	1	8	10179
65536	9	4	1	512	4	1	128	4	1	32	4	1	8	13841

Table 2. Sample results for regular rearrangeable networks composed of uniform elements E(8,32)

N	s	$u_{7,i}$	$v_{7,o}$	r_7	$u_{5,i}$	$v_{5,o}$	r_5	$u_{3,i}$	$v_{3,o}$	r_3	C
512	3	-	-	-	-	-	-	8	4	2	6
528	3	-	-	-	-	-	-	6	2	3	9
1024	3	-	-	-	-	-	-	8	2	4	12
4096	5	-	-	-	8	4	16	8	1	8	80
"	5	-	-	-	8	2	16	8	2	4	80
"	5	-	-	-	8	1	16	8	4	2	80
9088	5	-	-	-	7	1	41	7	1	6	215
65536	7	8	2	256	8	1	64	8	1	8	1792
"	7	8	1	256	8	2	32	8	1	8	1792
"	7	8	1	256	8	1	32	8	2	4	1792

N = 512 we have $u_{3,i}f_{3,i}$ = 128, and $u_{3,o}f_{3,o}$= 192. Thus, $u_{3,o}f_{3,o} < 2u_{3,i}f_{3,i} - 1$. This result illustrates the difference between nonblocking networks composed of uniform time-space elements and conventional nonblocking space-division networks, where $m \geqslant 2u - 1$. For some values of N there is more than one optimal structure (see, for example, N = 576).

Table 2 shows a sampling of results for regular rearrangeable networks composed of elements E(8,32). It is clear that for $N = 2^x$, where x is a positive integer, we have $u_{k,i}$ = 8, as was proven by Theorem 2. However, for N = 528 and N = 9088, $u_{k,i} \neq 8$. From Tables 1 and 2 we can see that path rearranging results in significant savings of hardware in comparison to strictly nonblocking networks. For example, the regular rearrangeable network having N = 65 536 requires about 8 times fewer uniform elements than the regular nonblocking network.

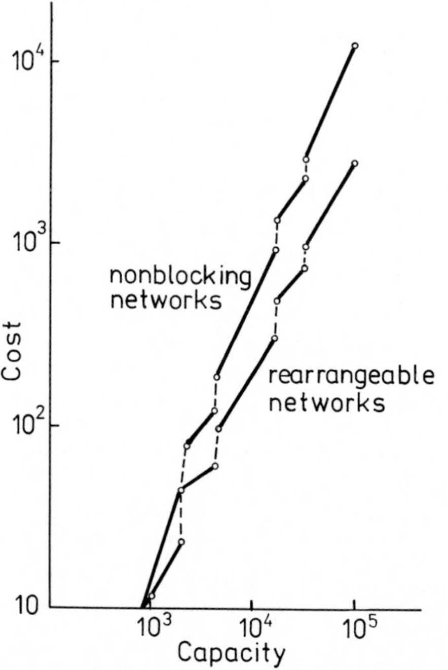

Fig.7. Cost versus capacity for irregular networks composed of elements E(8,32)

Figure 7 presents the dependence of cost on the capacity of irregular nonblocking and rearrangeable networks composed of elements E(8,32). It is seen that the jumps related to small increases of the capacity are smaller than for regular networks. Some characteristic features of irregular networks can be observed in Tables 3 and 4, which contain the parameters of chosen optimal networks. The notation used in these tables should be noted. If, for example, the value of the parameter $u_{1,i}$ is greater than n (i.e. the maximum capacity of one time-space element), then the middle stage of the switching network contains one-stage networks obtained by "parallel" connections of elements. Similarly, $u_{3,i}$ = 8 in Table 3 means that the stages adjacent to the middle stage are composed of modules, each containing two time-space elements.

The analysis of the tables of optimal irregular networks leads to the following conclusions. For such networks Theorems 3 and 4 do not hold. This means that sometimes the network having a greater number of stages contains fewer elements than the network of the same capacity and a smaller number of stages. The outer stages in multistage nonblocking networks usually consist of modules containing two time-space elements each. However, there are also exceptions, e.g. one version of the network having N = 16 896. For nonblocking irregular networks, in almost all cases, we have v = 1. In rearrangeable networks, modules containing parallelly connected switches are only present in the middle stage. In many cases, optimal rearrangeable networks have regular structures, i.e. the introduction of one stage modules does not lower the cost. Usually, an optimal rearrangeable network does not contain more stages than a nonblocking network of the same capacity. However, for example, for N = 65536 the situation is different.

6. CONCLUSION

Recent advances in semiconductor technology make it possible to construct switching networks using uniform time-space elements. The appropriate selection of structural parameters has a significant influence on the network's cost. In this paper two methods

Table 3. Sample results for irregular nonblocking networks composed of uniform elements E(8,32)

N	s	$u_{5,i}$	$v_{5,o}$	r_5	$u_{3,i}$	$v_{3,o}$	r_3	$u_{1,i}$	C
512	1	-	-	-	-	-	-	16	4
1536	1	-	-	-	-	-	-	48	36
2048	3	-	-	-	8	1	8	8	47
2560	3	-	-	-	8	1	10	10	100
"	3	-	-	-	7	1	12	12	100
"	3	-	-	-	6	1	14	14	100
"	3	-	-	-	5	1	16	16	100
"	1	-	-	-	-	-	-	80	100
4096	3	-	-	-	8	1	16	16	124
16896	5	6	2	88	8	1	22	22	1467
"	5	3	1	176	8	1	22	22	1467
65536	5	8	1	256	8	1	32	32	6544

Table 4. Sample results for irregular rearrangeable networks composed of uniform elements E(8,32)

N	s	$u_{7,i}$	$v_{7,o}$	r_7	$u_{5,i}$	$v_{5,o}$	r_5	$u_{3,i}$	$v_{3,o}$	r_3	$u_{1,i}$	C
512	1	-	-	-	-	-	-	-	-	-	16	4
576	3	-	-	-	-	-	-	6	2	3	6	9
"	1	-	-	-	-	-	-	-	-	-	18	9
768	1	-	-	-	-	-	-	-	-	-	24	9
2048	3	-	-	-	-	-	-	8	1	8	8	24
4096	3	-	-	-	-	-	-	8	1	16	16	64
16896	5	-	-	-	8	1	66	6	1	11	11	500
"	5	-	-	-	6	1	88	8	1	11	11	500
"	5	-	-	-	6	1	88	7	1	13	13	500
"	5	-	-	-	6	1	88	6	1	15	15	500
65536	7	8	2	256	8	1	64	8	1	8	8	1792
"	7	8	1	256	8	2	32	8	1	8	8	1792
"	7	8	1	256	8	1	32	8	2	4	8	1792

of optimization, i.e. the continuous and the discrete methods, have been presented. Some important results of optimization for nonblocking and rearrangeable networks are:

- regular networks should have as few stages as possible,
- in most practical cases the number of incoming channels per switch should be as great as possible,
- in some cases it is justifiable and/or necessary to construct networks having more incoming and outgoing channels than actually needed,
- irregular networks are, in many cases, more economical than regular ones; they often contain fewer switching stages,
- rearrangeable networks require significantly less hardware than networks nonblocking in the strict sense.

The last result shows that it may be economical to use more complicated software than for nonblocking networks in order to reduce the hardware. It can be seen that if rearrangement procedures are not used rearrangeable networks are quasinonblocking, i.e. the probability of blocking is comparable with the failure rate [2],[6]. This is due to the multiplicity of PCM links connecting switches at different stages. Thus, we can use structures obtained during optimization of rearrangeable networks as quasinonblocking ones. By applying simple control procedures these networks can offer nonblocking services for chosen terminals [9]. Such a solution can be of practical importance in ISDNs.

REFERENCES

[1] V.E.Beneš, "Mathematical Theory of Connecting Networks and Telephone Traffic", Academic Press, New York, 1965.

[2] P.Charransol, J.Hauri, C.Athènes and D.Hardy, "Development of a time division switching network usable in a very large range of capacities", IEEE Trans.Commun. vol.COM-27, no.7, pp. 982-988, 1979.

[3] C.Clos, "A study of non-blocking switching networks", Bell Syst. Tech. J., vol. 32, no.2, pp. 406-424, 1953.

[4] E.Hopner and M.A.Patten, "The digital data exchange - a space-division switching system", IBM J. Res. Develop., vol. 28, no.4, pp. 444-453, 1984.

[5] A.Jajszczyk, "Design concepts for switching networks composed of uniform time-space elements", Proc. 29th Int. Scientific Colloquium, Ilmenau, pp. 175-178, 1984.

[6] A.Jajszczyk, "Novel architecture for a digital switching network", Electronics Letters, vol.20, no.17, p. 683, 1984.

[7] A.Jajszczyk, "On nonblocking switching networks composed of digital symmetrical matrices", IEEE Trans. Commun., vol.COM-31, no.1, pp. 2-9, 1983.

[8] A.Jajszczyk, "Optimal structures of Benes' switching networks", IEEE Trans. Commun., vol. COM-27, no.2, pp. 433-437, 1979.

[9] A.Jajszczyk, "Priority rearrangements", Proc. 10th Int. Teletraffic Congress, Montreal, pp. 5.1/5/1-5.1/5/8, 1983.

[10] A.Jajszczyk, "Rearrangeable switching networks composed of digital symmetrical matrices", Proc. IEEE Mediterranean Electrotech. Conf., Athens, paper A4.11, 1983.

[11] L.B.Kovács, "Combinatorial Methods of Discrete Programming", Akademiai Kiado, Budapest, 1980.

[12] P.Odlyzko and S.Das, "Non-blocking rearrangeable networks with distributed control", Proc. Int. Conf. Commun., Amsterdam, pp. 294-298, 1984.

TELETRAFFIC ISSUES in an Advanced Information Society
ITC-11
Minoru Akiyama (Editor)
Elsevier Science Publishers B.V. (North-Holland)

STUDY OF BLOCKING FOR MULTISLOT CONNECTIONS IN DIGITAL LINK SYSTEMS

Georges FICHE (*), Pierre LE GALL (**), and Salvatore RICUPERO (*)

(*) CIT ALCATEL, LANNION, FRANCE
(**) CNET-DICET, FRENCH PTT, PARIS, FRANCE

ABSTRACT

A calculational method is presented for evaluating blocking probabilities for multislot calls in the time-division multistage switching networks of digital exchanges, which will be required to handle such calls in the future ISDN. The study shows the importance of the global peakedness factor resulting from mixing different classes of traffic within the same switching network. It explains how this factor can be used to define a reduced networks carrying equivalent average calls constituting ordinary (single-slot) Poisson traffic flows. And it demonstrates the increased importance of the stochastic dependence between switching stages in the case of multislot calls. Lastly, it is shown how to apply the reduced network method to evaluate, in a simple manner, the real blocking probabilities for the various classes of traffic.

1 - INTRODUCTION

Performance evaluation [1] as well as surveys on the grade of service [2] are more and more essential for large modern telephone exchanges offering more and more complex and varied facilities.

In the planned integrated services digital network (ISDN), digital exchanges will be required to establish MULTISLOT CALLS alongside ordinary single-slot calls.

Studies on such mixed traffic streams offered to a single group of circuits were performed some 20 years ago [3], and an algorithm for simplifying the numerical calculations was described recently by Roberts [4].

Mixed traffic handling in a multistage switching network (link system) with end-to-end path finding has also been analyzed, notably in [5].

This said, the calculational methods proposed to date do not permit simple evaluation of the impact of mixed traffic on blocking. Moreover, they assume stochastic independence between the various switching stages, an assumption first made by Jacobaeus [6].

In this paper, we show how the PEAKEDNESS FACTOR can be used to model the effects of traffic mixing ; and how it is possible to take into account the DEPENDENCIES between stages due to the existence of multislot connections and to the structures of large digital switching networks. The basic idea is to simplify analysis to that of ordinary (single-slot) Poisson traffic flows, obviating the need for special algorithms.
For this, we introduce the concept of EQUIVALENT AVERAGE CALLS carried by a network having a reduced number of paths, termed the REDUCED NETWORK, but without altering the offered traffic loads. It is then explained how the blocking probabilities in the real network can be simply calculated.

For preliminary calculations relative to the reduced network, use is made of the general method described in [8] and [9], which is valid for all switching networks.

After eliminating the effects of multislot connections on interstage dependencies, by translating the problem to the reduced network, we analyze the specific effects of the structures of digital switching networks, which tend to be large, have a small number of stages, and be heavily loaded.

Our results are also applicable to digital exchanges carrying only single-slot calls. In respecting grade-of-service specifications, it is necessary to consider the least-favorable case of "local" calls between lines or circuits connected at the same switch elements (e.g. at the same input and output switches of a TST network). Within the latter, each such local call occupies two time slots, even though it is an ordinary single-slot call.

2 - SINGLE CIRCUIT GROUP

2.1 - Basic assumptions and notation

We consider a group of N circuits offered x Poisson traffic streams labelled by an index (i = 1, 2, ..., x). For the ith stream, the offered traffic intensity is denoted a_i and is expressed as a number of calls ; the number of channels (time slots) used for each call is d_i; and the average call duration is equal to time unit. The latter duration is the same for all classes of traffic (any differences between classes would have no influence because each stream is Poissonian and can conserve its own time scale). Also, stationary conditions are assumed, meaning that the distribution of holding times has little importance, and the model is the lost call model.

The total offered traffic (expressed as a number of circuits) is :

$$M = \sum_{i=1}^{x} a_i d_i \qquad (1)$$

Its variance is :

$$V = \sum_{i=1}^{x} a_i d_i^2 \qquad (2)$$

And its peakedness factor is :

$$Z = V/M \qquad (3)$$

We denote by $P_k(N;M)$ the probabilities of finding k circuits occupied, and by B_i the call congestion for the ith traffic stream.

2.2 - Occupancy distribution function

J. Roberts [4] has given the following algorithm for numerically calculating probability P_k:

$$k.P_k = \sum_{i=1}^{x} a_i d_i P_{k-d_i} \qquad (4)$$

with : $P_{k-d_i} = 0$ if $k < d_i$.

P_0 is determined by the normalization formula below :

$$\sum_{k=0}^{N} P_k = 1$$

It can then be shown that ;

$$B_i = \sum_{k=N-d_i+1}^{N} P_k \qquad (5)$$

2.3 - Equivalent average calls and reduced circuit groups

Because all traffic is supposed Poissonian, the traffic mixing distribution takes a product form. To simplify, we can assume the circuit group to be of infinite size. This only change the normalization factor in the expression of $P_k(N;M)$, which becomes $P_k(\infty;M)$. We then have:

$$\sum_{k=0}^{\infty} P_k u^k = \prod_{i=1}^{x} e^{a_i(u^{d_i}-1)} = e^{\Phi(u)} \qquad (6)$$

Where :

$$\Phi_{(u)} = \sum_{i=1}^{x} a_i\left[(1+u-1)^{d_i}-1\right] =$$

$$\sum_{i=1}^{x} a_i\left[d_i(u-1)+d_i\frac{(d_i-1)}{2}(u-1)^2+\ldots\right]$$

Expanding the expression around u = 1.

We now introduce the global traffic cumulants (the first two of which are M and V) :

$$C_n = \sum_{i=1}^{x} a_i (d_i)^n \qquad (7)$$

It can be shown that :

$$e^{\Phi(u)} = e^{\Phi_0(u)} . e^{\Psi(u)} \qquad (8)$$

where :

$$\Phi_0(u) = \frac{M}{Z}(U^Z-1), \Psi(u) = \left[C_3 - \frac{V^2}{M}\right]\frac{(u-1)^3}{3!} + O\left[(u-1)^3\right] \qquad (9)$$

when u tends to 1.

The first two moments of $e^{\Psi(u)}$ are zero and hence the associated random number usually has a low value, oscillating around the origin. From (8) and (9), the circuit group can therefore be considered as virtually occupied by a single class of traffic made up of EQUIVALENT AVERAGE CALLS each occupying Z channels (with Z an integer).

We denote :

$$M_0 = M/Z , N_0 = N/Z \qquad (10)$$

defining a "REDUCED CIRCUIT GROUP" carrying equivalent average calls.

For equivalent average calls, the probability of finding $K = K_0.Z$ circuits occupied (with K_0 integer) is :

$$P_{k_0}(N_0;M_0) = \frac{\dfrac{M_0^{K_0}}{K_0!}}{1 + \dfrac{M_0}{1!} + \ldots \dfrac{M_0^{N_0}}{N_0!}} \qquad (11)$$

In time interval dt, the probability of arrival of a call finding this occupancy state is :

$$(M_0 dt).P_{k_0}$$

This is also the value found by an external observer paying no attention to the number of channels taken for each call, and for whom the offered traffic appears equal (in intensity and number of circuits) to M. For this observer, the preceding expression should be written :

$$(M.dt)(\frac{1}{Z} P_{k_0})$$

In other words, for integer K_0 :

$$P_k \simeq \frac{1}{Z} P_{k_0} \qquad (12)$$

Now, notice that $e^{\Phi_0(u)}$ defines a step function in which, for integer Z (>0) and N a multiple of Z :

$$P_{N-(y-1)Z} = P_{N-(y-1)Z-1} = \ldots. = P_{N-(y-1)Z-(Z-1)}$$

The purpose of function $e^{\Psi(u)}$ is to smooth the distribution with respect to $e^{\Phi(u)}$ without changing (12) for integer K_0. We can write with $n < Z$:

$$P_{N-(y-1)Z-n} \simeq P_{N-(y-1)Z}\,(K_y)^n$$

For n = Z (integer), this expression must be equal to $P_{N-yZ} = P_{N_0-y}$

Hence :

$$K_y = \left[\frac{P_{N_0-y}}{P_{N_0-(y-1)}}\right]^{1/Z} \qquad (13)$$

And also :

$$K_y = \left(\frac{N_0-y}{M_0}\right)^{1/Z} \qquad (14)$$

Finally, for an integer y (>0) not too large and an integer n (0<n<Z), we can write :

$$P_{N-yZ-n} \simeq P_N(K_1)^Z \ldots (K_y)^Z . (K_{y+1})^n \qquad (15)$$

When N is not an integer multiple of Z, we shall also use this aproximation. Formulas (11) and (12) then give the following expression for the TIME CONGESTION, taking into account (10) ;

$$P_N(N;M) \simeq \frac{1}{Z} E_{N_0}(M_0) \tag{16}$$

where $E_{N_0}(M_0)$ is the Erlang loss formula. Also, the CALL CONGESTION B for an equivalent average call is :

$$B = E_{\frac{N}{Z}}\left(\frac{M}{Z}\right) \tag{17}$$

This is Hayward's loss formula [7].

2.4 - Loss formulas

When di/z is not too large, (13) becomes:

$$K_i \simeq K = \left[\frac{P_{N_0-1}}{P_{N_0}} \right]^{1/Z} = \left(\frac{N}{M} \right)^{1/Z} \tag{18}$$

Formulas (5), (15), and (16) then give the loss value for the ith traffic stream :

$$B_i \simeq E_{\frac{N}{Z}}\left(\frac{M}{Z} \right).\left[\frac{1}{Z} \frac{K^{d_i}-1}{K-1} \right] \tag{19}$$

Table 1 shows the excellent accuracy of this formula compared with the results of calculations using algorithm (4). And our formula has the advantage of permitting fast, easy calculations.

TABLE 1

						Exact		Approximate	
a_1	d_1	a_2	d_2	M	Z	B_1	B_2	B_1	B_2
42.2	1	22	2	86.2	1.51	$6.1\ 10^{-4}$	$1.4\ 10^{-3}$	$5.7\ 10^{-4}$	$1.2\ 10^{-3}$
42.2	1	11	4	86.2	2.53	$2.3\ 10^{-3}$	$1.1\ 10^{-2}$	$2.2\ 10^{-3}$	$1.1\ 10^{-2}$
42.2	1	5.5	8	86.2	4.57	$4.9\ 10^{-3}$	$5\ 10^{-2}$	$4.9\ 10^{-3}$	$5.1\ 10^{-2}$
51	1	26	2	103	1.50	$1.3\ 10^{-2}$	$2.8\ 10^{-2}$	$1.3\ 10^{-2}$	$2.8\ 10^{-2}$
51	1	13	4	103	2.51	$1.5\ 10^{-2}$	$6.8\ 10^{-2}$	$1.5\ 10^{-2}$	$6.8\ 10^{-2}$
51	1	6.5	8	103	4.53	$1.5\ 10^{-2}$	$1.4\ 10^{-1}$	$1.5\ 10^{-2}$	$1.4\ 10^{-1}$

Table 1 - Mixing of two classes of traffic (on N = 120 circuits). Comparison of exact values and approximate values.

When N is an integer multiple of Z, expression (17) for B corresponds to the average blocking probability. If, however, N is not an integer multiple of Z, (17) does not give exactly the average blocking probability. Perturbations to the quasi-congested states result in a slight discrepancy from the exact result given by (19).

3 - SWITCHING NETWORK

We consider the multistage switching network of a digital exchange, operating as a link system in which the assumption of symmetry is valid [6]. The issue of stochastic dependence between the network's stages is examined further on.

This network carries a mixture of different classes of traffic, using the same assumptions and notation as for a single circuit group in paragraph 2

3.1 - The reduced network method

For establishment of a d-slot call, conventional point-to-point conditional selection involves looking for d free paths between the concerned input switch and output switch, out of N possible 'geographically" distinct paths.

Even when inter-stage stochastic dependencies exist, we assure almost symmetrical operation and selection for these N paths, which can be used indifferently for ordinary (single-slot) calls and multislot calls.

We denote respectively by M and Z the average value and the peakedness factor of the global traffic carried by the switching network.

We also assume that all point-to-point traffic streams have the same z value and the same characteristics. In [10], formula (160), it is shown that, for single-slot calls and taking the interstage dependencies into account, the probability of finding j channels (links) occupied per stage is :

$$P_j = P_0 . W(N;J) . \frac{M^j}{j!} \tag{20}$$

W(N;j)is the probability of being able to place j single-slot calls in the empty network independently of the nature of the random processes, with the random and independent draws taking place in proportion to the intensities of the partial traffic streams and respecting the symmetrical conditional selection rules.

For a mixture of different classes of traffics, (12) can be used to write, for any j which is an integer multiple of Z :

$$P_j \simeq P_0 . W(N;j) . \frac{(M)^{j/Z}}{(j/Z)!} \tag{21}$$

In view of the symmetry in hunting, W(N;j) virtually depends only on j and not on the distribution between the various classes of traffic. As the global random distribution is that of the equivalent average call, the latter has only (N/Z) paths available. It follows that :

$$W(N;j) = W\left(\frac{N}{Z} ; \frac{j}{Z}\right) \tag{22}$$

For a (point-to-point) series-parallel graph, in which all the paths are "geographically" disjoint, the above reasoning is strict. In the case of a "spider" graph, the analysis differs but only very slightly (because blocking takes place above all in the two end stages). Finally, formula (21) becomes :

$$P_j \simeq P_0 . W\left(\frac{N}{Z} ; \frac{j}{Z}\right) . \frac{(M/Z)^{j/Z}}{(j/Z)!} \tag{23}$$

even when j is not an integer multiple of Z. Thus, the concept of the equivalent average call can again be used to express the occupancy of internal links, and to translate the problem to the case of ordinary (single-slot) Poisson traffic streams carried by a reduced network in which the number of possible paths is (N/Z), without altering the loading of the internal links.

For a mixed point-to-point traffic stream, we define : $S_y(N;M)$ as the probability of finding exactly y free paths, where M is the global traffic intensity. $S_y(N;M)$ is the counterpart of $P_{N-y}(N;M)$ in the case of a circuit group (see paragraph 2)

We also denote :

$$M_0 = \frac{M}{Z}, \; N_0 = \frac{N}{Z} \tag{24}$$

by analogy with (10). And for the corresponding reduced network denote by : $R_y(N_0;M_0)$ the probability of finding exactly y free point-to-point paths. This is the counterpart for the reduced network of $S_y(N;M)$ for the real network.

If we write y_0 the integer part of y/Z then $R_{y_0}(N_0;M_0)$ is the counterpart of P_{N_0-y}

Formula (13) becomes :

$$K_{y_0} = \left[\frac{R_{y_0}(N_0;M_0)}{R_{y_0-1}(N_0;M_0)} \right]^{1/Z} \tag{25}$$

Lastly, formula (15) becomes for $y > 0$:

$$S_y(N;M) = \frac{1}{Z} R_{y_0} (K_{y_0+1})^{y-y_0 Z} \tag{26}$$

because of the coefficient (1/Z) in (12)

This smoothing approximation allows us to evaluate the point-to-point call congestion for each of the i classes of traffic :

$$B_i = \sum_{y=0}^{d_i-1} S_y(N;M) \tag{27}$$

We now write d_{0i} the integer part of di/Z

Taking account of (24) and (25), formulas (26) and (27) finally give :

$$B_i = \sum_{u=0}^{d_{0i}} R_u(N_0;M_0) \cdot \left[\frac{1}{Z} \frac{K_{u+1}^{v}-1}{K_{u+1}-1} \right] \tag{28}$$

with $v = Z$ for $u < d_{0i}$ and $v = d_i - d_{0i}Z$ for $u = d_{0i}$

As a first approximation, for di/Z not too large we can take

$$K_i \simeq K = (R_1/R_0)^{1/Z} \tag{29}$$

in which case:

$$B_i = R_0 \frac{1}{Z} \frac{K^{d_i}-1}{K-1} = B_1 \frac{K^{d_i}-1}{K-1} \tag{30}$$

analogous to (19).

In the case of a SINGLE-SLOT CALL, the point-to-point call congestion is :

$$B_1 \simeq \frac{1}{Z} R_0(N_0;M_0) \tag{31}$$

By simplifying the problem to a single class of traffic offered to a reduced network, formulas (30) and (31) allow us to take into account the effects of multislot calls on the stochastic dependence between stages.

3.2 - Calculation of blocking in the reduced network

It is now necessary to evaluate probability R_{y0}. This is quite straightforward, involving the conventional case of single-slot calls, offered to the reduced network. As a first step, we make the usual assumption of stochastic independence between the stages [6], and we introduce the generating function for the network :

$$G(s) = \sum_y R_y(N_0;M_0) \cdot s^y \tag{32}$$

Section III of [8] shows how to apply the Bernoulli distribution to occupancy in order to determine this generating function, and then conveniently extend the analysis to the case of arbitrary distributions at each stage.

For simplicity, we again assume that the N possible point-to-point paths are geographically independent of one another, referring to the discussion in paragraph 3.1. With the Bernoulli distribution at all stages, each path has a probability Q of being free. It follows that :

$$R_y(N_0;M_0) = \binom{N_0}{y} Q^y (1-Q)^{N_0-y} \tag{33}$$

The network is assumed to have E stages, with a loading p_i on the links of the ith stage ($i = 1, 2, \ldots, E$).

Denoting $q_i = 1 - p_i$, we have :

$$Q = q_1 \cdot q_2 \cdots q_E \tag{34}$$

Inserting this expression in (33), we next expand the resulting polynomial function of (p_1, p_2, ..., p_E) and then make the necessary substitutions to take into account the real occupancy distributions, using the method explained in [8]. This approach thus avoids the laborious manipulations of approaches based on combinatorial analysis.

In what follows, we apply our approach to the important case of a two stages link system.

4 - BLOCKING IN A TWO-STAGE LINK SYSTEM

We once again assume stochastic independence between the stages in the reduced network.

4.1. - "Exact" formulas

The Lee graph for paths from a point A to a point B is shown below in Figure 1 :

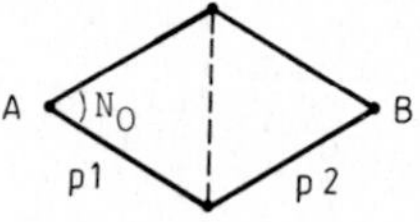

Figure 1 - Lee graph for a TST network (two-stage link system).

To ensure the required symmetry in hunting, it is assumed that a single channel (link) exists between two successive switches (nodes).

With the Bernoulli distribution at each stage, (33) can then be written :

$$R_y(N_0;M_0) = \binom{N_0}{y} \cdot (q_1 q_2)^y \cdot (p_1 + q_1 p_2)^{N_0-y}$$

Expanding the second term in brackets, we find :

$$R_y = \sum_{\lambda_1=y}^{N_0} \left[\binom{N_0}{\lambda_1} p_1^{N_0-\lambda_1} . q_1^{\lambda_1}\right] \left[\binom{\lambda_1}{y} p_2^{\lambda_1-y} . q_2^{y}\right]$$

To bring out the equivalent influence of each stage, the second term in brackets can also be written :

$$\binom{\lambda_1}{y} p_2^{\lambda_1-y} . q_2^y = \binom{\lambda_1}{y} p_2^{\lambda_1-y} . q_2^y \, (p_2+q_2)^{N_0-\lambda_1}$$

$$= \binom{\lambda_1}{y} \sum_{\lambda_2=y}^{N_0} \left[\binom{N_0}{\lambda_2} p_2^{N_0-\lambda_2} . q_2^{\lambda_2}\right] \frac{\binom{N_0-\lambda_1}{\lambda_2-y}}{\binom{N_0}{\lambda_2}}$$

From which it follows, with Bernoulli's model, that :

$$R_y = \sum_{\lambda_1=y}^{N_0} \sum_{\lambda_2=y}^{N_0} \left[\binom{N_0}{\lambda_1} p_1^{N_0-\lambda_1} . q_1^{\lambda_1}\right]\left[\binom{N_0}{\lambda_2} p_2^{N_0-\lambda_2} . q_2^{\lambda_2}\right] . N(y;\lambda_1,\lambda_2) \quad (35)$$

where :

$$N(y;\lambda_1,\lambda_2) = \binom{\lambda_1}{y} \frac{\binom{N_0-\lambda_1}{\lambda_2-y}}{\binom{N_0}{\lambda_2}} \quad (36)$$

In (35), it is now sufficient to replace each term in brackets by the real distribution. In practice, the ith stage corresponds to a group of circuits that must carry a traffic M_0 which is Erlang distributed. We can therefore make the substitution :

$$R_y = \sum_{\lambda_1=y}^{N_0} \sum_{\lambda_2=y}^{N_0} P_{N_0-\lambda_1} . P_{N_0-\lambda_2} . N(y;\lambda_1,\lambda_2) \quad (37)$$

with :

$$P_{N_0-\lambda_i} = \frac{M_0^{N_0-\lambda_i} / (N_0-\lambda_i)!}{1 + M_0/1! + \dots M_0^{N_0}/N_0!} \quad (38)$$

Expressions of the sort are presented in [5] for real networks.

Table 2a shows the very high accuracy of the reduced network method applied with the formula (28) to the case of a two-stage link system and N = 120-channel internal highways (groups of 120 links), as compared to the results obtained by using Roberts' algorithm.

TABLE 2

						Algorithm of Roberts		Reduced Network		
								Exact		Approximate
a_1	d_1	a_2	d_2	M	Z	B_1	B_2	B_1	B_2	B_2
48	1	3	4	60	1.6	$5.9\ 10^{-7}$	$1.8\ 10^{-5}$	$5.7\ 10^{-7}$	$2.4\ 10^{-5}$	$2.8\ 10^{-5}$
30	1	7.5	4	60	2.5	$2.5\ 10^{-5}$	$3.4\ 10^{-4}$	$5.4\ 10^{-5}$	$6.8\ 10^{-4}$	$7\ 10^{-4}$
57.6	1	3.6	4	72	1.6	$1.6\ 10^{-4}$	$3.1\ 10^{-3}$	$2.2\ 10^{-4}$	$4.1\ 10^{-3}$	$4.2\ 10^{-3}$
36	1	9	4	72	2.5	$1.3\ 10^{-3}$	$1.3\ 10^{-2}$	$2.3\ 10^{-3}$	$1.8\ 10^{-2}$	$1.6\ 10^{-2}$
40	1	10	4	80	2.5	$8.5\ 10^{-3}$	$6.4\ 10^{-2}$	$1.2\ 10^{-2}$	$7.7\ 10^{-2}$	$6.2\ 10^{-2}$
2a										2b

Table 2 : TST switching network with N = 120-channel internal highways. Comparison between calculation with the reduced network method and with Robert's algorithm.

4.2 - Approximate formulas

We now derive simpler, approximate formulas for use in numerical calculations.
(36) can be written :

$$N(y;\lambda_1,\lambda_2) = \binom{N_0}{y} \frac{\binom{\lambda_1}{y}}{\binom{N_0}{y}} \frac{\binom{\lambda_2}{y}}{\binom{N_0}{y}} \frac{\binom{(N_0-y)-(\lambda_1-y)}{\lambda_2-y}}{\binom{N_0-y}{\lambda_2-y}}$$

After a few transformations, and using (38), formula (37) becomes :

$$R_y = \binom{N_0}{y} P_{N_0-y} \sum_{\lambda=0}^{N_0-y} P_{N_0-y-\lambda} . H_y(\lambda) \quad (39)$$

with :

$$H_y(\lambda) = \sum_{\lambda'=0}^{\lambda} \frac{\binom{\lambda'+y}{y}}{\binom{N_0}{y}} . \frac{\binom{(\lambda+2y)-(\lambda'+y)}{y}}{\binom{N_0}{y}} \quad (40)$$

After additional transformations, we have :

$$H_y(\lambda) = \frac{1}{\binom{N_0}{y}^2} \sum_{\alpha=0}^{y} \binom{y}{\alpha} \binom{\lambda+1+y}{y+\alpha+1} \quad (41)$$

In practice, N_0 is large and y is small, permitting the following approximation :

$$H_y(\lambda) = \left(\frac{y!}{N_0^y}\right)^2 \sum_{\alpha=0}^{y} \binom{y}{\alpha} \frac{\lambda^{y+\alpha+1}}{(y+\alpha+1)!} \quad (42)$$

Similarly, we can write :

$$P_{N_0-y} \simeq E_{N_0}(M_0) . \frac{1}{p^y} \quad (43)$$

with, based on (24) :

$$p = \frac{M_0}{N_0} = \frac{M}{N} \quad (44)$$

Lastly, because $E_{N_0}(M_0)$ is usually very small, the following relationship holds :

$$\sum_{\lambda=0}^{N_0} \left(\frac{\lambda}{N}\right)^y . P_{N_0-\lambda} \simeq q^y \quad (45)$$

with q = 1 - p.

Also, expression (39) can be written in the following approximate form for y > 0 :

$$R_y(N_0;M_0) \simeq E_{N_0}(M_0) . \frac{\sum_{\alpha=0}^{y} \binom{y}{\alpha} \frac{(N_0 q)^{\alpha+y+1}}{(\alpha+y+1)!}}{\frac{(N_0 p)^y}{y!}} \quad (46)$$

where $E_{N_0}(M_0)$ is the Erlang loss formula.

For y = 0, formula (41) becomes : $H_0(\lambda) = \lambda+1$
(39) and (45) then give :

$$R_0(N_0;M_0) \simeq E_{N_0}(M_0) . (N_0 q+1) \quad (47)$$

This is the well-known Jacobaeus formula [6].

Table 2b shows the close correspondence between results obtained with the above approximation and formula (30) and those obtained with Roberts' algorithm.

Formulas (30), (46), and (47) can thus be used to calculate the individual loss values B_i, provided the structure of the switching network permits the assumption of stochastic independence between stages. However, as shown by the simulation results in Table 3a, this assumption is not very well respected in the case of digital networks. We now tackle this problem for the case of multislot call traffic, again using the reduced network approach.

4.3 - Stochastic dependence between stages

In digital (time-division) switching networks the elementary switches are large, the number of stages is small, and internal loads are high. This is incompatible with the assumption of stochastic independence between stages. To take into account the perturbations due to inter-stage dependencies, we use the MODEL WITH BALKING. For a transition to occur within a given stage from state $(N-\lambda-1)$ to state $(N-\lambda)$, the λ given links in the other stage capable of being used to set-up an additional call must not all be busy. Using the Palm-Jacobaeus formula, the probability of this can be written :

$$1 - \frac{E_{N_0}(M_0)}{E_{N_0-\lambda}(M_0)}$$

This multiplication factor is applied to the previous birth rate in the reduced network, and the substitution thus becomes :

$$P_{N_0-\lambda_i} \cdot D(\lambda_i, N_0) \qquad (48)$$

with :

$$D(\lambda, N_0) = \prod_{\lambda'=\lambda+1}^{N_0} \left(1 - \frac{E_{N_0}(M_0)}{E_{N_0-\lambda'}(M_0)}\right) \qquad (49)$$

This expression can be calculated using the recursive formula :

$$D(N_0, N_0) = 1$$

$$D(\lambda, N_0) = D(\lambda+1, N_0)\left(1 - \frac{E_{N_0}(M_0)}{E_{N_0-(\lambda+1)}(M_0)}\right) \qquad (50)$$

Table 3b shows the close correspondence between the results of calculation and simulation.

In multi-slot calls, the lowest sensitivity to the dependence phenomenon is noticeable and comes from the following :
Denoting

$$D(\lambda, N_0) \simeq \prod_{\lambda'=\lambda+1}^{N_0} (1 - p^{\lambda'})$$

where Bernoulli's approximation is used for not too high load values, as here only λ' low values are influential. (40) becomes :

$$H_y(\lambda) = \sum_{\lambda'=0}^{\lambda} \left[\frac{\binom{\lambda'+y}{y}}{\binom{N_0}{y}} . D(\lambda'+y, N_0)\right] . \left[\frac{\binom{(\lambda+2y)-(\lambda'+y)}{y}}{\binom{N_0}{y}} . D\big((\lambda+2y)-(\lambda'+y), N_0\big)\right] \qquad (51)$$

which allows formula (39) to be kept.

In practice, as N_0 is large, the formula analogous to (42) becomes :

$$H_y(\lambda) = \sum_{\lambda'=0}^{\lambda} \left[\left(\frac{\lambda'+y}{N_0}\right)^y D(\lambda'+y)\right] . \left[\left(\frac{(\lambda+2y)-(\lambda'+y)}{N_0}\right)^y . D\big((\lambda+2y)-(\lambda'+y)\big)\right] \qquad (52)$$

similary (39) becomes :

$$R_y(N_0, M_0) \simeq E_{N_0}(M_0) \frac{\left(\frac{N_0}{p}\right)^y}{y!} \sum_{\lambda=0}^{N_0} P_{N_0-\lambda} \cdot H_y(\lambda) \qquad (53)$$

taking into account (38), (48), (49), (52).

For y large enough, the low values of λ' and $(\lambda-\lambda')$ slightly influence (52) and consequently $D \simeq 1$. The independence stages is thus assumed again. In conclusion the dependencies between stages influence R_y only for y low values, while single-slot calls are strongly influenced by the stochastic dependencies between stages. It is not the case of multi-slot calls as long as the internal loads are not too high.

Finally let us remind that the dependencies between stages tend to seriously reduce the internal blocking rates.

TABLE 3

		Simulation		Reduced Network			
				Independence		Dependence	
M	Z	B_1 $(d_1=1)$	B_2 $(d_2=4)$	B_1	B_2	B_1	B_2
60	2.5	$2.6\ 10^{-5}$	$4.4\ 10^{-4}$	$5.4\ 10^{-5}$	$7\ 10^{-4}$	$4.6\ 10^{-5}$	$6.7\ 10^{-4}$
72	2.5	$6\ 10^{-4}$	$7.4\ 10^{-3}$	$2.3\ 10^{-3}$	$1.6\ 10^{-2}$	$1.4\ 10^{-3}$	$1.4\ 10^{-2}$
80	2.5	$2.7\ 10^{-3}$	$3.1\ 10^{-2}$	$1.2\ 10^{-2}$	$6.2\ 10^{-2}$	$4\ 10^{-3}$	$3.4\ 10^{-2}$
3a						3b	

Table 3 - Comparison between calculation and simulation

5 - LOCAL CALLS THROUGH THE SAME TERMINAL SWITCH

To complete this study of digital switching networks carrying multislot call traffic, we now examine the case of calls between two lines or circuits connected at the same terminal switch. The latter must then carry "local" calls each occupying two times d_i channels.

We consider the simple case of single-slot calls. The formulas for multislot calls can be derived quite easily.

Symmetrical path selection is again assumed. We denote by C the number of input switches (and output switches) in the network. And we denote by M the traffic in circuits offered to a single switch. Because calls are equally distributed, we can write :

$$M = x\left(\frac{C-1}{C}\right).1 + \frac{x}{C}.2 \qquad (54)$$

Where x is the traffic intensity in calls on the switch.

Hence :
$$x = M.\frac{C}{C+1} \qquad (55)$$

and because $M = a_1.1 + a_2.2$ it becomes :

$$a_1 = M\frac{C-1}{C+1}, \quad a_2 = \frac{M}{C+1} \qquad (56)$$

It is then possible to calculate blocking for the "local" calls and for "non-local" calls in the same way as for multislot calls in previous sections. Calculation of Z gives :

$$Z = \frac{V}{M} = \frac{C+3}{C+1} \qquad (57)$$

This result correctly reflects the relative influence of the number of switches.

6 - SOME PRACTICAL RESULTS

In conclusion to our study, you will find hereunder results showing coefficient K variations in formula (30) depending on opening N of TST network, and on coefficient of reduction Z for internal load value p = 0.7.

In brackets we give the corresponding blocking probability B_1 for ordinary calls.

These results are obtained by the exact calculation of section 4.3 taking into account the dependency between stages.

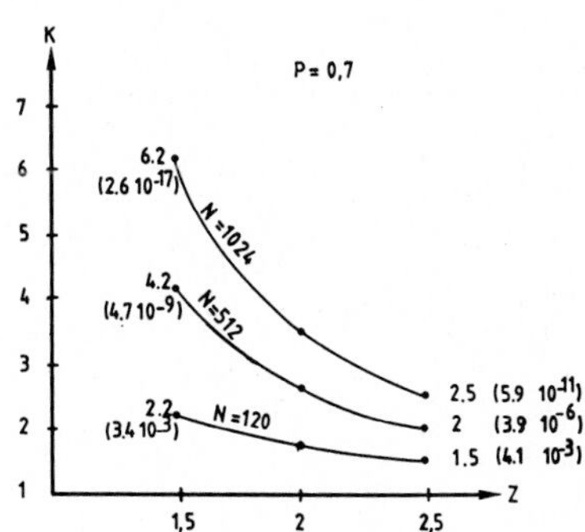

Figure 2 - K and B_1 in TST network, with p = 0.7.

Thus we show that the blocking characteristics for multislot connections can be calculated by the simple use of formula (30) applied to the reduced network whose parameters B_1 and K may be obtained by calculation or by simulation in the case of very high loads.

7 - CONCLUSION

We have shown how the REDUCED NETWORK METHOD resulting from the EQUIVALENT AVERAGE CALL concept, based on the PEAKEDNESS FACTOR Z, is applied to evaluate the real blocking probabilities for the multislot connections in the digital connection networks. As it allows the reference to single-slot connections, the general results already obtained on the networks can be used. The impact of the multislot calls on the ordinary traffic can thus be evaluated and the blocking probabilities derived for all types of traffic by using a simple relation-ship.

The increased importance of the dependencies between stages in the connection networks using large switching unit and a small number of stages has been stressed. It can be noticed that the phenomenon is less sensible for multislot connections than for ordinary connections. We have showed how to take into account the dependencies in the calculation.

Some practical results are given referring to a TST network such as those used in the French digital switching systems.

Finally, for the standards relating to the blocking of calls presented to a circuit group of a network handling mixed multislot connections, it can be referred to the blocking of ordinary single-slot calls in the reduced network defined by its parameters : N_0, Z, K.

References :

[1] G. Fiche "Method for evaluating traffic performance of the E10.B system" Commutation et Transmission n°3 - 1980.

[2] G. Fiche, M. Ruvoen "The E10 SYSTEM : functional evolution and quality of service." ISS 84 Florence

[3] L.A Gimpelson, "Analysis of Mixtures of wide and narrow band traffic", IEEE Trans. on Communication COM-13, N°3,p258 (Sept 1965)

[4] J.W Roberts, "Teletraffic models for the Telecom I integrated services network", Proc. 10 th Int. Teletraffic Congress, Montreal, 1983

[5] T. Saito, H. Inose, S. Hayashi, "Evaluation of traffic carrying capability ine one-stage and two-stage time-division networks handling data with a variety of speed classes", Proc. 9 th Int. Teletraffic Congress, Torremolinos, 1979.

[6] C. Jacobaeus, "A study on congestion in link systems", Ericsson Technics, N° 48, 1950, pp 1-68

[7] A.A Fredericks, "Congestion in blocking systems, a simple approximation technique", Bell Syst. Tech. J., Vol. 59, N°6, pp 805-827, 1980.

[8] P. Le Gall, "Méthode de calcul de l'encombrement dans les systèmes téléphoniques automatiques à marquage", Annales des Télécommunications, Paris, Vol. 12, N° 11, pp 374-386, Nov 1957.

[9] P. Le Gall, "Les Trafics téléphoniques et la sélection conjuguée en téléphonie automatique", Annales des Télécommunications, Paris, Vol. 13, N° 7-8, pp 186-207 ; N° 9-10, pp 239-253 ; N° 11-12, pp 278-301 ; 1958

[10] P. Le Gall, "Random processes applied to traffic theory and engineering", Commut et Electronique, SOCOTEL, Paris, N° 43 pp 5-51, Oct 1973 et ITC 7, Stockholm, 1973

TELETRAFFIC ISSUES in an Advanced Information Society
ITC-11
Minoru Akiyama (Editor)
Elsevier Science Publishers B.V. (North-Holland)

TRAFFIC DESIGN FOR TIME DIVISION WIDEBAND SWITCHING NETWORKS

Kohsoh MURAKAMI, Masafumi KATOH and Shunji ABE

FUJITSU LABORATORIES LIMITED
Kawasaki, Japan

ABSTRACT

This paper describes the traffic design for a time-division wideband switching network for high-speed bearer services more than 384kb/s including moving video. A key point in realizing wideband switching networks is to gain architecture efficient for switching variable bit-rate traffic and for integrating both point-to-point connection and broadcasting connection. In a conventional time division switching network, since higher bearer traffic requires multi-slot connection, the blocking probability is high. We propose a single slot connection method for all types of traffic. This paper shows that a single-slot connection can carry more traffic than the multi-slot connection and that broadcast services can be switched integrally with point-to-point establishment.

1. INTRODUCTION

Communication services are expected to progress from voice-only communication by telephone to image communication, which includes text, high-speed facsimile and moving video.

A wideband switching system is the heart of such enhanced integrated services digital networks (ISDN). However, there are many traffic problems, mentioned below, to be evaluated in realizing advanced services.

Although space division switching networks are adopted due to the limit of device technologies at present, time division networks will be required in practice and have become feasible through advances in high speed semiconductor technology.

This paper discusses the traffic design for time division wideband switching networks. In wideband communication networks, the switching system must provide the following functions for transparent transmission and switching.

(1) Balanced bi-directional end-to-end communication such as TV telephones.

(2) Multi-party balanced bi-directional communication, such as TV conferences.

(3) Unbalanced bi-directional center-to-end communication such as videotex, including moving video.

(4) CATV-like distribution services such as retransmission of broadcast programs.

There are many functions different from the 64kb/s network. These services imply inclusion of traffic at a variety of bit rates and require efficient establishment of both point-to-point connections and point-to-multiple connections.

In the conventional time division switching network, since higher bearer traffic requires multi-slot connection, which uses multiple internal time slots in a path establishment, the higher the bit rate, the larger the blocking probability, and also the larger the switching network control volume. To resolve these problems, it is necessary to introduce a switching method efficient for calls at variable bit rates.

This paper proposes a single-slot-connection method, which is more efficient than the multi-slot connection method for all types of bit rates. This paper also shows its traffic design results for time division wideband switching network based on traffic analysis.

2. GRADE OF SERVICE IN TIME DIVISION WIDEBAND SWITCHING NETWORKS

As mentioned above, basic functions required for time division wideband switching networks include switching various bit rate calls at a balanced grade of service, and establishment of point-to-multiple connections as well as point-to-point.

On the other hand, the probability distributions of interarrival time and service time of calls at various bit rates are not clear at present. But these probability distributions are believed to be similar to those for telephone and to have different mean value, because the main service is man to man. The loss system is assumed because wideband calls will be handled mainly in the circuit switching system.

There is an idea that a nonblocking switching network is necessary so as to confirm reservation connections for TV conferences. However, nonblocking switching is not economical in wideband switching network. So, the switching network should be economized by permitting a little blocking probability. In a wideband switching system, the common channel signaling method is essential for the administration of a variety of types of calls. Therefore, the blocking probability need not to be assigned to signaling trunks such as a multi-frequency sender. As a result, the blocking probability assigned to switching points, which is 0.001 at standard grade of service and 0.03 at overload, can be perfectly assigned to the switching network.

3. CONNECTION METHOD FOR TRAFFIC AT VARIOUS BIT RATES

3.1 Connection Method for Various Bit-Rate Calls

3.1.1 Problems of Switching for Various Bit-Rate Calls

The typical methods used to switch calls at various bit rates in a time division switching network follow:

(1) connection by a unit of lower bearer rate

(2) connection by a unit of higher bearer rate.

Method (2) is not economical when there are many lower bearer calls because lower bearer calls are converted to higher bearer calls.

In method (1) a higher bearer call is switched by using multi-time slots N. A few research papers on multi-slot connection show its traffic characteristics and show that the blocking probability for higher bearer calls is higher than for lower bearer calls and that large N is not allowed.[3] Further, path control volume for higher bearer calls increases in order to maintain time slot sequence integrity (TSSI). Therefore, method (1) is not economical when there are many higher bearer calls.

3.1.2 Single Slot Connection Method for Calls at All Bit Rates

For a wideband switching network where there are calls at various bit rates and these traffic ratios vary, the authors propose an efficient connection method, namely the single slot connection method. In this method, each bearer call is cyclically switched in time division mode by the period corresponding to its bearer rate. As a result, all the bearer rate calls are connected by a single slot during a time division switching period. An example of the connection method based on the above consideration is shown in Fig. 1. Several switching networks are combined, and in one of these networks, only one kind of call is switched for a period of its special bit rate. This method brings a little divided loss in switching network hardware when the number of bit rates increase. In practical phase are three or four bearer rates are standardized from a digital hierachy view point.

This method is expected to reduce the blocking probability and simplify switching control because any vearer call can be switched if there is one same-phase idle time slot in the 1st and 2nd links. In the input highway by this method, for a frame of a lower bearer call, multiple time slots of a higher bearer call must be periodic as shown in Fig. 2.

3.2 The Broadcast Connection Method

For a 1:M broadcast connection, the method that connects M paths independently is not efficient. In a time division switching network, the final-stage time switch can distribute the information to many subscribers on the same output highway. In this case, since the number of busy time slots on the internal link is smaller than on the output highway, the blocking probability

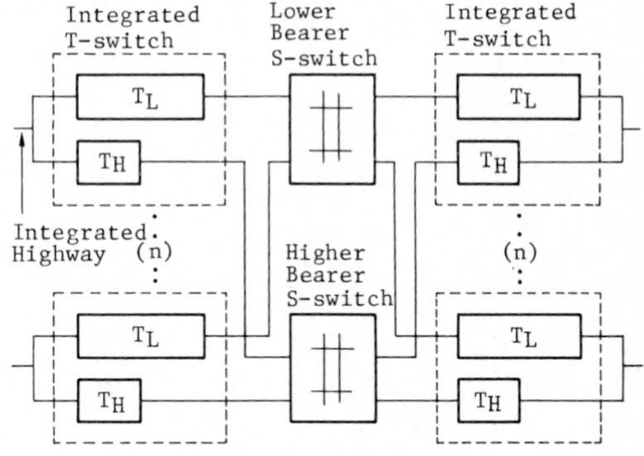

Fig. 1 Integrated single slot switching network.

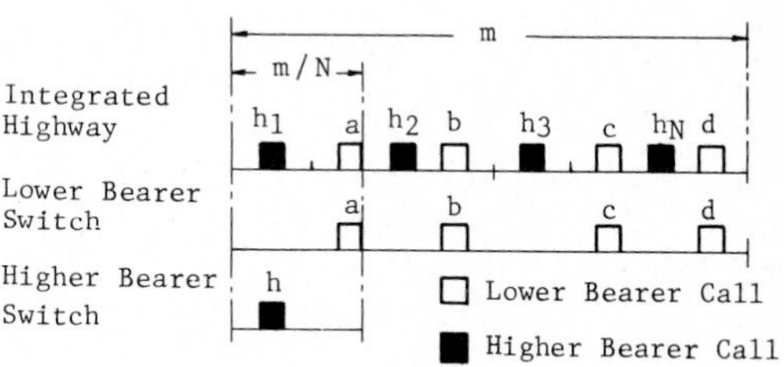

Fig. 2 Frame format of integrated single slot switching network.

for a 1:M broadcast call can be expected to be reduced, as mentioned in section 4.2. Consequently both 1:1 calls and 1:M calls can be expected to be efficiently switched in the same switching network.

4. TRAFFIC ANALYSIS

4.1 Performance Evaluation of Single Slot Connection

4.1.1 Analysis of Internal Blocking Probabilities for Single Slot Connection

Internal blocking probability for single-slot connection is analyzed and compared with that for multi-slot connection.

Fig. 3(a) shows a multi-slot connection model. A lower bearer call uses one time slot of a highway that has m time slots. A higher bearer call uses N time slots of the highway.

Fig. 3(b) shows a single-slot connection model. Lower bearer calls and higher bearer calls are separated at the entrance of the switching network. Lower bearer calls are switched in the lower bearer switching network having m time slots, and higher bearer calls are switched in the higher bearer switching network having m/N time slots. Each call uses one time slot during a time division switching period.

B_L and B_H denote the internal blocking probability of a lower bearer call and of a higher bearer call, in the multi-slot connection method. B_L and B_H are defined as the following conditional probabilities.

B_L = Pr{there is no same-phase idle time slot in the 1st and 2nd links of multiplexity m | there are one or more idle time slots in each link}

B_H = Pr{there are less than N same-phase idle time slots in the 1st and 2nd links of multiplexity m | there are more than N-1 idle time slots in each link}

To calculate B_L and B_H , following probabilities are defined.

$p_\ell(r)$: The probability that r time slots of the ℓ-th link are occupied

$Q_1(i,j)$: The probability that there is no same-phase idle time slot when i time slots of the 1st link and j time slots of the 2nd link are busy.

$Q_2(i,j)$: The probability that there are less than N same-phase time slots when i time slots of the 1st link and j time slots of the 2nd link are busy.

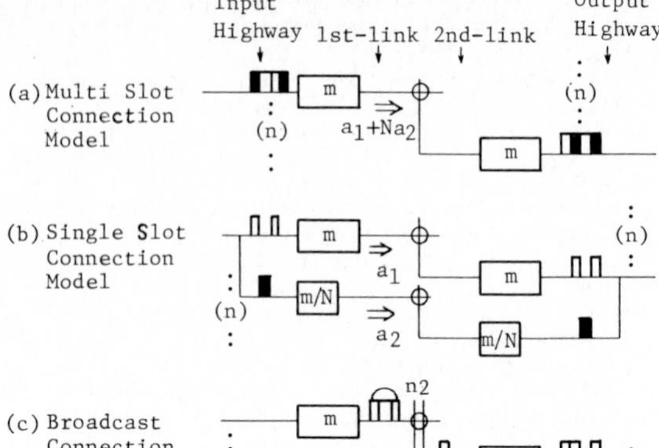

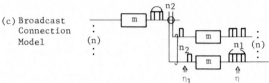

Fig. 3 Traffic models of various connections.

Then,

$$B_L = \sum_{i=0}^{m-1} \sum_{j=m-i}^{m-1} p_1(i)p_2(j)Q_1(i,j)$$

$$B_H = \sum_{\substack{m-i-j \leq N \\ i,j \leq m-N}} \sum p_1(i)p_2(j)Q_2(i,j). \tag{1}$$

If the selection of the time slot in the link is assumed to be random, $Q_1(i,j)$ and $Q_2(i,j)$ are

$$Q_1(i,j) = \binom{i}{m-j} \Big/ \binom{m}{j} \qquad i,j \leq m-1,\ i+j \geq m$$

$$Q_2(i,j) = \begin{cases} \displaystyle\sum_{k=0}^{(N-1)+i+j-m} \binom{m-i}{j-k}\binom{i}{k} \Big/ \binom{m}{j} & i,j \leq m-N,\ 0 \leq m-i-j < N \\ \displaystyle\sum_{k=0}^{N-1} \binom{i}{m-j-k}\binom{m-i}{k} \Big/ \binom{m}{j}. & i,j \leq m-N,\ m-i-j < 0 \end{cases} \tag{2}$$

Then the numbers of simultaneous calls of links are assumed to be independent and form a two-dimensional Erlang distribution. The probability $p(r_1, r_2)$ that a link is occupied by r_1 lower bearer calls and r_2 higher bearer calls is then

$$p(r1,r2) = \frac{a_1^{r_1}}{r_1!}\frac{a_2^{r_2}}{r_2!} \Bigg/ \sum_{r_2=0}^{[\frac{m}{N}]} \sum_{r_1=0}^{m-Nr_2} \frac{a_1^{r_1}}{r_1!}\frac{a_2^{r_2}}{r_2!} \tag{3}$$

The probability that r time slots of the link are occupied is then

$$p(r) = \sum_{r_2=0}^{[r/N]} p(r-N\cdot r_2,\ r_2). \tag{4}$$

Using equation (4), $p_\ell(i)$, $p_\ell(j)$ can be calculated. Note that $p_\ell(r)$ is normalized such that the sum of $p(r)$ $(r=0,1,\cdots,m-1)$ is equal to one when we calculate B_L and $p_\ell(r)$ is normalized such that the sum of $p(r)$ $(r=0,1,\cdots,m-N)$ is equal to one when we calculate B_H.

$\widehat{B}_L$ and $\widehat{B}_H$ denote the internal blocking probability of a lower bearer call and of a higher bearer call respectively, in the single slot connection method. These are defined as conditional probabilities as above mentioned about B_L.

As $p_\ell(r)$ in multi-slot connection, $x_\ell(r)$ and $y_\ell(r)$ denote the ℓ-th link occupancy in the lower bearer switching network and in the higher bearer switching network respectively, and $Q_3(i,j)$ and $Q_4(i,j)$ denote the probability that there is no same-phase idle time slot in each switching network. Then,

$$\widehat{B}_L = \sum_{i=0}^{m-1} \sum_{j=m-i}^{m-1} x_1(i)x_2(j)Q_3(i,j)$$
$$\widehat{B}_H = \sum_{i=0}^{m/N-1} \sum_{j=m/N-i}^{m/N-1} y_1(i)y_2(j)Q_4(i,j). \tag{5}$$

If the selection of time slot in the link is random,

$$Q_3(i,j) = \binom{i}{m-j} \Big/ \binom{m}{j}$$

$$Q_4(i,j) = \binom{i}{m/N-j} \Big/ \binom{m/N}{j}. \tag{6}$$

In the single-slot method, the numbers of simultaneous calls of the lower and higher switching network are independent and form an Erlang distribution. That is,

$$x_\ell(r) = \frac{a_1^{\,r}}{r!} \Big/ \sum_{r=0}^{m-1} \frac{a_1^{\,r}}{r!}$$
$$y_\ell(r) = \frac{a_2^{\,r}}{r!} \Big/ \sum_{r=0}^{\frac{m}{N}-1} \frac{a_2^{\,r}}{r!}. \tag{7}$$

4.1.2 Numerical Results

Figure 4 shows the internal blocking probabilities vs. total traffic offered under a balanced ratio of calls for the multi-slot connection and for the single-slot connection. When the bearer rate ratio N of higher bearer calls to lower bearer call is great, B_H is large. This is because the greater N is, the more difficult it is to find N same-phase idle time slots in the 1st and 2nd link. The greater N is, the greater $\widehat{B}_H$ is. The tendency of increased B_H ($\widehat{B}_H$) to traffic offered is weaker in the single-slot connection than in the multi-slot connection. This is because the occupancy of the higher bearer switching network in the single slot connection is near to half the total occupancy due to the separation of traffic offered to two types networks. If the blocking probability B_H ($\widehat{B}_H$) is permissible to be about 10^{-2}, the occupancy in the multi-slot connection for N = 16 can be risen to about 0.49 and the occupancy in the single-slot connection for N = 16 can be risen to about 0.84. Therefore, the single-slot connection is believed to be able to switch many more calls than the multi-slot connection.

In Fig. 5, the authors compare two bearers to three bearers for the single-slot switching network. The blocking probability of each traffic in the three bearer switching networks is lower than that in the two bearer switching networks. This is because the occupancy of the higher bearer switching network is lower in the three bearers than in two bearers due to separation of traffic offered to three networks. Therefore, even if the number of bearer rates is three or four, performance is believed to be by no means bad.

Figure 6 shows the internal blocking probabilities for single-slot and multi-slot connection (m = 256) when the total traffic offered is constant (160 erl) and traffic volume ratio of

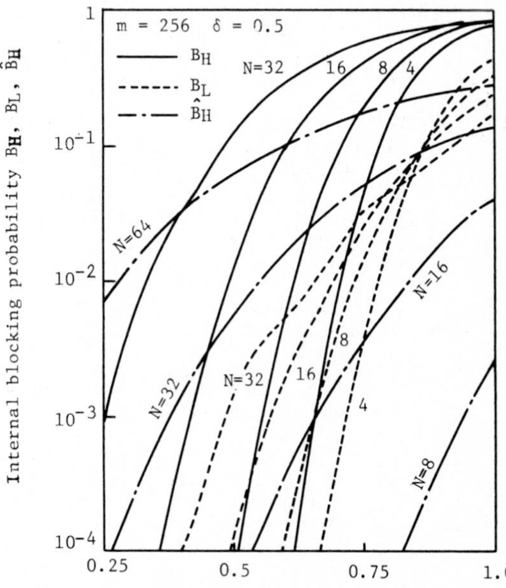

Fig. 4 Internal blocking probability to traffic offered.

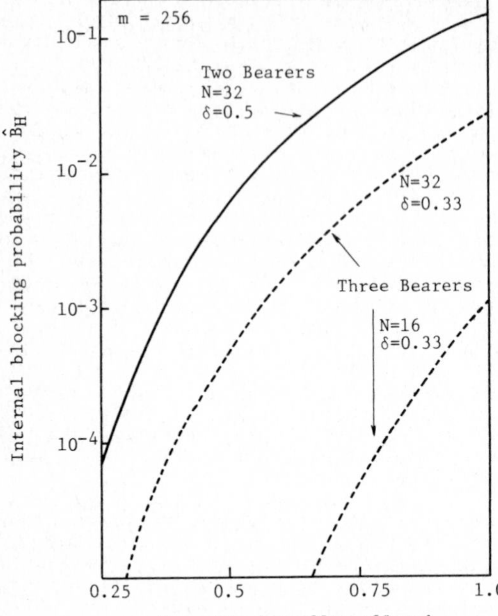

Fig. 5 Comparision between two bearers and three bearers.

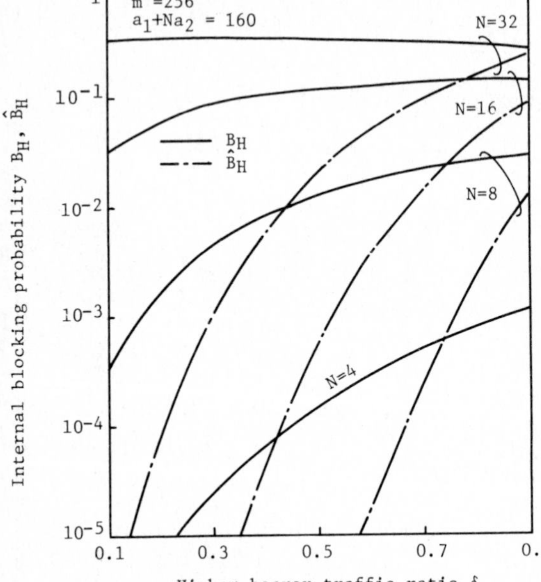

Fig. 6 Blocking probability vs. higher bearer traffic ratio.

higher bearer calls to both calls varies. The greater δ is, the greater is the blocking probability of a higher bearer call. $\hat{B}_H$ is much smaller than B_H. This fact also shows the single slot connection is thought to be able to switch many more higher bearer calls than the multi-slot connection.

4.2 Traffic Analysis of Broadcasting Connection

4.2.1 Internal Blocking Probability for Broadcasting Calls

The broadcast connection service in the time-division switching network is performed as shown in Fig. 3(c). The broadcast is performed by using the mechanism of the time switch. The broadcast from an input highway to a number of highways is executed by the multiple reading of the first-stage time switch memory. In the same way, broadcast to the output highway is executed at the final-stage time switch.

If the number of highways is n, the broadcast call must find each a time slot free on each of n highways. So, it is considered that the internal blocking probability is high. However, as the broadcast on the same output highway is performed by the final-stage time switch, it can reduce the internal link occupancy compared with that of output highway and result in reducing the internal blocking probability of the broadcasting call. This fact is expected to enable to connect both point-to-point connection (1:1 call) and point-to-multiple-connection (1:M call or broadcast) efficiently in the same time-division switching network. Following analysis will make this clear.

In Fig. 3(c), assume that the number n of highways in the 1st link equals the number of highways in the 2nd link and that the distribution number is balanced on each highway. Let n_1 be the number of distributing at the final-stage time switch, and let n_2 be the number of highways distributed by the first-stage time switch.

This broadcast call can be considered as a group call with the size $n_1 \times n_2$. The internal blocking probability of this call is defined as the probability of not choosing simultaneously the free time slot in n_2 highways.

Let η be the output highway occupancy. Assume that point-to-point and point-to-multiple-connection calls are occupied at a rate α : β in traffic volume. If it is assumed that all highways are used uniformly, the 1st link occupancy and the 2nd link occupancy are identical and is represented as η_1

$$\eta_1 = \left(\frac{\alpha}{\alpha + \beta} + \frac{\beta}{\alpha + \beta} \frac{1}{n_1} \right) \eta .$$

When it is assumed that the probability distribution of the number of simultaneous connections for each link is independent and forms Binomial distribution, the blocking probability of an arbitrary path between the 1st link and the 2nd link is represented by

$$\gamma = \{ 1 - (1 - \eta_1)^2 \}^m \tag{8}$$

Let B_1 and B_M be the blocking probabilities of point-to-point-connection calls and point-to-multiple-connection calls, respectively. The blocking probability B_1 is equal to γ.

Conversely, from the previous definition of the blocking probability for the broadcast call, the blocking probability B_M is given by the following equation.

$$B_M = 1 - (1 - \gamma)^{n2} \tag{9}$$

While it is natural to give a probability distribution to the group size (n1 x n2) of the broadcast call, it is difficult to do definitely so. In this paper, the internal blocking probability of the broadcast call is evaluated for the situation distributing situation broadcast call to all highways in 2nd links, i.e. $n_2 = n$.

4.2.2 Numerical Results

Figures 7 and 8 show numerical results of blocking probabilities B_1 and B_M for output highway occupancy and occupation ratio $\{\beta/(\alpha+\beta)\}$ of point-to-multiple-connection traffic to the total traffic.

In these numerical results, naturally, although the blocking probability B_M of the point-to-multiple-connection call is always larger than the blocking probability B_1 of the point-to-point-connection call, for increase of the number distributed by the final-stage time switch, blocking probability rapidly becomes lower. This causes the internal link occupancy to be reduced to the ratio $1/n_1$ of the occupation rate of point-to-multiple-connection calls. Although internal blocking probabilities are high in the range where the occupation ratio of point-to-multiple-connection traffic to the total traffic is small, in proportion to increase of this ratio, blocking probabilities rapidly become lower. Also for increase of time slot numbers, blocking probabilities rapidly become lower.

If the blocking probability is permissible to be about 10^{-2} the same order as 1:1 connection, from Fig. 7, the output highway occupancy can be risen to about 0.75 for n=16, m=32 and $n_1 = 2$. This satisfies the service grade of the same rank discussed in the previous section 4.1, although the number n1 is the small value. Further, as shown in Fig. 8, even if the occupation ratio of point-to-multiple-connection traffic to the total traffic is the small value 20%, the above service condition is completely satisfied for m=64 and $n_1 = 8$.

Therefore, it is possible that point-to-point and point-to-multiple-connection calls are very efficiently established in the same time-division switching network.

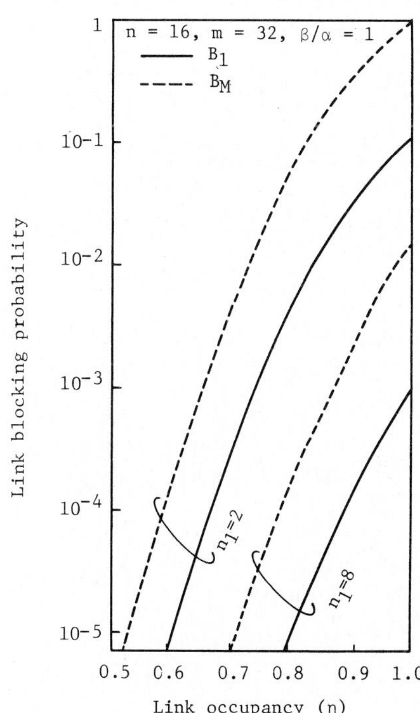

Fig. 7 Blocking probability VS. link occupancy.

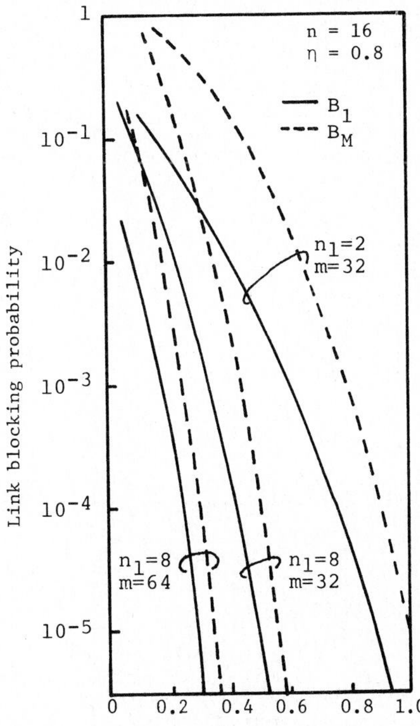

Fig. 8 Blocking probability VS. occupation ratio of 1:M traffic to total traffic.

only if the number of time slots in a highway is larger than 64, which is feasible by using high speed semiconductor technology.

5. DESIGN RESULTS FOR SWITCHING NETWORKS

In a switching network having calls at various bit-rates, important factors of the traffic characteristic are the bearer rate ratio N, the ratio δ of higher bearer traffic offered to the total traffic offered and the time division multiplexity m. This section shows the useful relationships of N, δ and m when the service condition is given.

Figure 9 shows the maximum occupancy under the practical condition B_H ($\widehat{B}_H$) = 10^{-2}, when lower bearer traffic offered is equal to higher bearer one (δ=0.5). The greater the time division multiplexity is, the more calls the switching network can switch. The tendency toward the permissible occupancy to the multiplexity is greater in the single-slot connection.

Figure 10 shows the N and δ relationship that is satisfied under the condition B_H ($\widehat{B}_H$) = 10^{-2} when the mean occupancy is 0.625. For the single-slot connection, the allowed N is larger than for the multi-slot connection. When δ is about 0.5, for the multi-slot connection, the maximum of N is 4 for m = 128, 8 for m = 256, and 16 for m = 512. That is, time division multiplexity that is 32 times N is necessary. Conversely, for the single-slot switching network, when δ is about 0.5, the maximum of N is 16 for m = 128 and 32 for m = 256. That is, the multiplexity that is only 8 times N is necessary. Currently, using a fast bipolar IC, memory can be read or written at 64Mb/s. Therefore , if switching network memories are constructed in 8 bit parallel, the switching can realize highway throughput of 512Mb/s. This fact shows that if two types of traffic are included, for example, 2Mb/s high speed data and 32Mb/s video, up to 70% video traffic can be carried under the time division multiplexity of 256.

6. CONCLUSION

Wideband switching networks must efficiently switch calls at various bit rates more than 384kb/s and establish not only point-to-point connection but also point-to-multiple connections. This paper proposed a single-slot connection method based on switching all bearer calls by using a single-slot during a time division switching period. Further, this paper showed that single-slot connection could carry more traffic than multi-slot connection and that broadcast connection could be handled integrally with point-to-point connection.

The blocking probability of higher bearer calls is not sufficiently small when there are many higher bearer calls, even by using the single-slot connection method. Further, the grades of service for higher bearer calls and for lower bearer calls are unbalanced. These problems must be resolved in the future.

REFERENCES

[1] L.A.Gimpelson, "Analysis of mixtures of wide-and narrow-band traffic",TEEE Trans.,COM-13,pp.258-266,1965.

[2] J.L.Lutton and J.W.Roberts, "Traffic performance of multi-slot call routing strategies in an integrated services digital network",Proc.ISS84,Florence,1984.

[3] S.Hattori, K.Murakami and M.Kato, "Some considerations on traffic characteristic of ISDN oriented switching systems",Paper of Tech.Group TGSE81-74,IECE Japan,1981.

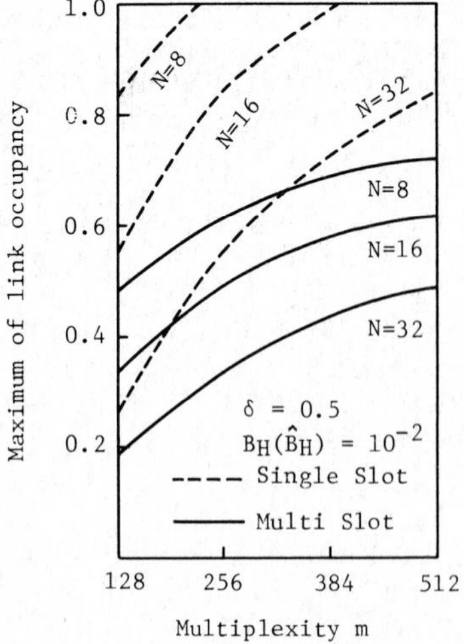

Fig. 9 Maximum of link occupancy.

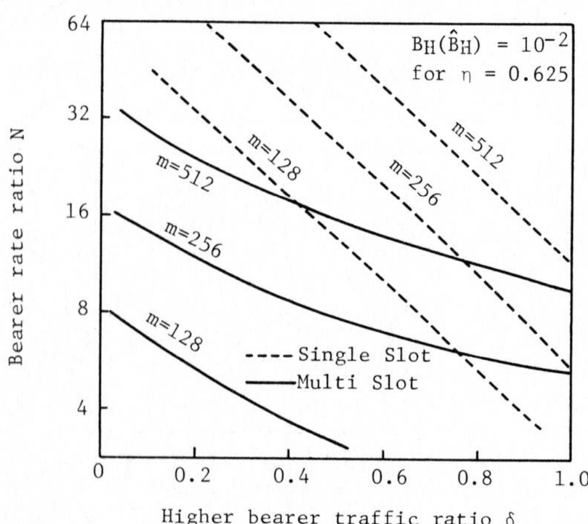

Fig. 10 Relation between δ and N.

TELETRAFFIC ISSUES in an Advanced Information Society
ITC-11
Minoru Akiyama (Editor)
Elsevier Science Publishers B.V. (North-Holland)

A TWO STAGE REARRANGEABLE BROADCAST SWITCHING NETWORK

G. W. RICHARDS

AT&T BELL LABORATORIES
INDIAN HILL, ILLINOIS

F. K. HWANG

AT&T BELL LABORATORIES
MURRAY HILL, NEW JERSEY

ABSTRACT

A new rearrangeable broadcast switching network has been invented which dramatically reduces the per outlet crosspoint cost. In its simplest form, the network has two stages. For larger networks, the switches in each stage can be replaced with two stage networks for a further reduction in crosspoint cost. The invention is based on an innovative approach of connecting each inlet channel to a multiplicity of first stage switches according to a predetermined connection pattern. The network grows easily and has simple path hunt and rearrangement algorithms.

1. INTRODUCTION

In recent years, various studies have been made of non-blocking multiconnection networks. The general case of multiconnection allows any number of inlets to be simultaneously connected to any number of outlets. Hwang has considered both non-blocking [1] and rearrangeable [2] solutions to this problem. Masson [3] along with Jordan [4] analyzed non-blocking and rearrangeable multiconnection networks that allowed any inlet to be connected to any or all outlets, simultaneously, but restricted each outlet to be connected to at most one inlet at a time. Thompson [5] considered this problem in the application of parallel processor intercommunication. Such networks are known by various names such as expansion, fan-out, generalized connection, broadcast, etc., and this paper addresses rearrangeable versions of these.

In the past, most multiconnection network studies [1,2,3,4] were based on three stage Clos networks and were concerned with the number of middle stage switches required to meet certain switching requirements. In the case of rearrangeable broadcast networks, an additional and substantially different approach has emerged [6,7,8,9]. These papers are similar in two important respects. First, the described networks have only two stages in their simplest forms. Second, inlet channels are multiply assigned to first stage switches according to various predetermined patterns. Alternately, these assignments may be viewed and manifested as particular crosspoint patterns in a single sparsely populated first stage rectangular switch.

These approaches are all based on the same fundamental concept. Namely, that any second stage switch can access up to a maximum number n_2 of any of the inlet channels. As will be seen the value of n_2 is determined by the multiplicity and pattern of inlet channel assignments on the first stage switches. These factors in conjunction with the total number of inlet channels determine the crosspoint utilization efficiency of the two stage network. Multistage networks improve on this efficiency for a sufficiently large number of inlet channels.

Assuming the efficacy of the two stage approach just described, we are left with the problems of synthesizing inlet assignment patterns, determining the corresponding values of n_2, and selecting various parameter values to minimize crosspoints (or to optimize with respect to some other variables). Much of this paper deals with the synthesis and analysis of the inlet assignment process, since these results most directly affect the practical utility of the networks in question.

The next section provides an overview and some fundamental aspects of the two stage rearrangeable design. Section 3 presents results on the synthesis and analysis of inlet assignment patterns. Section 4 discusses multistages and network cost growth rate. Path hunt and rearrangement algorithms are described in Section 5.

2. OVERVIEW OF THE NETWORK DESIGN

Our objective is to design a two stage rearrangeable broadcast network. Each stage is comprised of some number of rectangular switches and each switch in the first stage is connected by a single link to each switch in the second stage. To understand the approach better, we will initially limit our consideration to a network that has a single second stage switch.

We will assign some number n_2 of outlets to the second stage switch. We will also assign multiple appearances of inlets in some predetermined pattern to the various first stage switches. Assume that each inlet channel has the same number of multiple appearances, M. If there are N_1 inlet channels, then there will be a total of $M \times N_1$ inlets on the first stage switches.

This network will be rearrangeable if any n_2 of the N_1 inlet channels can be uniquely accessed on some n_2 first stage switches. By Hall's theorem of a system of distinct representatives [10], this will be true if for any $n \leq n_2$ there are at least n first stage switches containing appearances of any n inlet channels. If this condition holds, the configuration is a rearrangeable pairwise connection network and also a rearrangeable broadcast network if we make the standard assumption that the second stage rectangular switch can connect any inlet to any or all outlets simultaneously.

The extension from a single second stage switch to some arbitrary number is straightforward. Simply provide each first stage switch with a number of outlets equal to the number of second stage switches and connect each first and second stage switch with a single link. The fanout property of the first stage switches allows each second stage switch to access selected input channels independently of the activity of other second stage switches.

The connectivity function to be performed by any switch in either stage is the same as that for the entire network; i.e., to connect any inlet to any or all outlets simultaneously. Thus, for large networks, a further reduction in crosspoint count may be achieved by replacing switches with two stage networks.

Now consider a general two stage network. There are N_1 inlet channels and N_2 outlets. Each first stage switch has n_1 inlets and each second stage switch has n_2 outlets. The multiplicity factor is M. Thus, there are $\frac{M \times N1}{n_1}$ first stage switches and that same number of inlets on each second stage switch. Also, there are $\frac{N2}{n_2}$ second stage switches and that same number of outlets on each first stage switch.

It is easily verified that the number of crosspoints in the network is $N_1 N_2 [M(\frac{1}{n_1} + \frac{1}{n_2})]$. Obviously,

$$[M(\frac{1}{n_1} + \frac{1}{n_2})] \tag{1}$$

is the objective function we wish to minimize. To do this, we must understand the relationship of the three variables involved. This depends on the inlet assignment process(es), which we now discuss.

3. INLET ASSIGNMENT SYNTHESIS AND ANALYSIS

3.1 Inlet Assignment Pattern

The generation of the inlet pattern proceeds as follows. For the moment, we constrain N_1 to be the square of a prime number and limit M to be less than or equal to $\sqrt{N_1}$. We will generate a $\sqrt{N_1}$ by $M\sqrt{N_1}$ array of inlet channel numbers where the $\sqrt{N_1}$ numbers on a row represent inlet channels to be assigned to a given first stage switch. Thus, there will be $M\sqrt{N_1}$ first stage switches each with $n_1 = \sqrt{N_1}$ inlets.

The channel assignments in the array are defined as follows. The channel number of each input channel occurs exactly once in the first $\sqrt{N_1}$ rows of the matrix. A given channel number occurring in column c and row r in the first $\sqrt{N_1}$ rows also occurs M-1 additional times in column c, at rows given by

$$1 + (i-1)\sqrt{N_1} + [r+c(i-1) - i]_{\text{mod}\sqrt{N_1}}$$

for positive integers i from 2 through M. This assignment process effectively generates M sub-arrays of dimension $\sqrt{N_1}$ by $\sqrt{N_1}$, where each sub-array contains each channel number exactly once. The process also effectively rotates columns such that the positive slope diagonals (including wrap-around) of a given sub-array become the rows of the next sub-array. This means that, relative to the first sub-array, the rows of the other sub-arrays are obtained by selecting column entries that lie on lines with positive slope of 1,2,3 etc. We can eliminate the need for considering wrap-around of the positive slope lines by viewing the first sub-array as a torus, with the top connected to the bottom and the left connected to the right. If $\sqrt{N_1}$ is prime, then no pair of numbers will appear on the same row more than once throughout the entire array.

3.2 Bounds for n_2

What is the maximum value of n_2, given this method of assigning inlets? The answer is not known in general, but lower and upper bounds and some solutions have been established for certain cases [6].

A lower bound for n_2 is given by

$$M(M+1) - 1.$$

This bound implies that any number of inlet channels $I \leq M(M+1) - 2$ will appear on at least I switches.

Let S_j and S_k represent sets of first stage switches from two arbitrary sub-arrays. As stated above, no pair of inlet channel numbers appears on the same row more than once throughout the entire array. Thus, any switch in S_j will have exactly $|S_k|$ inlet channels in common with all of the inlet channels appearing in the set of S_k switches. And therefore, the number of inlet channels, which appear on switches in both S_j and S_k is given by $|S_j| \times |S_k|$. Thus, the maximum number of inlet channels, which can appear on particular sets of switches from all sub-arrays, is given by the product of the two smallest cardinalities of these switch sets.

Let S_1 and S_2 be the sub-array switch sets with the smallest and second smallest cardinalities, respectively. Let $|S_1|=|S_2|-d$, where $d \geq 0$. Since we seek a lower bound, we want the minimum number of total switches which will be exceeded by $|S_1| \times |S_2|$ or, equivalently, $|S_2| \times (|S_2|-d)$. The least number of switches obviously occurs when the other M-2 sub-arrays each contain $|S_2|$ switches, giving a total of

$$|S_1|+|S_2|+(M-2)|S_2|=M|S_2|-d.$$

Thus, we wish to find the minimum value of

$M|S_2|-d$ that satisfies

$$|S_2|(|S_2|-d) > M|S_2|-d. \qquad (2)$$

After some rearrangement we have

$$|S_2|(|S_2|-M) > d(|S_2|-1).$$

In all non-trivial cases $|S_2| \geq 1$ and it was given earlier that $d \geq 0$. Thus, in the above expression, $d(|S_2|-1) \geq 0$ implying that $|S_2|-M > 0$ or $|S_2| > M$. So, for integer values we have

$$|S_2| \geq M + 1. \qquad (3)$$

Solving expression (2) for d we find

$$d < \frac{|S_2|(|S_2|-M)}{|S_2|-1}.$$

Therefore,

$$M|S_2|-d > M|S_2|- \frac{|S_2|(|S_2|-M)}{|S_2|-1} = \frac{|S_2|^2(M-1)}{|S_2|-1}. \qquad (4)$$

Differentiating the right side of this expression with respect to $|S_2|$, we find a positive slope if $|S_2| > 2$. However, from expression (3) we know that $|S_2| \geq M + 1$, and since $M \geq 2$ (for non-trivial cases), the minimum value of the right side of expression (4) occurs at $|S_2| = M + 1$. Substituting in (4) we get

$$M|S_2|-d > M(M+1) - 1 - 1/M.$$

Or, for integer values we have

$$M|S_2|-d \geq M(M+1) - 1$$

which is the lower bound stated earlier. From expression (3) we have $|S_2| \geq M + 1$, and thus this bound obtains for prime $\sqrt{N_1} \geq M + 1$.

If $M = 2$, it is known that n_2 can never be greater than 5. And if $M = 3$, n_2 can never be greater than 13. All that is required to reach these upper bounds is for $\sqrt{N_1}$ to be at least 3 in the case $M = 2$, and to be at least 6 in the case $M = 3$. (Note that 6 is not a prime number. This will be discussed later.)

For any value of $M > 3$, the upper bound for n_2 increases as N_1 increases. However, this depends on which M of the $\sqrt{N_1}$ possible sub-arrays are selected for the inlet assignment array. For one known method of selecting these sub-arrays [6], an upper bound of n_2 (independent of N_1) is given by

$$\frac{2}{3}M^3 - 2M^2 + \frac{16}{3}M - 3. \qquad (5)$$

This upper bound was obtained from a geometric model that can be considered as an n×n array model with n tending to ∞ (no boundary effects). This bound remains an upper bound a fortiori for the array model since both use the same set of slopes.

Let us make some additional comments concerning the upper bound of n_2. We denote a group of r inlets as deficient if the total number of switches containing appearances of these inlets is less than r. Call r the deficiency number. For M given sub-arrays, any deficiency number minus one is an upper bound for n_2. And the minimum deficiency number minus one is the lower bound for n_2 for those M sub-arrays. Expression (5) is obtained by subtracting one from a set of deficiency numbers for a special pattern of selecting M sub-arrays. We conjecture these deficiency numbers are minimum (and therefore expression (5) a lower bound for n_2) for that pattern of selection. In fact, we conjecture these deficiency numbers are minimum for any selection of M sub-arrays, implying expression (5) a lower bound for n_2 for any selection of M sub-arrays. Additional support for this conjecture is provided by a result in [9] which shows that the lower bound for n_2 grows at least on the order of $M^{5/2}$ for large M. Other patterns of selecting sub-arrays may yield larger minimum deficiency numbers (which is desirable). Unfortunately, we know no general method of ascertaining deficiency numbers (minimum or otherwise) for all values of M. The matter is complicated further by the fact that boundary effects will occur if N_1 is not large enough.

For $M = 4$ and $\sqrt{N_1} = 61$, we have examined various patterns of selecting four sub-arrays and found a pattern with a minimum deficiency number of 50. This may not be a minimal value over all patterns but we believe that it is close. The purpose of discussing this example is to indicate that the maximum values of n_2 can be significantly larger than those given in (5). Estimates for values of $M > 4$ have not been attempted.

We can develop an expression for the upper bound of n_2 in terms of N_1 and M, using an approach similar to that in [11], where lower bounds were established for the number of crosspoints in concentrators. Let S_i represent a set of $n_2 - 1$ first stage switches. Let $I(S_i)$ represent the set of inlet channels which have appearances only on the S_i switches. If any n_2 inlet channels were to have all of their appearances on less than n_2 switches, the requirement of being able to access the n_2 channels on n_2 switches could not be met. Thus,

$$|I(S_i)| \leq n_2 - 1.$$

Assuming that each of the N_1 inlet channels appears on M different first stage switches and assuming n_1 inlets on each first stage switch, then summing over all possible selections of S_i, we have

$$\sum_{i=1}^{\binom{MN_1/n_1}{n_2-1}} |I(S_i)| \leq \binom{MN_1/n_1}{n_2-1} (n_2-1). \qquad (6)$$

Whenever a given inlet channel is a member of $I(S_i)$, it appears on exactly M of the S_i switches. Thus, there will be $\binom{MN_1/n_1 - M}{n_2 - 1 - M}$ selections of S_i for which a given inlet channel will be a member of $I(S_i)$. Therefore, we have

$$\sum_{i=1}^{\binom{MN_1/n_1}{n_2-1}} |I(S_i)| = \binom{MN_1/n_1 - M}{n_2 - 1 - M} N_1.$$

Substituting in expression (6) we get,

$$\frac{MN_1/n_1 - M}{n_2 - 1 - M} N_1 \leq \frac{MN_1/n_1}{n_2 - 1} (n_2-1).$$

After simplifying and rearranging we get,

$$(n_2-2)(n_2-3)\dots(n_2-M) \leq \frac{M}{n_1}[(MN_1/n_1 - 1), (MN_1/n_1 - 2)\dots(MN_1/n_1 - M + 1)]. \quad (7)$$

Thus, an upper bound for n_2 is given by a value of n_2 which satisfies expression (7) as an equality.

For the inlet assignment pattern described in the previous section, $n_1 = \sqrt{N_1}$, in which case expression (7) becomes

$$(n_2-2)(n_2-3)\dots(n_2-M) \leq \frac{M}{\sqrt{N_1}} [(M\sqrt{N_1} - 1), (M\sqrt{N_1} - 2)\dots(M\sqrt{N_1} - M + 1)]. \quad (8)$$

It is interesting to note that for $M = 2$, expression (8) becomes

$$n_2 \leq 6 - \frac{2}{\sqrt{N_1}}.$$

Thus, for $M = 2$ and finite N_1, the greatest integer value of n_2 is 5, as asserted earlier.

3.3 More on the Inlet Assignment Process

At this point, some additional comments should be made about the inlet assignment array. First, it should be noted that in addition to the $\sqrt{N_1}$ sub-arrays which can be generated by the above process, one more sub-array can be generated by interchanging the rows and columns of the first sub-array (or equivalently, any of the other sub-arrays). Thus, a total of $\sqrt{N_1} + 1$ sub-arrays can be generated.

Next, we consider the case where N_1 is not the square of a prime. The process is straightforward. Simply choose the next largest number that is the square of a prime, and leave blanks for unused inlet channel numbers when generating the array. If the number of blanks is large enough, some optimization can be done by choosing blanks so as to be able to eliminate entire rows in the array (and thus eliminate some first stage switches).

Earlier we restricted $\sqrt{N_1}$ to be a prime. When $\sqrt{N_1}$ is a composite number, the assignment pattern given in Section 3.1 plus the row-column interchange assures that f + 1 sub-arrays can be generated (where f is the smallest factor of $\sqrt{N_1}$), such that no pair of numbers appears in more than one row. Thus, when $M = 2$ or 3, N_1 may be the square of any number. When $M = 4$, N_1 may be the square of any odd number.

4. GROWTH RATE IN MULTISTAGE NETWORKS

It was mentioned earlier that the connectivity function to be performed by any switch in the network is the same as that for the entire network; i.e., to connect any inlet to any or all outlets simultaneously. Thus, for large networks, a potential exists for further reduction in crosspoints by replacing individual switches with two stage networks. And, of course, this process can be repeated to produce multistage networks.

The following summarizes results obtained in [7] concerning network growth rate for the sub-array method of assigning inlets. Let $G_s(N_1,N_2)$ denote the number of crosspoints needed for a multistage network with N_1 inlets, N_2 outlets, and s stages. We can show that $G_s(N_1,N_2)$ is linear in N_2; i.e., we can write $G_s(N_1,N_2) = N_2 \times G_s(N_1)$. Thus, we need to find expressions only for $G_s(N_1)$.

Suppose that n_2 can be approximated by M^k and that $G_s(N_1) \rightarrow G(N_1)$ for s, N_1 large. Then we can show that $G(N_1) \leq (\log N_1)^c$; where, for $k = 3$, c is approximately 1.51; for $k = 2.5$, c is slightly less than 2.

Suppose that n_2 is now approximated by $a \times M_k$. We can show that for $s = 2$ and $k = 2$,

$$G_2(N_1) = 2(1/a)^{1/2} N_1^{3/4},$$

for $s = 2$
and $k = 3$,

$$G_2(N_1) = 3(1/4a)^{1/3} N_1^{2/3},$$

and for $s = 3$
and $k = 3$,

$$G_3(N_1) = (8/a)^{1/2} N_1^{1/2}$$

for large N_1.

5. PATH HUNT AND REARRANGEMENT ALGORITHMS

5.1 Path Hunts

For the initial description, it will be assumed that there is only one second stage switch. The extension to multiple second stage switches is straightforward.

When an outlet requests a connection to an inlet, two possibilities may arise. One is that the requested inlet is already connected to the second stage switch. In this case, one simply operates the crosspoint in the second stage switch to connect the inlet and the outlet. If the requested inlet does not currently appear on the second stage switch, an attempt is made to find an idle first stage switch that has an appearance of the requested inlet. If this search is successful, the appropriate crosspoints are operated in both the first and second stage switches to connect the inlet and outlet. If this search is not successful, a rearrangement will have to be performed, as will be subsequently described.

In the generalized two stage network, each of the second stage switches has dedicated links to all first stage switches. Since each first stage switch is a non-blocking broadcast switch, a given second stage switch can hunt and have access to any inlet on a first stage switch, independent of the activity on the other second stage switches.

Since a multistage network can be viewed as a two stage network, where each stage is itself a multistage network, the path hunt strategy for a two stage network can be applied recursively for the multistage network. For example, in a three stage network, when an inlet request is made, the path is initially hunted by finding the inlet on an idle first stage switch. This will determine the appearances of the inlet on the middle stage switches and the path hunt for the last two stages then proceeds as described above.

5.2 Rearrangements

As with path hunts, rearrangements will be described initially for a two stage network having only one second stage switch. Extensions to generalize two stage networks and multistage networks are straightforward.

A rearrangement is required when a requested inlet is neither currently available at the second stage switch nor at an idle first stage switch. This means that all M first stage switches, which have an appearance of the requested inlet, are all being used to connect other inlets. Thus, one of these other inlet connections (denoted as level 1 connections) must be moved to another first stage switch. This will be immediately possible if any one of the level 1 channels is available on some other idle first stage switch.

If none of the level 1 channels can be accessed elsewhere on an idle first stage switch, then the search proceeds one level deeper. Now, an examination is made of those inlet channels (denoted as level 2) which are inhibiting the movement of level 1 channels. If a level 2 channel is available on some other idle first stage switch, it can be moved, thus making available an idle switch for a level 1 channel, which can now be moved, and thereby providing an idle switch for the originally requested channel.

This process is continued at successively deeper levels until a channel inlet is found that can be moved to an idle switch. The number of inlets which must be moved is equal to the level of depth reached in the search to find the first movable channel inlet. Thus, the algorithm determines (1) the minimum number of connections to be moved, (2) which connections are involved, (3) the order in which they are to be moved, and (4) the new connection configuration. It should also be noted that with this algorithm it is always possible to perform a "make before break" on a connection which must be moved; i.e., the inlet can be momentarily simultaneously accessed from both the old and new switches during the process of moving the connection. This is important in applications that cannot tolerate a loss of information resulting from a momentary open connection.

For multistage networks, the rearrangement process proceeds in a recursive manner similar to that described for path hunts.

8. REFERENCES

[1] F. K. Hwang, "Three-Stage Multiconnection Networks which are Nonblocking in the Wide Sense", B.S.T.J., Vol. 58, pp. 2183-2187, 1979.

[2] F. K. Hwang, "Rearrangeability of Multiconnection Three-Stage Networks", Networks, Vol. 2, pp. 301-306, 1972.

[3] G. M. Masson, "Upper Bounds on Fanout in Connection Networks", IEEE Transactions on Circuit Theory, Vol. 20, pp. 222-229, 1973.

[4] G. M. Masson and B. W. Jordan, "Generalized Multi-Stage Connection Networks", Networks, Vol. 2, pp. 191-209, 1972.

[5] C. D. Thompson, "Generalized Connection Networks for Parallel Processor Intercommunication", IEEE Transactions on Computers, Vol. 12, pp. 1119-1125, 1978.

[6] D. Z. Du, F. K. Hwang and G. W. Richards, "A Problem of Lines and Intersections with an Application to Switching Networks", Algorithms in Combinatorial Design Theory, 1985.

[7] G. W. Richards and F. K. Hwang, "A Two Stage Rearrangeable Broadcast Switching Network", to appear in IEEE Transactions on Communications, October 1985.

[8] G. M. Masson, "Binomial Switching Networks for Concentration and Distribution", IEEE Transactions on Communications, Vol. 9, pp. 873-883, 1977.

[9] R. W. Kufta and A. G. Vacroux, "Multiple Stage Switching Networks with Fan-Out Capabilities", IEEE Proceedings of the Computer Networking Symposium, Silver Spring, pp. 89-96, 1983.

[10] P. Hall, "On Representatives of Subsets", J. London Math. Soc., Vol. 10, pp. 26-30, 1935.

[11] S. Nakamura and G. M. Masson, "Lower Bounds on Crosspoints in Concentrators," IEEE Transactions on Computers, Vol. 12, pp. 1173-1179, 1982.

TELETRAFFIC ISSUES in an Advanced Information Society
ITC-11
Minoru Akiyama (Editor)
Elsevier Science Publishers B.V. (North-Holland)

THE ABSTRACT CHANNEL MODEL

Lihren WEY (Wei Li Ren)

Hunan Normal University,
Changsha, Hunan, The People's Republic of China

ABSTRACT

In this paper we generalize the channel model to abstract case. The results are obtained by compared with Shannon model. The theory of communication has developed extensively and gained enormous importance. We should not imagine the act of coding in too obvious a manner, thus we simplity the problem. When we doesn't care how to code, we speak of a abstract codes. To a certain extent, this theory is, of course, the basis of any special theory that the code is given. Our study may not have seemed to be too complete, but it had as purpose, at least partly, to establish those properties that we want to use in any special theory.

1. SHANNON PROBLEM

We will use the following notations:

(X, Bx) : message source
(Y, By) : message sink
U : signal source
V : signal sink

Let (X, Bx) be a measurable space, Bx is a B-field over X. Similarly, assume that (Y, By) is the measurable space, By is a B-field over Y. Let U and V be two sets of real numbers.

ACS = (E, F) is said a abstract communication systems with measure structure with regard to $X \cdot U \cdot V \cdot Y$, if

E = (X, Bx, Px(.)) is considered information source;

F = (U, P(v/u), V) is considered communication channel, where P(v/u) is the conditional distribution of V. Next,

f: X ⟶ U is called encoding

and

g: V ⟶ Y is called decoding.

If (f, g) is given, we define the measure m(x, y) on (X · Y, Bx · By) by

$$m(x \cdot y) = Px(x)P(g^{-1}(y)/f(x)) \qquad (1)$$

here

$$g^{-1}(y) = (v/g(v) = y)$$

The communication systems ACS is said to be determinate codes and is denoted ACS (f,g), (f,g) is given.

Let d(x y) be a measurable function. We should call d(x y) metric of distortion (MD). When we consider that ACS is a model of source sequance, we adopt the following convention. Let

$d(x \cdot B) = \min_x (d(x \cdot y)/y \in B,\ B \subset Y)$

such that

$\mathrm{Lim}_c P_x(x/x \in H(B, C)) = 0$

where

$H(B,c) = (x/x \in Bx,\ d(x \cdot B) \leq c)$

Finality we define the rule of reliability of ACS (f,g) by

$$\int_{x \cdot y} d(x \cdot y)\, dm \leq c \qquad (2)$$

Now let us consider Shannon problem: Under what conditions there are proper coding (f,g) for communication systems ACS such that (1), (2) hold.

In this paper, we shall give conditions of existence of dx and dy, but not straightway find f and g. dx and dy is called MD on X and Y, if for d(x · y) have

$$d(x \cdot y) = dx\,(x) - dy\,(y). \qquad (3)$$

Hence f and g is functions of (x, dx(x)) and (y, dy(y)) respectively. That is

$$f(x) = f(x,\ dx(x)) \qquad (4)$$
$$g(x) = g(y,\ dy(y)) \qquad (5)$$

2. MEASURE

We begin the investigation of the measure m(x · y) from (2). If for d(x · y) there exists a measure $m^*(x \cdot y)$ such that

$$c(m^*) = \min \int_{x \cdot y} d(x \cdot y)\, dm \qquad (6)$$

then (2) hold, so that $c(m^*) \leq c$

In order to obtain dx and dy, we must have a restriction for m(x · y). If Py(.) be given, we put

$$m(x \cdot y) \leq P\ x(x), \qquad x \in Bx \qquad (7)$$
$$m(x \cdot y) \geq P\ y(y), \qquad y \in By \qquad (8)$$

For example, if Px(x) Py(y) = 0, we have

$$m(x \cdot y) = cPx(x)Py(y) \qquad (9)$$

here, $c \in (1/x(x),\ 1/y(y))$. From this example we immediately conclude that if Px(.) Py(.) = 0 then (9) satisfies (7) (8).

Generally we have some measures m^n which satisfies (7) and (8).

On the other hand, if

$$m^n(X \cdot Y) \leq \text{ constant}$$

for all n and

$$m^n \longrightarrow m^*$$

By extended theory of Vitali-Hahn-Saks ([5], P.43), then m^* is measure on (X · Y, Bx · By). We wish to show that m^* satisfies (7), (8) and (6).

Let d be bounded continuous on X Y. Let Px(.) and Py(.) be finite measures. Then for (m^n) have

$$\left(\int_{x \cdot y} d\ \ dm^n \right). \qquad (10)$$

If there are a subsequence of (10) which converges to least upper bound (or greatest lower bound) of (10), written as

$$\int X \cdot Y \, d \, dm^* \qquad (11)$$

then m^* is called a optimization measure. It is known if the sequence (10) of real number is bounded. then (10) must contains a subsequence which converges to least upper bound (or greatest lower bound) of (10). (m^n) is called weakly converges and written as

$$m^n \xrightarrow{W} m^* \qquad (12)$$

As regards this matter we will use the following result.

Theorem 1: ([5]. P.196) let m^1, m^2 ... and m^* be finite measures on the Borel sets B (X · Y) of a metric space X · Y. For every $x \cdot y \in B(X \cdot Y)$ have

(i) $m^n(x \cdot y) \longrightarrow m^*(x, y)$

(ii) $m^*(d(x \cdot y)) = 0$

then

$$m^n \xrightarrow{W} m^*$$

We now proceed to show that m^* satisfies (7) and (8).

Theorem 2: Let X and Y be metric space. Let d is continuous bounded functions. Let Px(.) and Py(.) are complete and satisfies (7) and (8). If

$$m^n \xrightarrow{W} m^*$$

then m^* satisfies (7) and (8)

Proof: Since $m^n \xrightarrow{W} m^*$ by 5 (P.196), we have

$$m^n(X \cdot Y) \longrightarrow m^*(X \cdot Y)$$

and for every open subset $O \subset X \cdot Y$, have

$$\lim_n \inf m^n(O) \geq m^*(O) \qquad (13)$$

or for every closed subset $c \subset X \cdot Y$, have

$$\lim_n \sup m^n(c) \leq m^*(c) \qquad (14)$$

Let open set $O \in B(X)$, by (7) and (13), we have

$$PX(O) \geq \lim_n \inf m^n(O \cdot Y) \geq m^*(O \cdot Y) \qquad (15)$$

Since P(.) is complete, by (15), for any $x \in B(Y)$, we have

$$Px(x) = \inf (Px(O)/x \subset o \in B(X)) \geq \inf (m^*(O \cdot Y)/x \subset o \in B(X)) \geq m^*(x \cdot y)$$

Similarly let closet set $C \in B(Y)$, by (8) and (14), we have

$$Py(c) \leq \lim_n \sup m^n(X \cdot C) \leq m^*(X \cdot C) \qquad (16)$$

Since Py (y) is complete, by (16), for any $y \in B(y)$, we have

$$Y(y) = \sup (Py(c)/c \subset y \in B(Y)) \leq \sup (m^*(X \cdot c)/c \subset y \in B(Y)) \leq m^*(X \cdot Y)$$

Thus $m^*(x \cdot y)$ satisfies (7) and (8).

3. METRIC OF DISTORTION

In this section, we first give the definition of the support concept. The support of the measure m on $x \subset X$ is call measurable open neighborhood which has positive m-measure of X. The set of all support on X is called the support of m on X, denoted Sm(X).

Let m satisfy (7), (8) and let

$$d(x) - dy(y) \geq d(x \ y) \quad x \in Bx, \ y \in By \qquad (17)$$

hold. where $(dx, dy) \geq 0$. If

$$\int_{sm(X \cdot Y)} (d-(dx-dy)) \, dm = 0 \qquad (18)$$

We say that (dx, dy) is support of MD.

Theorem 3: Let Tx Bx be topology over X, and let $Ty \subset By$ be topology over Y. Let be continuous with respect to $Tx \cdot Ty \subset Bx \cdot By$. Then there exist functions Bx and By such that (dx, dy) is support of MD.

Proof: Let $(Xik \cdot Yik) \in Sm(X \cdot Y)$, k=1,2,..., ni(ni be finite). For any $x \in Bx$, we define

$$dx(x) = \sup_i (-d(x \cdot yi1) + \sum_{k=1}^{ni} d(X_{ik} \cdot Y_{ik}) - \sum_{k=1}^{ni-1} d(X_{ik} \cdot Y_{ik+1}))$$

By (17), for any y BY, we define

$$d(y) = \inf_{x \in Bx} (dx(x) + d(x \ y))$$

We imitate [6], then obtain that if d is bounded then dx is bounded. It follows that dy is bounded too. Next since

$$dx(x) \geq -d(x \cdot yi1) + d(xi1 \cdot yi1) + dx(xi1)$$

that is

$$dx(x) + d(x \cdot yi1) \geq d(xi1 \cdot yi1) + dx(xi1)$$

$$\inf_{x \in Bx} (dx(x) + d(x \cdot yi1)) = d(xi1 \cdot yi1) + dx(xi1)$$

Hence

$$dy(yi1) = d(xi1 \cdot yi1) + dx(xi1)$$

and the proof is complete.

REFERENCES

1 C.E. Shannon, "A. Mathematical Theory of Communication", B.S.T.J., 27, 379-423, 1948.

2 Guoding Hu and Shiyi Shen, "Some New Development of Shannon Theory," Advances in Math. (China), vol. 13, no. 4. 254-265, 1984.

3 Guoding Hu, Acta Math. Sinice, vol. 11. no. 3, 260-294, 1961.

4 Shiyi Shen, Annals of Math. (China), I(2), 226-234, 1980.

5 B.A. Robert, "Measure Integration and Functional Analysis," Academic Press Inc., 1972.

6 Lihren Wey, "The Dynamic Programming of Set-valued State," hunan Annals of Math., vol. 1.4, no. 1, pp66-73, 1984

TELETRAFFIC ISSUES in an Advanced Information Society
ITC-11
Minoru Akiyama (Editor)
Elsevier Science Publishers B.V. (North-Holland)

END-TO-END DIMENSIONING OF TRUNK NETWORKS A CONCEPT - BASED ON EXPERIENCE FROM THE DANISH NETWORKS

Chr.Asgersen

The Copenhagen Telephone Company

Copenhagen, Denmark

ABSTRACT

An end-to-end dimensioning concept for hierarchical trunk networks is presented. The dimensioning concept is developed for the Danish analogue network. Using the experience from this development an optimization method for a non-hierarchical digital network is elaborated. Finally the use of the Multi-Goal Theory is considered.

1. INTRODUCTION

The Danish trunk network existing during the analogue era was designed with extensive regard to traffic theory.

The reason for doing so was unsatisfactory service quality after the introduction of direct distance dialling in the early 1950'ies. Analyses showed that unsuitable dimensioning of the trunk network was one of the main reasons for the troubles.

Aiming at an improvement a new network concept was worked out basicly from viewpoints founded in traffic theory.

The first approach was the Cluster Engineering which was introduced in the Danish network about 1965.

During the next decade the Cluster Engineering concept was further developed. The result of this development was an end-to-end dimensioning concept for an entire hierarchical trunk network. This concept will be presented here.

Facing the challenge of designing a network for the digital age it was natural to use the experience from the analogue era. As a consequence of this experience it was decided to develope a dimensioning concept for a non-hierarchical modular structured trunk network, aiming at the digital network.

Derivation of formulas similar to the ones derived for the hierarchical network seemed to be impossible, however. In stead an optimization method was elaborated.

An outline of this optimization method will be presented.

Finally the use of the Multi-Goal Theory for Public Utilities is considered as a tool for taking the economics into account.

2. FRAMEWORK FOR THE DIMENSIONING

Modern networks require specific reliability plans. These plans form frameworks which include the grade of service standards for the traffic quality.

The grade of service standards are in general specified on a subscriber to subscriber basis and comprise parts covering faults etc. and parts saved for dimensioning of the network.

The part of the service standards at disposal for the dimensioning of the trunk network has to be distributed between contributions from the exchanges and contributions from the individual trunk groups in the network.

It is of the utmost importance for the performance of the network that the dimensioning rules for the individual trunk groups are elaborated in such a way that they adequately fit into the framework of the grade of service standards.

Therefore the dimensioning concepts, which I am going to describe, are elaborated on the basis of given point to point grade of service standards.

3. THE DIMENSIONING CONCEPT FOR ANALOGUE NETWORKS

The analyses of the service in the early direct distance dialling network showed that the basic problem was the large spread of the subscriber to subscriber congestion. Alternative routing was introduced with a varying number of hunting stages. The dimensioning was done ignoring the network architecture. The blocking versus the traffic formed a curve, having a very increasing shape. Often calls offered to the finals only, were completely quelled. Calls having access to a number of consecutively hunted thrunk groups on the other hand had an unnecessary low blocking rate.

In solving of the problem the papers by Mr.R.J.Wilkinson at the First International Teletraffic Congresses in Copenhagen (Litt.1) and the Fourth International Teletraffic Congress in London (Litt.2) were an important inspiration. Especially the London paper was used for modelling the network.

The dimensioning principle was what nowadays is denoted Cluster engineering.

The grade of service was related to a Cluster or "Final System" consisting of all the high usages having overflow to the same final - and the final itself.

This was an improvement of the engineering practise compared to the traditional method, where the grade of service standard was related to the final. By the traditional method the point to point blockings are haphazard results of the routes used through the network.

The Cluster Engineering as introduced in the Danish network is described in my paper to the Fifth International Teletraffic Congress in New York 1967 (Litt.3).

Experience with the Cluster Engineering as originally introduced did not allow adequately for dimensioning of the service protection groups.

Investigations on that occasion led to an extension of the Cluster Engineering method leading to a real end-to-end dimensioning method.

In the following I shall explain the end-to-end dimensioning concept elaborated.

The blocking between 2 exchanges connected by a direct high usage group can be expressed approximately as follows (fig.1):

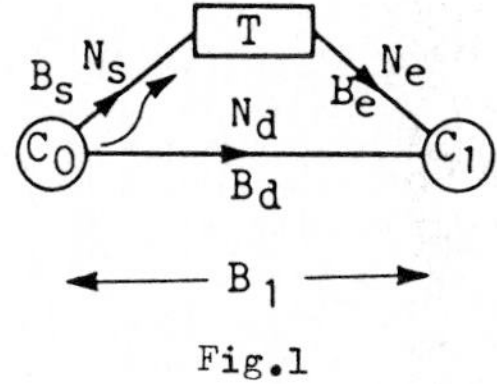

Fig.1

$$B_{o1} = B_d B_s + B_d B_e \quad (1)$$

B_{o1} = the blocking of the traffic from exchange C_o to exchange C_1

B_d = the engineering blocking for the direct high usage group N_d

B_s = the engineering blocking for the final trunk group N_s

B_e = the engineering blocking of the simple group N_e

If the grade of service standard for traffic between C_o and C_1 is given as the blocking B_1, we find:

$$B_{o1} = B_d B_s + B_d B_e = B_1 \quad (2)$$

Corresponding to equation (1) an equation can be derived for the case that no direct high usage group exist. The traffic in this case is offered to a service protection group N_p with overflow to N_s

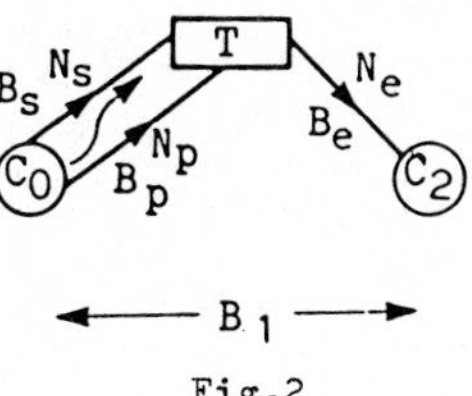

Fig.2

With reference to fig.2 it is easy to derive the following approximate equation:

$$B_{o2} = B_p B_s + B_e \quad (3)$$

B_p = the engineering blocking of the service protection group N_p

If we want to use the same grade of service standard B_1 for traffic between C_o and C_2 as we used for traffic between C_o and C_1, we find:

$$B_{o2} = B_p B_s + B_e = B_1 \quad (4)$$

From (2) and (4) we find:

$$B_d B_s + B_d B_e = B_p B_s + B_e \quad (5)$$

Using (5) B_p can be expressed by B_d, B_s, and B_e

$$B_p = B_d - \frac{B_e}{B_s}(1-B_d) \quad (6)$$

(6) shows that B_p cannot be chosen independently, if B_1 must be maintained.

(6) also shows that if we want to maintain the same B_1 for all pairs of exchanges subordinated to a transit exchange T_1 then we must choose the same value B_d for engineering of all the high usages.

This discovery raises the question of whether the traditional triangel optimization method for engineering high usage groups (Litt.4) imply a real optimization. In my opinion it does not. If more than one high usage group have overflow to the same final, the method does not secure that an optimal arrangement of the trunks is obtained, observing a given grade of service constraint. I have dealt with this problem in my contribution to the Seventh International Teletraffic Congress in Stockholm 1973 (Litt.5).

Continuing my explanation of the end-to-end dimensioning concept dealt with, I make use of the rule that the same grade of service value should be used for traffic between all exchanges subordinated to the same transit exchange. This implies the following expressions:

Case 1, (fig.3), a simple group between C_o and C_1:

$$B_{o1} = B_{e1} = B_1 \tag{7}$$

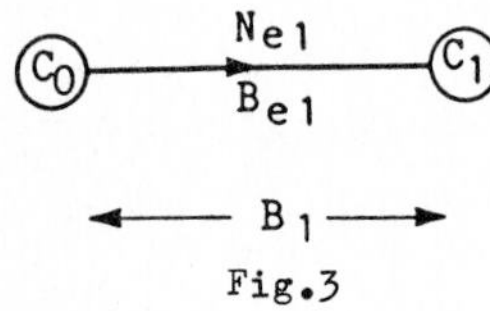

Fig.3

Case 2, (fig.4), a transit connection between C_o and C_1:

$$B_{o1} = B_e' + B_e'' = B_1 \tag{8}$$

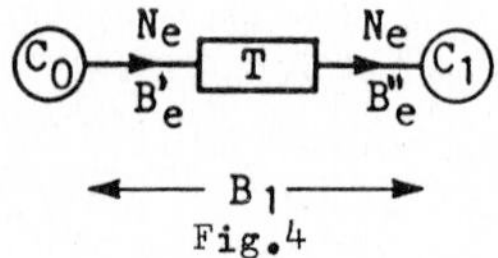

Fig.4

I will now do a trick: I claim that

$$B_e' = B_e'' = B_e \tag{9}$$

Then from (8) I find:

$$B_{o1} = 2B_e = B_1 \qquad B_e = \frac{B_1}{2} \tag{10}$$

From (7) together with (10) I find:

$$B_{e1} = B_1 = 2B_e \tag{11}$$

The equation (10) appears to be extremely valuable. This will be recognized by inserting (10) into (2):

(10) into (2) $\quad B_dB_s + B_d\frac{B_1}{2} = B_1$

$$B_d = \frac{1}{\frac{B_s}{B_1} + \frac{1}{2}} \tag{12}$$

In (12) B_d is expressed by B_1 and B_s

(12) can be converted to $B_s = B_1\left(\frac{1}{B_d} - \frac{1}{2}\right)$ (13)

In (13) B_s is expressed by B_1 and B_d

If (10) is inserted into (4) we find:

$$B_pB_s + B_e = B_1 \qquad B_pB_s + \frac{B_1}{2} = B_1$$

$$B_p = \frac{B_1}{2B_s} \qquad B_p = \frac{1}{\frac{2}{B_d} - 1} \tag{14}$$

The equations (7), (10), (12), (13), and (14) express the engineering values for all possible trunk groups connecting the exchanges subordinated to the same transit exchange. The engineering values can be calculated from the knowledge of the grade of service standard B_1 and a chosen value of B_d or B_s.

The idea of deriving the engineering values from the given grade of service can be extended to comprise all trunk groups in a hierarchic network.

The blocking for a connection as shown at fig.5 can be expressed by the following formula:

$$B_{oq} = B_{N1} + B_{N2} + \dots B_{Nq} \tag{15}$$

The formula expresses that the blocking from a given exchange C_o to another given exchange C_q with a close approximation, is equal to the sum of the blocking $B_{N1}, \dots, B_{Nq}$ for the circuit groups carrying the traffic from C_o to C_q.

If the grade of service value B_q is given, we find:

$$B_{N1} + B_{N2} + \dots + B_{Nq} = B_q$$

Fig.5

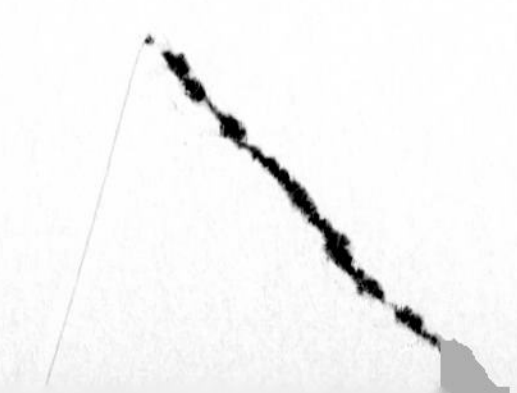

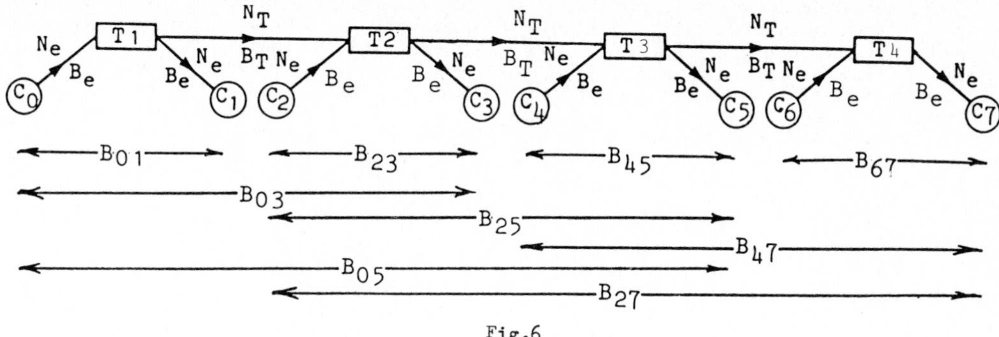

Fig.6

Fig.6 shows a hierarchical structured network with 4 transit exchanges:

Let us demand that

$B_{o1} = B_{23} = B_{45} = B_{67}$

Let us furthermore demand that

$B_{o3} = B_{25} = B_{47}$ and

$B_{o5} = B_{27}$

These claims cause the following formula:

$B_{o7} = B_e + (4 - 1)\, B_T + B_e$

or more general in the case of q transit exchanges "between C_o and C_q":

$$B_{oq} = B_e + (q - 1)\, B_T + B_e \qquad (17)$$

By the grade of service value B_q for the point to point connection given, we find:

$$B_e + (q - 1)\, B_T + B_e = B_q \qquad (18)$$

Inserting of (10) into (18) implies:

$(q - 1)\, B_T + B_1 = B_q$

$$B_T = \frac{B_q - B_1}{q - 1} \qquad (19)$$

One of the main ideas of this dimensioning concept is that the point to point blocking shall be the same, independent of the actual route for a given call and independent of whether a direct group exists or do not exist.

This claim can be used for deriving a formula for the engineering of high usage groups "embracing" a number of x transit exchanges (fig.7).

$$B_{dx} = \frac{B_1 + \frac{x-1}{q-1}(B_q - B_1)}{\frac{B_1}{B_d} + \frac{x-1}{q-1}(B_q - B_1)} \qquad (20)$$

It should be noticed that B_{dx} is given by the knowledge of the grade of service values B_1 and B_q as well as the engineering value B_d for a direct high usage between two exchanges subordinated to the same transit exchange.

For the direct trunk groups the number of circuits can be calculated direct on the basis of Erlang's B-formula. The number of circuits in finals must be calculated with regard to the peakedness for example according to the Equivalent Random Method (Litt.1).

The concept gives a simple and well defined correlation between the service standards and the engineering values, which makes it possible to get a clear picture of the congestion in the network.

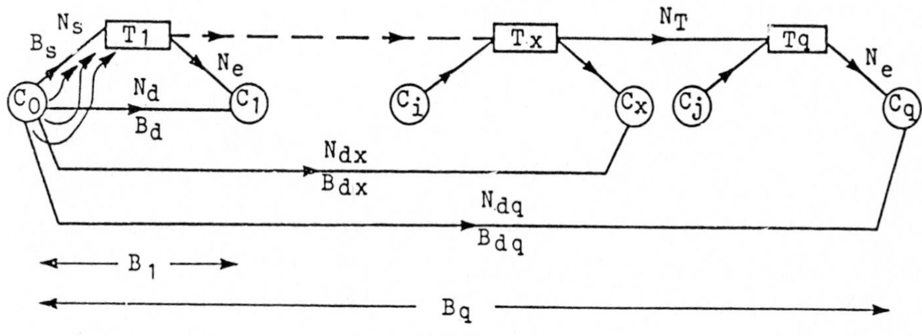

Fig.7

So far the costs are not included in the calculations. However, as mentioned in the introduction, I shall mention the Multi-Goal Theory as a tool for taking the economic demands imposed when dimensioning into account.

At this point I will mention that the dimensioning concept is fitted for use of a quasi-optimization method.

The quasi-optimization can be carried out as follows: The network is calculated on the basis of the service standards indicated and the B_d-value chosen. A certain configuration of high usages is chosen. Then the costs of the networks are calculated. The next step is changing some of the parameters and the configuration of high usages. Again the cost of the network is calculated. By series of revised calculations an economic optimization can be performed with close reference to the quality of service.

One of the features of the concept is that it is possible to decide, whether you want to put the major part at the blocking at the local part of the network, or you want to distribute the blocking in another way in accordance with the hierarchy of the network. A desirable distribution of the point to point blocking in the network can be obtained by varying the parameters in the formulas.

4. THE OPTIMIZATION METHOD FOR THE DIGITAL NETWORK

The digital network has some characteristics different from the ones of the analogue network. Here I shall only deal with characteristics important for the dimensioning problem.

Characteristics important for the dimensioning problem are among others.

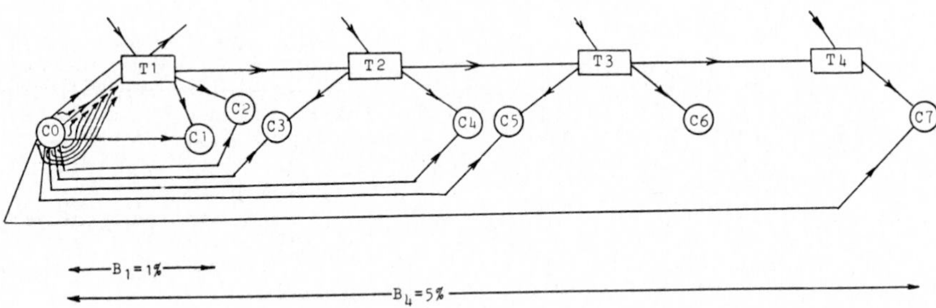

Fig.8

By measurements on networks at daily operation it is difficult to verify the grade of service values. This is due to the fact that a live network never is loaded in accordance with the number of circuits in operation. Therefore a number of simulations are done, aiming at verification of the concept. Overleaf I will show the results of one of the simulations: Fig.8 shows a hierarchical network which has been simulated. The results are indicated in fig.9.

Generally, the simulations show that the purpose of the dimensioning concept is achieved.

1. The network may not necessarily be hierarchically structured.

2. Trunk groups may be established by modules (e.g. PCM-systems of 24 or 30 circuits).

3. It must be accepted that a network node (a transit exchange) may be out of service e.g. for loading with new processor programs. This caused the need of connecting the local exchanges to more than one transit exchange.

From	To	Offered Load (Erlangs)	No. of Trunks	Blocking Aimed at	Calculation	Simulation
C0	C1	21.6	20	0.01	0.00912	0.00775
C0	C2	27.7	25	0.01	0.00923	0.00990
C0	C3	45.0	30	0.0233	0.02166	0.01904
C0	C4	29.3	20	0.0233	0.02138	0.01887
C0	C5	45.9	25 x)	0.0367	0.03389	0.03121
C0	C6	28.7	0 x)	0.0367	0.03454	0.03285
C0	C7	43,2	20	0.05	0.04566	0.04207

x) The traffic is offered to the service protection group.

Fig.9

The first approach of solving the dimensioning problem was to try to derive formulas similar to the ones derived for the hierarchical network. However, this seemed to be impossible, because of the complexity of the network.

Instead it was decided to create the network by a quasi-optimization method on the basis of the point to point blockings and cost parameters given, using a computer calculation system.

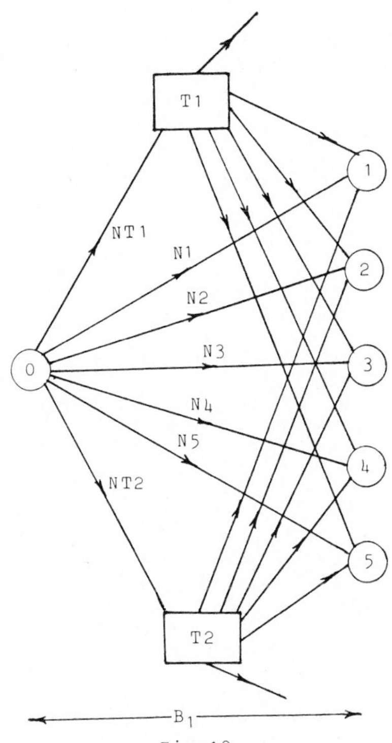

Fig.10

The idea of the computer calculation-system was to examine all possible networks which could serve the exchanges involved. "All possible" sounds ambitious. However, a limitation upwards of the number of modules of the individual potential trunk groups reduces "all possible" to "a limited number".

For each network the traffic flow and the blockings were calculated. All the cases, in which the grade of service was not kept, were abandoned. For all the other ones the costs were calculated and the cheapest chosen.

The calculations are made in two steps: At the first step the network between local exchanges and between local exchanges and their superior transit exchanges is designed.

At the second step the network between the transit exchanges is designed.

Fig.10 shows an example of a network pattern at local exchange level. Each local exchange is connected to 2 transit exchanges, the transit traffic is divided equally between them. A number of potential high usages exist. Each of the high usages may be established by a number of modules. A module may consist of p circuits. In this case p = 30 corresponding to a pcm system of 30 channels. Traffic blocked at the high usages overflow to the finals.

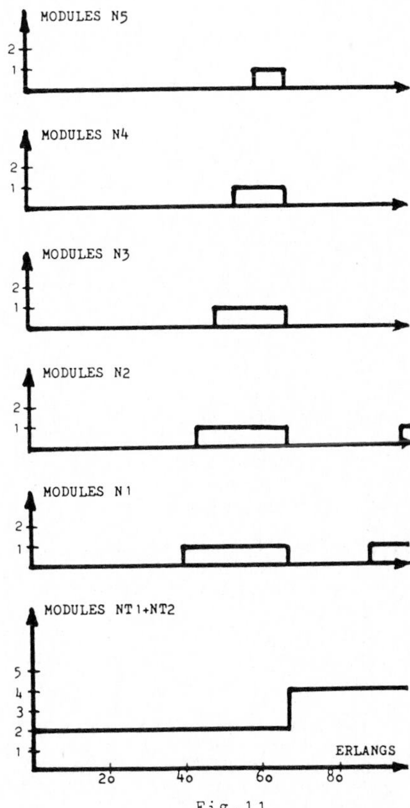

Fig.11

In fig.11 the results are depicted. At the x-axis the total traffic to the cluster is marked. These traffic amounts correspond to a development in time. By each amount of the total traffic the number of modules of the cheapest configuration is printed out (Fig.12).

It should be noticed that the optimal network pattern is varying dependent of traffic growth.

MODULE SIZE = 30

TRAFFIC OFFERED FROM C0 TO C1-C5:
10.0 9.0 8.0 7.5 7.0 PCT. OF TOTAL TRAFFIC

COST PROPORTION DIRECT ROUTE/TRANSIT ROUTE: 40000/70000

AT1	AT2	A01	A02	A03	A04	A05	NT1	NT2	N1	N2	N3	N4	N5	COST
8.00	8.00	1.60	1.44	1.28	1.20	1.12	1	1	0	0	0	0	0	280000.0
20.00	20.00	4.00	3.60	3.20	3.00	2.80	1	1	1	0	0	0	0	320000.0
22.00	22.00	4.40	3.96	3.52	3.30	3.08	1	1	1	1	0	0	0	360000.0
24.50	24.50	4.90	4.41	3.92	3.68	3.43	1	1	1	1	1	0	0	400000.0
27.00	27.00	5.40	4.86	4.32	4.05	3.78	1	1	1	1	1	1	0	440000.0
30.00	30.00	6.00	5.40	4.80	4.50	4.20	1	1	1	1	1	1	1	480000.0
34.00	34.00	6.80	6.12	5.44	5.10	4.76	2	2	0	0	0	0	0	560000.0
44.50	44.50	8.90	8.01	7.12	6.68	6.23	2	2	1	0	0	0	0	600000.0
49.50	49.50	9.90	8.91	7.92	7.43	6.93	2	2	1	1	0	0	0	640000.0
55.00	55.00	11.00	9.90	8.80	8.25	7.70	2	2	1	1	1	0	0	680000.0
61.00	61.00	12.20	10.98	9.76	9.15	8.54	2	2	1	1	1	1	0	720000.0
67.50	67.50	13.50	12.15	10.80	10.13	9.45	2	2	1	1	1	1	1	760000.0
76.00	76.00	15.20	13.68	12.16	11.40	10.64	3	3	1	0	0	0	0	880000.0
78.00	78.00	15.60	14.04	12.48	11.70	10.92	3	3	1	1	0	0	0	920000.0
86.50	86.50	17.30	15.57	13.84	12.98	12.11	3	3	1	1	1	0	0	960000.0
96.00	96.00	19.20	17.28	15.36	14.40	13.44	3	3	1	1	1	1	0	1000000.0

Fig.12

6. USE OF THE MULTI-GOAL THEORY

In Denmark the use of Moe's Principle is a tradition (Litt.6).

According to Moe's Principle the blocking at disposal for engineering purposes is distributed among the devices in a connection which carries traffic.

A rational distribution is done so that the improvement by adding one device to a group of devices is the same for all groups. Thus the blockings caused by the individual groups of devices are proportional to the costs.

However, this economic principle is only valid for a series of device-groups. If the network consists of parallel routes tied together by alternative routing, then the Moe's principle in its original version is not valid.

To meet the demand of a dimensioning concepts as presented here, Dr. Arne Jensen has developed the Multi-Goal Theory which will be subject to a contribution to this congress.

7. CONCLUSION

The dimensioning concept and the optimization method presented here appear as tools for planning in practice. The know-how fetched from traffic theory is partly hidden behind the computation programs. However, the development of the concepts would have been impossible without the contribution from the Teletraffic Congresses in the course of time.

REFERENCES

1. Wilkinson, R.J.: "Theories for Toll Traffic Engineering in the USA". The Bell System Technical Journal, March 1956.

2. Wilkinson R.J.: "Simplified Engineering of Single Stage Alternate Routing Systems" Paper presented at the Fourth International Teletraffic Congress in London, 1964.

3. Asgersen, Chr.: "A New Danish Traffic Routing and Plan with Single Stage Alternate Routing and Consistent Use of Service Protection Finals for First Routed Traffic to Tandem Offices". Congress book of the Fifth International Teletraffic Congress in New York, 1967, page 144/1 and Teleteknik, English Edition, 1967, No.1, page 11.

4. CCITT "Yellow Book, vol.II fascicle 11.3, Rec.E.522, page 67.

5. Asgersen Chr.: "Dimensioning of Trunk Networks According to a Uniform Overload Capacity Criterion". Congress book of the Seventh International Teletraffic Congress, Stockholm 1973, page 523/1 and Teleteknik English Edition 1973, No.2, page 33.

6. Jensen, Arne: "Moe's Principle", The Copenhagen Telephone Company, Copenhagen, 1950.

TELETRAFFIC ISSUES in an Advanced Information Society
ITC-11
Minoru Akiyama (Editor)
Elsevier Science Publishers B.V. (North-Holland)

Service Standards: Evolution and Needs

J. G. Kappel

Bell Communications Research, Inc.
Red Bank, New Jersey, USA

Abstract

The objective of this paper is to give an overview of telecommunications service standards in the USA: where we came from, where we are, new problems and areas for investigation. The main focus will be on Bell Operating Company (BOC) standards used directly for traffic engineering, and on performance monitored by their Public Utility Commissions.

I. Introduction

Increasing competition in the U.S. telecommunications industry heightens the needs for service standards investigations. How can a competing company degrade service without affecting the network as a whole? When capital constraints are imposed, it is important to know where and how service can be compromised with the least impact on customers.

Perhaps the most important technical factor is the introduction of digital switches. Virtually all components are sensitive to peak period loads, so new standards that minimize service risk are essential. Conversely, since service at average loads should be nearly perfect, a new look at other network criteria is needed.

To see where we should go, it is important to understand where we are and how we got there. this paper traces the development of traffic related service criteria in the former Bell System and discusses some of the key open issues.

II. Dial Tone Delay Standards

A great deal of attention has been devoted to dial tone delay standards. There are at least two important reasons for this.

(i) A customer who can't get dial tone can't place any kind of call.

(ii) Dial tone delay is measurable, and all Public Utilities Commissions rightly insist on monitoring it.

The earliest recorded standard (1919) was P.001 (one in a thousand blocking) using the Blocked Calls Held assumption for engineering of line finder groups and line switches in SxS and Panel [1]. This applied to the "busiest hour of the day". As yet there was no specific concept of busy season or of day-to-day variation affecting average service. Measurements of traffic intensity were crude: manual peg counts twice a month, augmented by periodic holding time studies. This was one good reason for the very stringent standard. There were at least two others.

(i) Dialing their own calls was a new chore for customers, and it was felt this should be made as easy as possible [1]

(ii) "If we consider that the selectors are sometimes held busy by 'shorts' or 'grounds' or by subscribers unwittingly (or wittingly?) leaving their receivers down, it becomes necessary to allow for some margin, and this margin is obtained by using the 1 in 1000 probability ..." [2] Mr. Baer was probably the first to recognize that Permanent Signal and maintenance usage were and would remain chronic problems in estimating effective line group loads.

In 1925 the standard was changed to P.01, because "it became apparent that a somewhat less liberal provision of equipment would not perceptibly degrade the service ..." [1]. Another reason, most likely [3], is that the trunk groups and internal SxS selectors were by then engineered to P.01, and it didn't seem logical to give perfect dial tone service if 3% to 4% of the calls could not complete dialing. The P.01 standard remained in effect until World War II.

Warren Turner[1] recounts how the standard was changed to P.02 by order of the War Production Board, the almost-perfect equivalence of P.02 to 1.5% delay over 3 seconds, and the reason why the standard was not changed back after the war. His major point is that shortage of capital and increased demand caused service much worse than P.02 in many offices during and just after the war. When service was improved to P.02 during the 1950s most customers were more than pleased. The P.01 objective had been 'forgotten', so why return to it? Apparently the Commissions saw no cause either, since the basic average busy season standard has remained 1.5% over 3 seconds delay for more than 40 years, and all dial tone speed (DTS) monitoring plans were built around this objective.

In common control (crossbar) offices it became clear that load vs. delay curves were much steeper than for SxS, and standards were proposed to protect customer service on the highest days. The first set of standards proposed in 1955 was the following [3]:

Engineering Time Frame	*% Dial Tone Delay over 3 secs.*
Average busy season busy hour (ABS)	1.5%
Average 10 high day busy hour (10 HD)	3% to 5%
High day busy hour (HD)	10%

This proposal was subject to many comments. R. I. Wilkinson observed that due to the steepness of the load-delay curves, it is extremely unlikely that any dial tone delays would occur outside the 10 high days. Thus an average for 10 days of 8%, and 50 days of no delay for a 60-day average busy season, would result in a 1.33% delay over 3 seconds, corresponding well to the 1.5% standard.

Thus the 10-high-day standard was set at 8% (5% for Panel and SxS, since mean delays were longer there than in crossbar). For several years there was no specific High Day standard, but in 1961 a 20% standard was added, to provide a limit on the worst service a customer could expect to receive. It was estimated that this change would increase overall CO equipment cost by less than 1% [3].

The seemingly innocent addition of a specific High Day threshold on service caused many more problems for traffic engineers and administrators than expected. First, service in crossbar and stored program control (SPC) machines was extremely difficult to predict, beyond the "knee" of the load/service curve [4], [5]. This meant that expected dial tone delay of 20% could easily become 50% or even 80%. Also, it was very difficult to provide guidelines for when to eliminate very unusual High Days from the annual database. It was stated [3] that "Act of God days" could be omitted, but the "highest day which regularly occurs" must be included. Given the statistics of peak period loads, the definition of 'regularly recurring' is very hard to quantify.

The necessity to set thresholds on peak period performance is even more evident in modern digital switching systems than it was in earlier common control systems [6] In most digital systems large, efficient line concentrators are a new source of dial tone delay, and the "knee" of the delay curve may occur at 90% occupancy or even higher. Little is known about customer response to long dial tone delays or to delays of more than 20% over 3 secs. Wilkinson's data studies show only that a Palm j=5 model works well for lower levels of delay, for originating-only service in SxS[7]. In the large digital concentrators, as in SxS, delays are subject to total call holding times, and most delays will be long delays - atypical of other common control systems.

With regard to customer perception of dial tone delay as a motivator for setting standards, three studies will be noted briefly.

1. In 1952 one of the BOCs was still suffering the effects of wartime shortages, and the return to a 1.5% over 3 seconds DTS standard was going to be costly. A special study was organized to determine customer attitudes toward dial tone delay. Nearly 150,000 observations of calls from 2000 customers were made, with the customers polled on their attitudes toward measured average dial tone delay levels [3].

Customer Response to Dial Tone Service	*Measured Percent Delay over 3 secs.*	*10 secs.*
Unsatisfactory	7.7	2.9
Noticeable	6.0	1.9
Never waited	3.8	1.3

On the basis of this some managers recommended an average busy season standard of 4%, with control objectives for the three highest days of each month. (The 10 HD concept had actually evolved from proposals to average service for the 3 highest days of the 3 highest months).

The study results were judged to be a conditioned-response phenomenon. Service considerably worse than 4% dial tone delay had been prevalent in this company for at least seven years, so 4% looked like good service to those customers who experienced it. Neither AT&T nor the other BOCs viewed this as grounds for changing the nationwide standard.

2. By 1975 SxS systems were being replaced much faster than the remaining systems in service were growing. Enough spare equipment was available to provide 'perfect' dial tone service in the remaining offices at low cost. We conducted a busy season study in two problem offices to assess the cost of "no dial tone" trouble report processing as a function of dial tone speed[6]. This showed that (a) customers often report dial tone delay when there is none, probably mistaking it for another type of trouble, (b) there is little noticeable difference between 0 and 1.5% delay, and (c) trouble report rates increase rapidly, beyond 2% delay. Unfortunately for us there was a mild winter that year, so no weather-related high traffic days were observed and we could not measure the effect of very severe delays.

3. Throughout the 1970's a number of controlled experiments were performed to assess the individual and combined effects of many kinds of service impairment: transmission loss, noise, cutoff calls, dial tone delay, post-dialing delay... A summary of results is given in[8]. Generally this report shows that moderate or even somewhat severe dial tone delay has less effect on a customer's attitude toward overall service than the other types of impairment. My own view is that

the results are not very relevant because of the controlled experimental conditions. Usually, customers were invited to participate in an interactive computer communications game. Dial tone delay may not have caused as much frustration as in other environments.

These experiments show no real reason to change the basic average busy season objective or its corollary, the ten-high-day busy hour objective. But the objectives, definitions and data screening methods for 'worst service the customer will see' (e.g. High Day, or once-a-year return period) are subjects that need to be re-examined. This is especially true because new overload control methods for SPC systems have served to dampen the effects of extreme loads on service [9], [10].

As noted, an area for study is the effect on customers of long dial tone delays inherent in the modern line concentrators. Basically, in a 400-server system under Erlang-C conditions, the "20% over 3 seconds" criterion would imply that 10% of the calls are delayed more than 10 seconds. Such delays could mean more than customer annoyance. They may well imply that emergency calls could not be completed, since the threshold at which these delays occur (almost 96% occupancy) is so close to system saturation that no traffic forecasting method could offer protection against near outages. And as yet there are no overload controls designed to protect customer service at the line group level.

III. End-to-End Blocking and Matching Loss

It is common practice to couple the concepts of overall blocking and internal link system matching loss. Standards for the latter come directly from the former, and they are still intimately tied together.

The (local area) end-to-end blocking standard evolved in much the same way as the dial tone delay standard, in the early years. When the Rorty (1903) and Molina (1907) tables were first published [11], trunk group blocking objectives were set at P.001 mainly because of the crude ways of estimating traffic. In the 1925 time frame it was observed that SxS selection stages could be engineered at P.01 with no real problems, and a 13% average equipment savings.

Since a local call through SxS equipment generally required three or four selection stages, including trunking between offices, the de facto standard for overall blocking or 'overflow' became 4%. Panel System District and Office frames were engineered accordingly. As crossbar systems development began in the early 1930's, E. C. Molina's investigations of the new matching loss phenomenon were aimed at estimating capacities in the 1% to 2% blocking range. Molina covered a full spectrum of probability distribution assumptions among linking stages in order to set bounds on possible capacities. The first crossbar offices were engineered for 1% blocking in each of the originating and incoming switching trains, with intraoffice trunks designed to the same level. Again, World War II shortages caused revision of the incoming matching loss standard to 2%, and it was never restored. However, as the retrial feature for outgoing or intraoffice trunk selection was soon introduced, it became evident that service levels of 0.5% or better could readily be maintained for the originating part of each connection. This meant that, with local trunks engineered at P.01, the 4% objective was still attained, at least in theory.

Automatic alternate routing of calls within metro areas did not change the objective. Tandem offices were engineered for 0.5% matching loss, so that the customer might expect up to 2.5% blocking on an alternate route, in addition to 2.5% or 3% matching loss. But since most of the calls were completed on a direct (high usage) route, 4% average blocking could still be maintained.

The overall 4% objective remains today, but there have been some complications. As W. S. Hayward reported in 1964 [12], the 4% standard was used as the initial intraoffice blocking objective in No. 1 ESS™. Later, when alternate routing within the No. 1 ESS switching train was introduced, for economic reasons, it became clear that overload service objectives were needed.

Systemwide data studies [13] showed that the average 'peaking factor' (HD to ABS ratio) for carried load in large offices was about 1.15 (for call volume, about 1.30). So it was decided to incorporate a "15% overload" response objective in the generic local switching systems requirements [14] then in preparation. Based on performance studies of SPC and crossbar systems, the outgoing, incoming and intraoffice objectives were set at 2%, 7% and 12% respectively.

These latter numbers were published as objectives rather than requirements, since it was recognized that (a) there were no human factors studies to back them up, and (b) the actual peaking factor could be much higher than 1.15, in any given office - especially one dominated by residential traffic [15]. But they served as a warning flag to designers and engineers that overload response had to be watched carefully in modern systems. This was familiar, of course, to those studying central processor capacity. But it was new to link systems analysts, accustomed to dealing with more well-behaved load/service functions[6].

IV. Trunking and the Toll Hierarchy

The early Rorty and Molina formulas were applied to trunking between manual offices, later to line finder groups, line switches and selector groups, as well as to trunking between local offices. By 1925 nearly everything was on a P.01 basis. Graded access groups to promote switching and trunking efficiency were introduced by 1910, and Wilkinson's tables later came into use [16], mostly based on 1% blocking, though P.005 was used in some stages if an average call used more than 4 stages of switching, to maintain the 4% end-to-end objective [3].

In the early 1900's most toll calling was on a 'call back' basis, because of the expensive facilities, without specific delay standards. From the 1920's until just after

™ Electronic Switching System No. 1: trademark of AT&T Technologies.

WWII average delay objectives known as T-speed standards were used. Typical standards were 8, 15, 30, 60 or 120 seconds average delay, depending on the importance and/or size of the route. When operator toll dialing through the new toll crossbar machine began, just after the war, P.01 was generally used in order to save operator work time and increase facilities in preparation for the introduction of nationwide toll dialing by customers.

To backtrack just a little, by 1938-39 alternate routing of local calls within major cities was beginning, using the new crossbar systems' intelligence. Crossbar tandem offices were in service soon after the first local crossbar offices. It was understood that the alternate route trunk groups would see 'worse than random' offered traffic, so that the Erlang or Molina assumptions did not quite apply. From empirical studies a 1.03 factor, later changed to 1.05, was applied to the offered load before using a P.01 table for local overflow routes [3].

The 1950s work of Bretschneider, Truitt, Wilkinson and others on the engineering of alternate routing systems is very familiar to ITC participants. Wilkinson's 1970 tables, which apply to local as well as toll trunking, included the effects of day-to-day variation as well as peakedness. For the first time the Blocked Calls Cleared model was used, for congestion in individual busy hours. At that time he said:

"We conclude that the Europeans [and T. C. Fry] were probably right in objecting to the use of the Poisson formula for estimating individual hour blocking. However, it is a happy circumstance that the early conservatism in trunk engineering adopted by the Bell System for rather different reasons should prove to have included just the right amount of extra capacity to allow for busy season day-to-day load variations..." Wilkinson provides evidence for this in [7].

C. Truitt and others [3] originally recommended a uniform P.03 standard for the engineering of final routes in the nationwide plan. With more than two-thirds of calls completed over high usage routes, it was expected that this would promote the long-term goal of making DDD service as good as local service. (In the interim many sparse routes were still T-speed engineered.) But it was recognized that the 1.05 "fudge factor" would not be adequate for finals in multi-alternate routing. Wilkinson began work on equivalent random theory in 1947 [3], but it was not easy to apply originally. Not until 1970 was it used universally, with full automation of traffic measurement processing and load forecasting.

Because of these difficulties and the lack of reliable measurements in some cases, a standard of P.01 on finals was recommended in 1965, to be fully implemented by 1970 [3]. But in parallel, during the 1960s, discount tariffs were introduced to promote higher usage of the network in off-hours. These created traffic peaks at previously unforeseen times, so the P.03 objective continued to be applied selectively for some years to trunk groups affected by the new tariffs. In the late 1970s, when P.01 was used almost universally, studies were made to determine if a return to the P.03 or even a P.05 objective would produce significant savings. It was found that revenue losses from poorer service [17] were about equivalent to savings in trunking, so the P.01 objective was retained.

V. Auxiliary Standards

It is an enigma of modern switching systems that the standards discussed in Sections II and III, the important customer service standards, are difficult to apply because they require load-service relations based on complex mathematical models of interacting service systems. But most switch components that are 'traffic engineered' are auxiliary elements such as service circuit groups or blocks of memory. The tables or curve-fitting procedures used for these are generally straightforward Erlang or Poisson functions. The process of tracking and forecasting traffic loads for these components is an ongoing problem, however.

Table I is taken from [14] and provides most of the BOC standards in use today. In addition, a 90% occupancy threshold is used for high day engineering of circuits such as digit receivers and senders, as well as key memory elements such as call progress and billing registers. This safeguard is to provide a cushion against forecast inaccuracy and is achieved at modest cost, in most modern systems. There is little of historical interest associated with most of these standards. They were usually arrived at by consensus judgement of traffic professionals, based on their a priori knowledge or studies of interactions under overload.

The most important 'hidden' or auxiliary standards, however, are those used for processor capacity estimation and engineering. In this context we may distinguish between capacity estimation, as applied to a single central processor, and 'engineering' as applied to determination of the number of minicomputers (or markers) needed in a parallel-processing environment. Modern systems usually employ distributed or serial call and administrative processing, with functions segregated, but the most time-consuming call setup functions may still be done by parallel processors.

In crossbar systems markers were the parallel processors, engineered based on average busy season data, but there is evidence [3] that a 'fudge factor' was used to protect peak period performance even before 1940. By the 1950's this was formalized in traffic practices, and in the early 1960's formal high day engineering was introduced. The problem here was not really what delay criterion to use for peak period engineering (anything from 200 ms. to 2 secs would admit very high occupancy), but how to estimate the traffic load. Markers were subject to delays from many sources, and were especially vulnerable to register and link frame congestion [18]. So the "knee" of the steep load/delay curve was very elusive. We knew when the early SPC systems were being designed that processor engineering was really not a standards problem, but a problem of characterizing performance. This has been well covered at prior ITC's. Modern systems with distributing processing and complex data communications among them offer a continuing challenge to performance analysts.

The most important advances of the 1970's were in overload control strategies [9], [10]. The paramount concern was to keep the processor handling as many effective or 'good' calls as possible, in any extreme overload condition. In this we succeeded very well, and capacity gains were also achieved. Ineffective attempt volumes were reduced and dial tone delay performance became somewhat more predictable in the "knee of the curve" region.

A challenge for the future is to characterize customer reaction to extreme overloads that are well controlled. 'Before and after' studies are very difficult to make in this context. But if we are to take full advantage of the capacity to be gained under controlled peak loads, we must ascertain what's really happening to customers when controls are invoked: how many calls are actually "lost", the real delays. Measurement systems today are not quite up to this task. When they are, it should be feasible to set new standards for peak period processor performance.

VI. Standards Needed

Table I shows standards for Extreme Value Engineering (EVE), with daily peak busy hour data processing assumed, as well as the conventional standards discussed, with time-consistent busy hour (TCBH) processing. As discussed in [19], [20], every effort was made to keep the standards consistent in order to ease the transition from TCBH to EVE. Average bouncing busy hour (ABBH), once a month return period (OAM), and Once a year return period (OAY) - later adjusted to Expected High Day (EHD) - coincide within a few percent in load or call volume to the traditional ABS, 10HD and HD time-consistent levels. EHD is intended be HD exactly, based on empirical formulas we had available [20]. So, few new service standards are needed for EVE.

For administrative and economic reasons, the BOCs recently decided that full conversion to EVE data processing is not warranted for modern digital systems. Left open for now is the question of whether EVE processing and standards will be needed for remote switching systems, digital loop carrier concentrators, and packet switching applications: either stand-alone or in the ISDN context. These latter cases seem natural EVE candidates, since the busy hour is not apt to be well determined and interacting components are relatively easy to specify.

The major missing numbers in Table I are for data switching and remotes. Actually, for remote concentrators an informal standard of P.005, with a 3-month return period (i.e., 4 exceedances of 0.5% blocking in a year) was established in the early 1970's to insure no noticeable overall service difference for customers on these switches [21]. For data access

Table I
Local Switching Systems Criteria

	Criterion Type	*ABSBH*	*THDBH*	*HDBH*	*ABBH*	*OAM*	*EHD*
Network Service							
Line-to-Trunk	Blocking (ML)	1%	-	2%	1%	-	2%
Trunk-to-Line	Blocking (FFM)	2%	-	#	2%	*	#
Line-to-Line	Blocking (ML)	2%	-	#	2%	*	#
Trunk-to-Trunk	Blocking (FFM)	2%	-	-	2%	-	-
Trunk-to-Trunk	Blocking (ML)	0.5%	-	2%	0.5%	-	2%
Dial Tone Delay	Delay (t>3")	1.5%	8%	20%	1.5%	-	20%
Dial Tone Delay	Average	0.6"	-	-	0.6"	-	-
Blocked Dial Tone	Delay (t>0)	-	-	#	-	-	#
IR Attachment Delay	Delay (t>3")	1.5%	8%	20%	1.5%	-	20%
Service Circuits							
Customer Digit Receivers	Delay (t>0")	-	-	5%	-	-	5%
Interoffice Receivers	Delay (t>0")	1%	-	-	1%	-	-
Interoffice Receivers	Delay (t>3")	-	-	0.1%	-	-	0.1%
Interoffice Transmitters	Blocking	-	1%	-	-	-	-
Ringing Circuits	Blocking	-	-	0.1%	-	-	0.1%
Coin Circuits							
Coin Control	Blocking	-	-	0.1%	-	-	0.1%
Overtime Announcement	Delay (t>0")	1%	-	-	1%	-	-
Announcement Circuits	Blocking	1%	-	-	1%	-	-
Tone Circuits	Blocking	1%	-	-	1%	-	-
Recorder Tone Circuits	Blocking	-	-	0.1%	-	-	0.1%
Conference Circuits	Blocking	0.1%	-	-	0.1%	-	-
Billing	Error	-	-	2%	-	-	2%

ML: Matching Loss (after retrials, where available).

FFM: First Trial Failure to Match.

* Where an OAM standard is desirable for specific systems applications, it should be derived based on the EHD and ABS standards and the HD/ABS distributions in Section 17, System Capacity.

\# HDBH and EHD standards are under study.

concentrators and public packet switched systems standards are in the proposal stages [22], and for packet switching in the ISDN context considerable CCITT/ITU work is in progress [23].

Enhanced services through ISDN open up a new array of standards problem. The major needs are for component peak period standards and for voice/data service tradeoff objectives under overload controls. This escalates the need for studies of customer reaction to service problems. Studies that characterize the load variability of remote switches are also urgently needed.

We will discuss how the missing numbers marked (#) in Table I can be filled in. These were the 7% and 12% blocking "objectives" for incoming and intraoffice matching loss under overload conditions. There are several considerations needed to decide whether these or alternative numbers should become firm requirements.

(i) Matching loss has been replaced by concentrator blocking, or "blocked dial tone", in modern digital systems. A blocked terminating call is now often coincident with a *long* dial tone delay, for an originating customer.

(ii) There are no studies or other specific evidence to indicate that a given level of blocking is more or less serious (evident to the customer) than the same level of (long) dial tone delay. But by the arguments and evolution tracked in Section II, it seems evident that dial tone delay should be the controlling standard. The load/service relations for long dial tone delay, or for terminating (incoming or intraoffice) blocking, are very steep. The system ranges from an equivalent 50 to about 500 full-access servers, with an unknown customer defection distribution - perhaps ranging from Wilkinson's J=5 assumption at low delay levels to Erlang-C when delays are excessive. Finite-source considerations are not very important for the largest systems, where 3000 to 5000 lines access 500 time slots.

(iii) Actual system designs are generally limited access, 2- or 3- stage concentrators, readily analyzed by Jacobaeus or Effective Availability methods (e.g. [24], [25]). Fortunately, because of their large access internally, the results look much like full-access load/service relations. So with reasonable assumptions in (ii), standardized functions may be used for analysis.

Figures 1, 2, and 3 show occupancy vs. blocking curves for three concentrator designs now in service. They cover equivalent server (time slot group) sizes of about 50, 100 and 400 respectively. Only Figure 3 is reproduced here, since the others are similar in form. The parametric curves are for total coefficients of variation (CV) ranging from 0 to 20%. The "0" curves include only source load variation, according to the analytical models used. The total variation may be considered day-to-day alone or combined day-day and group-group, depending on the application. For each curve a gamma distribution of load variation was assumed, consistent with TCBH observations of traffic on trunk groups [26].

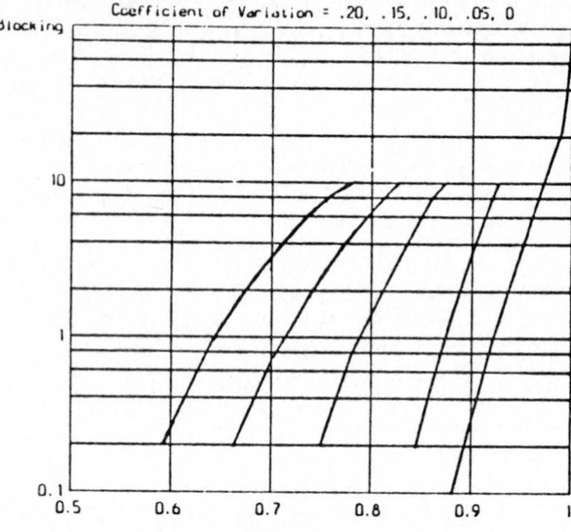

Fig. 3. Digital Line Concentrator Load vs. Service
Effective Access = 400

For example, in Figure 3 a 0.5% ABS objective with 10% CV would produce an adjusted capacity of .77 occupancy, down from .91 at 0% CV. To this we apply an EHD/ABS ratio similar to that given in [20] but modified by the gamma assumption, to estimate blocking under HD conditions. That is

$$EHD/ABS = 1 + 2.62 \cdot CV + 2.01 \cdot CV^2 = 1.282$$

and the corresponding blocking is about 20%, HD.

Table II shows results obtained from Figures 1, 2, and 3 by the straightforward method just given, as related to ABS blocking objectives of 1.5%, 1%, and 0.5%. As expected, the EHD service "blows up" when a 1.5% objective is assumed, especially for the most efficient concentrator design. A 1% objective might provide reasonable HD service protection except for the most efficient design, Figure 3, and a 0.5% objective appears to provide service protection for all cases, but too much so for the Figure 1 system. In Table III we draw one obvious compromise: varying the ABS objective with concentrator efficiency to maintain consistent HD objectives.

All of this assumes, of course, that we are willing to live with an as yet unquantified amount of service risk, inherent in engineering for 'expected' peak period loads. The statistics reported in [15] indicate that the *expected* HD-to-ABS ratio or

Table II
ABS Engineering with Adjustments for Traffic Variability

Equivalent *Group Size*	ABS *Objective*	Expected HD % Blocking at CV of *.05*	*.10*	*.15*	*.20*
50	1.5%	8	21	50	100
(Fig. 1)	1	5	15	29	48
	0.5	3	9	14	20
100	1.5	13	60	100	100
(Fig. 2)	1	9	30	70	100
	0.5	5	14	20	30
400	1.5	100	100	100	100
(Fig. 3)	1	37	100	100	100
	0.5	13	28	50	60

Table III

ABS Engineering Adjusted for Group Size and Traffic Variability
HD Objective = 20% Blocking

Equivalent *Group Size*	Adjusted ABS % Blocking at CV of *.05*	*.10*	*.15*	*.20*
50	(4.2)	1.4	0.7	0.5
100	(2.3)	0.8	0.5	0.4
400	0.7	0.4	0.3	0.3

increment can be estimated well by class of service, geographical region and time of day of the busy hour. But the year-to-year variation in these statistics remains large, and not all of it is due to Act of God days. Yet if we are willing to accept the type of "annually recurring HD" definition implied in this analysis, engineering based on adjusted average service objectives would be feasible.

Another approach is to redefine the peak period service problem as an outage problem. In the absence of good data on customer reaction to poor service, we might elect a criterion for engineering (e.g. 50% blocking) that is certain to cause severe service impairments, then specify that this level should occur no more than one hour in three years. Today's data processing systems can readily perform the required traffic projections.

VII. Acknowledgements

V. Bolotin provided the analysis of line concentrator service as part of a more general study. A number of former Bell System engineers provided historical information on standards.

REFERENCES

1. Turner, W.O. *Estimation of Requirements in Dial Telephone Central Offices,* Proc. Operations Research in Marketing, Case Inst. of Tech., Jan. 1953.
2. Baer, F.L. *The Computation of Quantities of Telephone Switching Equipment,* Paper presented to the Western Society of Engineers, Nov. 4, 1920.
3. Private communication or the author's interpretation of available notes.
4. Guess, H.A. and Kappel, J.G. *An Engineering Method to Account for Link Congestion Effects on Dial Tone Delay,* ITC8, Melbourne, Nov. 1976. Paper 136.
5. Farber, N. *A Model for Estimating the Real-Time Capacity of Certain Classes of Central Processors,* ITC 6, Munich, Sept. 1970. Paper 426.
6. Kappel, J.G. *Traffic Problems in Modern Digital Switches,* International seminar - Fundamentals of Traffic Theory, Moscow, June 1984.
7. Wilkinson, R.I. *Some Comparisons of Load and Loss Data with Current Traffic Theory,* ITC 6, Munich, Sept. 1970. Paper 521.
8. Kort, B.W. *Models and Methods for Evaluating Customer Acceptance of Telephone Connections,* GLOBECOM 83, Paper 20.6
9. Forys, L.J. *Performance Analysis of a New Overflow Strategy,* ITC 10, Montreal, June 1983. Paper 5.2-4.
10. Phelan, J.J.; Burkard, L.; Weekly, M. *Customer Behavior and Unexpected Dial Tone Delay,* ITC 10, Montreal, June 1983. Paper 2.4-5.
11. Wilkinson, R.I. *The Beginnings of switching Theory in the United States,* ITC 1, Copenhagen, 1955.
12. Hayward, W.S. *Traffic Design and Engineering of the Switching Network for the No. 1 Electronic Switching System,* ITC 4, London, 1964.
13. *LATA Switching Systems General Requirements,* Section 17 - System Capacity; a Bellcore publication.
14. *LATA Switching Systems General Requirements,* Section 11, Service Standards.
15. Bolotin, V.A. *Class of Service Analysis of Traffic Variations at Telephone Exchanges,* ITC 11.
16. Wilkinson, R.I. *The Interconnection of Telephone Systems - Graded Multiples,* BSTJ 10, Oct. 1931, pp. 531-564.
17. Liu, K.S. *Direct Distance Dialing - call Completed and Customer Retrial Behaviour,* ITC 9, Torremolinos, Oct. 1979. Paper 144.
18. Guess, H.A. *The Effect of Frame Load and Balance on Dial Tone delay in No. 5 Crossbar,* BSTJ, Dec. 1975.
19. Friedman, K.A. *Extreme Value Analysis Techniques,* ITC 9, Torremolinos, Oct. 1979.
20. Friedman, K.A. *Precutover Extreme Value of a Local Digital Switch,* ITC 10, Montreal, June 1983. Paper 1.4-1.
21. Gershwin, S.B. et al. *Peak-Load Traffic Administration of a Rural Line Concentrator,* ITC 7, Stockholm, June 1973.
22. Lee, J.I. and Armolavicius, R.J. *The Evolution of Traffic Standards,* ITC 10, Montreal, June 1983.
23. *CCITT Recommendations for ISDN,* Technical Session (7 papers), ICC '85, Chicago, June 23-26, 1985.
24. Jacobaeus, C. *A Study on Congestion in Link Systems,* Ericsson Technics, 1950, No. 48, pp.1-68.
25. Kuemmerle, K. *An Analysis of Loss Approximations for Link Systems,* ITC 5, New York, June 1967.
26. Wilkinson, R.I. *Nonrandom Traffic Curves and Tables,* June 1970. Distributed at ITC 6.

TELETRAFFIC ISSUES in an Advanced Information Society
ITC-11
Minoru Akiyama (Editor)
Elsevier Science Publishers B.V. (North-Holland)

GOAL THEORY FOR PUBLIC UTILITIES CONSISTENT WITH NATIONAL GOALS APPLIED TO THE TELE SECTOR

Arne Jensen

Technical University of Denmark
2800 Lyngby-Denmark

ABSTRACT

With this contribution it is intended to expand Moe's Principle (Arne Jensen(1950) and (1980)) to a general theory for a public utility. Both economic and qualitative demands, as well as national requirements, will become parts of the efforts to reach the goal. To maintain the position of the tele sector, it is a condition that a number of requirements are met, irrespective of their short-term financial advantages to the tele sector itself.

INTRODUCTION

The new general theory for Public Utilities formulated here includes Moe's Principle as a special case. It includes Christian Asgersen's requirements of upper limits for the total blocking of traffic and build on the Danish planning tradition using a two-level network.

Asgersen's results do not in practice deviate much from the ordinary dimensioning rules established by overloading requirements and by Moe's economic Principle.

As will be seen in Chapter I, the classic dimensioning rules do not deviate much from one another, either, except for their mathematical expression of levels of service. Their operabilities in the light of the technical economic possibilities are for this reason quite different.

The dimensioning rules for individual technical groups in the tele plant prove to remain improvement curves,

$$F_n(A) = A(E_n(A)-E_{n+1}(A)) = F_1,$$

such as presented in Moe's Principle.

But the norms F_1 - the level of service - are shown to depend on economic goals for administration and nation as well as subscribers' qualitative requirements to the individual traffic routes.

Dimensioning rules for direct circuit groups and protection groups can in this light be considered as part of superior safety views for teletraffic. Not only economic views, technical possibilities and blocking probabilities, but also the risk due to overloads or technical breakdowns become part of this general theory. An elaboration of these problems is taken up in the conclusion of the paper.

CHAPTER I
SOME CLASSICAL DIMENSIONING RULES

Molina used - as Erlang did earlier - the tail of the Poisson distribution in his dimensioning rule. The number of circuits was decided so that the tail was constant, e.g.

$$\sum_{i=n}^{\infty} P(i) = 0.02.$$

P.V.Christensen used an approximation for this tail, the "square-root law", $n = A+2,3\sqrt{A}$.

In engineering practice, the ability of the groups to meet overloading is often used as, e.g., in requiring blocking B = 0.07 at 25% overload.

In Moe's Principle the last supplementary "dollar" should have the same effect on reduced number of lost calls no matter where it was invested in the plant. That means $F_n = (A(E_n(A)- - E_{n+1}(A)) = F_1$, e.g., as in Denmark $F_1 = 0.05$ often used for the number of circuits.

In future, if you wish to split-up traffic and send it in parallels along different routes in an optimal way, you get (n+1-A(1-E))E = d. For d = 0.09 the results are given in Column "split".

Results show that the deviation between the various old and new rules applied is very modest.

The interesting issue left behind is how the service level as given by the constants for these rules depends on the required goals. This will be the central issues to be taken up.

A	Moe F_1=0.05 n	Molina 0.02 n	P.V.Christensen 2.3 n	Overload 25% B=0.07 n	Split d = 0.09 n	Blocking* Erlang in %
4.99	10	10.6	10.3	10	10.3	1.8
11.98	20	20	19.9	19	20	1.0
27.58	40	39.3	39.7	39	40	0.6
44.18	60	59	59.5	59	59	0.4
61.2	80	79	79.2	79	79	0.3
78.7	100	98.5	99.1	99	99	0.3
96.4	120	117.5	119	-	119	0.2

* along the Moe Curve calculated by Erlang's Formula.

CHAPTER II

FORMULATION OF GOALS FOR PUBLIC UTILITIES

In practice Public Institutions - to which a service sector such as telephony belongs - are subject to a special set of governing rules of a more or less formalized kind. At times such rules are just implied as a result of long stable traditions.

The Public Institution influences society's national product by its requirements of payments for services, by virtue of its "subsidies from and contributions to society", and by means of the service it grants. On top of this, its activities have an indirect effect on the national systems, e.g., the socalled multiplier effects.

The privileges enjoyed by Public Utilities are counterbalanced by the obligations to grant the service in question to all members of society, on equal terms and to some extent without regard to individual cost. Also, they are counterbalanced by the fact that the financial results of an institution, its price policy and the quality of its products are subject to public regulations, often in the form of minimum requirements. Therefore, we do not often talk about maximising the financial outcome of the public institution. On the contrary, the public institution has to maintain a certain economic - often linear weighted - balance between income and expense. Other economic restraints, such as staff policy and choice of technique, the draw on domestic and foreign manpower and products often have to be taken into consideration. The individual subscribers' minimum requirements of quality of the services offered on different routes have to be met. All these requirements take the shape of a long series of inequalities. Some to fulfill quality requirements, others are financially motivated. It may also be requirements to obtain a smooth functioning over time of other sectors in society.

The most essential issues of policies carried out within Public Utility, as, e.g., telephony - irrespective of society set-up - are aimed at an overall goal that tries to maximize the national product for society and the Public Utility as a whole under given constraints. In this way, the Public Utilities and institutions in a smooth way find their place in society. These requirements may include a predetermined profit for the Public Utility as well as demands for minimum contributions to society and clients and a maximum draw per unit of time on the society's different resources.

This influences both short-term and long-term investments, and often requires a smooth adaptation of the Public Utilities to changed economic and technical conditions.

Through the system of rates and service constants for the different groups of traffic, technical units and groups of workers, these requirements will influence the tele activity.

Decisive for the survival of the institutions is whether they directly or indirectly contribute positively to the national product.

An important draw in this picture is the overall subsidizing or transfers among the various sectors of society.

Even though the communication sector so far has done without subsidies, and has contributed its share to the society indirectly by meeting its requirements, you cannot in the long run forget transfers among sectors. Some are contributors. Others are recipients. The decision whether or not, and how, is made at a higher level within society.

It is characteristic of optimization under constraints, that the number of variables which management has to decide on is far less than the number of customer groups whose individual service demands have to be met.

This means that, as far as the optimal solution is concerned, the service will be just met in a limited number of cases, only. In fact, a number equal to the number of variables available to management. In far the most cases the requirements will be more than met. Or, said differently, only a few service requirements will establish norms for the different technical groups - one dominating all others in each technical group. Only one of the traffic routes it services will establish its norm for service. The service requirements of the other routes are more than met. Those traffic routes which in this way influence cost (and indirectly the technical organization and policy) get a special status in the planning.

As long as they find it profitable to go along with the too well-serviced groups, we get a stable organization. Not for long, though.

Are more-than-met quality norms to be or not to be credited as a contribution to the national product in our model is a good question.

To find the specific sets of norm-establishing traffic routes and their matching with the various technical groups could become a special mathematical task with directly, or indirectly political implications. First of all, it might in the hands of the experienced traffic engineer influence establishment of special routes for special traffic.

Before solving these problems for telenet in Chapter V on the basis of the general theory for Public Utilities formulated here, two classical problems in Chapters III and IV are taken up,

III Profit Maximization for a telephone company
IV Profit Maximization for a telephone company with quality constraints.

CHAPTER III

PROFIT MAXIMIZATION FOR A TELEPHONE COMPANY

Find the sizes $n = (n_1 \ldots n_M)$ for the M technical groups which service the N traffic routes with the offered traffic A erlang from the frequency of calls y with the holding time s, $A = y.s$, resulting in the probability of loss $E_n(A)$ and then the probability of getting through a group $\sigma = 1 - E_n(A)$.

The technical groups μ which service the route ν is given by $D = \{d_{\nu\mu}\}$ $\nu = 1,2 \ldots N$. $\mu = 1,2 \ldots M$, where $d_{\nu\mu}$ is 1 or 0 when used or not used.

The probability S_ν for getting through on route ν is then - on the assumption of independence among groups - given by for $\nu = 1,2 \ldots N$

$$S_\nu = \prod_{\mu=1}^{M} \sigma_\mu^{d\nu\mu} \tag{1}$$

$$\sigma_\mu = 1 - E_{n_\mu}(A._\mu) \ , \ A._\mu = \sum_{\nu=1}^{N} d_{\nu,\mu} A_\nu \tag{2}$$

Let the throughput of the traffic be called

$$v = A.S. \tag{3}$$

Let the price for the service on route ν per erlang be p_ν. The resulting revenue R will then be

$$R = \sum_{\nu=1}^{N} R_\nu, \quad R_\nu = p_\nu v_\nu \tag{4}$$

Let the cost C_μ for n_μ units in group μ be

$$C_\mu = b_{o,\mu} + n_\mu b_\mu , \tag{5}$$

$$C = \sum_{\mu=1}^{M} C_\mu = \sum_{\mu=1}^{M} (b_{o,\mu} + n_\mu b_\mu) \tag{6}$$

and

$$F_n(A) = A(E_n(A)-E_{n+1}(A)) \tag{7}$$

be the improvement function.

Operationally is used

$$\frac{\delta\sigma}{\delta n} \approx E_n(A)-E_{n+1}(A). \tag{8}$$

The problem is then to find the set n which maximizes the linear function of R and C as given by

$$\max_n (\alpha_1 R-\alpha_2 C-\alpha) \tag{9}$$

α_1 and α_2 depending on contribution and subsidies through, e.g., tax.

Using v = A.S,(4) and (6), this gives for $\mu = 1,2 \ldots M$

$$\sum_{\nu=1}^{N} \alpha_1 p_\nu \frac{\delta v_\nu}{\delta n_\mu} = \alpha_2 b_\mu \tag{10}$$

From (3)

$$\frac{dv_\nu}{d_{n_\mu}} = A_\nu - \frac{dS_v}{d_{n_\mu}} \tag{11}$$

Using (1), (2), (8), and (7) we get

$$\frac{\delta v_\nu}{\delta n_\mu} = \frac{A_v S_v}{\sigma_{n_\mu}^{d}{}_{v\mu}} \frac{\delta\sigma_{n_\mu}^{dv\mu}}{d_{n_\mu}} = d_{\nu\mu} \frac{v_\nu}{\sigma_{n_\mu}} \frac{F_{n_\mu}(A.\mu)}{A.\mu} \tag{12}$$

From (10) we then get for $\mu = 1,2 \ldots M$

$$\alpha_1 \frac{F_{n_\mu}(A._\mu)}{A._\mu \sigma_\mu} \sum_{\nu=1}^{N} d_{\nu\mu} p_\nu A_\nu S_\nu = \alpha_2 b_\mu \tag{13}$$

or

$$F_{n_\mu}(A.\mu) = \frac{\alpha_2}{\alpha_1} \frac{b_\mu}{R_\mu / A._\mu \sigma_\mu} \tag{14}$$

which can also be written for $\mu = 1, 2 \ldots M$

$$F_{n_\mu}(A._\mu) = \frac{\alpha_2}{\alpha_1} \frac{b_\mu}{\bar{p}_\mu} \tag{15}$$

where

$$\bar{p}_\mu = \frac{\sum_{\nu=1}^{N} d_{\nu\mu} p_\nu A_\nu S_n}{A._\mu . \sigma_\mu} = \frac{R_\mu}{A._\mu \sigma_\mu} \tag{16}$$

Since price generally follows marginal route cost.

$$p_\nu \sim \sum_{j=1}^{M} d_{\nu j} b_j = B(\nu)$$

within telephony, this in practice can be formulated like this, for $\mu = 1,2 \ldots M$

$$F_{n_\mu}(A._\mu) \sim F_1 \frac{b_\mu}{B_\mu} \tag{17}$$

where B_μ is the mean total marginal cost $B(\nu)$ for traffic routes ν passing group μ, that is $\mu = 1, 2 \ldots M$

$$B_\mu = \frac{\sum_{\nu=1}^{N} d_{\nu\mu} \left[\sum_{j=1}^{M} d_{\nu j} b_j\right] A_\nu}{\sum_{\nu=1}^{N} d_{\nu\mu} A_\nu} \tag{18}$$

CHAPTER IV

PROFIT MAXIMIZATION FOR A TELEPHONE COMPANY WITH QUALITY CONSTRAINTS

Using LaGrangean multiplier - the shadow prices -

$$\lambda_\nu, \quad \nu = 1,2 \ldots N,$$

for the quality requirement S_ν^0 to route ν, we can write this goal in the following way, for $\nu = 1,2 \ldots N$

$$\max_n \left[(\alpha_1 R-\alpha_2 C-\alpha)+ \sum_{\nu=1}^{N} \lambda_\nu y_\nu (S_\nu-\rho_\nu-S_\nu^0)\right] \tag{19}$$

$$\lambda_\nu \rho_\nu = o. \qquad \text{F.Zeuthen's Rule} \tag{20}$$

which means that

$$\sum_{\nu=1}^{N} \left[\alpha_1 p_\nu + \frac{\lambda_\nu}{s_\nu}\right] \frac{\delta v_\nu}{\delta n_\mu} = \alpha_2 b_\mu \tag{21}$$

Using (17), this gives for $\mu=1,2 \ldots M$

$$\alpha_1 \frac{F_{n_\mu}(A._\mu)}{A._\mu \sigma_\mu} \sum_{\nu=1}^{N} d_{\nu\mu} \left(p_\nu + \frac{\lambda_\nu}{s_\nu \alpha_1}\right) A_\nu S_\nu = \alpha_2 b_\mu \tag{22}$$

Using (16) and

$$\frac{\overline{\lambda}}{s}(\mu) = \frac{1}{A._\mu \sigma_\mu} \sum_{\nu=1}^{N} d_{\nu\mu} \frac{\lambda_\nu}{s_\nu} A_\nu S_\nu^0 \tag{23}$$

considering that $\lambda \neq o$, only, when $S_\nu = S_\nu^0$. We get in practice only a single term from the normgiving route. (22) can then as (15) be written, for $\mu=1,2 \ldots M$

$$F_{n_\mu}(A._\mu) = \frac{\alpha_2 b_\mu}{\alpha_1 (p_{\mu\theta} + \frac{1}{\alpha} \frac{\overline{\lambda}}{s}(\mu))} \tag{24}$$

or parallel with (17)

$$F_{n_\mu}(A_\mu) = F_1 \frac{b_\mu}{\left[B_\mu + \frac{1}{\alpha}\frac{\bar{\lambda}}{s}(\mu)\right]} \tag{25}$$

The strict solution by finding λ_ν can be reached by using the socalled relaxation method (see, e.g., Lasdon (1970)). It starts with renumbering ν and μ in $d_{\nu\mu}$ and use all $\lambda = o$ as initial values. After study of the biggest deviations from the required S^o_ν, it takes up some λ in the equation system and calculates S. This procedure is repeated and you reach the solution in a finite number of calculations.

In practice you will often go direct from the study of the deviations to the final discussion for the set of F to be used for the different technical groups. The set of F should be operable and that will often modify the preliminary requirements to the quality constraints.

It might help to use a modification of (17) as the initial set, by instead of the mean marginal cost B_ν for the passing traffic routes to use the maximum marginal cost

$$\underset{\nu}{\mathrm{Max}}\; d_{\nu\mu}\, B(\nu)$$

for the routes passing the group μ.

CHAPTER V

OPTIMIZING THE PUBLIC UTILITY FOR TELEPHONE SERVICE

Let the national product g be considered as a function of the throughput of served traffic $v_\nu = A_\nu S_\nu$ for the traffic route $\nu = 1, 2 \ldots N$.

Let the tax on income R and expenses C be t_1 and t_2. Let the use of national products be η_1, and the multiplier effect be η_2.

λ_σ is the Lanrangean multiplier to be used as a shadow price to get the required linear balance between R and C respected. We are then able to write the overall goal function for society and institution:

$$\underset{n}{\mathrm{Max}} \left[g - R(1-t_1) + (t_2 + \frac{\eta_1}{1-\eta_2})C + \lambda_\sigma(\alpha_1 R - \alpha_2 C - \alpha) + \right.$$
$$\left. + \sum_{\nu=1}^{N} \lambda_\nu y_\nu \,(S_\nu - \rho_\nu - S^o_\nu)\right] \tag{26}$$

$\lambda_\nu \rho_\nu = o,\ \nu = 1,2, \ldots N.$ F.Zeuthen's Rule (27)

which means that

$$\sum_{\nu=1}^{N} \left[\frac{\delta g}{\delta v} + (\lambda_o \alpha_1 - 1 + t_1)p_\nu + \frac{\delta_\nu}{s_\nu}\right] \frac{\delta v_\nu}{\delta n_\mu} =$$
$$= (\lambda_o \alpha_2 - t_2 - \frac{\eta_1}{1-\eta_2})\, b_\mu \qquad \mu = 1,2 \ldots M \tag{28}$$

Using (16) and (23)

and

$$\frac{\delta g}{\delta v}(\mu) = \frac{1}{A_{.\mu}\sigma_\mu} \sum_{\nu=1}^{N} d_{\nu\mu} \frac{\delta g}{\delta v} A_\nu S_\nu \tag{29}$$

(28) gives, see also (24) and (15), for μ = 1,2 ... M

$$F_{n_\mu}(A_{.\mu}) =$$
$$= \frac{\alpha_2\left(\lambda_o - \frac{t_2}{\alpha_2} - \frac{\eta_1}{1-\eta_2}\frac{1}{\alpha_2}\right) b_\mu}{\alpha_1 \left[\frac{\delta g}{\delta v}(\mu)\frac{1}{\alpha_1} + \left[\lambda_o - \frac{(1-t_1)}{\alpha_1}\right]\bar{p}_\mu + \frac{1}{\alpha_1}\frac{\bar{\lambda}}{s}(\mu)\right]} \tag{30}$$

or in parallel with (17) and (25)

$$F_{n_\mu}(A_{.\mu}) =$$
$$= F_1 \frac{b_\mu\left[\lambda_o - \frac{1}{\alpha_2}(t_2 + \frac{\eta_1}{1-\eta_2})\right]}{\left[\frac{dg}{\sigma v}(\mu)\frac{1}{\alpha_1} + B_\mu(\lambda_o \frac{1-t_1}{\alpha_1}) + \frac{1}{\alpha_1}\frac{\bar{\lambda}}{s}(\mu)\right]} \tag{31}$$

The mean influence of the technical groups on the national product has been taken into account when comparing (30) with (24) and the shadow price for deviations from the required economic balance for the Public Utility for tele service has influenced the service level of F, but the fundamentals in dimensioning rules remain the same as defined in Moe's Principle (Arne Jensen (1950)).

The norms in our tele plants are to be considered as a part of the general planning in the tele sector. That means that also effect outside the tele sector has to be taken into account even if it does not have direct influence on the own economy of the tele sectors. The individual influence of the different groups of traffic routes on standards and norms shows that a large part of our good service is the result of a rather few but important services which in all circumstances have to be met.

Variation in traffic around the clock and over seasons have not been considered here. I refer to Arne Jensen (1950), page 33 and Arne Jensen (1980) page 47 . It will add interesting results, but will not give major deviations from the above.

However, to what extent is communication service free as a consequence of investments being available for other purposes at another place and time during day and season is still an interesting question.

CHAPTER VI

SECURITY CONSIDERATIONS

To minimize cost, groups of direct circuits have been established. To protect overflow routes for direct access for fresh traffic, protecting circuits have been established. Dimensioning of these groups has often started, but not been fulfilled by using solutions to the triangle case. When we want security for at least some connections - even poor ones - there might be better possibilities.

It would only be natural to take the point of departure in the overall service from end to end for a traffic route and require that every single direct circuit should be able to meet that service. That is an improvement constant approximately equal to the sum of the constants for the groups the route is passing.

In my country, this might very well be $F_1 = 0.35$, not deviating much from the rules for direct groups, and in good accordance with a service level half that value $F_1 = 0.20$ for protection groups.

As we are in the middle of a technical revolution, it is important to take these preliminary security considerations up on par with other service considerations and to formulate an overall strategy.

Dimensioning rules for direct circuit groups and protection groups might be considered as a part of superior safety views. Not only lost calls and economics, but also risk due to overloading and technical breakdowns become part of our general theory for teletraffic.

In this connection, dimensioning of parallel independent traffic routes, with or without advanced cooperation, is going to be important. For that reason I have added the Column "Split" in the Table.

During this period of change, macro as well as micro considerations are coming into the limelight.

REFERENCES

Arne Jensen, "Moe's Principle. An Econometric Investigation Intended as an Aid in Dimensioning and Managing Telephone Plant" (1950). The Copenhagen Telephone Company. 155 pages.

Arne Jensen, "Traffic, Operational Research, Futurology. In Service of Research and Society". (1980). North-Holland Publishing Company, Amsterdam/New York/Oxford. 321 pages.

L.Lasdon, "Optimization Theory for Large Systems" (1970). McMillan.

TELETRAFFIC ISSUES in an Advanced Information Society
ITC-11
Minoru Akiyama (Editor)
Elsevier Science Publishers B.V. (North-Holland)

TELECOMMUNICATIONS POLICY AND TRAFFIC ENGINEERING THE INDUSTRY´S VIEWPOINTS

Christian JACOBAEUS

Telefonaktiebolaget L M Ericsson
Stockholm Sweden

ABSTRACT

The paper concerns itself with the interaction between telecommunications policy and traffic engineering specially from the viewpoint of industry. The industry can now offer equipment that can meet most of the policy requirement of the administrations (governments). This is true for both switching and transmission equipment. Also material for local plant have given the administrations much more freedom than before.

1. INTRODUCTION

The role of the telecommunications industry is to design and manufacture telecommunications equipment. The primary purchasers of the industry´s products are the telecommunications administrations. The latter´s customers are in turn the subscribers who need access to well functioning telecommunication plant for their business or private purposes.

It may well be said that telecommunications policy is ultimately a matter for the politicians. As it has proved that a country should have a single policy (on practical, technical, juridical grounds), the policy should also be formulated at government level. The government´s intentions must then be put into effect by the telecommunications administrations which, for that matter, may work on a national or regional basis. They may be privately owned, municipally-owned or state-owned. The administrations´ responsibility is to procure equipment and operate it in accordance with the country´s policy.

The administrations´ activities and operations have a very close and important link with the general subject of our congresses, telephone traffic theory and its applications. It is not the object of my address to illustrate this matter in greater detail. Otherwise it is self-evident, of course, that investments in equipment, at least as far as their traffic-dependent part is concerned, are determined by applications of traffic theory. Plant operation is also greatly assisted by traffic observations of different kinds; in actual fact traffic theory is now indispensable for this purpose.

My task in this context is to speak of the role of industry as regards telecommunications policy, with special regard for traffic theory and its applications. Those with a good memory perhaps recall that two years ago in Montreal I presented an invited paper in which I went through the various forms of equipment that industry had developed and could offer. I touched upon the extent to which telephone traffic theory was a necessary aid in design work. I shall here speak in rather more general terms about industry´s policy-creating or policy-facilitating role.

2. GENERAL POLICY-FACILITATING ROLE

It is obvious that, if industry can produce equipment that is inexpensive and well-adapted to its purpose, this facilitates the realization of a country´s policies. In the course of the years, furthermore, an enormous development towards cheaper and better systems and equipment has been forced ahead by the competition between different firms and between different technical solutions. At the same time a number of new traffic facilities have been introduced. We have got data communication on a large scale. A breakthrough for television-telephony is no longer a technical question and hardly an economic one either. The digital transmission systems also permit integration of different traffic facilities.

3. CONCENTRATION AND DECONCENTRATION

One of the spheres in which traffic theory plays a great role is in the location of exchanges. In small and medium-sized communities it is virtually self-evident that there shall be an exchange in the area. And where it should be placed is hardly any problem. In simple terms it may be said that the subscribers´ network must be as inexpensive as possible, i.e. the size and mutual positions of subscriber accumulations will determine the location of the exchande, naturally having regard to land availability and existing networks. In metropolitan networks the conditions are more complicated as they must have a larger number of exchanges to serve the area. The divisions between exchange areas will here be determined by many factors such as available exchange capacities, interest factors between different subscriber categories and different subscriber areas, costs of junction equipment between the exchanges, etc. Fortunately optimal solutions comprise fairly flat minima, so that moderate deviations can be made without appreciable increase of cost. In large cities the location of the transit centre(s) may also need special study.

The switching technique that has now been developed, as also new solutions for the subscriber network, have changed the entire approach to the problem.

Industry offers units remote from the larger exchanges, concentrators, of different capacities. In contradistinction to earlier attempts in this direction, with the aid of electronics the solutions are now economical. There is also a greater flexibility with regard to different changes in the social structure. A contributory reason for this is the tendency to build everything digitally. Switching and transmission are integrated right out to the subscriber´s telephone. The ISDN plans are an expression of this. Flexibility is facilitated also by the fact that the electronic equipment takes up little space and can be moved with a minimum of installation work.

The subscriber networks have also been made less expensive by new transmission systems. New ideas are being realized also for circuits to small places at medium distances.

The technique has, however, gone in the other direction as well. It is now possible to build exchanges with very great traffic capacity. This is of interest chiefly for large transit centres. The upper limit is hardly determined by considerations of traffic economy but by security aspects. Too large units are undesirable, as they are all too vulnerable to natural catastrophes, sabotage and war damage.

The complex represented by the size and location of exchanges and the form of the subscriber network constitutes an important field in the planning of telephone networks. This planning is based to a large extent on applications of traffic theory. We may expect that continued research will be necessary. But this perhaps concerns the administrations more than industry.

For the industrial products in the switching field the SPC exchanges have brought a change. The switching networks have very low congestion. The industry need generally make only relativly simple checks of the congestion. The interest from the traffic aspect has instead been concentrated to the central processors and their interworking. There are very complicated queuing problems to solve. With the high operating speeds offered by nodern electronics, however, the normal case here again becomes overdimensioning. In actual fact, accordingly, industry has attained a virtually ideal state as regards the switching equipment: congestion and waiting times are negligible. Dimensioning of the trunk circuits becomes the traffic-dependent part of the task. Governments and administrations have in the exchanges admirable means for realizing parts of their policy. The transmission systems have also become much cheaper in the course of the years.

4. OPERATIONAL DISTURBANCES

The telephone administrations obviously have a certain goal as regards operational reliability and performance. Since the beginning of the automatic era it is clear that telephone plant has become much more reliable. This has been accentuated by the development of the semiconductor technique. Extremely reliable are now the transmission systems on Atlantic cables and via satellites. The former were intended to operate without fault for at least 20 years, the latter at least 7 years. But clearly the networks must have adequate reserves to meet reduced transmission capacity when faults occur. Here traffic theory enters into the picture.

As regards switching equipment, a very careful balance is struck in the design stage between failure risks and standby equipment, usually by duplication. It is based on the components. A common specification is that the plant shall have at most 1 h breakdown in 40 years, regardless of the volume of traffic. And indeed the failure rate is generally independent of the volume of traffic. The electronic components exhibit no fatigue phenomena.

Another reason for the higher reliability is that administrations have acquired more efficient maintenance systems. The equipment has been designed not only for higher reliability but also for easier fault diagnosis, which makes faults simpler to find and remedy. The advances have often been achieved through collaboration between administrations and industry.

5. INTERNATIONAL ASPECTS

Telephone plant was originally regional but quite soon become national. Where the geographical conditions permitted, national networks in different countries were interconnected. With the emergence of Atlantic cables and satellites the world telephone network could finally be realized. This gave rise to new traffic theory problems. One such problem was that the busy hour occurs at different times in different zones. This made possible a better utilization of the satellites (over the 24 hours of the day). The speech channels in the satellites, furthermore, could be allotted certain destinations in a fixed and optional system. This has its basis in traffic theory.

As we all know, the conditions in the national networks must be such that they can work together in a fully adequate manner. Various technical conditions must be fulfilled. These are decided upon in CCITT´s committees. Certain requirements must, of course, also be placed on the networks from the traffic aspect. CCITT has, however, refrained from formulating these requirements, presumably because the economic consequences would be too burdensome for some countries.

6. SPECIAL BUSINESS NETWORKS AND EQUIPMENT

The chief sources of income for the administrations are business subscribers (taken in a broad sense). The units of business enterprises become increasingly often geographically dispersed across a country. Many, too, have their operations abroad. Special traffic facilities are often asked for. For speech communication there have for a long time been PBXs. For large customers with widespread activities these have also been interconnected in special networks. The latter may consist of privately owned or rented circuits or form part of the public network.Usually these various alternatives are used concurrently, the public network serving an overflow function.

The dimensioning of the junction network is a task for the traffic engineers.

In recent years, as is known, CCITT has worked on the development of a more universal communication system, especially for business subscribers. The system has been called ISDN (Integrated Switching Digital Network). Terminals of different kinds can be connected to the system and the circuits are digital, with two 64 kbit/s and one 16 kbit/s circuit to each terminal. The system allows transmission of speech and data and provides means for management information. When desired, furthermore, it has a supervisory and control function. It shall be so established that terminals can communicate on a global basis.

As yet there is limited experience of systems of this kind with many different forms of service integrated. But quite certainly they will face traffic engineers with new tasks.

I may also remind you that ISDN is to be dealt with in May 1986 in Belgium at a special seminar arranged by the Belgian Telephone Administration and the Belgian telephone industry.

7. THE INDUSTRY AND TELECOMMUNICATIONS TRAFFIC RESEARCH

Finally a few words about traffic research in industrial enterprises. As the pioneer in this field we reckon Erlang who was active in the first two decades of this century. He was associated with an administration, the Copenhaden Telephone Co. which for that matter had several prominent researchers among its staff. Erlang´s work had the effect, too, that industry started to take an interest in traffic research. This was needed in order to be able to understand the results of research and to be able to put them to proper use. The larger firms set up departments for traffic research. Independent research was widely done to solve proplems of particular interest. In cooperation with the system and design engineers solutions were found to problems of special interest for the firm´s range of products. In actual fact the traffic researchers acquired a great influence over, in particular, switching technique. A number of very important problems of a general nature have also found universally accepted solutions deriving from industry.

In the seventies and eighties industry recognized that the problems requiring a solution had been reduced in number. Research had provided industry with the answer to most of the questions posed in conjunction with design and system engineering. The result was a certain restriction of research activities in industry.

Traffic research, through its scientific nature has had fewer features of secretiveness than many other industrial activities. Researchers have been able to establish contacts with colleagues in other firms, as also with administrations and universities. ITC is a reflection of the fact that, in this as in other scientific fields, it has been possible to cooperate regardless of commercial and national interests. I think that to this fact must be ascribed the successes of ITC.

TELETRAFFIC ISSUES in an Advanced Information Society
ITC-11
Minoru Akiyama (Editor)
Elsevier Science Publishers B.V. (North-Holland)

Traffic Engineering in a New Competitive Environment

Walter S. HAYWARD, Jr.

AT&T Bell Laboratories
Holmdel, New Jersey 07733, U. S. A.

ABSTRACT

Over the past three decades, advances in technology have produced changes in the nature and structure of telecommunications. These, in turn, have raised questions about the suitability of the regulated monopoly environment as the best means of assuring growth in telecommunication services and effectiveness.

In the United States in particular, steps have been taken to establish a competitive environment. For the traffic engineer this has led to a very different traffic environment, requiring changes in traffic engineering practices and increasing the involvement of the traffic engineer in marketing and tariff making activities. Some examples are given of the challenging new traffic problems that are being encountered as the environment changes.

1. *INTRODUCTION*

Over the past two decades, technology has transformed telecommunications in ways affecting not only the nature of services and equipment but also the legal and political environment. Nowhere have these changes been as profound as in the United States. (A readable account for background is "Telephone" by John Brooks).[1] The direct effects of technology on traffic can be easily traced by reviewing the papers presented over the years to the ITC. The indirect effects which are moving telecommunications out of the regulated monopoly environment into a competitive environment have had less attention. These are the subject of this paper.

Traffic challenges, as they arise in the changing environment of the United States today, will be examined from three points of view, the interexchange carrier, IEC, the local exchange carrier, LEC, and the private network user. As will be seen in each of these areas, many of the assumptions previously underlying traffic engineering practices must be reexamined and new skills must be acquired.

Two comments are in order to get this paper in perspective. First, one of the products of a competitive environment is a reluctance to share special knowledge with competitors. Traffic engineering, when well done, can give a competitive edge. A slow-down in information flow may be one of the first noticeable effects of competition on traffic engineering. It has also made this paper more difficult to prepare. Second, the status of competition in the United States is still volatile. The changes set in motion by the divestiture of the Bell System are not yet completed. Consequently, the future telecommunications environment will almost certainly be different from today's.

2. *THE UNITED STATES 'NETWORK'*

The introduction of competition into the United States telecommunications environment has transformed a single telephone network into a group of networks interconnecting local offices that are operated by still-regulated, local exchange carriers. The latter provide a monopoly service, although inroads are being made by competition even here. In addition, separate networks serving non-telephone traffic are losing their functional differences and merging with telephony into telecommunications networks.

The Bell System, which was the predominant United States Carrier, is now divided into eight separate and independent companies. Seven of these are the owners of two or more operating companies, BOC's, which provide local service to former Bell System customers. The area served by each BOC is divided into LATAs, Local Access and Transport Areas. A LATA is intended to include a community of interest and may be a whole state or a large metropolitan area. Within one or more LATAs a BOC provides regulated monopoly service. The position of the "independent" or non-Bell operating companies is not as severely limited, but here too the local part of their business remains a regulated monopoly. In this capacity, both BOCs and independents will be referred to as Local Exchange Carriers, LECs.

The eighth company remaining after AT&T divestiture retained the name AT&T, and, as one of its functions, kept the business of interexchange carrier, IEC. This function has been expanded to cover not only the interstate carrier function, formerly filled by the Long Lines Department of AT&T of the Bell System, but also a newly defined class of interexchange traffic, that of inter-LATA, intra-state traffic.

More details about the inter-relationships of these new networks will be brought out in the following paragraphs.

3. *INTEREXCHANGE CARRIER TRAFFIC ENGINEERING*

3.1 *Forecasting*

Of all the changes brought about by competition, one of the most challenging to the traffic engineer is the higher volatility of demand that any one IEC will encounter. The most significant contributing cause is the customer's capability to choose which carrier to use on any call. While some customers will tend to maintain a consistent traffic pattern, some will change often. For example, customers with large traffic demands may use computing facilities to control traffic to take advantage of the most favorable parts of each IEC's offerings. As a result, traffic offered to a particular IEC is likely to shift quickly in response to competitive rate changes and new service offerings. The result of this higher uncertainty must be either a lowering of service objectives or a raising of capital requirements for each IEC.

The work reported by Franks et al.[2] or[3] and Coco et al.[4] describes a way of balancing the service and cost factors in choosing the best engineering course to follow. Figure 1. is based on the curves given in these papers. The ordinate is traffic capacity, which is related to investment; the abscissa is service, which is expressed in terms of entities not meeting service objectives. The curve relating the two can be found for a fixed degree of uncertainty for the difference between forecast loads and actual loads.

As shown in Figure 1, the operating curve moves to a higher level as the uncertainty of load forecasts increases. (The higher the uncertainty, the more equipment must be purchased to retain a given grade of service.) While it is evident that uncertainty is higher in a competitive environment, finding the exact location of the new curve will take time and a more stable environment than now exists. Meanwhile, estimates of new competitive services must include an element of chance which recognizes moves that might be made by competitors. Tariff plans that attract new business or redistribute traffic loads must also include consideration of moves by competitors. Also, the large and sophisticated users may well exploit such changes in ways that might not have been obvious to the IEC in the planning days.

3.2 *Grade of Service*

Not only are forecasts more uncertain but objective grade of service must also be reexamined. A call lost due to network blocking may be placed over another carrier's network, so the revenue penalty is higher. In fact, the penalty may be much higher than loss of a single call. It may be a lost account because the customer may change choice of preferred IEC in reaction to unsatisfactory service quality. The modern traffic engineer may well need to take a look at service with new, complex considerations given to hourly and daily variations. Consideration may also be extended to the kind of customer involved. The large business customer with large access trunk groups is much more sensitive to carrier blocking than the single line customer. Customized grade of service is a traffic challenge in both theory and measurement.

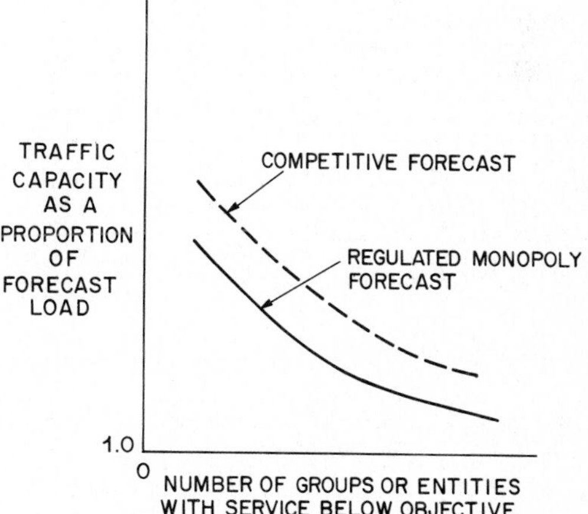

FIGURE 1
EFFECT OF COMPETITION ON TRADE-OFF BETWEEN UTILIZATION AND SERVICE

3.3 *Network*

The biggest changes in traffic engineering come at the interface between the networks of the IEC and LEC. Figure 2 is a skeleton illustration of such an interface. Here, two IECs are shown with connections to two exchanges in two LATAs. Many options are available beyond those shown. It should be remembered, however, the LEC is prohibited by law from carrying traffic between two LATAs - even when it serves local traffic in both. Interexchange carrier A is shown with direct trunks (through a non-switched facility point-of-presence, POP) to one of the exchanges in LATA 1 as well as with trunks to the access tandem for LATA 1. All "equal access" exchanges in the LATA can be reached through the access tandem. Interexchange carrier B is shown to have a POP in adjacent LATA 2. A great variety of configurations of internal trunking and switching within the IECs is seen to be possible. Since the economies and restrictions are different from those governing former trunk engineering and administration, the traffic engineer is faced with revising all engineering practices.

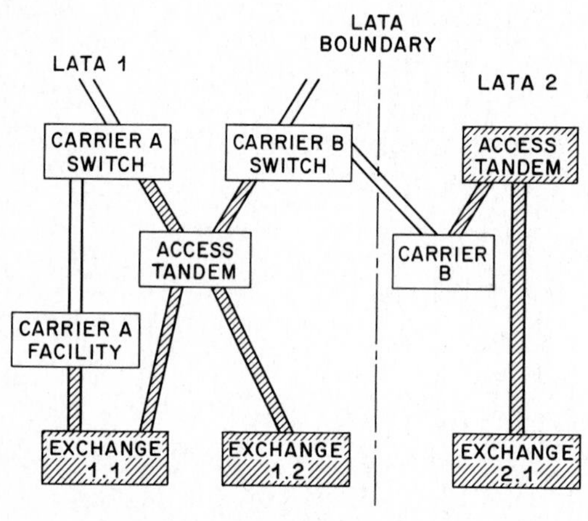

FIGURE 2
INTERFACES BETWEEN EXCHANGE CARRIERS AND INTEREXCHANGE CARRIERS

Provisioning of access trunks is made more complicated for both LEC and IEC because of the need for coordination between independent companies. A further complication arises from the kinds of tariffs that apply here. Regulators have perceived that the LECs would have problems if requests from IECs for access trunks were not controlled. Tariffs serve this purpose by requiring that the LECs charge the IECs for trunk additions and disconnects, for cancellation of orders, or for persistent overloading of access trunks by an IEC.

For regular use of the LEC LATA network, the IEC must pay an access charge per trunk, and, much more significantly, a charge must be paid per minute-of-use received from or sent to the LEC. The charge per minute is dependent on distance but is independent of traffic loads. There is, therefore, little financial incentive on the part of the IEC to route traffic so that the LEC may achieve trunk efficiency by large traffic volumes. At this point in an IEC's network, therefore, it is the tariff, rather than the actual LEC cost, which must be taken into account by the IEC in planning location of points of presence and planning choice of use of IEC trunks or LEC trunks. Of course, without knowledge of the IEC's plans, the LEC cannot respond as quickly to requests for network rearrangements by the IEC. Allowance is necessary for additional time lag in realizing trunk rearrangements.

Complications like these arise from restrictions which were imposed to protect the interface between regulated and unregulated carriers and to smooth the way for new competitors. The traffic engineer must keep a sharp lookout for these and other legal and tariff hurdles (or opportunities) that are quite different from those existing in a completely regulated environment. Perhaps a course in law should be added to the traffic engineering curriculum.

Finally, the traffic engineer may expect to find wider changes in traffic peakedness because of the choices available to customers.

3.4 *Switching System Traffic Engineering*

The problems brought up in the previous section have parallels in switching system traffic engineering. Switching system location and system capacity with complex routings (including new signaling procedures) are a couple of examples. In addition, carriers seek competitive advantage in variety and kind of services. Such things as large bandwidth, packet switching, and customer call control place large demands on switching system traffic capacity. New challenges come at an accelerating rate.

3.5 *International Traffic*

International telephone service, up to now, has been the responsibility of a single carrier within each country. From the United States, for non-telephone communications, several "record" carriers have provided international data and text communications services. With deregulation this separation of function between record and telephone has become less distinct. In addition, new interexchange carriers are entering the international telecommunications arena.

Traffic network design becomes more complex. The route of a call from the United States may involve one carrier for the within-country part, and another carrier for the international part. Since international telephone competition is in its early stages, it is not at all clear what the ultimate challenges to the traffic engineer are going to be. It is clear, however, that the involvement of additional carriers in series on a call raises traffic issues of grade of service, such as blocking and completion delay, as well as of routing and network management.

If other countries follow the trends of the United States, the problems can become very complicated indeed with a call traversing an LEC, a domestic IEC, an international IEC, a domestic IEC, and an LEC. Such a labyrinth is certainly technically achievable today; the challenge to the traffic engineer in each carrier is to maintain service and efficiency. If the engineer needs more work, the addition of international transit from one carrier to another will provide it. One is tempted to predict that traffic, operations, and revenue collection problems will be the restraining force against the increasing complexity that telecommunications technology can provide. These problems will need to be studied in CCITT so that international standards can be adopted.

3.6 *Traffic Measurement*

With the LEC and IEC functions separated, the full set of traffic measurements from end-to-end is no longer available to the IEC. The traffic engineer is challenged to evaluate service from incomplete data and to furnish as meaningful information as possible to the forecasting process.

4. *EXCHANGE CARRIER TRAFFIC ENGINEERING*

The LEC, in particular a Bell Operating Company, is restricted to regulated telecommunications within LATAs as described earlier. It is responsible for end-to-end service within the LATA and for handling traffic between end customers and the IECs.

4.1 *Forecasting*

The LEC engineer enjoys a somewhat better position in forecasting than does his IEC counterpart because, first, a monopoly condition persists for intra-LATA traffic and, second, the local distribution part of inter-LATA calls is only moderately affected by choice of inter-exchange carrier.

Two trends indicate that this position may be temporary. First, several companies want to compete for intra-LATA service, and second, large-traffic customers may elect to send their inter-LATA traffic directly to an IEC. The amount of this latter diversion is related to the amount of access charge that the IEC pays to reach its end customers. Here as elsewhere, the relationship between tariffs and traffic volumes has few precedents in the regulated environment; new traffic models must be found.

4.2 *Grade of Service*

To the extent that the LEC provides end-to-end service, many of the old grade of service criteria still apply. However the service given to the IECs in offering originating calls or completing terminating calls becomes a new subject for negotiation and concern. Because the IECs have their own goals in serving their customers they may be willing to pay higher prices for better service or want to pay lower prices for poorer service. The whole subject of calling-with-priority may have to be revisited for local completion traffic.

A further complication arises as a carrier feels that its traffic service data are proprietary. Ways must be found for making the measurements within a carrier's network and not relying on results from interconnecting carriers.

Adoption of standards appears to be one of the possible solutions to the dilemma. It is not clear at this time however, how such standards could be applied so that the public would benefit directly and not suffer later because of the stifling of competition among carriers.

4.3 *Network*

Planning of the LATA network is changed for the LEC traffic engineer by the volatility and new functions of points-of-presence. The IECs may change in location and traffic volume; plans must have built-in flexibility to accommodate these changes. As opposed to planning for intra-LATA network traffic, access network planning calls for a reactive mode.

The biggest challenge in this area is to establish a network which will provide low access cost to the large traffic customers. Such low costs will allow pricing that keeps traffic on LEC facilities for delivery to IEC points of presence. Again, precedents do not exist for an easy solution.

4.4 *Switching Systems*

There should be little change in the engineering of a local switching system; however the tandem switching systems are entirely new both in the position they occupy in the network and in the kind of traffic they must handle.

4.5 *Traffic Measurement*

In the past much of the traffic measurement for interstate traffic was based on samples of measured calls taken from billing records. This kind of information is proprietary for an IEC and is likely to become less and less available as the LECs and IECs progress in their adjustments to competition. A plan of traffic measurement must be worked out for each carrier without reliance on any connecting carrier.

5. *PRIVATE NETWORK TRAFFIC ENGINEERING*

An option open to every large customer is to design a private network. In doing this, portions of the IEC and LEC network, as well as parts of the services they offer, can be utilized. The private network engineer has more variables to consider. A PBX may well contain a computing facility for dynamic routing of traffic to take advantage of every private and public capability. What is accomplished is limited more by the imagination of the engineer rather than the limitations of computing. As ISDN becomes a reality, the options are likely to grow; the traffic engineer must have the background to proposed profitable arrangements and facilities to help the customer exploit to the fullest the competitive moves made by the carriers.

6. *OTHER ASPECTS OF TRAFFIC ENGINEERING*

6.1 *Operator Services*

In the Bell System divestiture, the operator forces were divided in two. The call handling group became part of the IEC and the number services group became part of the LECs. Since equipment and function are not so easily divided, it has been necessary to arrange for selling service between the two types of carriers. A new traffic problem is evident here in devising ways of providing service to "a customer's customer."

6.2 *Network Management*

The management of the aggregate of a number of interconnecting networks under overload is not only a complex traffic problem but a philosophical one as well. Can the proprietor of a thin IEC network rely on a robust IEC or on its connecting LECs to bail it out of a bad overload? A balance must be reached here between the economic realities and the need to protect the public in times of emergency. Definition of "emergency" is the starting place for solving this problem.

7. *CONCLUSION*

In all of the preceding discussion, several themes have come up again and again as the different engineering viewpoints were considered. First, the uncertainty with which the traffic engineer must work increases with competition. Second, the business moves of one's own company as well as the possible moves of competitors must be taken into account. Third, partial deregulation leaves the engineer with a complex number of factors that must be considered that do not exist in a completely deregulated environment. Fourth, grade of service takes on a different meaning to the traffic engineer.

Traffic engineers are coping with all of these changes. They are no strangers to uncertainty or the handling of complex situations. There is more work to be done than ever and no end is yet in sight.

REFERENCES

[1] John Brooks, "Telephone," Harper and Rowe, 1975, 1976, pp 91-93, 102-126

[2] R. L. Franks, H. Heffes, J. M. Holtzman, and S. Horing, "A Model Relating Measurement and Forecast Errors to the Provisioning of Direct Final Trunk Groups," ITC8 1976, Paper 133

[3] R. L. Franks, H. Heffes, J. M. Holtzman, S. Horing, E. J. Messerli, "A Model Relating Measurement and Forecast Errors to the Provisioning of Direct Final Trunk Groups, BSTJ Vol. 58, No. 2, pp 351, 378

[4] R. A. Coco, R. A. Farel, R. M. Potter, P. E. Wirth, "Relationships between Utilization, Service and Forecast Uncertainty for Central Office Provisioning", ITC 10 Montreal, 1983

TELETRAFFIC ISSUES in an Advanced Information Society
ITC-11
Minoru Akiyama (Editor)
Elsevier Science Publishers B.V. (North-Holland)

NEW DEVELOPMENTS OF TELECOMMUNICATIONS POLICIES IN JAPAN

Akiyoshi TAKADA and Tetsurô YAMAKAWA

Ministry of posts and Telecommunications
Tokyo, Japan

ABSTRACT

Telecommunications businesses provide telecommunications networks which form the foundation of the advanced information society. For the smooth materialization of such a society, it will be necessary to promote the wholesome development of the telecommunications businesses.

In order to achieve this end, on April 1 of this year, the Ministry of Posts and Telecommunications carried out a major reform of the telecommunications system in Japan. The major pillers of reform were placement of Nippon Telegraph & Telephone Public Corporation under private management, and introduction of the principle of competition into the telecommunications business.

Due to the enforcement of the new system, it has become necessary to interconnect the networks of those enterprises which are already in the business and those which will join the business in future. As a consequence, it has become necessary to examine new issues from the viewpoint of traffic engineering, such as routing and traffic control.

1. ADVANCED INFORMATION SOCIETY AND TELECOMMUNICATIONS

1.1 Changes in the Environment Surrounding Japan

Japan underwent remarkable economic and social development after W.W.II, and became a modern society and one of the advanced nations of the world. However, in the recent years, there have been marked changes in the situations surrounding it, and such situations have increased in severity. That is, domestically, there have been stagnation in the economy and changes in the social structure consequent upon the transition from the high economic growth to stable economic growth, and internationally, the effects of the high interest rates in the Unites States and accumulation of loans to developing countries have brought about multipolarization and instability.

Due to these changes in the environmental conditions, at present, Japan faces many difficult issues. These are for instance the need to achieve proper economic growth by utilizing the vitality of private management, coping with the rapidly progressing aging or maturing society, elimination of social and economic gaps between cities and local areas, and coping with the aggravating international trade and other frictions. If, toward the 21st century, Japan were to fulfill its obligations in the international society and to continue to enjoy its present prosperity, it will be necessary for it to properly cope with these issues.

1.2 Building of the Advanced Information Society and Telecommunications

Hopes are being placed on information as something which will play an important role in coping with these issues. It is undeniable that information has been necessary for the smooth operation of the society and economy up to the present. However, the rapid development of the electronics technology in the recent years has markedly enhanced its usefulness. And as a result, dependence by society on information has magnified to a degree uncomparable to the past. In the future, these tendencies are expected further to progress, leading to the materialization of a society in which smooth distribution of and access to the needed information will be secured for all of the members of the society including individuals, homes, enterprises, and administrative organs--that is, to the advanced information society.

As something which will allow us properly to cope with the various issues which Japan faces by utilizing information as a resource to substitute oil, so to speak, in the past high growth era, hopes are being placed on the advanced information society as one which makes possible the following:

i) Building of people-centered, rich national life

Through such means as home shopping, banking, security, and working at home, such society will enhance the convenience and efficiency in various spheres of people's living. As a result, the spare time created can be spend for creative activities by utilizing the rich information services provided through the videotex, etc. In addition, because in such a society the individuals and homes will be directly linked with the enterprises using networks, it will be easier for the latter to provide goods and services which are in coping with the former's individualized and divesified needs.

ii) Switchover to industrial structures which are more efficient and full of vitality

Due to the development of information and communication networks, it is expected that the so-called new-media-related, knowledge-intensive, high value added information industries such as the software industry and information services will develop. In addition, through efficient utilization of communication systems connecting enterprises in the same or different businesses. It can be expected that rationalization will be promoted in all areas of corporate activity such as production, distribution, sales, and business management.

iii) Promotion of independent development of local communities

By utilizing the information and communication networks, it becomes possible to decentralize the various social functions which had hitherto tended to concentrate in large cities to local regions. By so doing, such a society will solve such problems as depopulation and overpopulation, traffic congestion and housing shortage in large cities, and improve the amenity of local areas. In addition, by building up information and communication networks which have roots in the local community, it will improve local living and culture.

iv) Promotion of international solidarity and cooperation

Due to the advancement of international communications networks, it will be possible to eliminate international communication gaps, and build a stable international society which is based on the spirit of solidarity and cooperation.

In order to build such an advanced information society, what become indispensable are telecommunications networks to interconnect all of the members of the society such as individuals, homes, enterprises, and administrative organs. Today, telecommunications is therefore playing the role of leading the society in the formative stage of the advanced information society. And not only that, it is also expected to play a fundamental and nucleic role in the actual advanced information society. It can therefore be said that toward the smooth realization of such a society, telecommunications has further grown in importance.

2. NEW DEVELOPMENTS IN THE TELECOMMUNICATIONS POLICIES

2.1 Liberalization of the Telecommunications Businesses

The telecommunications businesses provide the telecommunications networks which form the foundation of the advanced information society. For the smooth materialization of such a society, it will be necessary to promote the wholesome development of telecommunications businesses. Based on such a recognition, the Ministry of Posts and Telecommunications carried out a major reform of the telecommunications system which had, as its main pillars, entrance by private enterprises into the telecommunications business and placement of Nippon Telegraph & Telephone Public Corporation under private management. And on April 1, 1985, the so-called three NTT reform laws were enforced: the Nippon Denshin Denwa Kabushiki Kaisha Law, the Telecommunications Business Law, and the Law Regarding the Revision and Adjustment of Laws Related to the above two laws.

In Japan, the telecommunications business had been uniterily operated by the state before W.W.II and by NTT Public Corporation (by Kokusai Denshin Denwa Co., Ltd. in the case of international communications) after the war. Through concentrated utilization of the limited amount of resources, NTT Public Corporation had answered the voluminous telecommunications-related demand from the post-war reconstruction period to the period of high economic growth, making Japan one of the foremost and advanced telecommunications nations in the world. However, now that backlog of demands has been eliminated and nationwide subscriber dialing service has been achieved, there has been less necessity to promote the concentrated utilization of resources through unitary operation. In addition, the telecommunications bausiness has been said as one in which natural monopolistic tendencies and technological uniformity exert strong influence, and these had been given as grounds for continuing the unitary operation. However, due to the advent of new media such as satellite communication and optical-fiber cables regarding which the concept of scale-merit does not necessarily apply, and to the progress of the interface technologies which enable the coexistence of a multiple number of networks, these problems are also about to be eradicated. Moreover, there has been a strong demand for pluralized operation due to such factors as the advent of various new media which is a result of progress of electronics technologies, advancement and diversification of users' needs, and activation of the telecommunications businesses in order to attentively cope with these needs. In view of these changes in the situation, the revisions effected in the system this time aim at promoting the smooth materialization of the advanced information society by establishing a pluralistic telecommunications system by introducing the principle of competition. (See Figure 1 and Table 1.)

2.2 The Basic Idea Governing the New System

The basic idea governing the new system is to activate and bring up the efficiency of telecommunications businesses by utilizing the vitality of private management. It also aimes at ensuring the public properties of communication businesses and promote the benefit of the people who are the ultimate users of communication services. The new system is made up of the following four pillars.

i) Ensuring of Flexibility

Telecommunications involves vanguard technology. Therefore, in order to enable enterprises to cope flexibly with technological progress and trends of demand in the future, the new system allows them to enter into any of the fields of telecommunication.

In addition, the telecommunication services are expected in the future to undergo diversification due to technological renovations and advancement in the demand for them. Therefore, the enterprises are striving to provide all kinds of communication services including telephone, telex, facsimile, VAN, and data communication. In this respect, in comparison to the former way of thinking of classifying telecommunication services into basic services and advanced services, the new system is believed to facilitate enterprises to cope appropriately with new developments; the recent trend of technological development from unitary services to integrated services.

The Telecommunications Business Law divides telecommunications businesses into Type I and Type II telecommunication businesses. It then

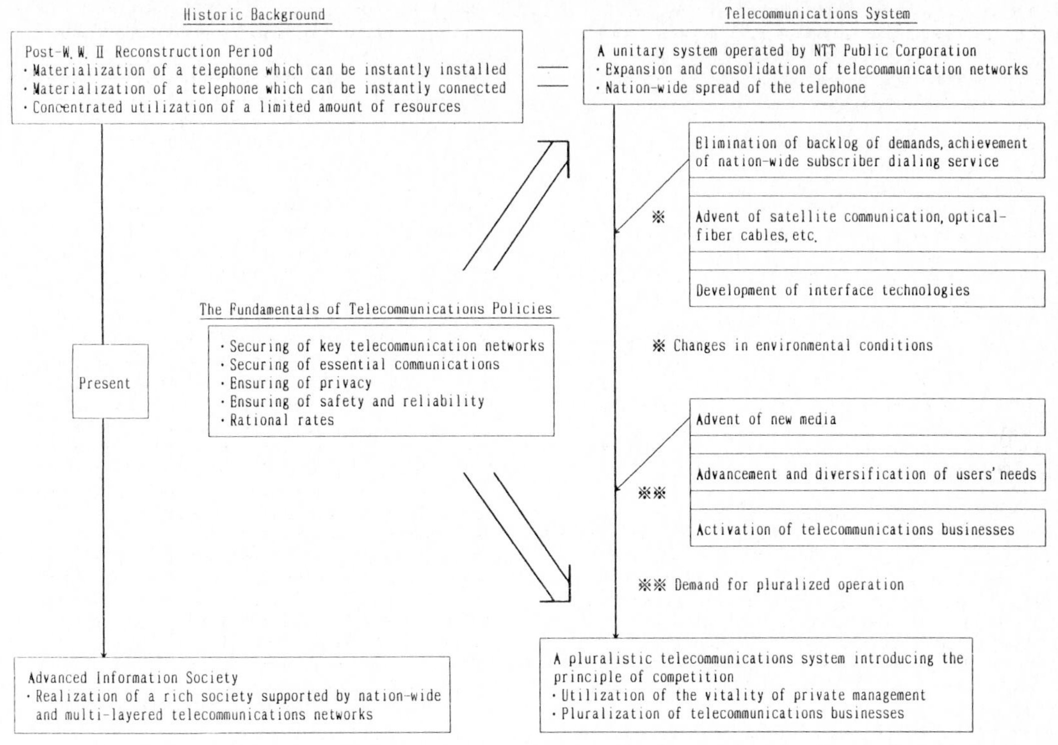

Figure 1. Revisions in the Telecommunications System

Table 1. Outline of the New Telecommunications System

	The Old System (Monopoly)	The New System (Competition)
Businesses operated by:	NTT Public Corp., KDD	Private enterprises ·Long-distance, large-capacity circuits ·Two-way CATV ·VAN
VAN business	Small and Medium Enterprises VAN only.	
Rates	Principal rates are to be determined by law (others to be approved)	Approval
VAN business	Approval not necessary	Approval not necessary
Services	To be determined by law	Approval (to cope with diversification by making the best use of the originality of private enterprises)
VAN business	Prohibited	Approval not necessary
Ensuring of privacy of communication, of transmission of important communication	To be determined by law	To be determined by law
Ensuring of services	Impartial service throughout the nation, as long as budget allows.	Provision of services (ensuring of the least necessary means of telecommunication)
Use of circuits	Strict restrictions on use by others, common use, and interconnections (guaranteeing of monopoly)	Approval not necessary in principle
Terminal Equipments	Direct management in principle (prohibition, in principle, of customer provision)	Liberalization (liberalization of customer provision of main telephones, multi-function and high-quality equipments through competition)

further divides the Type II businesses into General and Special Type II businesses. Figure 2 shows the positions of these different kinds of businesses in the form of a hierarchical structure.

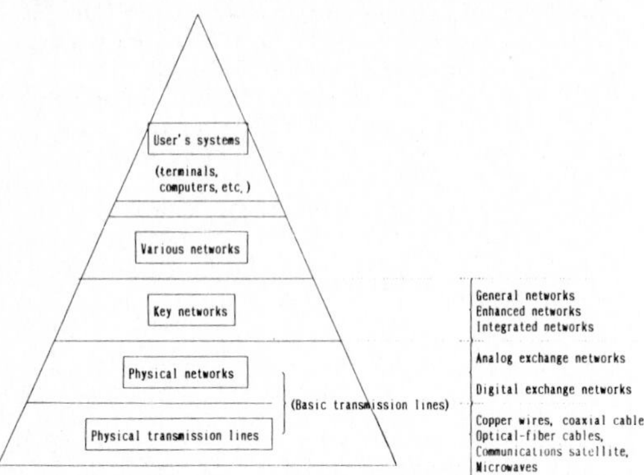

Figure 2. A Hierarchical Structure of Networks

(Source: a proposal made in "A Symposium on the Conception of an Integrated Data Communications Network.")

First, the lowest layer, which comprises the physical transmission lines, forms the infrastructure which provides the foundation for all of the telecommunication services. In the past, these lines were mainly made up of pair cables and coaxial cables. In the recent years however, optical-fiber cables which enable broadband, large-capacity, and economical transmission, and communications satellites which can cover the entire nation at once, have emerged.

The next layer comprises physical networks provided by the Type I telecommunications enterprises in a form which is closely related to and indivisible from the physical transmission lines set up by such enterprises. At present, the subscription telephone network, which is a nation-wide analog exchange network, has been consolidated. In the future however, due to the progress of the digital technology, the trend is expected to be toward an integrated digital network. In addition, due to the progress of the interface technology, it has become possible to interconnect a plural number of networks.

The third and fourth layers comprise functional networks operated by the Type II or Type I telecommunications enterprises and which are built on the physical networks operated by the Type I telecommunications enterprises. The so-called VAN which has diverse functions such as one enabling easy communication between different kinds of computers, belongs to this layer.

The third and fourth layers correspond to the Special and General Type II telecommunications businesses, respectively. In order to give objective clarity to the category of the Special Type II businesses, the law defines them as a) those which carry out businesses by covering many and unspecified persons, and b) those operating facilities of sizes larger than the one prescribed by cabinet order.

Finally, the fifth layer comprises the systems privately operated by individual enterprises, etc.

When these five layers of networks function organically, they mutually complement one another, and form the desired network society.

ii) Giving full play to the vitality of private management

The Type I telecommunications businesses, which provide telecommunication services by setting up their own telecommunications circuits and facilities, are placed under a permit system due to the following reasons. First, these are key businesses which construct and operate telecommunications networks which become foundations for providing telecommunications services which are indispensable to the people's daily living and industrial and economic activities. Secondly, as an industry, they are facility-intensive enterprises which require enormous amounts of facility investment, and their services are provided in a form closely related to and indivisible from the regions in which their facilities are set up (facility-bound).

In addition, in order to prevent excess facilities, their rates are not placed under a perfectly free competition market. As a consequence, the market rates will not necessarily and fully be formed. And because such rates must be public rates which are closely related to the people's daily living, and so that they should be "just and proper ones when compared with the proper cost under efficient management", they are placed under a permit system. By the same token, the Type I telecommunications businesses are obliged by law to provide their services in their own areas of business, so that users in such areas should be able to use the services with equality.

In this way, as public service enterprises, the Type I telecommunications businesses are placed under a minimum of restrictions. In fact, in order as much as possible to guarantee their free business activities, they are not placed under the usual restrictions binding on other public service enterprises such as restrictions on engaging in side businesses, and those concerning transferring of facilities.

Next, the Type II telecommunications businesses, which provide telecommunications services by borrowing telecommunications circuits and facilities from Type I enterprises, can relatively easily enter the business by simply setting up computers. And because this is a field which provides attentive services by coping with users' diverse needs, the market can be expected to expand if maximum respect is paid to the principle of free competition among enterprises. Therefore, the enterprises in this category are placed under a notification system in principle.

iii) Ensuring of public properties

The objective of the Telecommunications Business Law is to promote the wholesome development of the telecommunications businesses by utilizing the originality of private management. However, the telecommunications business has its own public properties based on its very nature as something dealing with communication. And because ensuring of these properties is also an objective of the Telecommunications Business Law, it lays down the measures necessary to achieve this end.

First, communication is indispensable if human beings were to carry on social life. And if they were to carry out free and frank communication with peace of mind, the most important thing is to secure the privacy of communication. For these reasons, the law prohibite third party state organs and private individuals to infringe on the privacy of communications being handled by telecommunications enterprises. At the same time, it also lays down the duty of individuals engaged in such businesses to protect the privacy and to take the necessary measures with respect to facilities and management.

In addition, as in the case of violation of privacy, interruption of service and outbreak of trouble due to accidents, etc., cause unexpected damage to users. Therefore, it is extremely important to ensure the safety and reliability of communication and to maintain wholesome networks. For these reasons, and from the standpoint of smooth provision of telecommunication services, the law lays down fixed technical standards for the telecommunication facilities of the particularly consequential Type I and Special Type II telecommunications businesses.

iv) Liberalization of circuit and terminal use

In the past, in order to ensure the unitary operation of telecommunications businesses by NTT Public Corporation and KDD, various restrictions had been imposed on the use of telecommunication circuits and terminal equipments. However, with the introduction of the principle of competition as a turning-point, the various legal restrictions have been abolished, and users are now able to use these services according to their own needs, and as freely and efficiently as possible.

In addition, with the liberalization of the use of terminal equipments, administrative functions such as laying down of technical standards for terminal equipments and compliance approval for the same which NTT had hitherto been carrying out were transferred to the state, so that enterprises may engage in fair competition among themselves.

3. PROMOTION OF NEW ENTRANCE AND ISSUES FOR TECHNOLOGICAL EXAMINATION

With the liberalization of the telecommunications businesses, it can be said that the telecommunications system has been consolidated for the time being toward the smooth materialization of the advanced information society. However, if telecommunications were truly to play a nucleic role in such a society, realistic measures and policies are needed which will make the most of the system's spirit. To achieve this end, the law encourages enterprises to enter into the various fields of the business. However, at the same time, it will be necessary to examine the various technological issues.

i) New entrance into long-distance, trunk line services

The enterprises which enter the business are expected to do so first in the field of long-distance, truck communication lines manily between Tokyo and Osaka where the telecommunication traffic is the most concentrated and the profitability is the highest. Already, several enterprises have filed for permission for starting Type I telecommunications businesses, and they plan to start leased circuit services from around 1986, and exchange services from around 1987. By interconnecting with NTT's existing local networks, the long-distance enterprises can provide wider services. The concrete form of such an interconnection is likely to be one which goes through the gateway switching offices as shown in Figure 3. When interconnecting the networks of different telecommunications enterprises in this way, it will be necessary to examine the technological issues such as are listed in Table 2. In particular, interesting issues have been raised from the viewpoint of traffic engineering and in connection with the appearance of a multiple number of businesses:

a) Changes in the traffic situations
b) Treatment of overflow traffic
c) Traffic control at the time of congestion

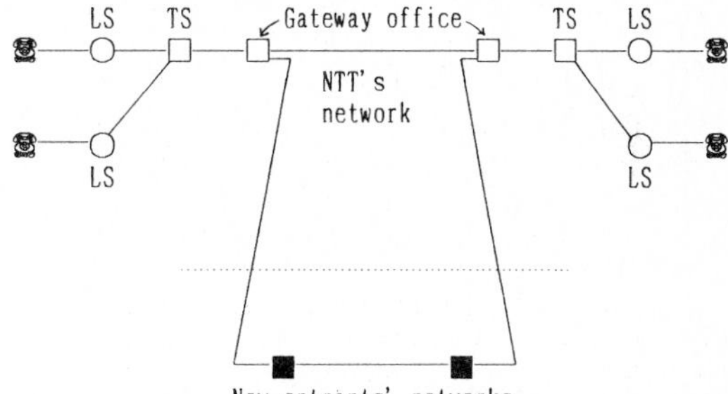

Figure 3. An Example of Interconnection

Table 2. Issues for Examination Regarding Interconnection

Items to be Examined
1. Forms of interconnection
2. Interface conditions
3. Provision of services
4. Design of Gateway facilities
5. Numbering plan
6. Charging plan
7. Maintenance and Testing method
8. Performance distribution

ii) New entrance into inter-city services

The Telecommunications Business Law also allows the entrance into the intra-city communication business. Some of the forms which such is expected to take are for wire broadcasting telephone companies and CATV companies to become Type I telecommunications businesses by interconnecting with inter-city networks, and entrance into car telephone and pocket-bell service businesses.

4. CONCLUSION

Today, the world is increasingly becoming interdependent in all of the areas of society and economy, and in the field of telecommunications also, various issues have emerged as a consequence of the progress of such interdependence. These are for example transborder data flow, distribution of such resources as satellite orbits and transmission frequencies, liberalization of service trade, and procurement of materials, etc.

These problems are developing on a world-wide scale, and among the Western advanced nations, they have emerged as problems related to liberalization of service trade between the United States and Europe which see the field of telecommunications as a strategic industry. And between the Western and the Eastern nations, they have emerged as problems associated with news agencies and satellite broadcasting in connection with the problem of free distribution of information. And between advanced nations and developing nations, they have emerged as problems of imbalance of information distribution and of north-south gap in the communications infrastructure.

By taking into consideration the progressing interdependence of the world, it is believed important for Japan also to deal comprehensively with these problems from a long-term perspective.

TELETRAFFIC ISSUES in an Advanced Information Society
ITC-11
Minoru Akiyama (Editor)
Elsevier Science Publishers B.V. (North-Holland)

TELECOMMUNICATIONS WORLDWIDE - A CHALLENGE FOR THE CCITT

Dipl.-Ing. Th. IRMER

Director CCITT, ITU Geneva

1. CCITT - more than its name actually reveals

When I was Chairman of the CCITT Study Group XVIII I believed that the CCITT, its framework and tasks were well-known to everybody. However, since January 1985, when I took office as Director of CCITT, I learned that my point of view on this matter was somewhat incomplete. The reasons are, of course, clear: as a Study Group Chairman I was mainly in contact with people directly involved in CCITT activities and therefore, aware of CCITT's tasks. But now having daily numerous contacts with other organizations, individuals and companies I realize that there is in fact a great lack of information as to the real nature of the CCITT, its potentials and constraints. Maybe the CCITT itself contributes to misunderstandings because of its denomination. When the CCITT was established in 1956 by merging CCIF and CCIT, the new denomination CCITT (CCITT = International Telegraph and Telephone Consultative Committee) was to indicate what its main tasks would be. However, during nearly 30 years, telecommunication technology changed rapidly - and so did CCITT's activities. Nowadays telegraphy plays only a minor role, but telephony is still important -and the technological evolution has created many other areas now to be covered by the CCITT such as telematics, data networks, optical fibres and, as ultimate but most complex area, the ISDN. At present, 15 Study Groups working in the field of telecommunication networks and services, are in charge of keeping existing Recommendations up-to-date or of drawing up new Recommendations covering new developments. Apart from these 15 Study Groups, which are in fact setting international standards by means of their Recommendations, there are presently 5 Specialized Autonomous Groups (GAS) drawing up Manuals to support mainly developing countries by supplying detailed information on important topics such as rural telecommunications, primary energy sources, traffic engineering, transition from analogue to digital networks, etc. - just to name a few of the topics covered so far. Finally, there is a World Plan Committee and the Regional Plan Committees, jointly administered by CCITT and CCIR in order to develop a general plan for the international telecommunication network facilitating a coordinated development of international telecommunication services. To promote their activities in their meetings, workshops and seminars on new developments are being held more frequently.

Exclusively for the 1985-1988 study period, two special CCITT bodies have been created: Special Study Group "S", charged with proposals for restructuring CCITT Study Groups and to improve working methods in general. The other group is the Preparatory Committee for the forthcoming World Administrative Telegraph and Telephone Conference in 1988 (PC-WATTC) to prepare this conference.

If sometimes even insiders do not quite know these areas of CCITT activities, one should not be surprised that even fewer people outside the organizations know the fundamentals on which CCITT activities are based. It is generally understood that participation in the CCITT is open to Administrations, Recognized Private Operating Agencies (RPOAs) and industrial and scientific organizations of all ITU Member countries. But how this cooperation actually works in practice, in particular within Study Groups in which all these organizations are represented, is seldom commonly known - and precisely this lack of information gives rise to many unreasonnable criticism. There is no doubt that no organization in the world is perfect and this holds also true for CCITT. But most of the criticism I have heard since my taking office - like for example CCITT standards are too complex, too late, not detailed enough and unequivocal - indicates a lack of information concerning the potentials on one hand and the constraints of CCITT on the other hand. In order to fill this gap and to promote a better understanding I would like to line out the circumstances under which CCITT performs its tasks. I shall restrict my deliberations to the work of the Study Groups, because this works forms the largest part of CCITT activities - it covers standardization of networks and services in the widest sense. This wide coverage requires close cooperation with CCIR, ISO and IEC, and it is clear that the more telecommunication technology is expanding, the higher is the degree of interworking and coordinating all standardization matters with these organizations.

2. The standardization scenario

The main task of CCITT Study Groups is to draw up Recommendations for the various areas allocated to them. In principle, these Recommendations are not binding to Administrations or RPOAs, and also each industrial or scientific organization is free to decide whether such Recommendations should be applied. In practice, however, this freedom is only a theoretical one: if there are CCITT Recommendations, they are nowadays applied throughout by all organizations involved in telecommunications; in fact, CCITT Recommendations may, therefore, be regarded as worldwide, genuine de facto telecommunication standards.

Why are CCITT Recommendations applied worldwide? There are mainly two reasons for it:

- Increasing complexity of networks, network elements (switching, transmission, etc.) and services (existing/enhanced/new services) calls for an increasing demand of standards - in the sound interest of users, industry and network providers. Each one of these three groups benefits from standardization but looses if no standards are drawn up at the appropriate time;

- the composition of CCITT Study Groups in which these three groups are represented ensures the coordination between users, industry and network providers. CCITT Recommendations are, therefore, up-to-date - they are taking account of modern technology and its potentials (submitted by industry), application-oriented technical and operational aspects (submitted by network providers) and user-related requirements (submitted by users). To establish a balance between the three groups (the views and interests of which are sometime naturally controverse) is difficult - and there are cases where a consensus could not be reached. It is, however, rewarding because, if standards are drawn up in such a way, all three groups are at an advantage. The ISDN-Recommendations are a typical example for such a consensus between these three groups which I will talk about later.

Reaching a consensus between the three groups is, however, becoming more and more difficult. By no means have the representatives of these groups in the various Study Groups become less cooperative - this is not at all the case, it is rather the contrary; not the people have changed their attitudes, but the conditions under which a consensus has to be reached.

As a matter of fact, over the years, the areas to be covered by Study Groups have increased steadily. Since 1972, each Plenary Assembly has approved a substantial number of new questions to be studied: from a total of 281 questions in 1972/76 it increased to 383 questions for the study period 1985-1988 - an increase of about 33% without any significant change in the organization and number of Study Groups in which workload increased therefore correspondingly.

This increase of questions is triggered mainly by two facts:

- apart from new topics generated by the progress of technology (e.g. telematics, optical fibres, ISDN) and which necessarily have to be studied and standardized, quite a few topics have been included which so far were subject to national, but not international standardization. An example is the entire area of digital trunk and local switching; up to some years ago, CCITT did not standardize switching equipment but only signalling systems crossing the borders. Standardization of signalling systems ensured international telephone service regardless of the type of switching equipment used in the various countries. The advent of SPC switching systems and the request by developing countries for standards of digital switching equipment has changed this situation completely: both software layout (CHILL) and design characteristics of trunk and local digital exchanges are now being standardized by CCITT, as well as many other topics which were previously regarded to be of national importance. In other words: CCITT standardization is no longer restricted to the genuine international part of networks and services; standards are increasingly drawn up in an "end-to-end" mode thus covering all parts of networks and services, respectively.

This trend to standardize more topics is superimposed by a trend towards higher complexity of the standards as such. For example, the Recommendations for coaxial cables are quite straight forward specifying very clearly the physical and geometrical parameters. Compared to that are the Recommendations for optical fibres more complex and more descriptive as a direct consequence of the more complex technology. The same holds true for the ISDN Recommendations, for example, the layout of the D-channel protocol. In other words: standardization of new technologies and services render Recommendations more complex, more details are necessary - and consequently it is becoming increasingly difficult to agree on such complex standards. The sharp increase in pages of the new Red Book (some 13,000 pages compared to some 7,000 pages of the Yellow Book) is, to a large extent, caused by the volume of many new Recommendations which are necessarily much more descriptive than the older ones because of their complexity.

Another factor contributing to the problems of standardization is the increasing number of participants in Study Groups meetings representing different views and defending different interests. Ten years ago, a big Study Group consisted of about 100 participants, if at all, representing the main Administrations and RPOAs as well as some big suppliers. Nowadays in some Study Group meetings, we have up to 350 participants. But not the mere increase of participants complicates the standardization process: they are representing in fact more network providers, more suppliers and more users - necessarily all holding different views on how standardization should be achieved. The increase in participation results to some extent from the widening of the scope of CCITT standardization which now attracts companies so far not represented (e.g. the semiconductor industry), or newly established companies entering the telecommunication market as newcomers. But also the network providers' camp is increasing: in many countries no longer one single network provider exists - now quite a few different and competing network providers are showing up instead. Also the users start to organize themselves in groups which raise their voices more and more. It is, therefore, not surprising that from a big country some 40 delegates are participating in a Study Group meeting today, compared to some 5 or 6 delegates some ten years ago. In addition, the increasing necessity to interwork with other international organizations entails an increase of participating organizations - and consequently also an increase of delegates from such organizations.

Both the increased complex workload and the different views of more and more participants certainly do not speed up the process of standardization - rather the contrary. But there are, on the other hand, some important reasons which categorically push for acceleration, even faster than during the past.

The most demanding pressure is generated by the speed of technological development. Standards for analogue technology have been consistently developed over some study periods - and for digital technology sometimes one study period is even considered to be too long. Recommendations just leaving the printers' are already under revision because the progress of technology provides better solutions - and this speed is likely to increase. The investments in research and new developments are tremendous, and everybody - be it industry or network provider - tries to push through his opinion, being fully aware of the fact that it bears financial consequences if his proposal fails to be a subject of standardization. This pressure is spurred by worldwide competition - a relentless and unmerciful process which has no room for philantropic considerations. The attempt to standardize videotex services is an instructive example - in my opinion, a true worldwide standard has failed to come through because each of the proposers of the different videotex systems believed that his system would eventually get the lion's share on this market.

And finally, political and regulatory influences are playing an increasing role in standardization, in particular in the area of telecommunication services. There must be necessarily different points of view between a single network provider operating all networks commonly, regulating all telecommunication services, and a number of network providers operating deregulated services in a highly competive environment. In 1988 the forthcoming World Administrative Telegraph and Telephone Conference (WATTC) will be the crucial test of how different views can be amalgamated and to which extent in particular the new telecommuniation services can be standardized.

The standardization scenario I have depicted here is very colourful indeed but it would be wrong to assume that, under such conditions, only very little progress could be achieved in doing the CCITT standardization work. As a practical example, I would like to refer to the setting up of the first set of ISDN Recommendations in the 1980-1984 study period. Although these Recommendations (I-Series Recommendations) are quite complex it has been posible to arrive at a consensus between all parties involved. Why? Because despite many divergent views on legal, technical and economic matters, the potentials and advantages of ISDN for user, manufacturer and network provider finally succeeded, and each of the three parties became aware of the fact that, failing a consensus, would be detrimental to each one of them.

3. The ISDN principles

ISDN Recommendations agreed on so far will not be discussed in this paper in detail. There have already been many publications dealing with the structure of these Recommendations, their contents and technical parameters. We will look into these Recommendations more generally, indicating the major principles laid down.

It should be noted that the agreement reached on these principles was constituted by a joint effort of many CCITT Study Groups. Study Group XVIII has, of course, been the focal point where most of the discussions took place and where the texts of most of the Recommendations were formulated. However, without close cooperation with literally more or less all the other CCITT Study Groups, this result would not have been achieved. At a time where there are many complaints about international standardization, the example of setting up the ISDN Recommendations within four years only clearly shows that such a complex and difficult task can be successfully completed - even in a worldwide operating standardization organization, provided that all parties involved are motivated and fully cooperative.

If we look into the ISDN Recommendation, a few main principles can be identified and summarized as follows:

- evolution
- modular interface characteristics
- service capabilities
- overall network aspects and functions.

We will investigate the main issues of these principles within the following paragraphs.

3.1 Evolution

In CCITT long and exhaustive discussions were held about the ways to establish an ISDN. In principle, each digital network - however, not necessarily a digital telephone network - can serve as a basis from which an ISDN may be developed. Finally, preference was given to the digital telephone network because it offers some advantages which other dedicated digital networks do not have. For technical and economic reasons has the analogue telephone network presently been converted in many countries into a digital telephone network. This digital telephone network is enhanced to an ISDN at 64 Kbit/s, often referred to as "narrowband ISDN". This ISDN accommodates all telecommunication services up to 64 Kbits, however, not services like sound programme and television as they require much higher bit rates than 64 Kbit/s. Consequently, it is assumed that an ISDN at bit rates above 64 Kbit/s ("broadband") will evolve from the ISDN at 64 Kbit/s.

The main point of the principle of evolution is that each network uses as much as possible the infrastructure of its preceding network so that mainly enhancements are necessary. These enhancements are kept at a minimum in order to provide for the economy of the evolving network at each stage.

3.2 Modular interfaces

Interfaces are essentials for the evolution of ISDN, both in the narrowband and broadband ISDN. Through these interfaces the services to be integrated are either entering or leaving the network, thus converting it to an ISDN. Taking into account the wide range of physical solutions for such interfaces, Recommendations for such ISDN interfaces are expected to be manyfold as well.

It should not be overestimated that this potential difficulty has indeed been overcome. This was only possible because a modular interface concept was developed and agreed on. The concept provides interfaces between the user installation and the network ("user/network interfaces") and between networks ("inter-network interfaces"). For each of these two classes of interfaces some sets of interface characteristics have already been or will be standardized. For the user/network interface, the basic access and primary level access have already been laid down in Recommendations. Both forms of access are covering a wide range of applications, however, represent only two different physical interfaces. It is hoped that, by covering at the same time a wide range of applications, this restriction of physical types of interfaces will be maintained when the higher order user/network interfaces is defined.

3.3 Service capabilities

Another focal point in the discussions was the definition of services to be integrated. The problem here can be seen in the fact that, with a few exceptions, most of the services are not yet defined to the extent necessary for providing the technical means for their integration. In addition it turned out during discussions that, in many countries all over the world, many diverging views exist on how to define telecommunication services because of the different regulatory principles.

At present, telecommunication services are either basic services and/or enhanced/value added services, at least in some countries. Basic services are the fully regulated and defined services like telephony, telex, Teletex; enhanced/value added services are offering additional features not provided by the basic services and their networks (such as message storage, code- and protocol conversions, etc.). This separation into basic and enhanced/value added services is understandable as it reflects to a large extent the evolution of networks and services. It is, however, more than questionable whether or not this separation should be maintained in the ISDN - in my opinion this would also mean to neglect in most cases the potential of ISDN. ISDN provides the chance of offering many enhanced/value added services (not all of them, only if sufficient demand exists) which are today necessarily offered by different networks to the customer as additional basic services within one and the same network. In other words: ISDN will increasingly blur the differences between basic and enhanced/value added services; if this is so, many of these services will have to be fully standardized, but here two obstacles appear: firstly, the separation into basic and enhanced/value added services is presently still continuing; secondly, it is not quite clear which of these services should be fully standardized.

For these reasons a new philosophy has emerged. One does not try any longer to define the services as such; instead, services will be defined according to their occupancy of the different levels in the OSI reference model. Following this approach telecommuication services are divided into two broad categories:

- Bearer services, characterized by a set of low layer attributes (attributes are classified as information transfer, access and general attributes), provide the user with the possibility of getting access to various forms of communications. The customer may choose any set of high layer protocols for his communication, however, the ISDN does not ascertain compatibility at these layers (4 to 7 of the reference model) between customers. Bearer services are providing the capability for the transmission of signals between user/network interfaces; an example of bearer service is a switched 64 Kbit/s circuit unrestricted service.

Teleservices, characterized by one set of low layer and one set of high layer attributes as well as operational and commercial attributes, provide the user with the possibility of getting access to various forms of applications. A user operating a specific application will not be prevented from using a terminal compatible with a given teleservice together with a bearer capability not recommended for this teleservice. Teleservices are providing the complete capability for communication between users of specified user terminals. (Examples of teleservices are telephony, Teletex, Videotex and message handling.)

Finally, supplementary services are supplementing or modifying a basic telecommunication service. A supplementary service should be offered together with a basic telecommunication service and could be the same for a number of telecommunication services.

3.4 Overall network aspects

The topics agreed upon under this principle are covering a wide range. The functional principles and reference models of the network should allow the definition and structure of protocols and procedures for communication in the ISDN. As to the service capabilities, principles and models have been developed in close relation with the OSI reference model. The network connection types are key elements in the ISDN functional architecture as they are supporting the bearer services, teleservices and telecommunication services mentioned above.

With the advent of ISDN, new plans for numbering, addressing and routing have to be considered. The numbering and routing plans developed for the dedicated networks so far are based on the requirements of a given service provided in a particular dedicated network, e.g. in the telephone network. A universal network like ISDN, consisting of a great number of different services, requires new principles which differ considerably from those established for the dedicated networks in the past.

To ensure a high performance of all services in an ISDN, performance objectives must be established. To a large extent, such objectives have already been the subject of Recommendations for digital telephone networks. However, when applying these Recommendations to an ISDN, it should be investigated whether or not the requirements of other non-telephony services are sufficiently covered by such Recommendations. For example, when drawing up the objectives for the bit error performance, the requirements of telephony were used as the basis for them since for telephony, the bit error rate is not too demanding. However, other services (e.g. data services) require a better performance if quality is not to suffer seriously. For this reason, the bit error performance in an ISDN has been set at a higher level than that which would be required for telephony alone. The same consideration holds true for other parameters determining the performance of a digital network.

4. Conclusion

Worldwide standardization in a rapidly developing area like telecommunications, and involving many different organizations at the right time and to the extent required, is a continuous and complex process. Because of its complexity not a single measure will improve this process but quite a few well-coordinated measures involving all organizations to participate in this process are needed. As an example, the unanimous agreement on the first set of the ISDN-Recommendations proves that a consensus can indeed be reached - if the appropriate measures are taken. Work will have to be carried out with a view to further elaborating the measures to be taken, and I can assure you that there will be no relaxation of effort - on the contrary.

TELETRAFFIC ISSUES in an Advanced Information Society
ITC-11
Minoru Akiyama (Editor)
Elsevier Science Publishers B.V. (North-Holland)

THE NEXT STEP FOR CCITT : ISDN TRAFFIC ENGINEERING

Géza GOSZTONY

BHG Telecommunication Works
Budapest, Hungary

ABSTRACT

In the 1980-84 study period CCITT telephone traffic engineering activity achieved results among others in compiling traffic requirements of SPC exchanges, in elaborating methods for the estimation of offered traffic and in improving alternate routing calculations with taking non coincident busy hours into account. Some problems of the new period /1985-88/ are also outlined including the reconsideration of the busy hour concept, advanced routing methods and service accessibility calculations. CCITT teletraffic engineering should find a general framework to approach ISDN grade of service problems. The main difficulties are the big number of parameters and the short introduction time for new services. Perhaps the division of service and network aspects could be a solution. The support given by ITC participants to CCITT is highly required in the future too.

1. INTRODUCTION

About two years ago, in the middle of a four years study period at the 10th ITC a detailed report was given about telephone traffic engineering activities of CCITT. Although the VIIIth Plenary of CCITT which endorsed the results of this period was only in October 1984, the main achievements could have almost completely been summarized then [GOSZ 83] .

Since the last ITC there have been no major structural changes in CCITT and just at the beginning of a new period there is hardly anything to review except some new problems. However there are at least two reasons which justify the presentation of this report.

- Traffic engineers from all over the world should be informed about the basic questions to be dealt with by CCITT, because the proper solution can be found only with their help.
- Traffic engineering is to undergo some important changes in the near future; the multi-service environment requires a reconsideration of the approach applied by CCITT at present.

This time no detailed description of aims and working methods of CCITT appears. Who is interested in it can find information in [GOSZ 83]. For ITC participants it is perhaps enough to know that CCITT studies refer to the pragmatic and practical application of traffic engineering from the aspect of world-wide telecommunication networks. The final aim is to unify the basic international traffic engineering practice.

CCITT work starts from contributions sent by participants representing operating, manufacturing and scientific organizations in answer to the so called Questions of the current Study Period. Discussions take place in regular meetings, results appear mainly in the so called "coloured" books. The VIIIth Plenary endorsed the Red Book.

Telephone traffic engineering studies are performed in Study Group II the structure of which appears in Table 1.1.

Table 1.1. Structure of CCITT Study Group II.

Working Party	Name
II/1	Operation and services
II/2	Human factors
II/3	Numbering, routing and interworking
II/4	Traffic engineering, forecasting, network planning
II/5	Quality of service, network management, mobile service
II/6	Availability, reliability

In Section 2 a summary of certain results is given which were not yet finalized at the time of preparation of the previous report. Section 3 surveys some important problems of the present period. Section 4 briefly reviews the basic traffic engineering problems of the multi-service ISDN environment. Proposals of how to approach ISDN grade of service /GOS/ questions can be found in Section 5. The last Section includes conclusions on the necessity of ITC-CCITT cooperation.

Considerations given below reflect also the personal views of the author, these are not necessarily the opinion of CCITT.

2. SOME RESULTS OBTAINED

This part of the paper should be regarded as a supplement to those results already mentioned at the 10th ITC.

2.1 TRAFFIC REQUIREMENTS OF SPC EXCHANGES

SPC exchanges have the capability to observe and adapt themselves to changing environment. The new Recommendation E.502 on traffic and operational requirements [CCIT 85]

- gives a detailed overview of all possible traffic streams of an exchange - see Fig.2.1,
- uses the rules of the CCITT MML /Man Machine Language/ to define the several types of traffic measurements /including scope, periodicity,etc./,

and as a result of present studies

- will cover real time analysis of measurements mainly for network management purposes,
- will probably be enhanced with network management decision making procedures and utilize the capabilities of CCS /Common Channel Signalling/ systems and networks.

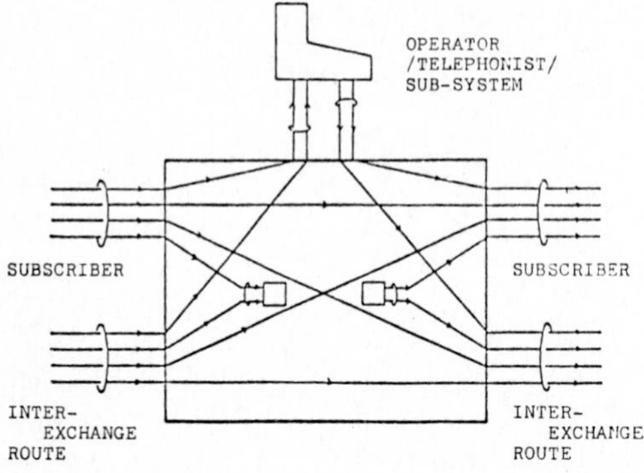

Fig.2.1 Main traffic flow diagram /From Rec.E.502/

Fig.2.1 serves as illustration, the detailed explanation included in the text of E.502 has not been reproduced here. This Recommendation will very likely form the prerequisite conditions of world-wide adaptive routing and the necessary uniform description of the procedures required.

2.2 TERMINOLOGY: A PERMANENT TASK

Following some important improvements the list of "Terms and definitions of teletraffic engineering" became Rec.E.600 [CCIT 85]. Although similar lists of definitions in the CCITT serve first of all the relevant studies, nevertheless a permanent transfer of terms into the International Electrotechnical Vocabulary /IEV/ takes also place and therefore there is a hope to arrive at an internationally acknowledged set of teletraffic definitions. Rec. E.600 will undergo permanent updating in the future.

2.3 OFFERED TRAFFIC, THE FIRST STEP

In Rec.E.600 traffic offered is defined as "the traffic that would be served by a pool of resources /assumed to be fully operative/ sufficiently large to serve that traffic without limitation by the finite size of that pool."

The environment of a given group of traffic handling resources has an effect on the traffic which is offered to the group. This effect is due to the changes in the average holding time and in the average interarrival time. These changes are caused by unsuccessful call attempts which penetrate into the network behind the considered group for some depth and by repeated attempts following unsuccessful ones. The offered traffic determined e.g. from the number of call attempts and the average holding time of calls carried by the group is valid only for the case if no changes took place either in the considered group /e.g. by augmentation/ or in the network where calls are directed to. Changes in the average number of unsuccessful calls effect offered traffic.

Considering that offered traffic is used in calculations mainly in a future oriented manner, the main difficulty can be summarized in the following way: how is it possible to determine the offered traffic in a future situation from measurement results referring to the present ?

Rec.E.501 on the estimation of traffic offered [CCIT 85] gives simple procedures for only route and for high-usage final-group arrangements with no-significant congestion. For only route circuit groups with significant congestion the following solution has been accepted:

"Let A_c be the traffic carried on the circuit group. Then on the assumption that augmentation of the circuit group would have no effect on the mean holding time of calls carried, or on the completion ratio of calls carried, the traffic offered to the circuit group may be expressed as

$$A = A_c \frac{1 - WB}{1 - B}$$

where B is the present average loss probability for all call attempts to the considered circuit group, and W is a parameter representing the effect of call repetition. Models for W are presented in Annex ... "

The model for W assumes the general validity of the

$$\beta = \frac{1}{1 - /1 - r/H}$$

formula./ β denotes the average number of call attempts per first call, r is the efficiency rate, H is the average perseverance. More about the above formula see in [GOSZ 85]./ It should be mentioned that now for the first time the effect of call repetition has explicitly been taken into account in CCITT. The study will continue.

2.4 SOME EXTRA ACTIVITIES

The NMDG /Network Management Development Group/ and the QSDG /Quality of Service Development Group/ have both been initiated by W.P.II/5 but live now a more or less independent life. The NMDG has been active for more than twelve years, the QSDG was formed in 1983. More than twenty operating companies take part in the first one, the members of the second count now about fifteen. These Groups are the practical results of the perception that in international telecommunications neither an effective network management, nor an improvement of quality can be achieved by isolated national actions.

The rapid growth and increasing complexity of the world-wide telephone network requires among others a better understanding of all sorts of traffic engineering also by those not directly involved in this activity. For this end the ITU Telecommunication Journal published special issues on network management /which can be regarded as real time engineering of traffic handling/ and on teletraffic problems. These special issues were initiated and taken care of by W.P. II/5 and W.P.II/4, respectively [TELE 84] .

A Handbook on quality of service, network management and network maintenance has also been prepared by Study Groups II and IV with the aim of summarizing knowledge and to support work in this field[CCIT 84].

3. OLD AND NEW PROBLEMS OF THE 1985-88 PERIOD

The list and presently valid numbering of Questions directly related to or having important common problems with traffic engineering appear in the Annex.

3.1 SERVICE ACCESSIBILITY

Fig.3.1 presents the hierarchy of concepts relevant to quality of service as they appear in Rec.G.106[CCIT 85b] . For detailed explanation see either the original text or [STRA 85].

Problems of reliability and availability /formerly studied by the dissolved CMBD/ belong now to Study Group II. Therefore the two important components of service accessibility: traffic capacity and availability can be studied in close cooperation.

Combined traffic and availability calculations are theoretically possible but there is a remarkable difference between the traffic and the failure process. Changes in the free-busy state of failure free traffic handling resources occur many times faster than changes in the operative-

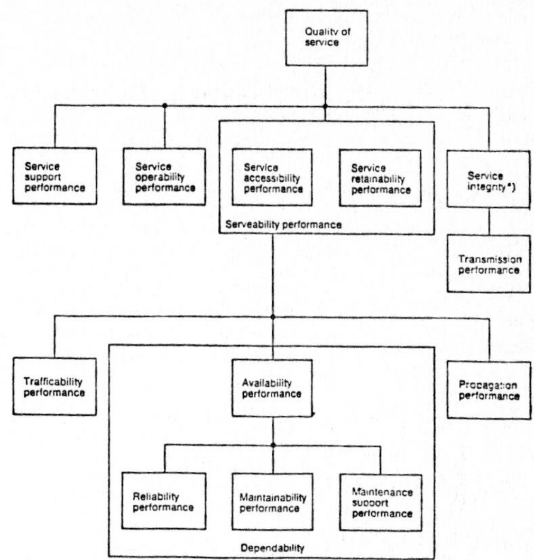

Fig.3.1 Performance concepts

failed state of the same resources. Therefore in spite of similarities e.g. in the mathematical /probabilistic/ modelling the practical aspects differ and make the combined approach difficult. From the users point of view there is no difference between a busy or a failed device: none of them can give access to the service.

3.2 NETWORK DESIGN AND ADVANCED ROUTING

Through the introduction of SPC exchanges and CCS systems sophisticated routing methods have become technically realistic. The new international routing plan /see Rec.E.171 [CCIT 85]/ prepared in the 1981-84 Study Period reflects current practices and recommends non-hierarchical routing. Several investigations have shown that more cost effective utilization of networks can be attained by the use of dynamic routings in which /time or state dependent / routing patterns are changed according to the traffic profile [FIEL 83, CAME 83]. The first implementations will very likely be realized in international networks and in the long distance part of big national networks.

International implementation requires the general guidance of CCITT. Relevant studies have already begun [CANA 85].

3.3 GRADE OF SERVICE VARIATIONS

Because of historical reasons and due to the success of the statistical equilibrium concept the parameters describing the congestion situation of a traffic handling system /e.g. loss probability, average delay/ are mainly understood in a simplified way: they refer to a single traffic value. GOS however should reflect the experience of the user including also the impact of traffic variations. The existence of variations is not questioned, several descriptions are possible, there is no generally accepted approach by CCITT.

The proper answer is even more difficult since GOS objectives are not and perhaps can not be defined in an exact way and have always been the result of common agreement and compromises.

CCITT has to answer two questions:

- how should traffic variations be reflected in GOS values /and calculations/?
- should all users /all call attempts/ of a given network enjoy the same GOS or if not what differences are tolerable?

3.3.1 TRAFFIC VARIATIONS

At present the GOS of most CCITT Recommendations are based on reference traffic values which are averages of certain high day busy hour traffics, see Rec.E.500 [CCIT 85]. The variation of high day traffic appears thus only through the mean i.e. in a smoothed way. There are however Rec.s which take day-to-day variations explicitly into account but different assumptions are used /normal distribution in E.500, gamma distribution in E.521/.

Both experience and simple examples show that GOS values derived from the mean of high day traffic underestimate the congestion found by customers. Let A be the traffic value to be used to calculate the B(A) congestion parameter. Let the high day busy hour traffic values have the density function f(A). For convex B(A) functions /which is normally the case/

$$B(M(A)) \leq \int B(A)\, f(A)\, dA$$

always holds. M(A) is the mean value of A.

The average congestion parameter is equal to or greater than the congestion at the average traffic level.

This average congestion can be seen by the user and this can simply be derived from congestion measurements.

Neither the formula nor the verbal statement above are new results, they have been well known for many years and have appeared in a structured form about twenty years ago e.g. in [ELLD 67]. What is missing is the CCITT standardization of this type of calculation.

To find a proper distribution for a given case or rather a generally applicable one requires detailed measurements and subsequent statistical analysis. The use of this distribution in a way outlined above is a matter of decision. Discussions prior to the decision should cover practical consequences of this change e.g. in the daily work and budgeting of a given operating company.

3.3.2 CLUSTER ENGINEERING

The majority of network planning procedures uses the blocking probability of the last choice route as GOS criterion. For historical reasons values about 0,01 are normally employed. In a network dimensioned in this way the overall GOS is much better, since only the traffic portion directed to the last choice route experiences congestion.

Those advocating the cluster approach of network dimensioning have the opinion that the 0,01 GOS value should apply to all traffic offered to the network considered and as a consequence the last choice route blocking probability could be increased and so the network would be cheaper. For traffic parcels entering the last choice route directly some service protection method /trunk reservation, etc./ is required. Studies have started. Possible solutions were already offered to CCITT for consideration, e.g. [NETH 85].

3.4 BUSY HOUR CONCEPTS IN PARALLEL?

The concept of the Time Consistent Busy Hour /TCBH/ and the load levels as defined in Rec.E.500 are results of an agreement and are necessary to have an unambiguous reference traffic for GOS calculations. Other busy hour concepts and other load levels could also be defined but in this case the relationship of the several reference traffics should also be clearly given. This relationship is the main problem, the relative benefits of measurement techniques to be used is of secondary importance [GOSZ 84].

In the 1981-84 Study Period the comparison of TCBH and the Average Daily Peak Hour method /ADPH/ has started with the aim of perhaps including ADPH in Rec. E.500 as a possible variant. The investigations which have been performed up to now were concentrated to ten day average load levels of circuit groups and have shown a rather good agreement. /The ADPH method measures carried traffic on a full hour basis, selects peak hours daily and applies the average over a number of days as reference traffic [PARV 85]/.

It is essential that the reference traffic /single value or perhaps distribution of high day busy hour traffics/ in a given case should be the same independently of the procedure of its determination. Applying different reference traffics as starting point the comparison of GOS values becomes difficult if not impossible.

3.5 TRAFFIC ENGINEERING OF COMMON CHANNEL SIGNALLING NETWORKS

The determination of the traffic capacity of CCS network is a new field CCITT embarked on in the 1985-88 Study Period. The structure of these networks is now under study by CCITT Study Group XI and the end-to-end performance of telephone networks will depend on how well signalling networks are dimensioned. CCS networks differ from speech networks: proposed and existing ones are packet switched, the type of traffic is similar to that in data networks, security requirements must be thoroughly taken into account etc. Traffic models and traffic measurements are needed to facilitate proper traffic engineering, special routing techniques should be defined and considered in network design.

4. THE CHALLENGES OF THE FUTURE

The user of tomorrow will live in a multiservice telecommunication environment. More and more different services will be offered to customers, and to serve the increased demand of information exchange /i.e. transfer, processing and storage/: communication and computer systems and techniques will be integrated [WEDL 83] [KOBA 83]. This future world will be realized by ISDNs. Independently of details two aspects seem to remain generally valid:

- the interworking of networks forming ISDNs will be required;
- proper balance of quality of service costs and income should be achieved.

CCITT is responsible for both aspects. Traffic engineering belongs to the second one.The systems which are necessary to satisfy the demand of information exchange will surely produce many traffic engineering problems.

Traffic engineering problems of new services and of old services served by new technologies have been dealt with by several authors [KATZ 83] [PAND 84]. Without the ambition to cover all questions some items to be considered are listed below.

Traffic engineering should take into account:

- the unique traffic behaviour associated with each of the new services /holding times, interarrival times, etc./
- the necessity to develop suitable traffic administration procedures /measurement, forecasting, etc./ to monitor the performance and to provide feedback,
- the increased importance of system and service protection e.g. by proper network management controls,
- the consequences if mixed traffic is carried by the network /mathematical modelling, sharing of resources, etc./.

As a consequence of this situation the well established traditional models of call generation, holding times should properly be enhanced if not entirely modified. But this is not the most important difficulty, since a continuous progress in teletraffic theory is quite normal.

4.1 THE MULTI-GOS PROBLEM

To solve GOS problems for a given service requires a clearly defined sequence of activities starting with the definition of GOS parameters up to the monitoring of performance. /See also Section 5./

The relevant tasks are well known for traditional services, they should however be reconsidered and performed repeatedly for each new service.

The rapid increase of the number of necessary GOS parameters together with the increased amount of traffic engineering work forms the first challange of the future.

4.2 THE PRESSURE OF TIME

"... the rapid proliferation of new network services and limited resources devoted to performance evaluations will seriously constrain the number of performance issues examined and may not permit a thorough treatment of performance objectives and standards. More reliance may be required on performance models than field assesment, due to the wide range of services to be evaluated. On the other hand, shorter intervals from service concept-to-introduction may permit little time for adequate performance modeling and pre-service testing. This situation would place more emphasis on field assesment and the establishment of end-to--end service standards based upon achieveable performance." [KATZ 83].

The time factor is the second challenge traffic engineering of our days is faced to.

4.3 RESPONSIBILITY FOR GOS

At present the number of companies offering conventional and new telecommunication services seems to increase. If so, quality of service - including GOS aspects - may have an increased importance since the choice by the customers between the companies involved will probably be based on service quality and costs. For big enterprises quality will be decisive while residential customers may estimate cost and features as more important.

However the increasing importance of quality aspects involves dangers. It is surely cheaper and more simple not to care too much for service quality. Customers and operating companies will surely discover yes-no type functional deficiences of a given equipment, network or service very soon, in most cases at the start. Quality problems of the better-worse type on the other hand are normally not evident at the beginning. The correction of quality impairments is not always possible later. What can be done if quality aspects were neglected? Is public opinion alone strong enough? The quality of the end--to-end service has "international" and "national" components. The responsibility is also shared in a similar way.

Since not only technology but in many countries also the structure of telecommunications /suppliers, operating companies, etc./ is rapidly changing the quality of service responsibility can be regarded being the third challenge for traffic engineers and traffic managers.

5. THE ISDN GOS PROBLEM

The way of how to approach the GOS questions of the multi-service ISDN by CCITT seems to be rather uncertain. This origins not only from the difficulties outlined in Section 4 but also from the decentralized distribution of GOS problems in CCITT.

5.1 INTERPRETATION OF GOS

Grade of service is defined by CCITT as follows:

"A number of traffic engineering parameters used to provide a measure of adequacy of plant under specified conditions these grade of service parameters may be expressed as probability of loss, probability of delay, etc.

The numerical values assigned to grade of sefvice parameters are called grade of service standards. The achieved values of grade of service parameters under actual conditions are called grade of service performances."
Rec.E.600 [CCIT 85].

From the CCITT point of view the plant is e.g. the world-wide PSTN /Public Switched Telephone Network/ or in some decades it will be composed of cooperating ISDNs. Assuming a point-to-point connection Fig.4.1 presents a simplified scheme of the plant representing the totality of traffic handling resources to be considered.

The plant is regarded to be adequate if it fulfills the quality requirements of the users and if it is profitable for the operating company. Adequacy is always the result of certain compromises.

GOS can often be described by a single parameter but in many cases several parameters are necessary /vector in the GOS space/. A given parameter may be a simple one, e.g. loss probability of a trunk group, or a composite one, e.g. a delay resulting from several delays caused by different groups of resources.

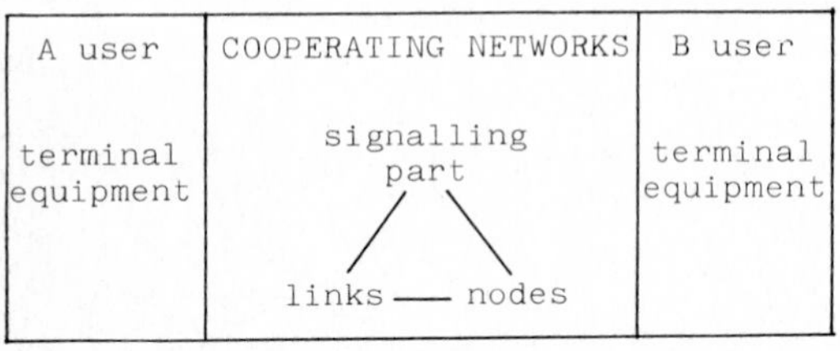

A user	COOPERATING NETWORKS	B user
terminal equipment	signalling part links — nodes	terminal equipment

Fig.4.1. The plant to be considered by CCITT

GOS performance is determined by two factors. One is the traffic /and availability/ situation which may be normal i.e. foreseen or extraordinary causing an overload state. The second is the resources in broad sense including traffic handling devices, their organization /e. g. routing strategies/, network management and overload control.

In the case of operational GOS the users' point of view dominates. This type of GOS is mainly system independent and the end-to-end aspect is emphasized. The design GOS refers mainly to the components of the network /Fig.4.1/, it is system dependent, economic factors have an important role in the GOS allocation procedure. Operating companies give priority to operational GOS, manufacturing companies are rather interested in design GOS. There is no well defined border between these two concepts.

GOS can be approached also from an overall traffic modelling side. The objective of relevant activities is to model the networks as a single entity by defining traffic flow, service and network resources related GOS parameters and their relationships.

Table 4.1 The GOS "process"

GOS related tasks	Relevance for CCITT
1. Definition of parameter/s/	high
2. Setting of objective/s/	high
3. Design and allocation	medium/high
4. Checking of performance	high
5. Feedback and actions	moderate

Table 4.1. shows the tasks which should be normally done in connection with GOS problems and gives an estimation of their relevance for CCITT. Definition is self explanatory. The setting of objectives assumes the definition of reference traffics. Design includes calculation procedures. Allocation with proper cost balancing is necessary if composite type parameters are considered. Traffic and GOS parameter measurement methods should be available to check performance.

5.2 THE PRESENT SITUATION

Up to now traffic engineering problems of different services have been studied by CCITT separately and in different Study Groups.

The CCITT traffic engineering Rec.s of the E.Series represent the operational GOS aspect. Besides GOS values for links and nodes /E.540 and E.543/ the end-to-end GOS also appears /E.541/ but without a definite numerical value. This deficiency is due to the difficulties of end-to-end GOS estimation. The design GOS aspect can be found e.g. in Rec.Q.504 on performance and availability design objectives of digital transit exchanges for telephony [CCIT 85c]. The main reason for producing this Rec.

was to have nodes of uniform quality characteristics in the network. GOS here is system dependent insofar current technology is assumed but no connection to any implementation appears.

Examples above refer to the telephone service, Study Groups II and XI are responsible for the E.and Q. Series, respectively. Unfortunately there is no clear framework of telephone GOS Rec.s which could be used as guideline in organizing future studies. Several Rec.s were prepared by Study Group VII including GOS parameters of circuit switched and packet switched connections /X.131, X.135, X.136, X.140 CCIT 85d/. Directly ISDN oriented GOS Questions are now examined by Study Groups II and XVIII.

5.3 POSSIBLE SOLUTIONS FOR THE FUTURE

Publications summarizing the traffic engineering aspects of ISDN and the related multiservice situation normally give examples of new services together with relevant GOS parameters.

The main objective of CCITT is to find a general approach to ISDN GOS problems which should be applied from the beginning of the GOS "process".

In the case of a single ISDN it seems to be clear that

- different services will require different GOS parameters or the same parameters but different objectives, and
- all services will be handled by the same network.

CCITT could restrict itself to the end-to-end approach and define GOS parameters for each service independently taking mainly the requirements of users /i.e. human factors/ into account. In this case only steps 1.,2. and 4. of Table 4.1 would be relevant. However this solution could hardly be looked upon as a traffic engineering activity in the original sense. Nevertheless this method could and will probably be applied mainly because of the pressure of time already mentioned.

The conventional way of considering all steps of Table 4.1 for each service independently including also design aspects seems not easy to be managed since the same network will be used for all services. This approach however will survive for a considerable time because separate networks for different services will exist.

The third way could be the separation of GOS aspects of the services and of the common network. The network GOS parameters could be defined as service independent and the service GOS parameters would be superposed service by service.
Network GOS parameters and values would not be independent of the type of connection /e.g. circuit switched and packet switched/. Using the nomenclature of the OSI /Open System Interconnection/ architecture, GOS problems of the lower four layers /physical, data link, network and transportation/ and those of the upper three /session, presentation and application/ would be handled separately.

There are many problems not mentioned here. The future evolution of the ISDN concept and the implemented ISDNs will definitely influence the GOS approach of CCITT and variants will coexist for some time. As a start GOS problems of non-voice services carried by the PSTN will be studied.

6. CONCLUSIONS

From the practical point of view the 11. ITC has been organized in a very suitable point of time. The CCITT Study Period has just begun, the questions to be answered are available and there is still time enough to contribute to the answers.

Nowadays CCITT is in the front line of the coordination of development in telecommunications. The retrospective standardization belongs to the past. Operating, manufacturing and scientific organizations contributing to CCITT take part in a big scale international project.

ITC is a necessary background for the GOS related activities of CCITT. The GOS questions cover a wide area from traffic measurement practice to sophisticated forecasting methods and from up-to-date network planning to exchange serveability problems. Many ITC participants work in one or other field of urgent CCITT interest, their results could and should form input to relevant studies.

To take part in CCITT work is not an easy task but the world-wide exchange of experiences taking place during discussions may be utmost interesting. To have an extra workload may certainly be not encouraging but offers a unique possibility to be up-to-date for the individual involved. The results of CCITT are beneficial internationally and also for organizations active in telecommunications. Taking this into account it seems justified to support CCITT and also those working for it.

ACKNOWLEDGEMENTS

During the preparation of this paper valuable suggestions and help were given by Ms.M. Ágostházi. The author is very indepted for her.

The support given by BHG Telecommunication Works and by the Hungarian Administration is highly acknowledged.

REFERENCES

[CANA 85] Canada: A routing method for the evolving telephone network - CCITT COM II-D5,1985 March.

[CCIT 84] CCITT: Handbook on quality of service,network management and network maintenance - ITU,Geneva,1984.

[CCIT 85] CCITT: International telephone network management and checking of service quality - Traffic engineering /Rec.s E.401-E.600/ - Red Book

Vol.II.,Fasc.II.3.,ITU, Geneva 1985.

[CCIT 85b] CCITT Recommendation: Concepts, terms and definitions related to availability and reliability studies - Red Book,Vol.III.,Fasc.III.1.,G.106, ITU,Geneva,1985.

[CCIT 85c] CCITT : Digital transit exchanges for national and international applications. Interworking of signalling systems /Rec.s Q.501-Q.685/- Red Book,Vol.VI.,Fasc.VI.5., ITU,Geneva,1985.

[ELLD 67] Elldin,A.: Dimensioning for the dynamic properties of telephone traffic - 5.ITC,New York,1967. pp. 25-36.

[GOSZ 83] Gosztony,G.: CCITT-a chance to unify the basic traffic engineering practice - 10.ITC,Montreal,1983.Paper 231/1-8.

[GOSZ 84] Gosztony,G.: The practicality of the busy hour concept and the CCITT approach - ITU Workshop on traffic engineering and forecasting, Athens, 1984,Nov.Doc.B5.

[GOSZ 85] Gosztony,G.: A general /rH / formula of call repetition: validity and constraints -11.ITC,Kyoto,1985.

[KATZ 83] Katz,S.S.: Introduction of new network services - 10.ITC,Montreal, 1983.Paper 111/1-12.

[KOBA 83] Kobayashi,K.: Strategic approaches to modern communications: "C and C" - 4.World Telecom.Forum, Geneva,1983.PartI,Paper III.1.3/1-11.

[NETH 85] Netherlands: Service protection methods in network design - CCITT COM II-D 23,1985 March.

[PAND 84] Pandya,R.N.,Robinson,W.R.: New services and their impact on traffic engineering - Telecom.J.,51.1984 June,pp.327-331.

[PARV 85] Parviala,A.: The stability of telephone traffic intensity profiles and its influence on measurement schedules and dimensioning - 11.ITC, Kyoto,1985.

[STRA 85] Strandberg,K.: CCITT quality of service concepts applied to telecommunication service planning - ICC-85,Chicago,1985.

[TELE 84] Telecom.J.,51.1984.Special issues on network management /Febr.,March/ and on teletraffic engineering /June, August/.

[WEDL 83] Wedlake,.J.O.: Customer services for the next decade - 4.World.Telecom.Forum, Geneva,1983.Part II, 2.6.1/1-8.

[CCIT 85d] CCITT: Data communication networks; transmission, signalling and switching, network aspects, ../Rec.s X.40 - X.180/ - Red Book Vol.VIII., Fasc.VIII.3., ITU,Geneva,1985.

ANNEX

TRAFFIC ENGINEERING RELATED QUESTIONS OF CCITT STUDY GROUP II FOR THE 1985-1988 STUDY PERIOD

Number	Title
22/II	Observations on the quality of international service
23/II	Network management
25/II	Terms and definitions of teletraffic engineering and international telephone operation
26/II	Traffic engineering of common channel signalling systems
27/II	Network design alternatives for international traffic
28/II	Methods for forecasting international traffic
29/II	Traffic models and measurements required to estimate traffic offered
30/II	Traffic models and measurements required for non-stationary traffic
31/II	Reference models for ISDN traffic engineering
32/II	Grade of service and performance criteria for international telephone exchanges under failure conditions
33/II	Traffic and operational requirements for SPC telecommunications exchanges
37/II	Service accessibility for telecommunication services
40/II	Allocation of accessibility and retainability objectives

TELETRAFFIC ISSUES in an Advanced Information Society
ITC-11
Minoru Akiyama (Editor)
Elsevier Science Publishers B.V. (North-Holland)

ITU/TETRAPRO TELETRAFFIC ENGINEERING TRAINING PROJECT

A.ELLDIN

Chairman ITC Training Working Party
Project Manager ITU/TETRAPRO

International Telecommunication Union

ABSTRACT

Training of traffic engineers within the developing countries is made possible by the ITU/TETRAPRO project by programmed courses, which preferably should be combined with on-the-job training. A basic course "Teletraffic Engineering" is available through the ITU/CODEVTEL sharing system. The course starts at B.Sc. level and gives comprehensive training in traffic engineering. Further specialized courses are being developed in advanced traffic engineering areas.

1. AIM OF PROJECT

The project aims at making it possible for developing countries to train their traffic engineers within the country. This will ensure that the countries can efficiently plan and operate their telecommunication networks.

2. PROCEDURE

Course material will be produced for the whole subject area, containing texts, exercises as well as instructions for the teacher, according to the CODEVTEL system. The instructor's guide will contain lesson plans, overhead pictures, solutions to exercises, tests with their solutions and scoring keys. The extensive and complete material will make it possible conduct the courses by the country's own teachers.

3. WHY COURSES AND NOT SEMINARS ?

Experience shows that seminars on more or less advanced subjects have very little impact on the traffic engineering skill in developing countries. The main reasons seem to be:
- The participants frequently lack sufficient background knowledge to assimilate the subject.(Wrong persons sent to seminar)
- Too short time to permit complete understanding of complicated matters. It is not enough time understand a new method and to learn to use it. (A seminar is generally 1, 2 or 3 weeks, the one-week seminar being the most common)

The courses are conducted at a lower speed and provide a better environment for learning and to individually practice a new method. The training becomes still more efficient if combined with on-the-job training. Further: It is easier for the administration to release its engineers for, say, one day per week than for continuous periods of several weeks and months.

4. COURSES

The course package comprises the following:

- TELETRAFFIC ENGINEERING

This course provides the general background for further advanced traffic engineering. The course starts at B.Sc. level and gives the theoretical background for all applications as well as their practical use. The training time is about 16 weeks. After this course and some practical work, the student can enter the specialized courses listed below.

SPECIALIZED COURSES:

- SUBSCRIBER AND TRAFFIC FORECASTING
- TRAFFIC MEASUREMENTS
- DIMENSIONING OF SWITCHING SYSTEMS
- DIMENSIONING OF TRUNK AND JUNCTION NETWORKS ("Network planning")
- TRAFFIC ENGINEERING PRACTICES AND MANAGEMENT

These courses will cover the latest and most advanced methods in their fields. Since the background knowledge is secured by the basic course "TELETRAFFIC ENGINEERING", these additional courses are estimated to take only 2 - 4 weeks each.

ELEMENTARY COURSE

This course concerns training of assistants to the traffic engineers. It is being developed by Siemens & Halske, Munich. The design of this course made it necessary of a close analysis of all traffic engineering work procedures.

5. FINANCING THE PROJECT

Initially it was planned that a number of experts should take part in the work and that it should be completed within 2 - 3 years. Difficulties arose however to find the money for carrying through the project, since ITU has no own funds for technical assistance. Fund were not available by UNDP and other donors. The work could however start late 1978 by the assignment of the project manager through contributions from Ericsson, Sweden, which company since then has supported the project financially.
Later on, the Government of the Netherlands provided an expert for 15 months, which made it possible to carry through a pilot course in India.
Further contributions were given by the Swedish Telecommunications Administration, by the British Telecom and by the Indian Posts and Telecommunications Adminstration by providing office facilities for the project. The latter organization also provided the training facilities for the pilot course.
The project has been supported by an Advisory Working Party which has met 7 times. The AWP was composed of members from supporting organizations and the ITC Training Working Party. as being the initiator of the project. Besides advising on extent, content and design of the courses, the individual members of AWP have given valuable contributions to the course material. The members of AWP have taken part in the work in kind.

The Elementary Course is being developed by Siemens & Halske, Munich, as a contribution to the United Nation's World Communication Year.

6. WORK ACCOMPLISHED SO FAR

The course material for TELETRAFFIC ENGINEERING is now available through the CODEVTEL Sharing System. The material consists of the following number of A-4 pages:

Text	1 000
Exercises and solutions	300
Instructor's guide:	
Lesson plans	200
Overhead transparances	750
Tests, answers, scoring keys	300

This type of programmed training requires much more material than ordinary text books used at universities. In fact, text and exercises are only about 40 % of the total material.The comprehensive course material will reduce the teacher's preparation time, but it is still assumed that he has traffic engineering experience.
A trial course has been conducted in India, at the Advanced Level Telecommunication Training Centre at Ghaziabad, outside New Delhi. The outcome of this course has influenced the final design of the course.

It is considered as important that the traffic engineers can present the numerical results of their work. Therefore, programmable pocket calculators have been used for the exercises. This is also a good introduction to the programming of more powerful computers.

A pilot course in TRAFFIC MEASUREMENTS will be conducted in Finland in the automn of 1985. It has been made possible by the assistance of the Finnish Aid Authorities. This course, as well as the other specialized courses are expected to be available in 1986.

7. THE FUTURE

The TETRAPRO project seems to become completed in 1986. New methods and new applications will apppear in the future. This will require up-dating and new courses may be desirable.

It is also desirable that the education in traffic theory and its applications at technical universities is improved.

8. FUTURE TASKS FOR ITC/TWP

The ITC Training Working Party should consider it as its task to advise the ITU on updating and extension of the TETRAPRO courses. It should further work for improved courses at universities on traffic engineering. If it is improved in the developed countries, the developing countries will certainly follow.

9. CONCLUSIONS

To extend a telecommunication network both money and trained staff is required. The financial problems can in many cases be solved, but without trained staff it will be impossible to run the networks efficiently. It generally takes longer time to train the staff than to install the new equipment. Training must therefore start early. In the preparations, skilled traffic engineering is needed early in the planning. The value of skilled traffic engineering cannot be overestimated. Mistakes in the dimensioning have severe economic effects, which only rich countries can afford to take. Developing countries must use their scarce resources in the best possible way.
The training must take its time and there are no short-cuts in learning this profession.

REFERENCES

A.Elldin, "Traffic Engineering in Developing Countries. Some Observations from the ESCAP Region." ITC 8, Paper No. 531, 1976.

and: Telecommunication Journal, vol. 44 (Sept. 1977, pp. 427 - 436.

A.Elldin, "Teletraffic Problems in Developing Countries", Ericsson Review, vol.54 (No. 4, 1977)

A.Elldin, "Report of ITC Training Working Party", ITC 9, 1979.

A.Elldin, "The Place of Traffic Engineering in the Planning Process", ITU Teletraffic Engineering Seminar, Istanbul, 5 - 16 May 1980

A.Elldin, "Training in Teletraffic Engineering", ITU Teletraffic Engineering Seminar, Istanbul, 5 - 16 May 1980

A.Elldin, "Report of ITC Training Working Party", ITC 10, Session 4.2, 1983,

A.Elldin, "Teletraffic Training - A MUST", Telecommunication Journal, vol. 51, (Aug. 1984) pp. 427 - 436.

J.Ernberg, "International Cooperation in Training Development. The ITU/CODEVTEL Project" Telecommunication Journal, vol. 45 (April 1978) pp. 154 - 161

CODEVTEL
= "Course Development in Telecommunications" is a universal ITU project. Countries in all parts of the world cooperate in producing courses on telecommunication subjects. The idea is that the same course (after adaptation) can be used in more than one country and that duplication work can be avoided if the countries exchange courses. The exchange is handled by the ITU Training Division and it is refered to as the ITU/CODEVTEL Sharing System. To make courses universally adaptable CODEVTEL provides also guidance on course standards and formats, as well as on the procedures for course development. The TETRAPRO courses will be available through this sharing system.
For further information on CODEVTEL, see ERNBERG below in the references!

ADDITION

The course material for Teletraffic Engineering has also been used in two courses conducted by international traffic engineering experts. In Dhaka, Bangladesh, a 12 week course was held in 1982 and another course was run in Amman, Jordan, comprising 8 weeks. The latter course used the material for the the first part of the Indian pilot course. The course was conducted in the last quarter of 1983.
Both courses gave valuable information for the design of the TETRAPRO course.

TELETRAFFIC ISSUES in an Advanced Information Society
ITC-11
Minoru Akiyama (Editor)
Elsevier Science Publishers B.V. (North-Holland)

"STORE-AND-FORWARD" FACILITTY IN TELEGRAPHIC NETWORK WITH OVERLOADED SUBSCRIBERS' LINES

Zbigniew DEC, Ryszard SOBCZAK,and Marian ZIENTALSKI

Institute of Telecommunication, Technical University
Gdańsk, Poland

ABSTRACT

There is influence of S&F facility on uneffective network traffic presented. Slightly decrement of this traffic and insignificant improvement of subscriber line's is showed. New S&F facility, connected with called subscriber, is proposed and is demonstrated possibility of significant improvement of overloaded subscriber accessibility.

1. INTRODUCTION

There are two reasons, which induced to evaluate "Store-and-Forward"/S&F/ facility on telegraphic network with overloaded subscribes' lines. The first is to knew if S&F facility decreases uneffective network traffic /traffic of repeated attemps to call/. The second is to find a proper criterion of facility allotment,when most of subscriber's needed new facility and could pay for them.

2. UNEFFECTIVE TRAFFIC

2.1. Network without S&F

Uneffective traffic model is build up founded following assumptions:

a/ subscriber will not resign of message sending,

b/ uneffective traffic arised from limited access to interexchange lines is to omit [1],

c/ duration of unsuccessful attempt to call is ε=0.2 of successful call duration,

d/ probability β_0 of subscriber's occupation is undepended on attempt number [2].

Ratio of uneffective traffic intensity to effective traffic intensity is called efficiency factor [1]. Figure 1 shows this factor according to occupation probability β_0. The range of β_0 come from fact, that called subscriber is overloaded.

2.2. Network with S&F

Traffic model for such a network is based on following additional assumptions:

a/ message delivery acknowlegement duration is α=2 times longer then attempt to call duration,

b/ elongation of call with S&F device is to omit,

Assumptions follow from call protocol [3]. Interesting efficiency factor is:

$$f_f = \beta_0 \alpha \varepsilon \frac{\beta_1}{1-\beta_1} \qquad 2.1$$

where β_1 is calling subscriber's occupation probability. Relative efficiency of put in S&F facility is equal:

$$\eta = \frac{f_f - f}{1+f} \qquad 2.2$$

Figure 2 presents this dependence for several β_0.

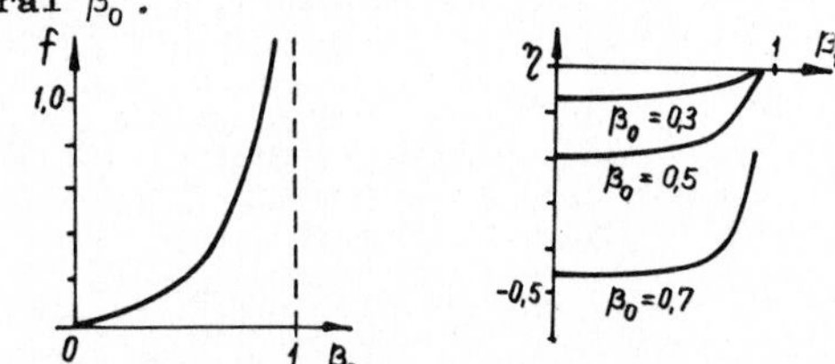

Fig.1 Efficiency factor

Fig.2 Relative efficiency factor.

2.3. Subscriber's line

For subscriber's line it is interesting to knew relative measure of line's load changes:

$$\eta_l = \frac{\beta_0 \alpha \varepsilon - f}{1+f} \qquad 2.3$$

Figure 3 illustrates above dependency.

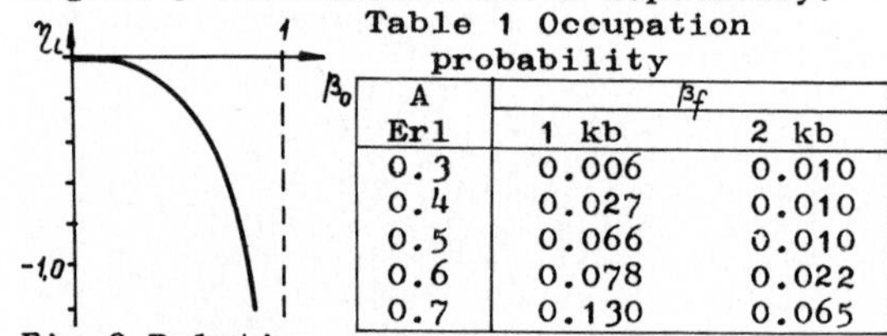

Fig.3 Relative measure

Table 1 Occupation probability

A	β_f	
Erl	1 kb	2 kb
0.3	0.006	0.010
0.4	0.027	0.010
0.5	0.066	0.010
0.6	0.078	0.022
0.7	0.130	0.065

3. CONCLUSIONS

S&F facility reduce uneffective traffic in insiquificant degree and doesn't improve subscriber accessibility. For these reasons it is offered new way to use S&F device resources. In new S&F facility buffer is connected with called subscriber and the occupation probability is much smaller. Simulation results for several incoming traffic intensity A /when outcoming traffic is also A/ contain table 1.

REFERENCES

[1] Z.Dec, R.Sobczak, M.Zientalski,"Służba "Zapamiętaj i Przekaż" w krajowej sieci telegraficznej", Prz.Telekom. R. 55 no 11 pp. 327-330, 1983

[2] P. Le Gall, "On a Theory of the Repetition of Telephone Calls", Ann. des Telecomm. vol. 24, no. 7-8, 1969

[3] "Store-and-Forward facilities in use in the swedish telex network", CCITT Study Group X, COM.X-86, 1979.

TELETRAFFIC ISSUES in an Advanced Information Society
ITC-11
Minoru Akiyama (Editor)
Elsevier Science Publishers B.V. (North-Holland)

LOW COST DESIGN MODIFICATION FOR IMPROVING REAL-TIME CAPACITY

Joe BRAND

Hitachi America, Ltd.
Norcross, Georgia, USA

A low cost design modification for significantly increasing the real-time capacity of a Stored-Program-Control (SPC) PBX is presented. The capacity increase is achieved with very little development effort and only a small increase in hardware cost. The method takes advantage of the fact that the majority of the processing involves a small fraction of the internal memory of the system. This trait is thought to be common for SPC switching systems [1].

For increasing the processor real-time capacity, two alternatives were considered. One involved increasing the processing speed (Alternative 1), and the other involved adding auxilliary processors and removing work from the central processor (Alternative 2). Using a queuing model calibrated with actual data, a capacity relationship for the central processor was determined as a function of both percent of work moved out of the central processor and percent increase in speed of the central processor. Although the idea of providing adequate capacity increase without auxilliary processors was initially met with skepticism, the model showed that either alternative would provide adequate capacity increase. The capacity increase for Alternative 1 results from three factors: (1) decreased time spent doing overhead work, (2) decreased time per call, and (3) increased allowable processor occupancy at maximum capacity. The third factor is a result of queuing theory when the capacity definition is based on maximum allowable queuing delays. Alternative 1 was chosen because it required less development effort and less hardware costs.

Advances in the microprocessor technology permitted up to a 50 percent increase in processor clock rate which suggested the alternative of increasing the real-time capacity by speeding up the processor. However, a 50 percent speed-up of the basic clock rate did not imply a 50 percent increase in effective processing speed because of the memory access speed limitation. In fact, without increasing the memory speed, only a marginal gain could be realized with a faster processor. Faster memory chips were available at lower density and higher cost. A large real-time processing capacity increase can be achieved by replacing the processor with the high-speed version and redesigning the memory boards. The redesign would involve replacing the memory chips with the high-speed, low density chips and optimizing other circuitry for speed. Because of the low density of the high speed memory devices, the number of memory boards required becomes excessive, escalating cost and physical space beyond acceptable limits.

A real-time analysis involving laboratory measurements showed that about 70 percent of the processing time involved less than one percent of the total memory. This amount of memory could easily fit on just one memory board even with the high-speed, low density memory devices. Thus, we found that we could replace a small percentage of the memory with high-speed, low-density, high-cost memory and achieve most of the benefits that would result from the less practical solution of replacing all the memory. The processor can retain most of the lower speed and lower cost memory while taking advantage of the higher speed of the memory used where most of the activity takes place. This solution is possible because of the asynchronous nature of the processor to memory allowing variable access speeds. With this approach, the resulting PBX can accommodate over two times as many lines (of equivalent traffic per line) as prior to increasing the processing speed. This increase is achieved at a very small relative cost.

The fixed organization of programs and data between high speed and low speed main memory as proposed here is analogous to the organization between main memory and file memory described in [2]. One difference is that here we are allocating to main memory only. Comparison of our memory organization with classical main-memory organizations, e.g., parallel, cache, etc. will be presented, showing why our method was chosen for our PBX application.

The results of this effort are offered to the ITC because they address the problem of increasing real-time capacity with a method shown to be very simple and inexpensive but effective. The method is thought to be of general applicability.

REFEREENCES

[1] Z. Koono, A. Shoda, and Y. Tokita, "A Distributed Control System for Electronic Switching Systems," International Switching Symposium, 1976

[2] K. Kusunoki and K. Yamamoto, "Optimum Program Design for Hierarchically Structured Memory in Electronic Switching System," Electronics and Communications in Japan, Vol. 59-A, No. 3, 1976

TELETRAFFIC ISSUES in an Advanced Information Society
ITC-11
Minoru Akiyama (Editor)
Elsevier Science Publishers B.V. (North-Holland)

PROCESSOR LOAD MODELLING

Peder J. EMSTAD

The Norwegian Institute of Technology
Trondheim, Norway

A TWO LEVEL MODEL

In a communication system a processor is part of the secondary resources engaged in administrating the primary resources carrying the subscriber generated traffic. In a well designed system the secondary system operation will not distort the desired behavior of the primary system in any noticable way.

Let the primary resource net be an open Jackson net without feedbacks. Node j in the net is assumed to be generating tasks according to a Poisson process with rate η_j and $\tilde{u}_j$ required instructions per task. Let there be m processors where processor i has a capacity e_i [instructions/sec]. The total arriving intensity to processor i is then

$$\lambda_i = \sum_{j \in I_i} \eta_j \tag{1}$$

I_i is the set of nodes feeding processor i.

We now model each processor as an M/G/1-model.

THE EFFECT OF DEPENDENT INPUT

Each node in the above Jackson network has a Poisson input and output process for primary system customers, but these processes are not independent. To study this dependence let the primary system be one $M/M/\infty$-system with arrival rate $\lambda/2$ and service rate μ. The secondary system is a one server delay system with n e d service time of mean $1/\gamma$. Tasks to the secondary system are generated when customers enter or leave the primary system. Under stationary conditions let

$$p(i,j) = \text{prob(there are I=i customers in the primary system and J=j tasks in the secondary)} \tag{2}$$

We find ($p(i,j) = 0$ for $i,j<0$):

$$-[\lambda/2+i\mu+\Delta(j)\gamma]p(i,j)+\lambda p(i-1,j-1)/2 +(i+1)\mu p(i+1,j-1)+\gamma p(i,j+1) = 0 \tag{3}$$

where $\Delta(j)=1$ if $j>0$, otherwise $\Delta(j)=0$.

We introduce the generating function

$$P(x,y) = \sum_{i=0}^{\infty} \sum_{j=0}^{\infty} p(i,j)x^i y^j \tag{4}$$

and find

$$\mu(y-x)y \frac{\partial P(x,y)}{\partial x} = [(\lambda/2+\gamma-\lambda xy/2)y-\gamma]P(x,y) +\gamma(1-y)P(x,0) \tag{5}$$

$P(x,0)$ has to be determined. We take the derivative of (5) with respect to y and set y=1.

$$\lambda E(J)-\mu E(I\cdot J)=\lim_{x\to 1} \frac{1}{x-1}[(\gamma-\lambda x/2-\lambda/2)e^{\frac{\lambda}{2\mu}(x-1)} -\gamma P(x,0)] \tag{6}$$

The left side of (6) must be finite which requires the pole for x=1 in the right side to be balanced by zeroes in the bracketted expression. This readily determines $P(x,0)$ to be

$$P(x,0) = (1-\frac{\lambda}{\gamma})e^{-\frac{\lambda}{2\mu}(1-x)} \tag{7}$$

Having determined $P(x,0)$, (5) can be solved using an approach similar to Lagrange's method for ordinary differential equations.

$$P(x,y) = (\gamma-\lambda)\cdot e^{-\frac{\lambda}{2\mu} + y\frac{\lambda}{2\mu}(1-y) + \frac{\lambda}{2\mu}yx} \cdot \{\frac{1}{-\lambda y^2/2-\lambda y/2+\gamma} - (1-y)\sum_{i=1}^{\infty} \frac{[\frac{\lambda}{2\mu}(1-y)]^i}{i!} \cdot \frac{(x-y)^i}{-\lambda y^3/2+(i\mu+\lambda/2+\gamma)y-\gamma}\} \tag{8}$$

(With x=1 and $\mu\to\infty$ we get the known p g f for number of jobs in a bulk arrival system with R=2 jobs in each bulk.)

The expected number of tasks in the secondary system is found to be

$$E(J) = \frac{3\lambda/2}{\gamma - \lambda} \tag{9}$$

This equals the expected number of jobs in the above mentioned bulk arrival system.

CONCLUSIONS

The above results suggest that the task generation process from a node with a substantial number of primary resources and arrival rate η should be one Poisson process with rate η. A task should be a lumping of subtasks at node arrival and departure instances.

REFERENCES

1 P.J. Emstad, "Multiprocessor System Load Analysis", Fjärde Nordiska Teletrafik-seminariet, Helsingfors May 11-13, 1982.

2 P.J. Emstad, "A Two Level Processor Load Model", Det 5. Nordiske Teletrafikkseminar, Trondheim June 5-7, 1984.

TELETRAFFIC ISSUES in an Advanced Information Society
ITC-11
Minoru Akiyama (Editor)
Elsevier Science Publishers B.V. (North-Holland)

ESSENTIAL SERVICE PROTECTION UNDER TRAFFIC OVERLOADS

J. J. PHELAN, G. L. Karel, and E. N. Barron*

AT&T Bell Laboratories
Naperville, Illinois, U.S.A.

ABSTRACT

One of the more sensitive issues in telephone administration is the provisioning of good service to vital lines during periods of traffic overload. Essential lines have usually been informally defined by local practice, but frequently include phones in fire and police departments, hospital operating rooms, telephone switching centers, some coin phones, etc. It is highly desirable that these phones receive essentially unimpaired service during traffic overloads, which may well be coincident with circumstances such as earthquakes, hurricanes, or other local catastrophes.

A novel strategy has been devised to protect the service to "essential" lines during traffic overloads, which, at the same time, avoids any significant impact on the total call-carrying capacity of stored-program-controlled switching systems. Other desirable attributes of a design for service protection for essential lines are that it responds rapidly and automatically to overloads, that it not place a significant overhead on the system, that it not unnecessarily degrade service to other lines, and that the controls be automatically reduced as the overload diminishes.

The Essential Service Protection (ESP) Feature implemented in the AT&T 1A ESS™ and 5ESS™ Switching Systems provides all of the above-mentioned attributes. In modern electronic switching systems, lines initiating a service request are first placed in a queue to wait to be served by the common control equipment. A very effective overload control strategy, the Improved Overload Strategy (IOS), to administer the flow of requests through this line service request queue, was first implemented in the 1ESS™ Switching System.[1] Building upon this IOS feature, ESP has been implemented using a network of queues and timing strategies. Simulation models have been used to predict the performance of the IOS/ESP strategy and to estimate parameters defined in its implementation. The generic software containing the ESP feature was introduced to operating 1A ESS Switches in early 1983. Difficult-to-obtain data on the performance of ESP in a large switching system under overload was, by chance, obtained during a minor earthquake on May 2, 1983, in California. An analysis of this system performance will be discussed.

The ESP strategy has also been adapted for implementation in the distributed architecture of the 5ESS Switching System. The distributed architecture allows effective control of partial or focused overloads near the periphery of the 5ESS Switch. A key element of the IOS overload control is the application of a Last-In-First-Out (LIFO) service discipline to the line origination queue. This discipline is very effective in controlling the impact of customer behavior on the switch.[1,2] However, it is possible for a service request to suffer an arbitrarily long waiting time in a LIFO queue. Thus, the LIFO queue is periodically searched for essential lines, and those found are moved to a second high-priority queue. If nonempty, this second queue is exhaustively served before the LIFO queue is served. A simulation of this ESP/IOS control strategy, under normal and overload conditions, has yielded expected queue length and delay distributions; these and other system performance results will be presented. This ESP/IOS strategy is now also in service in the 5ESS Switching System.

REFERENCE

[1] J. W. Borchering, L. J. Forys, A. A. Fredericks, and G. J. Hejny; "Coping With Overloads," page 78, Telephony, October 5, 1981.

[2] L. Burkard, J. J. Phelan, M. D. Weekly; "Customer Behavior and Unexpected Dial Tone Delay;" Proc. 10th Int. Teletraffic Congress 1983 (Montreal), 2.4-5.

---

* Present Address:
Department of Mathematical Sciences,
Loyola University,
Chicago, Illinois, U.S.A.

TELETRAFFIC ISSUES in an Advanced Information Society
ITC-11
Minoru Akiyama (Editor)
Elsevier Science Publishers B.V. (North-Holland)

SOME TRAFFIC ISSUES IN DIGITAL CROSS-CONNECT SYSTEMS

M. Segal and D. R. Smith

AT&T Bell Laboratories
Holmdel, New Jersey 07733 USA

ABSTRACT

Digital cross-connect systems allow for remote access to digital circuits for testing purposes and provide for cross connection of circuit legs remotely. Digital transmission facilities to various destinations are attached to frames of such systems via ports, thus, the frame of the cross-connect system plays the role of a distributing frame which can be activated remotely. For example, the Digital Access and Cross-Connect System (DACS), see [1,2,3,4,5,6] accepts up to 3048 DS0 circuits - 24 circuits for each of the 127 DS1 ports, with an additional port reserved for testing.

Multi-frame offices are often required because of the limited capacity of a single frame and ports are reserved for interframe connections or *ties*, sometimes via designated tandem frames.

Digital cross-connect systems are often used in special services applications [7] with circuit lifetimes measured in months. This results in a stream of connect and disconnect orders of circuits also known as the "churn" phenomenon [8].

Traffic models are presented to determine the number of ports that should be reserved for tying purposes. These models, resembling overflow systems [9], take into account the stochastic nature of the circuit activity, the spreading of same-destination facilities across frames, and their fill.

In a network setting, where the offices are equipped with cross-connect frames, structural and administration issues are raised and discussed.

REFERENCES

[1] J. R. Colton, "Cross-connections - DACS Makes Them Digital," AT&T Bell Laboratories Record, Vol. 58, No. 7, pp. 248-255, September 1980.

[2] R. C. Drechsler, "DACS Cross-connects - and That's Just the Beginning," AT&T Bell Laboratories Record, Vol. 58, No. 8, pp. 305-311, October 1980.

[3] J. R. Colton and A. J. Osofsky, "DACS Features and Applications," IEEE National Telecommunications Conference Record, Vol. 1, New Orleans meeting, November 29 - December 3, 1981, paper B1.1.

[4] R. P. Abbott and D. C. Koehler, "Digital Access and Cross-Connect System Architecture," IEEE National Telecommunications Conference Record, Vol. 1, New Orleans meeting, November 29 - December 3, 1981, paper B1.2.

[5] A. J. Cirillo, L. F. Horney and J. D. Moore, "DACS Microprocessor System," IEEE National Telecommunications Conference Record, Vol. 1, New Orleans meeting, November 29 - December 3, 1981, paper B1.3.

[6] L. C. Sweeney, "DACS in an Associated Company," IEEE National Telecommunications Conference Record, Vol. 1, New Orleans meeting, November 29 - December 3, 1981, paper B1.4.

[7] G. P. Ashkar, G. A. Ford and T. Pecsvaradi, "Reshaping the Network for Special Services," AT&T Bell Laboratories Record, Vol. 61, Vol. 7, pp. 4-10, September 1983.

[8] D. R. Smith, "A Model for Special-Service Circuit Activity," Bell System Technical Journal, Vol. 62, No. 10, pp. 2911-2934, December 1983.

[9] C. L. Monma and D. R Smith, "Probabilistic Analysis of Inter-Frame Tie Requirements for Cross-connect Systems," AT&T Bell Laboratories Technical Journal, Vol. 63, No. 4, pp. 643-664, April 1984.

TELETRAFFIC ISSUES in an Advanced Information Society
ITC-11
Minoru Akiyama (Editor)
Elsevier Science Publishers B.V. (North-Holland)

CHARACTERISTICS OF TELECOMMUNICATION SERVICES AND TRAFFIC

King-Tim KO

Telecom Australia Research Laboratories
Melbourne, Australia

The advance of new communication technology and market demands for new services have lead to an ever increasing array of available telecommunication services. The traffic characteristics and performance requirements of these new services vary considerably. Hence the network provider has great difficulty in

1. determining whether the service requirements are compatible with the facilities provided by the transit and end networks;
2. modelling the resultant network traffic pattern, and dimensioning the network for a specified quality-of-service.

To assist with the above problems, services can be classified according to the OSI Reference Model, or according to the requirements and characteristics of the services and the networks supporting them. While these approaches are generally accepted internationally, from the teletraffic engineer's point-of-view, services can be more conveniently divided into data and voice, which have different service requirements and traffic generating characteristics. Data networks are usually separate from voice networks because of their differing performance requirements which include throughput, delay and information integrity. In general, requirements for voice services are less demanding because their performance is assessed subjectively.

The assumptions of Poisson call arrival process and negative exponential call duration are considered to be adequate for voice traffic using a circuit-switched network. For data services, the network traffic pattern is the aggregated result of traffic profiles produced by the users of one or more available services. Three major categories of data transfer can be identified - bulk (e.g. file transfer), interactive (e.g. videotex) and short transfers (e.g. electronic fund transfer). Their general characteristics are :

(a) bulk transfer - high data volume, and connection via high speed line;
(b) interactive transfer - low data volume, long inactive periods, and terminal and remote host may follow a particular response time distribution for a given task;
(c) short transfer - low data volume, short session time, short and generally fixed message length, and a small number of dialog cycles per session.

Although some of the characteristics of the interactive and short transfers are similar, their demand for network resources are quite different. While the interactive transfers are generally more bursty, the short transfers have a more predictable behaviour, and impose greater demands on the signalling and switching functions of the network.

A telecommunication network can be dimensioned to the quality-of-service requirements once the network traffic pattern is characterized. The techniques of modelling the network traffic pattern and dimensioning circuit-switched networks are quite well developed. However similar techniques for packet-switched networks are still under intense investigation. For data networks, the assumptions of Poisson call arrival process, negative exponential call duration and geometrical message length are often used because of their mathematical tractability. These assumptions are generally considered to be valid when the number of users in the network is large and diverse.

Traffic modelling techniques for packet-switched networks are reported in [1] and [2]. A decomposition method is used in [1] where each of the queues in the network is considered individually. Some recent work on queueing theory give analytical results for particular cases of superposition of point processes [2]. The packet arrivals into a node are doubly stochastic, resulting from a superposition of call and packet arrival processes. A tractable approximation - an Interrupted Poisson Process, is used to model the packet arrival process. The first three moments of the packet arrival process are matched to those of the approximating process. The results obtained from the queueing analysis using the approximating process are extremely promising. Further research in this area has the potential of improving the model of traffic generated in a packet-switched network.

REFERENCES

[1] Rossister M.H., "Optimal analysis and design of a packet switching network", Proc. ICCC, Sydney, pp.427-433, 1984.

[2] Heffes H., "A class of data traffic processes - covariance function characterization and related queuing results",Bell Syst. Tech. J., Vol.59, No.6, pp.897-929, 1980.

TELETRAFFIC ISSUES in an Advanced Information Society
ITC-11
Minoru Akiyama (Editor)
Elsevier Science Publishers B.V. (North-Holland)

A SIMULATION STUDY ON A SINGLE-LINK, MULTIPLE-LAP PROTOCOL

Wolfgang FISCHER

Institute of Communications Switching and Data Technics
University of Stuttgart, Fed. Rep. of Germany

Within the recent study period a large amount of work has been done by CCITT in developping the recommendations for the Integrated Services Digital Network (ISDN) [1].

One of the most essential items of the ISDN is the specification of the user-network interface which is represented either by the basic access consisting of two B-channels with a transmission speed of 64 kbit/s each and a D-channel with 16 kbit/s or the primary rate access consisting of n B-channels (64 kbit/s each; in general n=30) and a D-channel with 64 kbit/s. Partitioning of the overall bandwidth of the transmission line into channels is achieved by time division multiplexing. The B-channels are used for the transmission of voice and data either separately and independently or in combination to achieve a larger bandwidth while the D-channel's main task is the transport of signalling information for the use of the B-channels. Besides this main task and in consideration of the relatively small amount of signalling information in the average of time, slow packetized data transfer will be done via the D-channel.

This paper deals with a simulation study of the layer 2 of the D-channel protocol. We shall concentrate on the basic access as the protocol for the primary rate access is only a subset of the former one. The reference configuration for the protocol is shown by Fig. 1.

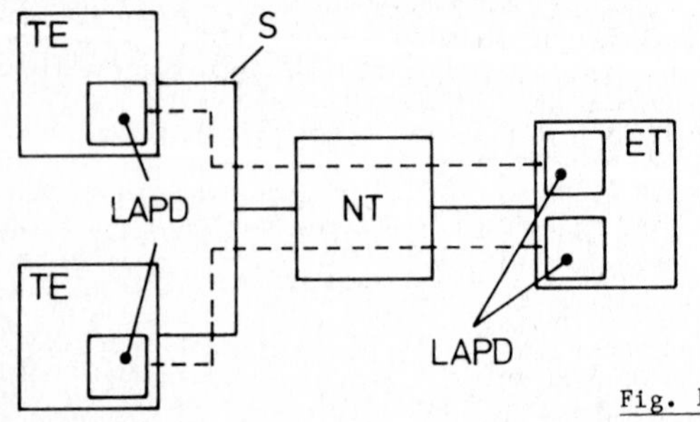

Fig. 1

A number of terminals (TE, Terminal Equipment) are connected to the network termination unit (NT) via the S-interface. Contention resolution is achieved by a CSMA/CD mechanism on the D-channel. This mechanism is controlled by the NT via a D-echo-channel.

For the case of basic access being considered here NT is transparent for the layers above layer 1. So signalling connections will exist between the terminals and the local exchange (ET, Exchange Termination).

Each signalling connection in layer 2 consists of a pair of LAPD processes in TE and ET respectively. LAPD is specified similarly as LAPB in X.25 level 2 [2] with a window size of 1. In general in every active TE only one LAPD process exists which implies that in ET there are as many LAPD processes as there are active TEs. All these connections are multiplexed onto a single D-channel. If they would produce a stationary load, results from throughput investigations for X.25 level 3 could be used [3] as multiplexing mechanisms are quite similar there.

What is essential in the D-channel protocol is not maximum throughput for signalling messages but short call setup times.

For each incoming or outgoing call a signalling connection has to be established before signalling message transfer is achieved. Especially in the case of an incoming call every terminal being able to accept that call will establish a logical link nearly at the same time before responding by means of signalling packets. So a very dynamic behaviour can be observed which is controlled by the layer 3.

The simulation study will concentrate on investigations of the call setup time which is influenced by

- number of TEs responding
- basic load of low-rate packetized data
- transmission speed of the D-channel
- processing delays
- error rate in the channel

References:

[1] CCITT, Draft Recommendations I.110 .. I.464 (1984)

[2] CCITT, Recommendation X.25 Yellow Book (1980)

[3] W.Dieterle, A Simulation Study of CCITT X.25 ...; Proceedings 10th ITC (1983)

TELETRAFFIC ISSUES in an Advanced Information Society
ITC-11
Minoru Akiyama (Editor)
Elsevier Science Publishers B.V. (North-Holland)

VIDEOTEX TELETRAFFIC IMPLICATIONS

Peter Farr

Coopers & Lybrand Services, Perth, Western Australia

ABSTRACT

Services such as videotex, facsimile, teletex and teleconferencing have been developed to meet business and social communication needs and will contribute a growing proportion of network traffic. The paper discusses the teletraffic issues which apply to the design of public and private videotex systems.

* * * *

Videotex is the modern method of communicating textual and graphical image information via the telephone network. It can bring worthwhile benefits in business, domestic and social applications. It is in use in 40 countries and the numbers of installations and users are growing rapidly.

Videotex will play an important part in the highly advanced information society. When videotex is frequently used as a transmission medium controlling goods and money flow, it will play a significant role in the socioeconomic activities of the modern world.

On the domestic scene, fast growth areas using videotex include electronic banking and shopping and down-line loading of programs into personal computers. In business and commerce, Closed User Groups represent the main market with significant use also occurring in electronic mail and data entry by third parties.

A turning point has been reached with the establishment of public videotex services by the Telecommunications Administrations in many countries. These serve a dual purpose: (i) a videotex service usable by anyone with access to a telephone; and (ii) a gateway to access videotex systems provided by private operators (Fig. 1).

A benefit to the users is that the cost of access is kept very low, and is often independent of the distance between the user and the videotex centre. Gateways serve important functions -

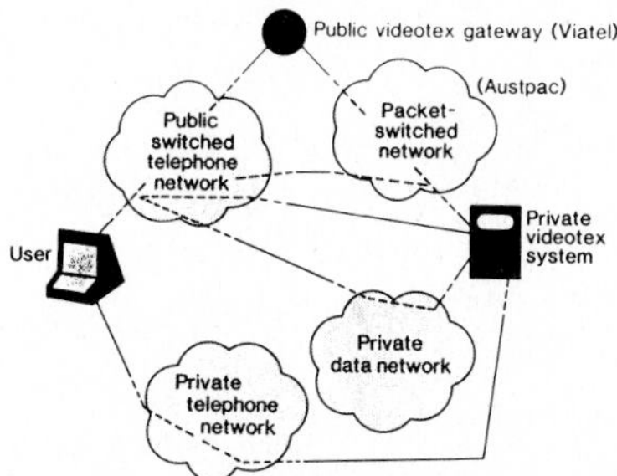

Fig. 1 Alternative Network Options

- They give private system operators indexes to their services on the public videotex system.
- They enable end users to switch easily between different private videotex systems without redialling.
- They allow short bursts of heavy traffic to be buffered.
- They offer a mechanism for centralised user billing.

The networking flexibility of videotex systems is one of their main advantages over conventional transaction-processing computer systems. Private videotex systems can be accessed over public and private telephone networks, over private and public data networks and over special videotex networks and gateways, both on a local and international basis.

The traffic characteristics of videotex "calls" need to be known in order to correctly dimension switching exchanges, multiplexors and transmission paths. From a system operator perspective, it is necessary to know how much traffic a central processor can handle at acceptable grades of service. An asset in handling this problem is that the system itself can maintain a statistical log of user sessions. In fact, a source of attraction to information providers and advertisers is the detailed market research which can be carried out via a videotex medium. This is possible by mapping the log of user sessions against user demographic profiles.

The economic viability of a videotex service is influenced by many factors, but the principle ones are -

- Ease of access and grade of service.
- Quantity and quality of information.
- The relative cost compared to obtaining the desired information from other sources.
- Cost and versatility of terminal services.

It may be observed that these factors, or similar ones, apply to various other products and services provided by telecommunications administrations.

Accordingly, the design of a videotex system, be it public or private, needs to take full account of the teletraffic issues. The data to be presented are recent figures from the Telecom Australia videotex service, VIATEL, and national videotex services provided by several large private operators.

TELETRAFFIC ISSUES in an Advanced Information Society
ITC-11
Minoru Akiyama (Editor)
Elsevier Science Publishers B.V. (North-Holland)

ON BOUNDS OF RANDOM NUMBER GENERATORS WITH FINITE CYCLE LENGTH FOR TRAFFIC SIMULATION

J. SÄGEBARTH

INSTITUTE FOR ELECTRONIC SYSTEMS AND SWITCHING, UNIVERSITY OF DORTMUND

GENERAL CONSIDERATIONS

This paper deals with properties of random-number-generators (RNGs) used in traffic simulations, and with new dimensioning methods for such generators with regard to the mutual independence of the generated random numbers (RNs) and the random events (REs) in the actual simulations, respectively. The methods presented, which can be only summarized here, will be treated in more detail in separate papers /1,2,3/. From the properties of RNGs with finite cycle length (CL) it can be concluded that the REs which are determined from the RNs produced by such a generator can only be mutually independent, if the number of events in the actual simulations does not exceed a certain limiting value.

In a first method, an upper limit is determined for the maximum number of independent random events (ire) for arbitrary generators with finite CL. This value ire depends only on the CL of the used generator but not on the chosen generating algorithm. In simulations with the usual assignments of RNs to REs, however, the actual admissible maximum number of REs can in most cases be considerably lower than this upper limit.

GENERATING ALGORITHMS

Usually a new RN y_k is a function of the preceeding r RNs, i.e.:

$$y_k = f(y_{k-1}, \dots, y_{k-r}) \mod m .$$

If a fast generating algorithm is desired, an algorithm of the following type is often used:

$$y_k = a\, y_{k-1} \mod 2^e.$$

In this algorithm n succeeding RNs always form a lattice after some simple transformations /1/. Based on the use of RNGs with a lattice structure, a generalized simulation model (GSM) which is commonly used can be defined as follows:

- o only neigbouring values of RNs can be comprised in a RE.
- o the sequence of each j-th selected RN is desired to be statistically independent.

For such GSMs, a second upper limit mnoc (maximum number of classes) can be derived /1/ which is considerably sharper than the limit ire. This limiting value mnoc is very suitable for dimensioning RNGs such that the independence of the REs in a simulation, and, in consequence, the accuracy of the simulation results are guaranteed.

SIGNIFICANCE OF STATISTICAL TESTS AND VALUES

Up to now statistical tests like POKER-test, RUN-tests etc. are the most used tools to qualify RNGs. Any statistical test represents a special application of the RNG which can be considered as a special simulation. As, however, the application of the RNs in such a test can differ considerably from the application in user simulations, it can easily be shown /1/ that, in general, such statistical tests are of little significance for qualifying RNGs.

Other methods often used for qualifying RNGs are based on statistical values like serial correlation of pairs. For mixed congruential RNGs with not interlocked hyperplanes it can be shown /1/ that the correlation factor can be adjusted to arbitrary values within a wide range, even to zero by means of an additive value. It is wellknown that this additive value has low significance to the maximum number of classes. This shows that statistical values like serial correlation of pairs are also not suited for qualifying RNGs.

CONCLUSION

To get simulations more reliable it is suitable to use a monitor watching the critical access distances. If each distance is known the value mnoc can be calculated. In user-simulations the minimal interval of RNs forming a RE must be greater than 1/mnoc. If this cannot be accomplished with the generator used the simulation must be carried out with a different generator which is dimensioned optimally for this application.

Any method which destroys the lattice, like scrambling RNs, coupled generators, etc., leads only to structures with uncalculable properties but does not lead to better results than an optimized generator with equivalent CL. Consequently, the user has lost a valuable tool for checking the accuracy of simulation results.

The presented methods enable an efficient implementation of RNGs in simulation programs and in complex universal simulation systems. They offer new facilities to get simulations more reliable.

REFERENCES

/1/ J. Sägebarth: Über Eigenschaften und Dimensionierung von rekursiven Zufallszahlengeneratoren und ihre Anwendung in der Verkehrssimulation. Ph. D. Thesis, University of Dortmund, 1985.

/2/ H. D. Ide, J. Sägebarth: On Properties of Random Number Generators and their Influence on Traffic Simulation. 10th ITC Montreal, 1983, and AEÜ 38, 3, 1984.

/3/ J. Sägebarth: Über Methoden zur Steigerung der Zuverlässigkeit bei der Simulation stochastischer Prozesse. 3rd GI/NTG Technical Conference, Dortmund, Oct. 1-3, 1985 (to be published).

TELETRAFFIC ISSUES in an Advanced Information Society
ITC-11
Minoru Akiyama (Editor)
Elsevier Science Publishers B.V. (North-Holland)

EVALUATION OF CORRELATED RANDOM SEQUENCES BASED ON THE BAYES-LAPLACE-STATISTICS OF MARKOV PROCESSES

Friedrich SCHREIBER   Martina HOMEYER

Technical University Aachen
Aachen, Federal Republic of Germany

ABSTRACT

1. It has been shown recently [1] that the unique solution of the basic statistical problem of the multinomial process which has been obtained as part of the so called Bayes-Laplace-statistics (BL-statistics) can be used to determine the a priori unknown transition parameters of Markov chains by proper evaluation of measured transition counts.

2. The work reported in [1] is continued in the present paper in order to find the **empirical stationary distribution function** of Markov chains with unknown transition parameters and of processes with correlated random sequences whose behaviour can be described by a Markov chain. For this purpose the BL-statistics is developed with respect to the following tasks.

2.1 First we consider a Markov chain with **two** nodes only, whose transition parameters are unknown. We measure n_o transitions **from** node 0 of which y_o lead to node 0 and n_o-y_o to node 1; and n_1 transitions **from** node 1 of which y_1 lead to node 1 and n_1-y_1 to node 0. It is shown that the posterior knowledge of the stationary probability P_o for the chain being in state 0 is uniquely described by a density $f(P_o|y_o,n_o;y_1,n_1)$ whose moments M_1 and M_2 are given by weighted sums of Gauss hypergeometric functions $F(a,b;c;z)$ [2] and express the **average stationary probability** $\overline{P}_o = M_1$ resp. the **relative error** when stating $\overline{P}_o$ by $d_o = [M_2/M_1^2-1]^{1/2}$.

2.2 Next we consider a Markov chain whose k+1 nodes (k=1,2,...) are interconnected by an arbitrary transition matrix with unknown transition parameters $p_{ji} = P(j|i)$; $i,j = 0,1,\ldots,k$. We may divide this chain in two parts, the left part consisting of the nodes $0,1,\ldots,r$ and the right part of the nodes r+1, r+2, ..., k. Then for any $r = 0,1,\ldots,k-1$ the left resp. right part can be attributed to node 0 resp. 1 of a 2-node Markov chain, called the "r-chain", and we may measure the number of transitions $n_o(r)$, $y_o(r)$ resp. $n_1(r), y_1(r)$ for every r-chain as defined in section 2.1. Using the results obtained in section 2.1 we may compute the average stationary probability $M_1(r) = \overline{P}_o(r)$ as well as the second moment $M_2(r)$ and then the desired **empirical stationary d.f.** $F_n(x)$ and its relative error $d_F(x)$ of the (k+1)-node Markov chain:

$$\left.\begin{array}{l} F_n(x) = M_1(r);\ d_F(x) = [M_2/M_1^2(r) - 1]^{1/2}; \\ r < x \leq r+1\ ;\ r = 0,1,\ldots,k-1. \end{array}\right\} \quad (1)$$

2.3 Finally we consider a **continuous random process** with an unknown stationary d.f. $F(x)$. In n+1 trials we measure and store the chronological vector $(x_t) = (x_o,x_1,\ldots,x_t)$ expressing the information which value x_{t+1} followed value x_t, $t = 0,1,\ldots,n-1$. After sorting the ordered vector $(x_r) = (x_o,x_1,\ldots,x_n)$, $x_{r+1} > x_r$ can be attributed to a (n+1)-node Markov chain with **one** transition only from each node except no transition from the "final node" belonging to the last measured value x_t, $t = n$. Since the transition counts $n_o(r)$, $y_o(r)$ and $n_1(r)$, $y_1(r)$ of the 2-node r-chain as defined in section 2.2 can be calculated for every $r = 0,1,\ldots,n-1$ we may obtain from Eq.(1) the empirical stationary d.f. $F_n(x)$ and associated error $d_F(x)$ as step functions with one step at each point x_r. This result should be compared with the empirical d.f. $F_n(x)$ in case of **independent** x-values [3].

3. The solution of section 2 for determining the stationary d.f. of Markov chains resp. of continuous processes with correlation effects is considered to belong to the class of objective solutions having principal importance within the BL-statistics. In order to make this solution available for practical work (e.g. for the evaluation of simulated teletraffic data) methods will be discussed for reducing the necessary computertime e.g. by collecting the measured x-values in intervals and/or by suitable approximations of the exact formulae for $F_n(x)$ and $d_F(x)$. Moreover it seems desirable to introduce here too an equivalent to the LRE-algorithm [4].

REFERENCES

[1] F.Schreiber, "The Bayes-Laplace-Statistic of the multinomial distribution and the measurement of unknown transient parameters of Markov chains". Proc. 5th Aachen Symposium "Mathematical Methods in Signal Processing", Aachen, Sept. 26-29, 1984 (to be published 1985 in AEÜ).

[2] M.Abramowitz and I.A. Stegun, "Handbook of mathematical functions." Dover Publ., New York, 1972.

[3] F.Schreiber, "The objective empirical distribution function and its error formulae." AEÜ 35 [1981], 473-480.

[4] F.Schreiber, "Time efficient simulation: the LRE- algorithm for producing empirical distribution functions with limited relative error." AEÜ 38 [1984], 93-98.

TELETRAFFIC ISSUES in an Advanced Information Society
ITC-11
Minoru Akiyama (Editor)
Elsevier Science Publishers B.V. (North-Holland)

ERIK BROCKMEYER AND THE TELETRAFFIC THEORY

Jens ARHNUNG
Copenhagen Telephone Company
Copenhagen, Denmark

Villy Baek IVERSEN
Technical University of Denmark
Lyngby, Denmark

Erik Brockmeyer (1901-1975) was employed by the Copenhagen Telephone Company (KTAS) from 1923 to 1971, and all along he worked with problems of teletraffic. (1) gives an account of his biography, publications, and contributions to the teletraffic theory.

In KTAS Brockmeyer was contemporary with such pioneers of teletraffic theory as F. Johansen, A.K. Erlang, P.V. Christensen, K. Moe, Arne Jensen, and E. Bjørn Christensen. His first reports deal with practical problems and traffic measuring problems. Later on he published more theoretical works, eg. an account of the application of probability theory to telephony, and the works mentioned below.

He was a member of the organizing committee of the First International Teletraffic Congress (ITC 1) in Copenhagen 1955. On this occasion he presented a review of traffic measuring methods in KTAS.

OVERFLOW THEORY

In 1954 Brockmeyer gave a solution to a key problem in teletraffic theory (2). Pure Chance Traffic Type 1 is offered to a finite trunk group and overflows to a second finite trunk group. Under the assumption of statistical equilibrium Brockmeyer gave explicit expressions for the elementary probabilities $p(j,k)$ of j trunks of the primary group and k trunks of the secondary group being occupied simultaneously. Furthermore he determined the overflow distribution, ie. the probability $\omega(k)$ that there will be k occupied trunks in the overflow group. Brockmeyer also presented algorithms for numerical calculations.

In 1937 L. Kosten gave the solution for an infinite overflow group. In 1956 J. Riordan derived the factorial moments of Kosten's model, and in 1976 R. Schehrer did the same for Brockmeyer's model. B. Wallström, C.E.M. Pearce and others have at previous Teletraffic Congresses generalized the Brockmeyer model in several ways.

ERLANG'S IDEAL GRADING

In 1948 the Copenhagen Telephone Company on the initiative of Arne Jensen published "The Life and Works of A.K. Erlang" (3). In this book Brockmeyer reviewed the mathematical works of Erlang. He also gave the theoretical basis for some of Erlang's works, i.a. Erlang's ideal grading (EIG), also referred to as Erlang's interconnection formula.

The assumptions for EIG are as follows. The offered traffic is Pure Chance Traffic Type 1. The system is in statistical equilibrium, and the lost calls are cleared. The number of trunks is n and the availability is k. The total traffic is equally divided among inlet groups, and there are as many inlet groups as there are ways of hunting k trunks out of the total n.

Practical applications of EIG has till now been very limited. This is partly due to a limited interpretation and understanding of the model, partly due to numerical problems, which are eliminated by the advent of computers and calculators.

EIG has several advantages over other grading formulas. The blocking probability is insensitive to the holding time distribution, and it gives the lower limit of the blocking probability for gradings with random hunting. Brockmeyer noticed that small gradings with sequential hunting may give smaller blocking. By intelligent hunting methods the minimal blocking is given by Erlang's multi dimensional loss formula. EIG is easily generalized to Engset traffic and to several traffic streams with individual availability.

Palm-Jacobæus formula is based on the same assumptions as EIG as concerns the number of inlet groups, but it does not take account of the actual state probabilities. Therefore it is only applicable for small blocking probabilities. EIG is also correct at high traffic levels, and grading calculations should be based on EIG, or modifications of this.

OTHER WORKS AND TOPICS OF DISCUSSION

Brockmeyer published several works on traffic measurements. For the demi-automatic exchanges of that time he introduced extreme value engineering based on the recording of the last choice traffic. He also gave some approximations for Erlang's B-formula.

The above-mentioned works of Brockmeyer will be the basis for a discussion on the state of the art in grading and overflow theory. Additional material will be distributed at the special interest group meeting.

REFERENCES

(1) J. Arhnung and V.B. Iversen, "Erik Brockmeyer and the Teletraffic Theory", Teleteknik, English edition, 1985. To appear.
(2) E. Brockmeyer, "The Simple Overflow Problem in the Theory of Telephone Traffic", Teleteknik, Vol. 5, pp. 361-374, 1954. In Danish. English edition published by the Copenhagen Telephone Company, April 1955. 15 pp.
(3) E. Brockmeyer, H.L. Halstrøm and Arne Jensen, "The Life and Works of A.K. Erlang. Copenhagen 1948. 277 pp.

TELETRAFFIC ISSUES in an Advanced Information Society
ITC-11
Minoru Akiyama (Editor)
Elsevier Science Publishers B.V. (North-Holland)

THE CONSTRUCTION OF EFFECTIVE ALGORITHMS FOR NUMERICAL ANALYSIS OF MULTILINEAR SYSTEMS WITH REPEATED CALLS

Sergey N. STEPANOV

Institute for Problems of Information Transmission
USSR Academy of Sciences Moscow, USSR

ABSTRACT

Our aim is to describe the general approach to constructing the numerical methods of probabilistic characteristics estimate of system with repeated calls. The approach being applied an upper bound of estimate error is calculated together with probabilistic characteristics estimate.

1. INTRODUCTION

As a rule,authors construct models with repeated calls according to the following traditional scheme.A servicing system with one or some incoming Poisson flows of primary calls is considered. Being refused servicing a subscriber repeats the call with a probability in exponentially distributed time.Service time of primary or repeated calls is exponentially distributed.These assumptions being fulfiled the model functioning is described by Markovian process with infinite number of states,which in most simple cases has a form (j,i),where j is the number of repeated subscribers and i is the number of busy lines.

2. METHOD DESCRIPTION

Estimate construction method is based on the property of strong decreasing of P(j,i)(probability of state (j,i)) with moving off j from its mean value or decreasing of i.Thus,it is possible to choose an area of states out of which the existence of process that describes the model functioning is almost equal to 0. The borders of the area (which we call further "reduced") may be found simply enough if a concrete model with repeated calls is considered proceeding from physical principles of system operating or after using some approximate technique. It is clear that if we take characteristics of Markovian process defined only on "reduced" state space as estimate of corresponding characteristics of initial model we should obtain a good approximation. The main part of the problem is to find the error of estimation.It can be solved if the Markovian process,defined on the "reduced" space of states,will majorize the Markovian process (or some of its components),described the initial model. It will be achieved by adding auxiliary fictitious calls into the "reduced" model.

Two types of estimates are to be considered.

2.1 Upper bounds

Let us substitute the initial space of states S (see Fig.1) by the space of states A that is given by m cutting levels on number of busy lines i and m cutting levels on number of repeating subscribers j.Let us denote by G a border of A. In G we include the states from which the initial process can move out of A after one step (on Fig.2,which shows "reduced" space of states A, we mark these states by *).Majorizing process in states $(j,i)\in A\setminus G$ is fuctioning by the usual way. If $(j,i)\in G$ then transition of the process into state with less number of busy lines due to ending of service time for one of busy line is impossible. It can be achieved by instantaneous addition to the system on service of the fictitious call.It will be shown that as a result of such substitution we obtain upper bounds for probabilistic characteristics of the initial model and find the calculating error in terms of probabilistic characteristics of process defined on "reduced" space of states.This method,compared with traditional one,when we take as "reduced" a space of states of rectangular type (see Fig.3) several times improves the accuracy of probabilistic characteristics calculation.

2.2 Effective estimates

Here majorizing process is also functioning as initial process in states $(j,i)\in A\setminus G$ but if $(j,i)\in G$ then as result of ending of service time for one of busy lines we immediately get into state (j-1, i). Such behaviour close to the natural behaviour of the initial process on the border. This allows to improve accuracy of estimates, introduced in 2.1 approximately in 10 times.However,the estimates obtained here can be either upper,or lower (it depends on situation) that impedes their error study.

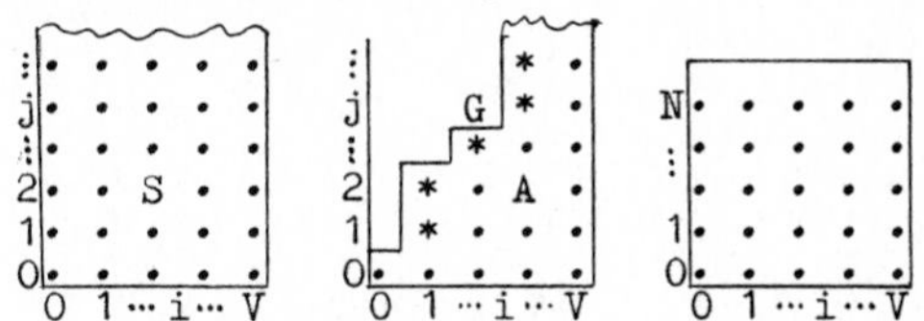

Fig.1 Fig.2 Fig.3

More detailed discussion of proposed method will be published later.

TELETRAFFIC ISSUES in an Advanced Information Society
ITC-11
Minoru Akiyama (Editor)
Elsevier Science Publishers B.V. (North-Holland)

CONGESTION ANALYSIS OF TELECOMMUNICATIONS NETWORKS UNDER NONSTATIONARY ARRIVAL CONDITIONS

M. Naim YUNUS

School of Mathematical Sciences
Universiti Sains Malaysia
Penang, Malaysia

INTRODUCTION

In teletraffic theory, the problem of congestion in a system experiencing time-dependent arrival rates (i.e. nonstationary arrivals which correspond to a nonhomogeneous Poisson distribution), $\lambda(t)$, and possibly time-dependent service rates, $\mu(t)$, has not been thoroughly investigated. This is mainly due to the difficulty in analysing mathematical models based on the above system. However, present day teletraffic is highly nonstationary due to several factors, such as social, economic, geographical and technological factors. Therefore the problem of determining time-dependent blocking probability is becoming more and more important in teletraffic engineering. This work will assume the service rate to be constant with mean one.

THE ERLANG LOSS FUNCTION

The Erlang Loss Function is a very well-known function and a lot is known about it and its computation is fairly easy. Therefore the possibility of trying to solve a difficult problem by using such a function is very tempting. Its derivation is based on the assumption of steady-state, that is the blocking probability doesn't change with time. However with the so-called modified offered load, the Erlang Loss Function can give fairly good approximations in the case of time-dependent blocking probability.

The first candidate for the modified offered load seems to be $\lambda(t)$, the arrival rate. This is not even the definition for offered load in the case of time-dependency, and the blocking probability it gives is wrong.

APPROACH 1

This approach firstly approximates the arrival rate with a series of step-functions, so that we have constant load over a small interval, where transience cannot be ignored. A possible modified offered load in this case is the transient offered load. The blocking probability obtained by it is not satisfactory. We can find another modified offered load at time t in any interval by using the first difference-differential equation of a set of such equations governing the probability distribution of the system in that interval. The first is used because it makes calculation much easier. However the time-dependent blocking probability function appears in it and we have to approximate it first before it can be used to find a more accurate blocking probability.

APPROACH 2

This approach considers time-dependent arrival rates directly. It utilizes the first difference-differential equation of another set of such equations for a system with continuous time-dependent arrival rates in the above manner. The modified offered load thus obtained involves integrals that could only be solved numerically for any t. Another approximation similar to the above would have to be made to this function too before it is usable.

CONCLUSION

In the above approaches the hard work, especially in the second one, is in the evaluation of the modified offered load at time t. Once this is completed the value is just used in the Erlang Loss Function and an approximation to the time-dependent blocking probability is obtained.

TELETRAFFIC ISSUES in an Advanced Information Society
ITC-11
Minoru Akiyama (Editor)
Elsevier Science Publishers B.V. (North-Holland)

DYNAMIC BEHAVIOR OF A COMMON STORE QUEUEING SYSTEM

H. R. van As*

IBM Zurich Research Laboratory
8803 Rüschlikon, Switzerland

1. INTRODUCTION

Traffic behavior of telecommunication systems is highly influenced by finite stores, which are normally shared by many traffic streams. In such an environment, the deficiencies caused by unbalanced arrival rates and priority schedules should be well understood.
The most important effects are : 1) storage domination by the high-rate traffic streams, and 2) the so-called priority deadlock in which high-priority traffic must be discarded owing to lack of store caused by backlogging low-priority traffic. These effects can be demonstrated most effectively by considering system dynamics as a response to a rectangular-shaped overload peak.

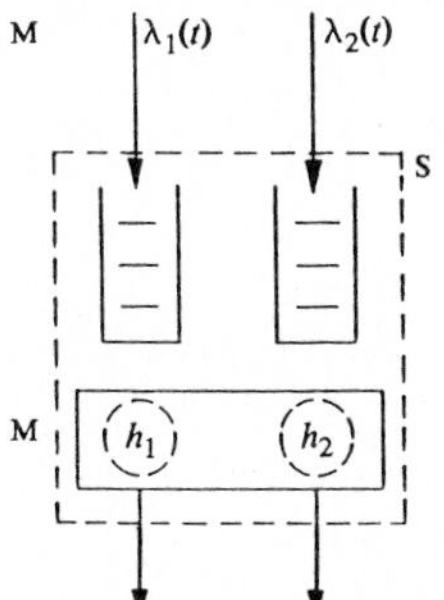

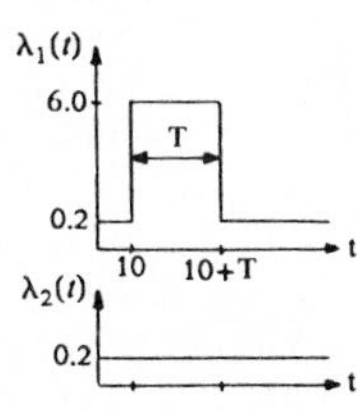

Fig. 1 Common store queueing system and overload peak specification for traffic stream 1

2. MODELING AND TRANSIENT QUEUEING ANALYSIS

Fig. 1 shows the traffic model consisting of a single-stage queueing system with one server and two traffic streams sharing a finite store, which includes the demand (packet or call) in service. Both random and nonpreemptive priority services are dealt with. Arrival and service processes are Markovian. The queueing analysis carried out was based on the calculation of the transient state probabilities by numerically solving the set of simultaneous Kolmogorov forward differential equations describing the transient system state process.

3. NUMERICAL RESULTS AND DISCUSSION

Fig. 2 depicts the mean system occupancies $E[X_1(t)]$ and $E[X_2(t)]$ of both traffic streams for different overload situations. To generate the overload peak, one of the two arrival rates is changed as specified in Fig. 1 . The common store queueing system has the capacity $S = 20$. The mean service time $h_1 = h_2 = 1$ is also the time unit.
Curves 0 , no priorities and an overload peak of traffic stream 1 :
1) Storage domination by the high-rate traffic stream 1, since the storage space becoming free is captured by the traffic streams in the proportion of the arrival rates.
2) Temporal surplus of the low-rate traffic stream 2 after disappearance of the overload peak : the acceptance rate for the low-rate traffic improves, while the backlogged high-rate traffic has not yet been worked off.
Curves 1 , overload peak of high-priority traffic stream 1 :
1) Decrease of $E[X_1(t)]$, since the store is filled up with backlogged low-priority demands.
2) Corresponding increase of $E[X_2(t)]$ and high temporal surplus after disappearance of the overload peak.
Curves 2 , overload peak of low-priority traffic stream 2 :
1) Almost complete storage occupation by the backlogging low-priority demands, since high-priority demands are given preference.
2) Hardly any high-priority demands are present, since once accepted they are almost immediately served.

Fig. 3 demonstrates a forthcoming priority deadlock. For long overloads of high-priority traffic, a backlog of low-priority demands builds up, and finally the store becomes completely useless :
the congested queueing system behaves like a loss system.

4. CONCLUSION

Both effects show that the continuity of traffic streams sharing storage can only be guaranteed, if some storage space is reserved for each traffic stream.

*) This work was done at the University of Siegen, W-Germany.

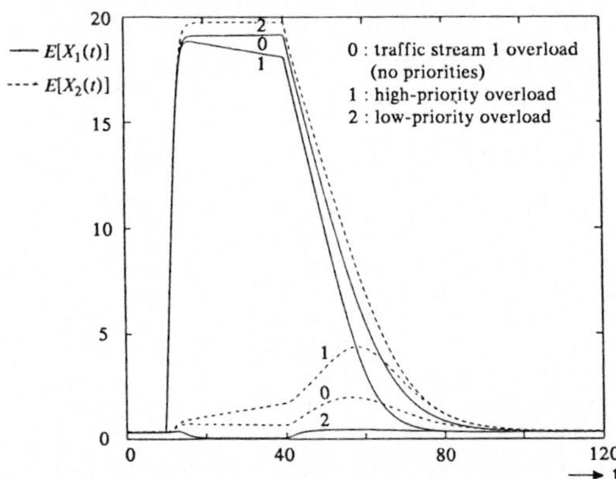

Fig. 2 Mean system occupancy for different overload situations

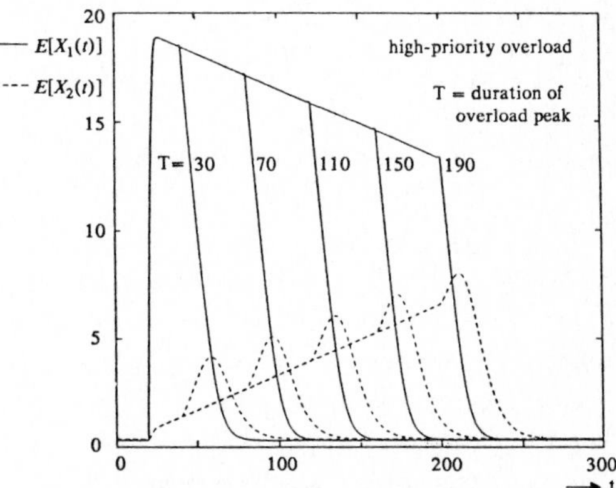

Fig. 3 Potential priority deadlock

TELETRAFFIC ISSUES in an Advanced Information Society
ITC-11
Minoru Akiyama (Editor)
Elsevier Science Publishers B.V. (North-Holland)

QUEUEING SYSTEMS WHERE ARRIVAL OR SERVICE TIMES ARE MARKOV-DEPENDENT

Gunnar Lind

Telefonaktiebolaget L M Ericsson

Stockholm, Sweden

1 INTRODUCTION

It is of interest to study the influence on queueing processes of dependencies between interarrival times and/or service times, i a in modelling the behaviour of telecommunications control systems.

In a contribution to ITC 10 the author formulated a class of queueing problems involving dependencies, which should allow fairly simple generalizations of classical analyses, where full independence assumptions are made. It was proposed to study one-stage systems under stationary conditions, where it is assumed that the sequence of interarrival times and that of service times are independent of each other, that one of them is a sequence of independent, identically distributed positive r v (random variables) (as in the classical theory) but that the other is a stationary sequence of Markov-dependent times. In the paper were given, i a, simple generalizations of the P-K formulae for the single server system having Poisson arrivals and general Markov-dependent service times with linear regression between successive service times.

Further research has since been done and in this short paper we will report on the analysis of the single server system where the interarrival times form a general stationary Markov sequence and the service times are independent of one another and of the interarrival times and exponentially distributed.

2 ASSUMPTIONS AND NOTATIONS

Let T_n = length of the interval between the nth and the (n+1)th arrival, X_n = nth service time and K_n = number of customers in the system immediately before the nth arrival (n=.., -1,0,1,..). The sequences (T_n) and (X_n) are independent of each other. (X_n) is a sequence of independent r v and $P(X_n \leqslant x) = 1-\exp(-\mu x)$, while (T_n) is a stationary sequence of Markov-dependent r v, completely described by $A(t) = P(T_n \leqslant t)$ and $A_1(t|y) = P(T_n \leqslant t | T_{n-1}=y)$. We will also use the alternative conditional distribution function $A_2(y|t) = P(T_{n-1} \leqslant y | T_n = t)$. We treat the absolutely continuous case with the probability density functions $a(t) = dA(t)/dt$, $a_1(t|y) = dA_1(t|y)/dt$ and $a_2(y|t) = dA_2(y|t)/dy$. Interpretations for cases where interarrival times are discrete or generally distributed should be obvious.

We denote: $E(T_n) = 1/\lambda$, $\rho = \lambda/\mu$=traffic offered.

3 SOLUTION

From the assumptions follows that (K_n, T_n) is a two-dimensional Markov process. We let $P(K_n=k, t<T_n \leqslant t+h)=r_k(t)h + o(h)$. Then we have $r_k=P(K_n=k)=\int_0^\infty r_k(t)dt$. Using the appropriate transition probabilities we get:

$$r_o(t)=\int_0^\infty \sum_{j=0}^{\infty} r_j(y)a(t|y)\sum_{l=j+1}^{\infty} q_l(y)dy$$

$$r_k(t)=\int_0^\infty \sum_{j=k-1}^{\infty} r_j(y)a(t|y)q_{j-k+1}(y)dy \quad (k=1,2,..)$$

where $q_s(y)=((\mu y)^s/s!)\exp(-\mu y)$

The solution is (k=0,1,..):

$$r_k(t)=(C(t))^k(1-C(t))a(t); r_k=\int_0^\infty r_k(t)dt$$

$$C(t)=\int_0^\infty a_2(y|t)\exp(-\mu y(1-C(y)))dy$$

This generalizes the classical case where $a_2(y|t)=a(y)$ and $C(t)=C$ for all t.

4 EXAMPLE

We use a four-point distribution for (T_{n-1}, T_n) with $E(T_n)=1/\lambda$, $Var(T_n)=1/\lambda^2$ and varying $r=Corr(T_{n-1}, T_n)$, and such that $r_k=\rho^k(1-\rho)$ (classical M/M/1 state distribution) if T_{n-1} and T_n are independent ($\Longrightarrow r=0$). This example gives quite simple calculations by which we can study how, for example, the mean waiting time varies when ρ and r are varied.

TELETRAFFIC ISSUES in an Advanced Information Society
ITC-11
Minoru Akiyama (Editor)
Elsevier Science Publishers B.V. (North-Holland)

OPTIMAL FEED-BACK CONTROL OF A STOCHASTIC SERVICE SYSTEM: A MARTINGALE APPROACH

Åke KNUTSSON

LM ERICSSON
Stockholm, Sweden

During an overload, the queue is long, and impatient customers drop out. That is a waste of customer time, of service-system resources, and of service-provider good-will. Before the queue will be too long, it may be better to throttle the input. Then the queue will decrease; but on the other hand, the risk of running empty of customers will increase. That is the dilemma. How to resolve it is outlined in this condensed paper.

1 THE SYSTEM DYNAMICS

The system, assumed to be Markovian, has one server, and the buffer has a limit, N-1. Let Q_t be the number of customers in the system at time t, $Q_t \varepsilon [0,N]$; let A_t be the number of arrivals during $(0,t]$; let I_t be the number of impatients dropping out during $(0,t]$; and let D_t be the number of served departures during $(0,t]$. Then

$$Q_t = Q_0 + A_t - I_t - D_t \qquad t \geq 0 \qquad (1a)$$

or in the form of a stochastic differential equation

$$dQ_t = dA_t - dI_t - dD_t \qquad (1b)$$

Let λ be the arrival intensity, let γ be a customer's drop-out intensity, and let μ be the server's capacity. Also, let $1(\cdot)$ denote the indicator function, which is unity when the argument is true, otherwise zero. Then the driving processes in (1b) have the following semi-martingale representations:

$$dA_t = \lambda u_t 1(Q_{t-} < N)\, dt + dM_t^A \qquad (2a)$$

$$dI_t = \gamma Q_{t-} dt + dM_t^I \qquad (2b)$$

$$dD_t = \mu\, 1(Q_{t-} > 0)\, dt + dM_t^D \qquad (2c)$$

where u_t is the feed-back control variable, the "throttle valve", $u_t \varepsilon [0,1]$. Let u_t depend on the present queue length only, $u_t = u(Q_{t-})$, then the Markov property is preserved M_t^A, M_t^I, and M_t^D are martingales relative to the σ-field generated by Q_s, $0 \leq s \leq t$.

2 THE SYSTEM ECONOMY

Assume a dropped-out customer will cost c_1 units, while a served customer will result in a gain of c_2 units. Then, at the final time, T, the expected total cost for control policy u is $C(u)$:

$$C(u) = E_u(c_1 I_T - c_2 D_T) =$$

$$= E_u \int_0^T (c_1 \gamma Q_{s-} - c_2 \mu 1(Q_{s-} > 0))ds \qquad (3)$$

The problem is to find a u that minimizes the cost function, $C(\cdot)$. The optimal control policy is denoted u^*.

3 THE OPTIMAL CONTROL POLICY

Guided by Brémaud, ref [1], p 202pp, we now apply a dynamic-programming procedure. Let $V(t,Q_t)$ denote the optimal cost-to-go from time t to the final time point, T, conditioned on the state of the Q-process at time t. Then

$$V(t,Q_t) = E_{u^*}\left\{ \int_t^T (c_1 \gamma Q_{s-} + c_2 \mu 1(Q_{s-} = 0))ds + \right.$$

$$\left. + V(T,Q_T) \mid Q_t \right\} \qquad (4)$$

"Itô-calculus" on the V-process results in

$$V(t,Q_t) = V(0,Q_0) + \int_0^t \left[\frac{\delta V(s,Q_s)}{\delta s} + \right.$$

$$+ (V(s,Q_{s-}+1) - V(s,Q_{s-}))\lambda 1(Q_{s-} < N)\, u_s +$$

$$+ (V(s,Q_{s-}-1) - V(s,Q_{s-}))(\gamma Q_{s-} +$$

$$\left. + \mu 1(Q_{s-} > 0)) \right] ds + martingale \qquad (5)$$

The Bellman-Hamilton-Jacobi equations for the problem are the following $N+1$ equations, where $q = 0,1,2,\ldots,N$:

$$\frac{\delta V(t,q)}{\delta t} + inf_{0 \leq u \leq 1} \left\{ (V(t,q+1) - V(t,q))\lambda 1(q < N) u_q \right\} +$$

$$+ (V(t,q-1) - V(t,q))(\gamma q + \mu 1(q > 0)) +$$

$$+ c_1 \gamma q + c_2 \mu 1(q = 0) \;=\; 0 \qquad (6)$$

It is sufficient for optimality, if there exists V and u such that (6) is satisfied. Using a conjecture by Högfeldt, ref [2], we nominate the following candidate:

$$V(t,q) = F(q) + (T-t)const \qquad (7)$$

We substitute (7) into (6), and eureka...., we find an optimal solution:

$$u_q = 1(F(q) > F(q+1)) \qquad (8)$$

The optimal control policy is stationary and has infinite horizon, as it depends neither on t nor on T. Also, it is of type 'bang-bang', the valve turns out to be an on/off switch.

REFERENCES

[1] P. Brémaud, 'Point Processes and Queues: Martingale Dynamics', Springer-Verlag: NewYork, 1981.

[2] P. Högfeldt, personal communication, 1984.

TELETRAFFIC ISSUES in an Advanced Information Society
ITC-11
Minoru Akiyama (Editor)
Elsevier Science Publishers B.V. (North-Holland)

TELETRAFFIC AND MANAGEMENT

Paolo de Ferra

S T E T - ITALY

Summary

Stimulated by a pressing technological evolution and by a competitive environment, the pace of innovation changes everywhere and increases its speed. Considerable opportunities, but connected with considerable burdens, are offered to both system suppliers and operating bodies. Certainly, the management will not ignore the advantages that can derive, at least in certain areas, from a positive convergence with the dynamic and vital current of the fast evolution. Considerable problems will arise, and also the teletraffic competences will be called to solve them. The key for obtaining timely and efficient solutions will be that of participation of all specialistic competences, teletraffic included, since the first phase of conception of systems, networks and services.

1.- INTRODUCTION

CCITT defines telecommunications traffic (teletraffic) as "a flow of attempts, calls and messages" /1/.

Management is defined as : "the conducting or supervising of something (as a business); esp: the executive function of planning, organizing, coordinating, directing, controlling and supervising any industrial or business project with responsibility for results" /2/.

For a telecommunications operating body the teletraffic is of course the main content of every activity, and its "raison d'être". The total volume of traffic flow can be considered a fundamental reference for estimating the size of the undertaking. Installation, operation and maintenance of the media required for the generation and flow of the telecommunication traffic are the main areas of activity for the management.

But teletraffic is also a science: indeed, a science with a particularly large number of branches. Its various fields include, for example:

- studies of the characteristics of the various types of traffic concerned in telecommunications;
- design of network models and definition of routing rules;
- definition of structures for assessing quality of service;
- study of the user's environment and behaviour;

and, more in general, definition of theories, models and methods for solving the manifold problems attached to telecommunications traffic.

If we consider this science from a certain point of view, there is no doubt that every one of its branches produces theories, models and methods that may have great importance as regards their applications. In other words, without any doubt, teletraffic science may be very readily contemplated by the management with a view to advantageous applications of its results. Indeed, in the everyday pursuit of the best compromise between cost and quality of service, a correct application of scientific results is certainly a sound way to guarantee the highest cost-effectiveness of networks and services.

But, from a different point of view, teletraffic science attains, through the deepness of its researches, levels of abstract conception and rational speculation that are quite sufficient to make it a basic science, with the dignity of full academic status, independently of the possibility of immediate applications of its results.

We can therefore say that teletraffic has two souls: a scientific soul, and an application-oriented soul. It seems correct to consider the scientific soul - the one that is cultivated in universities and in research centers - as the first one. The second one, concerning actual applications, is equally important and is cultivated mainly in industrial development laboratories and in telecommunications operating bodies.

But these two souls have never been completely separated: we can only observe that in certain bodies the scientific soul plays the major role, and, in other bodies, the applicational soul.

These two different souls, scientific and application-oriented, have been coexistent in teletraffic since its origins. Nevertheless, as happens to all living things, they have undergone a considerable evolution from their origins up to now. This paper is aimed at outlining this evolution, offering some comments and drawing some conclusions.

The paper falls into 5 sections, this being the first one. The second section is entitled "Teletraffic in the past". The third section: "Pressures for change" considers the radical evolution of telecommunications techniques and services that is now going on. This evolution already has considerable effects on teletraffic activities. Section 4, entitled "Teletraffic prospects", in particular refers to the evolution of the so-called "application-oriented soul" in the context of future foreseeable developments of telecommunications techniques, production media and policies. Finally, in section 5, some concluding remarks are given.

2.- TELETRAFFIC IN THE PAST

First of all, it seems important to point out a characteristic of the telecommunications world that was peculiar at the inception of teletraffic and is connected with the pace of technological evolution.

During the first decades of telecommunications as a public service, the technology changed very slowly. For example, the switching systems were mostly renewed about every twenty years. Every system was produced and installed within a few decades and was considered technically sound for at least forty years. In this framework, the size of R&D resources was kept small, and their cost proved low, particularly in comparison with the costs of production, operation and maintenance of the system.

As a consequence of this "slow pace" it was reasonable that a fairly long period might elapse between the availability of a system prototype and the actual application of the more or less ultimate version in public service. Indeed, development techniques and production techniques largely evolved independendently with respect to the optimization methods. It appeared acceptable during the process to allow long intervals, e.g. to the traffic experts, for introducing proper modifications.

After all, a certain separation between system designers and traffic experts (particularly network planners) was not a considerable disadvantage at that time.

At the beginning, teletraffic studies were oriented towards theoretical and modelling subjects, both for the above-mentioned reasons and because the stochastic processes were a new and complex matter. In fact, teletraffic - the beginning and end of all telecommunications calls - originates from processes of this type, characterized by random distribution on a probabilistic basis, and until then little studied.

The first important subject for study was the search for theories and models employing a sufficiently general approach to the problems of that time. A typical problem was the performance evaluation of systems with a stochastic type of service. When:

- intensity of the offered traffic,
- service procedures (e.g. loss or waiting system),
- system characteristics (e.g. full or limited availability)

were known, the question was to derive, with more or less approximate formulas, the grade of service in terms of loss or of waiting probability and of distribution of waiting times.

The same formulas and algorithms utilized to verify the system performances can, of course, be utilized to fix the dimensions of a system, i.e. to solve the reciprocal problem of evaluating the amount of resources required in order to provide a certain grade of service.

Along these general lines, the problems related to full availability links were first solved (A.K.Erlang /3/), followed by the gradings (G.F.O' Dell /4/, C.Palm /5/) and by the link systems (C.Jacobaeus /6/). So, engineers had at their disposal a number of formulas, graphs and tables, which were and will continue to be the fundamental tools commonly used to dimension and verify all single (circuit switched) links.

The availability of such tools laid down the bases for further developments, particularly in enabling the analysis to be extended not only to single links but also to ever larger and more complex systems. Obtaining the optimum condition of availability and the required number of selectors in a system with multiple selection stages and, in general, minimizing the number of crosspoints in the switching networks of telephone exchanges were typical applications.

A natural extension also occurred because of the progressive spreading of long distance direct dialling into larger areas. The increasing importance of transmission system costs made it more convenient to abandon the original, rigid routing policies and to use flexible routing rules depending on the state of engagement of the trunks. In these circumstances, the optimum conditions must be defined for the whole network, even in a wide area. With regard to this, the studies of A.Jensen /7/, Y.Rapp /8/, R.I.Wilkinson /9/, G.Bretschneider /10/, P.Le Gall /11/ and C.W.Pratt /12/ may be mentioned as important steps forward. They offered suitable theoretical and methodological tools for coping with the progressive extension and complexity of systems and problems connected with teletraffic.

The extension of the studies to complete systems also involves the use of more global and significant performance indexes, for a proper system dimensioning. With regard to this, for example, there is an increasing tendency to evaluate the network performance not only in terms of loss probability of final trunk groups, but also in terms of mean end-to-end blocking probability, as actually perceived by the users. At the same time, there is an increasing tendency to consider that sometimes the system must operate in degraded conditions (caused by overloads or failures) and to take account of these conditions also in the definition of the quality of service indexes.

In addition to the development of teletraffic as delineated above, it is worth mentioning another important development which has taken place at the same time : the development of computer science. This has deeply affected methods and instruments used in teletraffic calculations. The previous use of formulas, graphs and tables (calculations on paper) has been mostly replaced by the use of automatic calculation and operational research tools. Accordingly, entirely new programs, files, organizations and methods became necessary and were adopted.

This resulted in not only a processing capacity much greater than before, but also new capabilities. Among them, for example, a capability to forecast the future increase in traffic that is much more reliable than in the past. This fact not only allows obvious economic benefits in network planning but also makes other problems less difficult. Among them, for example, the problem of reducing the traffic peaks by means of tariff changes. In addition, the automatic evaluation of the required number of resources in almost real time is no longer unthinkable: the present and past values of the traffic may be processed, together with the state of the network, to provide automatically data relevant to extensions to be carried out within a certain period.

3.- PRESSURES FOR CHANGE

Several evolutionary steps, which are considerably influencing the development of teletraffic both as a science and as applications, have been taken in relatively recent times, regarding both telecommunications techniques and services. Mention may be made of the subjects related to:

- computer-controlled switching,
- integrated digital techniques,
- packet switching,
- satellite systems.

The expression "computer-controlled switching" seems inadequate to indicate the extent of the problems related to the application of computers as control equipment for switching exchanges, with a time-sharing approach. For the last twenty years the teletraffic experts have been compelled to devote more and more care to the research problems related to these systems, from the first mainframes performing all the logic functions of an exchange, to multiprocessing systems, and to the more recent systems organized with a multi-microprocessing approach to obtain a high degree of modularity. By now, an entire teletraffic branch is involved in studying the main parameters for evaluating the performances of these systems, in finding out suitable theories (e.g. the queuing theory) and in deriving suitable methods for solving the relevant problems. Several studies on these subjects testify their importance /13/, /14/.

But computers as control equipment of switching exchanges also have an impact on the network structure itself, because the same computers may also be used for a dynamic network management, i.e. to modify the routing rules from time to time. Modifications may be of a recurrent type, e.g. to follow seasonal traffic fluctuations, but dynamic network management is considered a particularly effective tool when a sudden overload arises or when a failure occurs. On these occasions, the interventions can follow two strategies: an increase in the resources (e.g. new routing schemes to escape critical areas) or a reduction in some traffic sources to prevent congestion of the network. On this subject, several notable contributions have been offered, e.g. by L.A. Gimpelson /15/. However, many problems have still to be solved. The subject is clearly delicate, in connection both with the possible consequences of the interventions (sometime even more detrimental than the phenomena that are to be prevented) and with the degree of timeliness (and automation) to be ascribed to the interventions themselves.

Besides, dynamic routing research problems will become increasingly complex as a consequence of the progressive introduction of integrated digital techniques in the network. The result will be a drastic reduction in the limits presently imposed on the maximum number of links and nodes that can be involved in a single connection. As a consequence, the hierarchical network structure itself could perhaps be suppressed in favour of more elastic structures that may allow greater economies and better quality: i.e. greater efficiency and greater adaptability to face emergencies. With this in mind, it is clear that further complex problems will have to be solved by teletraffic experts in the future.

The introduction of specialized networks for non-telephone services (with main reference to packet switched data networks) is another fairly recent source of research activities for teletraffic science. Various activities connected with traffic forecasting and network planning for the new services have assumed particular relevance and have prompted a large number of studies, on protocol evaluations, routing rules, flow control algorithms and so on. In addition, the common channel signalling network, connecting computer-controlled switching exchanges, is a particular case of packet-switched network, with peculiar characteristics requiring specific studies.

Furthermore, the introduction of satellite systems into the network is opening up new fields of activity. The peculiar characteristics of these systems may offer cheaper solutions to certain problems. Besides long-distance transmission, they may be used for switched connections (demand assignment) between points relatively closer but without other communication infrastructures (e.g. islands) or for an improved service availability in case of failures (e.g. using satellites as extra media in addition to the terrestrial network).

The above-mentioned list refers to the most important and recent new issues that the management has to face in this context, with a prospect of considerable technical and economic advantages. But, conversely, these opportunities create completely new problems and therefore require new branches of teletraffic activity.

It may be noted that the list is already long and complex. Nevertheless, it will lenghten in the next future: in fact, other important issues are near and already well recognized. Among them it is worth mentioning at least the ISDN, whose first experiments are now being carried out in several parts of the world.

ISDN does not only mean transparent digital connectivity at 64 kbit/s. This is only the first among five conceptual steps, the other four being /16/:

- ISDN including packet communications, with various multiservice customer interfaces;
- "Bearer" ISDN in the sense of Open System Interconnection (OSI) recommended by CCITT and ISO, convergence point for existing networks by means of gateways;
- ISDN with high-level OSI functions for communication-oriented application services: e.g. Message Handling System (MHS);
- "Intelligent" ISDN with high-level functions for application services in general: Value Added Services (VAS).

Obviously, each one of the previous conceptual steps will open up new and far-reaching problems for teletraffic: for example, new hybrid networks with packet and circuit switching capabilities, optimization of the gateways, optimum location, linkage and dimensioning of the resources for high level services, etc.

Moreover, the issues will become even more complex because in the ISDN framework also mobile media and reservation services have to be considered. Furthermore, ISDN will also include services with different bandwidths, from submultiples of 64 kbit/s to a very broad bandwidth (e.g. 34 Mbit/s). It will be a multiband network, where different services with different bandwidths will coexist.

It is evident that teletraffic will have to develop and extend its interests in the near future in order to play its part efficiently in solving these impending problems. These problems, as previously described, do not arise from scientific speculations but from precise prospects of technical and economic evolution of great interest for the management.

4.- TELETRAFFIC PROSPECTS

The following three fundamental subjects seem to stand out when taking an overall view of the teletraffic problems to be solved in the future:

- forecasts regarding the future pace of evolution of technologies, techniques and services;
- forecasts regarding the evolution of the production systems;
- forecasts regarding the evolution in telecommunications policies.

Regarding the development of technologies, techniques and services, there is no doubt that the evolution will continue to advance along its present lines at a steady rate. Telecommunications systems will progressively lose their present specific characteristics associated with single different services, and acquire general characteristics suitable for mixed applications in a "multiservice" environment.

Following this path, it is already possible to see the first prospects of "intelligent" multiservice terminals, equipped with various kinds of I/O devices to transmit and receive information: microphones, keyboards, scanners, television cameras, etc, and receivers, printers, displays, screens, etc. Regarding the network, the introduction of more and more "intelligent" resources is envisaged, for a progressively more sophisticated storage and processing of the signals. These resources could eventually be suitable for processing the signals to an extent that will allow recognition of the meaning of the signal itself. It is sufficient to mention recorded vocal messages, packet voice, speech recognition and synthesis. Regarding services, even the more traditional ones may evolve in very interesting ways. For example, the voice itself could be transmitted with different choices in terms of quality and tariff: with different bandwidths, with circuit or packet switching and, as an extreme case, through speech recognition and restitution at the far end via a choice of media: speech synthesizer, screen, printer, in the original language or even translated into a different one.

Of course, this is an extreme case, but it can give an idea of the increasing need for activities in the teletraffic sector. It will be necessary to solve more and more complex problems, both on account of the size of the systems and, mainly, because of the very numerous kinds of utilization of the resources. Among other things, new quality criteria, analytical models and simulation structures have to be developed to take into account the contemporary presence of services and sub-services of a different nature and value.

Moreover, new unexpected needs for activities in areas previously not related to teletraffic could arise. The control system for Computer Aided Manufacturing (CAM) could be a first example. Besides, this system has an outline similar to the scheme of the control systems for switching exchanges.

Regarding the evolution of the production apparatus, first of all it is worth mentioning how much the fast pace of technology has already changed the activity of development of new systems, in terms of methodology and consequently in terms of development times. This also may be considered as a consequence of the greater speed of innovation in technology, which shortens the period of technical and economic validity of the systems.

At the same time, the pressure to exploit the opportunities offered by new technologies, to either reduce the costs of the traditional systems or to offer new ones, becomes more and more urgent. As a consequence, economic competition grows among the systems suppliers. They continously have to compare the product portfolio with that of competitors, regarding not only present production but also, and especially, expected developments.

So, in the context of this open challenge, in terms of economic return, market shares, image, production costs etc, greater value may be attached to policies for extending markets by means of the best exploitation of the tecnological innovations, as regards both products and production media. This policy further encourages the speeding-up in the evolution of technology.

The increasing speed of innovation, together with the complexity of the new systems, requires larger and larger amounts of resources for the R&D process, and this trend will continue. At the same time, there is a steady decrease in the period of time available to improve new systems before their offer on the market.

Therefore, co-operation between technicians with different expertises becomes indispensable from the moment of conception of the new systems. System engineers have to work together with technicians of other fields of activity - technology, electronic design, software, marketing, reliability, quality control - and in general with the experts in every branch of activity needed to make the product successful. Among them, in the telecommunications field, the teletraffic experts cannot be ignored. Their specific technologies and know-how allow them both to evaluate the feasibility of new solutions (e.g. the new structure of a switching network, or the load factor of a computer) and to suggest completely new structures and ideas.

Cases of co-operation between traffic experts and system/network designers from the beginning of a project already exist, and in fact this kind of co-operation is spreading fairly fast. It is enough to consider the contributions offered by teletraffic experts to the conception of several recent systems such as: packet switched networks, multiprocessors, multiple access systems with collision detection for satellites and Local Area Networks (LAN), etc. But it may be predicted that this type of co-operation will spread dramatically and may prove to become the standard practice for future projects.

Finally, the following considerations concern the evolution of telecommunications policy, particularly regarding suppliers of services and the conflict between monopoly and liberalization.

It may be noted that, in different countries, various policies are nowadays practiced. It is enough to consider the variety of the more or less recent policies that are practiced in the United States, Canada, Japan, Great Britain, and a few other European countries. These policies try to meet the different needs that arise in the various countries in the world, which are so numerous and so different from one another (even if technical progress, and mainly the progress of telecommunications, are making this world smaller every day). Nevertheless, even if different policies are practiced, it is quite clear that on the one hand a complete liberalization of the entire sector appears extremely improbable, but on the other hand an evident trend towards a considerable reduction in the monopoly areas can be observed.

As a consequence, several bodies operating networks and services, even if traditionally acting in a protected market, will face direct competition at least in certain areas of activity. It is easy to forecast that, in certain areas, conditions and consequences will be similar to those already mentioned for the system suppliers. Incidentally, this situation can also rebound against manufacturers previously relying on considerable "captive markets".

Moreover, in the market there are also other tensions and causes for competition that have been detected for a long time but have not yet found full expression. They can bring about other consequences in repercussion. It is enough to consider, for example, the confluence, in the same market, of suppliers of telecommunications systems and services and of suppliers of systems and services for computers and office automation.

In this context, it is possible to forecast that also teletraffic experts working in the operating bodies (like their colleagues working in manufacturing firms) will be asked to enter into team activities with other experts: activities to be carried out promptly and efficiently, with the sole aim of winning and/or holding shares in a "not captive" market.

Among these activities, those connected with quality of service should be mentioned first. In fact, with the shrinkage of the monopoly areas, the quality of service is losing both its role as a constraint and its uniqueness. So, it is becoming an important parameter, which can assume different values to cope with the market requests, obtaining services and sub-services of different quality at different prices. The teletraffic experts may be asked to co-operate with marketing experts in the definition of company policies, by providing suitable instruments for the timely evaluation of the quality of service. In effect, it would be necessary to process in a short time, almost in real time, an increasing amount of information in order to verify the consequences of any given action on the service quality.

Moreover, it can be foreseen that the importance of the real time control of the efficiency of resources management will grow together with competition. As a consequence, the interdependence between traffic, reliability, maintainability and quality of service will increase, in order to ensure a cost-effective availability of service.

Considering the importance and complexity of these functions, this could be one of the first areas of application for the solutions offered by Artificial Intelligence, with particular reference to "Expert Systems".

5.- CONCLUDING REMARKS

In conclusion, after the preceding short notes, a few comments and observations may be made regarding the more suitable guidelines for the evolution of teletraffic in the future from a managerial point of view.

It has been pointed out that, since its inception, teletraffic has had two "souls": one scientific, the other application-oriented. These souls have always existed together in harmony. A connection with the process of innovation has existed since the inception of teletraffic, through its application-oriented soul. But, at first, it was a loose, secondary connection.

This may initially have offered some experts a certain peace of mind, a certain tranquillity for scientific researches and even for academic speculations, independently of the immediate applicational interests of the present industrial world.

But several powerful pressures for innovation have arisen, to modify the initial situation. At first they came from technological evolution, then from changed conditions in the process of system innovation, and finally from a clear political evolution, now oriented towards opening wide market areas.

Stimulated by a competitive environment, the pace of innovation changes and gathers speed, also within organizations which were previously accustomed to operate in more or less protected market sectors. Speeding-up of this kind began a certain time ago, but it will increase, even dramatically, in the near future.

In the face of these prospects, the management of all telecommunications organizations (manufacturers and operating bodies) will not ignore either the advantages that may derive from a positive convergence with the current of the fast evolution or the risks that it may run for non-convergence or for late convergence.

Therefore, it will be in the common interest to see that teletraffic expertise also sails in this current, which may be muddy but is certainly dynamic and vital.

In the same context, telecommunications have to face problems of great magnitude and complexity. For solving them, it has been shown that, from the initial stage of conception of systems, networks and services, a close and timely co-operation of all the essential expertises (with teletraffic among them) is needed. Indeed, this co-operation appears to be a very important key for the success of the same systems, networks and services.

These observations are certainly valid for the application-oriented soul of teletraffic. However, it would be reasonable to forecast that the great extent of the problems involved will prove highly attractive to the so-called "scientific soul" as well.

Acknowledgements

The author wishes to thank Messrs. A.Tonietti and A.Tosalli of CSELT and particularly Mr. R.Preti of STET for their valuable contributions and suggestions.

References:

/1/ - CCITT Yellow Book, Sup. No.7 (II.3).

/2/ - Webster's International Dictionary.

/3/ - A.K.Erlang: "Solution of some problems in the theory of probability of significance in automatic telephone exchanges"; The Post Office Electrical Engineers Journal(1917-18)

/4/ - G.F.O'Dell: "An outline of the trunking aspect of automatic telephony"; I.E.E. Journal (1927).

/5/ - C.Palm: "Calcul exact de la perte dans les groupes des circuits échelonnés"; Ericsson Technics (1943).

/6/ - C.Jacobaeus: "A study on congestion in link systems"; Ericsson Technics (1950).

/7/ - A.Jensen: "Moe's principle: an econometric investigation intended as an aid in dimensioning and managing telephone plant"; The Copenhagen Telephone Company (1950).

/8/ - Y.Rapp: "The economic optimum in urban telephone network problems"; Ericsson Technics (1950).

/9/ - R.I.Wilkinson: "Theories for toll traffic engineering in the USA"; Bell System Technical Journal (1956).

/10/- G.Bretschneider: "Die Berechnung von Leitungsgruppen fuer ueberfliessenden Verkehr in Fernsprechwaehlanlagen"; NTZ (1956).

/11/- P.Le Gall: "Les trafics téléphoniques et la selection conjuguée en téléphonie automatique"; 2.nd International Teletraffic Congress (1958).

/12/- C.W.Pratt: "The concept of marginal overflow in alternate routing"; 5.th International Teletraffic Congress (1967).

/13/- J.P.Buzen: "Computational algorithms for closed queueing networks with exponential servers"; Comm. ACM (1973).

/14/- F.Baskett et alii: "Open, closed and mixed networks of queues with different classes of customers"; J. ACM (1975).

/15/- L.A.Gimpelson: "Network management: design and control of communication networks"; Electrical Communication, (1974).

/16/- P.de Ferra et alii: "Objectives and results from the emerging ISDN standards"; 11.th International Switching Symposium (1984).

AUTHORS INDEX

INDEX OF PAPERS PRESENTED AT INTERNATIONAL CONGRESSES 1 THROUGH 10

INTRODUCTION

The International Teletraffic Congresses have been providing a forum for presentation of papers on a wide variety of teletraffic topics. The first Congress was held in Copenhagen in 1955. Congresses have been held about every three years since then. The most recent, ITC 11, was held in Kyoto, Japan in September, 1985 Although many of the papers have subsequently appeared in various publications, many have not and may be unknown to many workers in the teletraffic field. This index makes the information contained in the papers' titles more widely available.

Individuals who wish copies of particular papers may have difficulty in locating them. Copies of all ITC papers have been made available to major engineering libraries and might be found there. Otherwise the best probable source is the author of the paper. Publication of the proceedings in book form, as is being done for ITC 11, should ease the problem in the future.

In making this index, each paper has been given a number which consists of the congress number, session number, and paper number within the session. Because of variations among Congresses there are departures from this rule, but each paper has a unique identifying number. Liberties have been taken with spelling of certain words and with a few authors' names in order to be consistent in indexing. My apologies for mistakes that may have arisen.

Organization of the index is in three sections. The first lists papers in numerical order. The second lists authors and papers. The third lists key words chosen from the titles. For each key word all papers are listed with the key work highlighted. It was necessary to exclude certain words like "traffic", "circuit" and "switch" because so many papers contain these words that a listing for them is not very informative. (This means that a paper with a title such as "Traffic in Circuit Switching" would not appear at all in the keyword index.)

Finally, although much of the processing for this index was done by a personal computer, frequent human intervention was required. This has resulted in the occasional loss of pieces of data; it is hoped that the number of such lapses is small so that the usefulness of the index is not impaired.

Walter S. Hayward
October, 1985

INTERNATIONAL TELETRAFFIC CONGRESSES 1 TO 10
PAPERS INDEX

FIRST TELETRAFFIC CONGRESS, COPENHAGEN, DENMARK, JUNE 20-23, 1953

01-001 The Beginnings of Switching Theory in the United States--Wilkinson,R.I.
01-002 The Historical Development of the Theory of Probability in Telephone Traffic Engineering in Europe--Kosten,L.
01-003 Probability of Loss in Link Systems (In French)--Fortet,R.M.
01-004 On Equations of State for a Two-Stage Link System--Elldin,A.
01-005 Certain Delay Problems for a Full Availability Trunk Loaded by Two Traffic Sources--Cohen,J.W.
01-006 Some Aspects on the Variations in Traffic Intensity--Von Sydow,L.
01-007 Theories for Toll Traffic Engineering in the USA--Wilkinson,R.I.
01-008 Statistical Equilibrium Equations for Overflow Systems--Bech,N.I.
01-009 Optimal Use of Both-Way Circuits in Cases of Unlimited Availability--Taange,I.
01-010 Extension of the Erlang Formula in the Case Where Traffic in a Function of the Number of Busy Subscribers (In French) --Chaveau,J.
01-011 Some Recording Principles Applied in Time Zone Metering--Olsson,K.M.
01-012 On a Special Use of the Erlang Methods in Industry--Buch,K.R.
01-013 Probability Studies Applied to Telecommunications Systems with Storage--Wright,E.P.G. Rice,J.
01-014 On the Congestion in Gradings with Random Hunting--Elldin,A.
01-015 Probability Theory in Telephone Transmission--Jacobsen,B.B.
01-016 Consideration on the Significance of Random Sampling Process in Determinating the Magnitudes of Telephone Traffic--Rhode,K.
01-017 Measurements with Lamp Panel - A New Method for Traffic Studies--Jacobaeus,C.
01-018 A Survey of Traffic Measuring Methods in the Copenhagen Telephone Exchanges--Brockmeyer,E.
01-019 Traffic Simulations--Neovius
01-020 Application of Artificial Traffic Methods to Telephone Problems--Kosten,L.
01-021 Telephone Traffic Research with the Aid of the Swedish Traffic Analyzer--Ekberg,S.
01-022 An Analyzer for Telephone Traffic (In German)--Karlsson,S.A.
01-023 Adjustment of Traffic and Losses and Determination of the Overall Losses--Langer,M.
01-024 Analogies between the Congestion and Communication Theories--Syski,R.
01-025 The Applicability of Decision Theory in the Planning and Operation of Telephone Plant--Jensen,A.

SECOND INTERNATIONAL TELETRAFFIC CONGRESS, THE HAGUE, NETHERLANDS, JULY 7-11, 1958

02-001 Blocking Probabilities and Trunking Requirements in a Communications Grid--Berkowitz,R.S. Inose,H. Joseph,D.
02-002 Traffic and Blocking Problems in a Three Link Switching System--Van Den Bossche,M. Knight,R.G.
02-003 The Application of Boundary Curves to the Determination of the Capacity of Limited Access Groups (In German)--Bretschneider,G.
02-004 Some Recent Results on Traffic Theory--Cohen,J.W.
02-005 Estimates and Tests of Hypotheses Concerning Telephone Traffic (In French)--Fortet,R.M.
02-006 Calculation Method for Blocking in an Automatic Telephone System with Matching (In French)--LeGall,P.
02-007 The Trunking of Time Division Multiplex Electronic Telephone Exchanges--Harris,L.R.F., Ward,P.W.
02-008 A General Formula for the Calculation of Traffic Blocking in Telephone Networks (In German)--Karlsson,S.A.
02-009 Differentiation in Telephone Traffic Calculations (In German)--Langer,M.
02-011 Reciprocal Overflow Traffic--Rabe,F.W.
02-012 Congestion in Automatic Exchanges--Sanneris,A., Capello,F.
02-013 Markov Chains in Telephone Congestion Theory--Syski,R.
02-014 Some Basic Considerations Involved in the Calculation of the Amount of Storage Necessary in an Electronic Telegraph Switching System--Wright,E.P.G.
02-015 Loss Formulas for Homogeneous Gradings of the Second Order in Telephone Switching Employing Random Hunting--Kruithof,J.
02-016 Factors Influencing the Demand for Telephone Service--Leunbach,G.
02-017 The Use of Pure Chance and Smoothed Traffic Load Curves in Telephone Engineering--Bear,D.
02-018 Circuit Provision for Small Quantities of Traffic--Marrows,B.
02-019 Solving Equations of State in Digital Computers--Carlsson,S.G., Elldin,A.
02-020 The Influence of Telephone Rates on Local Traffic--Kraepelien,H.Y.
02-021 Statistical Supervision of Telephone Plant--Lind,G.
02-022 Artificial Traffic Trials on a Two-Stage Link System using a Digital Computer--Wallstrom,B.
02-023 Traffic Engineering and Administration of Line Concentrators--Hayward,W.S.,Jr.
02-025 The Economics of the Hiring of Private Wires in Great Britain - A Simulation Study--Jackson,R.R.P., Nickols,D.G.
02-026 The Use of Statistical Moments for Specifying Noise in Long Distance Telephone Circuits--Jacobsen,B.B.
02-027 Telephone Traffic and Matching Selection in Automatic Telephony (In French)--LeGall,P.
02-028A Contribution to the Study of Blocking in Link Systems with Matching--Fortet,R.M., Chastang,J.F.
02-029 Study of Load and Service Variations in Toll Alternate Route Systems--Wilkinson,R.I.
02-030 Statistical Methods for Administration of Dial Offices--Barnes,D.H.
02-031 Determination of the Traffic-Carrying Properties of Gradings with the Aid of some Derivative Parameters of Telephone Traffic Distribution Functions--Ekberg,S.
02-032 On a Certain General Type of Stochastical Problems in Telecommunications--Kosten,L.
02-033 The Use of Probability and Structure in Long Term Planning--Jensen,A.
02-034 Erlang's Formula over the Decade (In German)--Matkovic,V.
02-035 Approximate Equations for Two-Stage Links--Lotze,A.

THIRD TELETRAFFIC CONGRESS, PARIS, FRANCE, SEPTEMBER 11-16

03-002 Simulation of a Delay System with Single-Server and Special Priorities--Agard,J.
03-003 Assessment of Operations Reliability in Automatic Telephone Traffic--Ahlstedt,B.W.M.
03-004A Some Combinatory Properties of Connecting Networks--Benes,V.E.
03-004B The Evolution of Telephony--Bogaerts,B.W.M.
03-005 Some Statistical Data on the Reliability of Telephone Exchanges--Borel,G.
03-006 The Importance of Exact Methods for the Simulated and Approximated Calculus Intended for the Determination of Loss Probability --Bretschneider,G.
03-008 Maintenance of Telephone Plant from Statistical and Economic Aspects--Christiansen,U., Lind,G.
03-009 On Derived Markov Chains and Their Application to Queueing Theory--Cohen,J.W.
03-010 Simulation of Communication Networks with a Digital Computer--Dujack,R.L., Epstein,D.I., Robbins,C.

03-012 Traffic Measurements on an Alternative Network in Sweden--Ekberg,S., Taange,I.
03-013 On the Dependence between the Two Stages in a Link System--Elldin,A.
03-014 Long Time Measurements of Telephone Traffic--Elldin,A., Taange,I.
03-016 A Study of the Efficiency of Graded Multiple Delay Systems through Artificial Traffic Trials--Gambe,E.
03-017A Review of Simulations of Automobile Traffic in the USA--Gerough,D.L.
03-017B An Analog Computing Method for Investment Cost and Economic Studies Applicable to Telecommunications Development--Gillings,D.W.
03-018 The Results of Tests on Gradings Devised by the Method of G. F. O'Dell using an Artificial Traffic Machine--Harmston,A.T.
03-019 A Study of the Distribution of Traffic during the Busy Hour--Harmston,A.T.
03-020 Artificial Traffic Trials Using a Digital Computer--Hunter,D.G.N.
03-021 Measurement of Non-Random Telephone Traffic with Special Reference to Line Concentrators--Jones,A.W.
03-022 A Statistical Method for Charging the Telephone Calls--Karlsson,S.A.
03-023 On Operational Research Problems in Theory of Congestion--Kosten,L.
03-024 Minimum Costs of Multi-Stage Link Systems--DeKroes,J.L.
03-025 Some Recent Waiting Problems--LeGall,P.
03-026 Overflow Traffic--LeGall,P.
03-027 On the Determination of the Number of Circuits--Longley,H.A.
03-028 Loss Formula, Artificial Traffic Checks and Quality Standards for Characterizing One-Stage Gradings--Lotze,A.
03-029 Probability of Loss in Multi-Stage Selection on any Number of Stages (In French)--Martinez,R.
03-030 American and European Traffic Capacity Tables and Practices--Mina,R.R.
03-031A Electronic Traffic Measurement and Recording--Clement,M.A., Mina,R.R.
03-032 Bibliography Relating to the Application of Probability Theory to Telephone Systems and Similar Problems--Morris,N.H.G.
03-033 The Choice of Grade of Service Standards--Newstead,I.A.
03-034 Extension of Telephone Plant with Regard to the Value of Subscribers' Time--Rapp,Y.
03-035 On the Use of Markov Dependent Random Variables in Waiting-Time Theory--Runnenburg,J.Th.
03-036 More Accurate Calculation of Overflow Traffic from Link-Trunked Crossbar Group Selectors--Smith,N.M.H.
03-037 Estimation of Confidence Intervals for Telephone Traffic Parameters--Stormer,H.
03-039 Many Server Queues with Recurrent Input and Exponential Service Times--Takacs,L.
03-040 A Simplified Type of Gradings with Skipping--Einarsson,K.-A., Hakansson,L., Lundgren,E.
03-041A Alternative Routing in a Two Stage Link System, Congestion Theory and Simulated Traffic Studies--Wallstrom,B.
03-041B Some Traffic Characteristics of Communications Networks with Automatic Alternative Routing--Weber,J.H.
03-042 The Effective Accessibility as an Estimation of the Quality of the Mixing (In French)--Wendt,A.
03-043 An Exploration of Some Methods for Administering Final Groups in Alternate Route Systems--Wilkinson,R.I.
03-044 The Studies and Conclusions of the Working Party Set Up by the CCITT to Investigate More Accurate and Economical Methods of Traffic Determination for International Circuits--Wright,E.P.G.
03-045 Comments on Inter-Stage Correlation in Link Systems and on Communication No. 13 of Dr. Elldin--LeGall,P.
03-046 On the Condition of the Independence of the Steady-State Distribution from the Holding Time Distribution in Serving Systems --Kovalenko,I.

FOURTH INTERNATIONAL TELETRAFFIC CONGRESS, LONDON, UK, JULY 15-21, 1964

04-011 Some Aspects of Routing and Control in Communications Systems--Weber,J.H.
04-012 The Variability of the Proportion of Unsuccessful Attempts in Loss Systems--Descloux,A.
04-013 The Geometric Group Concept and its Application to the Dimensioning of Link Access Systems--Bridgford,J.N.
04-020 Study of the Congestion in a Loss System--Fortet,R.M., Grandjean,C.
04-021 Some Recursive Aids for Switching Network Blocking Computations--Van Bosse,J.G.
04-022 Modern Concepts Concerning Single-Stage and Two-Stage Switching Arrays as Derived from Exact Loss Calculations --Bretschneider,G.
04-023 Trunking and Traffic Principles of a PCM Telephone Exchange--Duerdoth,W.T., Walker,N.
04-024 Traffic Engineering of Communication Networks with a General Class of Routing Schemes--Segal,M.
04-026 On the Accuracy of Results from Certain Types of Counters Used for Traffic Supervising Purposes--Olsson,K.M.
04-031 Heavy Traffic Approximation in the Theory of Queues--Kingman,J.F.C.
04-032 A Note on Queueing Processes with Group Arrival--Cohen,D.
04-033 From Ballot Theorems to the Theory of Queues--Takacs,L.
04-034 The Derivatives of Erlang's B Formula--Akimaru,H., Nishimura,T.
04-035 A Measure of System Utility--Saporta,L.J., Toralballa,L.V.
04-036 A Practical Method of Calculation for Certain Types of Complex Common Control Systems--Bjorklund,M., Elldin,A.
04-037 Generalized Congestion Process--Syski,R.
04-040 Planning of Junction Network in a Multi-Exchange Area--Rapp,Y.
04-041 The General Plan for a Multi-Exchange Area--Rapp,Y.
04-042 The Reliability of Exchanges with Common Control--Stormer,H.
04-043 Some Economic Aspects of the Planning of Local Telephone Networks--DeBoer,J., Brouwer,J.F., DeFremery,F.
04-044 Network Planning Investigations Using an Electronic Computer--Freeman,A.H.
04-045 Reliability of Lattice-Interconnected Networks--Bryan,D.G, Knight,R.G.
04-046 Investigation of Subscribers' Complaints in some Exchange Areas of a Telephone Administration--Christiansen,U.
04-047 On a Class of Markov Processes Applicable to Control of Service Quality in Telephone Systems--Lind,G.
04-048 Telephone Prognostication from a Stochastic and a Structural Viewpoint--Matthiesson,P.C.
04-050 Digital Computer Simulation of Switching Systems Using the Algol Automatic Programming Language--Cole,A.C., Thomson,W.E., Williams,J.W.J.
04-051 Computer Simulation of a Private Wire Telephone Network--Shea,A.
04-052 Traffic Simulation According to Time-True Model - Methods and Results--Wagner,W.
04-053 Comparative Studies of Congestion Values Obtained in Grading Tested in Swedish Telephone Traffic Machine--Taange,I., Wikell,G.
04-054 Traffic Simulation Studies Using an Electronic Computer--Rubas,J.
04-055 Artificial Traffic Studies on Crossbar Link Systems Using a Digital Computer--Suzuki,T., Takagi,K.
04-056 Simulation Studies Respecting Delays with Common Signaling Channels--Lawn,M.
04-057 Application of Digital Simulation to Traffic Engineering in Large Communication Networks--Dujack,R.L., Weber,R.A.
04-070 A Method of Reducing the Probability of Blocking in a Three-Stage Link Connecting Network--Pollen,L.K.
04-071 The Distribution of Overflow Calls when Pure Chance Traffic is Offered to a Single Trunk--Harmston,A.T.
04-072 A Study of the Arrival Times of Calls During the Busy Hour--Harmston,A.T.
04-073 A Distribution Model for Telephone Traffic with Varying Call Intensity Including Overflow Traffic--Wallstrom,B.
04-074 Practical Application of Congestion Theory in Telephone Systems--Best,C.F.
04-075 Simplified Engineering of Single-Stage Alternative Routing Systems--Wilkinson,R.I.
04-076 A Solution to the Problem of Smooth Traffic--Mina,R.R.
04-077 Study of a Problem Concerning "Entraide" Traffic--Chastang,J.F., Grandjean,C.
04-080 A Traffic Variance Method for Gradings of Arbitrary Type--Lotze,A.
04-081 Traffic Design and Engineering of the Switching Network for the No.1 ESS--Hayward,W.S.,Jr.
04-083 Optimum Design of Switching Systems--Akimaru,H.
04-084 Trunking and Traffic Theory of the 5005 Crossbar System--Morris,N.H.G., Smith,J.L., Smith,J.W.Jr.

04-085 Certain Types of Cyclical Gradings for Random Hunting--Valenzuela,J.G.
04-086 A General Criterion to Evaluate the Quality of a Grading - Study of a Family of Gradings--Martinez,R.
04-090 Non-Parametric Aspects on the Grade of Service of Telephone Traffic--Molnar,I.
04-091 Duration of the Congestion State in Local Telephone Traffic--Ahlstedt,B.W.M.
04-092 Conclusions and Recommendations of CCITT XIII/5--Wright,E.P.G.
04-093 Rational Dimensioning of Telephone Routes--Karlsson,S.A.
04-094 Interconnection Systems--Karlsson,S.A.
04-095 The Traffic Variations and the Definition of the Grade of Service--LeGall,P.
04-100 Erlang Loss Tables for Variable Input Traffic--Smith,N.M.H.
04-103 Internal Traffic and its Effect on Congestion in Switching Systems--Botsch,D.
04-104 On the Congestion in TDM Systems--Huber,M.

FIFTH INTERNATIONAL TELETRAFFIC CONGRESS, NEW YORK, USA, JUNE 14-20, 1967

05-021 A Study of Traffic Variations and a Comparison of Post-Selected and Time-Consistent Measurements of Traffic--Povey,J.A.
05-022 The Dimensioning of Telephone Traffic Routes for Measured Integrated Peak Traffic--Karlsson,S.A.
05-023 Traffic Considerations for Demand Assigned Telephone Service Through a Satellite System--Anguera,F.
05-031 Dimensioning for the Dynamic Properties of Telephone Traffic--Elldin,A.
05-032 Reflections on Conditional Selection and Different Routing Methods (In French)--LeGall,P.L
05-033 The Concept of Marginal Overflow in Alternate Routing--Pratt,C.W.
05-041 Review of Recently Obtained Results in Queueing Theory--Cohen,J.W.
05-042 Poisson Service Systems with Priorities and Limited Waiting Capacity--Basharin,G.P.
05-043 On Combined Delay and Loss Systems with Non-Preemptive Priority Service--Wagner,W.
05-044 Priority Queues with Recurrent Input--Chang,W.
05-051 International Standardizing of Loss Formulas--Botsch,D.
05-052 Estimates of the Delay Distributions for Queueing Systems with Constant Server Holding Time and Random Order of Service --Durnan,C.J.
05-053 Variations of the Erlang, Engset and Jacobaeus Formulas--Joys,L.A.
05-054 Bernoulli Law Applied to Limited Groups of Inlets and Outlets--Kruithof,J.
05-055 New Graphic Methods for Practical Use of the Congestion Theory in Telephone Systems--Oettl,K.
05-056 Numerical Methods in the Use of Computers for Telephone Traffic Theory Applications--Szybicki,E.
05-057 Numerical Inversion of LaPlace Transforms with Applications to Queueing Theory--Wiener,S.
05-061 History and Development of Grading Theory--Lotze,A.
05-062 Exact Loss Calculations of Gradings--Bretschneider,G.
05-071 Some Further Studies on Optimal Grading Design--Hakansson,L.
05-072 Blocking in Practical Gradings with Random Hunting--Ward,P.W.
05-073 Measuring Interconnections with the Aid of a Traffic Machine--Rahko,K.
05-074 Simulation Methods on Telephone Gradings: Analysis of Macro-Structures and Micro-Structures--Cappetti,I.
05-075 Delay Systems with Limited Accessibility--Theirer,M.H.
05-082 On Automatic Traffic Control in Telephone Networks--Schwartzel,H.
05-083 Design of Service Systems with Priority Reservation--Faulhaber,G.R., Dunkl,P.
05-084 Automatic Traffic Overload Control in an Electronic Switching System--Hoover,E.S., Eckhart,B.J.
05-091 Number of Gates Required for Time Division Switching--Clos,C.
05-092 Non-Blocking and Nearly Non-Blocking Multi-Stage Switching Arrays--Kappel,J.G.
05-093 Optimum Link Systems--Lotze,A.
05-094 Design Tables for Optimum Non-Blocking Crosspoint Networks--Pollen,L.K.
05-101 Call Routing Strategies in Telecommunication Networks--Grandjean,C.
05-102 Comparison of Routing Methods in Small Link Systems by Simulation--DeBoer,J.
05-103 Combinatorial Solution to the Problem of Optimal Routing in Progressive Gradings--Benes,V.E.
05-111 Applications of Characteristic Functionals in Traffic Theory--Fortet,R.M.
05-112 Some Feedback Queueing Models for Time-Shared Systems--Coffman,E.G., Kleinrock,L.
05-113 Waiting-Times in Multi-Stage Delay Systems--Stormer,H.
05-114 On the Validity of the Particular Subscriber's Point of View--Descloux,A.
05-115 Waiting-Time Distributions for Networks of Delay Systems--Lee,L.
05-121 Recursive and Iterative Formulas for the Calculation of Losses in Link Systems of any Description--Bininda,N., Daisenberger,G.
05-122 An Analysis of Loss Approximations for Link Systems--Kummerle,K.
05-123 Two-Stage, Three-Stage, and Four-Stage Link Systems - A Comparison between Congestion results Obtained from Calculation and Simulation--Lundgren,E.
05-131 Artificial Traffic Studies on a Two-Stage Link System with Waiting--Gambe,E., Suzuki,T., Itoh,M.
05-132 Traffic Capacity of Two-Stage Switching Array with Graded Interstage Links--Rubas,J.
05-133 Entraide Networks Combined with Link Systems--Canceill,B., Gutierrez,D.
05-141 Optimal Design of Alternate Routing Systems--Schehrer,R.
05-142 Survey of "Congestions Studies in Telephone Systems with Alternate Routing"--Wallstrom,B.
05-143 A General Variance Theory Applied to Link Systems with Alternate Routing--Herzog,U.
05-144 A New Danish Traffic Routing Plan with Single Stage Alternate Routing and Consistent use of Service Protection Finals for First Routed Traffic to Tandem Offices--Asgersen,C.
05-145 Determination of Trunk Requirements, Alternate Route Networks--Blessum,R.B.
05-146 Some Practical Applications of Teletraffic Theory--Mina,R.R.
05-151 On the Accuracy of Measurements of Waiting-Times in the Single Server System with Arbitrary Distribution of Holding Times --Kosten,L., Ten-Brooke,A.M.
05-152 Alternating Renewal Processes, with Applications to some Single-Server Problems--Linhart,P.
05-153 A Problem Concerning the Accuracy of Traffic Measurements--Lind,G.
05-154 Some Different Methods in Using Markov Chains in Discrete Time Applicable to Traffic Simulations and Certain Accuracy Problems in this Context--Olsson,K.M.
05-161 A Study of Interconnecting Problems in a Unitized Link System--Burgess,J.A., Fisher,D.G.
05-162 Trunk Systems for PCM Exchanges--Duerdoth,W.T., Hughes,C.J., Bond,D.J.
05-163 TDM Link Systems with By-Paths for Non-Coincident Switching--Huber,M.
05-171 The Character and Effect of Customer Retrials in Intertoll Circuit Operation--Wilkinson,R.I., Radnik,R.C.
05-172 On the Consideration of Waiting-Times of Calls Departed from the Queue of a Queueing System--Wikell,G.
05-173 Customer Actions where Calls are Unsuccessful--Wikell,G.
05-181 Trunking and Traffic Aspects of the AEI No.18 REX System--Warman,J.B., Bear,D.
05-182 Traffic Combination in a General Purpose Switching Unit--Harland,G.
05-191 Load Variance and Blocking in Automatic Telephone Traffic. - Load Duration ,a Basis for Dimensioning. - Blocking Duration, A Yardstick of Traffic Quality--Ahlstedt,B.W.M.
05-192 Small Groups of Subscribers Traffic Unbalance--Rodriguez,A., Dartois,J.P.
05-193 The Balancing of Loads among Inputs to Loss and Delay Networks--Helly,W.
05-194 Probability Function and Trend of an Observed Series of Busy Hour Traffic Values--Carli,M.

05-201 Message Switching Traffic Formulas for Engineers--Robins,J.M.
05-202 Modular Engineering of Trunk Groups for Traffic Requirements--Levine,S.W., Wernander,M.A.
05-211 Statistical Performance Analysis of a Switched Communications Network--Katz,S.S.
05-212 Optimum Routing of Traffic in a Network of Local and Tandem Exchanges--Breary,D., Thomson,W.E.
05-213 Planning of Junction Network in a Multi-Exchange Area III. Optimum Types of Physical and Carrier Circuits--Rapp,Y.
05-214 Computer Techniques for Local Network Engineering--Gimpelson,L.A., Hinderliter,R.G.

SIXTH INTERNATIONAL TELETRAFFIC CONGRESS, MUNICH, FRG, SEPTEMBER 9-15, 1970

06-121 Traffic Engineering in the CCITT No.6 Signaling System--Wright,E.P.G.
06-131 Application of Traffic Theory in Industry and Administration--Jacobaeus,C.
06-132 Planning of Junction Network with Non-Coincident Busy Hours--Rapp,Y.
06-133 A Self-Optimising Network Model for the Long Term Planning of the UK Trunk Telecommunications Network--Breary,D.
06-134 Manual Service Criteria--Wikell,G.
06-135 On the Quality of Telephone Service in an Automatic Network and its Economical Expression--De Ferra,P., Masetti,G.
06-m136 Opportunities Provided by the Introduction of Decentralized Switching Stages into a Large Subscribers Network--Cappetti,I.
06-141 The Overflow Distribution for Constant Holding Time--Burke,P.J.
06-142 Trunk Engineering of Non-Hierarchical Networks--Katz,S.S.
06-143 Adaptive Routing Technique and Simulation of Communication Networks--Butrimenko,A.V.
06-144 A Traffic Routing Strategy Designed for Overload Protection Downwards the Hierarchy--Asgersen,C.
06-145 A Method for the Optimization of Telephone Trunking Networks with Alternate Routing--Caballero,P.A.
06-146 Network Dimensioning with Alternative Routing in a Multi-Exchange Area of a City--Arhnung,H.
06-147 On the Exact Calculation of Overflow Systems--Schehrer,R.
06-148 Program for Optimizing Networks with Alternative Routing--Valenzuela,J.G.
06-211 Analysis of Congestion in Small PABX's--Rubas,J.
06-212 Grade of Service for Full Available Trunk Groups with Faulty Trunks--Kilmontgowicz,A.
06-213 Comparative Evaluation of Gradings with the Aid of a Simulation Method--Miranda,G.
06-214 Engset Losses in a Full Availability Multiple Sequential Hunting, Improvement Function and Last Contact Traffic--Joys,L.A.
06-215 Lost Call Cleared Systems with Unbalanced Traffic Sources--Dartois,J.P.
06-216 Calculating Method for Call Congestion of Gradings for Random Routing whose Traffic has a Poisson Distribution--Jung,M.M.
06-217 Calculation of Two-Way Trunk Arrangements with Different Types of Traffic Input--Herzog,U.
06-221 Traffic in Connecting Networks when Existing Calls are Rearranged--Benes,V.E.
06-222 The Processes of Mutation: A General Model for Traffic Problems with or without Delay--Fortet,R.M.
06-223 Generalization of Kovalenko Theorem on the Invariance of Probability States of Service System with Respect to Service Time Distribution--Guseinov,B.T.
06-224 Loss Systems with Displacing Priorities--Katzschner,L.
06-225 Determination of the Powers of any given Matrix and Application on Discrete Markov Processes--Gatos,E., Keiser,F.
06-226 On Positive Recurrence for Denumerable Markov Chains--Delbrouck,L.E.N.
06-227 On Some Unresolved Problems in Mass Service (Queueing) Theory (In Russian)--Gnedenko,B.W.
06-231 On Analytical and Numerical Methods of Switching System Investigation--Basharin,G.P.
06-232 Application of Link Traffic Statistics to Estimating Link System Congestion--Pedersen,O.A.
06-233 About Multi-Stage Link Systems with Queueing--Hieber,L.J.
06-234 Traffic Problems in the Switching Network of Space Division Switching Systems of the PRX Type--DeBoer,J.
06-236 Exact Calculation of the Probability of Loss for Two-Stage Link Systems with Preselection and Group Selection--Lorcher,W.
06-241 The Number of Reswitching for a Three-Stage Connecting Network--Bassalygo,L.A., Grushko,I.I., Neiman,V.I.
06-242 Comparison between Measured and Calculated Values in a Link System with Overflow Facility--Fakhr El Din,M.E.
06-243 A Method to Formulate Blocking Functions of Network Graphs--Villar,J.E., Fontana,B.
06-244 Call Congestion in Link Systems with Internal and External Traffic--Bazlen,D.
06-245 The Structures of One-Sided Connecting Networks--Bassalygo,L.A., Grushko,I.I., Neiman,V.I.
06-311 Networks Containing Full Availability Groups and Ideal Grading Groups; Rearrangeable Link Networks (In Russian)--Lifshitz,B.S., Fidlin,Ya.V.
06-312 Generalized Priority Service and its Application in Analysis of One-Channel Data Transmission Systems--Dkhovny,I.M., Pankratov,V.I.
06-313 Analysis of a Non-Preemptive Priority Queueing System with Setup Times--Nakamura,G., Hashida,O.
06-314 A Simple Method of Calculating Queue Size Distributions of Priority Queueing Systems with Feedback--Enns,E.G.
06-315 One-Server Priority System with Arbitrary Service Time and Limited Waiting Room--Bocarov,P.P.
06-316 A Multi-Server Queueing System with Preemptive Priority--Brandt,J.G.
06-317 On a Multi-Server Queueing System with Constant Holding Time and Priorities--Langenbach-Belz,M.
06-321 On Memory Units with Finite Capacity--Cohen,J.W.
06-322 Delay Systems with Limited Availability and Constant Holding Time--Theirer,M.H.
06-323 Combined Delay and Loss Systems with Several Input Queues, Full and Limited Accessibility--Kuhn,P.
06-324 General Queueing Processes Treated by Means of a Time-Sharing Computer--Boettinger,R., Wallner,K.
06-326 A Single-Server Queueing System in which Service Time Depends upon Queue Size--Chang,W.
06-331 On Markovian Servers with Recurrent Input--Descloux,A.
06-332 A Multi-Server System with a Finite Number of Sources and Delayed Requests Served at Random--Segal,M.
06-333 The Queue as a Stopped Random Walk--Johnson,E.M.
06-334 Waiting-Time Prediction--Syski,R.
06-335 A Group of Servers Dealing with Queueing and Non-Queueing Customers--Pratt,C.W.
06-336 Study of a M/G/1 Delay Problem with Two Service Nodes--Van Bosse,J.G.
06-337 Queueing Systems with Priorities and Intervals of Saturation--Wagner,W.
06-341 Multi-Level Processor-Sharing Queueing Models for Time-Shared Systems--Kleinrock,L., Muntz,R.R.
06-342 Telephone Information Systems and Associated Traffic Theory Problems--Schwartzel,H.
06-343 Some Considerations about the Construction of a Communication Network--Kharkevich,A.D.
06-344 Optimal Dimensioning of a Satellite Network using Alternate Routing Concepts--Casey,J., Shimasaki,N.
06-345 Traffic Collection and Calculation in an International Telex Operation--Fleming,P.,Jr.
06-346 Computer Aided Traffic Engineering of the C-1 EAX Exchange using a Time-Shared Computer--Augustus,J.
06-347 Modification of the Input Process used in Classical Traffic Theory--Mina,R.R.
06-348 Choosing Proper Combination of Transmission Facilities in Non-Uniform Communication Networks (In German)--Oettl,K.
06-411 Simulation in Traffic Theory--Kosten,L.
06-412 Method for the Simulation of a Switching System having Non-Uniform Service Characteristics--Ret,M.
06-413 Traffic Model for the Simulation of Entire Exchanges--Dietrich,G.
06-414 An Approach of Real Feed-Back Type Telephone Traffic and its Effect on the Characteristics of Queueing Systems--Gosztony,G.
06-416 A Roulette Model for Simulation of Delay-Loss Systems--Rodriguez,A., De Los Mozos,J.R.
06-417 Accuracy in Measurements of the Mean of a Markov Process, Arbitrary Distribution of Intervals between Observations and Fixed Number of Observations or Fixed Measuring Period--Lind,G.
06-421 A Method of Estimating Dial Tone Delay in Telephone Switching Common Control Systems--Chen,R.W., Lee,L.
06-422 Comparison of Stratified and Full Simulation for a Large Telephone Exchange--Jolly,J.H.

06-423 Cyclic Queue with Changeover Times--Eisenberg,M.
06-424 Traffic Design and Engineering of Central Processing Systems in an Electronic Switching System--Suzuki,T., Nunotani,Y., Kaneda,O.
06-425 A GPSS-Simulation for the Study of the Capacity of a SPC Exchange Computer--Hakansson,L.
06-426 A Model for Estimating the Real-Time Capacity of Certain Classes of Central Processors--Farber,N.
06-427 Simulation of the Data Flow in Common Control Switching Systems--Hanglberger,P.
06-431 Human Factors in Telephone Systems and Their Influence on Traffic Theory, Especially with Regard to Future Facilities --Hayward,W.S.,Jr., Wilkinson,R.I.
06-432 On the Influence of Call Repetitions in Telephone Traffic Service--LeGall,P.
06-433 The Effect of Repeated Calls on Communication Systems--Shneps-Sneppe,M.
06-434 Repeated Calls with Limited Repetition Probability--Bretschneider,G.
06-435 Telephone Systems with Repeated Calls--Jonin,G.L., Sedol,J.J.
06-436 The Solution of the Repeated Call Problem - Full Availability Group Serving Limited Number of Sources(In Russian)--Fidlin,Ya.V.
06-441 Outlook to PCM Switching Systems and Their Calculation by Traffic Theory--Inose,H., Saito,T.
06-442 Calculation of the Blocking Probability in the Speech Path Network of the PDX Time Division Switching System by Trunk Exchanges --Jung,M.M.
06-443 Non-Blocking Switching in Integrated PCM Networks--Tanaka,M.
06-444 Traffic Aspects of Overall Control of a PCM Network--Bear,D.
06-511 What does ITC Expect from CCITT--Jensen,A.
06-512 What the CCITT Expects of Traffic Theory--Newstead,I.A.
06-521 Some Comparisons of Load and Loss Data with Current Traffic Theory--Wilkinson,R.I.
06-522 Some Grade of Service Problems--Longley,H.A.
06-523 Application of Traffic Forecasting Techniques to Circuit Provisioning--Leigh,R.B.
06-524 Analytical Models for Forecasting the Demand for Telephone Connections--Boehm,Erich
06-525 Parameters for the Evaluation of Extension Policies for Local Central Offices--Mossotto,C.
06-526 Recent Adaptive Trends in Telecommunications Traffic Engineering--Molnar,I.
06-527 Continual Checking of Route Loading in Telephone Traffic--Harva,M.
06-531 On the Extent of Routine Measurements in Telephone Plant. An Operational Research Approach--Elldin,A.
06-532 A Study of the Traffic Process Based on Measurements--Rahko,K.

SEVENTH INTERNATIONAL TELETRAFFIC CONGRESS, STOCKHOLM, SWEDEN, JUNE 12-20, 1973

07-131 Customer Line Use Characteristics--Barnes,D.H., Reutelhuber,W.O.
07-132 A Statistical Study of Telephone Traffic Data, with Emphasis on Subscriber Behavior--Myskja,A., Walmann,O.O.
07-133 An Analytical Model to Describe the Influence of the Repeated Call Attempts--Toledano,F., De Los Mozos,J.R.
07-134 On the Utilization and Observation of Telephone Traffic Completion Rate--LeGall,P.
07-135 Functional Algorithms of Automatic Telephone Systems and Automatic Answering Sets--Halatschew,W.II., Dimtschew,M.Iw.
07-136 On the Behavior of the Telephone Subscriber in the case of Blocking with Heavy Traffic (In French)--Pellieux,G.
07-137 Full-Availability Groups With Repeated Calls and Time of Advanced Service--Jonin,G.L., Sedol,J.J.
07-138 Extensive Traffic Investigations in a Local Telephone Exchange--Gaustad,O., Flo,A., Dadswell,R.L.
07-141 Statistical Problems in Teletraffic Theory--Gnedenko,B.W., Schneps-Schneppe,M.A.
07-142 Fluctuations of Load Measurements--Descloux,A.
07-143 A General Principle of Traffic Distribution--Solem,R.R
07-144 Test Calling: Experience, Theory, Prospects--Oberer,E., Riesz,G.W.
07-145 Measurement of Within-Busy-Hour Traffic Variations--Pedersen,O.A.
07-146 Peak-Load Traffic Administration of a Rural Line Concentrator--Gershwin,S.B., Laue,R.V., Wolman,E.
07-211 Tetrasim - A General Purpose Program for the Simulation of Telephone Systems--Rogenberg,T., Stenseth,B.
07-212 Simulation of the Group Selector Stage in the "ITT-8B" Exchange--Husstad,B., Thores,T.
07-213 A Simulation Study of a Network with Six Exchanges--Halvorsen,O., Kvase,H.
07-214 Some Methods for Improving the Efficiency of Simulation Programs--Nivert,K., Von Schantz,C.
07-215 Theoretical Aspects of Finite Source Input Process Simulation--Weisschuh,H.
07-216 An Extension of the Roulette Model to Simulate Speech and Signalling Networks--Villar,J.E., De Los Mozos,J.R.
07-221 Random Process Applied to Traffic Theory and Engineering--LeGall,P.
07-222 The Unserviceable Probability of an Interoffice Trunk Group--Lee,L.
07-223 On the Calculation of Full Available Groups with Offered Smoothed Traffic--Bachle,A.
07-224 Analysis of Traffic Processes Based on Data Obtained by the Scanning Method--Iverson,V.B.
07-225 Congestion as a Stochastic Process--Rahko,K.
07-226 Uncertainty Theory Applicable to Traffic Models--Mizuki,M.
07-227 Relations Between Redundance and Reliability in the Transmission and Processing of Information in Switching and Transmission Systems--Gatos,E.
07-231 On a Number of Reswitching in a Three-Stage Connecting Network--Bassalygo,L.A.
07-232 On the Influence of Hunting Mode and Link Wiring on the Loss of Link Systems--Bazlen,D., Kampe,G., Lotze,A.
07-233 A Comparison Between Classical Calculated and Artificially Measured Distributions in a Particular Link System--Michaux,J., DeRaedt,L.J., Jacquemijn,E.V.
07-234 Design of Efficient Link Systems with Concentration Stages--Kappel,J.G.
07-235 Comparison of Random Selection and Selection with a Fixed Starting Position in a Multi-Stage Link Network--DeBoer,J.
07-236 Inherent Load Balance in Link Systems with Random Slip Multiples--Johnson,E.M.
07-237 Congestion Studies in an Incomplete Link System--Loken,B.
07-238 Determination of Blocking Situations in the Three-Stage Connection Systems with Limited Number of Arrangements--Ershov,V.A., Lyashko,O.E.
07-241 Call Congestion and Distribution of Number of Engaged Outlets in a Given Split of Gradings with Random Routing--Jung,M.M.
07-242 Waiting-Time Distributions in Multi-Queue Delay Systems with Gradings--Kuhn,P.
07-243 Calculation of Simple O'Dell Gradings Using an Erlang Type Loss Formula --Joys,L.A.
07-244 An Iterative Method to Calculate Traffic Losses in Grading Systems--Skogan,E.
07-245 A Numerical Solution Method of Simplified Equations of State and Its Application to Grading Groups--Takagi,K.
07-246 An Investigation of the Effect of Alternative Hunting Rules in a Group of Code Receivers with Limited Availability--Kvaase,H., Lochen,R.C.
07-311 Semilattice Characterization of Non-Blocking Networks--Benes,V.E.
07-312 Nodal Switching Networks--Joel,A.E.Jr.
07-313 The Structure of Multi-Group Connecting Networks--Melik-Gaikazova,E.I.
07-314 Some Binary Connecting Networks and Their Traffic Characteristics--Neiman,V.I.
07-315 Bounds for the Growth of the Number of Contacts in Connecting Networks--Marcus,M.J.
07-316 Link Systems with Both-Way Connections and Outgoing Finite Source Traffic--Bazlen,D.
07-317 Point-to-Point Selection in Link Systems--Lorcher,W.
07-318 On the Complexity of a Concentrator--Pinsker,M.S.
07-321 Some Aspects of Queueing Theory--Cohen,J.W.
07-322 Total Waiting Time Distribution Function and the Fate of a Customer in a System with Two Queues in Series--Kramer,W.

07-323 Queueing Models for Buffer Memory in Store-and-Foreward Systems--Hashida,O., Fujiki,M.
07-324 Investigation of Queueing and Related Systems with the Phase Method--Schassberger,R.A.
07-325 An Approximate Model and Analysis for a Queue with Highly Clustered Arrivals--Rudin,H.,Jr.
07-326 A Teletraffic Problem Arising from the Design of Large Message Switches--Robins,J.M.
07-331 Telephone Traffic Model for Common Control Investigations--Dietrich,G.
07-332 Revision of the Busy Hour Concept in the Calculation of Common Control Telephone Exchanges--Nieto Nieto,A.M., Parra,J.
07-333 A Class of Common Control Circuits Systems: Single-Stage and Two-Stage Arrangements with Waiting--Bonatti,M., Kubanski,P.G., Popoli,P.
07-334 A Calculation Mewthod to Study in a Full Availability Group the Influence of Time-Outs--Rodriguez-Rodriguez,A., Fontana,B., Die-Fagoaga,G.
07-335 Common Control Performance of Switching Systems in Hierarchical Networks--Corsi,P., Ricciarelli,R., Casalotti,O.
07-336 Batch Switching Systems: Structures and Procedures--Garmash,V.A.
07-341 Full Availability One-Way and Both-Way Trunk Groups with Delay and Loss Type Traffic, Finite Number of Traffic Sources, and Limited Queue Length--Gosztony,G.
07-342 Dimensioning of Manual Operating Positions in PABX's--Rubas,J.
07-343 Computation of a Starlike Switching Network with a Large Number of Radial Rays--Dobrushin,R.L., Sukhov,Y.M.
07-344 A Queueing Problem in Data Transmission--Wollner,E.
07-345 Delay and Loss System with Early Departure Calls--Urmoneit,W.
07-346 Calculation of the Traffic Capacity of a Private Data Network--Fakhr El Din,M.E., Peiram,L.
07-411 Extension of the Equivalent Random Method to Smooth Traffic--Bretschneider,G.
07-412 Loss systems with Mixed Renewal and Poisson Inputs--Kuczura,A.
07-413 Service Protection for Direct Final Traffic in DDD Networks--Katzschner,L.
07-414 The Accuracy of the Equivalent Random Method with Renewal Inputs--Holtzman,J.M.
07-415 Message Switching Networks with Alternate Routing--Herzog,U.
07-416 On Arrival Processes for Overflow Calls--Jensen,E.
07-417 Loss Calculations in Certain Overflow Systems where the Holding Times in Successive Groups have Different Means--Wallstrom,B.
07-421 Optimization of Networks of Hierarchical Structure with Non-Coincident Busy Hours--Caballero,P.A., Diaz Lopez,F.
07-422 Calculation of the Optimum Number of Trunk Lines Based on Moe's Principle--Parviala,A.
07-423 Approximate Method for Determination of Overload Ability in Local Telephone Systems--Szybicki,E.
07-424 Optimization of Exchange Locations and Boundaries in Local Telephone Networks--Anderberg,M., Fried,T., Rudberg,A.
07-425 Network Efficiency and Network Planning Considering Telecommunication Traffic Influenced by Time Difference--Ohta,T.
07-426 Flexible Routing in the Global Communication Network--Nakagome,U., Mori,H.
07-427 Concepts of Optimality in Alternate Routing Networks--Harric,R.J.
07-431 A Continuum of Computer Processor-Sharing Queueing Models--Kleinrock,L., Hsu,J.
07-432 Analysis of Storage Hierarchy--Cohen,J.W.
07-433 Serving Strategies of Batch Arrivals in Common Control Switching Systems--Schwartzel,H.
07-434 Two-Stage Queueing System with Sampled Parallel Input Queues--Langenbach-Belz,M.
07-436 Control of Central Processor Load in a SPC System--Karlander,B.
07-437 An Approximate Calculation Method to Estimate Processor Capacity in SPC Switching Systems--Villar,J.E., Soto,O.G., Garcia-Higuera,J.A.
07-438 The Busy Period and the Probability that a Particular Queue is in Service in a Single Server Queueing System with N Queues, Arbitrary Priorities, and General Probablistic Interqueue Transition Matrix--Enns,E.G., Wignall,T.K.
07-441 Impact ot the Growth in Data Transmission on Network Planning--Kelly,P.T.F.
07-442 Traffic Calculations in Saturation Routing with Priorities--Grandjean,C.
07-443 Traffic Characteristics of the International Telex Calls--Hattori,N., Yamada,K.
07-444 Performance of Traffic Concentrators in Data Communication Systems--Fredrikson,G.
07-445 On One of the Approaches to the Synthesis of Structures on Information Exchange Systems--Pilschikov,E., Sharipova,M.M.
07-446 Simulation Experiments with GASP-II--Espvik,O.
07-511 On the Dependence between Traffic and Tariff--Elldin,A.
07-512 Computer Aided Forecasting for Point-to-Point Traffic Offered to an Automatic Toll Network--Masetti,G., Miranda,G.
07-514 Long-Range Planning of Hierarchical Trunk Networks with Rearrangement Costs--Katz,S.S.
07-515 Computing Aspects of a Traffic Prediction Model of a Telephone Exchange Network--Taylor,A., Batty,M.G.
07-516 Local Network Planning Computerized: Forecasting the Number of Subscriber Lines in Any Building as a Function of its Characteristics, Using Correlation Analysis--Parviala,A., Nurminen,M., Heiskanen,L.
07-521 DDD Network Optimization in Field Engineering - From Theory to Application--Lotze,A.
07-522 Standards for the Economic Provision of Direct Routes--Leigh,R.B.
07-523 Dimensioning of Trunk Networks According to a Uniform Overload Capacity Criterion--Asgersen,C.
07-524 Some Results on Simulation of Communication Network with the Usual and Constant Alternate Routing--Kharkevich,A.D.
07-525 A System for Telephone Traffic Measurement and Routing Analysis by Computer--Kyme,R.C
07-526 On Design Criteria for a Telephone Toll Network with Respect to Overload Protection--Miranda,G., Pallotta,P.
07-531 Some Theories of Telephone Traffic Distribution: A Critical Survey--Bear,D.
07-532 A Long Term Study of the United Kingdom Trunk Network--Breary,D., Gates,J.F.
07-533 The Operator Scheduling Problem - A Network Flow Approach--Segal,M.
07-534 Satellite Communication Traffic Demand Assignment Experiments--Pecchinotti,A.
07-536 A Traffic Prediction Model for a Telephone Exchange Network--Bear,D., Seymour,C.A.
07-537 Reliability Prediction in Telephone Systems Engineering--Strandberg,K.
07-541 Economic Considerations in the Development of Traffic Data Gathering Programs--Birch,L.A., Cole,J.B.
07-542 A Sensitivity Study of Traffic-Parameter Estimation Procedures Used for Engineering Trunk Groups--Berger,M., Neal,S.R.
07-543 VEMAUS - A General Purpose Program System for Editing and Evaluating and Analyzing the Results of Traffic Load Measurements --Daisenberger,G.
07-544 Measurement of Subscriber Reaction to Unsuccessful Call Attempts and the Influence of Reasons of Failure--Evers,R.
07-611 On Traffic Engineering Studies in the CCITT--Pratt,C.W., Taange,I.
07-612 Research - Administration - Development - Planning. Problems related to the Cooperation between ITC and CCITT--Jensen,A.
07-613 An Onlooker's Reflections on Recent Grade of Service Deliberations by the International Telegraph and Telephone Consultative Committee (CCITT)--Molnar,I.
07-621 Network Management in the United States - Problems and Progress--Macurdy,W.B.
07-622 An Approach to the Network Management--Lazarev,W.G.
07-623 Traffic Routing and Control in a Network--Rahko,K., Erke,T.
07-624 Network Management Strategies--Wean,O.
07-631 Evaluation of PCM Toll Switching Networks with Partial Access Pulse Shifters--Inose,H., Saito,T., Yanagisawa,Y.
07-632 A Quasi-Non-Blocking TDM Switch--Duerdoth,W.T., Seymour,C.A.
07-634 On a Queueing Problem in TDMA Networks--Lajtha,G., Mazgon,A.
07-635 An Analysis of Mixtures of Multiple Bandwidth Traffic on Time Division Switching Networks--Enomoto,O., Miyamoto,H.
07-636 Trunking Studies for a TDM Terminal Exchange--Manera,L., Pellegrini,R.

EIGHTH INTERNATIONAL TELETRAFFIC CONGRESS, MELBOURNE, AUSTRALIA, NOVEMBER 10-17, 1976

08-121 "E.P.G." - A Tribute to E.P.G. Wright--Gimpelson,L.A., Rice,J.

08-122 On the Conditional Blocking Probabilities in a Link System for Preselection--Michaux,J.
08-123 Functional Algorithm and Structural Switching of Special Automatic Telephone Information Equipment--Dimchev,M.I., Halachev,V.I., Ivanov,P.D.
08-124 On Erlang's Formula for the Loss System M/G/K.--Cohen,J.W.
08-125 General Telecommunications Traffic without Delay--LeGall,P.
08-126 The Influence of a Preceding Selector Stage on the Loss of Gradings--Bazlen,D., Lorcher,W.
08-131 Structural Properties of Connecting Networks--Neiman,V.I.
08-132 Multi-Hour Engineering in Alternate Route Networks--Eisenberg,M.
08-133 A Model Relating Measurement and Forecast Errors to the Provisioning of Direct Final Trunk Groups--Franks,R.L., Heffes,H., Holtzman,J.M., Horing,S.
08-134 Traffic Models for the Traffic Service System (TSS)--Augustus,J.
08-135 Equivalent High Usage Circuits and Their Application in Analysis & Design of Local Networks--Employing Alternate Routing Harrington,J.S.
08-136 An Engineering Method to Account for Link Congestion Effects on Dial Tone Delay--Guess,H.A., Kappel,J.G.
08-141 Some Practical Problems of the Traffic Engineering of Overloaded Telephone Networks--Gosztony,G., Honi,G.
08-142 A Method for the Calculation of Traffics Offered to an Alternative Routing Arrangement--Sagerholm,B.
08-143 Dimensioning of Traffic Routes According to the EERT-Method and Corresponding Methods--Rahko,K.
08-144 Dimensioning of Alternative Routing Networks Offered Smooth Traffic--Rubas,J.
08-145 Computations with Smooth Traffics and the Wormald Chart--Nightingale,D.T.
08-146 Modular Engineering of Junction Groups in Metropolitan Telephone Networks--Farr,J.P.
08-211 Forecasting Telephone Traffic in the Bell System--Blair,N.D.
08-212 An Economic Forecasting Model of the Demand for International Voice Telecommunication from Canada--Khadem,R.
08-213 Traffic Forecasting with an Inadequate but Improving Data Base--Turner,W.M., Willett,R.B.
08-214 Use of Computer Programs for Trunk Forecasting--Gitten,L.J.
08-221 Yngve Rapp - A Memoir--Jacobaeus,C.
08-222 A Mathematical Model for the Long Term Planning of a Telephone Network--Bruyn,S.
08-223 Computer Aided Planning of the Telephone Network of Rural Areas--Caballero,P.A., Sole-Gil,J., Knutsen,K., Hansen,I.
08-224 Hypothesis of a Toll Network with Separate Routes for First-Choice and Overflow Traffic--Diano,G., Pallotta,P., Masetti,G.
08-225 A Nodal Route Switching Network Composed of Standard Modules--Grimbly,J.L.C., Smith,J.L.
08-227 Computer Aided Study for the Determination of the Structure of the Algerian Long Distance Network--Belhadj,A., Caballero,P.A.
08-231 A Queueing System with Time-Outs and Random Departures--Wallstrom,B.
08-232 Queueing Model with Regular Service Interruptions--Fujiki,M., Murao,V.
08-233 Approximations for Certain Congestion Functions in Single Server Queueing Systems--Delbrouck,L.E.N.
08-234 Queue Discipline NIFO for Tree Structured Generalized M/G/1-Queueing System & its Application (NIFO = Nearest IN First Out) --Egenolf,F.J.
08-235 Approximate Formulae for the Delay in the Queueing System GI/G/1--Kramer,W., Langenbach-Belz,M.
08-236 Analysis of Complex Queueing Networks by Decomposition--Kuhn,P.
08-237 Provision of Signalling Equipment According to a Delay Criterion--Johnsen,S., Smith,J.L.
08-241 Traffic Data - The Need, Nature and Use--O'Shaughnessy,J.J.
08-242 Extreme Value Engineering of Small Switching Offices--Barnes,D.H.
08-243 Cost Effectiveness of Traffic Measurements--Chin,Y.M.
08-244 Traffic Measurements and the Grade of Service--Erke,T., Rahko,K.
08-245 Accuracy Requirements Concerning Routine Traffic Measurements with Regard to Service Level Objectives in Telephone Network and to Certain Error & Cost Factors--Parviala,A.
08-251 Graded Delay Systems with Infinite or Finite Source Traffic and Exponential or Constant Holding Time--Kampe,G., Kuhn,P.
08-252 Waiting-Time Distribution in Link Systems--Haugen,R.B., Osterud,H.
08-253 A Network Flow Model Analysis of a Multi-Queue Operator Service System with Priority--Liu,J., Kettler,D.A.
08-254 Accuracy of Observed Mean Values of Waiting-Times, Queue Lengths & Proportion of Delayed Calls in Markovian Many-Server Systems --Olsson,K.M.
08-255 Turfing - A Queueing Model for the Work Backlog of a Telephone Repairman/Installer--Segal,M.
08-256 A Calculation Method for Link Systems with Delay--Villen,M.
08-261 The Total Network Data System--Buchner,M.M., Hayward,W.S.,Jr.
08-262 Busy Hour Traffic Variations Determined from Continuous Traffic Measurements--Leigh,R.B., Little,A.J.
08-263 The Rigorous Calculation of the Blocking Probability and its Application in Traffic Measurement--Mina,R.R.
08-264 A Study of Subscriber Dialed Trunk Traffic in the Short Term--Cole,A.C., Macfadyen,N.W.
08-265 Customer Line Usage Studies--Hartman,M.G.
08-266 An Improvement in Traffic Matrices Calculation--Marin,J.I.
08-311 Operational Research Methods in Traffic Engineering--Fujiki,M.
08-312 On General Point Processes in Teletraffic Theory with Applications to Measurements and Simulation--Iversen,V.B.
08-313 Applications of Processing State Transition Diagrams to Traffic Engineering--Gerrand,P.
08-314 Decomposition Techniques for Evaluating Network Reliability--Fratta,L., Montanari,U.
08-315 Uncertainty Model for An M/M/I System--Mizuki,M.
08-316 Some Applications to Telephone Traffic Theory Based on Functional Limit Laws for Cumulative Processes--Lindberger,K.
08-317 Average Number of Disjoint Available Paths--Timperi,G.
08-321 Comparison of Calculated and Simulated Results for Trunk Groups with Repeated Attempts--Gosztony,G.
08-322 On the Interaction Between Subscribers and a Telephone System--Myskja,A., Aagesen,F.A.
08-323 The Configuration Theory: The Influence of Multi-Part Tariffs on Local Telephone Traffic--Kraepelien,H.Y.
08-324 Some Traffic Characteristics of Subscriber Categories & Influence from Tariff Changes--Bo,K., Gaustad,O., Kosberg,J.E.
08-325 Experimenting with the Effect of Tariff Changes on Traffic Patterns--Cohen,G.
08-331 A General Purpose Blocking Probability Calculation Program for Multi-Stage Link Systems--Kodaira,K., Takagi,K.
08-332 Probability of Blocking in Non-Hierarchical Networks--Vestmar,B.J.A.
08-333 Some Applications of Quadratic Programming to the Calculation of Traffic Matrices--Nivert,K., Von Schantz,C.
08-334 An Approximate Method of Calculating Delay Distributions for Queueing Systems with Poisson Input and Arbitrary Holding Time Distributions--Bear,D.
08-335 On Point-to-Point Losses in Communications Networks--Butto,M., Colombo,G., Tonietti,A.
08-336 Transformed Probability Distributions of Indetermined Form: Calculation Methods for Moments of Arbitrary Order and their Application in Queueing Systems--Schreiber,F.
08-337 Traffic Engineering with Programmable Pocket Calculators--Bretschneider,G.
08-341 Automatic Fundamentals for the Calculation of Traffic Distribution in Telephone Networks--Daisenberger,G.
08-342 The Effects of Non-Uniform Traffic Distribution in Switching Networks--Hofstetter,H.
08-343 Peakedness in Switching Machines: its Effect and Estimation--Heffes,H., Holtzman,J.M.
08-344 A Mathematical Model of Telephone Traffic Dispersion in Some Australian Metropolitan Networks--Dunstan,A.W.
08-345 Analysis of Traffic Flows on Subscriber Lines Dependent of Time and Subscriber Class--Evers,R.
08-411 Evaluation of Reliability and Serviceability in Communication Systems with its Applications to Network Planning--Mori,H., Hiroichi,T.
08-412 A Study of Software Reliability--Andersson,H., Peiram,L., Strandberg,K.
08-414 Effects of Faults on the Grade of Service of a Telephone Exchange--Kaniuk,G., Smith,J.L.

08-415 On the Influence of Certain Typical Equipment Faults on Grade of Service--Jensen,E., Toledano,F.
08-416 Effectiveness Characteristics of Partly Disabled Device Groups--Andersson,H., Kjell,S.
08-421 Some Formulae Old and New for Overflow Traffic in Telephony--Pearce,C.E.M., Potter,R.M.
08-422 On Higher Order Moments of Overflow Traffic Behind Groups of Full Access--Schehrer,R.
08-423 Mean and Variance of the Overflow Traffic of a Group of Lines Connected to One or Two Link Systems--DeBoer,J.
08-424 The Accuracy of Overflow Traffic Models--Freeman,A.H.
08-425 Behavior of Overflow Traffic and the Probabilities of Blocking in Simple Gradings--Kosten,L.
08-426 Cyclic Overflow of Calls Offered to Subgroups of Full-Availability Groups of Lines--Jung,M.M., DeBoer,J.
08-427 Correlation Induced in Traffic Overflowing from a Common Link--Wilson,K.G.
08-431 Statistical Problems in Simulation of Telecommunications Traffic Systems, Some Analytical Results--Olsson,K.M.
08-432 An Interactive Simulation System for Data Communication Network Design - ICANDO---Ono,K., Yoshiyori,U.
08-433 Subcall-Type Control Simulation of SPC Switching Systems--Dietrich,G., Salade,R.
08-435 ENTRASIM; A Real-Time Traffic Environment Simulator for SPC Switching Systems--Gruszwcki,M.
08-436 Traffic Studies of Message Switching Computers--Thuan,L.D., Bogner,R.E.
08-441 Factors Influencing the Call Completion Ratio--Riesz,G.W.
08-442 Statistical Aspects in Detecting Critical Traffic Conditions--Miranda,G., Tosalli,A.
08-443 Concent - An Aid to the Business Management of Telephone Networks--Marlow,G.O'H.
08-445 Individual Circuit Measurements--Graves,R.D., Pearson,D.A.
08-511 Optimization of Telephone Networks with Alternative Routing and Multiple Number of Route Group Channels--Tsankov,B.
08-512 A Comparison of System and User Optimized Telephone Networks--Harris,R.J.
08-513 A Generalization of Takagi's Results on Optimum Link Graphs--Van Bosse,J.G.
08-514 A Method for Determining Optimal Integer Numbers of Circuits in a Telephone Network--Berry,L.T.M.
08-515 An Optimal Dimensioning of a Certain Local Network--Yechiali,U.
08-516 A Probabilistic Model for Optimization of Telephone Networks--Krishnan Iyer,R.S., Downs,T.
08-521 Data Communications Through Large Packet Switching Networks--Kleinrock,L., Kamoun,F.
08-522 Evaluation of Traffic Characteristics of Some Time-Division Switching Networks for Data With Plurality of Speeds--Inose,H., Saito,T., Wakahara,Y., Fukushima,Y.
08-523 Measurement and Analysis of Data Traffic Originated by Display & Teletypewriter Terminals in a Teleprocessing System --Pawlita,P.F.
08-524 Modeling & Analysis of Store-And-Forward Data Switching Centers with Finite Buffer Memory and Aknowledgement Signalling--Bux,W.
08-525 Probability of Loss of Data Traffics with Different Bit Rates Hunting One Common PCM-Channel--Katzschner,L., Scheller,R.
08-531 Traffic Engineering in Developing Countries. Some Observations from the ESCAP Region--Elldin,A.
08-532 International Network Planning - Two Case Studies. The Pan-African Telecom Network and the Middle East & Mediterranean Telecom Network--Engvall,L.
08-533 Integration of Communication Satellites into International Networks; Planning & Economy, Traffic Engineering--Pernau,W., Urmoneit,W.
08-534 Structuring of Telephone Network with Alternative Routing but with No Hierarchization of Centers--Chapuis,R.J., Hofer,C., Perrin,J.L.
08-541 Point-To-Point Selection Versus Point-to-Group Selection in Link Systems--Lotze,A., Roder,A., Thierer,G.
08-542 Wide-Sense Non-Blocking Networks, and Some Packing Algorithms--Smith,D.G., Rahman Khan,M.M.
08-543 Blocking Probabilities for Connecting Networks Allowing Rearrangements--Hwang,F.K.
08-544 Investigations on Folded and Reversed Link Systems--Lotze,A., Roder,A., Thierer,G.
08-545 Some Blocking Formulae for Three-Stage Switching Arrays, with Multiple Connection Attempts--Morrison,M., Fleming,P.,Jr.
08-546 Design of Mixed Analog and Digital Switching Networks--Harland,G.
08-547 PPL - A Reliable Method for the Calculation of Point-to-Point Loss in Link Systems--Lotze,A., Roder,A., Thierer,G.
08-549 Congestion in a Switching Network with Rearrangement--Boyd,J.C., Hunter,J.M.
08-611 Traffic Calculations in SPC Systems--Villar,J.E.
08-612 Investigation on the Traffic Behavior of the Common Control in SPC Switching Systems--Weisschuh,H., Wizgall,M.
08-613 The Waiting-Time Distribution for Markers and Other Peripheral Devices in SPC Switching Systems--Gerrand,P., Guerrero,A.
08-621 Methods of Estimating Central Processing System Traffic Performances in SPC - Electronic Switching Systems--Itoh,M., Nunotani,Y., Ueda,T., Okada,K.
08-622 Traffic Problems Within Multi-Processor SPC Systems--Bradley,I.A., McTiffin,M.J.
08-623 Priority Models for Communication Processors Including System Overhead--Herzog,U.
08-624 Analytical Study of Feedback Effects on Processor Traffic in SPC Systems--Haugen,R.B., Jensen,E., Sanchez-Puga,M.J.
08-631 Studies on the Probability of a Called Subscriber Being Busy--Lind,G.
08-632 A Non-Constant Failure Rate Distribution in the Conversation Time of Telephone Calls--Nombela,L.F., Pillado,J.M.
08-633 The Variance of Observations on Markov Chains--Songhurst,D.J.
08-634 Statistical Design of Load Balance Traffic Studies--Pedersen,O.A.
08-635 Call Generation, Holding Times & Load for Individual Telephone & Telex Subscriber Lines--Anderberg,M., Wikell,G.
08-636 A Modified Chi-Squared Test for Computer Simulations of Telephone Switching Networks--Peterson,M.M.
08-641 Report of the Working Party for the Documentation of First Six International Teletraffic Congresses--DeLange,S.J.
08-642 Reporting of the Working Party on Teletraffic Training--

NINTH INTERNATIONAL TELETRAFFIC CONGRESS, TORREMOLINOS, OCTOBER 16-24, 1979

09-110 Long-Term Planning in Telephony--Jensen,A.
09-111 New Methods for Administrative Area and Local Entity Usage Forecasting--Blair,N.D., Hayward,W.S.,Jr.
09-112 Causal Models for Forecasting Telephone Services by Market Sectors--Bohm,E., Wacker,W., Thornton,P.
09-113 The Sequential Projection Algorithm: A New and Improved Traffic Forecasting Procedure--David,A.J., Pack,C.D.
09-114 Implementing the Sequential Projection Algorithm in an Operating Telephone Company--Harris,R.H., Sharkey,R.E.
09-115 Planning and Forecasting Using a Simulation Model--Khadem,R., Schulzki,A.
09-116 A Growth Rate Forecast System for Intertoll Traffic--MacDiarmid,A.N.
09-117 Application of an Empirical Bayes Technique to Telephone Traffic Forecasting--Pandya,R.N., Robinson,W.R.
09-121 Traffic Routing and Trunk Forecasting with a Computer Based System--Bezdek,P.V.
09-122 Some Measurements of the Traffic Between Subscribers in Small Geographical Areas--Dahlstrom,P., Nivert,K., Sjostrom,L.E.
09-123 Comparison of Different Methods for the Calculation of Traffic Matrices--Debiesse,J.L., Matignon,G.
09-124 Traffic Distribution for a Base Unit Remote Switching Unit Entity--Hert,E.de
09-125 A Traffic Forecast Technique for Network Planning--Lee,L.
09-126 A Mathematical Model for Traffic Matrices Calculation--Marin,J.I.
09-127 A Study of Traffic Distribution Models with Reference to the Polish Telex Network--Palmowska,K.
09-130 Teletraffic Theory and its Practical Applications: A Veteran's View on the Present Situation--Jacobaeus,C.
09-131 Characterizing Traffic Variations--Bolotin,V.A., Prokopovich,V.S.
09-132 Further Studies on Lost Call Telephone Traffic--LeGall,P.
09-133 A Renewal Theoretic Approach to the Solution of Traffic and Reliability Problems--Gebraegs,H., Holfe-Isphording,U.
09-134 Estimating Peakedness From Arrival Counts--Holtzman,J.M., Jargerman,D.L.
09-135 The Effect of Holding Times on Loss Systems--Iversen,V.B.
09-136 On a New Class of Queueing Models for Distributed Environment--Kleinrock,L.

09-137 Modeling of Non-Stationary Traffic Process--Myskja,A.
09-138 Perturbed Markovian Queues in Teletraffic--Syski,R.
09-141 On the Observation and Augmentation of Subscriber Lines with High Probability of Being Busy--Biot,J., Massant,J.
09-142 Observations of User Behavior and Service Quality for a Long Distance Telecommunications Network Under Sustained Traffic Overload--Farr,J.P., Waldron,E.H.J.
09-143 Studies on Traffic Processes of Single Subscribers and Groups of Subscribers--Lind,G.
09-144 Direct-Distance-Dialing Call Completion and Customer Retrial Behavior--Liu,K.S.
09-145 Statistical Observation of Repeat Attempts in the Arrival Process--Macfadyen,N.W.
09-146 Percentage Estimation of Busy Lines in Subscribers Selections--Pillado,J.M.
09-147 Recent Observations of Subscriber Behavior--Roberts,J.W.
09-148 Effect of Subscriber Behavior and Traffic Abnormalities on the Design of SPC Exchanges--Roy-Choudhury,P.K., Shukla,M.N., Wadhwa,T.R.
09-210 Traffic Data Acquisition Systems--Connell,J.
09-211 ISTC: Integrated System for Traffic Measurements and Control--Asher,A.
09-212 The Sapart Network. Auxiliary System Engineering for Planning and Analysis of Telephone Networks--Gabarda,S., Garcia,J.L., Gimeno,F.J.
09-213 Evolution of the Total Network Data System--Grantges,R.F., Riesz,G.W., Snyder,B.E.
09-214 Automatic Message Supervision System (AMSS)--Herzberg,M.
09-215 Traffic Measurements for Telephone Network Planning--Rahko,K.
09-216 An Integral Observation on an Automatic Telephonic Network, Plan Application and Resultant--Sanchez Garrido,F.
09-217 Optimal Planning and Processing of Teletraffic Measurements--Warfield,R.E.
09-221 Observations of the Quality of Service of International Traffic by Means of a Minicomputer--Bodard,H., Guerineau,J.P.
09-222 Stochastic Models for Service Evaluation Systems--Descloux,A.
09-223 Analysis of Traffic--Girolami,A., Cerasoli,L.T.
09-224 Measurement and Analysis of Subscriber Traffic Rates--Gregory,E.W., Songhurst,D.J.
09-225 A Traffic Measuring Device Based on the Concept of State Transition--Kraushaar,J.M., Mina,R.R.
09-226 Effective Network Administration in the Presence of Measurement Uncertainty--Neal,S.R.
09-227 An Analog-Digital Method for Measuring Parameters and Probability Distributions of the Telephone Traffic--Szczepanski,J.
09-231 Reliability of Combined Measurement of Mean Holding Time and Traffic Load Based on Counting Calls and Sampling Seizures States --Daisenberger,G.
09-232 Optimization of MonteCarlo Simulation--Jobmann,K.
09-233 Accuracy of Estimating Congestion Quantities from Simulation Results--Kosten,L.
09-234 The Distribution of Measured Traffic with Poisson Call Arrivals--Milne,C.B.
09-235 Significance of Time Tagging in Each Simulated Sequence--Mizuki,M.
09-236 Accuracy of Measurements on Markovian Service Systems--Peiram,L.
09-237 Determination of Confidence Limits for the Estimates of Congestion--Rubas,J., Warfield,R.E.
09-238 Improved Simulation by Application of the Objective Bayes Statistics--Schreiber,F.
09-239 Confidence Limits for the Expected Telephone Traffic in Simulation Models Using Arma-Models--Stordahl,K., Sollie,B.H., Damsleth, E.
09-241 Derivatives of Wilkinson Formula & Their Application to Optimum Design of Alternative Routing Systems--Akimaru,H., Tokushima,H., Nishimura,T.
09-242 A Telephone Network Performance Model--Berry,L.T.M.
09-243 The Accuracy of Carried Traffic Measurements with Applications to Supervision of Overflow Systems--Lindberger,K.
09-244 A Moment Method for the Analysis of Telephone Traffic Networks by Decomposition--Manfield,D.R., Downs,T.
09-245 Estimation of Traffic Service in a System of Full-Availability Trunk Group with Mutual Overflow--Miernik,J.
09-246 The Variance of Time Congestion in the Final Route of an Alternative Routing System--Olsson,K.M.
09-247 The Equivalent Non-Random Method and Restrictions Imposed on Renewal Overflow Systems by the Specification of a Finite Number of Overflow Traffic Moments--Potter,R.M.
09-248 Estimation of Service Effectiveness of Overflow Traffic Component Flows in Overflow Systems with Full-Availability Trunk Groups --Rucinski,A., Klimontowicz,A.
09-249 On Individual Losses in Overflow Systems--Wallstrom,B., Reneby,L.
09-310 CCITT Practice - A Permanent Challenge to Teletraffic Engineering--Gosztony,G.
09-311 Observations of Extreme Value Statistics in Small Switching Offices--Barnes,D.H.
09-312 Performance Index Describing the Degradation of the Grade of Service in Service-Integrated Systems with Overload--Evers,R., Mrozynski,G.
09-313 Extreme Value Analysis Techniques--Friedman,K.A.
09-314 Extreme Value Control and an Additive Seasonal Moving Average Model for the Evaluation of Daily Peak Hour Traffic Data --Lehtinen,P.
09-315 The Role of Traffic Engineering in the Environment of a Regulated Monopoly Telephone Service--Mina,R.R., Kraushaar,J.M.
09-316 Spacing Theory in Some Traffic Topics--Nombela,L.F.
09-317 Traffic Engineering of a Remotely Controlled Switching Systems--Stone,R.C.Jr., Sheft,J.A.
09-318 Measures of Effectiveness Performance Applied to Telecommunication--Strandberg,K.
09-320 Some Effects of U.S.Interstate Message Tariff Changes on Traffic Levels and Distributions--Auray,R.R.
09-321 Intertoll Traffic Versus Tariffs Evolution--Bravo,V., Gomez,M.
09-322 Mathematical Model of Telephone Subscriber--Gechtman,V., Biran,D.
09-323 Frequency Distribution of Subscribers by Their Monthly Local Traffic--Kraepelien,H.Y.
09-324 Improved Utilization of the Telephone Network Through Introduction of Differentiated Rates--Stordahl,K., Bakken,T.
09-325 The Influence of Degree of Telephone Development and Tariffs on Telephone Traffic--Todorov,P., Mirski,K.
09-326 Productivity Based Planning Model for Teleglobe--Werner,M.
09-331 Traffic Capacity of Highly Loaded Link Systems--Elldin,A.
09-332 Blocking in Three-Stage Digital Switching Network--Gallassi,G.
09-333 Traffic Behavior of Reversed Link Systems with Short Path Possibilities--Hofstetter,H., Weber,D.
09-334 Link System Effective Availabilities for Point-to-Point Selection--Pedersen,O.A.
09-335 The Loss of One-Sided Link Systems with Short Path Connections--Roder,A.
09-336 On the Calculation of Full Access Groups for Internal and External Traffic with Two Types of Sources--Schehrer,R.
09-337 The Method of Loss Calculation in the Class of Eight-Stage Link Systems--Stastny,M.
09-338 Loss Calculation of Subscriber Link Systems--Thierer,G.
09-339 Reduction of the Rank of the System of Linear Equations to be Solved for Exact Calculation of the Loss Probability of Link Systems--Urmoneit,W.
09-341 A Study of Non-Blocking Rearrangeable Switching Networks--Clos,C.
09-342 Superior Channel Graphs--Hwang,F.K.
09-343 A Mathematical Model of Path Searching in General Type Switching Networks--Kocay,L.T.
09-344 Worst-Case Analysis of Some Algorithms for Rearranging Three-Stage Connections Networks--Kubale,M.
09-345 TDM Versus SDM Switching Arrays - A Comparison--Lotze,A., Rothmaier,K., Scheller,R.
09-346 New Architecture and the Traffic Aspects of Switching Network Using Microprocessors--Okada,K., Imase,M., Ichikawa,H.
09-347 Design of Economic PCM-Arrays with a Prescribed Grade of Service--Rothmaier,K., Scheller,R.
09-348 Generalized Optimal Channel Graphs for SLIP Connections in a Class of Link Systems--Takagi,K.
09-349 Link Systems - Access to Outgoing Routes Connected to a Number of Switches--Walker,E.

09-411 An Approximation for the Delay Distribution of the M/G/N Queue and an Exact Analysis for the M/E2/N Queue with Random Service --Bear,D.
09-412 Two Queues in Series with Non-Overlapping Service Times--Boxma,O.J.
09-413 Single Server Queues with General Interarrival and Phase-Type Service Time Distribution Computational Algorithms--Bux,W.
09-414 Some Transform Approaches to the Computation of the Normalization Constant in Queueing Networks--Eklundh,B., Stavenow,B.
09-415 The Output of Multi-Server Queueing Systems with Poisson Input--Everitt,D.E., Downs,T.
09-416 Analysis of Delays by Counting Events--Jolley,W.M.
09-417 Analysis of Post-Call Activity in Queueing Systems--Jolley,W.M., Harris,R.J.
09-418 On the Delay-Ratio Queue Discipline--Segal,M.
09-421 Dimensioning Trunk Groups for Digital Networks--Elsner,W.B.
09-422 Traffic Considerations of a New Generation of Telephone Set--Fleming,P.,Jr.
09-423 Traffic Consideration of Digital Class 5 Offices--Fleming,P.,Jr.
09-424 Optimization of Digital Local Networks Taking Reliability Aspects Into Account--Holmberg,G., Lindberg,P.
09-425 Use of Stochastic Models in Choosing Network Synchronization Methods, Clock Stabilities and lastic Buffer Sizes--Marlevi,A.
09-426 On a Full Digital Long-Distance Network--Mazei,U., Miranda,G., Pallotta,P.
09-427 Reliability of Service Integrated Communication Networks and Its Dependencies of Traffic Load and Traffic Distribution --Mrozynski,G.
09-430 Traffic Problems in Computer Networks--Frank,H.
09-431 A Traffic Study on the Satellite Channel Access Algorithm CPODA--Aagesen,F.A.
09-432 Analysis of Shared Storage in a Computer Communication Network--Korner,U.
09-433 Treatment of Some Important Aspects of Data Networks Planning--Lera,E., Smith,E.A.
09-434 Evaluation & Comparison of Dialogue Traffic Characteristics in Teleprocessing Systems Based on Measurement Results-Pawlita,P.F.
09-435 Parameters and Communication Efficiencies in the Modeling of a Packet Switching Network--Soto,O.G., Martinez Miguez,L.
09-436 Some Results from a Simulation of Queueing Networks with Dependent Service Times--Vrana,C.
09-441 Specific Call Congestion of Multi-Flow Bulk Traffic in Loss Systems--Bastiani,S., Kubanski,P.G.
09-442 High Level Data Link Control - Traffic Considerations--Bux,W., Truong,H.L.
09-443 On Networks with Generalized Processor Sharing (and a New Property of Erlang's B Formula)--Cohen,J.W.
09-444 Performance Characteristics for Hierarchically Organized Multi-Processor Computer Systems with Generally Distributed Processing Times--Herzog,U.
09-445 A Drop and Throttle Flow Control (DTFC) Policy for Computer Networks--Kamoun,F.
09-446 Congestion Control in Packet Switched Networks by Gradual Restrictions of Virtual Calls--Matsumoto,J., Mori,H.
09-447 Evaluation of Traffic Carrying Capability in One-Stage and Two-Stage Time-Division Network Handling Data with a Variety of Speed Classes--Saito,T., Inose,H., Hayashi,S.
09-511 Traffic Properties of Multi-Stage Waiting Systems--Fujihashi,C., Gambe,E.
09-512 Traffic Analysis of an Overloaded Telephone Network with Both Centralized and Distributed Control--Haugen,R.B., Fredrikson,G.
09-513 Busy Period Distribution of a Single Server Queueing System to Which Traffic is Offered via Clock-Pulse Operated Gate-Jung,M.M.
09-514 Analysis of Switching System Control Structures by Decomposition--Kuhn,P.
09-515 On the Use of Digital Approximation for Modeling Multi-Processor Systems--Nilsson,A.A., Stewart,W.J., Seraj,J.
09-516 Some Queueing Models of AXE-1--Rapp,D.
09-517 An Approximate Model for a Conventional Telephone Exchange--Salles,J.M.
09-518 Traffic Analysis of a Class of Distributed SPC Systems--Wilen,M., Morales,G.
09-519 A Static-Priority Queue with Time Slicing--Walke,B.
09-521 A General Exchange Simulator for Capacity Studies--Addie,R., Codsi,G., Gerrand,P., Rubinstein,M.
09-522 Large ESS Traffic Engineering Methods--Chambers,M.D., Stockert,A.A., Kappel,J.G.
09-523 A Characterization of Traffic Variability for SPC Systems--Forys,L.J., Zucker,H.
09-524 Simulation Language Palm--Jonin,G.L., Sedol,J.J.
09-525 E12 - System-Software Performances--Lavergne,R., Demard,M.
09-526 A Special Processor for Fast Simulation of Queueing Networks--Lehnert,R.
09-527 An Approach to Flexible Modeling and Simulation for Control Processor Analysis--Soto,O.G., Garcia Higuera,J.A., Berzosa,C.D., Martinez Miguez,L.
09-528 The Flow of Control Data Within the Analog Concentrator of the System IFS--Wizgall,M., Kramer,W., Mayer,J.
09-531 Statistical Methods for Detecting Overloads in a Trunk Group--Bostica,P., Butto,M., Tonietti,A.
09-532 Comparison of Regulation Methods for Traffic Overloads in SPC Systems--Briccoli,A.
09-533 A Time-Series Transfer Model for the Relationships Between Registers and Junctors in a Telephone Exchange with Applications to Traffic Control--Damsleth,E., Sollie,B.H., Bakken,T.
09-534 Progressive Method of Overload Control for Processors--Garcia-Higuera,J.A., Berzosa,C.D.
09-535 Optimal Control and Call Acceptance in a SPC Exchange--Schoute,F.C.
09-536 Dynamic Processor Overload Control and Its Implementation in Certain Single-Processor and Multi-Processor SPC Systems --Somoza,P., Guerrero,A.
09-537 Call Handling and Control of Processor Load in a SPC - A Simulation--Wilding,K., Karlstedt,T.
09-541 Maintenance Interval for a Full Availability Group of Devices in a Loss System--Boer,J.
09-542 The Performance of Various Gradings Under Severe Load Unbalance--Das,P.K., Smith,D.G.
09-543 Observation of the Effect of Killer Trunks on Grade of Service--Fernandez,F., Aguilar,M.S.
09-544 Improved Network Administration Process Utilizing End-to-End Service Considerations--Katz,S.S.
09-545 Balanced Skip Multiples and Their Wiring Algorithms--Kodaira,K., Ueda,T.
09-546 Study of Traffic Imbalance in Subscriber Matrices--Morales,G., Crespo,A., Rafart,M.G.
09-547 A Report on Telephone Traffic Variation in Japan--Shikiba,E., Mihara,T., Mizuashi,K., Kaiba,H.
09-548 Traffic Management in Congested Telephone Networks--Venkatram,M.C.
09-611 Variable Communication Network Design--Akiyama,M.
09-612 Dynamic Routing of Calls in the Local Telephone Network; Performance of Different Routing Algorithms--Bean,A., Szybicki,E.
09-613 An Economic Method for Improving Telephone Network Serviceability--Chan,W.C.
09-614 Feedback Methods for Calls Allocation on the Crossed Traffic Routing--Hennion,B.
09-615 Administrative Concepts in an Advance Routing Network--Lavigne,M.E., Barry,J.R.
09-616 Improved Trunk Engineering Algorithm for High Blocking Hierarchical Networks--Rau,R.N.
09-617 Optimal Allocation of Telephone Cable Drums to Construction Sites - A Cutting Stock Problem with a Generalized Objective Function--Resh,M., Yechiali,U.
09-618 Planning of Three Level Telecommunication Network for More than One Busy Hour--Tsankov,B.
09-621 Optimizing Telephone Network with Alternate Routes: An Integer Programming Solution--Basu,K.K.
09-622 Some Results on the Design of Tree Structured Communication Networks--Camerini,P.M., Fratta,L., Maffioli,F.
09-623 International Network Optimization Methods--Engvall,L.
09-624 Distribution of the Grade of Loss in a Long Distance Network--Gavassuti,M., Scavo,G.G.
09-625 Network Design Taking Into Account Breakdowns and Traffic Overloads--Guitoneau,S., Camoin,B.
09-626 Determining Grades of Service in a Network--Harvey,C., Hills,C.R.
09-627 End-to-End Connection Probability - The Next Major Engineering Issue?--Horn,R.W.
09-628 An Optimum Strategy for Improving Network Traffic Throughput with Resources Constraints--Sallai,G.
09-631 Telephone Network Behavior in Repeated Attempts Environment - A Simulation Analysis--Grillo,D.
09-632 Network Management Systems in the United States--Haenschke,D.G.

09-633 Congestion Mechanisms and Overload Control in Telephone Networks--Hashida,O., Nakajima,S., Kazuhiko,O., Shinohara,M.
09-634 An Analytical Model for Traffic Flow in a Network with Both Centralized and Distributed Control--Helvik,B.E., Skogan,E.
09-635 Service Protection for Special Traffic Flows--Hopmans,A.C.M., Van der Ven,R.
09-636 Modeling for Automatic Network Management: An Adaptive Time-Recursive Parameter Estimator and Traffic Predictor--Knutsson,A.
09-637 System Approach to Telecom Network Management--Sinha,N.K.

TENTH INTERNATIONAL TELETRAFFIC CONGRESS, MONTREAL, CANADA, JUNE 8-15,1983

10-111 Introduction of New Network Services--Katz,S.S.
10-112 Teletraffic Models For The Telecom 1 Integrated Services Network--Roberts,J.W.
10-113 Modeling and Performance of a New Integrated Switching System for Voice and Data with Distributed Control--Denzel,W., Weiss,W.
10-114 Integrating Voice and Data in a Switching Node -- A Comparison of Strategies--Crogier,M.W., Pandya,R.N., Doshi,R.
10-115 Subscriber Repeat Attempts, Congestion, and Quality of Service: A Study Based on Network Simulation--Songhurst,D.J.
10-116 An Approach to Integrated Voice and Data Switch Modeling--Liu,Feng-Kuang
10-117 A Numerical Algorithm for the GI/G/K Queue--Everitt,D.E., Downs,T.
10-118 Lost Traffic in a Service Integrated TDM Switching System with Separated and Common Used Channel Areas--Mrozynski,G.
10-121 Relationships between Utilization, Service, and Forecast Uncertainty for Central Office Provisioning--Coco,R.A., Farel,R.A., Potter,R.M., Wirth,P.E.
10-122 Switching Capacity Relief Model: Theoretical Development--Chow,K.-Y., David,A.J., Ionescu-Graff,A.M.
10-123 Application of Multi-Dimensional Scaling to Grouping of Toll Areas and Proposition of a Forecasting Method--Ueda,T.
10-124 A Forecasting Model to Set Up Multi-Season Traffic Matrices--Passeron,A., Eteve,E.
10-125 Minimum-Cost Multi-Year Trunk Provisioning--Kashper,A,, Pack,C.D., Varvaloucas,G.C.
10-126 Long Term Growth Trends in International Telecommunications Traffic: an Intelsat Study--Parthasarathy,R, Casey,J., Troutman,P.
10-127 An Econometric Model of the Demand for Trunk Calls in the Netherlands During 1950-1981--Knottnerus,P., Velding,W.E.
10-132 A Strict Priority Queueing System with Overload Control--Eisenberg,M.
10-133 Approximate Analysis of Markovian Queueing Networks with Priorities--Schmitt,W.
10-134 A Feedback Queue with Overload Control--Wallstrom,B.
10-135 Delay Time Analysis in Multi-Microprocessor Systems--Satoh,S., Tanabe,S., Oguchi,N.
10-136 Description of Interprocessor Communication by Queueing Breakdown Analysis--Giglmayr,J.
10-141 Precutover Extreme Value Engineering of a Local Digital Switch--Friedman,K.A.
10-142 Estimation of Reference Load From Daily Traffic Distributions--Rubas,J.
10-143 Novel Concepts for the Measurement of the Mean and Variance Traffic Parameters and Their Application--Kraushaar,J.M.
10-144 Microprocessor and Eve in the Traffic Measurement and Administration Problems. An Empirical Specification of G.O.S.-Nombela,L.F.
10-145 Engineering Telecommunications Networks through Integration of Differences in Day-to-Day Traffic Variation Profiles Among Circuit Groups--Kodaira,K., Harada,Y.
10-146 Development of Traffic Theory for SPC Exchanges and New Teleservices Based on Measured Data--Lehtinen,P., Martikainen,O.
10-147 A Study of Traffic Flow in Relation to CCITT Recommendations for the Measurement of Telephone Traffic Flow, Call Attempts and Seizures--Cole,A.C.
10-212 An Approach to Traffic Analysis of Chronically Overloaded Networks--Guerrero,A., De Los Mozos,J.R.
10-213 Cost 201: A European Research Project. A Flexible Procedure for Minimizing the Costs of a Switched Network Taking into Account Mixed Technologies and End-to-End Blocking Constraints--Dressler,J., Gomes,J.A.C., Mantel,R.J., Mepuis,J.M. Sara,E.J.
10-214 Performance Evaluation of Private Networks with Automatic Route Selection and Call Queueing--Girard,A., Taillon,V.G.
10-215 A Dynamic Planning Method for Telecommunication Networks and Its Performance Evaluation for District Trunk Networks --Camerini,P.M., Fratta,L., Maffioli,F., Bonatti,M., Gallassi,G.
10-216 Optimal Dimensioning of Circuit Switched Digital Networks--Berry,L.T.M.
10-217 End-to-End Dimensioning of Hierarchical Telephone Networks and Traffic Combination--LeGall,P.
10-221 Circuit-Switched Communication Networks Analysis -- Call Delay Point of View --Lubacz,J.
10-222 Design and Implementation of the Evolving Italian Toll Network: Methodologies and Computer Tools--Filippini,M., Mazzei,U., Miranda,G.
10-223 Optimization of Portuguese Trunk Telephone Network--Nordeste,P.M.N.
10-224 The Suboptimal Alternate Routing Practice with Non-Coincident Busy Hours--Parviala,A.
10-225 A Computer Procedure for Optimal Dimensioning of Evolving Toll Networks Under Constraints--Gavassuti,M., Giacobbo-Scavo,G., Trimarco,U.
10-226 Some Aspects of the Dimensioning and Optimization of Digital Networks in Both-Way Circuits--DeMiguel,M., De Los Rios,F.J.
10-227 "Killer Routes" and "Killer Numbers" in Telephone Networks--Simcha,Becker, Meir,H.
10-231 CCITT -- A Chance to Unify the Basic Traffic Engineering Practice--Gosztony,G.
10-232 Statistical Modeling of the Reliability of SPC Software--Downs,T.
10-233 Cost 201: A European Research Project: A Procedure for Minimizing the Cost of a Transmission Network Under Service Availability Constraints in Failure Conditions--Lindberg,P., Mocci,U., Tonietti,A.
10-234 The AGT Toll Network Reliability Objectives--Lee,L.
10-236 Evaluating Network Survivability Through Exposure--McMenamin,P.G., Fitzpatrick,G.J.
10-237 Terminal Reliability of a Computer Communication Network with a Maximum Allowed Delay--Shatwan,F., Smith,D.G.
10-238 The Evolution of Traffic Standards--Lee,J.I., Armolavicius,R.J.
10-239 Optimum Distribution of Congestion in a National Trunk Network. An Economical Evaluation as a Base for Grade of Service Standards--Nivert,K., Gunnarsson,R., Sjostrom,L.E.
10-241 Bell System Traffic Usage by Class of Service--Bolotin,V.A., Kappel,J.G.
10-242 Measurement of Repeat Call Attempts in the Intercontinental Telephone Service--Lewis,A., Leonard,G.
10-243 Application of Bayesian Methods to Teletraffic Measurement and Dimensioning--Warfield,R.E., Foers,G.A.
10-244 Time Efficient Simulation: A Simple Algorithm for Producing Empirical Distribution Functions with Predetermined Constant Error --Schreiber,F.
10-245 Customer Behavior and Unexpected Dial Tone Delay--Burkard,L., Phelan,J.J., Weekly,M.D.
10-246 On Properties of Random Number Generators and Their Influence on Traffic Simulation--Ide,H.D., Sagebarth,J.
10-247 Comparison of Bayesian and Sampling Theory Estimates of Parameters Based on Simulation Results--Poo,G.S.
10-312 Distributed Access Workload Performance Sensitivity--Arthurs,E., Stuck,B.W.
10-313 Token-Ring Performance: Mean-Delay Approximation--Bux,W., Truong,H.L.
10-314 Models of Demand Access of Satellite Communication Channels--Emstad,P.J.
10-315 End-to-End Delay in Local Area Networks--Kiesel,W.M.
10-316 Traffic Evaluation of Ambre: A Multi-Services Local Network--Boyer,A., Dupuis,A., Pitie,J.-M., Romoeuf,L.
10-317 Throughput of a Satellite Channel Communication--Pujolle,G., Spaniol,O.
10-321 The Benefits of Dynamic Non-Hierarchical Routing in Metropolitan Traffic Networks--Field,F.A.
10-322 Intercity Dynamic Routing Architecture and Feasibility--Ash,G.R., Kafker,A.H., Krishnan,K.R.
10-323 Dynamic Routing for Intercity Telephone Networks--Cameron,W.H., Regnier,J., Galloy,P., Savoie,A.M.
10-324 The Overload Performance of Engineered Networks with Non-Hierarchical and Hierarchical Routing--Akinpelu,J.M.
10-325 DNHR: A New SPC/CCIS Network Management Challenge--Haenschke,D.G., Kettler,D.A., Oberer,E.
10-326 Grade of Service of a Dynamic Call Routing System--Regnier,J., Blondeau,P., Cameron,W.H.
10-327 Centralized Routing Based on Forecasts of the Telephone Traffic--Karstad,T., Stordahl,K.
10-328 The Switch Planning System for the Dynamic Non-Hierarchical Routing Network--David,A.J., Farber,N.

10-332 Time-Out Management in Packet Switching Networks--Van As,H.R.
10-333 Traffic Characteristics of PCR Method for CCITT Signalling System No. 7--Watanabe,Y., Ikeda,Y.
10-334 User Behavior in Teleprocessing Networks: Analytical Models Based on Empirical Data--Pawlita,P.F., Suedhofen,H.-D.
10-335 Congestion Control in Packet Switching Computer Communication Networks--Korner,U.
10-336 A Simulation Study of CCITT X.25: Throughput Classes and Window Flow Control--Dieterle,W.
10-337 Evaluation of Preemptive Queueing Strategies with Overhead and their Applicability in Data Networks--Goerg,C.
10-338 Modeling and Performance Evaluation of a Packet Switching Module in an X.25 Network--Soto,O.G., Martinez-Miguez,L., Niemegeers,I.G.
10-339 A Flexible, High Performance Multi-Processor for Data Network Simulation--Barel,M.
10-341 User Traffic Characteristics on the Public Packet Switched Network and their Influence on Network Performance--Kaminuara,K., Nakamura,M.
10-342 Analysis of HDLC Normal Response Mode with Full-Duplex Transmission -- A New Calculation Method and Simulation--Raith,T., Truong,H.L.
10-343 An Analytical Model of a Class of Adaptive Virtual Call Routing Procedures--Whitehead,M.J.
10-344 Analysis of a User Communication Path in a Store and Forward Network--Soto,O.G., Gutierrez,D., Martinez Miguez,L.
10-345 A Simulation Study of HDLC-ABM with Selective and Non-Selective Reject--Goeldner,E.-H., Truong,H.L.
10-346 Approximate Methods for Multi-User-Class System Analysis--Burakowski,W.
10-412 Approximation For Networks Of Queues--Whitt,W.
10-413 Queueing Systems With A Finite Number Of Sources And Internal Traffic--Kallenberg,P.J.M.
10-414 Queueing Analysis Of An Arrival-Driven Message Transfer Protocol--Manfield,D.R., Tran-Gia,P.
10-415 On The Overflow Processes From The Ph+Ph/M/S/K Queue--Machihra,F.
10-416 Queue Problems With A Two-Dimensional State Space--Cohen,J.W.
10-417 Blocking And Delays Conditional On The Number Of Arrivals In A Time Period--Halfin,S.
10-418 Queue With Two Priorities And Feedback, Joint Queue Length Distribution And Response Time Distributions For Specific Sequences --Fontana,B.
10-421 Estimation Of Traffic Distribution With Shortage Of Data--Tome,F.M., Cunha,J.A.
10-422 Some Aspects Of Trunk Traffic Forecasting--Mirski,K.
10-423 Differentiated Rates - Effects Of Its Introduction In A Developing Country--Conde,R., Carvalho,P.C.
10-424 Telephone (TV And Radio) Usage In Developing Countries--Engvall,L.
10-425 Administration Of Traffic Forecasting For Sparse Rural Telephone Networks--Farr,J.P., Waldron,E.H.J.
10-426 Report Of Training Working Party--Elldin,A.
10-43A1 Routing And Trunking Network Evolution When Introducing Local Digital Switching--Burattin,A., Mazzei,U.
10-43A2 A Uniform Approach To The Analysis And Optimization Of Circuit-Switched Communication Networks--Pioro,M.P.
10-43A3 Optimization Of Alternative Routing Networks With Regard To Existing Equipment And Modularities--Hedberg,I.
10-43A4 Effects Of Digitalization On The Optimum Structure Of Junction Network--Caballero,P.A., Silva,J.M.
10-43A5 A Sensitivity Study On Network Planning--Okazaki,H., Miyamoto,H.
10-43A6 Performance Evaluation Of DA/SS-TDMA Satellite Systems With Wide Bandwidth Traffics--Butto,M., Giovanni-Scarati,C.
10-43A7 Traffic Engineering For Automatic Call Distribution Networks--DeSousa,P.J.T.
10-43B1 Optimum Design Of Alternative Routing Systems With Constrained Individual Call Losses--Akimaru,H., Takahashi,H., Ikeda,T.
10-43B2 Modular Engineering And Full Grouping Theorems For Alternate Routing Networks--Sallai,G., Dely,Z.
10-43B3 Comparison Of Network Dimensioning Models--Harris,R.J.
10-43B4 Simple Modular Algorithm For Dimensioning Digital Links--Combes,M.C., Horn,F., Kerneis,J.
10-43B5 Allocation Of Trunks Over Diversity Routes--Ekvall,D., Lee,L.
10-43B6 Optimization Of A Telecommunication Network Based On Adjustable Access Probabilities--Jansons,A.
10-43B7 A Method For Calculation Of Traffic Carried On Network Paths--Dziong,Z.
10-44A1 Analysis Of Multi-Cellular Mobile Radiotelephone Systems With Loss--Everitt,D.E., Macfadyen,N.W.
10-44A3 General Solution Of The Equilibrium Equations Of A Non-Preemptive Multi-Priority Queue System--Daisenberger,G., Teufel,B.
10-44A4 On Queueing Systems With Clocked Operation And Priorities--Jans,H.
10-44A6 Asymptotic Analysis Of M(T)/M(T)/1, The Time Dependent M/M/1 Queue--Massey,.W.A.
10-44A7 Multiple Channel Occupation Period In A Packet Switching System--Krzysztof,P.
10-44A8 An Analysis Of Teletraffic Generated By A Large Number Of Markovian Sources--Weiss,A.
10-44A9 Approximate Analysis Of Separable Queueing Networks--Peiram,L.
10-44B1 Simple Approximation Techniques for Congestion Functions for Smooth and Peaked Traffic--Sanders,B., Haemers,W.H., Wilcke,R.
10-44B2 A Three Moments Method for Traffic in a Network Based on a Coxian Input Process--Guerineau,J.P., Labetoulle,J., Lebourges,M.
10-44B3 Generalized Peakedness of Teletraffic Process--Eckberg,A.E.
10-44B4 Distribution Models for Quasi-Stationary Random Traffic--Myskja,A.
10-44B5 Some Analytical Aspects of the Peakedness Concept--Van-Doorn,E.A.
10-44B6 The Loss Probability Measurement Accuracy for the Subscriber Line with Repeated Calls--Naumova,E.O.
10-44B7 Tail Probabilities Through Use of Extreme Order Statistics--Johnsen,S.
10-511 Gradings in Link Systems--Jacobaeus,C.
10-512 Switching Networks with Unreliable Elements--Garmash,V.A., Shor,L.A., Abramovich,A.I.
10-513 An Application of Information Theory to Switching Network Design--Tanaka,Y., Akiyama,M.
10-514 Blocking Calculation Method for Digital Switching Networks with Step-by-Step Hunting and Retrials--De-Los-Mozos,J.R., Bucheister,A.
10-515 Priority Rearrangements--Jajszczyk,A.
10-516 Some Studies of Generalized Optimal Channel Graphs--Takagi,K.
10-517 On Two Control Algorithms for Rearrangeable Switching Networks--Bassalygo,L.A., Neiman,V.I.
10-518 On the Traffic Capacity and Loss of PCM Switching Arrays as a Function of Internal Traffic and Path Allocation Modes --Rothmaier,K., Scheller,R.
10-519 Blocking Probability of Folded Switching Networks with Respect to Their Capacity Stages--Hofstetter,H., Weber,D.
10-5110 Telecommunication Systems and Traffic Theory--Jacobaeus,C.
10-521 Subcall-Oriented Modeling of Overload Control in SPC Switching Systems--Tran-Gia,P.
10-522 Analysis of Overload Control Schemes for a Class of Distributed Switching Machines--Doshi,B.T., Heffes,H.
10-523 Adaptive Overload Control for an SPC Exchange--Schoute,F.C.
10-524 Performance Analysis of a New Overload Strategy--Forys,L.J.
10-525 Average Response Times in an MG/1 Queue with General Feedback and Priorities--Villen,M.
10-526 Analyzing the Effectiveness of Audible Ringing--Forys,L.J., Zucker,H.
10-527 Analysis of a Traffic Shaping Scheme--Ronningen,L.A.
10-528 On a Delay-Loss System for Overload Protection--Schehrer,R.
10-531 Analysis of Individual Traffic Characteristics for Queueing Systems with Multiple Poisson and Overflow Inputs--Matsumoto,J. Watanabe,Y.
10-532 Approximating Parcel Blocking Via State Dependent Birth Rates--Fredericks,A.A.
10-533 Simple Approximations of Overflow System Quantities for Additional Demands in the Optimization--Lindberger,K.
10-534 Limits and Asymptotes of Overflow Curves--DeBoer,J.
10-535 On Individual and Overall Losses in Overflow Systems--Reneby,L.
10-536 Some Methods for Approximation of Congestion--Rahko,K.
10-537 A Generalization of the Brockmeyer System--Pearce,C.E.M.

INTERNATIONAL TELETRAFFIC CONGRESSES 1 TO 10
AUTHOR INDEX

INTERNATIONAL TELETRAFFIC CONGRESSES 1 TO 10
TITLE KEYWORD INDEX